Review of Progress in
# QUANTITATIVE NONDESTRUCTIVE EVALUATION
Volume 6A

A Continuation Order Plan is available for this series. A continuation order will bring delivery of each new volume immediately upon publication. Volumes are billed only upon actual shipment. For further information please contact the publisher.

Review of Progress in
# QUANTITATIVE NONDESTRUCTIVE EVALUATION

Volume 6A

Edited by
**Donald O. Thompson**
Ames Laboratory (USDOE)
Iowa State University
Ames, Iowa

and
**Dale E. Chimenti**
Materials Laboratory
Air Force Wright Aeronautical Laboratories
Wright–Patterson Air Force Base
Dayton, Ohio

PLENUM PRESS • NEW YORK AND LONDON

The Library of Congress has cataloged the first volume of this title as follows:

Review of progress in quantitative nondestructive evaluation—Vol. 1- — New York: Plenum Press, 1982-
   v.: ill.; 28 cm.
   Annual.
   Vols. 2-   published in 2 pts.: A and B.
   Vol. 1-   are the Proceedings of the 8th, 1981-   Air Force/Defense Advanced Research Projects Agency Symposium on Quantitative Nondestructive Evaluation.
   ISSN 0743-0760 = Review of progress in quantitative nondestructive evaluation.
   1. Non-destructive testing—Congresses. I. United States. Air Force. II. United States. Defense Advanced Research Projects Agency. III. Air Force/Defense Advanced Research Projects Agency Symposium on Quantitative Nondestructive Evaluation.
TA417.2.R48                   620.1'127—dc19                   84-646699
Library of Congress                    [8502]                   AACR 2 MARC-S

ISBN 0-306-42584-X

First half of the proceedings of the Thirteenth Annual Review of
Progress in Quantitative Nondestructive Evaluation,
held August 3-8, 1986, in La Jolla, California

© 1987 Plenum Press, New York
A Division of Plenum Publishing Corporation
233 Spring Street, New York, N.Y. 10013

All rights reserved

No part of this book may be reproduced, stored in a retrieval system, or transmitted in any form or by any means, electronic, mechanical, photocopying, microfilming, recording, or otherwise, without written permission from the Publisher

Printed in the United States of America

PREFACE

This volume (Parts A and B) contains the edited papers presented at the annual Review of Progress in Quantitative Nondestructive Evaluation held at the University of California (San Diego) in LaJolla, August 3-8, 1986. The Review was organized and sponsored by the Center for NDE at Iowa State University and the Ames Laboratory, in cooperation with the Office of Basic Energy Sciences, USDOE, and the Materials Laboratory at Wright-Patterson Air Force Base. Approximately 400 attendees, a new record, representing various government agencies, industry, and universities participated in the technical presentations, poster sessions, and discussions. This Review, with its wide-ranging interchange of technical information, stands as one of the most comprehensive in the field of NDE research and engineering.

In order to present the reader with a more useful document, we have organized the symposium papers in these Proceedings by subject rather than by the order of presentation at the Review. Topical subject headings have been selected under which the large majority of papers would reasonably fall. Here, again, we have revised the format used in former years to accommodate an evolving focus of interest in the field. These categories cover a broad spectrum of research in NDE and encompass activities from fundamental work to early engineering applications. In the following paragraphs we offer a brief summary of the research presented in these Proceedings.

Considering the profound influence on research--indeed, on our lives--of the digital computer, it was with great interest that we heard from Prof. Robert S. Engelmore of Stanford University on the prospect of thinking machines. An active area of research that is already finding application under controlled conditions, artificial intelligence holds forth the promise of a new paradigm in the person-computer interaction. While most of Prof. Engelmore's examples were taken from the medical field, the audience could easily substitute similar situations from inspection science.

A respectable number of papers in this volume deals specifically with developments in the basic methods. These are collected in a single chapter on Generic Techniques-Fundamentals. This chapter contains papers on ultrasonics, eddy current, thermal wave methods, and acoustic emission. In each case, fundamental or generic aspects of the techniques are emphasized. Also included is a section on tomography and magnetic resonance techniques.

Chapter 2 combines papers on imaging, microscopy, inversion, and reconstruction, a grouping that reflects the close physical and mathematical relationship among these topics. Several inversion papers concentrate on eddy current data, discussing a variety of approaches to this problem.

Following a trend established last year, the importance of continuous monitoring or in-process control to NDE is acknowledged in the third chapter on Sensors and Probes. In two sections, Ultrasonics and Electromagnetics, are collected papers on the development of improved or entirely new means of extracting useful information on the condition of the system being probed.

Chapter 4 properly highlights as a separate division the work combined in the previous volume with sensor development. Efforts in Image Analysis, Signal Processing, and Artificial Intelligence experienced a substantial growth in the last year. In particular, further development of the analysis and processing of images seems assured considering the strong general interest in providing inspectors with pictorial representations of a nondestructive test. A special session (the first in this Review series) on artificial intelligence for NDE closes this chapter.

Part A of this volume ends with a chapter devoted to engineering aspects of NDE. Several large-scale automated inspection systems--ultrasonic, x-ray, and electrothermal--are discussed in a series of papers. The important question of inspection reliability, for both the nuclear and aerospace industries, is treated in the final section of this chapter.

To accommodate the rapid growth of papers on materials properties, we have followed last year's format and organized the entire Part B of this volume around the materials theme. Chapter 6 contains research reports on Advanced Composites, headed by invited papers from this year's minisymposium on ceramic composites. The length and diversity of this chapter attests to the increasing importance of composite materials in many critical applications requiring high strength-to-weight ratios. To underscore the explosion of interest in this topic, only three papers related to composites appeared in volume one of this series.

Reflecting the value of cross fertilization between NDE and other areas, the number of papers in Electronic Materials and Devices has risen steadily. The work reported here has now begun to encompass techniques new to the field. These papers are collected in Chapter 7.

The last chapter of contributed papers, Materials Characterization, is the second largest in the volume and contains the majority of papers in this Part of the Proceedings. As in the previous volume, this chapter is further subdivided into five sections in order to separate related, but distinct, topics. The first section collects those papers which treat general properties of materials, and the second contains work on acousto-elasticity, stress and texture. The third section presents results on cracks and deformation, while ferromagnetic materials are the subject of the fourth section. Closing Chapter 8 is a section on weldments and bonds.

Following what, by now, must be called a tradition, the editors have included the paper presented at the informal evening problem session along with an edited transcript of the accompanying discussion as a part of the Proceedings. Unified Life Cycle Engineering is a design concept which seeks, through modeling and simulation, to promote an information-rich design environment and establish considerations such as manufacturability and inspectability as equal partners with performance in the system design process. Dr. Harris M. Burte, Chief Scientist of the Air Force Materials Laboratory, was joined by a panel of experts in discussions which amplified and extended the evening's presentation. This paper and the group discussion are incorporated as Chapter 9 of this volume.

The organizers of the Review wish to acknowledge the assistance provided by several organizations and many people. They include the Ames Laboratory and Iowa State University as well as the agencies noted in the first paragraph of the Preface. They are especially grateful to Dr. R. Engelmore of Stanford University for his excellent keynote lecture on artificial intelligence and expert systems and to Dr. Harris Burte of the Air Force Materials Laboratory for his discussion of Unified Life Cycle Engineering as noted above. A number of people helped to organize sessions. They include: J. Achenbach (Northwestern University), R. Beissner (Southwest Research Institute), A. Berens (University of Dayton Research Institute), R. Green (Johns Hopkins University), A. Loos (Virginia Polytechnic Institute), J. Opsal (Thermal Wave, Inc.), Y. Rajapakse (Office of Naval Research), W. Scott (Naval Air Development Center), L. Shambaugh (Pratt & Whitney), H. Wadley (National Bureau of Standards), and K. Wickramasinghe (IBM, T. J. Watson Research Center). These "special" sessions contained several invited, overview papers for which the organizers are grateful. The organizers also wish to acknowledge the assistance of the chairpersons who managed the sessions and the many participants who contributed to the meeting through both session and hall-way discussions. The organizers are particularly indebted to Mrs. Diane Harris for her logistics management of the Review, to Ms. Linda Dutton for her preparation of Review materials and assistance at the meeting, and to Ms. Elizabeth Bilyeu for her preparation of this manuscript.

         Donald O. Thompson
         Ames Laboratory, USDOE
         Iowa State University

         Dale E. Chimenti
         Materials Laboratory
         Air Force Wright Aeronautical
          Laboratories
         Wright-Patterson AFB

CONTENTS

VOLUME 6A

KEYNOTE ADDRESS

    Artificial Intelligence and Knowledge Based Systems:
Origins, Methods and Opportunities for NDE................... 1
R. S. Engelmore

CHAPTER 1: GENERIC TECHNIQUES--FUNDAMENTALS

Section A: Ultrasonics

    UK Developments in Theoretical Modeling for
Ultrasonic NDT............................................... 21
A. Temple

    On the Direct and Inverse Elastic Wave Scattering
Problem to Characterize Damage in Materials.................. 37
J. D. Achenbach, D. A. Sotiropoulos, and H. Zhu

    Time Dependent Pulse Propagation and Scattering
in Elastic Solids: An Asymptotic Theory...................... 45
A. N. Norris

    Ultrasonic Surface and Bulk Wave Interaction with
Fluid-Saturated Porous Solids................................ 51
M. J. Mayes, P. B. Nagy, L. Adler, B. P. Bonner, and
R. Streit

    Interactive Diffraction of a Plane Longitudinal
Wave by a Pair of Coplanar Central Cracks in an
Elastic Solid................................................ 59
Y. M. Tsai

    Ultrasonic Scattering by Planar and Non-Planar
Cracks....................................................... 69
S. K. Datta and A. H. Shah

    Acoustic Wave Scattering from a Circular Crack:
Comparison of Different Computational Methods................ 79
W. M. Visscher

    3-D Modeling of Ultrasonic Scattering from
Inter-Granular Stress Corrosion Cracks....................... 87
J. D. Achenbach and D. E. Budreck

CHAPTER 1: (CONTINUED)

Application of Models for IGSCC Inspection.................... 93
T. A. Gray, R. B. Thompson, B. P. Newberry,
J. D. Achenbach, and D. Budreck

Plate Modes Generated by Emats for NDE of Planar
Flaws......................................................... 101
S. K. Datta, R. E. Schramm, and Z. Abduljabbar

Reflection of Bounded Acoustic Beams from a
Layered Solid................................................. 109
A. K. Mal and T. Kundu

Application of Forward and Inverse Scattering
Models for Ultrasonic Waves in Complex Layered
Structures.................................................... 117
P. R. Smith, L. J. Bond, D. T. Green, and C. Chaloner

Finite Element Studies of Transient Wave
Propagation................................................... 125
M. Sansalone, N. J. Carino, and N. N. Hsu

Modeling Ultrasonic Waves Using Finite
Difference Methods............................................ 135
L. J. Bond, N. Saffari, and M. Punjani

## Section B: Eddy Current

Eddy Current Response to Three-Dimensional Flaws
by the Boundary Element Method................................ 145
R. E. Beissner and J. H. Hwang

On Boundary Integral Equation Method for Field
Distribution Under Cracked Metal Surface...................... 153
V. G. Kogan, G. Bozzolo, and N. Nakagawa

Applications of the Volume Integral Technique to
Modeling Impedance Changes due to Surface Cracks.............. 161
W. S. Dunbar

Eddy Current Induction by a Coil Near a Conducting
Edge in 2D.................................................... 169
S. K. Burke

3-D Eddy Current Nondestructive Testing Modeling
for Surface Flaws............................................. 177
R. Grimberg, M. Mayos, and A. Nicolas

Eddy-Current Probe Interaction with Subsurface
Cracks........................................................ 185
J. R. Bowler

Recent Studies in Modeling for the A. C. Field
Measurement Technique......................................... 193
D. H. Michael and R. Collins

Three Dimensional Finite Element Modeling..................... 201
N. Ida

CHAPTER 1: (CONTINUED)

A Computational Model for Electromagnetic
Interactions with Advanced Composites.......................... 211
H. A. Sabbagh, T. M. Roberts, and L. D. Sabbagh

### Section C: Thermal Waves

Fundamentals of Thermal Wave Physics........................... 217
J. Opsal

Classical and Quantum Mechanical Aspects of
Thermal Wave Physics........................................... 227
A. Mandelis

Temporal Behavior of Modulated Reflectance
Signal in Silicon.............................................. 237
A. Rosencwaig, J. Opsal, and M. W. Taylor

Modulated Reflectance Measurement of Reactive-Ion
and Plasma Etch Damage in Silicon Wafers....................... 245
W. L. Smith and P. Geraghty

Picosecond Transient Thermoreflectance:
Time-Resolved Studies of Thin Film Thermal Transport........... 253
G. L. Eesley

Probing Through the Gas-Solid Interface with
Thermal Waves: A Study of the Temperature
Distribution in the Gas and in the Solid....................... 263
L. J. Inglehart, J. Jaarinen, P. K. Kuo, and
E. H. Le Gal LaSalle

Reflection-Mirage Measurements of Thermal Diffusitivity........ 271
C. B. Reyes, J. Jaarinen, L. D. Favro, P. K. Kuo,
and R. L. Thomas

Coating Thickness Determination Using Time
Dependent Surface Temperature Measurements..................... 277
J. C. Murphy, and L. C. Aamodt, and G. C. Wetsel, Jr.

Nondestructive Characterization of Coatings on
Metal Alloys................................................... 285
G. C. Wetsel, Jr., J. C. Murphy, and L. C. Aamodt

Thermal Wave Techniques for Imaging and
Characterization of Materials.................................. 293
L. D. Favro, P. K. Kuo, and R. L. Thomas

### Section D: Acoustic Emission

Numerical Calculations of Acoustic Emission.................... 301
J. A. Johnson

Point-Source/Point-Receiver Materials Testing.................. 311
W. Sachse and K. Y. Kim

Thermally Induced Acoustic Emission in
Homogeneous Metals and Composites.............................. 321
J. M. Liu

# CHAPTER 1: (CONTINUED)

Internal Monitoring of Acoustic Emission
in Graphite-Epoxy Composites Using Imbedded
Optical Fiber Sensors........................................... 331
K. D. Bennett, R. O. Claus, and M. J. Pindera

Application of Pattern Recognition Techniques
to Acoustic Emission from Steel and Aluminum.................. 337
M. A. Friesel

Digital Acquisition and Analysis of Acoustic
Emission Signals for Crack Site Initiation Studies............ 345
B. A. Barna, J. A. Johnson, and R. T. Allemeier

Effect of Temperature and Heat Treatment
on Crack Growth Acoustic Emission in 7075 Aluminum............ 353
S. L. McBride and J. L. Harvey

A Semi-Adaptive Approach to In-Flight
Monitoring Using Acoustic Emission............................ 361
C. M. Scala

Real-Time Aircraft Structural Monitoring
Using Acoustic Emission....................................... 371
S. Y. Chuang

## Section E: X-Ray, CT, and NMR

Quantitative Analysis of Real-Time Radiographic
Systems....................................................... 379
M. D. Barker, R. C. Barry, R. A. Betz, P. E. Condon,
and L. M. Klynn

Spatial Resolution of Linear Array Xenon
Ionization Chamber X-Ray Detectors............................ 389
J. W. Eberhard

Engineering Tomography: A Quantitative NDE Tool............... 401
R. L. Hack, D. K. Archipely-Smith, and W. H. Pfeifer

A Linearization of Beam-Hardening Correction
Method for X-Ray Computed Tomographic Imaging
of Structural Ceramics........................................ 411
E. Segal, W. A. Ellingson, Y. Segal, and I. Zmora

Photon CT Scanning of Advanced Ceramic Materials.............. 421
B. D. Sawicka and W. A. Ellingson

Applications of Film Tomography Technique for QNDE............ 433
A. Notea

NDE of Polymer Composites Using Magnetic
Resonance Techniques.......................................... 441
W. A. Bryden and T. O. Poehler

Proton NMR Imaging of Green State Ceramics.................... 449
L. B. Welsh, S. T. Gonczy, R. T. Mitsche, L. J. Bauer,
J. Dworkin, and A. Giambalvo

## CHAPTER 1: (CONTINUED)

### Section F: Other Techniques

Optical Generation of Coherent Ultrahigh Frequency Surface Waves .................................................. 459
S. Gracewski and R. J. D. Miller

Use of a Chirp Waveform in Pulsed Eddy Current Crack Detection .................................................. 467
R. E. Beissner and J. L. Fisher

A New Method for the Measurement of Ultrasonic Absorption in Polycrystalline Materials ..................... 473
H. Willems

A Single Transducer Broadband Technique for Leaky Lamb Wave Detection ........................................ 483
P. B. Nagy, W. R. Rose, and L. Adler

Simultaneous Measurements of Ultrasonic Phase Velocity and Attenuation in Solids ............................ 491
N. K. Batra and P. P. Delsanto

## CHAPTER 2: IMAGING, MICROSCOPY, INVERSION AND RECONSTRUCTION

### Section A: Imaging and Microscopy

Shear Wave Imaging of Surface and Near-Surface Flaws in Ceramics ................................................. 501
J. D. Fraser, C. S. DeSilets, and B. T. Khuri-Yakub

A Real-Time SAFT System Applied to the Ultrasonic Inspection of Nuclear Reactor Components ........... 509
T. E. Hall, S. R. Doctor, and L. D. Reid

On the Ultrasonic Imaging of Tube/Support Structure of Power Plant Steam Generators ..................... 519
J. Saniie and D. T. Nagle

Eddy Current Imaging for Material Surface Mapping ............. 527
E. J. Chern and A. L. Thompson

Imaging of Advanced Composites with a Low-Frequency Acoustic Microscope ............................................ 535
J. D. Fraser, C. S. DeSilets, and B. T. Khuri-Yakub

Acoustic Microscopy Using Amplitude and Phase Measurements .................................................. 543
P. Reinholdtsen and B. T. Khuri-Yakub

Acoustic Microscopy: Materials Art and Materials Science ............................................................. 553
R. S. Gilmore, R. E. Joynson, C. R. Trzaskos, and J. D. Young

### Section B: Inversion and Reconstruction

Ultrasonic Inversion: A Direct and an Indirect Method ......... 563
C. A. Chaloner and L. J. Bond

## CHAPTER 2: (CONTINUED)

Imaging of Flaws in Solids by Velocity Inversion .............. 573
J. K. Cohen, N. Bleistein, and F. G. Hagin

Characterization of Flaw Shape and Orientation
Using Ultrasonic Angular Scans ............................... 585
D. K. Hsu, S. J. Wormley, and D. O. Thompson

Reconstruction of the Electromagnetic Wavefield
from Scattering Data .......................................... 595
J. H. Rose

Uniform Field Eddy Current Probe: Experiments
and Inversion for Realistic Flaws ............................. 601
J. C. Moulder, P. J. Shull, and T. E. Capobianco

Inversion of Eddy Current Data and the
Reconstruction of Flaws Part 1: Acquisition of Data ........... 611
J. C. Treece, J. M. Drynan, J. A. Nyenhuis, and C. D. Beaman

Inversion of Eddy Current Data and the
Reconstruction of Flaws Part 2: Inversion of Data ............. 619
L. D. Sabbagh and H. A. Sabbagh

## CHAPTER 3: SENSORS AND PROBES

### Section A: Ultrasonics

A Pulsed Laser/Electromagnetic Acoustic
Transducer Approach to Ultrasonic Sensor Needs for
Steel Processing .............................................. 627
G. A. Alers and H. N. G. Wadley

Development and Comparison of Beam Models for
Two-Media Ultrasonic Inspection ............................... 639
B. P. Newberry, R. B. Thompson, and E. F. Lopes

Transducer Models for the Finite Element Simulation
of Ultrasonic NDT Phenomena ................................... 649
R. L. Ludwig, D. Moore, and W. Lord

Transducer Radiation Modeling for Ultrasonic
Inspection Purposes ........................................... 657
D. D. Bennink, A. L. Mielnicka-Pate, D. O. Thompson,
and R. B. Thompson

Technique for Generation of Unipolar Ultrasonic Pulses ........ 667
D. O. Thompson and D. K. Hsu

Acoustic Transducers and Lens Design for
Acoustic Microscopy ........................................... 677
C-H. Chou and B. T. Khuri-Yakub

### Section B: Electromagnetics

Field Mapping and Performance Characterization
of Commercial Eddy Current Probes ............................. 687
T. E. Capobianco

CHAPTER 3: (CONTINUED)

Design and Characterization of Uniform Field
Eddy Current Probes.................................................. 695
P. J. Shull, T. E. Capobianco, and J. C. Moulder

Assessment of Eddy Current Probe Interactions with
Defect Geometry and Operating Parameter Variations............ 705
W. D. Rummel, B. K. Christner, and D. L. Long

Effects of Shielding on Properties of Eddy Current
Probes with Ferrite Cup Cores....................................... 713
S. N. Vernon and T. A. O. Gross

Pickup Coil Spacing Effects on Eddy Current
Reflection Probe Sensitivity........................................ 721
T. E. Capobianco and K. Yu

Eddy Current Probe Performance Variability.................... 727
G. L. Burkhardt, R. E. Beissner, and J. L. Fisher

Capacitive Arrays for Robotic Sensing.......................... 737
M. Gimple and B. A. Auld

Inductive Sensor Arrays for NDE and Robotics................. 745
A. Rosengreen and A. J. Bahr

Optical Range Finder.............................................. 751
G. Q. Xiao, D. B. Patterson, and G. S. Kino

The Effect of Oxygen on the Ion-Acoustic
Signal Generation Process......................................... 759
F. G. Satkiewicz, J. C. Murphy, J. W. Maclachlan, and
L. C. Aamodt

CHAPTER 4: IMAGE ANALYSIS, SIGNAL PROCESSING AND AI

       Section A: Image Analysis and Signal Processing

Application of Additive Regional Kalman
Filtering to X-Ray Images in NDE............................... 767
J. P. Basart, Y. Zheng, and E. R. Doering

An Image Segmentation Algorithm for
Nonfilm Radiography.............................................. 773
Z. W. Bell

Applications of Digital Image Enhancement
Techniques for Improved Ultrasonic Imaging of
Defects in Composite Materials.................................. 781
B. G. Frock and R. W. Martin

An Improved Defect Classification Algorithm Based on
Fuzzy Set Theory................................................. 791
M. Carkhuff and S. S. Udpa

Frequency-Modulated (FM) Time Delay-Domain Thermal
Wave Techniques, Instrumentation and Detection: A Review
of the Emerging State of the Art in QNDE Applications......... 799
A. Mandelis

## CHAPTER 4: (CONTINUED)

An *A Priori* Knowledge Based Wiener Filtering Approach to Ultrasonic Scattering Amplitude Estimation.................. 807
S. Neal and D. O. Thompson

Signal Processing of Leaky Lamb Wave Data for Defect Imaging in Composite Laminates...................... 815
R. W. Martin and D. E. Chimenti

A Highly Interactive System for Processing Large Volumes of Ultrasonic Testing Data...................... 825
H. L. Grothues, R. H. Peterson, D. R. Hamlin, and K. S. Pickens

Ultrasonic Flaw Detection Using a Time Shifted Moving Average......................................... 831
D. A. Stubbs and B. Olding

An Update on Automatic Positioning, Inspection, and Signal Processing Techniques in the RFC/NDE Inspection System........ 839
R. T. Ko, W. C. Hoppe, D. A. Stubbs, D. L. Birx, B. Olding, and G. Williams

Multiparameter Methods with Pulsed Eddy Currents.............. 849
C. V. Dodd and W. E. Deeds

An Efficient Technique for Storing Eddy Current Signals....... 855
S. S. Udpa

### Section B: Artificial Intelligence

Image Processing and Artificial Intelligence for Detection and Interpretation of Ultrasonic Test Signals....... 863
K. S. Pickens, J. C. Lusth, P. K. Fink, K. K. Palmer, and E. A. Franke

An Expert System for Ultrasonic Materials Characterization and NDE....................................... 871
M-S. Lan and R. K. Elsley

Development of an Expert System for Ultrasonic Flaw Classification.......................................... 879
L. W. Schmerr, Jr., K. E. Christensen, and S. M. Nugen

SIIA: A Knowledge-Based Assistant for the SAFT Ultrasonic Inspection System........................... 889
R. B. Melton, S. R. Doctor, T. T. Taylor, and R. V. Badalamente

An AI Approach to the Eddy Current Defect Characterization Problem....................................... 899
L. Udpa and W. Lord

## CHAPTER 5: NDE SYSTEMS AND RELIABILITY

### Section A: System Design and Performance

Design and Operation of a Dual-Bridge Ultrasonic Inspection System for Composite Materials..................... 907
D. C. Copley

CHAPTER 5: (CONTINUED)

Using a Squirter to Perform Pulse-Echo Ultrasonic
Inspections of Gas Turbine Engine Components:
The Pros and Cons............................................. 915
D.A. Stubbs

An Apparatus for the Automated Nondestructive
Testing of Electro-Explosive Devices Using the
Electrothermal Transient Technique........................... 925
H. L. Jacoby, K. S. Pickens, J. F. Tyndall, S. P. Clark,
R. A. Baker, and R. H. Peterson

An Automated Real-Time Imaging System for
Inspection of Composite Structures in Aircraft................ 933
D. R. Hamlin, B. M. Jacobs, R. H. Peterson, and
W. R. Van der Veer

System of Inspection Assisted by Microprocessor............... 943
J. L. Arnaud, M. Floret, and D. Lecuru

A Semi-Automatic System for the Ultrasonic
Measurement of Texture........................................ 951
S. J. Wormley and R. B. Thompson

Section B: System Reliability

Model-Based Ultrasonic NDE System Qualification
Methodology................................................... 957
T. A. Gray, R. B. Thompson, and B. P. Newberry

Crack Sizing by the Time-of-Flight Diffraction Method,
in the Light of Recent International Round-Robin Trials,
(UKAEA, DDT and PISC II)...................................... 967
G. J. Curtis

Results of the Phase I Reliability Test on the
RFC/NDE Eddy Current Station.................................. 977
R. T. Ko

Analysis of the RFC/NDE System Performance
Evaluation Experiments........................................ 987
A. P. Berens

Application of the RFC/NDE System Testing Results............. 995
C. G. Annis, Jr. and T. Watkins, Jr.

Status of Advanced UT Systems for the Nuclear Industry........ 1003
M. M. Behravesh, M. J. Avioli, G. Dau, and S-N. Liu

The Detection of Cracks Under Installed Fasteners
by Means of a Scanning Eddy-Current Method.................... 1013
D. Harrison

Real Defects for Verification of Ultrasonic
Testing Models................................................ 1019
J. R. Bower

## VOLUME 6B

## CHAPTER 6: ADVANCED COMPOSITES

### Section A: Properties

Toughness and Flaw Responses in Nontransforming
Ceramics: Implications for NDE.................................. 1023
B. R. Lawn and C. J. Fairbanks

NDE of Fiber and Whisker Reinforced Ceramics................... 1033
D. B. Marshall

Continuum Modeling of Ultrasonic Behavior in
Fluid-Loaded Fibrous Composite Media with
Applications to Ceramic and Metal Matrix Composites........... 1047
A. H. Nayfeh and D. E. Chimenti

Multiparameter Ultrasonic Evaluation of
Ceramic Matrix Composites...................................... 1057
R. W. Reed and R. F. Murphy

Detection of Ultrasonic Waves Propagating in
Boron/Aluminum and Steel/Lucite Composite Materials........... 1065
S. Huber, W. R. Scott, and R. Sands

Interface Effects on Attenuation and Phase
Velocities in Metal-Matrix Composites.......................... 1075
S. K. Datta, H. M. Ledbetter, Y. Shindo, and A. H. Shah

Ultrasonic Dispersion in Fluid-Coupled Composite
Plates......................................................... 1085
D. E. Chimenti and A. H. Nayfeh

Measurement of Ultrasonic Wavespeeds in Off-Axis
Directions of Composite Materials.............................. 1093
L. H. Pearson and W. J. Murri

Effects of Reflection and Refraction of
Ultrasonic Waves on the Angle Beam Inspection of
Anisotropic Composite Material................................. 1103
S. I. Rokhlin, T. K. Bolland, and L. Adler

Evaluation of Anisotropic Properties of
Graphite/Epoxy Composites Using Lamb Waves.................... 1111
W. R. Rose, S. I. Rokhlin, and L. Adler

Acoustic Wave Reflection from Water/Laminated
Composite Interfaces........................................... 1119
A. H. Nayfeh

Guided Interface Waves......................................... 1129
E. Drescher-Krasicka, J. A. Simmons, and H. N. G. Wadley

### Section B: Defects

Analytical Treatment of Polar Backscattering
from Porous Composites......................................... 1137
J. Qu and J. D. Achenbach

CHAPTER 6: (CONTINUED)

Porosity Characterization in Fiber-Reinforced
Composites by use of Ultrasonic Backscatter................... 1147
R. A. Roberts

Ultrasonic Characterization of Porosity in
Composite Materials by Time Delay Spectrometry............... 1157
M. D. Fuller and P. M. Gammell

Ultrasonic Characterization of Cylindrical
Porosity - A Model Study...................................... 1165
S. M. Nair, D. K. Hsu, and J. H. Rose

A Morphological Study of Porosity Defects in
Graphite-Epoxy Composites..................................... 1175
D. K. Hsu and K. M. Uhl

Evaluation of Porosity in Graphite-Epoxy Composite
by Frequency Dependence of Ultrasonic Attenuation............ 1185
D. K. Hsu and S. M. Nair

Defect and Damage Characterization in Composite
Materials..................................................... 1195
I. M. Daniel, S. C. Wooh, and J. W. Lee

Characterization of Impact Damage in Composites............... 1203
C. F. Buynak and T. J. Moran

Ultrasonic Techniques to Produce Damage Profiles
Through the Thickness of a Sample............................ 1213
W. J. Murri, K. W. Bailey, B. W. Sermon, and T. J. Todaro

Estimating Residual Strength in Filament
Wound Casings from Nondestructive Evaluation
of Impact Damage.............................................. 1221
E. I. Madaras, C. C. Poe, Jr., W. Illg, and J. S. Heyman

Photoacoustic Microscopy of Ceramics Using
Laser Heterodyne Detection.................................... 1231
H. I. Ringermacher and C. A. Kittredge

Imbedded Optical Fiber Sensor of Differential
Strain in Composites.......................................... 1241
M. Reddy, K. D. Bennett, and R. O. Claus

Ultrasonic Signal Analysis of Composite
Structures Using the Entire Waveform......................... 1247
W. D. Brosey

Section C:  Processing

Processing of Thermoplastic Matrix Composites................. 1257
A. C. Loos and P. H. Dara

Dynamic Viscosity Measurements of Fluids Employing
Resonance Characteristics of Piezoelectric
Element Vibrating in the Shear Mode........................... 1267
M. Bujard, B. R. Tittman, L. A. Ahlberg, and
F. Cohen-Tenoudji

CHAPTER 6: (CONTINUED)

Nondestructive Evaluation of the Curing of Resin
and Prepreg Using an Acoustic Waveguide Sensor................ 1277
R. T. Harrold and Z. N. Sanjana

Process Monitoring of Polymer Matrix Composites
Using Fluorescence Probes...................................... 1287
B. Fanconi, F. Wang, and R. Lowry

Dynamic Dielectric Analysis for Nondestructive
Cure Monitoring and Process Control........................... 1297
D. E. Kranbuehl, S. E. Delos, M. S. Hoff, M. E. Whitham,
L. W. Weller, and P. D. Haverty

CHAPTER 7: ELECTRONIC MATERIALS AND DEVICES

Section A: Electronic Materials and Devices

Atomic Force Microscopy: General Principles and
a New Implementation.......................................... 1307
G. M. McClelland, R. Erlandsson, and S. Chiang

Scanning Differential Contrast Microscopy..................... 1315
G. S. Kino, C-H. Chou, T. R. Corle, and P. C. D. Hobbs

Nondestructive Submicron Dimensional Metrology
Using the Scanning Electron Microscope........................ 1327
M. T. Postek

Thermal and Plasma Waves in Semiconductors.................... 1339
J. Opsal

Thermal Wave Characterization of Coated Surfaces.............. 1347
J. Jaarinen, C. B. Reyes, I. C. Oppenhiem, L. D. Favro,
P. K. Kuo and R. L. Thomas

Nondestructive Characterization of Polishing
Damage in Silicon Wafers Using Modulated
Reflectance Mapping........................................... 1353
W. L. Smith, S. Hahn, and M. Arst

Measurement of Semiconductor Transport Properties
Using Scanned Modulated Reflectance........................... 1361
F. A. McDonald, D. Guidotti, and T. M. DelGiudice

Room Temperature Photo-Luminescence Imaging of
Dislocations in GaAs Wafers................................... 1369
D. Guidotti, H. J. Hovel, M. Albert, and J. Becker

CHAPTER 8: MATERIALS CHARACTERIZATION

Section A: Properties

Applications of NDE to the Processing of Metals............... 1377
J. F. Bussiere

NDE Methods for Determination of Thermal History
and Mechanical Properties of Al-Li Alloys..................... 1395
D. J. Bracci, P. Garikepati, D. C. Jiles, and O. Buck

CHAPTER 8: (CONTINUED)

Nondestructive Characterization of Aluminum Alloys............ 1403
S. Razvi, P. Li, K. Salama, J. H. Cantrell, Jr., and
W. T. Yost

Ultrasonic Attenuation Measurement by Backscattering
Analysis..................................................... 1411
P. B. Nagy, D. Rypien, and L. Adler

The Effect of Nonspherical Pores and Multiple Scattering
on the Ultrasonic Characterization of Porosity................ 1419
J. H. Rose

Ultrasonic Reflection from Rough Surfaces in Water............ 1425
J. H. Rose and D. K. Hsu

Surface Roughness Effects in Porosity Assessment
by Ultrasonic Attenuation Spectroscopy........................ 1435
P. B. Nagy, D. V. Rypien, and L. Adler

Ultrasonic NDE of Green-State Ceramics by
Focused through Transmission.................................. 1443
R. A. Roberts

The Influence of High Temperature Creep on
the Ultrasonic Velocity in Alloy 800H......................... 1453
H. Willems

Ultrasonic Determination of Recrystallization................. 1463
E. R. Generazio

Materials Characterization by Ultrasonic Attenuation
Spectral Analysis............................................. 1475
R. L. Smith

Radiographic Detection of 100 A Thickness Variations
in 1-um-Thick Coatings on Submillimeter-Diameter
Laser Fusion Targets.......................................... 1485
D. M. Stupin

Section B: Acoustoelasticity, Stress, and Texture

A Comparison of Predictions and Measurements of the
Acoustoelastic Response of a Textured Aggregate............... 1495
G. C. Johnson and W. C. Springer

The Role of Texture Development and Dislocations in
Acoustoelasticity During Plane Deformation.................... 1505
G. C. Johnson, T. E. Wong, and S. E. Chen

The Use of Ultrasonics for Texture Monitoring in
Aluminum Alloys............................................... 1515
A. V. Clark Jr., A. Govada, R. B. Thompson,
J. F. Smith, G. V. Blessing, P. P. Delsanto,
and R. B. Mignogna

Influence of Plate Wave Characteristics on the Angular
Dependence of Ultrasonic Velocities........................... 1525
R. B. Thompson, J. F. Smith, and S. S. Lee

CHAPTER 8: (CONTINUED)

Ultrasonic Separation of Stress and Texture
Effects in Polycrystalline Aggregates......................... 1533
P. P. Delsanto, R. B. Mignogna, and A. V. Clark, Jr.

Illustration of Texture with Ultrasonic Pole Figures.......... 1541
J. F. Smith, R. B. Thompson, D. K. Rehbein, T. J. Nagel,
P. E. Armstrong, and D. T. Eash

An Ultrasonic Technique for Axial Bolt-Stress
Determination................................................ 1549
A. C. Holt, B. Cunningham, G. C. Johnson, and D. Auslander

Measurement of Thermal Stress in Railroad Rails
Using Ultrasonic SH Waves.................................... 1559
D. T. MacLauchlan and G. A. Alers

Ultrasonic Characterization of Residual Stress
and Texture in Cast Steel Railroad Wheels.................... 1567
A. V. Clark, H. Fukuoka, D. V. Mitrakovic, and J. C. Moulder

Ultrasonic Studies of Stresses and Plastic
Deformation in Steel During Tension and Compression.......... 1577
J. Frankel and W. Scholz

Uniaxial Stress Effects on the Low-Field
Magnetoacoustic Interactions in Low and Medium
Carbon Steels................................................ 1585
D. Utrata and M. Namkung

### Section C: Cracks and Deformation

Compressive Stress Effects on the Ultrasonic
Detection of Cracks in Welds................................. 1593
T. A. Siewert and R. E. Schramm

Determining Crack Tip Shielding by Means of
Acoustic Transmission and Diffraction Measurements........... 1601
O. Buck, D. K. Rehbein, and R. B. Thompson

Ultrasonic Attenuation as a Trace of Growth of
Fatigue Crack in Steels...................................... 1609
Z. Pan

Resolution of Closely-Spaced Machining-Damage-Induced
Surface Cracks in Ceramics................................... 1617
L. R. Clarke, B. T. Kuri-Yakub, and D. B. Marshall

Evaluation of Strain-Induced Surface Changes
by Optical Correlation....................................... 1625
N. S. Chang and W. L. Haworth

NDE Positron Study of Cu and Cu-Al Alloys
Thermally Charged with Hydrogen and Deformed................. 1633
Y. Pan and J. G. Byrne

CHAPTER 8: (CONTINUED)

### Section D: Ferromagnetic Materials

On-Line Inspection of Ferromagnetic Materials
Using Magnetic NDE Methods.................................... 1641
R. Ranjan, O. Buck, and R. B. Thompson

Residual Leakage Field Modeling............................... 1651
Y. S. Sun, S. S. Udpa, and W. Lord

The Computation of Fields and Signals due to
Ferromagnetic Anomalies....................................... 1659
H. A. Sabbagh and L. D. Sabbagh

Magnetomechanical Effect in Iron and Nickel
Polycrystals During Cyclic Loading............................ 1665
P. Ruuskanen and P. Kettunen

Improved Interpretation of the Downhole Casing
Inspection Logs for Two Strings of Pipes...................... 1673
S. G. Marinov

Investigation of the Microstructural Dependence
of the Magnetic Properties of 4130 Alloy
Steels and Carbon Steels for NDE.............................. 1681
D. C. Jiles and J. D. Verhoeven

Electromagnetic Nondestructive Evaluation of
Surface Decarburization on Steels:
Feasibility and Possible Applications......................... 1691
M. Mayos, S. Segalini, and M. Putignani

### Section E: Weldments and Bonds

Ultrasonic Characterization of Diffusion Bonds................ 1701
B. J. Hosten, L. A. Ahlberg, B. R. Tittmann, and J. Spingarn

Origin of Spurious Ultrasonic Echoes in Stainless
Steel Piping with Weld Overlay................................ 1707
D. S. Kupperman

In-Process Ultrasonic Evaluation of Spot Weld Quality......... 1715
S. I. Rokhlin, S. Meng, and L. Adler

Ultrasonic Detection of Weld Bead Geometry.................... 1723
N. M. Carlson and J. A. Johnson

Sizing Canted Flaws in Weldments Using Low-Frequency EMATS.... 1731
R. E. Schramm and T. A. Siewert

Ultrasonic NDE of Tubing Pinch Welds.......................... 1737
D. K. Rehbein, D. K. Hsu, R. B. Thompson, and T. A. Jones

Ultrasonic Evaluation and Imaging of Tube Closure Welds....... 1747
G. H. Thomas, J. R. Spingarn, and S. Benson

NDE Characterization of Metallic Interfaces................... 1755
D. D. Palmer, D. K. Rehbein, J. F. Smith, and O. Buck

CHAPTER 8: (CONTINUED)

Ultrasonic NDE of Integrally Fabricated Turbine Rotors........ 1763
T. A. Gray, R. B. Thompson, and F. J. Margetan

Thermal Wave Evaluation of Kapton-Laminated Copper
Diffusion Bonds.............................................. 1773
J. Jaarinen, L. D. Favro, P. K. Kuo, R. L. Thomas,
C. M. Woods, and G. L. Houston

Inspection of Bonded Assemblies Using Roller-Type
Ultrasonic Probes without Coupling Fluid..................... 1779
D. Lecuru

An Ultrasonic F-Scan Inspection Technique for the
Detection of Surface Preparation Variances in
Adhesively Bonded Structures................................. 1787
J. B. Nestleroth, J. L. Rose, D. Lecuru, and E. Budillon

CHAPTER 9: UNIFIED LIFE CYCLE ENGINEERING

Unified Life Cycle Engineering: An Emerging
Design Concept............................................... 1797
H. M. Burte and D. E. Chimenti

Discussion................................................... 1811

ATTENDEES........................................................ 1827

CONTRIBUTORS INDEX............................................... 1849

SUBJECT INDEX.................................................... 1853

ARTIFICIAL INTELLIGENCE AND KNOWLEDGE BASED SYSTEMS:

ORIGINS, METHODS AND OPPORTUNITIES FOR NDE

Robert S. Engelmore

Knowledge Systems Laboratory
Computer Science Department
Stanford University

INTRODUCTION

The title of my paper covers a lot of territory, so my remarks will tend to be broad rather than deep. Some of the topics I will cover are:

- What is AI?
- A brief history of AI.
- What is Knowledge Engineering?
- What are Expert Systems?
- Why should you care?
- Examples of Expert Systems.
- Opportunities for NDE.
- Selecting appropriate problems.
- Criticisms of Expert Systems.
- Looking ahead.

WHAT IS ARTIFICIAL INTELLIGENCE?

Asking for a definition of Artificial Intelligence (AI) is like the old story of the blind men describing an elephant (Fig. 1). You can get many different definitions, depending on one's point of view. Some people would describe AI as general symbolic, i.e., non-numeric, computation. The field of expert systems, which receives a lot of attention in the media and is frequently equated there (erroneously) with AI, is certainly part of the story. Advanced robotics and Vision are major areas of research within the realm of AI. Programs have been developed that "understand" what they see, in the sense of generating a high-level description of a video image or an aerial photograph. Natural Language understanding is another major area of AI research. Some people regard the activity of understanding actual written or spoken language as the fundamental problem of AI. There's also the definition that says AI is the leading edge of computer science. In other words, as complex problems are solved successfully, they are no longer regarded as AI; they're considered instead to be part of some other field, like chemistry or mechanical engineering. Therefore, the domain of AI could be defined as the residue of unsolved problems in computer science.

Let me try to make some sense of AI by describing it in terms of different types of computing. In Fig. 2 we have a two-by-two matrix

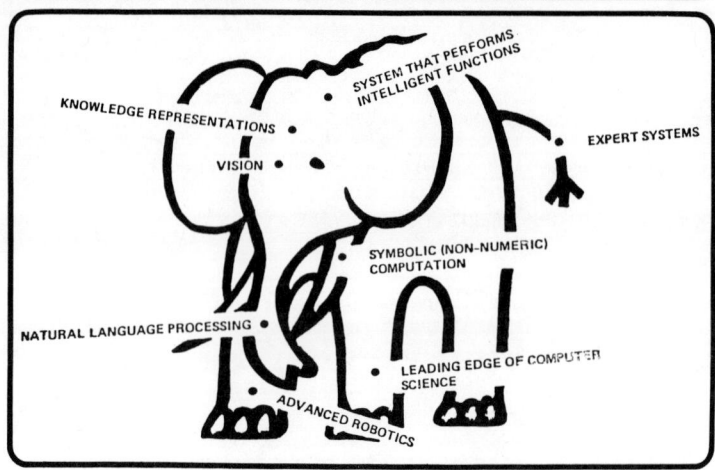

Figure 1. Views of Artificial Intelligence.

## Type of Information

|  | NUMERIC | SYMBOLIC |
|---|---|---|
| **ALGORITHMIC** | TRADITIONAL SCIENTIFIC CALCULATIONS | DATA PROCESSING |
| **HEURISTIC** | COMPUTATION-INTENSIVE APPLICATION WITH HEURISTIC CONTROL (MANIPULATORS) | ARTIFICIAL INTELLIGENCE |

Type of Processing

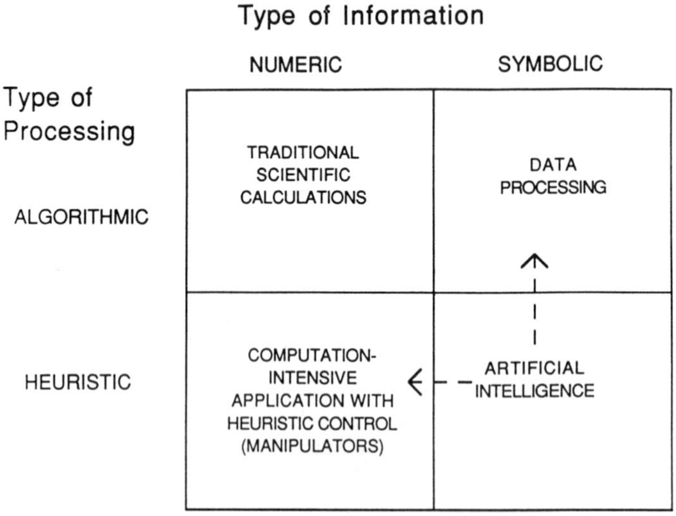

Figure 2. Types of Computing.

in which the type of processing is listed in the rows and the type of information, numeric or symbolic, with which the processing deals, is in columns. Numeric algorithmic[1] processing deals with the traditional scientific or engineering calculations with which we are all familiar. Algorithmic processing also applies to symbolic data. Retrieving information from bibliographic data bases, where the data are not numbers but text streams, is an example of this kind of activity. Here, too, the techniques for retrieving that information are very clearly defined in an algorithm, and the process is always done the same way.

---

[1] I use the term "algorithmic" in the sense of a fixed computational procedure that will always solve a well-defined problem having a well-defined solution.

On the bottom row is a type of processing called heuristic. Heuristics are techniques which you might think of as rules of thumb, used to solve problems by using judgmental or experiential knowledge rather than mathematical formulae. A corollary is that one is not guaranteed to find the best solution, or even any solution, but using heuristics usually works, and can be a very efficient way to reach a solution that is good enough. The term was originally coined by the mathematician Georg Polya, who defined heuristics as the art of good guessing. An example of heuristic computing with numeric informatin would be a program that must explore a large space of possible solutions, where each candidate involves extensive numeric computation to generate and/or test that solution. (Imagine all the conformations of a complex organic molecule, searching for the one conformation that best explains a set of x-ray crystallographic data.) Heuristics could be used here to guide the search, avoiding unlikely classes of solutions or enforcing the presence of certain partial solutions.

Artificial Intelligence finds its niche in the bottom right corner of the figure: heuristic symbolic computing, where the data are largely symbolic and the problems are "ill-structured", requiring heuristic techniques to reach a solution. AI makes contributions to the neighboring quadrants of the matrix. In the area of numeric heuristic processing, AI has provided techniques for searching large solution spaces efficiently. With respect to symbolic algorithmic processing, AI has linked natural language query techniques with data base retrieval, so that one can request information, using the sort of imprecise, ambiguous language that we tend to use in ordinary conversation, and let the "intelligent front end" infer what the user really asked for.

Although AI itself is primarily a field of academic research, it has spawned several subfields that are concerned with practical applications (see Fig. 3). The subfield of knowledge-based systems, which deals with building systems containing large amounts of knowledge for specific tasks, will be discussed further below. Natural language understanding is just what the name implies. In the subfield of vision and robotics, the emphasis is on building sensors and effectors for performing intelligent action. The fourth subfield is concerned with providing the hardware and software environments for building systems that manipulate and reason about symbolic information.

There are now a number of commercial companies that specialize in one or more of these subfields. Teknowledge, Inc. (Palo Alto), Intellicorp (Mt. View), Carnegie Group (Pittsburgh) and Inference (Los Angeles) are the four major AI companies that are developing knowledge-based systems and generic tools for their customers. Artificial Intelligence Corporation (Waltham) develops natural language understanding systems, and one of its products is marketed by IBM. Xerox, Symbolics, LMI and Texas Instruments are marketing computer systems specifically for AI applications.

A BRIEF HISTORY OF AI

Figure 4 (for which I thank Teknowledge for an earlier version) shows some of the main events in the history of AI. The term "Artificial Intelligence" was coined almost exactly 30 years ago by John McCarthy who has continued to be one of the field's chief theoreticians. During the first decade of its history, which we call the conception phase, the emphasis in AI research was on generality, i.e., finding general problem-solving techniques for solving symbolic problems. Two areas that provided the desired complexity for exploring and testing these general techniques were chess and symbolic logic. An early program, called LT (Logic Theorist), could prove theorems from the Principia Mathematica by Russell and Whitehead. One of the first AI programming languages, called IPL, was developed and used in the late 1950's, and GPS (General Problem Solving) was developed at Carnegie-Mellon in the early 1960s. McCarthy developed the Lisp programming language during this period, also.

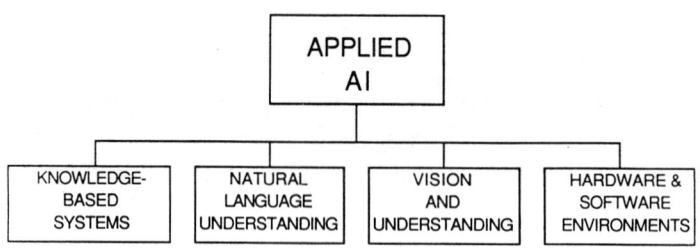

Figure 3. Subfields of Artificial Intelligence.

From about 1966 to about 1973, there was a redirection away from the idea of generality. It was clear by then that one could get just so far with general (i.e., weak) problem solving methods. Most of the successes that researchers could point to were what we would today call toy problems. On the other hand, there were two projects in the mid-to-late 60s that were attacking very significant scientific problems.

One of them was the Heuristic Dendral project at Stanford, a collaborative effort of computer science, chemistry and genetics. The problem is the interpretation of mass spectrometric data, reasoning from data (a table of the masses of the molecular fragments and their relative abundance) and knowledge of analytic chemistry and mass spectrometry to derive the chemical structure of the molecule under investigation. Without going into details, let me just say that the project, which was active for over 15 years, was very successful. One measure of its success was the number of technical papers -- on the order of fifty -- that were published in refereed chemistry journals (not AI journals). The program achieved a level of expertise of a fairly advanced graduate student in chemistry. A commercial product came out of part of this work -- a program called CONGEN for generating all molecular structures from the stoichiometric formula plus various structural constraints. Dendral was one of the very first of what we now call Expert Systems.

4

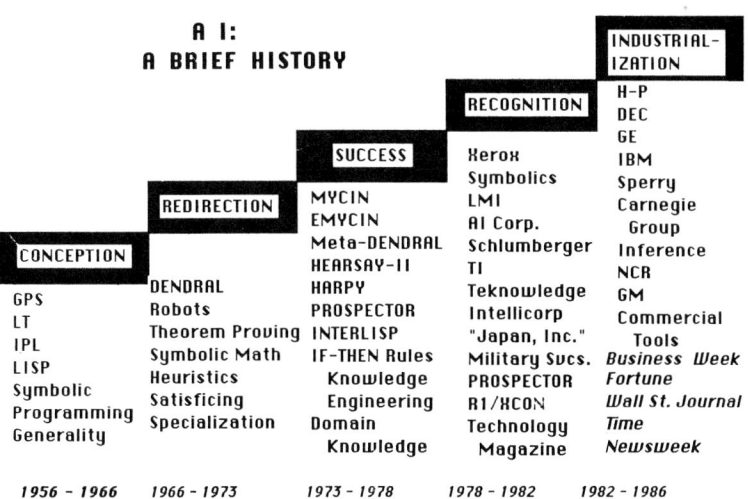

Figure 4. A Brief History of AI.

Another landmark program during this period was a program, developed at MIT, for solving mathematics problems symbolically, e.g., solving indefinite integrals. That work led to a system called MACSYMA, which is now used routinely by a large number of scientists and engineers.

The major theme during the era of redirection was the emphasis on achieving high levels of performance by using lots of specialized knowledge rather than to rely on general purpose techniques. During the next five years, this idea caught on, and there were a number of successful knowledge-based systems developed. MYCIN is by now perhaps the best known of the classic Expert Systems, and I will discuss it in more detail later. An important spinoff of the MYCIN effort was EMYCIN, the first tool for building rule-based expert systems. Meta-Dendral was a program that learned from examples the kind of knowledge that was needed for the Dendral program. It was clear early on that a major bottleneck in developing Expert Systems is the acquisition of all the knowledge that the program needs to exhibit expert performance. Meta-Dendral was one of the first systems capable of learning new knowledge automatically.

Another major research area during this period of "Success" was the Speech Understanding Program, initiated by DARPA. Two systems developed at CMU -- Hearsay II and Harpy -- achieved at least moderate performance in understanding normal speech, in a quiet room, in near real time. But perhaps more importantly, these projects produced some very interesting problem solving frameworks which have since been used in other applications. The whole field of knowledge engineering was spawned in this five-year period, when we began to understand how to develop these knowledge-based systems on something other than an ad hoc basis.

For the next four years, starting about 1978, a number of different organizations recognized the potential value of this technology, and launched their own internally sponsored AI projects. Xerox's Palo Alto Research Center had already been active in the field, and built one of the first Lisp machines that became a commercial product. Schlumberger was one of the first non-computer companies to see the value of expert systems in data interpretation, and set up their own research center. Carnegie-Mellon and Digital Equipment teamed up to produce the first expert system that went into commercial use and demonstrated real value. That was R1, a system for configuring the components of VAX computers. VAX systems are highly modular and can be ordered pretty much a la carte. Not all components are compatible, however, and there are many implicit interdependencies among components. Thus configuring one of these systems is a non-trivial task. R1, renamed XCON when it was transferred to DEC, was used on over 1000 orders per week by the end of 1983. Stanford University spun off two companies, Intelligenetics (now called Intellicorp) and Teknowledge, that specialized in knowledge engineering. The public at large became aware of this technology in the early 80s, with lots of publicity appearing in such places as Business Week, Wall Street Journal, Time Magazine.

In 1981, the Defense Science Board recognized AI as one of the most promising investments for the military. AI soon became a buzzword throughout the military services, and it seemed like all new proposal requests contained a requirement that any software must contain some artificial intelligence! Two years later DARPA,. which had been nurturing the field since the early 1970s, stepped up its support significantly when it initiated the $100M/yr Strategic Computing Program with AI as one of the core technologies.

During the past four years we have witnessed a period of increasing industrialization, and hence an increased emphasis on the applied side of AI. More new AI companies, like Carnegie Group and Inference Corp. have sprung up and are doing well, and many of the familiar names in computing, like IBM, Sperry, Digital Equipment, and Texas Instruments, now have a large investment of people and money in AI research and development. We're still a little too close to current events to identify all the milestones of the current period, but certainly one must include the arrival of low-cost workstations and PC-based knowledge systems, making the technology affordable to a large segment of the industrial and commercial sectors.

To summarize, the first decade emphasized search, i.e., the idea that you can solve any problem by search in a space (usually very large) of possible solutions and use a variety of general problem solving techniques to search efficiently. The second decade emphasized the use of application-specific knowledge, and demonstrated the power of this idea in a number of relatively small systems. The third decade, I believe, has been mainly a time of consolidation and a combining of a number of different types of problem-solving methods in order to deal with the complexity of real-world problems.

What lessons have we learned? Certainly one important lesson is that knowledge powers the solution to complex problems (it's better to be knowledgeable than smart is another way to put it). Another is that knowledge can be expressed in relatively small chunks, and that a knowledge-based system can consist of 100 or 1000 or 100,000 such chunks. And we've also learned that the key tasks in building these systems are mining the necessary knowledge and molding it so it can be used in an actual running system.

WHAT IS KNOWLEDGE ENGINEERING?

These tasks have given rise to the field called knowledge engineering, which is concerned with the acquisition, representation and use of symbolic knowledge to solve problems that normally require human attention.[2] The products of knowledge engineering are called by many names: knowledge-based expert systems, knowledge-based systems, expert systems, knowledge systems. For the purposes of this talk they are all synonymous, and I'll use the most common term, expert systems.

WHAT ARE EXPERT SYSTEMS?

We can describe expert systems on several dimensions to help distinguish them from other kinds of computer programs. One dimension has to do with methodology. Expert systems are members of the family of AI programs, and as I mentioned earlier, they emphasize symbolic information and heuristic processing. A second dimension is quality: expert systems are expected to exhibit expert-level performance in their respective application domains. A third dimension is the design dimension: Expert systems should be easy to modify or augment, and their behavior should be understandable. Finally, on the implementation dimension, a key element of expert systems is that the knowledge base is kept separate from the inference procedures, as shown in Fig. 5. In thinking about designing an expert system, one should always keep in mind the two parts: a knowledge base, which you may think of as a global data base containing facts, rules, relations, etc., and an inference engine which interprets the knowledge and controls the problem solving procedure according to some strategy. This design is in sharp contrast with most computer programs, in which the knowledge of the domain is embedded directly into the program.

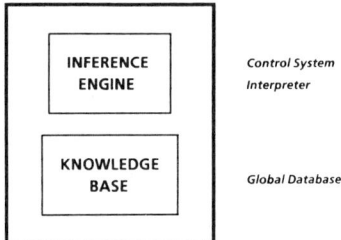

Figure 5. Separation of Knowledge and Inference Mechanism.

---

[2]The question naturally arises, what sort of problems that normally require human attention could or should a machine solve? Many of us like to think that all tasks can be automated up to but not including our own. As time goes on that threshold keeps moving up.

WHY SHOULD YOU CARE?

Two practical reasons for building expert systems is that they can reduce costs and/or increase the quality of many decision-making tasks. Expertise is always a scarce resource, and costly problems can arise when novices are left to solve problems on their own. Another well-known dictum is that crises and experts are almost never in the same place at the same time. Moreover, experts are not permanent: they change jobs or retire, eventually, and quality of performance drops. Expert systems can provide the expertise that is typically unavailable to less experienced personnel when or where it's needed. Moreover, experts should not have to spend valuable time on routine decision-making or bookkeeping tasks, and expert systems can be used to automate such tasks. During times of crisis, record keeping is often inaccurate or incomplete. Expert systems are able to keep consistently accurate and complete records of their actions.

EXAMPLES OF EXPERT SYSTEMS

MYCIN is a rule-based system for medical diagnosis and therapy. It was developed at Stanford about twelve years ago. It started as a Ph.D. thesis by Ted Shortliffe, who also has an M.D. degree and is now on the faculty at Stanford in the Department of Medicine. MYCIN is one of the landmark expert systems, not only because it solves an interesting problem, but also because it does it in a very clean way. By separating the knowledge from the program that uses the knowledge, MYCIN is easy to understand, replicate, and augment, and its knowledge base can be used for purposes other than diagnosis, e.g., teaching.

Why did he choose to develop this particular expert system? Shortliffe, of course, was interested in medicine. He found some doctors at Stanford who are world class experts in the diagnosis and treatment of various kinds of infectious diseases, and he focused his attention on the problem of diagnosing infecting organisms in blood and meningitis infections.

The idea behind MYCIN is that you would like to be able to make a diagnosis and prescribe an effective antibiotic treatment, using test results and information about the patient as supplied by the doctor. You would also like to be able to do it early in the treatment when you don't necessarily have all the laboratory data available. Moreover, non-expert physicians tend to overuse antibiotics, or use them irrationally, and MYCIN is intended to counteract such practice. Finally, MYCIN was intended to be an active medium for distributing expertise in the use of antibiotics, as well as the diagnosis of infectious diseases, to a broad community of practitioners who normally don't have access to experts as you might find at a university hospital.

Figures 6 and 7 show how MYCIN's diagnosis and therapy recommendations might look to a user. The diagnosis stage usually consists of identifying certain kinds of infections, and for each of these infections, identifying particular organisms that are responsible for them. It tends to be exhaustive, not just finding the first thing that might explain the data, but to find all of the possible explanations of the patient's symptoms.

The next phase, therapy, recommends a set of drugs that will cover for each of the five items in the diagnosis. In this case, MYCIN recommends some two antibiotics, prescribes the dosage, and includes comments for modification of dosage in case of certain complications such as renal failure.

## How MYCIN looks to the user: Diagnosis

Infection-1 is Cystitis
    &lt;Item 1&gt; Pseudomonas-Cepacia [Organism-5]
    &lt;Item 2&gt; Citrobacter-Diversus [Organism-4]
    &lt;Item 3&gt; E-coli [Organism-3]
Infection-2 is Upper-Respiratory-Infection
    &lt;Item 4&gt; Listeria [Organism-2]
Infection-3 is Bacteremia
    &lt;Item 5&gt; Enterococcus [Organism-1]

Figure 6.  Example of a MYCIN Diagnosis.


## How MYCIN looks to the user: Therapy recommendation

[REC-1] My preferred therapy recommendation is as follows:
In order to cover for items <1 2 3 4 5>:
Give the following in combination:
  1: Kanamycin
    Dose: 750 mg (7.5 mg/kg) q12h IM (or IV) for 28 days
  Comments: Modify dose in renal failure
  2: Penicillin
    Dose: 2,500,000 units (25000 units/kg) q4h IV for 28 days

Figure 7.  Example of a MYCIN Therapy Recommendation.


    The Knowledge base in MYCIN consists of about 400 diagnostic rules and about 50 therapy rules.  A typical diagnosis rule is shown in Fig. 8.  One thing to notice is that it's symbolic.  There are no numbers (with one exception) in it.  It has the typical "if-then" format.  The conditions have to do with the site of the culture and so forth.  So, this is symbolic data.  Moreover, the conclusion is also symbolic.  It has to do with the identity of an organism that would produce such data. The one number appearing in the rule is what is generally called a certainty factor.  It's similar to a probability measure, although not used exactly that way.  MYCIN has a calculus for dealing with these certainty factors so that when you chain together several rules, each of which may embody uncertain knowledge, in a line of reasoning, one can calculate an overall certainty of the conclusion.

**MYCIN DIAGNOSIS RULES**

**IF** The site of the culture is blood
The gram stain of the organism is gramneg
The morphology of the organism is rod, and
The patient is a compromised host,

**THEN** There is suggestive evidence (0.6) that the identity of the organism is Pseudomonas-Aeruginosa.

Figure 8. Example of a Diagnosis Rule in MYCIN.

How does MYCIN actually use these? The rule in Fig. 8 is a typical chunk of knowledge. You can see that the conclusion of this rule is a conclusion about the identity of an organism, and it's clear that this is one of the things one needs to know in order to be able to prescribe any drugs. So, one of the sub-goals of the program is to identify all of the organisms. In order for this rule to fire, each of the clauses in the IF portion must be true. The truth of the first clause, "if the site of the culture is blood," may be determined by simply asking the user, or by inferring that fact from other rules whose conclusion part says that the site of the culture is blood. To evaluate these rules, their IF portion must be evaluated, so the procedure is recursive. We call this procedure "backward chaining" because we start from a goal and work backwards by first finding the rules that conclude the goal, then generating sub-goals from the clauses on the condition side of the rule, and so forth.

Figure 9 shows a therapy selection rule. One can include justifications to the rules. These are not used by the program, but they help establish confidence in the program by indicating where these rules come from.

**MYCIN: Therapy Selection Rules**

**IF** You are considering giving chloramphenicol, and
The patient is less than 1 week old

**THEN** It is definite (1.0) that chloramphenicol is contraindicated for this patient.
[Justification: Newborn infants may develop vasomuscular collapse due to an immaturity of the liver and kidney functions resulting in decreased metabolism of chloramphenicol.]

Figure 9. Example of a Therapy Selection Rule in MYCIN.

One of the most powerful aspects of MYCIN is its ability to explain its line of reasoning, particularly for physicians who are understandably suspicious of using computers for such high level decision making. It is absolutely essential that the system provide not only an answer but also the ability to explain why it reached the conclusion it did. So MYCIN was developed to create confidence by letting the physician ask it several types of questions (see Fig. 10).

MYCIN frequently asks questions that seem a little baffling to the physician; they just don't seem to be relevant to the task at hand. So he or she might ask why that question was asked, and MYCIN would explain that it's trying to evaluate a particular rule. If the physician types "Why?" again, meaning "Why are you trying to evaluate that rule?", MYCIN replies that it's trying to evaluate that rule to establish the truth of a condition in a higher level rule. By asking why repeatedly you can work right back up the chain to what the ultimate goal of the consultation is.

After the program has made its recommendations, one may ask "How (did you arrive at that conclusion)?" and MYCIN will explain its line of reasoning in terms of the rules that were invoked. MYCIN can even handle a question of the form, "Why didn't you consider something else?" as shown in Fig. 11. The rules in MYCIN have a simple format, which makes it relatively easy to translate and display them in fairly standard English.

MYCIN also was significant in that it gave rise to a whole family of other systems, as shown in Fig. 12. EMYCIN is really not an expert system in itself, but is just the inference part of MYCIN, i.e., MYCIN without the medical knowledge. Because of that separation, one can add new knowledge in related domains or totally different domains and produce different expert systems. And that's what has happened over and over in the past few years. This family tree continues to grow and spread beyond what's shown in the figure. For example, Texas Instruments sells a product called "The Personal Consultant" which is a direct descendant (if not a twin sister) of EMYCIN.

## How does MYCIN create confidence in the user?

- Answering "Why?" (Why did you ask me that?)
- Answering "How?" (How did you arrive at that conclusion?)
- Answering "Why not X?" (Why did you not consider X?)

MYCIN's simple rule format and friendly explanations in "English" are the key.

Figure 10. Questions that can be asked of MYCIN.

## MYCIN Explanation

**User:** Why didn't you consider Streptococcus as a possibility for Organism-1?

**MYCIN:** The following rule could have been used to determine that the identity of Organism-1 was Streptococcus:
Rule 33.
But Clause 2 ("the morphology of the organism is Coccus") was already known to be false for Organism-1 so the rule was never tried.

Figure 11. Example of an Explanation in MYCIN.

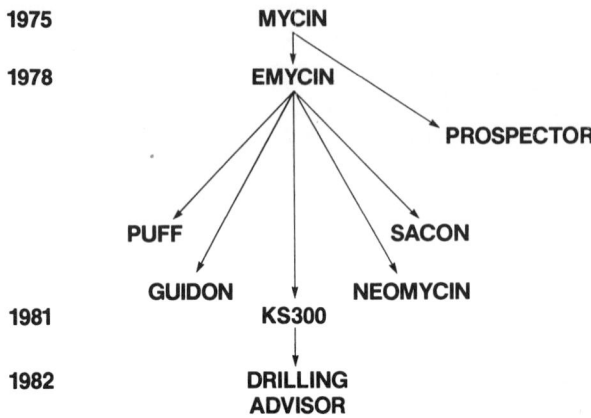

Figure 12. The MYCIN Family Tree.

I won't have time to talk about all of the systems shown in the figure, but I'll mention a few. PUFF (PULmonary Function diagnosis) analyzes data collected by a pulmonary function machine at Pacific Medical Center in San Francisco. It interprets the data and prints out a report, in standard format. About 80 percent of the time, the report says what the expert interpreter would have written himself, and he just signs his name at the bottom.

GUIDON is a system for teaching students about the medical knowledge that is contained in MYCIN. GUIDON also led to a reformulation of MYCIN, called NEOMYCIN, which made explicit the distinction between the domain knowledge (the facts, rules, relations, etc.) and the problem-solving strategy knowledge, and that became the basis of other intelligent tutoring systems.

Let me discuss one more expert system with which I was involved. This application took EMYCIN completely outside the area of medicine and into the domain of structural engineering. By adding knowledge about structural analysis, we created a small prototype system called SACON (Structural Analysis CONsultant). SACON advises engineers on the use of a large structural analysis program called MARC. The advice consists of a high level analysis strategy to use along with a list of specific recommendations. SACON was developed in a very short time, less than six months, and was basically a demonstration that this technology could work in other areas as well.

Figure 13 shows how SACON would look to the user after he or she answers questions about the type of structure (shape, material composition) and the type of analysis desired (material behavior, accuracy, etc.). SACON concludes its consultation with two kinds of recommendations. One is something called the "analysis class". In this case the recommended analysis class is "general inelastic", which was one out of about 36 different kinds of analysis classes that he could choose from. The second kind is a list of specific recommendations about certain input parameters that are provided for the program.

Since this was a concept demonstration, we didn't actually go as far as to generate the actual input data to the MARC program in the form of a tape or punch cards or whatever, but that could have been done straightforwardly. (I know of at least one other similar sort of expert system, a consultant for a seismic data analysis system, that does generate the input data.)

Figure 14 shows a typical rule in SACON, having to do with the type of material in the substructure, the non-dimensional stresses which are calculated by the system, loading cycles and so forth, and then concluding that fatigue must be considered as one of the stress behaviors.

The key ideas that came out of SACON were (a) complex computer systems require consultants to help non-expert users, and (2) an expert system can be an effective replacement for computer manuals and local "wizards." Perhaps this is an appropriate place to allay the fear that these expert systems are putting people out of work. In the case of SACON, there were a very small number of wizards who thoroughly understood the MARC program. They spent a lot of time on the phone talking to customers instead of doing the R&D work that was their primary responsibility. For them, an automated consultant was a way to save them time and let them focus on more productive tasks. This situation is typical.

### How SACON looks to the user:

The following analysis classes are relevant to the analysis of your structure: general-inelastic.

The following are specific analysis recommendations you should follow when performing the structure analysis:

1 Activate incremental stress—incremental strain analysis.
2 Model non-linear stress-strain relation of the material.
3 Solution will be based on mix of gradient and Newton methods.
4 User programs to scan peak stress at each step and evaluate fatigue integrity should be used.
5 User programs to scan stresses, smooth, and compare with allowable stresses (with appropriate safety factors) should be used.
6 User programs to scan deflections, calculate relative values, and compare with code limits, should be called upon.
7 Cumulative strain damage should be calculated.

Figure 13. Example of SACON Output.

## SACON RULE

**IF:** The material composing the substructure is one of metal,
The analysis error (in %) that is tolerable is between 5 and 30,
The non-dimensional stress of the substructure is greater than .9, and
The number of cycles the loading is to be applied is between 1000 and 10,000,

**THEN:** It is definite that fatigue is one of the stress behavior phenomena in the substructure.

Figure 14. Example SACON Rule.

OPPORTUNITIES FOR NDE

Based on analogy with other areas of science and engineering, I can see a number of different opportunities to use expert systems technology in NDE.

Data Interpretation

A typical problem in this category is, given the location and size of a defect, the stress to which the piece is to be subject, and perhaps some other physical and/or environmental data; is the defect serious? This is clearly an area where judgmental, experiential knowledge plays a major role.

Signal Characterization

Everyone knows that measuring devices can produce erroneous data. It is common, for example, to detect artifacts in signals, induced by the geometry of the object under test or by the environment. Signal peaks may be spurious, transducers can generate false signals, etc. Each source of data has a relative likelihood of credibility. Characterizing signals requires the application of a large number of heuristics.

Control of Data Collection

Intelligent control of how data are collected can often compensate for collecting large quantities of data which must be stored and later culled through at great cost in time. For example, one might use long wavelengths for routine scanning, switching to shorter wavelengths only where closer inspection is required because of a higher load in that area, or detection of an unusual signal, or because of historical data, etc.

Integrated Sources of Knowledge

Defects in a material part sometimes can be corrected on the manufacturing line by experienced line operators who seem to have an almost magic ability to twiddle the right knobs and correct the problem. For problems that are new or especially complex (e.g., when the "disease" is due to the interaction of multiple failures), the analytical skill of well trained engineers or scientists is required. Knowledge based systems can integrate multiple sources of knowledge and achieve better problem solving performance than any one source of expertise. This has been demonstrated many times, but a classic case is the expert system called PROSPECTOR, which combines the knowledge of several expert geologists, each of whom is a specialist in one or more models of ore deposits. In 1980, PROSPECTOR became the first expert system to achieve a major commercial success in predicting a large molybdenum deposit in Washington state.

Inspection Plan Designer

Given a new part to be fabricated, how should it be inspected? Where should you look for the most likely occurrence of defects? How do you totally cover the volume of interest, which involves selecting angles, the appropriate transducer (shear waves vs longitudinal waves, e.g.)? How do you quantify the size and type of a defect once it's located? The seriousness of a defect depends on the applied load--a smooth sphere inclusion is not as serious as a defect near a sharp edge, which can crack. One may need two or more transducers in a pitchcatch relationship, or one may need a different technique, e.g., x-rays instead of or in addition to ultrasonics. Clearly, there are numerous issues that must be considered, and experiential knowledge, rather than textbook procedures, can make the difference between an excellent and a mediocre inspection plan.

Intelligent Front End to Simulation Models

Quantitative simulations of the behavior of structures under various loading conditions provide critical information to NDE analysts. If your work is like many other areas of engineering analysis, you now have simulation packages that are at least a decade old, contain hundreds of options that must be specified before submitting a run. Moreover, the program itself contains tens of thousands of lines of code, the original authors have disappeared, no one understands why or how the program works in its entirety, and the documentation is both too cumbersome to read and understand, out of date, and doesn't tell you what you really want to know. Each organization has a small number of wizards to whom you go in order to learn how to set up a simulation that applies to your specific problem.

This is an excellent area for using expert systems to help nonexperts, and there have been a number of successful demonstrations of intelligent front ends to complex software packages. I've had direct experience with such systems in the areas of structural analysis and seismic data processing. The IFE can conduct a dialogue with the user, querying him or her about the specific application in the familiar terms of the domain. The conclusions drawn by the IFE are then translated into data processing terms -- what values go in which fields of which records. The analyst no longer need be a data processing specialist.

Expert System - Simulation Hybrid

Carrying the idea of combining a reasoning program with a numerical simulation one step further, imagine a closed-loop system where an expert system generates a hypothesis about the type and location of a defect, sets up and runs a simulation program to explore the consequences of that hypothesis, and finally interprets and uses the results of the simulation to generate alternate hypotheses.

These are just speculations. I'm certainly a novice in your field and I'm just going on the basis of experience of similar applications in other fields.

SELECTING AN APPROPRIATE APPLICATION

Here are some guidelines in selecting an appropriate application for applying this technology:

1. The solution to the problem primarily involves symbolic, not numerical reasoning. If there's a way to do it algorithmically using a numerical method, you should do it that way. Don't rely on expert systems.

2. Since we are talking about expert systems, there has to be an expert. I think that's been overlooked, that somehow, these expert systems will magically produce expert performance. They are based on knowledge obtained from experts.

3. It has to be a problem where somebody cares about the solution. The eventual fielded system has to either save people time or money or increase productivity.

4. The problem needs to be bounded in scope or you'll never finish. We have some of our own heuristics about that. If a human expert practitioner can solve the problem in a few minutes, it's too easy. If it takes the expert a week or more it's likely to be too difficult. So a problem that takes an expert a few hours to a few days to solve is appropriately bounded.

5. There should be a vocabulary of the field which is not unbounded. There should be no more than a few hundred terms that would completely describe all the objects and relationships that exist in the area.

6. The experts should agree with one another on the solutions to the problems. When the experts disagree among themselves, there will be a major problem in establishing the credibility of the system you develop.

7. Numerous test cases should be available. Experts generally cannot tell you what they know directly, but can impart that knowledge indirectly by solving lots of problems. Thus, at present, expert systems are developed by analyzing cases.

8. Combinatorial problems are often likely candidates. That's one area where it's difficult for humans to perform very well. We are generally poor enumerators when there are a large number of possible solutions to a problem, whereas machines are very methodical about that.

9. Finally, you ought to be able to show progress in an incremental way so that your sponsor doesn't have to wait for two or three years before anything comes out the other end of the tunnel.

CRITICISMS

Critics of expert systems have frequently pointed out that the knowledge in them is shallow, that they don't exhibit any common sense, and that some problems that would be obvious to you can't be solved by the expert system because it only knows what's in its knowledge base; if it requires using some piece of knowledge that's not in it, it fails. And finally, it doesn't improve with experience. You run the same system day after day after day and it solves the same problems but it doesn't get any smarter.

These are all the areas that we are aware of and they are all areas of research. Rather than say that we have been overselling the systems, I would say that they are not oversold; they are just merely underdeveloped.

LOOKING AHEAD

I'd like to conclude with some personal remarks on where I believe the field is heading over the next 5 to 10 years.

Knowledge Acquisition

Knowledge acquisition is a well-known bottleneck in the development of practical expert systems that require very large knowledge bases. There are two major problems here. One is the time-consuming process of entering thousands of chunks of knowledge into a knowledge base. At present we do this more or less manually, one chunk at a time. Moreover, we have to run the evolving system through hundreds to thousands of cases in order to verify the knowledge acquired thus far and to see what's missing. The second problem is that knowledge bases always have a structure to them; that is, the objects in the knowledge base have particular interrelationships that are important to solving the problem, or at least to solving the problem efficiently. Eliciting that structure is a design activity that requires currently creativity and a lot of handcrafting. That's why we now have to rely on experienced (and therefore scarce) knowledge engineers to build knowledge systems. Thus every new system is built from scratch, and there's no economy of scale.

I think the research now going on at Stanford and other places will eventually widen, if not totally break, this bottleneck. Meta-DENDRAL, now ten years old, demonstrated that one can acquire new knowledge from examples. More recently, one of our Ph.D. students developed a program, called RL, that uses a rough mode, or half-order theory, of the domain in order to guide a systematic search through a space of plausible concept definitions and associations. Preliminary results show because of the rules and metarules learned. Another mode of learning, learning by analogy, is a promising area for constructing KBs. To be successful here, we have to learn how to find the appropriate analogies and to use the analogies correctly. I wouldn't hold my breath waiting for breakthroughs in this area, but I'd keep my eye on the work of Doug Lenat and his colleagues at MCC. I refer you to his recent article on this topic in the Winter, 1986 issue of AI Magazine. Lenat is embarked on a ten-year project to build an encyclopedic knowledge base, structured so that one can search efficiently for analogies and exploit them when direct knowledge of a particular domain is incomplete.

Multiprocessor Architectures

I believe we will see an increasing amount of activity in both the hardware and the software areas, aimed at speeding up the execution of large expert systems. On the hardware side, it's clear that we continue to move in the direction of higher performance and lower cost uniprocessors, and it's also pretty clear that we will see more and more examples of multiprocessor machines with multi-instruction, multidatapath architectures on the market over the next decade. The real challenge is discovering how to program these beasts to fully realize their potential for parallel symbolic computation. My own feeling is that we're going to have to take the low road on this for a while, looking at specific applications, or classes of applications, finding the parallel problem solving tasks that are inherent in that domain, appropriate language in which to implement the system and then a specification for a machine architecture that can support the whole activity -- in other words, the multi-system level approach we're taking on our own advanced architectures project. Only after a few successes will we be able to step back and see what generic progress we've made. I don't see any breakthroughs here for at least another few years, but I think we can see an order of magnitude speedup very soon simply from using several tens of processors and reasonably intelligent techniques for knowledge and data base management. In five years I expect to see two orders of magnitude speed-up, and in ten years at least three (but that's based only on faith that a few geniuses will emerge to crack this problem).

Workstations and Distributed, Interacting Systems

Powerful workstations for symbolic computation are rapidly getting down to a price where we can start thinking about them like we now think about terminals, in other words an ordinary piece of office equipment that you might put on every worker's desk. When this happens, I think we'll see a quantum jump in the level of work performed by executives, managers, administrators and secretarial personnel. High-resolution, multiwindow workstations will be as commonplace as the telephone (and in fact the telephone may be just one component of it). Having N times as much processing power as we now have in our centralized time-shared systems will have a payoff not so much in the quantity of computer use as in the quality. The key will be in the sophistication of the user interface that will be permitted by the enhanced processing power of the workstation. Even we executives will be able to understand how to retrieve a vital piece of information that we can't describe exactly, or get help with a planning or budgeting task, or make marginal notes on an on-line document, without having to ask someone for help or read a manual, because our workstation can represent and utilize a model of what we know and don't know, and draw inferences from our often ambiguous commands. After all, the Macintosh can trace its roots to some AI ideas in the early 70s. More recent work on Intelligent Agent, Knowledge-based programming and in language understanding is likely to lead to these simple-to-use, intelligent job aids --"White Collar Robotics" as Business Week described the field in its February 10th issue.

REFERENCE

1.  B. G. Buchanan and E. H. Shortliffe, Rule Based Expert Systems, (Addison-Wesley, Reading, MA, 1984).

DISCUSSION

Bill Karp, Westinghouse: I would like to make a comment and pose a question. At Westinghouse, we have been in the business of artificial intelligence as it impacts the field of NDE. The comments that you made with regard to paying attention to just exactly how you define and identify your expert application in fact is absolutely critical and very often is not a specific individual.

Just to give you a feel for where I'm coming from, we developed some automated systems for the interpretation of eddy current data. Our experts were, in fact, our scientists and engineers that put the concepts together. To get this thing really to work, we found out near the end of the program that if we had taken advantage of the experts in the field, specifically in the field of the inspectors, right up front in the planning of the program, things would have gone much more smoothly. So, identification of expert or the experts is absolutely critical.

R. Engelmore: Yes, it really is and they have to be identified early. Also, it requires quite a bit of time for the expert to participate in such a project and that can be a real problem. Experts are, by their nature, scarce. They are always needed elsewhere.

We found that you usually can't convince an expert's boss to release him for a task like this for much time, so you have to jump over his boss and go to his boss's boss who has a little broader or longer-term view of the goals of the organization and is willing to tell the expert's boss to make this person available.

Bob Green, Johns Hopkins: Since even experts or wizards make mistakes, what consideration is given to false positives or redundancy in these programs? And can we expect malpractice suits against computers?

R. Engelmore: It's a good question that comes up all the time. Yes, these expert systems are going to make mistakes because they are based on the knowledge of humans, who make mistakes. One should not necessarily expect any better performance than you are going to get from humans.

I'd like to think, at least for medical system applications and perhaps some other critical ones, that these expert systems will be regarded as providers of second opinions. It will still be the responsibility of the primary physician or the primary user of that system to make the final decision. I think that's the only way it's going to make sense for a while.

The other side of the coin is (again, with analogies to medicine), when these kinds of systems which are shown to perform very well are available out in the field, will there be malpractice suits for people who don't use them?

Dick Berry, Lockheed: One of the things that we are seeing more and more is that the military is interested in getting automatic accept-reject systems built-in for many of the routine inspections that are carried out in the military in ammunition and a variety of other things. Has AI made any inroads in helping to solve this problem?

R. Engelmore: I'm not aware of any inspection systems, using AI, that are actually in use today, but there may very well be some prototype efforts going on now, and there may be people in this room who could answer that question.

Ward Rummel, Martin Marietta: We have one at Martin.

Steve Huber, NADC: When you mentioned about the "no improvement" experience, it comes to mind immediately: Why isn't the feedback built in? For example, if a doctor makes a diagnosis and the patient dies, he tells the system, "The patient died," and therefore the system would refine its statistics. Has that been done? And if it has been done, why has it failed so far?

R. Engelmore: There actually have been some attempts to do just that. There's a program at Stanford called "RX", which looks over a large medical data base and tries to make some interesting statistical correlations that people have not made before, thereby learning new knowledge from past experience. So that's one attempt. But there haven't been very many such efforts to do that.

Now the whole idea of machine learning is a big research topic, getting programs to actually develop knowledge bases, either learning from examples or learning from textbooks, or learning by sort of looking over the shoulder of the practicing expert as he solves the problem, and keeping records of what he or she does. That's a research area now, and our lab for one is actively involved in it.

# UK DEVELOPMENTS IN THEORETICAL MODELING FOR NDT

Andrew Temple

Theoretical Physics Division, Harwell Laboratory
Didcot, Oxon OX11 ORA, United Kingdom

INTRODUCTION

Non-destructive inspection is widely used to ensure that engineering structures such as railway rails, bridges, nuclear reactor pressure vessels, offshore oil platforms, airplane airframes and so on contain no unacceptably large defects. Such defects, if they were present in the structures, could cause failure under certain applied loads. Generally, the most serious defects are cracks which occur during manufacture, either in castings or in welds, or during service due to cyclic loads and environmental attack. The non-destructive inspections are carefully designed to be capable of detecting these crack-like defects. I am concerned here only with ultrasonic inspection techniques and consider the recent developments which have taken place in modeling them. Some of the modeling work has been driven by requirements of inspection of pressurized water reactors, some by inspection requirements foreseen for fast reactors. All the developments have application to some aspects of inspection of all reactor types: fast reactors, pressurized water reactors and gas cooled reactors; and indeed to the inspection of many other engineered structures too.

If the response from the defect can be calculated then it is possible to examine the inspection strategies and procedures and demonstrate that they are adequate for the purpose for which they are intended. Thus the technique, either proposed or already in use, can be validated with a modeling approach. This will be particularly useful once the models themselves have been tested against experiments and shown to give satisfactory agreement.

I concentrate on those areas which are most recent. The consolidating work done on modeling of pulse-echo and Tandem inspections is not included since this is reviewed elsewhere [1]. Similarly, work done on Time-of-Flight Diffraction [2] is not included. I emphasize the physical features of defects such as: defect roughness; material ingress; the effects of compressive stress; the effects of crack size; and the implications of variations in the reliability of ultrasonic testing due to these effects for structural integrity.

Models designed to understand some particular physics are usually concerned with the direct scattering problem and this is also true of models used for validation of inspection procedures. The direct scattering problem can be expressed in the form: 'given this particular incident wave and this specific scatterer what is the scattered wave at some particular point or points?'. Another type of model is concerned with data interpretation in which solutions to the inverse scattering problem are required. The inverse scattering problem is: 'given this output at a known position of the receiver, and what we believe to be the input from the transmitter, what is the scattering object or objects responsible for the transformation?'. The direct scattering problem for elastic waves can be very hard for certain ranges of parameters, usually those of most relevance from a practical point of view, but the inverse problem is rigorously more difficult. Needless to say, most work has been carried out to date on the direct scattering problem. Early attempts at carrying out inversion of data to obtain information about the scatterer have made use of the Born approximation [3,4] and recent progress in the UK is being reported [5].

Results of the modeling are used to guide experiments, aid in design of scanners, optimize inspection techniques, and extrapolate experimental results. A review of the modeling techniques most widely used is given in [6].

INSPECTION TECHNIQUES

The diffraction coefficients for high frequency ultrasonic waves incident on a slipping crack, i.e., one modeled by boundary conditions of continuity of normal stress and displacements, have been calculated recently and compared with those for a stress-free crack [7]. The results are for backscatter and demonstrate that the signal diffracted from a lubricated crack is lower than that from a stress-free crack, typically by between about 6 dB and 10 dB. For compression waves the backscattered signal is always lower for the slipping contact crack compared with that from a stress-free crack whereas for a shear SV-wave (relative to the crack) there are some angles for which the signal is larger from the slipping crack.

FINITE DIFFERENCE RESULTS

Finite difference results have the capability of providing answers to questions about scattering which are not convolved with transducer responses. I have chosen the effects of crack size and ultrasonic frequency to illustrate developments with this approach to obtaining the scattered signal. Figure 1a, taken from [8] shows experimental measurements of the conversion factors of compression waves to Rayleigh waves for saw cuts of various depths and for grazing incidence. The amplitude is a strong function of both frequency and crack size. Figure 1b shows similar results obtained with finite difference calculations.

In Fig. 2, results of finite difference calculations of shear wave reflection from the mouth of a surface breaking crack are presented as a function of the ratio of crack depth to wavelength [9]. Various angles of incidence and observation are represented on Fig. 2. The angles are all measured from the normal to the plate surface and the receiver is on the same side of the crack as the transmitter. The results in Figs. 1a and 1b are for slots of width, typically 4 mesh spacings or greater (~0.25 mm at 5 MHz) whereas those in Fig. 2 are for a slot width much less than one mesh spacing (~0.03 mm at 5 MHz) achieved by using a novel difference scheme around the crack. Again the non-monotonic nature of the reflection coefficient is very striking.

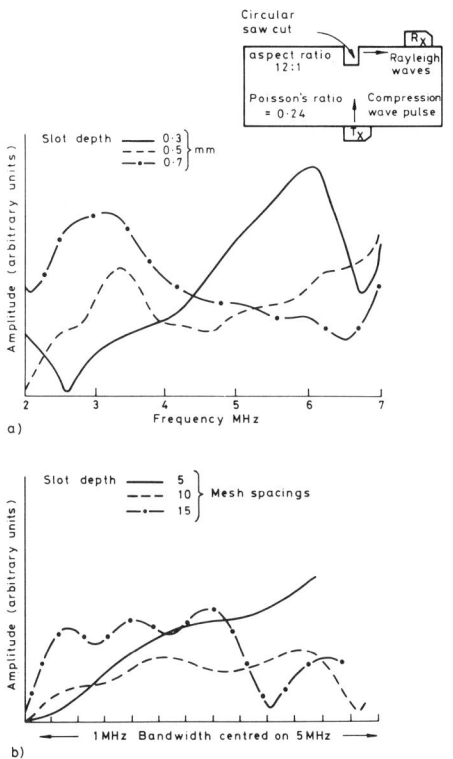

Figure 1. a) Experimental results; b) Finite difference calculations. (From Bond and Saffari 1983).

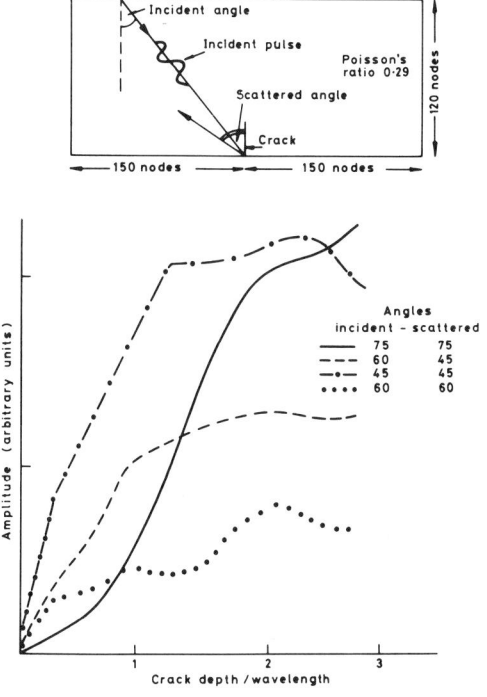

Figure 2. Shear wave amplitudes for a 2 cycle sine-wave input pulse. (After Harker, 1984).

## INTERGRANULAR ATTACK

Another interesting development is a model of the scattering from regions of degraded material such as inter-granular attack. A model of this has been developed [10]. This, together with recent experimental work [11] suggests that ultrasonic inspection would be a viable alternative to eddy currents for the inspection of steam generator tubing in all reactor types. Ultrasound has been used successfully to inspect the ferritic heat exchanger tubes of the SNR fast breeder reactor at Kalkar, in the Federal Republic of Germany, and the same technique is also being advanced for steam generators of pressurized water reactors [12].

The material considered in the model is nickel with a region of degradation comprising disbonding of the grains. This is represented by two parameters $\varepsilon$ and $\bar{D}$ with $\varepsilon$ a wholly metallurgical parameter dependent on the shape of the grains and $\bar{D}$ a parameter depending on metallurgical features such as heat treatment, alloy content, grain boundary orientation and on the chemical nature of the corroding environment. The parameter $\bar{D}$ has dimensions of length and is a measure of the effectiveness of the above collection of features in promoting intergranular attack. Based on results of O'Connel and Budiansky [13] expressions can be obtained for the speed of propagation of compression or shear waves in such degraded material. The time taken for a short pulse to travel across from one tube-wall to the other and back again will be different in corroded material compared with sound material. This time difference is a function of the depth of attack and is typically a few tens of nanoseconds. A comparison of experimentally observed values in simulated inter-granular attack [11], and theoretical predictions [14] with plausible values of the two parameters $\varepsilon$ and $\bar{D}$ are given in Fig. 3 for compression waves. More experimental work is required in order to fix the parameters $\varepsilon$ and $\bar{D}$ but even so the model is clearly capable of qualitative and quantitative prediction.

## THE EFFECTS OF COMPRESSIVE STRESS ON ULTRASONIC SIGNAL AMPLITUDES FROM CRACKS

There has been concern that real fatigue cracks under compressive stress may be transparent to ultrasound. This would reduce the ultrasonic signals and hence impair the reliability of detection correct sizing of such defects. A model, originally due to Haines [15] has been extended by Temple [16] to investigate this problem. The model of Haines was for normal incidence and so did not involve mode conversion whereas the extended model is for arbitrary angles of incidence and takes account of mode conversions. The behavior of a fatigue crack under compressive stress with varying parameters is shown in Fig. 4. Whapham et al [17] measured the amplitude of Time-of-Flight Diffraction signals from fatigue cracks under compressive stress with results shown in Fig. 5 confirming the validity of the model. The results shown are for ultrasonic center frequencies of 6 MHz and for a defect with a root mean square roughness of 1.5$\mu$ in a material representing steel with a flow pressure of 1200 MPa.

In all cases the overall decrease in ultrasonic signal amplitude tends to saturate at high values of applied compressive stress. The theory and experiment agree very well on this value. We see, for example, that the maximum decrease in signal of Time-of-Flight with compression waves, is likely to be in the region of 10 to 12 dB compared with the signals from a fully open crack. This applies to frequencies round about 5 MHz. As the frequency is lowered for a given defect, then

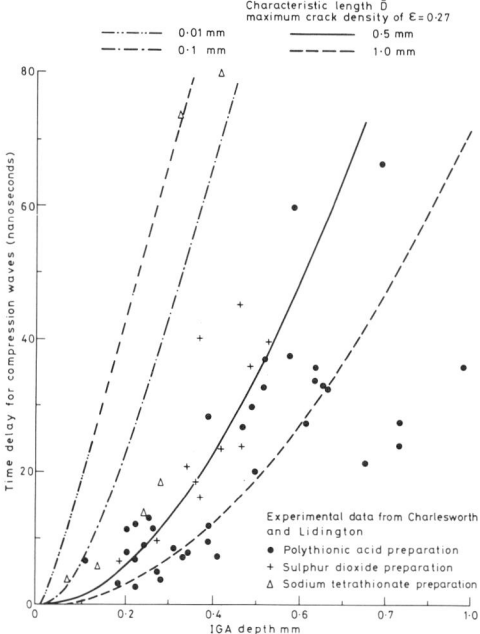

Figure 3.  Theoretical predictions for the time delay of compression waves in material containing IGA.

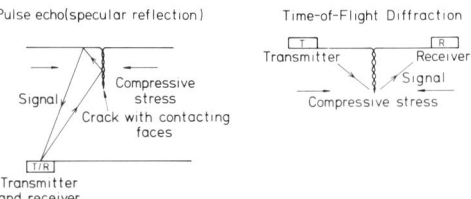

Figure 4a.  Geometry of pulse-echo technique employing specular reflection and Time-of-Flight Diffraction. The contacting faces of tight crack under compressive stress will reduce the signal amplitudes compared with those from an open crack.

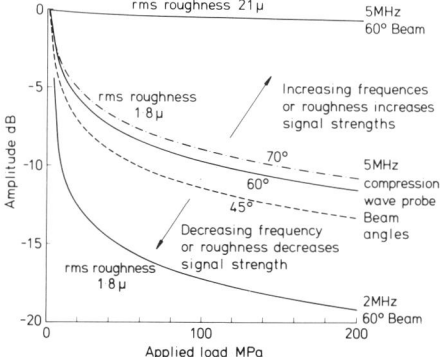

Figure 4b.  The effects of compressive stress and crack roughness on the amplitude of reflected pulse-echo or diffracted compression wave signals.

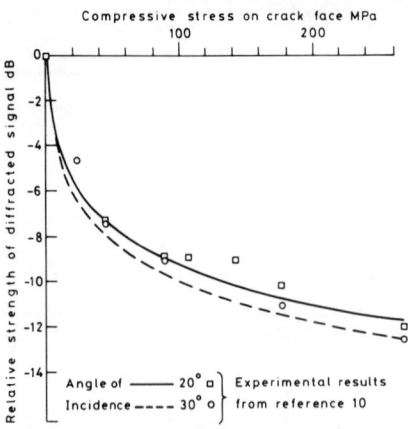

Figure 5. Comparison of results predicted by Temple 1984 with experimental mesurements by Whapham 1985. The theoretical curves are for a surface rms roughness of 1·5 µ, the flow pressure of the material is 1200 MPa, the frequency of the ultrasonic compression waves is 6 MHz.

the signal reflected or diffracted decreases. For shear SV-waves the signal loss is greater and saturates typically around -20 dB. Whether or not this will seriously hinder the successful detection and accurate sizing of cracks under compressive stress will depend on the signal to noise level in the material.

Thompson and Fiedler [18] considered a model in which the usual stress free boundary condition on the faces is replaced by a coupling between the crack faces created by a massless spring. By allowing an exponential rate of change of spring constant with distance from the tip, Thompson and Fiedler achieved good agreement with experiments on ultrasonic reflection and transmission at normal incidence on fatigue cracks. Kirchhoff theory for the scattering was used and predicted the observed variation in crack length with frequency. Achenbach and Norris [19] considered both normal and shear tractions across the interface. By allowing the conditions of traction continuity and displacement discontinuity to vary between opening and closing stresses nonlinear equations are obtained as boundary conditions.

Finite difference solutions to the scattering of elastic waves in two dimensions by a partially closed crack have been presented [20]. The most striking difference between the ultrasonic signals from open and partially closed cracks is that those from the crack with contacting faces are much more complex, being a superposition of many diffractions. The relative amplitudes of the signals depend in detail on the many parameters representing the contacting faces.

Ingress of a material such as liquid sodium can affect the ultrasonic signal in much the same way as compressive stress: reducing it below the level expected from an open, unfilled crack. Work on this has been part of the work on inspection of fast reactors in the UK for some years. The theoretical prediction was that cracks or width less than 10 µ could be difficult to detect if filled with liquid sodium. Experimental tests of this prediction with glycerine as a substitute for sodium confirmed the theoretical predictions provided roughness on the faces of the crack and the frequency filtering effect of the transducer shoe were taken into account [21,6].

# THE EFFECTS OF DEFECT ROUGHNESS ON ULTRASONIC AMPLITUDES

In Fig. 6 a plane compression wave is incident on a rectangular, rough defect, at angle θ, to the normal. The surface of the defect is rough in only one direction in this example and is described by a root mean square deviation from flatness of 0.12 mm, with a root mean square gradiant of 0.18. The defect itself is 30 mm wide and 10 mm high with a thickness, between mean planes, of 0.3 mm. The roughness is assumed to exhibit a Gaussian distribution of heights. Results for the far field scattered displacement field as a function of polar angle $\theta_2$ and azimuthal angle $\theta_3$ are shown in Fig. 7, using elastic Kirchhoff theory [22]. The results are for the amplitude of compression waves. Compared with a reflection from the same size smooth defect, the scattered peak amplitude is about 30 dB lower which is similar to that observed [23].

Ogilvy [24] has carried out a variational calculation for acoustic scattering from a rough defect. By substituting the Kirchhoff approximation into the variational principle, as a trial field, estimates are obtained of whether Kirchhoff over - or under-estimates the scattered field. The variational and Kirchhoff amplitudes agree for both backscattered and specularly reflected amplitudes from smooth cracks but for rough cracks the Kirchhoff approximation gives a larger signal amplitude than the variational approach in the specular direction. For smooth cracks the elastic Kirchhoff approximation always underestimates the backscattered signal compared with the high frequency asymptotic solution [25]. Ogilvy's variational principle therefore gives a lower bound to the scattered amplitude. Derivation of a complementary variational principle would give an upper bound. The ratio between the amplitude calculated using the variational principle with the Kirchhoff approximation as a trial field and that obtained using just the Kirchhoff approximation itself is given in Fig. 8 as a function of the two parameters describing the roughness of the defect. The two parameters used are the root mean square deviation of the defect surface from a plane and the correlation length along the surface of the defect. In this case, the defect is not corrugated but has the same roughness in two orthogonal directions in its mean plane (isotropic roughness) and hence only a single correlation length. The results presented in Fig. 8 are for an observation point lying in the plane containing the incident wavevector.

The elastic Kirchhoff calculations for scattering from rough defects have been used [26] to explain some of the differences in the correct detection rate of defects in the PISC II exercise [27]. Roughness of defects raises the level of diffuse scatter to levels similar to, or in excess of the backscattered amplitude at non-specular angles of incidence thereby enhancing defect detectability for a given defect through-wall size. If the defect detectability averaged over teams is plotted against the through-wall size of the cracks in the PISC-II exercise, then experimentally it seems that a rough defect with a through-wall size in the range 5 to 10 mm will be correctly detected by about 50% of the teams using ASME inspection procedures with a sensitivity of 20% DAC (distance amplitude correction) [28]. However, for a smooth crack, the defect through-wall size giving 50% chance of correct detection by the same inspection procedure was found to be in the range 15 to 20 mm, that is a shift in through-wall size of about 10 mm. Ogilvy's Kirchhoff calculations suggest that roughness on the faces of the defect can account for a shift of about 7 mm in the through-wall size of defects which the ASME procedures will detect 50% of the time. These calculations are for the roughness of the defect faces and do not include any effects due to non-sharp defect edges.

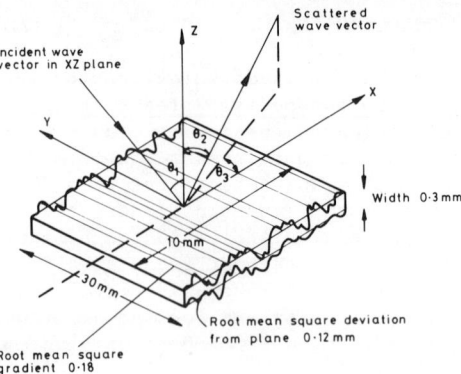

Figure 6. Geometry of calculation of the scattering from a rectangular rough defect.

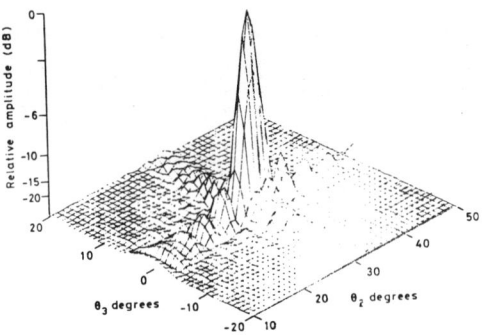

Figure 7. Scattered compression wave amplitude from an incident 10 MHz compression wave at 30° to the normal. The material has a Poisson's ratio of 0·29. The rectangular defect is characterized by root mean square deviations from smoothness of 0·12 mm and a width of 0·3 mm. The root mean square gradient of the surface is 0·18. (After Ogilvy, 1985)

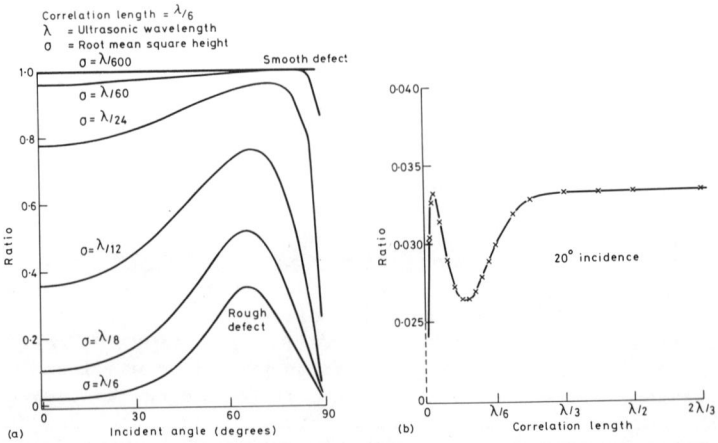

Figure 8. Ratio of amplitude calculated using the variational principle with the Kirchhoff approximation as a trial field to the amplitude calculated from Kirchhoff alone. The scattered amplitudes are in the specular direction

## ANISOTROPIC MEDIA AND CLADDING EFFECTS

The cladding layer of a pressurized water reactor and variations in its thickness can lead to sizable variations in measured signal amplitudes if no corrections are made for it [23]. Corrections for the cladding layer were carried out in setting up the Harwell Time-of-Flight inspection equipment used in the Defect Detection Trials and PISC II [29] based on phase and group velocities calculated for elastic wave propagation through anisotropic, yet homogeneous material. Experimental verification of the necessary corrections for the cladding layer has been carried out [30]. As well as calculating phase and group velocities for single crystal pure materials, the model allows the reflection coefficient for plane waves incident on the boundary between two anisotropic materials to be calculated. The two materials are assumed to be bonded together with continuity of normal and tangential stresses and displacements across the interface. The materials can be of different symmetry and can have any relative orientation.

The cladding layer on a pressurized water is one example where anisotropic media are important and another is in the various designs of liquid metal cooled fast reactors. A ray tracing model has been developed and applied to pulse-echo inspections [31] and to inspection of welds in austenitic steel [32]. Figure 9 shows some results for the way in which the ultrasonic energy passes through a typical austenitic V-weld as a transducer scans across the top of the weld. For comparison the material outside the weld region is taken to be isotropic, and so does not deviate the path of the ultrasound. The weld is modeled as long grains of transversely isotropic material with a long grain axis which follows the gradients set up in the cooling weld and is, therefore, a curved path. The results show the difficulty which may be experienced in practice in interrogating certain regions of the weld in the search for defects and demonstrates the relative effectiveness of the different wave modes. SH-waves are the most efficient, but are harder to generate and detect, while SV-waves are the most unsatisfactory. Compression waves are almost as good as SH-waves, especially at angles around 45° to the surface normal.

## THE RELIABILITY REQUIRED OF INSPECTION TO ENSURE STRUCTURAL INTEGRITY

The ultimate aim of inspection is to ensure structural integrity. Modeling can also be used to specify the level of reliability required of an ultrasonic inspection. Probabilistic fracture mechanics models have been used to prescribe what the effects of different levels of inspection reliability will have on the integrity of, say, a pressure vessel of a pressurized water reactor (PWR). In this approach, the failure of the vessel is governed by those defects from the initial defect population which were not successfully detected and repaired following manufacture, and which are either of sufficient size to cause failure or which grow to such a size. A function $B(a)$ is taken to represent the chance that cracks of through-wall size a will remain in the vessel. A schematic diagram of this function is shown in Fig. 10. Initially, for small acceptable defects, $B(a)$ is unity; then there comes a region of decreasing likelihood that defects will remain in the vessel - this is due in part to the decreasing likelihood that big defects would be created in the first place and, second, to the increased likelihood of successful detection, correct classification and repair; finally, there is an asymptote to which $B(a)$ tends for large defects. The asymptote represents factors beyond the capability of the non-destructive testing technique to detect and size defects accurately. An example of such a factor would be gross human error

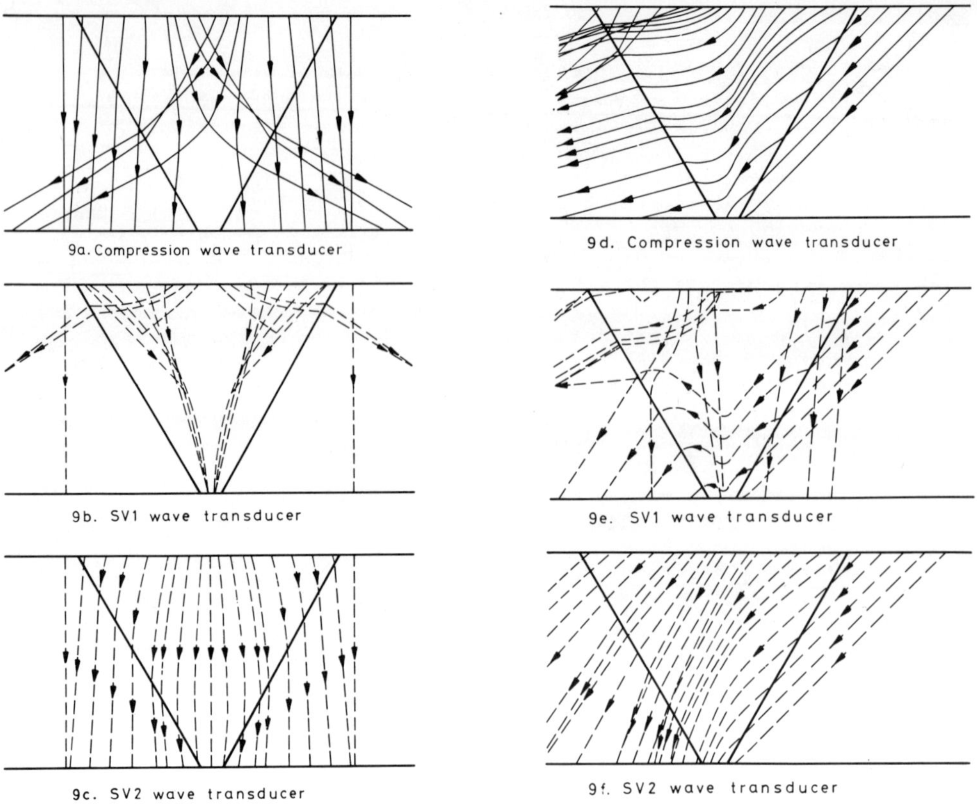

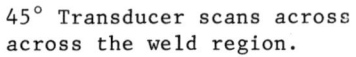

| 9a. Compression wave transducer | 9d. Compression wave transducer |
| 9b. SV1 wave transducer | 9e. SV1 wave transducer |
| 9c. SV2 wave transducer | 9f. SV2 wave transducer |

Figure 9.  0° Transducer scans across the weld region.    45° Transducer scans across across the weld region.

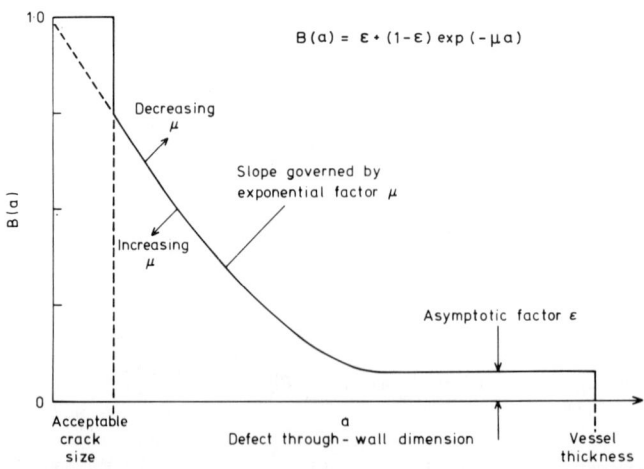

Figure 10.  Schematic view of the various regions of the function $B(a)$.

such as omitting an inspection altogether. It is expected that this asymptote will have a low likelihood of occurrence such as between $10^{-3}$ and $10^{-4}$. Defects so large that the vessel leaks or fractures into several parts will not go unnoticed so $B(a)$ becomes zero at this point.

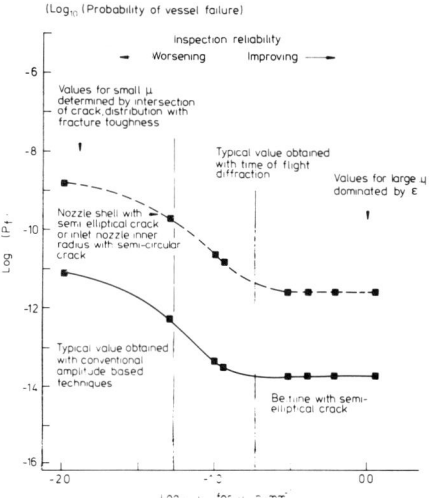

Figure 11. Implications of inspection reliability for structural integrity.

Figure 11 shows the results of some calculations of predicted failure rates per reactor year of a PWR pressure vessel as a function of the quantity μ in B(a), (see Fig. 10). This parameter represents in essence the capability of the ultrasonic inspection to distinguish between critical and non-critical defects. Large values of μ indicates that the inspection reliability quickly reaches the asymptote on Fig. 10 and hence the technique distinguishes easily between acceptable or rejectable defects. Figure 11 shows the predicted variations in vessel failure rate for a fixed asymptote of $10^{-3}$ and as a function of μ [33].

CONCLUSION

Modeling has played a useful role in developing the capability of ultrasonic inspection, in understanding some of the difficulties posed by physical characteristics of real defects and components, and in demonstrating how inspection techniques can be applied most effectively. More work remains, the properties of defects considered here have not all been explored in the detail necessary nor have all properties which affect the ultrasonic response been studied. More models of the physics of interaction of ultrasound with defects are required and eventually the aim should be to incorporate these interaction models into inversion algorithms, expert systems for guiding ultrasonic inespectors, or into models of complete inspections (including perhaps the human being) so that inspection reliability can be validated.

ACKNOWLEDGEMENTS

I am grateful to L. J. Bond, N. Saffari, R. F. Cameron, A. H. Harker, J. A. Ogilvy, J. P. Charlesworth and B. H. Lidington, for permission to make use of their published results. All the figures are original and are based on figures in Harwell reports for which the United Kingdom Atomic Energy Authority holds the copyright. I am grateful to the UKAEA for permission to publish these. All the models developed at Harwell make use of the Harwell Subroutine Library [34].

# REFERENCES

1. R. K. Chapman, J. M. Coffey, "A theoretical model of ultrasonic examination of smooth flat cracks", Review of Progress in Quantitative Nondestructive Evaluation 3A, ed. by D. O. Thompson and D. E. Chimenti, (Plenum Press, New York, 1984), 151-162.
2. J. A. G. Temple, "Time-of-Flight Inspection: Theory" OECD/IAEA Specialist Meeting on Defect Detection and Sizing, Ispra, Italy, and Nuclear Energy 22(5) 335-348, (1983).
3. J. H. Rose, "Inverse scattering at long wavelengths", Proceedings of DARPA/AFML Review of Quantitative NDE, (Boulder, 1981), (Plenum Press, 1981).
4. J. H. Rose, J. L. Opsal, "Inversion of ultrasonic scattering data" Proceedings of DARPA/AFML Review of Quantitative NDE, (Boulder, 1981), (Plenum Press, 1981).
5. C. A. Chaloner, L. J. bond, "Ultrasonic inversion: a direct and an indirect method" Review of Quantitative NDE, this volume, (1986).
6. J. A. G. Temple, "Developments in theoretical modeling for ultrasonic NDT" post SMiRT Seminar, Varese, Italy, August 1985, and to be published as a chapter in Non-Destructive Examination in Relation to Structural Integrity ed. by R. W. Nichols, G. Dau, S. Crutzen, Elsevier Applied Science, (1985).
7. R. K. Chapman, "Ultrasonic scattering from smooth flat cracks: exact solutions for a liquid-filled semi-infinite crack" CEGB report NWR/SSD/86/0020/R (1986).
8. L. J. Bond, N. Saffari, "Crack characterization in turbine disks" Review of Progress in Nondestructive Evaluation 3A, ed. by D. O. Thompson and D. E. Chimenti, (Plenum Press, New York, 1983), 251-262.
9. A. H. Harker, "Numerical modeling of the scattering of elastic waves in plates", Journal of Nondestructive Evaluation 4(2) 89-106, 1984).
10. J. A. G. Temple, "Calculations of the feasibility of ultrasonic inspection of steam generator tubing in presurized water reactors" report AERE-R.10980 H.M.S.O. London (1983).
11. J. P. Charlesworth, B. H. Lidington, "The use of ultrasonic techniques for the detection of intergranular attack" NDT 85, ed. P. J. Emerson, G. Oates, (Emas, Warley, West Midlands, UK, 1985).
12. M. Lodder, C. Broere, J. Reinhoudt, "Ultrasound and tube integrity of steam generators" Nuclear Europe Journal of the European Nuclear Society, 6, 18-19, (1986).
13. R. O'Connell, B. Budiansky, "Viscoelastic properties of fluid saturated cracked solids", J. Geophys. Research 82, 5719-5735, (1977).
14. J. A. G. Temple, "Estimating the depth of intergranular attack by ultrasonic time delay: a theoretical analysis" to be published (1986).
15. N. F. Haines, "The theory of sound transmission and reflection at contacting surfaces" CEGB report RD/B/N4744 (1980).
16. J. A. G. Temple, "The effects of compressive stress and crack morphology on Time-of-Flight Diffraction signals" Int. J. Pres. Ves. & Piping 19 185-211, (1985).
17. A. D. Whapham, S. Perring, K. L. Rusbridge, "Effects of stress on the ultrasonic response of fatigue cracks" AERE-R10854 (1985).
18. R. B. Thompson, C. J. Fiedler, "The effects of crack closure on ultrasonic scattering measurements", Quantitative Nondestructive Evaluation 3A, ed. D. O. Thompson and D. E. Chimenti, (Plenum Press, New York, 1986), 207-215.

19. J. D. Achenbach, A. N. Norris, "Specular reflection by contacting crack faces", Review of Progress in Quantitative Nondestructive Evaluation 3A, ed. D. O. Thompson and D. E. Chimenti, (Plenum Press, New York, 1983), 163-173.
20. M. Punjani, L. J. Bond, "Scattering of plane waves by a partially closed crack", Review of Progress in Quantitative Nondestructive Evaluation 5, ed. D. O. Thompson and D. E. Chimenti, (Plenum Press, New York, 1986).
21. R. S. Crocker, B. S. Gray, private communication, (1983).
22. J. A. Ogilvy, "Ultrasonic wave scattering from rough defects: the elastic Kirchhoff approximation" report AERE-R.11866 H.M.S.O. London, (1985).
23. D. B. Langston, R. Wilson, "The UKAEA defect detection trials: application of some results to model validation and assessment of technique capability" CEGB report TPRD/B/0465/R84 (1984).
24. J. A. Ogilvy, "An estimate of the accuracy of the Kirchhoff approximation in acoustic wave scattering from rough surfaces" accepted for publication in J. Phys. D. (Applied Physics), (1986).
25. J. M. Coffey, R. K. Chapman, "Application of elastic scattering theory to the quantitative prediction of ultrasonic defect detection and sizing", OECD/IAEA Specialist Meeting on Defect Detection and Sizing, Ispra, Italy, and Nuclear Energy 22(5) 319-333, (1983).
26. J. A. Ogilvy, "A theoretical study of the effects of surface roughness on defect detection, with reference to the PISC-II results presented at the PISC Symposium, Varese, Italy, October, 1986.
27. S. Crutzen, "PISC exercises: looking for effective and reliable inspection procedures" Nuclear Engineering and Design 86, (1985), 197-218.
28. S. Crutzen, P. Jehensen, R. Nichols, M. Stephens, "The PISC-ii project: initial conclusions regarding the procedures used in the round-robin tests" NDT International 18(5) 243-249, (1985).
29. G. J. Curtis, B. M. Hawker, "Automated Time-of-Flight studies of the defect detection trial plates 1 and 2" Brit. J. NDT 25(5) 240-248, (1983).
30. J. P. Charlesworth, J. A. G. Temple, "Ultrasonic inspection through anisotropic cladding", Proceedings of Periodic Inspection of Pressurized Components Institute of Mechanical Engineers, London C142/82 pps 117-124, (1982).
31. J. A. Ogilvy, "Identification of pulse-echo rays in austenitic steel", NDT International 18(2) 67-77 (1984).
32. J. A. Ogilvy, "Ultrasonic beam profiles and beam propagation in an austenitic weld using a theoretical ray tracing model" accepted for publication in Ultrasonics 1986 and report AERE-TP. 1156 (1985).
33. R. F. Cameron, J. A. G. Temple, "Quantification of the reliability required of non-destructive inspection of PWR pressure vessels", Nuclear Engineering and Design 91, 57-68, (1985).
34. Computer Science and Systems Division, "Harwell Subroutine Library" AERE-R.9185, H.M.S.O, London, (1985).
35. M. J. Hopper, "TSSD Typesetting system for scientific documents" AERE-R.11160, H.M.S.O., London, (1983).

DISCUSSION

Mr. J. D. Achenbach, Northwestern: When you mentioned intragranular attack, you said it was a random distribution of the bonding of the grain. You mean that's a different kind of damage there than stress corrosion crack, where the damage still seems to be very much in a plane or in some branches of that plane?

Mr. Andrew Temple, Harwell: It's like a lot of stress corrosion cracks essentially.

Mr. Achenbach: All over the place?

Mr. Temple: I think that's fair, yes. You see two morphologies of intragranular attack. One is with fingers which look more like conventional cracks and the other form is where a whole region is just degraded as though it was a pile of grains pushed together with very little bonding left between the grains.

Mr. Achenbach: And the time-of-flight method?

Mr. Temple: As the wave propagates through this region, it slows down. If you like, the material has less and less modulus because it's not bonded, so the waves travel slower through that region, and you see a time difference in the time of travel through the region compared with propagation through good material.

You don't get the reflection from the interface with the good material because what you see there is something which deteriorates gradually from good material to poor material, so you don't have an interface in the conventional sense and you get a very poor reflection. You can't just measure the reflection from the interface to get the depth, but allowing the wave to propagate through the material, if you have time difference between that and good material, then you can measure the depth of attack.

Mr. Achenbach: The attenuation might be useful as an additional feature?

Mr. Temple: Yes, certainly.

From the Floor: When you do your wave tracing model, do you calculate the amplitude of the wave or only the propagation path?

Mr. Temple: The amplitude and the phase.

From the Floor: And will you concede that the mode conversion, when the wave hits the interface, that the shift of (inaudible) compressional and you also do (rays inaudible) by the shear and compressional, things like that?

Mr. Temple: Yes and no is the answer. We can do two things. We can either consider it to be material which consists of discrete regions, each region of which has a unique set of axes associated with it. Each region is like a single crystal of anisotropic material, and then the answer to your question would be yes, we would consider the amplitude of all the rays that go into that region and discard some or keep some, according to some fairly arbitrary rules.

Also, we can do a different situation in which the material is inhomogeneous in the sense that, if you like, it's as though it was a single crystal, but the axes rotate smoothly as you move along the direction in which you are going, in which case we take very small steps through the material, but you are not treating an interface in each stage then. You don't get a reflection coefficient.

From the Floor: Thank you.

Mr. Shambaugh: One more question.

From the Floor: Just to follow-up on this anisotropic behavior, is it frequency dependent, the mode conversions as well as the (Barkwitz) changes?

Mr. Temple: It depends which limit you are in. If you are in a high-frequency limit where the interface is essentially planar, then no, and that's the situation we would like to be in. We would like to treat it as a high-frequency limit.

From the Floor: Do you get any false indications when you go to anisotropic materials?

Mr. Temple: You mean because of grains or...

From the Floor: Yes.

Mr. Temple: Oh, yes. If you have a random assortment of anisotropic grains, you get a lot of noise back, and that's one of the things we would like to calculate and haven't yet done.

ON THE DIRECT AND INVERSE ELASTIC WAVE SCATTERING PROBLEM

TO CHARACTERIZE DAMAGE IN MATERIALS

J.D. Achenbach, D.A. Sotiropoulos and H. Zhu

The Technological Institute
Northwestern University
Evanston, IL 60201

INTRODUCTION

In this paper we consider a region of damage as a region of the material in which the elastic constants have substantially smaller values than in the undamaged state. The reduction of the elastic constants may be due to void distributions, zones of plastic deformation, microcracking and other generally continuous distributions of inhomogeneities which have a deleterious effect on the stiffness and presumably also the local strength of the material.

If the zone of damage is localized, it is equivalent to a low moduli inhomogeneity in an otherwise homogeneous elastic solid. In that case the damaged region can be detected and its effect on the overall strength of the body can be assessed by ultrasonic methods, as discussed in general terms in this paper.

From the mathematical point of view this work is concerned with scattering by an inhomogeneity which is contained in an infinite isotropic elastic material. The inhomogeneity is of general shape, it is elastic anisotropic and it is the scatterer of incident plane elastic waves. The integral representation of the scattered field is obtained for relatively large wavelengths as compared with the size of the scatterer, and at large distances from the scatterer. For a given scattered field the inverse problem is subsequently formulated as a nonlinear optimization problem. Its solution gives the location of the centroid of the scatterer, its force vector and moment tensor. In addition, the interaction energy between the material and the inclusion is obtained for a complementary static stress state. This latter result has direct relevance to failure conditions in the material under service conditions.

FORMULATION OF THE DIRECT PROBLEM

An anisotropic elastic inhomogeneity of general shape and of volume V is contained in an unbounded homogeneous isotropic elastic solid. The density and elastic constants of the inhomogeneity are $\rho^*$ and $C^*_{ijk\ell}$, respectively, while those of the host material are $\rho$ and $C_{ijk\ell}$ where in terms of Lame's elastic constants:

$$C_{ijk\ell} = \lambda \delta_{ij}\delta_{k\ell} + \mu(\delta_{ik}\delta_{j\ell} + \delta_{i\ell}\delta_{jk}) \tag{1}$$

The position vector of a point inside the inhomogeneity is given by $\vec{\zeta}$, while $\vec{x}$ is the position vector of a point outside the inhomogeneity. The origin of the Cartesian coordinate system is chosen to coincide with the centroid of the inhomogeneity, i.e.,

$$\int_V \zeta_i \, dV_\zeta = 0, \quad i = 1,2,3. \tag{2}$$

A plane time-harmonic displacement wave, whose spatial dependence is defined by $u_i^I(\vec{x})$, is incident upon the inhomogeneity. The components of the scattered displacement field, $u_i^S(\vec{x})$, in the frequency domain may then be written as [1,2]

$$u_i^S(\vec{x}) = \int_V G_{im}(\vec{x}-\vec{\zeta}) f_m(\vec{\zeta}) dV_\zeta - \int_V \frac{\partial G_{im}}{\partial \zeta_k} m_{mk}(\vec{\zeta}) dV_\zeta \tag{3a}$$

where

$$G_{im}(\vec{x}-\vec{\zeta}) = G_{im}^T(\vec{x}-\vec{\zeta}) + G_{im}^L(\vec{x}-\vec{\zeta}) \tag{3b}$$

$$G_{im}^T(\vec{x}-\vec{\zeta}) = \frac{1}{4\pi\rho\omega^2} \left( k_T^2 \frac{e^{ik_T R}}{R} \delta_{im} + \frac{\partial^2}{\partial x_i \partial x_m} \frac{e^{ik_T R}}{R} \right) \tag{3c}$$

$$G_{im}^L(\vec{x}-\vec{\zeta}) = -\frac{1}{4\pi\rho\omega^2} \frac{\partial^2}{\partial x_i \partial x_m} \frac{e^{ik_L R}}{R} \tag{3d}$$

$$f_m(\vec{\zeta}) = \omega^2(\rho^* - \rho)(u_m^I + u_m^S), \quad m_{mk}(\vec{\zeta}) = (C_{mk\ell j}^* - C_{mk\ell j}) \frac{\partial(u_\ell^I + u_\ell^S)}{\partial \zeta_j} \tag{3e,f}$$

In (3a)-(3f), $\omega$ is the angular frequency of the fields, $R = |\vec{x}-\vec{\zeta}|$ and $k_L = \omega/c_L$, $k_T = \omega/c_T$, $c_L$ and $c_T$ being the longitudinal and transverse wavespeeds, respectively, in the host material. Repeated indices imply summation.

It may be shown that under the assumptions

$$(k_T a) \left.\frac{\partial f}{\partial \zeta_i}\right|_{\vec{\zeta}=0} a \ll f(0), \quad \left.\frac{\partial^2 f(\vec{\zeta})}{\partial \zeta_i \partial \zeta_j}\right|_{\vec{\zeta}=0} a^2 \ll f(0) \tag{4,5}$$

$$(k_T a)^2 \ll 1, \quad r = |\vec{x}| \gg a, \tag{6,7}$$

Eqs.(3a)-(3f) reduce to

$$u_i^S(\vec{x}) = G_{im}(\vec{x}) F_m + \frac{\partial G_{im}}{\partial x_k} M_{mk} \tag{8a}$$

$$\frac{\partial G_{im}}{\partial x_k} = -\frac{k_L}{4\pi\rho c_L^2} \frac{e^{ik_L r}}{r} A_{imk}^L + \frac{k_T}{4\pi\rho c_T^2} \frac{e^{ik_T r}}{r} [A_{imk}^T + \gamma_k \delta_{im}(i - \frac{1}{k_T r})] \tag{8b}$$

$$A_{imk} = -\gamma_i \gamma_m \gamma_k (i - \frac{1}{kr}) - \frac{1}{kr}[1 + \frac{3}{kr}(i - \frac{1}{kr})][\gamma_i \delta_{mk} + \gamma_m \delta_{ik} + \gamma_k \delta_{im} - 5\gamma_i \gamma_m \gamma_k] \tag{8c}$$

(here $k = k_T$ and $k = k_L$ for $A_{imk}^T$ and $A_{imk}^L$, respectively). Also

$$F_m = \omega^2[(\rho^* - \rho)(u_m^I + u_m^S)]_{\vec{\zeta}=0} V \tag{8d}$$

$$M_{mk} = \left[(C^*_{mk\ell j} - C_{mk\ell j})\frac{\partial(u_\ell + u^S_\ell)}{\partial \zeta_j}\right]_{\vec{\zeta}=0} V, \tag{8e}$$

where $G_{im}(\vec{x})$ is given by Eqs.(3b,c,d) with $\vec{\zeta} = 0$, and $\vec{\gamma} = \vec{x}/r$ with $r = |\vec{x}|$. It was also assumed that $m_{mk}(\vec{\zeta})$ of Eq.(3f) satisfies Eqs.(4,5) as well. Equation (8a) represents the scattered field as the sum of two terms. The first term is the field at $\vec{x}$ due to forces located at the centroid of the inhomogeneity and of strength $F_m$ (force vector). The second term is the field at $\vec{x}$ due to double forces at the centroid of the inhomogeneity and of strength $M_{mk}$ (moment tensor). Previous studies [3,4] assumed that ka << 1 and kr >> 1. This is a more restrictive assumption than Eqs.(6,7). Consequently, their expression for the scattered field does not include the full expressions for $G_{im}$ and $\partial G_{im}/\partial x_k$ as given by Eqs.(3c,d;8b,c), since they assumed 1/kr = 0.

Having formulated the direct problem, the solution of the inverse problem may now be obtained as will be shown in the next section.

METHOD OF SOLUTION OF INVERSE PROBLEM

A nondestructive test yields data for the scattered field. Of course, the physical properties of the host material are known. The purpose of the test is to "characterize" the inhomogeneity, ie., to determine its location as well as its volume and elastic constants, or more preferably to determine its effect on the overall strength of the component. It has been shown in an earlier study [4] that one cannot determine independently the volume and elastic constants of the inhomogeneity, for long wavelengths such as considered in this study.

In this section, Eqs.(8) are used to determine the location of the centroid of the inhomogeneity, i.e. $\vec{x}$, as well as the force vector $F_m$ and the moment tensor $M_{mk}$. The method of solution is described in what follows.

At first, we redefine $M_{mk}$ of Eq.(8e) as

$$M_{mk} = C^*_{mk\ell j} \varepsilon_{\ell j}\Big|_{\vec{\zeta}=0} V \tag{9}$$

where

$$C^*_{mk\ell j}\varepsilon_{\ell j} = (C^*_{mk\ell j}-C_{mk\ell j})\frac{\partial(u^I_\ell + u^S_\ell)}{\partial \zeta_j}. \tag{10}$$

Equation (10) is the necessary and sufficient condition for the equivalency of an inhomogeneity and an inclusion[5]. The reason Eq.(9) is used will be explained later. Substitution of Eq.(9) in Eq.(8a) gives

$$u^S_i(\vec{x}) = G_{im}(\vec{x}) F_m + \frac{\partial G_{im}}{\partial x_k} C^*_{mk\ell j}\varepsilon_{\ell j}\Big|_{\vec{\zeta}=0} V. \tag{11}$$

Equation (11) is the partial equivalent representation of the scattered displacement. The vector-matrix form of Eq.(11) is

$$\{u^S(x)\}_{6\times 1} = [L^*(\vec{x})]_{6\times 18}\{P^*\}_{18\times 1}, \qquad (12)$$

where, considering its real and imaginary parts, the scattered displacement vector $\{u^S\}$ has 6 components. Similarly, the force vector and the integrated strain tensor that are included in $\{P^*\}$ add up to a total of 18 components, since $\varepsilon^*_{\ell j}$ has 6 complex independent components.

The ith equation of Eq.(12) in complex form is

$$\{u^S_i\} = [G_{i1}\ G_{i2}\ G_{i3}\ Q_{i11}\ Q_{i112}\ Q_{i13}\ Q_{i22}\ Q_{i23}\ Q_{i33}]\{P^*\}, \qquad (13a)$$

where

$$Q_{i11} = (\lambda+2\mu)G_{i1,1} + \lambda(G_{i2,2} + G_{i3,3}), \quad Q_{i12} = 2\mu(G_{i1,2} + G_{i2,1}) \qquad (13b,c)$$

$$Q_{i13} = 2\mu(G_{i1,3} + G_{i3,1}), \quad Q_{i22} = (\lambda+2\mu)G_{i2,2} + \lambda(G_{i1,1}\ G_{i3,3}) \qquad (13d,e)$$

$$Q_{i23} = 2\mu(G_{i2,3} + G_{i3,2}), \quad Q_{i33} = (\lambda+2\mu)G_{i3,3} + \lambda(G_{i1,1} + G_{i2,2}) \qquad (13f,g)$$

$$\{P^*\} = [F_1\ F_2\ F_3\ e_{11}\ e_{12}\ e_{13}\ e_{22}\ e_{23}\ e_{33}]^T, \quad e_{ij} = \varepsilon^*_{ij}\bigg|_{\zeta=0} V \qquad (13h,i)$$

Equation (12) therefore defines a set of 6 nonlinear equations for 21 unknowns (3 components of vector $\vec{x}$, and 18 components of vector $\vec{P}^*$). This is an underdetermined system of nonlinear equations. To obtain a unique solution the system has to be determined or overdetermined. Since the components of vector $\vec{P}^*$ are independent of $\vec{x}$, additional equations can be obtained from other observation points. For the present problem the data $u^S_i$ is needed at 4 different observation points. We then have 24 equations and 21 unknowns.

The first step to solve the overdetermined system of nonlinear equations is to eliminate the unknown vector $\vec{P}^*$ and to subsequently solve for $\vec{x}$. However, an inspection of Eq.(12) reveals that the left inverse of $[L^*(\vec{x})]$ does not exist (rows more than columns). Thus to solve for $\vec{P}^*$ one needs to consider the data in 3 observation points simultaneously. This gives

$$[u^S(\vec{x}^1)\ u^S(\vec{x}^2)\ u^S(\vec{x}^3)]^T_{18\times 1} = [A^*(\vec{x}^1)]_{18\times 18}\{P^*\}_{18\times 1}. \qquad (14)$$

Equation (14) can be formally inverted:

$$\{P^*\} = [A^*(\vec{x}^1)]^{-1}[u^S(\vec{x}^1)\ u^S(\vec{x}^2)\ u^S(\vec{x}^3)]^T. \qquad (15)$$

Substitution of Eq.(15) in Eq.(12) for the 4th observation point gives

$$\{u^S(\vec{x}^4)\}_{6\times 1} = [L^*(\vec{x}^4)]_{6\times 18}[A^*(\vec{x}^1)]^{-1}_{18\times 18}[u^S(\vec{x}^1)\ u^S(\vec{x}^2)\ u^S(\vec{x}^3)]^T_{18\times 1}. \qquad (16)$$

The solution of the inverse problem has therefore been reduced to the solution of Eq.(16). This is a system of 6 nonlinear equations and 3 unknowns $(x^1_1, x^1_2, x^1_3)$, since the relationship between $\vec{x}^1, \vec{x}^2, \vec{x}^3, \vec{x}^4$ is known.

To solve the nonlinear optimization problem defined by Eq.(16) the following 6 residuals, $g_i(\vec{x})$, are defined

$$\{g(\vec{x}^1)\} = \{u^S(\vec{x}^4)\} - [L^{*}(\vec{x}^4)][A^{*}(\vec{x}^1)]^{-1} [u^S(\vec{x}^1)\ u^S(\vec{x}^2)\ u^S(\vec{x}^3)]^T . \quad (17)$$

The residuals $g_i(\vec{x}^1)$ are now minimized in the least squares sense with respect to $\vec{x}^1$, i.e., we seek

$$\underset{\vec{x}^1}{\text{Min}} \sum_{i=1}^{6} g_i^2(\vec{x}^1) . \quad (18)$$

Equations (17,18) define a nonlinear least squares problem. Its solution is obtained through use of a modification of the Levenberg-Marquardt algorithm as outlined in the User-Guide for MinPack [6]. The solution so obtained is the unknown $\vec{x} = (x_1^1, x_2^1, x_3^1)$.

In performing the initial numerical calculations it was found that the solution was identical to the initial guess used for $\vec{x}^1$, when Eq.(9) was not used. This happened because the Jacobian of $L(\vec{x}^4)[A(\vec{x}^1)]^{-1}$ was zero. Redefinition of $L(\vec{x}^4)$ and $[A(\vec{x}^1)]^{-1}$ to $L^{*}(\vec{x}^4)$ and $[A^{*}(\vec{x}^1)]^{-1}$ through use of Eq.(9) gave the correct solution $\vec{x}^1$. Substitution of $\vec{x}^1$ in Eq.(13h) gives $\{P^*\}$. In turn, Eq.(9) gives $M_{mk}^*$.

NUMERICAL RESULTS

Several numerical experiments were performed to test the validity of the method of solution of the inverse problem. One of the calculations was carried out for a spherical inclusion of an isotropic material, with $\nu^* = \nu = 0.3$, $\lambda^* - \lambda = -.5\lambda$. The mass density of the inclusion was taken equal to that of the host material, which implies that $F_m \equiv 0$. The incident displacement field was taken as a longitudinally polarized wave. At the position of the inclusion the incident wave was considered as a plane wave of amplitude $u_o$ and wavenumber $k_L$, propagating along the $x_3$ direction.

To obtain the position of the centroid of the inhomogeneity, together with the components of the moment tensor, $M_{mk}$, the scattered displacement field is needed at 3 observation points. For the present purpose, synthetic data was obtained by using the solution to the direct problem given by Eq.(8). Several values of $k_L a$ (a is the radius of the sphere) were considered, the largest being $k_L a = 0.6$. By using the quasi-static values of the strain inside the inclusion, the components of the moment tensor $M_{mk}$ of the spherical inhomogeneity were found to be linear functions of $k_L a$. Subsequent use of Eq.(8) provided the synthetic data for the inverse problem. The actual values of $|u^S|/u_o$ are very small. This does not affect the mathematical process of inversion, but it does of course have implications for the practical applications of the method, as discussed briefly in the concluding section of this paper.

The synthesized data was used to solve the inverse problem as explained in the preceding section. The position of one of the observation points was obtained as well as the moment tensor of the inhomogeneity. The match with the actual values was found to be excellent. Numerical tests were also performed on the stability of the solution. It was established that errors in the scattered data cause errors of the same order in the location of the inhomogeneity and in the average of the components of the moment tensor. Therefore, the solution of the inverse problem is stable.

STRENGTH CONSIDERATIONS

In this study, scattering data from a nondestructive test was used as input, and the inverse problem was subsequently solved to obtain the position of the centroid of the inhomogeneity and its moment tensor. It is, however, of practical importance to relate the inverse problem solution to an actual service situation, for example, to a static state of stress. For this reason, the incident field for the nondestructive test should be such that in the limit of zero frequency there will be a nonzero applied stress. This is achieved by redefining the incident displacement field for an incident longitudinal wave propagating in the z-direction as

$$u_z^I = \frac{u^o}{ik_L} e^{ik_L z} . \qquad (19)$$

Then

$$\lim_{k_L \to 0} \sigma_{ij}^I = \sigma_{ij}^o = C_{ij33} u^o . \qquad (20)$$

Analogously, and also motivated by the fact that for a spherical inhomogeneity the moment tensor is linearly proportional to frequency for small frequencies, we define

$$M_{mk}^S = \lim_{k \to 0} \frac{M_{mk}}{ik} \qquad (21)$$

where $M_{mk}^S$ is the low frequency limit obtained from the $M_{mk}$ solved for in the inverse problem.

It can be shown [7] that the interaction energy, $\Delta W$, i.e., the change in the total potential energy (Gibbs free energy) caused by the presence of the inhomogeneity, is

$$\Delta W = \frac{1}{2} C_{ijk\ell}^{-1} \sigma_{k\ell}^o M_{ij}^S . \qquad (22)$$

This interaction energy can be computed from the solution of the inverse problem based on data from a nondestructive test. For a static service situation defined by a far field $\Sigma_{ij}^o$, the interaction energy $(\Delta W)_{ss}$, of the actual service situation, is related to $\Delta W$ by

$$(\Delta W)_{ss} = \frac{\sigma_{ij}^o C_{ijk\ell}^{-1} \sigma_{k\ell}^o}{\Sigma_{ij}^o C_{ijk\ell}^{-1} \Sigma_{k\ell}^o} \Delta W . \qquad (23)$$

Therefore, for a service situation, $(\Delta W)_{ss}$ can be calculated via Eq.(23) and the results obtained in this study from the solution of the inverse problem. It may be expected that the magnitude of $(\Delta W)_{ss}$ has direct relevance to failure of the material.

CONCLUDING COMMENTS

The success of the proposed inverse method depends on the availability of suitable low frequency scattering data. Appropriate signal processing of experimental time-domain measurements will give frequency domain data. In the low-frequency range it may, however, be necessary to fit the displacement data with a curve of the general form $Ck_L^2$, to improve the accuracy. It should also be noted that for an inhomogeneity of lower stiffness than the host material, an incident plane wave produced by a regular transducer will generate a very small scattered field. This is primarily due to the geometrical attenuation caused by the term a/r. The displacements tend to be particularly small in the plane perpendicular to the ray connecting the point of observation and the centroid of the inhomogeneity. The scattered displacement magnitude can, however, be amplified considerably by the use of a focused transducer. But even then only the radial displacement component may be useful. In that case the number of observation points would have to be tripled to make up for the lack of transverse displacement data. This would have to be done in any event if a water-bath configuration would be used.

ACKNOWLEDGMENT

This work was supported by the Office of Naval Research under Contract N00014-85-K-0401 with Northwestern University.

REFERENCES

1. A.K. Mal and L. Knopoff, "Elastic Wave Velocities in Two-component Systems," J. Inst. Maths. Applics. 3, 376 (1967).
2. J. E. Gubernatis, E. Domany and J.A. Krumhansl, "Formal Aspects of the Theory of the Scattering of Ultrasound by Flaws in Elastic Materials," J. Appl. Phys. 48, 2804 (1977).
3. J. E. Gubernatis, J.A. Krumhansl and R. M. Thomson, "Interpretation of Elastic-wave Scattering Theory for Analysis and Design of Flaw-characterization Experiments: The Long-Wavelength Limit," J. Appl. Phys. 50, 3338 (1979).
4. W. Kohn and J. R. Rice, "Scattering of Long-wavelength Elastic Waves From Localized Defects in Solids," J. Appl. Phys. 50, 3346 (1979).
5. J.D. Eshelby, "The Determination of the Elastic Field of an Ellipsoidal Inclusion, and Related Problems," Proc. Roy. Soc. Lon. A241, 376 (1957).
6. J. More et al, "User Guide for MINPACK-1," Argonne National Lab. Rep. ANL-80-74, Argonne, IL (1980).
7. T. Mura, "Micromechanics of Defects in Solids," Martinus Nijhoff Publ. (1982).

# TIME DEPENDENT PULSE PROPAGATION AND SCATTERING IN ELASTIC SOLIDS; AN ASYMPTOTIC THEORY

A.N. Norris

Rutgers University
Department of Mechanics & Materials Science
P.O. Box 909
Piscataway, NJ 08854

## INTRODUCTION

Predictive modeling of ultrasonic pulse propagation in elastic solids is usually formulated in the frequency domain. Tractable solutions can then be obtained by using, for example, the powerful technique of geometrical elastodynamics and ray theory for wavefront propagation [1]. Recent advances [2,3] allow us to incorporate the finite pulse width by means of Gaussian profiles. However, a more realistic model should also include the fact that the pulse is of limited duration and therefore spatially localized in all directions. This paper outlines a theory for pulses in the form of a localized disturbance with a Gaussian envelope. The theory is valid if the associated carrier wavelength is short in comparison with typical length scales encountered in the solid. The method provides results explicitly in the time domain without the necessity of intermediate FFTs required by frequency domain methods. Applications to pulse propagation in smoothly varying inhomogeneous media, interface scattering and edge diffraction are discussed. The present theory contains an extra degree of freedom not explicitly considered before, i.e., the temporal width or duration of the pulse. An extensive treatment of the related problem for the scalar wave equation can be found in reference 4.

## MATHEMATICAL THEORY

For the sake of generality, the theory is presented for a smoothly varying, inhomogeneous, isotropic elastic medium. The equations of motion are

$$(\lambda u_{k,k})_{,i} + [\mu(u_{i,j}+u_{j,i})]_{,j} - \rho u_{i,tt} = 0, \qquad (1)$$

where the elastic moduli $\lambda$, $\mu$ and the density $\rho$ can vary with position. In Eq. (1), $u_i$ is the displacement in the $x_i$ direction, subscript j denotes the derivative in the $x_j$ direction and the summation convention is assumed. Also, the subscript t means the time derivative. The key to the present theory is the assumption that the displacement is of the form

$$u_i(\underset{\sim}{x},t) = V_i(\underset{\sim}{x},t) e^{i\omega\phi(\underset{\sim}{x},t)}, \qquad (2)$$

where $\omega$ is the central or characteristic frequency of the disturbance. In the standard time harmonic theory, it is implicitly assumed that $\phi(\underset{\sim}{x},t) =$

$\Phi(\underset{\sim}{x})\cdot t$, but this is not the case here. The frequency $\omega$ is assumed to be large, so that when (2) is substituted into (1), we can treat (1) as a sequence of distinct asymptotic equations. The first such equation comes from the terms that are of order $\omega^2$:

$$(\lambda+\mu)V_k\phi_k\phi_i + \mu V_i\phi_k\phi_k - \rho\phi_t^2 V_i = 0 \qquad (3)$$

This is the eikonal equation of elastodynamics, and can be simplified by writing $\underset{\sim}{p} = \nabla\phi$, $\underset{\sim}{V} = \underset{\sim}{V}^{(1)} + \underset{\sim}{V}^{(2)}$, where $\underset{\sim}{V}^{(1)}$ and $\underset{\sim}{V}^{(2)}$ are the parts of $\underset{\sim}{V}$ parallel and perpendicular to $\underset{\sim}{p}$, respectively. Then (3) becomes

$$(c_L^2 p^2 - \phi_t^2)\underset{\sim}{V}^{(1)} + (c_T^2 p^2 - \phi_t^2)\underset{\sim}{V}^{(2)} = 0, \qquad (4)$$

where $c_L$ and $c_T$ are the longitudinal and transverse wave speeds, $c_L^2 = (\lambda+2\mu)/\rho$ and $c_T^2 = \mu/\rho$. Thus, either $\phi_t^2 = c_L^2 p^2$ and $\underset{\sim}{V}$ is parallel to $\underset{\sim}{p}$, or $\phi_t^2 = c_T^2 p^2$ and $\underset{\sim}{V}$ is perpendicular to $\underset{\sim}{p}$. The former corresponds to a longitudinal pulse and the latter a transverse pulse.

The next equation in the asymptotic sequence is the transport equation for the pre-exponential amplitude. Let the pulse center be at $\underset{\sim}{\bar{x}}_0$ at time $t = 0$. The subsequent path of the center is at $\underset{\sim}{\bar{x}}(t)$ where $\underset{\sim}{\bar{x}}(t)$ is the solution to the ray equation defined by the eiconal equation. The rays are just straight lines in homogeneous media, but in general satisfy

$$\dot{\bar{x}}_i = c(\underset{\sim}{\bar{x}}(t))e_i(t) \qquad (5)$$

$$\dot{e}_i = e_i e_j \frac{\partial c}{\partial x_j} - \frac{\partial c}{\partial x_i} \qquad (6)$$

where $\underset{\sim}{e}(t)$ is the unit direction vector of the ray and the overdot denotes the total time derivative along the ray. The speed $c$ in (5) and (6) can be either $c_L$ or $c_T$, and the system is completely described by the initial conditions $\underset{\sim}{\bar{x}}(0) = \underset{\sim}{\bar{x}}_0$ and $\underset{\sim}{e}(0) = \underset{\sim}{e}_0$, the initial ray direction. The transport equation can be solved along the ray in the form.

$$\underset{\sim}{V}(\bar{x}(t),t) = V_0 \frac{c(\bar{x}(t))}{c(\bar{x}_0)} \left[\frac{(\rho c^2 \det \underset{\sim}{A})(0)}{(\rho c^2 \det \underset{\sim}{A})(t)}\right]^{1/2} \qquad (7)$$

where $\underset{\sim}{V} = V\underset{\sim}{e}(t)$ for longitudinal motion and $\underset{\sim}{V} = V\underset{\sim}{n}(t)$ for transverse motion, where $\underset{\sim}{n}$ is a unit vector orthogonal to $\underset{\sim}{e}(t)$. An equation for the rotation of $\underset{\sim}{n}(t)$ about $\underset{\sim}{e}(t)$ can be obtained, but we defer the details until a later publication. The matrix $\underset{\sim}{A}(t)$ is a complex, 3x3 matrix which satisfies

$$\dot{A}_{ij}(t) = e_i \frac{\partial c}{\partial x_k} A_{kj} + c^2(B_{ij} - e_i e_k B_{kj}) \qquad (8)$$

$$\dot{B}_{ij}(t) = \frac{-1}{c} \frac{\partial^2 c}{\partial x_i \partial x_k} A_{kj} - \frac{\partial c}{\partial x_i} e_k B_{kj} \qquad (9)$$

Initial condition on $\underset{\sim}{A}$ and $\underset{\sim}{B}$ must be specified. It turns out that the combination $\underset{\sim}{B}\underset{\sim}{A}^{-1} \equiv \underset{\sim}{M}(\bar{t})$ is equal to the matrix of second derivatives of $\phi$ along the ray, i.e.,

$$M_{ij}(t) = \frac{\partial^2 \phi}{\partial x_i \partial x_j}(\underset{\sim}{\bar{x}}(t),t) \qquad (10)$$

It follows from the eikonal equation that $\dot{\phi} = 0$ along the ray, or $\phi(\bar{x}(t),t) = \phi_0$, a constant which can be taken as zero without loss of generality. Also, the gradient of $\phi$ along the ray, $\nabla\phi(\bar{x}(t),t) = \underline{p}(\bar{x}(t),t)$ is equal to $\underline{e}(t)/c(\bar{x}(t))$. Therefore, a local Taylor series expansion of $\phi(\underline{x},t)$ about the ray position up to second order in distance gives the paraxial approximation, $\phi \cong \phi_p$,

$$\phi_p(\underline{x},t) = \frac{1}{c} \underline{e}(t) \cdot (\underline{x}-\bar{\underline{x}}(t)) + \frac{1}{2} M_{ij}(t)(x_i-\bar{x}_i(t))(x_j-\bar{x}_j(t)) \tag{11}$$

Equations (2), (7) and (11) define what we call a Gaussian wave packet or GWP for brevity. The GWP will be localized in space if and only if the imaginary part of $\underline{M}(t)$ is positive definite. It can be shown that if $\underline{M}(0)$ satisfies this requirement, then (8) and (9) will automatically generate an $\underline{M}(t)$ that has this property. Without loss of generality, we can specify $\underline{A}(0) = \underline{I}$, the identity matrix. Then the subsequent evolution of the GWP is completely prescribed by the initial parameters $\bar{\underline{x}}_0$, $\underline{e}_0$ and $\underline{M}(0)$.

PROPERTIES OF GAUSSIAN WAVE PACKETS

The key quantity in the theory is the 3x3 complex matrix $\underline{M}(t)$ which describes the Gaussian envelope of the packet about the pulse center $\underline{x} = \bar{\underline{x}}(t)$. The matrix $\underline{A}(t)$ is like the ray tube area of classical geometrical optics. However, since $\underline{M}(0)$ is assumed to have a positive definite imaginary part, and $\underline{A}(0) = \underline{I}$, the subsequent values of $A_{ij}(t)$ as determined from (8) and (9) are in general complex. Thus, $\det(\underline{A})$ is really a complex ray tube area. In fact, it can be shown that $\det(\underline{A})$ is always non-zero, even at the geometrical caustics and loci of classical geometrical optics. Thus, the GWP solution has no singularities. This is one of its most advantageous features; it obviates the necessity of patching up the solution near caustics and loci. The price paid is actually very little in comparison with what one has to do in classical geometric optics to describe the propagation of a ray bundle of time harmonic form. The real set of equations for the ray tube cross-section, the curvature matrix, is a simple projection of the 3x3 complex Eqs. (8) and (9) onto a very similar 2x2 system. The extra algebra in the present theory is really quite trivial.

One example of quite general significance is for a material in which the gradient of $c(\underline{x})$, which is either $c_L$ or $c_T$, is constant. A more generally inhomogeneous medium can be considered by splitting it up into regions of constant $\nabla c$. It is well known in classical geometrical optics that the rays become arcs of circles and the real ray tube cross-section curvatures can be obtained explicitly. Similar explicit solutions can be obtained for all of the GWP parameters, including $\underline{M}(t)$. A trivial special case of interest is the homogeneous medium, $c$ = constant. Then, $\underline{e}(t) = \underline{e}_0$, $\bar{\underline{x}}(t) = \bar{\underline{x}}_0 + ct\underline{e}_0$, and

$$\underline{M}(t) = \underline{M}(0) [\underline{I}+c^2 t\ \underline{P}\underline{M}(0)]^{-1} \tag{12}$$

where $\underline{P}$ is the projection matrix $(\underline{I}-\underline{e}\underline{e}^T)$ onto the plane perpendicular to the ray direction. The amplitude follows from Eq. (7), where now

$$\left[\frac{\det \underline{A}(0)}{\det \underline{A}(t)}\right]^{1/2} = [\det(\underline{I}+c^2 t\ \underline{P}\underline{M}(0))]^{-1/2} . \tag{13}$$

Various limiting cases of pulses are obtained by taking appropriate values for the initial complex matrix $\underline{M}(0)$. (1) A very short, thin initial pulse results by letting $\text{Im}(m_s(0)) \to \infty$, where $m_s(t) \equiv e_i(t)e_j(t)M_{ij}(t)$ is the element that determines the length of the pulse in its direction of propagation. (2) Similarly, if $\text{Im}(m_s(0)) \to 0$, a very long pulse is obtained. This limit is equal to the case of a time harmonic Gaussian beam, (3) The orthogonal width of the GWP depends upon the 2x2 submatrix $\underset{\sim}{M}_n = \underset{\sim}{P}\underset{\sim}{M}$, where $\underset{\sim}{P} = (\underset{\sim}{I} - \underset{\sim}{e}\underset{\sim}{e}^T)$. Thus, an initially plane pulse corresponds to $\underset{\sim}{M}_n \to 0$, (4) Finally, a GWP of zero initial wavefront curvature, or a point source, is obtained by letting $\text{Im}(\underset{\sim}{M}_n(0)) \to \infty$.

## REFLECTION AND TRANSMISSION OF GWPs

When a GWP strikes a surface of material discontinuity, e.g., a free surface or the surface of an inclusion, it splits up into reflected and transmitted GWPs, in the same way that a classical ray does. The precise formulation of the interface jump conditions is beyond the scope of the present article, but the general procedure is in the same spirit as, for example, time harmonic geometrical elastodynamics [1]. The properties, i.e., amplitude V(t) and envelope matrix M(t) of the transmitted and reflected GWPs are completely determined by the material properties of the interface in the neighborhood of the point where the central ray of the incident GWP strikes the interface. The subsequent propagation of the split GWPs is described by the same theory as before.

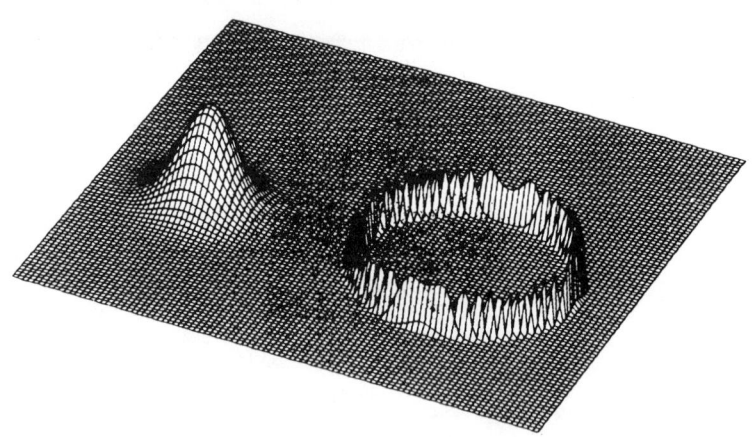

Fig. 1. The incident GWP at $t = 0$. The circular rim defines the region of slower speed, $c_1/c_0 = 1/4$.

These ideas are made apparent by a specific illustration. The sequence of four figures shows the scattering picture for a pressure wave in an acoustic medium incident upon a region of lower wave speed. The model is two-dimensional for simplicity of presentation. The first figure shows the incident GWP at time $t = 0$. The GWP is indicated by its envelope function, which we define as $(\text{ReV})|e^{i\omega\phi}|$. The edge of the "inclusion" is depicted by the circular rim. The initial pulse is circular but broadens in the orthogonal direction as it approaches the interface. The reflection and transmission process occurs between Figs. 2 and 3. Both the reflected and transmitted GWPs are visible in Figs. 3 and 4. Note that the reflected GWP appears as a depression. This is due to the choice of the envelope function, and illustrates that the amplitude V has gone through a 180° phase shift, as we would expect for a soft target, since V defines the displacement in the direction of propagation. This effect would not be apparent for an envelope function like $|Ve^{i\omega\phi}|$.

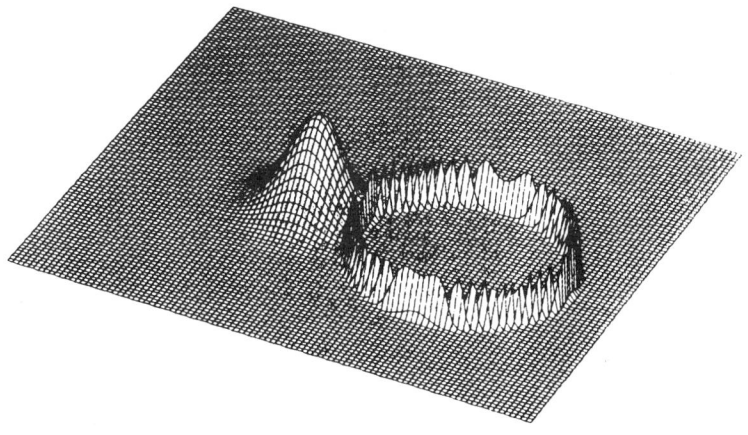

Fig. 2. The incident GWP at t = 1, shortly before it strikes the interface.

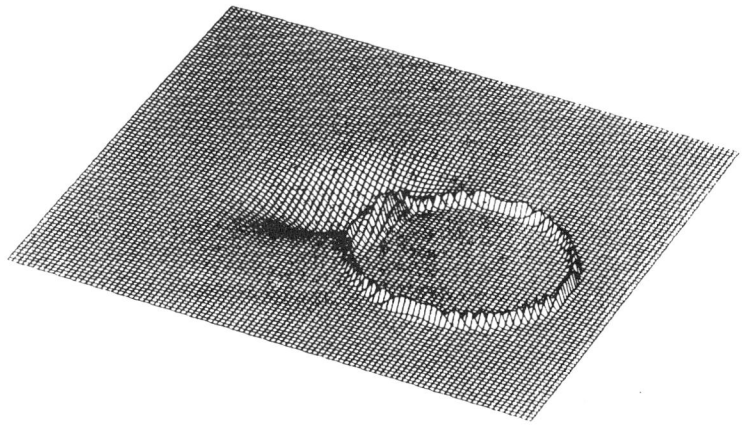

Fig. 3. The reflected and transmitted GWPs at t = 2.

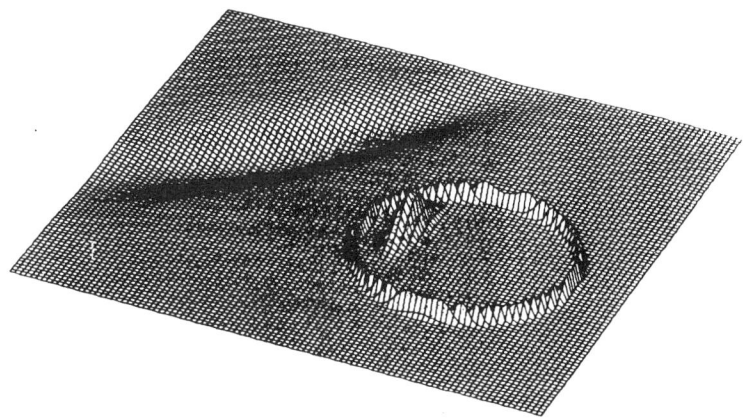

Fig. 4. The transmitted and reflected GWPs at t = 3. Note the significant effects due to the interface curvature, causing focusing and defocusing, respectively.

## EDGE DIFFRACTION AND CRITICAL ANGLE REFLECTION

When a GWP strikes the edge of a crack there is a diffraction effect similar to that for a plane wave [1]. Energy is diffracted in all directions away from the edge and mode conversion into both body waves and surface waves on the crack faces occurs. The specific form and amplitude of the diffracted waves depends critically on which part of the GWP strikes the edge. Thus, the diffraction is maximum if the central ray hits the edge. This case is most similar to plane wave diffraction theory. However, in general, the edge is not reached by the central ray, but by a point in the waist of the pulse. The incident field at the crack tip behaves locally like an evanescent wave. This can be treated by an extension of the Geometrical Theory of Diffraction (GTD) [1] that allows for complex angles of incidence. We do not give any details of the procedure here, but note that it does not present any conceptual difficulties.

Critical angle Rayleigh wave phenomena occur when ultrasound is incident at a liquid-solid interface at the angle which couples the acoustic wave to the pseudo-Rayleigh wave in the solid. The most significant effect is the beam displacement phenomenon which has been explained by Bertoni and Tamir [5]. Their model considers a CW beam of Gaussian profile. Using their approach, we have treated the same problem for transient GWP incidence. The solution can be obtained in closed form, and has the advantage of being fully time dependent. It contains both the specularly reflected GWP and a surface leaky Rayleigh wave part.

## CONCLUSIONS AND FUTURE APPLICATIONS

The GWP idea offers a simple procedure for modeling pulse propagation and scattering in solids and fluids. It is particularly suitable to piecewise continuous media, where interface reflection/transmission and edge diffraction can be treated by well-known techniques. The main advantage of using GWPs is that they are explicitly time dependent: no transforms are necessary. They also have many physically important features built in. For example, it can be shown that a GWP conserves energy. In the area of ultrasonics in composites, GWPs may be very useful in helping NDE researchers and practitioners understand the ways in which pulses propagate. Future research will focus on the propagation of GWPs in anisotropic materials, with possible application to fiber and ply reinforced composites.

## REFERENCES

1. J.D. Achenbach, A.K. Gautesen and H. McMaken, _Ray Methods for Waves in Elastic Solids_, Pitman, Boston (1982).

2. R.B. Thompson and E.F. Lopes, "A Model for the Effects of Aberrations on Refracted Ultrasonic Fields," _Review of Progress in Quantitative NDE5_, D.O. Thompson, D.E. Chimenti, Eds., Plenum Press, NY (1986).

3. M.M. Popov, "A New Method of Computation of Wave Fields Using Gaussian Beams," Wave Motion, 4, 85-97 (1982).

4. A.N. Norris, B.S. White and J.R. Schrieffer, "Gaussian Wave Packets in Inhomogeneous Media with Curved Interfaces" (unpublished).

5. H.L. Bertoni and T. Tamir, "Unified Theory of Rayleigh-Angle Phenomena for Acoustic Beams at Liquid-Solid Interfaces," Appl. Phys., 2, 157-172 (1973).

ULTRASONIC SURFACE AND BULK WAVE INTERACTION WITH FLUID-SATURATED

POROUS SOLIDS

M. J. Mayes, P. B. Nagy*, and L. Adler

Department of Welding Engineering
The Ohio State University
Columbus, Ohio 43210

B. P. Bonner and R. Streit

Lawrence Livermore National Laboratory
Livermore, California 94550

INTRODUCTION

There has recently been considerable interest in the acoustics of fluid-saturated porous media. The most interesting peculiarity, the existence of two compressional waves, was first predicted by Biot [1] in 1956. It was not until 1980 that Plona [2] experimentally observed the additional slow compressional bulk wave in a water-saturated porous solid composed of sintered glass spheres. Even more recently, Feng and Johnson [3] have extended the Biot theory to numerically predict the velocities of a new surface mode as well as the expected pseudo Rayleigh and pseudo Stoneley modes at fluid-porous solid (i. e. fluid/fluid-saturated porous solid) interface. The expected modes are either below the fluid velocity, so that energy will not leak into the fluid at all, or just above it so that energy leaks at such a high angle, in which case simple phase-matching techniques are not applicable for excitation and detection. In a recent paper [4], the authors suggested the application of a slightly corrugated periodic surface, since, in this case every surface mode becomes somewhat "leaky" in certain directions at particular resonant frequencies. In this work we are presenting experimental results of surface wave velocity measurements, using this periodic surface technique. The surface wave velocities are compared to bulk results obtained by a new wedge technique.

BULK WAVE EXPERIMENT

There are three different bulk modes in a fluid-saturated porous solid: (i) shear wave with the same velocity as in the drained skeleton frame, (ii) fast compressional wave with velocity higher than both the fluid and frame longitudinal velocities, and (iii) slow compressional wave with velocity lower than both the fluid longitudinal and frame shear velocities.

---

*Permanent Address: Applied Biophysics Laboratory, Technical University
Budapest, Hungary

Plona [2] was the first to observe Biot's propagating slow wave at ultrasonic frequencies. He used water-saturated plates made of sintered glass beads, and distinguished the three modes by their different transit time through the sample as he changed the angle of incidence.

We used a somewhat different "wedge" technique to separate the through-transmitted signals by their different propagation directions according to the law of refraction. Fig. 1 shows the received fast wave signal which is bent toward the thinner edge of the sample and arrives earlier when transmitted through the thicker part of the wedge. Fig. 2 shows the received slow wave signal which is bent toward the thicker edge of the sample and arrives later when transmitted through the thicker part of the wedge. The materials used in our experiments were Eaton Products EP Brand Porous Structures of Grades 15 and 55 (grades indicate nominal pore size in microns). The porous structures are manufactured by cementing glass beads and their total void volume is about 30% for all grades. The results of the bulk wave velocity measurements are shown in Table 1.

TABLE 1.  Results of the bulk wave velocity measurements

| MATERIAL | SLOW WAVE [m/s] | FAST WAVE [m/s] | SHEAR WAVE [m/s] |
|---|---|---|---|
| EP Grade 15 | 820 | 2860 | 1480 |
| EP Grade 55 | 840 | 2920 | 1540 |

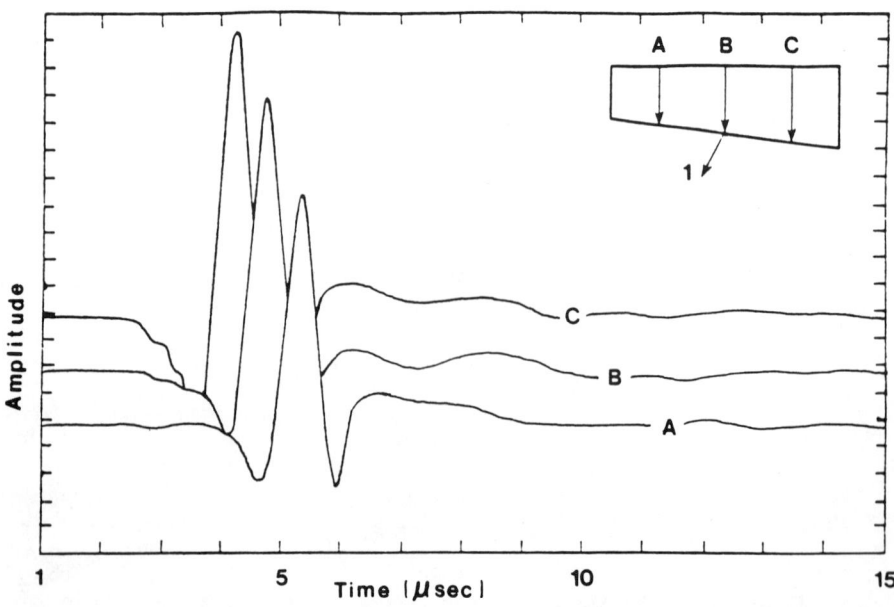

Fig. 1.  Fast wave arrivals through different sample thicknesses.

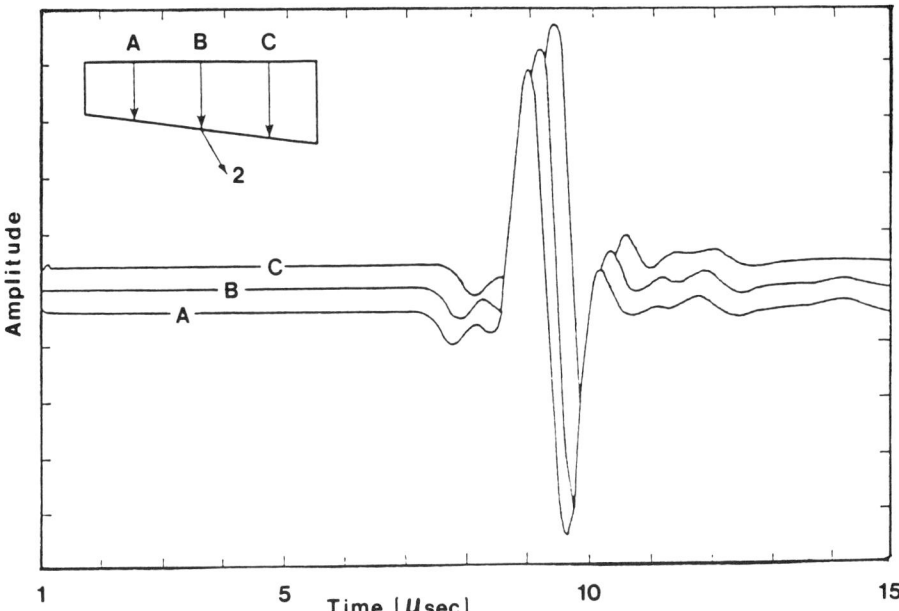

Fig. 2. Slow wave arrivals through different sample thicknesses.

SURFACE WAVE EXPERIMENT

At a fluid-isotropic solid interface, two different types of surface waves may exist. The true surface wave is the so-called Stoneley wave with a velocity lower than all of the bulk wave velocities in the neighboring media. The second wave is a Rayleigh type with a velocity higher than that of at least one of the bulk velocities of the two media, and this "leaky" or "pseudo" surface wave will be attenuated by leaking energy into the medium of lower velocity.

If the leaky surface velocity is higher than the sound velocity in the fluid, which is true for most cases except for a few solid materials such as plexiglas, the so-called Rayleigh angle phenomenon occurs due to mode conversion at the angle of incidence where the bulk wave in the fluid is phase-matched to the surface wave propagating along the interface. The reradiation of the leaky surface wave will occur at the Rayleigh angle, which can be detected in many different ways [5-7], such as observing the Schoch displacement, beam splitting on a schlieren image or by detecting the increased backscattering signal using mechanical scanning. The surface velocity is easily calculated from the Rayleigh angle using Snell's law.

If the surface wave does not leak into the fluid under ordinary conditions, special measures must be taken to induce mode coupling between the surface wave and the fluid bulk wave. One solution is to corrugate the surface with a series of periodic grooves. In this case, the condition of phase matching is expressed by the law of diffraction:

$$\frac{f}{c_s} - \frac{f \Lambda \sin\theta_i}{c_f} = m, \text{ where } m = 0, \pm 1, \pm 2 \ldots \quad (1)$$

Here f is the frequency, $\Lambda$ is the periodicity of the surface, $c_s$ and $c_f$ are the velocities of the surface wave and the fluid wave, respectively, and $\Theta_i$ is the mth order direction in which the surface wave will leak energy into the fluid as it propagates along the surface. By this technique, which was originally developed by Jungman, Adler and Quentin [8], we insonify the sample at normal incidence ($\Theta_i = 0$) and determine the surface velocity from the first order (m = 1) coupling condition of Eq. 1, $c_s = f_1 \Lambda$.

The presence of mode coupling can be detected by the sharp drops at certain frequencies in the reflectance of the interface. Fig. 3 shows the spectrum of the reflected broadband signal from a plexiglas sample which has about the same shear and longitudinal velocities as the shear and fast compressional wave velocities in the water-saturated porous samples. The two strong minima at 1.7 and 2.4 MHz correspond to the true Stoneley and pseudo Rayleigh velocities of about 1070 and 1500 m/s, respectively. The details of the periodic surface technique have been described in detail by Jungman et al. [9] and will not be repeated here.

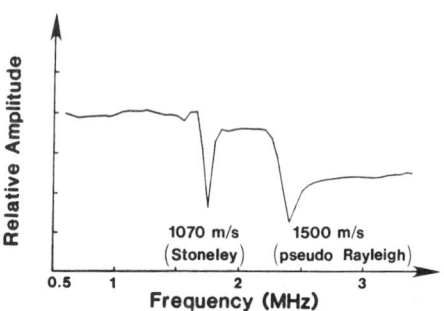

Fig. 3. Reflected spectrum from a water-plexiglas surface of 635 μm periodicity and 85 μm groove depth.

The above nomenclature often causes confusion. In most cases, the shear velocity of the solid is considerably higher than the sound velocity in the liquid, therefore the pseudo Rayleigh velocity is just a little higher than the true Rayleigh velocity of the free solid surface, and the Stoneley velocity is slightly below the fluid velocity. In this rather unusual case of plexiglas or porous solid, the shear velocity is lower than the fluid velocity, therefore the pseudo Rayleigh velocity is slightly higher than the fluid velocity and the Stoneley velocity is just below the Rayleigh velocity of the free solid surface.

FLUID-POROUS SOLID INTERFACE

In a recent study [3], Feng and Johnson introduced a numerical approach to predict surface modes and calculate their velocities at a fluid-porous solid interface. According to their predictions, a water-satured fused glass bead sample can exhibit three different surface modes: (i) pseudo Rayleigh wave with a velocity between the fluid velocity and the shear velocity of the porous solid, (ii) pseudo Stoneley wave with velocity higher than the slow bulk wave velocity in the saturated solid b ut lowe than both the shear velocity and the fluid velocity, and (iii) the true Stoneley wave with velocity lower than the lowest bulk velocity, i.e. the slow wave velocity. The detailed analysis shows that depending on the stiffness of the frame and surface conditions (open or closed pores), one, two or all three of these modes can appear on the interface.

The above introduced periodic surface technique was used to observe the different surface modes on the water-saturated porous samples. As an example, the deconvolved spectrum of the reflected signal for a grade 15 sample with periodic grooves at 635 µm spacing is shown in Fig. 4. The two higher frequency minima in the reflectance correspond to 1200 and 1500 m/s, indicating the presence of both pseudo Stoneley and pseudo Rayleigh waves. The third, less pronounced, minimum below these frequencies was found in many, but not all cases, and it corresponds to approximately 860 m/s. Figs. 5 and 6 show further examples of the reflected spectrum for different periodicities and groove depths.

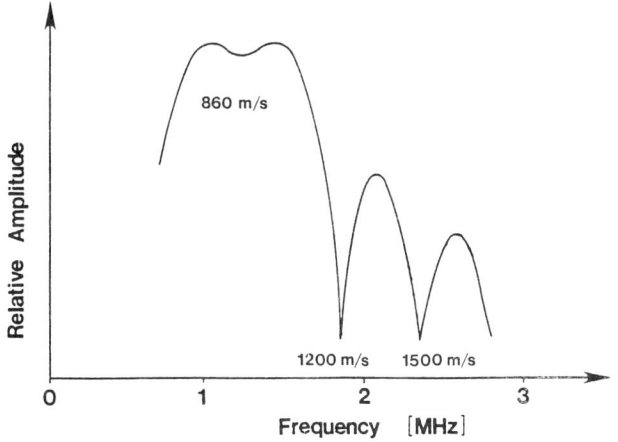

Fig. 4. Deconvolved spectrum of the reflected signal from a water-saturated porous sample of 635 µm periodicity and 320 µm groove depth.

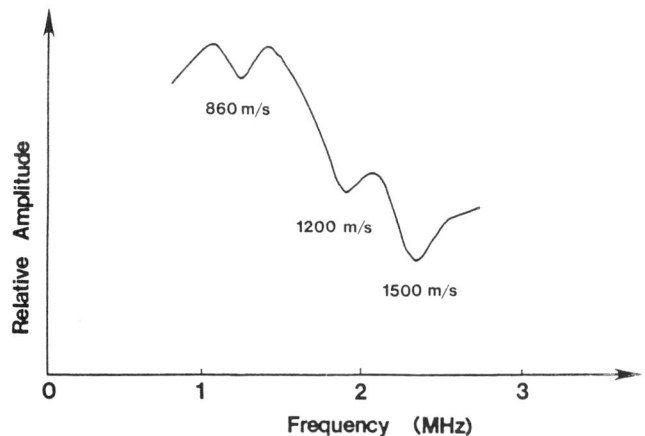

Fig. 5. Deconvolved spectrum of the reflected signal from a water-saturated porous sample of 635 µm periodicity and 180 µm groove depth.

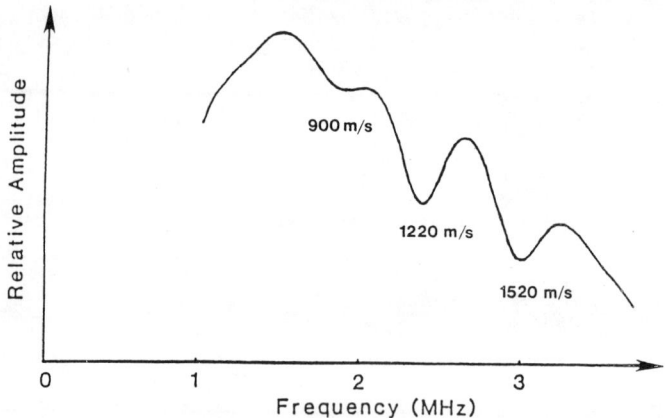

Fig. 6. Deconvolved spectrum of the reflected signal from a water-saturated porous sample of 508 μm periodicity and 254 μm groove depth.

The two higher surface modes were easily identified in all cases. The lowest minimum seems to be very close to the slow compressional wave, and we could not find it in other porous materials such as natural rocks where the slow bulk mode was missing, too. This clearly indicates that the lowest mode is associated somehow with the slow compressional wave. It might be attributed to a weak bulk slow wave propagating at grazing angle. We have found, in isotropic materials, that a bulk wave parallel to the surface will also produce a slight minimum. Otherwise, this minimum might be identified as the above mentioned true Stoneley wave with velocity just below the slow compressional wave speed. This surface mode is predicted only for closed, but not for open pores at the surface, therefore this would contradict Feng and Johnson's results [3] unless the pores are proved to be at least partially closed.

Table 2 summarizes our surface wave velocity results. Although these velocities are obviously pore size dependent, we found the differences between grades 15 and 55 to be smaller than the spread of the experimental data for different groove periodicities and depths, therefore we averaged the results for different grades as well.

TABLE 2. Experimental results for different surface modes

| SURFACE MODE | 1st | 2nd (pseudo Stoneley) | 3rd (pseudo Rayleigh) |
|---|---|---|---|
| Average Velocity | 870 m/s | 1210 m/s | 1530 m/s |
| Spread | 4.5% | 3.9% | 2.8% |

CONCLUSIONS

Our measurements provide experimental verifications of the existence of different surface modes for fluid/fluid-saturated porous solid interfaces. The results are in good qualitative agreement with theoretical predictions.

The accuracy of this method will depend on the precision with which the periodic grooves can be machined into the samples, proper time gating, and distortion of the surface mode by the periodic nature of the interface.

ACKNOWLEDGEMENT

This work was supported by the United States Department of Energy through the Lawrence Livermore National Laboratory, under Contract W7405-ENG.#48.

REFERENCES

1. M. A. Biot, J. Acoust. Soc. Am. 28, 168 (1956) and 28, 179 (1956).
2. T. J. Plona, Appl. Phys. Lett. 36, 259 (1980).
3. S. Feng and D. L. Johnson, J. Acoust. Soc. Am. 79, 906 (1983) and 79, 915 (1983).
4. M. J. Mayes, P. B. Nagy, L. Adler, B. P. Bonner and R. Streit, J. Acoust. Soc. Am. 79, 249 (1986).
5. A. Schoch, Acustica 2, 18 (1952).
6. H. L. Bertoni and T. Tamir, Appl. Phys. 2, 157 (1973).
7. W. G. Neubauer, J. Appl. Phys. 44, 48 (1973).
8. A. Jungman, L. Adler and G. Quentin, J. Appl. Phys. 53, 4672 (1982).
9. A. Jungman, L. Adler, J. D. Achenbach and R. Roberts, J. Acoust. Soc. Am. 79, 1025 (1983).

# INTERACTIVE DIFFRACTION OF A PLANE LONGITUDINAL WAVE BY A PAIR OF COPLANAR CENTRAL CRACKS IN AN ELASTIC SOLID

Y. M. Tsai

Department of Engineering Science & Mechanics and
Engineering Research Institute,
Iowa State University, Ames, Iowa 50011

## INTRODUCTION

The problem of the diffraction of a plane longitudinal wave by a penny-shaped crack has been solved using the techniques of Hankel transforms [1]. The interaction of elastic waves with a Griffith crack has been investigated for a range of values of the wave frequency [2]. More recently, approximate formulas have been derived for the problem of diffraction of elastic waves by two coplanar Griffith cracks in an infinite elastic medium [3].

The scattering of a plane longitudinal wave at a pair of coplanar central cracks in an elastic solid is investigated in the present work. The wave is harmonic in time, impinging normally on the crack surfaces, which are assumed to remain separated during small deformations of the solid. The interaction between the cracks affects not only the scattering of the wave, but also the severity of the cracks.

The diffraction problem forms a three-part mixed boundary problem and is solved by the method of Fourier transform. The singular integral equation involved in the problem is solved using the techniques of the finite Hilbert transform. The results are simplified through the process of contour integrations.

The stress intensity factors at the inner and outer crack tips are obtained in exact expressions. The parameter functions involved in the expressions are solved by an iteration process. The maximum values of the stress intensity factors are investigated in terms of the frequency factor, the ratio of the crack separation distance to the crack length and the value of Poisson's ratio. The motion of the crack surfaces is also investigated as a function of the wave frequency, the material constants, and the distance between the cracks.

## FORMAL SOLUTION

A two-dimensional infinite elastic solid is assumed to contain a pair of central cracks situated along the x-axis. The cracks are located symmetrically with respect to the vertical y-axis. A plane harmonic wave is moving in the positive direction of the y-axis and impinges normally on the crack surfaces. The scattered wave field generated by the cracks must satisfy the equations of motion for a homogeneous, isotropic solid.

The boundary conditions for the scattered wave field at y = 0 can be written as [1,4]

$$\sigma_{xy} = 0, \text{ all } x \tag{1}$$

$$\sigma_{yy} = -P_0 e^{i\omega\tau}, \text{ in } L_2 \tag{2}$$

$$v = \begin{cases} \phi(x), & \text{in } L_2 \\ 0, & \text{in } L_1 \text{ and } L_3 \end{cases} \tag{3}$$

where the interval $L_2$ is $a \leq x \leq b$; $L_1$ $0 < x < a$; $L_3$; $b < x < \infty$. The distances a and b, respectively, locate the inner and outer tips of the cracks. The unknown-crack-shape function $\phi(x)$ is to be determined later. The quantities A and B are determined after Eqs. (9) and (11) have been satisfied. In terms of the results for A and B, the normal stress at y = 0 can be written as

$$\sigma_{yy}/M = -\sqrt{\frac{2}{\pi}} \int_0^\infty \cos(sx) s\, \bar{\phi}_c \, ds$$

$$+ \sqrt{\frac{2}{\pi}} \int_0^\infty \cos(sx) s Q(s,\omega) \bar{\phi}_c \, ds \tag{4}$$

$$Q = \frac{(2s^2 - k_2^2)^2 - 4s^2\alpha\beta}{2s\alpha(k_2^2 - k_1^2)} + 1, \quad M = \mu/(1-\nu). \tag{5}$$

$$\alpha^2 = s^2 - k_1^2, \quad \beta^2 = s^2 - k_2^2, \quad k_1 = \omega/c_1, \quad k_2 = \omega/c_2$$

The omission of the time factor $\exp(i\omega\tau)$ is understood in the above expressions and in the later analyses; $c_1$ and $c_2$ are, respectively, the dilatational and shear wave speeds. To satisfy the conditions in Eqs. (2) and (3), the cosine transform of $\phi(x)$ is written as

$$\bar{\phi}_c = \frac{1}{s} \int_a^b h(t^2) \sin(ts)\, dt . \tag{6}$$

In terms of the parameter function $h(t^2)$, the integration over s in the second term on the right-hand side of Eq. (4) can be simplified by using the techniques of contour integration [1,5,6]. The results after integration are written in the following form:

$$\int_0^\infty Q(s,\omega)\cos(sx)\sin(st)\,ds$$

$$= (-i)2(1-\nu)k_2 \left\{ \int_0^{1/k} \frac{(2\xi^2-1)^2}{2\xi(1/k^2-\xi^2)^{1/2}} \cos(xk_2\xi)\sin(tk_2\xi)\,d\xi \right.$$

$$+ \int_0^1 2\xi(1 - \xi^2)^{1/2}\cos(xk_2\xi)\sin(tk_2\xi)d\xi\Big\}$$

$$= L\cos(xk_2\xi)\sin(tk_2\xi)$$

$$= (-i)2(1 - \nu)k_2L_0\cos(xk_2\xi)\sin(tk_2\xi) \tag{7}$$

where $k = c_1/c_2$. Both operator $L$ and $L_0$ operate over the variable $\xi$.

In terms of Eqs. (4), (6), and (7), the boundary condition in Eq. (2) leads to a singular integral equation in $L_2$ as follows:

$$\sqrt{\frac{2}{\pi}} \int_a^b \frac{th(t^2)dt}{t^2 - x^2} = \frac{P_o}{M} + \sqrt{\frac{2}{\pi}} L \int_a^b h(t^2)(\sin(tk_2\xi)dt \cos(xk_2\xi). \tag{8}$$

The techniques of the finite Hilbert transform [7] are applied to Eq. (8) to convert it into the following integral equation:

$$h(t^2) = \sqrt{\frac{2}{\pi}}\frac{P_o}{M}\left(\frac{t^2 - a^2}{b^2 - t^2}\right)^{1/2} + C_1\left[(t^2 - a^2)(b^2 - t^2)\right]^{-1/2}$$

$$+ \left(\frac{2}{\pi}\right)^2 L \int_a^b h(\eta^2)\sin(\eta k_2\xi)d\eta \left[\frac{1}{2} I_o(t,\xi)\right.$$

$$+ \left.\left(\frac{t^2 - a^2}{b^2 - t^2}\right)^{1/2} I_1(k_2\xi)\right] \tag{9}$$

$$I_1 = \int_a^b \lambda[(b^2 - \lambda^2)(\lambda^2 - a^2)]^{-1/2}\cos(k_2\xi\lambda)d\lambda \tag{10}$$

$$I_o = k_2\xi \int_a^b \ln\frac{[(b^2 - t^2)(\lambda^2 - a^2)]^{1/2} + [(t^2 - a^2)(b^2 - \lambda^2)]^{1/2}}{[(b^2 - t^2)(\lambda^2 - a^2)]^{1/2} - [(t^2 - a^2)(b^2 - \lambda^2)]^{1/2}}$$

$$\sin(k_2\xi\lambda)d\lambda. \tag{11}$$

The boundary conditions in Eq. (3) require $h(t^2)$ to satisfy the following equation:

$$\int_a^b h(t^2)dt = 0. \tag{12}$$

This condition determines the value of $C_1$ in Eq. (9). For later convenience in determining the stress intensity factors at the crack tips, the integral equation is also written as

$$h(t^2) = -\sqrt{\frac{2}{\pi}}\frac{P_o}{M}\left(\frac{b^2 - t^2}{t^2 - a^2}\right)^{1/2} + C_2\left[(t^2 - a^2)(b^2 - t^2)\right]^{-1/2}$$

$$- \left(\frac{2}{\pi}\right)^2 L \int_a^b h(\eta^2) \sin(\eta k_2 \xi) d\eta \left[\frac{1}{2} I_o(t,\xi)\right.$$

$$\left. + \left(\frac{b^2 - t^2}{t^2 - a^2}\right)^{1/2} I_1(k_2 \xi)\right] . \qquad (13)$$

The value of $C_2$ is also determined by satisfying the condition in Eq. (9).

## STRESS INTENSITY FACTOR

The stresses are singular at both the inner and outer crack tips. The intensity of the singularity depends upon the wave frequency and the distance between the cracks. The normal stress on the crack plane can be calculated from Eq. (4) in terms of Eqs. (6), (7), and the parameter function $h(t^2)$. The function is to be solved from Eqs. (9), (12), and (13).

The integral equation (9) is solved using the method of successive approximations. The first two terms on the right-hand side of Eq. (9) are considered as the first order solution, subject to the condition of Eq. (12). The results can be written as

$$\sqrt{\frac{\pi}{2}} \frac{M}{P_o} h_1(t^2) = \left(\frac{t^2 - a^2}{b^2 - t^2}\right)^{1/2} + \frac{b^2 Y_1}{[(b^2 - t^2)(t^2 - a^2)]^{1/2}} \qquad (14)$$

$$Y_1 = a^2/b^2 - E(\delta)/K(\delta) \qquad (15)$$

where $K(\delta)$ and $E(\delta)$ are, respectively, the complete elliptic integrals of the first and the second kind with the modulus $\delta = (1 - a^2/b^2)^{1/2}$.

If the first approximation is substituted into Eq. (6) the second approximation becomes

$$\sqrt{\frac{\pi}{2}} \frac{M}{P_o} h_2(t^2) = \left(\frac{t^2 - a^2}{b^2 - t^2}\right)^{1/2} (1 - i\varepsilon_1)$$

$$+ b^2 Y_2 [(b^2 - t^2)(t^2 - a^2)]^{-1/2} - i G_1(t,k_2) \qquad (16)$$

$$\varepsilon_1 = d_{o1} + Y_1 d_{o2} \qquad (17)$$

$$Y_2 = Y_1 (1 - i\varepsilon_1) + i[e_{o1} + Y_1 e_{o2}]/K(\delta) \qquad (18)$$

$$G_1(t,k_2) = T_{o1}(t,k_2) + Y_1 T_{o2}(t,k_2) \qquad (19)$$

$$d_{o1} = 2 \, do \, \frac{b}{c} (1 - a^2/b^2) L_o \int_o^{\pi/2} \cos^2\psi \, \sin(\bar{k}\bar{\eta}\xi) \frac{d\psi}{\bar{\eta}} \int_o^{\pi/2} \cos(\bar{k}\bar{\lambda}\xi) d\theta$$

$$d_{o2} = 2d_o \frac{b}{c} L_o \int_0^{\pi/2} \sin(\bar{k}\bar{\eta}\xi) \frac{d\psi}{\bar{\eta}} \int_0^{\pi/2} \cos(\bar{k}\bar{\lambda}\xi) d\theta$$

$$\bar{\eta} = \frac{b}{c} [1 - (1 - a^2/b^2)\sin^2\psi]^{1/2}$$

$$\bar{\lambda} = \frac{b}{c} [1 - (1 - a^2/b^2)\sin^2\theta]^{1/2}$$

$$d_o = 2(1 - \nu)\bar{k}\left(\frac{2}{\pi}\right)^2 \frac{b}{c}$$

$$T_{o1} = d_o \frac{b}{c} (1 - a^2/b^2) L_o I_o(t,\xi) \int_0^{\pi/2} \cos^2\psi \sin(\bar{k}\bar{\eta}\xi) d\psi/\bar{\eta}$$

$$T_{o2} = d_o \frac{b}{c} L_o I_o(t,\xi) \int_0^{\pi/2} \sin(\bar{k}\bar{\eta}\xi) d\psi/\bar{\eta}$$

$$b\, e_{o1} = \int_a^b T_{o1}(t,k_2) dt; \qquad b\, e_{o2} = \int_a^b T_{o2}(t,k_2) dt. \tag{20}$$

The frequency factor is $\bar{k} = \omega c/c_2$, where the crack length is $c = b - a$.

The process of approximation can be continued to any higher-order solution. The n-th order solution can be written as

$$\sqrt{\frac{\pi}{2}} \frac{M}{P_o} h_n(t^2) = \left(\frac{t^2 - a^2}{b^2 - t^2}\right)^{1/2} (1 - i\,\varepsilon_{n-1})$$
$$+ b^2 Y_n[(b^2 - t^2)(t^2 - a^2)]^{-1/2} - i\, G_{n-1}(t,k_2). \tag{21}$$

The function $G_{n-1}(t,k_2)$ has no effects on the stress intensity factors and is presented in the next section in conjunction with the crack surface displacement. In the process of approximation, there are recurrent functions for $n \geq 1$ as follows

$$b\, G_{n1}(\xi) = \int_a^b T_{(n-1)1}(t,k_2) \sin(k_2 \xi t) dt$$

$$b\, G_{n2}(\xi) = \int_a^b T_{(n-1)2}(t,k_2) \sin(k_2 \xi t) dt$$

$$T_{n1}(t) = d_o L_o G_{n1}(\xi) I_o(t,\xi)$$

$$T_{n2}(t) = d_o L_o G_{n2}(\xi) I_o(t,\xi)$$

$$d_{n1} = 2 d_o L_o G_{n1}(\xi) \int_0^{\pi/2} \cos(\bar{k}\bar{\lambda}\xi) d\theta$$

$$d_{n2} = 2 d_o L_o G_{n2}(\xi) \int_0^{\pi/2} \cos(\bar{k}\bar{\lambda}\xi) d\theta$$

$$b\, e_{n1} = d_o L_o G_{n1}(\xi) \int_a^b I_o(t,\xi) dt$$

$$b\, e_{n2} = d_o L_o G_{n2}(\xi) \int_a^b I_o(t,\xi) dt \cdot \qquad (22)$$

The nondimensional functions in Eq. (21) can now be written in the following forms:

$$\varepsilon_n = \sum_{j=0}^{n-1} (-i)^j [(1 - i\,\varepsilon_{n-j-1}) d_{j1} + \gamma_{n-j}\, d_{j2}] \qquad (23)$$

$$\gamma_n = \gamma_1 (1 - i\,\varepsilon_{n-1}) - \sum_{j=0}^{n-2} (-i)^{j+1} [(1 - i\,\varepsilon_{n-2-j}) e_{j1}$$

$$+ \gamma_{n-j-1}\, e_{j2}]/K(\delta) \qquad (24)$$

The value of $\varepsilon_n$ vanishes if the subscript n is equal to or less than zero.

If the solution $h_n(t^2)$ is substituted into Eq. (4), the normal stress on the crack plane for $x > b$ can be determined in terms of Eqs. (6) and (7). The singular part of the normal stress for $x > b$ is obtained in closed forms as follows:

$$\frac{M}{P_o} \sigma_{yy}\Big|_{singular} = (1 - i\varepsilon_{n-1}) \left(\frac{x^2 - a^2}{x^2 - b^2}\right)^{1/2} + \frac{b^2 \gamma_n}{[(x^2 - a^2)(x^2 - b^2)]^{1/2}} \cdot$$

$$(25)$$

The stress intensity factor $K_b$ at the outer crack tip $x = b$ is calculated and normalized in the following form:

$$\frac{K_b}{K_I} = \left[\frac{2}{(1 - a/b)(1 - a^2/b^2)}\right]^{1/2} \left|(1 - i\varepsilon_{n-1})\left(1 - \frac{a^2}{b^2}\right) + \gamma_n\right| \qquad (26)$$

where the associated, stress intensity factor for a single crack is $K_I = p_o(\pi c/2)^{1/2}$.

The stress intensity factor $K_a$ at the inner crack tip $x = a$ is determined using the second form of solution in Eq. (13). The solution has the following form:

$$\sqrt{\frac{\pi}{2}} \frac{M}{P_o} h_n(t^2) = - \left(\frac{b^2 - t^2}{t^2 - a^2}\right)^{1/2} (1 - i\,\varepsilon'_{n-1})$$

$$+ b^2 \gamma_n' [(t^2 - a^2)(b^2 - t^2)]^{-1/2} - i\, G_{n-1}'(t, k_2) \tag{27}$$

$$\gamma_1' = 1 - E(\delta)/K(\delta), \quad \varepsilon_1' = -d_{o1}' + \gamma_1' d_{o2}$$

$$d_{o1}' = 2d_o \frac{b}{c} \left(1 - \frac{a^2}{b^2}\right) L_o \int_0^{\pi/2} \sin^2\psi \sin(\bar{k}\bar{\eta}\xi) \frac{d\psi}{\bar{\eta}} \int_0^{\pi/2} \cos(\bar{k}\bar{\lambda}\xi)\, d\theta$$

$$T_{o1}' = d_o \frac{b}{c} \left(1 - \frac{a^2}{b^2}\right) L_o\, I_o(t, \xi) \int_0^{\pi/2} \sin^2\psi \sin(\bar{k}\bar{\eta}\xi)\, d\psi/\bar{\eta}$$

$$b\, e_{o1}' = \int_a^b T_{o1}'(t, k_2)\, dt \tag{28}$$

$$\varepsilon_n' = -\sum_{j=0}^{n-1} (-i)^j [(1 - i\, \varepsilon_{n-j-1}) d_{j1}' - \gamma_{n-j}' d_{j2}]$$

$$\gamma_n' = \gamma_1'(1 - i\varepsilon_{n-1}') + \sum_{j=0}^{n-2} (-i)^{j+1} [1 - i\varepsilon_{n-2-j}') e_{j1}'$$

$$- \gamma_{n-j-1}' e_{j2}]/K(\delta) \tag{29}$$

The "primed" quantities on the right-hand sides of Eq. (29) are obtained from Eq. (22) by placing primes on all the quantities whose second subscripts are one. The exact expression for the singular part of the normal stress for $x < a$ is obtained as follows:

$$\frac{M}{P_o} \sigma_{yy}\bigg|_{singular} = (1 - i\varepsilon_{n-1}') \left(\frac{b^2 - x^2}{a^2 - x^2}\right)^{1/2} - \frac{b^2 \gamma_n'}{[(b^2 - x^2)(a^2 - x^2)]^{1/2}} \tag{30}$$

The normalized, stress intensity factor has the following form:

$$\frac{K_a}{K_I} = \left[\frac{2}{(1 - a/b)(1 - a^2/b^2)a/b}\right]^{1/2} \left|(1 - i\varepsilon_{n-1}')\left(1 - \frac{a^2}{b^2}\right) + \gamma_n'\right| \tag{31}$$

The normalized, stress intensity factors at both the inner and outer crack tips are functions of the frequency factor $\bar{k}$, the distance ratio $a/b$, and Poisson's ratio. The recurrent expressions are systematic and convenient for numerical calculations to any higher-order approximations.

Numerical calculations were carried out for $\nu = 0.3$ and for various values of $\bar{k}$ and $a/c$. The approximation process converges rapidly; very accurate results are obtained for $n = 5$, which is the highest order approximation for all the results presented. The normalized, stress intensity factors at the inner and outer crack tips are shown, respectively, in Figs. 1 and 2. It can be seen from the curves that the wave frequency and the crack separation distance have much larger effects at the inner crack tip than at the outer tip. The higher values at the inner tip indicate

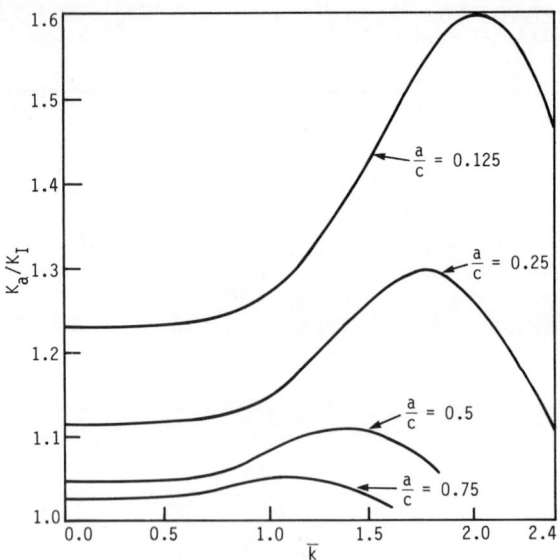

Fig. 1. Normalized stress intensity factor at the inner crack tip.

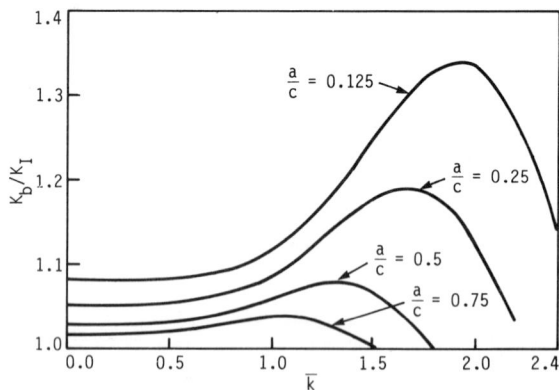

Fig. 2. Normalized stress intensity factor at the outer crack tip.

that the crack growth tends to start from the inner tip, which is consistent with a recent finding on the fracture of specimens containing double cracks under static loading [8].

The maximum values of the stress intensity factors depend upon the value of the frequency factor and the ratio of the crack separation distance to the crack length (see Figs. 1 and 2). The maximum value occurs at a higher frequency for a smaller value of $a/c$. If the crack separation distance decreases, the maximum value of the stress intensity factor increases. For $a/c = 0.125$, the maximum $K_a$ has 60% increase over the associated, single-crack stress intensity factor $K_I$.

Crack Surface Displacement

The normal displacement of the crack surface is calculated from Eqs. (3) and (6) for x in $L_2$ as follows:

$$v = \phi = \sqrt{\frac{2}{\pi}} \int_0^\infty \bar{\phi}_c \cos(sx)ds = \sqrt{\frac{\pi}{2}} \int_x^b h(t^2)dt . \qquad (32)$$

Substituting Eq. (21) yields, after some calculations, the following crack shape function:

$$\frac{M\phi}{p_o b} = (1 - i\varepsilon_{n-1})\left[E(\delta,\phi_x) - \frac{a^2}{b^2}F(\delta,\phi_x)\right]$$

$$+ \gamma_n F(\delta,\phi_x) - i\sum_{j=0}^{n-2}(-i)^j(1 - i\varepsilon_{n-2-j})\int_x^b T_{j1}(t,k_2)dt$$

$$- i\sum_{j=0}^{n-2}(-i)^j \gamma_{n-1-j}\int_x^b T_{j2}(t,k_2)dt \qquad (33)$$

$$\phi_x = \sin^{-1}[(b^2 - x^2)/(b^2 - a^2)]^{1/2} . \qquad (34)$$

The functions $F(\delta,\phi_x)$ and $E(\delta,\phi_x)$ are, respectively, the elliptic integrals of the first and the second kind of the amplitude $\phi_x$ and the modulus $\delta$. The last two summation terms represent the integration of the function $G_{n-1}(t,k_2)$ in Eq. (21).

The crack shape function is calculated for various values of $\bar{k}$ and a/c. For a/c = 0.125, the calculated crack shape normalized by the displacement at the center of the associated static crack surface $\phi_M$ is shown in Fig. 3 for four different values of $\bar{k}$. The crack surface reaches its maximum opening at $\bar{k} = 2$. The surface of the crack is not symmetrical; the crack opening is larger near the inner tip than the outer tip.

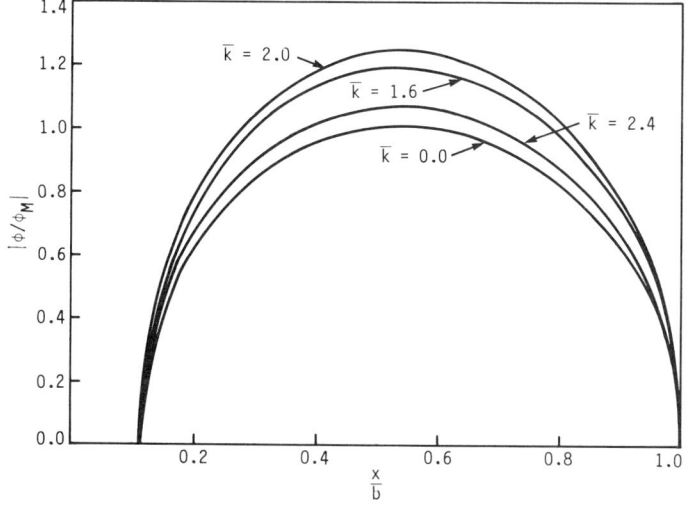

Fig. 3. Normalized crack shape for $\frac{a}{c} = 0.125$.

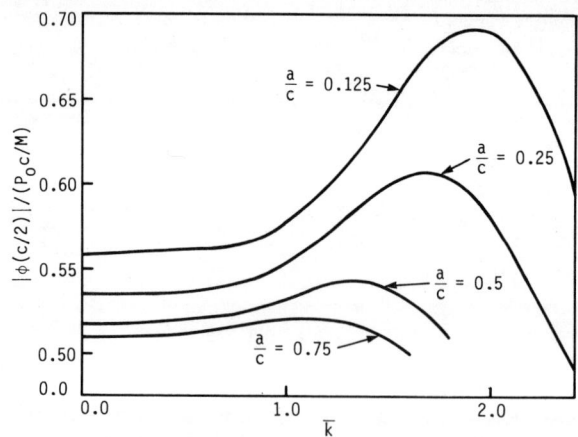

Fig. 4. Normalized crack center displacement.

The normalized value of the magnitude of the surface displacement at the crack center is shown in Fig. 4. The normalized displacement is a function of the frequency factor and the ratio $a/c$. For $a/c = 0.125$, the maximum crack opening occurs at $\bar{k} = 2$ and is 25% larger than the associated maximum static displacement. The frequency at which the maximum crack opening occurs increases with decreasing value of $a/c$. The magnitude of the maximum crack surface displacement also increases if the value of the ratio between the crack separation distance and the crack length decreases.


ACKNOWLEDGMENT

This research is supported by the Engineering Research Institute of Iowa State University.


REFERENCES

1. I. A. Robertson, Proc. Cambridge Philos. Soc., 63, 229 (1967).
2. A. K. Mal, Int. J. Eng. Sci., 8, 763 (1970).
3. D. L. Jain and R. P. Kanwal, Int. J. Solids Struct., 8, 961 (1972).
4. Y. M. Tsai, Int. J. Solids Struct., 9, 625 (1973).
5. Y. M. Tsai and H. Kolsky, J. Mech. Phys. Solids, 15, 263 (1967).
6. H. Lamb, Philos. Trans. R. Soc. London, A 203, 1 (1904).
7. M. Lowengrub and K. N. Srivastava, Int. J. Eng. Sci. 6, 359 (1968).
8. W. H. Lin, M.S. Thesis, Iowa State University, Ames (1985).

# ULTRASONIC SCATTERING BY

# PLANAR AND NON-PLANAR CRACKS

Subhendu K. Datta

Department of Mechanical Engineering
and Cooperative Institute
for Research in Environmental Sciences
University of Colorado
Boulder, CO 80309, U.S.A.

Arvind H. Shah

Department of Civil Engineering
University of Manitoba
Winnipeg, Canada R3T 2N2

## INTRODUCTION

Problems of elastic wave scattering by surface-breaking and near-surface cracks are of considerable current interest for ultrasonic nondestructive evaluation. Ultrasonic scattering by planar cracks near or at the free surface of a semi-infinite elastic homogeneous medium has been studied theoretically by many authors. References to recent papers on this subejct can be found in [1], [2] and [3]. Some experimental works on surface-breaking normal planar cracks have also appeared ([4] - [6]).

Ultrasonic scattering by surface-breaking planar and branched cracks of arbitrary orientation is the subject of this investigation. To our knowledge this problem has not received much attention in the literature. An approximate solution that is valid at low frequencies was presented in [7] for SH wave diffraction by a canted surface-breaking planar crack. Subsequently, a hybrid finite element and eigenfunction technique was used in [8] to study SH wave diffraction by planar surface-breaking canted crack.

In this paper we use the same hybrid technique as in [8] to study the scattering of in-plane body and surface waves by canted planar and normal surface-breaking branched cracks. We focus our attention to the near-field. Numerical results are presented for the vertical surface displacement amplitudes near the base of the crack.

## FORMULATION AND SOLUTION

Consider a homogeneous, isotropic, and linearly elastic medium with a near-surface inhomogeneity of arbitrary properties and shape as shown in Figure 1. Assume the displacement $u(x,y,t)$ at a point $P$ to be time-harmonic of the form $u(x,y)e^{-i\omega t}$,

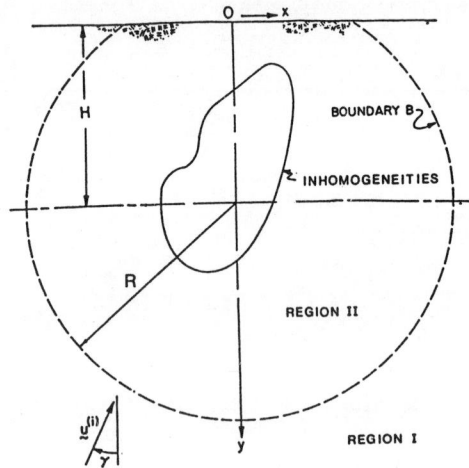

Fig. 1 Geometry of a near-surface inhomogeneity

where $\omega$ is the circular frequency. Then $u$ satisfies the equation of motion in $y > 0$ (at points not on the crack)

$$\mu \nabla^2 u + (\lambda + \mu) \nabla \nabla \cdot u = -\rho \omega^2 u \qquad (1)$$

where $\lambda, \mu$ are Lamé constants, $\rho$ the mass density and the factor $e^{-i\omega t}$ has been dropped.

The solution of (1) can be expressed in terms of the longitudinal and shear wave potentials, $\phi$ and $\psi$, in the form

$$u = \nabla \phi + \nabla \times (\psi e_z) \qquad (2)$$

Furthermore, in a homogeneous half-space, $\phi$ and $\psi$ can be expressed in infinite series of multipolar potentials as [9],

$$\phi = \sum_{n=-\infty}^{\infty} (a_n \phi_n^p + b_n \phi_n^s)$$

$$\psi = \sum_{n=-\infty}^{\infty} (a_n \psi_n^p + b_n \psi_n^s) \qquad (3)$$

where expressions for $\phi_n^p, \psi_n^p, \phi_n^s$ and $\psi_n^s$ can be found in [9]. The coefficients $a_n$, $b_n$ are found by satisfying the appropriate boundary conditions.

The representation (3) is not useful for satisfying the boundary conditions on the crack surface. For this reason, a different representation is needed in this near-field region. In this paper, the region inside the fictitious boundary $B$ (Fig. 1) is divided into finite elements having $N_I$ number of interior nodes and $N_B$ number of boundary nodes.

For the finite element representation in region II, the energy functional is taken to be

$$F = \frac{1}{2} \int \int_{R_{II}} (\sigma \cdot \epsilon^* - \rho \omega^2 u \cdot u^*) dx dy - \frac{1}{2} \int_B (P_B \cdot u_B^* + P_B^* \cdot u_B) ds \qquad (4)$$

where " * " indicates complex conjugate and $\sigma$ and $\epsilon$ are column vectors defined as

$$\sigma = \{\sigma\} = (\sigma_{xx}, \sigma_{yy}, \sigma_{xy})^T \tag{5}$$

$$\epsilon = \{\epsilon\} = (\epsilon_{xx}, \epsilon_{yy}, \epsilon_{xy})^T . \tag{6}$$

Superscript "$T$" denotes transpose. The $P_B$ and $U_B$ represent the traction and displacement on $B$, respectively.

It is assumed that the displacement field within the $j^{th}$ element is represented in terms of the shape functions $L_j(x,y)$ and elemental nodal displacements $\{q_j^e\}$ as

$$u^e = \sum_{j=1}^{N_e} L_j q_j^e \tag{7}$$

where each $q_j^e$ has two components $u_{xj}$ and $u_{yj}$ along the $x$ and $y$ directions, respectively. The $N_e$ represents the number of nodes in each element.

The $\sigma_{ij}^e$ and $\epsilon_{ij}^e$ are computed by substituting Eq. (7) into strain-displacement relations and these, in turn, into the stress-strain relations. Using these in Eq. (4), we get

$$F = q_I^{*T} S_{II} q_I + q_I^{*T} S_{IB} q_B + q_B^{*T} S_{BI} q_I$$
$$+ q_B^{*T} S_{BB} q_B - q_B^{*T} P_B^{(1)} - P_B^{*T(1)} q_B \tag{8}$$

in which $q_I = q_I{}^{(2)}, q_B = q_B{}^{(2)}, P_B{}^{(1)} = P_B{}^{(2)}$ and the elemental impedance matrices $S_{ij}$ are defined as

$$[S^e] = \int\int_{R_e} ([B^e]^T [D][B^e] - \rho_e \omega^2 [L]^T [L]) dx dy. \tag{9}$$

In Eq. (9),

$$[B^e] = \begin{bmatrix} \frac{\partial}{\partial x} & 0 \\ 0 & \frac{\partial}{\partial y} \\ \frac{\partial}{\partial y} & \frac{\partial}{\partial x} \end{bmatrix} \begin{bmatrix} L_1 & 0 & L_2 & .. \\ 0 & L_1 & 0 & .. \end{bmatrix} = [N][L].$$

Note that $[L]$ is a 2 x $2N_e$ matrix.

For an isotropic material $[D]$ is given by

$$[D] = \begin{bmatrix} \lambda_e + 2\mu_e & \lambda_e & 0 \\ \lambda_e & \lambda_e + 2\mu_e & 0 \\ 0 & 0 & \mu_e \end{bmatrix}$$

where $\lambda_e$ and $\mu_e$ are the Lamé's constant.

To find the constants $a_n, b_n$ appearing in (3) and the nodal displacements in region II, it is necessary to use the continuity of displacement and traction on $B$. This is discussed in the following.

The incident displacement fields will be assumed to arise from the incident plane $P$ and $SV$ waves, and their reflections from the free surface $y = 0$. The case of incident Rayleigh waves will also be considered.

Let us suppose that in the absence of the crack the free field is the sum of the incident and reflected fields, that is

$$u_j^{(0)} = u_j^{(i)} + u_j^{(r)} \qquad (j = 1, 2). \tag{10}$$

For the Rayleigh wave $u_j^{(0)}$ is the associated displacement.

The total field outside $B$ then is

$$u_j = u_j^{(S)} + u_j^{(0)} \qquad (j = 1, 2), \tag{11}$$

where $u_j^{(S)}$ is given by (2) and (3).

Using (2) and (3), the displacements at the nodes on $B$ can be written as

$$\{q_B^{(S)}\} = [G]\{a\} \tag{12}$$

where $[G]$ is a $2N_B \times 2N_B$ matrix formulated in [10] and vector $\{a\}$ is

$$\{a\} = [a_1, ..., a_{N_B}, b_1, ..., b_{N_B}]^T .$$

Similarly, using (2) and (3) in the stress strain relation, the traction at the nodes on $B$ can be expressed in the form

$$\{\sigma_B^{(S)}\} = [F]\{a\} \tag{13}$$

where $[F]$ is also a $2N_B \times 2N_B$ matrix defined in [10].

To express $\{\sigma_B^{(S)}\}$ in terms of $\{q_B^{(S)}\}$, we use the expression for the virtual work done on the boundary $B$, which is

$$\delta\pi = \int_B \{\delta q^{*(1)}_B\}^T \{\sigma_B^{(1)}\} d\Gamma \tag{14}$$

where superscript (1) denotes the total field in region $I$ (outside $B$).

Because of the continuity of displacements and traction on $B$, we have

$$q_B^{(1)} = q_B^{(2)} = q_B^{(0)} + q_B^{(S)} \tag{15}$$

$$\sigma_B^{(1)} = \sigma_B^{(2)} = \sigma_B^{(0)} + \sigma_B^{(S)} \tag{16}$$

where superscript (2) denotes the total field in region II.

Substituting (12), (13), (15), and (16) in Eq. (14), and noting that $\delta q_B^{(1)} = \delta q_B^{(S)}$, we obtain from eq. (14)

$$\delta\pi = \{\delta a^*\}^T \{P_B^{(1)}\} \tag{17}$$

where $P_B^{(1)}$ is given by

$$\{P_B^{(1)}\} = [\bar{R}]\{a\} + \{P_B^{(0)}\} \tag{18}$$

Here
$$[\bar{R}] = \int_B [G^*]^T [F] d\Gamma \quad (19)$$

and

$$\{P_B^{(0)}\} = \int_B [G^*]^T \{\sigma_B^{(0)}\} d\Gamma. \quad (20)$$

Substituting Eq. (12) into Eq. (8) and taking the variation, we obtain a set of simultaneous equations which may be written in matrix form as

$$\begin{bmatrix} S_{II} & S_{IB} G_I \\ G^{*T} S_{IB}^T & G^{*T} S_{BB}^G \end{bmatrix} \begin{bmatrix} q_I \\ a \end{bmatrix} = \begin{bmatrix} -S_{IB} q_B^{(0)} \\ -G^{*T} S_{BB} q_B^{(0)} + P_D^{(0)} \end{bmatrix}. \quad (21)$$

Using Eqs. (18) and (21), we obtain

$$[G^{*T}(S_{BB} - S_{IB}^T S_{II}^{-1} S_{IB} G^* - \bar{R}]\{a\} =$$
$$-G^{*T}(S_{BB} - S_{IB}^T S_{II}^{-1} S_{IB}) q_B^{(0)} + P_B^{(0)}. \quad (22)$$

In Eq. (22), the generalized coordinates $\{a\}$ are the only unknowns. Therefore, $\{a\}$ can be evaluated. Once $\{a\}$ are known, the near and far displacement and stress fields can be determined.

## NUMERICAL RESULTS AND DISCUSSION

In this paper, the boundary $B$ enclosing the interior region is not a complete circle, and so the potentials $\phi_n^p, \psi_n^p$, and $\phi_n^s, \psi_n^s$ cannot be expanded in cylindrical wave functions as was done in [1]. So the integrals giving these potentials and their derivatives were evaluated numerically for every node on $B$. The details are discussed in [11].

The hybrid method is employed to study scattering by P, SV and Rayleigh waves by three types of surface breaking cracks: a vertical crack (Fig. 2 with $\alpha = 90°$), a 45° inclined crack (Fig. 2 with $\alpha = 45°$), and a vertical branched (Y) crack (Fig. 3).

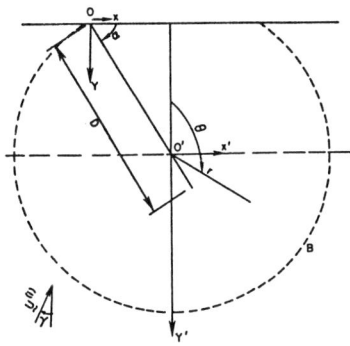

Fig. 2. Geometry of a planar crack

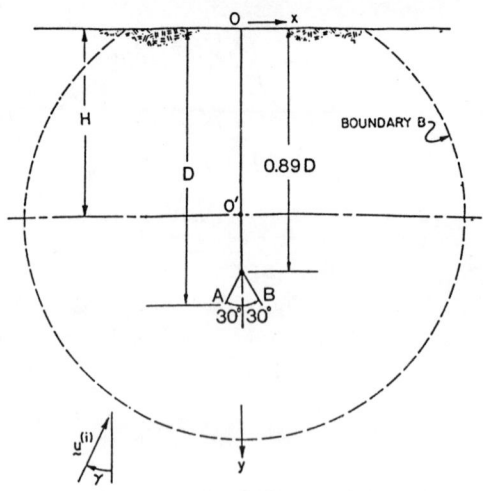

Fig. 3. A branched crack

Stress intensity factors at the tips of the cracks were calculated and for the particular case of a planar surface-breaking normal crack they were found to agree well with the results of [12].

The surface displacements at $y = 0$ are calculated by using (3) in (2) after $\{a\}$ are calculated. Normalized values of $u_y^{(S)}$ are presented in Figures 4-8. For each type of crack mentioned above five cases of incident waves were considered: plane P wave incident at 0° and 45°, plane SV wave incident at 0° and 45°, and finally Rayleigh wave. Some representative results are shown here.

Figures 4 and 5 show the scattered vertical surface displacement amplitudes for a Rayleigh wave incident from the left on a normal planar and branched crack. It is seen that there are large differences in the forward direction between the two cases as the frequency becomes large. In the backward direction, however, the differences are not very significant. Figures 6 and 7 show the results for an incident SV wave moving vertically as well as at 45° to the vertical. Large differences are found for vertical incidence, but not in the other case. Finally, in Figure 8 is shown the case of a Rayleigh wave incident on a canted crack. This figure is to be contrasted with Fig. 4. The large contrast shown clearly distinguishes a canted crack from a normal crack.

CONCLUSION

Model calculations of elastic wave scattering by surface-breaking planar and non-planar cracks have been presented. These calculations show that near-field surface displacements due to scattering by planar and branched cracks are quite different even when the branches are small. Also, it is found that signatures of normal and canted cracks are very dissimilar. These characteristic differences can be used to discriminate between the various cases.

Although the results presented here are for homogeneous medium, the technique can be generalized to study cracks in a composite medium. These are presently under investigation and will be reported later.

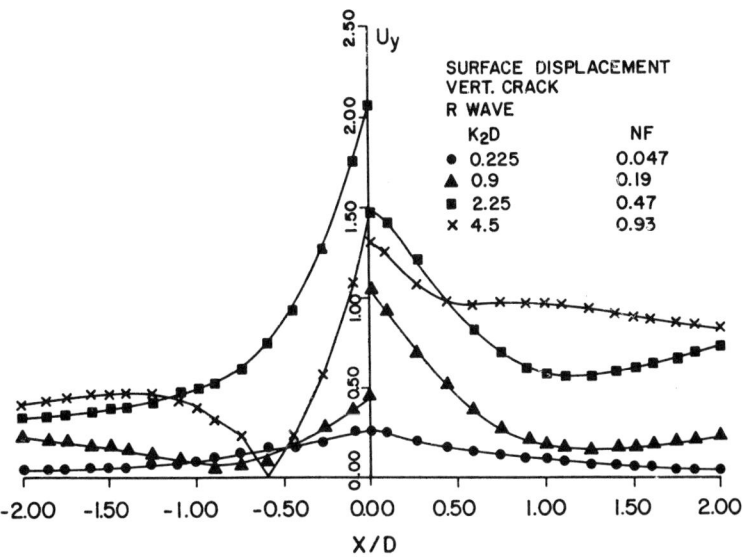

Fig. 4. Scattered vertical surface displacement amplitude due to normal planar crack.

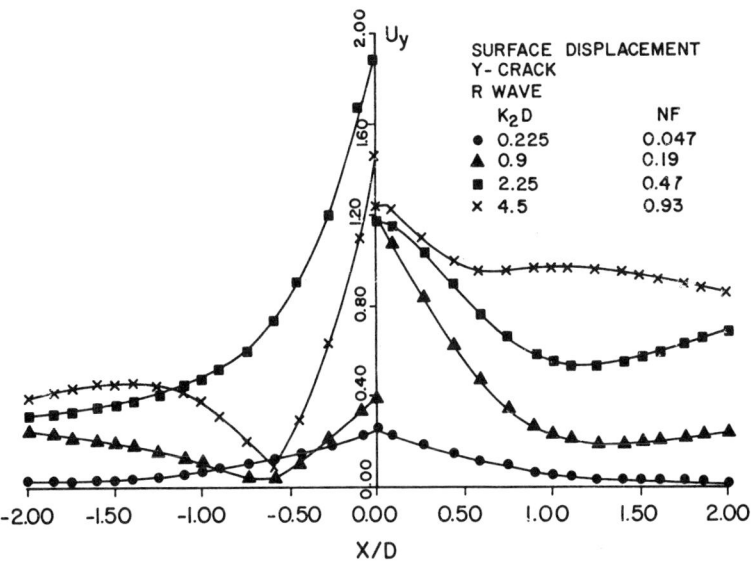

Fig. 5. Scattered vertical surface displacement amplitude due to normal branched crack.

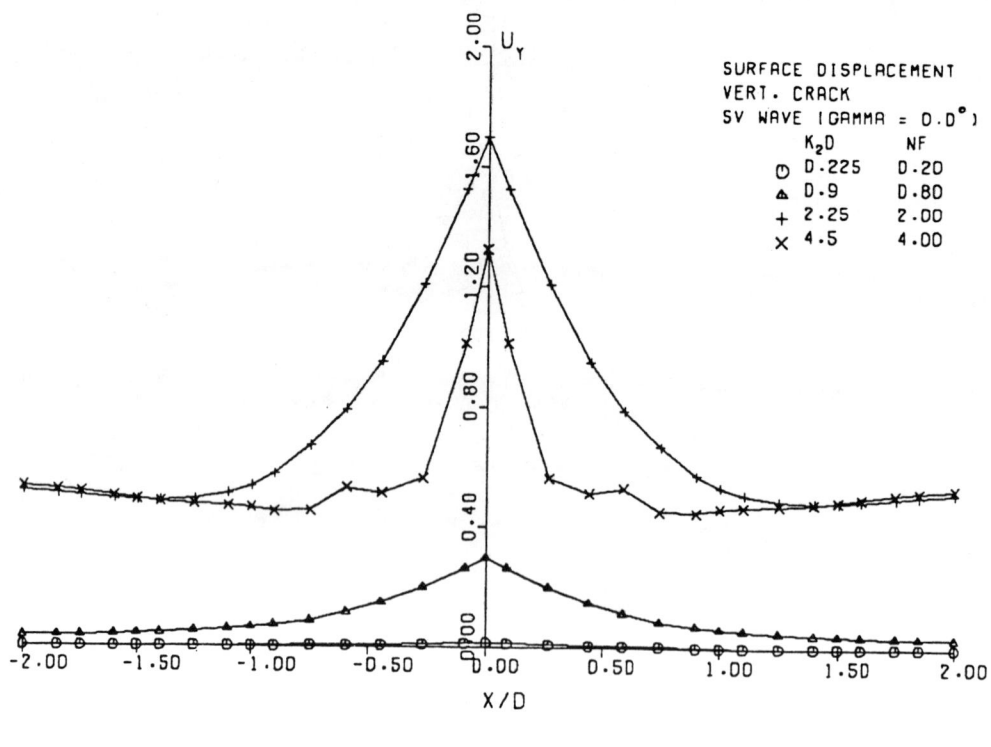
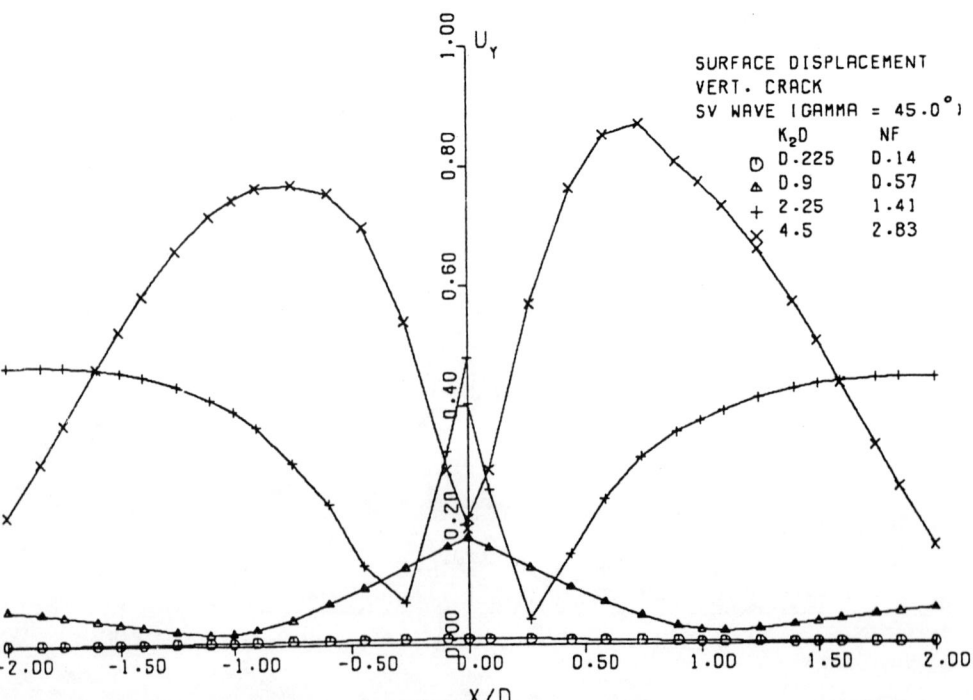

Fig. 6. Scattered vertical surface displacement due to vertical crack for incident SV waves

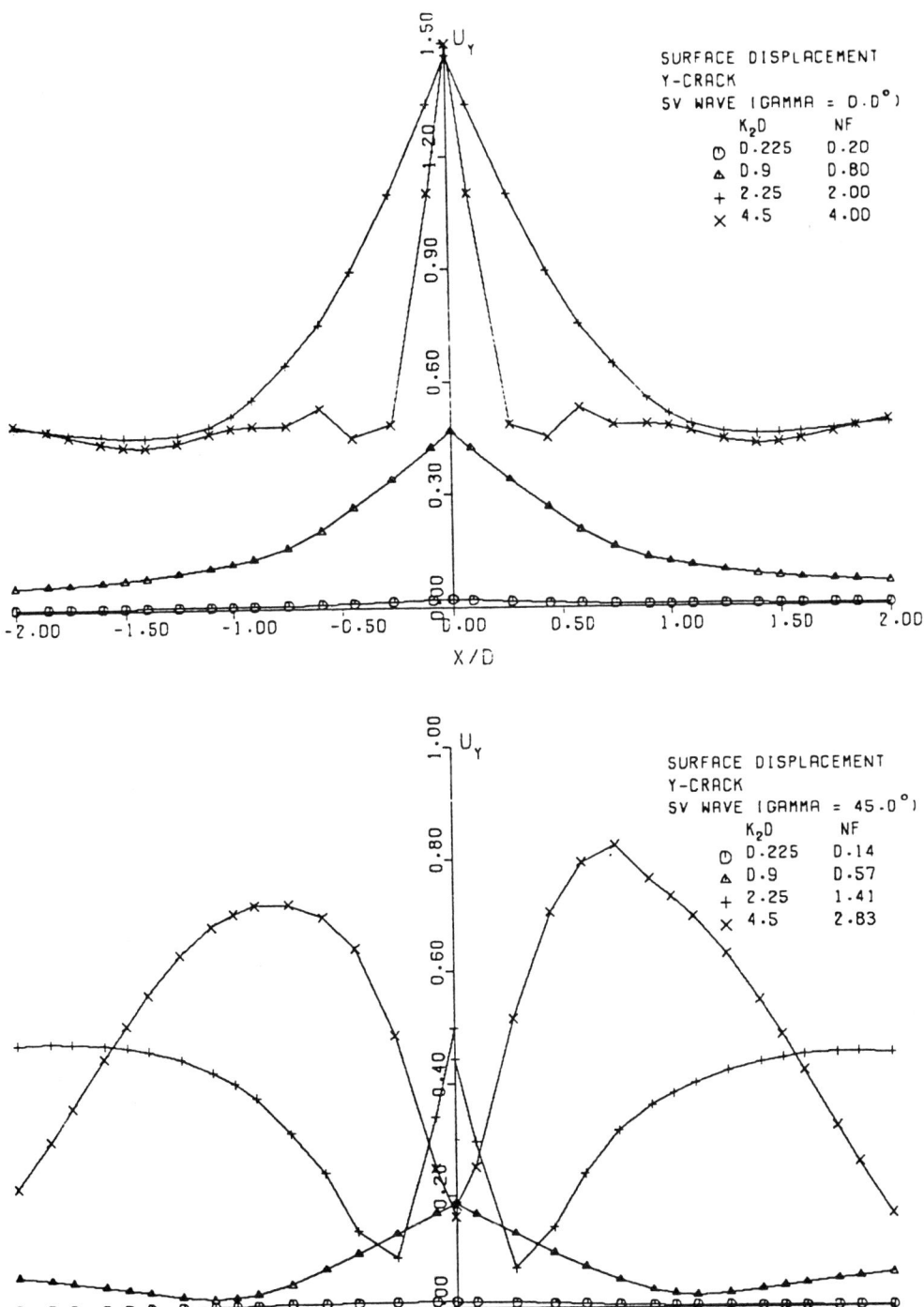

Fig. 7. Scattered vertical surface displacement due to Y crack for incident SV waves

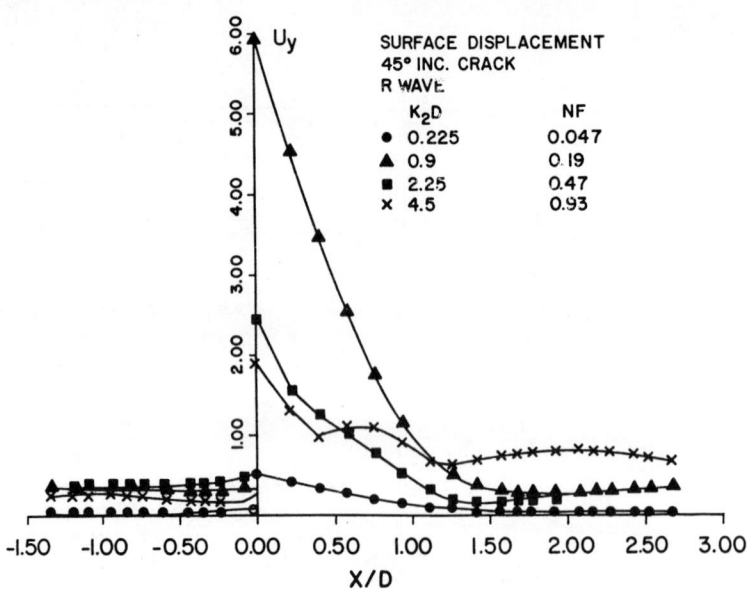

Fig. 8. Scattered vertical surface displacement due to 45° inclined crack for incident R wave

## ACKNOWLEDGEMENT

Results presented here were partly supported by grants from the Office of Naval Research (N00014-86-K-0280), the National Science Foundation (CEE81-20536) and the Natural Science and Engineering Research Council of Canada (A-7988).

## REFERENCES

1. A.H. Shah, K.C. Wong and S.K. Datta, Wave Motion 7, 319 (1985).
2. J.D. Achenbach, Y.C. Angel and W. Lin, in: "Wave Propagation in Homogeneous Media and Ultrasonic Nondestructive Evaluation," G.C. Johnson, ed., American Society of Mechanical Engineers, New York (1984).
3. J.H.M.T. van der Hijden and F.L. Neerhoff, J. Acoust. Soc. Am. 75, 1694 (1984).
4. M. Hirao, H. Fukuoka and Y. Miura, J. Acoust. Soc. Am. 72, 602 (1982).
5. C.H. Yew, K.G. Chen and D.L. Wang, J. Acoust. Soc. Am. 75, 189 (1984).
6. R. Dong and L. Adler, J. Acoust. Soc. Am. 76, 1761 (1984).
7. S.K. Datta, J. Appl. Mech. 46, 101 (1979).
8. S.K. Datta, A.H. Shah and C.M. Fortunko, J. Appl. Phys. 53, 2895 (1982).
9. S.K. Datta and N. El-Akily, J. Acoust. Soc. Am. 64, 1692 (1978).
10. A.H. Shah, Y.F. Chin and S.K. Datta, to be published.
11. Y.F. Chin, "Scattering of Elastic Waves by Near-Surface Inhomogeneities," M.S. Thesis, Department of Civil Engineering, University of Manitoba (1985).
12. J.D. Achenbach, L.M. Keer and D.A. Mendelsohn, J. Appl. Mech. 47, 551 (1980).

ACOUSTIC WAVE SCATTERING FROM A CIRCULAR CRACK: COMPARISON OF DIFFERENT
COMPUTATIONAL METHODS

William M. Visscher

Theoretical Division, MS B262
Los Alamos National Laboratory
Los Alamos, NM 87545

INTRODUCTION

This work was motivated by a disagreement between the results obtained from two computations of scattering of an axially incident elastic p-wave on a circular crack. One calculation, using the method of Mal [1], involving the direct solution of the Helmholtz integral equation for this case, shows the total cross-section oscillating with a considerable amplitude about $\sigma_{tot} = 2\pi a^2$ as a function of $k_R a$ with period $\pi$, where $k_R = 2\pi/\lambda_R$ is the Rayleigh surface wavenumber. Another calculation, [2] using MOOT, in which the elastic displacement near the crack is expanded in regular spherical eigenfunctions of the elastic wave equation, agrees with the first calculation reasonably well up to $k_R a = 10$ or so, but thereafter

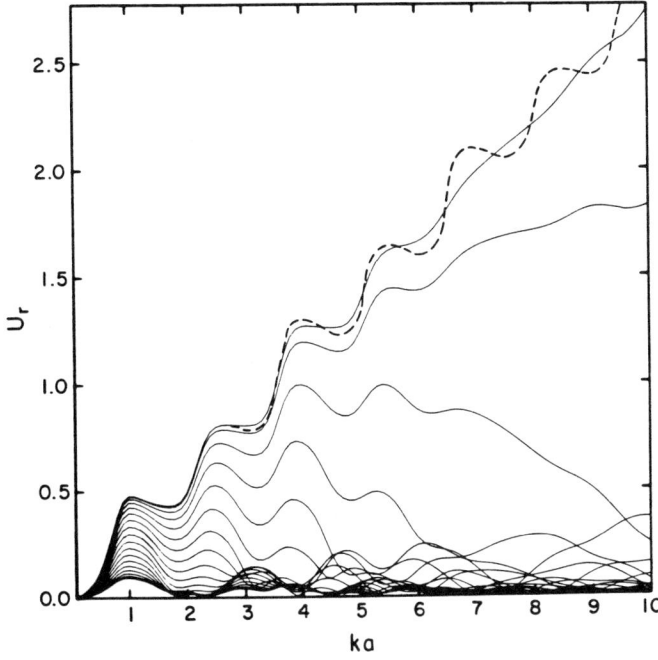

Fig. 1. Elastic wave scattering from a circular crack. The solid lines are the p-p backscattering amplitudes from a crack oriented broadside (top) to edge-on (bottom) in 5° intervals, computed using MOOT with spherical eigenfunctions of the elastic wave equation as basis functions. The dashed line is obtained from Mal's solution for the axisymmetric (broadside) case. The oscillations in the Mal solution (thought to be quite accurate) continue to large ka, while the oscillations in the MOOT results damp rapidly. From Opsal and Visscher [2].

the oscillations in $\sigma_{tot}$ rapidly disappear. Figure 1 contrasts the different results.

We thought that perhaps the reason for this discrepancy was that the basis for the MOOT expansion ($j_n$(kr) and its derivatives) was inappropriate; in fact, we mistakenly stated that it is not complete on $0 < kr < ka$ (it _is_ complete; see 9.1.86 in ref. [3]), and that the difference might be ameliorated by a different choice of basis.

A simple system on which to test this speculation is the scalar wave incident on a circular crack. The wave function $\phi$ satisfies

$$(\nabla^2 + k^2)\phi = 0 \quad , \tag{1}$$

asymptotic scattering conditions, and certain boundary conditions (BC's) on the crack surface C. The crack is shown on Fig. 2; it is a mathematical crack (zero thickness) in the xy plane with radius a.

The simplest BCs to impose on $\phi$ would be Dirichlet ($\phi = 0$ on C) or Neumann ($\phi,_n = 0$ or C, where $\phi,_n = \vec{\nabla}\phi \cdot \hat{n}$). The scattering can be obtained for these cases by a variety of methods. The T-matrix of Waterman has been obtained for both Dirichlet and Neumann BCs [4]. The Helmholtz integral equation has been solved for Dirichlet BCs and axial incidence [5], and MOOT has been applied to this case, with two different choices for the basis set [5].

Unfortunately, though, all these methods give results (for the Dirichlet case; not all have been worked out for Neumann BCs) which agree with one another; in particular, for large ka no oscillations appear in the scattered amplitude. This is a reflection of the fact that for large ka and Dirichlet BCs $\phi,_n$ on C approaches a <u>constant</u> (independent of $\rho = \sqrt{x^2+y^2}$) [5].

In contrast, the elastic wave case illustrated in Fig. 1 has oscillations in the scattered amplitude caused by resonance modes (drumhead vibrations) which are standing surface waves on the crack surface (this is why the oscillations in Fig. 1 have roughly period $\pi$ in $k_R a$).

The reason for this difference is that the Helmholtz equation (1) admits no surface wave solutions with either Dirichlet or Neumann BCs, and without surface waves one can't get standing waves on C and one won't get resonance oscillations in the scattered amplitude. Our model is just <u>too</u> simple to exhibit the effect we wish to study.

A solution to this problem is to change the BCs to mixed boundary conditions (MBCs)

$$\phi + \gamma\phi,_n = 0 \quad \text{on C} \quad , \tag{2}$$

which admits, with (1), a solution

$$\phi(x,y,z) = e^{i\vec{\kappa}\cdot\vec{\rho}-\gamma z} \tag{3}$$

with $\kappa^2 = k^2 + \gamma^2$. Equation (3) describes a surface wave if the surface is $z = 0$, $\gamma > 0$, and the incompressible fluid occupies the upper half-space. If we solve the crack problem with the BCs (2), one expects to see resonances corresponding to standing surface waves on the crack surface.

The MBCs however, complicate the mechanics of solving the scattering problem considerably. The T-matrix method can no longer be applied, because a feature of the method which is essential to its application to cracks, the symmetry of the Q-matrix, no longer holds (or at least has not been demonstrated).

The Helmholtz integral equation method, too, becomes much more difficult. The Helmholtz integral equation is

$$\phi(r) = \phi_0(r) - \int_C \{G(r,r')\phi,_{n'}(r') - G(r,r'),_{n'}\phi(r')\}dS' \quad , \qquad (4)$$

for r outside the crack C, with $G(r,r') = e^{ikR}/4\pi R$, $R = |\vec{r}-\vec{r}'|$. For axial incidence, $\phi_0(r) = e^{ikz}$, and in order to solve (4) for $\phi(r)$, r on S, one considers $\phi_+(r)$, and $\phi_-(r)$, which are $\phi(\rho,+0)$ and $\phi(\rho,-0)$ respectively. It can be shown that

$$G(r,r'),_z = -\frac{\text{sgn}(z-z')}{4\pi\rho}\delta(\rho-\rho')$$

for z, z' small, so that (4), with (2), yields

$$\bar{\phi}(\rho) = 1 + \gamma^{-1} \int_{C+} G(\rho,\rho')\bar{\phi}(\rho')dS' \quad , \qquad (5)$$

with $\bar{\phi} = \frac{1}{2}(\phi_+ + \phi_-)$ and C+ = top surface of crack. Equation (5) can be solved for $\bar{\phi}(\ell)$, which, when inserted in (4), will give the even (in z) part of $\phi(r)$.

In order to obtain an equation for $\hat{\phi} = \frac{1}{2}(\phi_+ - \phi_-)$, which, when plugged into (4) will give the odd part of $\phi(r)$, one needs to differentiate (4) with respect to z before letting z → ±0. This yields

$$-\gamma^{-1}\hat{\phi}(\rho) = ik - \int_{C+} G,_{zz}(\rho,\rho')\hat{\phi}(\rho')dS' \quad , \qquad (6)$$

with

$$G,_{zz} = \frac{d^2G}{dz^2}\bigg|_{z=z'=0} \quad . \qquad (7)$$

Equation (6) is a much nastier one than (5), because (1) has a $|\vec{\rho}-\vec{\rho}'|^{-3}$ singularity. Although it turns out that this is no problem in principle (the singularity is integrable, and one can replace the surface integral with a "principal value" integral by omitting a small circle around $\rho' = \rho$), it is a serious one in practice because it drastically worsens the convergence of the Fourier integrals with which it is natural to represent (7).

This leaves us only MOOT with which to compute acoustic scattering from a crack with MBCs.

## Moot

We will now briefly sketch the method of optimal truncation (MOOT), as applied to circular flat cracks. It will be clear that it is applicable to calculation of a scattering from any isolated flaw.

The idea is to expand $\phi$ in truncated sets of eigenfunctions of the Helmholtz operator (1) independently in each of the regions I, II, and III shown on Fig. 2. Then integrate the square of the residual (the amount by which the BCs or matching conditions fail) on the surfaces $S_\pm$ and $C_I$. Thus

$$I = \int_C \{|\phi_I + \gamma\phi_{I,n}|^2 + |\phi_{II} + \gamma\phi_{II,n}|^2\} dS$$

$$+ \int_{S+} \{|\phi_0 + \phi_{III} - \phi_I|^2 + \beta/k^2 |\phi_{0,n} + \phi_{III,n} - \phi_{I,n}|^2\} dS$$

$$+ \int_{S-} \{|\phi_0 + \phi_{III} - \phi_{II}|^2 + \beta/k^2 |\phi_{0,n} + \phi_{III,n} - \phi_{II,n}|^2\} dS \quad , \quad (8)$$

where $\phi_I(r) = \sum_{n=1}^{N} a_n \phi_n(r)$, $\phi_{II}(r) = \sum_{n=1}^{M} b_n \phi_n(r)$ ,

and $\phi_{III}(r) = \sum_{\ell=0}^{\ell_{max}} c_\ell h_\ell^{(1)}(r) Y_\ell^0(\cos\theta)$, $\phi_0(r) = e^{ikz}$

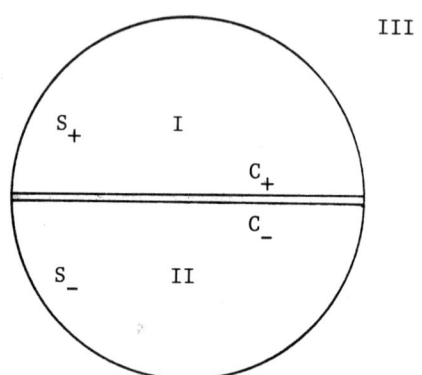

Fig. 2. The circular crack on the xy plane. $S_\pm$ are the upper and lower hemispheres surrounding $C_+$, the top and bottom surfaces of the circular crack.

$\beta$ is a dimensionless constant we take to be $\min(1,(ka)^{-2})$. Varying it by an order of magnitude either way has little effect on results. Clearly $I \geq 0$, with equality attained if and only if $\phi_I$, $\phi_{II}$, $\phi_{III}$ comprise an exact solution of the scattering problem with $\phi_0$ incident. The functions $\phi_n(r)$ are any convenient set of solutions of $(\nabla^2 + k^2)\phi_n = 0$; they need not be mutually orthogonal. The truncation limits N, M, $\ell_{max}$ are mostly dictated by the value of ka we consider. Although there is in principle no reason they can't be different, we will take $N = M = \ell_{max} + 1$.

Now I is a bilinear form in $\alpha_n = \{a_n, b_n, c_n\}$, which we wish to minimize. Thus

$$\frac{\partial I}{\partial \alpha_n^*} = 0$$

is a set of 3N linear inhomogeneous equations for the 3N unknowns a, b, c, with coefficients which are integrals of pairwise products of $\phi_0$, $\phi_n$, and

$Y_\ell^0(\cos\theta)$ on $C_+$ and $S_+$. The matrix of the coefficients can be readily inverted (at least if N is not too large), and the solution for $\alpha_n$ obtained.

So MOOT is uniquely specified except for choosing $\phi_n$, the set of N independent solutions of (1) with which $\phi$ in the upper and lower hemisphere is represented. We will choose two sets, and compare the results. The first choice will be

$$\phi_n = j_n(kr) Y_n^0(\cos\theta) , \qquad (9)$$

in analogy with the set used in [2] to compute elastic wave scattering from the circular crack. The second choice will be

$$\chi_n = J_0(p_n\rho) \begin{array}{c} \sin q_n z \\ \cos q_n z \end{array} , \qquad (10)$$

where $p_n a$ are the roots of $J_0(x)$ and of $J_0'(x)$, and $p_n^2 + q_n^2 = k^2$. Most of the $q_n$'s are imaginary. Both (9) and (10) comprise complete sets as $N \to \infty$; the question we wish to address here is "which set will closely approximate the correct answer with the least labor?"

Numerical Considerations

In the case of Dirichlet BCs the solution for $k \to 0$ is for r on C

$$\phi,_n(\rho) = -2/\pi\sqrt{a^2-\rho^2} , \qquad (11)$$

and this inverse square root singularity at the crack edge is presumably preserved for all k. For the mixed BCs (2) the behavior of $\phi,_n$ and consequently also of $\phi$ is undoubtedly also singular at $\rho = a$, but we don't know the nature of the singularity. If $\phi,_n$ for MBC (and consequently also $\phi$) behaves like (11), then the integrals on C in I will contain logarithmic divergent terms, presumably cancelling one another. Since we don't know the nature of the singularity, however, we will proceed as if there were none, and let the results tell us what it is.

Most of the integrals which are the coefficients of the bilinear form (8) must be performed numerically, which we do by Gauss-Legendre quadrature with 50 points (on the interval $0 < \rho < a$ for the C-integrals; on the interval $0 < \cos\theta < 1$ for the S-integrals).

We will show results of calculations for a variety of choices of $\ell_{max}$, up to 24, and for values of ka up to 14. For these values of $\ell_{max}$ 50 in the Gauss-Legendre quadrature is more than adequate; whether $\ell_{max} = 24$ is sufficient for ka = 14 can be judged from the results.

RESULTS

In Fig. 3 is shown the value of Re$\phi$ on the top surface of the crack as a function of $\rho$ and $\ell_{max}$ computed with MOOT using a spherical basis. The phase of $\phi$ has been adjusted here so that it is real in each case at $\rho = 0$. This is for ka = 10; $\phi$ does not approach its true value until $\ell_{max} \geq 15$. Even for ka = 0 $\phi$ has 3 nodes in $0 < \rho < 1$, and one always needs $\ell_{max} \geq 15$ or so for accurate results.

Figure 4 shows $I/I_0$ and $4\pi$ Im $f(0)/k\sigma_{TOT}$ (the optical theorem ratio) for this system. $I/I_0 = 0$ for an exact solution. It doesn't vanish, but seems to be decreasing as $\ell_{max}$ increases as if the MOOT solution is trying, with slow success, to accommodate a singularity (Fig. 3 shows a discontinuity) in $\phi$ at $\rho = 1$. The optical theorem ratio should be unity; it is about 0.98 and increasing at the largest $\ell_{max}$.

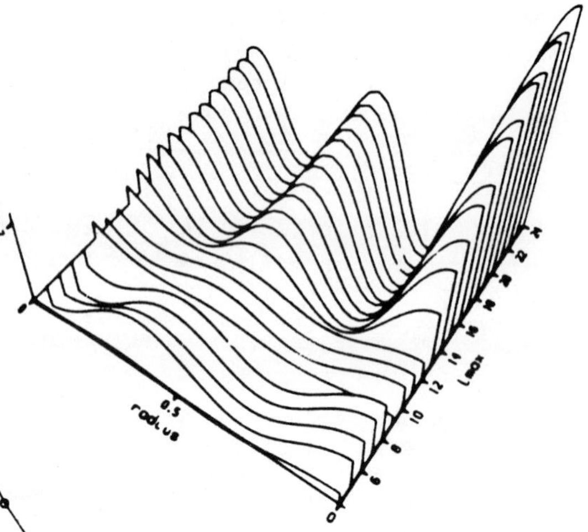

Fig. 3. Pressure Re$\phi$ on the top surface of a circular crack of unit radius caused by an axially incident wave with $ka = 10$. As $\ell_{max}$ increases, $\phi$ seems to converge nicely, except at $\rho = 0$. But the importance of $\phi(0)$ is diminished by the fact that $\phi(\rho)$ is always weighted with $\rho d\rho$. $\phi(\ell)$ begins to resemble its true value at $\ell_{max} \sim 15$. This figure was computed using spherical basis functions.

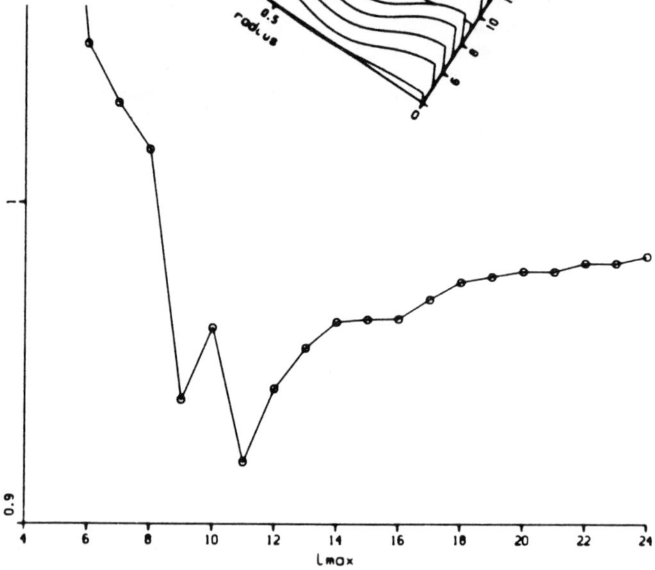

Fig. 4. Integrated residual $I/I_0$ (left ordinate scale) and optical ratio (right ordinate scale) for the system described in Fig. 3. $I_0$ is $I$ with only $\phi_0 \neq 0$.

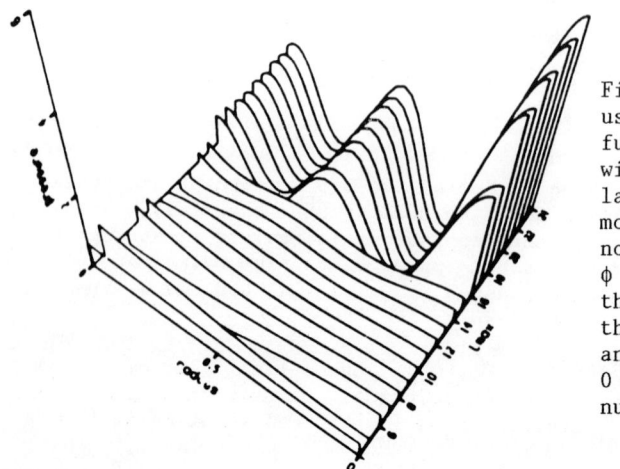

The next two figures illustrate the same quantities for the cylindrical basis set. The results are similar.

Fig. 5. Same as Fig. 3, but using cylindrical basis functions. This $\phi$ agrees with that of Fig. 3 for large $\ell_{max}$. The relatively more sudden change from noise to nearly the correct $\phi$ at $\ell_{max} \sim 15$ is caused by the fact that at that point the number of nodes and antinodes in $J_0(p_n\rho)$ in $0 \leq \rho \leq 1$ coincides with the number in the correct $\phi(\rho)$.

Fig. 6. Same as Fig. 4, but for cylindrical basis functions.

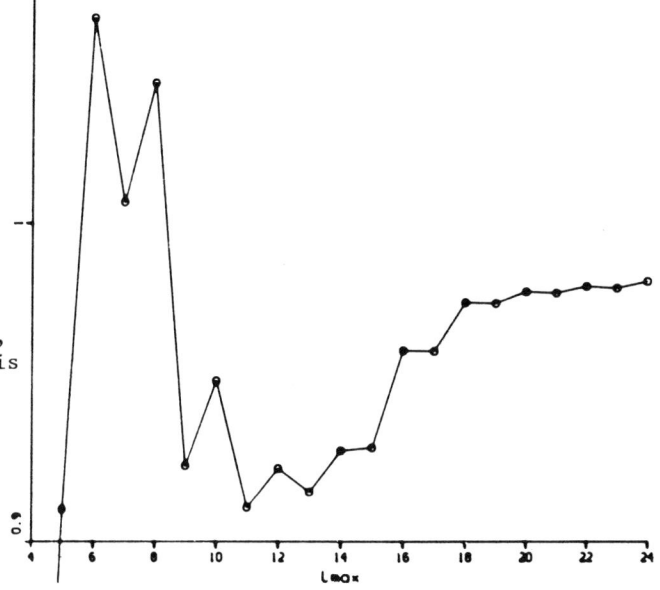

The final series of figures shows how some of the same quantities vary as ka goes from 0.5 to 14 for the circular crack with spherical basis functions ($\ell_{max}$ = 24) and with cylindrical basis functions ($\ell_{max}$ = 23). The residual integral plots indicate the trustworthiness of the calculation. Figures (7) and (9) are in close agreement (notice the different vertical scales).

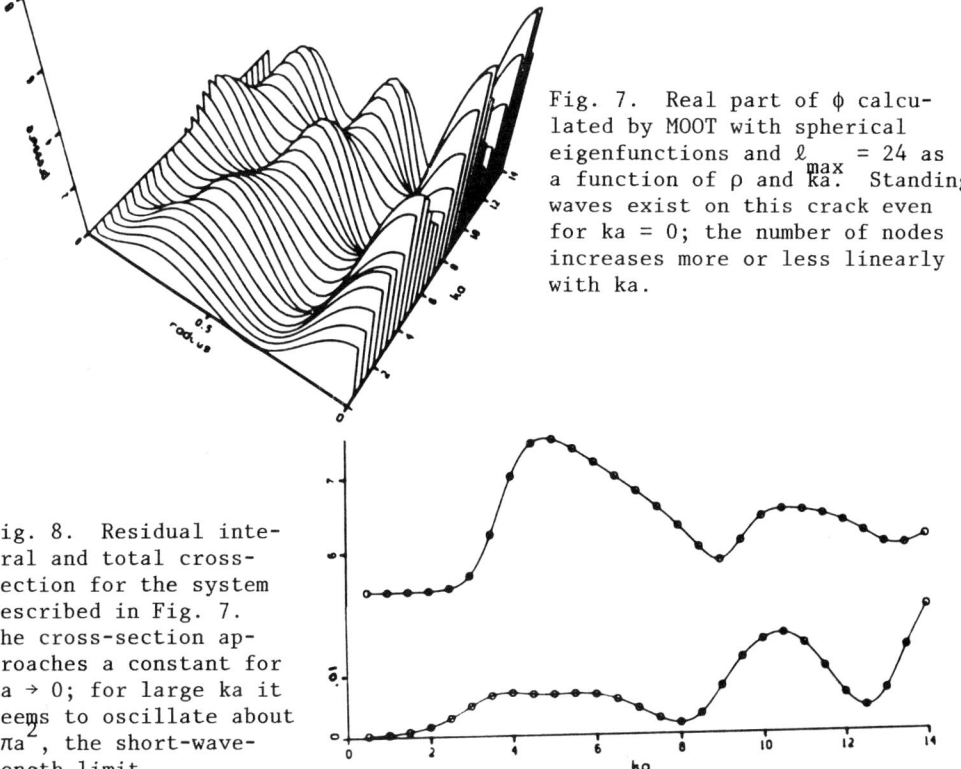

Fig. 7. Real part of $\phi$ calculated by MOOT with spherical eigenfunctions and $\ell_{max}$ = 24 as a function of $\rho$ and ka. Standing waves exist on this crack even for ka = 0; the number of nodes increases more or less linearly with ka.

Fig. 8. Residual integral and total cross-section for the system described in Fig. 7. The cross-section approaches a constant for ka → 0; for large ka it seems to oscillate about $2\pi a^2$, the short-wavelength limit.

CONCLUSIONS

Our results indicate that our original speculation, that the discrepancy of Fig. 1 was caused by inadequacy of the spherical basis set, was wrong. In application to the present test problem, in fact, the spherical basis set works better than the cylindrical one does. Both are quite capable, with the same truncation limit $\ell_{max}$ = 24, of accurately describing the pressure (analog of the crack-opening-displacement in the elastic wave scattering case) at least up to ka = 14, when the pressure has 5 nodes in $0 < \rho < a$.

Fig. 9. Same as Fig. 7, but with a set of 23 cylindrical basis functions. After a different vertical scale is taken in to account, Figs. 7 and 9 are in close agreement.

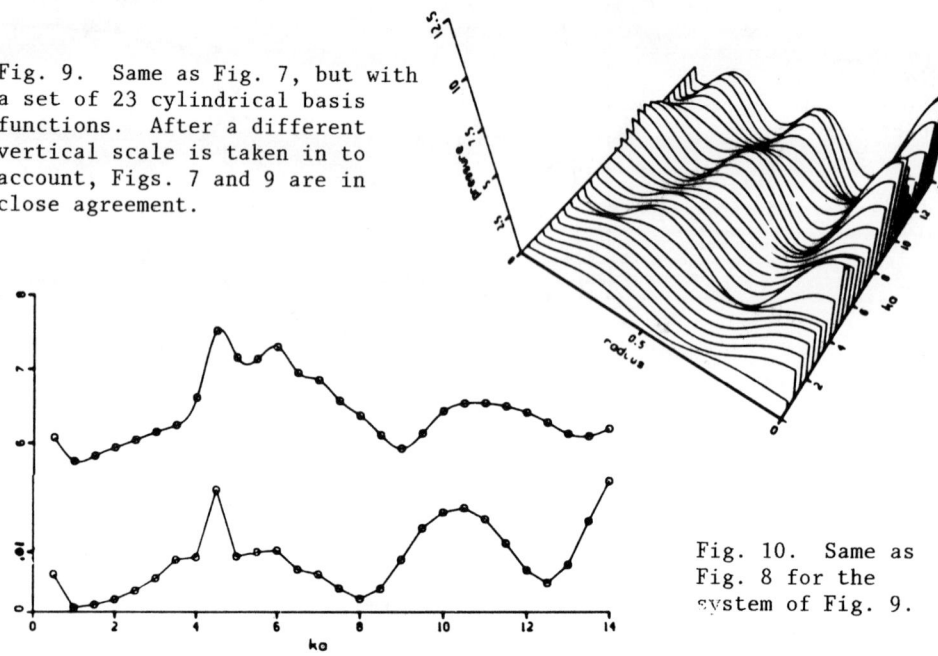

Fig. 10. Same as Fig. 8 for the system of Fig. 9.

The original question then returns: if it is not due to a bad basis set, what <u>does</u> cause the difference between the two results on Fig. 1? Discounting the possibility that Mal's method yielded wrong results here, one is forced to the conclusion that $\ell_{max}$ was not large enough in the MOOT calculation reported in [2]. A rough estimate, obtained from the results of the present scalar MBC problem, of the minimum $\ell_{max}$ required for a given ka >> 1, is

$$\ell_{max} \gtrsim 1.5 ka . \tag{17}$$

The largest value of $k_R a$ shown in Fig. 1 is $k_R a$ = 21.4 ($k_\rho a$ = 10); the criterion (17) indicates that in order to insure accuracy to this value of ka one should take $\ell_{max} \sim 30$. The $\ell_{max}$ used in the MOOT calculation of [2] was only 20. It may be repeated with larger $\ell_{max}$ to see if this conjecture is correct.

REFERENCES

1.  A. K. Mal, Int. J. Eng. Sci. <u>8</u>, 381 (1970).
2.  Jon L. Opsal and William M. Visscher, J. Appl. Phys. <u>58</u>, 1102 (1985).
3.  M. Abramowitz and I. A. Stegun, "Handbook of Mathematical Functions," (National Bureau of Standards, 1964).
4.  Gerhard Kristensson and P. C. Waterman, J. Acoust. Soc. Am. <u>72</u>, 1612 (1982).
5.  William M. Visscher, unpublished.

# 3-D MODELING OF ULTRASONIC SCATTERING FROM INTERGRANULAR STRESS CORROSION CRACKS

J. D. Achenbach and D. E. Budreck

The Technological Institute
Northwestern University
Evanston, IL 60201

INTRODUCTION

   This paper describes the derivation of a three-dimensional model for scattering of elastic waves by an intergranular stress corrosion crack (IGSCC). The model is based on a geometrical abstraction of the IGSCC. The crack has a main stem and left and right branches. The transducer is on the upper surface and the crack breaks the lower surface of a flat plate. The beam of transverse wave motions radiated by the transducer insonifies the crack under an angle of 45° with the lower face of the plate. Figure 1 shows the plane of symmetry of the configuration. The current profile of the beam is Gaussian, but the use of other beam profiles is possible. The position of the center of the transducer can be varied in the plane of symmetry of the crack. This allows the simulation of sizing techniques which require probe motion. The backscattered field has been computed by using the Kirchhoff approximation for the crack-opening displacements of the individual components of the branched crack. The backscattering by the main stem and the branches has been analyzed and superimposed to provide the total backscattered field in the frequency domain. Numerical results are presented.

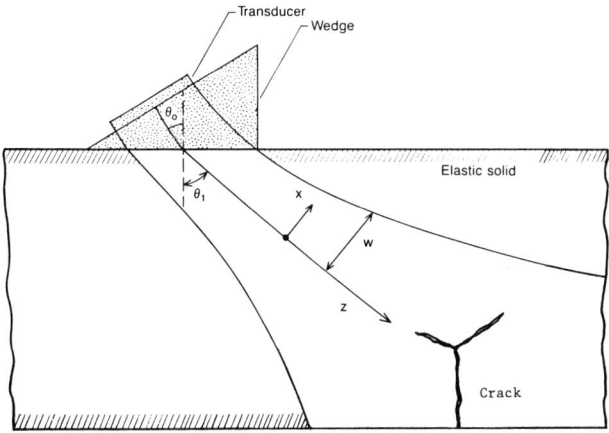

Fig. 1   Configuration in the plane of symmetry of the transducer and the crack.

## MATHEMATICAL MODELLING

*Crack Model.* The 3-D model presented herein is an extension of the 2-D model of Thompson et al, see [1]. The lengths and widths of the vertical stem and the branches, as well as the angular orientations of the two branches, were considered as variables. The mathematical crack was constructed by patching together three half-ellipses in the manner depicted in Fig. 2b.

*Beam Model.* In the description of the transducer field we follow the model proposed by Thompson and Lopes [2], which assumes that the beam profile is Gaussian in all cross-sectional planes. In this model the displacement field of a beam of transverse wave motion takes the form

$$u_i = A\, e^{-i\phi}\, d_i\, e^{ik_T z}\, e^{-r^2(1/w^2 - ik_T/2R_c)}, \qquad (1)$$

where $|\underline{d}| = 1$ and $\underline{d} \perp \underline{u}$, $z$ is the direction of propagation, and $k_T = \omega/c_T$ is the wavenumber, $c_T$ being the velocity of transverse waves. The third exponential of (1) defines the Gaussian nature of the beam, and the first is simply a phase factor which is premultiplied by an amplitude factor. The term $\exp(-i\omega t)$, where $\omega$ is the angular frequency, has been suppressed.

## AULD'S FORMULA

To characterize the back-scattered field we will employ Auld's formula, as opposed to the usual representation integrals for stresses or displacements. There are two advantages to the use of Auld's formula. Firstly, it gives us a field quantity (here and henceforth called Auld's parameter) which can be measured (electrically) external to the elastic body. Secondly, the integration involved in Auld's formula is somewhat simpler, since the integrand is free of Green's functions. For the derivation of Auld's formula the reader is referred to [3]. For back-scattering by a flat crack lying in the $x_1 x_2$-plane the formula takes the form

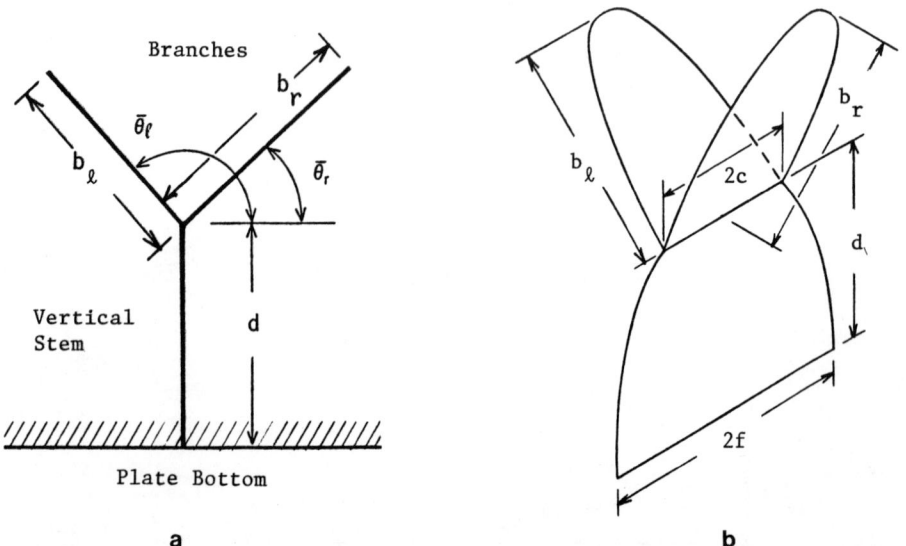

Fig. 2 Model of the IGSCC: (a) side view, (b) skew view.

$$\delta\Gamma = \frac{i\omega}{4P} \int_A \sigma_{3i}^{tr} \Delta u_i n_3 dA , \qquad (2)$$

where $\delta\Gamma$ is Auld's parameter, and P is the electrical power incident on the transducer. Also, $\sigma_{ij}^{tr}$ denotes the transducer stress field, i.e. the stress field that would exist in the absence of the crack, and $\Delta u_i$ denotes the crack-opening displacement. The integral is carried out over the insonified surface with unit outward normal $n_3$.

## CALCULATION OF CRACK-OPENING DISPLACEMENT

The utility of (2) lies in our ability to first calculate $\Delta u_i$, the crack-opening displacement. To do so for the stated problem would involve the solution of a complicated singular integral equation over the crack surface, which for the geometry of our branched crack would necessitate the use of a numerical scheme, such as the boundary element method. Alternatively, we can make an intelligent estimate of the crack opening displacement with far less computational effort by making use of a few simple assumptions, each of which will be briefly discussed below.

*Assumptions.* Firstly, we will ignore the effect of the top surface of the plate. In doing so we can predict only those first-arrived signals that have interacted either solely with the crack or with the crack and the lower surface of the plate. Secondly, we will assume that the beam behaves locally like a plane wave. This assumption will greatly simplify both the evaluation of Auld's formula and the description of the beam after reflection from a traction-free surface. For example, the assumption implies that (for an angle of incidence of 45°) the incident beam is reflected as another beam of transverse motion, without mode conversion.

Lastly, and related to the previous assumption, in calculating $\Delta u_i$ it will be assumed that the field on the insonified crack face is locally the same as on an insonified traction-free plane, while on the shadow side the field is assumed to vanish. This assumption is known as the Kirchhoff approximation, and amounts to taking as $\Delta u_i$ the sum of the incident and reflected waves on an insonified traction-free plane.

*Ray Cases.* In order to apply the Kirchhoff approximation to the branched crack, we first consider the ways in which the segments of the crack can be insonified: either directly by the transducer, or after one reflection off the bottom surface of the plate. These paths of insonification are denoted as "ray paths". In considering the crack as consisting of three segments, i.e., two branches and the vertical stem, the crack's insonification consists of six different "ray cases". However, it may be assumed that the right branch is not insonified by wave motion reflected from the bottom surface of the plate, since this branch is shadowed by the left branch (see Figs. 1 and 2a). Hence, the six ray cases are further reduced to five. To each of these ray cases we apply the Kirchhoff approximation. For any one ray case we write the crack-opening displacement as

$$\Delta u_i = U_T(x_1,x_2) [d_i^T + R_L^T d_i^{rL} + R_T^T d_i^{rT}] e^{ik_T p_1^T x_1} , \qquad (3)$$

where we have affixed a local $x_1 x_2$ coordinate system to the plane of the crack segment. The premultiplying function contains the amplitude, phase, and Gaussian nature of the beam. The three terms in square brackets represent the successive contributions from the incident T-wave, and the reflected L- and T-waves. Explicit expressions for the longitudinal and transverse reflection coefficients, $R_L^T$ and $R_T^T$, can be found elsewhere, see for example [4].

TRANSDUCER STRESS FIELD

Since we have five crack opening displacements, $\Delta u_i$, Auld's parameter $\delta\Gamma$ of (2) will consist of contributions from five separate integrals. Each of these integrals can, however, be further broken into two integrals, since the transducer stress field $\sigma_{ij}^{tr}$, which appears in (2), consists of two terms:

$$\sigma_{ij}^{tr} = (\sigma_{ij}^{tr})^d + (\sigma_{ij}^{tr})^r . \tag{4}$$

Here the d-superscript term denotes the field emanating directly from the transducer, and the r-superscript term denotes the field generated by one reflection off the lower surface of the plate. These two fields are calculated from their respective displacement fields, giving at $x_3 = 0$

$$(\sigma_{ij}^{tr})^{d,r} = ik_T \mu [d_i^T p_j^T + d_j^T p_i^T]^{d,r} U_T^{d,r}(x_1,x_2) e^{ik_T (p_1^T)^{d,r} x_1}, \tag{5}$$

where in carrying out the differentiation we have operated only on the plane-wave parts of the displacement fields.

EVALUATION OF THE INTEGRALS

In evaluating the integrals of (2), we make further use of the assumption that the incident beam may be locally approximated by a plane wave, by assigning the Gaussian function $U_T(x_1,x_2)$ of (3) and (5) its value at the centroid of the relevant crack segment. In doing so, the evaluation of the integrals of (2) is reduced to the evaluation of integrals of the form

$$I = ik_T \int_A e^{ik_T \gamma x_1} dx_1 dx_2 , \tag{6}$$

where $\gamma$ can take on any of the following forms, depending upon the ray case:

$$\gamma = \pm 2\sin\theta_T, \quad \pm(\cos\theta_T - \sin\theta_T), \quad \pm(\cos\theta_T + \sin\theta_T). \tag{7}$$

Here $\theta_T$ is the angle of incidence of the incoming beam of transverse motion measured from the outward normal of the particular crack segment.

The area integrals (6) over the half-elliptical branches can be converted to line integrals, as shown in Ref.[5]. The integration can subsequently be carried out rigorously using the result of [6]. For a branch of length b and width 2c (see Fig. 2.), there results

$$I_{Branch} = \frac{c}{\gamma}[2 - \pi\{H_1(k_T \gamma b) - iJ_1(k_T \gamma b)\}] - \frac{2c}{\gamma} , \tag{8}$$

where $H_1$ and $J_1$ denote the first order Struve and Bessel functions, respectively, both of the first kind. Using the same approach, the evaluation of (6) over the truncated half-elliptical main stem can be approximated as the difference between integrations over two half-ellipses.

NUMERICAL RESULTS

Numerical results will now be given for a crack with the following configuration, see Fig. 2: $\theta_\ell = 120°$, $\theta_r = 60°$, $b_\ell = b_r = .15$ cm, $d = .25$ cm, $2f = 1.0$ cm, $2c = .96$ cm. The magnitude of a dimensionless version of Auld's parameter, $|\overline{\delta\Gamma}|$, is plotted versus the dimensionless shear wavenumber, $k_T d$. The plots are essentially frequency response functions of the branched crack.

In Fig. 3 we show the contributions to the response function of two of the five ray cases. In these two plots the beam centerline coincides with the crack mouth M, i.e. the beam is being directed towards the mouth of the crack. There is considerable difference in amplitudes in the two figures, since one ray case is much closer to specular reflection than the other.

In Fig. 4a we show the combined contributions of all five ray cases, giving us the total backscattered response function. A composite plot of the two ray cases corresponding to corner reflection looks essentially like Fig. 4a without the wiggles. The wiggles in Fig. 4a are due to scattering by the branches.

In Fig. 4b we show the effect of moving the transducer over the upper plate face, so that the center of the beam will be directed to points a distance of $\sqrt{2} \cdot \overline{x}$ from the crack mouth M. This simulation of probe motion is essential to crack sizing techniques.

To conclude we mention the utility of the complex-valued frequency response functions, whose norms have been plotted in Fig. 4a,b. By multiplying the complete response functions by the Fourier transform of an incident time-domain waveform, and then taking the inverse Fourier transform of the product, there results the time-domain back-scattered pulses. Thus one can create synthetic experimental data with which to

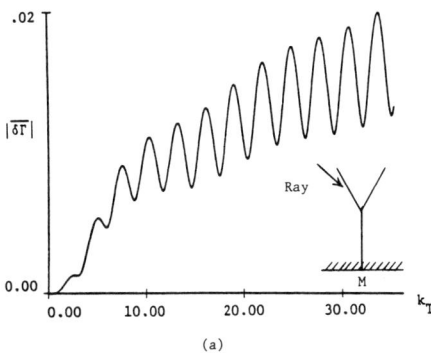 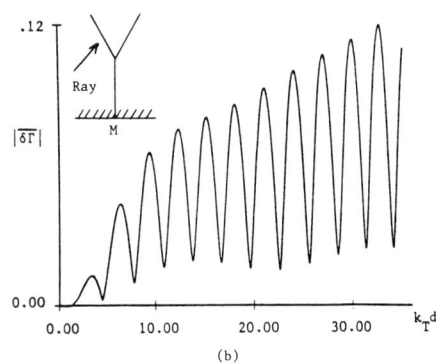

Fig. 3  $|\overline{\delta\Gamma}|$ versus $k_T d$ for (a) direct insonification of left branch, and (b) insonification of left branch by reflected wave.

compare to actual experimental back-scattered pulses. The details of this procedure have been carried out for the two-dimensional case in Ref.[1]. Furthermore, for an unknown crack geometry, measured data can be contrasted with model results in a method to infer the geometrical parameters of the crack.

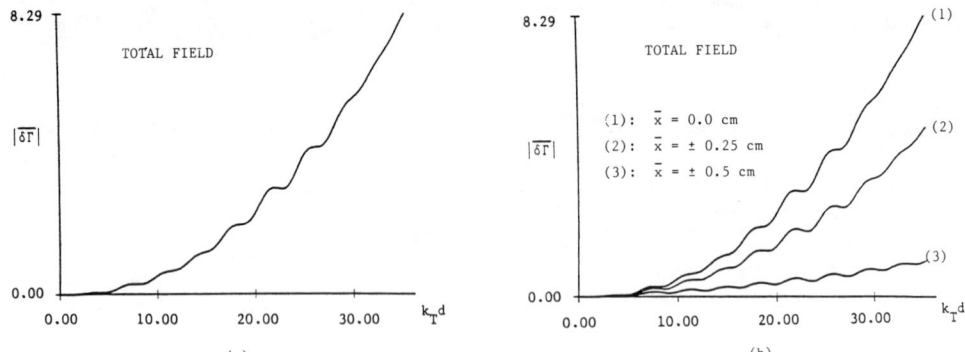

Fig. 4   $|\overline{\delta\Gamma}|$ versus $k_T d$ for (a) total backscattered field, and (b) total backscattered field with simulated probe displacement.

ACKNOWLEDGMENT

This work was partially sponsored by the Electric Power Research Institute, under subcontract SC-85-089 from Ames Laboratory, and partially supported by the US Army Office of Research under Project DAAG29-84-K-0163.

REFERENCES

1. R.B. Thompson, T.A. Gray, J.D. Achenbach and K.Y. Hu, "Ultrasonic Scattering from Intergranular Stress Corrosion Cracks: Derivation and Application of Theory", EPRI NP-3822, January 1985, Electric Power Research Institute, Palo Alto, Cal.
2. R.B. Thompson and E.F. Lopes, "The Effects of Focussing and Refraction on Gaussian Ultrasonic Beams", Journal of Nondestructive Evaluation, Vol. 4, No. 2, 1984.
3. B.A. Auld, "General Electromechanical Reciprocity Relations Applied to the Calculation of Elastic Wave Scattering Coefficients", WAVE MOTION, 1, 1979, p. 3.
4. J.D. Achenbach, Wave Propagation in Elastic Solids, Amsterdam/New York: North Holland Publishing Co., 1973, p. 179.
5. J.D. Achenbach, A.K. Gautesen and H. McMaken, Ray Methods for Waves in Elastic Solids, Boston: Pitman Publishing, Inc. 1982, pp. 243-244.
6. N.W. McLachlan, Bessel Functions for Engineers, 2nd Ed., London: Oxford University Press, 1961, p. 77.

APPLICATION OF MODELS FOR IGSCC INSPECTION

T. A. Gray, R. B. Thompson and B. P. Newberry

Ames Laboratory, USDOE
Iowa State University
Ames, IA  50011

and

J. D. Achenbach and D. Budreck

Department of Civil Engineering
Northwestern University
Evanston, IL  60201

INTRODUCTION

Ultrasonic detection and sizing of intergranular stress corrosion cracks (IGSCC) in nuclear reactor cooling systems is a difficult practical problem due to the complicated geometry of these defects and to the variety of other reflectors (e.g., welds) which produce competing ultrasonic indications. The use of models of scattering from such defects can help in improving physical insight into ultrasonic scattering from IGSCC's and may ultimately be of use in defining inspection protocols and signal processing algorithms which can lead to improved inspection reliability and discrimination between IGSCC's and other geometrical reflectors. This paper will discuss the application of a model of ultrasonic scattering from a simple Y-shaped crack based upon the Kirchhoff approximation. In this model, the ultrasonic beam is approximated by a Gaussian profile which includes the effects of diffraction and allows calculation of the full ultrasonic radiation pattern which may be used, for example, to simulate a scanned inspection. Included in this paper will be a brief description of the model and a presentation of simulated IGSCC results. Comparisons of the model to experimental measurements will then be addressed followed by an application to the problem of IGSCC sizing based upon the dB-drop and PAT (pulse-arrival-time) techniques.

MODEL SUMMARY

The basis of the IGSCC model is an electromechanical recriprocity relationship which relates the ultrasonic fields in the vicinity of the flaw to the signals that could be received in a practical inspection[1]. Initially based upon a 2-D scattering model, recent work has extended the previously reported results[2] to allow the crack to have finite length[3]. The model 3-D Y-shaped crack is illustrated in Fig. 2(b) of Ref. 4. The main stem of the crack is a truncated semi-ellipse and the branches are semi-ellipses. The inspection configuration, shown in Fig. 1, is 45-degree shear

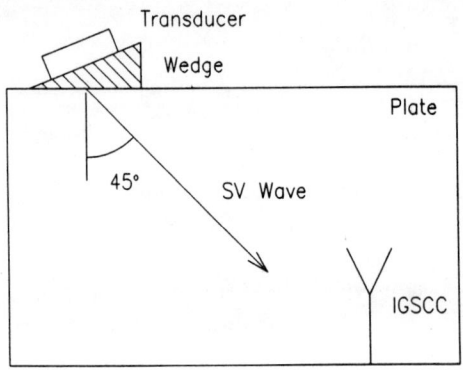

Fig. 1. Inspection configuration for Y-crack model.

wave backscatter. The ultrasonic beam is represented by a Gaussian beam model[5] which includes the effects of diffraction (beam spread) and allows computation of the full fields in order to simulate a scanned inspection. Scattering from the simulated IGSCC is described by a Kirchhoff approximation in which the ultrasonic fields which illuminate and scatter from the branches and stem are computed according to the Gaussian beam theory. The resulting model can be expressed in the following dimensionless form[3]:

$$\frac{HP\delta\Gamma}{2(U_o T_{o1})^2 \mu c_T f} = \frac{ik_T H}{8f} \iint_{A^+} \frac{\Delta u_i [(\tau_{i2}^{tr})^d + (\tau_{i2}^{tr})^r]}{(U_o T_{o1})^2 \mu} n_2 dx_1 dx_3. \qquad (1)$$

The various terms in Eq. 1 are:

- $H$ = plate thickness;
- $P$ = electrical power incident on the transducer;
- $\delta\Gamma$ = variation of electrical reflection coefficient due to presence of a flaw;
- $U_o$ = maximum displacement of transducer face;
- $T_{o1}$ = wedge/plate ultrasonic transmission coefficient;
- $\mu$ = shear modulus;
- $c_T$ = shear velocity;
- $f$ = surface breaking half-length of crack stem;
- $A^+$ = illuminated surface of crack;
- $k_T$ = shear wavenumber;
- $\Delta u_i$ = crack opening displacement;
- $\tau_{ij}$ = stress fields induced by the transducer with no flaw present, superscript refers to illumination by direct or reflected rays;
- $n_j$ = unit vector normal to crack surface.

In the current model implementation, the crack opening displacement and stress fields are approximated by assuming that the local illuminating fields on the branches or stem of the crack can be replaced by a plane wave whose amplitude is defined by the Gaussian beam theory evaluated at a user-specified point on the crack element. This represents a simplification of the actual scattering phenomena, but allows fast computation time, since the model, as indicated in Eq. 1, requires a double integration over the crack face. For plane waves, the integrals can be evaluated analytically.

Equation 1 is, in effect, an "impulse response" spectrum which is independent of the frequency characteristics of a particular transducer. To simulate a signal that might be measured in a practical situation, Eq. 1 must be convolved with an appropriate transducer frequency response. To illustrate this, it is convenient first to rewrite Eq. 1 as

$$\delta\Gamma = N \left[ \frac{(2f) \cdot c_T^2}{\omega^2 H \cdot A} \right] \left[ \frac{\rho c_T \omega^2 (U_o T_{o1})^2 A}{P} \right] \qquad (2)$$

where N denotes the dimensionless quantity of Eq. 1 and A is the cross-sectional area of the ultrasonic beam. Note that the final term in Eq. 2 is a dimensionless power conversion efficiency term representing the acoustic power transmitted into the bulk of the plate per unit incident electrical power for a specific transducer.

This term can be determined by a reference measurement, e.g., the reflection from a corner of a plate identical to that containing the IGSCC. This reference signal can be expressed as

$$\delta\Gamma_R = D \left[ \frac{\rho c_T \omega^2 (U_o T_{o1})^2 A}{P} \right] \qquad (3)$$

where D is a term to account for diffraction loss (beam spread). Equation 3 can be solved for the efficiency factor which is then inserted into Eq. 2 to yield

$$\delta\Gamma = N \cdot \delta\Gamma_R \frac{(2f) c_T^2}{D \omega^2 H \cdot A} \qquad (4)$$

Experimentally, one measures a voltage rather than a reflection coefficient. However, since the proportionality between the two is the same for both the scattering and reference measurements, $\delta\Gamma$ and $\delta\Gamma_R$ in Eq. 4 may be interpreted as the measured signals.

Some means for computing the diffraction term D and the ultrasonic beam area A in Eq. 4 must be determined. In the following, the latter factor is approximated as the refracted elliptical "footprint" of the beam in the plate. Thus,

$$A = \pi a^2 \frac{\cos\theta_1}{\cos\theta_0} \qquad (5)$$

where a is the probe radius and $\theta_{0,1}$ are the angles of the beam in the wedge and in the plate relative to normal to the plate surface. The diffraction loss term D is approximated in an <u>ad hoc</u> manner by a formula which is valid for circular piston probes at normal incidence to the sample surface[6],

$$D(s) = 1 - e^{-i2\pi/s} (J_0(2\pi/s) + iJ_1(2\pi/s)) \qquad (6)$$

where

$$s = 2 \left( \frac{z_0 \lambda_0}{a^2} + \frac{z_1 \lambda_1}{a^2} \right) \qquad (7)$$

$\lambda_{0,1}$ and $z_{0,1}$ being the ultrasonic wavelength and path length in the wedge and the solid, respectively. It should be noted that Eqs. 5 through 7 as implemented in Eq. 4 are not expected to be quantitatively correct in this application to IGSCC inspection but should introduce roughly the correct frequency dependence for a specific transducer selection.

EXPERIMENTAL COMPARISONS

To test the validity of Eq. 4 as implemented using Eqs. 5-7, experiments were performed on a Ti-6Al-4V plate containing four approximately semi-elliptical EDM slots with depths ranging from 0.038 to 0.152 cm (0.015 to 0.060 in.) and surface-breaking lengths of 0.140 to 0.358 cm (0.055 to 0.141 in.). Measurements were performed in immersion using a 1.27 cm (0.50 in.) diameter unfocussed 5 MHz transducer oriented at a 19.2° angle relative to the sample surface to couple into a 45° shear wave in the plate. A corner reflection from the edge of the plate was used as the reference signal. A representative comparison of model predictions to experimental measurements is shown in Fig. 2, in which the frequency domain result of Eq. 4 has been Fourier transformed into the time domain. This figure compares results for the 0.114 cm (0.045 in.) deep slot. Qualitatively, the agreement in pulse shape, including the amplitude of the tip signal (the small precursor to the main signal feature) relative to the crack mouth signal, is quite good. Unfortunately, the signal amplitudes differ by an order of magnitude. Results from the other three slots were similar, except that for the 0.038 cm (0.015 in.) deep slot, neither the model nor experimental waveforms showed a distinct tip signal. Since the ultrasonic wavelength in the solid is approximately 0.064 cm (0.025 in.), the tip signal would not be expected to be resolved in time from the mouth signal. Investigations are currently under way to identify the cause of the amplitude discrepancy between model and experiment.

APPLICATION TO IGSCC SIZING

As an example of the practical application of the IGSCC model, we will illustrate its use by predicting the performance of two typical IGSCC sizing methods - the dB drop and pulse-arrival-time (PAT) techniques. The former method attempts to size a crack by first positioning a probe to achieve the maximum signal from the crack and then repositioning the probe to either side of the defect such that the signal is reduced by a specified amount. The defect height estimate is the distance between the latter two probe positions. The PAT method begins, once again, by maximizing the crack signal. For a typical IGSCC, this position will correspond to the probe position which directs the ultrasonic beam axis toward the crack mouth (corner trap). Next, the probe position is varied to maximize the crack tip signal. The crack height is then simply related to the time delay between these two maximized signals[2]. It is known that the dB-drop technique cannot work for a surface-breaking defect, since the primary signal feature in that case is the crack mouth reflection[2]. Thus the dB-drop technique only provides a measure of the ultrasonic beam width. The PAT method, on the other hand, does typically provide accurate height estimates. These statements have been verified by experimental tests[7].

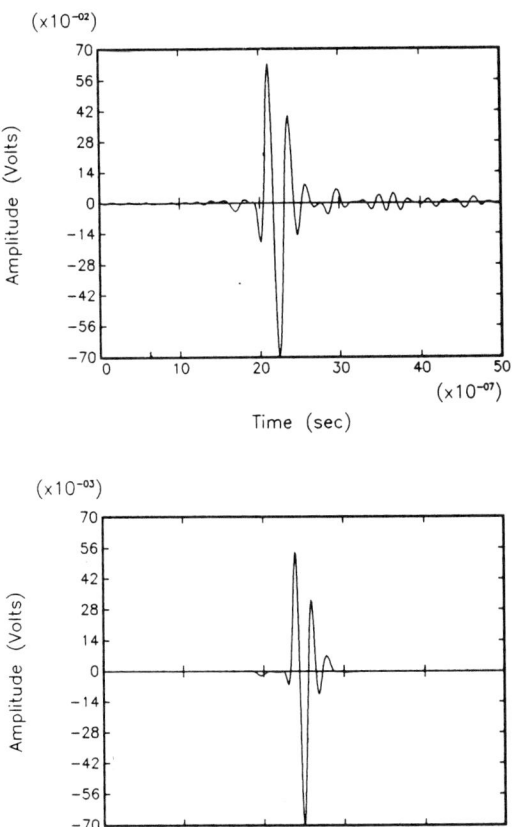

Fig. 2. 45° shear wave signals from a 0.045 in. deep semi-elliptical defect. (Top = experiment, bottom = model)

To simulate the sizing performance, the model was exercised for a set of branched cracks ranging in total depth from 0.318 to 2.286 cm (0.125 to 0.90 in.) for which the angle between branches was varied from zero to 90° and the ratio of stem height to crack height was varied from 0.5 to 0.99. The results of all simulations are shown in Figs. 3 (dB-drop) and 4 (PAT). The dB-drop method shows no correlation between estimated and actual IGSCC height, while the PAT technique exhibits an essentially perfect capability. Similar results were obtained previously using a 2-D Y-crack model[2], except that the PAT method showed some scatter in the estimated height for simulated IGSCC's of identical height but different "topologies" (e.g., different

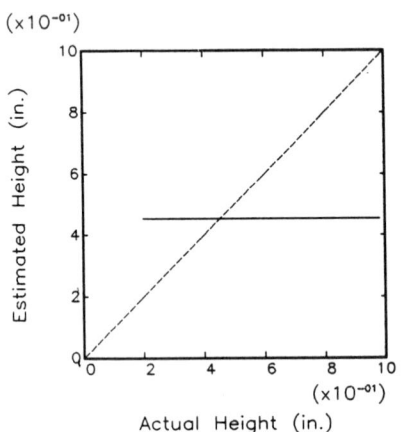

Fig. 3. Predicted sizing performance for dB-drop technique (solid line).

branch angles). The improvement in model accuracy illustrated in Fig. 4 is due to the ability in the 3-D model to approximate the crack opening displacement using the ultrasonic fields incident at an arbitrary user-specified point on the crack element (stem or branch). In the 2-D model, these fields were evaluated at the crack centroid. Thus, the signal from a branch, for example, would be maximized when the beam axis is directed toward the branch's centroid rather that toward the tip. Choosing the user-specified point to be the crack tip in the 3-D model causes the signals to be maximized when the beam axis strikes the tip, as is observed experimentally.

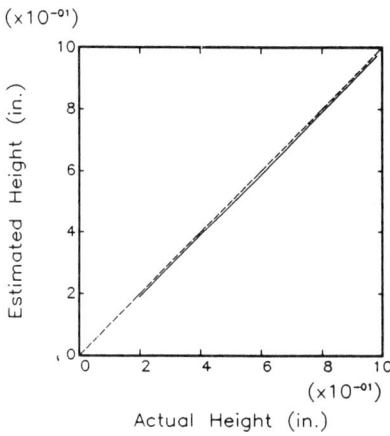

Fig. 4. Predicted sizing performance for P.A.T. technique (solid line).

SUMMARY

This report has described the implementation and application of a model for predicting the ultrasonic response from IGSCC's. There is still the need, however, for further model refinement, particularly to determine the source of an order of magnitude discrepancy between measured and model predicted signal amplitudes. Other model considerations which are warranted include development and implementation of more accurate ultrasonic beam models and scattering theories. Further practical considerations include the possibility of modeling the response from "geometrical" reflectors, such as a nearby weld, and the need to incorporate surface finish effects to allow simulation of inspection through cladding, etc. The future goal of this work is to develop a model-based tool for use in the design of new inspection techniques and configurations, the verification of their capability, the selection of alternative approaches, and the validation of existent techniques. The ability to synthesize simulated waveforms could also be of considerable value in training programs. Such a capability could thus offer a significant economic benefit in that it would allow NDE systems to be designed and evaluated using predictive approaches, rather than empirical methods relying upon extensive sample preparation and experimentation.

ACKNOWLEDGMENT

This work was sponsored by the Electric Power Research Institute under research project RP2687-1.

## REFERENCES

1. B. A. Auld, Wave Motion, 1, 1979, pp 3-10.
2. R. B. Thompson, T. A. Gray, J. D. Achenbach and K. Y. Hu, 1983, EPRI Report NP-3822.
3. R. B. Thompson, T. A. Gray, J. D. Achenbach and D. Budreck, final report for EPRI contract RP2687-1 (in press).
4. J. D. Achenbach and D. E. Budreck, these proceedings.
5. R. B. Thompson and E. F. Lopes, J. Nondestr. Eval., 4, 1984, pp. 107-123.
6. R. B. Thompson and T. A. Gray, J. Acoust. Soc Am. 74, 1983, pp. 1279-1290.
7. G. J. Dau and M. Behravesh, in Effective Nondestructive Examination for Pressurized Components, R. Nichols and G. J. Dau, eds., Applied Science Pub. Ltd. (in press).

PLATE MODES GENERATED BY EMATS FOR NDE OF PLANAR FLAWS*


S. K. Datta

Department of Mechanical Engineering and CIRES
University of Colorado
Boulder, CO 80309

R. E. Schramm

Fracture and Deformation Division
National Bureau of Standards
Boulder, CO 80303

Z. Abduljabbar

Department of Mechanical Engineering
King Saud University
P.O. Box 800
Riyadh, Saudi Arabia

INTRODUCTION

The theory of SH-wave generation by electromagnetic-acoustic transducers (EMATs) and the representation of the waveform in propagating plate modes was developed in [1]. Since then several theoretical and experimental studies have been reported that deal with SH-wave scattering by normal flaws (cracks) and its use for sizing these cracks [2-10].

In this paper we have studied theoretically and experimentally SH-wave scattering by planar canted cracks in a plate. Generation of various plate modes by a transducer, their interaction with a canted crack, and then reception of the scattered signal by a receiver have been analyzed. Calculations have been based on a hybrid finite element and modal expansion technique. Predicted scattered waveforms due to canted cracks of different lengths are compared with experiments performed on slotted plates. This background narrows the experimental search for the best transducer geometry to obtain a simple signal most useful for flaw detection and sizing. The NDE parameter used in some recent studies has been the ratio of the amplitudes of signals back-scattered from and transmitted through a flawed region in a plate [9,10].

---
*Contribution of the National Bureau of Standards; not subject to copyright in the United States.

## GOVERNING EQUATIONS AND SOLUTION

Consider a canted crack in a plate as shown in Fig. 1. For the purpose of the present method of analysis, the plate is divided into three regions, $R_1^-$, $R_2$, and $R_1^+$ as shown. It is assumed that the plate has the same homogeneous uniform properties in $R_1^\pm$ but may have different properties in $R_2$. Also, $R_2$ may have any number of cracks or inhomogeneities.

For SH motion with nonzero displacement $w(x,y,t)$ in the z-direction only, the equation satisfied by w is,

$$\frac{\partial^2 w}{\partial x^2} + \frac{\partial^2 w}{\partial y^2} + k_2^2 w = 0, \quad (x,y) \in R_1^\pm \tag{1}$$

where $k_2$ is the wavenumber and we have assumed that $w(x,y,t) = w(x,y)e^{-i\omega t}$.

In $R_1^\pm$, w satisfies the boundary conditions

$$\frac{\partial w}{\partial y} = 0 \text{ on } y = 0, H \tag{2}$$

Solution to equation (1) satisfying (2) admits an infinite set of propagating and nonpropagating waves of the form,

$$w_n = \cos(n\pi y/H) \, e^{\pm i k_n x}, \quad n = 0, 1, 2, \ldots \tag{3}$$

where $k_n = \sqrt{k_2^2 - n^2\pi^2/H^2}$. It was shown in [1] that the waves generated by an EMAT placed on the surface $y = 0$ can be expressed in terms of the plate modes (3) as, (see [3])

$$w(x,y) = \frac{ik_2 MD}{2\rho v_2 H} \frac{N_c I_e B_o}{\omega W} \sum_{n=0}^{\infty} \frac{\varepsilon_n F}{k_n} \cos\frac{n\pi y}{H} e^{\pm i k_n x}, \tag{4}$$

where:
- $\rho$ = density of the plate.
- $v_2$ = shear wave speed in the plate,
- $W$ = width of the transducer,
- $I_e$ = eddy current induced in the plate,
- $N_c$ = number of turns in the coil,
- $B_o$ = normal component of the applied magnetic field,
- $F = \dfrac{\pi \sin(k_n D/4)}{2(k_n D/4)} \dfrac{\sin[(MD/4)(k_n - 2\pi/D)] \exp(iM\pi/4)}{M \sin[(D/4)(k_n - 2\pi/D)]}$
- $M$ = number of magnet pairs
- $D$ = magnet spacing
- $\varepsilon_n = \begin{cases} 1, & n = 0 \\ 2, & n > 0. \end{cases}$

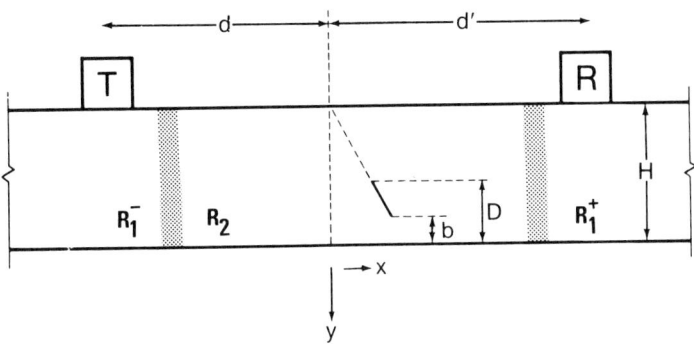

Fig. 1  Cross section of transmitter and receiver (T and R) transducers on the surface of a plate containing a canted flaw. The plate is divided into three regions: $R_1^\pm$ = unflawed, $R_2$ = flawed. The z-coordinate is normal to this drawing.

Now for the laboratory configuration shown in Fig. 2 with the transmitter and receiver placed at y = H at a distance d apart, the normalized receiver signal as a function of time can be expressed approximately as

$$S(d,t) = \sum_{\ell=1}^{L} W_\ell \sum_{n=0}^{N} F_{\ell n}(d,t)\, A_{\ell n}^{T}\, A_{\ell n}^{R}\, \sin(\omega_\ell t - k_{\ell n} d) \tag{5}$$

when there is no scatterer present. The definitions of various quantities above are:

$\ell$ = frequency index,

n = mode index,

$W_\ell$ = weight assigned to each frequency component,

N = highest order plate mode allowed to propagate,

$F_{\ell n}(d,t)$ = envelope factor,

$A_{\ell n}^{T}$ = normalized transmitter array factor,

$A_{\ell n}^{R}$ = normalized receiver array factor,

$k_{\ell n}$ = $\sqrt{(\omega_\ell/v_2)^2 - n^2\pi^2/H^2}$ .

Figure 2 shows the receiver signals at different values of d. The transducers were EMATs with 16 pairs of magnets (M) with a spacing of 3.9 mm (D). The frequency was 454 kHz (2.85 × $10^6$ rad/s). The test material was A516 steel. Also shown in this figure are the signals computed according to Eq. (5) with L = -3, $\omega_1$ = 2.639, $\omega_2$ = 2.827, and $\omega_3$ = 3.016, all in $10^6$ rad/s. The weights assigned on the basis of experimental spectral analysis were $W_1 = W_3 = 0.6$, $W_2 = 1.0$, and the pulse envelope $F_{\ell n}$ was taken to be a triangle as shown in Fig. 3. The shear wave velocity in the plate was $v_2$ = 3.17 km/s. Qualitatively, there is good agreement between the computed and measured signals, although there are quantitative differences.

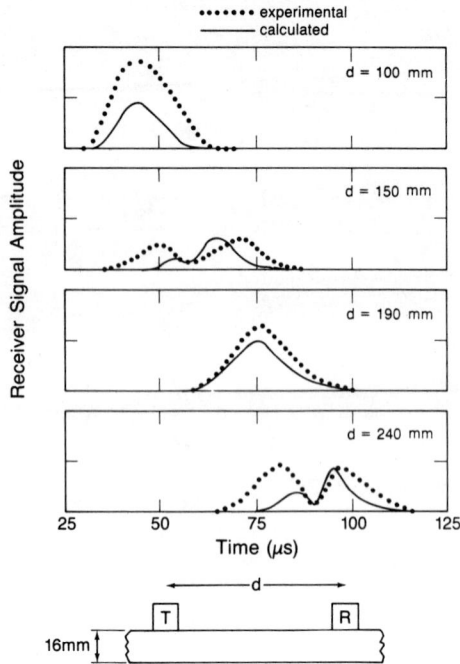

Fig. 2  Signal characteristics of an SH-wave transmitted along an unflawed plate. This compares the experimental and calculated signal shape at several transducer separations. The rf carrier is omitted here and only the signal envelope is shown.

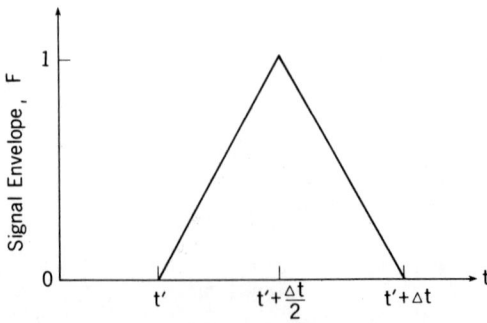

Fig. 3  Simple triangular envelope imposed here on the calculated signal. The true pulse shape is largely due to the duration of the gated rf tone burst (5 cycles here), and the transducer size and configuration. $t'$ is the arrival time of the pulse front, and is a function of the plate modes since they each travel down the plate at different speeds as determined by the angle of their wave fronts. $\Delta t$ is the pulse duration; the experimental value in this study was 20 µs.

Once the reflection and transmission coefficients are obtained [8] for a particular frequency, then the reflected and transmitted fields can be calculated at any point. In particular, for the experimental setup considered here (Fig. 1) we obtain the reflected or transmitted pulse at a point as

$$S(d,d',t) = \sum_{\ell=1}^{L} W_\ell \sum_{m=0}^{N} \sum_{n=0}^{N} F_{\ell mn}(d,d',t) A_{\ell m}^{T} A_{\ell n}^{R} (-1)^{m+n} |S_{\ell nm}|$$
$$\times \sin(w_\ell t - k_{\ell m} d - k_{\ell n} d' - \alpha_{\ell nm})$$

where the parameters are the same as in Eq. (5) except:

m = incident mode index
n = scattered mode index

$$S_{\ell nm} = |R_{\ell nm}| e^{i\alpha_{\ell nm}^{r}} \quad \text{for the reflected field}$$

$$= |T_{\ell nm}| e^{i\alpha_{\ell nm}^{t}} \quad \text{for the transmitted field.}$$

R and T are the reflection and transmission coefficients with phase shifts on scattering of $\alpha^r$ and $\alpha^t$, respectively.

The experimental conditions were the same as noted above except that now there was a sawcut canted at 30° to the surface normal to simulate the flaw shown in Fig. 1. On the test plate, b=0 (the sawcut was on the bottom surface) while the calculation for the scattering coefficients assumed b/H=0.1 (buried crack). To compare the theoretical predictions with experimental results, three different angular frequencies $\omega_1$, $\omega_2$, $\omega_3$ as given before were chosen. For each of these frequencies there were five propagating modes (N = 4). Computed pulse shapes are compared with observed ones for the different flaw sizes in Figs. 4-7. At some receiver locations the agreement is reasonably good. However, at some other locations there is only qualitative agreement.

The discrepancies between the computed and observed pulse shapes may arise for several reasons:

1. Finite transducer size;
2. Actual flaws not buried, but surface-breaking;
3. Frequency content of the actual pulse continuous with a finite bandwidth;
4. Inhomogenous plate with non-flat surfaces;
5. Gated pulse shape.

Even with the simplifying assumptions made in the theoretical computations, there is a remarkable qualitative agreement with observations. Improving on items 2, 3, and 5 above would be especially important in increase this agreement.

As an aid to experimental NDE, these calculations will serve to guide the selection of transducer locations. With proper placement, it is possible to obtain a signal with maximum amplitude, simplest shape, and least sensitivity to minor changes in test geometry to increase probability of detection and sizing reliability. Flaw size, orientation, and location through the depth all influence mode interaction. This gives the possibility that detailed analysis of signal configuration will yield much information on flaw character.

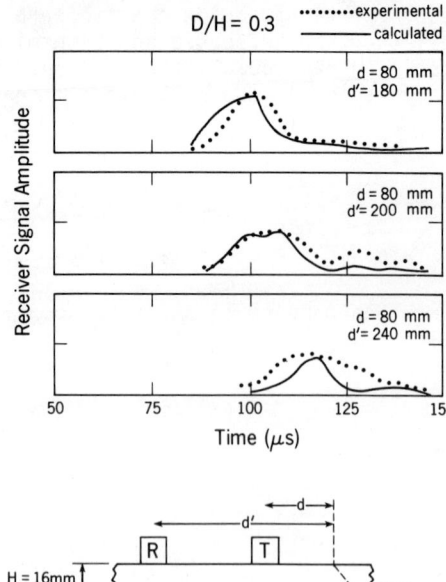

Fig. 4    Reflected signal envelopes from a 30° canted flaw.  D/H = 0.3.

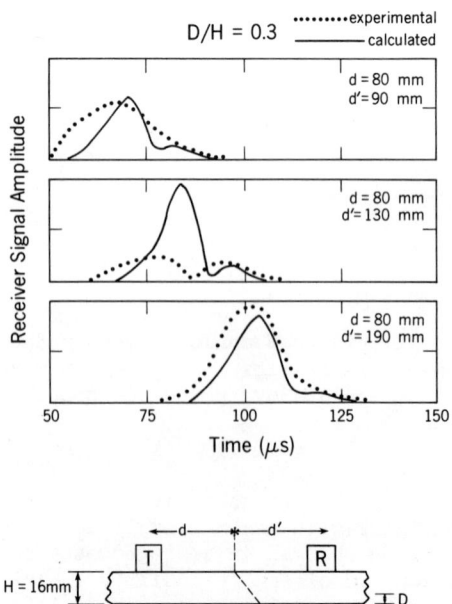

Fig. 5    Transmitted signal signal envelopes from a 30° canted flaw.  D/H = 0.3.

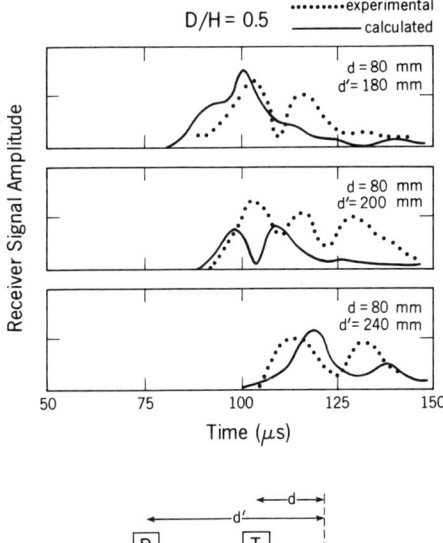
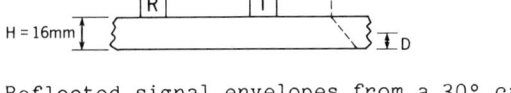

Fig. 6  Reflected signal envelopes from a 30° canted flaw. D/H = 0.5 for calculations and D/H = 0.6 for experiment.

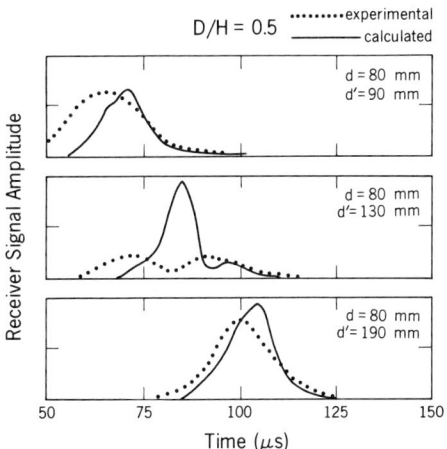
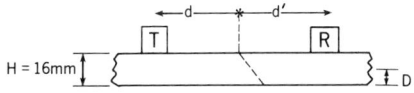

Fig. 7  Transmitted signal envelopes from a 30° canted flaw. D/H = 0.5 for calculations and D/H = 0.6 for experiment.

ACKNOWLEDGMENT

Theoretical work reported here was supported in part by grants from the National Science Foundation (CEE78-24179, CEE81-20536) and the Office of Naval Research (N00014-86-K-0280). The experimental work was supported by the Naval Sea Systems Command, (N00167-86-WR6-0038).

REFERENCES

1. C. F. Vasile and R. B. Thompson, J. Appl. Phys. 50, 2583 (1979).

2. S. K. Datta, C. M. Fortunko, and R. B. King, in: "1981 Ultrasonics Symposium Proceedings," B. R. McAvoy, ed. IEEE, New York (1981).

3. C. M. Fortunko, R. B. King, and M. Tan, J. Appl. Phys. 53, 3450 (1982).

4. C. M. Fortunko and R. E. Schramm, Welding J. 61, 39 (1982).

5. Z. Abduljabbar, S. J. Datta, and A. H. Shah, J. Appl. Phys. 54, 461 (1983).

6. C. M. Fortunko and R. E. Schramm, J. NDE 3, 155 (1983).

7. Z. Abduljabbar and S. K. Datta, in: "Numerical Methods for Transient and Coupled Problems," R. W. Lewis, E. Hinton, P. Betess, and B. A. Schreffer, eds., Pineridge Press, Swansea (1984).

8. Z. Abduljabbar, "Diffraction of Horizontally Polarized Shear Waves in a Plate," Ph. D. thesis, University of Colorado, Boulder, 1983.

9. R. E. Schramm and T. A. Siewert, in: "Review of Progress in Quantitative Nondestructive Evaluation, 5B," D. O. Thompson and D. E. Chimenti, eds., Plenum Press, New York (1986), p. 1705.

10. R. E. Schramm and T. A. Siewert, "Sizing Canted Flaws in Weldments Using Low-Frequency EMATs," these proceedings.

# REFLECTION OF BOUNDED ACOUSTIC BEAMS FROM A LAYERED SOLID

A. K. Mal

Mechanical Aerospace and Nuclear Engineering Department
University of California, Los Angeles, California  90024

and

T. Kundu

Department of Civil Engineering and Engineering Mechanics
University of Arizona, Tuscon, AZ 85721

INTRODUCTION

It is well known that when a bounded beam of acoustic waves is incident on a fluid-solid interface at certain critical angles, the reflected beam is significantly distorted and displaced due to the interference between specularly and nonspecularly reflected waves. Measurement and analysis of the reflected field can be used to estimate certain near surface elastic properties of the solid by means of several alternative nondestructive experimental arrangements [1,2]. In most of these experiments the interface generated leaky waves play a significant role.  Thus a good understanding of the interface phenomena is a prerequisite to the design of experiments for their practical applications.

A detailed examination of the leaky waves and of their influence on the reflected field was carried out in [3] for a uniform half space model of the solid.  The problem of the multilayered solid consisting of one or more isotropic elastic layers bonded to a homogeneous half space has been considered in a number of papers [4-6]. The multilayered plate with the bottom surface free of traction has also been considered in a more recent paper [7].  These and other recent studies have shown that in presence of layers, the interface phenomenon exhibits many features which, in general, can not be extrapolated from the simple half space [3] or thin layer approximations [4].  The primary reason for this is the presence of dispersive guided waves which can propagate along the layers with relatively low spatial attenuation.

In this paper we discuss the interface related wave phenomena in a solid consisting of a top layer (InSb) which is bonded to a homogeneous substrate (Si) by means of a relatively thin and low velocity bonding material (epoxy).  We also consider the same problem for a plate obtained by removing the substrate as a reasonable model of the layered solid containing a large debond at the epoxy-Si interface.  We present dispersion curves for both models in a wide frequency range and show that the epoxy layer, however thin, has a strong influence on the phase velocity of the Rayleigh waves in the half space model. In addition, we calculate the V(z) curve or the AMS for each model at a fixed frequency and show that the presence of the debond significantly alters their oscillatory character.

THEORY

Let a cartesian coordinate system be located at the fluid-solid interface with the y-axis directed into the solid (Fig.1). We assume, for simplicity that all field variables describing the wave motion in the medium are independent of z, so that the mathematical problem is two dimensional.

We first consider the guided wave problem in the solid in absence of the fluid. The general theory of Rayleigh wave propagation in a multi-layered half space can be found in the standard seismological literature [8]. The theory of Lamb type wave propagation in multilayered plates appears to have received less attention in the literature except in the case of the uniform (i.e., single layered) plate which has been studied in great detail [9]. The dispersion or secular equation for a multi-layered plate has been discussed in [7]. The general form of this equation for both the half space and the plate is

$$D(k, \omega) = 0 \qquad (1)$$

where k is the wavenumber of the guided waves, $\omega$ is the circular frequency and $D(k, \omega)$ is the dispersion function. For a homogeneous half

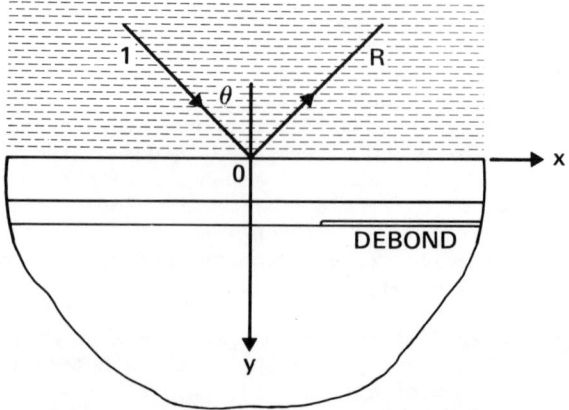

Figure 1. Geometry of the problem.

space or a plate, explicit expressions for D can be easily obtained. In presence of layers, it is difficult if not impossible to derive such closed form expressions. However, D can be expressed as an element of a 4x4 matrix which is itself the product of a number of other 4x4 matrices. The number of these matrices is equal to the number of the layers and the general method is based on the so called Thomson Haskell matrix method [8]. The elements of the individual matrices are certain functions of $\omega$, k and the properties of the layers. Calculation of the wavenumber k or, equivalently, the phase velocity $v_p = \omega/k_p$ requires a numerical treatment of the matrices. It is well known that a direct numerical evaluation of the matrix product becomes unstable at higher frequencies when evanescent waves are present within the layers. In the seismological literature, submatrix manipulation of the Thomson Haskell matrices and an alternative formulation called the reflectivity method have been used to avoid this so called precision problem. A discussion of this and other related issues can be found in [10].

We have used the submatrix (also called the delta matrix) manipulation in calculating the dispersion functions for both the half space and the plate models. We have developed computer codes for the calculation of the roots of the dispersion equations. The codes appear to be extremely efficient and yield all roots at a given frequency in double precision arithmetic. These codes have been used to calculate the phase velocities of the guided waves in the two models. The results of these calculations will be presented and discussed in the next section.

The next problem of interest is the calculation of the plane wave reflection coefficient from the interface as a function of the incident angle. It has been shown [3,4,6,7] that the general form of the reflection coefficient is

$$R(k,\omega) = \frac{D(k,\omega) + i\omega q(k)G(k,\omega)}{D(k,\omega) - i\omega q(k)G(k,\omega)} \qquad (2)$$

where k is the horizontal wave number of the incident acoustic waves,

$$q(k) = \omega \rho_f / \nu_f \qquad (3)$$

$$\nu_f = (k_f^2 - k^2)$$

$$k_f = \omega/\alpha_f$$

and $\rho_f$, $\alpha_f$ are the density and acoustic wave speed in the fluid. The function $G(k,\omega)$ is an element of the same product matrix from which the dispersion function D is derived. If the angle of incidence is $\theta$ then

$$k = k_i = k \sin\theta \qquad (4)$$

It can be seen from equation (2) that if the phase velocity of the guided waves is greater than $\alpha_f$ and if

$$\theta = \arcsin(\alpha_f/v_p) \qquad (5)$$

then the reflection coefficient becomes -1, resulting in a critical phenomenon. For a given frequency there may be several critical angles corresponding to the number of roots of the dispersion equation (1).

We now consider the reflection of a bounded acoustic beam incident at a critical angle $\theta$. Assuming that the incident beam has a Gaussian profile of width 2b at the fluid solid interface, the incident and the reflected potentials in the fluid may be expressed in their fourier integral forms

$$\phi_i(x,y,\omega) = (b_o/2\sqrt{\pi}) \int_{-\infty}^{\infty} \exp\{g(x,y,k,\omega)\} \, dk \tag{7}$$

$$\phi_r(x,y,\omega) = (b_o/2\sqrt{\pi}) \int_{-\infty}^{\infty} R(k,\omega)\exp\{g(x,-y,k,\omega)\} dk \tag{8}$$

where

$$g(x,y,k,\omega) = -\{(k_i-k)b_o/2\}^2 + i(k_i-k)x + i\nu_f y \tag{9}$$

and
$$b_o = b \sec\theta$$

Clearly, the calculation of the reflected field requires numerical evaluation of the integral (8). Although the significant contribution to the integral comes from a small interval of k near $k_f$, the integrand is not very well behaved due to rapid variations in R near the critical point and to the dense oscillations of of the sinusoidals. We have developed a special quadrature code [11] based on a modification of the original work of Clenshaw and Curtis [12]. The code appears to be extremely efficient in evaluating this type of integrals and is adaptive with excellent accuracy control features.

If the incident beam is convergent as in the case of the acoustic microscopy experiment, then it is decomposed into a finite number of narrow beams at the critical angles corresponding to the operating frequency of the microscope. The reflected waves from each of these beams are calculated separately as described above. In addition, the reflection of the normally incident beam is also considered as a special case. These reflected fields are then propagated through the buffer rod of the microscope and the resulting normal displacements are superposed to give the transducer response. The V(z) curve is obtained by changing the focal distance z of the lens from the interface. Details of these calculations can found in [6]. Results for the two layered solid are presented in the next section.

RESULTS AND DISCUSSIONS

The properties of the two layered half space are given in Table 1. In the absence of any debonding the possible guided waves in the solid are dispersive Rayleigh waves which decay exponentially in the substrate. The calculated dispersion curves in the frequency range 0-100 MHz are presented in figure 2 for the model of Table 1 with a 5 m thick epoxy layer. It can be seen that there are three possible modes of propagation in this frequency range with cutoffs at 15 and 30 MHz. As expected the phase velocity of these higher modes is 6 mm/μsec (the shear wave velocity in the substrate) at the cutoffs and approaches the Rayleigh wave velocity in top layer (2.09 mm/μsec) at the high frequency limit. The phase velocity of the fundamental mode equals the Rayleigh velocity in the substrate (5.41 mm/μsec) at zero frequency and approaches 2.09 mm/μsec at high frequencies.

Table 1. Properties of Component Materials

| Material | Thickness $h$ (micron) | P-wave vel. $\alpha$ (km/sec) | S-wave vel. $\beta$ (km/sec) | Rayleigh Wave vel. $v_R$ (km/sec) | Density $\rho$ (gm/cc) |
|---|---|---|---|---|---|
| InSb | 8 | 3.78 | 2.29 | 2.09 | 5.77 |
| Epoxy | 5 | 2.20 | 1.10 | 1.03 | 1.20 |
| Silicon | $\infty$ | 9.30 | 6.00 | 5.41 | 2.33 |

Clearly, the low velocity epoxy layer has a strong influence on the velocity of the fundamental mode, which decreases rapidly to about 1.4 mm/μsec and then increases rather slowly toward its limiting value of 2.09 mm/μsec.

In order to further examine the influence of the bonding layer on the phase velocity of the fundamental mode, we calculated its values for different thicknesses of this layer. The results are shown in figure 3. It can be seen that as the thickness of the layer decreases, the decrease in the phase velocities becomes less rapid and that they approach the top curve (no epoxy layer). However, even the presence of a 1 m epoxy layer has a substantial influence on the fundamental mode.

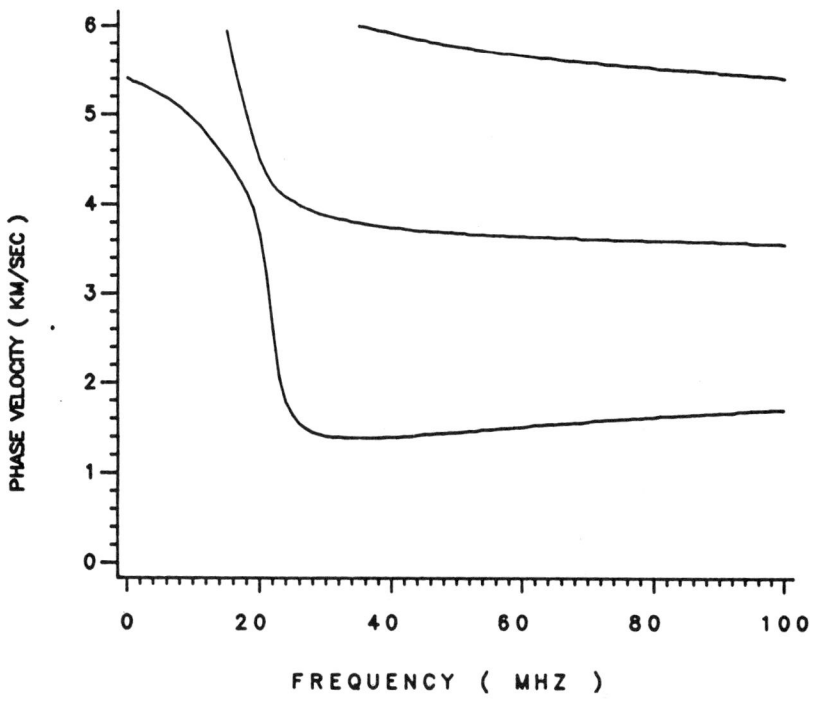

Figure 2. Phase velocity of the three Rayleigh modes in the half space model of Table 1.

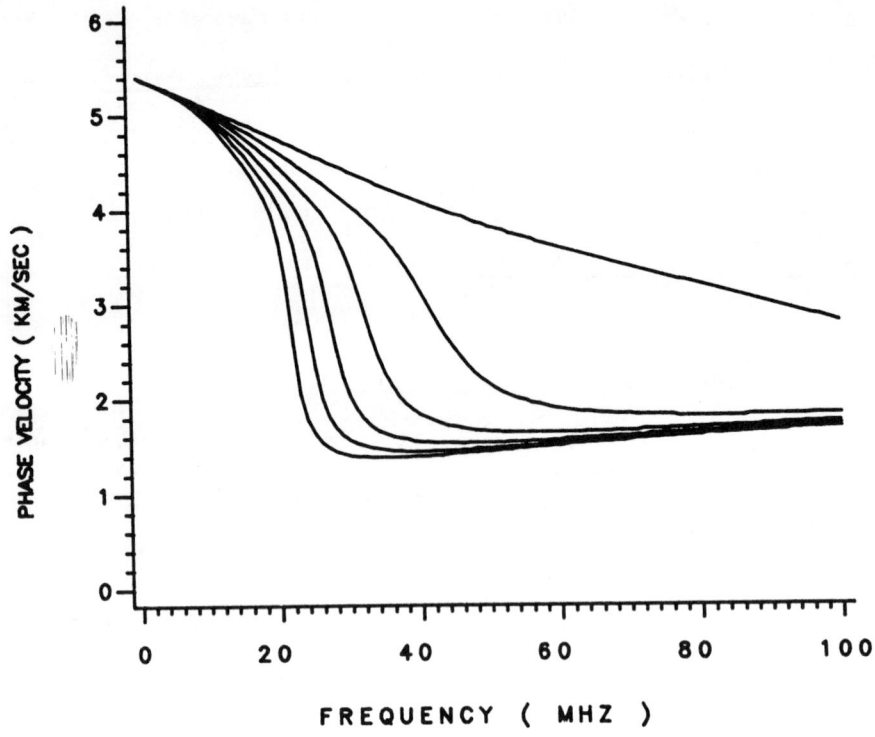

Figure 3. Phase velocity of fundamental Rayleigh modes for half space model with different thicknesses of the epoxy layer. Uppermost curve is for no epoxy layer, next lower for 1 μm thick layer, etc.

In Figure 4 we present the calculated phase velocity of the Lamb waves which can propagate in the two layered plate obtained by removing the substrate from the half space model. In this case the phase velocity of the fundamental mode is almost flat in the entire range of frequencies and all velocities approach 2.09 mm/μ sec at high frequencies as expected. It should be noted that in contrast to a homogeneous plate, the waves in the layered plate can not, in general be classified into symmetric and antisymmetric motions.

Finally the calculated V(z) curves at 60 MHz for the two layered solid in presence and in absence of the substrate are shown in figure 5. Only the total contribution from all the critically reflected beams are shown; their individual contributions have been presented in [7]. It can be seen that the oscillations in the curves in absence of debonding are of a significantly different nature than those in presence of debonding. For the half space model, two periodicities can be detected; the short period oscillations are due to leaky wave radiation from the fundamental mode Rayleigh waves, while the longer period ones are due to leakage from critically refracted P-waves propagating in the substrate. In the case of the plate, the oscillations are entirely due to the first higher mode of the Lamb waves; the fundamental mode does not produce any leaky waves in the fluid.

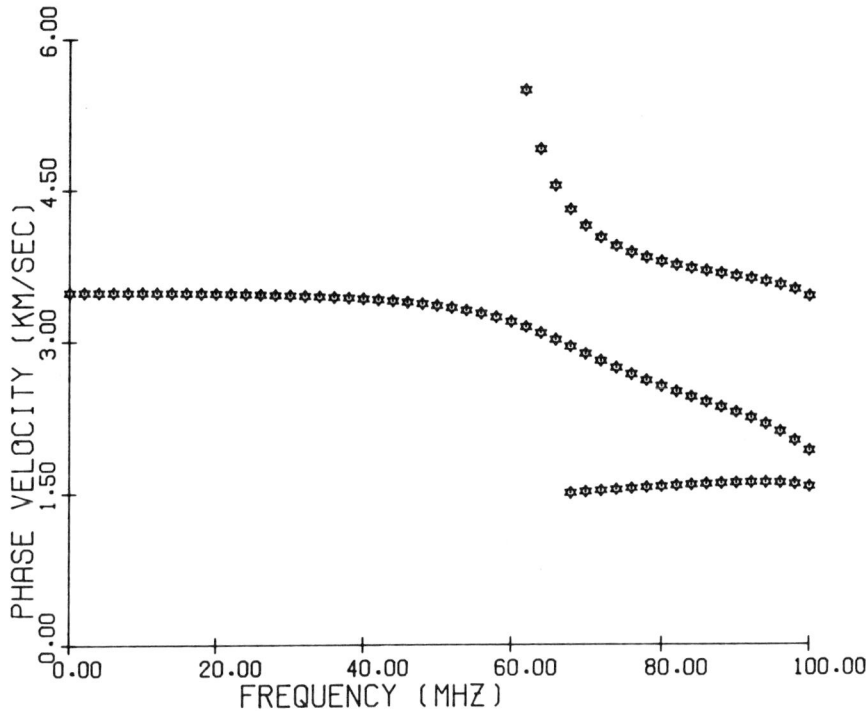

Figure 4. Phase velocity of Lamb waves in a two layer plate obtained by removing the Si substrate from the model in Table 1.

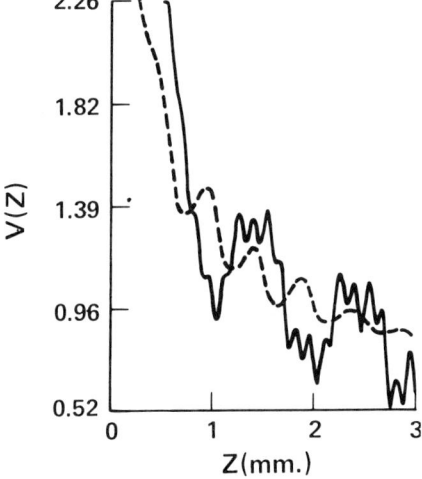

Figure 5. Computed V(z) curves for the two layered model of Table 1 (solid) and the two layered plate in absence of the substrate (dashed) at 60 MHz.

## ACKNOWLEDGEMENT

This research was partially supported by the National Science Foundation under Grant MEA-82-19592.

## REFERENCES

1. Y. Bar-Cohen and D. E. Chimenti, Review of Progress in Quantitative NDE, 3B, 1043 (1984).
2. R. D. Weglein, R. F. Wilson and S. D. Benson, Review of Progress in Quantitative NDE, 4A, 271 (1985).
3. H. L. Bartoni and T. Tamir, Applied Physics, 2, 157 (1973).
4. A. H. Nayfeh, D. E. Chimenti, L. Adler and R. L. Crane, J. Appl. Phys., 52, 4985 (1981).
5. D. B. Bogy and S. M. Gracewski, Int. J. Solids and Structures, 20, 747 (1984).
6. T. Kundu, A. K. Mal and R. D. Weglein, J. Acoust. Soc. Am., 77, 353 (1985).
7. T. Kundu and A. K. Mal, Int. J. Engr. Sc. (in press).
8. K. Aki and P. G. Richards, Quantitative Seismology, Theory and Methods, Freeman, San Francisco (1980).
9. N. Vasudevan and A. K. Mal, J. Appl. Mech., 52, 356 (1985).
10. T. Kundu and A. K. Mal, Wave Motion, 7, 459 (1985).
11. P. -C. Xu and A. K. Mal, Wave Motion, 7, 235 (1985).
12. C. W. Clenshaw and A. R. Curtis, Numer. Math., 2, 197 (1960).

APPLICATION OF FORWARD AND INVERSE SCATTERING MODELS FOR

ULTRASONIC WAVES IN COMPLEX LAYERED STRUCTURES

P.R. Smith and L.J. Bond

NDE Centre, Department of Mechanical Engineering
University College London, England

D.T. Green and C.A. Chaloner

R & D Centre, Royal Ordnance Explosives Division
Westcott, England

INTRODUCTION

In circumstances where multi-layered structures play a critical role in the performance of manufactured components, a thorough and detailed inspection procedure is required. However the construction of such objects is such that it is not easy to provide an adequate non-destructive test to monitor the state of the structure. It is possible to employ x-ray techniques, but on these layered (or cylindrically symmetric) structures many exposures are required over a range of positions, which is both expensive and time consuming. It is therefore attractive to use an ultrasonic NDT system. A further important motivation for using ultrasound is its sensitivity to a large number of material parameters.

The scattering of ultrasonic waves in complex layered structures which combine a range of materials can provide useful data about the state of the structure. However, such signals are difficult to interpret. It is desirable to recover quantitative information from the scattered field, and this requires an inverse solution method for an appropriate forward model. In this paper we investigate the application of an inverse scattering algorithm, for the reconstruction of depth-dependent acoustic impedance, using data derived from a simulated pulse-echo experiment.

Models for the forward scattering problem of pulsed ultrasonic waves in cylindrical structures have been provided by finite difference methods in cylindrical coordinates, and other models for waves in layers [1,2]. Systems which consider layers of steel, rubber and perspex have been produced. Given reasonable agreement between theory and experiment for the forward problem, attention has turned to consider the inverse problem, in particular, the solution of the inverse scattering problem for the acoustic impedance profile. Here we apply the impediography method [3] to the inversion of the scattered field from a layered system, for pulse-echo measurements. Synthetic data are derived from a layered model based on that of Scott and Gordon [4] in which the effects of multiple scattering (reverberation) are included. The accuracy of the inversion algorithm is tested in the presence of band-limited noise and frequency dependent absorption.

The importance of the technique as a method for characterising defects in adhesive bonds will be illustrated and discussed.

FORWARD MODELS: FINITE DIFFERENCE

For many acoustic and elastic wave propagation scattering problems no analytical solutions are available. For solving such problems a range of numerical modelling techniques have been employed [1], including those which use finite difference (FD) methods. These latter methods are particularly well suited to the study of pulsed waves and interactions in the mid-frequency scattering regime. A comparison between the explicit FD methods and those based on T-matrix formulations is given in Table 1. [2]. Practical FD modelling is restricted (by present computing capability) to two spatial dimensions and time, and models have been developed in Cartesian, cylindrical and spherical coordinates [2,5].

Table 1. Comparison of 'T'-matrix and explicit methods for modelling elastic wave phenomena

| Problems | T-matrix | Finite difference |
| --- | --- | --- |
| $ka$ range | $0 < ka < 10$ | $0.1 < ka < 20$ |
| Field region | near and far | near (limited by available computers) |
| Dimensions | 2-D. and 3-D. | 2-D. (3-D. possible, but limited by available computers) |
| Defect shape | wide range of types possible | |
| Mode conversion | explicitly included | |
| Short pulse | impractical | very good |
| Multiple scatter | good | 2-D. very good |
| Wave types | good range | shear, compressional and Rayleigh |

The FD models can be formulated, in many cases, as initial value problems; where the displacement field at all grid points is specified at the start of the computation, and the numerical model then calculates the displacement field at all future times.

When complex layered structures are considered it is necessary to define a set of system parameters which here include: (i) pulse spectral content, (ii) layer thickness, (iii) material elastic properties, and (iv) interface conditions. Structures of interest could be composed of rubbers, metals and filled polymers.

An example of pulsed wave propagation into a three layered system which includes Voigt-type damping is shown in Figure (1). The input pulse is a smoothed delta function applied across the surface with a cosine bell weighting function, as has been used elsewhere [5]. The observed wave

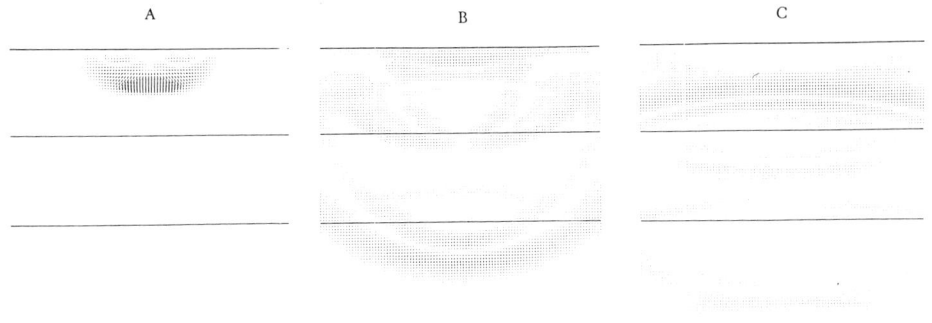

Fig. 1. Numerical visualisation of waves for a strip pulse source on a three layered plate structure. Figures (1a)-(1c) represent a time sequence, and the amplitude and direction of the displacement field is shown by lines originating from the 2-D lattice.

field is seen to be similar to that given by square ultrasonic transducers which include both plane and edge wave components [6]. The interaction of curved wave fronts with layers is further complicated in that mode-conversion is then important.

Data from these and similar numerical studies provide solutions to the forward scattering problem for relatively complicated scattering objects. However many inversion schemes require some form of analytical forward model. For such schemes to be easily implementable a scattering approximation is often used, and this aspect is now considered.

FORWARD MODEL: ANALYTIC

Since we are dealing with layered structures whose material properties only change at the interfaces between layers (to a good approximation), we assume that the media are uniformly homogeneous. At an interface between the acoustic semi-infinite media i and j the reflection (R) and transmission (T) coefficients for plane waves of normal incidence are given, as functions of the impedances ($Z_i$, $Z_j$), by

$$R_{ij} = (Z_j - Z_i)/(Z_i + Z_j)$$
$$R_{ji} = -R_{ij}$$
$$T_{ij} = 1 + R_{ij}$$
$$T_{ji} = 1 - R_{ij}$$

Scott and Gordon [4] have derived an iterative procedure for generating the R and T coefficients of a wave A(t), where t=time, normally incident at an arbitrary number of uniform layers. The result is given in the form of reflection and transmission spectra (as functions of the frequency w) incorporating the effects of reverberation within the layers. These spectra also depend on the R coefficients at each interface, and the

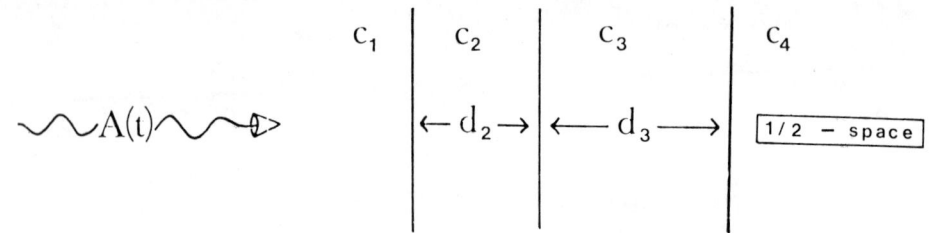

Fig. 2. Schematic of four component layered structure with source pulse incident from the left.

characteristic frequency ($w_i$) of each layer, defined as the ratio of the sound speed in layer i ($c_i$) to the width of layer i ($d_i$).

Here we consider a four component layered structure (three interfaces) as depicted in Figure (2). Our intention is to simulate the layered system: couplant/rubber/adhesive/perspex, which is representative of a variety of commonly occuring structures. Using the iterative formulae derived by Scott and Gordon the reflection coefficient for this object ($R_{14}(w)$) may be evaluated algebraically as

$$R_{14} = [R_{12}+T_{12}T_{21}(R_{23}+R_{34}\exp(-2ia))\exp(-2ib)/F]A(w) \qquad (1)$$

where

$a = w/w_3$

$b = w/w_2$

$F = 1 + B\exp(-2ia) + C\exp(-2ib) + D\exp(-2i[a+b])$

$B = -R_{34}R_{32}$

$C = -R_{23}R_{21}$

$D = -R_{21}R_{34}$

and $A(w)$ is the Fourier transform of the signal $A(t)$. Equation (1) constitutes the forward model which is used to generate synthetic pulse-echo data.

Having obtained $R_{14}$ explicitly (as opposed to a numerical recursion relation) it is illuminating to analyse the expression further. In an experiment we might hope to measure the quantity $20\log_{10}[R_{14}(w)]$, and therefore examination of equation (1) indicates that the shape of this function, in the frequency domain, will be modulated by the positions of turning values in $|F|^2$. These turning points are given by satisfying both $a=n\pi/2$ and $b=m\pi/2$, where n,m are integers. Thus in terms of the measured frequency $f=w/2\pi$, turning points are located at

$$nw_3/4 = f = mw_2/4 \qquad (2)$$

This expression is then an extension of the familiar quarter wavelength condition: $f = nw_i/4$, which applies to a single layer i, and defines a condition for maximum (or minimum) reflection. Eliminating f from equation (2) yields

$$n/m = w_2/w_3 \qquad (3)$$

which has some interesting consequences. The equation (3) is clearly only satisfied for certain values of n,m thereby giving rise to a set of 'allowed' harmonics of the single layer, quarter wavelength condition. Since these 'allowed' features are functions of the dimension and property of the constituent materials, it may be possible to distinguish between a variety of structures by applying the condition (3). A more detailed analysis of the structure of equation (1) would be fruitful, but is not within the scope of the present work.

## INVERSE SCATTERING ALGORITHM

The impediography equation has been used extensively in the field of seismic processing [7] to recover the impedance profile, as a function of acoustic travel time ($t_a$), from backscattered signals. If the depth is measured in x then $t_a$ is defined as

$$t_a = \int_0^x dx'/c(x')$$

where $c(x')$ is the acoustic velocity profile. The impedance profile $Z(t_a)$ is given in terms of the impulse response $I(t_a)$ as

$$Z(t_a) = Z(t_a=0)\exp(2\int_0^{t_a} I(t_a')dt_a') . \qquad (4)$$

Note that in a real experiment the impulse response is not measured directly, but would be found after deconvolution of the source pulse. The impediography equation (4) is based on the Born approximation, and therfore only includes single scattering phenomena.

Since the synthetic data are derived from a multiple scattering model we do not expect equation (4) to reconstruct the impedance profile exactly. Although algorithms are available to take into account these multiple reflections [8], they are computationally far more expensive to implement than the impediography method. If the technique employed here is adequate then a relatively fast automated testing procedure can be envisaged.

## INTERROGATION OF PERSPEX/RUBBER INTERFACE

In this section we simulate a pulse-echo experiment performed on a layered system consisting of couplant, rubber, adhesive and perspex. We consider three cases:
    (i)   Normal adhesive (perfect bond)
    (ii)  Contaminated adhesive (partial bond)
    (iii) No adhesive (de-bond), i.e. air gap.
The parameter being changed between the objects (i)-(iii) is the acoustic impedance of the third material (i.e. the second layer). The rubber layer is 1.64mm. thick (so that the characteristic frequency = 1MHz.), and the adhesive layer is 0.85mm. thick (so that the characteristic frequency = 2MHz.). In cases (ii) and (iii) the layer widths remain the same, although the characteristic frequency of layer 2 will change. The impedance profiles for (i)-(iii) are shown in Figures (3a)-(3c) respectively,

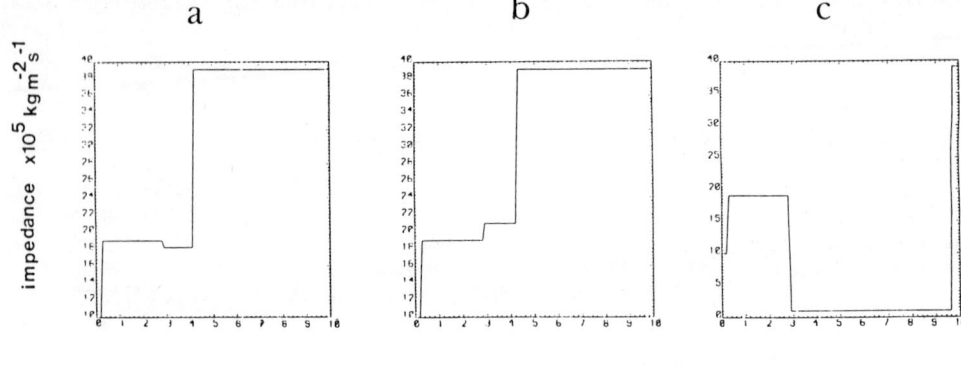

Fig. 3. Impedance profiles for cases (i)-(iii) are depicted in (3a)-(3c) respectively.

plotted as a function of travel time. The synthetic spectra for (i)-(iii) are shown in Figures (4a)-(4c) respectively, in the frequency range 0-20 MHz. Each spectrum has been artificially attenuated with increasing frequency by modulation with a Gaussian profile, to simulate an absorption process.

The reconstructed impedance profiles are shown in Figures (5a)-(5c) for cases (i)-(iii) respectively. Also shown (Figures (5d)-(5f)) are the reconstructed profiles after addition of noise in the frequency range 5-15 MHz. at a level of 50% of the signal amplitude. Each profile is clearly distinguishable from the others, and the important features are recovered even in the presence of quite intense noise. In terms of the precise features, case (iii) is probably the least accurate reconstruction and this is most likely due to the highly reverberating first layer (rubber).

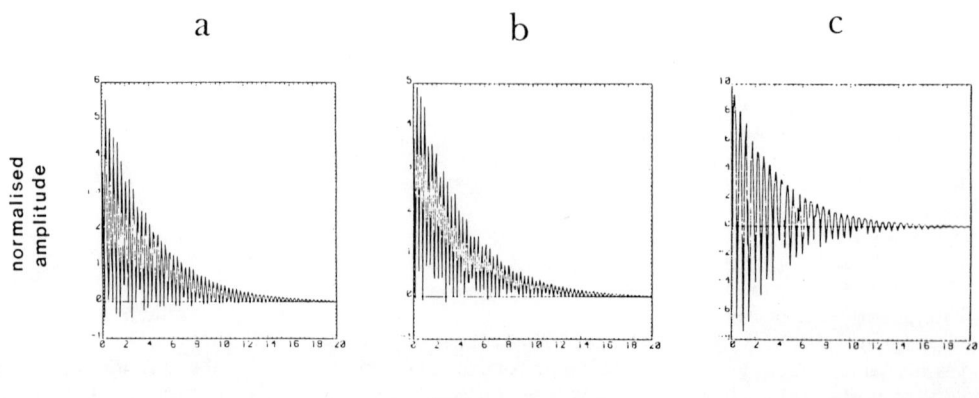

Fig. 4. Synthetic spectra for cases (i)-(iii) are depicted in (4a)-(4c) respectively.

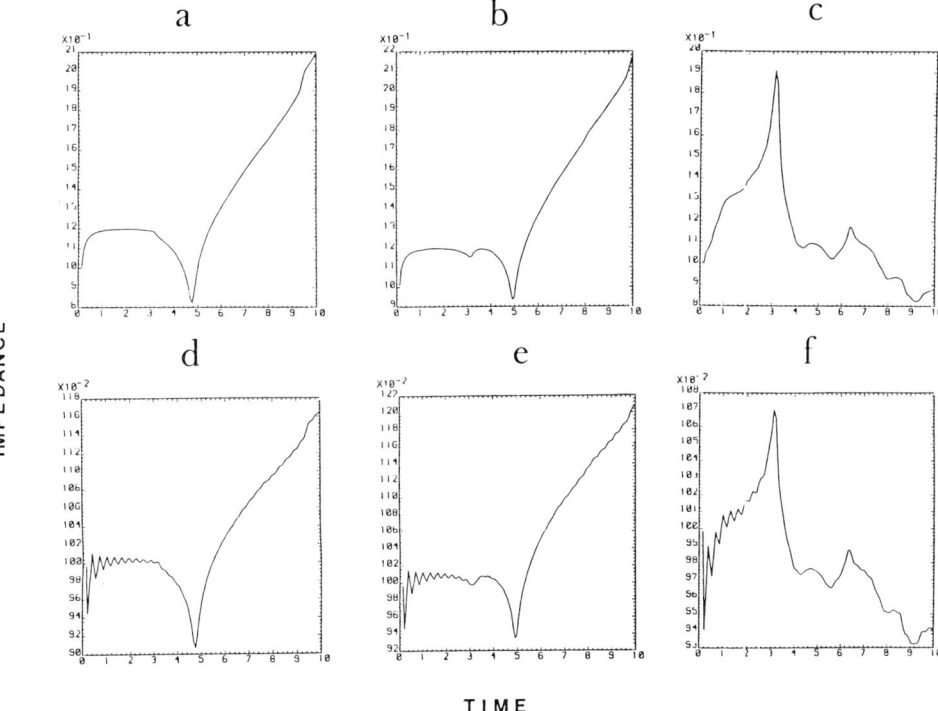

Fig. 5. Reconstructed impedance profiles for cases (i)-(iii) Noise free: (a)-(c). 50% noise in the range 5-15 MHz. (d)-(f).

CONCLUSIONS

For the complex layered components considered, our synthetic study has demonstrated that the impediography method is a simple attractive technique for interrogation of the perspex/rubber interface. In a pulse-echo experiment this inspection procedure exploits phase changes in the back-scattered signal, which result after reflection from remote layers. We have deliberately investigated a structure with large impedance variations, to test the reliability of the reconstruction procedure for a wide range of materials. Instead of an operator making an experienced appraisal of such phase changes, the method described here provides a quantitative analysis of the reflected signal with easily interpretable results. Another attractive feature of the method is that it is does not rely on amplitude measurements, although this would become important if a calibrated reconstruction were required.

ACKNOWLEDGEMENTS

PRS performed this work with the support of the Royal Ordnance Explosives Division and the Procurement Executive of the Ministry of Defence, UK.

# REFERENCES

1. L.J. Bond, Methods for the computer modelling of waves in solids, in "Research Techniques in NDT, vol. 6" pp 107-150, ed. R.S. Sharpe, Academic Press, (London, New York), (1982).

2. L.J. Bond, M. Punjani and N. Saffari, Review of some recent advances in quantitative ultrasonic NDT, IEE Proceedings, Part A. **131**, 265 (1984).

3. K.A. Berteussen and B. Ursin, Approximate computations of the acoustic impedance from seismic data, Geophys., **48**, 1351 (1983).

4. W.R. Scott and P.F. Gordon, Ultrasonic spectrum analysis for nondestructive testing of layered composite materials, J. Acoust. Soc. Am., **62**, 108 (1977).

5. R.J. Blake, L.J. Bond and A. Downie, Advances in numerical studies of elastic wave propagation and scattering, Review of Progress in QNDE, vol. 1, ed. D.O. Thompson and D.E. Chimenti, Plenum Publishing Corp. (1982).

6. A.J. Hayman and J.P. Weight, Transmission and reception of short ultrasonic pulses by circular and square transducers, J. Acoust. Soc. Am., **66**, 945 (1979).

7. G. Kunetz, Quelques exemples d'analyse d'enregistrements sismiques, Geophys. Prosp., **11**, 409 (1963).

8. J.A. Ware and K. Aki, Continuous and discrete inverse-scattering problems in a stratified elastic medium. I. Plane waves at normal incidence, J. Acoust. Soc. Am., **45**, 912 (1969).

# FINITE ELEMENT STUDIES OF TRANSIENT WAVE PROPAGATION

Mary Sansalone, Nicholas J. Carino, and Nelson N. Hsu

National Bureau of Standards
Gaithersburg, Maryland

## INTRODUCTION

The National Bureau of Standards (NBS) has been working to develop a nondestructive test method for heterogenous solids using transient stress waves [1-5]. The method is referred to as the impact-echo method. The technique involves introducing a transient stress pulse into a test object by mechanical impact at a point and measuring the surface displacement caused by the arrival of reflections of the pulse from internal defects and external boundaries. Successful signal interpretation requires an understanding of the nature of transient stress wave propagation in solids containing defects. A primary focus of the NBS program is on using the finite element method to gain this understanding.

The purpose of this paper is to show the versatility and power of the finite element method for solving stress wave propagation problems, and to provide background information about the finite element program that was used to carry out the NBS studies. To achieve this purpose, this paper illustrates the use of the method to solve the following three problems: 1) stress and displacement fields produced by transient point impact on the surface of an elastic plate; 2) the interaction of transient stress waves with a planar disk-shaped void within an elastic plate; and, 3) stress fields produced by an ultrasonic transducer radiating into an elastic solid. The finite element results are compared to exact Green's function solutions for a point source on an infinite plate, experimentally obtained surface displacement waveforms for point impact on a plate containing a planar flaw, and photoelastic pictures of the stress fields produced by an ultrasonic transducer radiating into silica.

## BACKGROUND

The finite element method is a general numerical technique for obtaining approximate solutions to the partial differential equations that arise from boundary value problems. The method involves dividing a continuum into a finite number of discrete parts - the finite elements. The discretized representation of the continuum is referred to as the finite element model. For stress analysis, the behavior of each element is described by a set of assumed functions which represent the variation of displacements within that element. Variational (or energy) principles are used to formulate force-displacement equations for the elements. These element equations are then used to construct the global equations which describe the behavior of the entire continuum. Solution of these global equations gives the displacements at points in the element [6].

An explicit, two-dimensional (axisymmetric or plane strain), finite element code (DYNA2D), developed at Lawrence Livermore National Laboratory for solving finite-deformation, dynamic contact-impact problems [7-9], was used to perform the studies discussed in this paper. An input generator (MAZE) [10] was used to create the finite element model. A mini-computer with a virtual operating system, 8 MBytes of memory, and a floating point processor were used to carry out the analyses.

In DYNA2D, a continuum is divided into constant strain triangular and quadrilateral elements. Higher order elements (e.g., linear strain, quadratic strain) are not available in DYNA2D because they are computationally much more expensive in wave propagation applications than the use of constant strain elements. For a particular element type, the accuracy of the finite element solution is partly determined by element size. In wave propagation problems, the optimum element size depends on the geometry of the continuum and on the time-history of the dynamic loading.

In dynamic finite element analysis, numerical integration of the equations of motion must be carried out; DYNA2D uses the central difference method to perform this integration. The central difference method requires a small time step for numerical stability. This is not a drawback because wave propagation applications require the use of very small time steps to obtain an accurate solution. Numerical stability requires that the time step, h, meets the following criterion:

$$h \leq h_{max} = L/C_p \qquad (1)$$

where L = shortest dimension of the element; and, $C_p$ = P-wave velocity in the material. In DYNA2D, the time step is taken as 0.67 $h_{max}$ unless the user specifies some other value. During an analysis, results are stored in data files at a time interval specified by the user.

A dynamic finite element analysis generates a large amount of data; therefore, an efficient and versatile post-processor is essential. An interactive, graphic, post-processor (ORION) [11] was used to process the results of the analyses. ORION allows the user to look at the results of an analysis in a variety of different ways. Full-field contour or vector plots or time-history plots can be obtained for a number of different parameters, including displacements, velocities, accelerations, stresses, and strains. Examples of several different types of data display are shown in this paper.

PLATE RESPONSE TO POINT IMPACT

Initial finite element studies were carried out using a solid plate so that the solutions could be compared to exact Green's function solutions for point impact on an infinite plate [3,4,12]. An axisymmetric, linear elastic analysis was performed for point impact on the surface of a 0.5-m thick, 1.5-m diameter, unsupported plate (Poisson's ratio of 0.2 and a P-wave velocity of 4000 m/s). The force-time history of the impact was modeled as a half-cycle sine curve with a duration of 25 microsec. The force was applied as a uniform pressure over the two elements at the center of the top surface of the plate.

Figure 1 illustrates two useful forms of data display: a vector plot of displacements, and a contour plot of stresses. In the vector plot, the magnitude and direction of the average nodal displacement of each element is indicated by a vector. Darker areas in the vector plot indicate areas of larger displacements; thus, the various waves can be easily identified. The right side of Figure 1(a) shows the vector displacement field through a cross-section of the plate 125 microsec after the start of the impact. At 125 microsec, the P-wavefront arrives at the bottom of the plate. The position of the P- and S-wavefronts are indicated on the left side of the figure. Figure 1(b) shows a contour plot of minimum principal (compressive) stress 125 microsec after the start of the impact. Since a state of pure shear stress is equivalent to a state of equal biaxial tension and compression, a plot of minimum principal stress also shows the variation of stresses in the S-wave. The letters A through E indicate the relative magnitude of the stresses. The position of the P- and S-waves are indicated on the figure. The magnitude of the stresses in the spherical P-wave are maximum near the centerline of the plate. The stresses in the S-wave are small at the center of the plate and become larger along rays located at increasing angles from the center of the plate. Near the surface of the plate, the large amplitude stresses caused by the Rayleigh (R) wave interfere with those produced by the S-wave making it difficult to separate the stresses caused by each wave.

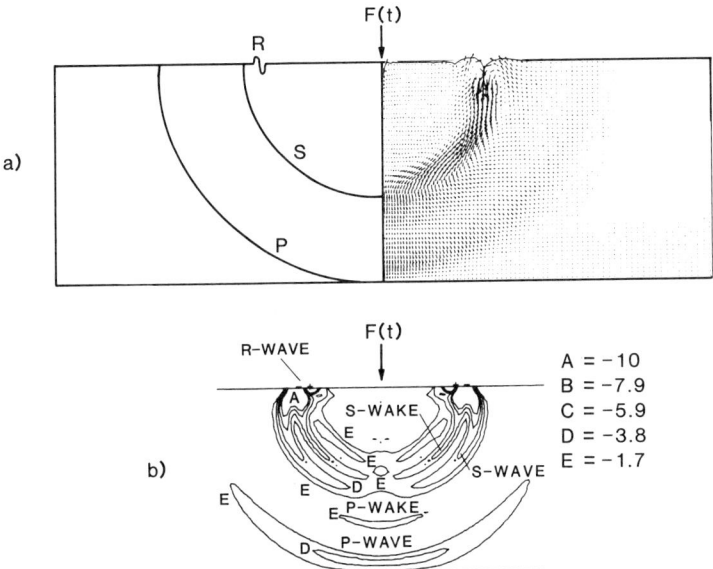

Fig. 1. Displacement and stress fields within a solid plate: a) vector plot of displacements and the locations of the wavefronts; and, b) minimum principal stress contour plot.

The observed pattern of displacements and stresses in the P- and S-waves are similar to those expected based on the displacement fields produced by a harmonic point source [13]. However, in addition to the P- and S-wave radiation patterns, Figure 1(b) shows that in the region between the P- and S-waves there is a stress contour that resembles that in the P-wave; this disturbance is the "P-wake." In addition, there is a region of nonzero stresses trailing the S-wave that resemble the contours in the S-wave; this is the "S-wake." These wakes do not exist in the far-field solution for a harmonic point source. They were a significant finding of the finite element analysis and helped to explain previously unexplained features of surface displacement waveforms produced by a transient point source.

Figure 2 shows a comparison of a surface displacement waveform obtained from a Green's function solution (Figure 2(a)) with a surface displacement waveform obtained from the finite element analysis (Figure 2(b)). The test configuration and duration of the impact were the same for both solutions; in each case, the displacement was calculated 0.05 m away from the impact point. There is excellent agreement between the two waveforms. The only discrepancy appears following the R-wave, where the finite element waveform exhibits some spurious oscillations due to excitation of the zero-energy modes of the finite elements [7]. In this case, the zero-energy modes are excited by the element distortion caused by the rapid, large changes in displacement that occur in the R-wave. This "numerical ringing" does not affect the echo pattern due to the multiply reflected body waves.

The agreement between the waveforms shown in Figure 2 was achieved only after a correct finite element model was constructed. Convergence studies were carried out to determine the optimum element size for the constant strain quadrilaterals and the dynamic loading functions used in the linear elastic, plate analyses. The criterion for convergence was agreement between finite element displacement time-histories obtained at points on the top and bottom surfaces of a plate and the waveforms obtained at the same points by the Green's function solution for an infinite plate. For 0.25- to 0.5-m thick plates subjected to impacts with contact times of 25 to 30 microsec, rectangular elements with dimensions on the order of 0.02 times the plate thickness were found to give accurate results.

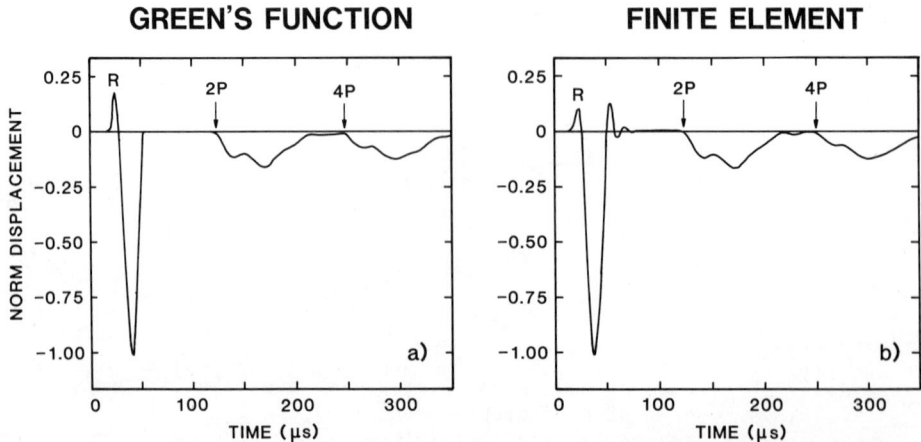

Fig. 2. Surface displacement waveforms: a) Green's function solution; and b) finite element analysis.

## PLATES CONTAINING FLAWS

The next problems that were analyzed were point impacts on plates containing planar flaws [3,5]. The main focus of these studies was to determine the effects on stress and displacement fields in a plate and on surface displacement waveforms caused by diffraction of waves at the edges of a flaw. The capability of the finite element method to model diffraction at the edges of a flaw was established by comparing surface displacement waveforms obtained from analyses of an aluminum plate containing a flat-bottom hole to experimentally obtained waveforms. These comparisons are shown in Refs. [3,5]. Excellent agreement was found between the finite element and experimental waveforms. Thus, the finite element method was used to study various flaw geometries and impact conditions.

As an example, an analysis was performed for point impact on a 0.5-m thick, 1.5-m diameter plate (Poisson's ratio of 0.2 and P-wave velocity of 4000 m/s) containing a disk-shaped void. The void was 0.01-m thick and 0.1 m in diameter; it was located 0.25 m below the top surface of the plate. The contact time of the impact was 20 microsec. Figures 3(a)-(d) show contour plots of the minimum principal stresses in the plate at 65, 80, 95, and 125 microsec after the start of the impact.

The P-wavefront arrives at the flaw at 62.5 microsec, and is incident upon the edge of the flaw at 64 microsec. At 65 microsec, the stress contours in Figure 3(a) show that diffraction of the P-wave is occurring; the diffracted P-wavefront is beginning to emerge from the edge of the flaw. The diffracted P-wave forms a toroid with a circular cross-section. Rays emanating from the impact point and intersecting the edges of the flaw delineate the shadow zone beneath the flaw.

Figure 3(b) shows the stress contours at 80 microsec. At this time, the front of the reflected P-wave is just overlapping the front of the direct S-wave. The diffracted P-waves have overlapped in the center region of the plate, above and below the flaw. This figure clearly illustrates how diffraction causes large stresses to penetrate into the shadow zone.

At 95 microsec (Figure 3(c)), the P-wave has completely passed the flaw. The P-wave is formed by the overlapping of diffracted P-waves in the shadow zone and the original P-wave outside of the shadow zone.

Figure 3(d) shows the stress contours at 125 microsec. The various waves that are present at this time can be vividly seen in these contours. The P-wave reflected from the surface of the flaw has arrived at the top surface of the plate and the stress contour associated with this reflected wave is clearly seen. The S-wave produced by mode-conversion of the P-wave incident upon the flaw has traveled approximately 60 percent of the distance from the flaw to the top surface of the plate; the stress contour of this wave is also evident. Reflection and diffraction of the S-wave by the flaw have occurred, the P-wake has reformed below the flaw, and the P-wavefront is nearing the bottom surface of the plate.

Compare Figure 3(d) with Figure 1 which showed the stress field in a solid, 0.5-m thick plate at 125 microsec. In the plate containing the flaw, the stress pattern is much more complicated because of the interaction of the waves with the flaw. Notice that the stress field produced by the P-wave is similar to the stress field in the solid plate because diffraction allowed the P-wave to penetrate the shadow zone.

Displacement waveforms recorded near the point of impact at the top surface of a plate containing a flaw consist of displacements caused by waves reflected and diffracted from the flaw and waves reflected from the bottom surface of the plate and subsequently diffracted by the flaw. The relative importance of the effects caused by each of these phenomena on displacement waveforms depends on the flaw geometry and the test conditions. Studies were carried out to better understand how waveforms are affected by these

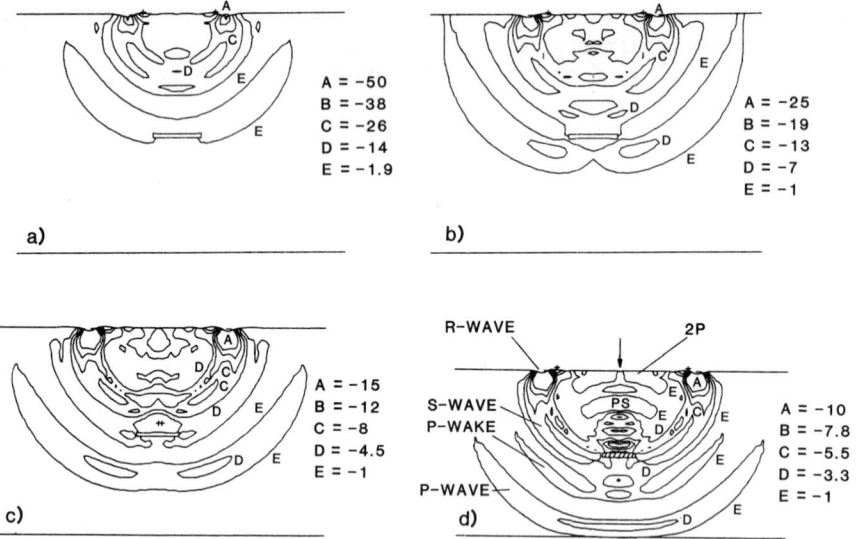

Fig. 3. Minimum principal stress contour plots occurring in a plate containing a disk-shaped flaw at various times after the start of the impact: a) 65; b) 80; c) 95; and, d) 125 microsec.

parameters. The variables examined included: the contact time of the impact; the diameter and depth of a flaw; and the test configuration, that is, the point where the displacement was recorded. The results of these studies are discussed in Ref. [3].

## ULTRASONIC TRANSDUCER FIELD

To show the versatility of the finite element method for studying wave propagation in solids, results obtained from an axisymmetric analysis of an idealized, 25.4-mm diameter transducer radiating into silica are presented. The input pulse produced by the transducer was simulated by applying a uniform pressure over a 25.4-mm diameter area at the center of a 140-mm diameter, 70-mm deep cylinder. The time-history of the pressure loading is shown in Figure 4. The loading is approximately a single cycle, damped sine curve; the duration of the loading was 1 microsec.

Figure 5(a) shows a stress contour plot of maximum shear stress obtained 11.5 microsec after the start of the input pulse. The P- and S-waves are labeled in this plot and letters A through E indicate the relative magnitude of stresses in the waves. The transducer can be though of as a multitude of point sources, each generating P- and S-waves and wakes. The radiation patterns produced by all these point sources are superimposed to produce the radiation pattern for the transducer. Notice that both the P- and S-waves are composed of regions of stresses directly below the transducer. In addition, the S-wave is also composed of a toroidal ring which is centered about the edge of the transducer. The largest stresses in the P-wave are an order of magnitude larger than those in the S-wave. This is in contrast to the stress pattern produced by a single point source (Figure 1(b)) in which the stresses in the S-wave are approximately the same or larger than those in the P-wave.

Figure 5(b) shows a photoelastic visualization 11.5 microsec after the start of an input pulse produced by a 1 MHz, broadband transducer radiating into silica [12]. The photoelastic technique relies on the difference in principal stresses (or shear stress). Therefore, Figure 5(b) can be compared directly to the shear stress contour plot that was shown in Figure 5(a). In Figure 5(b), the higher intensity of the illuminated P-wave indicates that the stresses are much higher in the P-wave than in the lower intensity S-waves. This is in agreement with the variation of stress contours in Figure 5(a). The agreement between the stress fields obtained by the finite element method and that obtained from photoelasticity is excellent.

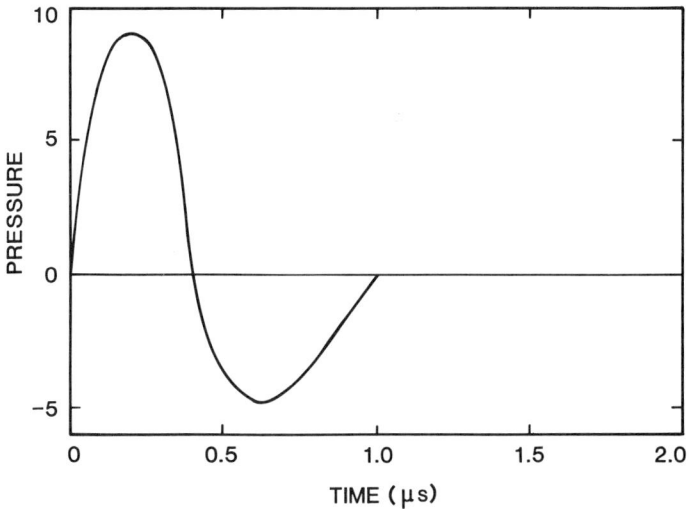

Fig. 4. Time-history of pressure loading for an idealized 25.4-mm diameter broadband transducer.

The results obtained from this finite element analysis can be used for much more than just visualization of the radiation pattern. Quantitative full-field information about stresses and displacements can be obtained, and finite element waveforms recorded on the top surface of a test specimen can be compared to experimentally obtained pulse-echo waveforms. A variety of input pulses can be modeled. Radiation patterns in specimens containing flaws can be studied. In particular, studies of standard reference blocks, such as the flat-bottom hole specimens, can be carried out.

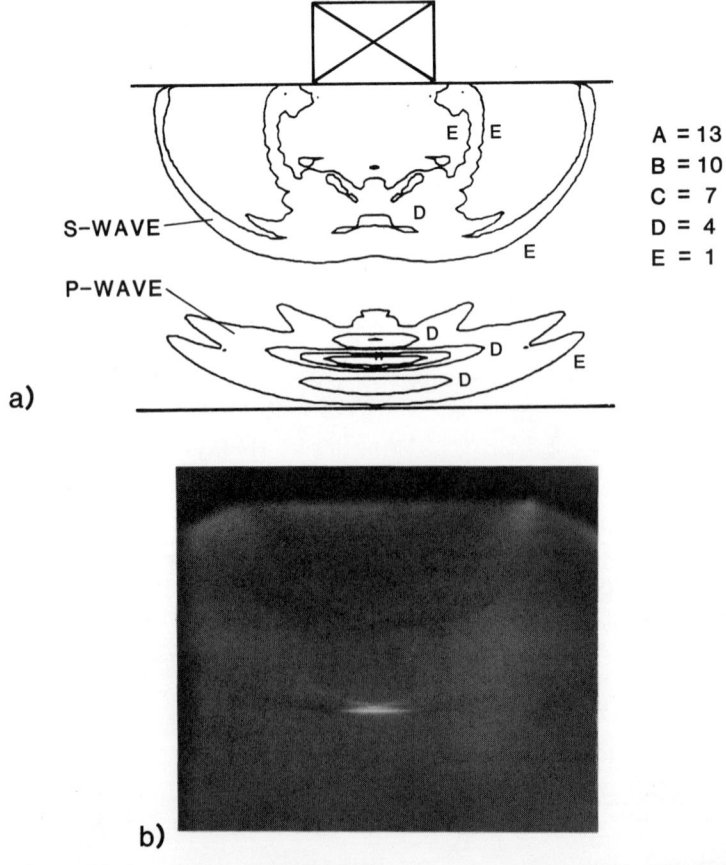

Fig. 5. Stress field produced by a broadband transducer radiating into silica: a) finite element shear stress contour plot; and, b) photoelastic visualization.

SUMMARY

These numerical studies were presented to illustrate the power of the finite element method as a tool for studying wave propagation in bounded solids subjected to arbitrary applied loads, boundary conditions, and containing flaws. In each of the examples presented in this paper, excellent agreement was obtained between finite element results and existing Green's function solutions for an infinite plate or experimental results. It is hoped that the work discussed in this paper and in Refs. [3-5] offers convincing evidence of the versatility of the finite element method for solving a wide variety of wave propagation problems for which there are presently no other solutions.

REFERENCES

1. N. J. Carino, M. Sansalone, and N. N. Hsu, J. Am. Conc. Inst. __83__, 199 (1986).
2. N. J. Carino, M. Sansalone, and N. N. Hsu, Flaw Detection in Concrete by Frequency Spectrum Analysis of Impact-Echo Waveforms, in: "International Advances in Nondestructive Testing," W. J. McGonnagle, ed., Gordon & Breach Science Publishers, New York (1986).
3. M. Sansalone and N. J. Carino, "Impact-Echo: A Method for Flaw Detection in Concrete Using Transient Stress Waves," NBSIR 86- National Bureau of Standards, Gaithersburg, MD (1986).
4. M. Sansalone, N. J. Carino, and N. N. Hsu, ASME J. of App. Mech., submitted (Jan., 1986).
5. M. Sansalone, N. J. Carino, and N. N. Hsu, ASME J. of App. Mech., submitted (June, 1986).
6. R. H. Gallagher, "Finite Element Analysis, Fundamentals," Prentice-Hall, Englewood Cliffs (1975).
7. J. O. Hallquist, "A Procedure for the Solution of Finite-Deformation Contact-Impact Problems by the Finite Element Method," UCRL-52066, Lawrence Livermore Laboratory (1976).
8. G. J. Goudreau and J. O. Hallquist, Comp. Meth. in App. Mech. and Eng., __33__, 725 (1982).
9. J. O. Hallquist, "User's Manual for DYNA2D - An Explicit Two-Dimensional Hydrodynamic Finite Element Code with Interactive Rezoning," Lawrence Livermore Laboratory (1984).
10. J. O. Hallquist, "User's Manual for MAZE: An Input Generator for DYNA2D and NIKE2D," Lawrence Livermore Laboratory (1983).
11. J. O. Hallquist, "User's Manual for ORION: An Interactive Post-Processor for the Analysis Codes NIKE2D, DYNA2D, and TACO2D," Lawrence Livermore Laboratory (1983).
12. N. N. Hsu, "Dynamic Green's Function of an Infinite Plate - A Computer Program," NBSIR 85-3234, National Bureau of Standards, Gaithersburg, MD (1985).
13. R. Roderick, Radiation Pattern from a Rotationally Symmetric Stress Source on a Semi-Infinite Solid, in: Ph.D. Thesis, Brown University, Providence (1951).
14. W. Sachse, N. N. Hsu, and D. G. Eitzen, Visualization of Transducer Produced Sound Fields in Solids, in: IEEE Ultrasonics Symposium Proceedings (1978).

MODELING ULTRASONIC WAVES USING FINITE

DIFFERENCE METHODS

L. J. Bond, N. Saffari and M. Punjani

University College London
Department of Mechanical Engineering
Torrington Place, London, WC1E 7JE  United Kingdom

INTRODUCTION

Models based on explicit finite difference methods are used to study pulsed elastic wave propagation and scattering. These models were first developed in seismology and have now been developed and applied to SAW electronics and ultrasonic NDT.

The region under consideration is divided into a 2-D grid of points. The initial conditions require that the displacements at two initial time levels are specified. For each point the time development of the displacement field is then calculated in sequence. The models employ difference forms of the continuum equations which provide nodal formulations for each class of point, e.g., body node, 90 degree corner node, etc. The nodal formulations are then the building blocks from which models are constructed.

Table 1 lists the references to initial conditions and nodal formulations used in modelling systems which introduce slots into the stress-free surface of an elastic solid, and on the interface of an elastic solid with a viscous fluid.

Table 1 - References to various algorithms used in the models

| Algorithm | Reference |
|---|---|
| Solid interior points | Alterman and Rotenberg [1] |
| Horizontal and vertical free surfaces | Ilan and Loewenthal [2] |
| 90° corner | Ilan [3] |
| 270° corner | Ilan et al [4] |
| Absorbing boundaries | Reynolds [5] |
| Input pulse: C-pulse | Ilan et al [6] |
| SV-pulse | Punjani [7] |
| Solid-fluid interface | Saffari [8] |
| Solid-solid interface with partial closure | Punjani [7] |

## LINE COMPRESSION WAVE INTERACTION WITH A 270° WEDGE

A classical problem is the interaction of a line impulse of compression waves with a 270° wedge.

A series of vector plots are shown for the case of a 60° incident pulse as figure 1. Alternatively, the surface displacements were recorded at all points along the horizontal and vertical surfaces at one time step, figure 2. The magnitude of the mode-converted Rayleigh wave component was then considered. This problem was studied as the angle of incidence was varied from +90° to -90°. The conversion coefficients thus obtained were compared with experimental data due to Gangi and Wesson [9], figure 3. They are seen to be in good agreement.

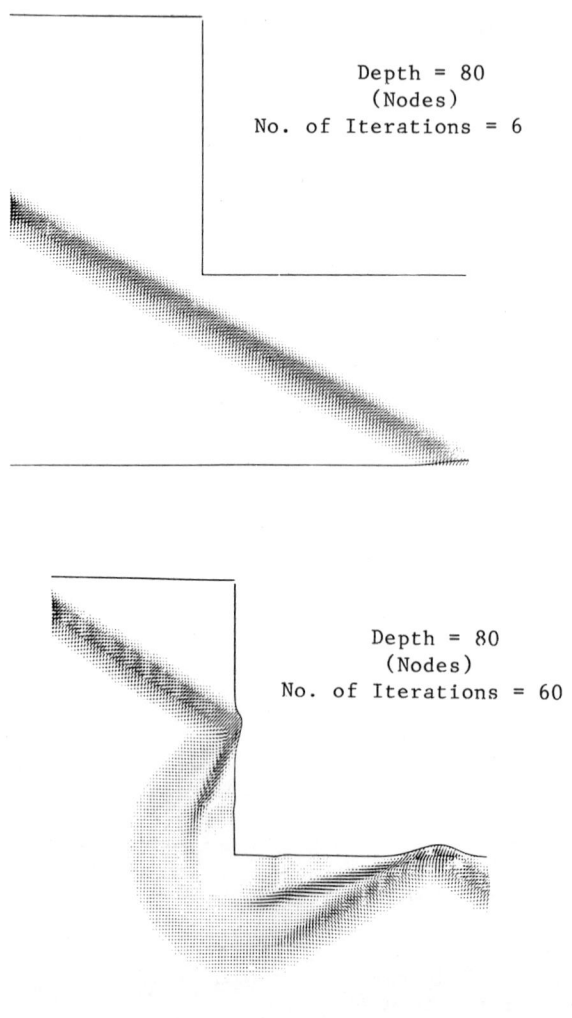

Figure 1. Vector plots of the interaction of a compression impulse with a 270° wedge, incident at 60°.

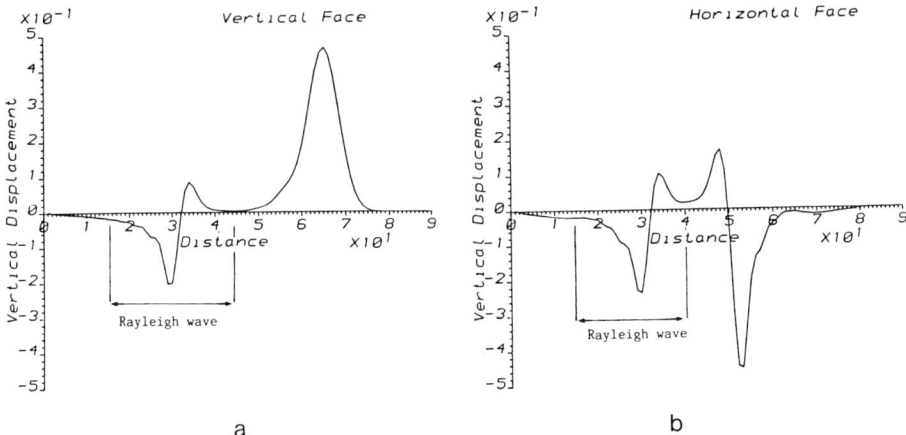

Figure 2. Displacements along the vertical (a) and horizontal (b) surfaces of the wedge at one time step.

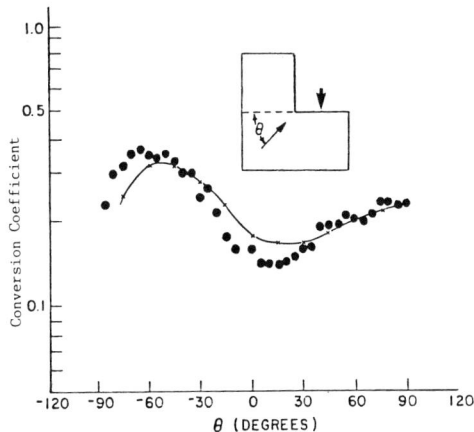

Figure 3. Numerical C-to-R conversion coefficients (solid line) com-compared to the experimental results (dots) of [9].

COMPRESSION WAVE INTERACTION WITH A VERTICAL STEP

The vertical step is the simplest feature which provides frequency dependent scattering in the mid-frequency regime.

A line impulse of compression waves was normally incident from below on a vertical up step. A series of vector plots are shown in figure 4. At a receiver point located on the free surface to the right hand of the step, the vertical component of displacement was recorded with time, figure 5. The receiver was sufficiently distant from the step to allow the various scattered components in the wave field to separate. The Rayleigh wave component was gated and its spectrum deconvolved with that of the input pulse, figure 6. The position of the peak was measured for a number of models with different step depths and an average value of 0.74 found. The position of this peak is found to be in good agreement with that predicted by an interference model proposed in [8].

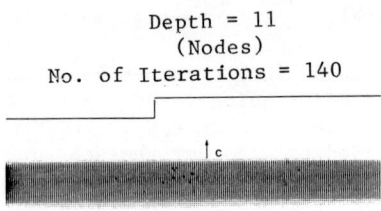

Figure 4. Vector plots for the interaction of a compression impulse with a step.

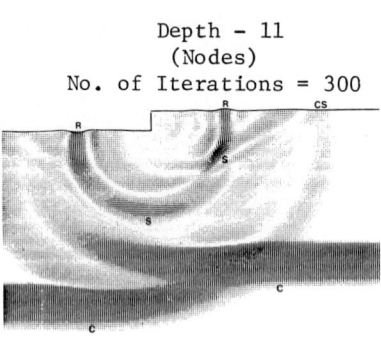

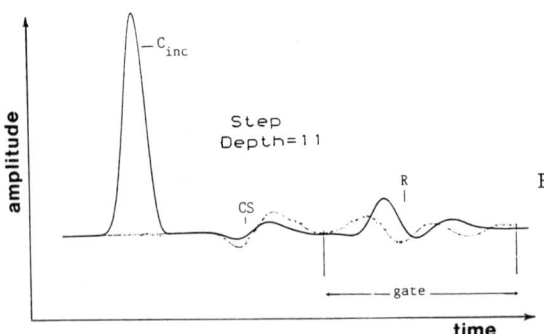

Figure 5. Vertical displacements at a receiver point located on the free surface.

Figure 6. Deconvolved spectrum of the mode-converted Rayleigh wave on the free surface.

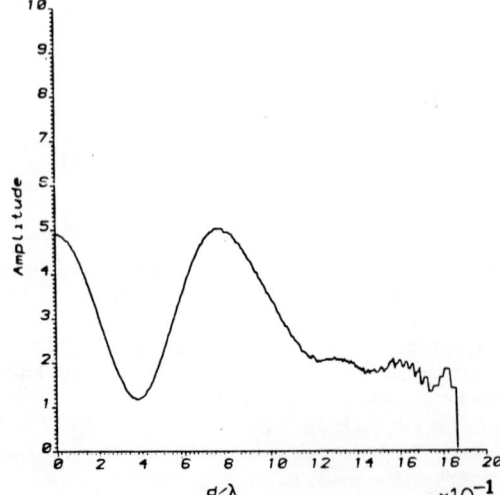

## SCATTERING OF A COMPRESSION WAVE BY A CLOSED CRACK

In a previous paper [10] the study of the scattering of transient plane compressional and shear vertical waves by a broken crack with regions of partial closure and vacuum, was reported. Here, the previous work is extended to consider the scattering by a crack with continuous partial closure (a partially closed perfect crack). In contrast to the complicated broken crack, the geometry of the partially closed perfect crack is simple. The boundary condition for partial closure is, physically, continuity of traction but not of displacement. The constants $\underline{k}$ $Nm^{-3}$ of this condition are stiffness (spring) coefficients.

Figures 7a and 7b, displacement vector and magnitude plots, respectively, show the diffraction of a plane compression wave by a crack with $\underline{k} = 0$. In this figure, the well-known diffraction pattern of scattered body and surface waves is clearly observed. Figures 8a and 8b show the diffraction of a plane compression wave by a crack with $k_1 = k_2 = 0.5 \times 10^7$ $Nm^{-3}$. It is evident in this figure that the crack becomes almost transparent to the incident wave. The scattering, which is seen to be weak, still consists of outgoing circular body waves plus scattered waves due to the continuous partial closure.

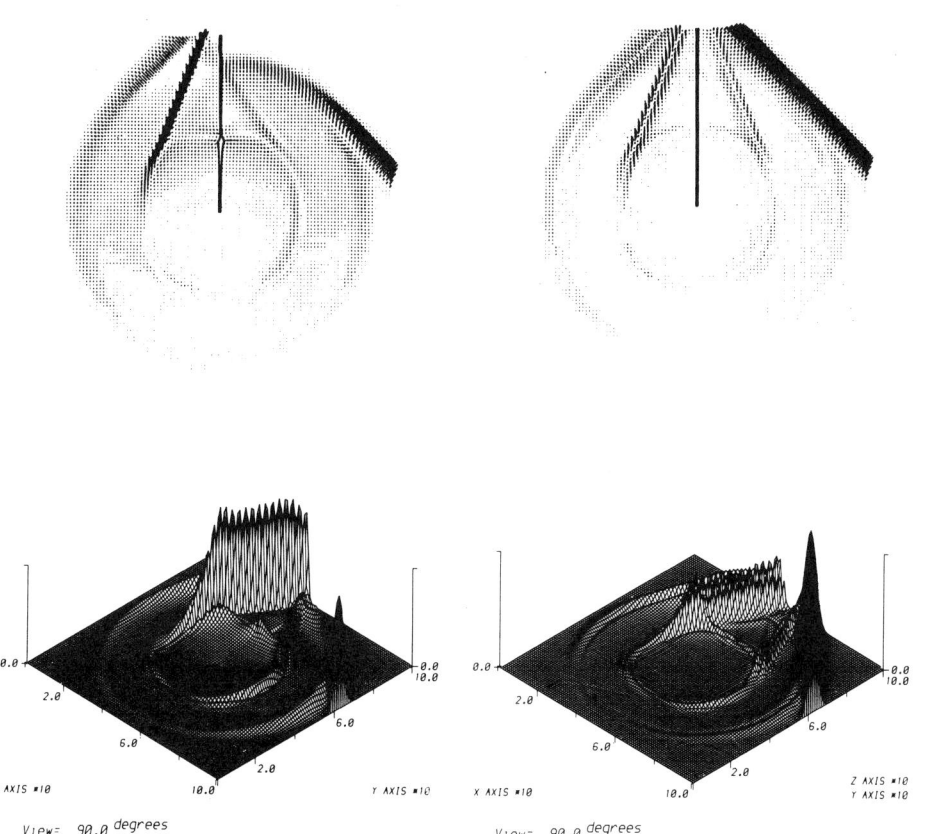

Figure 7. Diffraction of a compression wave by a crack ($\underline{k} = 0$); angle of wave incidence is 45°.

Figure 8. Diffraction of a compression wave by a crack ($k_1 = k_2 = 0.5 \times 10^7$ $Nm^{-3}$); angle of wave incidence is 45°.

COMPRESSION WAVE INTERACTION WITH VERTICAL SLIT BELOW THE SURFACE OF A HALF-SPACE

A line compression wave was normally incident on a vertical slit from below. The depth of the slit below the free surface was a traction of the incident wavelength (0.1 to 0.2$\lambda$), and the length being up to about a wavelength. A series of vector plots are shown in figure 9. When the displacements are plotted at selected points along the free surface for models with a range of depths and lengths, figure 10, the mode-conversion phenomena are found to be related to the void dimensions. It is seen that the amplitude of the Rayleigh wave component, $R_T$, is a function of void depth and the surface-skimming shear wave, $S_{HS}$, is a function of both depth and length.

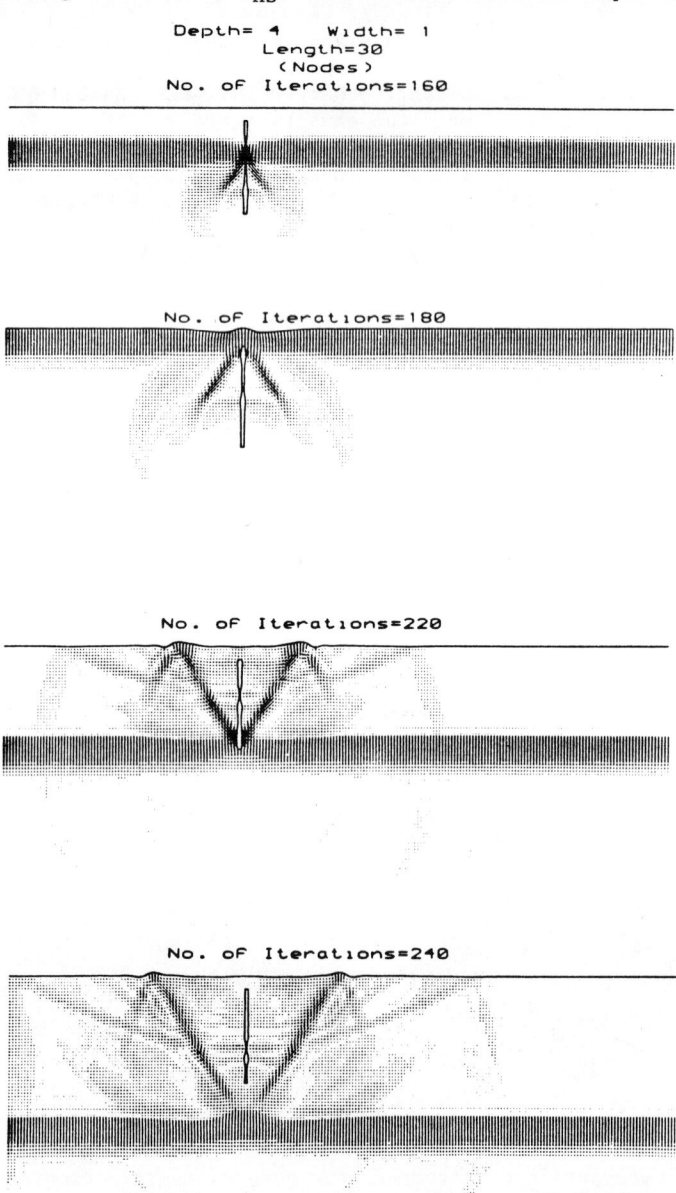

Figure 9.  Vector plots for the interaction of a compression impulse with a vertical near-surface slit.

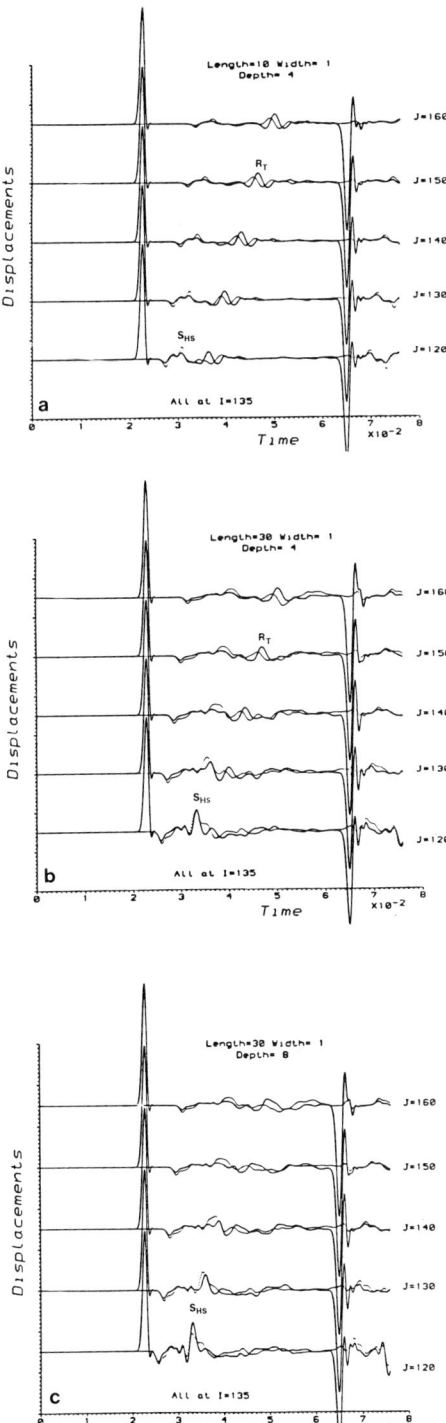

Figure 10. Displcement versus time plots at different receiver points along the free surface for three different slits: (a) length 10, depth 4; (b) length 30, depth 4; (c) length 30, depth 8 - the slit is at J = 80.

## LEAKY RAYLEIGH WAVE INTERACTION WITH PERPENDICULAR SUB-SURFACE SLIT

A line compression wave was incident on a solid-fluid interface, from the fluid side, at the Rayleigh critical angle. On the interface, and decaying into the solid, a leaky Rayleigh wave was thus excited which is shown interacting with a near-surface perpendicular slit, figure 11. Part of the energy is reflected which travels back along the the interface as a leaky Rayleigh wave. The vertical displacement due to this reflection wave is sampled on the curve for the frequrncy dependent reflection coefficient obtained, figure 12. The reflection coefficients for two slits with different depth (d) to length (l) ratios are shown  These are compared with those for Rayleigh wave interactions with slots on free surfaces calculated in [11]. They are seen to be similar.

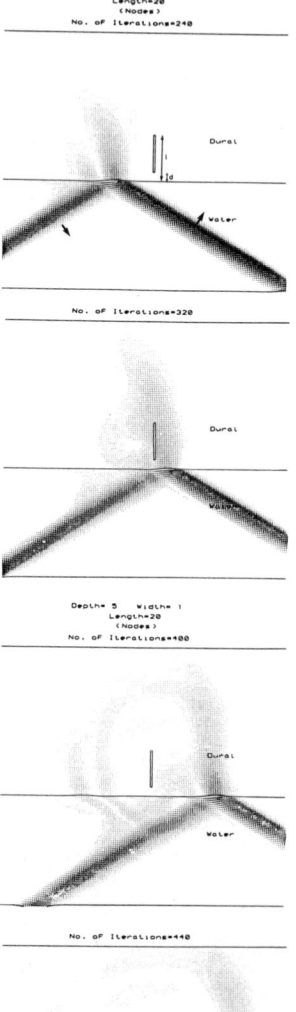

Figure 11. Vector plots for the interaction of a leaky Rayleigh wave pulse with a near-surface slit.

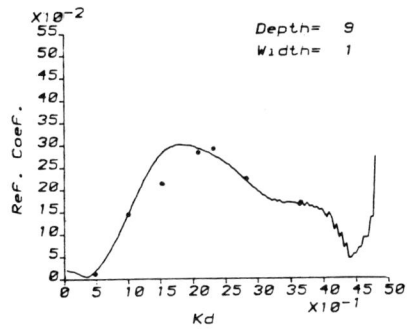 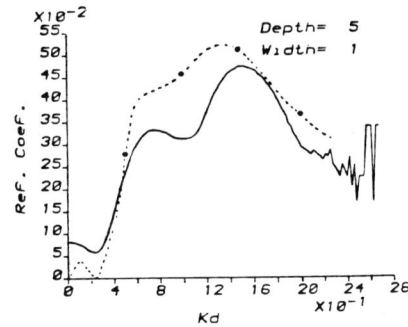

Figure 12. Leaky Rayleigh wave reflection coefficients for near-surface slits of different d/l ratios (solid lines) compared with the results of [11] (dots).

## SCATTERING OF A SHEAR-VERTICAL WAVE ON A HALF SPACE WITH A SLOT

A new source function, which consists of a Ricker-type amplitude spectral distribution, has been developed to model a plane shear-vertical wave at a half space for wave incidence from 0 to 90°. The details of the analytic synthesis of this shear-vertical Ricker-type wavelet for a numerical scattering model have been presented in [7].

Figure 13 shows a sequence of displacement fields which arise when a transient plane shear-vertical wave interacts with a slot perpendicular to the surface of a half space. In this figure, the angle of wave incidence is 45° relative to the half space, and the incident wave is traversing the half space from left to right.

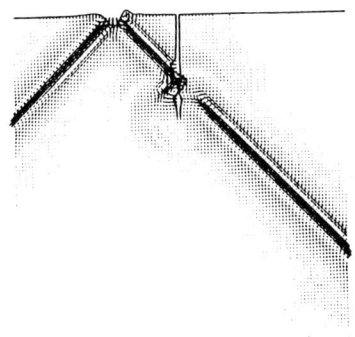 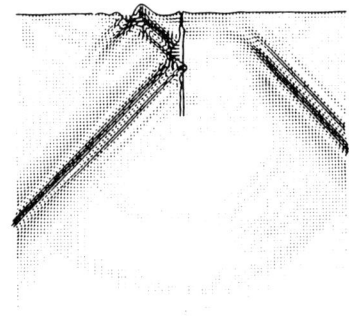

Figure 13. Scattering of a transient plane shear-vertical wave by a slot perpendicular to a half space; displacement field and contour plot of magnitude of displacement field.

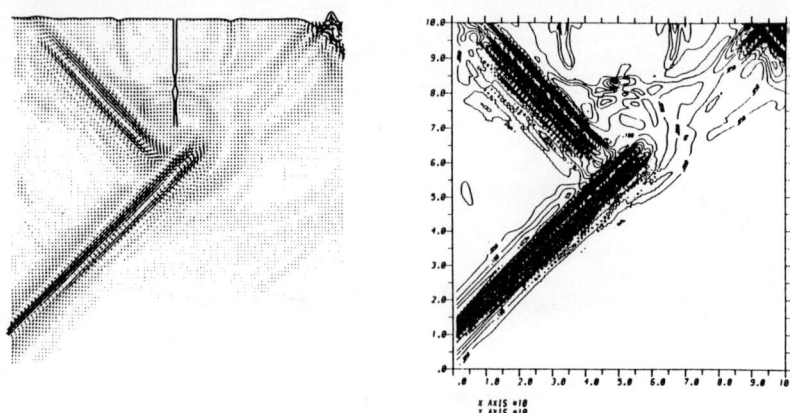

Figure 13 (continued)

ACKNOWLEDGEMENT

This work was performed with the support of the Procurement Executive, Ministry of Defence, UK.

REFERENCES

1. Z.S. Alterman and A. Rotenberg, Bull. Seism. Soc. Am.,59(1),347-368 (1969)
2. A. Ilan and D. Loewenthal, Geophys. Prosp. 24, 431-453 (1976).
3. A. Ilan, J. Comp. Phys. 29, 389-403 (1978).
4. A Ilan, A Ungar and Z S Alterman, Geophys. J Royal Astr Soc 43, 727-742 (1975).
5. A C Reynolds, Geophys 43 (6), 1099 - 1110 (1978).
6. A Ilan, L J Bond and M Spivack, Geophys. J Royal Astr Soc 57, 463 - 477 (1979).
7. M Punjani, PhD Thesis, University College London (1986).
8. N Saffari, PhD Thesis, University College London (1986).
9. A F Gangi and R L Wesson, J of Comp Phys. 29, 370 - 388 (1978).
10. M Punjani and L J Bond, Review of Progress in QNDE, Vol 5, D O Thompson and D E Chimenti, Eds, 61 - 70, Plenum Press, New York (1986).
11. J D Achenbach and R J Brind, J of Sound and Vib 76(1), 43 - 56 (1981).

EDDY CURRENT RESPONSE TO THREE-DIMENSIONAL FLAWS BY THE BOUNDARY ELEMENT METHOD

R. E. Beissner and J. H. Hwang

Southwest Research Institute
San Antonio, Texas 78284

INTRODUCTION

In planning an inspection procedure, or in designing parts with flaw detectability as a design goal, it is essential that the engineer have available some form of model for estimating the probability of flaw detection. In the past this need has been met, with varying degrees of success, by relying on experience in the inspection of similar parts, sometimes supplemented by experimental testing. With the rapid advances in computer technology in recent years, it is now feasible to consider replacing, or at least enhancing, such practices with predictions based on numerical simulation of the flaw detection process [1].

For eddy current NDE, numerical simulation requires the solution of Maxwell's equations as a step toward predicting the response of a probe to the presence of a flaw [2]. In general this poses a very difficult problem because flaw, part and probe geometries are not amenable to analytic treatment. This means that computer simulation of eddy current flaw detection will almost always require the numerical solution of Maxwell's equations in a complex geometry. To further complicate matters, one needs not just a single calculation of probe response, but a multitude of such calculations, one for each probe position as it is moved over the surface of the part [1,3]. Finally, it is important to note that the inspection simulation problem is inherently three-dimensional because, at a general point in the course of a scan, the induced eddy current field is not symmetrical with respect to the flaw position.

In this paper we describe a hybrid approach to the three-dimensional simulation of an eddy current inspection. Our approach makes use of the boundary element method (BEM) [4] for solving the boundary integral form of Maxwell's equations for the current density and tangential magnetic field on the surface of a flaw in a known incident field. Incident field data, i.e., the current density and magnetic field in the material in the absence of a flaw, are provided by analytic solutions [5,6] for simple part geometries, or by an additional boundary element calculation if the part geometry is complex. Probe response is then calculated by means of the reciprocity theorem [7], with receiver field data provided again by analytic or boundary element calculations for the unflawed part.

By formulating the problem in this way it is possible to separate the calculation of flaw surface fields from calculations of incident fields and receiver response. The formal solution for the flaw surface field is the product of a solution matrix, which depends only on the flaw geometry and skin depth, and a column matrix representation of the incident field. Thus, to simulate a scan of the eddy current probe, only one calculation of the solution matrix is required. Changes in probe response caused by changes in probe position are then completely determined by calculations of the transmitter and receiver fields for the unflawed part. This makes for an efficient simulation of flaw signals as a function of position in a scan pattern for the purpose of determining the probability of flaw detection.

The theoretical elements that comprise this hybrid approach are developed elsewhere [8] and are therefore reviewed only briefly in the next section. The principal purposes of this paper are to provide illustrations of inspection simulations for a simple geometry, and to discuss extensions of the method to more complex, three-dimensional applications.

## THEORY

The reciprocity theorem [7], which is given by (1), shows that the flaw signal can be expressed as an integral over the flaw surface of a vector product of certain fields which are labeled here with subscripts T and R.

$$\Delta Z = \frac{1}{I^2} \int_{S_F} [\vec{E}_R \times \vec{H}_T - \vec{E}_T \times \vec{H}_R] \cdot \vec{dS} \qquad (1)$$

The T fields are those produced on the flaw surface when coil T, the transmitter or induction coil, is activated. The R fields are those that would be produced in an unflawed part if the receiver coil were activated. If we expand the vector product in the integrand, we see that only the tangential components of the T and R fields are involved. The reciprocity theorem therefore tells us that we need only the tangential components of the T and R fields on the flaw surface to determine the response of an eddy current probe. The boundary element method (BEM) is a numerical procedure for calculating these tangential fields.

Before describing the BEM as used here, it is first necessary to introduce a class of functions called dyadic Green's functions [9]. Physically, these functions relate the electric and magnetic fields at an arbitrary point $\vec{x}$ within a conductor to the current in the conductor at another point $\vec{x}'$. In general, the Green's functions must be dyads, or, equivalently, tensors, in order to satisfy the boundary conditions at the surface of the conductor. Also, except in very simple geometries, dyadic Green's functions are complicated functions of position that cannot be expressed in terms of simple analytic functions. However, if the Green's dyads are known for a particular geometry in the unflawed conductor it is possible to simplify the calculation of flaw surface fields. So, for the present, let us assume that the dyadic Green's functions for the unflawed conductor are known.

In this case, starting with Maxwell's equations, it is possible to develop a set of coupled integral equations that involve only the tangential components of the fields on the flaw surface. The result is [8]

$$\frac{\Omega c}{4\pi} q_i(\vec{x}) = q_i^o(\vec{x}) + \sum_{j=1}^{3} \int_{S_F} [T_{ij}^E(\vec{x},\vec{x}')q_j(\vec{x}') + U_{ij}^E(\vec{x},\vec{x}')h_j(\vec{x}')]dS$$

(2)

$$\frac{\Omega c}{4\pi} h_i(\vec{x}) = h_i^o(\vec{x}) + \sum_{j=1}^{3} \int_{S_F} [T_{ij}^H(\vec{x},\vec{x}')h_j(\vec{x}') + U_{ij}^H(\vec{x},\vec{x}')q_j(\vec{x}')]dS$$

with $\vec{q}(\vec{x}) = \vec{n}(\vec{x}) \times \vec{j}(\vec{x})$, $\vec{h}(\vec{x}) = \vec{n}(\vec{x}) \times \vec{H}(\vec{x})$, where $\vec{n}(\vec{x})$ is the normal to the flaw surface at the point x, and $\vec{j}$ and $\vec{H}$ are the current density and magnetic field intensity, respectively. In these equations the kernels U and T are determined by the dyadic Green's functions, $\Omega_c$ is the solid angle subtended by the conductor at the point $\vec{x}$, and $q_i^o$ and $h_i^o$ are the tangential components of the unperturbed fields, i.e., the fields that would exist at the point x if no flaw were present. We assume, for the present, that the unperturbed fields can be calculated by another method. The simplification that results from use of the dyadic Green's functions is that the integrals are over points on the flaw surface only. If we had used simpler Green's functions that do not satisfy the boundary conditions on the surface of the conductor, then additional integrals over the surface of the conductor would appear on the right sides of these equations.

The boundary element method is a numerical technique for solving integral equations of this form. To develop an approximate set of algebraic equations, the surface of the flaw is divided into surface elements as shown in Figure 1, and each element is defined by a number of nodal points around the periphery of the element. Integration over the element is accomplished by expressing the fields inside the element in terms of their values at the nodal points, and evaluating the resulting integral by double Gaussian quadrature. The end result is a set of simultaneous algebraic equations for the fields at the nodal points, which can be written in matrix form. The solution for the tangential fields on the flaw surface is therefore of the form

$$\begin{bmatrix} q \\ h \end{bmatrix} = Z^{-1} \begin{bmatrix} q_o \\ h_o \end{bmatrix}$$

(3)

where q and h are column matrices containing the components of the vectors $\vec{q}$ and $\vec{h}$ at each node. An important property of this solution, which is demonstrated elsewhere [8], is that the inverse matrix, which we will call the solution matrix, is a function only of the flaw geometry and skin depth. It is independent of the unperturbed field and is therefore independent of probe geometry and position. The vector containing $q_o$ and $h_o$ is, on the other hand, independent of the flaw geometry and depends only on the probe configuration and position. The final solution therefore involves a factor that depends only on the flaw and another factor that depends only on the probe. For a given flaw geometry and skin depth, this means that we need compute the solution matrix only once to determine the probe response as a function of probe position and/or configuration, which simplifies the simulation of an inspection as a function of scan pattern and probe geometry. Examples of such applications are given in the next section.

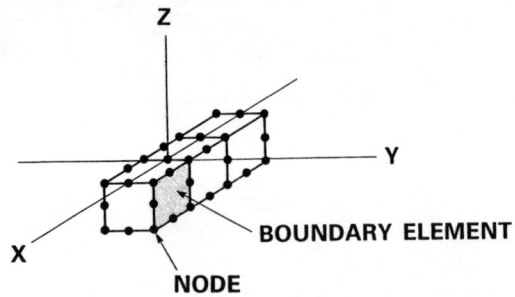

Fig. 1. Discretization procedure for the solution of integral equations by the boundary element method

FLAW SIGNAL PREDICTIONS

We now consider a simple example that illustrates the simulation of an eddy current inspection. In this case the flaw is a cube, about 8 mils on an edge, located about one skin depth below a plane surface as shown in Figure 2. We choose the skin depth to be large compared to flaw dimensions because, as we have shown elsewhere [8], this allows us to uncouple the electric and magnetic field solutions. Also, because the flaw is a skin depth below the surface, we use the infinite medium Green's dyad to calculate current densities on the flaw surface, and thus ignore effects of the plane surface in the BEM solution. To further simplify the calculation we also assume that the magnetic field is unperturbed by the presence of the flaw. It should be noted that none of these approximations are essential to the application of the boundary element method; they are introduced only as a way of saving computer time in these illustrative examples.

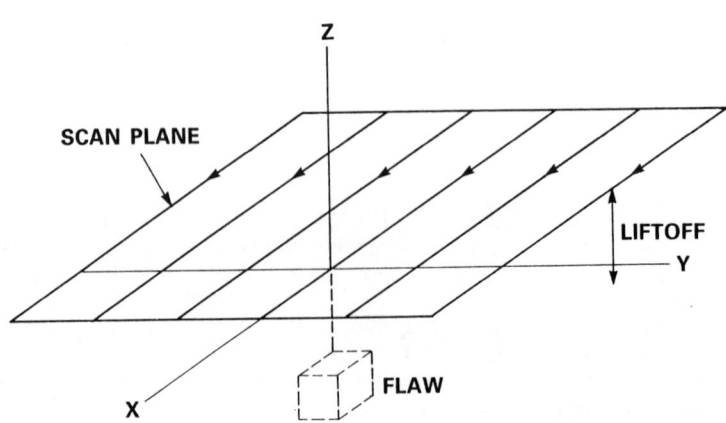

Fig. 2. Geometry for the calculation of eddy current flaw signals

Our illustrations are plots of the magnitude of the probe impedance as different probes are scanned over a raster pattern like that shown in Figure 2. To provide data on the unperturbed fields and the $E_R$ and $H_R$ fields of (1) we use a one-dimensional Fourier transform model equivalent to that of Dodd and Deeds [5], and, when necessary, a two-dimensional generalization of that theory [6].

Figure 3 shows the absolute value of the complex impedance of an absolute probe as a function of probe position; the flaw is located at the center of the pattern and the response shows the expected symmetry for a circular coil over a cubic flaw. Another calculation for an absolute probe of smaller diameter produces a similar pattern (not shown), the only difference being that the signal is better localized, as one would expect. The results of a third calculation show that an asymmetric signal is obtained when we use a separate receiver located adjacent to the transmitter and displaced in the positive X direction of Figure 2.

When the receiver coil is rotated so that its axis is parallel to the surface of the conductor and along the X axis, we see a different type of asymmetric signal as shown in Figure 4. Finally, in the fifth calculation of this series, still another asymmetric pattern (not shown) is obtained when the receiver coil axis is parallel to the Y axis of Figure 2.

The flaw model used for these examples is, of course, a very simple one and several approximations have been introduced to reduce computation time. Still, the results should serve as a first illustration of what can be done with a single boundary element calculation in the simulation of eddy current flaw detection. To apply the method to more practical problems the next steps we must take are to remove the various approximations we have introduced and extend the program to the treatment of more realistic part and flaw geometries.

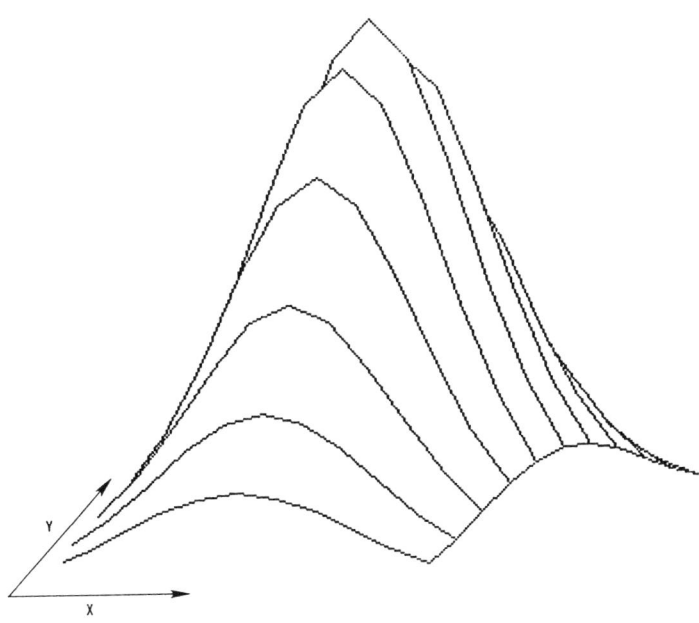

Fig. 3. Flaw signal magnitude for an absolute probe in the geometry of Figure 2

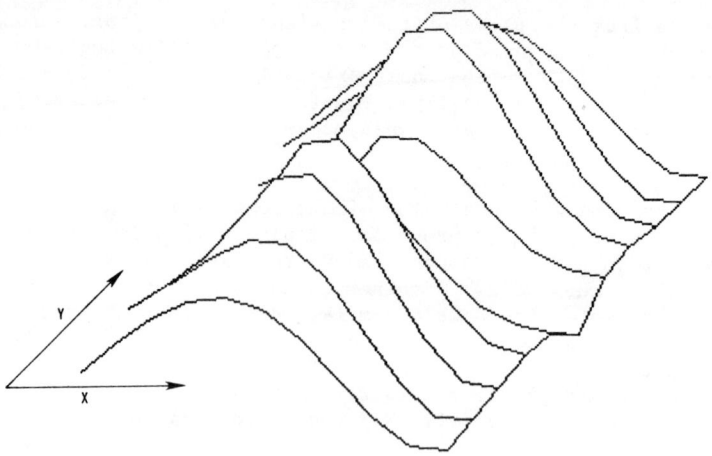

Fig. 4. Flaw signal magnitude for a probe with a separate receiver coil with axis in the X direction

EXTENSIONS OF THE MODEL

An important special case is the surface crack in a half-space or a flaw near a plane surface. Because the Green's dyads are known for the half-space [10], this class of problems can be handled in much the same way as the infinite medium calculations just illustrated. All we need do is replace the infinite medium kernels with half-space kernels in the existing code; there is no need to introduce boundary elements on the plane surface because boundary conditions on that surface are automatically satisfied through the use of the half-space Green's dyads. We expect that this version of the program will be useful mostly for studies of the effects of probe and flaw geometry on flaw detectability under ideal, flat-surface conditions.

If the part to be inspected has a complicated shape, then the calculation of appropriate Green's dyads becomes so complex as to be prohibitive. In such cases it becomes necessary to introduce boundary elements on the surface of the part, as well as the flaw, and solve for both the unperturbed and flaw surface fields by the boundary element method. It is still possible to present the solution for flaw surface fields as the product of a solution matrix times an incident field vector, but in this case the incident field is the field in air, rather than the unperturbed field in the conductor. Modification of the program to treat this, the most general case, will require considerable effort. However, development of such a code is considered feasible, and its implementation should be practical with existing computer capabilities.

Our plans for the immediate future therefore call for completion of the half-space version of the boundary element model, and, over a somewhat longer period of time, development of a general, complex geometry code. We also believe it would be advantageous to explore various approximations, such as those used in the calculations presented here, which could greatly reduce computational requirements while providing accuracy adequate for most purposes.

CONCLUSION

Our early experience with the boundary element method indicates that it should prove to be an efficient approach to modeling complex, three-dimensional eddy current problems. Because the desired solution can be expressed as the matrix product of a flaw-dependent factor and a probe-dependent factor, the formulation presented here appears to be well suited to the simulation of eddy current scanning operations and the evaluation of candidate probe designs.

ACKNOWLEDGEMENT

This work was sponsored by the Center for Advanced Nondestructive Evaluation, operated by the Ames Laboratory, USDOE, for the Air Force Wright Aeronautical Laboratories/Materials Laboratory under Contract No. W-7405-ENG-82 with Iowa State University.

REFERENCES

1. R. E. Beissner, "Predictive Models and Reliability Improvement in Electromagnetic Nondestructive Evaluation," Review of Progress in Quantitative NDE, Vol. 5A, D. O. Thompson and D. E. Chimenti, eds., Plenum, New York (1986).
2. W. Lord, "An Overview of Numerical Models for Eddy Current NDT Phenomena," these proceedings.
3. G. L. Burkhardt and R. E. Beissner, "Probability of Detection of Flaws in a Gas Turbine Engine Component Using Electric Current Perturbation," Review of Progress in Quantitative NDE, Vol. 4, D. O. Thompson and D. E. Chimenti, eds., Plenum, New York (1985).
4. T. A. Cruse and F. J. Rizzo, eds., "Boundary-Integral Equation Method: Computational Applications in Applied Mechanics," ASME Proc. AMD-Vol. 11 (1975).
5. C. V. Dodd and W. E. Deeds, "Analytical Solutions to Eddy-Current Probe-Coil Problems," J. Appl. Phys. $\underline{39}$, 2829 (1968).
6. R. E. Beissner and M. J. Sablik, "Theory of Eddy Currents Induced by a Nonsymmetric Coil Above a Conducting Half-Space," J. Appl. Phys. $\underline{56}$, 448 (1984).
7. B. A. Auld, "Theoretical Characterization and Comparison of Resonant-Probe Microwave Eddy-Current Testing with Conventional Low-Frequency Eddy-Current Methods," in "Eddy Current Characterization of Materials and Structures," ASTM STP 722, G. Birnbaum and G. Free, eds., American Society for Testing and Materials, Philadelphia (1981), p. 332.
8. R. E. Beissner, "Boundary Element Model of Eddy Current Flaw Detection in Three Dimensions," J. Appl. Phys. $\underline{60}$, 352 (1986).
9. C. T. Tai, "Dyadic Green's Function in Electromagnetic Theory," International Textbook, Scranton (1971).
10. R. E. Beissner, "Analytic Green's Dyads for an Electrically Conducting Half-Space," J. Appl. Phys. $\underline{60}$, 855 (1986).

ON BOUNDARY INTEGRAL EQUATION METHOD FOR FIELD DISTRIBUTION UNDER

CRACKED METAL SURFACE

V. G. Kogan, *G. Bozzolo, and N. Nakagawa

Ames Laboratory and Center for NDE
Iowa State University

INTRODUCTION

Two approaches to the problem of the AC magnetic field distribution inside conducting media are currently used. One consists of solving the quasi-stationary Maxwell differential equations subject to certain boundary conditions. Another one is based upon certain boundary integral equations that involve only the field components at the boundary surface [1,2]. Aside from some numerical advantage in treating unknowns of a lower dimension, the second method fits better the common formulation of the eddy current NDE, which aims at determination of the total impedance of a metal part (or its change due to a defect). The impedance is related to the integrated energy flux through the metal surface and as such it can be expressed in terms of the field components at the surface exclusively.

We consider in this report only the two-dimensional (2-D) case. Having little to do with real situations, this case is nevertheless instructive. Analytical development can be carried in some 2-D sitatuions "almost to the end" as opposed to the general 3-D configurations. Solvable 2-D problems may serve as test cases for more involved and less transparent general approaches.

INTEGRAL EQUATION

Let us consider the external magnetic field uniform in space and harmonic in time: $B_z = B_0 \exp(-i\omega t)$. The metal part is a long (infinite) "rod" in the z direction with the surface parallel to the external field. The cross-section of such a part is shown in Fig. 1. The external field in this configuration remains uniform for any shape of the cross-section, thus making the external problem "solved". The internal problem, therefore, consists of solving the 2-D Maxwell equation

$$(\nabla^2 + k^2) B(x,y) = 0, \quad k = (1+i)/\delta, \tag{1}$$

with $\delta$ being the skin depth. At the boundary contour $B(x,y)=B_0$.

---

*Now at NASA Lewis Research Center, Cleveland, OH

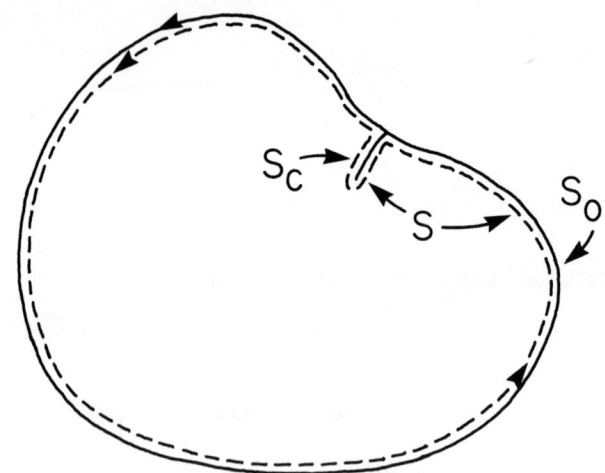

Fig. 1. Cross-section of a long metal part. $S_o$ is the "uncracked" surface. Surface S consists of $S_o$ and of the crack's faces, $S_c$.

Consider now a Green's function, $G(\vec{r},\vec{r}_o)$, which satisfies

$$(\nabla^2+k^2)G(\vec{r},\vec{r}_o) = -4\pi\delta(\vec{r}-\vec{r}_o) \quad (2)$$

under certain boundary conditions, and integrate the combination [Eq. (1)xG - Eq. (2)xB] over the volume $d^3\vec{r}$). After applying the Green's theorem one obtains

$$\int d\vec{s}\cdot(G\nabla B - B\nabla G) = 4\pi B(\vec{r}). \quad (3)$$

Here $d\vec{s}$ is the area element directed as the external normal to the metal surface. This equation gives the field $B(\vec{r})$ everywhere inside the metal in terms of the magnetic field at the surface (which is constant, $B_o$, in our geometry) and of its normal derivative (which is proportional to the tangential component of the electric field: $\partial B/\partial n = \sigma E_t$ with $\sigma$ being the metal conductivity). We note that Eq. (3) holds irrespective of what specific Green's function is chosen (e.g., what boundary conditions are imposed upon G). The problem would have been solved, had the Green's function for a particular metal shape under the condition <u>G=0 at the surface</u> been known. In this case, Eq. (3) reduces to

$$B(\vec{r}_o) = (B_o/4\pi)\int d\vec{s}\cdot\nabla G. \quad (4)$$

The trivial example of such a case is the half-space y>0 filled with metal. The known solution, $B(y)=B_o\exp(iky)$, can be also obtained from Eq. (4) using the Green's function

$$G/i\pi = H_o(kR) - H_o(kR_1), \quad R^2 = (x-x_o)^2 + (y-y_o)^2, \quad R_1^2 = (x-x_o)^2 + (y+y_o)^2 \quad (5)$$

where $H_o$'s are Hankel functions of the first kind and of the zero order, and $R_1$ is the distance between $\vec{r}_o=(x_o,y_o)$ and the image of $\vec{r}$: $(x,-y)$.

Another example of a known Green's function is that of the 90° corner:

$$G/i\pi = H_o(kR) - H_o(kR_1) - H_o(kR_2) + H_o(kR_3), \qquad (6)$$

$$R_2^2 = (x+x_o)^2 + (y-y_o)^2, \quad R_3^2 = (x+x_o)^2 + (y+y_o)^2,$$

and R, $R_1$ are defined in Eq. (5). One easily verifies that G=0 at both corner faces which coincide with positive x and y axes. Equation (4) now yields after some algebra [1]:

$$\frac{B(x,y)}{ikB_o} = y \int_o^{x/y} \frac{dv}{\beta} H_1(ky\beta) + x \int_o^{y/x} \frac{dv}{\beta} H_1(kx\beta), \quad \beta^2 = v^2+1. \qquad (7)$$

Differentiate this with respect to y and take the limit y→0 to obtain:

$$\sigma E_x(x,0) = ik^2 B_o \int_o^x H_o(ku) du. \qquad (8)$$

We now define the total surface impedance as the time average over the period of the integrated Pointing vector

$$Z = \int \vec{E} \times \vec{B}^* \cdot d\vec{s}/2 \qquad (9)$$

where the integral is taken over the metal surface with the normal directed into the metal. One can easily verify with the help of Maxwell equations that

$$\text{Re}Z = \int \vec{j} \cdot \vec{E}^* dv/2, \quad \text{Im}Z = -\omega\mu \int dv |B|^2/2, \qquad (10)$$

with the current density $\vec{j} = \sigma \vec{E}$. Thus, ReZ is the total dissipation power in the metal averaged over the period, while ImZ is proportional to the magnetic energy stored. It is worth noting that ReZ>0, while ImZ<0 [3].

For the corner, the total Z diverges. However, the difference, $Z-Z_o$, can be evaluated with $Z_o$ being the total impedance of the plane surface ("unfolded corner").

$$Z - Z_o = B_o \int_o^\infty [E_x(x) - E_o] dx. \qquad (11)$$

Here the electric field on the plane surface (or far from the corner's edge), $E_o = ikB_o/\sigma$. The integrand of (11) is

$$E_x(x) - E_o = -kE_o \int_o^\infty H_o[k(x+t)] dt \qquad (12)$$

155

(this is verified with the help of Eq. (8) and using the identity $\int_0^\infty H_o(t)dt=1$). Substitute (12) in (11) and introduce polar coordinates $x=r\cos\phi$, $t=r\sin\phi$ to do the double integration. The result turns out to be a real number: $Z-Z_o = -B_o^2/\pi\sigma$. One can normalize this on the dissipation power of 1 cm² of plane surface, $\text{Re}z_o=\text{Re}(E_oB_o/2)=B_o^2/2\sigma\delta$, to obtain

$$(Z-Z_o)/\text{Re}z_o = -(2/\pi)\delta = -0.637\delta. \tag{13}$$

One can say that the dissipative part of the corner impedance is depleted with respect to the plane surface as if the corner is "$2\delta/\pi$ shorter". This result has been obtained numerically by Kahn [1].

In most situations, the Green's function that vanishes at the metal part's surface is difficult to construct. One takes then $G=i\pi H_o(kR)$ (the simplest possible G), places the point $\vec{r}$ at the surface (where $B(\vec{r})=B_o$) to obtain from (3) an integral equation for the tangential component of the electric field:

$$\sigma\int ds G(\vec{s},\vec{s}_o)E_t(\vec{s}) = B_o(4\pi + \int d\vec{s}\cdot\nabla G) \tag{14}$$

($d\vec{s}$ is directed along the external normal, while $ds$ is a scalar surface element). This can be solved numerically for $E_t$, thus making it possible to evaluate the total impedance (9) as well as the distribution $B(\vec{r}_o)$ (if the latter is needed) by substituting $E_t$ in Eq. (3).

Let us turn now to the situation where the exact Green's function vanishing at the whole surface of the metal part, is unknown, while it is known for a surface "close" to the actual one. An example is the case of a tight crack normal to the plane metal surface (we take this example for simplicity, although the same approach can be used for other crack shapes). One can utilize then the Green's function (5) of the plane surface (y=0). Integral (3) splits in two: one over the plane surface (over x from $-\infty$ to $\infty$), and another one over the crack's face. The term $G\nabla B$ does not contribute to the integral over the plane surface, while $B\nabla G$ in this integral yields the unperturbed by the crack field $B_o\exp(iky)$. In the integral over the crack faces, the contribution of $B\nabla G$ cancels out ($B=B_o$ along the crack, $\nabla G\cdot d\vec{s}$ has opposite signs on two crack faces). One obtains

$$B(\vec{r}_o) = B_o e^{iky} - \frac{i\sigma}{2}\int_o^d E_y(y)[H_o(kR)-H_o(kR_1)]dy. \tag{15}$$

Here d is the crack depth, R and $R_1$ are defined in (5) (set crack's location as x=0), and $E_y(y)$ is the electric field at the right crack face $x = +0$. An integral equation for this quantity is obtained using the boundary condition $B = B_o$ at the crack face:

$$B_o(1-e^{iky}) = -\frac{i\sigma}{2}\int_o^d E_y(y)\{H_o(k|y-y_o|)-H_o[k(y+y_o)]\}. \tag{16}$$

This can be solved numerically for $E_y(y)$. The solution has been given by Kahn [4] (though he used $H_o(kR)$ as Green's function and, therefore, had to integrate over the metal surface in addition to the crack faces).

Given $E(0,y)$ at the crack face, we can proceed in evaluation of the impedance change, $\Delta Z$, due to the crack:

$$\Delta Z / B_o = \int_0^d E_y(0,y) dy + \frac{1}{2} \int_{-\infty}^{\infty} [E_x(x,0) - E_o] dx. \tag{17}$$

One could show by a direct evaluation of $E_x(x,0)$ using Eq. (15) for the magnetic field, that the second contribution to $\Delta Z$ in (17) is equal to $-B_o \int_0^d E_y(0,y) \exp(iky) dy$, i.e., it is expressed in terms of the quantity $E_y(0,y)$ we already know. It is instructive, however, to see that this result is a direct consequence of the general reciprocity theorem.

The theorem states that at a given frequency, any two solutions of the Maxwell equations (which could correspond to different boundary conditions) satisfy the identity

$$\int_S (\vec{E}_1 \times \vec{B}_2 - \vec{E}_2 \times \vec{B}_1) \cdot d\vec{s} = 0, \tag{18}$$

if the whole integration surface $S$ is situated in a region where the material parameters ($\sigma$ in our case) corresponding to these two situations are the same. Let us call $\vec{E}_1, \vec{B}_1$ the inside fields for the uncracked metal piece shown in Fig. 1. Let $\vec{E}_2, \vec{B}_2$ be the fields in the same piece with a crack. Then choosing the surface $S$ as shown in the figure, we apply theorem (18). We further separate the surface $S$ into the "uncracked" piece $S_o$ and the crack face, $S_c$, to obtain

$$\int_{S_o} (\vec{E}_2 - \vec{E}_1) \times \vec{B}_o \cdot d\vec{s} = \int_{S_c} \vec{E}_2 \times \vec{B}_1 \cdot d\vec{s}. \tag{19}$$

We have used here that both $\vec{B}_2$ and $\vec{B}_1$ are equal to the constant $\vec{B}_o$ at the surface $S_o$.

Turning to the case of a closed crack in the plane surface, we obtain from (19) the above mentioned result:

$$-\int_{-\infty}^{\infty} [E_x(x,0) - E_o] dx = 2 \int_0^d E_y(+0,y) e^{iky} dy. \tag{20}$$

(Note: $\vec{B}_1 = B_o \hat{z} e^{iky}$, $\vec{B}_o = B_o \hat{z}$). Equation (17) now yields the impedance change due to the crack:

$$\Delta Z = B_o \int_0^d E_y(y)(1 - e^{iky}) dy. \tag{21}$$

Once again, we see that the total impedance change due to the crack is expressed in terms of the field $\vec{E}_t$ at the crack face. Thus, the whole problem reduces to that of finding $E_y(y)$ at the crack.

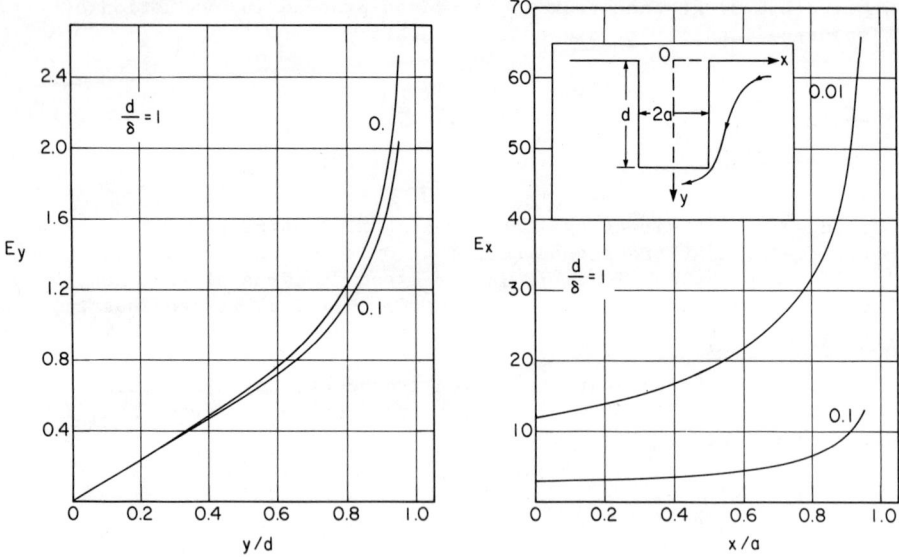

Fig. 2. Electric fields $\text{Re}E_y(y)$ and $\text{Re}E_x(x)$ at the surfaces of an open crack for $d/\delta=1$. The configuration and the coordinates are given in the insert. The numbers by the curves indicate the ratio of the width to the skin depth, $2a/\delta$. For simplicity, the field $B_0$ and the conductivity $\sigma$ are set equal to unity.

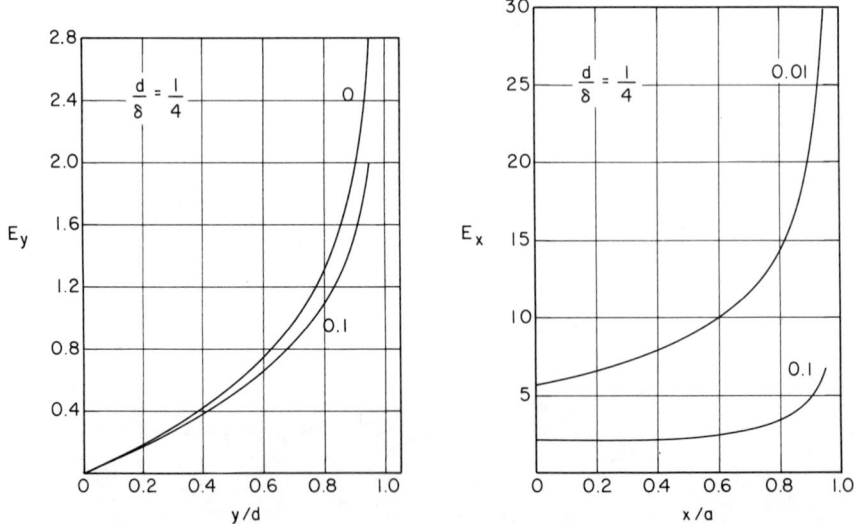

Fig. 3. Electric fields $\text{Re}E_y(y)$ and $\text{Re}E_x(x)$ at the surfaces of an open crack for $d/\delta=1/4$. The numbers by the curves indicate the ratio of the width to the skin depth, $2a/\delta$.

With essentially the same argument, one can deduce from the reciprocity relation (18) for any closed two dimensional crack

$$\Delta Z = \sum_{n=1}^{2} \int ds E_t(\vec{s},n)[B_o - B_o(\vec{s})]/2 \qquad (22)$$

where n=1,2 corresponds to two opposite crack faces, $\vec{s}$ is the point at the crack, $B_o(\vec{s})$ is the "unperturbed" by the crack magnetic field (at the crack's location), and $E_t(\vec{s},n)$ is the tangential to the crack electric field at the side n. (Note that in general the field $E_t$, at two sides of the crack, are not necessarily equal in magnitude as in the case of a plane crack normal to the plane metal surface).

Equation (22) can be applied to evaluate the impedance change due to a closed crack in the 90° corner. In this situation, the field $B_o(\vec{s})$ is given in Eq. (7), where x and y take their values at the crack's location. The tangential component of the electric field $E_t(\vec{s},n)$ at the crack faces is found by solving an appropriate integral equation (similar to Eq. (16)). Numerical work on this problem is in progress.

The same approach can be exploited for open 2D cracks in the plane surface. Using the Green's function (5), one obtains the magnetic field distribution:

$$B(\vec{r}) = B_o e^{iky} - \frac{\sigma}{4\pi} \int_{S_c} ds\, E_t(\vec{s}) G(\vec{s},\vec{r}) - \frac{k^2 B_o}{4\pi} \int dA_o G(\vec{r},\vec{r}_o). \qquad (23)$$

To derive this expression from Eqs. (3) and (5), we added and subtracted $B_o \int \Delta G \cdot d\vec{s}$ over the crack's opening ("crack's mouth"). This allows us to extract the contribution of the unperturbed field $B_o \exp(iky) = B_o \int_{-\infty}^{\infty} dx (\partial G/\partial y)_{y=0}$. The part $B_o \int \nabla G \cdot d\vec{s}$ over the crack faces complemented with the same integral over the "crack mouth", results in the last term in Eq. (23). For points $\vec{r}$ at the crack faces ($\vec{r}=\vec{s}_o$), $B=B_o$ and Eq. (23) yields an integral equation for the tangential electric field $\vec{E}_t$ at the crack faces and at the crack bottom. The preliminary numerical solutions for $ReE_y(y)$ (at the faces) and $ReE_x(x)$ (at the bottom) are given in Figs. 2, 3, and 4 for three different ratios of the crack depth d to the skin depth $\delta$. It is worth noting that the real part of the electric field $E_y(y)$ at the crack faces depends rather weakly upon the crack width. In fact, for $d/\delta=4$ within the accuracy of about 10%, $ReE_y(y)$ are the same for $2a/\delta=0$, 0.01, and 0.1. The derivation of the impedance change due to open cracks along with details of the numerical procedure and accuracy estimates will be published elsewhere.

ACKNOWLEDGEMENT

This work was supported by the Center for Advanced Nondestructive Evaluation, operated by the Ames Laboratory, USDOE, for the Air Force Wright Aeronautical Labortories/Materials Laboratory under Contract No. W-7405-ENG-82 with Iowa State University and the NSF University/Industry Center for NDE at Iowa State University.

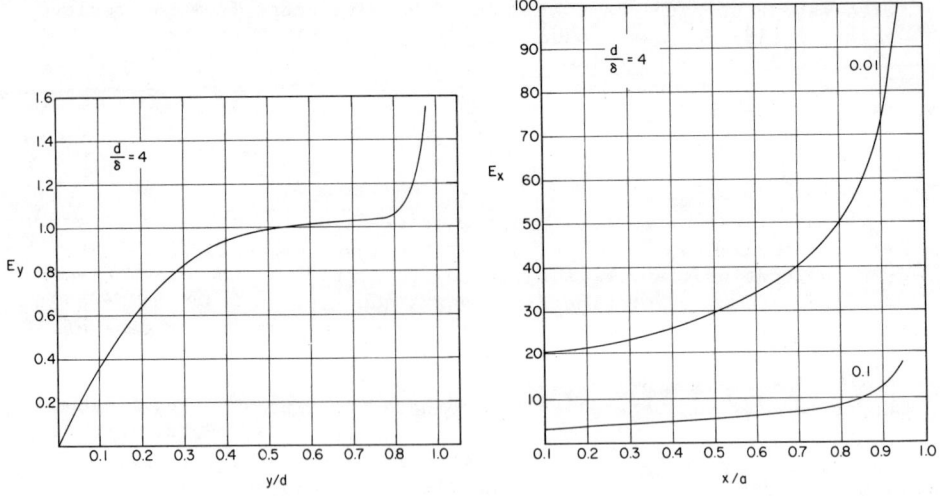

Fig. 4. Electric fields $ReE_y(y)$ and $ReE_x(x)$ at the surfaces of an open crack for $d/\delta=4$. The numbers by the curves indicate the ratio of the width to the skin depth, $2a/\delta$. Within about 10% accuracy, $ReE_y(y)$ for values of $2a/\delta=0$, 0.01, and 0.1 are the same.

REFERENCES

1. A. H. Kahn, R. Spal, and A. Feldman, J. Appl. Phys. <u>48</u>, 4454 (1977).
2. R. E. Beissner, J. Appl. Phys. <u>60</u>, 352 (1986).
3. L. D. Landau and E. M. Lifshitz, "Electrodinamics of continuous media", Pergamon, NY, 1975, p. 67.
4. A. H. Kahn, <u>Review of Progress in Quantitative NDE, 1</u>, D. O. Thompson and D. E. Chimenti, Eds., (Plenum Press, NY, 1982), p. 369.

APPLICATIONS OF THE VOLUME INTEGRAL TECHNIQUE TO MODELING IMPEDANCE
CHANGES DUE TO SURFACE CRACKS

W. Scott Dunbar

Department of Metallurgical Engineering
University of British Columbia
Vancouver, B.C., Canada

INTRODUCTION

The volume integral technique solves the electromagnetic diffusion equation for the electric field vector E in the presence of a flaw of conductivity $\sigma_f$ in a host medium of conductivity $\sigma_h$. Given the electric field, the impedance change in a probe coil may be computed. The technique has been used to model electromagnetic data collected in geophysical exploration programs [1].

THEORETICAL BACKGROUND

Using Maxwell's equations and ignoring displacement currents, the partial differential equation for the electric field E due to a source (probe coil) current density J° is found to be

$$\nabla^2 E - \gamma^2 E = i\omega\mu J° \tag{1}$$

where $\gamma^2 = i\omega\mu\sigma$, $\omega$ is the frequency of the source, $\mu$ is the magnetic permeability, $\sigma$ is the conductivity and $i = (-1)^{1/2}$.

For a source J° distributed over a volume V, the solution to Equation 1 is

$$E°(r) = \int_V G(r,r') \cdot J°(r')dV \tag{2}$$

where $G(r,r')$ is the Green's tensor for Equation 1; i.e., the solution of the equation

$$\nabla^2 G - \gamma^2 G = i\omega\mu I\delta(r,r') \tag{3}$$

where I is the identity matrix and $\delta(r,r')$ is the Dirac delta function. In three dimensions, Green's tensor contains nine components $G_{ij}(r,r')$. $G_{ij}(r,r')$ is the electric field component $E_i$ at r due to a point source of current density of unit magnitude oriented in the j direction at the point r'. Green's tensor may be derived for homogeneous and layered half spaces [1, 2].

To model flaws of conductivity $\sigma_f$ in a host medium of conductivity $\sigma_h$, the conductivity $\sigma$ is written as $\sigma = \sigma_h + \delta\sigma$ where

$$\delta\sigma = \begin{matrix} 0 & \text{outside flaw} \\ \sigma_f - \sigma_n & \text{inside flaw} \end{matrix} \qquad (4)$$

The constitutive equation relating current density I to electric field E therefore becomes

$$J = \sigma E + J° = (\sigma_h + \delta\sigma)E + J° . \qquad (5)$$

When the above equation is used to re-derive Equation 1, the result is

$$\nabla^2 E - \gamma_h^2 E = i\omega\mu(J° + \delta\sigma E) \qquad (6)$$

for which the solution is

$$E(r) = E°(r) + \int_V \delta\sigma G(r,r') \cdot E(r') dV \qquad (7)$$

where Equation 2 has been used. Equation 6 shows that the flaw is modeled as an equivalent distribution of dipole current sources of strength $\delta\sigma E$. The volume integration in Equation 4 extends only over the region where $\delta\sigma$ is non-zero i.e., the flaw volume.

An integral equation may be derived to compute the electric field due to the dipole distribution (the scattered field). Assuming $\delta\sigma = \sigma_f - \sigma_n$ is constant in V (this is not a necessary assumption). Equation 7 may be rearranged to give

$$E°(r) = E(r) - (\sigma_f - \sigma_h) \int_V G(r,r') \cdot E(r') dV . \qquad (8)$$

Thus, knowing the incident electric field E°, the scattered electric field E which is equivalent to the presence of the defect may be computed. Given the scattered field, the corresponding impedance change in the probe coil may be computed, as discussed in the next section.

The discretization and solution of Equation 5 begins by dividing the volume of the flaw into N volume elements, within which the scattered electric field is assumed constant. (More complicated variations of E could be assumed). Equation 8 is then written

$$E°(r_j) = E(r_j) - (\sigma_f - \sigma_h) \sum_{k=1}^{N} \left[ \int_{V_k} G(r_j, r') dV_k \right] \cdot E_k \qquad (9)$$

where $r_j$ is a position vector and $E_K$ denotes the electric field vector at the centroid of the Kth element. A system of equations of order 3N in $E_k$ is formed by letting $r_j$ be the centroid of each of the volume elements $V_j$, $1 \leq j \leq N$:

$$E° = (I-C)E = AE \qquad (10)$$

where E° is a vector of length 3N composed of incident field valves, I is the identity matrix and

$$C_{jk} = (\sigma_f - \sigma_h) \int_{V_k} G(r_j, r') dV_k \qquad (11)$$

is a 3 by 3 matrix formed by integrating the Green's tensor over the volume element $V_k$.

The details of the discretization of the flaw, the integration of the Green's tensor and the solution of the system of equations are described by Dunbar [3]. Generally the matrix A is asymmetric and fully populated.

## CALCULATION OF THE IMPEDANCE CHANGE

The change in coil impedance due to the scattered electric field E is given by Faraday's law

$$\Delta Z = \frac{-i\omega}{I} \int_S B \cdot n \, dS \tag{12}$$

where B is the magnetic field in the coil due to the scattered current density in the flaw volume and I is the current in the probe coil. The vector n is the unit normal to the coil surface S. For a flat-lying coil whose axis lies along the Z coordinate axis directed into the host, n = (o,o,-1) and

$$\Delta Z = \frac{i\omega}{I} \int_S B_z \, dS \tag{13}$$

The above equation was used to compute impedance changes.

An even simpler method of computing impedance changes would be to use the formula for $\Delta Z$ derived by means of the reciprocity theorem:

$$\Delta Z = \frac{1}{I^2} \int_{V_f} (\sigma_f - \sigma_h) E^\circ \cdot E \, dV \tag{14}$$

where $V_f$ is the flaw volume [4]. The author is grateful to Dr. S. Burke of the Aeronautical Research Laboratories in Australia for pointing this out.

## EXAMPLES

Two examples are presented:

1) Rectangular surface slot in aluminum plate.
2) 'Realistic' surface crack in aluminum plate.

In each example, the flaw was assumed to occur in a semi-infinite half space. This assumption is valid since the skin depth in each example is shallow; i.e. the coil frequencies are sufficiently high that the host may be assumed to be a half space.

The geometry and discretization of the slot of Example 1 are shown in Figure 1. Table 1 lists the model and coil parameters for this example. The coil is of rectangular cross section with an air core so that the incident electric field may be calculated by the Dodd and Deeds [5] formulation. The impedance change profile along the positive y axis is shown in Figure 2. This profile is similar in form and magnitude to that calculated by Auld et al [6] by another method for a similar geometry. This is regarded as partial verification of the program.

The geometry and discretization of the 'realistic' surface crack of Example 2 is shown in Figure 3. Table 1 lists the model parameters for this example. The same coil as in Example 1 was used. The impedance change profile along the positive y axis is shown in Figure 4. As expected, it is similar to the impedance profile of Example 1, but lower in magnitude.

In both Examples 1 and 2, the impedance change profile is symmetric about the x axis. The results in Figures 2 and 4 were found to be sensitive to the amount of discretization of the flaw, particularly along the y axis. This is to be expected in these cases but implies that, in general, different discretizations should be attempted so that confidence in the results can be established.

TABLE 1

**Model Parameters – Examples 1 and 2**

**Coil**

Inner radius: 2.0 mm
Outer radius: 10.0 mm
Height: 5.0 mm
Lift-off: 0.1 mm
Frequency: 200 kHz
Number of turns: 1000

**Test Specimen**

Conductivity: $2.6 \times 10^7$ S m$^{-1}$

**Defect**

Example 1
Conductivity: 0.0 S m$^{-1}$
Length: 4.0 mm
Depth: 1.0 mm
Width: 0.2 mm

Example 2
Conductivity: 0.0 S m$^{-1}$
Length: 4.0 mm
Depth: 1.0 mm
Width: 0.0 mm at ends; 0.2 mm maximum

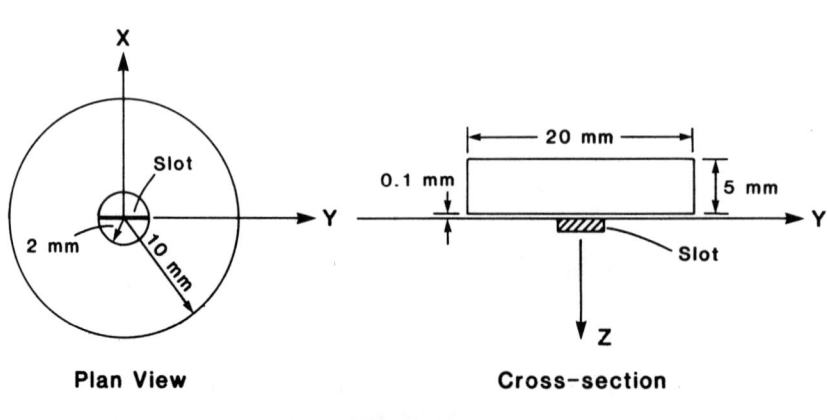

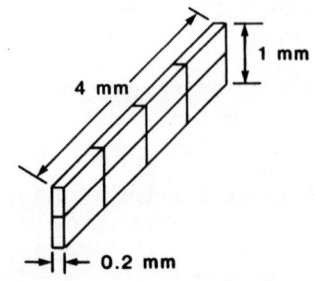

FIGURE 1. Example 1. Plan view, cross-section and discretization

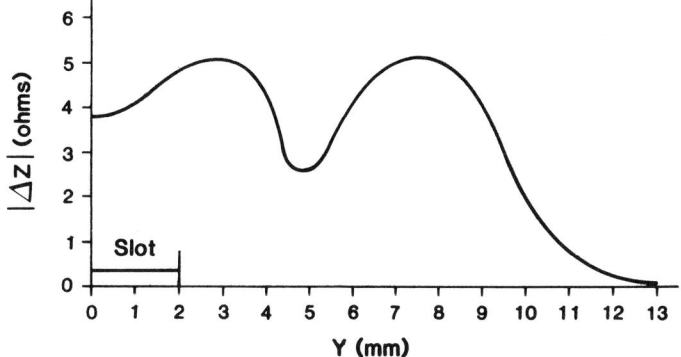

FIGURE 2. Example 1. Impedance change profile along y axis

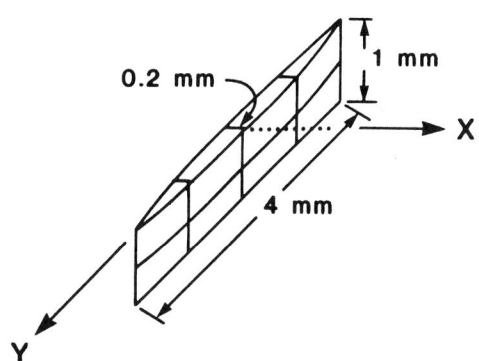

FIGURE 3. Example 2. Discretization

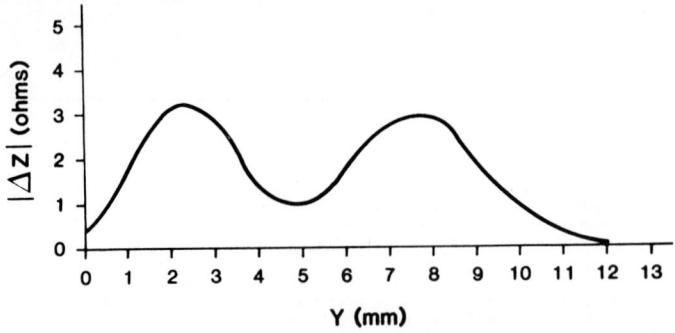

FIGURE 4.  Example 2.  Impedance change profile along y axis

Very little time was required to set up the data for each model. Although the method is economical with respect to computer storage, computing times are long. Most of the computer time is spent evaluating Hankel transforms associated with the incident field, $E^\circ$, and the magnetic field, B, in the coil. Use of Equation 14 for the impedance calculation would undoubtedly lower computing time requirements.

VOLUME INTEGRAL-FINITE ELEMENT HYBRID

For some geometries (e.g. near a corner) it is impossible to derive an analytical solution for the incident electric field or for Green's tensor. An alternative is the finite element method. However, in order to model the boundary conditions at infinity, the mesh of a finite element model must be extended out to large distances where the electric field may be assumed negligible. A hybrid finite element technique, which uses the Green's tensor of the volume integral method to satisfy the boundary conditions at infinity, is described in this section.

The system of equations associated with a finite element model is

$$KE = S \tag{15}$$

where K is a 'diffusion matrix', E is a vector of electric field components at the nodes of the finite element mesh and S is a source vector. The matrix K is sparse, banded and symmetric. The vector S is zero everywhere except in the elements comprising the source (probe coil).

Equation 15 may be partitioned as follows:

$$\begin{bmatrix} K_{vv} & K_{vb} \\ K_{bv} & K_{bb} \end{bmatrix} \begin{bmatrix} E_v \\ E_b \end{bmatrix} = \begin{bmatrix} S_v \\ S_b \end{bmatrix} \tag{16}$$

where the subscripts v and b denote the interior and boundary respectively of the finite element mesh as shown in Figure 5. The first equation in the system 16 may be written

$$K_{vv}E_v = S_v - K_{vb}E_b \;. \tag{17}$$

Within the finite element mesh and on its boundary, the volume integral method may be used to compute $E_b$ by means of Equation 8 (using the notation of Equation 10).

$$E_b = E_b^\circ + CE_v \quad . \tag{18}$$

Substituting Equation 18 into Equation 17 gives

$$(K_{vv} + K_{vb}C)E_v = S_v - K_{vb}E_b^\circ \tag{19}$$

which may be solved for Ev. Equation 18 may then be used to compute the electric field in the probe if it lies outside the finite element mesh.

Owing to the presence of the matrix C, the matrix $K_{vv} + K_{vb}C$ is asymmetric and tends to be fully populated. However, it is likely to be of relatively small dimension compared to K.

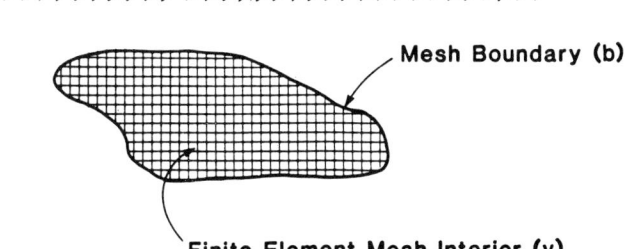

FIGURE 5. Exterior and interior regions of a hybrid mode.

Lee et al [7] proposed an iterative scheme that avoids the inversion or decomposition of asymmetric matrices. The iteration proceeds as follows:

1) Assume a value for $E_b$, say $E_b^*$,
2) Solve for $E_v$ from Equation 17,
3) Compute new $E_b$ from Equation 18,
4) If the difference between $E_b$ and $E_b^*$ is greater than a prescribed tolerance, set $E_b^* = E_b$ and go to 2. Otherwise quit.

Lee et al [7] found that when the conductivity contrast between the interior and exterior regions is greater than 1000, the rate of convergence is extremely slow or divergence can occur. This might pose a problem in some applications. However, a similar iterative algorithm is used in the stress analysis of solids. The iteration proceeds by adjusting displacements at the boundary between the exterior and interior regions. When the ratio of the elastic moduli of the interior and exterior regions exceeds a critical value, slow convergence or divergence occurs. The remedy is to reformulate the problem in terms of stresses at the boundary between the interior and exterior regions. Thus, if one analogizes the electric field E with displacements, derivatives of E (the magnetic field) with stresses and conductivity with elastic modulus, the problem should be reformulated in terms of the magnetic field when slow convergence occurs. This is somewhat speculative and needs to be tested.

ACKNOWLEDGEMENTS

This research was supported by the Canadian Defence Research Establishment, Pacific Division. (DREP) under Contract Serial No. 8SB84-00574 to Barrodale Computing Services Ltd. of Victoria, B.C.

REFERENCES

1. G. W. Hohmann, Three Dimensional Induced Polarization and Electromagnetic Modeling, Geophysics, 40, 309-324, (1975).
2. P. Weidelt, Electromagnetic Induction in Three Dimensional Structures, Journal of Geophysics, 41, 85-109, (1975).
3. W. S. Dunbar, The Volume Integral Method of Modelling Eddy Current Test Data, Report to Canadian Defence Research Establishment, Pacific. Contract NO. 8SB84-00574, (1986).
4. T. G. Kincaid, M. V. K. Chari, Z. J. Csendes, K. Fong and R. O. McCary, Two Approaches to Solving the Inversion Problem for Eddy Current NDE, in "Proceeding of the DARPA/AFML Review of Progress in Quantitative Nondestructive Evaluation", (Thousand Oaks, CA, 1980).
5. C. V. Dodd and W. E. Deeds, Analytical Solutions to Eddy Current Probe Coil Problems, Journal of Applied Physics, 39, 2829-2838, (1968).
6. B. A. Auld, S. Ayter, F. C. Muennemann and M. Riaziat, Eddy Current Signal Calculations for Surface Breaking Cracks, in "Review of Progress in Quantitative Nondestructive Evaluation, 3", D. O. Thompson and D. E. Chimenti, eds., (Plenum Press, New York, 1984).
7. K. H. Lee, D. F. Pridmore and H. F. Morrison, A Hybrid Three Dimensional Electromagnetic Modeling Scheme, Geophysics, 46, 796-805, (1981).

EDDY CURRENT INDUCTION BY A COIL NEAR A CONDUCTING EDGE IN 2D


S.K. Burke

Aeronautical Research Laboratories
Department of Defence
Melbourne, Australia.


INTRODUCTION

Difficulties often occur in eddy current inspection when a defect is situated close to an edge. This is due to the large signal arising from the edge ('edge-effect') which tends to obscure any signals coming from the defect[1,2]. To understand the edge-effect in eddy current NDI in more detail, the induction of eddy currents in a conducting 90° edge by a two-dimensional coil is considered in this paper. Firstly, the impedance of a two-dimensional coil in the vicinity of a conducting quarter-space is calculated numerically using a dual boundary-integral-equation (BIE) method and the results compared with experiment. The behaviour of the induced currents (vector potential) in the vicinity of the edge is then examined using the numerical (BIE) results and the results of an analytical approximation [3] valid in the limit of small skin-depth.

NUMERICAL CALCULATION OF COIL IMPEDANCE

The calculation of coil impedance was performed for the case shown in Fig. 1. A two-dimensional coil consisting to two identical rectangular windings is located near a uniform quarter-space ($x \geq 0$, $y \leq 0$) with conductivity $\sigma$ and magnetic permeability $\mu_0$. The coil windings carry equal and opposite alternating current $\pm Ie^{i\omega t}$, the centres of the windings are separated by a distance 2D and the windings have a turn density N/(4WT). The base of the coil is parallel to the top surface of the conductor, the coil 'liftoff' is H and the coil centreline is situated at a distance XC from the edge. In this two-dimensional idealization, the coil and quarter space are infinite in length normal to the X-Y plane; the coil windings may be visualized as a superposition of line-current sources.

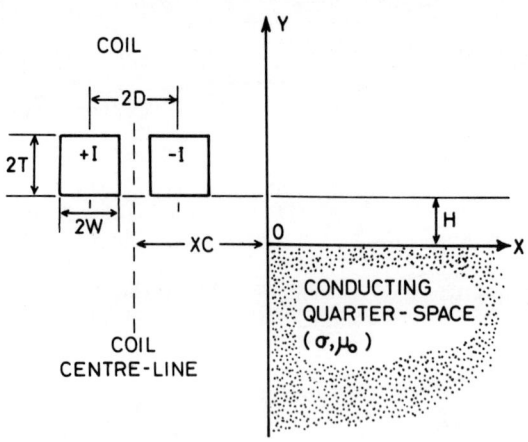

Fig. 1. Two-dimensional coil conducting quarter-space. Here XC is defined as positive when the coil is over the conductor.

The calculation of coil impedance closely follows the BIE method described by Kahn[4], who reduces the problem of obtaining the quasi-static electric and magnetic field to that of solving a pair of simultaneous integral equations for the z component of the magnetic vector potential (A) and its normal derivative ($\partial A/\partial n$) on the boundary of a conductor. These equations are,

$$\tfrac{1}{2} A(s) - \int_B ds' \frac{\partial G(s,s')}{\partial n'} A(s') + \int_B ds' G(s,s') \frac{\partial A}{\partial n'}(s') = A^\circ(s) \qquad (1)$$

$$\tfrac{1}{2} A(s) + \int_B ds' \frac{\partial g(s,s')}{\partial n'} A(s') - \int_B ds' g(s,s') \frac{\partial A}{\partial n'}(s') = 0 \qquad (2)$$

where

$$G(s,s') = -(2\pi)^{-1} \ln |s-s'|, \quad g(s,s') = (2\pi)^{-1} K_o (k|s-s'|), \quad k = (i\omega\mu_o\sigma)^{\tfrac{1}{2}}$$

and $A^\circ(s)$ is the (known) incident vector potential on the boundary. The integrations are over the conductor boundary (B) and the normal derivative is defined to be outward from the conductor. The coil impedance per unit length, Z, can be expressed in terms of A and $\partial A/\partial n$ obtained from the solution of equations 1-2 as follows,[5]

$$Z = \frac{i\omega}{\mu_o I^2} \int_B ds \left[ A(s) \frac{\partial A^\circ(s)}{\partial n} - A^\circ(s) \frac{\partial A(s)}{\partial n} \right] + Z_o \qquad (3)$$

where $Z_o$ is the coil impedance (per unit length) for the coil in isolation.

A system of coupled integral equations such as eq.1-2 can only realistically be solved by numerical methods and a variety of numerical techniques of greater or lesser sophistication may be used. All of these methods, however, rely on reducing the coupled integral equations to a set of simultaneous linear algebraic equations. The method which is adopted is similar to that employed by Kahn [4], using a point matching scheme but with triangular hat basis functions rather than triple-pulse hat basis functions.

To implement this scheme, a grid of variable size was used. Close to the edge and in the immediate vicinity of the coil a fine mesh was employed, whereas at large distances from the edge a coarse grid was used. A grid of intermediate width was used elsewhere and no grid point was placed on the edge itself. For satisfactory convergence it was necessary to extend the grid to large distances from the edge. The diagonal matrix elements and matrix elements involving $G(s,s')$ were calculated from exact analytical results. Accurate series and asymptotic expansions were used to calculate the off-diagonal matrix elements involving $g(s,s')$. Recalling that A and $\partial A/\partial n$ are phasors (complex) it proved convenient to code the system of simultaneous equations as COMPLEX variables and a Nag library routine [6] (f04adf) was used to solve this system numerically. The coil impedance was then calculated via eq.3, approximating the integral by using the trapeziodal rule. No significant change in the solution was found on increasing n above 200 points for a typical set of grid points.

## EXPERIMENTAL VERIFICATION

The validity of these calculations was tested experimentally by measuring the impedance of a 'two-dimensional' coil in the vicinity of a 90° conducting edge. A rectangular coil of length 40.1cm was wound with N=158 turns of 'litz' wire. The coil dimensions (see Fig.1) were: 2D=1.02cm, 2W=0.84cm and 2T=0.95cm. As the coil had a large aspect ratio (40) it was essentially two-dimensional[3]. The inductance of the coil in isolation ($L_0$) was 3.50mH, in resonable agreement with the theoretical value of 3.87mH calculated for a two-dimensional coil of this size.

The experiments were performed on a large aluminium alloy block (1m x 1m x 11cm) with a resistivity of 5.4 $\mu\Omega$ cm. The coil impedance was measured, using an HP-4192A low-frequency impedance analyser, as the coil was translated along the top-surface of the block and across the edge with the base of the coil maintained at constant Y=H=1.3mm (refer Fig.1.) Data were collected at frequencies of 25kHz and 1kHz, corresponding to skin-depths ($\delta=\sqrt{2}/|k|$) of 0.74mm and 3.7mm respectively. The results are shown in Figures 2-3 and are in excellent agreement with the (parameter-free) numerical calculations.

## THE EDGE EFFECT AND INDUCED FIELDS

The most significant feature of these results is that, while the coil reactance increases monotonically as the coil is translated across the edge, the resistance peaks before decreasing. This is the so-called 'edge-effect'.

This enhancement of the coil resistance can be traced to a concentration of the induced fields at the edge itself, analogous to the concentration of charge near a sharp edge in electrostatics. This is illustrated in Fig. 4 where the variation of the magnetic vector potential on the conductor surface, obtained from the numerical results for the 25kHz data set, is shown for selected values of coil center translation (XC). Two calculated curves are also shown here. The first represents the results of an approximate analytical calculation[3] for the vector potential on the conductor surface to first order in $\delta$, $A^*(1-i)\delta A''+..$ This approximation is in very good agreement with the numerical results for Re(A), the poorer agreement with Im(A) reflects the fact that the next term in the expansion for Im(A) is of order $\delta^2$ whereas the next term for Re(A) is of order $\delta^3$ for small $\delta$. It should be noted that the vector potential in this small $\delta$ limit has a singularity of order -1/3 at the edge. The second set of curves in Fig. 4 represent the vector potential for the coil on an infinite half-space and are the result of an exact analytical calculation.[7]

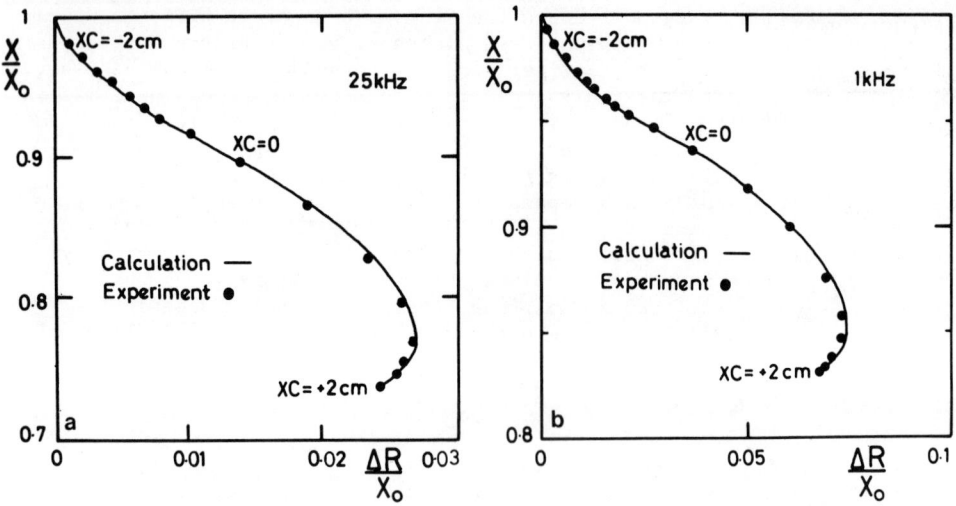

Fig. 2. Edge-effect loci in the normalized impedance plane for frequencies of (a) 25kHz and (b) 1 kHz. The solid circles are the experimental results and the curves are the results of the numerical calculation. The coil reatance (X) and change in coil resistance ($\Delta R$) are normalized to the isolated coil reactance ($X_o = \omega L_o$) following the usual convention.

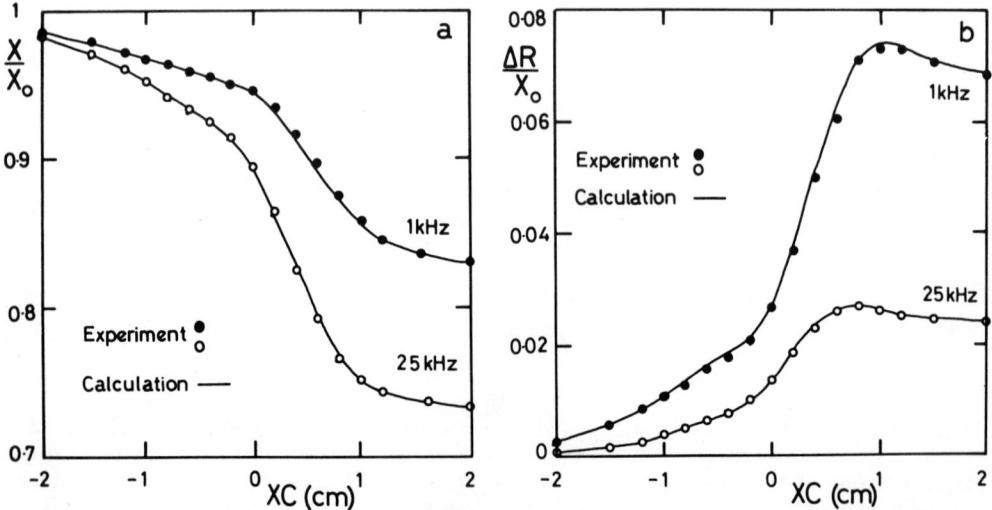

Fig. 3. Normalized coil reactance (a) and resistance (b) as a function of coil translation (XC) near a 90° edge for frequencies of 25kHz and 1kHz. The circles are the experimental data and the solid lines are the results of the numerical calculation.

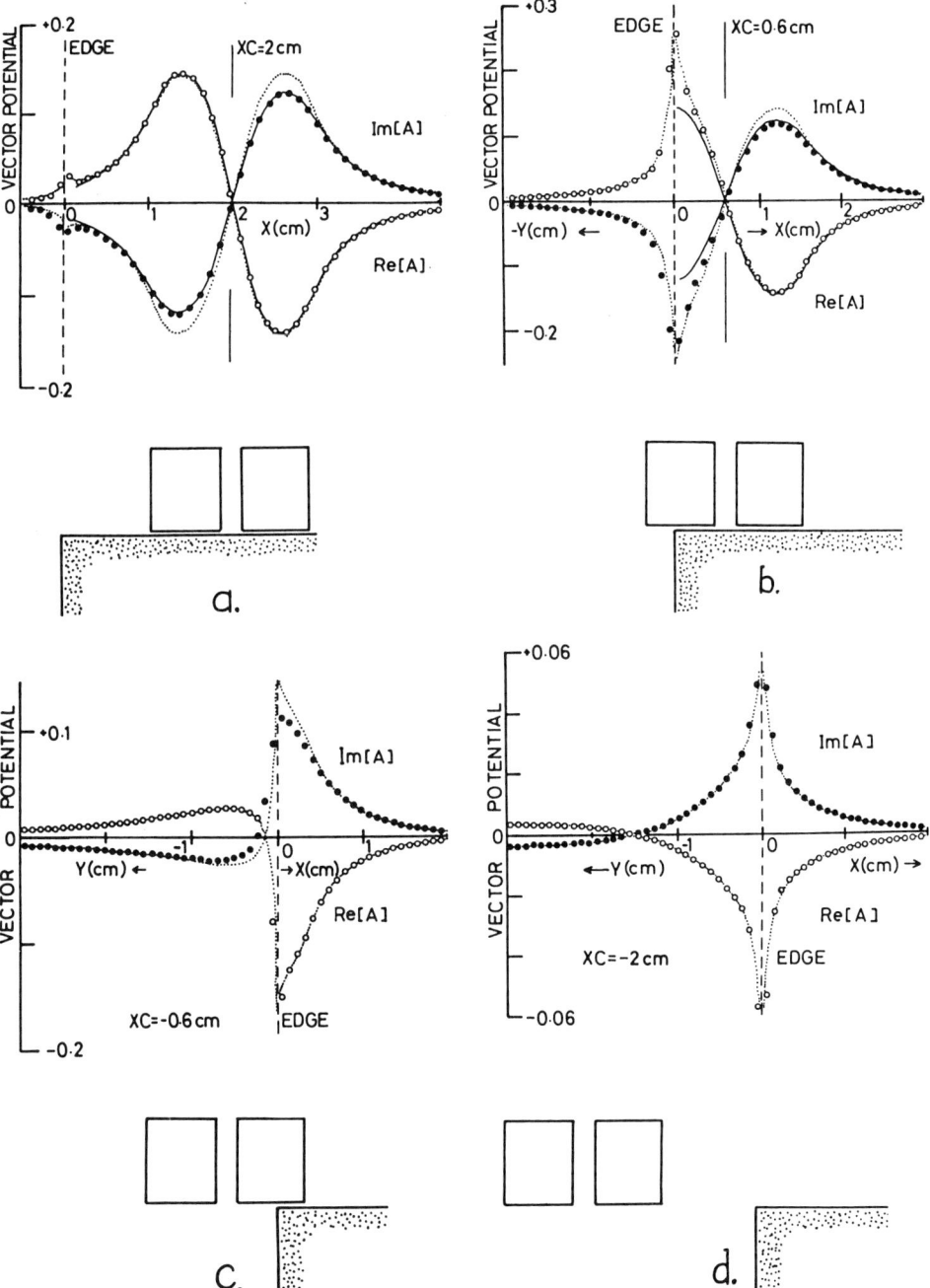

Fig. 4. Calculated vector potential on the surface of the aluminium alloy block. Real (o) and Imaginary (•) parts of A obtained numerically for the 25kHz data-set are plotted as a function of surface position for coil centre positions (XC) of (a) 2cm (b) 0.6cm (c) -0.6cm and (d) -2.0cm. Also shown are the results of the small skin-depth approximation for a 90° edge (dotted line) and the exact analytical calculation the coil on a half-space (solid line). Note the changes in scale.

The variation of vector potential on the conductor surface as the coil is translated across the edge may be described as follows. For XC=2cm, the coil is resting on the block at some distance from the edge and the vector potential on the top surface (X>0) is indistinguishable from that obtained from the exact calculations for a 2D coil in a half-space [7], except for a small increase close to the edge. This edge contribution increases significantly in size as the coil is moved closer to the edge of the block. This is illustrated in Fig. 4b where XC=0.6cm and the coil resistance is close to its maximum value. At this position the left-hand coil winding is centred on the edge, the incident vector potential at the edge is thus a maximum and a large edge response could be expected. The vector potential on the top surface approaches the half-space values for X>XC (i.e. beyond the coil centre) but is significantly larger near the edge. As the coil is translated across the edge and off the block, the vector potential decreases rapidly. When the right-hand coil winding is centred on the edge (XC=-0.6cm) an enhanced edge contribution is observed, corresponding to the weak 'shoulder' seen in the coil resistance-position curves in Fig. 3b. The tendency for the vector potential (or induced current) to be concentrated near the edge is well illustrated in Fig.4d, where the coil is distant from the edge (XC=-2cm).

To account for the variation of coil impedance with position it is convenient to turn to an alternative expression [8],

$$Z = \frac{i\omega}{I} \frac{N}{4WT} [\iint_{(+)} A \, dxdy - \iint_{(-)} A \, dxdy] \qquad (6)$$

which relates the impedance to the integral of the vector potential over the coil windings (denoted (+) and (-)), rather than using the surface integral formulation of equation 3. This expression, taken with the fact that Im(A) decays smoothly out from the conductor surface, implies that the peak in the coil resistance is a direct consequence of the enhancement of Im(A) at the conductor edge. The coil reactance, however, shows no such peak and increases rapidly as the coil is translated across the edge. This difference in behaviour is most easily understood by considering the small $\delta$ expansion for A outside (and on) the conductor [3],

$$A = A' + (1-i)\delta A'' + \ldots \qquad (7)$$

$A'$, the vector potential in the perfect conductor limit ($\delta=0$), is purely real and is <u>zero</u> on the conductor surface. The second term, which consists of equal and opposite real and imaginary parts, was shown for selected values of XC in Fig 4. Re(A) at the coil windings, which determines the coil reactance is dominated by $A'$ and any edge contributions from $A''$ show up as a weak 'shoulder' in the reactance-position curve.

These results for a 2D coil are relevant to those 3D geometries in which a significant proportion of induced current flows parallel to a conducting edge, as for example is the case for a pancake coil resting flat on the conductor surface and with the coil centre not too close to the edge.

CONCLUSION

The impedance of a two-dimensional coil in the vicinity of a homogenous conducting quarter-space was calculated using a dual boundary-integral-equation method. The results were in excellent agreement with a series of experiments performed using a rectangular coil of large aspect ratio which approximated to a 2D coil. The behaviour of the vector-potential on the conductor surface was considered in detail and the maximum in coil resistance was traced to the concentration of A near the edge.

REFERENCES

1. B. C. Bishop and N. T. Goldsmith, Non-Destructive Testing Australia 21 2 6-8 (1984).
2. D. J. A. Williams, J. P. Tilson and J. Blitz, Effects of the Edges of Samples on Eddy-Current Flaw Evaluation, in Proceedings of Fourteenth Symposium on Nondestructive Evaluation, (San Antonio, Texas, 1983).
3. S. K. Burke, J. Phys. D: Appl. Phys. 18 1745-60 (1985).
4. A. H. Kahn, J. Res. National Bureau of Standards 89 47-54 (1984).
5. B. A. Auld, Theoretical Characterization and Comparison of Resonant Probe Microwave Eddy-Current Testing with Conventional Low Frequency Eddy-Current Methods, in Eddy-Current Characterization of Materials and Structures, ASTM STP 722, eds., G. Birnbaum and G. Free, (American Society for Testing and Materials, Philadelphia, 1981), 332-347.
6. Nag Fortran Library MK 11, Numerical Algorithms Group, 256 Banbury Rd., Oxford, OX2 7DE UK.
7. Induction by a current filament above a half-space is treated by many authors (e.g. J. A. Tegopoulos and E. E. Kriezis, "Eddy Currents in Linear Conducting Media. Studies in Electrical and Electronic Engineering 16" (Elsevier, Amsterdam, 1985), p. 89ff). Results for a coil of finite cross-section are obtained by superposition.
8. C. V. Dodd and W. E. Deeds, J. Appl. Phys. 39 2829-38 (1968).

3-D EDDY CURRENT NONDESTRUCTIVE TESTING

MODELING FOR SURFACE FLAWS

R. Grimberg*, M. Mayos*, and A. Nicolas**

*IRSID, 78105 St. Germain En Laye Cedex, France
**ECL, Dpt. Electrotechnique, UA CNRS 829
BP 163, 69131 Ecully Cedex, France

INTRODUCTION

In eddy current non-destructive testing, numerical models have been used for a long time to optimize the method. Those models solve Maxwell's equations and compute the coil impedance. Many 2-D programs have been realized:
- for axisymmetrical geometries, with axisymmetrical or very small flaws, to model tube sheets of a steam generator or wires [1,2].
- for an E-shape sensor which generates a one dimensional field to scan a plane surface [3].

Unfortunately, a lot of actual problems can't be solved with such approximations. Two important examples for industrial NDT in steel plants can be mentioned:
- transverse cracks on steel slabs shown in Fig. 1;
- long cracks on wires.

To study such cases, a 3-D model is necessary because, even if a 2-D computation gives interesting results for a test piece without defect, the symmetry is broken by the presence of a flaw.

Full 3-D programs have been realized [4,5], but are not used for ferromagnetic materials and are very expensive in CPU time (they are implemented on vectorial computers, Cray I or Cyber 205). The aim of our study is to realize a 3-D program for small skin depth problems which can be performed on a mini-computer (HP 9000, VAX 780) with a "reasonable" CPU time. For ferromagnetic materials, the frequencies used in NDT are generally at least 10 kHz: so, the skin depth $\delta$ is very small and negligible compared to the other dimensions of the problem. For example,

for ordinary steel, $\sigma = 10^7 \, \Omega^{-1} \cdot m^{-1}$ $\mu r = 100$
for $f = 10$ kHz, and $\delta = 0.2$ mm.

This paper presents a specific formulation using a surface impedance condition for small skin depth.

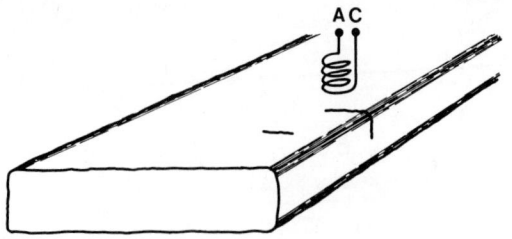

Figure 1. Transverse cracks on steel slab.

THE SURFACE IMPEDANCE BOUNDARY CONDITION

To reduce the computation time, only the exterior magnetic field will be computed. In air, curl $\vec{H}$ is zero so that $\vec{H} = -\,\mathrm{grad}\,\phi$, where $\phi$ is a scalar potential. We must then find a boundary condition at the surface of the conducting material which contains all the information about the metal.

The surface impedance is well known in acoustic and electromagnetic wave propagation. It has been used for eddy currents modeling [6] with a finite element resolution. In this work, the surface impedance leads to a boundary condition for a boundary integral equation (BIE) resolution. Some hypotheses are necessary but are not restrictive for many eddy current NDT problems. They are:

- $\delta$ is small compared to the other geometrical data.
- Each variable ($\vec{E}, \vec{H}, \vec{B}, \vec{J}, I,...$) is sinusoidal.
- The radius of curvature of the interface is much larger than $\delta$.
- In the local coordinate system (s1, s2, zn), with the previous hypothesis and in the conducting material, $\vec{H}$ (resp. $\vec{E}, \vec{J}, \vec{B}$) decreases with depth according to equation (1):

$$\vec{H}(s1, s2, zn) = \vec{H}o(s1, s2)\exp\left(-\frac{1+j}{\delta}zn\right) \qquad (1)$$

The physical meaning of the surface impedance condition is to replace, as in [6], the eddy currents in the metal by a surface current density $\vec{J}s$, given by

$$\vec{J}s = \frac{\sigma\delta}{1+j}\vec{E}ot \qquad (2)$$

in which $\vec{E}ot$ is the tangential electric field on the surface and

$$z_s = \frac{1+j}{\sigma \delta} \qquad (3)$$

is the surface impedance.

The air-metal interface relationship between the tangential fields in the air $\vec{E}t$ and $\vec{H}t$ becomes:

$$\vec{E}t = z_s \vec{H}t \wedge \vec{n} \qquad (4)$$

Relation (4) is introduced in Maxwell's equations. The final result is expressed in air in terms of the total scalar magnetic potential $\phi$:

$$\vec{H} = -\vec{\text{grad}}\ \phi. \qquad (5)$$

The impedance boundary condition is then obtained:

$$\frac{\partial \phi}{\partial n} = \frac{1+J}{\delta \mu r} \phi \qquad (6)$$

With this boundary condition at the surface of the metal, Maxwell's equations can be solved outside the conducting parts only.

RESOLUTION BY A BIE METHOD

The total exterior magnetic field $\vec{H}$ is computed in two steps:

$$\vec{H} = \vec{H}o + \vec{H}i \qquad (7)$$

where $\vec{H}i$ is the induced magnetic field and $\vec{H}o$ the magnetic field generated by the eddy current probe alone in the air, without any conducting material. $\vec{H}o$ is calculated analytically (using Biot-Savart's law for example). Curl $\vec{H}i$ is zero in the air and $\vec{H}i = -\vec{\text{grad}}\ \phi i$, where $\phi i$ is the induced magnetic potential and $\Delta \phi i = 0$.

$\phi i$ is the unknown function and Laplace's equation, with the surface impedance boundary condition, is solved in 3-D by a classical BIE method (we have implemented the surface impedance boundary condition in the BIE package 'PHI3D' [8] developed in the electrical engineering department of Ecole Centrale de Lyon).

Why do we choose the induced potential as the unknown function? The variation of the magnetic field due to the flaw is very weak. Computing an induced potential instead of a total value gives the best accuracy because the flaw modifies only the induced field.

RESULTS OF THE FIELD COMPUTATION

An example of the results of the field computation obtained to validate the method is presented. The best piece is a steel cylinder (conductivity $10^7\ \Omega^{-1}.m^{-1}$, permeability 100) whose geometry is shown on Fig. 2. The probe is a single turn coil, fed with 10 kHz alternating current, centered on the axis of the cylinder. Two symmetries are used for the 3-D computation. Figure 3 shows the mesh of a quarter of the test piece. This mesh has 115 nodes and 32 second order quadrilateral elements. On each node, the unknown function is the scalar potential $\phi i$, so that there are as many unknowns as nodes. The result of the comparison between the 3-D computation and an axisymmetric program is shown on Fig. 4, Hn being the normal magnetic field on a radius of the top surface of the cylinder.

The good agreement between the two computations shows that the surface impedance boundary condition, with a BIE resolution, is well adapted to model small skin depth problems and can be used to acknowledge a lot of NDT phenomena.

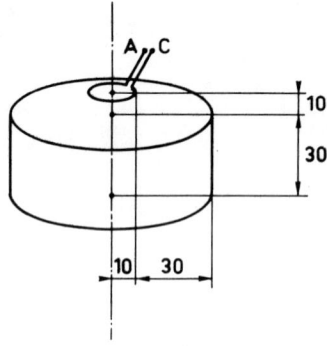

Figure 2. Geometry of the test piece.

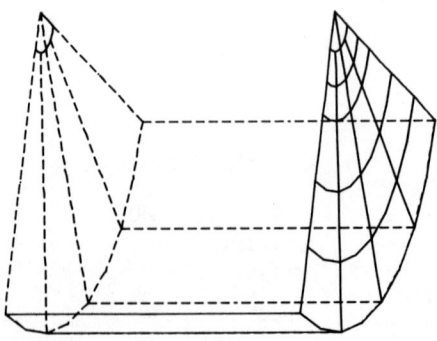

Figure 3. 3-D mesh of a quarter of cylinder.

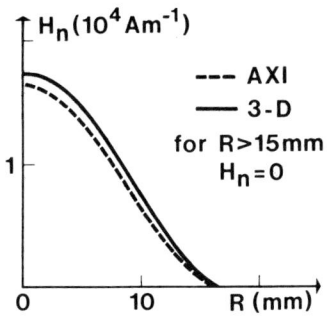

Figure 4. Comparison of 3-D and axisymmetrical results.

## ADVANCES IN IMPEDANCE COMPUTATION AND FLAW MODELING

Impedance Computation

Different methods are generally used to compute the probe impedance: energy calculation, theorem of reciprocity, flux calculation. The first method is used by [3 and 5] but it appears to be interesting only when the exterior induced field is zero [3]. With our method, only the energy dissipated in the metal is easy to compute, but it does not lead to impedance values. The theorem of reciprocity will be programmed because it is well adapted to our computation (the impedance is calculated from the values of the fields on the surface of the crack and those values are easy to obtain from our computation). In order to simulate separate functions probes, a flux calculation is essential because the two other methods do not take into account the geometry and the position of the receiver. So the flux calculation is the most interesting way to compute the sensor impedance, but unfortunately, the most expensive in CPU time.

Flaw modeling

The hypothesis of large radius of curvature is not true for corners, flaws or cracks. According to [9], the error due to a corner can be neglected. Figure 5 shows the current lines on a perfect conductor with a flaw, computed with 'phi3d'. For a real crack, we are developing a special crack element to take into account the influence of the crack on the surface of the metal, without computing the magnetic field inside the crack.

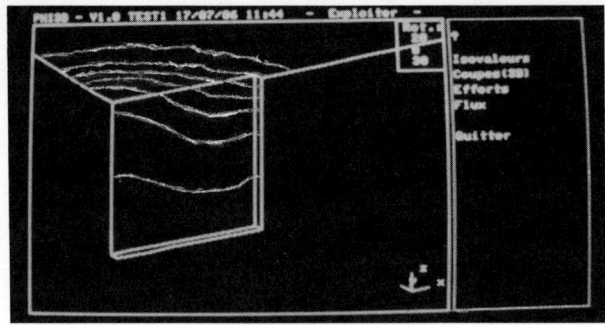

Figure 5. Current lines on a perfect conductor with a flaw.

CONCLUSION

The surface impedance boundary condition has been presented for small skin depth eddy current problems encountered in NDT phenomena. The model was tested and compared to an axisymmetrical program, showing good agreement. Different methods are tested to compute the probe impedance. The flux calculation seems to be the best one to simulate the largest range of sensors, especially separate functions probes. The model will be used on a test object with a crack and compared with experimental data to carry out the model of the crack.

ACKNOWLEDGEMENTS

The authors would like to thank Q. S. Huang and H. Mansir (ECL) for many useful technical discussions and L. Kranbuehl and L. Nicolas (ECL), authors of 'phi3d'.

REFERENCES

1. R. Palanisamy, W. Lord, "Prediction of eddy current probe signal trajectories", IEEE Trans. on Mag., Vol. MAG-16, (September, 1980), 1083-1085.
2. C. V. Dodd, W. E. Deeds, W. G. Spoeri, "Optimizing defect detection in eddy current testing", Material evaluation, (March, 1971), 59-63.
3. N. Burais, A. Foggia, A. Nicolas, J. C. Sabonnadiere, "Electromagnetic field formulation for Eddy Current calculations in non-destructive testing systems", IEEE Trans. on Mag., Vol. MAG-18, (November, 1982), 1058-1061.
4. A. Bossavit, J. C. Verite, "The TRIFOU code: solving the 3-D eddy current problem by using H as state variable", IEEE Trans. on Mag., Vol. MAG-19, (November, 1983), 2465-2470.
5. N. Ida, "Development of a 3-D eddy current model for nondestructive testing phenomena", Review of Progress in Quantitative NDE 3, (Santa Cruz, 1983), 547-554.

6. T. W. Preston, A. B. J. Reece, "Solution for 3 dimensional eddy current problems: the T-O method", IEEE Trans. on Mag., Vol. MAG-18, (March, 1982), 486-491.
7. C. Vassalo, "Electromagnetisme classique dans la matiere", Dunod Universite, (1980), 58-65.
8. L. Kranbuehl, L. Nicolas, A. Nicolas, "A graphic interactive package for boundary integral equations", IEEE Trans. on Mag., Vol. MAG 21, (November, 1985), 2555-2558.
9. S. Ratnajeevan, H. Hoole, C. J. Carpenter, "Surface impedance models for corners and slots", IEEE Trans. on Mag., Vol. MAG-21, (September, 1985), 1841-1843.
10. R. Grimberg, "Modelisation tridimensionnelle en CND par courants de Foucault de defauts de surface", internal report IRSID MCA-SG 965, (Juillet, 1986).
11. R. Grimberg, H. Mansir, J. L. Muller, A. Nicolas, "A surface impedance boundary condition for 3-D nondestructive testing modeling", INTERMAG 1986, IEEE Trans. on Mag., Vol. MAG-22, (Setember, 1986).

EDDY-CURRENT PROBE INTERACTION WITH SUBSURFACE CRACKS

John R. Bowler

Department of Physics
University of Surrey
Guildford
Surrey
GU2 5XH
United Kingdom

INTRODUCTION

Electric current will flow around on open crack in a conductor and give rise to very abrupt variations in the field. If the crack has a negligible opening it acts as a surface barrier where the field is virtually discontinuous. Effectively the crack is then equivalent to a layer of current dipoles with the dipole orientation normal to the surface and pointing upstream. An integral equation for the dipole density has been derived for an idealised subsurface crack using the Green's function method [1]. Numerical solutions have been found by assuming a piecewise constant dipole density and satisfying boundary conditions on the crack at a finite number of points. Here we shall develop the theory further, making use of a knowledge of the dipole distribution for a given incident field, to calculate probe impedance changes $\Delta Z$, due to subsurface cracks.

An advantage of the present approach is that the unperturbed incident field may be found quite independently of the scattering problem. Here we consider axially symmetric probes and use both analytical and finite element methods to calculate the fields in the absence of defects. There are well-known closed form expressions for the field of a cylindrical air-cored probe whose axis of symmetry is normal to the surface of a conductor [2]. However the usual analysis is incomplete. By taking the derivation a stage further the air-cored probe field is given by integral expressions containing Struve functions. We also determine $\Delta Z$ for a probe with a ferrite core and a probe with a ferrite core and ferrite shield. The unperturbed fields in these cases being calculated using a general purpose two-dimensional finite element code [3].

FIELD EQUATIONS

As is common in scattering problems, we formally write the total electric field as the sum of an incident and a scattered field. Thus, suppressing the time harmonic phase factor $e^{-i\omega t}$, the electric field in air (j = 1), and in a half space conductor (j = 2) is

$$\underline{E}_j(\underline{r}) = \underline{E}_j^i(\underline{r}) + \underline{E}_j^s(\underline{r}) \qquad j = 1,2 \qquad (1)$$

$\underline{E}_j^i$ being the field in the absence of the defect and $\underline{E}_j^s$ the scattered field.

Assuming that the scatterer is a virtually closed crack at an open surface $S_o$ completely embedded in the conductor, the scattered field may be expressed as [1],

$$\underline{E}_j^s(\underline{r}) = i\omega\mu_o \int_{S_o} \underline{G}_j(\underline{r},\underline{r}') \cdot \underline{p}(\underline{r}') \, dS' \qquad (2)$$

where $\underline{p} = \hat{n}p$ is the current dipole density on $S_o$ and $\hat{n}$ is a unit vector normal to $S_o$ (figure 1). $\underline{G}_j$ is a half-space dyadic Green's function for a source in the conductor [4]. (2) assumes that the discontinuity in the tangential field at the crack $\Delta E_t$, may be written as

$$\Delta E_t = -\frac{1}{\sigma} \nabla_t p \qquad (3)$$

where $\nabla_t$ is the gradient tangential to $S_o$, $\sigma$ being the electrical conductivity. The magnetic field is assumed to be continuous at $S_o$. For a known dipole distribution the field anywhere in the conductor can be found using (2) or the jump in the field at the defect can be determined from (3). Essentially the introduction of p reduces a three-dimensional vector field problem to one of finding a surface scalar distribution with a corresponding reduction in the computation needed.

The dipole density can be found by applying the condition that the normal component of the total electric field at the crack is zero. Thus

$$\hat{n} \cdot \underline{E}(\underline{r})^{\pm} = 0 \qquad (4)$$

where the $\pm$ sign refers to limiting values on either side of $S_o$. The

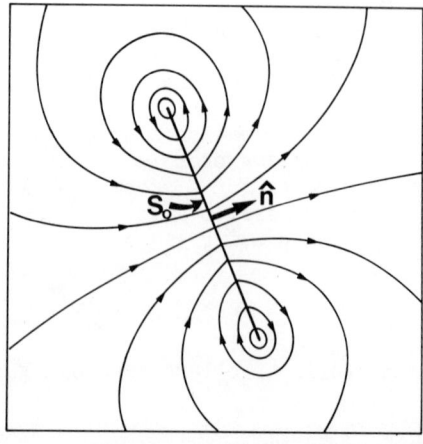

Figure 1. Scattered electric field schematic

crack faces actually acquire a small surface change in an eddy current field and hence a charging current flows normal to the crack but in the quasi-static limit this current component is negligible. Combining (1) (2) and (4) gives an integral equation for p that may be solved numerically [1] using standard boundary integral methods [5].

A relationship between the dipole density and the probe impedance change due to a crack can be derived using a reciprocity theorem. Thus

$$I^2 \Delta Z = \int_{S_o} \underline{E}^i(\underline{r}) \cdot \underline{p}(\underline{r}) \, dS \tag{5}$$

where the integration is over the surface $S_o$ and I is the probe current. This power balance relationship is a generalisation of a result obtained by Burrows for point scatterers [6].

## Incident Field

In order to determine $\Delta Z$, the incident field $E^i$ must be specified for a given source. Expressions for the field due to cylindrical air-cored probes of annular cross-section are well-known from the work of Dodd and Deeds [2], but here a modified equation is used. The electric field of an annular coil may be found by a superposition of solutions for a circular current filament [7], often referred to as a "delta function coil", or one may take a more fundamental approach and integrate over the source region using the appropriate Green's function. Either procedure leads to a radial integral that has often appeared in the literature without being evaluated in terms of standard functions. To define this integral, suppose $\rho$ is the radial source coordinate, and $a_1$ and $a_2$ the external and internal coil radii respectively, then we find [8]

$$\int_{a_2}^{a_1} \rho \, J_1(\kappa\rho) \, d\rho = a_1^2 \, \chi(a_1\kappa) - a_2^2 \, \chi(a_2\kappa) \tag{6}$$

where $J_1$ is a first order Bessel function and

$$\chi(s) = \frac{2\pi}{s} [J_1(s) H_0(s) - J_0(s) H_1(s)] \tag{7}$$

$H_0$ and $H_1$ being Struve functions. The electric field in a half-space conductor is then given by

$$\underline{E}_2^i(\rho, z) = \frac{i\omega\mu_0 I N \hat{\phi}}{(a_1 - a_2)b} \int_0^\infty \frac{1}{\kappa + \gamma} \sinh(b\kappa) [a_1^2 \chi(a_1\kappa) - a_2^2 \chi(a_2\kappa)]$$

$$J_1(\kappa\rho) \, e^{\gamma z - \kappa h} \, d\kappa \tag{8}$$

with $\gamma = (\kappa^2 - i\omega\mu_0\sigma)^{1/2}$, where the root with a positive real part is used. 2b is the axial length of the coil, h is the height of the centre of the coil above the conductor, $\hat{\phi}$ is an azimuthal unit vector and N the number of turns.

To compare an air-cored induction coil with shielded and non-shielded ferrite cored probes, we have calculated the azimuthal field in each case using a finite element package and checked the air-cored probe results from the finite element calculations against equation (8). The accuracy is mesh-dependent but in the conductor was found to be better than 0.2%.

For each of the cases examined the coil geometry was the same (figure 2) and the same ferrite core parameters were used in modelling both shielded and non-shielded probes. The ferrite was assumed to have linear material properties, a relative permeability of 220 and a conductivity of $10^{-7}$ S.m$^{-1}$. A coil current density of 1 A mm$^{-2}$ was assumed, at a frequency of 1 MHz giving a skin depth in the test material of 0.55mm. These values correspond roughly to test conditions suitable for detecting small surface or near surface defects in nickel alloys.

Contour diagrams of the ferrite probe of the azimuthal fields are shown in figure 3. Clearly the shield has the effect of confining the field and locally increasing its intensity, particularly near the surface of the test piece immediately below the coil (figure 4a). However, below about 0.2mm from the surface, the non-shielded probe produces a greater field intensity (figure 4b), with obvious consequences for the relative sensitivity to subsurface defects.

PROBE RESPONSE

The dipole density was found as before [1], using the moment methods [5], with $S_o$ divided into n square patches and p approximated by assuming it to have a constant value $p_\alpha$, over each patch ($\alpha = 1, n$). The field equations (1) and (2) are then used to get a matrix equation for $p_\alpha$ by demanding that (4) is satisfield at n matching points at the centre of each patch. Finaly the impedance perturbation is calculated from a discretised version of (5). Thus

$$\Delta Z = \frac{A}{I^2} \sum_{\alpha = 1, n} E^i_\alpha p_\alpha \qquad (9)$$

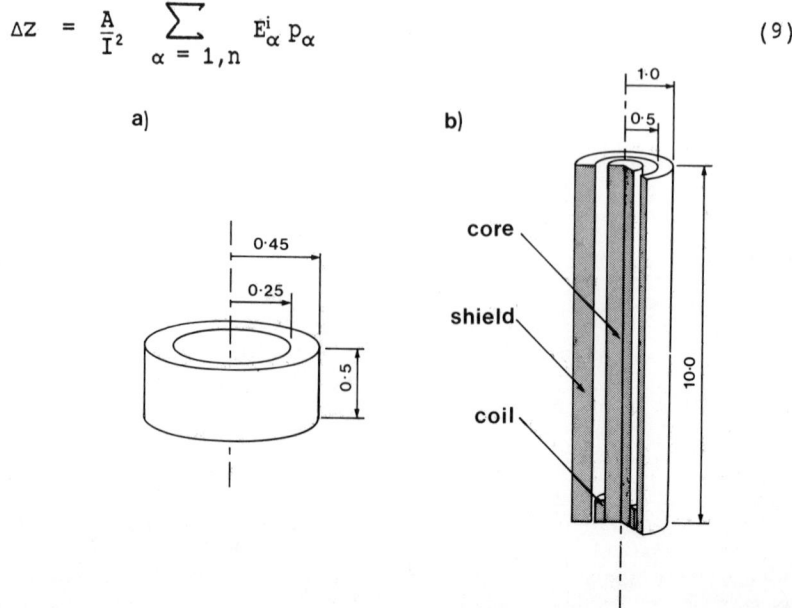

Figure 2. (a) coil, (b) shielded probe. Dimensions in mm.

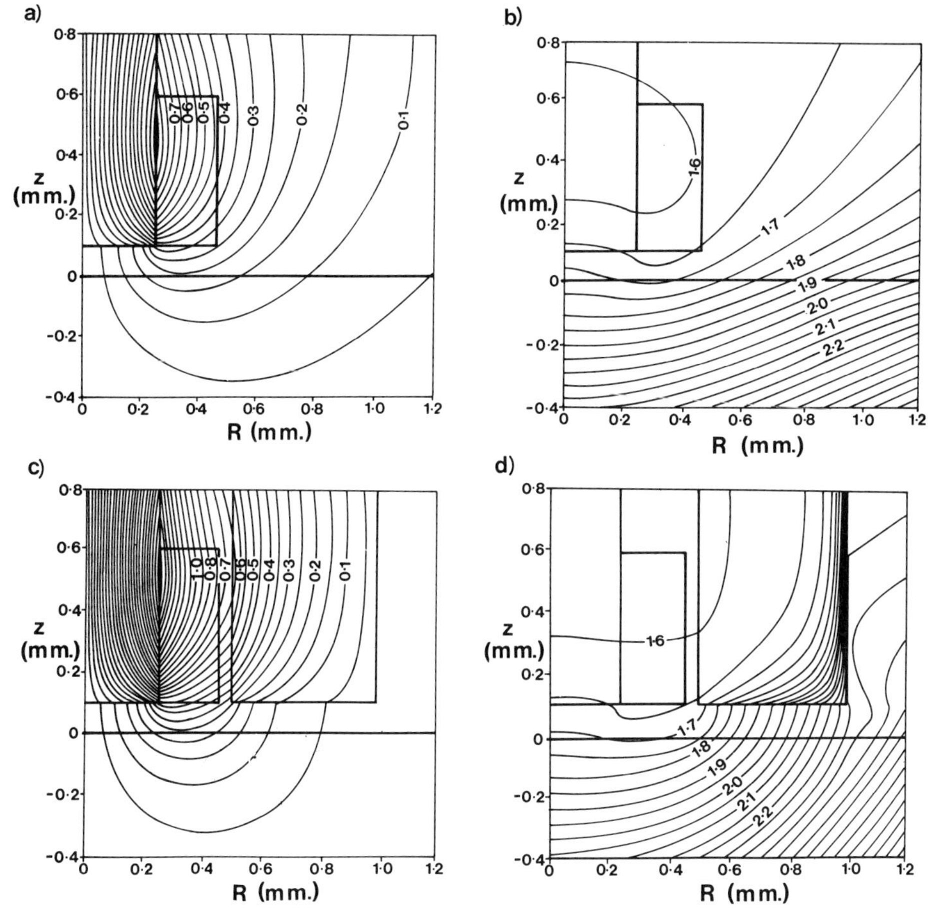

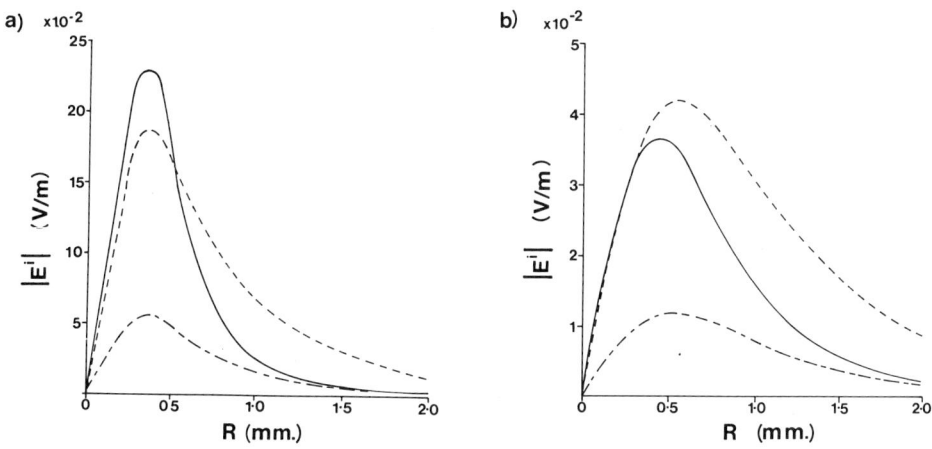

Figure 3. Electric field contours (a) modulus ($Vm^{-1}$) (b) phase (rads.) referred to source current - nonshielded ferrite probe. (c) modulus and (d) phase contours - shielded probe.

Figure 4. Modulus of electric field. Solid line - shielded probe, dashed line - nonshielded probe, long and short dashes - air-cored probe. (a) at the surface of the conductor (Z = 0) (b) below the surface (Z = -0.4mm).

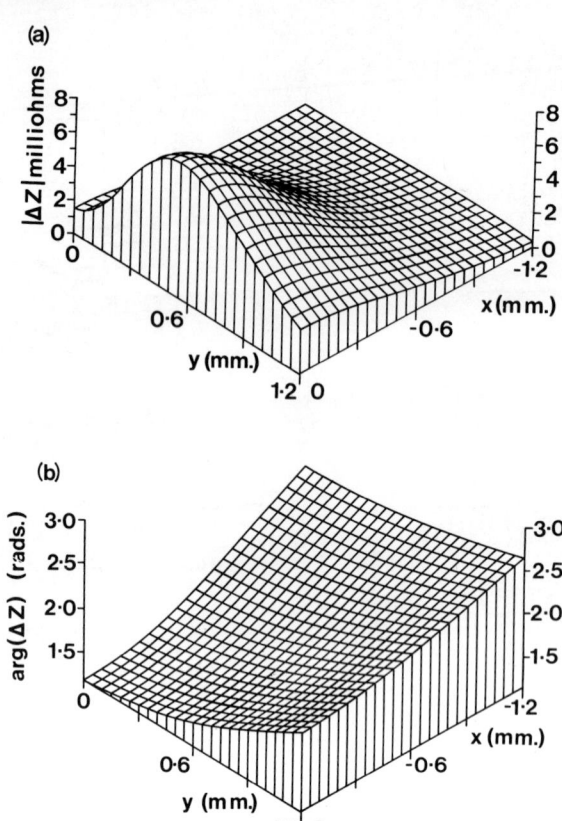

Figure 5. Impedance variations, nonshielded ferrite cored probe. Lift off 0.1mm. Crack in y-z plane.

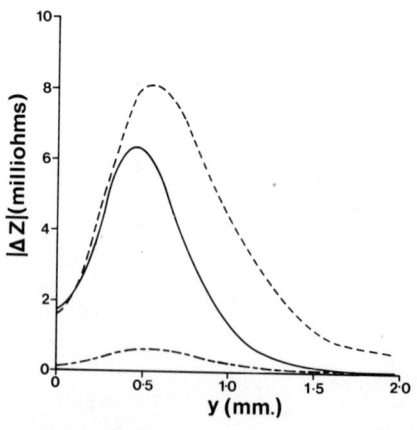

Figure 6. Impedance variations for transverse scans (x = 0). Solid line - shielded probe, dashed line - nonshielded ferrite probe, long and short dashes - air-cored probe.

where A is the patch area and $E^i_\alpha = \hat{n}.E^i(\underline{r}_\alpha)$, is the normal component of the electric field at the matching point $\underline{r}_\alpha$.

In the present formulation discretisation of the governing equations leads to an overestimate of the dipole density and hence an overestimate of $|\Delta z|$. These errors arise, at least partly because p is not accurately represented by a piecewise constant distribution, especially at the edge of the crack. One can always increase the number of patches to reduce these errors but only at the cost of greater computer time. As a compromise a 20 x 10 patch array was chosen. Based on sample calculations using a larger array, the error in $|\Delta z|$ for the standard patch format was estimated to be 5%.

A crack was modelled as a rectangular subsurface scattering object in a plane normal to the surface of the test material. Its dimensions are $\delta \times \delta/2$, the longer side being parallel to the surface, at a depth of $0.4\delta$ ($\delta = \sqrt{2/\omega\mu_0\sigma}$). Figure 5 shows impedance variations with displacement for the unshielded ferrite-cored probe assuming a 40 turn coil winding. Where the probe is located directly above the centre of the crack, at $(x,y) = (0,0)$, $|\Delta z|$ has a saddle point and the overall variation exhibits the familiar double-hump pattern. Also at the saddle point we have a phase minimum related to the defect depth and skin depth $\delta$ (figure 5b).

Figure 6 shows variations in $\Delta z$ for transverse scans, with the axis of the probes in the plane of the defect. Changes in the air-cored probe impedance are smallest and the nonshielded ferrite-cored probe gives the largest variation. This is consistent with the comparative incident field intensities in the defect region (figure 4). Although the shielded probe response is smaller than that of the nonshielded ferrite probe, the results indicate that it has a better resolution even for subsurface defects.

CONCLUSION

The usual approach in modelling eddy current-defect interactions is to calculate the electromagnetic field and then determine the probe response from the results. It is then possible to see in detail what the scattered field is like. However, one is often only interested in finding the probe response to a narrow crack, in which case it is simpler just to calculate the jump in the field of the crack or the corresponding dipole density before evaluating $\Delta z$. By using a combination of boundary integral and finite element methods we have shown that this can be accomplished for realistic probe structures.

ACKNOWLEDGEMENT

The author would like to thank Dr D J Harrison (R.A.E. Farnborough, U.K.) for helpful comment.

The work was carried out with the support of the Procurement Executive, Ministry of Defence, U.K.

REFERENCES

1. J. R. Bowler, Review of Progress in Quantitative Nondestructive Evaluation Vol. 5A, D. O. Thompson and D. E. Chimenti, Eds. Plenum, New York (1986).
2. C. V. Dodd and W. E. Deeds, J. Appl. Phys. $\underline{39}$, 2829 (1968).
3. A. G. A. M. Armstrong and C. S. Biddlecombe, I.E.E.E. Trans. Mag. Vol. Mag-18, No.2, (1982).
4. P. Weidelt, J. Geophys. $\underline{41}$, 85 (1975).
5. R. F. Harrington, "Field Computation by Moment Methods", Macmillan, New York (1968).
6. M. A. Burrows, "A Theory of Eddy Current Flaw Detection". Ph.D Thesis, University of Michigan (1964).
7. P. Hammond, I.E.E. $\underline{514S}$, 508, (1962).
8. I. S. Gradshteyn and I. M. Ryzhik, "Tables of Integrals Series and Products", 4th Edition, A.P. (1965) p.683.

RECENT STUDIES IN MODELING FOR THE A.C. FIELD MEASUREMENT TECHNIQUE


D.H. Michael and R. Collins

London Centre for Marine Technology
University College London
London, United Kingdom

INTRODUCTION

Theoretical modeling studies associated with the a.c. field measurement technique have been a major feature of work at UCL for almost the last decade [1]. In that technique, the objective is to establish a spatially uniform current flow in the surface of a metal and to use this to interrogate a defect such as a surface-breaking crack by directing it broadside on to the crack and measuring the perturbations in the surface voltage distribution which it produces (Figure 1a). We have described the first major result of the theoretical studies as an unfolding theory. It shows that when the electrical skin depth $\delta$ is small compared with the defect size, the field problem posed by the interception of a uniform current flow by a surface-breaking crack is the plane potential problem shown in Figure 1c and d. The descriptive name was adopted because the problem domain is that formed by conceptually sectioning the material in the plane of the crack, as in figure 1b, and unfolding the crack plane about the surface edge BC to make it coplanar with the metal surface. This approach has been successfully used to solve the field problems associated with cracks of various forms, for example elliptical, circular arc, rectangular and triangular, and it has enabled us to incorporate the influence of crack aspect ratio on the readings of an a.c.f.m. instrument such as the Crack Microgauge [2]. Earlier review papers [1,3] give numerous examples of the comparison between theory and experiment.

The explicit assumption that the skin depth is negligibly small in this theory implies that perturbations in the field produced by the corner of the surface edge BC and the crack interior edge BRC are negligible also, since they are assumed to occur over a length scale comparable with the skin depth. This view has been confirmed by deeper studies of the corner and edge fields [4] which show the solutions in those regions to be passive and adapting to the conditions imposed on them by the global solution which is determined by the unfolding approach. Implicit also in the unfolding theory as it is depicted in Figure 1 are the assumptions that the crack and metal surfaces are plane, that the crack faces are electrically insulated from each other and that the crack does not penetrate the plate in which it is formed. The purpose of this paper is to outline some current work at UCL which is concerned with modeling phenomena such as these by further development of the unfolding approach.

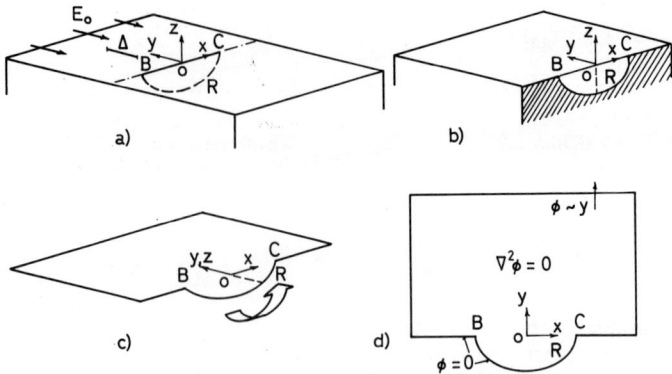

Figure 1. Unfolded field problem for a surface-breaking crack in the thin skin limit.

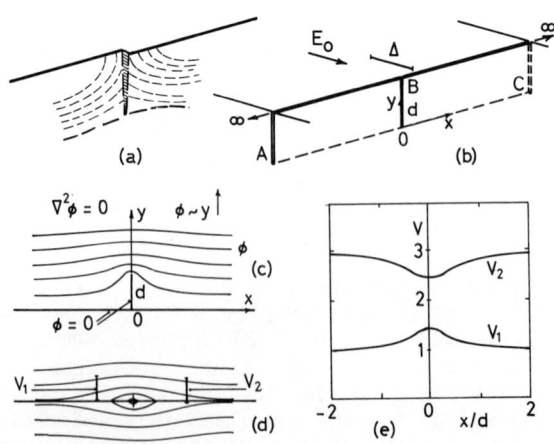

Figure 2. The line-contact problem. (a) facet formation, (b) model, (c) unfolded problem, (d) surface potentials, (e) nature of signals $V_1$ and $V_2$.

## LINE CONTACTS AND OVERLAPPING CRACKS

An important circumstance in which the faces of a crack can be in electrical contact arises during the growth and coalescence of fatigue cracks at the toe of a weld in a steel structure [5]. Cracks initiate at various locations along the weld and grow towards each other on slightly different planes under repeated load. The cracks may overlap slightly and the bridge between them can shear leaving a facet which maintains electrical contact between the faces even when the crack is 'open' (Figure 2a). A model of this phenomenon occurring on a crack of uniform depth is shown in Figure 2b and electrical contact is maintained across the crack faces along the line OB, the line contact. We wish to know how this feature will affect the field on the metal surface and the influence it will have on the readings of voltage given by a probe contacting the metal surface in its vicinity. Although this problem is a complicated three-dimensional problem with singular behaviour expected along the line contact, it is made immediately tractable by the unfolding approach and the solution is remarkably simple. Since AOBC is a plane of symmetry, the potential $\phi$ must be constant on AOC and must have the same constant value along OB. Without loss of generality this constant may be set at zero and the unfolded field problem is thus as shown in Figure 2c. It is mathematically identical to the problem of plane irrotational flow past a normal barrier with fluid stream lines analogous to lines of constant electric potential $\phi$. In terms of a complex variable $z = x + iy$ with origin at the bottom of the line contact, the solution is

$$\phi = \text{Im}(z^2 + d^2)^{1/2} \quad (1)$$

Figure 2c shows the lines of constant potential in the unfolded plane, Figure 2d illustrates the bunching of potential lines on the metal surface brought about by the line contact. From this solution the voltage readings given by a probe of length $\Delta$ (see Figure 2d) when aligned along the y axis and moved in the direction x may be deduced. The presence of the line contact produces a characteristic increase in the reference signal $V_1$ taken just before one leg of the probe crosses the crack and a decrease in the cross-crack signal $V_2$ as shown in Figure 2e. These features have been confirmed by experiments on models of the line contact phenomenon and the results of this study have been successfully applied to real line contact situations on both Tee butt welds and tubular welded joints, where the predicted character of the reference and cross-crack signals is again observed [5]. A system of line contacts disposed periodically in the x direction also has a straightforward solution and this too has been employed in interpreting measurements made on tubular joints so as to describe the influence of neighbouring line contacts.

As a development of this work, the problem associated with two overlapping cracks of uniform depth and shown in Figure 3 has also been considered [6]. In this example the separation of the two cracks is assumed to be small compared with the crack depth but large compared with skin depth so that corner effects do not intrude. Formulation of the unfolded field problem in this case is achieved by considering the metal to be sectioned on two planes which continue each crack so as to produce the domain shown in Figure 3b. The shaded regions are voids with no counterparts in the real space and the boundary conditions to be imposed on them involve both the potential and the potential gradients. These conditions are required to achieve analytic continuation across those lines which become contiguous when the problem domain is refolded and to give the required antisymmetric character to the solution. The unfolded field problem shown in Figure 3b has been programmed and solved by finite difference methods - other approaches are of course possible. Figure 3c shows an example of the potential lines generated by the solution for the

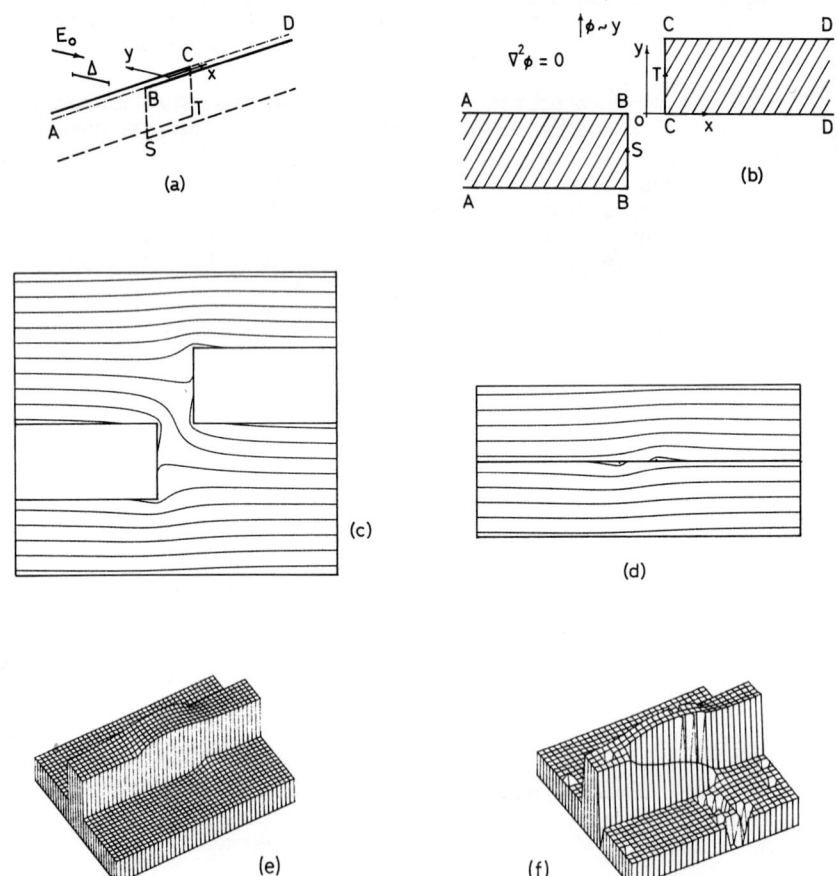

Figure 3. (a) Overlapping cracks of uniform depth, (b) unfolded domain, (c) potential lines, (d) surface potential lines, (e) perspective of predicted probe signal, large overlap, (f) experimental probe signal.

case when the overlap is equal to the crack depth, and Figure 3d the character of the potentials on the metal surface derived from it. A perpective of the probe signal predicted by the model for a situation where the overlap is large is shown in Figure 3e. The interesting antisymmetrical features of this plot are clearly reproduced in experiments on overlapping spark-eroded notches, the results of which are plotted in figure 3f. For cases where the overlap is small compared with crack depth, the antisymmetry remains but the plateau in the signal at the centre of the overlap is replaced by a trough, and in the limit of zero overlap the solution reproduces that for the line contact problem.

CORNER CRACKS

For cracks on corners we seek to direct a uniform current along the edge and towards the crack; figure 4 shows a simple symmetric arrangement. It is well understood, however, that in the case of very thin skin currents flowing along corners without cracks, the current density rises as the corner is approached so that the interrogating current is not uniform and the surface field is no longer a plane Laplacian. This behaviour presents an obstacle to the extension of the unfolding method. It is also well known, however, that the concentration of current density at the edge can be alleviated by rounding off the corner [7] and we have found it possible in practice to achieve a sensibly uniform current density around the edges of a mild steel block by virtue of such a slight rounding. It was therefore considered worthwhile to use the unfolding approach and the plane Laplace problem it produces as a first approximate model in cases where the radius of curvature of the rounded edge is large compared with skin depth

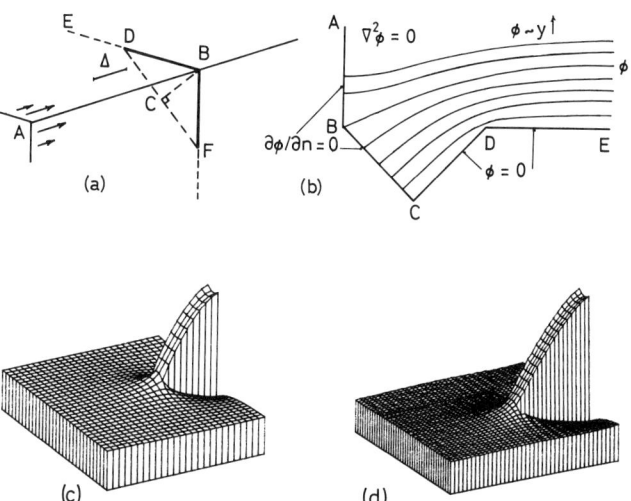

Figure 4. (a) Symmetrical corner crack, (b) unfolded approximation and potentials of solution, (c) perspective of predicted probe signal, (d) experimental probe signal.

while still small compared with the crack dimension. Figure 4b shows the unfolded field problem, (by symmetry only one half of one face need be considered) together with potential lines from its solution obtained by using the Schwarz-Christoffel transformation to map the boundary ABCDE on to the real axis. Potential lines on one quarter of the metal surface in the vicinity of the crack may be visualized by covering over the triangular part of Figure 4b. As with the previous section, the final two elements of this figure compare the predicted probe signals (Figure 4c) with those obtained experimentally for a particular combination of probe size and crack depth (Figure 4d). A more detailed exposition and comparison are given in a recent paper [8]. Despite its acknowledged approximate nature, the model proves to be very satisfactory. Asymmetric corner cracks have also been treated but with the resulting boundary value problem solved numerically.

CURVED SURFACES

The objective in unfolding a surface field is to produce a Laplace problem in a single plane and it has been possible to do this in the examples considered so far because of the plane surfaces involved and the straight edges about which the unfoldings can be made. The treatment of curved surfaces involves a further development in which one part of the field is mapped into the required plane by a suitable coordinate or conformal transformation. An example involving the latter is the crack of circular arc form part way through a cylindrical rod, as in Figure 5a. The two two-dimensional Laplace problems involved here are on the crack plane labelled z and on the surface plane $s = p + iq$ which can be considered to be wrapped around the rod with the coordinate $q = a\theta$. These fields join

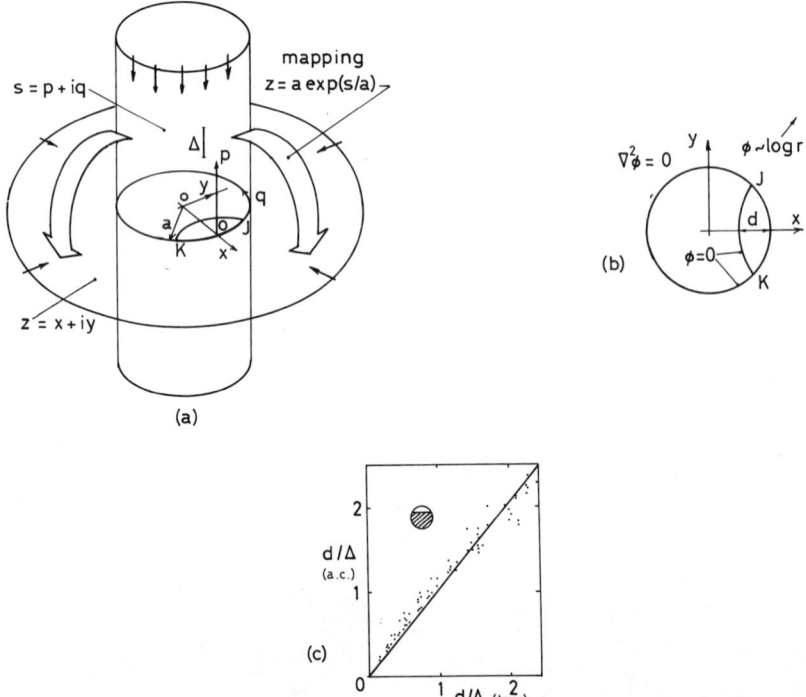

Figure 5.   Circular arc crack on a circular rod, (a) mapping on to the z plane, (b) plane field problem, (c) measurements of centre-line depth interpreted through this model.

along the curved crack surface edge KJ. A single Laplace problem in the z plane is generated in this case by mapping the s plane with the conformal transformation z = a exp(s/a), which maps the rod surface into the region in the z plane outside the circular section containing the crack (Figures 5a, b). The incident uniform flow on the rod surface is mapped by this transformation into a sink flow towards the origin in the z plane and the transformed plane potential problem is thus that of sink flow into the lunate boundary at constant potential (Figure 5b). This can be readily solved by standard techniques, the Karman-Trefftz transformation for example, and the potential distribution so found is then mapped back on to the cylindrical rod surface to enable probe signals to be calculated. The model has been tested experimentally by measuring cross-crack and reference signals with probes of various lengths on a series of notches of different depths cut on mild steel rods of various radii. For convenience, these notches had straight lower boundaries; the method of course can model any combination of curvatures of the rod and crack lower boundary. Agreement between the real depth and the depth interpreted from the probe signal data using this model is excellent as shown in Figure 5c.

Examples of the use of a coordinate transformation to generate the plane potential problem arise from consideration of the effects of surface pits. This work began by considering pits of spherical cap form [9] but it has recently been extended to pits of arbitrary axisymmetric shape, examples of which are conical, cylindrical and annular [10]. The effect of the coordinate transformation there is to map the curved surface so that it occupies the plane circular region inside the rim and thus continues the surface plane.

PENETRATING CRACKS

Consider the situation shown in Figure 6a where the crack has grown so that it penetrates the plate in which it is formed. We wish to know how the surface fields on both faces of the plate are affected by this penetration. Unfolding in this case involves the two surface edges BG and CF and

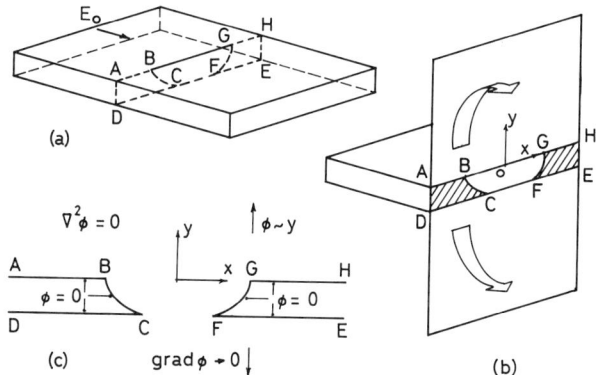

Figure 6. Unfolded problem for a crack penetrating a plate of finite thickness.

the procedure is perhaps better described as folding the upper surface plane up about BG and the lower surface down about CF so that both become coplanar with the crack plane, as in Figure 6b. The boundary conditions to be applied at infinity in the lower half of this z plane depend on the way in which the lower surface of the plate can communicate electrically with the upper. If the plate is a laboratory specimen of finite size and the field inputs are arranged symmetrically at two of its ends, then the upper and lower surfaces would both have a uniform current sheet flowing on them. If the specimen is very large, however, and is such that the lower surface communicates electrically with the upper only through the penetrating crack, then there is the possibility of leakage of the field on to the lower face but there will be no net current flow at infinity on this face. The problem shown in Figure 6c is for the latter situation. The field problem which arises when the plate has negligible thickness compared with crack length is very simple and the solutions on the two surfaces may be shown to involve the normal barrier solution in equation (1). For arbitrary plate thickness and arbitrary crack shape, the problem has been solved by application of the Schwarz Christoffel mapping with the crack profile approximated by a polygonal boundary. Theoretical and experimental results are in good agreement with rectangular and circular arc penetrating slots. A paper on this work which also deals with penetrating holes is in preparation.

ACKNOWLEDGEMENT

It is a pleasure to acknowledge the contributions made by Riaz Haq, Richard Leetham, Martin Lugg, Maureen McIver, Andrzej Niemiro, David Parramore, Huai-Min Shang and Rowland Travis to the work described in this review paper.

REFERENCES

1. R. Collins, W.D. Dover and D.H. Michael, The use of a.c. field measurements in non-destructive testing in "Research techniques in non-destructive testing", vol. VIII, R.S. Sharpe, ed., Academic Press, London (1985).
2. Inspectorate Unit Inspection Ltd, Sketty Hall, Swansea, UK.
3. R. Collins, D. Mirshekar-Syahkal and D.H. Michael, The mathematical analysis of electromagnetic fields around surface flaws in metals in "Review of progress in quantitative non-destructive evaluation", Vol. 2B, D.O. Thompson and D.E. Chimenti, eds., Plenum, New York (1983).
4. D.H. Michael, R.T. Waechter and R. Collins, Proc. Roy. Soc., A 381, 139 (1982).
5. W.D. Dover, G. Glinka and R. Collins, Automated crack detection and monitoring of crack shape evolution in tubular welded joints, in Proc. Int. Conf. on NDT in fitness for purpose assessment of welded constructions, Paper 11, The Welding Institute, London (1984).
6. In preparation.
7. J.D. Cockroft, Proc. Roy. Soc., A 122, 533 (1929).
8. H.M. Shang, R. Haq, R. Collins and D.H. Michael, to be submitted for publication.
9. R. Collins, D. Mirshekar-Syahkal and D.H. Michael, Proc. Roy. Soc., A 393, 159 (1984)
10. R. Collins, D.H. Michael, D. Mirshekar-Syahkal and H.G. Pinsent, Journal of NDE 5, (1985)

# THREE DIMENSIONAL FINITE ELEMENT MODELING

Nathan Ida

The University of Akron
Electrical Engineering Department
Akron, Ohio 44325

## INTRODUCTION

Numerical modeling of electromagnetic nondestructive testing problems has become in the last few years common enough to be considered an integral part of the engineering design practice. In particular, eddy current modeling activity has been on the increase because of the benefits afforded by such models and their relative simplicity and availability. This is certainly true for two dimensional and axisymmetric modeling where both general [1] and NDT models [2] exist. In many such cases, little more than a small computer and a reasonably trained person are needed.

Three-dimensional models for general field problems and for NDT applications present a totally different situation. The formulation of Maxwell's equations is in itself more complex and the type of formulation has important implications. Formulations ranging from single and double scalar potentials [3] through magnetic vector potentials [4] to combinations of scalar and vector potentials [5] abound. Far more significant than the choice available is the fact that these formulations are not necessarily equivalent and their applicability is not universal.

This work is intended as an outline to 3-D eddy current modeling as it relates to nondestructive testing. Because of their importance, the work deals with some of the most important problems involved in the solution of 3-D eddy current problems and to appropriate methods to overcome some of them. As it so often happens with numerical calculations, no method is universal or problem independent. Yet, the methods outlined here are in general applicable and, with proper development promise to make the solution of 3-D electromagnetic problems easier, faster and less expensive.

## METHODS OF FORMULATION

Formulation of a field problem involves two steps: 1) representation of Maxwell's equations in a form suitable solution and , 2) formulation of the resulting field equations in a form suitable for solution using a numerical method.

The first step starts with the simplification of Maxwell's equations. Simplification in this case means the use of the equation in a reduced form.

For low frequency problems, within conducting media, the displacement currents are neglected as are free charge densities. Linearity may also be assumed in many cases as well as uniform current distributions within source regions. At this stage one obtains an equation of the following form

$$\nu \nabla \times \overline{B} = \sigma \overline{E} \tag{1}$$

where the material properties are assumed to be linear (and isotropic) and the excitation sinusoidal. The equation can now be solved in terms of A or the equations can be rewritten in terms of any of a number of functions from which the field can be calculated. The most common form of formulation is the use of the magnetic vector potential [4]. This, potential, being a vector does not reduce the problem size but the magnetic vector potential is continuous over material boundaries and therefore it allows inclusion of complex material interfaces and boundaries without the need to specify interface conditions. The general 3-D problem takes the following form under the assumptions in Eq. (1).

$$\nabla \times \nabla \times \overline{A} = -\overline{J}_s + j\omega\sigma\overline{A} \tag{2}$$

In regions void of current sources, the field can be obtained from a scalar potential. This is extremely attractive since the solution involves the calculation of a single scalar at each node of the mesh as compared to three components of a vector. This has been attempted [3] but found to have two serious drawbacks 1) large errors due to cancellation of terms and 2) it cannot be used in current regions. To overcome the first problem, the use of two scalar potentials was suggested [3] while the second can only be treated by a vector potential. The natural outcome of this was the use of a vector potential in current regions and a scalar potential in the rest of the solution region [5]. This approach seems to be optimal but the interface between vector regions and scalar regions has not yet been solved satisfactorily. Among the many methods suggested are the coupling of the magnetic vector potential with the magnetic scalar potential [7] and the electric vector potential and the magnetic scalar potential [5] to mention but two. Another method is to solve directly for H or B [6].

The second step, that of the numerical formulation is usually quite simple once the field equations have been defined in terms of any particular function. One particularly simple method is the use of Galerkin's method [6]. More often, an energy functional and variational methods are used for the formulation [8]. The choice here depends on the form of the original equations and user preference.

APPLICATION OF THE MODEL

The size of 3-D problems and the computer resources needed to solve these problems are by far the biggest difficulty in their effective application to solution of realistic problems. There are four areas that need to be addressed in order to make a 3-D formulation more attractive from the user point of view. These are: 1) size of problem, 2) computer resources, 3) accuracy of solution and 4) display. Depending on the formulation used, the requirements may vary but, in general all of these must be improved to ensure maximum efficiency in solution and simplicity in model application.

While the formulation is important and has some effect on the problem size, there is an urgent need for methods to reduce the computer time needed. The types of improvements that can be implemented are as are: a) use of faster computers b) use of infinite elements c) hybrid solution method and d) use of more efficient solvers.

COMPUTER RESOURCES

The use of larger, faster computers, although being a "brute force" method is quite effective in improving solution times and handling larger problems. While it may not be very appealing from an intellectual point of view, the trend in computer technology seems to favor this approach. The reason for this is in the fact that both vector and parallel computers, the fastest architectures available today, require the problem to be solved using a minimum of simple repetitive operations. The basic ideas of algorithmic development has changed drastically. Instead of trying to reduce the number of operations and/or memory usage, one is trying to recast the problem such that simple, standard operations such as dot products can be used. In particular, branching and single, scalar operations should be avoided on vector machines. Parallel machines also require a high degree of parallelism in the algorithm itself, a feat that can only be achieved with simple algorithmic constructs. To understand the importance of this approach to solution of field problems it is useful to consider the figures in Table 1. Here, a large eddy current problem was solved on a VAX 11/780, an IBM 3033 and a CYBER 205. The transfer from one machine to another is not straightforward but the improvements are significant. More important, one can reasonably assume that future machines should bring these numbers down. In this respect, consider the figures in the first column in Table 2. Here, four problems were solved on a personal computer (IBM AT). The first 3 are axisymmetric problems while the fourth is three dimensional. New computer systems like the Multiple Instructions Multiple Data (MIMD) and vector machines [9] could have a significant impact on numerical computation.

INFINITE ELEMENTS

Part of the reason for the size of electromagnetic field problems is the fact that the field only decays to zero at infinity. The model requires the discretization of large volumes of space in order to reduce the influence of boundaries, artificially located at some finite distance. The natural

Table 1. Solution times for a large eddy current problem (12,513 unknowns, 3360 elements) for different computers. 1, 2 and 4 are in 32 bit mode, 3 is in 64 bit mode.

|   | Sol. Time CPU | Sol. Time CLock |
|---|---|---|
| 1 VAX 11/780 | 21 Hrs | 82 Hrs |
| 2 IBM 3033 | 6.5 Hrs | 14 Hrs |
| 3 CYBER 205 | 29 Min. | 31 Min. |
| 4 CYBER 205 | 21 Min. | 23 Min. |

Table 2. Solution times of four problems on an IBM AT with an ICCG Solver.

|   | Sol. Time | No. Iter. | Nodes | Band. | Elem. |
|---|---|---|---|---|---|
| 1. (2-D) | 32 Sec. | 6 | 55 | 7 | (T) |
| 2. (2-D) | 37 Sec. | 6 | 55 | 7 | (Q) |
| 3. (2-D) | 1.75 Hrs | 16 | 1927 | 49 | (Q) |
| 4. (3-D) | 2.17 Hrs | 58 | 735 | 147 | (H) |
| 4. (3-D) | * 1.37 Hrs | 31 | 735 | 147 | (H) |

\* With Preconditioning
T-Triang., Q-Quadrilat., H-Hexahedral elements

solution to this problem is to use the so called "infinite finite elements". Infinite elements are in practice a simple extension to normal finite elements. Instead of using polynomial shape functions over an element, one could use a decaying function instead. This function is usually obtained by simply multiplying the normal shape function by an exponentially decaying factor or some other similar function [10]. This in effect allows the field to decays to zero over a single element which now spans an infinite region. Although the principle is simple, in practice one tries to match the decay of the field to that of the shape functions. This is quite inaccurate and requires extensive experimentation until a correct decay parameter is found.

HYBRID SOLUTION

This alternative method [11] is based on the fact that many 3-D testing geometries consist of a two-dimensional or axisymmetric geometry with a small region that is three-dimensional (i.e. a defect or inclusion in a conducting material). This allows the use of a two-dimensional or axisymmetric solution as the boundary values for a closely truncated three-dimensional mesh with better accuracy than can be obtained by the use of infinite elements. In effect, one assumes that the 3-D portion of the problem is so small as to affect the solution very little outside its immediate vicinity.

Since very accurate solutions in 2-D and axisymmetric geometries can be generated at little computational cost it would be of advantage if such solutions could be used as the far field solution. This can only be done if the disturbance in the field created by the three dimensional effect is local and affects the far field values very little. While this is not the case in general electromagnetic field problems it is quite common in NDT applications. In these applications one can solve for the field in the sample without the defect and then use the calculated values as boundary conditions for a full three-dimensional calculation of the sample with the defect but with a mesh truncated close to the region of interest. The method also allows the analysis of a variety of defects in the same sample without the need to recalculate the boundary conditions.

To illustrate this approach a relatively large 3-D problem was analyzed using a 3-D solution and a hybrid solution. The 3-D mesh consists of 3,360 elements and 12,513 variables with a bandwidth of 336. The truncated mesh consists of 1,200 elements, 4,758 variables and has a bandwidth of 216. An axisymmetric mesh with 3,000 elements and 3,146 nodes was used with some of its nodes in locations identical to the 3-D mesh. From the values calculated with the axisymmetric mesh (without the defects) the boundary values on the surface of the 3-D mesh are set. The components of the magnetic vector potential at each boundary that cannot be found from such a calculation are left unspecified. Figure 1 shows the solution obtained using this method and the experimental impedance plane trajectory. The truncated mesh solution and the full 3-D solution are practically the same. Although these results indicate that the two 3-D meshes yield the same results it cannot be concluded that the full 3-D mesh is adequate to describe the problem. In fact, the comparison to the experimental curve in Figure 1 indicates that the meshes are far too coarse. The results only indicate that the same result can be obtained with a far smaller mesh and that the effect of the defect on the far field is minimal. Perhaps even more dramatic than the solution itself is the improvement in solution time. This is shown in Table 3 and shows a factor of about 10 in CPU time and about 9.5 in total (clock) time.

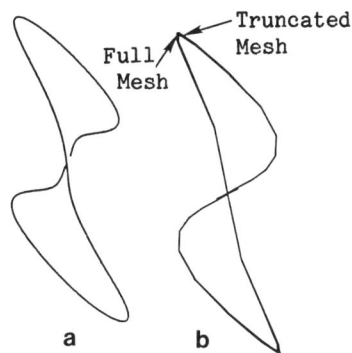

Fig. 1. Experimental and 3-D (full and truncated mesh) impedance plane trajectories.

Table 3. Solution times on a VAX 11/780 for the full 3-D and truncated 3-D meshes.

|              | Full Mesh    | Truncated Mesh |
|--------------|--------------|----------------|
| Unknowns     | 12,513       | 4,758          |
| Elements     | 3,360        | 1,200          |
| Bandwidth    | 336          | 216            |
| Matrix Size  | 12,513 X 336 | 4,758 X 216    |
| CPU Time     | 21 Hrs       | 2.3 Hrs        |
| Clock Time   | App. 82 Hrs  | 8.6 Hrs        |

Results are for 24 probe positions

Table 4. Memory requirements for four different problems with direct and iterative methods of solution.

|               |       | Choleski | ICCG   | Nodes | Band. | Elem. |
|---------------|-------|----------|--------|-------|-------|-------|
| Prob. 1       | (2-D) | 385      | 440    | 55    | 7     | (T)   |
| Prob. 2       | (2-D) | 385      | 550    | 55    | 7     | (Q)   |
| Prob. 3       | (2-D) | 94,423   | 19,270 | 1927  | 49    | (Q)   |
| Prob. 4       | (3-D) | 108,045  | 22,275 | 735   | 147   | (H)   |

T-Triangular (3 nodes), Q-Quadrilateral (4 nodes), H-Hexahedral (8 nodes)

SOLUTION METHOD

The accepted method of solving linear systems of equations is by some form of Gauss Elimination. In many cases, the Choleski decomposition method is used because of its reduced number of operations. Elimination methods are particularly convenient since they are direct methods and are relatively insensitive to the conditioning of the matrices. This is important in 3-D eddy current problems because the matrices are not as well behaved as in 2-D problems and 3-D magnetostatic problems. This approach has a severe limitation due to the fact that the size of the matrix depends on the bandwidth of the mesh. Some improvement can be obtained through the use of special mesh generators [12] or by designing a mesh such that the original node and element numbering is minimal. The problem of mesh bandwidth is aggravated in the so called frontal methods [13]. Here, one seeks to minimize both bandwidth and frontwidth [13] both of which may not always be done independently.

On the other hand, iterative methods of solution offer significant advantages in terms of storage requirements. The memory needed for the solution of a problem does not depend on the bandwidth but rather on the basic structure of the mesh. Thus, for finite element problems, the number of memory locations needed for the solution is closely related to the number of neighboring nodes at each node. This is both smaller than the bandwidth (by orders of magnitude in many cases) and remains constant within a mesh regardless of the size of the mesh. Table 4 shows the memory requirements for some 2-D and 3-D problems using Choleski decomposition and an iterative method. Although the method itself has not been defined yet, the requirements are about the same for most iterative methods. The figures in Table 4 are for the matrix itself and no auxiliary arrays are reflected in these figures. In the case of the iterative method, both the indexing array and the coefficient matrix are included. The memory requirements in Table 4 should be viewed as approximate because the actual usage may be different depending on the storage scheme.

Among the various iterative methods available, the most promising is the Incomplete Choleski Conjugate Gradient (ICCG) method [14]. The method is applicable to magnetostatic problems and eddy current problems where it takes the form of a Bi-Conjugate method, specially developed for the solution of complex systems of linear equations. The ICCG method was never popular for 2-D or small 3-D problems because the great advantage in storage requirements is often offset by the significant increase in the number of operations required for solution. In moving probe problems, an additional aspect plays a significant role: The need for resolution of the system of equations for each probe position. In a direct method this is done by an elimination step followed by a number of backsubstitution steps - one for each probe position. In iterative methods, this cannot be done since one has to start with an approximation to the solution and the intermediate steps are all lost. The fact that the previous solution can be used as a good estimate to the current solution improves matters but does not eliminate the basic need for resolution from scratch.

For large problems, the picture changes considerably in favor of the ICCG method (or any other iterative method). In such large application the savings in memory and disc access time more than compensates for the increased number of operations and the solution is considerably faster. Iterative methods offer other advantages as well. The most obvious is independence of roundoff errors. At the same time, ill conditioned systems are more difficult to solve if not impossible. Some methods of preconditioning can be used to either render the problem solvable or to reduce the number of iterations needed for solution.

In order to get some feel for this type of solution, Table 4 summarizes the solution times and number of iterations needed for the solution of some relatively small 2-D problems and a small 3-D eddy current problem. It should be noted that, in the 3-D case, a significant difference is achieved by preconditioning of the system.

CALCULATION OF PROBE IMPEDANCES

A situation unique to NDT applications arises in moving probe calculations. In axisymmetric geometries the impedance of coils is calculated by integrating around the coil's cross section. This method is applicable for absolute and differential probes or for any combination of coils. In 3-D problems, the flux around the coil is not constant circumferencially and therefore this method cannot be used. In particular, the calculation of differential impedances is complicated. The only correct way of calculation is through calculation of stored and dissipated energies in the solution region from which inductance and resistance can be calculated. These are global quantities and therefore, differential impedances cannot be calculated (i.e. zero impedance would correspond to zero applied current, not to identical conditions in both coils). One possible solution is to calculate the impedance of a single coil and subtract subsequent probe position impedances such as to correspond to the location of the second coil in a differential pair [15].

MESH AND OUTPUT DISPLAY

Display of mesh data is extremely important for the purpose of checking the correctness of the input data. This is particularly so for 3-D problems where one would like to ensure correct data before hours of CPU time are lost. A variety of methods, none completely satisfactory are used. Cross sectional displays, hidden line removal algorithms and rotations are only a few possible approaches. An example to the importance of mesh display can be seen in Fig. 2, where a simple mesh without and with hidden lines removed is presented. Similarly, meaningful display of output data is important. In NDT applications, one is sometimes content with an impedance plane plot. If flux plots are needed, only cross sectional plots can be displayed and even then, some cross sections may be more useful than others. A general cross section, on which the field is aligned arbitrarily is quite useless for field representation since one would have to use a 2-D plane to display the three components of the field.

In addition to display of data, and more significant from the user point of view is the generation of input data for the model. Since the user is likely to spent a significant portion of the total time needed to model a problem in this initial step, it is important that efficient methods for pre- and post-processing be developed. Particularly attractive for this purpose are personal computers and graphics workstations. These highly interactive computers allow the user to have total control of the geometry definition process and therefore design a correct mesh in a relatively short period of time. One such preprocessor has been developed by the author for 2-D and axisymmetric mesh generation and is currently being extended for 3-D applications. Figure 3 shows a sequence taken from the design of a simple mesh. The process starts with the definition of an outline of the geometry (Fig. 3a). Then, the geometry is partitioned according to the materials or structures within it (Fig. 3b). This general definition can be altered at any stage. Next, material properties and boundary conditions are entered interactively. The basic mesh in Fig. 3b is then subdivided into any number of elements and the obtained mesh displayed either on a raster display (Fig. 3c) or on a plotter (Fig. 3d). Parts of the mesh can also be displayed selectively. The user can go back to any step he desires and change any of the data generated with simple commands. The extension of this process to 3-D problems should free the user from most of the hassle associated with mesh generation and allow him to concentrate on the problem itself.

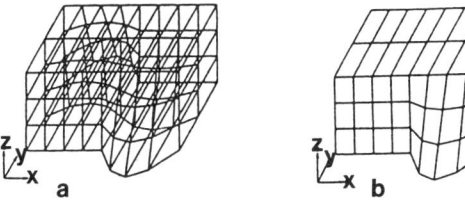

Fig. 2. A simple mesh displayed with all lines shown (a) and with hidden lines removed.

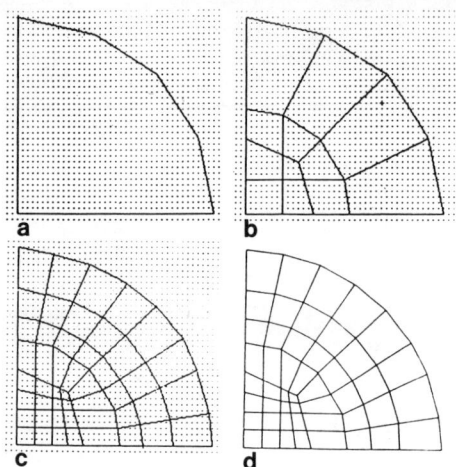

Fig. 3. Interactive design of a simple mesh. a) Outline of geometry, b) subdivision into blocks, c) display of mesh and d) plot of mesh.

CONCLUSIONS

The need for modeling of electromagnetic field problems does not need justification by itself. The effort required and the expense of such models is another mater. This impediment to the development and use of 3-D models can be overcome by proper development of the basic tools already available. The size of problems solvable can be increased significantly by the use of faster computers, more efficient solution algorithms and the use of special methods, including infinite elements to reduce the computer resources required. New formulations should be inspected carefully for better compatibility with the newer computer architectures. It is hoped that in the future, with the new parallel machines the difficulty of solving realistic 3-D electromagnetic field problems will be reduced to a level similar to that existing today in 2-D and axisymmetric modeling.

REFERENCES

1. P. P. Silvester and M. V. K. Chari, "Finite element solution of saturable magnetic field problems", IEEE Transactions on Power Apparatus and Systems, Vol. 89, 1970, pp. 1642-1651.
2. R. Palanisamy and W. Lord,"Finite element modeling of electromagnetic NDT phenomena," IEEE Transactions on Magnetics, Vol. MAG-15, No. 6, November 1979, pp. 1479-1481.
3. C. W. Trowbridge, "Three-dimensional field computation," IEEE Transactions on Magnetics, Vol. MAG-18, No. 1, pp. 293-297, January 1982.
4. N. Ida and W. Lord, "A finite element model for 3-D eddy current NDT calculations," IEEE Transaction on Magnetics, Vol. MAG-21, No. 6, Nov. 1985, pp. 2635-2643.
5. T. W. Preston and A. B. J. Reece, "Solution of 3-dimensional eddy current problems: the T - method," IEEE Transactions on Magnetics, Vol. MAG-18, No. 2, pp. 486-491, March 1982.
6. P. Hammond, "Use of potentials in calculation of electromagnetic fields,"IEE Proceedings, Vol. 129, Part A, No. 2, pp. 106-112, March 1982.

7. J. Simkin and C. W. Trowbridge, "Which potential? - a comparison of the various scalar and vector potentials for numerical solution of the nonlinear Poisson problem," Rutherford Laboratory, 1978, RL-78-009/B.
8. M. L. Brown, "Calculation of 3-dimensional eddy currents at power frequencies," IEE Proceedings, Vol. 129, Part A, No. 1, pp. 46-53, Jan. 1982.
9. N. Ida "Electromagnetic field modeling on Supercomputers," IEEE Transactions on Magnetics, Vol. MAG-21, No. 6, November 1985, pp. 2490-2494.
10. P. Bettess, "Infinite elements," International Journal for Numerical Methods in Engineering, Vol. 11, (1977) pp. 53-64.
11. N. Ida, "Efficient treatment of infinite boundaries in electromagnetic field problems," accepted for publication in COMPEL, (Ireland).
12. N. Ida, "A mesh generator with automatic bandwidth reduction for 2-D and 3-D geometries," Presented at the EM-COMP Conference, Pittsburgh, PA., December 13-14, 1984, to be published in Mathematics and Computers in Simulation.
13. E. G. Thompson and Y. Shimazaki, "A frontal procedure using SKYLINE storage," International Journal for Numerical Methods in Engineering, 15, 1980, pp. 889-910.
14. J. Ortega and E. Poole, "Incomplete Choleski Conjugate Gradient on the CYBER 203/205," Supercomputer Applications Symposium, Oct. 31 - Nov. 1, 1984, (Purdue University and Control Data Corporation)
15. N. Ida, "Alternative approaches to the numerical calculation of impedance," submitted for publication in IEE Proceedings, Part A.

# A COMPUTATIONAL MODEL FOR ELECTROMAGNETIC INTERACTIONS WITH ADVANCED COMPOSITES

Harold A. Sabbagh, Thomas M. Roberts and L. David Sabbagh

Sabbagh Associates, Inc.
2634 Round Hill Lane
Bloomington, IN 47401

## INTRODUCTION

Composite materials in the form of fiber-reinforced matrix materials as, for example, graphite-epoxy, are being increasingly used in critical structures and structural components because of their high strength-to-weight ratio. In order to assess the integrity of these structures, it is necessary to employ suitable methods for quantitative nondestructive evaluation (NDE). One such method uses electromagnetics (eddy-currents), but the problem is that composite materials are inherently anisotropic, which means that much of the classical eddy-current technology and design procedures are not applicable. In this paper we compute fields by applying a rigorous model of electromagnetic interactions with graphite-epoxy composites, which is based on a continuum approach. In this approach the graphite fibers produce a macroscopic conductivity tensor that has different conductivities in the directions parallel and transverse to the fibers.

In [1] we described a method for computing electromagnetic fields within composite media, which is based on a matrix form of Maxwell's equations in Fourier space. By using this approach we computed the tensor Green's function for a plane-parallel slab of graphite-epoxy. With the Green's tensor in hand, we showed how one can compute the fields interior to the slab, due to abritrary sources of excitation; we gave as an example the field due to an infinite current sheet parallel to the slab. In [2] we extended the computations to include the fields due to planar, circular current-loops that are also parallel to the slab. These results simulate some of the effects one might encounter when dealing with circular coils in the presence of a composite slab.

The fundamental relationship between the field at level $z$, due to an impressed current source at $z'$, is, in Fourier-transform space, $(k_x, k_y)$:

$$\tilde{e}(k_x, k_y, z) = \int \tilde{\bar{\bar{G}}}(k_x, k_y, z; z') \cdot \tilde{j}^{(i)}(k_x, k_y, z') dz', \qquad (1)$$

where $\tilde{e}(k_x, k_y, z)$ is the transform of the transverse field vector at $z$, $\tilde{\bar{\bar{G}}}(k_x, k_y, z; z')$ is the transform of the Green's tensor, with source point at $z'$ and field point at $z$, and $\tilde{j}^{(i)}(k_x, k_y, z')$ is the transform of the transverse components of the impressed current source at $z'$. The computation of $\tilde{j}^{(i)}(k_x, k_y, z')$ for circular current sheets is discussed in Appendix C of [2].

## COMPUTATION OF FIELDS WITHIN THE SLAB

When $\tilde{\tilde{G}}$ and $\tilde{j}$ are substituted into (1), we obtain the transverse Fourier transform of the transverse field vector at any level, $z$, within the slab. Then, upon taking the inverse Fourier transform, say by using the Fast Fourier Transform (FFT) algorithm, we get the fields in physical space, at level $z$. We have done this for a number of circular current distributions, including a filamentary loop, and have found that the results are qualitatively similar in all cases; hence, we will display results of the filamentary loop, only. The frequency of excitation is $10^6$ Hz, and the parallel and transverse conductivities are, respectively, $\sigma_p = 2 \times 10^4$ S/m, $\sigma_t = 100$ S/m. Hence, the anisotropy ratio is 200. The slab thickness is 1.27 cm (0.5 inches), and the current loop is 2.54 mm (0.1 inch) above the slab. The radius of the current loop is 0.5 inch (1.27 cm).

Before going into the anisotropic problem, we illustrate in Figure 1 the field induced into an isotropic medium (with a conductivity of $2 \times 10^4$ S/m) at a depth of 0.05 inch (0.127 cm). The isotropic nature of the response is clearly apparent; if we were to look vertically downward we would see a circular response region. Each pixel is a square, whose side is 0.05 inch. Thus, the response region has a diameter of about 1.0 inch, which is the diameter of the current loop. Therefore, the result agrees with our intuition.

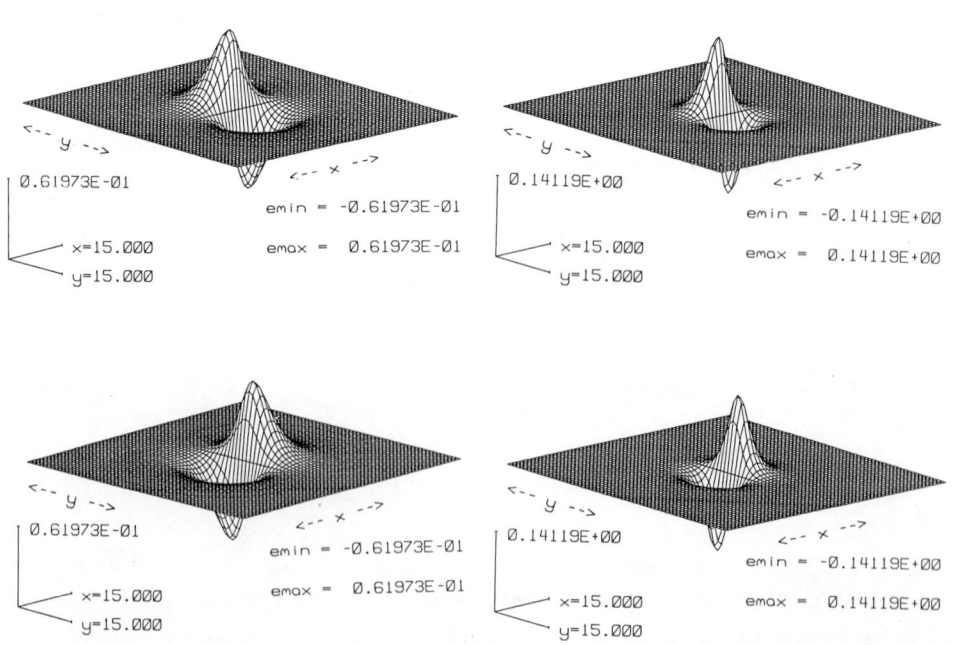

Fig. 1 Field induced into isotropic slab by a circular current loop. Left-hand column: real part; right-hand: imaginary part. Top row: $x$-component; bottom: $y$-component. (Same arrangement for Figs. 2, 3)

The situation in an anisotropic material is changed dramatically, however. In this case the fibers will 'guide' the field, so that it will die out much less rapidly in the $x$-direction (along the fibers) than in the $y$-direction. This is illustrated quite clearly in Figure 2, where the complex values of the $x$- and $y$-components of the electric field at a depth of 0.05 inch are shown. The response region in this figure is highly elongated in the $x$-direction, when viewed from directly above. The $x$-component of the induced electric current field is obtained by multiplying the $x$-component of the electric field by $\sigma_p$, which is equal to $2 \times 10^4$ S/m, and the $y$-component of the current is given by the product of the $y$-component of electric field with $\sigma_t$ (100 S/m). Therefore, the eddy-currents do not flow in the usual circular paths of an unbounded isotropic medium, as suggested by Figure 1, but, rather, flow in highly elongated quasi-elliptical paths. The degree of eccentricity of the paths depends upon the degree of anisotropy, as measured by the ratio, $\sigma_p/\sigma_t$.

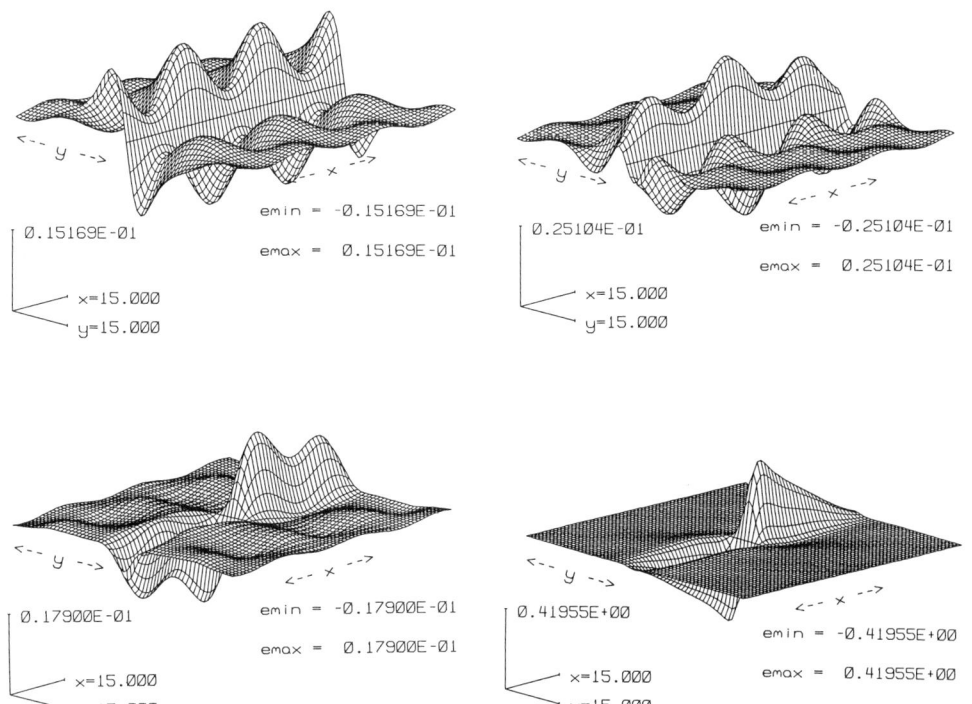

Fig. 2 Field induced into composite (anisotropic) slab by a circular current loop. Depth = 0.05 inch.

In many applications it is important to know how rapidly the induced field dies out with depth into the slab. In an anisotropic medium there is no unique skin-depth, because the conductivity varies with direction of the electric field. Therefore, the problem must be handled numerically in most cases. We present in Figure 3 model calculations of the induced electric field at a depth of 0.4 inch within the slab, under the same conditions as above. Upon comparing this figure with Figure 2, we draw the following conclusions: the field magnitude is reduced by about an order-of-magnitude, the field is much more spread out in the $y$-direction and is very uniform in the $x$-direction. These results are consistent with the notion of diffusion in isotropic media; they are an obvious manifestation of the filtering-out of the higher spatial frequencies, $k_x$, $k_y$, with depth, and the crowding of the spatial-frequency spectrum toward the origin. In addition, we note that the $x$-component of the electric field dies out much more rapidly with depth than does the $y$-component. This is due to the fact that in the principal axis coordinate system the $x$-components 'sees' a much larger conductivity, $\sigma_p$, than does the $y$-component. This supports our statement that there is no unique skin-depth in a single layer of graphite-epoxy.

Model computations of the type presented here can be very useful in setting up eddy-current experiments in graphite-epoxy and interpreting the results.

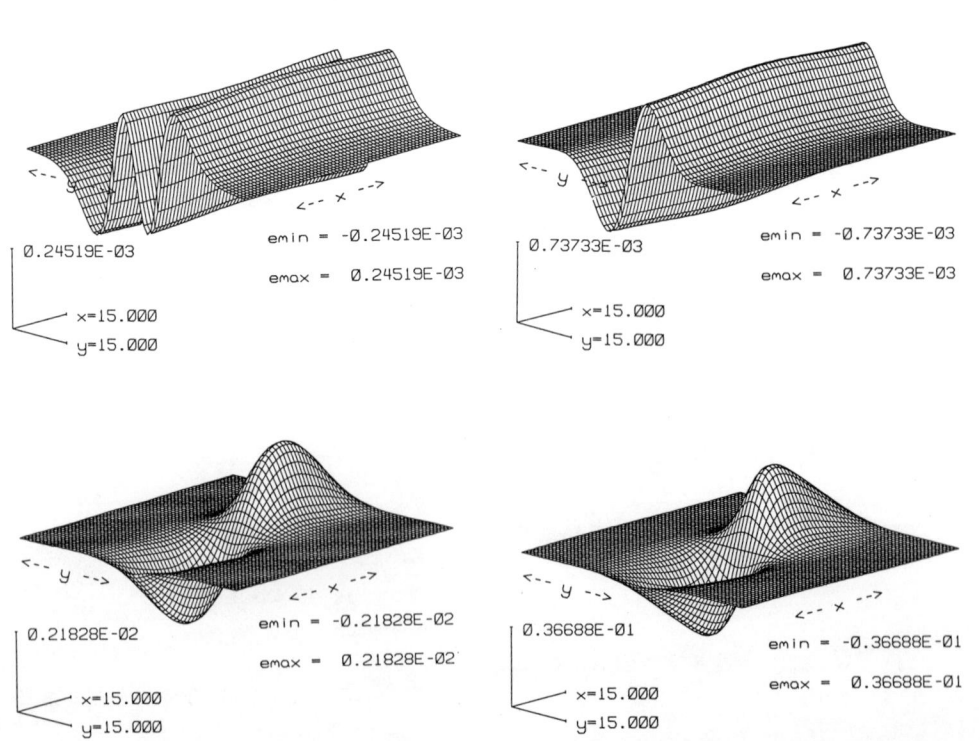

Fig. 3 Field induced into composite (anisotropic) slab by a circular current loop. Depth = 0.40 inch.

ACKNOWLEDGEMENT

This work was supported by the Naval Surface Weapons Center, White Oak Labs, Silver Spring, MD under contract No. N60921-85-C-0046 with Sabbagh Associates, Inc.

REFERENCES

[1] Thomas M. Roberts and Harold A. Sabbagh, *Review of Progress in Quantitative Nondestructive Evaluation*, 5B, D. O. Thompson and D. E. Chimenti, eds., (Plenum Press, New York, 1986), pp. 1105-1111.

[2] Harold A. Sabbagh, L. David Sabbagh and Thomas M. Roberts, ``An Eddy-Current Model for Three-Dimensional Nondestructive Evaluation of Advanced Composites'', (Naval Surface Weapons Center, Silver Spring, MD 20903-5000 technical report NSWC TR 85-304, 28 July 1985).

FUNDAMENTALS OF THERMAL WAVE PHYSICS

Jon Opsal

Therma-Wave, Inc.
47734 Westinghouse Drive
Fremont, CA 94539

INTRODUCTION

In this talk, we present the basic elements of thermal wave physics with a particular emphasis on the propagation and scattering of thermal waves. The most significant aspects of thermal waves in terms of their potential for materials characterization will be shown using simple examples and a minimum of mathematical analysis. Although most practical applications require a full 3-dimensional analysis for quantitative interpretation, much of the physics can be qualitatively understood in terms of a simpler 1-dimensional treatment appropriately modified to account for 3-dimensional effects. Following a prescription for the rigorous 3-dimensional analysis we discuss some of the implications in terms of the simpler modified 1-dimensional model. To emphasize the practical aspects of thermal wave physics we next describe a highly sensitive optical method for detecting thermal waves that is nondestructive and capable of making quantitative measurements of material properties. Finally, to illustrate the rapid evolution of this field, we conclude by presenting some recent results on semiconductors which are a combination of thermal and electronic effects and potentially significant for their sensitivity to surface conditions.

THEORY

The mathematics of thermal waves begins with the assumption of linear proportionality between the heat flux, $J$, and the temperature gradient, $\nabla T$,

$$J = -\kappa \nabla T \tag{1}$$

in which the thermal conductivity, $\kappa$, is assumed to be independent of variations in the temperature, $T$. In most thermal wave experiments, the induced sample temperature excursions are less than 10 °C and $\kappa$ is, therefore, essentially constant. The next step towards a description of thermal waves is to invoke energy conservation which in differential form is the equation of continuity

$$\nabla \cdot J + \partial(\rho C T)/\partial t = f(r,z)Q(t) \tag{2}$$

where $\rho C$ is the volume specific heat expressed in terms of the density, $\rho$, and mass specific heat, C, both of which are also assumed to be independent of variations in T. The right side of Eq. (2) represents the rate of heat input per unit volume with a spatial distribution, $f(r,z)$, and a time dependent factor, $Q(t)$. The thermal wave equation is then obtained by assuming the sinusoidal time dependent form for $Q(t)$,

$$Q(t) = Qe^{-i\omega t} \qquad (3)$$

That is, from Eqs. (1)-(3) we have

$$\nabla^2 T + q^2 T = -f(r,z)Q/\kappa \qquad (4)$$

which is the basic equation of thermal wave physics [1]. Equation (4) describes waves propagating with a complex wave vector q given by

$$q = (1 + i)(\omega \rho C/2\kappa)^{1/2} \qquad (5)$$
$$= (1 + i)/\mu,$$

where $\mu$ is the thermal diffusion length. Thermal waves are, therefore, critically damped becoming insignificant after but a few thermal diffusion lengths from their point of origin and, consequently, potentially useful for localized defect detection and depth profiling.

One of the complicating features of any wave equation is the source distribution, in this case $f(r,z)$, which must be specified prior to obtaining a complete solution. One can, however, learn a lot about thermal waves without having to study their dependence on sources. Since the thermal wave equation is linear, the general solution for an arbitrary source can be formulated as a linear superposition of basis functions obtained from known sources. A particularly useful representation comes from the 1-dimensional $\delta$-function distribution, $\delta(z-z')$. Replacing the spatial distribution $f(r,z)$ in Eq. (4) with $\delta(z-z')$, leads to the wave equation for the 1-dimensional thermal wave green's function, $T_g(z,z')$,

$$d^2 T_g/dz^2 + q^2 T_g = -\delta(z-z')Q/\kappa \qquad (6)$$

which gives the response at z due to a $\delta$-function source of strength $Q/\kappa$ at $z'$. Multiplying the solution of Eq. (6) by an arbitrary distribution $f(z')$ and integrating over $z'$ then gives the complete solution for an arbitrary source, $f(z')Q/\kappa$, i.e.,

$$T(z) = \int dz' \, T_g(z,z')f(z'). \qquad (7)$$

Thus, the thermal wave problem, like any other wave problem, has been reduced to finding the solution of a wave equation with a singular (but known) source.

The next question then is how to solve for $T_g(z,z')$. We do this by first assuming that there are no boundaries that will reflect thermal waves; that is, we solve for the infinite medium green's function, $g(z,z')$. As depicted in Fig. 1, a source at $z'$ will give rise to plane thermal waves propagating to the right and to the left of $z'$. For $g(z,z')$ we then expect

$$g(z,z') = \frac{iQ}{2q\kappa} e^{iq|z-z'|} \qquad (8)$$

which can be verified by direct substitution. When boundaries are present we then add to $g(z,z')$ plane wave solutions of the homogeneous wave equation (Eq. 6 without a source) in such a manner that the boundary conditions are satisfied. That is, for $T_g(z,z')$ we have

$$T_g(z,z') = g(z,z') + A(z')e^{iqz} + B(z')e^{-iqz}, \quad (9)$$

where A and B are coefficients dependent on $z'$ and determined by the boundary conditions.

To illustrate the green's function method of solution in a multilayered medium, we consider the single layer of thickness d shown in Fig. 1. The left boundary defines the origin of the z-axis and the source is shown at $z'$. As implied in the figure, the source gives rise to the response $g(z,z')$ which then interacts with the boundaries at $z = 0$ and $z = d$.

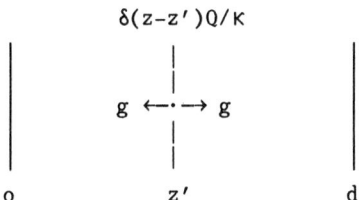

Fig. 1 Pictorial depiction of thermal wave response to a δ-function source.

For the boundary conditions we assume the ideal conditions, i.e., continuity of heat flux and temperature. At $z = 0$ we have for the heat flux, $J_o$, and temperature, $T_o$,

$$iJ_o = q\kappa [ A - (g_o + B) ] \quad (10)$$

and

$$T_o = A + (g_o + B) \quad (11)$$

which can be combined into the single equation

$$\frac{iJ_o}{T_o} = -q\kappa \left[ \frac{1 - R_o}{1 + R_o} \right] \quad (12)$$

with

$$R_o = \frac{A}{g_o + B} \quad (13)$$

and

$$g_o = g(0,z') . \quad (14)$$

The point here is that $R_o$ is the reflection coefficient for thermal waves propagating to the left, incident on the boundary at $z = 0$ and therefore a known (or at least calculable) source-independent property of the medium. At $z = d$ we similarly obtain

$$\frac{iJ_d}{T_d} = q\kappa \left[ \frac{1 - R_d e^{-2iqd}}{1 + R_d e^{-2iqd}} \right] , \quad (15)$$

where

$$R_d = \frac{B}{g_d + A} \quad (16)$$

and
$$g_d = g_o(-z')  \qquad (17)$$

with $R_d$ the reflection coefficient from the boundary at $z = d$ and therefore also a known source-independent property of the medium. Using Eqs. (14) and (16) we then have for A and B,

$$A = \frac{R_o(g_o + R_d g_d)}{1 - R_o R_d} \qquad (18)$$

and

$$B = \frac{R_d(g_d + R_o g_o)}{1 - R_o R_d} \qquad (19)$$

which essentially completes the solution of the problem. Once A and B are known, we have $T_g(z,z')$ inside the layer. By applying the boundary conditions again, we can obtain $T_g(z,z')$ in the regions $z < 0$ and $z > 0$. For the multilayered problem, this procedure is then repeated for each layer in which sources are present. Although straightforward, implementing the solution requires a fair amount of careful bookeeping.

EXAMPLES

Let's consider first the half-space problem obtained by setting $d = \infty$, $R_d = 0$ and assuming zero conductivity in the region $z < 0$, i.e., $R_o = 1$. The thermal wave green's function in this case is then

$$T_g(z,z') = \frac{iQ}{2q\kappa} \left[ e^{iq|z-z'|} + e^{iq(z+z')} \right]. \qquad (20)$$

This solution can be thought of as arising from two localized sources; the first at $z'$ and the second an image source at $-z'$ introduced to satisfy the boundary condition of zero heat flux at $z = 0$. When $z' = 0$, the two sources, of course, produce identical responses; but as $z'$ moves away from the surface, the first term becomes dominant with $T_g(z,z')$ approaching the simpler infinite medium green's function, $g(z,z')$, when $z' >> \mu$. In order to appreciate the role of the source distribution on the thermal wave response, let's assume the exponential form as often occurs in practice, $f(z') = \alpha \exp(-\alpha z')$, where $\alpha$ is an absorption coefficient. Performing the integration indicated in Eq. (7) we find for $T(z)$,

$$T(z) = \frac{i\alpha Q}{2q\kappa} \left[ \frac{e^{iqz} + e^{-\alpha z}}{\alpha - iq} + \frac{e^{iqz} - e^{-\alpha z}}{\alpha + iq} \right]. \qquad (21)$$

In the limit of weak absorption, $\alpha\mu << 1$, the heating is essentially uniform on the scale of a thermal diffusion length, and the resulting temperature profile is therefore the same as the source and independent of the thermal conductivity. Also in this limit, the frequency dependence is $1/\omega$ and the phase is $\pi/2$. One can, of course, obtain this limit directly from the continuity equation, Eq. (2), by simply assuming a weak spatial dependence in T, that is, by setting $\nabla T = 0$. Although the response is not really a thermal wave, this limit does enable one to make accurate measurements, for example, of the optical absorption in partially transparent materials.

In the other extreme, $\alpha\mu >> 1$, the spatial dependence is now set by the thermal wave vector, q. That is, the response is a propagating thermal wave with an amplitude proportional to $1/\sqrt{\omega}$ and a phase of $\pi/4$. Obviously, one can use frequency dependence and phase to detect surface absorption that might occur in an otherwise weakly absorbing homogeneous material. This is well known and has been employed extensively in spectroscopic studies. More germaine to this talk, however, is the potential of thermal waves for imaging defects and depth profiling. Thus, to study thermal wave interactions in the purest sense, we simply assume the strong absorption limit, keeping in mind that finite absorption effects need to be included in any quantitative treatment of the problem.

As an illustration of the sensitivity of thermal waves to changes in material parameters, we now consider a layer with a thermal wave vector $q_1$ and thickness d on a half-space with a thermal wave vector $q_2$. Assuming that all of the energy is absorbed at z = 0, we find for the surface temperature, T(0),

$$T(0) = \frac{iQ}{q_1 K_1} \left[ \frac{q_1 K_1 - i q_2 K_2 \tan q_1 d}{q_2 K_2 - i q_1 K_1 \tan q_1 d} \right]. \quad (22)$$

Varying the layer thickness from d = 0 to d = ∞ causes T(0) to increase or decrease (depending on $q_1 K_1 / q_2 K_2$) from $iQ/q_2 K_2$ to $iQ/q_1 K_1$ as one would expect. To see the effect of a thin layer, ($d \ll \mu_1$) on T(0) we expand Eq. (22) to first order in $q_1 d$ with the result

$$T(0) = \frac{iQ}{q_2 K_2} \left[ 1 + i q_1 d \left( \frac{q_1 K_1}{q_2 K_2} - \frac{q_2 K_2}{q_1 K_1} \right) \right]. \quad (23)$$

For a thin layer, we see that the magnitude and phase have the same linear dependence on $d/\mu_1$ and, therefore, provide identical information about the material. Depth profiling, that is, for this simple example extracting d and $\mu_1$ from a measurement of T(0), requires higher frequencies than assumed in this approximation. Equivalently, this corresponds to expanding Eq. (22) to higher order in $q_1 d$. Carrying out the expansion to second order, in fact, brings in terms that only affect the phase of T(0), thus providing the additional information needed for depth profiling.

GENERALIZATION TO 3-DIMENSIONS

Up to this point, we've assumed in the analysis a 1-dimensional source in Eq. (4). In practice, this is never quite the case, and it is therefore necessary to consider the effects of sources with lateral dimensions comparable to or smaller than a thermal diffusion length. Since our 1-dimensional description of thermal waves has been in terms of plane waves, it is only natural to also seek a solution for the 3-dimensional problem in terms of plane waves. That is, let's assume for the source term, f(r,z) in Eq. (4), the spatial Fourier transform representation

$$f(r,z) = \int d^2 q_r \, e^{i q_r \cdot r} f(q_r, z) . \quad (24)$$

Then by linear superposition we also have for T(r,z) the representation,

$$T(r,z) = \int d^2 q_r \, e^{i q_r \cdot r} T(q_r, z) \quad (25)$$

where $T(q_r,z)$ satisfies the 1-dimensional wave equation,

$$d^2T/dz^2 + q_z^2 T = -f(q_r,z')Q/\kappa \qquad (26)$$

and $q_z$ is the z-component of the thermal wave vector,

$$q_z = (q^2 - q_r^2)^{1/2}. \qquad (27)$$

Formulated in this manner, the solution of the 3-dimensional thermal wave problem is simply a superposition of solutions to 1-dimensional problems in which the thermal wave vector, q, is replaced by the z-component $q_z$ defined above. Although $q_r$ is an integration variable and not constant, one can always find an effective value that approximately describes the thermal wave without having to actually carry out the integration. Suppose, for wxample, that the lateral profile of the source is always slowly varying on the scale of a thermal diffusion length. Then, $q_r$ is effectively zero relative to q, and the thermal waves are 1-dimensional propagating in the z-direction. In the other limit where the source dimensions are much smaller than a thermal diffusion length, the effective lateral wave vector depends on the average source radius, a, and the distance, r, away from the source. Near the source where r/a << 1, we have $q_r \sim 1/a$, and away from the source where r/a >> 1, we have $q_r \sim (2r/a)(1/a)$. This limit is equivalent to setting $\omega = 0$ which means that $q_z = iq_r$. Thus, the surface temperature as given in Eq. (22) with $iq_r$ replacing $q_1$ and $q_2$ shows a stronger dependence on thermal conductivity as compared to the 1-dimensional limit. Of course, one does not get something for nothing; all of the phase information is lost in this low frequency limit.


DETECTION AND NEW DIRECTIONS

Thermal waves are of practical significance since they can easily be detected and measured. There are presently several techniques in use [2], the most prevalent based on optical methods. Here, we describe a relatively old phenomenon, photoreflectance, but one that is recent in its application to thermal waves [3]. Since the complex refractive index of most materials depends on temperature, a modulated temperature will induce a corresponding modulation in the refractive index and, consequently, a modulation in the optical reflectivity. If this modulated temperature is due to the absorption of energy from an intensity modulated pump laser, then the resulting modulated reflectance is called photoreflectance [4].

To see how this refractive index modulation affects an optical probe beam, we consider the reflectance from a half-space. Letting R denote the unperturbed reflectance, then in terms of the complex index of refraction, $n + ik$, we have

$$R = \frac{(n-1)^2 + k^2}{(n+1)^2 + k^2}; \qquad (28)$$

and applying a small perturbation $\Delta n + i\Delta k = T_0(dn/dT + idk/dT)$ we find for the thermal wave-induced modulated reflectance,

$$\Delta R/R = \operatorname{Re}\left(\frac{4T_0}{(n+ik-1)(n+ik+1)}(dn/dT + idk/dT)\right) \qquad (29)$$

In deriving this expression we have tacitly assumed that the refractive index modulation is slowly varying with respect to the optical wavelength in the material. This is generally valid, since thermal wavelengths are typically much longer than optical wavelengths and, therefore, the modulated reflectance essentially measures the surface temperature, $T_o$. One particularly significant feature of this detection scheme is the high degree of localization it affords. By focusing the pump and probe beams onto the same spot, measurements with better than 1 μm spatial resolution are possible which is especially important in semiconductor applications.

In semiconductors, however, there are electronic effects that can complicate the interpretation of a modulated reflectance measurement. As discussed by Opsal [5], an optical pump beam will also generate an electron-hole plasma wave similar in many respects to the thermal wave. Thus, we need to include in the modulated reflectance any significant plasma effects. One of the simplest and one that depends linearly on the plasma density, N, is the optical Drude effect given by [6]

$$dn/dN = -\lambda^2 e^2/(2\pi n m c^2) \quad (30)$$

$$dk/dN = -(k/n)(dn/dN) \quad (31)$$

which is valid for silicon when relaxation effects in the plasma are negligible. Assuming a probe wavelength, $\lambda = 633$ nm, we also have $k \ll n$ and Eq. (29) (with T replaced by N) to a good approximation then reduces to

$$\Delta R/R = \frac{-2\lambda^2 e^2}{\pi n(n^2-1)mc^2} N_o \quad . \quad (32)$$

To evaluate Eq. (32) we use for the electron's charge $e = 4.8 \times 10^{-10}$ esu, the velocity of light $c = 3.0 \times 10^{10}$ cm/sec, the effective mass $m = 0.15 m_o$ where the bare electron mass $m_o = 9.1 \times 10^{-28}$ gm, and the index of refraction $n = 3.9$. With these values we have $\Delta R/R \sim -10^{-22} N_o$ so that a plasma density $N_o = 10^{18}/\text{cm}^3$ implies a modulated reflectance $\Delta R/R = -10^{-4}$ which is of the same order as but of opposite sign to the expected thermal wave-induced modulated reflectance [3].

Another optical effect to consider is that due to having a spatially nonuniform perturbation on the refractive index. Following the analysis of Aspnes and Frova [7] we have for the modulated reflectance

$$\Delta R/R = \text{Re}\left(\frac{4\langle N\rangle}{(n+ik-1)(n+ik+1)}(dn/dN + idk/dN)\right) \quad (33)$$

with $\langle N\rangle$ the weighted average of $N(x)$

$$\langle N\rangle = -2iK\int dx\, N(x)\, \exp(2iKx) \quad (34)$$

and where K is the electromagnetic wave vector in the material,

$$K = (2\pi/\lambda)(n+ik) \quad . \quad (35)$$

We first note that Eq. (33) reduces to the same form as Eq. (29) in the limit that the variation in $N(x)$ is slow compared to the spatial variation of the probing optical beam. Next, in the other extreme where $N(x)$ goes to zero beyond $x = \delta$ and $|2K\delta| \ll 1$, we have

$$\langle N\rangle = -2iK\delta N_1 \quad (36)$$

where $N_1$ is the unweighted average of $N(x)$. For silicon with $k \ll n$, we then obtain for the modulated reflectance

$$\Delta R/R = \frac{16\lambda nke^2}{(n^2-1)^2 mc^2} N_1 \delta \qquad (37)$$

which we note is positive and opposite in sign to the normal Drude effect of Eq. (32). Using for the extinction coefficient $k = .025$, we have that this latter surface-type effect is equal to the bulk-type effect of Eq. (32) when $N_1 \delta = N_0 \times 10^{-4}$. That is, for $\delta = 100$ Å this would require $N_1 = 10^{20}/\text{cm}^3$. Such a magnitude is possible if, for example, there are trapping sites at the surface ($\sim 10^{14}/\text{cm}^2$) that can effectively pin some fraction of the plasma at the surface. However, we should point out that the effect becomes much more significant if k differs from the value we've assumed here. In fact, increasing k in a thin layer near the surface by an order of magnitude would dramatically affect the modulated reflectance while leaving the dc reflectance essentially unchanged. Thus, near surface lattice damage which may be insignificant in a normal optical reflectivity measurement could be readily observed in modulated reflectance through its effects on k (in addition to any plasma wave propagation effects).

To conclude we present a simple multiple trapping model as a mechanism for producing the nonuniform spatial effects on the optical properties discussed above. Under conditions of intense illumination with above band gap light we create electron-hole pairs which, in addition to diffusing and eventually recombining, can be trapped into available surface states. Trapped electrons will pin holes at the surface and, conversely, trapped holes will pin electrons. Assuming that the intrinsic dangling bond surface states trap electrons which could then be reemitted into bulk states with some characteristic time $\tau_1$, we would expect for the modulated component of the pinned hole density a frequency dependence of the form $(1 - i\omega\tau_1)^{-1}$. If, in addition to these intrinsic states, there are defect or damage related states which can trap either electrons or holes with a characteristic emission time $\tau_2$, then we similarly expect their dependence to be of the form $(1 - i\omega\tau_2)^{-1}$. Since the total number of carriers pinned at the surface depends on the net charge that's been trapped, we would then have for the modulated component of the total number of pinned carriers

$$N_p = \gamma N_0 [(1 - i\omega\tau_1)^{-1} + \beta(1 - i\omega\tau_2)^{-1}] \qquad (38)$$

where $\beta$ is the ratio of the amount of charge trapped into defect surface states to the amount trapped into intrinsic surface states. Also in Eq. (38), $\gamma$ is a constant of proportionality between the total number of pinned carriers and the number of photogenerated carriers left in the bulk, $N_0$. If the intrinsic and defect surface states trap charge of opposite sign, then $\beta$ is taken to be negative. Furthermore, since the Drude effect does not depend on the sign of the charge, we also take the absolute value of the real part of Eq. (38) while retaining the imaginary part to preserve phase.

For the modulated reflectance we then have including these trapping effects,

$$\Delta R/R = [\Delta R/R]_0 |1 - \alpha[(1 - i\omega\tau_1)^{-1} + \beta(1 - i\omega\tau_2)^{-1}]| \qquad (39)$$

where $[\Delta R/R]_0$ is the bulk-type Drude effect given by Eq. (32)

$$[\Delta R/R]_o = \frac{-2\lambda^2 e^2}{\pi n(n^2-1)mc^2} N_o \qquad (40)$$

and $\alpha$ is a constant, $\alpha = \gamma[8\pi n^2 k\delta/\lambda(n^2-1)]$. Also in Eq. (39), as we did in Eq. (38), we take the absolute value of the real part of the quantity in square brackets while keeping the imaginary part as it is. This preserves the charge independence of the Drude effect while maintaining the relative phase difference between the bulk and surface contributions to the modulated reflectance.

There are some interesting predictions of this phenomenological model that we should emphasize. The most apparent is that $\Delta R/R$ can increase with increasing modulation frequency from a minimum of $[\Delta R/R]_o |1 - \overline{\alpha[1 + \beta]}|$ to a maximum of $[\Delta R/R]_o$ whenever $\alpha|1 + \beta| \leq 1$. This behavior is consistent with all of our measurements [8] on silicon samples for which the diffusion coefficient is large enough to ensure that the bulk density $N_o$ is not decreasing significantly with increasing frequency. More interesting, however, are the possibilities when $\beta$ is a time dependent quantity. That is, if the defect surface states are somehow modified in time by the photogenerated plasma either through recombination or trapping processes, then one can expect to observe a time dependence in $\Delta R/R$ that increases, decreases, or does both with time depending on the initial value of $\beta$, the emission times, and the modulation frequency. If in a p-type sample the signal decreases with time, one could perhaps expect an increasing signal in an n-type sample since the defect states, if due to the doping, would trap charge of the opposite sign. In general we expect the time dependent effects to go away at sufficiently high modulation frequencies. These, as well as other effects have been observed, as discussed in the talk by Rosencwaig, et al [8].

REFERENCES

1. A. Rosencwaig, in VLSI Electronics: Microstructure Science, Vol. 9, edited by N. G. Einspruch, Academic, Orlando (1985) p. 227.
2. A. C. Tam, Rev. Mod. Phys. 58, 381 (1986).
3. A. Rosencwaig, J. Opsal, W. L. Smith and D. L. Willenborg, Appl. Phys. Lett. 46, 1013 (1985).
4. E. Y. Wang, W. A. Albers and C. E. Bleil, in II-VI Semiconducting Compounds, edited by D. G. Thomas, Benjamin, New York (1967) p. 136.
5. J. Opsal, these proceedings, see also, J. Opsal and A. Rosencwaig, Appl. Phys. Lett. 47, 498 (1985).
6. L. A. Lompre, J. M. Liu, H. Kurz and N. Bloembergen, Appl. Phys. Lett. 44, 3 (1984).
7. D. E. Aspnes and A. Frova, Solid State Commun. 7, 155 (1969).
8. A. Rosencwaig, J. Opsal and M. W. Taylor, these proceedings.

CLASSICAL AND QUANTUM MECHANICAL ASPECTS OF THERMAL WAVE PHYSICS

Andreas Mandelis

Photoacoustic and Photothermal Sciences Laboratory
Department of Mechanical Engineering  University of Toronto
Toronto, Ontario M5S 1A4, CANADA

INTRODUCTION

The ability of thermal waves to perform non-destructive depth-profiling studies in materials with spatially variable thermal/thermodynamic properties has been exploited mostly qualitatively so far. The lack of appropriate general theoretical models in the literature has been largely responsible for the near absence of quantitative depth-profiling, especially in media with large thermal property variations within depths on the order of the thermal wavelength. As a result of mathematical difficulties, theoretical treatments have been essentially confined to discrete, multilayered solid structures with constant thermal and thermodynamic properties within each thin layer [1,2]. Furthermore, Afromowitz et al. [3] have applied discrete Laplace transformations to the heat conduction equation to treat the production of the photoacoustic signal in a solid with continuously variable optical absorption coefficient as a function of depth, however, the thermal parameters of the solid were assumed constant. Thomas et al. [4] calculated the Green's function for the three-dimensional heat conduction equation describing thermal wave propagation in a thermally uniform solid with a subsurface discontinuity ("flaw"). More recently, Jaarinen and co-workers [5,6] used Finite Difference and Inverse methods for thermal wave depth-profiling of samples with spatially variant thermal properties from measurements of the surface temperature distribution. Aamodt and Murphy [7] very recently used vector/matrix methods to calculate thermal wave responses from discretely layered samples. These authors further considered the case of continuously varying thermal properties as the limit of infinitely thin layers.

In this work the thermal wave propagation problem in generalized continuously non-homogeneous media is approached through the formal analogy that exists between classical and quantum wave fields. It is shown that the thermal wave field Hamiltonian is nondissipative irrespective of the spatial dependence of the relevant thermal/thermodynamic properties of the system. The classical mechanical Hamiltonian can be shown to be that of a harmonic oscillator in the temperature potential field. Solutions to the classical mechanical problem lead to a thermal ray description which is rigorously valid only at material depths large compared to the thermal wavelength. Thermal wave propagation at any depth, however, is shown to be rigorously describable by a drastic step: quantization of the thermal wave field Hamiltonian [8].

HAMILTON-JACOBI FORMULATION OF THERMAL WAVE PHYSICS

The temperature field T(x) in a continuously or discretely non-homogeneous medium subject to harmonic optical or otherwise surface excitation at angular frequency $\omega_o$ is given by the Fourier-Helmholtz equation. For simplicity we will consider the one-dimensional case, with the three-dimensional problem constituting a straight forward extension of the fundamental concepts developed herein:

$$\frac{d}{dx}[k(x)\frac{d}{dx}T(x)] - \omega_o\rho(x)c(x)T(x) = 0 \qquad (1)$$

$k(x)$, $\rho(x)$, and $c(x)$ are the spatially variant thermal conductivity (W/m °K), density (kg/m$^3$), and specific heat (J/kg °K), respectively, of the medium. The Lagrangian function corresponding to Eq. (1)

$$L = \tfrac{1}{2}k(x)[\frac{dT(x)}{dx}]^2 + \tfrac{1}{2}i\omega_o T(x)^2 \ . \tag{2}$$

Upon defining the generalized coordinate $q_T = T(x)$ and momentum $p_T = \dfrac{\partial L}{\partial (dT/dx)} = k(x)\dfrac{dT(x)}{dx}$, the classical mechanical Hamiltonian can be written:

$$H(x, T, p_T) = p_T[\frac{dT(x)}{dx}] - L \tag{3}$$

$$= \frac{p_T^2}{2k(x)} - \frac{i}{2}\omega_o \rho(x) c(x) T^2 \ . \tag{4}$$

This form of the Hamiltonian is not appropriate, however, for use in the consideration of thermal wave dynamics, because it is an explicit (non-cyclic) function of x. A canonical transformation is thus required [9] such that both q and p will be constants of the motion. Using the following parametric transformations

$$\zeta = \frac{1}{J}\int_0^x [\frac{\rho(y)c(y)}{k(y)}]^{1/2} dy \tag{5}$$

$$\tau \equiv [k(x)\rho(x)c(x)]^{\frac{1}{4}} T(x) \tag{6}$$

and

$$J \equiv \frac{1}{L}\int_0^L [\frac{\rho(y)c(y)}{k(y)}]^{1/2} dy \tag{7}$$

the Hamiltonian can be written in its Hamilton-Jacobi representation

$$H(\tau, \frac{\partial W}{\partial \tau}) = \alpha \equiv E \ , \tag{8}$$

where $\alpha$ (and E) is a constant of the motion, corresponding to the total generalized energy of the thermal wave field. $W(\tau, \alpha)$ is Hamilton's characteristic function. As a consequence of the fact that H is cyclic in x in the coordinate system ($\tau, p_\tau$), the generalized momentum can be written [9];

$$p_\tau = \frac{\partial W}{\partial \tau} = \text{constant} \ . \tag{9}$$

Eqs. (4) - (9) now yield the complete functional form of the canonical Hamiltonian

$$H(\tau, p_\tau) = \tfrac{1}{2}J p_\tau^2 + \tfrac{1}{2}K\tau^2 \tag{10}$$

with

$$K \equiv -i\omega_o J \tag{11}$$

as the Thermal-wave Harmonic Oscillator (THO) spring constant. The effective mass $m \equiv J^{-1}$ is subject to a restoring, conservative force $F = -K\tau$ generated by the effective harmonic potential field $V(\tau) = \tfrac{1}{2}K\tau^2$. The combination of Eqs. (8) - (10) yields [9]

$$\tau = \frac{\partial W}{\partial \alpha} = \zeta + \beta , \qquad (12)$$

where $\beta$ is a constant of integration, which represents the constant generalized coordinate in the canonical system, and can be determined from the initial or boundary conditions to the problem. Eq. (12) gives the general solution to the Hamilton-Jacobi problem:

$$\zeta + \beta = \frac{\partial W}{\partial \alpha} = \frac{1}{J}\int \frac{d\tau}{\sqrt{\frac{2}{J}[\alpha - V(\tau)]}} . \qquad (13)$$

Eq. (13) can, in principle, be integrated and turned "inside out" to give $\tau = \tau(\zeta)$ or, using the parametric transformations (5) - (7), to yield the desired solution in the form $T = T(x)$ upon imposition of appropriate boundary conditions.

The frequency of the THO spatial oscillation may, however, be obtained without explicitly solving Eq. (13), via the use of the action-angle variable

$$I_\tau = \oint p_\tau d\tau = \oint [\frac{\partial W(\tau, \alpha)}{\partial \tau}]d\tau . \qquad (14)$$

The integral $I_\tau$ can be evaluated between 0 and $2\pi$, and its conjugate generalized angle variable coordinate $v_\tau$ representing the spatial frequency can be calculated:

$$v_\tau = \frac{\partial H}{\partial I_\tau} = \frac{1}{2\pi}(KJ)^{\frac{1}{2}} . \qquad (15)$$

Now the angular frequency can be written as

$$\Omega_\tau \equiv 2\pi v_\tau = \pm \frac{(1-i)}{L}\int_0^L [\frac{\omega_o \rho(y)c(y)}{2k(y)}]^{\frac{1}{2}} dy \qquad (16)$$

$$\equiv \pm \frac{(1-i)}{L}\int_0^L a_s(\omega_o, y)dy \qquad (17)$$

where $a_s(\omega_o, y)$ is the local thermal diffusion coefficient of the Rosencwaig-Gersho theory [10] at depth y in the medium.

Thermal Ray Limit of the Hamilton-Jacobi Theory

The Hamilton-Jacobi formulation of the thermal wave problem leads to an equation for the trajectory of propagating thermal waves at any point in space. The trajectory is given by the direction of the momentum $\vec{p}_\tau$. Upon combination of Eqs. (8) - (10) one finds

$$\vec{p}_\tau = \nabla_\tau W \qquad (18)$$

which is the rule for the construction of constant phase surfaces. Assuming a *slow* variation of the medium thermal diffusivity

$$\alpha_s(x) = \frac{k(x)}{\rho(x)c(x)} \tag{19}$$

in space, particularly over distances of the order of one thermal wavelength, the exact Eq. (18) reduces to the approximate form

$$(\nabla L)^2 = \alpha_{so}/\alpha_s(x) , \tag{20}$$

where $\alpha_{so}$ is a constant reference thermal diffusivity, and $L = a_s(\omega_o, x)x$. Eq. (20) is the *eikonal equation of thermal ray mechanics* and is entirely equivalent to the eikonal equation of geometrical optics with $\alpha_{so}/\alpha_x(x)$ replaced by the square of the index of refraction [11]. Surfaces of constant L define the thermal wave fronts. The thermal ray trajectories are everywhere perpendicular to the wave fronts and can thus be determined from Eq. (20). Another compact way for obtaining thermal ray trajectories is offered by the thermal ray variational Fermat's principle, which is simply stated in the form

$$\delta \int \frac{ds}{\alpha_s(x)} = 0 \tag{21}$$

where ds is the incremental length of the thermal ray trajectory.

The concepts presented in this section put in perspective the methods available for recasting well known Classical Mechanical ray formulations into the thermal wave problem and exploiting the considerably abundant expertise which has been developed with the former treatments (e.g. light ray analyses) to attack thermal ray problems. Typical applications of the thermal ray concept to-date can be found in the works by Burt [12-14], Bennett and Patty [15], and Mandelis et al. [16]. The main features of these applications are a) good to excellent agreements with experiments [12-16]; b) reduction of computational labor over numerical integration [12-14]; c) the requirement of only few thermal rays to calculate the entire trajectory picture [12-14]; d) applicability to many types of subsurface geometries [12-16]; e) success in rederiving the Rosencwaig-Gersho model [10] from thermal ray interference principles [15,16]; and f) the ability to measure interferometrically thin solid film thicknesses [16]. The most important disadvantage of the thermal ray concept is the approximate nature of the eikonal equation (20). The short wavelength approximation (or slow spatial variation of $\alpha_s(x)$ involved therein) restricts the applicability of the thermal ray concept, since most problems of practical interest involve thermal property gradients at material depths shorter than, or on the order of, one thermal wavelength (e.g. thermal wave imaging of microelectronic materials [17]). A mathematical statement of this restriction is that the eikonal equation (20) *cannot* rederive the Fourier-Helmholtz eqn (1) with purely algebraic manipulations. It can be shown, however, that Eq. (1) can be recovered from thermal ray mechanics through quantization of the thermal ray field.

## QUANTUM THEORY OF THERMAL WAVE PHYSICS

A complete analogy to the conventional quantum theory can be drawn from the previous classical formulations upon replacing all classical variables of the Hamilton-Jacobi theory with thermal wave quantum mechanical operators:

$$\tau \rightarrow \hat{\tau} = \tau \tag{22a}$$

$$p_\tau \rightarrow \hat{p}_\tau = -i\hbar \frac{\partial}{\partial \tau} \tag{22b}$$

$$H \rightarrow \hat{H} = i\hbar \frac{\partial}{\partial \zeta} . \tag{22c}$$

The constant $ƀ$ is the thermal wave equivalent to Planck's constant whose units are W °K/m$^2$. An eigenfunction solution is assumed to the operator eq. (22c).

$$\hat{H}\psi = iƀ\frac{\partial \psi}{\partial \zeta} \tag{23}$$

Eqs. (10) and (23) yield the canonical coordinate-dependent "Schrödinger equation" of thermal wave quantum mechanics:

$$-ƀ^2(\frac{J}{2})\frac{\partial^2}{\partial \tau^2}\psi(\tau, \zeta) + V(\tau)\psi(\tau, \zeta) = iƀ\frac{\partial}{\partial \zeta}\psi(\tau, \zeta) \tag{24}$$

where $V(\tau) = \frac{1}{2}K\tau^2$. Separation of variables in the form

$$\psi(\tau, \zeta) = \phi(\tau)\exp(-iE\zeta/ƀ) \quad , \tag{25}$$

together with the requirement for $ƀ$ to be a complex constant of the form $ƀ = (1 + i)|ƀ|$, in order for the energy eigenvalues to be real and positive, results in the following coordinate-independent Schrödinger equation

$$|ƀ|^2 J\frac{d^2}{d\tau^2}\phi(\tau) + (\omega_o J)\tau^2\phi(\tau) = 2iE\phi(\tau) \quad . \tag{26}$$

Upon definition of a new variable

$$z = (4\omega_o/|ƀ|^2)^{\frac{1}{4}}\tau e^{-i\pi/4} \quad , \tag{27}$$

eq. (26) can be written in terms of a parabolic wave Weber-Hermite equation [18]

$$\frac{d^2}{dz^2}\phi_n(z) + (n + \frac{1}{2} - \frac{z^2}{4})\phi_n(z) = 0 \tag{28}$$

with n = positive integer. The more general case with n = real has been treated elsewhere [8]. Eq. (28) admits solutions expressed in terms of Hermite polynomials

$$\phi_n(z) = 2^{-(n/2)} e^{-\frac{z^2}{4}} H_n(z/\sqrt{2}) \tag{29}$$

with the eigenvalues

$$E_n = (n + \frac{1}{2})\omega_o^{\frac{1}{2}} |ƀ| J \quad . \tag{30}$$

Writing eq. (16) in terms of eqs. (5) - (7) and substituting in eq. (30) yields:

$$E_n = (n + \frac{1}{2})|ƀ||\Omega_\tau| = (n + \frac{1}{2})ƀ\Omega_\tau \quad . \tag{31}$$

Now eq. (31) can be used to interpret $ƀ$ in terms of the generalized energy of thermal wave packets (thermions!), as the constant ratio of the energy to the angular frequency of such wave packets:

$$E = ƀ\Omega_\tau = bv_\tau \quad . \tag{32}$$

231

Eq. (32) can be used in the quantum mechanical wave packet sense [9] to provide the deBroglie thermal wavelength

$$\lambda_{th}(\zeta) = \frac{\hbar}{p_\tau} = 2\pi[\frac{k(\zeta)}{\omega_o \rho(\zeta) c(\zeta)}]^{1/2} \tag{33}$$

The proportionality relation between $\hbar$ and $\lambda_{th}$ is analogous to that observed between Planck's constant of the quantum theory of light and the wavelength of the optical radiation. It can be further shown that this relationship is consistent with the correspondence principle of quantum mechanics [8]: in the limit of $\hbar \to 0$, i.e. $\lambda_{th} \to 0$, the thermal wave Schrödinger's equation (24) becomes identical to eq. (8) of the classical Hamilton-Jacobi theory.

## Expectation Functions and Ehrenfest's Theorems

The most important application of thermal wave quantum mechanics is its ability to calculate expectation functions for various macroscopic observables. It is especially valuable in its handling the derivation of expressions in the near-field range, i.e. for material depths shallower than, or of the order of, one thermal wavelength. In this limit of great practical importance the eikonal equation (20) is not valid and the temperature field may thus be difficult or impossible to derive from the Fourier-Helmholtz eq. (1) via conventional means.

### A. Potential Energy of the THO

In terms of quantum mechanical observable notation we can write:

$$<V(z)>_n = \int_{-\infty}^{\infty} \psi_n^*(z, \zeta) z^2 \psi_n(z, \zeta) dz \tag{34}$$

where

$$\hat{H}\psi_n(z, \zeta) = E_n \psi_n(z, \zeta) \ . \tag{35}$$

Using the orthogonality property of the Weber functions [18] and the normalization condition for the eigenfunctions $\psi_n$

$$\int_{-\infty}^{\infty} \psi_n^*(z, \zeta) \psi_m(z, \zeta) dz = \delta_{nm} \tag{36}$$

eq. (34) gives [8]

$$<z^2>_n = <V(z)>_n = 2n + 1 = \tfrac{1}{2}E_n \ . \tag{37}$$

Eq. (37) shows that, for any value of n, the average potential energy is half of the total generalized energy per cycle of oscillation, a result familiar from the classical and quantum mechanical harmonic oscillators.

## B. Temperature Field

The expectation function for T(x) can be found from

$$<z>_{n,m} = \int_{-\infty}^{\infty} \psi_n^*(z, \zeta) z \psi_m(z, \zeta) dz \tag{38}$$

$$= \begin{cases} (\frac{n+1}{2})^{\frac{1}{2}} G_n^*(\zeta) G_{n+1}(\zeta) & ; \quad m = n+1 \\ (\frac{n}{2})^{\frac{1}{2}} G_n^*(\zeta) G_{n-1}(\zeta) & ; \quad m = n-1 \\ 0 & ; \quad m \neq n \pm 1 \end{cases} \tag{39}$$

where

$$G_n(\zeta) = \exp(-iE_n\zeta/\hbar) \quad . \tag{40}$$

For the purpose of obtaining an expression for the temperature field which is consistent with direct solutions to the macroscopic Fourier-Helmholtz eq. (1) in the limit of constant k, ρ, and c, the particular eigenmodes n = 0, m = 1 must be chosen, so that

$$T(x) \equiv <T(x)>_{0,1} . \tag{41}$$

It can be shown that after some algebraic manipulation [8] the expectation function (41) for the temperature field becomes:

$$T(x) = \frac{Q(x)}{k(x)\sigma_x(\omega_o, x)} \exp[-\int_0^x \sigma_s(\omega_o, y) dy] \tag{42}$$

where $\sigma_s(\omega_o, x) \equiv (1+i) a_s(\omega_o, x)$, and

$$Q(x) \equiv Q_o [\frac{k(x)\rho(x)c(x)}{k(0)\rho(0)c(0)}]^{\frac{1}{4}}$$

$$\times \{1 + (\frac{e^{-i\pi/4}}{4\omega_o^{1/2}}) [\frac{k(0)}{\rho(0)c(0)}]^{\frac{1}{2}} [\frac{d}{dx} \ln(k\rho c) \mid_{x=0}]\}^{-1} \quad . \tag{43}$$

In eq. (43) $Q_o$ is the constant heat flux at the material surface. In the limit of k, ρ, and c constant, eq. (42) reduces immediately to a simple well-known form [1], as expected.

## C. Ehrenfest's Theorems

These theorems can be easily formulated using thermal wave quantum mechanical operator algebra in the form of commutation relations (See eqs. (22)):

$$[\hat{\tau}, \hat{p}_\tau] = i\hbar \tag{44}$$

$$[\hat{\tau}, \hat{H}] = \hbar J \frac{\partial}{\partial \zeta} \tag{45}$$

$$[\hat{p}_\tau, \hat{H}] = -\hbar\omega_o \tau J \ . \tag{46}$$

It can thus be shown [8] that:

$$<\frac{d}{d\zeta}\tau(\zeta)>_{n,m} = \frac{d}{d\zeta}<\tau(\zeta)>_{n,m} \tag{47}$$

$$= \frac{1}{F^{\frac{1}{4}}(\zeta)}[J<p_\tau(\zeta)>_{n,m} + (\frac{d}{d\zeta}F^{\frac{1}{4}}(\zeta))<\tau(\zeta)>_{n,m}] \tag{48}$$

where

$$F(\zeta) \equiv k(\zeta)\rho(\zeta)c(\zeta) \ . \tag{49}$$

The validity of eq. (47) originates in the fact that the potential field $V(\tau)$ for the THO is harmonic and does not involve terms higher than second order. For $F$ = const., eq. (48) reduces to the Ehrenfest theorem

$$<p_T>_{n,m} = k\frac{d}{dx}<T>_{n,m} \tag{50}$$

in agreement with the classically expected result.

If F (which is the square of the material effusivity) is not constant, one may use the relations [8]

$$<\frac{d^2}{d\zeta^2}\tau(\zeta)>_{n,m} = \frac{d^2}{d\zeta^2}<\tau(\zeta)>_{n,m} = J\frac{d}{d\zeta}<\Pi_\tau(\zeta)>_{n,m} \quad , \tag{51}$$

where

$$\Pi_\tau(\zeta) \equiv \frac{1}{F^{\frac{1}{4}}(\zeta)}[p_\tau(\zeta) + \frac{1}{J}(\frac{d}{d\zeta}F^{\frac{1}{4}}(\zeta))\tau(\zeta)] \tag{52}$$

is the effective generalized thermal momentum. Eqs. (47) and (48) can give the following Ehrenfest's theorem:

$$\frac{d}{d\zeta}<\tau>_{n,m} = \frac{1}{J^{-1}}<\Pi_\tau>_{n,m} \ . \tag{53}$$

Here $J^{-1}$ plays the role of a generalized mass for the medium. Differentiation of $\Pi_\tau(\zeta)$ and use of eq. (51) finally yields the equation of motion of the thermal wave heat centroid in the presence of a harmonic generalized potential $V(\tau)$ and for general forms of $F(\zeta)$:

$$\frac{d^2}{d\zeta^2}<\tau(\zeta)>_{n,m} + [\Omega_\tau^2 - F^{-\frac{1}{4}}(\zeta)\frac{d^2}{d\zeta^2}F^{\frac{1}{4}}(\zeta)]<\tau(\zeta)>_{n,m} = 0 \ . \tag{54}$$

An equation similar in structure to (54) has been derived by Morse and Feshbach [Ref. 19, Eq. (6.3.22)] in connection with the eigenvalue problem of the Liouville equation. That derivation formally establishes the equivalence between the thermal wave quantum mechanical approach, eq. (54), and the original Fourier-Helmholtz eq. (1), itself a special case of the Liouville problem.

CONCLUSIONS

1. The Hamilton-Jacobi thermal *ray* formalism shows complete analogy with classical ray optics. It is strictly valid for thermal probing at material depths large compared to the thermal wavelength. Its main advantage over other, more specialized theories is the ability to treat materials with continuously varying thermal/thermodynamic properties. A few applications of thermal rays have emerged.

2. Quantum mechanical thermal *wave* formalism gives solution to the thermal wave temperature field problem for generalized media, with continuous or discontinuous variations in thermal/thermodynamic properties. Applications are anticipated in non-destructive depth profiling of materials with large degrees of inhomogeneity at depths within one thermal wavelength and rapidly varying thermal diffusivities/effusivities.

REFERENCES

1. J. Opsal and A. Rosencwaig, J. Appl. Phys. *53*, 4240 (1982).
2. S.D. Campbell, S.S. Yee, and M.A. Afromowitz, IEEE Trans. Biomed. Engn. *BME-26*, 220 (1979).
3. M.A. Afromowitz, S.P. Yeh, and S.S. Yee, J. Appl. Phys. *48*, 209 (1977).
4. R.L. Thomas, J.J. Pouch, Y.H. Wong, L.D. Favro, P.K. Kuo, and A. Rosencwaig, J. Appl. Phys. *51*, 1152 (1980).
5. J. Jaarinen and M. Luukkala, J. Phys. (Paris) *44*, C6-503 (1983).
6. H.J. Vidberg, J. Jaarinen and D.O. Riska, in "Inverse Determination of the Thermal Conductivity Profile in Steel from the Thermal Wave Surface Data" Res. Inst. Theor. Phys. Univ. Helsinki, Preprint # HU-TFT-85-38 (1985).
7. L.C. Aamodt and J.C. Murphy, J. Appl. Phys. (in press).
8. A. Mandelis, J. Math. Phys. *26*, 2676 (1985).
9. H. Goldstein, *in* "Classical Mechanics", Addison-Wesley, Reading, MA, (1965).
10. A. Rosencwaig and A. Gersho, J. Appl. Phys. *47*, 64 (1976).
11. D. Marcuse, *in* "Light Transmission Optics", Van Nostrand, New York (1982).
12. A. Burt, Proc. IEEE Ultrasonics Symp. 815 (1981).
13. A. Burt, J. Phys. (Paris) *44*, C6-453 (1983).
14. A. Burt, Proc. 4th Intern'l Topical Meeting on Photoacoustic, Thermal and Related Sciences Tech. Digest *MA10* (1985).
15. C.A. Bennett and R.R. Patty, Appl. Opt. *21*, 49 (1982).
16. A. Mandelis, E. Siu, and S. Ho, Appl. Phys. *A33* 153 (1984).
17. J. Opsal, A. Rosencwaig, and D.L. Willenborg, Appl. Opt. *22, 3169 (1983)*.
18. E.T. Whittaker and G.N. Watson, *in* "A Course of Modern Analysis", Cambridge Univ. Press, Cambridge (1963).
19. P. Morse and H. Feshbach, *in* "Methods of Theoretical Physics", McGraw-Hill, New York (1953).

TEMPORAL BEHAVIOR OF MODULATED REFLECTANCE SIGNAL IN SILICON

Allan Rosencwaig, Jon Opsal and Michael W. Taylor

Therma-Wave, Inc.
47734 Westinghouse Drive
Fremont, CA 94539

INTRODUCTION

In another paper in these proceedings, Opsal [1] discusses the origin of the modulated reflectance signal observed in silicon under the experimental conditions employed in the Therma-Probe system [2]. These experimental conditions are described in the paper by Smith, Hahn and Arst in these proceedings [3]. Table I lists the major differences between our type of modulated optical reflectance experiments and the more conventional photoreflectance experiments [4-10].

TABLE I: Comparison between our modulated reflectance experiments and conventional photoreflectance experiments.

| CONVENTIONAL PHOTOREFLECTANCE | MODULATED OPTICAL REFLECTANCE |
|---|---|
| PUMP BEAM < 1 $W/cm^2$<br>PROBE BEAM < 1 $mW/cm^2$ | PUMP BEAM > $10^4$ $W/cm^2$<br>PROBE BEAM > $10^3$ $W/cm^2$ |
| BEAM SPOTS ~ mm | BEAM SPOTS ~ 1 µm |
| PUMP MODULATION < 1 kHz | PUMP MODULATION ~ MHz |
| PROBE λ SCANNED<br>SPECTROSCOPY-CRITICAL POINTS<br>→ BAND STRUCTURE | PROBE λ FIXED<br>TRANSPORT PROPERTIES OF<br>THERMAL & PLASMA WAVES |
| NON-SEMICONDUCTORS;<br>SIGNAL FROM THERMOREFLECTANCE | NON-SEMICONDUCTORS;<br>SIGNAL FROM THERMOREFLECTANCE |
| SIGNAL IN SEMICONDUCTORS;<br>FROM ELECTROREFLECTANCE | SIGNAL IN SEMICONDUCTORS;<br>FROM THERMOREFLECTANCE<br>AND DRUDE EFFECTS |
| NONLINEAR IN PUMP INTENSITY<br>NO TEMPORAL BEHAVIOR | LINEAR IN PUMP INTENSITY<br>TEMPORAL BEHAVIOR<br>-LASER "ANNEALING" EFFECT- |

Of particular interest are the facts that: 1) in our modulated reflectance experiments the main source of signal appears to come from thermoreflectance and Drude effects from the electron-hole plasma [1,11] rather than the electroreflectance effects [12] that tend to dominate the photoreflectance signal; and 2) that the modulated optical reflectance signal appears to exhibit a temporal or laser "annealing" behavior under continuous laser irradiation whereas no such temporal behavior has been observed in photoreflectance.

As described in the paper by Opsal [1], the Drude portion of the modulated reflectance signal in a semiconductor like silicon may be written as the sum of three terms:

$$\Delta R/R = [\Delta R/R]_0 |1 - \alpha[(1 - i\omega\tau_1)^{-1} + \beta(1 - i\omega\tau_2)^{-1}]| \qquad (1)$$

where the first term is the Drude term arising from the <u>free</u> photogenerated carriers; the second term is the Drude term from those photogenerated carriers that are temporarily pinned or trapped at the surface by intrinsic (i.e., dangling-bond) surface states: and the third term is the Drude term from the photogenerated carriers that are temporarily pinned or trapped at the surface by extrinsic (i.e., defect or impurity related) surface states. Expressions and definitions for the various terms and symbols in Eq.(1) are given in [1].

Continuous laser irradiation is assumed to have no effect on the intrinsic dangling-bond surface states as long as the laser intensity is non-damaging. However, it is possible that even non-damaging laser irradiation may electronically modify the extrinsic or defect-related surface states through the presence of a high density of photogenerated carriers. Thus we may hypothesize that while $\alpha$ in Eq.(1) may be time-independent, $\beta$ may be time-dependent. Our results, as discussed below, indicate that $\beta \to 0$ as $t \to \infty$ under continuous intense laser irradiation.

EXPERIMENTAL RESULTS

Our experiments on incoming bare silicon wafers show that typical modulated reflectance signals, $\Delta R/R$, at 1 MHz lie in the range of $0.3-2 \times 10^{-3}$. Our data indicate that the higher signal levels are indicative of the small amount of residual damage present in the surface regions of the wafers that result from the chemipolishing and scrubbing processes used in the final stages of the manufacture of the wafers [3]. As described by Opsal [1], theoretical calculations predict that the modulated reflectance signal should generally increase with the extent of damage, in the form of surface states or of lattice disorder in the near surface region of the silicon wafer. These theoretical predictions are in agreement with experimental results on both incoming wafers and on ion-implanted wafers [3,13]. In Fig. 1 we see that the lowest signal is for an incoming wafer that has been thermally annealed and has several hundred Angstroms of thermal oxide grown on it. This wafer would be expected to have the least amount of damage, both from the thermal annealing and from the consumption of any damaged silicon region by the growth of the oxide film. Our experiments on ion-implanted wafers [13,14] show that the modulated reflectance signal increases monotonically with the dose of implanted ions into the wafer. Furthermore when an implanted wafer is thoroughly annealed the $\Delta R/R$ signal decreases dramatically. All of these results indicate that the modulated reflectance signal tends to increase with the amount of damage in the near surface region of the wafer.

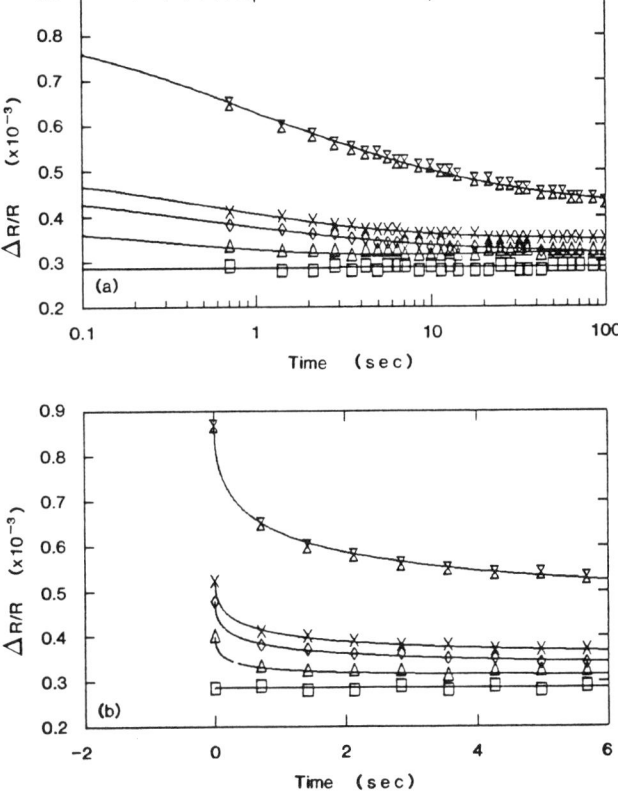

Fig. 1 Modulated reflectance as a function of time on a number of different silicon samples: □ - Si with 300 Å thermal-oxide film, △ - p-Si 5 Ω-cm (#1), ◇ - p-Si 50 Ω-cm, ✕ - p-Si 5 Ω-cm (#2), ✖ - p-Si 5 Ω-cm (#3, 3 x polishing pressure). The lower plot in this figure and in all subsequent plots of temporal data shows the behavior near t = 0.

Returning to Fig. 1, we see that except for the wafer with the thermal-oxide film, all other incoming "bare" wafers tend to exhibit a temporal dependence of the signal, that is, the magnitude (and often the phase) of the ΔR/R signal changes if the pump and probe laser beams are allowed to continually illuminate the same spot on the silicon wafer. Of considerable importance is the fact that the time-dependent curves can be fit with high accuracy to an expression of the form,

$$f(t) = A\{1 + B\exp(Ct)\mathrm{erfc}(\sqrt{Ct})\} \quad , \quad (2)$$

thus indicating that this laser-induced temporal effect is a result of a diffusion process. This particular solution of the the diffusion equation [15] corresponds to having a source term which decays exponentially in time as exp(-Ct) and should therefore approximate the removal of the defect surface states from the illuminated region that we believe is occuring in our measurements. We will henceforth refer to this laser-induced temporal effect on the ΔR/R signal as a laser annealing effect. Using Eq.(2) we are then able to predict the t = ∞ value of the ΔR/R signal and this value is indicated in Table II for the various wafers. We note that the t = ∞ signal as well as the t = 0 signal are lowest for the silicon wafer with the least amount of damage, that is, the wafer with the 300 Å of thermal oxide film. On the other hand, both the t = 0 and t = ∞ signal levels for the other wafers vary considerably and these variations appear to be related to the amount of near-surface damage present and to the amount of intrinsic disorder from the as-grown dopant concentrations. Thus a heavily doped (not shown) p-Si wafer with 0.01 Ω-cm resistivity exhibits the highest t = ∞ signal. We attribute the increased t = ∞ signal for the heavily doped 0.01 Ω-cm wafer to the effects of the high doping concentration on the the electron-hole diffusivity D (with the understanding of course that the recombination lifetime τ and the surface recombination velocity S may also be affected).

TABLE II: Actual t = 0 and predicted t = ∞ values for the modulated reflectances shown in Fig. 1.

| Sample | ΔR/R at t = 0 (×10$^{-3}$) | ΔR/R at t = ∞ (×10$^{-3}$) |
|---|---|---|
| Si with 300 Å thermal-oxide film | 0.285 | 0.286 |
| p-Si 5 Ω-cm (#1) | 0.397 | 0.326 |
| p-Si 50 Ω-cm | 0.477 | 0.312 |
| p-Si 5 Ω-cm (#2) | 0.523 | 0.345 |
| p-Si 5 Ω-cm (#3) (3 × polishing pressure) | 0.867 | 0.404 |

However if we compare the 5 Ω-cm (#1) wafer with the 50 Ω-cm wafer, we see that their t = ∞ signals are essentially identical, indicating that a change in doping concentration by a factor of 10 does not noticeably change the intrinsic ΔR/R signal when the doping concentration is low. Nevertheless these two curves have different t = 0 values, possibly indicating different amounts of near surface damage.

In fact if we compare three 5 Ω-cm wafers (#1,2 and 3) that have received different polishing treatments with wafer #1 receiving that polishing treatment that is considered to produce minimal damage while wafer #3 received the most damage, it is clear that the t = 0 signal is especially sensitive to the presence of this damage while the t = ∞ signal is less so. We are able to account for this observation by assuming that chemipolishing damage results in both some lattice disorder and, more significantly, in the presence of additional extrinsic, i.e., impurity or defect surface states. Since our laser beams are generating temperatures of only ~ 10°C, we would not expect any laser annealing effect on the lattice disorder damage. However, as we postulate [1], it may be possible that extrinsic non-dangling bond surface states may be altered, by charge neutralization, bond reconfiguration or promotion into another state such as a bulk defect state, by the presence of the photogenerated electron-hole plasma. In particular the energy for the alteration of the extrinsic surface states may come from a local electron-hole recombination event. This may be analogous to the photogeneration of recombination defects in amorphous silicon.[16-18] At any rate we would thus expect that while the t = 0 signal is indicative of the presence of both extrinsic surface states and lattice disorder, the t = ∞ signal would be mostly due to the amount of lattice disorder present, with most of the extrinsic surface states "annealed" out by the laser beams. This explanation is consistent with the observation that silicon wafers with a thermally grown oxide film exhibit no laser annealing effect. It is well known that a good $Si/SiO_2$ interface will have only dangling bond interface states and that these would not be altered by the presence of an electron-hole plasma or by local electron-hole recombination events. On the other hand extrinsic, i.e. impurity or defect-related, surface states may be altered by the electron-hole plasma and by electron-hole recombination events.

Similar results are obtained with epitaxial silicon wafers. Although epitaxial films are usually of higher crystalline quality than as-grown bulk Si, these films can exhibit high concentrations of extrinsic surface defect states from the hydrogen that is incorporated into the surface layers during the epitaxial growth reaction.

Thermal Effects

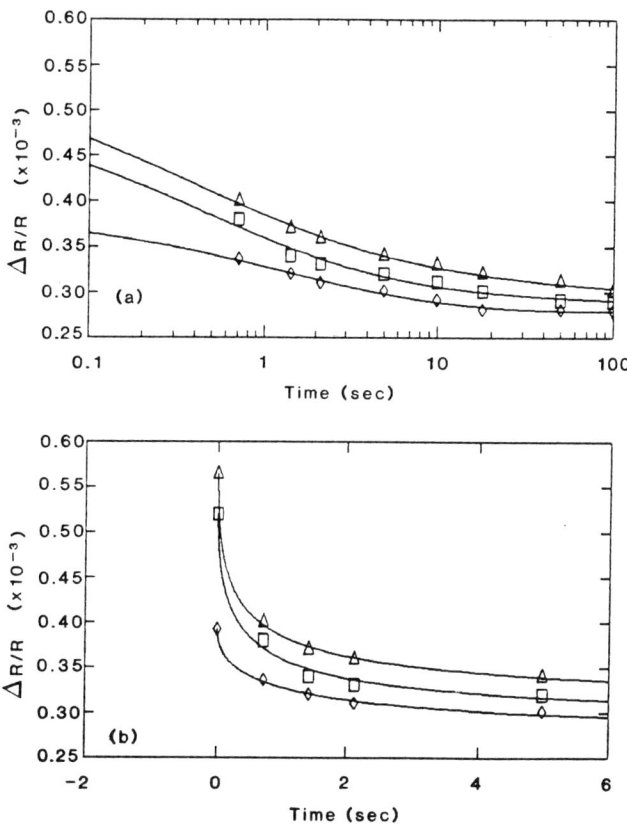

Fig. 2   Modulated reflectance as a function of time showing effects of heat treatment on a 0.21 Ω-cm epitaxial Si sample: ◊ - after 30 minutes at 200°C, □ - 72 hours after thermal treatment, Δ - before heat treatment.

In Fig. 2 we show the temporal behavior of the ΔR/R signal for a 0.21 Ω-cm epitaxial Si Film before and after a thermal anneal at a fairly low temperature of 200°C. We note that while such low temperatures appear to have little noticeable effect on the t = ∞ signal, this is not true for the t = 0 signal. In fact after only 30 minutes at 200°C, the t = 0 signal decreases substantially. However, of even greater interest is the fact that if the wafer is then reexamined after another 72 hours at room temperature in air, most of the original t = 0 signal returns. Other experiments on various bulk and epitaxial wafers have confirmed these findings. Thus moderately low temperature thermal annealing in air tends to have the same effect on the extrinsic surface states as does the laser annealing. This indicates that the activation energy for neutralizing or transforming these extrinsic surface states is indeed quite small, in keeping with the concept that the laser annealing effect is a result of an electron-hole recombination phenomenon. The fact that the thermal annealing of these states is reversible is also in keeping with the diffusion aspects of the laser annealing effect.

Effect of Ion Implantation

In Fig. 3 we show how the t = 0 and t = ∞ ΔR/R signals vary with boron and arsenic implant into Si wafers. Of considerable importance is the observation that the t = ∞ signal increases rapidly with increasing implant dose, thus indicating that the major portion of the ΔR/R signal arises from lattice disorder.

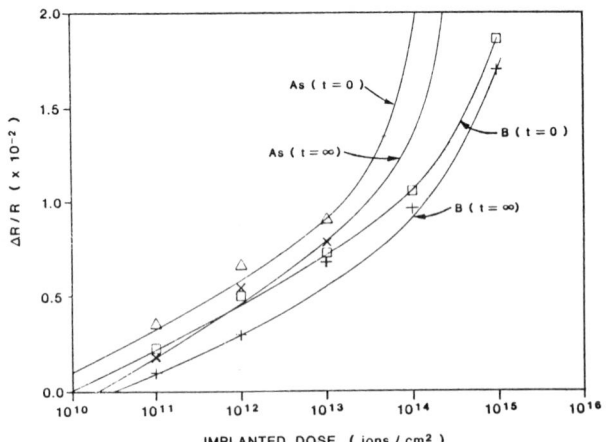

Fig. 3 Modulated reflectance signal as a function of implanted dose at t = 0, and t = ∞.

Although some of the implant-generated damage is due to extrinsic surface states, the data in Fig. 3 indicate that the contribution to the total signal from these implant-generated extrinsic surface states does not increase as rapidly as does the contribution from the lattice disorder itself.

As we noted before, a Si wafer with a thermal oxide film exhibits no laser annealing behavior. However, this is not true of such a wafer which is then implanted. The implantation process disrupts the Si/SiO$_2$ interface and allows the creation of extrinsic interface states associated with either localized structural defects or with implanted dopant ions and other impurities at the Si/SiO$_2$ interface. Thus, similar t = 0 and t = ∞ data are obtained for implants on oxidized wafers as on bare wafers.

However, implanted oxide coated wafers exhibit quite different behavior from implanted bare wafers after a high temperature anneal. While bare wafers will initially exhibit no temporal dependence or laser annealing behavior immediately after the high temperature anneal, this behavior will manifest itself a few days later. This result is similar to that observed for unimplanted epitaxial films or polished bulk wafers. However, wafers that have been implanted through a screen or gate oxide permanently lose all evidence of any laser annealing behavior after a high temperature thermal anneal. This indicates that once the high temperature post-implant thermal anneal heals the $Si/SiO_2$ interface, the extrinsic interface states are either removed or acquire too high an activation energy to be altered by the laser beams. This is consistent with our prior observations of no temporal dependence in Si wafers having a good thermal oxide film.

Duration of Laser Annealing Effect

We have investigated the duration of the laser annealing effect by first irradiating a fairly large area on a Si wafer until its modulated reflectance signal reaches its steady state condition (i.e. the t = ∞ level). We then perform some quick checks on the signal from this region as a function of time. These quick checks are performed in a manner so as to minimize as much as possible any additional laser annealing effect from the checks themselves. Our results indicate that the signal slowly comes back to its initial t = 0 level in a diffusive fashion. However, whereas the laser annealing effect occurs within minutes, the return of the signal to its initial t = 0 level usually takes several days.

This is similar to the results obtained for thermal annealing, thus indicating that the annealed state is in both cases a metastable state that will in time revert back to its original extrinsic surface state at room temperature.

Spatial Extent of Annealing Effect

Of considerable interest is the spatial extent of this laser annealing effect. This experiment is performed by illuminating a micron sized spot on the silicon wafer for 60 seconds until a reasonably annealed state is achieved, then rapidly moving a set distance from this location and recording the t = 0 signal at the new location. By repeating this procedure many times we are able to map out the spatial extent of the annealing effect after a 60 second exposure at a given site. The data obtained indicate that in Si wafers with high levels of structural damage such as heavily implanted wafers, the annealing effect appears to be confined to the irradiated site. However for wafers that have little or no structural damage such as starting Si wafers or epitaxial films, the laser annealing effect appears to extend 50-100 μm beyond the initial illuminated point after 60 seconds. This is most surprising since three-dimensional calculations readily show that neither the DC nor the AC thermal waves and electron-hole plasma waves extend much beyond 5-10 μm from the illuminated spot.

These data indicate that the laser annealable surface states either diffuse away from the illuminated area under their own concentration gradient, or that a neutralyzing factor diffuses out from the illuminated area. This neutralyzing factor may be a trapped charge from the photogenerated carriers that can hop or diffuse a considerable distance away from the irradiated area provided the Si crystal has little or no structural damage.

SUMMARY

We have presented laser-induced modulated reflectance data on crystalline silicon, epitaxial silicon, and ion-implanted, but unannealed, silicon which in general is dependent on the time of exposure to the pump and probe beam illumination. This effect, which is reversible, appears to be a new phenomenon, certainly for crystalline and epitaxial silicon, and related to the presence and temporal evolution of electronic surface states. To support our conjecture that electronic surface states are involved, we have shown data on samples which have undergone different surface treatments as well as a number that have been ion implanted. This temporal behavior, which we have termed a laser annealing effect, is present to varying degrees in all samples and appears to depend on the amount of surface damage. In addition, the effect completely disappears for samples with a thermally grown oxide as well as on samples that have been implanted through a screen or gate oxide and then subjected to a high temperature anneal. Finally, our data on bare silicon wafers having little surface damage shows that the laser annealing effect can extend well beyond the area of illumination indicating that a surface diffusion process may be taking place.

In conclusion, we want to emphasize that this laser annealing effect is reproducible and appears to be well correlated with surface conditions. Thus we believe that, in addition to its potential for providing more information about electronic surface states in silicon (and perhaps in other semiconductors as well), the temporal behavior of the laser-induced modulated reflectance can be used in practice as a nondestructive method for semiconductor surface characterization.

REFERENCES
1. J. Opsal, this proceedings.
2. Therma-Probe systems are a product of Therma-Wave, Inc. Fremont, CA.
3. W.L. Smith, S. Hahn and M. Arst, this proceedings.
4. E.Y. Wang, W.A. Albers, and C.E. Bleil, in II-VI Semi- Conducting Compounds, ed. Thomas (Benjamin, New York, 1967) p. 136.
5. J.G. Gay and L.T. Klauder, Jr., Phys. Rev. 172, 811 (1968).
6. W.A. Albers, Phys. Rev. Lett. 23, 410 (1969).
7. R.E. Nahory and J.L. Shay, Phys. Rev. Lett. 21, 1569 (1968).
8. N.G. Nilsson, Solid State Commun. 7, 479 (1969).
9. F. Cerdeira and M. Cardona 7, 879 (1969).
10. D.E. Aspnes, Solid State Commun. 8, 267 (1970).
11. A.M. Bonch-Bruevich, V.P. Kovalev, G.S. Romanov, Ya.A. Imas, and M.N. Libenson, Sov. Phys. Tech. Phys. 13, 507 (1968).
12. B. O. Seraphin, in Semiconductors and Semimetals, ed. Willardson and Beer, (Academic, New York, 1972), p. 1.
13. W.L. Smith, A. Rosencwaig, and D.L. Willenborg, Appl. Phys. Lett. 47, 584 (1985).
14. W.L. Smith, A. Rosencwaig, D.L. Willenborg, J. Opsal and M.W. Taylor, Solid State Tech. p. 85, January, 1986.
15. H.S. Carslaw and J.C. Jaeger, Conduction of Heat in Solids, (Clarendon, Oxford, 1959) p. 70.
16. D.L. Staebler and C.R. Wronski, Appl. Phys. Lett. 31, 292 (1977).
17. D.L. Staebler and C.R. Wronski, J. Appl. Phys. 51, 3262 (1980).
18. B. Aker and H. Fritzsche, J. Appl. Phys. 54, 6628 (1983).

MODULATED REFLECTANCE MEASUREMENT OF REACTIVE-ION

AND PLASMA ETCH DAMAGE IN SILICON WAFERS

W. Lee Smith

Therma-Wave, Inc.
47734 Westinghouse Drive
Fremont, CA 94539

Patrice Geraghty

Advanced Micro Devices
Sunnyvale, CA 94088

INTRODUCTION

Reactive ion etching (RIE) and plasma etching (PE) are vital processes for the attainment of densely-packed, micron-scaled structures for VLSI integrated circuits. However, it is widely recognized that undesirable modifications of semiconductor or insulator materials may accompany the use of these dry etch processes. For example, during the RIE process, samples are exposed to high energy ions, UV photons and x-rays, all of which can result in radiation damage [1] in the form of non-annealable structural defects in gate oxide or $Si/SiO_2$ interface regions, deep level traps or surface states. In terms of IC device performance, these effects cause transistor threshold voltage shifts, poor subthreshold performance, increased junction leakage or decreased capacitor charge retention time, degradation of minority carrier lifetime, barrier shifts in Schottky diodes and reduction of the integrity of trench isolation structures [2,3]. Contamination from sputtering of the chamber parts or of the oxide mask may add to these problems. In addition, polymer material that may be deposited from carbon-containing gases has been found to create oxygen-induced stacking faults [4]. All of the above can lead to significant yield reduction.

A number of techniques have been developed to study the effects of RIE or PE on single crystal silicon. MOS capacitors can be fabricated and capacitance-voltage plots generated to obtain threshold shifts or interface state densities [5]. Generation lifetimes can be measured. Deep level transient spectroscopy can be used to determine trap levels in silicon, but in this technique wafers are subjected to temperature extremes (100 - 300 °C) during the measurements [6]. Alternatively, a decorative, wet chemical etch such as a Seeco [7] or Wright [8] etch can be used to highlight dislocations in silicon. These etches are, however, destructive in nature. Channeling-mode Rutherford back scattering, reflected high energy electron diffraction, electron spin resonance and measurement of the current-voltage characteristics of Schottky diode structures [9] have also been employed to study radiation damage in silicon. Each of the above methods is either destructive to the examined wafers or lacks the speed or ease of use necessary to constitute an effective, practical method for detecting RIE or PE damage during dry etch process development or during the integrated circuit manufacturing process.

Recently it has been shown that measurement of laser-induced modulated optical reflectance can nondestructively detect the radiation damage in Si due to RIE or PE with good sensitivity [10,11]. This method also measures etch uniformity over the wafer surface and can detect the presence of deposited polymer material on micron-scale device features on patterned wafers. In the present study, we have used this method to measure RIE damage at the bottom of 6-micron deep trench structures as a function of RIE process parameters such as dc bias and pressure. This parametric study allows the RIE processs variables to be adjusted to minimize damage to the silicon surface.

EXPERIMENTAL PROCEDURE

Sample Preparation

The trench samples used are p-type <100> silicon substrates with $n^+$ buried layer and $n^-$ epi silicon as illustrated in Fig. 1. The masking material for the silicon trench etch is 4% P-doped CVD oxide etched in a commercial oxide etcher. Trench openings are 1.2 microns wide, with one 10-micron wide trench surrounding the die. This wider trench is used for the modulated reflectance measurements. Prior to silicon etch, samples are cleaned in RCA solution and dipped in 10:1 HF.

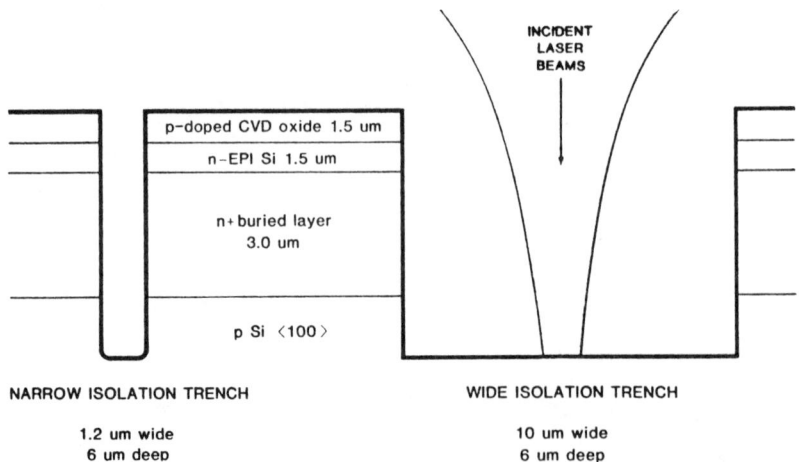

Fig. 1: Diagram of the structure of the trench wafer samples. Also shown is the measurement laser beam focused on the wider trench bottom.

Etch Apparatus and Process

The main etch step is a low pressure, medium power process with which 80% of the trench is etched. As will be described below, two of the

parameters varied in the experiments to determine their effect on damage
are the relative proportion of the constituent gases and the total flow in
this etch step. A "clean up" etch step is incorporated to remove the
damaged Si produced by the "main" step while completing the etch. As
experimental parameters in this "clean up" step, the dc self-bias voltage
is varied between -60 and -250 volts and the pressure between 15 and 100
mTorr to determine their effect on silicon damage.

Evaluation Technique

The apparatus employed is the commercial Therma-Probe inspection
system (Therma-Wave, Inc., Fremont, CA), which is described in the article
by Smith, Hahn and Arst in these proceedings [12]. In this apparatus,
thermal waves are generated and detected by two low-power laser beams
auto-focused to a 1-micron spot diameter on the wafer surface. Absorption
of light from an acousto-optically modulated (1 MHz) Ar-ion "pump" laser
generates thermal and plasma waves within the surface region of the silicon
down to an effective depth of about three microns. These waves are then
detected by the HeNe "probe" laser through the pump-induced, 1 MHz
modulation of the sample reflectivity at the wavelength of the probe laser.
As described in [13-16], the effects of the thermal and plasma waves on the
reflectivity of silicon result in a net modulated reflectivity signal that
is very sensitive to the presence of disorder or defects in the surface
region of the wafer. The measurement method is similar to that employed
for ion implant monitoring [17,18] and polishing damage characterization
[12] in which damage to the silicon is detected by the Therma-Probe system.
Significant characteristics of this method are the use of low-power laser
beams for noncontact, nondestructive measurements, the use of a 1-micron
probe spot allowing measurements to be made on patterned, product wafers
as well as test wafers, the speed of the measurements (3-7 minutes) and
the high sensitivity allowing extremely low levels of damage to be measured.

For all the measurements presented in the next section, a Therma-Probe
scan of 50-micron length is made along the bottom of a wide trench (Fig. 1)
on each sample wafer. The average value of the modulated reflectance,
$\Delta R/R$, signal from each wafer is displayed on the following graphs. Longer
(diameter) scans and whole-wafer contour maps are also available in this
method.

RESULTS AND DISCUSSION

Dependence on Bias Voltage

Figure 2 illustrates the results of varying the dc self-bias voltage
from -60 to -250 V in the "clean up" step of the silicon etch process. The
purpose of this step is to remove the first few hundred Angstroms of
silicon, damaged by the previous high-bias main etching step, from the
bottom of the trench while rounding out the trench bottom corners by
operating at a slightly higher pressure. The results, in general agreement
with previous work by Pang et al. [19], show that the level of RIE damage
as measured by the modulated reflectance technique increases markedly with
increasing bias voltage above a certain threshold bias voltage, in this
case about -120 volts. The data points in Fig. 2, the light and dark dot
symbols representing measurements on two separate runs in the same etcher
but several weeks apart, illustrate the reproducibility of the damage
produced by the etch process and measured by the Therma-Probe system. Note
that at low bias voltage, the measured $\Delta R/R$ signals are very close to
$30 \pm 3 \times 10^{-4}$, the value generally obtained on non-damaged silicon in
other studies [12].

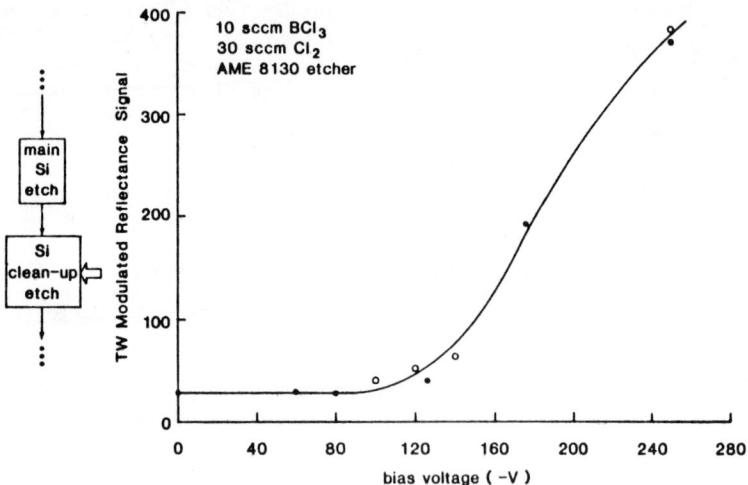

Fig. 2: Measured modulated reflectance signal, indicating the level of RIE-induced damage, versus bias voltage for the "clean up" step of a silicon etch process. The light and dark dot symbols denote data recorded on two separate runs weeks apart in time.

## Dependence on Pressure

The effect of varying pressure in the "clean up" silicon etch step is shown in Fig. 3. As the pressure is increased under constant flow conditions, the mean free path of the reactive ions is shortened, and the mechanisms of etch become more chemical in nature, rather than physical (sputtering). Thus damage to the crystal lattice should lessen. This is reflected in our data, where the general trend shows a decrease in crystal damage with increasing pressure. The anomalous data point at a pressure of 15 mTorr may be due to an inadequate supply of reactant gas at such low operating pressure.

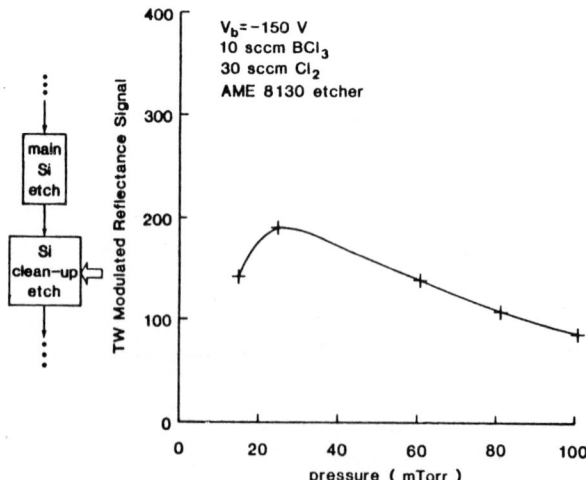

Fig. 3: Modulated reflectance signal versus pressure in the "clean up" step of a silicon etch process.

## Dependence on Etch Chemical Composition

The effect of etch chemistry on crystal damage has also been investigated. Figure 4 shows the dependence of the $\Delta R/R$ signal on the percentage of $BCl_3$, in a $BCl_3/Cl_2$ mixture at two different values of total flow rate. The substantial variation of the indicated damage in these data is surprising and will be further investigated.

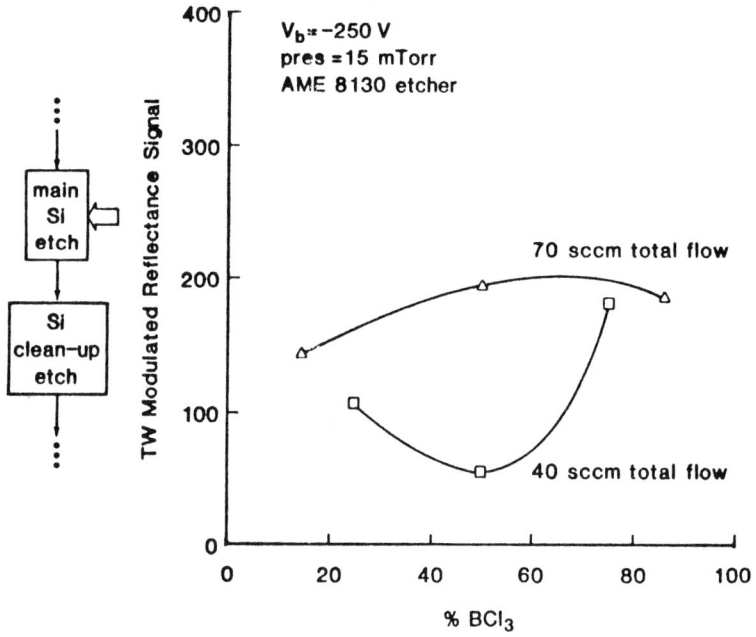

Fig. 4: Modulated reflectance signal versus percentage of $BCl_3$ in a $BCl_3/Cl_2$ mixture, at two values of total flow rate, in the "main step" of a silicon etch process.

## Modulated Reflectance Signal vs. Process Step

The above results have illustrated the capability of modulated reflectance measurements to provide a quick, nondestructive means for optimizing the silicon etch process to minimize damage. Since device wafers can be used, another application of this technique is as a routine monitor of critical process steps that may contribute to lower yields. Figure 5 illustrates the $\Delta R/R$ signals obtained at eight process steps in the trench-formation sequence: starting material, after oxide etch, after pre-silicon-etch clean, etc. Any deviation from a standard set of $\Delta R/R$ signals can be immediately observed and the process problem rectified before device yields are impacted. For example, wafers can be recleaned after oxide etch if not "in spec". Any drift in the equipment or process that affects the level of silicon damage can be identified immediately rather than later at the nonsalvagable stage of wafer probe.

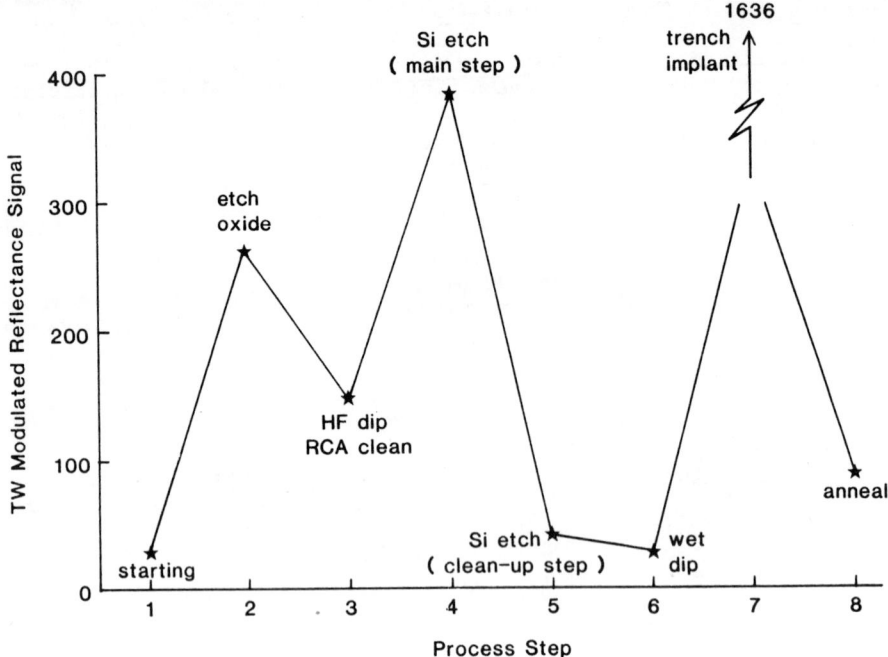

Fig. 5: Modulated reflectance signal measured at the bottom of 10-micron wide trenches on product wafers at eight steps in the trench-formation process sequence. Nondamaged silicon is characterized by a $\Delta R/R$ signal of $\simeq 30 \times 10^{-5}$.

CONCLUSIONS

In this work we demonstrate that RIE-induced damage to silicon wafers is readily detected and quantified using a modulated reflectance technique. This method is rapid, noncontact, nondestructive, and has 1-micron spatial resolution. No special patterns or devices need to be fabricated or test structures created; rather, actual device wafers can be used. Measuring along the bottom of 10-micron wide trenches, RIE damage versus etcher bias, pressure, flow and gas composition can be measured. These parametric results enable adjustment of process etch parameters to minimize damage. We also demonstrate the use of modulated reflectance measurements to monitor process and equipment stability at critical steps in the RIE trench etch sequence.

REFERENCES

1. S.W. Pang, D.D. Rathman, D.J. Silversmith, R.W. Mountain and P.D. DeGraff, J. Appl. Phys. 54, 3272 (1983).
2. S. Pang, Solid State Tech., p. 294, April (1984).
3. J. Dielemn and F.H.M. Sanders, Solid State Tech. p. 191, April (1984).
4. R. Ogden, R.R. Bradley and B.E. Watts, Phys. Status Solidi 26, 135 (1974).
5. L.M. Ephrath and R.S. Bennett, J. Electrochem. Soc. 129, 1822 (1982).
6. H. Matsumoto and T. Sugano, J. Electrochem. Soc. 129, 2823 (1982).
7. F. Seeco d'Aragona, J. Electrochem Soc. 119, 948 (1972).
8. M. Wright Jenkins, Proc. of ECS Meeting, Washington DC, 1976, p. 63.
9. S.J. Fonash, Solid State Tech., p. 201, April (1985).
10. W. L. Smith and A. Rosencwaig, Bull. Amer. Phys. Soc. 31, 273 (1986).
11. Application Note 200.02, Therma-Wave, Inc., March 1986.
12. W.L. Smith, S. Hahn and M. Arst, this proceedings.
13. A. Rosencwaig, J. Opsal, W.L. Smith and D.L. Willenborg, Appl. Phys. Lett. 46, 1013 (1985).
14. J. Opsal and A. Rosencwaig, Appl. Phys. Lett. 47, 498 (1985).
15. J. Opsal, this proceedings.
16. A. Rosencwaig and J. Opsal, this proceedings.
17. W.L. Smith, A. Rosencwaig and D.L. Willenborg, Appl. Phys. Lett. 47, 584 (1985).
18. W.L. Smith, A. Rosencwaig, D.L. Willenborg, J. Opsal and M.W. Taylor, Solid State Tech., p. 85, January (1986).
19. S.W. Pang, C.M. Horwitz, D.D. Rathmon, S.M. Cabral, D.J. Silversmith and R.W. Mountain, Proc. Electrochem. Soc. 83-10, 84 (1983).

# PICOSECOND TRANSIENT THERMOREFLECTANCE:

# TIME-RESOLVED STUDIES OF THIN FILM THERMAL TRANSPORT

Gary L. Eesley

Physics Department
General Motors Research Laboratories
Warren, MI 48090

The advent of new and sophisticated material growth processes (molecular beam epitaxy, chemical vapor deposition and ion sputter deposition) has produced new exotic materials such as amorphous alloys and compositionally modulated structures [1]. The atomic level structure of these materials can be proved by techniques such as x-ray diffraction. The electrical and thermal transport properties are also used to characterize these materials, which are usually deposited as thin films onto supporting substrates. Although the substrate may be electrically isolated from the film, complete thermal isolation is more difficult to achieve and thermal transport measurements are complicated.

A variety of optical, non-contact techniques have been developed to measure the thermal diffusivity of thin films. For instance, pulsed photo-thermal radiometry can be used to measure the thermal diffusivity of free standing metal films of known thickness [2]. This technique uses short laser pulses to heat the sample surface, followed by a time-resolved measurement of the black-body radiation to determine the surface cooling rate. Another method, photothermal deflectometry, has been used to determine thermal properties by measuring the optically-induced thermoelastic deformation of the surface or the heating induced refractive index gradient in a gas above the surface [3,4]. More recently, Rosencwaig et al. [5] have shown that a modulated thermoreflectance measurement is an equally sensitive method which can yield the same information as the deflection techniques. All of these techniques have been demonstrated with relatively low time resolution measurements, and in order to determine supported thin film thermal properties a knowledge of the substrate thermal properties is required. As a result, the accuracy of determining the film thermal diffusivity will depend on the precision to which the substrate thermal properties are known. The thermal impedance of the film-substrate interface can also complicate the determination of the film thermal properties. Although the theoretical treatments of heat flow in a simple multilayer system are correct, they only account for heat flow across ideal boundaries where there is an abrupt change in thermal properties [6-9]. In reality, many interfacial boundaries are characterized by strains, contamination and chemical interactions which can alter the thermal properties relative to the bulk values of the constituent materials. Detailed knowledge of the interfacial properties can be difficult to obtain and modelling of the heat flow may not be accurate.

As a solution to this problem, we have developed a technique which uses ultra-short laser pulses to generate transient thermoreflectance (TTR) signals which correlate with the transport of heat away from the film surface. Time-resolved measurement of these signals can be used to determine the thermal diffusivity of a supported film, independent of the substrate. In the following we will review the time domain analysis of thermal transport relevant to our technique. We will show how TTR can be used to study the thermal impedance of interfaces, in addition to measuring the thermal diffusivity of single element metal films. We will also show that on the picosecond timescale some interesting deviations from the standard heat flow equation can occur.

## THERMAL TRANSPORT: TIME DOMAIN

In the classical theory of heat conduction, the heat flux J is directly proportional to the temperature gradient. For the case of one dimensional heat flow, this relationship can be expressed as

$$J(z,t) = -K\frac{\partial T(z,t)}{\partial z} \quad , \tag{1}$$

where K is the thermal conductivity and T is the temperature. When this flux is combined with the energy conservation equation

$$C\frac{\partial T(z,t)}{\partial t} = -\frac{\partial J(z,t)}{\partial z} + P(z,t) \quad , \tag{2}$$

we obtain the standard parabolic heat diffusion equation

$$C\frac{\partial T(z,t)}{\partial t} = K\frac{\partial^2 T(z,t)}{\partial z^2} + P(z,t) \quad , \tag{3}$$

where $P(z,t)$ is a source term and C is the material heat capacity.

We are interested in the time response of the temperature increase produced by an incident laser pulse. In the case of metals, the optical radiation will be exponentially attenuated as a function of distance into the metal. If we assume a Gaussian shaped optical pulse of duration $\tau$, and intensity I, the source term in Eq. (3) becomes

$$P(z,t) = I(1-R)\alpha R e^{-\alpha z} e^{-(t/\tau)^2} \quad , \tag{4}$$

where R and $\alpha$ are the metal reflectivity and absorbtivity, respectively. Substituting this heating source term into Eq. (3) and solving for the surface temperature as a function of time results in the curve shown in Fig. 1. This result represents the general response of a metal to an optical heating pulse with duration short compared to the time required for the heat to diffuse out of the heated volume. Here we have asssumed a 4 psec wide heating pulse, and the horizontal axis is labelled in delay time units; that is, time relative to the arrival of the heating pulse at the surface. Since the heating pulse has a finite duration, the zero time delay is defined to occur at the peak of the heating pulse intensity.

In thermal transport measurements, the decay time of the temperature is the quantity of interest. This decay time is directly related to the thermal diffusivity of the material, which is defined as

$$\kappa = K/C \quad . \tag{5}$$

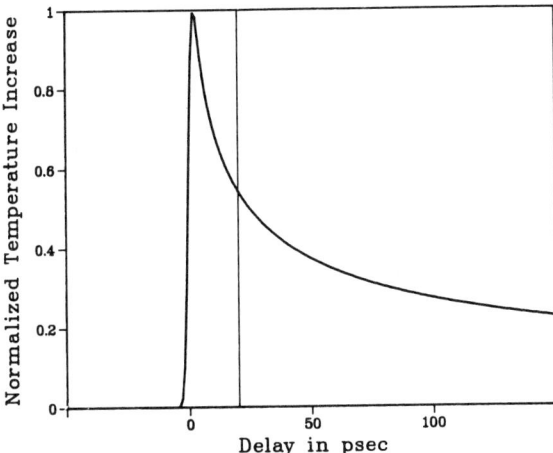

Figure 1. Calculation of the normalized temperature increase of the surface of a metal.

If the optical power is deposited at the surface of the medium, instead of in the heating depth $1/\alpha$, then $\kappa$ would solely determine the decay time of the surface temperature. In reality, however, the finite heating depth must be accounted for in determining the temperature decay profile. As a result, the decay time depends on $\alpha$ as well. This is understandable since we are measuring the transport of heat away from the surface, and this occurs in a time on the order of $10/\kappa\alpha^2$. In Fig. 1, a thermal diffusivity of $2.3 \times 10^{-5}$ m$^2$/sec and a heating depth of $1/\alpha = 14$ nm were used.

Our discussion so far has only dealt with the macroscopic nature of thermal transport described by Eq. (3). Some interesting phenomena can be described if we consider the microscopic details of light absorption and thermal transport. In metals, the thermal conduction is dominated by the free electrons which absorb a fraction of the incoming optical pulse. The electrons which absorb the light thermalize very rapidly with the surrounding electrons and then cool by transferring energy to the metal lattice via electron-phonon scattering. This scattering process proceeds simultaneously with the diffusion of thermal energy down the temperature gradient. In Eq. (3), the thermal conductivity is that of the electrons, whereas the heat capacity used is that of the lattice. This is because the electronic thermal conductivity is much larger than that of the lattice. As a result of the electron-phonon coupling, the lattice heat capacity is used since it is much larger than that of the electrons. Thus the thermal diffusion is a coupled process, involving both the electrons and the lattice ion cores in a metal.

The validity of Eq. (3) in describing thermal transport really depends on the timescale we are considering. Clearly the electron-lattice temperature equilibration requires a finite amount of time. In metals this time is on the order of one picosecond ($10^{-12}$ sec). Thus for heating pulsewidths of tens of picoseconds or longer, Eq. (3) describes both the electron and lattice temperatures. When the optical heating pulsewidth is comparable to or shorter than this equilibration time, then we would expect a nonequilibrium temperature difference to exist between the electrons and the lattice. The fact that the electrons can be elevated to a temperature above the lattice results from the relatively small electron heat capacity.

Under these conditions the thermal transport and cooling process must be modeled by a two-temperature system of coupled differential equations, describing the electron temperature $T_e$ and the lattice temperature $T_i$

$$AT_e \frac{\partial T_e}{\partial t}(z,t) = K\frac{\partial^2 T_e}{\partial z^2}(z,t) - G(T_e-T_i) + P(z,t) \quad , \quad (6)$$

and

$$C_i \frac{\partial T_i}{\partial t}(z,t) = G(T_e-T_i) \quad . \quad (7)$$

In Eqs. (6) and (7), A is the electronic constant of heat capacity (linear in temperature, $T_e$), $C_i$ is the lattice heat capacity and G is the electron-phonon coupling constant.

This nonequilibrium situation was postulated and modelled theoretically nearly thirty years ago [12]. Only recently have we been able to observe this phenomenon using picosecond pulsed lasers [13,14]. This nonequilibrium heating can be observed in both semiconductors and metals. With the advent of femtosecond ($10^{-15}$ sec) pulsed lasers, time-resolved measurements of hot electron transport and electron-phonon relaxation are possible. On these ultrashort timescales, Eq. (6) predicts that hot electrons can be generated which transport heat with a thermal diffusivity much larger than that of the equilibrium transport in Eq. (3). That is, $K/AT_e \gg K/C_i$, for temperature excursions of several tens of degrees.

A solution to the coupled equations (6) and (7) can also be found by the method of finite differences, and results for the metal copper are shown in Fig. 2. We have used a heating pulsewidth of 5 psec and a pulse energy of 0.5 nJ, and during this time we see that the electron temperature does exceed that lattice temperature by a few degrees [14]. The use of substantially shorter pulses results in a larger electron-lattice temperature mismatch, and the equilibration time exceeds the laser pulsewidth. Such studies are currently in progress using 80 fsec visible light pulses [15].

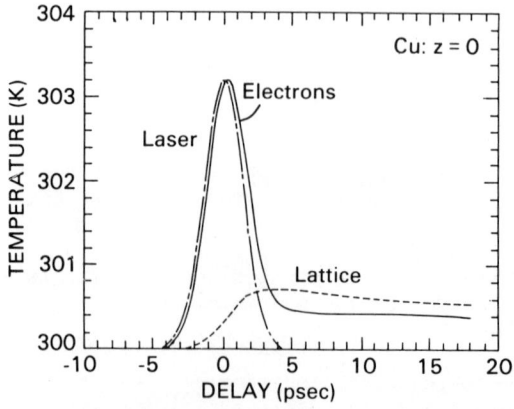

Figure 2. Calculation of the nonequilibrium heating of copper electrons. The laser heating pulse is centered at zero time delay.

The thermal transport processes we have discussed thus far have been based upon the heat flux of Eq. (1). When this flux is combined with the energy conservation equation, a heat diffusion equation is obtained. This equation predicts the instantaneous propagation of a thermal disturbance throughout the medium (see the Greens function in [10]). Despite this unrealistic result, these diffusion equations are quite accurate in modelling thermal transport in most situations. However, it has been shown that in situations involving transient heating at very low temperatures, the classical diffusion theory breaks down. Thermal transport takes the form of wave propagation with a finite velocity. This phenomena is referred to as second sound, in that phonon heat pulses have been observed to propagate macroscopic distances in a manner analogous to phonon acoustic pulses [16].

This deviation from the diffusion model can be accounted for by postulating a new heat flux equation

$$J + \tau_r \frac{\partial J}{\partial t} = -K \frac{\partial T}{\partial z} \quad . \tag{8}$$

When this equation is combined with the energy conservation equation (2), a hyperbolic differential equation is obtained

$$\frac{1}{v^2} \frac{\partial^2 T}{\partial t^2} + \frac{1}{\kappa} \frac{\partial T}{\partial t} = \frac{\partial^2 T}{\partial z^2} + \frac{P}{K} + \frac{\kappa}{v^2} \frac{\partial P}{\partial t} \quad . \tag{9}$$

In these equations, a propagation velocity v appears which is related to the thermal diffusivity $\kappa$ by the relation

$$v^2 = \kappa/\tau_r \quad , \tag{10}$$

where $\tau_r$ is the thermal carrier relaxation time (electron-phonon relaxation time in conductors).

It is clear that in the limit of zero relaxation time, the propagation velocity becomes infinite and Eq. (9) reduces to the diffusion equation. Alternatively, if transient heating occurs on a timescale which far exceeds $\tau_r$, then one would expect the diffusive nature of heat transport to apply. At very low temperatures (~1K) in solids, when $\tau_r$ is very large, the wave nature of thermal transport would become more apparent. Experiments have confirmed this phenomenon in solid helium and NaF crystals, where phonon mean free paths are comparable to the sample dimensions [17,18].

Beyond these rare situations, it is not clear that Eq. (9) is applicable or even necessary to describe thermal transport. Nevertheless there exists a core of literature on this subject, and solutions to the hyperbolic heat flow equation under a variety of conditions are well documented [19]. With the recent advances in generating ultrashort laser pulses for annealing and melting applications, it is possible that the wave nature of thermal propagation will be an important feature to consider in modelling heat transport at early times. In cases where film thicknesses are less than carrier mean free paths and heating pulsewidths are comparable to carrier scattering times, wave propagation at elevated temperatures may be observable. So far such observations have not been made, and the relevance of this form of transport is either overlooked or dismissed. It is a serious undertaking to generate realistic solutions to Eq. (9), and no attempt to do so will be given in this work.

## TRANSIENT THERMOREFLECTANCE

Transient thermoreflectance (TTR) is a technique which uses two synchronous picosecond laser pulses to measure thermal diffusion. The first pulse produces ultra-fast heating to peak temperatures on the order of 10K above ambient. The second pulse has a variable delay with respect to the heating pulse, and it is used to measure the thermally-induced change in surface reflectivity ($\sim 10^{-5}$/K). For small temperature deviations the reflectivity change is linear in temperature, and a temperature profile over several hundred picoseconds can be measured [11,20]. The penetration depth of visible light in a metal is approximately 20 nm, and thermal diffusion out of this region occurs in a few hundred picoseconds. Therefore, for film thicknesses of 100 nm or greater, the TTR measurement can be completed before substrate effects become important.

Since the heating depth is small compared to the diameter of the illuminated surface, a one-dimensional heat flow model can be fit to our measurements. Our fitting routine solves Eq. (3) by the method of finite differences. The routine involves a two parameter fit with the thermal diffusivity and a constant scaling factor as the free parameters. The accuracy of the fit is sensitive to the value of the optical absorption coefficient of the material, since we are monitoring the flow of heat out of the optical heating depth. We determine the complex refractive index of our samples by measuring the ratio of s-polarized to p-polarized reflectivity for both 30° and 70° angles of incidence. We fit these

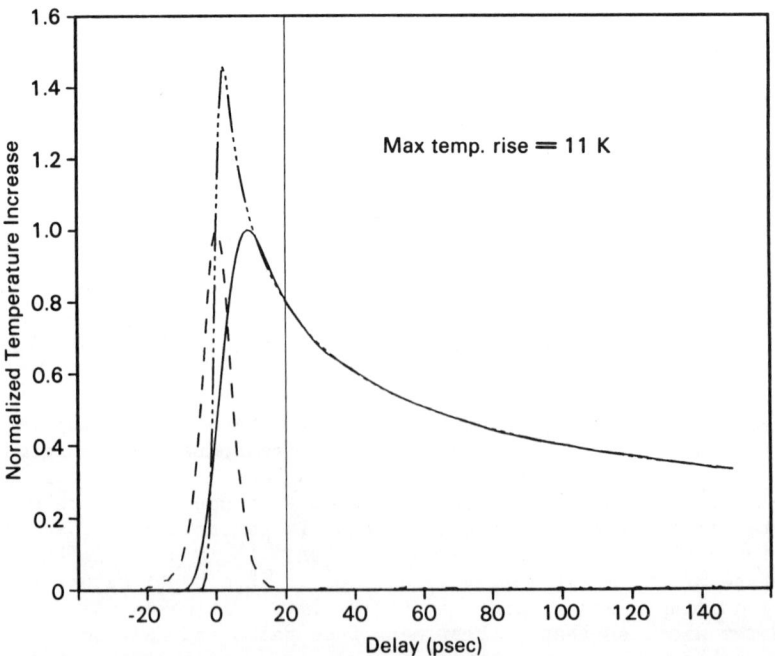

Figure 3. Best fit solution of the heat conduction equation to the normalized TTR measurement (solid line) of single crystal Ni. The cross correlation of the heating and probing pulses (long dash line) is also shown.

measurements to the Fresnel reflection formula and determine the real(n) and imaginary (k) parts of the metal refractive index. The imaginary part is then used to calculate the absorbtivity from the relation $\alpha = 4\pi k/\lambda$, where $\lambda$ is the heating laser wavelength.

The validity of our approach for measuring the thermal diffusivity was initially tested by TTR measurements on bulk single-crystal nickel (110). Figure 3 shows the normalized TTR signal versus the time delay of the probing pulse relative to the heating pulse. Superimposed on the data is the best fit solution of Eq. (3) at z=0, calculated using a heating pulsewidth of 4 psec (full-width-at-half-maximum). The mismatch between the fitted solution and the data at early times is a result of convolution effects in the measured data which are not accounted for in the calculated temperature profile. These effects are only important at early times and we commence our fitting routine after a time delay of 20 psec. Thus the actual details of the heating pulse shape are not important [11]. The measured cross correlation of the heating- and probing-pulse is also shown in Fig. 3 to demonstrate the effective time resolution of our measurement and establish the zero time delay position. The thermal diffusivity fitted by our model is $\kappa = 2.1 \times 10^{-5}$ m$^2$/sec, compared with the literature value of $2.23 \times 10^{-5}$ m$^2$/sec [21]. The mean square error per data point is $5 \times 10^{-4}$.

We have used TTR to measure the thermal diffusivity of deposited single element films and compositionally modulated Ni-Ti and Ni-Zr films [11]. We find that the thermal diffusivity of the modulated metal films is substantially smaller than that measured for the constituent single element films. This indicates that the interface between two different metallic layers can alter the thermal transport in a direction perpendicular to the film plane. Our goal is to use TTR to measure the thermal impedance of interfaces by fitting our results to a one-dimensional heat flow model which includes the interfacial boundary condition [9]

$$K_1 \frac{\partial T}{\partial z}1(1^-,t) = K_2 \frac{\partial T}{\partial t}2(1^+,t) = \frac{1}{\rho}[T_2(1^+,t) - T_1(1^-,t)] \quad . \quad (11)$$

$K_1$ and $K_2$ are the thermal conductivities of the metals on either side of the interface located at z=1. The thermal impedance of the interface is $\rho$.

We are currently implementing a fitting procedure which solves a heat flow equation appropriate to each single element region and coupled by the boundary condition in Eq. (11). We will use measurements of thermal diffusivity in single element films as input parameters to characterize the constituents in single interface, bilayer metal films. Our fitting routine will then determine the interfacial thermal impedance by fitting solutions of the coupled heat flow equations to the TTR measurement on the bilayer film.

Although the fitting procedure is not yet operational, we have performed a series of measurements of thermal transport across single metal-metal interfaces. The samples used for these measurements were fabricated to consist of a 30 nm Ni cap over a 300 nm metal underlayer of either Cu, Mo, Ti or Zr. The samples were prepared in a dual source magnetron sputter deposition system with a base pressure of $10^{-7}$ Torr. Silicon wafers were used as substrates under the simultaneously operating, shuttered sputter sources. Depostion rates were held constant at 0.5 nm/sec and the sputtering atmosphere was $2 \times 10^{-3}$ Torr of argon.

A graphic example of the degradation in thermal transport produced by the presence of a single interface is shown in Fig. 4. TTR measurements are shown for both single element Ni and Ti films and for a Ni-Ti bilayer film. (All films are produced with the same sputtering conditions.) The decrease in thermal transport due to the metal-metal interface is clearly demonstrated and this effect is also observed in the other bilayer films [22].

Tabulated in Table 1 are the results of fitting our TTR measurements to a single heat flow equation (3) and determining an effective thermal diffusivity of the bilayer film. We have included measurements of the film optical properties as well. Figure 5 shows the TTR measurements for all four bilayer films investigated. The thermal diffusion is faster in the Ni-Cu film, followed by the Ni-Mo bilayer film. This ordering would be expected since the thermal diffusivities of Cu and Mo are larger than Ni, Ti and Zr.

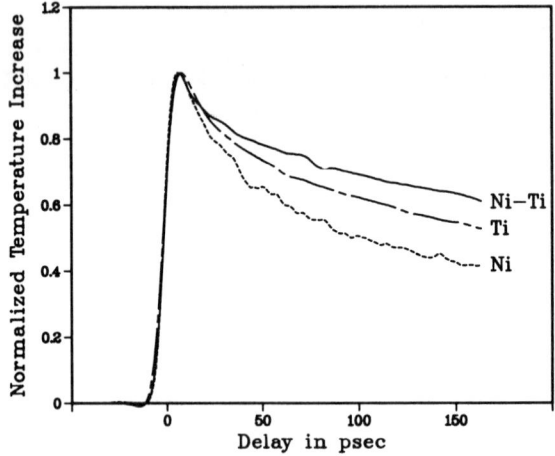

Figure 4. TTR measurements from single element Ni and Ti films, and from a single interface Ni-Ti bilayer film.

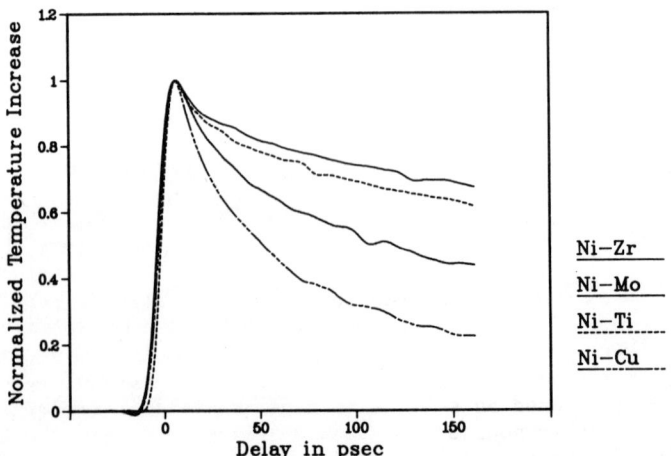

Figure 5. TTR measurements from different bilayer films.

A somewhat unexpected result is observed in the relative thermal diffusion of the Ni-Ti and Ni-Zr bilayer films. In this case we find that the diffusion in the Ni-Ti film is faster than in the Ni-Zr film, even though the thermal diffusivity of Zr is larger than that of Ti (see Table 1). We believe this reversal in trend may be attributed to the fact that the thermal impedance of the Ni-Zr interface is larger as a result of a larger atomic lattice mismatch and the ensuing higher degree of interfacial disorder. The relative lattice mismatch of the bilayer samples is calculated from literature values of nearest-neighbor lattice constants [23], and the results are also contained in Table 1. We find that the trend in degradation of thermal diffusion due to a metal-metal interface is correlated with the relative lattice mismatch of the metal constituents.

Table 1. The measured optical refractive index (n+ik), the absorptivity ($\alpha$), the best fit thermal diffusivity ($\kappa$), and the lattice mismatch for both single element and bilayer films studied.

| Sample |  | n | k | $\alpha$ ($\times 10^7$ m$^{-1}$) | $\kappa$ ($\times 10^{-6}$ m$^2$/sec) | Lattice Mismatch (%) |
|---|---|---|---|---|---|---|
| Ni | single element | 1.80 | 3.71 | 7.4 | 4.4 | - |
| Mo |  | 3.07 | 3.84 | 7.6 | 12.5 | - |
| Ti |  | 2.34 | 3.28 | 6.5 | 1.5 | - |
| Zr |  | 2.13 | 3.31 | 6.6 | 2.5 | - |
| Ni-Cu | bilayer film | 1.73 | 3.70 | 7.3 | 32.0 | 2.8 |
| Ni-Mo |  | 1.80 | 3.85 | 7.6 | 6.1 | 9.2 |
| Ni-Ti |  | 1.63 | 3.54 | 7.0 | 0.42 | 16.1 |
| Ni-Zr |  | 1.81 | 3.80 | 7.5 | 0.33 | 27.3 |

The real atomic structure of our samples is currently being investigated by x-ray diffraction and the details will be presented in a later publication. We expect this analysis to provide more accurate mismatch values and a quantitative correlation with the interfacial impedance may be possible. What is clear at this point is that even between two metals, interfacial thermal impedance has a dramatic effect on thermal transport.

CONCLUSIONS

We have discussed the time domain analysis of the one-dimensional heat conduction equation, with the purpose of interpreting our picosecond time-resolved measurements of thermal diffusion in thin metal films. Our measurement technique is generally nonperturbative and provides a means of determining thin film thermal diffusivity independent of the supporting substrate. We have shown that the ability to do this is important because of the degradation in thermal transport produced by the presence of an interface. We expect that an additional advantage of ultrafast time-resolved measurements is the ability to measure interface thermal impedances in novel thin film structures.

We have shown that in the case of ultra-short heating, interesting nonequilibrium thermal transport can occur. This regime of thermal transport is important to fundamental studies of ballistic electron transport and electron-phonon relaxation in metals and semiconductors. The advent of laser heating sources in the femtosecond time regime may also require that we alter our conventional view of thermal transport as being diffusive in nature, at least over dimensions comparable to carrier mean free paths. Such dimensions are becoming more relevant as new growth techniques are being used to fabricate compositionally modulated materials with repeat distances on the atomic scale.

# REFERENCES

1. *Synthetic Modulated Structures*, Edited by L. L. Chang and B. C. Giessen, Academic Press (New York, 1985).
2. W. P. Leung and A. C. Tam, Opt. Lett. $\underline{9}$, 93 (1984).
3. J. Opsal, A. Rosencwaig and D. L. Willenborg, Appl. Opt. $\underline{22}$, 3169 (1983).
4. L. J. Inglehart, K. R. Grice, L. D. Favro, P. K. Kuo and R. L. Thomas, Appl. Phys. Lett. $\underline{43}$, 446 (1983).
5. A. Rosencwaig, J. Opsal, W. L. Smith and D. L. Willenborg, Appl. Phys. Lett. $\underline{46}$, 1013 (1985).
6. M. Vaez Iravani and H. K. Wickramasinghe, J. Appl. Phys. $\underline{58}$, 122 (1985).
7. J. Baker-Jarvis and R. Inguva, J. Appl. Phys. $\underline{57}$, 1569 (1985).
8. J. Opsal and A. Rosencwaig, J. Appl. Phys. $\underline{53}$, 4240 (1982).
9. D. L. Balageas, J. C. Krapez and P. Cielo, J. Appl. Phys. $\underline{59}$, 348 (1986).
10. J. H. Bechtel, J. Appl. Phys. $\underline{46}$, 1585 (1975).
11. C. A. Paddock and G. L. Eesley, J. Appl. Phys. $\underline{60}$, 285 (1986).
12. M. I. Kaganov, I. M. Lifshitz and L. V. Tanatarov, Sov. Phys. -JETP $\underline{4}$, 173 (1957).
13. G. L. Eesley, Phys. Rev. Lett. $\underline{51}$, 2140 (1983).
14. G. L. Eesley, Phys. Rev. B$\underline{33}$, 2144 (1986).
15. R. W. Schoenlein, W. Z. Lin, J. G. Fujimoto and G. L. Eesley, in the conference proceedings of Topical Meeting on Ultra-fast Phenomena, Snowmass, CO (June 1986).
16. C. C. Ackerman and R. A. Guyer, Annals of Phys. $\underline{50}$, 128 (1968).
17. C. C. Ackerman and W. C. Overton Jr., Phys. Rev. Lett. $\underline{22}$, 764 (1969).
18. T. F. McNelly, S. J. Rogers, D. J. Channin, R. J. Rollefson, W. M. Goubau, G. E. Schmidt, J. A. Krumhansl and R. O. Pohl, Phys. Rev. Lett. $\underline{24}$, 100 (1970).
19. D. E. Glass, M. N. Ozisik, D. S. McRae and B. Vick, J. Appl. Phys. $\underline{59}$, 1861 (1986).
20. C. A. Paddock and G. L. Eesley, Opt. Lett. $\underline{11}$, 273 (1986).
21. *American Institute of Physics Handbook*, (McGraw-Hill, New York, 1972).
22. C. A. Paddock, G. L. Eesley and B. M. Clemens, presented at the International Quantum Electronics Conference, San Francisco, CA (June 1986).
23. C. Kittel, *Introduction to Solid State Physics*, 4th Edition, Ch. 1, Table 5 (John Wiley and Sons, Inc., New York (1971).

PROBING THROUGH THE GAS-SOLID INTERFACE WITH THERMAL WAVES:

A STUDY OF THE TEMPERATURE DISTRIBUTION IN THE GAS AND IN THE SOLID

L.J. Inglehart

The Johns Hopkins University
Center for Nondestructive Evaluation
Baltimore, MD 21218

J. Jaarinen

The University of Helsinki
Department of Physics
Siltavuorenpenger 20 D
SF-00170 Helsinki 17, Finland

P.K. Kuo

Wayne State University
Department of Physics
Detroit, MI 48202

E.H. Le Gal La Salle

Laboratoire D'Optique Physique
Ecole Superieure de Physique et de Chimie
10, Rue Vauquelin
75231 Paris Cedex 05 France

INTRODUCTION

Thermal wave nondestructive evaluation is one of the few NDE methods applicable to the characterization of ceramic and composite materials. As a near surface characterization technique, it has been shown to be useful for metals and semiconductors and has been developed for use with those materials. Limited work has been done with the technique on ceramics, with little effort to investigate the resulting fundamental differences which can occur in the case of insulating materials.

In this work we investigate the case of thermally insulating materials, in particular to describe the ac temperature distribution near the gas-solid interface. We make use of the mirage effect [1], or optical beam deflection (OBD) detection method, which permits probing the temperature distribution locally on both sides of the sample-gas interface and provides spatially resolved information about the local temperature gradients. For this purpose, we choose a material which absorbs infrared radiation, transmits visible radiation, and has a low volume expansion coefficient, fused silica. By scanning the heat source relative to the probe beam and sample surface while varying the experimental parameters, chopping frequency, probe distance to sample surface, heating beam power, probe and heating beam diameters, we have obtained a description of the ac temperature distribution near the sample surface both in the gas and in the solid.

Probing inside the solid has been explored as a sensitive method to obtain optical absorption coefficients for materials which are both weakly absorbing [2] and strongly absorbing [3,4]. Recent experiments in semiconductors [5] have provided information about near surface transport.

In this paper, we relate our experimental results to a three dimensional model assuming a homogeneous, isotropic bulk solid with a surface source [6]. We find that our results may be explained when the thermal diffusivity of the probed media is low, and the chopping frequency of the heat source is low. This condition occurs for many of the materials which are presently of interest and can be studied effectively with thermal methods (see Table I).

Table 1. Material studied. The literature value of the thermal diffusivity and the calculated value of the thermal diffusion length for each is given at a frequency of 100 Hz.

| Material | Thermal Diffusivity $\frac{cm^2}{s}$ | Thermal Diffusion Length (cm) |
|---|---|---|
| Cu | 1.19 | .062 |
| Al | .980 | .056 |
| SiC (hot pressed) | .4-.6 | .044 |
| Ga | .307 | .031 |
| Air | .300 | .031 |
| GaAs | .302 | .031 |
| $Si_3N_4$ (hot pressed) | .1-.16 | .023 |
| $Al_2O_3$ | .080 | .016 |
| Stainless Steel | .040 | .011 |
| Glass (Fused Silica) | .008 | .005 |
| Plastic | .002 | .003 |

THEORY

We consider a model which has been developed [6,7] to include coatings (a three layered system, gas-coating-solid) and which may also include subsurface absorption. However, for simplicity, we consider here only surface absorption and a bulk material and find that there is good qualitative agreement between the theoretical results and the experimental results. We assume a gaussian heating beam and a line probe beam. A complete description must include subsurface absorption; probe beam effects have been addressed in [7].

We start by solving the time dependent heat diffusion equations for the periodic solutions for the temperature in each region to be probed (see [6,7] for details). Once the temperature in each region is determined, the deflection of the probe beam, which is the measured effect, may be related to the gradient of the temperature distribution integrated along the probe beam. These steps may be carried out in momentum space ($\lambda$), and then a Fourier transform may be performed to give a result in real space. When this is done, the total deflection has the form:

$$\phi(x,z,t) = 1/n \; dn/dT \; \exp(i\omega t) \; \nabla \int_0^\infty F(\lambda,z) \cos(\lambda x) \, d\lambda \qquad (1)$$

where n is the index of refraction, T is the temperature, t is the time, $\omega$ is the angular frequency, z is the distance of the center of the probe beam to the surface, and x is the transverse distance of the probe beam away from the source. The functional form of $F(\lambda,z)$ depends on absorption parameters, heating beam size, distance of the probe to the sample, thermal reflection coefficients (fmr the case of coated substrates), and whether the probe beam is in the gas or the solid.

For our work here we consider only the solutions in the direction normal to the sample surface, $\phi_n$, for which $\nabla \to \partial/\partial z$. In this case, for probing in the gas, we find that $F(\lambda,z)$ is:

$$F(\lambda,z) = P/(\pi^2 \kappa_s b_s) \; \exp(-\lambda^2 R^2/4) \; \exp(-b_g z) \qquad (2)$$

where P is the heating beam power absorbed, $\kappa_s$ is the thermal conductivity of the solid, R is the heating beam radius at 1/e, and $b_{s,g}$ is given by

$$b_{s,g} = (\lambda^2 - i\omega/\alpha_{s,g})^{1/2} \qquad (3)$$

where s and g refer to the solid and gas respectively, and $\alpha$ is the thermal diffusivity. For the case of probing in the solid, the expression given in (2) includes the term with the heating beam parameter, while the term $b_g$ in the exponential is replaced by $b_s$. For each case, probing in the air and the solid, we have the convolution of two exponentials with a term involving the ratio of the thermal diffusivities of the gas and the substrate.


EXPERIMENT

The experiments were performed using mirage effect detection with a focused HeNe probe beam and a chopped, focused $CO_2$ heat source (both beams were ~ 100 μm at 1/e for most experiments unless otherwise noted). A quadrant detector was used to monitor the probe beam deflections and was maintained with a null dc voltage during the experiments where the deviation of the probe from the center of the detector was significant. The voltage from the detector was put into a vector lock-in amplifier set to give magnitude and phase. Signals for both $\phi_n$ and $\phi_t$ were measured; however, we report here only measurements for $\phi_n$, the deflection normal to the surface.

The sample was a slab of fused silica, 3 x 5 x 10 mm, and was studied by focusing the heat source at the surface and positioning the probe beam at a distance z from the surface and at a distance x from the heat source. The signals were monitored as the distance between the source and the probe beam was varied (transverse offset) for a given frequency and probe beam height, z. This measurement of the normal and transverse deflection signals as a function of transverse offset was done for different z positions, different frequencies, different heating beam powers, and different probe and heating beam diameters. By varying the transverse offset of the beams, the profile of the temperature distribution can be probed and studied.

RESULTS AND DISCUSSION

Probing in the Gas

In Fig. 1, we have plotted the result of the theoretical calculation for the magnitude of the normal deflection for the case of materials with decreasing thermal diffusivity. For each of the theoretical and experimental plots made as a function of beam transverse offset we show data plotted from the center of the heat source to one side only, since the curve is symmetric about zero transverse offset. Notice that for materials with small diffusivities a minima and second maxima occur as the transverse offset increases. This behavior has been reported previously [6] for the case of a glass material with a coating, but occurs for all of the materials we have studied in Table 1 with diffusivities less than air, independent of any coatings. The main peak width is determined by the heating beam dimensions. The side dip and second peak depend on the diffusivity of the probed material, the height of the probe beam, chopping frequency, and surface characteristics, such as roughness.

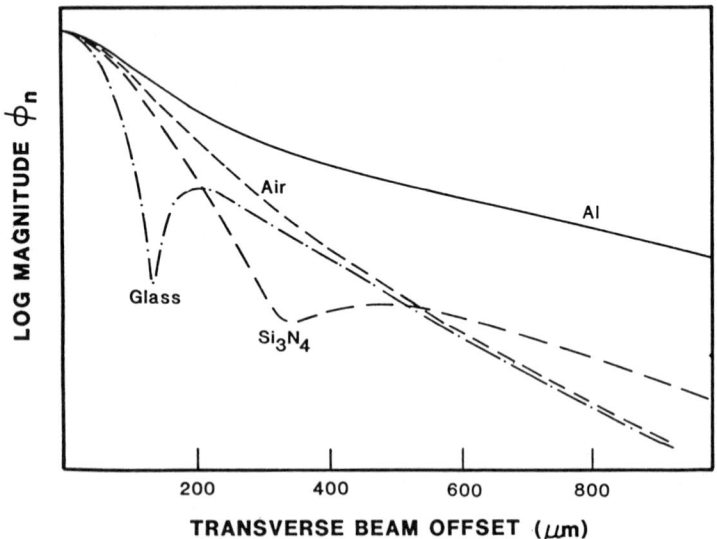

Fig.1. Theoretical magnitude of $\phi_n$ versus transverse offset for Aluminum, Air, $Si_3N_4$, and Glass at frequency 100 Hz, probe height of 100 μm from surface in air, 100μm diameter heating spot size.

In Fig. 2 we show the dependence of the transverse offset signal on probe beam height, where we have used the diffusivity of glass in the calculation, and a frequency of 100 Hz. Comparing with Fig. 1, we see that the magnitude of the $\phi_n$ signal in Fig. 2(A) has no minima when the probe beam is far away from the surface and that there is a slight minima which increases as the probe beam moves toward the surface. However, in contrast to the dependence on diffusivity in Fig. 1, the probe beam height dependence produces a shift in the minima both toward the source (transverse offset=0) and to larger amplitude. The phase curves for the same parameters with the addition of one extra height, shown in Fig. 2(B), have a 180 degree phase reversal which occurs with the minima in the magnitude, also reported in [6].

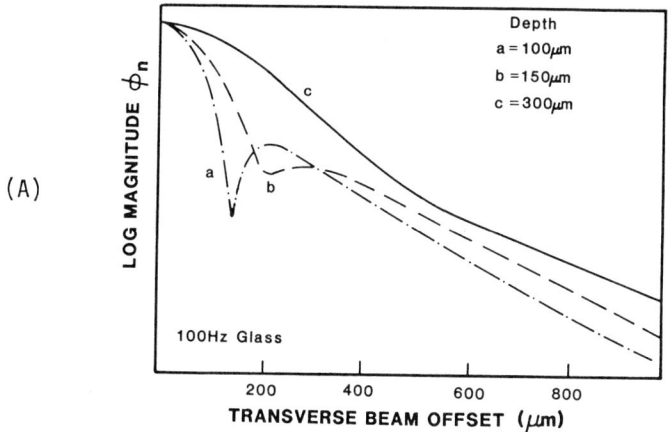

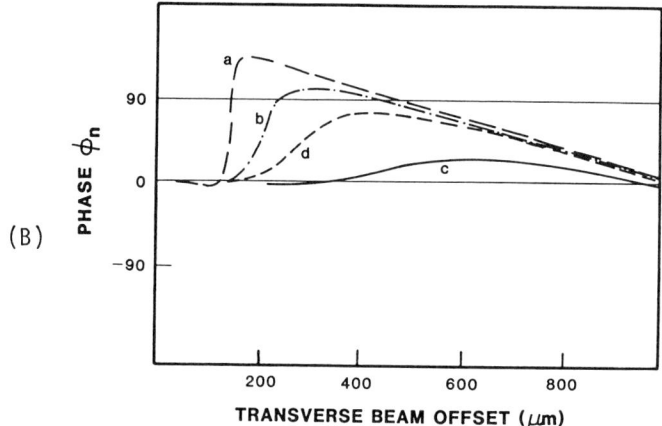

Fig. 2. (A) Theoretical magnitude of $\phi_n$ versus transverse offset for glass, at three probe beam heights: (a) 100 μm, (b) 150 μm, and (c) 300 μm; frequency 100 Hz.
(B) Theoretical phase of $\phi_n$ for four probe beam heights: (a) 100μm, (b) 150 μm, (c) 300 μm, and (d) 200 μm, for the same parameters as (A).

We plot the experimental results for $\phi_n$ magnitude and phase in Fig. 3, for the same frequency as shown in Fig. 2, for comparison. In this experiment we are probing in the air above the sample surface. The qualitative comparison with the theoretical calculation is good. The dependence with height shows the same trends in both magnitude and phase. We point out that, when the probe beam is very near the sample surface, there is an inversion in the phase for both the calculated and experimental phases.

In addition, we have found that changing the frequency to higher values produces the same smoothing effect as the theory predicts and that moving the probe beam closer to the surface produces more pronounced dips, also predicted by the theory. Changing the heating beam diameter has the effect of spreading the heat over a larger area, without affecting the behavior of the minima and second maxima. Changing the diameter of the probe beam does appear to improve the resolution of the shape of the curves, giving the appearance of more tightly narrowed peaks. Experiments at higher heat

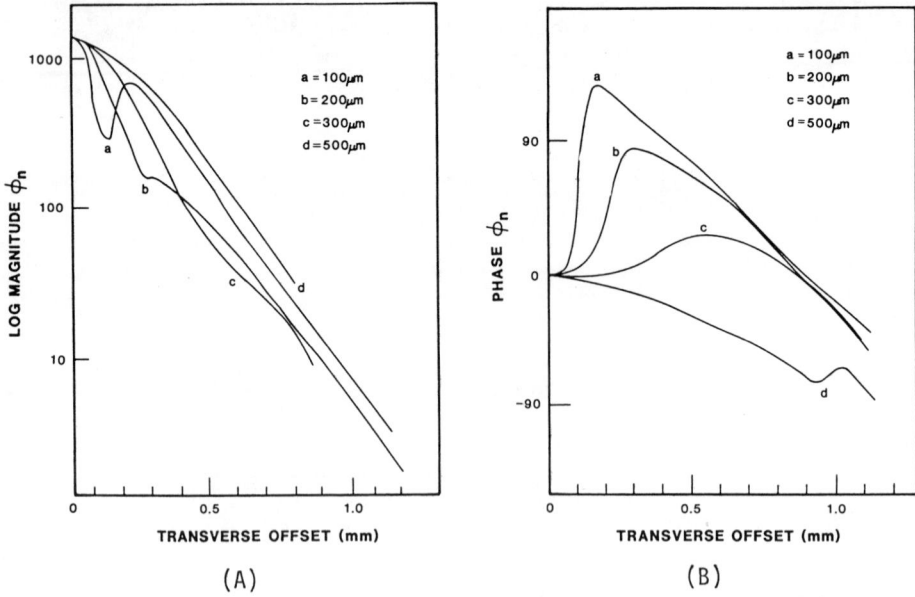

Fig. 3. Experimental results for the magnitude (A) and phase(B) of $\phi_n$ for a chopping frequency of 100 Hz, at different probe beam heights in air: (a) 100 μm, (b) 200 μm, (c) 300 μm, (d) 500 μm.

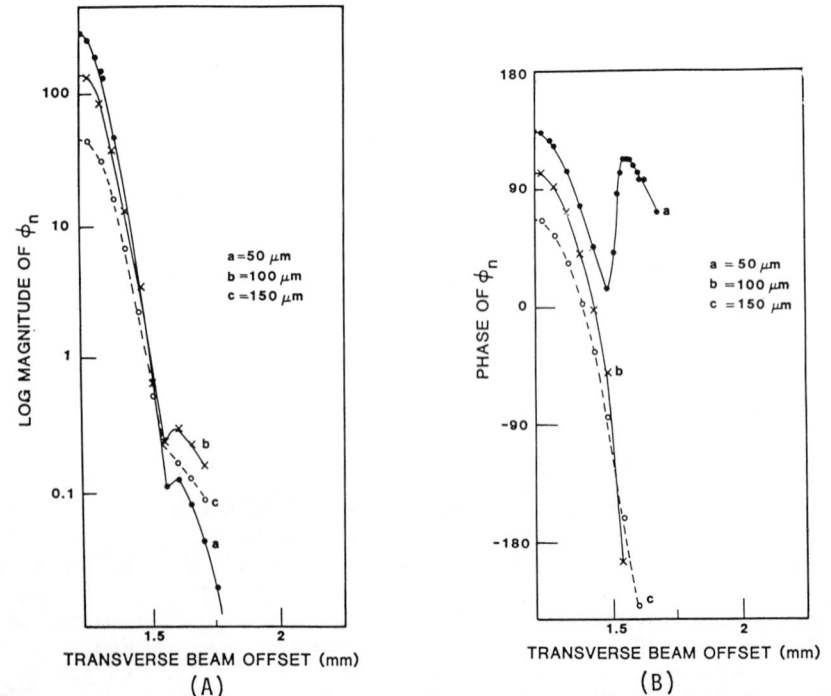

Fig. 4. Magnitude (A) and Phase (B) of $\phi_n$ versus beam transverse offset probing in the solid for different probe beam depths from the surface (z): (a) 50 μm, (b) 100 μm, (c) 150 μm. Frequency is 100 Hz, probe beam diameter is 50 μm.

source powers have shown that a saturation in the central peak occurs which may be related to an increase in thermal radiation being emmitted by the surface.

Probing in the Solid

We present experimental results for probing in the solid in Fig. 4. The amplitude of the signal is larger, as it should be due to the increased temperature coefficient of the solid (dn/dT), and we see that there is also a minima in the magnitude. We find that the minima and second peak occur at distances slightly greater than two diffusion lengths in the solid below the surface. The phase of the signal in the solid has a very different behavior from that in the gas. Near the surface, the phase has a sharp minima, and away from the surface it decreases rapidly.

We have also studied the depth dependence of the signal measured at the maximum of the magnitude of $\phi_n$, both in the air and in the solid for different frequencies, plotted in Fig. 5. The amplitude of the signal reaches its maximum value in the solid when the probe beam is entirely in the solid and at a distance below the surface equal to the diameter of the probe beam. The fast decay at higher frequencies is expected in both media and is expected to be more rapid in the solid than in the gas.

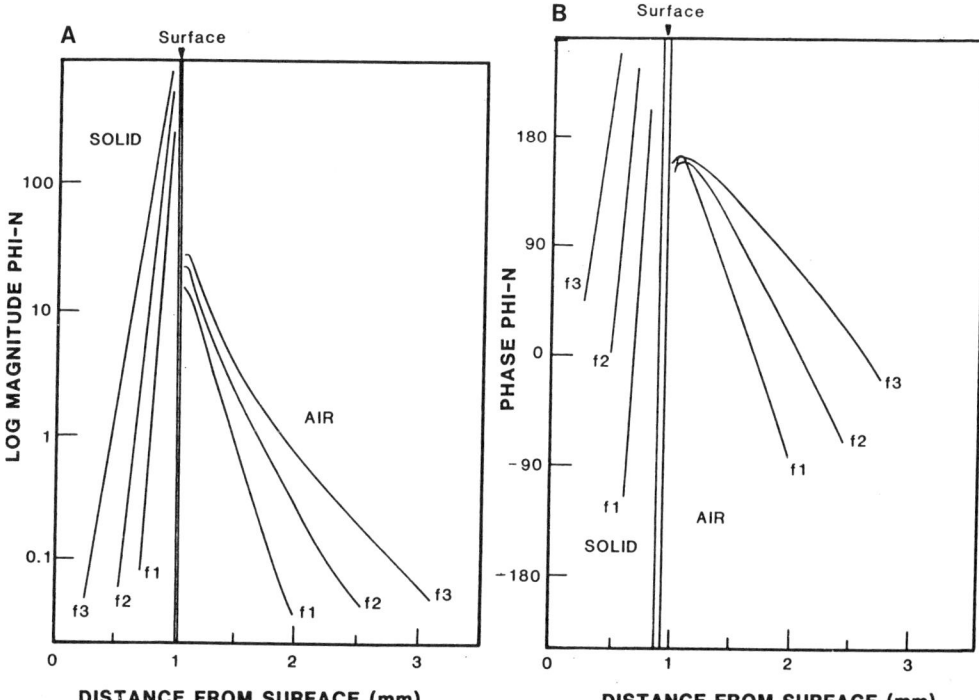

Figure 5. Magnitude (A) and Phase (B) of the maximum value of $\phi_n$ versus probe beam distance to the surface (z), both in the gas and in the solid, for three frequencies: f1=300 Hz, f2=100 Hz, and f3=40 Hz.

We are presently studying the application of this technique to measurements of surface roughness, porosity, and near surface damage. Advantages of the measurement are that it provides spatially resolved information and both magnitude and phase information, and does not require that the probe beam touch the sample surface.

ACKNOWLEDGEMENTS

This research was supported in part by the National Bureau of Standards, Office of Nondestructive Evaluation, and in part by the Wayne State University Institute of Manufacturing Research. We would especially like to thank Christie Van Nguyen for her assistance with the figures.

REFERENCES

1. A. C. Boccara, D. Fournier, and J. Badoz, Appl. Phys. Lett., 36, 130 (1980).
2. A. C. Boccara, D. Fournier, W. Jackson, and N. Amer, Opt. Lett. 9, 377 (1980).
3. G. C. Wetsel, Jr. and S. A. Stotts, Appl. Phys. Lett. 42, 931 (1983).
4. G. C. Wetsel and S. A. Stotts, J. de Phys. 44, 215 (1983).
5. D. Fournier, C. Boccara, A. Skumanich, and N. M. Amer, J. Appl. Phys. 59, 787 (1986).
6. W. B. Jackson, N. M. Amer, A. C. Boccara, and D. Fournier, Appl. Opt. 8, 1333 (1981).
7. P. K. Kuo, E. D. Sendler, L. D. Favro, and R. L. Thomes, to be published in Can. J. Phys., 1986.

# REFLECTION-MIRAGE MEASUREMENTS OF THERMAL DIFFUSIVITY

C.B. Reyes, J. Jaarinen, L.D. Favro,
P.K. Kuo, and R.L. Thomas

Department of Physics
Wayne State University
Detroit, MI 48202

## INTRODUCTION

The experimental technique for the measurement of thermal diffusivity using the mirage effect, or optical probe beam detection of thermal waves in opaque solids has been described elsewhere. [1-3] This is carried out by scanning the probe beam relative to the heating beam with a constant height, h. The separation, $x_o$, of the two points on either side of the center of such a scan where the phase of the transverse deflection signal reaches ninety degrees effectively measures the thermal wavelength, $\lambda = 2(\pi\alpha/f)^{1/2}$ in the solid. The determination of the thermal diffusivity, $\alpha$, is accomplished by plotting this separation versus the inverse square root of the frequency. It has been shown theoretically [4,5] that the ratios of the slopes of such plots correspond, in the low frequency limit, to the ratios of the actual diffusivities of the solid. The numerical constant which relates the thermal diffusivity to the slope depends on the value of h. The previous measurements [1-3] of $\alpha$, carried out using the "skimming" optical probe beam technique (see Fig. 1), were found to be in reasonable agreement with nominal values calculated from handbook data, provided that the slopes of the plots were set equal (heuristically) to $(1.0 \pi\alpha)^{1/2}$. Careful examination of the theory [5], however, shows that when h is negligibly small compared to the other characteristic lengths in the experiment, the slope should be $(1.4 \pi\alpha)^{1/2}$. In this paper we present the results of reflection-mirage measurements on a number of these same samples. In this technique (see Fig. 2) the probe beam is incident at an angle of a few degrees from the sample surface and reflects from the surface into the position sensor. Utilization of this reflection experimentally provides a much smaller value of h than can be achieved by "skimming". To assess the contribution to the deflection of the beam from the thermally induced surface curvature we have made measurements on the same specimen at both atmospheric pressure and at greatly reduced pressure. In Fig. 3 we show the comparison between theory and experiment for the dependence of the mirage phase signal on the heating-probe beam separation at the two pressures. The same parameters were assumed, except for the diffusivity of the gas at low pressure, which was calculated from that at atmospheric pressure using the pressure ratio. It can be seen from Fig. 3 that good agreement is obtained at both pressures, indicating that the effect of the curvature of the surface is small in these experiments.

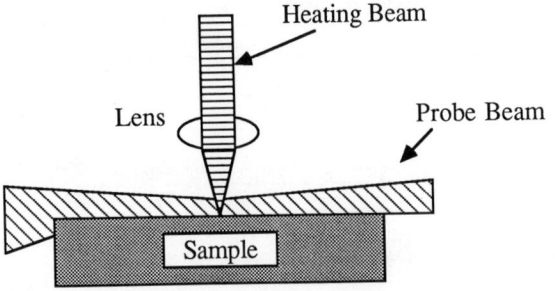

Fig. 1  Schematic diagram of "skimming" optical beam technique.

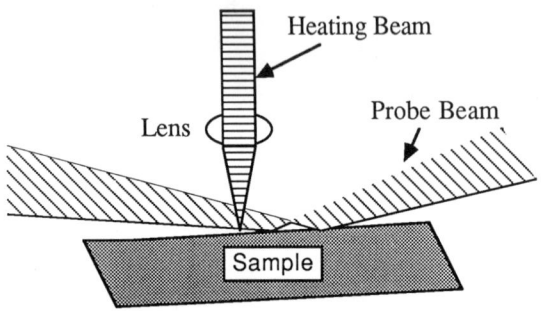

Fig. 2  Schematic diagram of "bouncing" optical beam technique.

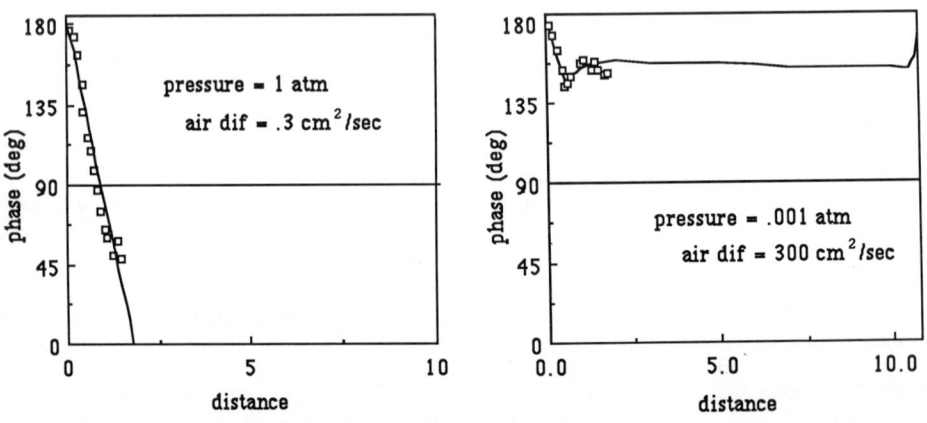

Fig. 3  Comparison between theory and experiment for the dependence of the mirage phase signal on the heating-probe beam separation at the two pressures.

## EXPERIMENTAL RESULTS

In Figs. 4 and 5, we present plots of the parameter $x_o$ as a function of $f^{-1/2}$ for several pure single crystal elements, obtained from reflection mirage measurements on these samples. In Fig. 6 our measured diffusivities for pure elements at room temperature are compared with nominal values as calculated from the values of thermal conductivity, heat capacity and density given in the Handbook of Chemistry and Physics [6], using the relation $(1.4\,\pi\alpha)^{1/2}$. In Table 1 we summarize the diffusivity results for these elements.

It can be seen from Fig. 6 and Table 1 that this method of measuring diffusivities gives reasonable agreement with nominal values for most of the samples studied, over a wide range of diffusivity (1.29 cm$^2$/s to 0.045 cm$^2$/s). It is expected that the agreement will improve with the use of the full set of experimental data (rather than just $x_o$) in a least-squares fit to the theory.

## CONCLUSIONS

This direct measurement of thermal diffusivity has several advantages. It is totally contactless and only involves measurements of length and time, each of which can be made with a high degree of accuracy. Also, the measurement is inherently local ( 1 mm) and only one surface of the sample needs to be accessible, so that measurement of non-uniform or small samples can be made. Furthermore, the diffusivity can be measured in different orientations on anisotropic materials, such as composites [6] by simply rotating the probe beam direction. A disadvantage is that the sample surface must be reasonably flat, so that the probe beam can skim the surface at a small value of h. Furthermore, the detailed dependence of the measurement on the beam sizes and probe beam height have not been studied. A more detailed theoretical and experimental study of these questions is in progress.

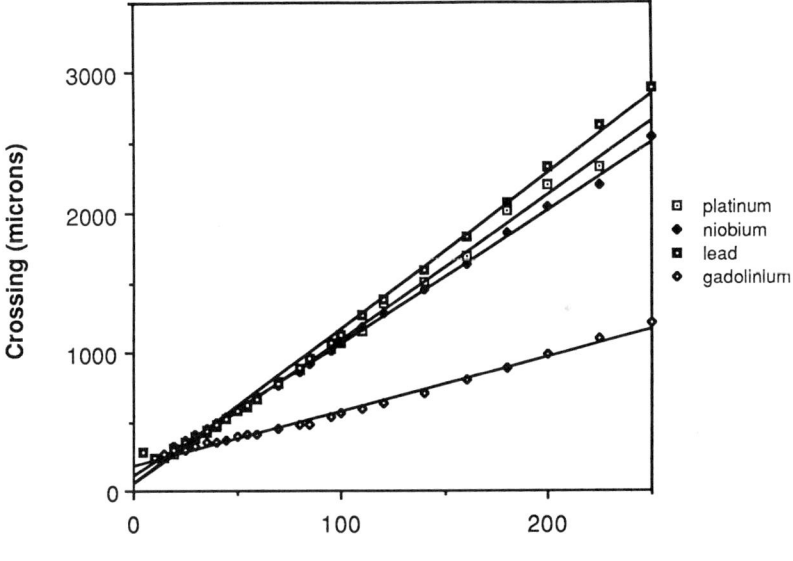

Fig. 4  Experimentally observed frequency dependence of the parameter $x_o$ for several samples of pure elements.

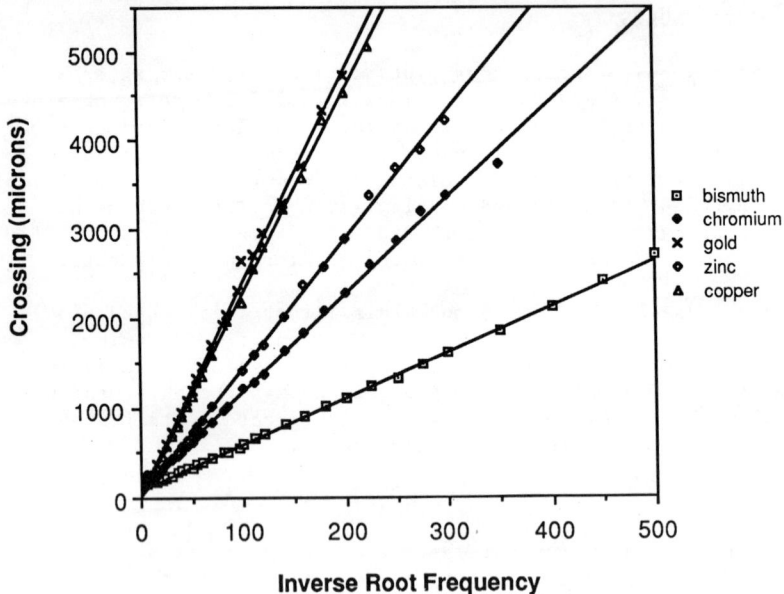

Fig. 5  Experimentally observed frequency dependence of the parameter $x_o$ for additional samples of pure elements.

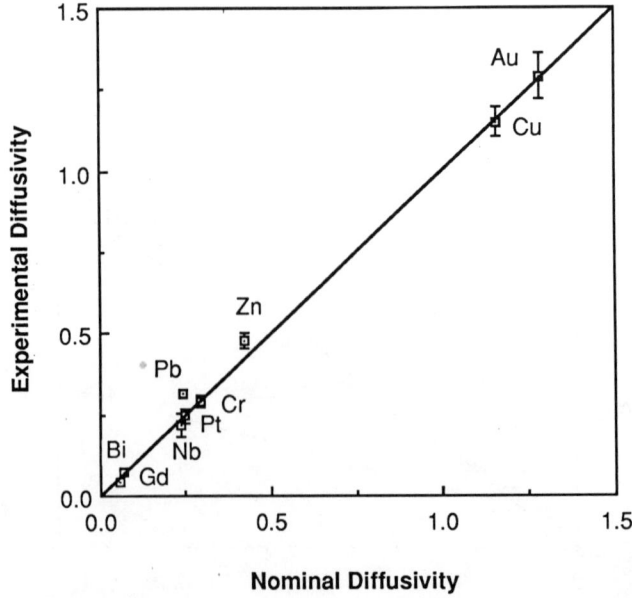

Fig. 6  Measured diffusivities of pure elements plotted versus literature values.

Table 1. Experimental Values of Thermal Diffusivities

| Material | Thermal Diffusivity ($cm^2/s$) | |
| --- | --- | --- |
| | This work | Nominal |
| Au | 1.29 | 1.28 |
| Bi | 0.072 | 0.044-0.095 |
| Cr | 0.290 | 0.292 |
| Cu | 1.15 | 1.16 |
| Gd | 0.045 | 0.056-0.059 |
| Nb | 0.219 | 0.238 |
| Pb | 0.315 | 0.240 |
| Pt | 0.245 | 0.250 |
| Zn | 0.479 | 0.420 |

ACKNOWLEDGEMENTS

This work was sponsored by ARO under Contract No. DAAG 29-84-K-0173, and by the Wayne State University Institute for Manufacturing Research. J. Jaarinen also acknowledges the support of the Finnish Cultural Foundation.

REFERENCES

1. R.L. Thomas, D.S. Kim, L.D. Favro, P.K. Kuo, C.B. Reyes, and Shu-Yi Zhang, Review of Progress in Quantitative NDE, Vol. 5B, edited by D.O. Thompson and D. Chimenti (Plenum, New York, 1986), 1379.

2. P.K. Kuo, M.J. Lin, C.B. Reyes, L.D. Favro, R.L. Thomas, D.S. Kim, Shu-yi Zhang, L.J. Inglehart, D. Fournier, A.C. Boccara, and N. Yacoubi, Can. J. Phys. 64(in press).

3. P.K. Kuo, L.J. Inglehart, E.D. Sendler, M.J. Lin, L.D. Favro, and R.L. Thomas, Review of Progress in Quantitative NDE, Vol. 4B, edited by D.O. Thompson and D. Chimenti (Plenum, New York, 1985), 745.

4. P.K. Kuo, E.D. Sendler, L.D. Favro and R.L. Thomas, Can. J. Phys. 64 (in press).

5. J. Jaarinen, C.B. Reyes, I.C. Oppenheim, L.D. Favro, P.K. Kuo, and R.L. Thomas, Review of Progress in Quantitative NDE, Vol. 6, edited by D.O. Thompson and D. Chimenti (Plenum, New York, 1987), (this volume).

6. Handbook of Chemistry and Physics, 61th Ed., 1980, Chemical Rubber Press (West Palm Beach).

COATING THICKNESS DETERMINATION USING TIME DEPENDENT SURFACE TEMPERATURE
MEASUREMENTS

J. C. Murphy, L. C. Aamodt, and G. C. Wetsel, Jr.[*]

The Johns Hopkins University,
Applied Physics Laboratory
Johns Hopkins Road, Laurel, Md. 20707

INTRODUCTION

Thin film coatings are used to protect metal alloys from oxidation. To be effective they must have adequate thickness and be well bonded to the metal substrate. This is especially important for metals subject to high temperatures or to highly oxidizing environments. As a consequence, there is a need for a non-destructive method for evaluating coating thickness, bonding, and other coating properties.

In this paper we report measurements made on nickel based superalloys (IN 738) coated with an aluminide coatings (AEP 32) using a dynamic thermal method of measurement. In addition, we describe an analysis which could be a convenient means of monitoring the thermal parameters of thin film coatings.

The three samples reported on here were in the form of circular cylinders coated on one end and on the cylinder side. The other end was uncoated. These cylinders were designated as samples DC#12, DC#13, and DC#15 and have the following dimensions: (a) DC#12, 12.1 mm in diameter x 10.5 mm long, with a nominal coating thickness of 38-51 µm; (b) DC#13, 12.1 mm in diameter x 9.6 mm long, with a nominal coating thickness of 25-38 µm; and (c) DC#15, 12.1 mm diameter x 10.5 mm long, with a nominal coating thickness of 56-64 µm. Measurements of the coated and the uncoated ends were made on each sample.

THERMAL MODEL/PHOTOTHERMAL RADIOMETRY

Figure 1 shows the thermal model and terminology that we use. A pulsed light source illuminates the thin coating and is partially reflected. The absorbed light heats the opaque coating surface and heat diffuses into the coating. At the coating-metal interface, a thermal mismatch occurs if the thermal effusivities of the coating and substrate are not identical. [Thermal Effusivity(E) = $\sqrt{(\rho \kappa C)}$ where $\rho$ is the density, $\kappa$ is the thermal conductivity, and C is the thermal capacity.] When there is no thermal mismatch, heat flows uninterrupted into the sample bulk; otherwise a portion of the heat is reflected back toward the sample surface, while the remainder flows into the metal substrate.

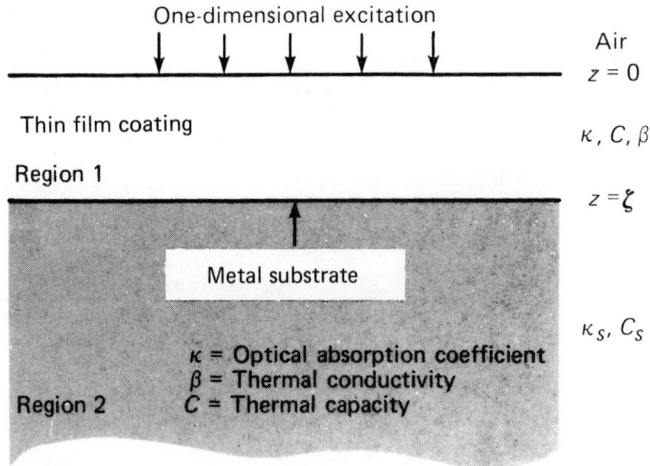

Fig. 1  Thermal model used in this paper - thin coating of thickness $\zeta$ on a thermally thick substrate; planer geometry.

If the thermal bonding of the coating to the metal is good, the magnitude of this reflection is determined by the ratio of the thermal effusivities of the coating and the base metal. If the thermal bond is not good, e.g., a void exists, a large heat reflection can be produced by the thermal barrier associated with the disbonding. For partial bonding, the amount of heat reflection lies between these extremes.

The time, $\tau$, that it takes for the reflected heat to return to the sample surface depends upon the thermal diffusivity ($= \kappa/\rho C$) of the coating, the coating thickness, and the distribution of heat generated in the coating (if the coating is not perfectly opaque).

The effect of heat reflection can be seen in Fig. 2. The particular curve identified as $E_b/E_f=1$ is the decay pattern of the surface temperature when there is no thermal mismatch between the coating and the metal substrate (or if the coating were thermally thick). The drop in surface temperature is caused by thermal diffusion into the coating bulk. The initial decay pattern corresponds to diffusion in the coating and is the same for all thermal mismatches. At longer times, the transit time, $\tau$, is approached and the surface temperature decay pattern deviates from the decay pattern of the thick specimen. The direction of the deviation depends upon whether the metal is a better or worse conductor than the coating, while the magnitude of the deviation depends upon the magnitude of the thermal effusivity ratio. Since the temporal surface temperature decay pattern is affected by coating thickness and by bonding quality, a study of this pattern potentially provides a means for determining both of these quantities.

We use photothermal radiometry to measure surface temperature. The general features of this method are seen in Fig. 3. Details of the method can be found in References 1-3 for both CW and pulsed illumination.

IR radiometry theoretically measures the temperature at a point on the sample surface; in actuality, it measures the temperature integrated over a small area on the sample surface determined by the spatial resolution of the detector. The surface radiance of the heated sample (see Fig. 4a) varies with specimen temperature and IR wavelength. The incremental change in radiance with changes in specimen temperature is shown in Fig. 4b. This example corresponds to the experiment reported here where a change in

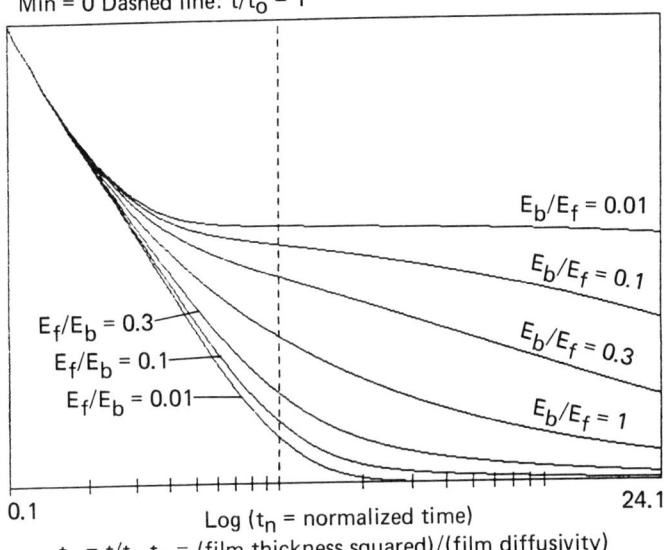

Fig. 2 Theoretical time dependent surface temperature following an instantaneous pulse for several ratios of coating and substrate thermal effusivity.

specimen temperature is induced by modulated laser heating and the resultant change in IR emission is monitored. While the radiance, L(t), is proportional to the fourth power of temperature, $L(t) = \varepsilon(\lambda) \sigma T^4$, the change in radiance with temperature, $\Delta L(t)$, is linear with the change in surface temperature, $\Delta T$, for small $\Delta T$, i.e., $\Delta L(t) \approx 4 T^3 \Delta T$. For homogeneous materials (no coating) the time dependent signal obtained from the IR detector is proportional to the excess surface temperature,

$$T_s(z=0,t) = (\pi \kappa \rho C)^{-1/2} \int_0^\infty \exp[-\beta z - z^2/4\alpha t] \, dz$$

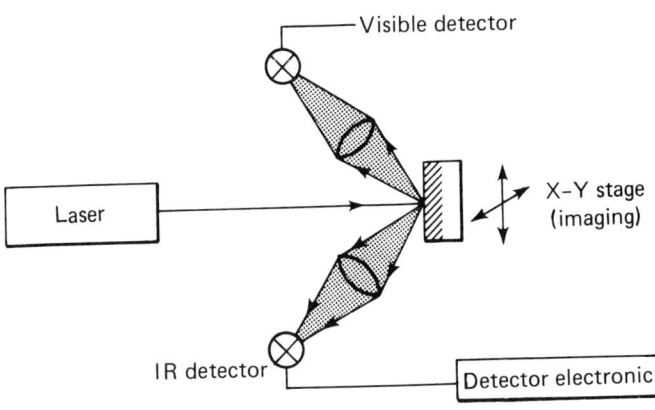

Fig. 3 Experimental IR radiometer arrangement used to obtain temperature data.

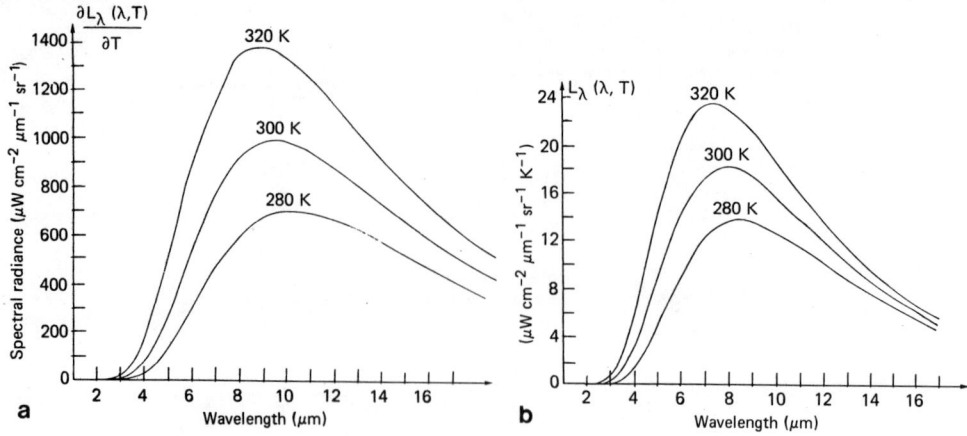

Fig. 4. a) Spectral radiance of a blackbody. b) First partial derivative of spectral radiance of a blackbody with respect to temperature.

where $\beta$ is the optical absorption coefficient, $\alpha$ is the thermal diffusivity, and C is the thermal capacity. The first exponential factor is the heat-source profile (with depth) in the sample and the second factor introduces the diffusion time, $z^2/4\alpha$ needed for the heat to flow from a depth z to the sample surface.

EXPERIMENTAL RESULTS

Using the experimental set-up of Fig. 3, we have measured the thermal response under pulsed heating of the three samples described above. The laser was a 30 mJ Nd:Yag laser with a pulse width of approximately $7\times10^{-8}$ sec (Fig. 5), which is short compared with the thermal response of the coated and uncoated samples. The IR detector used was a $LN_2$ cooled HgCdTe detector with a 20 MHz bandwidth and data was stored in a Data Precision Model 6000 transient analyzer with a 100 MHz digitizing rate.

Figure 6 shows the thermal response for the coated ends of the three samples while Fig. 7 shows the corresponding curve for the three uncoated ends. The IR emission from the uncoated samples peak at essentially the same time while the emission from the coated samples shows distinct delays which are correlated with coating thickness. The different response patterns for coated and uncoated samples is also evident in Fig. 8 (for DC#13). This result is typical of the response patterns for the other samples, with the coated sample peaking later than the uncoated sample.

One factor affecting these results is the coating structure and topography which results in non-uniform coating properties and thickness.

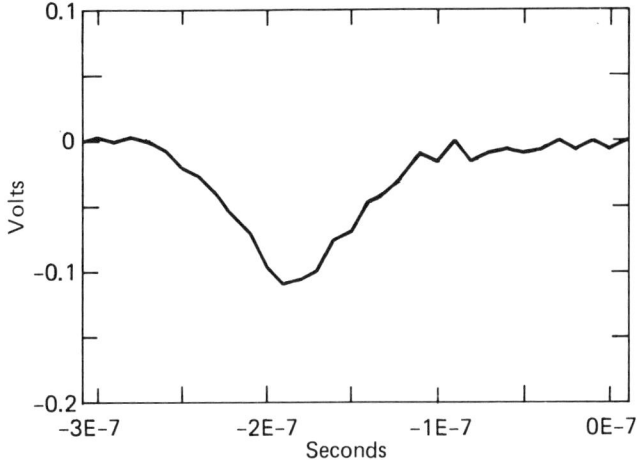

Fig. 5  Intensity of a 30 mJ Nd:Yag laser pulse used to heat samples.

Profilometer scans show variations ranging up to 25% of the nominal coating thickness across the specimen with occasional variations even greater. Coating nonuniformity is evident in Figure 9a,b (for DC#12) which shows an apparent porous, "spongy" texture of the coating surface. These figures are two optical microscope views of the coating surface focused respectively at the top and bottom visible layer of the coating. The estimated texture height from this measurement is 15 microns.

The "porous" appearance of the coating surface suggests that this is the origin of the delay of the thermal peak in the coated samples. Incident radiation is absorbed within the porous structure, and IR emission would emerge from a range of sites. Diffusion processes would also be affected with a more complex process related to the reduced dimensionality of the layers occuring. Quantitative treatment of these issues requires more analysis, which is in progress.

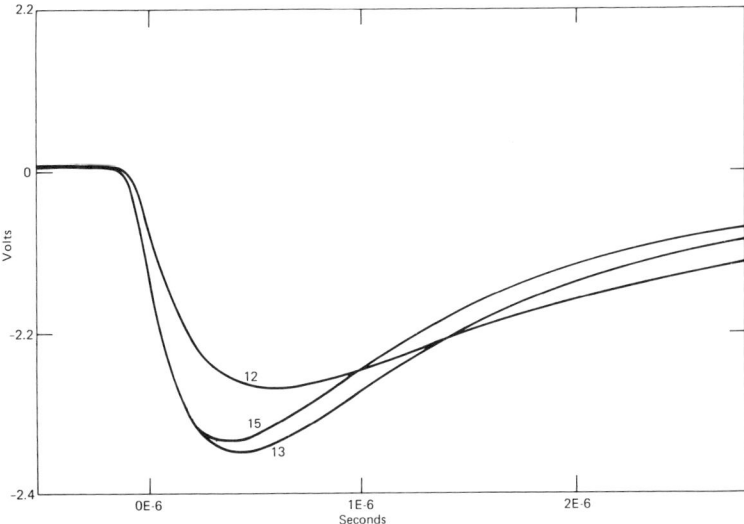

Fig. 6  IR thermal response pattern of three IN 738 specimens DC 12, 13, 15 (Alpak) coating. Nominal thickness 2.0, 1.5, 2.5 mils.

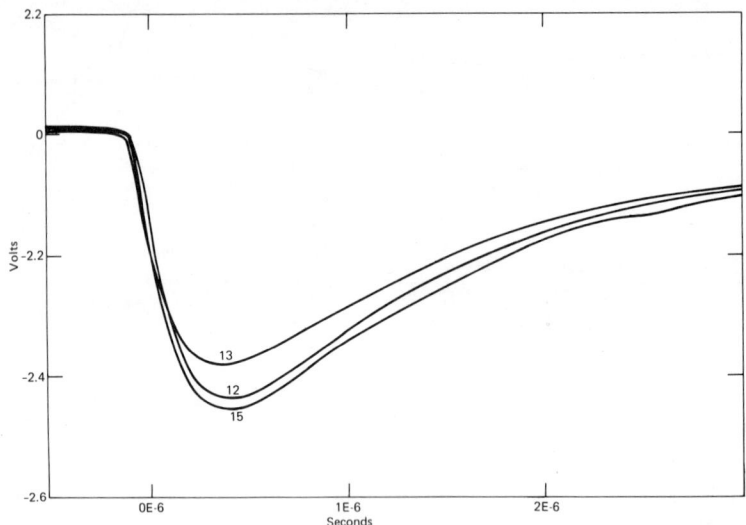

Fig. 7   IR thermal response pattern of three uncoated specimens of IN 738.  DC 12, 13, 15.

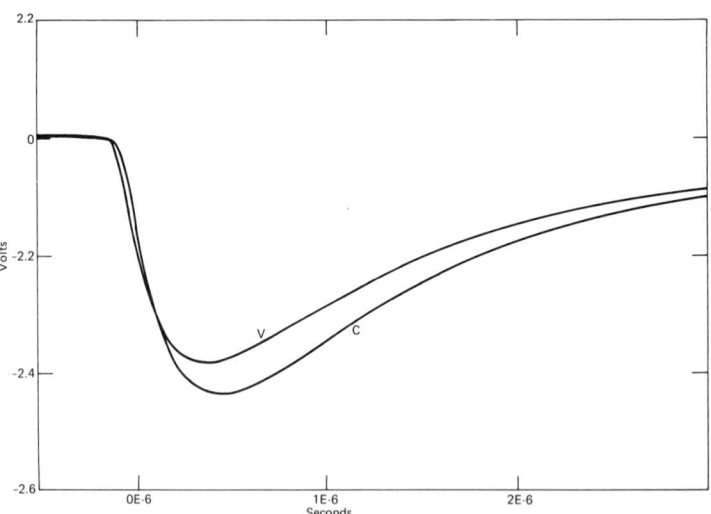

Fig. 8   Thermal response pattern of specimen DC#13 showing coated and uncoated response.  This result was typical of the response pattern for the other specimens.

ANALYSIS OF HEAT FLOW IN THIN FILMS

The surface temperature of a homogeneous uncoated material is modified when a thin film coating is applied.  If the coating is opaque or near opaque, the optical absorption coefficient of the coating governs the source profile of heat generation rather than the optical absorption coefficient of the sample.  If the substrate is thermally thick, its thermal properties affect the thermal mismatch at the coating-substrate interface, however, the time for heat diffusion back to the coating surface is determined by the thermal diffusivity of the coating and not the substrate diffusivity.  When the sample is porous and specimen structure need be considered, a statistical approach must replace the deterministic approach used here.  The present analysis does not accommodate this case.

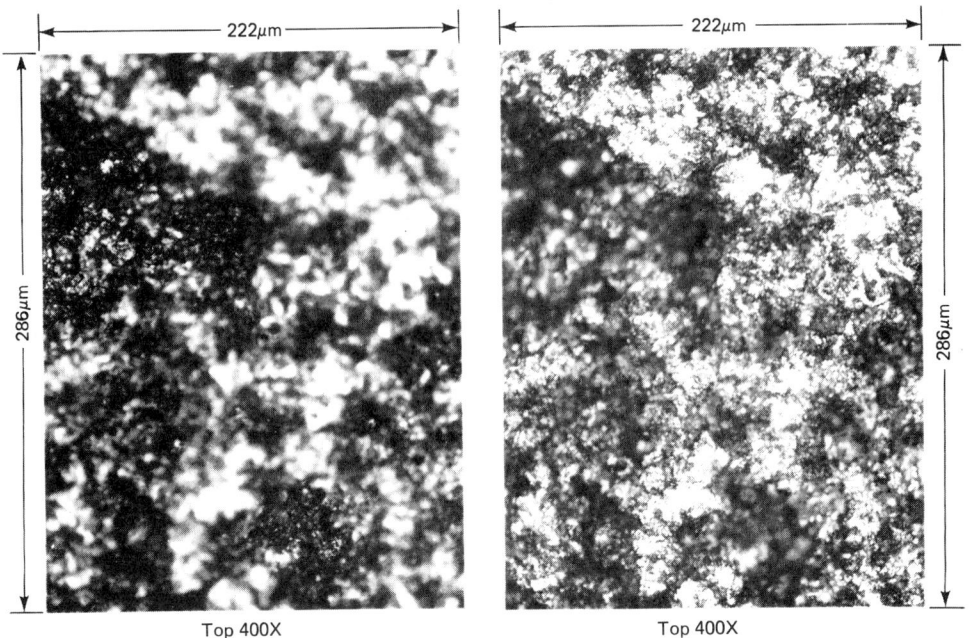

Fig. 9  Two optical microscope views of the coating surface focused at the top and bottom visible layers of the coating, respectively. Estimated texture height, 15 microns.

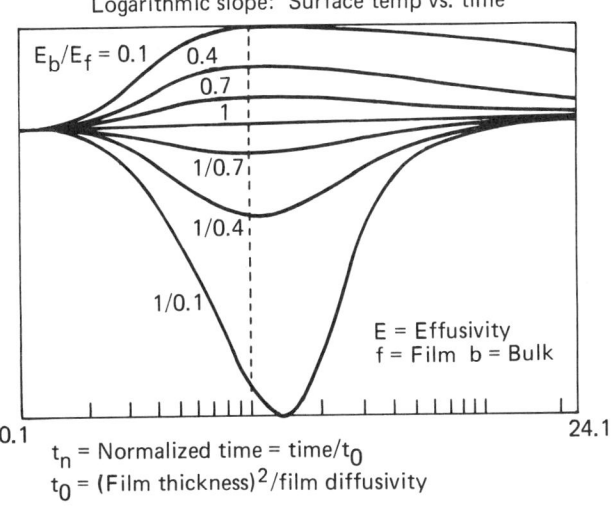

Fig. 10  Logarithmic derivative of the temperature patterns in Figure 1 vs. time. The curves peaks (or troughs) in the vicinity of the time, $t = \zeta^2/\alpha$.

The excess surface temperature of an almost opaque coated material heated by a spatially uniform temporal pulse is given by the expression,

$$T_s(z=0,t) = (\pi\rho\kappa C)^{-1/2} \int_0^\zeta \exp[-\beta z] \{\exp(-z^2/4\alpha t) + \sum_{\ell=1}^\infty (-\Gamma)^\ell [\exp(-z/4\alpha t) + \exp(z^2/4\alpha t)] \exp(\ell\tau_n/t) \} dz$$

where $\tau = \xi^2/\alpha$. $\zeta$ is the coating thickness, $\alpha$, the coating thermal diffusivity, $\beta$, the coating optical absorption coefficient, $\Gamma = [1 - E_s/E_c] / [1 + E_s/E_c]$, and $E_s$ and $E_c$ are the sample and coating thermal effusivities, respectively.

The sum in the equation represents successive thermal reflections between the upper and lower coating surfaces. The temperature decay patterns obtained from this expression for a fully opaque coating are shown in Fig. 2. The decay patterns in this figure are different for different thermal mismatches, but to ascertain the coating thickness or thermal mismatch from this pattern would involve a difficult task of curve fitting. A more convenient method way to obtain this information is to measure the logarithmic derivative of this decay (Fig. 10). As seen in this figure, the derivative's peak or trough always occurs close to the time, $t = \tau_n$, from which the sample thickness can be obtained if the coating thermal diffusivity is known. It is also evident from this figure that a thermal approach to characterizing film thickness requires the thermal mismatch between coating and sample to be significant.

ACKNOWLEDGEMENTS

This work has been jointly supported by the Army Research Office and the Naval Sea Systems Command under Contract No. N00024-85-C-5301 (ARO Nbr. 21066-MS).

*Permanent address: Department of Physics, Southern Methodist University, Dallas, TX 75275

REFERENCES

1. P. E. Nordal and S. O. Kanstad, Physica Scripta, 20, 659 (1979).
2. G. Busse, Infrared Phys., 20, 419 (1980).
3. P.-E. Nordal and S. O. Kanstad, Infrared Phys., 25, 1/2, 295, (1985).

NONDESTRUCTIVE CHARACTERIZATION OF COATINGS ON METAL ALLOYS*

G. C. Wetsel, Jr.†, J. C. Murphy, and L. C. Aamodt

The Johns Hopkins University
Applied Physics Laboratory
Laurel, MD 20707

INTRODUCTION

There is a need to nondestructively evaluate coatings on metal alloys that must endure high-temperature or highly oxidizing environments, such as for aircraft turbine blades. Such coatings are opaque and of the order of 25 μm in thickness. Characteristics of interest include the uniformity and thickness of the coating, which must be sufficient to protect the substrate from oxidation at high temperatures.

In this paper we report the results of measurements on samples of IN 738 alloy substrates covered by AEP 32 coatings. The samples were in the forms of cylindrical reference standards and actual aircraft turbine blades. Characterization methods included photothermal-optical-beam-deflection (PTOBD) imaging [1], optical microscopy, and surface profilometry, with the principal method being the first.

RESULTS

A turbine blade is shown in Fig. 1. The coating has been abrasively removed to the right of the arrow exposing the substrate. Optical micrographs of 1.14 mm × 0.889 mm areas on each side of the boundary indicated by the arrow in Fig. 1 are shown in Fig. 2. The substrate, shown in Fig. 2a, exhibits inclusions typical of metal alloys and scratches (presumably put there by the abrasion process). The coating, shown in Fig. 2b, exhibits a porous, sponge-like appearance. Evidently, the coating is nonuniform.

A transverse PTOBD amplitude contour map of a 1 mm × 1 mm area near the boundary between coated substrate and exposed substrate indicated by the arrow in Fig. 1 is shown in Fig. 3. The boundary is clearly visible as a dark arc across the center of the figure. The exposed substrate is in the upper part of Fig. 3; the coated substrate is in the lower part of Fig. 3. This photothermal image indicates that the coated substrate has thermal inhomogeneities on a scale of 0.1 mm, whereas the exposed substrate is relatively homogeneous.

In Fig. 4, a profilometer scan across the boundary between the exposed substrate and the coated substrate is shown. The surface roughness of the

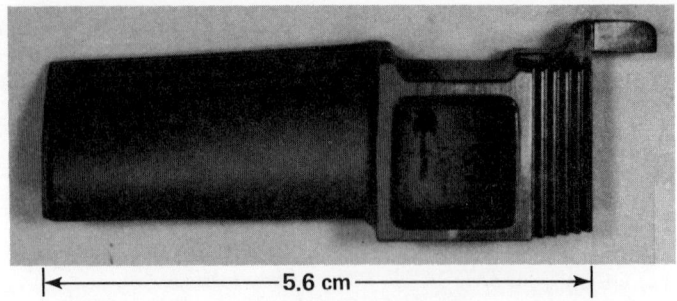

Fig. 1   Photograph of turbine blade:  IN 738 alloy substrate, AEP 32 coating.

coating can be seen to be greater than the surface roughness of the substrate.

Three standard samples of cylindrical shape were studied: (a) DC#12, 12.1 mm in diameter × 10.5 mm long, with a nominal coating thickness of 38-51 μm; (b) DC#13, 12.1 mm in diameter × 9.6 mm long, with a nominal coating thickness of 25-38 μm long; (c) DC#15, 12.1 mm diameter × 10.5 mm long with a nominal coating thickness of 56-64 μm. Optical micrographs of a 572 μm × 445 μm area of the coating surface are shown for each sample in Fig. 5.

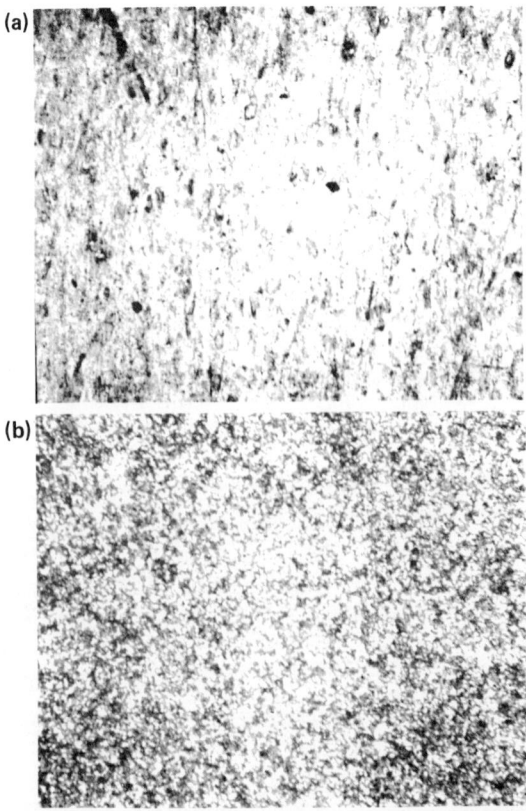

Fig. 2   Optical micrographs of 1.14 mm × 0.889 mm area on turbine blade: (a) substrate, (b) coating.

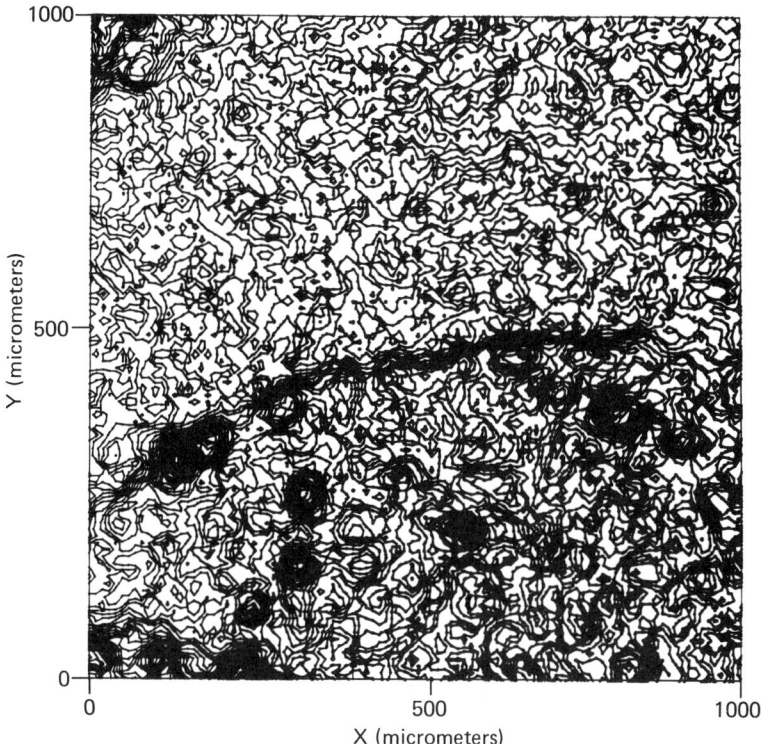

Fig. 3  Transverse PTOBD amplitude contour map of area near substrate-coating interface on turbine blade, f = 2 kHz.

Normal PTOBD amplitude contour maps of a 500 µm × 500 µm area of the coating on the three cylindrical samples are shown in Figs. 6, 8, and 9. In each case, the photothermal images indicate substantial thermal inhomogeneity. The transverse PTOBD amplitude contour map of Fig. 7 should be compared with Fig. 6. The area imaged in Fig. 7 is shifted along the x axis to the left of the imaged area of Fig. 6 by 150 µm. Generally, the thermal inhomogeneities that are so prominent in the normal component of

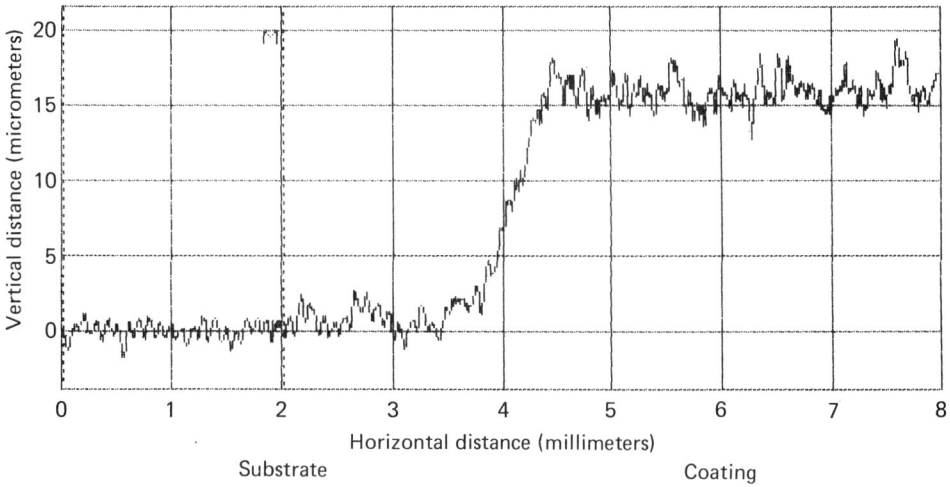

Fig. 4  Profilometer scan across substrate-coating interface on turbine blade.

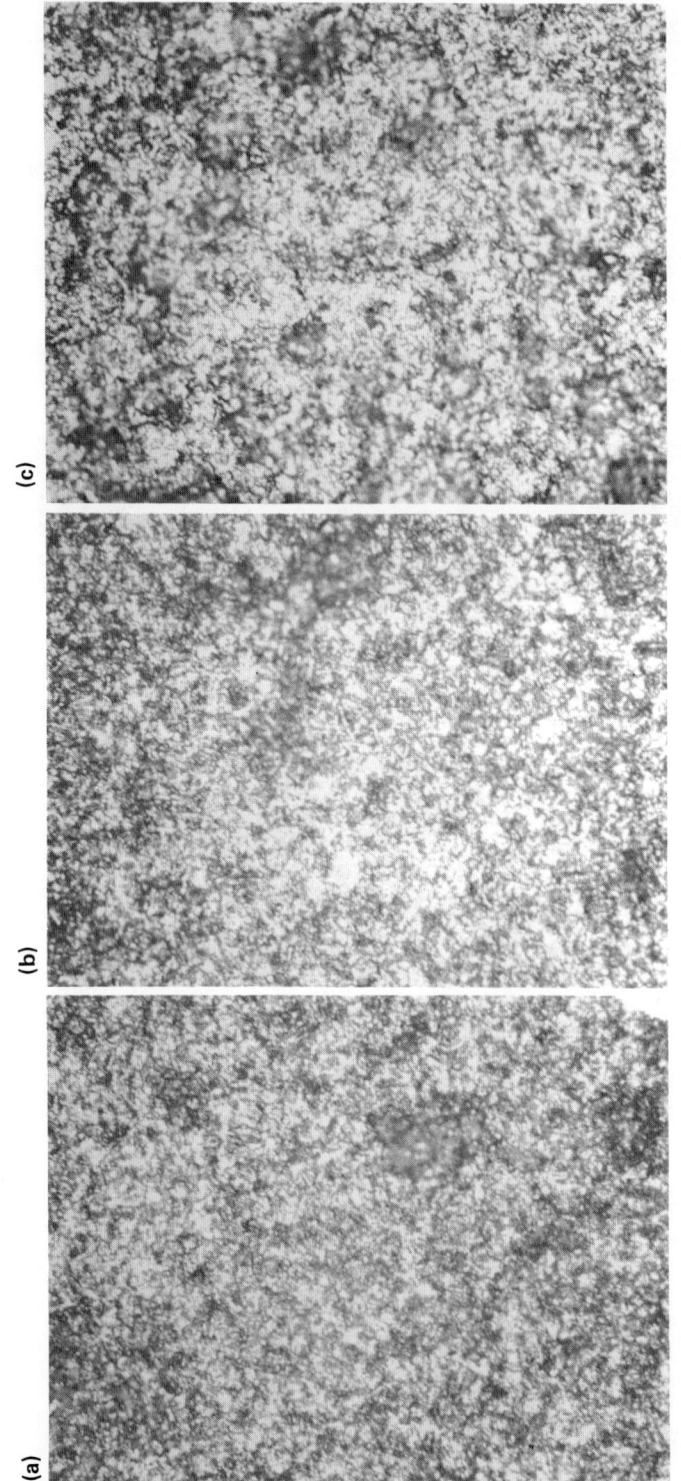

Fig. 5 Optical micrographs of 572 μm × 445 μm area of coating on samples: (a) No. 12, (b) No. 13, (c) No. 15.

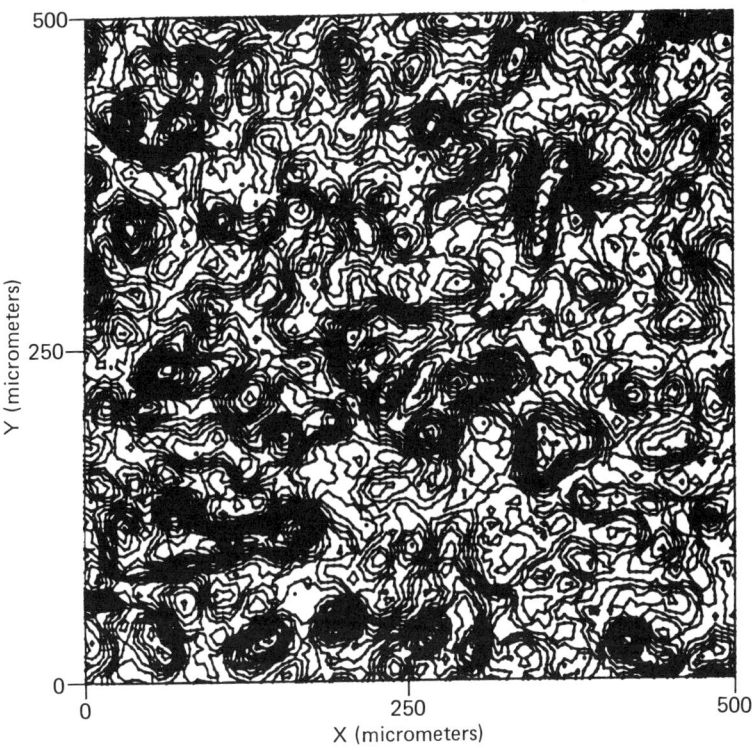

Fig. 6  Normal PTOBD amplitude contour map of 500 $\mu$m $\times$ 500 $\mu$m area of coating on sample No. 12, f = 2 kHz.

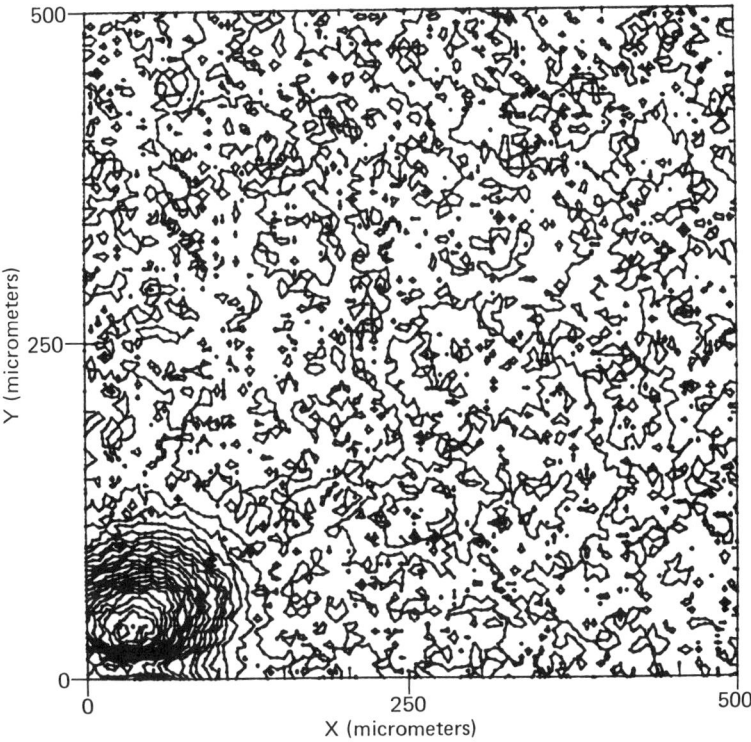

Fig. 7  Transverse PTOBD amplitude contour map of 500 $\mu$m $\times$ 500 $\mu$m area of coating on sample No. 12, f = 2 kHz.

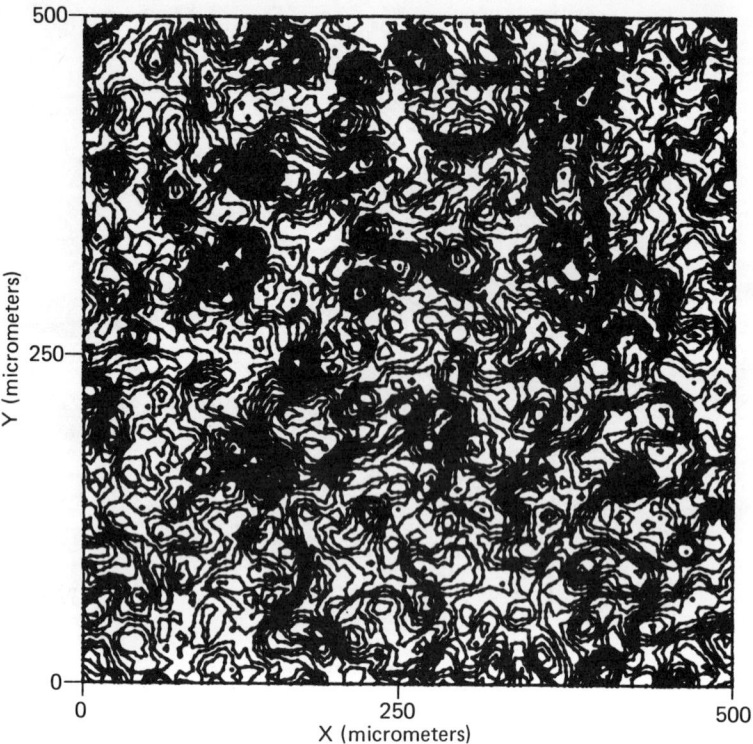

**Fig. 8** Normal PTOBD amplitude contour map of 500 $\mu$m $\times$ 500 $\mu$m area of coating on sample No. 13, f = 500 Hz.

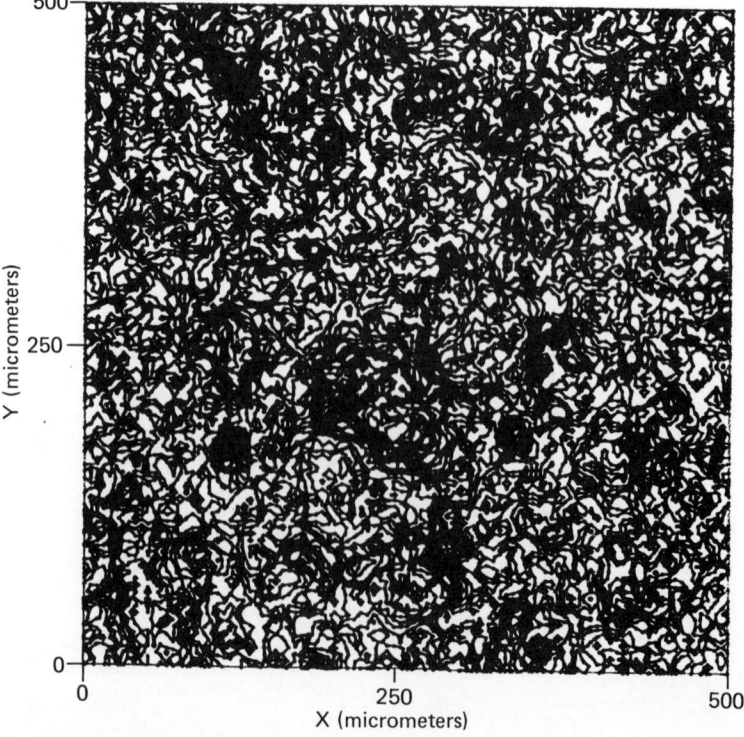

**Fig. 9** Normal PTOBD amplitude contour map of 500 $\mu$m $\times$ 500 $\mu$m area of coating on sample No. 15, f = 500 Hz.

beam deflection of Fig. 6 are not prominent in the transverse component of beam deflection of Fig. 7. The "hot spot" shown near the origin in Fig. 7 is an area not included in the area imaged in Fig. 6.

DISCUSSION

The results of the PTOBD imaging measurements clearly show that in each sample studied the AEP 32 coatings are substantially inhomogeneous. The lateral dimensions of the inhomogeneities may be several times as large as the nominal coating thickness. It is not known at this time if the observed inhomogeneities are due to variations in the coating composition, variations in the coating thickness, variations in heat transfer from coating to substrate (an indication of bonding variation), or all of the above. Surface profilometry reveals that surface topography variations can be of the order of 25% of the nominal coating thickness or greater. Optical micrographs reveal coating variations on a scale consistent with the other two methods of measurement. These coating variations, particularly the thickness variations, are serious potential failure hazards.

\* Supported in part by the Air Force Program in Quantitative Nondestructive Evaluation.
† Permanent address: Department of Physics, Southern Methodist University, Dallas, Texas 75275

REFERENCES

1. G. C. Wetsel, Jr., J. W. Maclachlan, J. B. Spicer, and J. C. Murphy, "Nondestructive Evaluation and Materials Characterization Using Photothermal-Optical-Beam-Deflection Imaging," <u>Review of Progress in Quantitative Nondestructive Evaluation</u>, Vol. 5A, Ed. by D. O. Thompson and D. E. Chimenti, pp. 713-719, Plenum, New York (1986).

# THERMAL WAVE TECHNIQUES FOR IMAGING AND CHARACTERIZATION OF MATERIALS

L.D. Favro, P.K. Kuo and R.L. Thomas

Department of Physics
Wayne State University
Detroit, MI 48202

## IMAGING

Thermal wave imaging is proving to be a useful technique for the nondestructive evaluation (NDE) of subsurface features of opaque solids. This imaging is achieved with various intensity-modulated heat sources, such as laser or particle beams, and with various detectors, such as microphones, ultrasonic transducers, infrared detectors, and laser probes. The authors have recently reviewed these techniques and their application to NDE [1]. Common to the techniques is the fact that they each involve the interaction of a highly damped thermal wave with surface or subsurface thermal features. They also have in common the fact that the source is localized. The techniques differ in that the detectors may be local or non-local to a greater or lesser degree. For example, the focused infrared detector is a local point temperature detector; the mirage effect laser probe is a line detector; and the microphone is an area detector. The presence of the localized source gives all of these methods the potential for high spatial resolution. The symmetry of the non-locality of the detector, however, may seriously limit detection of particular kinds of flaws. For some of the detection schemes, comparisons between experiment and theory for imaging of flaws with simple geometry (planar cracks, cylindrical or spherical inclusions) are straightforward, and good agreement has been achieved in most cases. Other schemes, such as piezoelectric detection, are more complex in nature. For example, the details of the conversion from thermal energy to acoustic energy may involve several different processes, and no quantitative three-dimensional theoretical model yet exists to assess the relative importance of these processes. In this section we give a brief review of the principles of imaging in the extreme near field, followed by descriptions of three selected experimental thermal wave imaging techniques, including geometrical considerations, signal-to-noise considerations, and examples of NDE applications. For descriptions of other thermal wave imaging techniques the reader is referred to the literature [1].

The resolution of any microscope depends on its ability either to localize the illumination at the scatterer (typical of scanning microscopes) or to localize the region of the scatterer to which the detector is sensitive (such as in conventional microscopes, which focus different points of the scatterer on different detectors). In microscopes which use lenses as focusing elements to achieve this localization, the localization is limited by the wavelength of the radiation being focused (the Rayleigh criterion). For example, the resolution of high-quality optical microscopes is limited to about $\lambda/2$, where $\lambda$ is the wavelength. On the other hand, if it is possible to localize, say, the source to much better than a wavelength, and to bring it into close (compared to a wavelength) proximity to the scatterer, resolution many times better than a wavelength is possible. This situation is perhaps best described as "the extreme-near-field limit" [1-3]. This limit was achieved with an optical microscope in 1984 by Pohl et al [4] and in the macroscopic domain using microwaves in 1972 by Ash and Nichols [5]. In the case of thermal wave microscopes, the sources are normally lasers or particle beams, which can be focused to dimensions which are small compared to the thermal wavelength. Therefore, it is much easier to satisfy the small source, close proximity criteria for thermal wave microscopy than it is for optical microscopy. Thermal wave imaging generally falls into the category of scanned microscopy with a localized source. Thermal waves by nature are heavily damped, dying out in distances of the order of a wavelength (typically a few microns to a few mm) or less. This appears to preclude the use of thermal wave lenses. However, it also means that the contrast of thermal wave images is dominated by scatterers located within a fraction of a thermal wavelength from the source. Thus, by varying the thermal wavelength one can vary the region of the specimen which contributes to the image. Another consequence of the heavily damped nature of these waves is that thermal wave imaging is especially well suited for the nondestructive evaluation (NDE) of near (submicron to a few mm) subsurface defects in opaque solids. This region is extremely important because very small flaws near the surface often lead to catastrophic failure.

Thermal wave microscopes take on a number of different forms with corresponding advantages and disadvantages, depending on the nature of the source and the detector. Gas-cell detection [6-12] uses a microphone to detect the pressure variations in an enclosed volume of gas as a focused, intensity-modulated laser beam is scanned over the surface of a sample. Conceptually, this can be thought of as imaging with a point source and an area detector (the acoustic wavelength is typically larger than the dimensions of the cell, so that the microphone responds equally to all parts of the cell). For this type of thermal wave microscope, because the source is a laser and the solid is normally opaque, the source is localized on the surface to a region whose lateral dimensions are limited by the optical wavelength. The proximity of the source to the scatterers is determined by the depth of the scatterers, and thus the resolution is limited by that depth. An advantage of this imaging technique is that it lends itself readily to theoretical analysis. This ease of theoretical analysis originates in the planar symmetry of the detection scheme, which allows one to utilize plane-wave scattering theory [11]. This same symmetry can be a disadvantage, however, since it precludes the detection of an important class of defects, closed vertical cracks. [12]

A block diagram of the gas-cell technique is shown in Fig. 1. An acousto-optic modulator, controlled by the audio reference signal from the lock-in amplifier, is used to provide a modulated heating beam from the Ar-ion laser. This beam is focused through the window of the gas-cell (shown schematically in Fig. 2) to the surface of the sample to which the cell is attached. A miniature hearing-aid microphone, coupled to the fixed volume of air trapped between the sample surface, the cell walls and the wax gasket (see Fig. 2), is used to convert the area integral of the ac surface temperature of the surface into a corresponding electrical signal for phase-sensitive detection by the lock-in amplifier. Since the microphone is attached rigidly to the sample, if stepping motors are used to scan relative to the focal point of the heating beam, signal-to-noise considerations suggest that it is advisable not to monitor the microphone signal during the resulting accelerations and decelerations of the sample. Usually, the data are taken for one line (x-scan) by stepping the position of the focal spot with the sample stationary (see Fig. 1). The sample's y-position is then stepped between lines to complete the area scan for imaging. A disadvantage of this technique is the limited range (audio) of modulation frequencies and hence the limited range of thermal wavelengths. Perhaps a more serious disadvantage is that it involves contact with the sample surface. The quality of the gasket seals also plays an important role in maximizing the signal-to-noise ratio.

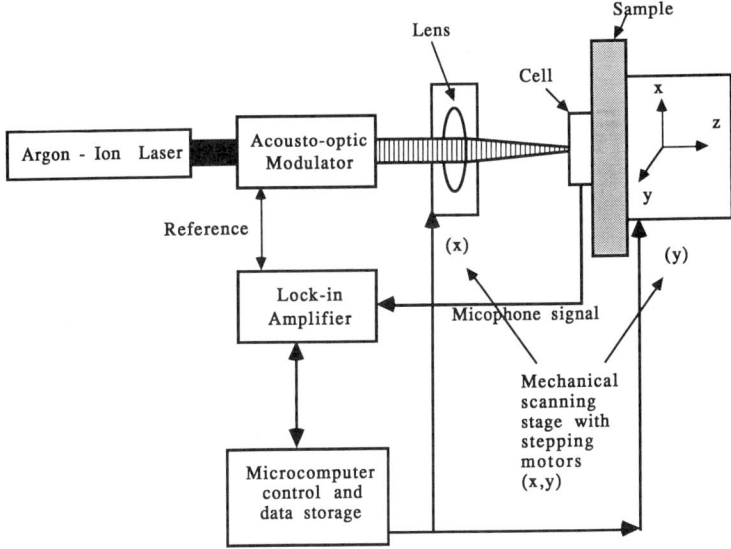

Fig. 1    Block diagram of the gas-cell method for thermal wave imaging.

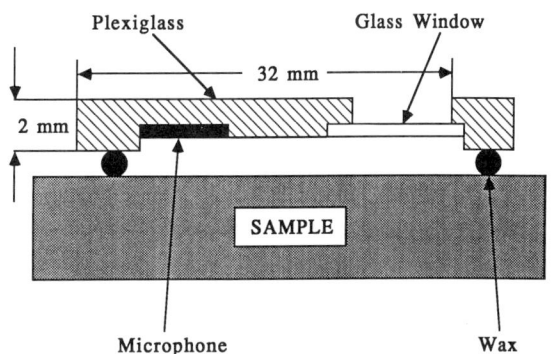

Fig. 2    Schematic diagram of the gas cell.

A thermal wave imaging technique which avoids the disadvantage of surface contact, and which reduces the symmetry of the detection scheme, utilizes the mirage effect or optical beam deflection (OBD) detection [13-16]. Again, the localization is achieved by focusing the source, but the detection is along a line rather than over an area. This method, in addition to being contactless, can operate over a much wider frequency range (1 Hz to a few hundred kHz). In contrast to the gas cell technique, which cannot detect vertical cracks, the mirage effect provides an excellent method for imaging such cracks [12,16]. An obvious disadvantage to this technique is the requirement of flat or cylindrical sample surfaces. Furthermore, there are the practical difficulties of maintaining two laser positions, as well as the height of the probe beam, during the scan. Also, the line symmetry of the detector means that a theoretical analysis requires, at the very least, considerations of scattering with cylindrical waves, and hence is considerably more difficult than that for gas-cell imaging.

A block diagram of the mirage effect technique is shown in Fig. 3. The heating beam is an intensity-modulated Ar-ion laser, and the probe beam is a HeNe laser which skims the sample surface with its (ac) position monitored synchronously by a quad-cell position detector and lock-in amplifier. Either the normal component or the transverse component of the ac deflection of the probe beam can be monitored by selecting the appropriate combinations of the four segments of the position sensor. A narrow band optical filter is used to prevent scattered (modulated) Ar-ion radiation from reaching the position sensor. For imaging, the positions of both beams are fixed during the x-y scan of the sample. In selecting the probe laser, it is important to achieve minimum intensity noise in the operating frequency range

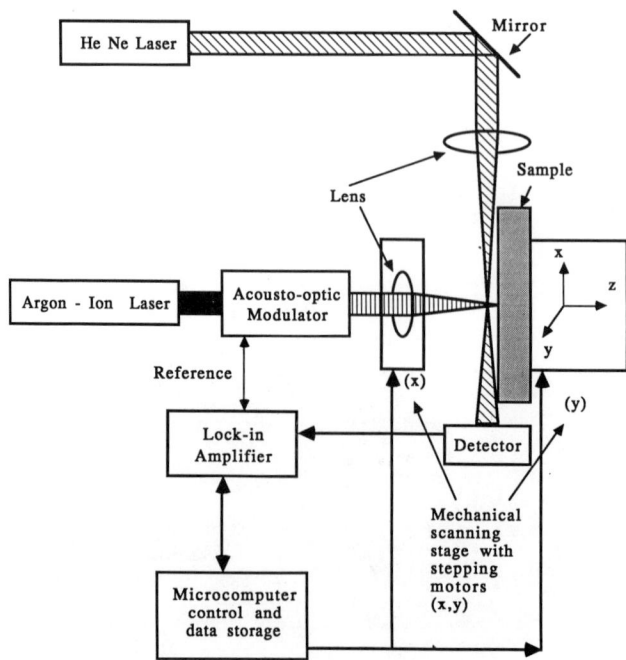

Fig. 3   Block diagram of the mirage effect method for thermal wave imaging.

and to have good pointing stability. Alignment of the beams relative
to one another, the position sensor, and the sample surface is
achieved with appropriate micrometer driven translation stages and
tilt tables, and in mounting these components to the optical table,
careful attention must be paid to the avoidance of relative motions in
the operating frequency range.

Another contactless thermal wave imaging technique is
photothermal radiometry[17-20], in which variations in the local
surface temperature are measured with a focused infrared (IR)
detector. The source is again a modulated laser beam, focused on the
surface. In this case, one has a point source and a point detector,
with the ultimate resolution in the extreme-near-field limit being
provided by the optical, rather than the IR, focal spot because of the
shorter optical wavelength. An advantage of this technique is that,
in contrast to both the gas-cell and mirage effect techniques, it does
not depend upon heat flow from the solid to the air and therefore
does not have the complications associated with the additional phase
delays and magnitude changes of the signal due to the presence of the
air. This also permits the application of the technique in vacuum. A
disadvantage of the technique is that variations in the emissivity of
the surface can obscure the thermal wave image. However, the
theoretical analysis is facilitated by the absence of complications
associated with the presence of the air.

A block diagram of the photothermal radiometry technique is shown
in Fig. 4. A Ge lens is used to collect the 3-12 micrometer
black-body radiation (modulated by the varying sample temperature in
the presence of the heating beam) on a suitable IR detector. An
inexpensive version is to use a pyroelectric IR detector. Such
detectors also have reasonable sensitivity at low modulation
frequencies (< 40Hz), a frequency region which is quite useful for
deep probing of the subsurface region of the sample. Cooled IR photon
detectors (e.g. HgCdTe) are more expensive but have very fast
response times and better sensitivity. If one combines that feature
with a suitable scanned array or IR camera and operates in the time
domain instead of the frequency domain (pulsed source, rather than
modulated source), considerable improvements in imaging speed can be
achieved [21].

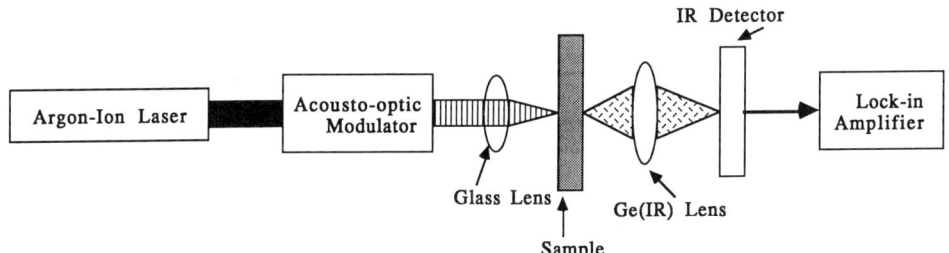

Fig. 4    Block diagram of the photothermal radiometric method for
thermal wave imaging.

## CHARACTERIZATION OF MATERIALS

The mirage effect method for thermal wave imaging (Fig. 3) can also be utilized for the characterization of thermal properties of pure and coated materials [22-26]. The experimental technique is to measure the transverse deflection of the (stationary) probe beam as the heating beam is scanned across the sample surface at right angles to the probe beam with a constant probe beam height and with the sample held stationary. In a pure material, to a good approximation, the separation of the two points on either side of the center of such a scan where the phase of the transverse deflection signal reaches ninety degreees effectively measures the thermal wavelength in the solid. The determination of the thermal diffusivity is then accomplished by plotting this separation versus the inverse square root of the frequency. More detailed agreement can be obtained, and the technique can be extended to the case of coated surfaces, by comparison to three dimensional thermal diffusion calculations of the mirage effect signal for this geometry. The reader is referred to two papers [25,26] in this volume for further details of the method and the accompanying theoretical developments. Several other papers in this volume and other volumes of this series describe thermal wave measurements in the time domain which have been used to measure thermal diffusivities of bulk materials and thin films.

## DISCUSSION AND CONCLUSIONS

The general principles of thermal wave imaging have now been established experimentally and theoretically. At least for the simpler detection schemes it is possible to carry out detailed calculations of images for subsurface defects having simple geometries, and quantitative agreement with experiments has been obtained. All of these thermal wave techniques are capable of producing high resolution images under the right experimental conditions. Different detection techniques have different degrees of symmetry, and this symmetry, or lack of symmetry, can be exploited for the detection of flaws with particular geometries. A number of the imaging schemes involve the generation and detection of acoustic waves in addition to the thermal waves. Interpretation of such images is necessarily more complicated, but such microscopes present the possibility of making high resolution images with both thermal waves and acoustic waves using the same instrument. It should be noted that as compared to optical, and even acoustic imaging, thermal wave imaging is very young, with the first images having been obtained just a few years ago. Clearly, much more instrumental development will occur. A primary objective of such development will undoubtedly be to improve the imaging speed. This probably will be accomplished through some form of parallel processing, rather than point by point scanning. The potential for production applications for NDE of thermal wave imaging would then be established.

## ACKNOWLEDGEMENTS
This work was supported by ARO under contract No. DAAG 29-84-K-0173, and by the Institute for Manufacturing Research, Wayne State University.

## REFERENCES

1. R.L. Thomas, L.D. Favro, and P.K. Kuo, Can. J. Phys. $\underline{46}$ (in press).
2. L.J. Inglehart, K.R. Grice, L.D. Favro, P.K. Kuo and R.L. Thomas, Appl. Phys. Lett. $\underline{43}$, 446 (1983).

3. L.J. Inglehart, M.J. Lin, L.D. Favro, and R.L. Thomas, Proceedings of the 1983 IEEE Ultrasonics Symposium, edited by B.R. McAvoy (IEEE, New York, 1983), p. 668.
4. D.W. Pohl, W. Denk and M. Lanz, Appl. Phys. Lett. 44, 651 (1984).
5. E.A. Ash and G. Nichols, Nature 237, 510 (1972).
6. P.K. Kuo, L.D. Favro, L.J. Inglehart, R.L. Thomas, and M. Srinivasan, J. Appl. Phys. 53, 1258 (1982).
7. Y.H. Wong, R.L. Thomas, and G.F. Hawkins, Appl. Phys. Lett. 32, 538 (1978).
8. G. Busse, Appl. Phys. Lett. 35, 759 (1979).
9. M. Luukkala and A. Penttinen, Electron. Lett. 15, 326 (1979).
10. R.L. Thomas, J.J. Pouch, Y.H. Wong, L.D. Favro, P.K. Kuo, and A. Rosencwaig, J. Appl. Phys. 51, 1152 (1980).
11. P.K. Kuo and L.D. Favro, Appl. Phys. Lett. 40, 1012 (1982).
12. K.R. Grice, L.J. Inglehart, L.D. Favro, P.K. Kuo, and R.L. Thomas, J. Appl. Phys. 54, 6245 (1983).
13. A.C. Boccara, D. Fournier, and J. Badoz, Appl. Phys. Lett. 36, 130 (1980).
14. J. C. Murphy and L. C. Aamodt, J. Appl. Phys. 51, 4580 (1980).
15. L.C. Aamodt and J.C. Murphy, J. Appl. Phys. 54, 581 (1983).
16. M.J. Lin, L.J. Inglehart, L.D. Favro, P.K Kuo, and R.L. Thomas, in Review of Progress in Quantitative Nondestructive Evaluation, Vol. 4B, Ed. D.O. Thompson and D. E. Chimenti, Plenum, New York, 1985), 739.
17. P.E. Nordal and S.O. Kanstad, Phys. Scr. 20, 659 (1979).
18. G. Busse, Appl. Opt. 21, 107 (1982).
19. M. Luukkala, A. Lehto, J. Jaarinen, and M. Jokinnen, Proceedings of the 1982 IEEE Ultrasonics Symposium, edited by B.R. McAvoy (IEEE, New York, 1982), p. 591.
20. D.P. Almond, P.M. Patel, and H. Reiter, J. de Physique Colloque C6, 491 (1983).
21. W.N. Reynolds, Proceedings of the 4th International Topical Meeting on Photoacoustic, Thermal, and Related Sciences, Esterel, Quebec 1985, Technical Digest, Paper WB1.1.
22. R.L. Thomas, L.J. Inglehart, M.J. Lin, L.D. Favro, and P.K. Kuo, Review of Progress in Quantitative NDE, Vol. 4B, edited by D.O. Thompson and D. Chimenti (Plenum, New York, 1985), 859.
23. R.L. Thomas, L.D. Favro, D.S. Kim, P.K. Kuo, C.B. Reyes, and Shu-Yi Zhang, Review of Progress in Quantitative NDE, Vol. 5B, edited by D.O. Thompson and D. Chimenti (Plenum, New York, 1986), 1379.
24. P.K. Kuo, C.B. Reyes, L.D. Favro, R.L. Thomas, D.S. Kim, and Shu-Yi Zhang, Review of Progress in Quantitative NDE, Vol. 5B, edited by D.O. Thompson and D. Chimenti (Plenum, New York, 1986), 1519.
25. C.B. Reyes, J. Jaarinen, L.D. Favro, P.K. Kuo, and R.L. Thomas, Review of Progress in Quantitative NDE, Vol. 6, edited by D.O. Thompson and D. Chimenti (Plenum, New York), to be published.
26. J. Jaarinen, C.B. Reyes, I.C. Oppenheim, L.D. Favro, P.K. Kuo, and R.L. Thomas, Review of Progress in Quantitative NDE, Vol. 6, edited by D.O. Thompson and D. Chimenti (Plenum, New York), to be published.

# NUMERICAL CALCULATIONS OF ACOUSTIC EMISSION

John A. Johnson

Idaho National Engineering Laboratory
EG&G Idaho, Inc.
Idaho Falls, Idaho 83415

## INTRODUCTION

A computer program [1] which solves the partial differential equations for sound propagation numerically is applied to the study of problems in acoustic emission. The program uses finite difference techniques to calculate sound fields due to distributions of sources in complex geometries in two dimensions. The potential to handle more complex geometries and to model more realistic sources is the main advantage of this type of calculation over the analytic calculations. The main disadvantage of the numerical technique is the cost of obtaining results since a large main frame computer or supercomputer is required.

In this paper the fields due to a simple source in a planar geometry are calculated and compared to the analytic results using the methods developed by Pao and his collaborators [2,3]. The purpose of these calculations is to test the feasibility of the code for calculations of acoustic emissions and to determine the limitations by comparing it with the Green's function or generalized ray theory results. Further calculations in more complex geometries which cannot be handled by the Green's function method are planned, including the effect of sources near or at the tip of a crack and the effect of geometric distortions due to plastic strain in a sample in a tension test.

## METHOD OF CALCULATION

In the finite difference method, the partial differential equations describing the system are approximated by finite differences in both space and time. The physical system to be analyzed is divided up into small quadrilateral zones. The four zones surrounding the grid point labeled (I,J) are shown in Fig. 1 with their respective stresses. The size of the zones should ideally be less than about 1/10 of the shortest wavelength of interest. Economic considerations may require some compromise of this ideal.

The stress in each zone is calculated from the integrated strain rate in that zone, which is determined by the velocities of the points at the four corners of the zone. Then the acceleration of the points of the mesh

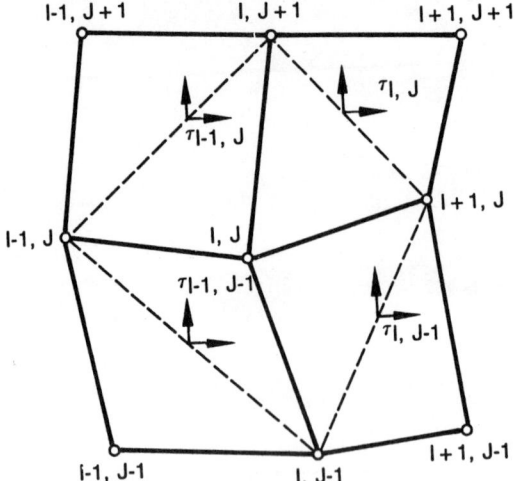

Fig. 1.     around a mesh point in a finite difference grid.

are calculated from the stresses around that point and integrated in time using a time-centered formulation to find new velocities and displacements from the previous values. Boundaries are represented by special values of the stresses or velocities corresponding to the desired physical situation. For example, a free boundary is modeled by having zero stress on the sides of a zone on the boundary. In this paper the acoustic emission on the surface is modeled by putting a time dependent stress on the sides of the relevant zones.

COMPARISON TO GREEN'S FUNCTION METHOD

The Green's function method, also known as the theory of generalized rays, was first applied by geophysicists for the study of stress waves generated by earthquakes. Rays from the source are traced to the locations of the receiver. The transient response of each ray is calculated from the inverse Laplace transform of a combination of integrals which depend on the source, receiver, and ray path, including mode conversions at the boundaries of the plate [2,3]. The solution is then the sum of the transients for each individual ray which arrives during the time of interest.

The Green's function analysis uses an exact heaviside time function for a point force. However, such a forcing function in a finite difference scheme would contain high-frequency components in both space and time which would produce waves of very short wavelengths. Such short wavelengths would be too small for any mesh and would result in numerical noise. Thus the forcing function must be approximated by a function that is bandwidth limited and of finite size. The bandwidth is chosen so that the equivalent minimum wavelength is ten times larger than the zone size:

$$F = F_o (1 - e^{-t^2/T^2})$$

This function has a 20 dB bandwidth of:

$$f = \frac{\sqrt{\ln 10}}{\pi T} = 0.48/T$$

Similarly the spatial width of the forcing function in the analytic calculations is zero, i.e., the spatial part is a delta function and acts at a point. Again the numerical scheme does not allow such a distribution and requires the source to be spread over several zones.

These limitations of the numerical finite difference method are not really all that deleterious since in real experimental situations the sources, and certainly the receivers, do have limited bandwidth. Some sources, however, are point-like in space on the scale of the discretization used here. A cracking inclusion in a metal, for example, could have dimensions much smaller than 0.2 mm, the zone size in this calculation.

Modeling the forcing function for the finite difference method results in some differences between the results of the numerical method and the Green's function method. The rise and fall times of the calculated displacements in the Green's function calculations can be zero, corresponding to the input step function force. In the numerical calculation the rise and fall times are limited by the bandwidth of the input force. A second effect is to lower the displacement amplitude in the numerical calculation, especially when the Green's function result is a step displacement. The finite source size has a similar effect of smearing out the wave form since the time of arrival of a pulse is different, depending on what part of the source it came from. This also tends to reduce the peak displacement amplitudes when compared to the Green's function solutions.

The Green's function solutions can be extended to finite size and limited bandwidth sources using standard filtering and integration techniques.

Sample Acoustic Emission Problem

In order to test the method, a point normal heaviside force on the surface of a plate is modeled using finite difference methods. A schematic of the problem is shown in Fig. 2. The exact solution to this problem can be calculated using Green's functions and has been verified experimentally [2-4]. Thus the results from the finite difference calculations can be compared and verified with confidence.

Ceranoglu [5] has presented in graphical form the results of just such a problem. The geometry and material properties in this study have been chosen to match those used in his calculation. Ceranoglu's data are in normalized nondimensional form in both time and displacement. To simplify the comparison, the longitudinal wave speed and the plate thickness have been chosen to have equal magnitudes (6 mm/$\mu$s and 6 mm) so that the normalized time and the real time in microseconds are the same. The shear wave speed was then chosen to be equivalent to that used by Ceranoglu, who used a longitudinal-to-shear wave sound speed ratio of $\sqrt{3}$ (3.46 mm/$\mu$s).

The numerical calculation actually requires the bulk and the shear moduli and the density. For a density approximately that of steel (8.0 g/cm$^3$), the required bulk and shear moduli to obtain sound speeds above are 1.60 and 0.96 Mbars respectively.

The normalized nondimensional displacement response used by Ceranoglu is determined by multiplying the actual displacement response by

$$\text{Nomalized Displacement} = \frac{\pi \mu h^2}{F_o} \times \text{displacement}$$

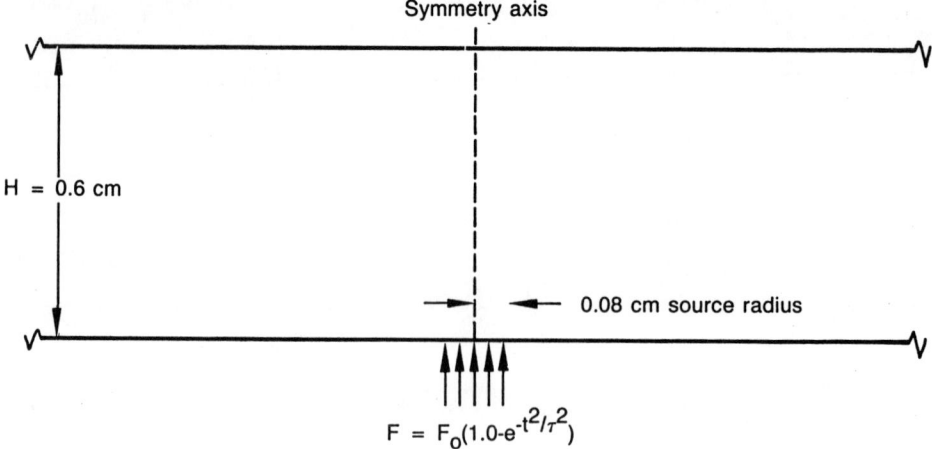

Fig. 2. Acoustic emission in a plate due to a source on the surface.

where h is the plate thickness, $\mu$ is the shear modulus, and $F_0$ is the strength of the force. In the finite difference calculation the point source is modeled by a uniform normal stress of radius 0.8 mm and strength $10^{-4}$ Mbars. The force is the area of this circle times the stress, or $2.0 \times 10^6$ dynes. Then the final conversion factor for determining the normalized displacement is:

Normalized displacement = $0.54 \times 10^6$ × displacement in cm

The mesh consisted of zones of 0.2 mm on a side. The height of the mesh was chosen to be 6 mm, as noted above. The radial size of the mesh was chosen so that reflections from the far boundary would not return to points in a radius of 12 mm during the time of interest (5 µs). The zone size of 0.2 mm then requires that the minimum wavelength in the problem be greater than about 2.0 mm or 10 times the zone size. In Table 1 the 20 dB bandwidth and the shear wavelength in mm and in zone lengths are given for various values of the risetime used in the calculation. From this table, the forcing function with a risetime of 0.32 µs should be adequate for the a mesh with zones 0.2 mm on a side.

Comparison of the Finite Difference and Green's Function Calculations

The finite difference program results are shown in graphical form in Figs. 3 through 6. The left side of the plots is the axis of symmetry,

Table 1. Forcing Function Bandwidth and Shear Wavelength for several values of Forcing Function Rise Time.

| Rise Time Micros | Bandwidth MHz | Wavelength mm | Zone Lengths |
|---|---|---|---|
| 0.45 | 1.1 | 3.2 | 16.0 |
| 0.32 | 1.5 | 2.3 | 11.5 |
| 0.14 | 3.4 | 1.0 | 5.0 |

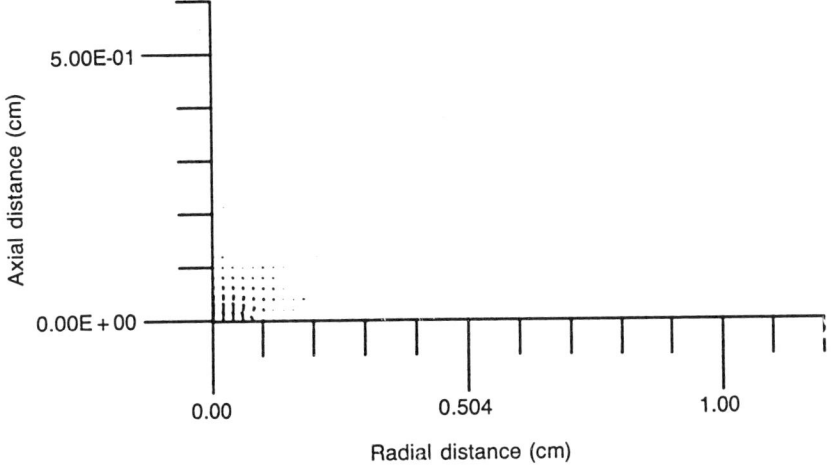

Fig. 3. Vector field plot at 0.2 μs. The left axis is the axis of symmetry. The horizontal axis is the radial distance.

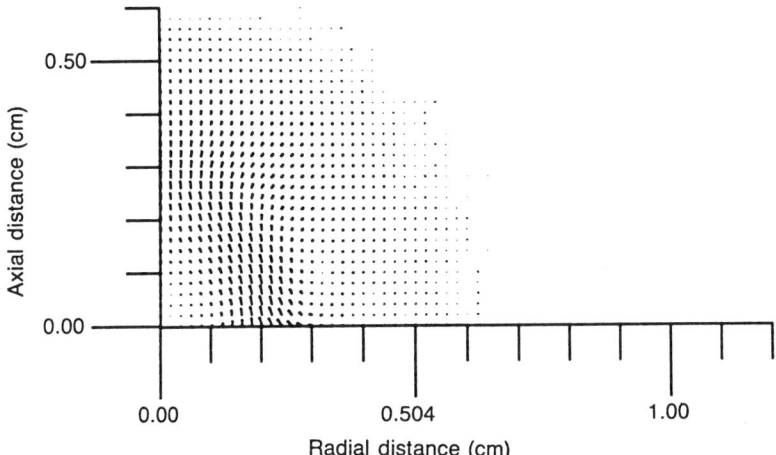

Fig. 4. Velocity field plot at 1.0 μs.

corresponding to the center of Fig. 2. Each plot is a snapshot of the velocity vector field at various times. In Fig. 3 the field at 0.2 μs is confined to the lower left corner of the grid, next to the source. In succeeding figures waves propagate out from the source on the axis. The longitudinal wave has reached the top of the plate while the shear wave forms an arc at 1.0 μs in Fig. 4. Along the radial (horizontal) axis the arc is distorted by the Rayleigh surface wave which has a slightly slower sound speed than the shear wave. The shear wave has reached the top of the plate in Fig. 5 and can be seen reflecting off the top in Fig. 6 forming a second arc.

For comparison with the Green's function calculations, plots of the displacement as a function of time can be made at any point in the grid. Ceranoglu [5] has shown a plot of the displacement versus time for several

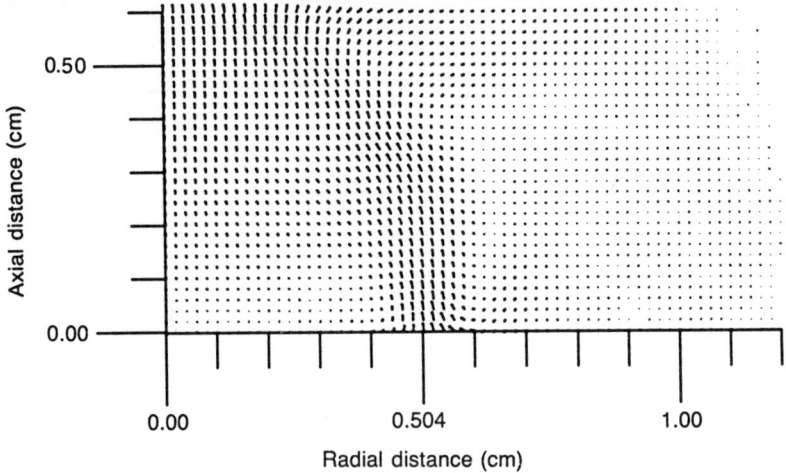

Fig. 5. Velocity field plot at 2.0 µs.

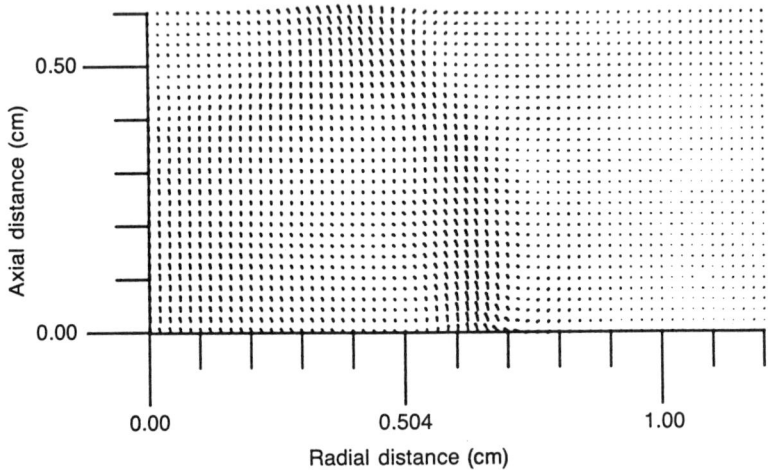

Fig. 6. Velocity field plot at 2.4 µs.

points in the plate. In Fig. 7 one of his plots is reproduced, giving the displacement on the bottom of the plate (same side as the source) at a radial distance of twice the thickness of the plate. This corresponds to a radius of 12 mm in the finite difference calculations.

In Figs. 8 and 9 the radial displacements versus time for two different source rise times (0.32 and 0.14 µs) are shown. Similar plots for the axial displacements are displayed in Figs. 10 and 11. These plots are smoothed versions of the displacements given in Fig. 7, actually corresponding to the convolution of the displacements from the Green's function calculations and the source time function, integrated over the source function. In all the plots the small longitudinal wave arrives at 2.0 µs. The normalized amplitude for the radial component is approximately 0.036, corresponding to $0.67 \times 10^{-7}$ Mbars, more than twice that shown in Figs. 8 and 9. The dip at about 3.8 µs, marked with an R

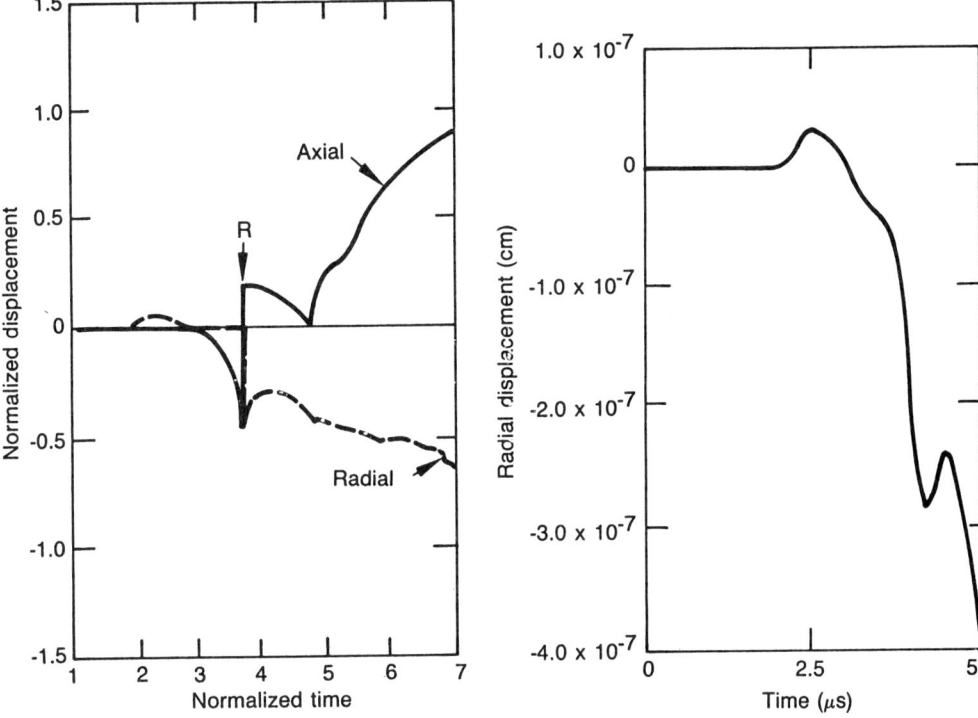

Fig. 7. Green's function calculation of the displacement versus time for a point on the same side of the plate at the source, at a radial distance of twice the thickness of the source.

Fig. 8. Radial displacement calculated using finite difference methods for a source risetime of 0.32 μs.

in Fig. 7, corresponds to the arrival of the Rayleigh wave. This, and the rise after 4.0 μs, is reproduced in a smoothed fashion in Figs. 8 and 9.

The axial component of the displacement in the finite difference calculations also follows the Green's function results. A very small longitudinal component arrives at 2.0 μs, followed by the Rayleigh wave component. The positive excursion in Fig. 7 just after the Rayleigh wave arrival has an amplitude of about 0.15 normalized units, corresponding to $2.8 \times 10^{-7}$ Mbar. From the finite difference method values of $1.4 \times 10^{-7}$ and $1.6 \times 10^{-7}$ are obtained in Figs. 10 and 11.

In summary, the effect of the finite rise time and finite source size in the numerical calculations is a smoothed version of the Green's function results with significantly lower amplitude. The plots with the faster rise times more closely reproduce the sharp features of the Green's function results as expected.

CONCLUSION

Finite difference calculations of acoustic emission on the surface of a plate have been shown to be equivalent to the Green's function or generalized ray theory results within certain limitations resulting from a reduced frequency bandwidth and finite source size in the numerical calculations. The next steps include simulating a buried source of

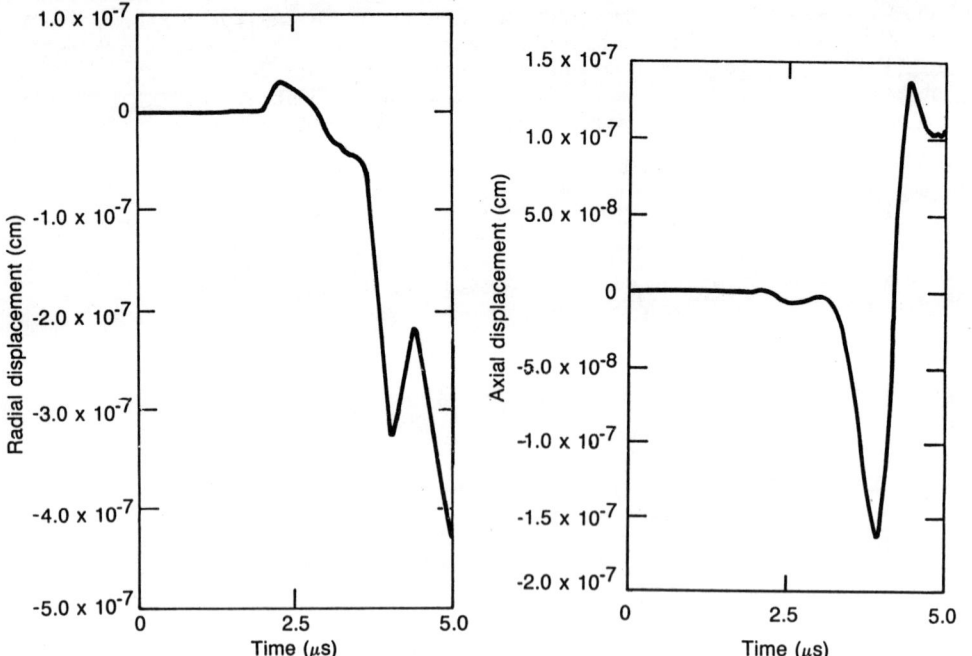

Fig. 9. Radial displacement for a source risetime of 0.14 μs.

Fig. 10. Axial displacement for a source risetime of 0.32 μs.

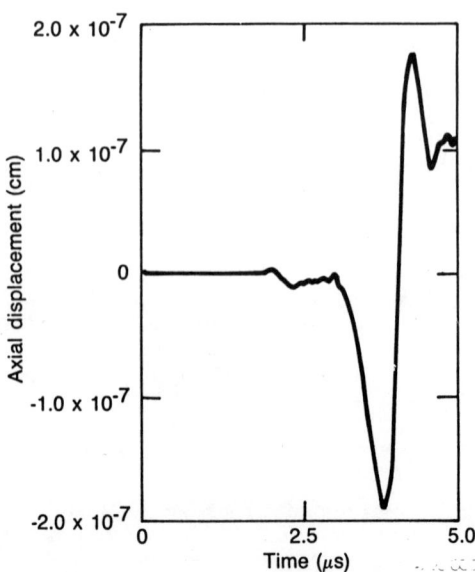

Fig. 11. Axial displacement for a source risetime of 0.14 μs.

acoustic emission and comparing the results to the analytic theory and to a model of more complex geometries that the analytic theory is not capable of handling. These include sources near or at a crack tip with the complex boundary conditions of the crack and accounting for large strains in the plate, similar to what might be encountered in tension tests of ductile specimens.

ACKNOWLEDGMENTS

This work is supported by the U. S. Department of Energy, Office of Energy Research, Office of Basic Energy Sciences under DOE Contract No. DE-AC07-76ID01570.

REFERENCES

1. J. A. Johnson, "Numerical Calculations of Ultrasonic Fields: Transducer Near Fields", J. of NDE 3, 27 (1982).
2. Pao, Gajewski, and Ceranoglu, "Acoustic Emission and Transient Waves in an Elastic Plate", JASA 65, 96 (1979).
3. A. N. Ceranoglu, "Acoustic Emission and Propagation of Elastic Pulses in a Plate", Ph.D. Thesis, Cornell University (1979).
4. J. E. Michaels, T. E. Michaels, and W. Sachse, "Applications of Deconvolution to Acoustic Emission Signal Analysis", Materials Evaluation 39, 1032-1036 (1981).
5. A. N. Ceranoglu, op. cit., 137.

POINT-SOURCE/POINT-RECEIVER MATERIALS TESTING


Wolfgang Sachse and Kwang Yul Kim

Department of Theoretical and Applied Mechanics
Cornell University, Ithaca, New York - 14853 U S A


INTRODUCTION

   Conventional measurements in the ultrasonic testing of materials, when used as the basis of a materials characterization procedure, typically rely on one or two piezoelectric transducers operating as source and receiver, attached to a specimen to launch and detect ultrasonic waves in the object to be characterized. Measurements of signal arrival time (or velocity) and amplitude (or attenuation), possibly as a function of frequency, are then correlated with the composition and the macro- and micro-structure of the material, which may include voids, flaws and inclusions distributed through a region of the material. While relative measurements of the time-of-flight and ultrasonic amplitudes do not present extraordinary measurement challenges, absolute measurements do. It is unfortunate that absolute quantities are often required since they are difficult to obtain reliably with a conventional piezoelectric transducer-based ultrasonic system. For this reason, a considerable effort over the past decade has been undertaken to develop and improve non-contact methods for generating and detecting ultrasonic signals in materials. However, a limiting factor of all the existing non-contact measurement systems is the care required for their use and their reduced sensitivity in comparison to those utilizing piezoelectric transducers.

   Over the past several years notable advances have been made in the development of quantitative acoustic emission methods in which the signals emitted by a source in or on a structure are detected by one or more receiving transducers at the surface of the structure. Knowing the transfer characteristics of the sensors and the appropriate dynamic Green's functions of the structure, the temporal and spatial characteristics of the emission source can be recovered from the detected signals by signal processing techniques [1, 2]. An essential assumption in such a measurement procedure is that the material encompassing the source and receiver points is homogeneous, isotropic, elastic and non-attenuative. While the Green's functions for an anisotropic material can, in principle, be computed, none appear to have been published. It is known that the propagation of acoustic emission signals from a source to a receiver point is influenced not only by the geometry of the specimen but also by the material's macrostructure specified in terms of its size, shape and for composite materials, the ply configuration. In addition, the wave propagation is strongly affected by the specimen material's microstructure, including its anisotropy, heterogeneity, elastic, inelastic and viscous properties, and its wave attenuation characteristics. To determine these variables it is proposed to utilize an ultrasonic testing system analogous to that used in exploration geophysics, consisting of a well-characterized point source and point receiver. This is shown schematically in Figure 1. If both the source and receiver possess known temporal characteristics, then the principal advantage of a point-source, point-receiver system over a conventional ultrasonic system is that quantitative ultrasonic measurements are possible. Although the geometric characteristics of the wave propagation may be more complex than those for the case in which plane waves are used, these effects can be accounted for in the detected signals to recover the material-related wavespeeds and attenuation properties. Because the excitation and detection regions are small, the required specimen surface preparation is minimal and furthermore, specimens of arbitrary geometry can be easily tested, in particular those which are neither planar nor flat.

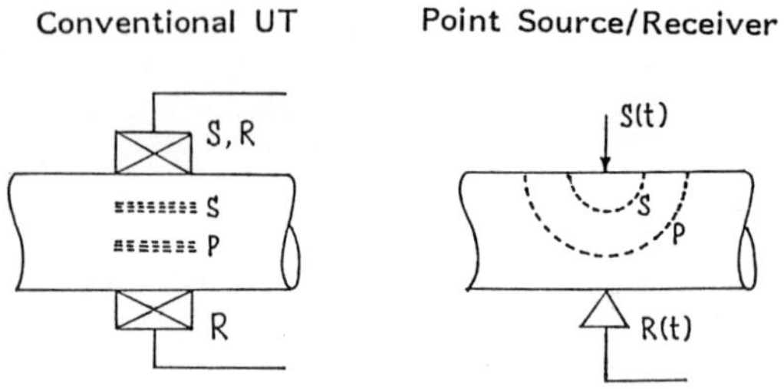

Fig. 1 - Conventional ultrasonic and point-source/point-receiver testing configurations.

WAVE PROPAGATION, SYSTEM CHARACTERISTICS AND SIGNAL ANALYSIS

The basis of a point-source/point-receiver testing system is a source and a receiver whose temporal and spatial transfer characteristics are known and a theoretical basis by which the measured signals can be identified, interpreted and processed to recover the characterisitcs of the propagating medium. The term "point" refers ideally to a signal source or detection region whose lateral dimension is much smaller than the effective wavelength of the highest frequency component of interest in the measured signal. This wavelength will also be much shorter than any dominant dimension of the specimen.

## Wave Propagation

The analysis of transient elastic waves between a point source and a point receiver in a bounded structure is discussed in several papers [c.f. 3]. The displacement signals $u_k$ detected at a receiver location, $\underline{x}$, in a structure from an arbitrary source $f(\underline{x}',t)$ located at $\underline{x}'$ having source volume V can be written as a sum of contributions due to a monopole, dipole and higher-order terms. The monopole source contribution to the signal can be rewritten compactly as

$$u_k^{(m)}(\underline{x},t) = F_j(t) * G_{jk}(\underline{x}/\underline{x}',t) \qquad (1)$$

where $F_j(t) = \int_V f_j(\underline{x}',t)\,dV$ is the total force acting throughout the source volume and $G_{jk}(\underline{x}/\underline{x}',t)$ is the dynamic Green's function of the structure. It is a function of time only for a point source. For all the monopolar sources utilized in this paper, the force is always a force normal to the surface of the specimen, i.e., $F_z$. The dipole contribution to the signal can be rewritten compactly as

$$u_k^{(d)}(\underline{x},t) = M_{ij}(t) * G_{jk,i}(\underline{x}/\underline{x}',t) \qquad (2)$$

where $M_{ij}(t) = \int_V x_i' f_j(\underline{x}',t)\,dV$ represents the moment tensor of the source and $G_{jk,i}(x/x',t)$ is the spatial derivative of the Green's function. This representation is used to model many thermoelastic sources [4].

A solution to the forward problem can be readily computed for a specimen of flat plate-like geometry. That is, given the thickness of the specimen, the material's longitudinal to shear wavespeed ratio and the source/receiver separation, the dynamic Green's functions appearing in Eqs. (1) and (2) can be found using one of the available algorithms [3, 5]. If the measurement system includes a point receiving transducer whose output voltage signal is related to the input displacement signal by the sensor's transfer function, R (t), then the results given by Eqs. (1) and (2) must be further convolved with this transfer function to obtain the output voltage signals of the system corresponding to each of the excitations.

An important example is shown in Figure 2. This is the normal displacement signal computed for a vertical force source acting on a plate specimen and detected at the epicenter point on the back surface of the plate directly under the source point. This example was obtained for a 7090 Al/SiC composite 1.884 cm thick whose longitudinal and shear wavespeeds were 0.708 cm/μsec and 0.368 cm/μsec, respectively, with zero wave attenuation. Thus, this result represents the behavior expected for an ideal material. The signals corresponding to other types of source and receiver configurations can be computed similarly. It is seen that the arrivals of both the P- and S-waves can be easily identified. Hence, even though the force was applied normal to the specimen surface, both longitudinal and shear wave modes are excited. The identification of the wave arrivals is possible, provided that the source/receiver separation is not larger than about 10h - 15h, where h is the thickness of the plate. It is also clear from the waveforms shown in the figure that while the wave arrivals are readily identifiable, the signals characteristically possess a "tail"; that is, each wave is geometrically dispersed as it propagates through the specimen. This dispersion is unrelated to the properties of the medium and its presence in a signal must be removed if the correct material-related frequency characteristics of a measured signal are to be determined.

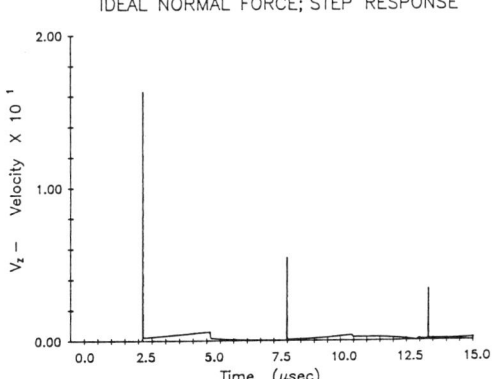

Fig. 2 - Ideal epicentral velocity signal; Step normal force excitation; Zero attenuation.

Additional insight is gained by considering the evaluation of the displacement signals in the frequency domain. This has recently been carried out for the case of a viscoelastic medium [6]. It is found that the Fourier phase function of the P-wave arrival of the velocity signal is given by

$$\phi(\omega) = kL + A/kL \qquad (3)$$

where k is the wavenumber, L the source/receiver separation and A is a correction factor approximately equal to 2. The first term of the phase function is identical to that derived for plane waves [7]. The amplitude corresponding to the first shear wave arrival can be processed similarly and the phase functions of other source types are expected to exhibit a similar form [8].

Source and Receiver Characteristics

The ideal point source and point receiver with perfect impulse response can only be approximated with real sources and receivers. However, in order to achieve acceptable signal-to-noise ratios, it is possible to use transducers with a finite aperture provided that the generated and detected acoustic fields in the specimen are uniform and resemble those expected from a point source and point receiver. Equally important is that the transduction characteristics of the source and receiver are known. This includes both the primary and secondary quantities being generated and detected. The temporal transfer characteristics of both source and receiver must be known a priori or determined in a calibration experiment; they must possess an appropriate frequency response relative to the material property being investigated. There are, however, many measurement situations in materials whose viscous dispersion and wave attenuation is sufficiently low so that only a measurement of the arrival of a signal is required and a complete characterization of the sensor is not needed. In these cases, conventional piezoelectric point sensors sensitive to the wave motions normal to the surface of the specimen can be used. With such a sensor the measurement is simplified since no special surface preparation of the specimen or critical transducer alignment with the specimen surface is required. Obviously, in all cases the source and the receiver must possess an adequate signal-to-noise ratio to permit signal identification and subsequent processing operations to be performed reliably. A discussion of various point-sources and receivers and their operating characteristics is given in Ref. 9.

Signal Identification and Waveform Analysis

According to the convolution equations (Eqs. (1) and (2)) for a source of known type and time function and for a specified source-receiver separation, only the time-dependence of the input source function and output displacement signals is required to invert these equations to recover the dynamic Green's function corresponding to the particular testing geometry and specimen material [10]. If the measurement system consists of a source whose excitation is an impulse or a Heaviside step and the receiver is a high-fidelity displacement or velocity sensor, then the detected signals will correspond directly to the dynamic Green's function of the specimen, thus requiring no further signal processing. This observation emphasizes the advantage of using a source and a receiver possessing ideal characteristics.

Once the Green's function has been determined, the ultrasonic wavespeeds can be recovered by identifying the arrival times of the P- and S-wave signals. If the instant of excitation is known, only the first arrivals of these signals need to be determined. When this is not known, the arrival of other signals propagating through the specimen are needed. The 3P signal corresponding to the longitudinal wave propagating three times through the specimen is identified in the actual waveform shown in Figure 3. The longitudinal and shear wavespeeds of the material can be recovered from the measured arrival times according to the formulae shown in the figure.

The frequency-dependence of a particular wave arrival is determined by properly windowing that portion of the waveform containing the signal amplitude and transforming this amplitude to obtain the Fourier phase function of the signal. Once this function is found, it is substituted into Eq. (3) which, in turn, is solved numerically to obtain a solution for the dispersion relation of the wave in the material. Once the dispersion relation has been found, the phase and group velocities, c $(=\omega/k)$ and v $(=\partial\omega/\partial k)$, can be evaluated.

Since the computed, ideal waveform corresponds to the propagation in a non-attenuative material, the attenuation of either the P- or the S-wave amplitude in a real material can be determined by making a comparison of the measured and computed waveform amplitudes of the corresponding wave arrivals. It follows that the frequency dependence of the attenuation can be determined by processing the windowed signal amplitude in the frequency domain to form its magnitude spectrum V (f) and evaluating

$$\alpha(f) = 20 \log_{10} [V(f) / V_{ref}(f)] \quad (4)$$

where $V_{ref}(f)$ refers to the magnitude spectrum of the wave velocity amplitude of the non-attenuated, ideal signal.

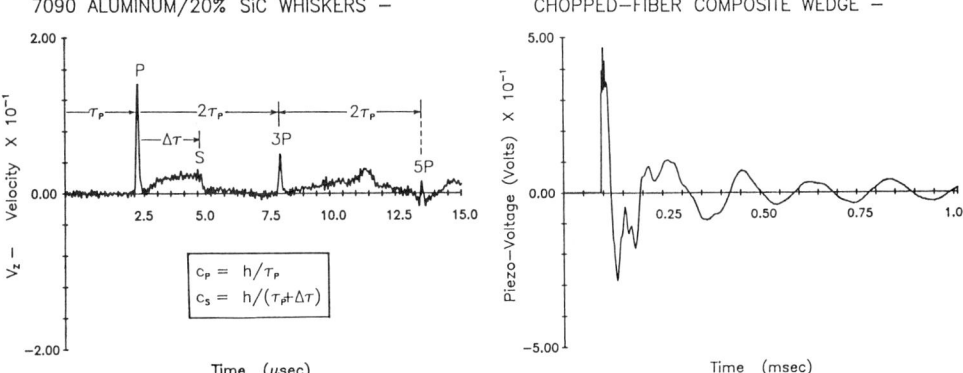

Fig. 3 - Measured epicentral velocity signal; Step source. (Wave arrivals are indicated).

Fig. 4 - Ultra-absorptive chopped fiber composite; Step source; Piezoelectric transducer detection.

MEASUREMENTS

The measurement system utilizing a point-source/receiver resembles a system used to make quantitative acoustic emission studies. The exception is the presence of the input source element which may or may not have a sensor attached to it in order to generate a synchronization pulse with the excitation signal. The normal velocity signal at epicenter corresponding to a step excitation on a specimen of a 7090 Al/SiC metal-matrix composite was shown in Figure 3. Measurement of the arrivals of the P- and S-wave amplitudes leads at once to the recovery of the longitudinal and shear wavespeed values, $c_p$ = 0.700 cm/μsec; $c_s$ = 0.380 cm/μsec. These are within 3% of the values determined in a conventional ultrasonic measurement.

Since in these measurements only the time of arrival of a particular wave is required, an uncalibrated piezoelectric point transducer can often be used. The waveform obtained in a highly attenuative, chopped fiber/epoxy, wedge-shaped specimen having a non-uniform layer of another material on one side is shown in Figure 4. The detection region of the specimen was left unprepared and hence the piezoelectric transducer with the excess couplant exhibited considerable ringing. However, even for this unfavorable testing situation, the first P-wave arrival is easily detected and can be used to determine an effective longitudinal wavespeed value for this material.

To determine the orientation dependence of wavespeeds in a sample, an array of transducers is required. In the simplest configuration, the receiving elements are located equi-distant about the source point. The sensors may be on either side of the sample, but they should be within 10h - 15h of the source so that the first P- and S-wave arrivals can be clearly identified in the detected signals. An example of the results of waveform measurements made in a specimen of graphite-epoxy comprised of 32 plies whose layup was at (+/- 45°) is given in Figure 5. Shown are the wavespeeds of the P-wave in various directions of the material. To obtain this result, the fracture of a capillary was used as a monopolar source with eight point piezoelectric sensors placed at various angles about it. The time of the first arrival was measured in each of the detected waveforms and the wavespeed was computed by dividing the arrival time into the source/receiver separation.

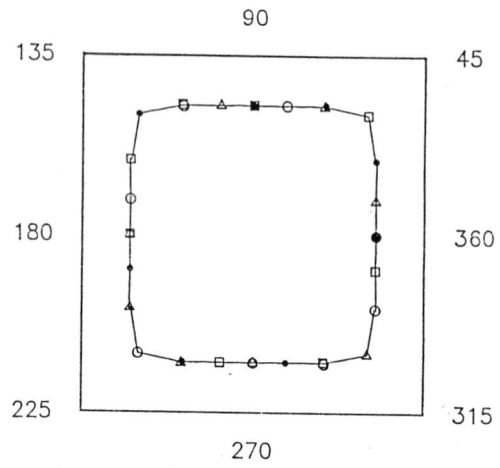

Fig. 5 - Wave velocity surface in 32-ply Graphite/Epoxy; Layup; (±45°).

The graphite-epoxy specimen possesses a four-fold symmetry. This can be determined from inspection of the detected signals or from knowledge of the material's fabrication. Recognizing this symmetry, it is possible to generate additional pseudo-points by projecting each of the measured data values in directions oriented at 180, 90, -90 degrees to those measured. As the results of Figure 5 demonstrate, the twenty-four additional points all lie on the same wavespeed surface. This finding verifies the consistency of the measurement results.

Application of the Fourier phase analysis method for determining the dispersion relation and the frequency-dependent phase velocity of the longitudinal wave in a 6061 Al/SiC metal-matrix composite specimen is shown next. The signal resulting from a capillary fracture source detected at epicenter with a piezoelectric point transducer whose response approximated a velocity sensor is shown in Figure 6(a). Also indicated is the windowed, first arrival of the P-wave signal. From the magnitude spectrum it is found that the signal contains little energy above 8 MHz reflecting the frequency response of the transducer and amplifier used to detect the signals. Because the low-frequency correction is only significant at frequencies below 0.5 MHz, it is omitted from the dispersion relation of the derived phase velocity shown in Figure 6(b). It is seen from the latter that the phase velocity between 3 and 10 MHz is nearly constant at 0.690 cm/μsec. At lower frequencies there is a decrease to lower wavespeed values which is due principally to the response of the piezoelectric transducer used to make the measurements and the omission of the low-frequency correction.

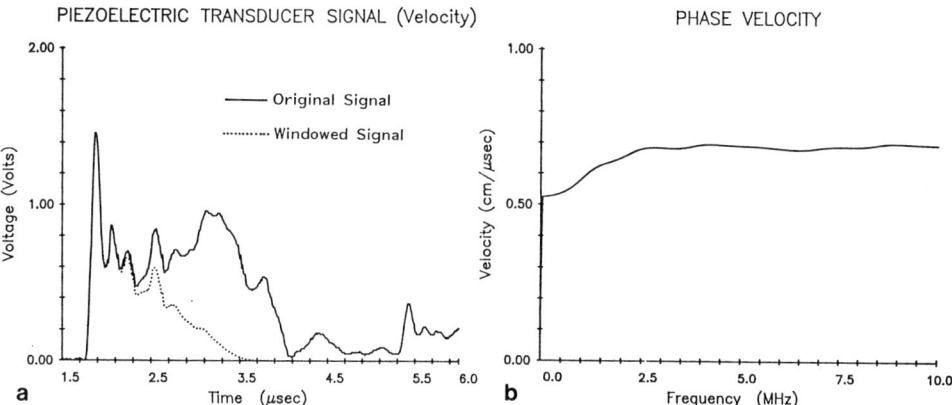

Fig. 6 - Wave dispersion measurement in a metal-matrix composite (a) Original, P-wave windowed signal; (b) Derived phase velocity.

An example of an attenuation measurement is shown in Figures 7(a)-(b). The velocity signal resulting from a step force applied in a 7090 Al/SiC metal-matrix composite specimen was shown in Figure 3. In the procedure, the first P-wave is windowed and compared to the computed response for the ideal case of a non-attenuating material shown in Figure 2. The Fourier magnitude spectra of the measured and ideal P-wave amplitudes are shown in Figure 7(a), while the result obtained from applying Eq. (4) is shown in Figure 7(b). In this example, only a relative measure of the attenuation of the longitudinal wave is determined since the vertical scale in Figure 3 was not calibrated absolutely and the magnitude of the force drop of the source used to generate the signal in this experiment was not measured.

In the waveforms detected in extremely absorptive materials, only the lowest frequencies are able to propagate and, hence, an unambiguous

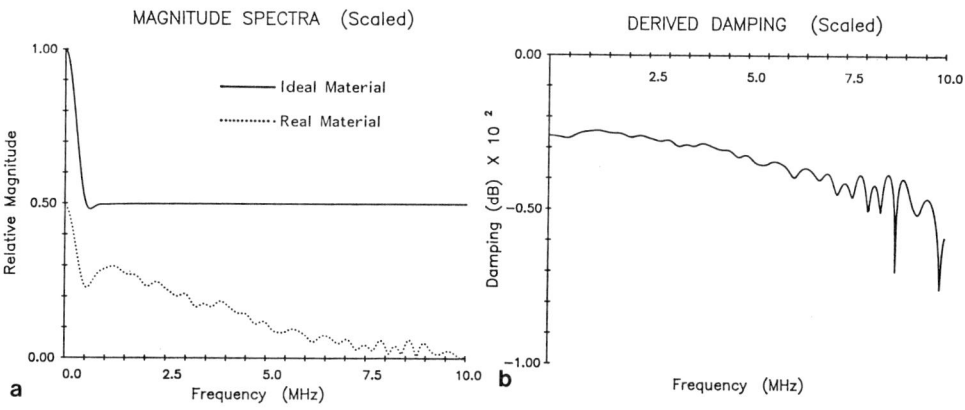

Fig. 7 - Wave attenuation measurement in a metal-matrix composite (a) Magnitude spectra of ideal, real signals; (b) Derived attenuation.

identification of the particular wave arrivals may be difficult. In such cases it may be advisable to choose an epicentral testing configuration with a sufficiently thick specimen so that the separation of the P- and S-wave arrivals is distinct in the detected signals. In cases in which only the lowest frequency components of the signal are propagated, it may also be necessary to consider other sensors to detect the signals.

The few examples shown here were used to illustrate the various signal analysis procedures described in the previous section. Numerous additional examples obtained in a variety of different materials are contained in a full length paper [9].

CONCLUSIONS

The components and characteristics of a point-source/point-receiver material testing system have been described by which the ultrasonic wavespeeds and attenuation can also be determined as a function of frequency in a variety of materials. The method utilizes a source and receiver whose transduction characteristics are known or can be determined in a calibration experiment. The measurements require a minimal amount of surface preparation and they can be made on specimens which are neither planar nor flat. Information regarding the propagation characteristics of both longitudinal and shear wave components is possible from a single waveform. It is also possible to select an excitation source whose time characteristics result in high energies at low frequencies which facilitates measurements in ultra-absorptive materials. It was demonstrated that while the characteristics of the wave propagation are more complex than those for plane waves, the existence of a theory of transient elastic waves permits a proper interpretation of the detected signals, provided that appropriate and calibrated point sources and receivers are utilized to make the measurements and the source/receiver separation is known. Results of several experiments were shown in which a composite material's longitudinal wavespeed can be recovered from the detected waveforms. It was demonstrated that by using an array of sensors, the wave velocity surface of a material can also be determined.

A procedure was also described by which the frequency-dependent wavespeeds and attenuation can be determined from the detected signals. The wavespeeds are recovered from an analysis of the Fourier phase functions of the normal velocity amplitudes corresponding to the arrival of either the longitudinal or shear wave signals. An analysis of the Fourier magnitude spectra of these signal amplitudes are compared to the magnitude spectra of the corresponding wave amplitudes for an ideal, non-attenuating specimen. With the continued development of non-contact point-sources and receivers, this measurement technique shows great promise as a powerful tool for characterizing micro- as well as macro-structural features of a large number of materials under a variety of measurement conditions.

ACKNOWLEDGEMENTS

We acknowledge the valuable discussions we have had with R. L. Weaver and the specimen materials we have received from D. Divecha. This work was supported in part by the Mechanics Division (Dr. Y. Rajapakse) of the Office of Naval Research and by the Solid and Geo-Mechanics Program (Dr. K. Thirumalai) of the National Science Foundation. Use of the facilities of the Materials Science Center at Cornell University which is supported by a grant from the National Science Foundation is also acknowledged.

# REFERENCES

1. N. N. Hsu, J. A. Simmons and S. C. Hardy, "An Approach to Acoustic Emission Signal Analysis - Theory and Experiment", *Materials Evaluation*, Vol. 35, No. 10, 100-106 (1977).

2. Y. H. Pao, "Theory of Acoustic Emission", in *Elastic Waves and Non-destructive Testing of Materials*, Y. H. Pao, Ed., AMD-Vol. 29, Am. Soc. Mech. Engrs., New York (1978), pp. 107-128.

3. A. N. Ceranoglu and Y. H. Pao, "Propagation of Elastic Pulses and Acoustic Emission in a Plate: Part I. Theory; Part II. Epicentral Response; Part III. General Responses", *ASME J. Appl. Mech.* 48, 125-147 (1981).

4. C. B. Scruby, "Laser Generation of Ultrasound in Metals", in *Research Techniques in Nondestructive Testing*, Vol. 5, R. S. Sharpe, Ed., Academic Press (1985), pp. 281-327.

5. N. N. Hsu, "Dynamic Green's Functions of an Infinite Plate - A Computer Program", Report NBSIR 85-3234, National Bureau of Standards, Gaithersburg, MD (November 1985).

6. R. L. Weaver, "Frequency Dependence of Generalized Ray Arrivals at Epicenter in a Viscoelastic Plate". In preparation.

7. Y. H. Pao and W. Sachse, "On the Determination of Phase and Group Velocities of Dispersive Waves in Solids", *J. Appl. Phys.*, 48, 4320-4327 (1978).

8. R. Weaver and W. Sachse, "Viscoelastic Generalized Rays: Theory and Experiment". In Preparation.

9. W. Sachse and K. Y. Kim, "Applications of Ultrasonic Point-Source/Point-Receiver Measurements". Submitted for publication.

10. J. E. Michaels, T. E. Michaels and W. Sachse, "Applications of Deconvolution to Acoustic Emission Signal Analysis", *Materials Evaluation*, 39, No. 11, 1032-1036 (1981).


# DISCUSSION

Mr. Marvin Hamstad, University of Denver: Wolfgang, what kind of fiber volumes were you using typically in these samples? And would you comment on the effect of fiber volume on this approach?

Mr. Sachse: In the graphite/epoxy specimens?

Mr. Hamstad: Yes. Or even the aluminum composites.

Mr. Sachse: The graphite/epoxy specimens typically had a fiber volume ranging from 63-67%. The metal-matrix aluminum composites contain about 20% silicon carbide.

Mr. Hamstad: Could you comment on what the effect would be if it was, say, 60 percent?

Mr. Sachse: I don't know the answer so I'll answer your question with a plea. If someone could supply us with samples to test, we would be delighted to test them and return them, unharmed. With the small number of specimens we only recently obtained, we had no way of investigating this. But I have no way of knowing the answer to your question. I have only two small specimens that I obtained only recently.

From the Floor: I was interested in the use of the capacitive transducer. Was it a standard one; did it suffer from sensitivity limitations, or have you developed a new type of sensor?

Mr. Sachse: It was developed by my colleague, Dr. Kim. A special feature about this capacitive transducer--I have not seen another like it--is that it is self-aligning with the surface of the specimen. This is achieved by having the electrode resting on a small ball. Thus one only needs to polish a very small section of the sample. We have several versions of this transducer with the smallest element about 1.2 millimeters in diameter. In using this sensor, one only needs polish or have a small, smooth surface of that dimension available on the specimen.

In use, the plate is brought in contact with the sample, and the ball permits a self-alignment of the moving electrode with the sample. A very fine micrometer adjustment is used to set the gap.

From the Floor: What was the sensitivity of the transducer?

Mr. Sachse: The answer depends on the gap dimension, the bias voltage and the charge amplifier sensitivity. In our system the charge amplifier sensitivity was 0.25 V/pf, so when we use a gap of 5 µm and a bias supply of 30 volts, we have a sensitivity of approximately 100 mV/Å. You can see from the waveforms we obtained that they are very good.

# THERMALLY INDUCED ACOUSTIC EMISSION IN HOMOGENEOUS METALS AND COMPOSITES

John M. Liu

Materials Evaluation Branch (R34), Naval Surface Weapons Ctr.
White Oak Laboratory, Silver Spring, MD  20903-5000

## INTRODUCTION

Both nonuniform heating in a homogeneous material and uniform heating in an inhomogeneous material produce local stresses. Inhomogeneous materials include polycrystals with anisotropic grains, two or more phase materials and composites. The thermally generated stresses can potentially induce acoustic emission via microscopic deformation and dynamic stress relieving mechanisms. It is anticipated that the ability to follow the history and characteristics of acoustic emission would be useful as a research tool for the study of microscopic deformation mechanisms, and could serve as a basis for establishing accept/reject criteria during thermal proof-testing of these inhomogeneous materials and composites. The current work reports some results on acoustic emission detected by a simple system equipped with energy processing capabilities, during thermal cycling in anisotropic, polycrystalline alumina, silicon carbide whisker reinforced aluminum, and a continuous graphite fiber reinforced epoxy.

## EXPERIMENTAL PROCEDURE

The metal and ceramic specimens studied were 0.5 in. diameter, 6 in. long rods. The graphite epoxy specimen was a uniaxially reinforced plate of approximately 8 in. long, 2 in. wide, and 0.2 in. thick. Three approaches for heating were attempted. The first was a laboratory furnace which allowed heating one end of a ceramic specimen, while the acoustic emission sensor was mounted on the other end. The sensor end of the specimen was cooled with a clamp-on, water circulating device. The furnace was modified for D.C. operation to minimize electromagnetic noise. The second approach to specimen heating was by means of a 500 watt heat lamp. Both the acoustic emission sensor and the thermocouple were screened off from direct heating, and the ceramic and metal specimens were supported by strings to minimize potential noise generated by mechanical fixtures. The acoustic emission sensor was a commercial piezoelectric uinit, with a resonant frequency of 800 KHz, and was coupled to the specimen via a viscous fluid. Thermal cycling of the specimens was achieved by periodically inserting and withdrawing the specimen from the furnace, or by turning on and off the radiant heating lamp. Cooling was achieved by forced convection. The third approach to heating was by means of a C.W. Ar laser. This was focused on the surface of the graphite/epoxy specimen by a lens of 50 in. focal length. Thermal cycling was achieved by simply blocking off the laser beam periodically.

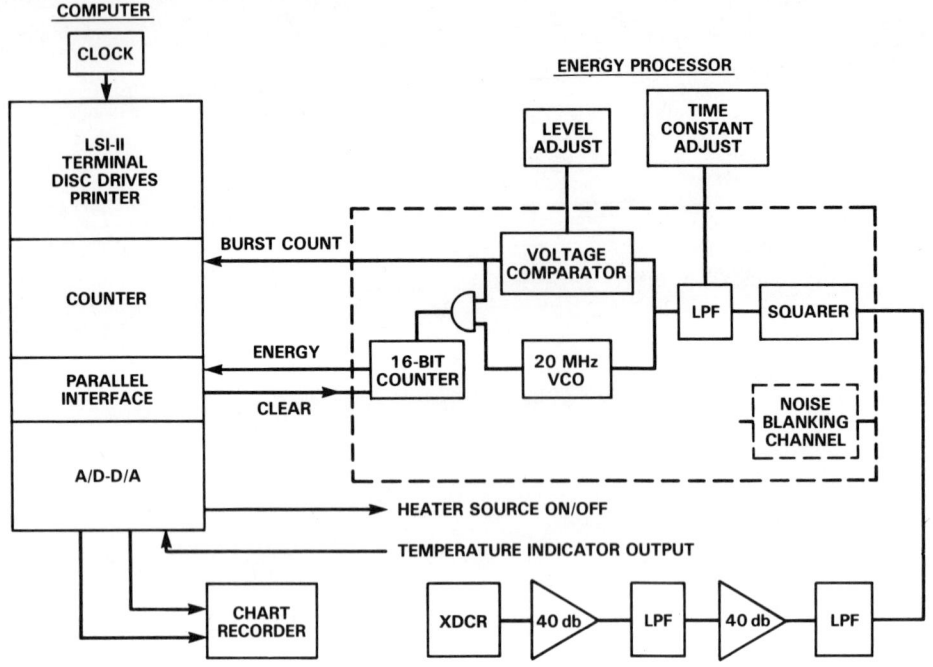

Fig. 1. Schematic of the acoustic emission measurement system.

The acoustic emission measurement system is shown in Fig. 1. Besides the usual amplification and low pass filtering, this system was equipped with an "energy" processor. Briefly, this allowed the voltage sensed by the transducer to be integrated in time, and was achieved by the squaring and voltage control oscillator circuitry. Acoustic emission burst type signals above a presetable level were converted to "energy counts" which were accumulated as thermal cycling proceeded. The number of bursts and the "energy" associated with them were measured and stored, controlled by a program implemented in a LSI-11 microcomputer. The output from the computer during an experiment were data tables as well as x-y recorder displays of 2 acoustic emission or temperature parameters versus time.

This system was calibrated using ring down signals generated by a conventional ultrasonic transducer facing the acoustic emission sensor, excited with square wave voltage waveforms. Figure 2 shows that, for continuous excitation, there was a linear relationship between the frequency output of the "energy" processor and the power output from a R.M.S. voltmeter. Furthermore, the number of energy counts was related approximately linearly with the square of the excitation voltage as shown in Fig. 3. This indicated that the energy counts did approximate the strength or the energy associated with bursts type acoustic emission signal. Only those bursts with voltage levels above 200 μV at the transducer before amplification were recorded in our experiments.

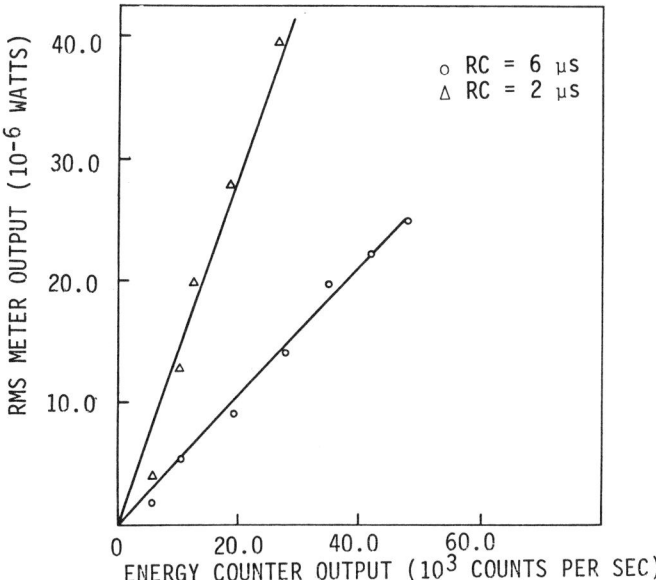

Fig. 2. Relationship between the output from the energy processor and a R.M.S. voltmeter for continuous wave excitation.

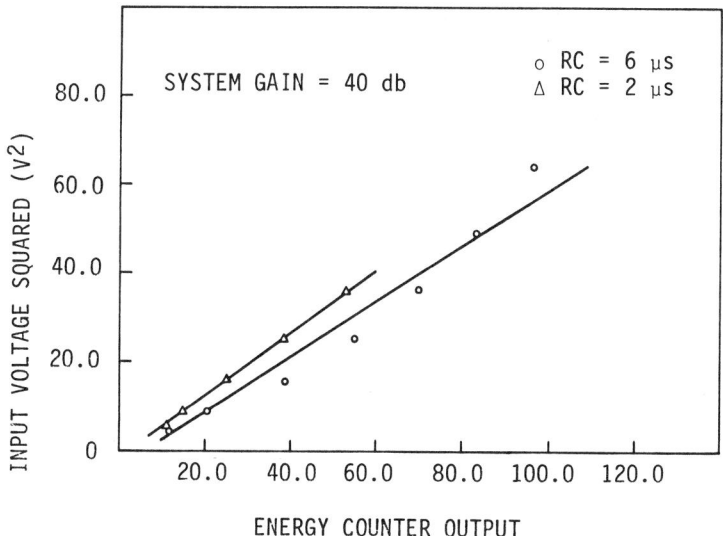

Fig. 3. Relationship between the square of excitation voltage and the energy count for burst type voltage waveform.

## RESULTS

The number of bursts and their associated energy during thermal cycling of alumina in the temperature range 300-900°F are shown in Fig. 4 and 5, respectively.

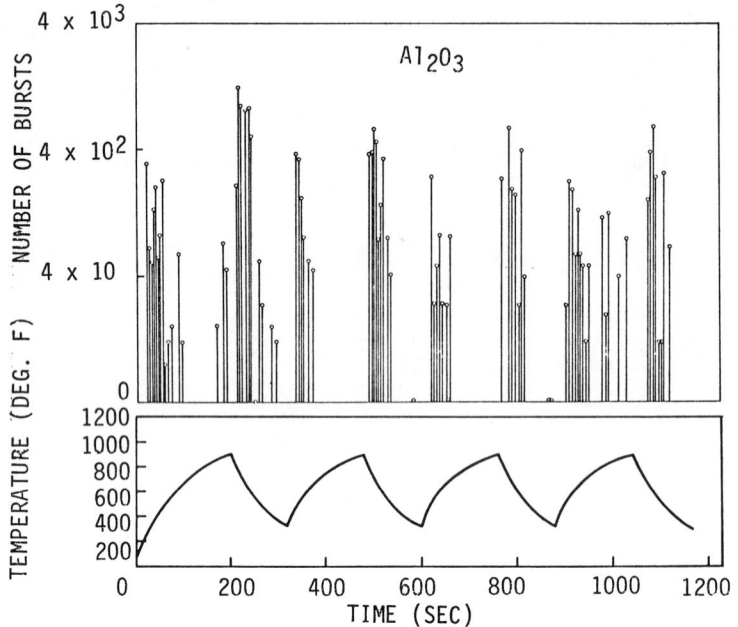

Fig. 4. Number of bursts during thermal cycling of alumina at 300-900°F.

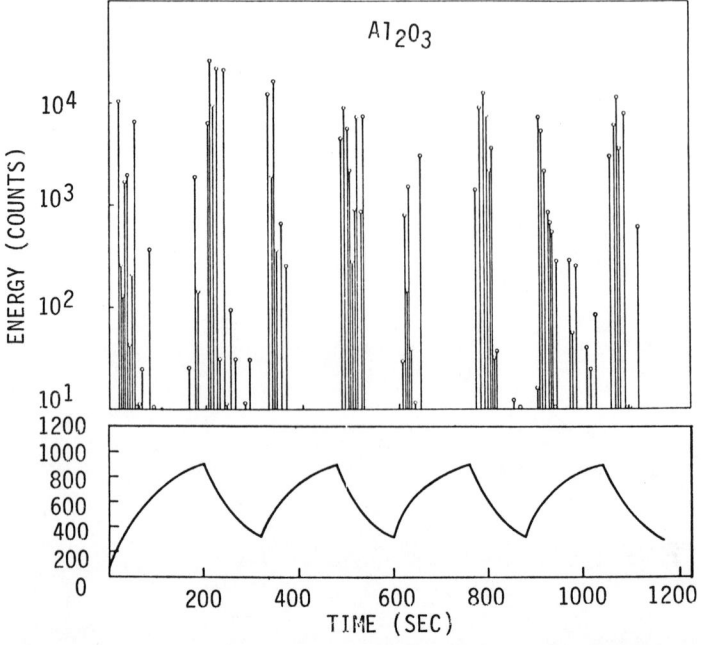

Fig. 5. Energy counts associated with bursts in Fig. 4.

The acoustic emission parameters were accumulated every four seconds during the tests. It appears that most emission occurred when the heating and cooling rates were highest. The laboratory furnace was used to obtain these relatively high temperature results. When radiant heating (heat source was approximately 1 foot from the specimen) was used, the range of specimen temperature, as sensed by the thermocouple at one end of the specimen, was lowered to 300---200°F for heating times comparable to those observed when the laboratory furnace was used. The number of bursts and their associated energy during these low level thermal cycling tests are shown in Fig. 6 and 7 for alumina, and in Fig. 8 and 9 for silicon carbide reinforced aluminum.

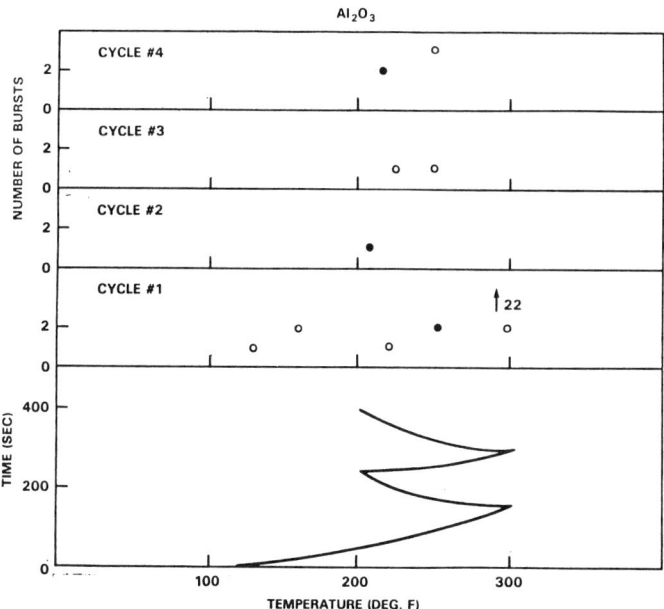

Fig. 6. Number of bursts of acoustic emission in alumina during low level radiant heating and cooling.

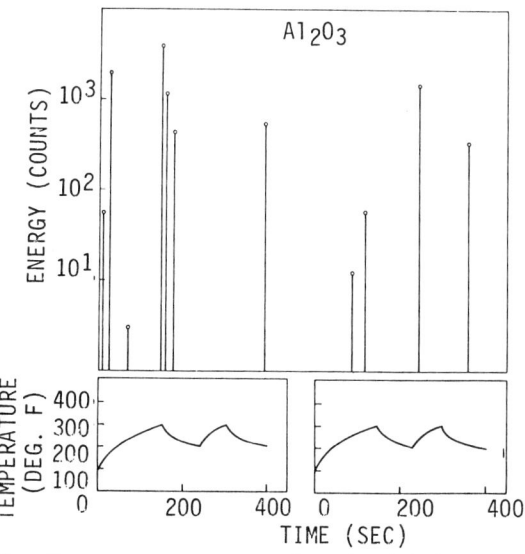

Fig. 7. Energy counts associated with bursts in Fig. 6.

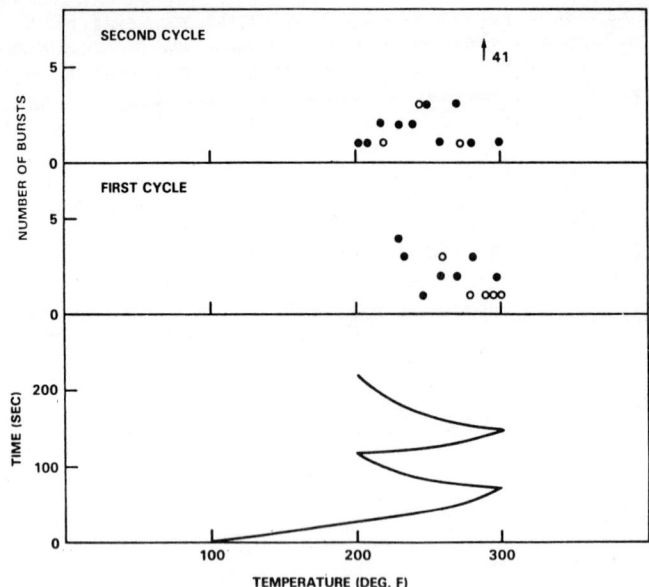

Figure 8. Number of bursts of acoustic emission in silicon carbide whisker reinforced aluminum during low level radiant heating and cooling.

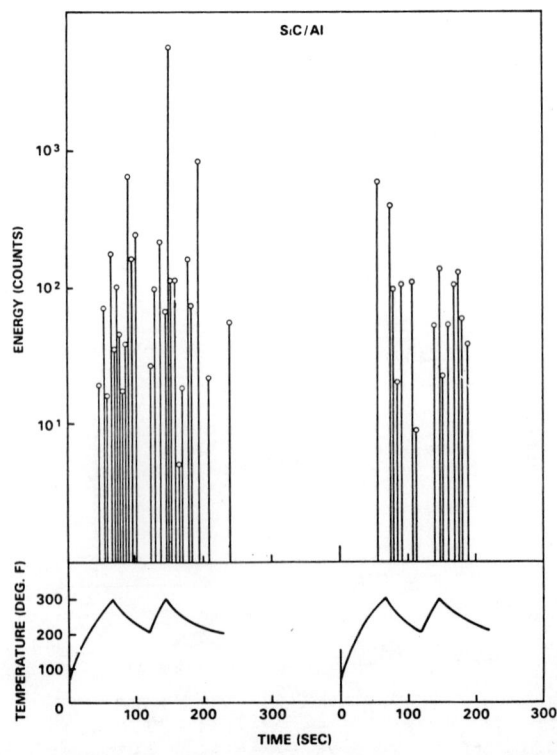

Fig. 9. Energy counts associated with bursts in Fig. 8.

A comparison between the acoustic emission characteristics for alumina and silicon carbide reinforced aluminum during these four thermal cycles is shown in Table I. It appears that the number of emissions was more numerous in the aluminum composite, and the emitted energy was higher during the cooling part of the cycles. Since the total number of bursts in alumina was smaller, the energy per unit burst was larger. The acoustic emission energy generated in the graphite/epoxy composite when heated by the C.W. Ar laser is shown in Fig. 10. Apparently, emission was recorded during both the heating and the cooling part of the cycles.

TABLE I. SUMMARY OF ACOUSTIC EMISSION ACTTIVITIES IN ALUMINA AND CARBIDE REINFORCED ALUMINUM

| MATERIAL | NUMBER OF BURSTS | | ENERGY (COUNTS) | |
|---|---|---|---|---|
| | HEATING | COOLING | HEATING | COOLING |
| $Al_2O_3$ | 33 | 5 | 8620 | 1293 |
| SiC/Al | 17 | 93 | 1426 | 9539 |

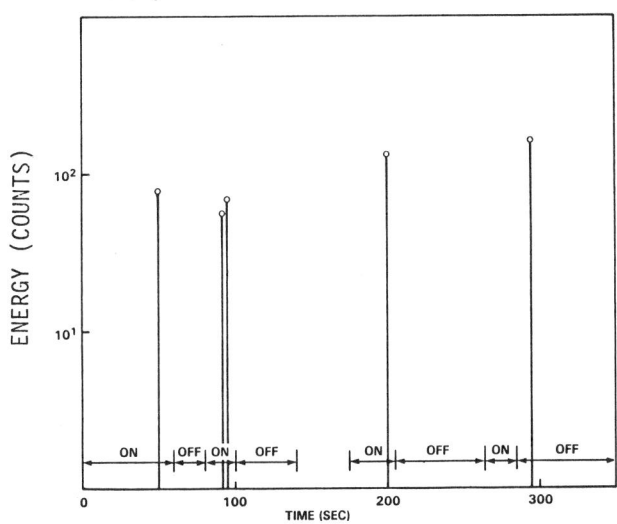

Fig. 10. Laser generated acoustic emission in graphite/epoxy.

DISCUSSION

There is considerable discussion in the literature on thermal stress generation in anisotropic ceramics [1], metals with ceramic inclusions [2], and composites [3]. Based on the known coefficients of thermal expansions and anisotropic elastic properties, the thermal strains and stresses generated by a given temperature rise or fall can be estimated in theory. In general, two considerations must be given when attempting to interpret the observed acoustic emission when a given piece of material is subjected to thermal cycling. The first is for the initial microstructural state of the specimen, which depends on the manufacturing process for the material. It was shown by Evans [4] that starting from the high temperature at which creep deformation should readily reduce internal stresses to zero, the cooling down of alumina to room temperature during the manufacturing process could produce grain boundary cracks as a result of the anisotropic thermal and elastic properties. Thermoelasticity theory can also be used to show that in the as manufactured state, there was compressive stresses inside the silicon carbide particles, and the aluminum matrix yielded plastically and hardened [5]. This initial state was, of course, a function of any heat treatment subsequent to the manufacturing process.

The second consideration for interpreting our acoustic emission results is for the thermoelastic response and stress emission mechanisms consistent with the initial state of the materials. The acoustic emission observed in alumina during thermal cycling could be indicative of the microscopic growth of the initial grain boundary cracks. It is interesting to note that the ranges of temperataure over which acoustic emission was observed coincided with those over which quenching induced strength degradation in alumina was reported [6], the latter presumably the result of the propagation of pre-existing microcracks [7]. Our data in Table I suggest that compared to alumina, silicon carbide reinforced aluminum had more emission sources of low energy. The larger energy per unit burst observed in alumina appears to be consistent with the interpretation that these were associated with microcrack propagation, while the emissions observed in silicon carbide reinforced aluminum were associated with lower energy, microscopic events.

It has to be noted that considerable thermal gradient existed in the experiments, since the temperatuare on the specimen surface must have been different from its interior, which was measured at one end of the specimen. Thermal gradient effects may have been, in part, the cause of thermally induced low level acoustic emission being observed in conventional, wrought, aluminum 6061 alloys. Other sources of thermally induced acoustic emission in this homogeneous material, however, cannot be excluded at present. We also note that the time required for the specimen interior to reach the peak temperature was longer in alumina than in the metallic composites as a result of the lower thermal diffusivity in the former. Thermal gradients became significantly larger in the laser heating experiments for graphite/epoxy, and a detailed thermoelastic stress analysis for the locally heated composite may be required for a proper interpretation of our acoustic emission data.

SUMMARY

1. Thermally induced acoustic emission in alumina occurred in the temperature range in which strength degradation following quenching was reported, and the propagation of pre-existing grain boundary cracks was suggested.

2. Bursts emitted were more numerous in silicon carbide reinforced aluminum than in alumina during thermal cycling in 80-300°F.

3. The averaged energy per burst was higher in alumina than in silicon carbide reinforced aluminum in the same temperature range.

4. CW laser heating should provide a potentially well defined source for acoustic emission excitation in a locallized region in composites.

ACKNOWLEDGEMENT

Most of the work reported was performed while author was with the Department of Materials Science and Engineering, State University of New York at Stony Brook. He wishes to thank Prof. F. P. Chiang, Department of Mechanical Engineering, State University of New York at Stony Brook, for making available to him the argon laser.

REFERENCES

1. J. E. Blendell and R. L. Cable, J. Amer. Ceramic Soc., 65, 174 (1982).
2. T. A. Hahn and R. W. Armstrong, Thermal Expansion 7, pp. 195, Plenum Press (1982).
3. M. W. Hyer, Dept. Mechanical Engineering, U. Maryland, College Park, MD, Private communication.
4. A. G. Evans, Acta Metall. 26, 1845 (1978).
5. R. J. Arsenault and R. M. Fisher, Scripta Met. 17, 67 (1983).
6. R. W. Davidge and G. Tappin, Trans. Brit. Ceramics Soc. 66, 405 (1967).
7. D. P. H. Hasselman, J. Amer. Ceramic Soc. 52, 600 (1969).

INTERNAL MONITORING OF ACOUSTIC EMISSION IN GRAPHITE-EPOXY

COMPOSITES USING IMBEDDED OPTICAL FIBER SENSORS

K. D. Bennett and R. O. Claus

Dept. of Electrical Engineering
Virginia Polytechnic Institute and State University
Blacksburg, VA  24061

M. J. Pindera

Dept. of Engineering Science & Mechanics
Virginia Polytechnic Institute and State University
Blacksburg, VA  24061

INTRODUCTION

The monitoring of acoustic emission (AE) is an important technique for the nondestructive characterization of strained materials because time and frequency domain analyses of AE events yield information about the type, geometry, and location of defects, as well as how material failure may occur. The quantitative interpretation of AE event signatures is critically dependent upon the faithfulness of the acoustic transduction and signal processing system in reproducing localized stress wave amplitude as a function of time. Although the usual sensor for acoustic emission is the piezoelectric transducer, several investigators have considered the application of interferometric optical sensing techniques which offer good spatial resolution and frequency response [1,2]. These techniques typically focus one beam of a modified Michelson interferometer to a small spot on the surface of a specimen and measure the time-dependent normal component of surface displacement at the location of that spot.

This paper reports the self-referenced interferometric optical detection of acoustic emission in graphite-epoxy composites using an optical fiber waveguide imbedded directly within the composite matrix. In particular, it discusses the operation of the fiber sensor system and the measurement of AE events produced by both graphite fiber breaking and matrix cracking.

OPTICAL FIBER SENSING OF MATERIAL PROPERTIES

Optical fiber waveguides, originally designed for high-speed long-distance telecommunications applications two decades ago, have been applied to a variety of sensing problems within the past ten years. Due to the inherent similarity between 100-micron-diameter unjacketed glass-on-glass optical fibers and individual graphite fibers in graphite-epoxy composites, a number of investigators have considered the particular use of optical fibers imbedded between composite laminae as sensors of composite material properties [3-5]. The status of such research at this time is that the effects of strain and temperature, or other observables which produce strain or temperature, integrated along the length of the sensor fiber in the composite can be determined using simple methods which measure optical phase, intensity, polarization, wavelength of modal properties. Spatial resolution and discrimination between strain and temperature effects may be obtained using several more complicated fiber sensing methods [6,7], and localized strain tensor quantities may be determined in addition by both presuming accurate models of the applied stress and knowing the photoelastic and mechanical properties of the imbedded fiber [8].

Modal Domain Sensing

The modal domain sensing system shown in Figure 1 was used for the detection of acoustic emission described below. Here, 633 nm light from a helium-neon laser travels through a glass optical fiber with an 850 nm cutoff wavelength. At the output of the fiber, the speckle pattern produced by the mutual interference between the far field distributions of the four propagating core modes is spatially filtered and detected. If the fiber is mechanically perturbed, the individual mode contributions to the far field intensity function change, the speckle pattern shifts, and the detected signal varies.

Several authors have considered similar mode-mode beating effects in fibers for quasi-static and sinusoidal strain loading [9,10]. For slowly varying loads, a simple model of phase modulation [11] may be used to demonstrate the sensitivity of the fiber to different strain components. For example, if temperature variations are neglected, the total phase change produced by stress alone may be written in general for a single-mode step-index fiber as

$$\Delta\phi = \frac{2\pi L}{\lambda} \left\{ \epsilon_1 - \frac{n^2}{2} \left( P_{11} + P_{12} \right) \epsilon_r + P_{12}\epsilon_1 \right\}, \tag{1}$$

where L is the length of the sensing section of the fiber, $\lambda$ is the optical guide wavelength, $\epsilon_1$ and $\epsilon_r$ are the longitudinal and radial strain, respectively, n is the core index, and $P_{11}$ and $P_{12}$ are the photoelastic constants of the silica fiber [8]. For a multimode fiber the phase of each mode varies in a slightly different way so the total output speckle pattern is complicated, especially if the number of modes in the fiber is large.

The behavior of the phase of a single mode in (1) is an appropriate model for the instantaneous modulation produced by an acoustic wave with a wavelength much larger than the fiber diameter. The different acoustic wavelengths of the event not only modulate the fiber differently but also drive plate mode vibrations of the composite which also contribute to the fiber modulation [12].

EXPERIMENT

Optical fiber was imbedded between the two center plys in eight-ply symmetric cross-ply composite prepreg laminates and cured in a standard heated-platen press. The edges from which fibers emerged were insulated with paper during cure to reduce the embrittlement of the adjacent fiber coating, and protective quick-cure polymer tabs were applied to the fiber-matrix joints soon after cure to minimize fiber breakage. The specimens were then saw-trimmed in width to approximately 6.5 cm.

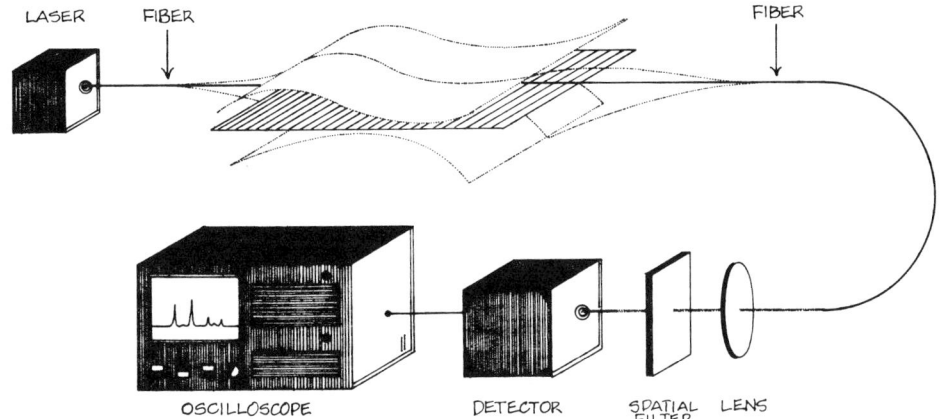

Figure 1. Optical fiber modal domain sensing system.

The specimens were axially loaded in a computer-controlled load frame using special grips designed to securely hold the specimen without cracking the imbedded fiber sensors. As the specimens were loaded, the optical fiber sensor output was monitored using a self-triggerable digital storage oscilloscope. The detection system recorded burst events at the same times that audible acoustic events were noted, as well as smaller amplitude events which were inaudible. Each of the four specimens tested failed at about 65,000 pounds of load and before the internal fiber sensors failed due to load.

Typical events recorded for the same composite specimen at different times during loading are shown in Figures 2 and 3. Analysis of this data indicates that the acoustic emission event in Figure 2 is due to the cracking of the composite matrix material. This is evidenced by the fast inital risetime of less than twenty microseconds, a primary emission followed by induced plate mode vibrations, and cascading emissions related to subsequent strain release. The very different signature shown in Figure 3 is typical of an event caused by graphite fiber breakage in that it consists of a single relatively long duration pulse of slow risetime and does not exhibit following events.

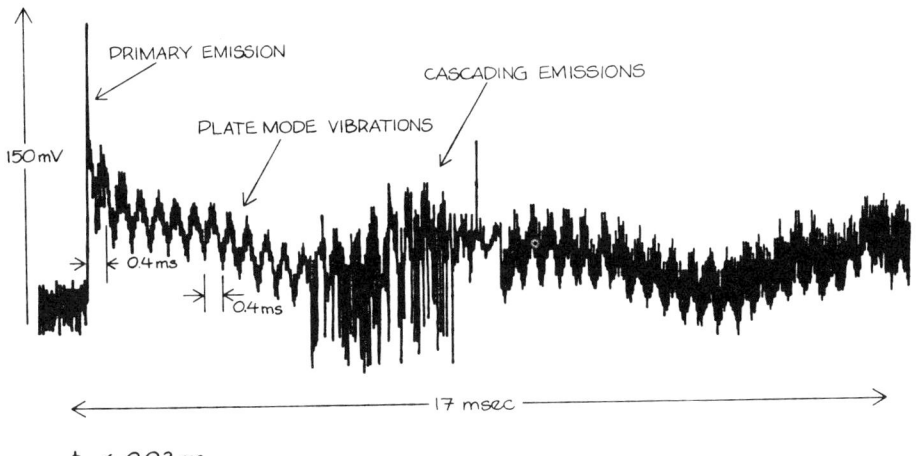

Figure 2. Emission due to matrix crack.

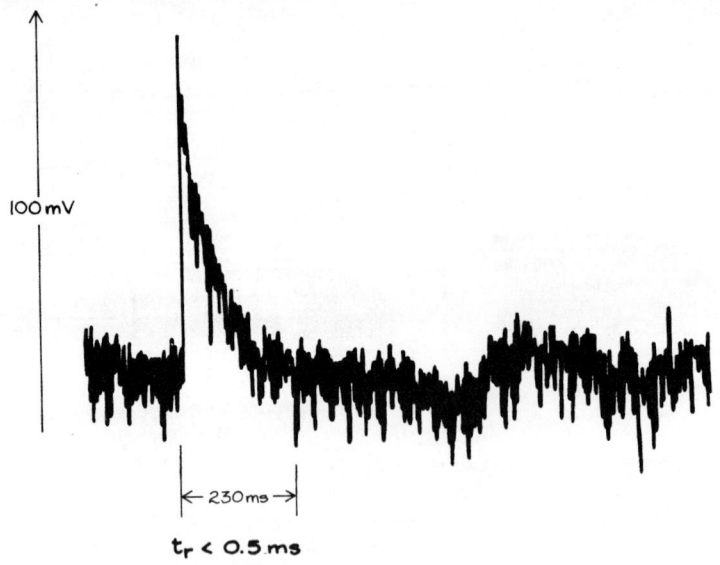

Figure 3. Emission due to composite fiber breakage.

Events recorded simultaneously using both the optical fiber and conventional "acoustic emission" piezoelectric transducers were also compared. The piezoelectric transducer was attached to the specimen in the center of one side, directly above the imbedded fiber. Typical comparison data is shown in Fig. 4.

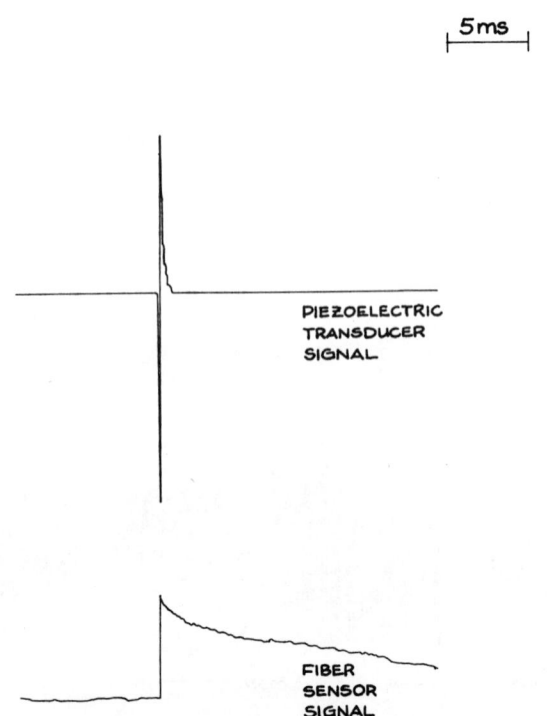

Figure 4. Comparison of acoustic emission events detected by piezoelectric transducer (top) and imbedded optical fiber sensor (bottom).

RESULTS AND CONCLUSIONS

The measurements reported in this paper indicate that acoustic emission events in graphite-epoxy composite materials may be observed using imbedded optical fibers employed in a multimode self-referenced interferometric detection system which is simple and inexpensive. The sensitivity of this method to the amplitude of the acoustic waves generated by the acoustic events, as well as the frequency response of the fiber sensor system, allow the discrimination between those events generated by graphite fiber breakage and internal matrix cracking. Quantitative agreement between the occurrance times of events recorded by the fiber sensor and acoustic emission piezoelectric transducers is good although the frequency responses differ due to the poor dc-response of the piezoelectric and the integrated acoustic response of the entire imbedded fiber. Extensions of this method include the location of the emission source in the two planar directions of the composite by triangulation and improvement in the spatial resolution of the sensor by selectively coating sections of the fiber.

ACKNOWLEDGEMENTS

The authors thank J. S. Heyman, B. W. Brennan, G. Meltz and E. G. Henneke for suggestions and comments and N. K. Shankarananayan and K. T. Srinivas for laboratory assistance. This work was supported by Simmonds Precision and the NASA Langley Research Center.

REFERENCES

1. C. H. Palmer and R. E. Green, Appl. Opt. 16, 2333 (1977).
2. R. A. Kline, R. E. Green and C. H. Palmer, J. Acoust. Soc. Am. 64, 1633 (1978).
3. B. W. Brennan, W. B. Spillman and J. R. Lord, Proc. 3rd Ann. SEM Conf. on Hostile Env. and High Temp. Measurements (Cincinnati, OH), March 1986.
4. J. R. Dunphy, G. Meltz and R. M. Elkow, Proc. 1986 ISA Conference, pp. 145-149.
5. R. O. Claus, B. S. Jackson, and K. D. Bennett, Proc. SPIE 566, 60 (1985).
6. G. Meltz and J. R. Dunphy, Proc. SPIE 566, 159 (1985).
7. R. O. Claus and J. C. Wade, J. Nondestructive Evaluation 4, 23 (1984).
8. B. Culshaw, Optical Fibre Sensing and Signal Processing (Peter Peregrinus, Ltd., 1984).
9. S. A. Kingsley, Electron. Lett. 14, 419 (1981).
10. M. R. Layton and J. A. Bucaro, Appl. Opt. 18, 666 (1979).
11. R. O. Claus, K. D. Bennett and K. T. Srinivas, Proc. Nondestructive Characterization of Materials Symposium (Montreal, Canada), July 1986.
12. K. D. Bennett and R. O. Claus, Proc. IEEE Region 3 Conf. (Richmond, VA), April 1986, pp. 95-98.

APPLICATION OF PATTERN RECOGNITION TECHNIQUES

TO ACOUSTIC EMISSION FROM STEEL AND ALUMINUM

Mark A. Friesel

Engineering Physics Department
Battelle, Pacific Northwest Laboratory
Richland, Washington   99352

INTRODUCTION

Much of the interest in using acoustic emission to monitor structural components is a result of the ability of the technique to detect growing cracks. The greatest problem standing in the way of applying this technique is the presence of signals from innocuous sources, which can make identification of crack-produced emission difficult, especially in those circumstances when cracks must be detected in real time and without other reliable correlation parameters. The test results presented here suggest an approach to eliminate spurious signals from acoustic emission data, which may be applicable to real time analysis and amenable to a variety of monitoring methods.

Taken at face value, some of the better work in acoustic emission waveform analysis suggests severe difficulties in the path of pattern recognition applications. Theoretical surface response to a number of microscopic source models was derived by Scruby, using a half-space Green function [1], and some of the results verified experimentally. The surface displacement produced by the various sources is quite similar when viewed through the response of a single sensor, and it is doubtful whether source differences could be identified in application, especially in the presence of background noise. In addition, for a crack and sensor on the same surface of a thick specimen, most of the energy is in the Rayleigh wavemode, so detection of the relatively low energy P and S modes may require saturation of the recording instrumentation.

As described here, these apparent limitations are partially overcome due to the way microscopic mechanisms combine to form a macroscopic event. In the limits where the macroscopic and microscopic source processes are similar, other techniques may be used to extend the usefulness of the signal features.

ZB-1 PRESSURE VESSEL TEST

The first set of data to be presented was obtained by hydraulically loading a medium size steel pressure vessel, designated ZB-1. This project was funded by the U. S. Nuclear Regulatory Commisison [2]. Three flaws were produced by notching and fatigue cycling a steel patch, which was then welded into a side of the vessel, and an unflawed patch of degraded steel was welded in on the opposite side of the vessel. Signals were detected through two foot long waveguide sensors made of 308 welding rod mounted by drilling and tapping the vessel wall. Along with waveforms, source location and other standard acoustic emission parameters were obtained.

Source location, load position (the location of the event on the load cycle), signal amplitude, and operator interpretation of the test conditions were used to identify the source of detected events. The waveforms of Figure 1 are typical of signals from crack growth in the flawed notches of the insert. The obvious characteristics of these signals were the distinctness and consistency of the three-pulse pattern. The first two pulses in the signals are caused by a longitudinal wavegroup, a second group traveling with approximately the shear wave speed, while the third pulse is an internal reflection of the first. The top two signals in Figure 1 originated at flaws in the outside of the vessel on the same side of the wall as the sensor, and therefore, are probably were the response of the rod to an incident Rayleigh wave. Waveforms from crack growth in the weld around the degraded steel patch, illustrated in Figure 2, show the same characteristic pattern. The lower Figure 1 signal is composed of a small number of similar response patterns and is always associated with a flaw on the inner wall of the vessel.

The type of signal associated with crack growth was not producible by any recognized noise source. Electronic noise would not produce the three-pulse response as it does not pass through the waveguide, and so does not have the waveguide characteristics as seen in Figure 3. Other noise signals whose sources are known, principally noise from oxide fracture and (probably) noise from a loose manhole cover, also occurred on the pressure vessel. Surprisingly, the waveguide response cannot be discerned in signals from these sources either. Figure 4 shows waveforms of events known to be caused by oxide fracture on the inside of the vessel wall. Oxide fracture is itself a brittle fracture process, and one would therefore suppose that the microsource process is the same as metallic fracture. The explanation

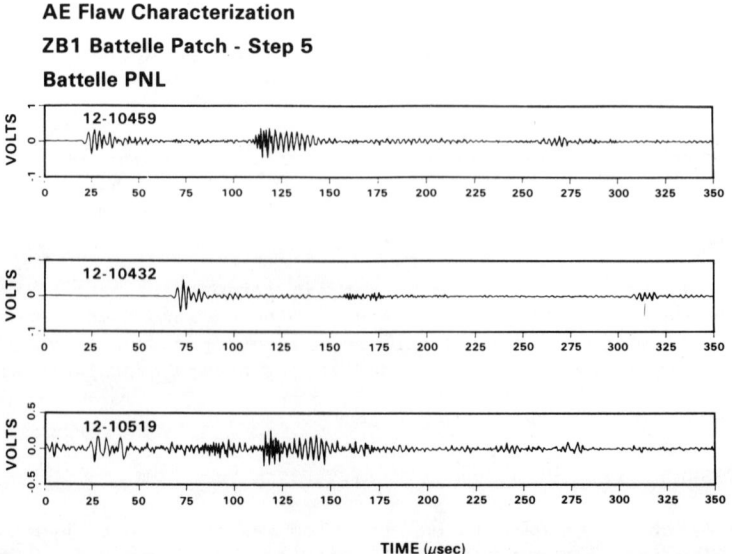

Fig. 1. Crack-Growth Acoustic Emission.

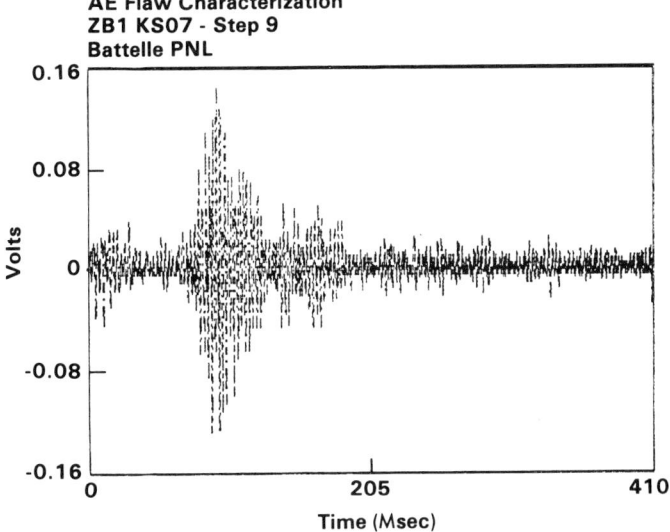

Fig. 2. Crack-Growth Acoustic Emission from Weld Region.

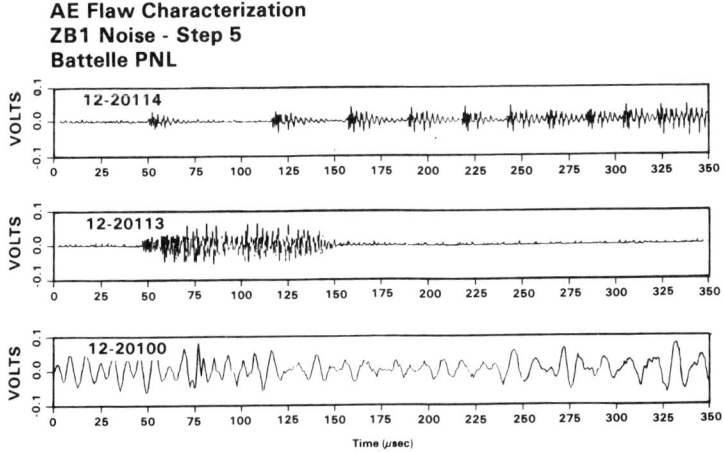

Fig. 3. Electronic Noise.

of the observed signal characteristics appears to lie in the occurrence of oxide fracture as a macrosource process composed of a large number of sequential microscopic events. Signals ascribed to the manhole cover are similarly extended and lack an easily identifiable waveguide response, although verification of this source is rendered more difficult by the absence of accurate source location due to the origin of these events being on an endcap of the vessel.

A simple pattern recognition scheme was devised which produced correct source identification in nearly 10,000 sequential waveforms of greater than 95%. Classification was carried out by creating templates using the envelopes of crack-related signals and cross-correlating test signals with the templates. If the maximum correlation coefficient exceeded a set cut-off value, the signal was classified as crack-related and otherwise discarded. Of the test data, over 900 randomly chosen signals were verified directly by examination and correlation with source location and load position

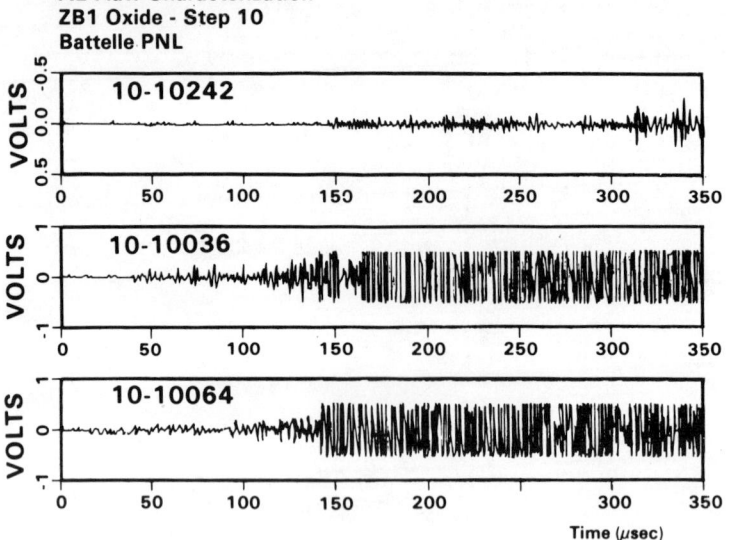

Fig. 4. Oxide Fracture

information. Missed signals included some events from the inside wall flaw for which no template was used, and signals of such low amplitude that their characteristics were obscured by the resolution of the digitizing instrument.

Electronic noise cannot be so easily discarded if waveguide sensors are not used, but the signal characteristics of macroscopic events should be detectable by surface-mounted sensors. If so, this concept could enable a fairly large number of noise sources to be relatively easily discerned in a variety of applications.

SIMULATED JOINT SPECIMENS

A set of tests were performed for the NADC under vastly different conditions than those described above. The NADC tests used thin aluminum simulated joint specimens with surface-mounted sensors. The specimens were flat plates, four inches wide and about 14 inches long; one specimen being .75 inch and the other .15 inch thick. Both specimens contained a large hole at one end, through which a pin was inserted to attach the specimen to the test frame. The specimens were cyclically stressed in two phases. The first phase was to produce and record fretting acoustic emission alone from rotation of the specimen about the pin. For the second phase, the specimen was notched near the edge of the hole and a fatigue crack grown and emission recorded. During this second phase, fretting was reduced by polishing and lubricating the test joint.

Since fretting is caused by friction between two surfaces, the emission from fretting should be a macro-event, comprised of a number of microsources, each microsource being the breaking of a single microweld or the release of a single catch point on the microscopically rough surfaces. Since crack growth should be a single microscopic event, as seen in the ZB-1 vessel data, acoustic emission from cracking should be discernable on the basis of its relatively short duration compared to the fretting signals.

Typical waveforms for crack growth and fretting signals taken from the .75 inch thick specimen are shown in Figure 5, where the signal from the fretting portion of the test is obviously of much greater duration than the crack-growth event, and does not show a clearly defined initial peak. The fretting characteristics shown in the figure are also found in fretting signals generated during the crack growth phase of the test.

The case is not so clear-cut for emission from the 0.15 inch thick specimen. Figure 6 compares crack-related acoustic emission to a waveform obtained during the fretting phase, and although the fretting signal is again somewhat more extended than the crack signal, the distinguishing characteristics found in previous macrosource emission are becoming less evident. Increasing similarity of the fretting signal characteristics to the proposed microsource characteristics of crack growth waveforms may result from the reduced surface area contributing to fretting events, due, in turn, to reduction in specimen thickness. During the crack-growth phase of the test, emission was found which was attributed to fretting on the basis of the position of the event on the load curve. This emission went one step further, in that it was visually indistinguishable from crack-growth emission.

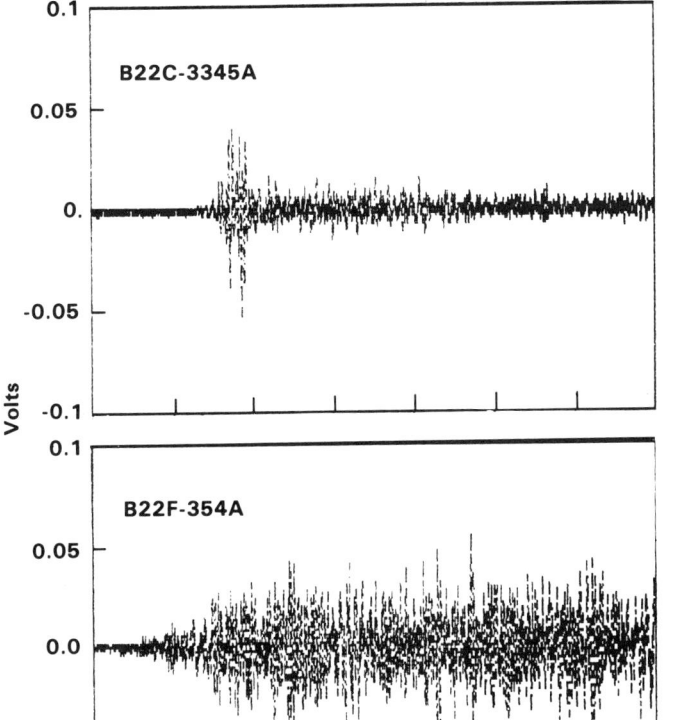

Fig. 5. Crack Growth (top) and Fretting (bottom) from 0.75 Inch Thick Aluminum Specimen.

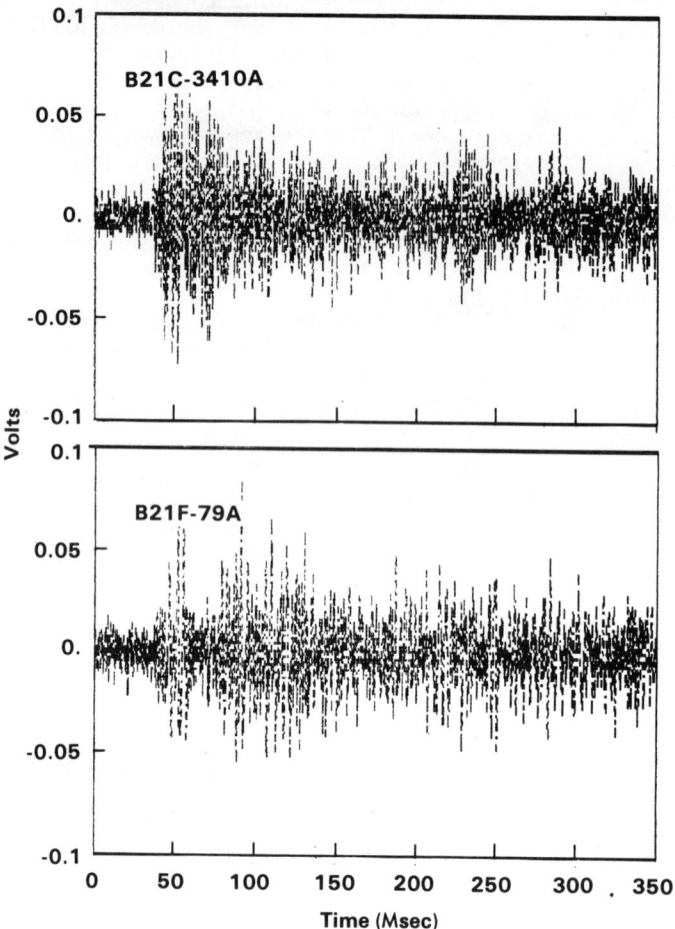

Fig. 6. Crack Growth (top) and Fretting (bottom) from 0.15 Inch Thick Aluminum Specimen.

Correct source identification of about 85% or better was obtained using either frequency or autocorrelation features, combined with statistical analysis techniques, which were not derived from or correlated with specific source processes. The problem which arose was that particular features and feature values which worked well on a particular data set (specimen type and sensor location) did not work well on other data, allowing analysis only after the data sets were obtained and requiring retraining of the algorithm. The difficulty with this phenomenological approach of simply applying available features is that there is no way to determine the ultimate cause of the values obtained, whether the cause is geometrical or actually some fundamental source-related process. The statistical techniques used, however, illustrated that overlapping two-category data can be separated, providing that good features are available.

CONCLUSION

An approach to identifying acoustic emission from crack-growth from among types of noise now begins to present itself. Insofar as crack-growth emission is due to a single microsource, characterization of the sensor response to such a signal should allow the elimination of most macrosource-type processes, and the lower limits of discrimination extended by characterizing crack growth stohastically to account for random variations in the data and possibly, small geometric effects.

For any set of features, it may be possible to imagine a noise process which renders those features useless for source discrimination, but by combining the most successful approaches, it may eventually be possible to discern acoustic emission caused by a growing crack from all of the commonly occurring noise sources.

REFERENCES

1. C. B. Scruby, "Quantitative Acoustic Emission Techniques", AERE-R 11261, AERE Harwell, Oxfordshire, England; July 1984.
2. P. H. Hutton, R. J. Kurtz, R. A. Pappas, J. F. Dawson, L. S. Dake, and J. R. Skorpik, "Acoustic Emission Results Obtained from Testing the ZB-1 Intermediate Scale Pressure Vessel", NUREG/CR-3915, PNL 5184, R5, 65; September 1985.

DIGITAL ACQUISITION AND ANALYSIS OF ACOUSTIC EMISSION

SIGNALS FOR CRACK SITE INITIATION STUDIES

Basil A. Barna, John A. Johnson, and Randy Allemeier

Idaho National Engineering Laboratory
EG&G Idaho, Inc.
Idaho Falls, Idaho 83415

INTRODUCTION

In the study of elastic-plastic fracture mechanisms it is necessary to determine accurately the time and location history of crack initiation and to discriminate between the fracture mechanisms of ductile and brittle rupture. In order to accomplish this, a real-time digital acquisition system for analysis of acoustic emissions has been configured, and some preliminary results are described in this paper. The goals for this system are to locate crack initiation sites with an accuracy of 200 μm or better and to respond to events separated in time by less than 100 ms.

HARDWARE IMPLEMENTATION

A block diagram of the system is shown in Fig. 1. A CAMAC-based PDP 11/73 is used to control the overall process. The A/D convertors are 8 bit flash digitizers that can be operated at up to a 32 MHz sampling rate. As currently configured the system has capability for digital storage of signals on up to four independent channels with the first channel acting as the timing reference for analysis of the data. Once armed, the system constantly digitizes the output of all active channels until the receipt of a "stop" trigger generated by the signal level in channel 1 crossing a predetermined threshold. After receipt of the stop trigger a specified number of pre- and post-trigger samples are transferred to computer memory and the system is cycled and armed for the next event. In addition to storing the digitized signal received by the transducer, the software also reads the system clock and records the time at the beginning of the data transfer. To decrease the "dead time", all the data are stored in computer memory. This way the time when the system is not able to acquire data due to the storing of the previous event's data is much shorter than if the data were stored on disk with a much slower access time. The actual dead time of the system depends on the number of channels which are active and the length of each data set. A typical dead time is 65 ms for a four channel set-up collecting 161 points of data per channel. With the existing software a total of 32 Kbytes of data can be collected before filling the memory.

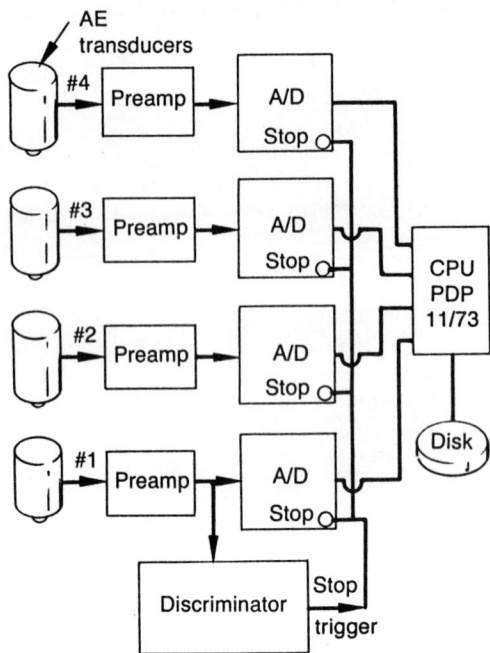

Fig. 1. System block diagram.

## LASER SOURCE CALIBRATION EXPERIMENTS

After the system was assembled and debugged, its performance and accuracy were determined. The first series of these experiments used two transducers from a design by Proctor [1] affixed to a steel fracture specimen 6 mm thick. A Nd-YAG laser beam was then used to generate simulated AE signals on the edge of the specimen between the two transducers. This configuration is shown schematically in Fig. 2. Twenty-one data sets were collected with a spacing of 0.1 mm between spot positions.

Since the path length between source and transducer is a function of the source location, the signals detected in each channel will show differing time delays that depend on source location. Examples of the type of data collected are shown in Figs. 3a and 3b. The data were then analyzed for location of the source by iteratively solving the equation that describes the time of arrival differences between the active channels. With the laser path on an exposed surface of the specimen for this first set of experiments, it was possible to simplify the experiment and only use two channels of data to calculate the source positions. In the general case it is necessary to utilize all four channels for unambiguous location in three dimensional space.

After calculation of the location from the measured time delays, the calculated source locations are plotted to show their relative positions and to give a visual impression of the measurement accuracy. A typical plot from the laser calibration experiments is shown in Fig. 4. Each circle represents a calculated source location. Where two circles of differing

Laser Source Calibration

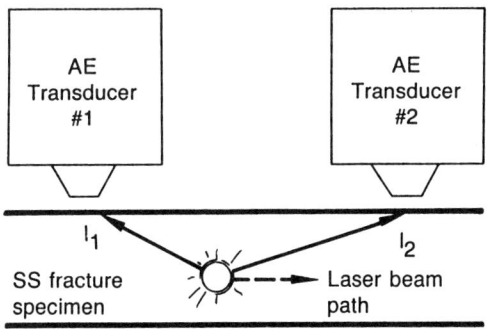

Fig. 2. Laser Generated AE for Calibration.

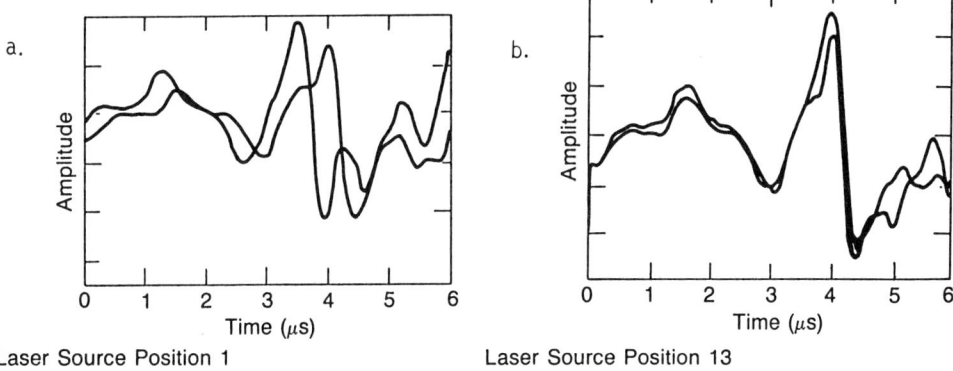

a. Laser Source Position 1

b. Laser Source Position 13

Fig. 3. Digitized data at two laser spot positions separated by 1.2 mm.

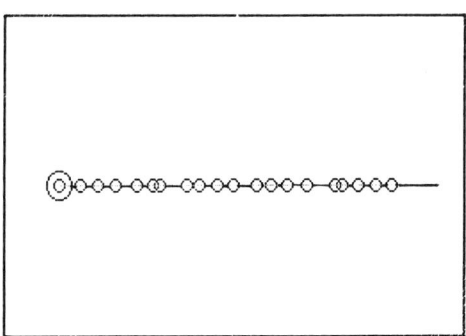

Fig. 4. Graphical output of the locations calculated from the laser calibration data. Each circle represents a source point and the solid horizontal line represents the 2.0 mm path of the laser spot.

radii but with a common center are shown the calculated positions were the same since the measured time delays were the same. The solid line on the plot shows the 2.0 mm path over which the laser spot was moved. The total calculated path from the time delays was 1.76 mm, 12% less than the true path. The most likely sources of this error are the finite size of the laser spot and measurement errors in locating the transducer positions. However it is worthy of note that in all but one case the system was able to discriminate between sources separated by 0.1 mm.

TITANIUM FRACTURE EXPERIMENTS

After the calibration experiments, data were acquired with the system using three-point-bend titanium fracture specimens that provided actual acoustic emissions for evaluation. The intent of these preliminary experiments with actual fracture specimens was to determine how the system would respond to real events. The rapidity and pattern of real event generation, spurious electrical and mechanical noise, and the spectra of the acoustic emissions all affect system performance and the parameters used to control the data collection.

Two specimens were examined with different pre-crack sizes. The first had a 25.4 mm electrical-discharge machined notch from which a small fatigue crack was grown before the test. The specimen was then instrumented with two AE transducers as shown in Fig. 5. As a simplification for these preliminary tests it was assumed that the crack initiation sites in this brittle material would originate on the pre-crack tip, so again only two transducers were used.

In addition, a moire grid was attached to the surface to monitor the strain field at one end of the pre-crack. Future experiments will attempt to correlate the strain field recordings with the acoustic emission data, but the purpose at this stage was to evaluate the two systems with respect to electrical and mechanical compatibility. An example of the typical moire output and the three-point-bend setup is shown in Fig. 6.

During the test the specimen was stressed until catastrophic failure with the crack completely cleaving the specimen. The AE data recording was begun at a load of 500 pounds and continued until failure. The digitized output of a typical AE event is shown in Fig. 7. As in the laser source experiments, the time shift between the two signals is indicative of the AE source location.

After the test the measured time shifts for each event were used to predict a crack initiation site for each event. Since some of the data recorded were due to electrical noise or other spurious sources, the data from the two channels were cross-correlated and only events with a correlation value above a reasonable threshold were analyzed. The results are shown in Fig. 8. Each circle again represents a calculated crack initiation site with the solid line indicating the 25.4 mm extent of the pre-crack. As can be seen, all but one of the events lie within the extent of the pre-crack and the small deviation in the extreme right hand event is most likely due to true crack growth beyond the original size or to an error in measuring the location of the transducers.

A similar series of data was collected on a second sample with a much smaller pre-crack size of 2.4 mm. If the AE system was prone to error a significant number of events would be mislocated outside the region defined by this small pre-crack. Such was not the case as the results in Fig. 9 demonstrate. The solid line now represents a 2.4 mm pre-crack. For this

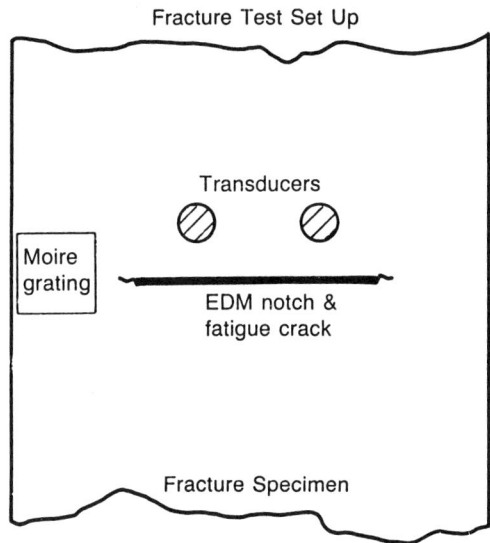

Fig. 5. Fracture test setup showing the locations of the transducers, pre-crack, and moire grating on the titanium fracture specimen.

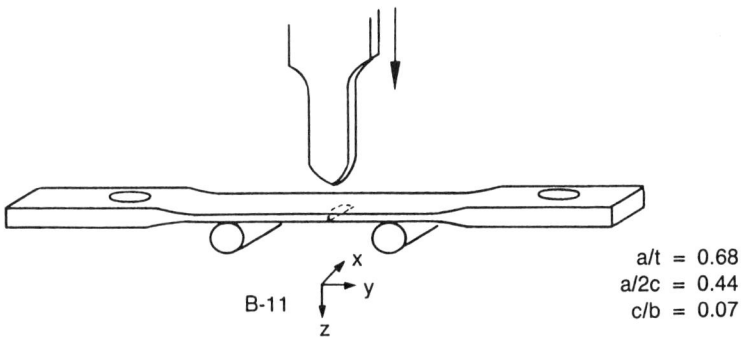

X displacement field
f = 2400 l/mm

Fig. 6. Three-point-bend configuration and typical moire images of the strain field during the fracture tests.

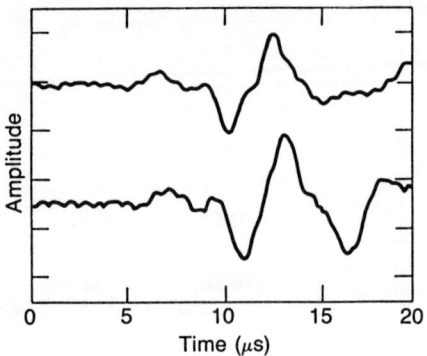

Sample B12 Event 25

Fig. 7. Digitized output of a typical fracture event during the three-point-bend tests.

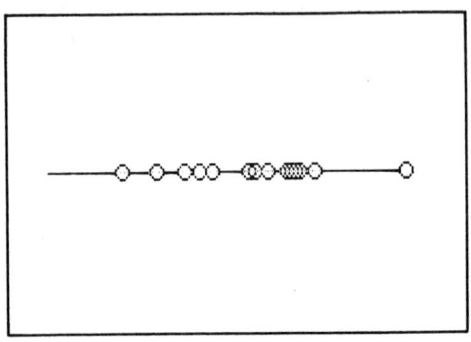

Fig. 8. Calculated source locations for sample with pre-crack size of 25.4 mm.

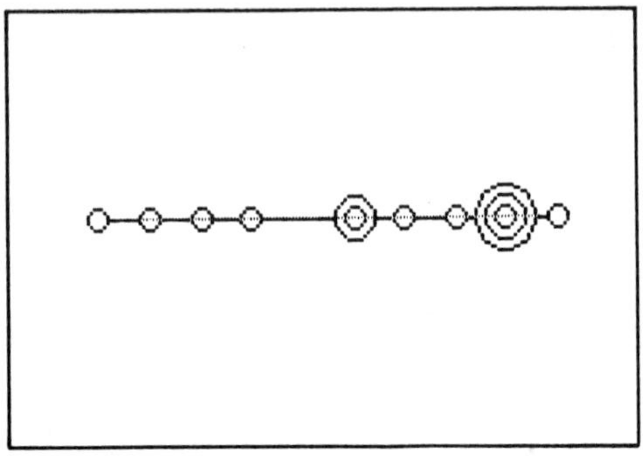

Fig. 9. Calculated source locations for sample with pre-crack size of 2.4 mm.

experiment several of the events share a common location which is due to the smaller pre-crack which extends through only 12 resolution elements, compared to more than 120 for the larger pre-crack.

Some of the interesting and potentially useful features of the data resulting from the fracture tests are the time history of the events and the pattern of occurrence. As the stress on the sample with the 25.4 mm pre-crack was increased, the AE activity would begin on one side of the crack and then jump to the opposite side. In addition there were areas of the crack front that were quiet until catastrophic failure. Table 1 summarizes the time of occurrence and calculated location of the valid events recorded during the test. The fifteen events cover a time span of approximately 102 s immediately preceding catastrophic failure. The location value is referenced to an origin at the center of the crack.

CONCLUSIONS

A real-time digital acoustic emission data acquisition system has been configured and used to collect preliminary data from both laser calibration experiments and actual fracture specimens of titanium. The calibration

Table 1. Titanium Sample with 25.4 mm Pre-crack

| Time of Event (min:sec) | Location (mm) |
|---|---|
| 43:23.983 | -4.5 |
| 43:27.716 | 3.6 |
| 43:28.250 | 2.3 |
| 43:35.600 | -7.1 |
| 43:54.166 | 7.0 |
| 43:54.633 | -2.4 |
| 43:54.833 | 5.0 |
| 44:03.850 | 5.3 |
| 44:46.683 | -0.4 |
| 44:49.266 | -7.1 |
| 44:59.866 | 13.7 |
| 45:03.950 | 5.3 |
| 45:04.733 | 2.6 |
| 45:05.183 | 5.6 |
| 45:06.283 | 6.0 |

experiments show a worst-case resolution between adjacent sources of 200 μm. It is anticipated that subsequent experiments and refinements will show a system capability of somewhat better resolution between 100 and 200 μm. The system has also shown apparent sensitivity to actual brittle fractures in a sample of titanium in that the locations of the calculated fracture sites are consistent with the known pre-crack extent before significant crack growth. System response times with the configuration used in these experiments are on the order of 65 ms.

Further experiments are now required to verify these preliminary results. An error analysis is required to quantify the sensitivity of the location calculation to errors in the measured transducer positions and the effects of the number of channels used as input to the calculation. Once the sources of error are further identified and quantified, the source location information that the system provides can be correlated with a time history of the free surface strain field surrounding the crack observed using moire techniques. These correlations will provide insight into the dynamics of the fracture event. Future work will also attempt to extend the analysis capability beyond source location to characterizing the type of fracture generating the emission.

ACKNOWLEDGMENTS

This work is supported by the U. S. Department of Energy, Office of Energy Research, Office of Basic Energy Sciences under DOE Contract No. DE-AC07-76ID01570.

REFERENCE

1. T. M. Proctor, "An Improved Piezoelectric Acoustic Emission Transducer", JASA, 71, 1163 (1982).

# EFFECT OF TEMPERATURE AND HEAT TREATMENT ON CRACK GROWTH ACOUSTIC EMISSION IN 7075 ALUMINUM

S.L. McBride and J.L. Harvey

Royal Military College of Canada
Kingston, Ontario, Canada

## ABSTRACT

The acoustic emission activity due to crack advance in 7075 aluminum alloys has been found to vary with both temperature and heat treatment. An increase in temperature or overaging of material in the -T6 condition each reduces the acoustic emission activity by changing the probability of occurrence of acoustic emission events and lowering their amplitudes. These observations suggest that the acoustic emission source mechanism is not inclusion fracture but rather is a property of the ductile matrix.

## INTRODUCTION

Investigations of 7075-T6 and -T651 aluminum undergoing tensile or cyclic loading have led several investigators to the conclusion that the fracture of brittle intermetallic inclusions is the source of burst acoustic emissions in these materials [2, 3, 4, 5, 6]. It has been shown by examination of fatigue crack fracture faces that inclusion fracture is observed when burst emissions occur. Indeed McBride et al (1981) showed that both the presence of brittle intermetallic inclusions and a matrix material with high enough yield strength are required to produce fracture-related burst acoustic emissions during crack growth.

Recent observations by our laboratory have shown that the number of detected acoustic emissions is less than 1% of the number of fractured inclusions observed thus casting some doubt on inclusion fracture as the fundamental source mechanism (McRae, 1984). To further investigate the source mechanism in aluminum alloys we present results on the effect of varying bulk mechanical properties of 7075-T651 aluminum on the acoustic emission behaviour during fatigue crack growth. The variation of bulk mechanical properties was accomplished by selection of the ambient crack growth temperature or by overaging the material.

It will be demonstrated that both the amplitude and probability of occurrence of burst acoustic emissions resulting from crack growth are dependent on the bulk mechanical properties suggesting that the source mechanism is not inclusion fracture but rather results from matrix deformation or microcracking within the plastic zone.

## EXPERIMENTAL

### Specimens

All specimens were machined from the aluminum alloy 7075-T651. In the T651 condition, 7075 aluminum has a yield strength of 540 MPa and an elongation to failure of 11% at 24°C. The bulk mechanical properties (yield strength, ultimate tensile strength, elongation to failure, toughness etc) are temperature dependent (Aluminum Standards and Data 1984). In general the yield strength and ultimate tensile strength decrease with increase in temperature while the toughness and elongation to failure increase. Below the peak aging temperature (120°C for 7075-T651 aluminum) these changes are reversible while at or above the peak aging temperature the material overages resulting in irreversible changes in the bulk mechanical properties. Such overaging reduces the strength of the alloy and affects the crack propagation behaviour which becomes much more complex as a result of irreversible microstructural changes. In this work the effect of bulk property changes on the acoustic emission resulting from crack growth in 7075-T651 aluminum will be studied. These changes will be accomplished in two ways (a) change from room temperature of the ambient crack growth temperature to reversibly change the mechanical properties and (b) overaging of the alloy at 190°C to produce irreversible changes in the bulk mechanical properties by changing the microstructure.

The single-edge notched (SEN) fatigue specimens (Fig. 1) were manufactured from a 4.7 mm thick sheet of Alcoa 7075-T651 aluminum with the longest dimension (tensile axis) parallel to the rolling direction. A stress raising side edge notch, 0.3 mm wide and 10 mm deep was machined in the specimen at the location indicated using a Buehler Isomet low speed metallurgical saw. This specimen geometry is used to allow the acoustic emission sensor to be air cooled at room temperature while the fatigue crack is maintained at a different temperature by the surrounding environmental chamber.

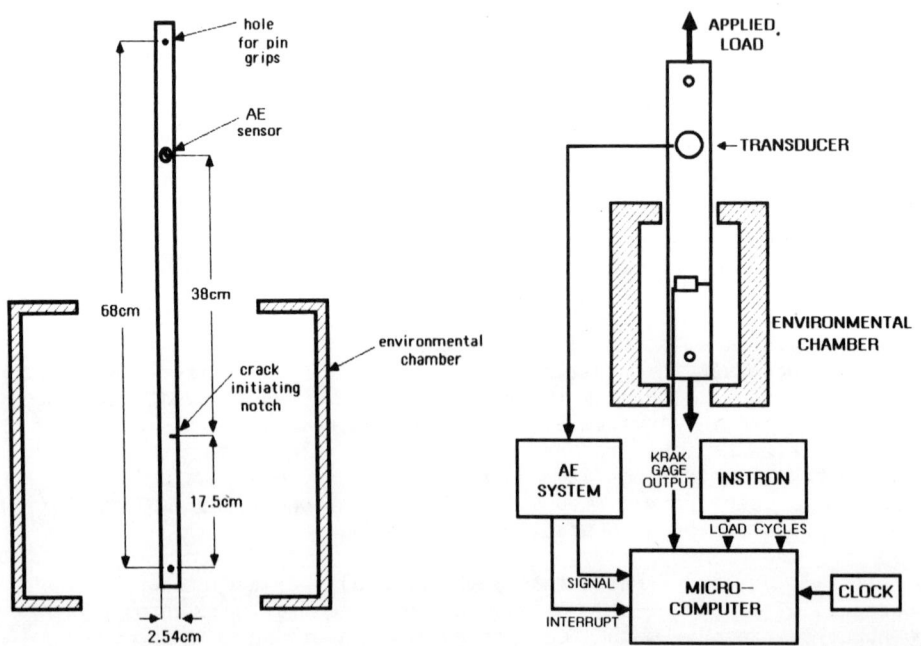

Fig. 1. The single-edge notched fatigue specimen used here.

Fig. 2. Block diagram of experimental set up used here.

Table 1a  Measured yield strength of 7075-T651 aluminum at the indicated crack growth temperature for each of specimens used to study the effect of ambient crack growth temperature on the acoustic emission behaviour.

| SPECIMEN | AMBIENT CRACK GROWTH TEMPERATURE ($^\circ$C) | YIELD STRENGTH (MPa) |
|---|---|---|
| T1 | -35 | 568 |
| T2 | -19 | 559 |
| T3 | 22 | 540 |
| T4 | 37 | 531 |
| T5 | 64 | 517 |
| T6 | 92 | 503 |
| T7 | 120 | 448 |
| T8 | 120 | 448 |
| T9 | 138 | 409 |

Table 1b  Measured yield strength of the overaged 7075-T651 aluminum specimens studied here. These yield strength measurements were made at room temperature which was the ambient crack growth temperature used in the study of the effect of overaging.

| SPECIMEN | AMBIENT CRACK GROWTH TEMPERATURE ($^\circ$C) | YIELD STRENGTH (MPa) |
|---|---|---|
| O1 | 22 | 540 |
| O2 | 22 | 522 |
| O3 | 22 | 478 |
| O4 | 22 | 438 |
| O5 | 22 | 391 |

Table 1 lists the ambient crack growth temperature and yield strength for each of the specimens studied here.

Specimen Loading

A schematic diagram of the experimental set-up is shown in Fig. 2. An Instron Model 1123 Universal Testing Instrument (maximum capacity of 25 kN) was used to fatigue the samples using the approximately triangular waveform produced by this machine under constant amplitude cyclic loading conditions. In this work the fatigue specimens were clamped in custom designed pin grips which minimized friction related noises by means of the teflon liners placed between the specimen and grip surfaces. Set screws tightened to the edges of the specimens were used to minimize movement of the specimen relative to the grips. In addition, Molykote 41 grease was used to quieten potentially movable contacting surfaces such as the loading pin and bolt hole to further reduce noise. Each sample was fatigued under tension-tension cyclic loading with an R-factor of 0.5 and a maximum load of 3.8 kN. These loads were chosen to produce a crack growth rate in the range of 0.1 microns to 0.2 microns per load cycle. The cyclic loading frequency of approximately 0.5 Hz was achieved using a crosshead speed of 5mm/min. The ambient crack growth temperature was controlled using an Instron Model 3111 Environmental Chamber which could maintain a constant temperature throughout the range -40$^\circ$C to +300$^\circ$C by oven heating or gaseous $CO_2$ cooling as required.

## Data Recording and Analysis

During fatigue crack propagation, the amplitude of each acoustic emission signal is required for analysis along with the specimen load, number of load cycles, time and crack length at which the signal occurred. These parameters were recorded using the data acquisition system described elsewhere (Pollard, 1981).

When triggered by an acoustic emission signal, the data acquisition computer records a data file containing each of the required parameters. The acoustic emission sensor output is preamplified, logarithmically amplified and envelope-followed prior to being read by the data acquisition system. The acoustic emission sensor used was a Bruel and Kjaer Model 8313 transducer with a resonant frequency of 200 kHz. The sensitivity of the sensor-couplant system was maintained constant throughout each fatigue experiment by air cooling the sensor attached to the specimen as illustrated in Fig. 1.

Crack length was measured using the 4-lead measurement of the electrical resistance of an epoxy-backed aluminum foil which was attached to the specimen surface in the path of the growing crack (Paris and Hayden, 1979). This device uses a constant current supply and provides a change in output voltage in the range 0 to 10 volts which is proportional to crack length for crack lengths in the range 0 to 10 mm. This output, along with the Instron load cell output are each recorded by the data acquisition system via an A/D converter at the time of occurrence of each acoustic emission signal.

The data analysis system (Pollard, 1981) provided X-Y plots and distribution functions of recorded parameters. With this system, it is possible to select signals which occurred under specified conditions by windowing on up to three chosen parameters, such as load of occurrence, crack length or minimum signal amplitude. To maintain consistency throughout all fatigue experiments reported here, the following conditions were met:

(a) the portion of crack growth used is that for which there is a linear relation between the number of acoustic emission events and crack length (crack length 1.5 mm to 4.5 mm).
(b) the system sensitivity (sensor and couplant) is constant throughout, to within 0.2 dB as measured using the helium gas jet calibration method.
(c) the detection threshold for acoustic emission signals is selected to be 0.1 mV at the preamplifier input.
(d) the minimum load at which acoustic emission signals are recorded is set at approximately 80% of the maximum cyclic load.
(e) crack face rubbing noises were removed from the acoustic emission data after thorough verification via criteria (a) and (d).

Using these conditions, the recorded data is that which results from fracture related events resulting from crack growth. The use of a specimen-sensor-couplant system with identical sensitivity and a common amplitude threshold for all experiments permits the variation of acoustic emission behaviour with material and environmental parameters to be properly established.

## RESULTS AND DISCUSSION

### Effect of Ambient Temperature

Figure 3 shows the cumulative acoustic emission amplitude distributions obtained from crack growth at the ambient temperatures indicated. Each result is normalized for 10 mm$^2$ of increase in crack face area. From the data of Figure 3 it is evident that the shape of each of the amplitude distributions is similar and that increasing ambient crack growth temperature reduces the acoustic emission activity by reducing the amplitude of the events (lateral displacement of the cumulative amplitude distribution to the left) and also reduces the number of emissions of a given amplitude (downward vertical displacement of the cumulative distribution). Since the data in Fig. 3 are presented in a logarithmic format these displacements correspond to multiplication factors in a linear presentation.

To separately investigate the effect of temperature on the amplitude and probability of occurrence of acoustic emission events, a least squares fitting was carried out between the empirical fit to the room temperature data and the cumulative amplitude distributions recorded at each of the other temperatures (Fig. 3). The best fit was found by vertical and lateral translation of each distribution to achieve superposition with the ambient room temperature result. This process can be described algebraically by the equation

$$(y-y_o) = a(x-x_o)^3 + b(x-x_o)^2 + c(x-x_o) + d \qquad (1)$$

where

a,b,c,d are constants defining the best fit to the room temperature cumulative distribution ($x_o$, $y_o$ = 0)
$x_o$ and $y_o$ are the lateral and vertical displacements required to obtain the best least squares fit for the logarithmic data presentation

Figure 4 shows the variation of the <u>linear</u> multiplication factors with ambient crack growth temperature. The multiplication factor $X_o$ (which is derived from $x_o$ and describes the <u>relative amplitude</u> of acoustic emission signals at different ambient temperatures) is seen to decrease linearly with increasing ambient temperature in the temperature range -40°C to +100°C. Above 100°C the rate of decrease of acoustic emission amplitude with increasing temperature is more rapid. Below 40°C the acoustic emission amplitude appears to decrease slightly as the ambient temperature decreases. The linear multiplication factor $Y_o$ (which is derived from $y_o$ and describes the effect of temperature on the relative probability of emissions) decreases with increasing ambient temperature for temperatures within the range -40°C to +100°C. Above 100°C the rate of decrease of probability of emission with increasing temperature is more rapid.

The data of Figure 4 plotted as a function of bulk yield stress are shown in Figure 5. Also shown in Figure 5 are the best least squares linear fits for $X_o$ and $Y_o$ as a function of yield stress. Both the relative amplitude of acoustic emission signals ($X_o$) and their probability of occurrence ($Y_o$) are each seen to vary approximately linearly with yield stress throughout the temperature range studied.

The variation in relative event amplitude ($X_o$) with material yield strength is seen to be approximately linear (Figure 5). This result is inconsistent with fracture of $Mg_2Si$ inclusions being the actual source of acoustic emission but suggests rather that the strength of the matrix material is the operative parameter. The role of $Mg_2Si$ inclusions in the

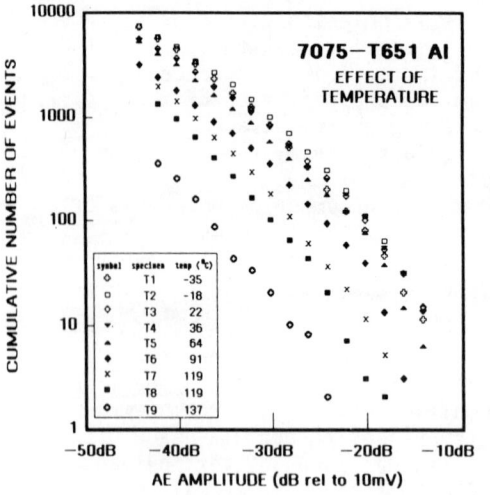

Fig. 3. Cumulative acoustic emission amplitude distributions at the indicated ambient crack growth temperatures.

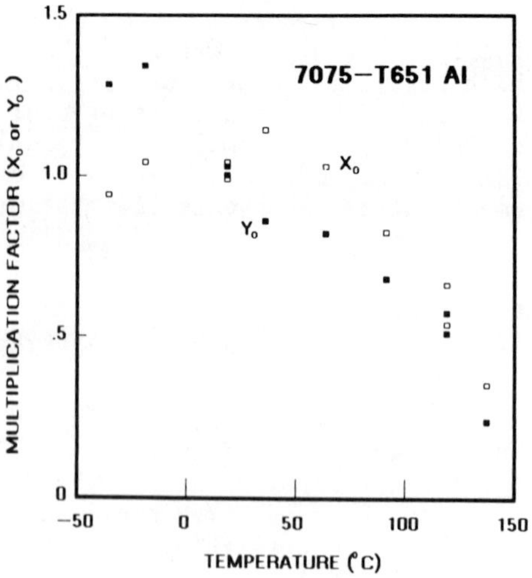

Fig. 4. The variation of the factors $X_0$ and $Y_0$ with ambient crack growth temperature.

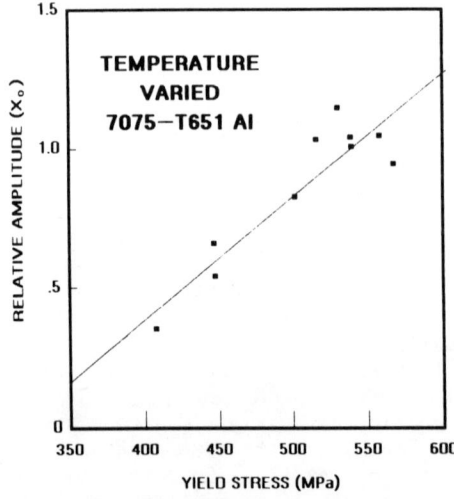

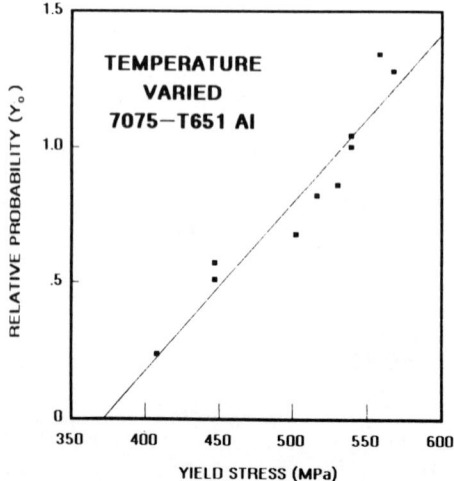

Fig. 5b. Variation of the factor $Y_0$ with matrix yield stress at the ambient crack growth temperature.

Fig. 5a. Variation of the factor $X_0$ with matrix yield stress at the ambient crack growth temperature.

acoustic emission source mechanism is indisputable, however, due to the remarkable agreement between shapes of the acoustic emission amplitude distribution and the shape of the size distribution of the $Mg_2Si$ inclusions (McBride, Maclachlan and Paradis, 1981). We propose that the acoustic emission source is triggered by the inclusion fracture. The resulting change in local stress then causes sudden plastic deformation or microcracking of the matrix material which links the crack tip to the fracturing inclusion. Such a mechanism would explain the role of both inclusion fracture and matrix yield stress in determining the acoustic emission event amplitude.

Effect of Overaging

Figure 6 shows the cumulative acoustic emission amplitude distributions obtained from crack growth at room temperature in 7075-T651 material which has been overaged at a temperature of 190°C. The degree of overaging of each sample is defined here by the bulk yield strength measured at room temperature after heat treatment. From these data it is evident that the shapes of each of the amplitude distributions are similar and that overaging results in the displacement of the amplitude distributions relative to one another. As in the previous section this displacement can be defined by equation 1 as lateral and vertical displacements ($x_o$, $y_o$) in the logarithmic presentation or as derived multiplication constants ($X_o$, $Y_o$) in the linear presentation.

To separately investigate the effect of overaging on the amplitude and probability of occurrence of acoustic emission events a least squares fitting was carried out between the empirical fit to the room temperature data and the cumulative amplitude distributions recorded for each heat treatment (Figure 6). The procedure carried out is identical to that used to represent the effect of temperature via equation 1.

Figure 7 shows the multiplication factors $X_o$ and $Y_o$ plotted as a function of yield stress for each of the overaged samples. The relative amplitude of acoustic emission signals ($X_o$) is seen to vary approximately linearly with change in yield stress. The relative probability of occurrence of acoustic emissions ($Y_o$) has an average value of 1.5 times

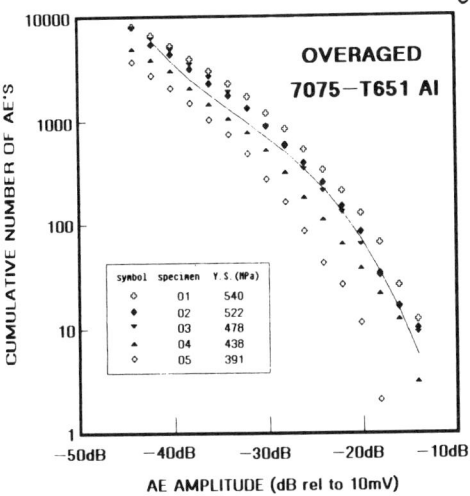

Fig. 6. Cumulative amplitude distribution of acoustic emission signals for 7075 T-651 aluminum overaged at 190°C.

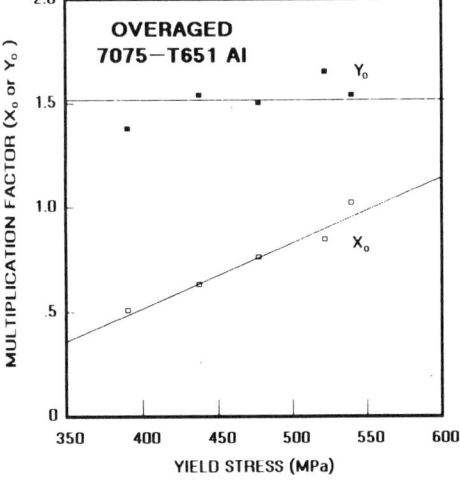

Fig. 7. The variation of $X_0$ and $Y_0$ with yield stress for the overaged 7075-T651 aluminum.

that obtained for 7075-T651 aluminum and is independent of specimen yield stress in the range 380 MPa to 540 MPa. Hence overaging has increased the probability of occurrence of acoustic emission ($Y_o$) by 50% independent of the degree of overaging as measured by the material yield strength. The relative amplitude of acoustic emission signals ($X_o$) varies with overaging in a manner similar to that observed for the variation with temperature.

CONCLUSIONS

For crack growth in 7075 aluminum alloys both the amplitude and probability of occurrence of burst acoustic emissions depend on the bulk mechanical properties of the matrix material. In particular the acoustic emission amplitude increases linearly with yield strength suggesting that the acoustic emission source is a matrix effect and not the sound of fracture of brittle inclusions. Possible acoustic emission sources are localized yielding or microcracking within the plastic zone. Such effects are observed to take place in the matrix material between the macrocrack and fractured inclusions.

ACKNOWLEDGEMENTS

Supporting funds were provided by the Department of National Defence, Canada (ARP 3610-208 and FE 329 F072 D6792). The project was monitored by Mr. W.R. Sturrock of the Materials Section Defense Research Establishment Pacific, Victoria, British Columbia.

REFERENCES

1. Aluminum Standards and Data (1984), The Aluminum Association, Inc., December 1984, 8th edition, 37.
2. R. Bianchetti, M. A. Hamstad, and A. K. Mukherjee (1976) "Origin of Burst-Type Acoustic Emission in Unflawed 7075-T6 Aluminum", J. Test. Eval., 4. 313-318.
3. S. M. Cousland, and C. M. Scala (1981) "Acoustic Emission and Microstructure in Aluminum Alloys 7075 and 7050", Metal Science 15, 609-614.
4. L. J. Graham, and W. L. Morris (1975) "Sources of Acoustic Emission in Aluminum Alloys", Proc. ASNT 35th National Fall Conf., Atlanta, GA, American Society of Nondestructive Testing.
5. C. R. Heiple, and S. H. Carpenter (1982) "Changes in Acoustic Emission Peaks in Precipitation Strengthened Alloys with Heat Treatment", J. Acous. Emis., 1, 4, 251-262.
6. S. L. McBride, J. W. Maclachlan and B. P. Paradis (1981) "Acoustic Emission and Inclusion Fracture in 7075 Aluminum Alloy", J. Nondes. Eval., 2, 35-41.
7. K. I. McRae (1984) "A Microscopic Examination of the Acoustic Emission Source Mechanism for Fatigue Crack Growth in 7075-T6 Aluminum", MSc Thesis, Royal Military College, Kingston, Canada.
8. P. C. Paris, and B. R. Hayden, "A New System for Fatigue Crack Growth Measurement and Control", Proceedings of the A.S.T.M. Symposium on Fatigue Crack Growth, (Pittsburgh, October, 1979).
9. M. D. Pollard (1981) "A Data Recording and Analysis Apparatus for Crack Growth Monitoring", MSc Thesis, Royal Military College, Kingston, Canada.

# A SEMI-ADAPTIVE APPROACH TO IN-FLIGHT MONITORING USING ACOUSTIC EMISSION

C.M. Scala

Aeronautical Research Laboratories
P.O. Box 4331,
Melbourne 3001, Australia

## INTRODUCTION

Acoustic emission (AE) is one of several promising non-destructive inspection (NDI) techniques for assessing defect growth in aircraft. Its potential for monitoring defects on a continuous basis, in situ, has stimulated various laboratories to carry out extensive research and application programmes in AE. Much of this work is applicable to structural testing of aircraft (1) and may later be applicable for use as part of a damage tolerance approach for aircraft stuctural management.

In any AE application, a major problem is to discriminate between AE from spurious sources and from (say) crack growth (1-5). Frequency discrimination and source location are common initial steps in AE source discrimination. Two distinct approaches have evolved for subsequent processing (6). The first approach, described variously as fundamental, deterministic or non-adaptive, involves developing quantitative relationships between defect parameters and AE (6, 7); the second, labelled as empirical, stochastic or adaptive, involves acquiring large quantities of data and seeking empirical correlations between defect types and AE data (6, 8, 9, 10, 11).

This paper briefly reviews existing procedures for distinguishing between AE sources in aircraft structures. Limitations in the applicability of frequency discrimination and source location techniques are addressed, and the failings of the fundamental and empirical approaches are shown. A semi-adaptive approach is proposed, which combines the results of research on AE sources, sensors and wave propagation (using calibration studies) to predict features of AE waveforms from crack growth in different locations in a structure. The semi-adaptive approach is applied to the processing and analysis of AE waveforms, detected during full-scale fatigue testing of a Mirage aircraft (5, 12, 13), to distinguish between spurious sources and AE from fatigue crack propagation in bolt holes in the aircraft's aluminium alloy wing-spar.

## PROCESSING AND ANALYSIS OF AE DATA

Many AE sources are present in an aircraft during flight or fatigue testing. Hence, considerable signal processing and analysis are required if AE from fatigue crack propagation is to be distinguished from spurious AE sources.

### Initial Treatment Of Extraneous Sources

AE from fatigue crack propagation will have frequency components up to many megahertz while vibration signals most often occur below 100 kHz. Thus, many of the latter signals can be removed by filtering. In addition, modern equipment is usually adequately protected against transient electromagnetic interference, which is usually at very high frequencies.

Source location systems are used to locate any AE sources within, and reject extraneous sources from outside, a region of interest. Differences in arrival times of elastic waves at an array of sensors are measured, a wave speed is assumed and algorithms are used to calculate the location of an AE source. Some location systems assign co-ordinates to the source while others assign it to a zone within the region of interest.

Source location is most readily achieved in simple structures where the mode of wave propagation is known (14, 15). However, problems can be expected in complex 3-D structures where wave propagation occurs as a complex combination of longitudinal, shear and Rayleigh waves. The type of wave detected by each sensor in the array will depend on the magnitude of the AE source, the relative locations of source and sensor, and the effects of boundaries on the presence of reflected or mode-converted waves. There will also be errors in the location of an internal source, of the order of the source depth (6). Clearly, an extensive sensor array, complicated algorithms and time-consuming computation are involved, if AE source location is to be achieved in an irregular 3-D structure.

Guard sensors are used in combination with a sensor array to minimize the effects of sources extraneous to a region of interest. Noise sources external to the array are identified and a guard is mounted between noise and array; AE events reaching the guard before the array are rejected. Alternatively, rejection is controlled by measuring differences in time-of-flight between guard and array sensors. The effectiveness of guard sensors depends on factors such as the wave speed assumed for calculation of time-of-flight, how many guard sensors are used and the activity of any extraneous sources. Our experience has shown that (i) highly active extraneous sources (which are often difficult to locate and identify) exist in aircraft applications, (ii) large numbers of guard sensors are needed to control noise sources, and (iii) assignment of a suitable wave speed in a complex component is difficult (5, 16) - additional AE signal processing will always be needed to distinguish between AE signals from different sources.

Limitations Of Fundamental Or Empirical Approaches

Three elements determine the characteristics of a detected AE signal: the source mechanism, the wave propagation path in the structure between source and sensor, and the sensor itself. In principle, if the effects of wave propagation and detection can be deconvoluted from the AE signal, the complete character (type, magnitude and time dependence) of an AE source can be obtained by measurements at six independent sensors (6).

This fundamental approach appears relatively straight-forward when an 'ideal' sensor (having a calibrated, wide-band response to a simple, physical parameter (6, 7, 17) is used for the AE measurements and the structure under investigation approximates to a half-space (or perhaps to a plate) (6). In a recent fundamental study of fatigue crack growth in a compact tension specimen of an aluminium alloy (18), multi-sensor measurement of directivity patterns of longitudinal waves allowed the discrimination of microcrack AE events associated with fatigue crack propagation and shear sources consistent with fretting between a loading pin and the specimen. However, this work is incomplete as a wider range of source types (particularly crack face rubbing) must be characterized. Even then, the measurement of features like directivity patterns will not be possible in complex structures.

A completely different approach ignores fundamental information and relies on adaptive learning principles. Empirical pattern recognition is applied to extract the features of possible AE sources from training sets of data (8-11). Source discrimination in an arbitrary data set is then undertaken using an appropriate combination of features to achieve an 'acceptable' error rate. Application of this statistical approach has not been successful because (i) well-defined training data sets from likely sources are not always available for analysis (e.g. AE from crack face rubbing), and (ii) propagation - and sensor-related rather than source-related features tend to be extracted from the training sets. Although it is possible to distinguish between vastly different sources such as crack-related AE and mechanical impacts, hydraulic noise and electrical transients, the differences between sources such as cracking and fretting are more subtle (10, 19). Thus a modified approach to data analysis in an arbitrary application is required.

## Development Of A Semi-Adaptive Approach

The availability or usability of a priori information determines the degree of adaption that can be employed in signal processing problems (11). Clearly, a semi-adaptive approach is needed to distinguish between AE sources in complex aircraft structures, given limitations in both the availability and usability of information for a non-adaptive approach and the problems in implementing an adaptive approach.

The semi-adaptive approach we have developed is based on comparing features extracted from pattern recognition analysis (5, 19) of AE waveforms from an aircraft, with features predicted for possible sources. These features have been identified as a result of our research on AE sensors, wave propagation (modelled in irregular structures using calibration studies) and AE sources during fatigue crack propagation in aircraft aluminium alloys.

AE sensors. Our research programme has highlighted the need for careful selection of sensors so that sensor characteristics do not dominate pattern recognition features extracted from AE waveforms. A sensor with small aperture and wide-band response is important for pattern recognition, while characteristics such as high sensitivity, ease of coupling and robustness are always useful (17).

Calibration studies. Calibration studies have been used extensively in the determination of AE sensor characteristics and in some deconvolution studies to yield source characteristics (6, 7) but their use in pattern recognition analysis has been neglected. A feature of our semi-adaptive approach is the use of calibration measurements, undertaken using a simulated AE source, to predict features of an AE source appropriate to different combinations of test-piece, sensor and source location (where accessible). Propagation features are also obtained, to discriminate between subtly different sources (like bolt fretting and fatigue crack propagation), when they occur at different locations in a structure. We have found the fracture of pencil lead (7) to be a convenient, inexpensive and reproducible source for calibration studies in the laboratory and field (16, 19).

AE sources. Our AE studies of aircraft aluminium alloys indicate that inclusion fracture in the crack tip plastic zone is a likely AE source during fatigue crack propagation (5). The risetime, duration, autocorrelation characteristics and the distribution of load cycle positions of AE signals have been identified as useful pattern recognition features. Inclusion fracture is expected to occur at the peak of a load cycle during fatigue crack propagation, and may also occur on a positive load gradient (depending on loading history). A variability in AE activity from cycle to cycle is also expected, related to any inhomogeneity in the size and spatial distribution of the inclusions.

The results of other researchers (see below) suggest that other possible sources could occur at a range of different load cycle positions, depending on factors like the loading conditions used (load limits, non-axiality in load application etc), the specimen configuration and microstructural considerations. For example, AE from fretting of a bolt in a hole was observed on negative load gradients for fixed load limit cycling of bolted plates (10) and at other load cycle positions for misaligned specimens. AE from crack face rubbing has usually been observed on positive load gradients, below the mean load (20 - 22), being <u>repetitive</u> for many hundreds of cycles. Results of a recent study (23) suggest that AE from crack face rubbing could also occur at the peak of a load cycle under certain experimental conditions. However, the results of this study were difficult to evaluate, because of the possible effects of an inhomogeneous size distribution of inclusions on the AE detected during a limited number of load cycles.

## AE MONITORING OF MIRAGE

### Background Information

Some bolt and rivet holes situated in the bottom flange of the aluminium alloy wing-spar of a Mirage aircraft were known to be fatigue critical. An NDI technique was sought to enable continuous monitoring of these holes to be conducted. Interference fit bolts were used on the forward side of the flange; close tolerance bolts and rivets were used on the rear side of the flange, the five bolt holes closest to the mainframe having interference fit stainless steel bush inserts (24). Thus, AE appeared a promising technique for continuous monitoring of any defects in the wing-spar, because the application of more conventional NDI techniques would have required factory disassembly of the wing.

Aeronautical Research Laboratories (ARL) contracted Battelle Pacific Northwest Laboratories to install AE equipment on a full-scale fatigue test of the Mirage aircraft, conducted at a Swiss aircraft factory (5, 12). Two wings were monitored by AE, wing RH79 and wing RH56 which replaced wing RH79 (5).

Battelle mounted a six-sensor array on the lower surface of the bottom flange of wing RH56, for use in the time-of-flight source location of AE events. Two of the sensors were used to divide the spar into zones along the spar axis, while the four other sensors were used to distinguish between the forward and rear sides of the lower flange. Rayleigh wave detection was assumed, and AE events were either located into one of sixteen zones (labelled 4 to 19) or discarded (Fig 1).

The sixteen zones defined by time-of-flight criteria extended beyond the bottom flange of the wing-spar to surrounding regions (e.g. parts of the wing-skins and top flange of the spar). Nine guard sensors were used to minimize the likelihood of detecting AE events from these regions. Three guard sensors were coupled to a repair patch on the cracked rear wing-skin (Fig 2), four on the web between the top and bottom flanges and two on the lower surface of the bottom flange of the spar (Fig 2).

The guard and array sensors were made by Battelle especially for AE source location and comprised in essence a 1.7 mm radius air-backed disc. A similar type of sensor (located at W (Fig. 2)) was installed by Battelle for collection of waveform data but was rarely used. Amplified AE signals were digitized in a Biomation 1010 transient recorder with a capacity of 4096 10-bit words and sample interval set at 0.1μs. Each digitized waveform was recorded on 9-track magnetic tape. The number of the zone in which the event was located and time in seconds relative to the start of the test were also recorded for each AE event.

### Experimental Conditions for Semi-Adaptive Approach

Carefully designed AE experiments were undertaken on wing RH56 at the Swiss aircraft factory in August 1984 to allow application of the semi-adaptive approach to AE waveform processing and analysis (16). The existing data acquisition system was modified as follows. Valpey Fisher pinducers, wide-band sensors comprising a thin piezoelectric disc and delay-line matched backing, were used for waveform collection at either W1 or A1 (Fig 2). (The pinducers were particularly suited to pattern recognition measurements (17)). Information on the load cycle position of AE events was obtained by recording data from four strain gauges mounted along the length of the spar.

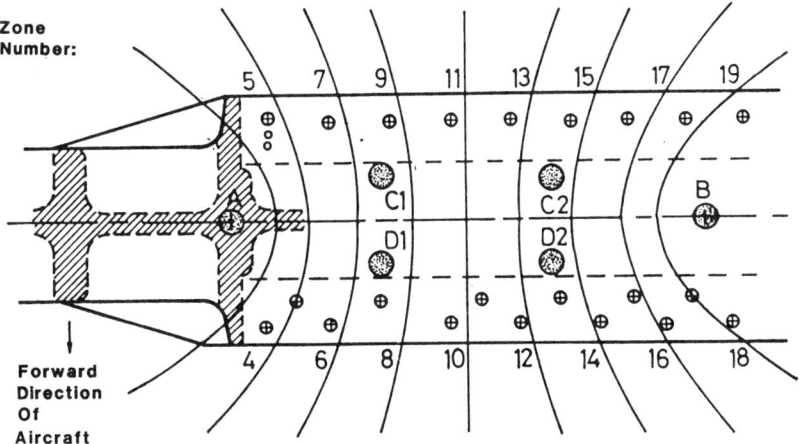

Fig. 1 Lower surface of bottom flange of Mirage wing RH56 showing
(i)  ◉ AE sensor array (A, B, C1, C2, D1 and D2) used for source location,
(ii) ⊕ bolt holes and ○ rivet holes,
(iii)   the zones numbered 4 to 19.

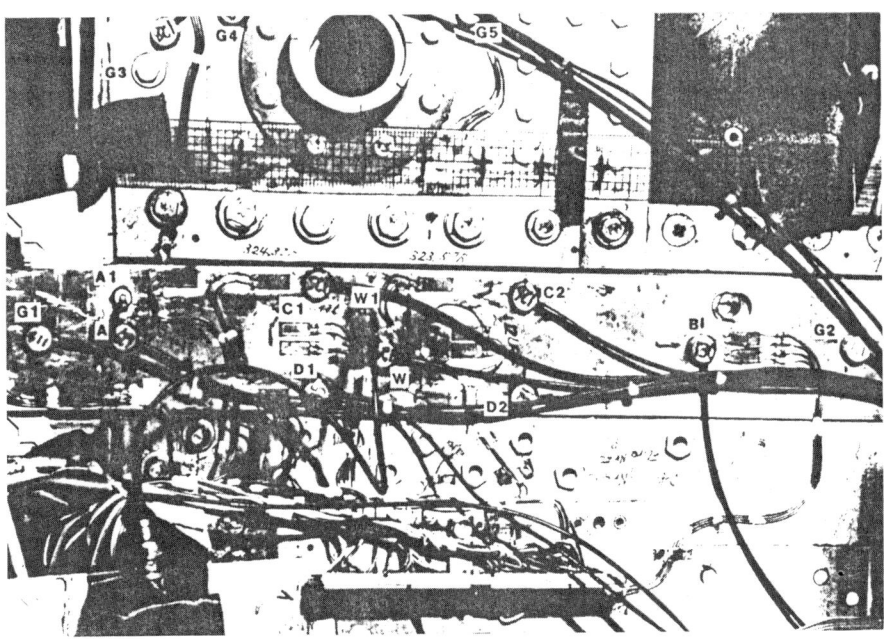

Fig. 2 Location of AE sensors on the lower surface of Mirage wing RH56: AE sensor array (A, B, C1, C2, D1, D2), waveform sensors W, W1 and A1, guard sensors G1 and G2 on the bottom flange of the wing-spar, and guard sensors G3, G4, and G5 on a patch on the wing-skin.

Calibration measurements were also undertaken on wing RH56, using fracture of a 2H 0.3 mm Pentel lead as a simulated source. The pinducers were characterized in situ, propagation features were determined for waves generated at the various bolt/rivet holes and propagating to the waveform sensors at A1 and W1, and the effectiveness of the zone location system was assessed (16).

The pencil lead source was used in supplementary calibration studies at ARL, to determine propagation characteristics for AE generated at a single bolt hole in a thick cylinder and at bolt and rivet holes in a Mirage wing-spar without wing-skins or bolts/rivets.

Implementation of Semi-Adaptive Approach

The semi-adaptive approach was applied to the processing and analysis of the specialist AE waveforms from the Mirage full-scale fatigue test. Some examples of its use (viz. for the elimination of extraneous sources and the detection of fatigue crack propagation) will now be given, emphasis being placed on evaluation of waveforms assigned to zones 10 and 11. However, the procedure is representative of that involved for any monitored zone (19).

Elimination of Sources Extraneous to the Bottom Flange. Results from our calibration studies (see earlier) were used to define risetime/autocorrelation function criteria for AE waveforms from the zones in the bottom flange of the wing-spar. A waveform was rejected, as originating from outside the region of interest in the bottom flange, if its risetime was greater than 100 μs or if the first minimum in its autocorrelation function occurred at a lag of greater than 21. 77% of the waveforms in zone 10 were rejected but only 25% of the waveforms in zone 11.

The guard system was clearly satisfactory for zone 11, but allowed a significant number of extraneous sources to be located in zone 10 (probably because signals from a fairing attachment from the main-frame to the nearby wing-skin were not guarded out).

Detection of the AE from fatigue crack propagation. AE waveforms were identified as originating from inclusion fracture in the crack tip plastic zone during fatigue crack propagation, provided that the following three requirements were satisfied: (i) the waveforms occurred intermittently at the peak of a load cycle or on a positive load gradient, (ii) the waveform features lay within the range predicted by calibration studies, (iii) such features were not obtained for waveforms detected elsewhere on the load cycle.

AE waveforms satisfying all the above criteria, were assigned by the location system to zone 11 at both waveform sensor locations ($A_1$ and $W_1$). Furthermore, these events were not repetitive, as would be expected for AE from crack face rubbing, but occurred infrequently, in keeping with the expected crack propagation rates (25). (Risetime characteristics of a sample of AE waveforms detected at A1 and assigned to zone 11 are shown in Fig 3, while a typical test flight used in the Mirage fatigue test, showing the occurrence of a peak load, zone 11 event, is illustrated in Fig 4).

No AE waveforms with characteristics corresponding to inclusion fracture were assigned to zone 10.

Incorrect Assignment of Zone 11 Events. It was found from the calibration measurements that the characteristics of the peak load signals assigned to zone 11 corresponded to AE from fatigue crack propagation in the wing-spar at the zone 9 bushed bolt hole. In addition different features were obtained for calibration measurements undertaken at other likely sources of cracking and fretting for which AE could possibly be assigned to zone 11. For example, signal risetimes detected at A1 for the pencil lead source applied to the edge of the bolt in the zone 11 bushed hole corresponded to those obtained for decreasing load gradients and load minima during fatigue testing (Fig 3). Calculations based on the spar geometry confirmed

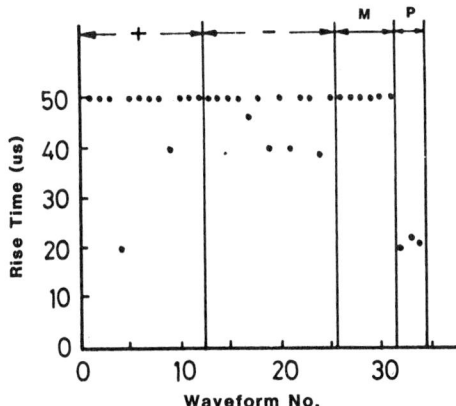

Fig. 3  Risetimes and load cycle position (+ positive load gradient, - negative load gradient, M load minimum, P load peak) are given for a sample set of waveforms assigned to zone 11. The waveform set was obtained during 35 test flights with the waveform sensor at A1 and an amplifier gain of 60dB(16), and had been subjected to preliminary processing to reject AE from extraneous sources. Note that all P load signals have a short rise-time, that all M or - load signals have a long risetime and that an occasional + load signal has a short risetime.

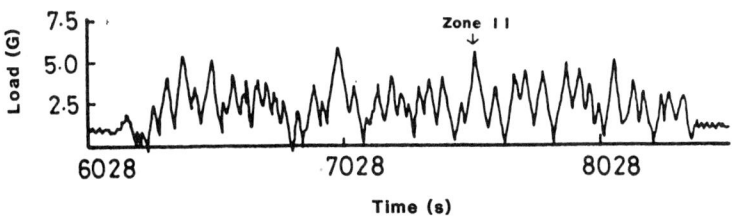

Fig. 4  A typical test flight in the Mirage fatigue test showing a peak load AE event (assigned to zone 11) which satisfied the criteria for inclusion fracture AE. (Only one such event occurred on the test flight despite the occurrence of several isolated peak loads).

that zone 9 events were assigned to zone 11 when first-arrival longitudinal rather than Rayleigh waves, were of sufficient amplitude to be detected by the array sensors A and B.

Confirmation of AE predictions. Metallography following completion of the full-scale testing of Mirage wing RH56 showed that several large cracks had propagated in the wing-spar at the zone 9 bolt hole during the fatigue test. (The largest crack had a depth of 6 mm.) No significant cracks (larger than about 0.5 mm in depth) were found in the zone 11 bolt hole or in the bolt holes which could have contributed to zone 10 emission. Thus the detection of inclusion fracture waveforms from zone 9 (incorrectly assigned to zone 11) and their absence in zone 10 is explained.

## CONCLUSIONS AND FUTURE WORK

A powerful range of signal processing and analysis procedures are now available to distinguish between different AE sources. Sensor arrays and guard sensors can be used to eliminate many extraneous sources and to obtain a preliminary estimate of AE source location. Fundamental characteristics of AE sources can be obtained in simple structures. In addition, the semi-adaptive approach we have developed is applicable in more complex structures such as aircraft - in the Mirage application, this material/science-based approach was successfully used to distinguish between AE from fatigue crack propagation and other AE sources detected during the full-scale fatigue test.

Further research on AE sources and sensors would allow a reduction in the degree of adaptation which was required in the Mirage analysis and would also facilitate future in-flight AE monitoring.

AE sensors. Suitable AE sensors now exist for short term AE monitoring - the pinducer was successfully utilized in the semi-adaptive approach developed for the Mirage analysis, while the NBS conical sensor (26) was recently used for quantitative AE measurements (18). Unfortunately neither of these sensors is easily coupled to a structure (see Fig. 2 for the pinducer), and the present NBS design is not robust. Hence long-term (or in-flight) continuous AE monitoring would require development of a modified sensor design.

Amplitude of AE signals. In principle, signal amplitude appears a useful feature for distinguishing between different AE sources. For example, in a recent study of 7050 aluminium alloy (21) the signal amplitudes associated with crack face rubbing extended to much higher levels than those associated with peak load crack propagation. However, recent theoretical results (27) suggest that the presence of a macrocrack significantly amplifies the amplitude of AE from a microcrack, i.e. the amplitude of AE from inclusion fracture in the crack tip plastic zone will depend on the relative sizes of the inclusion and the crack and their separation distance. Additional studies are required to clarify these effects, particularly for quantitative AE measurements.

Other source-related features. Further studies on the load cycle dependence of different AE sources are needed. The approach used in this paper was based on extending results from a few studies carried out during simple load cycling, but the load cycle occurrence of AE events during random load cycling needs to be rigorously investigated before it can be applied with complete confidence to the discrimination of different AE sources.

## ACKNOWLEDGEMENTS

The considerable contribution to the Mirage project by S.J. Bowles, R.A. Coyle and I.G. Scott (Aeronautical Research Laboratories) and by P.H. Hutton, D.K. Lemon and J.R. Scorpik (Battelle Pacific Northwest Laboratories) is gratefully acknowledged.

## REFERENCES

1. I.G. Scott, The status of in-flight monitoring using acoustic emission, presented at the N.Z. NDTA 10th annual Symposium, Wellington, (August 1986)
2. S.L. McBride and J.W. Maclachlan, J. Acoustic Emission 3, 1 (1984).
3. C.R. Horak and A.F. Weyhreter, Materials Evaln. 35 (5), 59 (1977).
4. W.M. Pless, C.D. Bailey and J.M. Hamilton, Materials Evaln. 36(4), 41 (1978).
5. C.M. Scala, R.A. Coyle and S.J. Bowles, An analysis of acoustic emission detected during fatigue testing of an aircraft, in: "Review of Progress in Quantitative Nondestructive Evaluation," Ed. D.O. Thompson and D.E. Chimenti, Vol. 4B, 709 Plenum Publishing Co., New York (1985).
6. C.B. Scruby, Quantitative acoustic emission techniques, in: "Research techniques in nondestructive testing," Ed. R.S. Sharpe, Vol. VIII, 141 Academic Press Inc., London (1985).

7. N.N. Hsu and S.C. Hardy, Experiments in acoustic emission waveform analysis for characterization of AE sources, sensors and structures, in: "Elastic waves and non-destructive testing of Materials," Ed. Y.H. Pao, AMD 29, 85 (1978).
8. D.R. Hay, R.W.Y. Chan, D. Sharp and K.J. Siddiqui, J. Acoustic Emission 3, 118 (1984).
9. P.H. Hutton, D.K. Lemon, R.B. Melton and P.G. Doctor, Develop in-flight acoustic emission monitoring of aircraft to detect fatigue cracks, in: Review of Progress in Quantitative Nondestructive Evaluation," Ed. D.O. Thompson and D.E. Chimenti, Vol. 1, 459 Plenum Publishing Co., New York (1982).
10. L.J. Graham and R.K. Elsley, J. Acoustic Emission 2, 47 (1983).
11. J.M. Richardson, R.K. Elsley and L.J. Graham, Pattern Recognition Letters 2, 387 (1984).
12. I.G. Scott, Monitoring crack growth during fatigue testing of aircraft by means of acoustic emission, in :"Proc. 4th Pan Pacific Conf. on Nondestructive Testing," Paper AE4, Sydney (1983).
13. C.M. Scala and R.A. Coyle, AE waveforms detected during fatigue testing of a Mirage aircraft (in preparation, 1986).
14. C.B. Scruby, J. Acoustic Emission 4, 9 (1985).
15. H.J. Rindorf, Bruel and Kjaer Technical Review 2 (1981).
16. C.M. Scala, Report on an overseas visit, Materials Tech. Memo 388, Aeronautical Research Labs., Melbourne (1984).
17. C.M. Scala, J. Acoustic Emission 2, 275 (1983).
18. C.B. Scruby, G.R. Baldwin and K.A. Stacey, Int. J. Fract. 28, 201 (1985).
19. C.M. Scala and R.A. Coyle, NDT International 16, 339 (1983).
20. T.C. Lindley, I.G. Palmer and C.E. Richards, Mater. Sci. Engng. 32, 1 (1978).
21. C.J. Kim and J. Weertman, Crack extension acoustic emission during fatigue crack growth, in : "Microstructural characterization of materials by non-microscopical techniques," Ed. N. Hessel Anderson et al., 349, RISØ (1984).
22. T. Ohira, T. Kishi and R. Horiuchi, Acoustic emission analysis of fatigue crack propagation in 7049 Al-Zn-Mg, Proc. Acoustic Emission Conf., Bad Nauheim, 241 (1980).
23. G. Weatherly, J.M. Titchmarsh and C.B. Scruby, Mater. Sci. and Technol. 2, 374 (1986).
24. J.Y. Mann, A.S. Machin and W.F. Lupson, Improving the fatigue life of the Mirage IIIO wing main spar, Structures Report 398, Aeronautical Research Labs., Melbourne (1984).
25. N.T. Goldsmith, Private Communication, Aeronautical Research Labs, (1985).
26. T.M. Proctor, J. Acoustic Emission 1, 173 (1982)
27. J.D. Achenbach, K.I. Hirashima and K.I. Ohno, J. Sound and Vibration 89, 529 (1983).

REAL-TIME AIRCRAFT STRUCTURAL MONITORING USING ACOUSTIC EMISSION

S. Y. Chuang

General Dynamics
Fort Worth, TX. 76101

INTRODUCTION

Advanced design of aircraft structures may incorporate areas inaccessible for in-service inspection. It is important to monitor high stress intensity substructures on aircraft to assure the integrity of aircraft structures and to decrease life-cycle cost of advanced military aircraft. Acoustic emission (AE) provides an ideal means for real-time structural monitoring. Acoustic emission is the stress wave generated by rapid release of energy from localized sources within a stressed material. Acoustic sensors combined with an effective signal processor can be used to detect acoustic emission events related to structural damages and serve as a real-time nondestructive evaluation (NDE) tool.

In-flight AE monitoring of airframe components had been conducted on Air Force C-5A and C-135 transport aircraft [1,2]. RCAF and RAAF are currently investigating applications of AE monitoring for their inventories of trainer/fighter aircraft [3-5]. AE was also considered by USAF for monitoring F-105 fatigue-critical areas, but was not implemented due to short remaining operation service life of F-105 fleet [6]. The feasibility of AE monitoring for modern high performance fighter aircraft is the subject of this investigation.

SCOPE

AE monitoring system for use in high performance fighters must be light-weight and small in size. It must be capable of filtering out in-flight noise and must be able to monitor composite as well as metallic structures. In order to develop an effective system that fits the requirements, research efforts have been directed toward the following areas: (a) on-board sensors evaluation, (b) damage signature and background noise characterization, and (c) optimization of noise filtering techniques.

The program started with evaluation of on-board sensors. State-of-the-art acoustic sensors can be constructed from piezoelectric ceramic, polymer film, or fiber optics. Polymer film, such as polyvinylidene fluoride (PVDF), has recently been used as transducers for many sound applications [7]. Flexibility is a major advantage of polymer film sensors. Placed in critical locations on the airframe, they can withstand flexing without cracking. PVDF performed as well as conventional PZT sensors in detecting AE events in composite structures, but was not as sensitive for use in aluminum components [8].

Characterization of AE signals due to various damage modes in composites as well as in aluminum alloys were also investigated by performing static tension tests and fatigue tests [9]. In this report, recent results from in-flight structural noise measurements of a fighter aircraft and a component durability test are presented.

IN-FLIGHT STRUCTURAL NOISE MEASUREMENTS

Acoustic sensors were installed at several structural locations of a fighter aircraft for investigating in-flight structural noise. Configuration of AE sensors on the aircraft is shown in Fig. 1. Hardman fast setting Double/Bubble epoxy was used to mount the sensors. Acoustic Emission Technology (AET) Corporation's AC-175L resonant response sensors were used. The sensors are of differential type construction and are completely shielded. A preamplifier (AET Model 140B) was mounted close to each sensor to provide a 40 dB amplification of output signals from the sensor. An analog airborne magnetic tape recorder (AR-700) is used to record the output signals from the preamplifier directly.

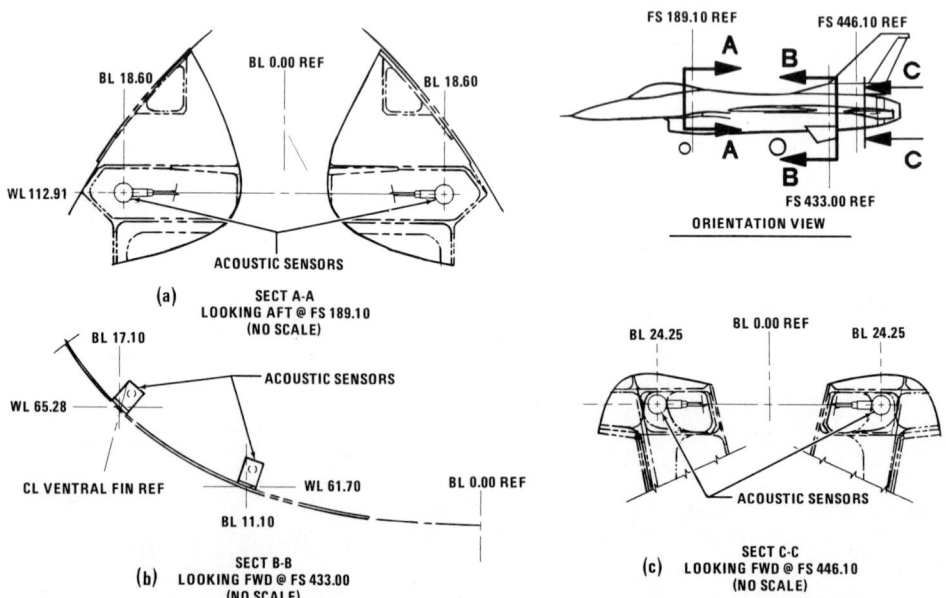

Fig. 1. Acoustic Sensor Configuration for In-Flight Structural Noise Measurement

Prior to flight test, the aircraft was subjected to an extensive flight loads calibration ground test. During the flight loads calibration, acoustic emission was real-time monitored by using the onboard sensors and preamplifiers. AE signals were analyzed using AET-5000 acoustic emission system. The system is a 8-channel microcomputer based AE system. It can process signals from each sensor channel independently according to the following AE parameters: ringdown counts, event counts, event duration, amplitude, rise-time, and energy. Distribution displays of peak amplitude, event duration, rise-time, energy, and slope (amplitude/rise-time) versus events are also provided. The system can be used for real-time AE monitoring with extensive capability for data discrimination based upon characteristic of AE parameters. AE data measured by AET-5000 were recorded in a 15 M-bytes hard disk for post-processing data analysis. A Sangamo magnetic tape recorder was also used in conjunction with AET-5000 for recording analog AE waveforms. Thirty-three test conditions were investigated. For each test condition, the applied load was ramped up from zero to 100% and returned to zero load, and it was performed for four times. AE data were recorded during the entire test procedure. All observed acoustic noise was transient in nature, as observed in the typical "ringdown" patterns shown in Fig. 2 (a), (b), and (c). Electrical transients usually have only a couple of ringdown counts per event, as shown in Fig. 2(d). Only very few electrical transients were observed. By gating the ringdown count parameter in AET-5000, noise due to electrical transients can be easily discriminated. AE data were characterized according to event rate, peak amplitude, and event duration parameters.

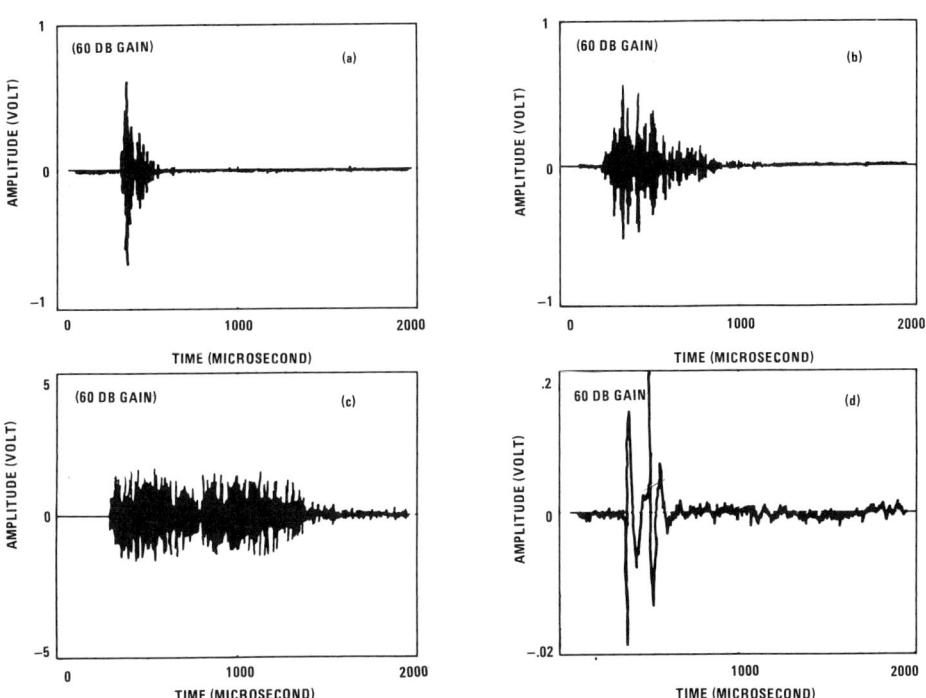

Fig. 2. Typical Acoustic Transients (a, b, and c) and Electrical Transient Waveform (d) During Ground Test.

Most of AE events occurred during the first run of the test. There is a qualitative correlation between the total number of AE events recorded by a sensor and the maximum microstrain measured at the location of the sensor during the test. Those AE events have amplitudes ranged from 35 to 75 dB and have durations less than 300 microseconds. Obviously, those AE events are stress induced structural noises, including structural rubbing and fretting.

Flight tests of the aircraft are currently in progress. In-flight AE measurements are performed concurrently with other dedicated flight tests. Two channels of AR-700 magnetic tape recorder with frequency response from 10 KHz up to 250 KHz are available for AE measurements. A CEC VR-3600 tape drive is used to play back AE signals into the AET-5000 AE system for data analysis. A waveform analyzer, ANALOGIC DATA-6000, is also used for AE waveform characterization. A flight test tape containing one flight hour of AE data has been analyzed. Preliminary results show that most of acoustic transient wave forms are similar to those observed during ground tests. In Fig. 3, AE event rate (A) which was recorded by the sensor located at bulkhead FS 446 is compared with various flight parameters: the stress measured near the location of the sensor (B), altitude (C), Mach number (D), and normal acceleration (E). It is clearly shown that AE event rate is associated with structural loading, as is produced during high-G maneuvers or rapid changes in aircraft speed, altitude and stress.

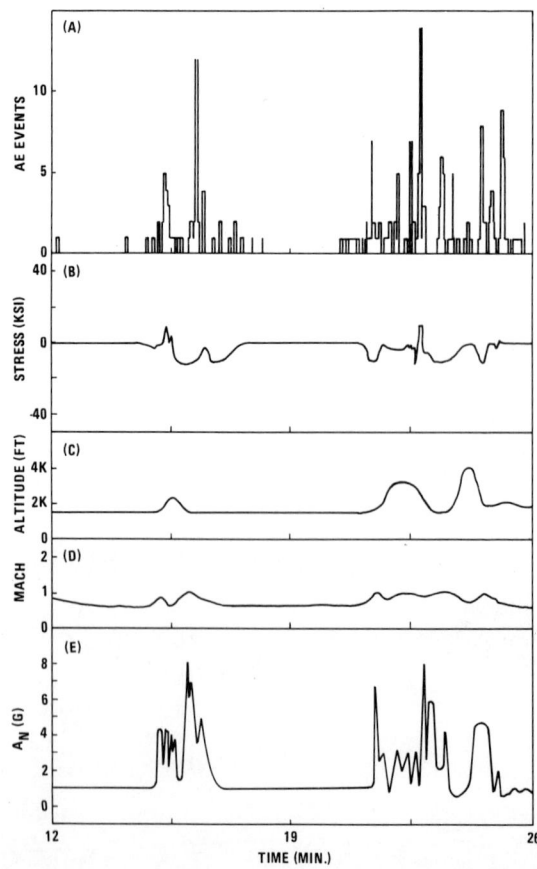

Fig. 3. Correlation of AE Event Rate with Various Flight Parameters.

COMPONENT DURABILITY TEST

A bulkhead test component was fatigue loaded with F-16 flight load spectrum for 16,000 equivalent flight hours of durability test. Two AE sensors were mounted on the test component at high stress locations as shown in Fig. 4. The AET-5000 system was used for real-time monitoring. It started monitoring at 3,000 flight hours. The test data were analyzed on-line. The energy parameter, which is a function of amplitude and duration parameters, provides a convenient means to identify changes in AE characteristic during the fatigue test. The energy distribution graphs after a 500 equivalent flight hours of test were printed out for comparison. Before 8,000 flight hours, energy distribution of AE events showed very little change. It has a sharp peak at about 82 dB (relative energy unit). After 8,000 flight hours, the AE spectrum started to change gradually as the test progressed. It has a major broad distribution centered at 82 dB and also a high energy peak at 125 dB. The energy distribution graphs for the period between 7,500 to 8,000 flight hours and the period between 14,500 to 15,000 are shown in Fig. 5 (A) and (B) respectively.

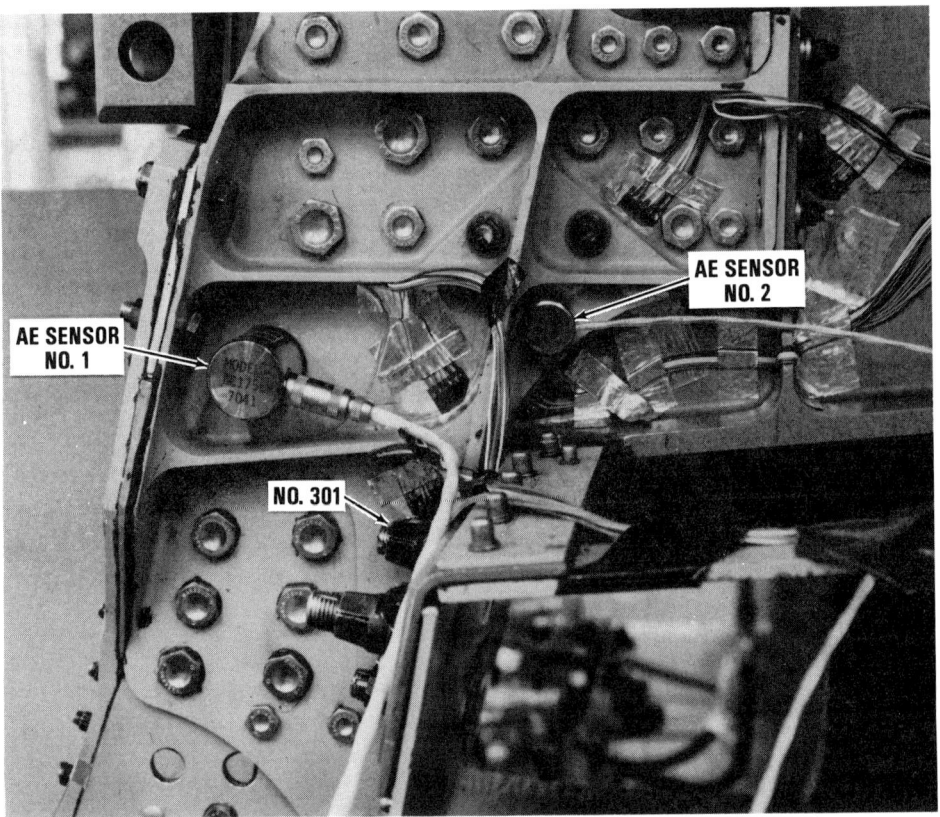

Fig. 4. Acoustic Sensor Configuration On A Bulkhead Test Component.

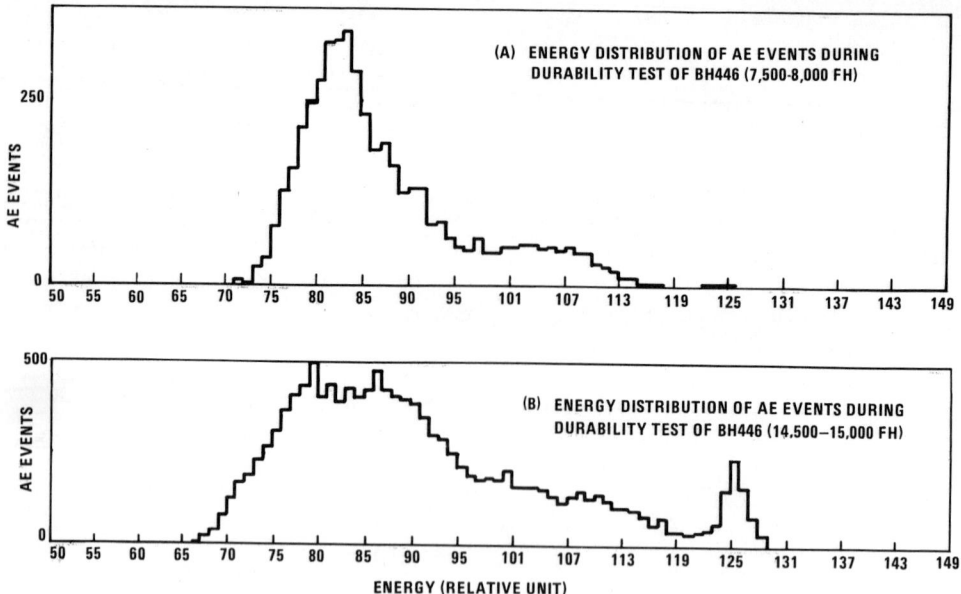

Fig. 5. Energy Distributions of AE Events During Durability Test of A Bulkhead Component: (A) 7,500 - 8,000 Flight Hours; and (B) 14,000 - 15,000 Flight hours.

A linear source location (time-of-flight) program was used in a later stage of the test in an attempt to determine the location of AE sources. NDE techniques, including eddy current, ultrasonic, fluorescent penetrant, surface replication, and x-radiography were used to inspect the test component after 16,000 equivalent flight hours of durability test. Two fatigue cracks were found inside a fastener hole (#301 hole shown in Fig. 4) close to the armpit. The largest crack length was about 0.17 inch. The location of the fastener hole where fatigue cracks were found coincided with the high AE activity zone detected by the time-of-flight source location technique. By viewing the AE monitoring data, the initiation of the fatigue cracks was estimated to begin at about 8,500 flight hours of test.

CONCLUSION

In conclusion, the results of ground test and flight test have shown that airframe structural noise comprises both acoustic and electrical transients. The electrical transient can be filtered out by gating ringdown count parameter. Acoustic transient noise is most likely due to structural rubbing and fretting. Acoustic transient event rate is correlated with structural loading and is not too high for practical operation of conventional AE systems. The result of the bulkhead component test has shown that AE energy distribution is a convenient parameter to examine sudden changes in AE characteristic during fatigue loading. The time-of-flight source location technique is useful for monitoring a localized area in a complex structure. However, fretting signals and crack growth signals can not be distinguished by simple AE parametric analysis and spatial filtering. Additional advanced filtering techniques are needed to separate fretting signals from crack growth signals in a high fretting noise environment.

REFERENCE

1. C. D. Bailey, "Acoustic Emission for In-Flight Monitoring of Aircraft Structure", Mater. Eval. 34, 165-171(1976).
2. M.E. Mizell and W. T. Lundy, Jr., "In-Flight Crack Detection System for the C-135 Lower Center Wing Skin", ISA Proc. 22nd Int. Instr. Symposium, San Diego, CA., p.259f, 1976.
3. S. L. McBride, "Canadian Forces In-Flight Acoustic Emission Monitoring Program", Proc. of the ARPA/AFML Rev. of Progress in Quantitative NDE, July 17-21, 1978.
4. I. G. Scott, "In-Flight Fatigue Crack Monitoring Using Acoustic Emission", Presented at the 13th Symposium on Nondestructive Evaluation, San Antonio, 1981.
5. P. H. Hutton and J. R. Skorpic, "In-Flight Crack Monitoring Using Acoustic Emission", 26th Int. Instr. Symposium, Seattle, WA., May 1980.
6. J. Rodgers, "PRAM Project-The F-105 Acoustic Monitoring", Final Report, USAF PRAM Program, July 1979.
7. H. Sussner, "The Piezoelectric Polymer PVF2 and Its Applications", 1979 Ultrasonic Symposium Proceedings, 1079, pp 491-498; Institute of Electrical and Electronic Engineers, New York, N.Y.
8. S. Y. Chuang and F. H. Chang, "On-board Acoustic Sensor Development", Engineering Research Report No. 2379, General Dynamics, Fort Worth, Texas, 1984.
9. S. Y. Chuang and F. H. Chang, "Characterization of Acoustic Emission from Gr/Ep Composites", Engineering Research Report No. 2445, General Dynamics, Fort Worth, Texas, 1985.

QUANTITATIVE ANALYSIS OF REAL-TIME RADIOGRAPHIC SYSTEMS


M. D. Barker, R. C. Barry, R. A. Betz, P. E. Condon,
and L. M. Klynn
Nondestructive Testing Technology Laboratory
Lockheed Research and Development Division
Palo Alto, Ca.


INTRODUCTION

Radiographic inspection is an essential tool in the nondestructive evaluation of devices such as solid rocket motors which must work properly when fired. The advent of real-time radiographic (RTR) inspection systems has dramatically improved the throughput and coverage of these inspections over film-based techniques. The RTR inspection, however, is only as sensitive as the system used. Qualitative measures of image quality which were originally developed for film-based inspection have been applied to these real-time systems. While these qualitative measures are useful, there is a clear need to develop quantitative measures which are more appropriate for the RTR inspection systems which do not rely on a subjective judgement and which can be used to indicate the cause of problems in the system.

A program to develop quantitative image quality indicators (IQI) was initiated in response to problems occurring over time in a group of 16 MeV solid rocket motor inspection systems originally put into service in 1976. The goal of the program was to develop a system of IQI's which would be analyzed digitally to indicate the quality of the radiographic images being obtained and which would be utilized to indicate which component of the RTR system was not working correctly. The components of these systems are shown schematically in Fig. 1. The parameters of these systems which require regular monitoring are (1) the size of the x-ray focal spot of the Linatron accelerator, (2) degradation in the imaging chain which consists of the x-ray to light conversion screen, mirror, lens, camera and electronics, and digital image processor, and (3) degradation in the CRT monitor used as the main image display device. The progress made to date on this project is reported in this paper.

QUANTITATIVE IMAGE QUALITY INDICATORS

One of the problems with the ASTM E-142 qualitative plaque and wire penetrameters [1] is that typically both the contrast and spatial resolution limits of the system are being measured at the same time. Figure 2 illustrates the detectability of an object depending on the object contrast and size. For film-based inspection systems where the spatial resolution is significantly better than that required to resolve the holes or wires in the penetrameter, the detectability as measured with the penetrameters is close to the contrast asymptote in Fig. 2. However, for RTR systems, the penetrameters are typically probing the response of the system in the region indicated by the circle in Fig. 2. Thus the spatial resolution and contrast sensitivity are intermixed in the reading of the penetrameter sensitivity. It is often important to separate the effects of system resolution and sensitivity, especially when troubleshooting the system.

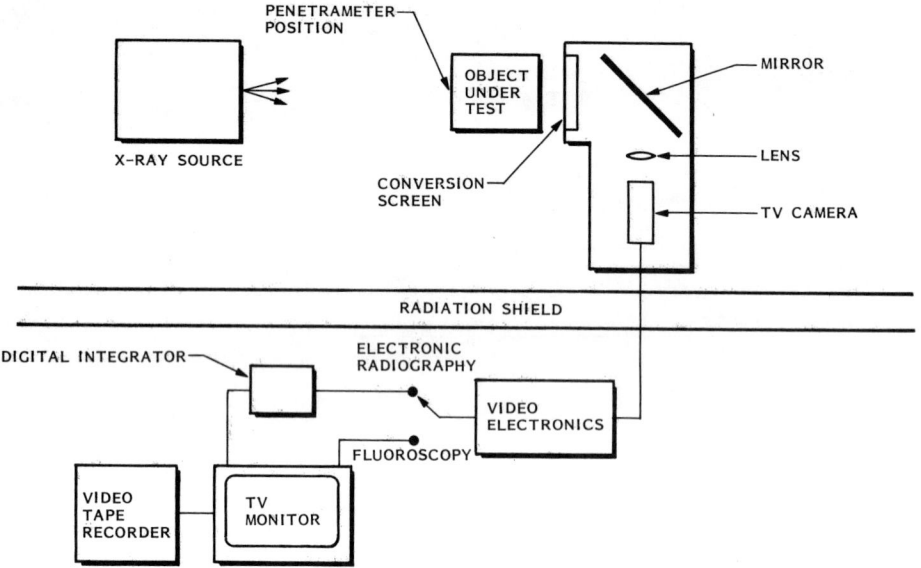

Fig. 1 Schematic of the high energy radiography inspection system

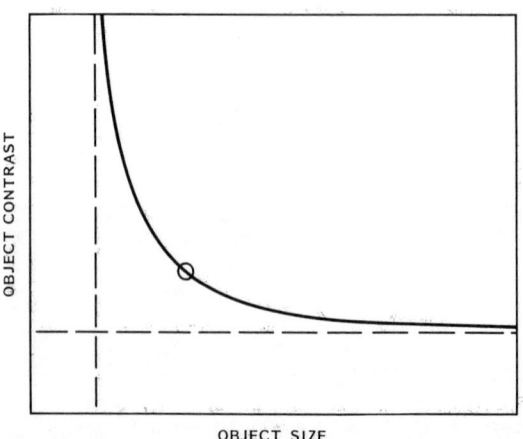

Fig. 2 Illustration of the inter-relationship of contrast sensitivity and spatial resolution in object detectability. Qualitative penetrameters typically probe the response of the system in the region indicated by the circle.

A new system for image evaluation has been developed which independently measures the contrast sensitivity by utilizing large low contrast objects, i.e., a step tablet with shims, and the spatial resolution by using small high contrast objects, i.e., a bar pattern. The system utilizes the digital information in the images to obtain quantitative information. This evaluation system involves using optical IQI's to measure the camera performance and x-ray IQI's to analyze the x-ray system performance. The x-ray source focal spot size is monitored by using geometric magnification of the bar pattern, while the intrinsic resolution of the x-ray detection system is monitored with the bar pattern at a magnification of one. In addition, a simple method has been developed to measure the sensitivity of the CRT monitors used to display the radiographic images.

To facilitate testing the efficacy of the x-ray IQI's, all experiments were performed at low energy (around 100 kV). The materials and thicknesses used for the final IQI's will be appropriate for the 16 MV Linatron energy needed for the solid rocket inspection systems. The concept and computations required are independent of the x-ray energy range actually used.

One serious problem to overcome in evaluating the digital information in the images is the image shading. With a camera based system, a uniform input does not yield a uniform image after passing through the optics and camera imaging chain; there is typically a bright center with the signal level falling off toward the edges. In addition, electronic shading correction is included in the camera control unit to flatten the signal prior to digitization of the image. This electronic shading has linear (ramp) and quadratic (parabola) components. In evaluating images, the changes in signal level due to shading must be distinguished from changes in signal level due to changes in the input. Thus, the image must be corrected for the shading. This can be done in one of two ways. One is to subtract an image acquired with a uniform input from the image acquired with the actual object. This method removes the image shading, but requires the acquisition of two images where the object is changed between the two images. In addition, subtraction adds to the image noise. In the 16 MeV solid rocket motor inspection systems, it is too time consuming to obtain subtracted images of this type. An alternative is to correct the shading mathematically. A region with known uniform input can be fitted using a two-dimensional quadratic Taylor's series expansion. The true signal is then the difference between this fit and the measured signal value.

In order to perform the quadratic fit to the image, the system must provide: (1) digital signal values; (2) access to the signal value at any given pixel; (3) computer processing of the digital information, and (4) a region of interest cursor display (not required, but helpful). To make use of the quadratic fit, particular IQI's were designed. First the contrast sensitivity IQI will be discussed, then the spatial resolution IQI will be described.

Contrast Sensitivity Measurement

The contrast sensitivity was measured using a step tablet with shims placed in the center of each step as shown in Fig. 3a. The contour of the image for a given step, including the effects of shading, is illustrated in Fig. 3b. The thickness of the steps should be chosen to measure the sensitivity at two different x-ray intensity levels so that the sensitivity over the dynamic range of the system can be determined. The shims should be a small fraction of the step thickness, in the range of 0.5% to 8%.

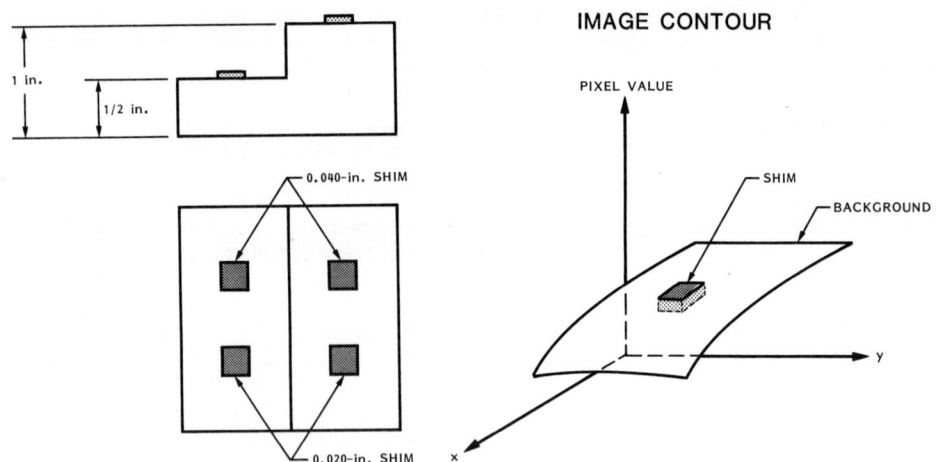

Fig. 3.  (a) Aluminum step tablet used to measure system resolution, Steps of 0.5" and 1.0" Al were used with 0.02" and 0.04" thick shims. (b) Contour of the digitized image in the region surrounding a shim. This contour indicates the presence of image shading.

For the low energy tests of this method, an aluminum object with 1/2" and 1" step thicknesses was used with shim thicknesses of 2% and 4% for the 1" step, and 4% and 8% for the 1/2" step. The regions outside and inside of each shim were simultaneously fit with the two-dimension quadratic Taylor's series expansion

$$p_o = a + b x + c y + d x^2 + x y + f y^2. \quad (x,y) \text{ outside shim region,}$$

$$p_i = p_o + S_1 \quad (x,y) \text{ inside shim region.}$$

The difference between the inner and outer regions, $S_1$ is then the image signal change due to the presence of the shim with the effects of image shading removed.

The image noise, $\sigma$, is estimated from the reduced chi-squared, $\chi^2/\text{NDF}$, of the fit as

$$\sigma = (\chi^2/\text{NDF})^{\frac{1}{2}}$$

$$\chi^2 = \sum_{ij} (p_{ij} - p(x_i, y_j))$$

where $p_{ij}$ are the actual values and $p(x_i\ y_j)$ the Taylor expression evaluated at pixel location $x_i$, $y_j$ and NDF is the number of degrees of freedom, i.e., the number of pixels included in the fit minus the number of fitting parameters. The contrast sensitivity is then defined to be

$$\text{CS} = S / [(\%T)\sigma]$$

where S and $\sigma$ are the shim signal and image noise as defined above and %T is the thickness of the shim expressed as a percentage of the step thickness. Note that, just as with the qualitative plaque penetrameters, the CS value will be different for different step thicknesses and material types. The larger the numerical value of CS, the better the image sensitivity.

Table 1. Typical contrast sensitivity, CS, measurement compared to ASTM E-142 penetrameter, PENE, observations.

| kV | mA | CS | PENE | CS | PENE |
|----|----|-----|------|-----|------|
| 80 | 1 | 0.9 | 2-4T | 0.6 | 2-4T |
| 80 | 2 | 1.3 | 2-4T | 0.8 | 2-4T |
| 80 | 4 | 1.7 | 2-4T | 1.1 | 2-4T |
| 110 | 4 | 1.9 | 2-2T | 1.6 | 2-2T |

Table 1 gives the results of several measurements taken using this new quantitative contrast sensitivity method. Images of ASTM E-142 plaque penetrameters were taken with the same camera settings for each x-ray kV and mA setting, and the plaque sensitivity, noted in Table 1, was measured by a trained observer. X-ray images of both the new and plaque IQI's are given in Fig. 4a and b. Subtracted images have been shown to improve the visibility for reproduction. Several trends are worth noting from these data. As the mA is increased, the CS value also increases and shows discrimination between the mA settings which is not available from the qualitative penetrameter results. When the kV is increased, the CS value increases by a larger amount for the thicker step and once again shows more discrimination than is available with the plaque penetrameters.

Repeated tests of this method have demonstrated good consistency between repeated measurements, between measurements taken with shims of different thicknesses, and between measurements taken with and without subtraction. The method is easily automated and takes only a few seconds of computer time to obtain quantitative results. Work is continuing on correlating the quantitative methods to penetrameter observations and on establishing ranges of CS which indicate acceptable image quality.

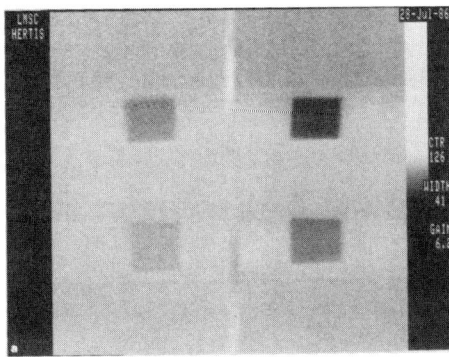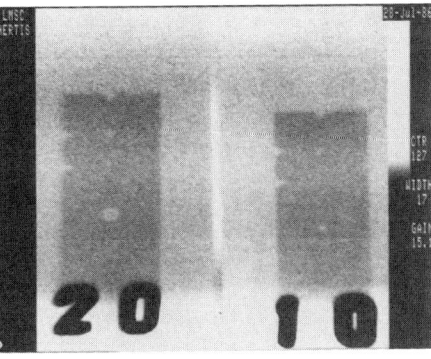

Fig. 4. (a) X-ray image obtained with the contrast sensitivity phantom at 110 kV, 4 mA. The 1" step is on the left with the 0.5" step on the right half of the image. (b) X-ray image obtained using the aluminum step tablet and 2% aluminum penetrameters. For both (a) and (b) a sugtracted image is shown to improve visibility for reproduction.

Spatial Resolution Measurement

The spatial resolution was measured using a bar pattern. A two-dimensional quadratic Taylor's series expansion was fitted to the data in a region containing an integral number of line pairs following the concepts of Ref. 2. The fit is able to follow the image shading, but has no component which can fit the bar modulations so long as two or more line pairs are included in the region. Thus, the amplitude of the bar pattern modulation, A, contributes to the reduced chi-squared of the fit as

$$\chi^2/NDF = A^2 + \sigma^2 = (A')^2$$

The square root of the reduced chi-squared, $A'$, is a measure of the modulation amplitude in the image. This measure is plotted against the line pairs per mm (lp/mm) of the bar pattern, as shown in Fig. 5. The value of $A'$ increases until it is dominated by the image noise and then becomes flat. A straight line fit, on log-log paper, is constructed of the decreasing $A'$. The limiting resolution of the system has been defined as the point where the constructed line intersects the image noise. The values obtained by this method are very close to those which were selected by trained observers looking at the x-ray image. Figure 5 shows the modulation for a geometric magnification of 1.75 taken with 1.5 mm and 0.4 mm nominal focal spot sizes.

These data can be used to separate the effects of x-ray focal spot size and limiting resolution of the detection system. To make this separation, a uniform detector response and uniform focal spot intensity were assumed. This results in a system modulation transfer function (MTF) composed of sinc functions:

$$MTF(f) = \text{sinc}\,(\pi f a/m)\,\text{sinc}\,(\pi f s (m-1)/m)$$

where f is the spatial frequency in cycles per mm, a is the effective system

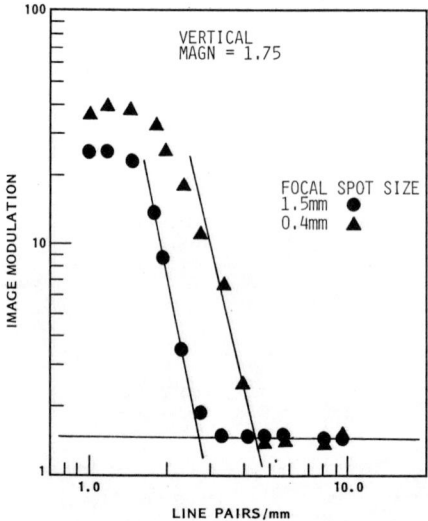

Fig. 5. The square root of the chi-squared or $A'$, labelled Image Modulation, is shown plotted against the line pairs per mm of the object.

resolution element in mm, s is the effective focal spot size in mm, and m is the geometric magnification of the image. While this approximation oversimplifies the actual MTF, it gives a useful estimate of the relative resolution and focal spot size which will indicate when changes have occurred in the system.

The limiting resolution, $f_o$, as measured with the bar pattern, is equated to the spatial frequency at which the MTF equals zero. For a magnification near unity, the zero crossing due to the focal spot size can be neglected, thus

$a = m / f_o$, for m = 1.

To measure the focal spot size, the zero crossing due to the system resolution must occur at a higher spatial frequency than that due to the focal spot size. Thus,

$s = m / ( (m-1) f_o )$, if a $f_o/m < 1$.

This typically holds true for magnifications of 1.5 or greater, but must be checked for the particular system under investigation.

Using these expressions, the effective system resolution element and the effective focal spot size were estimated in the horizontal and vertical directions for a Siefert 160/10 x-ray tube with dual focal spots of 0.4 mm and 1.5 mm nominal size. These results are given in Table 2. This method of measuring system resolution is easily automated and gives numerical estimates of the system resolution and focal spot size which will be quite useful to detect changes in the system performance.

Table 2. Resolution and focal spot measurements taken with two different spot sizes using the new quantitative method. Part (a) gives the effective resolution elements, a, of the system measured with a magnification of 1. Part (b) gives the focal spot size, s, measured with a geometric magnification of 1.75. All measurements are in mm.

|  |  | Nominal Spot Size (mm) 0.4 | 1.5 |
|---|---|---|---|
| Part (a) | Horizontal | 0.33 | 0.33 |
|  | Vertical | 0.31 | 0.31 |
| Part (b) | Horizontal | 0.71 | 1.3 |
|  | Vertical | 0.51 | 0.83 |

CRT Evaluation

The light output of a CRT screen is a monotonic function of the voltage applied to the control electrode of the CRT. For large voltages, the output saturates at a maximum value; for very low control voltages, the output bottoms out at a small, but non-zero background level. In the middle range of control voltages, the output is approximately a linear function of the control voltage. As the tube ages, the maximum brightness decreases, the background level increases, and the slope of the function in the middle range decreases. The middle range slope is a measure of the brightness responsivity of the CRT.

The contrast control on a CRT monitor adjusts the gain of the amplifier that drives the CRT control electrode. The brightness control adjusts the offset of the input-output function in the middle range by changing the bias voltage on another electrode of the CRT. To evaluate the CRT fairly, these controls must be set to some standard condition.

In our evaluation procedure, the CRT monitor is driven with a staircase function, while the screen brightness is measured with a portable light meter held in direct contact with the CRT screen. The staircase function drive produces a vertical bar step pattern on the screen. To set a standard condition of the contrast knob, the drive signal on the control electrode is adjusted with an oscilloscope to 7 volt steps. To set the brightness control, the measured brightness of the brightest bar is adjusted to 50 ft-L. This value was chosen because it is the minimum specified brightness of new CRTs. The brightness of the other bars in the pattern is then measured and recorded. Data for a new CRT monitor and for a very old one are shown in Fig. 6. The brightness responsivity (slope) of the old tube is clearly less than that of the new tube. Also, the maximum brightness of this old tube had degraded so much that 50 ft-L could not be obtained. Data will be acquired on tubes of intermediate age. These data will be used to select a slope value at which the tube should be retired from service.

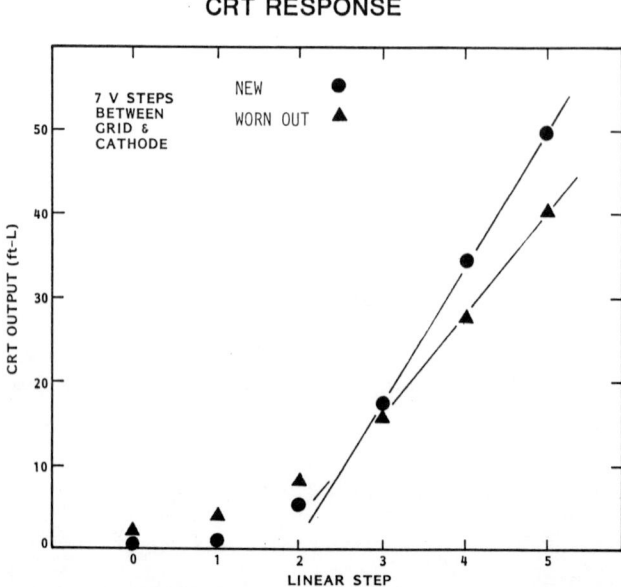

Fig. 6  Output brightness as a function of a staircase driving voltage for both a new and a very old CRT.

SUMMARY

A new method has been developed to quantitatively extract information on the spatial resolution, contrast sensitivity, image noise, and focal spot size from real-time radiography images on a routine basis. The method requires simple image quality indicators and computer calculations. It is used for x-ray and optical images to allow for trouble shooting to determine which component of the system is not operating to the required standard. The method is easily automated and provides greater discrimination than ASTM E-142 qualitative penetrameters without need for subjective evaluation of images. In addition, a method has been developed to monitor the performance of CRT display systems using a simple procedure and a hand-held light meter.

This method has been tested for low energy x-rays, and will soon be employed at high energy to monitor the performance of 16 MeV solid rocket motor inspection systems. The method shows great promise for future quantitative evaluation of a wide variety of radiographic inspection systems.

REFERENCES

1. Annual Book of ASTM Standards, Part 11: Metallography; Nondestructive Testing, ASTM, Philadelphia (1982)

2. R. T. Droege and M. S. Rzeszotarski, Med. Phys. 12, 721 (1986).

SPATIAL RESOLUTION OF LINEAR ARRAY XENON

IONIZATION CHAMBER X-RAY DETECTORS

Jeffrey W. Eberhard

General Electric Corporate Research and Development
1 River Road
Schenectady, New York 12345

INTRODUCTION

One of the most important characteristics of any imaging system is spatial resolution. This report describes a custom linear array Xenon ionization chamber detector developed by GE for sub-millimeter spatial resolution applications. The suitability of such a detector for high resolution digital imaging of turbine blades in both Digital Fluoroscopy and Computed Tomography modes is demonstrated.

In X-ray systems, spatial resolution is determined primarily by the effective source aperture, the effective detector aperture, and the imaging geometry. This report describes the measured spatial resolution characteristics of a digital X-ray Inspection Module (XIM) developed for airfoil inspection by General Electric under Air Force Contract F33615-80-C-5106. The XIM is one of a series of nondestructive inspection devices developed under the Integrated Blade Inspection System (IBIS) Program. A more general description of the IBIS XIM is given in [1].

Blades are a critical component of aircraft gas turbines. These parts undergo high temperature thermal cycling and very high stress. In order to meet these requirements, modern superalloy materials are widely used. As a result, critical flaw sizes are quite small. The major goal of the XIM system is to find flaws in turbine blades as small as 0.010 in linear dimension and assess their impact on blade performance. This report describes the spatial resolution characteristics of the XIM system.

SYSTEM HARDWARE CONFIGURATION

The XIM system is designed to inspect turbine blades with approximately 0.010" resolution. These blades fit into a part envelope 3 inches in diameter and 12 inches high. In order to achieve reasonable inspection rates, it is very desirable for both the X-ray beam and the detector to span the 3 inch part diameter. Therefore, a fan beam inspection configuration with a linear array detector is required. With this configuration, a DF image data set can be generated by scanning the part vertically past the linear array detector, and a CT image data set can be generated by rotating the part at the appropriate height for the CT slice. The data acquisition configuration for these two imaging modes is depicted in Figure 1. In the XIM, the data is handled in pipeline fashion, and the completed image is available essentially as soon as the data acquisition is complete.

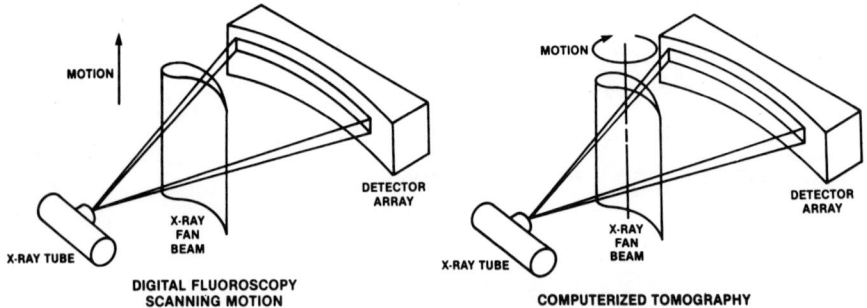

Fig. 1. DF and CT Data Acquisition

Fig. 2. XIM Detector

The X-ray source in this system is a Phillips industrial X-ray tube which operates from 230 kilovolts peak (KVP) to 420 KVP. This range of energies is required to provide adequate penetration of several inches of nickel based superalloy. The absorptivity contrast at these energies is still quite sufficient to provide high quality images of small flaws. The fan angle of the tube is approximately 30 degrees. Since spatial resolution is a key concern, the tube is typically operated in small focal spot mode (1.5 mm spot size). Large focal spot mode (4.5 mm) is also available, and comparisons of results in the two modes are described in the Results Section.

The detector is essentially the key element in the system, since it provides the required spatial resolution. In order to provide 0.010" resolution, the sampling theorem [2] requires that individual detector elements must be spaced on 0.005" centers. To cover the 3" diameter of the inspection field of view, at least 600 detector elements must be provided. This number of elements implies a system of substantial complexity, both in the detector itself, and in the associated data acquisition elements.

In addition to providing a large number of closely spaced individual elements, the detector must have high quantum efficiency for effective imaging, and the reliability of the detector must be good for use in the factory. In order to meet these requirements simultaneously, a Xenon ionization chamber linear array X-ray detector was developed by General Electric for use in the XIM system. This report summarizes some of the key characteristics of that detector which affect spatial resolution.

The requirement for a 600 channel linear detector array in a 3 inch space led to the choice of a Printed Circuit Board (PCB) implementation of a multi-element Xenon ionization chamber. In order to achieve the required spacing with sufficient accuracy, advanced photolithography techniques are required. This technology was developed at General Electric Corporate Research and Development early in the program on a prototype 64 element detector, [3]. The 600 element XIM detector was developed next. The detector board forms one plate of a parallel plate capacitor which resides in a pressure vessel. A high voltage plate is attached above the detector plate, and high pressure Xenon gas is the ionization medium. The finger spacing is uniform over the region contained in the pressure vessel, then it fans out to the connectors on the edges of the board.

A second major factor driving the use of PCB technology is the requirement of bringing 600 leads out of the Xenon pressure vessel. With the PCB, this can be achieved by simply passing the circuit card through the rear pressure vessel flange and making cable connections to flat ribbon cable. This is a major simplification over alternate connector techniques. A photograph of the 600 element XIM detector board mounted in the pressure vessel is shown in Figure 2.

In order to achieve comparable resolution in the vertical direction for DF imaging, a resolution of 10 mils with data taken in 5 mil steps is again required. The spatial resolution is achieved by collimation. Two tungsten blocks thick enough to attenuate the incident beam by a factor of 1000 are spaced 10 mils apart in front of the ionization chamber detector. This aperture defines the slice thickness in CT imaging.

The length of the detector elements in the X-ray beam direction and the X-ray absorption properties of the Xenon dielectric [4] determine the quantum efficiency of the detector. The effective energy of an X-ray tube operated at 420 KVP is approximately 250 keV. The quantum efficiency of the XIM detector at this energy is about 70%. In order to achieve this efficiency, a Xenon pressure of 75 atmospheres is required.

The imaging requirements leading to the detailed detector specification have been discussed. Spatial resolution is the major constraint driving the design. The detector is a 600 element linear array ionization chamber, implemented in PCB technology. The individual elements are spaced on approximately 5 mils centers. Horizontal resolution is on the order of 10 mils for a Xenon pressure of 75 atmospheres. Tungsten collimators limit the vertical slice height to 10 mils, and data is taken in 5 mil steps. The measured spatial resolution of this detector is the major topic of this report.

SPATIAL RESOLUTION CONSIDERATIONS

The spatial resolution of an X-ray inspection system is determined primarily by the focal spot size of the X-ray source, the detector element width, and the position of the part relative to the source and the detector. Suppose SS is the width of the focal spot, DIA is the width of the target, D1 is the distance from the focal spot to the target, and D2 is the distance from the target to the detector. The width of the penumbra (region where any part of the beam is blocked by the target) at the detector plane is given by

$$X2 = SS * (D2/D1) + (1 + D2/D1) * DIA. \tag{1}$$

This expression consists of two terms, one due to the focal spot size SS and one due to the width of the target DIA. For a point target (DIA = 0), the projected size of the focal spot at the detection plane is simply SS*(D2/D1), which is otherwise known as the unsharpness of the image due to focal spot size [5]. For a point source (SS=0), the shadow of the target on the detector plane is (1+D2/D1)*DIA, which is a magnified image of the target with magnification M=(1+D2/D1).

The system magnification (i.e., the distance ratio D2/D1) and the detector element width are determined by the required system resolution. This is done by specifying the target size which must be resolved, DRES. In the XIM system, DRES is chosen to be 10 mils. (The focal spot size is determined once the X-ray source is specified. At this point, we know SS and specify DIA=DRES=10 mils in Equation 1.) For an optimum system, the focal spot size at the detection plane is set equal to the shadow of the target at the detector plane.

$$(1 + D2/D1) * DRES = (D2/D1) * SS \qquad (2)$$

The result is given by

$$D2/D1 = DRES/(SS-DRES); \quad M = SS/(SS-DRES). \qquad (3)$$

If the focal spot size is smaller than the target width, (D2/D1)*SS is always smaller than (1+D2/D1)*DIA and the system designer has one additional parameter at his disposal to aid in system optimization.

The detector element spacing is determined by the required spatial resolution and the sampling theorem [2], which requires two detector measurements in each resolution cell. Since the spatial resolution is specified at the part, the system magnification must be taken into account as well in specifying detector element width. The result is

$$\text{ELEMENT SPACING} = (M * DRES)/2. \qquad (4)$$

The vertical resolution and the CT slice thickness are determined by the collimator height. The normal choice is

$$\text{COLLIMATOR HEIGHT} = M * DRES. \qquad (5)$$

The measured magnification in the XIM system is 1.15.

The horizontal system resolution is affected by factors besides system geometry. The primary factor involved in the XIM system is the interaction of incident X-rays with the Xenon detection medium. As incident X-rays ionize the Xenon molecules, both secondary electrons and secondary X-rays are created. These secondaries travel away from the primary detection site before they interact with the Xenon again, and the distance they travel affects the spatial resolution. This distance is determined primarily by the energy of the incident X-rays and the density of the Xenon gas. In short, the detection process is not a strictly localized process, and spreading is therefore introduced into the width of the system resolution function. If the distance the secondaries travel is greater than the detector element spacing, crosstalk between elements will be observed. The spatial resolution of the detection process is described in more detail in [6].

Resolution of an optical system is defined [7] as the minimum separation of two adjacent points that is detectable by the system. Resolution of photographic emulsion is expressed as the number of line pairs per millimeter that can be distinguished. Since the XIM system is essentially a filmless X-ray imaging system, the spatial resolution should also be characterized by the number of line pairs per millimeter that can be distinguished. Spatial resolution is, therefore, characterized by measurement of the Modulation Transfer Function (MTF). The MTF for x LP/mm is defined by

$$MTF(x) = (AL(x) - AP(x))/(AL(0) - AP(0)) \tag{6}$$

where AL(x) is the average signal amplitude through the lead lines of a standard resolution gauge at spatial frequency x, and AP(x) is the equivalent quantity for plastic lines. The denominator normalizes the results to the dc value, so the MTF ranges in amplitude from 0 to 1.

Spatial resolution as defined above clearly requires the ability to separately identify adjacent objects in the image. In certain imaging applications, detection of an object of a certain size is sufficient, and resolution of two closely spaced objects is not required. THE ABILITY TO DETECT AN OBJECT OF A CERTAIN SIZE IS A SUBSTANTIALLY EASIER TASK THAN ACHIEVING RESOLUTION OF COMPARABLE SIZE. As shown below, with the XIM system it is possible to detect objects of 1 or 2 mil diameter, even though the measured spatial resolution is between 10 and 14 mils. Therefore, it is critically important, when comparing imaging systems, to make the comparison based on the same criterion.

EXPERIMENTAL RESULTS

The measured spatial resolution characteristics of the XIM system are presented in this section. Under typical conditions, the X-ray tube was operated with a peak voltage of 320 kilovolts, and the focal spot size was 1.5 mm (60 mils). The magnification was 1.15. The voltage on the Xenon detector was 1000 V, and the Xenon pressure was around 1050 psig. Both horizontal and vertical spatial resolution were measured as a function of various parameter values around these nominal settings. If a specific parameter is not mentioned in the description below, it can be assumed that its value is the default mentioned in this paragraph.

The horizontal resolution characteristics are presented first. Figure 3 shows the measured horizontal MTF for the default values specified in the previous paragraph. (Slightly better resolution may be achieved for other parameter values). The graph shows a MTF of 5.4% at 2.8 LP/mm. At 2.9 LP/mm, the data no longer resolves 5 plastic lines and 4 lead lines. Therefore, the horizontal resolution of the detector is quoted at 2.8 LP/mm. At this spacing, each structure in the gauge (lead or plastic line) has a width of .179mm = 7 mils. However, in order to resolve (differentiate between) two structures of this width, they must be separated by 14 mils. Hence the resolution is 14 mils.

Horizontal resolution is determined both by spreading of secondary radiation and by the interelement spacing. Tam's model of secondary spreading [6] predicts a Full Width at Half Maximum of 6 mils for the spreading of secondaries in Xenon gas at a density of 1 gm/cm**3 for a typical 320 KVP tube spectrum. In conjunction with a finger width of 5.8 mils and a projected focal spot size (Eq. 1) of 9.1 mils, this implies a horizontal resolution of 10.2 mils. Uncertainties in the shape and size of the focal spot and the spreading of the secondary radiation probably account for the discrepancy with measured results. In any case, it is clear from this measurement that for the X-ray energies and Xenon pressures involved, closer spacing of the individual detector elements would not be fruitful.

The measured horizontal resolution function using the large spot of the X-ray tube is shown in Figure 4. Here, the MTF is 7.8% for a line spacing of 1.3 LP/mm. The difference compared to Figure 3 is strictly due to the much larger size of the focal spot in this measurement (4.5 mm instead of 1.5 mm). The result is generally consistent with the standard model of system resolution width [5] in which the total resolution width is given by the square root of the sums of the squares of the X-ray source width and the detector element width (including secondary radiation effects).

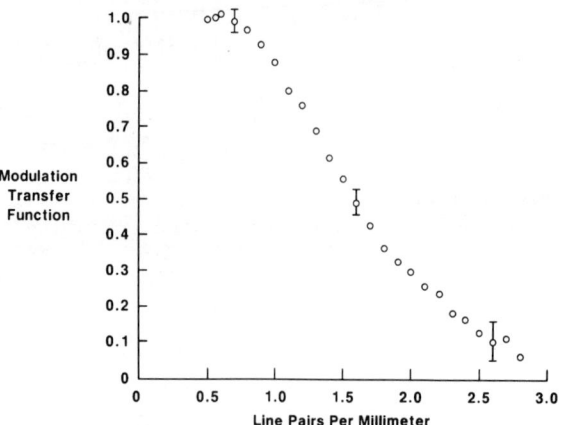

Fig. 3. Horizontal Resolution Small Spot

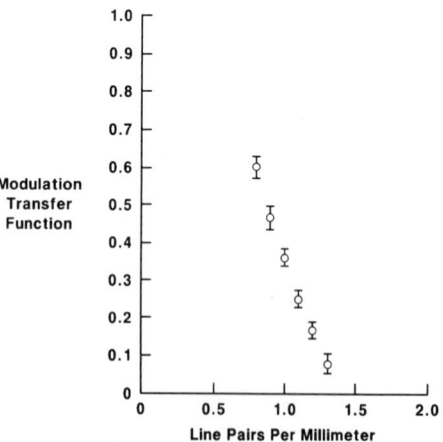

Fig. 4. Horizontal Resolution Large Spot

The measured vertical resolution MTF is shown in Figure 5. Here, 4.0 LP/mm data is resolved with an MTF of 10.2%. Each structure has a width of 0.125 mm = 4.9 mils, and the quoted resolution is 9.8 mils. The vertical resolution in the system is essentially determined by the collimator opening. If a better resolution in this direction were required, it could be achieved by narrowing the jaws of the collimator. However, the resulting reduction in X-ray flux would reduce the available signal to noise ratio.

The measured vertical resolution function using the large spot of the X-ray tube is shown in Figure 6. The MTF is 9.3% for a line spacing of 3.7 LP/mm. Even though the focal spot size is 4.5 mm, the measured resolution is only slightly different from the case of a 1.5 mm focal spot size. The 10 mil tungsten vertical collimator limits the effective spot size dramatically, and prevents any significant degradation in resolution.

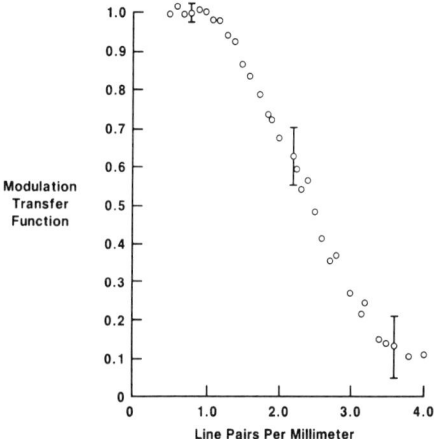

Fig. 5. Vertical Resolution Small Spot

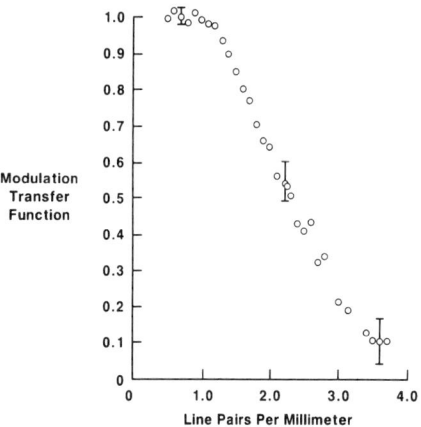

Fig. 6. Horizontal Resolution Large Spot

This result is, of course, significantly different than the horizontal resolution case, where no collimator is available to reduce the effective spot size.

The effect of Xenon pressure on system performance is important and somewhat surprising. Detector signal amplitude as a function of Xenon pressure is shown in Figure 7 for 2 individual elements of the 600 element XIM detector. These channels are representative of detector performance in general. Signal amplitude in the 2 elements varies slightly, but the shape of the curve is the same in the two cases. Signal amplitude has a broad maximum around 800 psig, and drops rather quickly above 950 psig. Nonetheless, system resolution continues to improve at least up to 1000 psig, as shown in Figure 8. Improvement in spatial resolution with pressure is expected since the density of the Xenon is increasing rapidly with pressure in the 600 to 1000 psig range, and the range of secondary radiation in the ionization chamber decreases with increasing density. The observed signal amplitude behavior is more difficult to explain, however, it is consistent with increased ion recombination rates and lower ion mobilities at higher gas densities. A more detailed explanation requires further investigation.

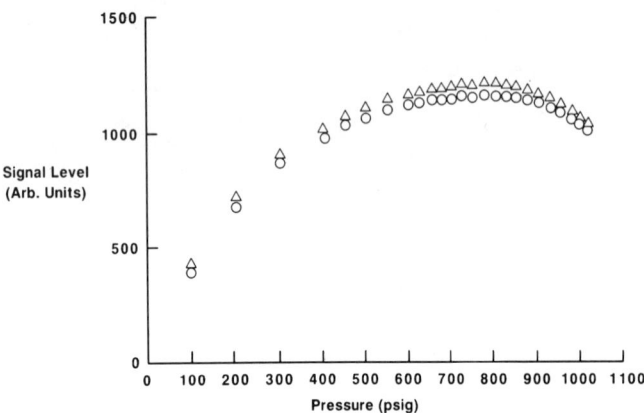

Fig. 7. Detector Signal vs. Xenon Pressure

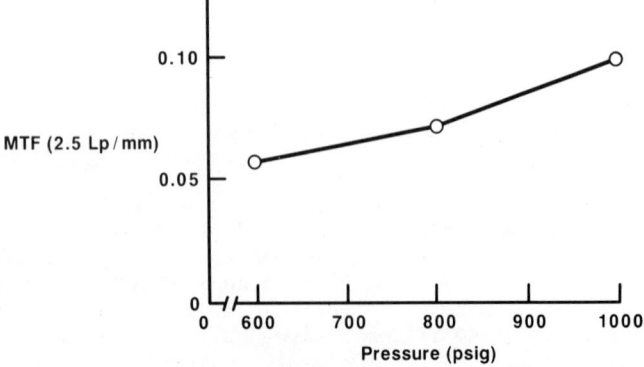

Fig. 8. System Resolution vs. Xenon Pressure

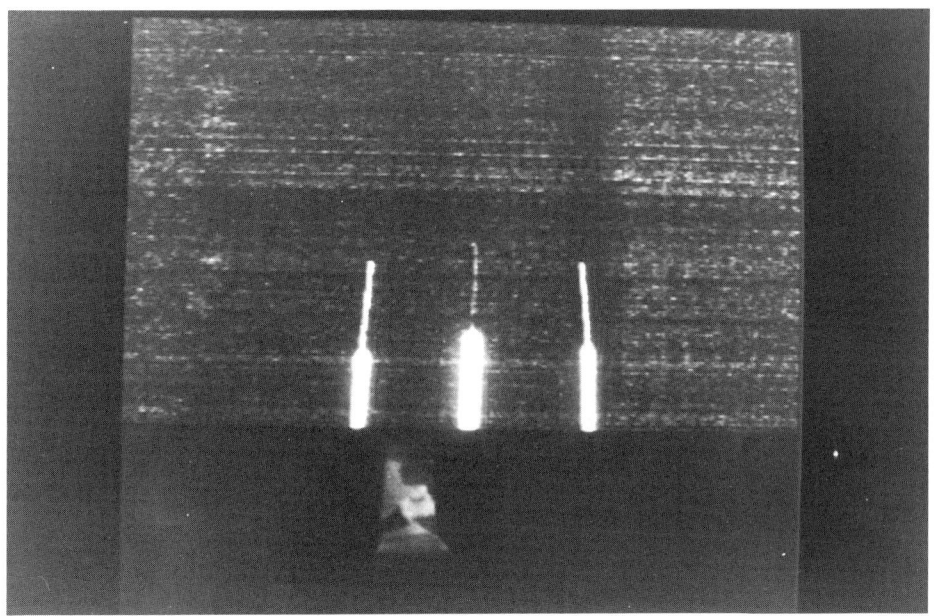

Fig. 9. DF Image of 2 Mil Diameter and 1 Mil Diameter Tungsten Wires

The spatial resolution properties of the XIM system were described in the last few paragraphs. The system, however, is capable of DETECTING much smaller objects. This capability is presented in Figure 9, where a DF image of 3 tungsten wires encapsulated in glass is displayed. The two outer wires are 2 mils in diameter and the center wire is 1 mil in diameter. All three wires are clearly visible in the image, though the 1 mil wire is near the detectability threshold. These images demonstrate the necessity of properly specifying the characteristics of an imaging system in order to meet the necessary performance requirements without overspecifying and adding unnecessarily to the cost.

CONCLUSIONS

The XIM system was specifically designed for turbine blade imaging applications. Typical blades and both DF and DT images are shown in [1]. The XIM has been remarkably successful in meeting its blade inspection goals, and two systems are now installed and operating in GE aircraft engine manufacturing plants. The spatial resolution is clearly satisfactory for this challenging application, and the custom GE Xenon ionization chamber detector is a key element in the success.

ACKNOWLEDGEMENTS

The author would like to acknowledge the contribution of N.R. Whetten to the development of the first experimental printed circuit board Xenon detector. D.S. Steele carried the development through the pre-prototype and prototype stages, and is now responsible for the detector effort at GE Aircraft Engine Business Group in Evendale, Ohio. C.R. Trzaskos was responsible for detector fabrication.

REFERENCES

1. D. W. Oliver, J. Brown, M. K. Cueman, J. Czechowski, J. Eberhard, J. Eng, R. Joynson, J. Keaveney, R. Koegl, R. Miller, K. Silverstein, L. Thumhart, R. Trzaskos, T. Kincaid, H. Scudder, C.R. Wojciechowski, L. Howington, I. Ingram, R. Isaacs, L. Meyer, J. Portaz, J. Schuler, J. Sostarich and D. Steele. XIM: X-ray Inspection Module for Automatic High Speed Inspection of Turbine Blades and Automated Flaw Detection and Classification, in "Review of Progress in Quantitative NDE, 5", eds. D. O. Thompson and D. E. Chimenti, (Plenum Press, New York, 1986), p. 817.
2. R. N. Bracewell, The Fourier Transform and its Applications, Ch. 10 (McGraw-Hill Inc., New York, 1978).
3. D. S. Steele, Private communication, (1979).
4. E. Storm and H. Israel, Photon Cross sections from 0.001 to 100 MeV for Elements 1 through 100, Los Alamos Scientific Laboratory Report LA-3753, Clearinghouse for Federal Scientific and Technical Information, National Bureau of Standards, U. S. Department of Commerce, (Springfield, Virginia 22151, 1967).
5. Lawrence E. Bryant, Technical Editor, Radiography and Radiation Testing, Section 6, High-Energy Radiography, "Nondestructive Testing Handbook, 2nd Edition, $\underline{3}$" (American Society for Nondestructive Testing, U.S.A., 1985).
6. K. C. Tam, Spatial Resolution in High Pressure Xenon X-ray Detectors, Rev. Sci. Instrum. $\underline{64}$, (1983), 1771-1776.
7. J. Thewlis, "Concise Dictionary of Physics", (Pergamon Press, Oxford, England, 1979).


DISCUSSION

From the Floor: Maybe I misunderstood, but did you say that in the image, you wanted to make the contribution of the source and the magnified image of the target the same? Is there some reason for that?

Mr. Eberhard: Yes. The best way to look at that is to assume you have two objects separated by a gap of width equal to the system resolution, DRES. You then choose the distances D1 and D2 such that the penumbras of the two objects just touch at the detection plane. This is the limiting case in which you can resolve the two objects. It also corresponds to the situation where the contribution of the source and the magnified image of the target are the same.

Mr. John Goss: In the CT scans I've seen, the cracks in the material are basically a gray smear across the image. If I take a photograph of the crack and I take a look at the CT scan I have to do a lot of inference to try to figure out where the crack really was.

Mr. Eberhard: The detectability depends on a lot of factors. The major point is that you can typically detect things in a system before you can resolve them. For objects smaller than the size of the system resolution function, the CT image of the object is typically larger than the object itself, so determination of size and position are difficult. Also, if two small cracks were present in the image, you can detect the presence of the cracks before you can resolve the presence of two separate objects.

Mr. Goss: Well, if the two cracks were parallel and, say, 28 mils apart, could I distinguish between the two?

Mr. Eberhard: In this system, if they are 28 mils apart and you have sufficient contrast to see the cracks, you should be able to tell that there are two of them.

From the Floor: Were you seeing cracks with CT or with Digital Fluoroscopy?

Mr. Eberhard: The resolution measurements presented here were taken in Digital Fluoroscopy mode. We have certainly seen small flaws in CT as well, but we have not yet attempted to make quantitative measurements.

Mr. Oliver: Jeff, there is the experience with the CT gauge. We made a gauge with a very small hole in the middle of it. This was done by taking a rod, cutting it in half, grinding it flat, putting a tiny hole in it, and then gluing it back together under pressure so that the interface was no more than several 10,000ths of an inch thick. All we could see was the interface. The CT method can be very sensitive to cracks. But it hasn't been qualified.

From the Floor: What is the resolution you can get if you improve your system by using a microfocus X-ray source?

Mr. Eberhard: In the current geometry, the problem is still that you have finite detector aperture effects and you still have the spreading of the secondary radiation in the detector. Therefore, as the spot size decreases, these detector effects will quickly become the key contributor to the width of the system resolution function, and I wouldn't expect a substantial improvement from using a microfocus source.
    On the other hand, if you switch to a high geometric magnification geometry, or you collimate the detector instead of using an array, you should be able to do much better.

Mr. Oliver: Last question.

From the Floor: We have seen delaminations in some composite structures down around 5 mils with conventional CT systems.

ENGINEERING TOMOGRAPHY: A QUANTITATIVE NDE TOOL

Richard L. Hack, Donna K. Archipley-Smith, and William H. Pfeifer

PDA engineering, 1560 Brookhollow Drive
Santa Ana, CA 92705

## BACKGROUND

The development and application of advanced materials, whether composite, metal matrix or ceramic, has progressed to a point where qualitative non-destructive inspection of components is no longer sufficient. As confidence in the validity of material properties increases, structures utilizing these advanced materials will be designed without the excessive safety factors characteristic of earlier structures utilizing the same materials. While this trend has the advantage of economizing on the use of the advanced, expensive materials, it underscores the need to quantify the flaw structure of advanced material components so that accurate, flawed material thermostructural response can be predicted and so that a quantified accept/reject criteria for a given component can be established.

Engineering tomography is an overall engineering tool and plan to achieve the quantified accept/reject criteria previously mentioned. Engineering tomography should be viewed as a collection of 4 subcomponents.

- Obtain quantitative digitized NDE data (i.e. X-ray CT, digital ultrasonics, magnetic resonance imaging) of a component to locate and identify flaw structure.

- Establish material property data base with degraded material properties as a function of the physical measured property obtained with the quantitative NDE inspection.

- Construct a 3-dimensional finite element model of the component including flaws (location and size) noted with the quantitative NDE technique.

- Conduct a finite element analysis on the flaw laden model integrated with the degraded material property data base to obtain a defective component response and ultimately quantify the acceptability or rejectability of the component.

As a whole, engineering tomography should prove to be a very powerful quantitative NDE tool. All of these steps need to be addressed to make engineering tomography a reality. The first step, quantifying flaws and the effects of flaws, is the focus of this paper.

Recent progress in the field of NDE has resulted in a number of techniques that go a long way to achieving the quantitative inspection required; techniques such as x-ray computed tomography (CT), advanced and digital ultrasonics (UT), magnetic resonance imaging (MRI) and advanced X-ray fluoroscopy all go a long way to improving our quantitative understanding of flaw structure.

EXPERIMENTAL APPROACH

One approach to developing quantitative accept/reject criteria is shown schematically in Figure 1. At the present time, PDA is funding an IR&D program to correlate CT data, density and mechanical/ thermal properties of advanced composites.

Computed tomography appears to have the potential of ultimately "driving" a thermostructural analysis if quantitative density data can be provided (1). Recent PDA studies indicate that quantitative CT density correlations are possible (2). This paper describes additional progress made in quantitively assessing the density of various composites by CT and describes the calibration procedures necessary to achieve meaningful

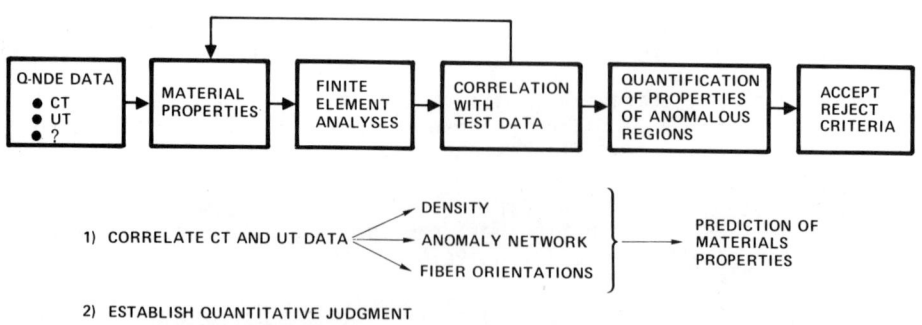

Figure 1. Quantitative NDE Accept/Reject Flow Chart

correlations between CT numbers and absolute density. Selected correlations of microstructural observations and CT images are presented as are studies concerning dimensional precision and resolution.

## CT NDE PROCEDURE

Computed Tomography scanning (also known as Computerized Axial Tomography - CAT Scanning) is an X-ray imaging technique. CT is based on the principle that radiation directed through a given volume of material will be absorbed to some degree by the material. The amount of absorption will be dependent on certain characteristics of the material including atomic number and physical density. CT images produced are of cross-sectional slices through the object which show the internal distribution of the X-ray attenuating properties of the material. CT procedures involve taking sophisticated measurements of radiation absorption and utilizing these measurements to obtain material characterization information.

The general procedure undertaken for a CT examination is outlined in Reference 3.

## DENSITY CALIBRATIONS

Critical to the CT examination process is the formulation of a mathematical relationship between measured absorption data and absolute material densities. <u>Only with this step accurately completed</u>, can CT NDE provide the desired level of quantitative material characterization.

The absorption/density relationship is a variable known to be dependent upon: the CT system, component or assembly geometry, component or assembly material (atomic number), scanning parameters (slice thickness, scan time, voltage, etc.), and CT system calibration. The effects of these parameters have been studied extensively in the medical community and can usually be accounted for by using proper calibration techniques.

The calibration process typically used by PDA to establish or verify the absorption/density relationship is described below:

- Preparation of the component for scanning includes positioning several calibration control rods within the component (typically aligned with the centerline).

- Control rods (normally numbering from 4 to 6) are materials of identical atomic number which span a range of densities and include at least one material similar to the actual component (i.e. similar in type of construction and absolute density).

- Accurate physical material density measurements are obtained for each control material.

- The absorption/density relationship is then defined by the control rod absorption magnitudes derived from the CT examination and the measured control densities, Figure 2.

Control specimens are usually prepared from actual composite hardware in order to provide absolute density standards during CT scanning. Refer to reference 3 for further details on density calibration.

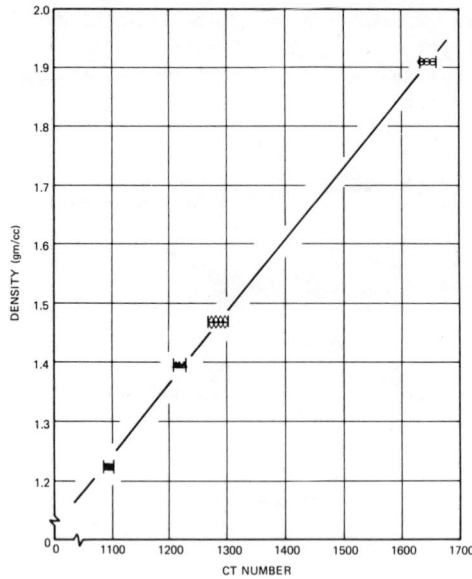

Figure 2. CT Number Versus Absolute Density

EXPERIMENTS: DESCRIPTION AND RESULTS

The following points were addressed in the current research described herein.

1. Effect of material composition on CT number/absolute density linear relationship. The validity of the CT number/absolute density relation is critical in the NDE of components with more than one material.

2. The precision of dimensional measurements. Direct dimensional measurements of complicated components via CT is a natural and powerful extension of the unique capabilities of CT.

3. The resolution and detactability limits of the CT system. Minimum flaw resolution and minimum flaw detection of a system will determine its applicability as a quantitative NDE tool.

PDA designed and fabricated a test article to address all three of these points. Called the Resolution Phantom, the test specimen consisted of a 6 inch (approx.) diameter by 1 inch thick graphite disk (Union Carbide AGSR; density = 1.60 g/cc nominal). A series of holes were drilled around the perimter and at other select spots to accept a variety of polymeric, carbon, and other low atomic weight materials. Other holes were precisely located for dimensional and resolution investigations.

1. Material composition effects on the CT number/absolute density relationship

Figure 3 describes the resolution phantom and the variety of polymeric, carbon and other materials utilized. Note that all the materials are comprised of relatively low and similar atomic number materials (i.e. all have components with atomic numbers of 9 or less). The lone exceptions are the glass filled epoxies (silicon has Z = 14).

The compositions and densities of all the materials were precisely determined prior to scanning. The materials were then inserted into the graphite disk and the assembly inserted into the scanner. Parameters for the X-ray source were 140 KV, 140 mA, and 3 seconds exposure. A CT number/absolute density calibration curve was created from the data based upon the carbon and graphite disk only.

To assess the effect of material composition on the materials relation to the CT number/absolute density curve, the CT number for each specimen was extracted from the scan data. The specimen CT number was then plotted against its measured density on the CT number/absolute density curve created from the carbon/graphite data. The results are shown in Figure 4. Note that all the materials with the exception of the glass/epoxy materials

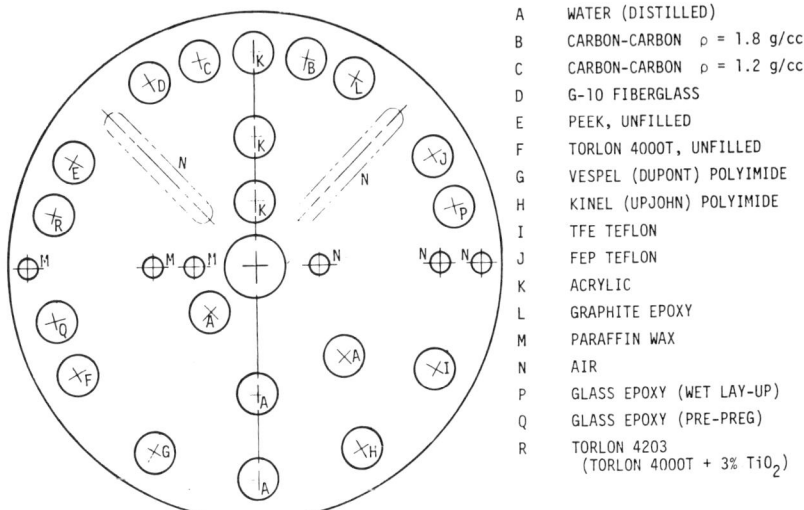

Figure 3. Resolution Phantom: Material Description

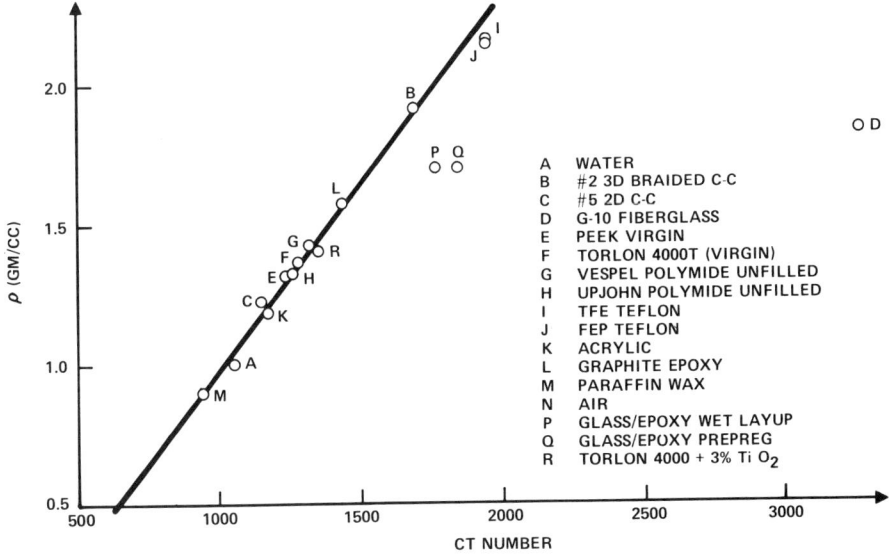

Figure 4. Resolution Phantom Material Study. CT Number Versus Absolute Density

405

fit very nicely upon this curve. The contribution of the silicon (Z = 14) in the glass/epoxy materials is felt to be the major cause for these materials not correlating with the data; the attenuation of the X-rays is much higher in the glass/epoxy than the material density would dictate due to attenuation being a function of the atomic number to the fourth power (approx.) and the glass epoxy having a higher effective atomic number. All the other materials have similar effective atomic numbers and hence fall very close to the line.

## 2. Precision dimensional measurement

Figure 5 shows the dimensions between points used in the dimensional measurement study. The basic procedure for measurement entails counting the number of pixels between points and determining the characteristic pixel[1] size based upon the scanning matrix size and the field of view of the CT system (both quantities are user defined on the GE9800 system). Counting the number of pixels between points is readily available due to the digital nature of the CT system and due to the CT number data being stored on a pixel by pixel basis. The characteristic dimension of the pixel is given by

$$d_{pixel} = \frac{\text{field of view size}}{\text{matrix length (or width)}}$$

The dimensional studies on the resolution phantom utilized a 7.09 inch (18 cm) field of view and a 512 x 512 matrix yielding pixels of .0138 x .0138 inches (.035 x .035 cm).

Definition of an edge was difficult owing to partial volume effects[2]. We defined an edge to be the pixel that had a CT number closest to the average CT number between the two materials in question (i.e. if air has a CT number of 0 and graphite a CT number of 1500, the edge is defined as the pixel with a CT number closest to 750).

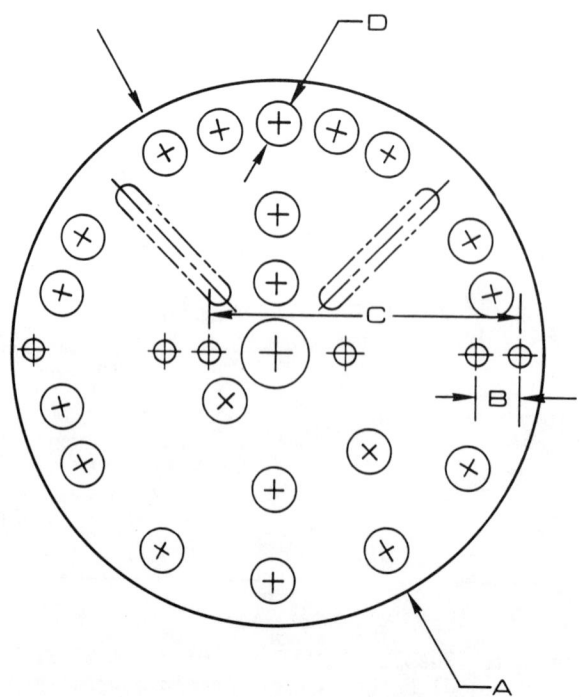

Figure 5. Dimensional Study

1. Pixels: Picture elements or segments of the CT image. The number of pixels in a scan slice is equal to the matrix, i.e. a 512 x 512 matrix has a total of 262,144 pixels.
2. Partial volume effect: Due to the finite size of a pixel, the CT system must average the properties within the pixel. Hence, at the transition between two or more materials, the CT number is weighted average of the materials.

Using this definition of an edge, measurements between the specific points on the resolution phantom were made. The results are summarized in Table 1.

Note the variance between CT measured and physically measured dimensions. We would expect a possible variance of .007 inches (half a pixel); variances of less than .007 are due to complimentary positioning of the edge pixels with respect to edge (i.e. the percentage of material 1 in one pixel is equal to the percentage of material 2 in the pixel at the other end of the desired dimension).

Table 1. Dimensional Variance CT Versus Physical Measurements

- CT PIXEL SIZE: 0.0138 INCHES (512 x 512; 18 CM FOV)

| LOCATION | DESCRIPTION | PHYSICAL | CT | VARIANCE |
|---|---|---|---|---|
| A | OUTER DIAMETER | 5.975 | 5.982 | 0.007 |
| B | HOLE: CENTER TO CENTER (AIR TO AIR) | 0.500 | 0.500 | 0.000 |
| C | HOLE: CENTER TO CENTER (PARAFFIN TO AIR) | 3.500 | 3.495 | 0.005 |
| D | HOLE: DIAMETER OF ACRYLIC SPECIMEN | 0.470 | 0.469 | 0.001 |

3. Resolution/Detectability

Spatial resolution is defined in the CT context as the ability to distinguish two small high contrast objects located a small distance apart. By comparison, detectability is the ability to observe a difference or the effects of a difference between two high contrast objects but not necessarily distinguish it.

A series of small holes was drilled in the resolution phantom as shown in Figure 6. The presence of a 80 drill hole (.013 inches diameter) is detectable with the GE9800 system (.014 pixel size) as noted in Figure 7a. The visualization of the holes is improved with a change to gray scale colors and by applying image sharpening convolutions away from the GE9800 as shown in Figure 7b. Note the inherent danger of not being able to resolve the hole; if its presence was not known, one would be hard pressed to say that the flaw was actually a hole or just a low density region.

The hole triplets were intended to check the resolution. However, the inherent spacing between holes proved to be too thin in virtually all cases and was not a definitive test although the presence of a wall between the holes was detected. Figure 7 clearly resolves a circular hole of diameter .028 inches diameter.

CONCLUSIONS

The experimental program undertaken has provided the following conclusions:

- Materials of similar effective atomic number can be directly correlated with the CT number/absolute density relationship.

- CT systems can be utilized to measure dimensions to an accuracy of 1/2 pixel. Improved accuracy can be achieved in post-scanning image processing by pixel multiplication and interpolation.

- The GE9800 system can detect flaws on the order of one pixel size. The resolution studies indicated 4.028 inch diameter hole limit to resolution.

In general X-ray CT can be used as a quantitative NDE tool for both flaw detection and dimensional studies. One can quantitatively characterize a flaw which is the first step in developing engineering tomography as a quantitative NDE tool.

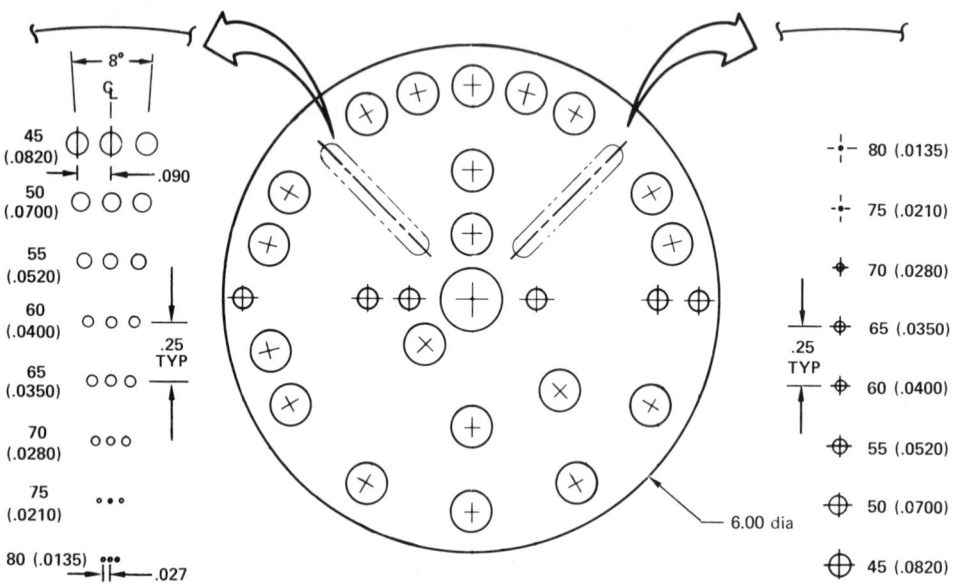

Figure 6. Resolution Phantom: Geometric Design

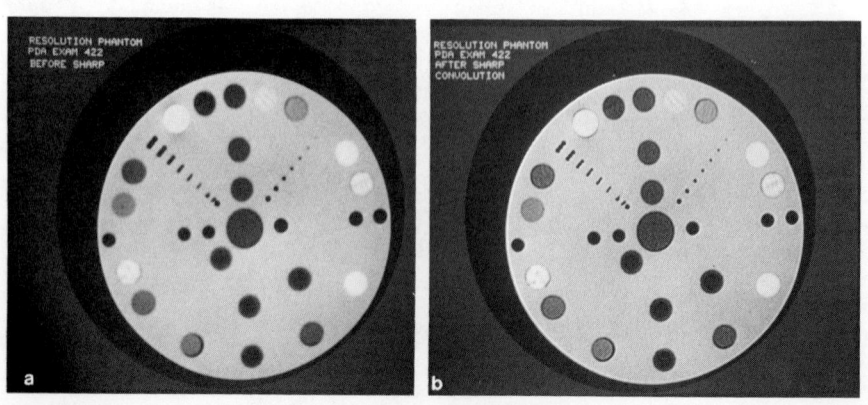

Figure 7. CT Images a) Without Enhancement
b) With Enhancement.

ACKNOWLEDGEMENTS

The authors would like to acknowledge the efforts and assistance of Mike Davidson and Terry Linn in the construction, data acquisition and CT imaging utilized in this study. Their efforts were critical to the successful completion of this research effort.

REFERENCES

1. Pfeifer, W.H., "Development of Improved Non-Destructive Evaluation Techniques for Carbon-Carbon Exit Cones", 1982 JANNAF Rocket Nozzle Technology Subcommittee Meeting, Monterey, CA.; CPIA Publication 367.
2. Crose, J.G., Pfeifer, W.H., Kipp, T.E., Kochendorfer, P.E. and J. Jortner, "Nozzle Attachment Technology Program", Final Report PDA-TR-1828-02, under contract F04611-83-C-0010 (AFRPL, Edwards Air Force Base, CA.); October 1985 (to be published).
3. Pfeifer, W.H., "Computed Tomography of Advanced Composite Materials", Advanced Composite Conference 2-4 December 1985, Dearborn, Michigan. Sponsored by American Society for Metals, Paper 8521-011.

DISCUSSION

Mr. A. Notea (Israel Institute of Technology): In terms of the slide where you see very nicely the delaminations, about 25 slides back -- the thickness of the delamination is far, far off the real thickness betweeen plies. You go back to the same phenomena I have been talking about previously. To correct for it, you have to go back into the memory of the computer, extract the numbers, and use a completely different algorithm. You cannot use the conventional element volumes to extract the real size of the flaw.

Mr. R. L. Hack (PDA Engineering): What he's talking about is cylinder with the involute plies and some noticeable delaminations. You can't use CT to really accurately predict the size of that delamination. You can detect that there is a delamination, but because of partial volume effects, you tend to spread the lower density over quite a few pixels and hence, you can't really acheive a good, accurate measure of that defect.

However, we have been able to see and measure defects on other parts that have more contrast than this part with better accuracy.

Mr. Notea: Another point is you should not mix so many materials in the same phantom because there is a memory. Between every point in the tomograph, there's the knowledge about the entire image itself due to the crossover of the beams. And once you mix so many together, it doesn't behave as in a practical object where you have only two or three.

Mr. Hack: I concur with that also. We have since been scolded for doing that, but I still feel it's valuable. They do line up very nicely, even in such a huge part.

From the Floor: On that involute cylinder, how big was that? And what was the pixel size?

Mr. Hack: The diameter of the cylinder was about eight inches. So the pixel size would have been approximately 15 to 20 mills.

Mr. Robert E. Green: Since this is the last talk on the CAT scan with composites, I'd like to ask a question.

Do you or does anyone else here have any thoughts on doing CAT scans inside an autoclave in the processing of composite material?

Mr. Hack: I think we actually have. It's a matter of now trying to arrange for somebody to let us use their CAT scanner inside of an autoclave. (Laughter).

Mr. Green: I mean outside the autoclave.

Mr. Hack: You have a problem there, and I'll let -- Bill Pfeifer is in the back. I'll let him make a comment.

I feel you're going to have a problem with penetration through the autoclave itself. The medical systems have a problem primarily because they utilize low-energy x-rays and they won't penetrate heavy steels or most other metals. A large high energy system, maybe like G.E.'s XIM System, if it was scaled up, might be capable of doing that.

Do you have any comments, Bill?

Mr. Bill Pfeifer (PDA Engineering): Yes. Only that we have done some sensitivity studies to be able to determine loss of contrast by running the experiment through the aluminum cylinder.

In the petroleum industry, of course, they are looking at two-phase fluid flow through rock cores in a pressurized system. In the medical system, such as the 9800, you can get away with a certain thickness of aluminum. In the inspection of nozzles, you can go through some higher atomic number materials, also. But you've got a filter in the system now, so you have to pay attention to the physics and the atomic number of the material being penetrated.

Mr. Hack: But as far as a means of monitoring the process in the autoclave, yes, we are aware that it would be a tremendous tool to do that.

Mr. Oliver: Another comment on that is if you look at the contrast-to-noise ratio in an image, there is an optimum attenuation Alpha L product for that quantity, and if you have a high Z wall and then a low-density material on the inside, it's hard to get enough low-energy protons to get the contrast sensitivity that you want.

A LINEARIZATION BEAM-HARDENING CORRECTION METHOD FOR X-RAY COMPUTED

TOMOGRAPHIC IMAGING OF STRUCTURAL CERAMICS*

E. Segal**, W. A. Ellingson, Y. Segal[+], and I. Zmora[++]

Materials and Components Technology Division
Argonne National Laboratory
Argonne, Illinois  60439

INTRODUCTION

Computed tomographic (CT) imaging with both monochromatic and polychromatic x-ray sources can be a powerful NDE method for characterization (e.g., measurement of density gradients) as well as flaw detection (e.g., detection of cracks, voids, inclusions) in ceramics. However, the use of polychromatic x-ray sources can cause image artifacts and overall image degradation through beam hardening (BH) effects [1]. Beam hardening occurs because (i) x-ray attenuation in a given material is energy dependent and (ii) data collection in CT systems is not energy selective.  Without an appropriate correction, the BH effect prevents the establishment of an absolute scale for density measurement.  Thus, quantitative density comparisons between samples of the same material but of different geometrical shape becomes unreliable [2].

Many different correction approaches are employed in medical CT systems to eliminate or reduce the BH effect.  These range from the early "water bag" approach (i.e., prepatient beam filtering) to a dual-energy approach [3,4] and correction of the image after reconstruction [5,6]. The intensive correction effort undertaken for medical CT systems has reduced the BH for tissue and tissue-like material to less than a few Hounsfield units or to tenths of a percent.  However, BH problems still exist for higher density bone and bone-like material.

For many industrial components made of relatively high-density materials, the BH effect is considerably greater than encountered in medical applications, but very little has been done to cope with this problem [7]. Rather, the BH effect is avoided in many industrial CT systems by using monochromatic isotope sources [8,9].  The main disadvantages of isotope-source CT systems are the low intensity (which leads to longer

---

*Work supported by the U.S. Department of Energy, Fossil Energy, Advanced Materials Development Program, and Conservation and Renewable Energy, Ceramic Technology for Advanced Heat Engines Project, under Contract W-31-109-Eng-38.
**Visiting Scientist, Technion, Israel Institute of Technology, Haifa, Israel.
[+]Technion, Israel Institute of Technology, Haifa, Israel.
[++]Elscint Corporation of America, Boston, Massachusetts.

image data acquisition times) and the stringent safety measures required to protect personnel. The purpose of this paper is to present a linearization BH correction method applicable to the CT examination of ceramic materials with a polychromatic x-ray source.

PRINCIPLE OF LINEARIZATION BH CORRECTION

It has been recognized for some time [10] that the nonlinear CT image reconstruction process can be linearized if the material being scanned can be assumed to be homogeneous. This linearization process can be mathematically explained as follows: The intensity, $I(x)$, of a polychromatic x-ray beam after penetrating a homogeneous material to a depth x is given by

$$I(x) = \int S(E) e^{-\mu_L(E)x} dE, \qquad (1)$$

where $S(E)$ is the spectrum of the polychromatic source and $\mu_L(E)$ is the total linear attenuation coefficient (i.e., the total of the photoelectric, compton, and Rayleigh components). The polychromatic x-ray beam can be represented by an equivalent monoenergetic x-ray beam, by first substituting an effective total linear attenuation coefficient, $\mu_{L(eff)}(x)$:

$$I(x) = \int S(E) e^{-\mu_L(E)x} dE = I_o e^{-\mu_{L(eff)}(x)x}, \qquad (2)$$

where

$$I_o = \int S(E) dE \qquad (3)$$

and $\mu_{L(eff)}(x)$ is the total effective linear attenuation coefficient obtained over the energy spectrum of interest. Having obtained $\mu_{L(eff)}(x)$, one can refer to the attenuation vs. energy plot and obtain an equivalent monoenergetic photon energy. Figures 1 and 2 illustrate this process for the case of $Si_3N_4$ and a typical polychromatic x-ray spectrum. Figure 1 shows the x-ray spectrum for a Siemens Somatom DR-H CT scanner operated at 125 kV. Figure 2 is a plot of the total linear attenuation coefficient as a function of photon energy for dense and green $Si_3N_4$ and for a fluorinated hydrocarbon, Freon TF. At each of 100 points on the x-ray energy spectrum curve, the relative flux was multiplied by the total linear attenuation coefficient at that energy. The weighted average of these 100 values is $\mu_{L(eff)}$; this average was 0.901 for dense $Si_3N_4$. This total effective linear attenuation coefficient is independent of the depth of penetration, and thus the total attenuation becomes a linear function of x. From the linear attenuation curve for dense $Si_3N_4$ (Fig. 2) and the $\mu_{L(eff)}$ value of 0.901, the equivalent monoenergetic photon energy is found to be 60.6 keV.

Figure 3 shows this effect graphically by comparing the uncorrected attenuation coefficients with the corresponding $\mu_{L(eff)}$ values for two $Si_3N_4$ densities. Note that the energy dependence of the uncorrected linear attenuation coefficient has a thickness dependence which is significant at specimen sizes of engineering interest (e.g., > 1 cm). From Fig. 3, the BH correction value for specimens of different thickness can be determined. The relationship can be put into the CT reconstruction algorithm as a polynomial or as a look-up table. In order to calculate the BH correction values, one has to know accurately (1) $\mu_L$ as a function of energy for the material being studied, at photon energies relevant to commercial scanners (20-150 keV); and (2) the spectrum of the x-ray head. For medical CT imaging, extensive calculations of $\mu_L(E)$ have been done, and an accuracy of better than 0.5% has been claimed [12].

The $\mu_L(E)$ values presented here for $Si_3N_4$ compounds were calculated from attenuation coefficients for the corresponding elements [13]. The claimed accuracy of the attenuation coefficients of ref. 13 is better than 1% for the relevant energy range.

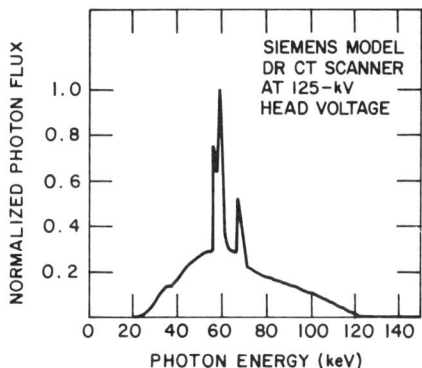

Figure 1. Polychromatic X-Ray Spectrum of Siemens DR-H CT Scanner Operated at 125 kV [11].

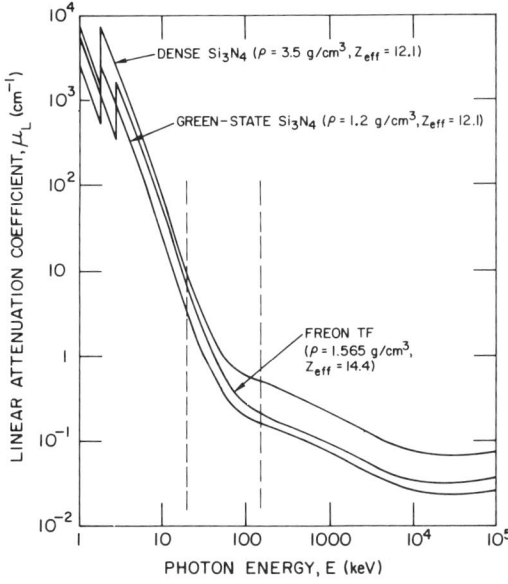

Figure 2. Total Linear Attenuation Coefficient for Dense and Green-State $Si_3N_4$ and Freon TF.

413

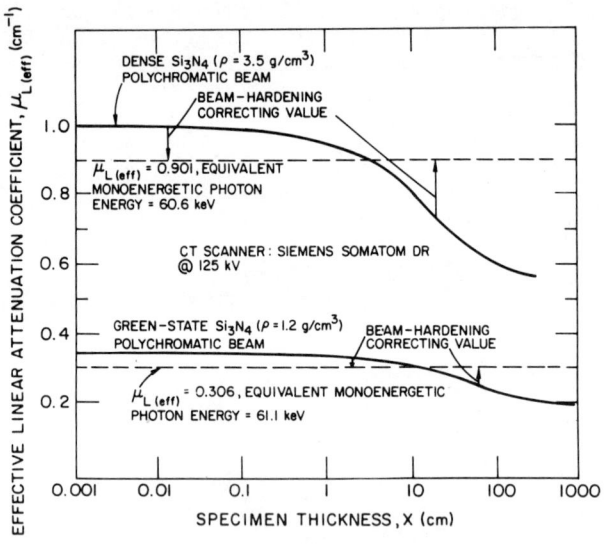

Figure 3. Comparison between the Attenuation Coefficient Uncorrected for X-Ray Polychromaticity and the Total Effective Linear Attenuation Coefficient Used in the BH Correction Process.

IMPLEMENTATION OF THE LINEARIZATION

Implementation of this linearization correction requires knowledge of the type of detector being used, the spectrum of the x-ray head, and the composition of the material being studied, as well as access to the raw detector data. Several excellent references [14] are available on CT detectors and we will not discuss detection here. In order to evaluate the accuracy of the effective linear attenuation coefficient method for a known x-ray spectrum and a homogeneous material, a theoretical calculation was completed and compared with an experimental measurement on a green-state $Si_3N_4$ specimen ($\rho$ = 1.995 g/cm$^3$) with dimensions of 5.7 x 4.3 x 3.1 cm. Figure 4 shows a comparison between the experimental data and theoretical calculations based on the x-ray head spectrum shown in Fig. 1. The excellent agreement between the experimental and theoretical results demonstrates that the BH effect can be calculated for ceramic materials. Figure 4 also shows how severe the BH effect can be.

The linearization BH correction method for ceramic materials was further experimentally verified with an Elscint Excel 2002 second-generation medical CT scanner. Access to the normalized detector data for this scanner was obtained. An approximate energy spectrum, S(E), was used to represent the polychromatic source. Freon TF was chosen as the test material because this fluid has a mass density ($\rho$ = 1.565 g/cm$^3$) and an electron density ($z_{eff}$ = 14.4) close to those of both green and dense $Si_3N_4$ (see Fig. 2). The test specimen was a 53-mm-diameter thin-walled polyethylene bottle filled with Freon TF and placed in the CT machine so as to produce a circular cross-sectional image. Figure 5 shows a plot of the uncorrected nonlinear attenuation and the linearization correction obtained by using $\mu_{L(eff)}$ at equivalent monoenergetic photon energy (60.6 keV). The nonlinear polynomial-curve coefficients were empirically established during tests on the machine.

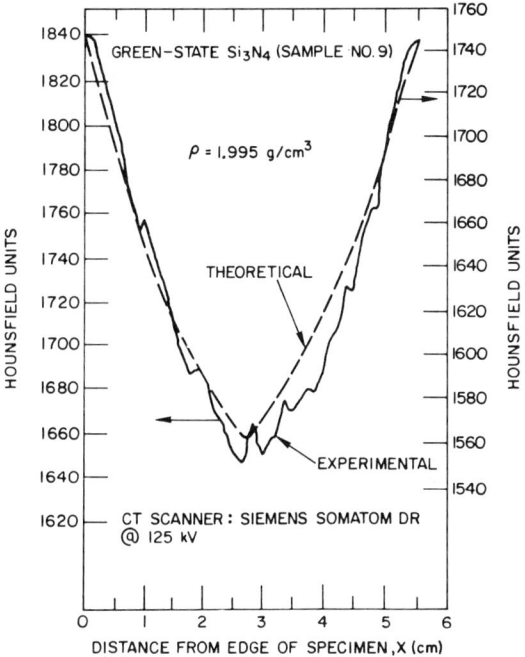

Figure 4. Comparison between Theoretically Calculated BH Effect and Experimentally Measured BH Effect for a Green-State $Si_3N_4$ Specimen.

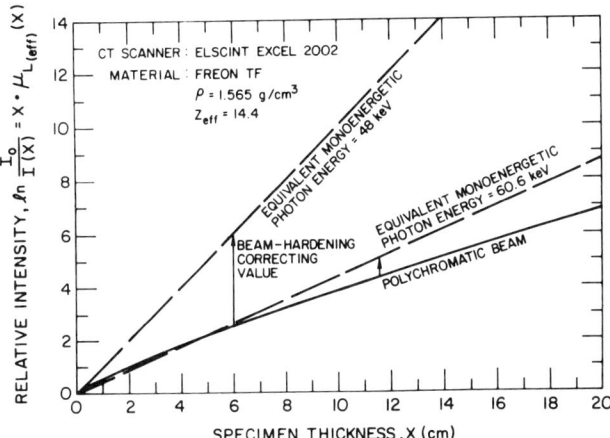

Figure 5. Comparison of Theoretically Derived Uncorrected Linear Attenuation with Corrected Linear Attenuation for Estimated X-Ray Spectrum of Polychromatic Source from Elscint Excel 2002 CT Scanner.

Figure 6 shows the Freon TF CT image obtained with a standard "water equivalent" BH correction. The BH effect is about 10%. Figure 7 shows the CT image obtained when the linearization BH correction was implemented. In this case the BH was reduced to <1%.

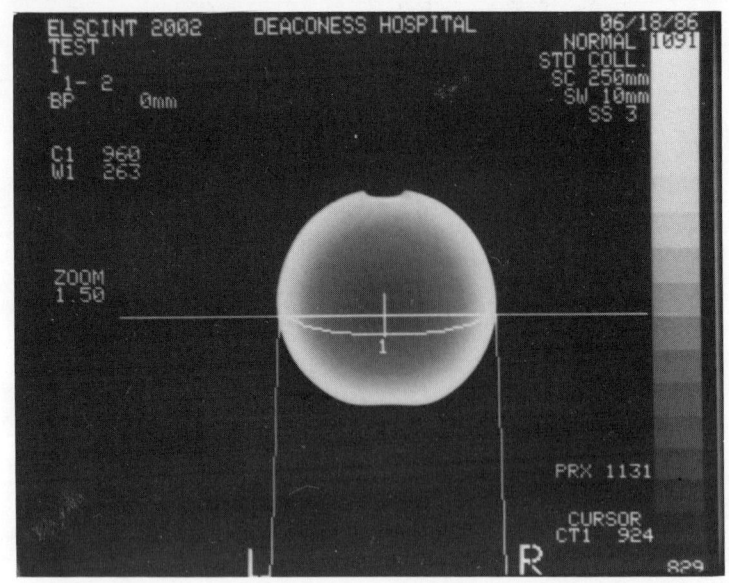

Figure 6. CT Scan (10-mm Slice) of 53-mm-Diameter Polyethylene Bottle Filled with Liquid Freon TF, with Water BH Correction. BH effect is $\sim 10\%$.

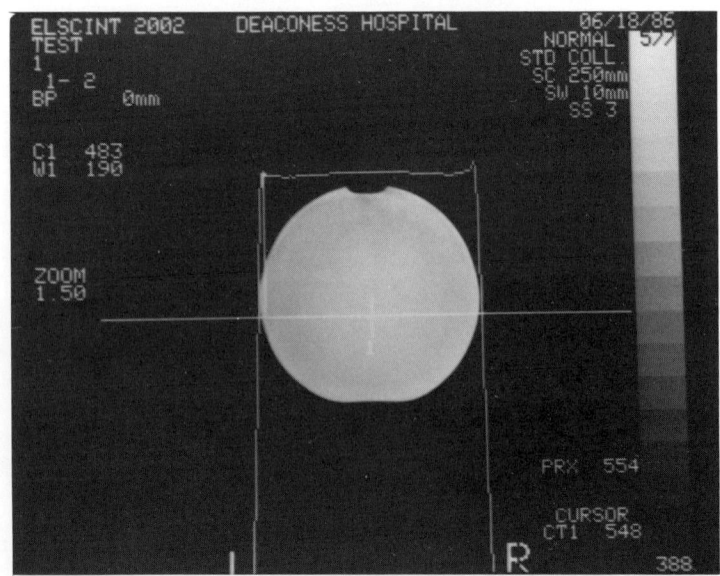

Figure 7. CT Scan (10-mm Slice) of Same Specimen Shown in Fig. 6, with Linearization BH Correction. BH effect is <1%.

CONCLUSIONS

The results presented here show that a linearization BH correction procedure which takes into account the material composition of the specimen and the x-ray spectrum of the CT scanner can reduce the BH effect to less than 1%. Further reduction of the BH effect to the 0.1% level may not be possible, as scattering effects are present. Theoretically, a special BH calibration should be performed for the material of interest and for each density of this material. This presents a problem for ceramic components, as uniform ceramic calibration blocks may be difficult to produce. It would be very useful if the material mass-density/electron-density trade-off could be established so that calibrations could be done on known homogeneous substances such as the liquid Freon used in these experiments.

REFERENCES

1. E. Segal and W. A. Ellingson, Beam Hardening Correction Methods for Polychromatic X-Ray CT Scanners Used to Characterize Structural Ceramics, to be published in Proc. 2nd Int. Symposium on Nondestructive Characterization of Materials, Montreal, Canada, July 21-23, 1986.
2. P. Rüegsegger, Th. Hangartner, H. U. Keller, and Th. Hindeling, J. Comput. Assisted Tomography 2(2), 184-188 (1978).
3. R. A. Brooks and G. Di Chiro, Phys. Med. Biol. 21(3), 390-398 (1976).
4. P. B. Dunscombe, D. E. Katz, and A. J. Stacey, Br. J. Radiol. 57, 82-87 (Jan. 1984).
5. D. J. Moh, G. Zheng, and B. Eddleston, Br. J. Radiol. 58, 873-880 (Sept. 1985).
6. G. T. Herman and S. S. Trivedi, "A Comparative Study of Two Post-reconstruction Beam Hardening Correction Methods," Medical Image Processing Group Technical Report No. MIPG76 (Feb. 1983).
7. M. D. Silver, Identification and Removal of Density Artifacts in CT Images of C-C Components, presented at Composites in Manufacturing Conference, Westin Bonaventure Hotel, Los Angeles, CA, Jan. 14-16, 1986.
8. B. D. Sawicka and W. A. Ellingson, Photon CT Scanning of Advanced Ceramic Materials, these proceedings.
9. P. Reimer and J. Goebbels, Mater. Eval. 41, 732-737 (May 1983).
10. J. A. Randmer, T. J. Koller, and W. P. Holland, X-Ray Sources and Controls, in: Radiology of the Skull and Brain, Volume 5, "Technical Aspects of Computed Tomography," T. H. Newton and D. G. Potts, eds., C. V. Mosby Co., St. Louis (1981), pp. 4058-4095.
11. "CT Scanner Information Booklet," Siemens Corporation, Medical Engineering Group, Erlangen, FRG.
12. M. E. Phelps, E. J. Hoffman, and M. M. Ter-Pogossian, Radiology 117, 573-583 (Dec. 1975).
13. J. H. Hubbel, Int. J. Appl. Radiat. Isot. 33, 1269-1290 (1982).
14. P. Haque and J. Stanley, Basic Principles of Computed Tomography Detectors, in: Radiology of the Skull and Brain, Volume 5, "Technical Aspects of Computed Tomography," T. H. Newton and D. G. Potts, eds., C. V. Mosby Co., St. Louis (1981), pp. 4096-4103.

# DISCUSSION

Mr. Ron Stripe, Lawrence Livermore: It seems to me that unless you know a priori the object size very accurately, that you are gaining a free variable from nowhere. You don't have the thickness or the internal structure of your object. How can you deconvolve the actual density on your radiograph or your C.T. from whether it's beam hardening or actual geometry effects? Also you looked at that first one, and you saw where you have that very light halo. How do you know, if you didn't know what was in your object, that that's not actually internal stress?

Mr. Segal: Well, here I know it isn't an actual internal structure because I have taken a liquid.

Mr. Stripe: But you have a priori knowledge of exactly what it contains.

Mr. Segal: Yes. So here I wanted to be sure that I am correcting the beam hardening and only the beam hardening.

Mr. Stripe: But the effect of beam hardening is a function of the thickness that the beam goes through, and if you don't know that thickness because you are trying to measure--that's one of the variables you are trying to evaluate. It seems you get back to the

Mr. Segal: This was your first question. If you are taking this picutre (Fig. 3), we really should know the dimension of the sample. It depends on the geometry. But if we are implementing our correction at the stage of the normalized measure data, the correction will be independent of the thickness or geometry of the sample. The normalized measured data, or the preprocessed data

$$\ln\left(\frac{I_o}{I(X)}\right) = X \cdot \mu_{L(eff)}(X)$$

is represented by the curved line in Fig. 5. At this stage we transfer this curve to a straight line. A straight line implies that the absorption coefficient $\mu_{L(eff)}(X)$ is constant, i.e., a monochromatic beam.

Mr. Green, Johns Hopkins: Could you describe briefly, again, your specimen that you used for this?

Mr. Segal: My specimen was a plastic bottle filled with Freon T. F.

Mr. Green: And it's cylindrical, like a glass? And looking down?

Mr. Segal: Yes to both questions. It is quite heavy so there is some flatness.

Mr. Green: But the thickness is constantly going down the length. It's round?

Mr. Segal: It's just a plastic bottle.

Mr. William Friedman, Standard Oil: It seems like most of the beam hardening occurs in perhaps the outer 1 centimeter. Would it be possible to filter the beam so that you remove that soft component? And then get rid of most of these problems?

Mr. Segal: Yes. This was done for the human body. The patient was surrounded by a water bag. What it does, really, is it reduces the beam hardening from the edges where the slope is higher. The equivalent here is surrounding the sample by a cylinder of the same material. Then most of the beam hardening will be taken out. However you will still have the central part.

This procedure reduces the beam hardening and also causes the beam hardening to be the same for all directions. Usually in a cylindrical object, the beam hardening will be different along different lines. However, it does not eliminate beam hardening.

Mr. R. Morris, Los Alamos National Laboratory: How did you derive your polychromatic beam hardening curve, experimentally or from first principles?

Mr. Segal: From first principles.

Mr. Morris: From tabulated cross-section values?

Mr. Segal: Yes.

Mr. Oliver: Just as a comment, the XIM system which Jeff described earlier used corrections essentially identical to yours. The corrections were made by taking real data from step blocks experimentally, externally, and that whole procedure can be automated because the correction can be put in essentially as a look up table. It is indeed dependent on material, but it is independent of thickness.

Mr. Segal: But here, measured data of the attenuation coefficient for ceramics is a real problem, due to the difficulty in preparing constant density samples having different thicknesses.

# PHOTON CT SCANNING OF ADVANCED CERAMIC MATERIALS

B.D. Sawicka

Atomic Energy of Canada Limited
Chalk River Nuclear Laboratories
Chalk River, Ontario, Canada K0J 1J0

W.A. Ellingson

Argonne National Laboratories
Argonne, IL 60439, USA

## INTRODUCTION

Advanced ceramic materials (e.g. $Si_3N_4$, $ZrO_2$, SiC, $Al_2O_3$) are being developed for high temperature applications in advanced heat engines and high temperature heat recovery systems [1]. Although fracture toughness has been a constant problem, advanced ceramics are now being developed with fracture toughnesses close to those of metals [2]. Small size flaws (10-200 μm), small non-uniformities in density distributions (0.1-2%) present as long-range density gradients, and porous regions which can be seen as localized areas of slightly lower density, are critical in most ceramics. The need to detect these small flaws is causing a significant effort to be devoted towards nondestructive evaluation. Detection of "defects" such as those noted in engineering ceramics has presented problems for conventional non-destructive evaluation methods [3].

The use of computed tomographic (CT) imaging provides a means of obtaining accurate two-dimensional attenuation mappings of cross sections through an object, from which density variations can be obtained, within the limits of the contrast of the CT image [4-6]. CT imaging is undoubtedly going to play a crucial role in the development of reliable ceramic materials. Since small density gradients and small defects are to be observed, CT for industrial ceramics must provide images that have very good density resolution (i.e., high contrast, low noise) and be free from artifacts. While medical CT scanners with X-ray sources can be applied in some cases, especially for low-density materials, they have some drawbacks (see Sec. 2). Another possibility is to use isotopic sources. Sources can provide photons at various energies, and, with the choice of appropriate energy, are suitable to study both low and high-density ceramics. If monoenergetic sources are used, the images are free from beam hardening (BH) effects and artifacts. For multi-energy sources proper BH corrections can be easily introduced, since the energy spectra of isotopic sources are well known.

Preliminary CT evaluations of a selection of engineering ceramic components have been reported earlier [7], and it has been demonstrated that the use of isotopic sources in CT could play a key role in the development of reliable engineering ceramic materials. In the present paper we present photon source, low-noise CT scans performed for two large ceramic objects. The high contrast images obtained demonstrate the ability of the technique to detect and to quantify small density variations within ceramic tiles, as well as to detect internal cracks. The CT analyses are compared with the results of low-kV contact radiographic images.

PHOTON CT SCANNING FOR CERAMICS

Characterization of many ceramics (especially in the green state) can be achieved using medical scanners with polychromatic radiation. However, the use of these scanners has its drawbacks. One problem is that medical scanners use X-ray sources that have an energy which may be too low for large objects of high atomic number and/or density. Secondly, the polyenergetic radiation should be corrected for the so-called beam hardening (BH) effect [5,8,9], which might be a problem, especially if the energy spectrum of the X-ray sources is not well known. Thirdly, the geometry in medical scanners is usually fixed, and not necessarily well matched to specific industrial applications.

A CT image provides an accurate two-dimensional map of the X-ray (photon) attenuation in a cross-section of an object. This corresponds to a map of density in the measured cross section, but the two are not equal. If information about densities is needed, it is necessary to transfer the attenuation data into the density data using the known mass attenuation coefficients of the compound being studied. The measured intensity, I, for photons of energy E, is related to the linear attenuation as:

$$\mu \cdot x = \ln I_0/I \qquad (1)$$

where  $\mu$ is the linear attenuation coefficient (in cm$^{-1}$) for a given material at energy E,
$x$ is the distance (in cm) that the beam travels through the material,
$I_0$ is the measured unattenuated photon intensity,
$I$ is the intensity of transmitted photons.

The linear attenuation coefficient $\mu$ is related to the mass attenuation coefficient, $\mu_m$ (at energy E, i.e., for a monoenergetic source), as

$$\mu_m \cdot \rho = \mu. \qquad (2)$$

The absorption coefficients are strongly energy dependent, but are known for all elements in a wide energy range [10,11] and therefore can be calculated for all materials of known compositions. For a fixed energy:

$$\mu_m \cdot \rho \cdot x = \ln I_0/I, \text{ and } \rho = (\ln I_0/I) \cdot \mu_m \cdot x. \qquad (3)$$

For a CT scan obtained with a monoenergetic source, the absolute values of the density can be obtained from measured attenuation data in a straightforward way, using the above formulas and the corresponding values of $\mu_m$ for the materials. A photon source of energy close to 1.2 MeV (e.g. Co-60) is especially convenient, because for this energy the mass absorption coefficients are the same (within 5-10%) for all elements except hydrogen, and therefore, the density data can be determined from such scans quite accurately even without advance knowledge of the chemical composition of the object.

To obtain a density map from an attenuation CT map measured using a polyenergetic source (e.g., X-rays), a transformation function has to be constructed, which integrates over all photon energies in the source and properly corrects for beam hardening (BH) effects [8,9]. The BH effects are especially strong for dense, high atomic number objects imaged by low-energy spectrum. BH correction is possible if the energy spectrum is well known, as is the case, e.g., for polyenergetic sources such as Ir-192. For CT scans obtained with X-ray sources, the transfer from attenuation data to the density data is complicated and therefore is not made in most cases, and instead, the CT images are presented relative to the standards of known densities.

A requirement for the application of CT to ceramics is that the CT images have to be measured with high contrast (i.e., low noise), since otherwise small density gradients and small flaws would not be detected. The density can be measured within each pixel of the image with an accuracy equal to the noise of the image. The pixel-to-pixel statistical noise, which determines the density resolution, depends on several factors [12]. The contrast of the CT image (i.e., the inverse of the noise) for fixed scan parameters depends on the value of the absorption coefficient (and therefore the energy of the source) and the intensity of the radiation transmitted through the object (and therefore the source intensity and the sample thickness). Thus, for industrial CT applications one would like to both optimize the photon (X-ray) energy and to increase the beam intensity. Since X-ray sources are generally more intense, the second requirement often results in X-ray sources being preferred over isotopic sources. However, for dense ceramics, especially large objects, the energy of the X-ray source may be too low to penetrate the object, in which case isotopic sources are superior.

The optimum energy choice is defined by the requirement that $\mu \cdot x = 2$, which comes directly from the formulas for the CT noise [12]. This yields an energy of about 1.25 MeV for 10 cm thick objects of both alumina ($\rho$ = 3.95 g/ccm) and zirconia ($\rho$ = 4.65 g/ccm) - i.e., the photon energy of a Co-60 source. In comparison, the use of an energy of 60 keV, which is about the effective energy of a 120 kV X-ray tube, decreases the signal-to-noise ratio (for the same scan parameters, source intensity and scan time) by a factor of 10 in the case of alumina, and by many ($\sim 8$) orders of magnitude in the case of zirconia. For a 10 cm alumina object this increase in noise can be compensated for by using a source 100 times as intense, which is possible for an X-ray source in comparison to an isotopic source. However, for a 10 cm zirconia object a 60 keV source is simply inadequate. The lowest practical energy is about 150 keV (this corresponds to a tube voltage of ~300 kV) and even then the intensity must be a factor of $10^4$ greater in comparison to the isotopic source, to compensate for the energy mismatch; for 100 keV the required increase in the intensity is $10^8$, and for 60 keV, $\sim 10^{16}$. For less dense materials, the energy optimization is not as important since the signal-to-noise ratio is not strongly energy dependent over a wide energy range (provided the objects are not too small, i.e. >~1 cm). For example, for a 10 cm thick graphite object ($\rho$ = 1.8 g/cm$^3$), the optimum energy is about 350 keV, but the noise conditions do not worsen greatly by the use of energies between 60 keV - 1.2 MeV.

EXPERIMENTAL METHODS AND OBJECTS FOR STUDY

Two 146 mm x 146 mm densified $Al_2O_3$ tiles ($\rho$ = 3.93 g/cm$^3$), one 6 mm thick and one 23 mm thick, see Fig. 1, were used in this study.

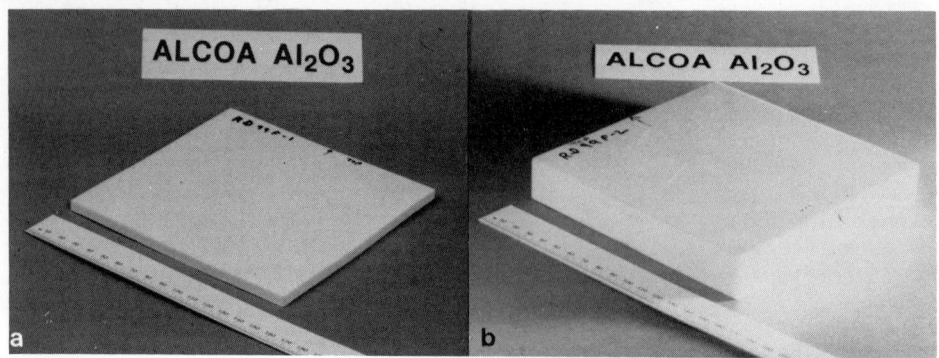

Fig. 1: Photograph of sintered $Al_2O_3$ ceramic tiles used for the tests. Tile sizes are 146 mm x 146 mm, with thicknesses of (a) 6 mm and (b) 23 mm.

CT tests were conducted with a first generation (translate-rotate, single detector) scanner built and operated at the Chalk River Nuclear Laboratories. The scanner consists of a gamma-radiation source (12 Ci Co-60), a CsF detector and associated electronics, a mechanical unit that rotates and translates the test object relative to the source-detector axis, and a computerized data acquisition and control system. The gamma-ray beam is collimated equally from both detector and source sides. The spatial resolution of the system is defined by the FWHM (full width at half maximum) of the beam, which can be changed to match specific applications. The performance of this scanner and its parameters have been described in detail elsewhere [13,14].

The CT images were obtained using projections sampled twice per beam width, so that the pixel (spatial element of the image) size was half of the spatial resolution, with the spatial resolution equal to the FWHM of the beam. For each image, 224 rays (beams) separated by one half of the beam width were measured for each of 360 projections. The projections were spaced 0.5° apart and spanned 180°. Two of the images were obtained using a spatial resolution of 1.92 mm FWHM (i.e., 0.96 mm pixel size), and another one with a FWHM of 1.42 mm (i.e., 0.7 mm pixel size). The CT 'slice' thickness was 1.9 mm (i.e., the FWHM of the beam in the plane of the image). The images were reconstructed using the filtered back-projection method of linearly interpolated projections with a Ram-Lak filter [6,15], and then represented in matrices of 256x256 pixels (a 224x224 image plus borders), with the pixel size equal to the ray spacing. The pixel-to-pixel statistical noise ranged from 0.5% to 1%. The use of a Co-60 photon source, which has a spectrum close to monoenergetic, assured that the images are free from beam hardening effects.

Additional non-destructive examination of the same tiles was performed by means of low-kV contact radiographic imaging, using a Picker 110 Hot-Shot X-ray imaging system. The system was operated at 70 kV and 8 mA, with a 30 inch source-to-film distance, type M-8 Ready Pack film, and a 0.002 inch Pb screen to reduce radiation scatter; the resulting film densities were 3.2.

## EXPERIMENTAL RESULTS: 'THIN' TILE

Low-kV radiographs of the 6 mm thick ('thin') tile (see Fig. 2) suggested that this tile had three regions with either somewhat lower density or a change in thickness. Thickness measurements showed no appreciable change in dimension and thus the change in optical film density indicated regions of lower density of the ceramic. From the radiographs one cannot determine the absolute density to quantify the density decrease in the three regions, nor can one determine how far they extend across the tile.

CT scans for this tile were made in two planes, one parallel and one normal to the large (146 mm x 146 mm) face. Fig. 3 shows a scan taken parallel to the large face at the midplane. To expose details in density variation, this image is presented using a thresholded scale, so that the dynamic range of the scale covers only about 2% of the nominal density (i.e., densities below 3.85 g/cm$^3$ are presented as black, above 3.93 g/cm$^3$ - white, and the densities in between these limits are in various tones of gray). The three lower density regions (A, B and C) observed in the radiograph are clearly observed.

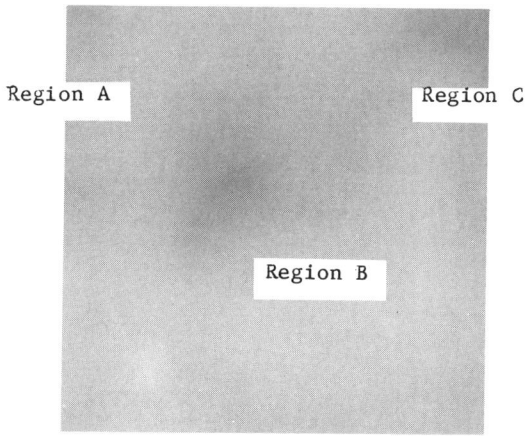

Fig. 2: Low-kV contact radiographic image of the 6 mm thick ('thin') tile.

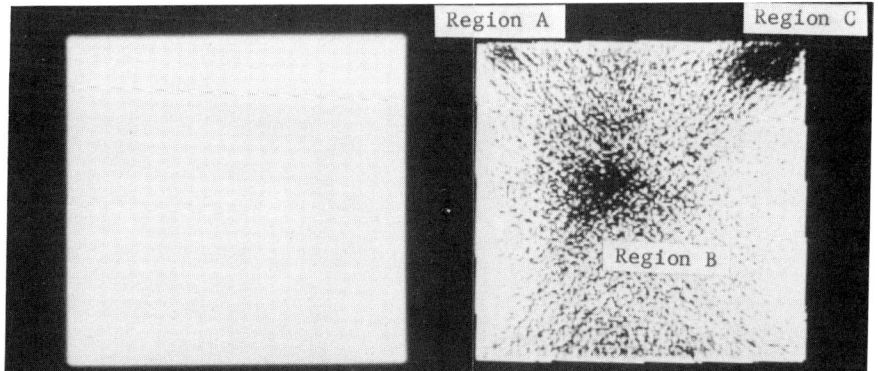

Fig. 3: CT image through the midplane of the 6 mm thick tile. The image is presented using two different linear scales: covering the whole range of densities from 0 g/cm$^3$ to 3.93 g/cm$^3$ (left image) and thresholded to encompass about 2% of the average specimen density.

To better illustrate the spatial distribution of the densities disclosed in the CT image, density profiles were plotted. Fig. 4 shows some examples of the density profile graphs, plotted along horizontal lines crossing regions A, B and C. The scale in the profiles covers a range within 20% of the maximum density (~3.3-4.1 g/cm$^3$). The three (A,B,C) regions are seen as corresponding dips in the graphs, from which the size of these regions can be determined, as well as the value of the density within each pixel. At the bottom of Fig. 4, are shown density profiles across the tile outside of the A, B, C regions. The vertical profile (the lowest graph) shows a scattering of the data around the average value, which reflects the statistical pixel-to-pixel noise of the image. In the horizontal profile, apart from the noise, one can also see that the density in this direction is not uniform and varies slightly but steadily from the edges to the center.

The density within each pixel can be determined to the limit imposed by the noise of the image. One can also determine the density averaged over an area enclosing a number of pixels, which can be used to decrease the error in the density determination. The average density obtained from the CT image in various locations of the tile outside of the A, B, C regions is 3.92-3.93 g/cm$^3$, which agrees perfectly with the nominal density of the tile determined by weighing. In regions A, B and C, the average densities determined from the CT image over an area of 10-20 pixels were 3.87 g/cm$^3$, 3.81 g/cm$^3$ and 3.84 g/cm$^3$, respectively, with an uncertainty of less than ±0.01 g/cm$^3$. The three low-density regions have therefore a density that is lower than the nominal density of the tile by, correspondingly, 1%, 3%, and 2%. Very close to the A, B, C regions the density is 3.91 g/cm$^3$, slightly below the nominal density of the tile. Because the tile material is chemically uniform, one concludes that the A, B and C regions indicate porous regions, with an average porosity of about 1%, 3% and 2%, for the three regions.

To examine the density variation through the thickness of the tile a cross-sectional CT scan was made across region B, along the scan line shown in Fig. 4, middle image. For this scan the pixel size was 0.7 mm. Fig. 5 shows the image of this cross section (6 mm x 146 mm), and the density profile along a horizontal line through the center of the image. The data show that the low-density B region extends throughout the thickness of the sample and is not just in the midplane displayed in Figs. 3,4.

EXPERIMENTAL RESULTS: 'THICK' TILE

Visual examination of the 23 mm thick tile showed significant cracking (see Fig. 1b). Not known was the depth to which the cracking extended, or whether the cracking was caused by a significant density gradient in the specimen. Before CT imaging this tile, low-kV contact radiographic images had been obtained (Fig. 6). Again cracks are detected, but the depths are unknown. A CT image was obtained at the midplane of the 23 mm thick tile, as for the 6 mm thick tile. Fig. 7 shows the CT image using two different gray scales, one that covers the full density range and another one that covers about 10% of the maximum density (i.e., densities below 3.7 g/cm$^3$ are black, above 3.95 g/cm$^3$ are white). The density profiles are shown in Fig. 8.

From the CT image, densities in various regions of the tile were calculated as before. The density is generally higher at the edges of the tile, 3.93 g/cm$^3$, and decreases towards the centre, forming two large, well separated areas of lower density (3.75 g/cm$^3$ at the rims of these

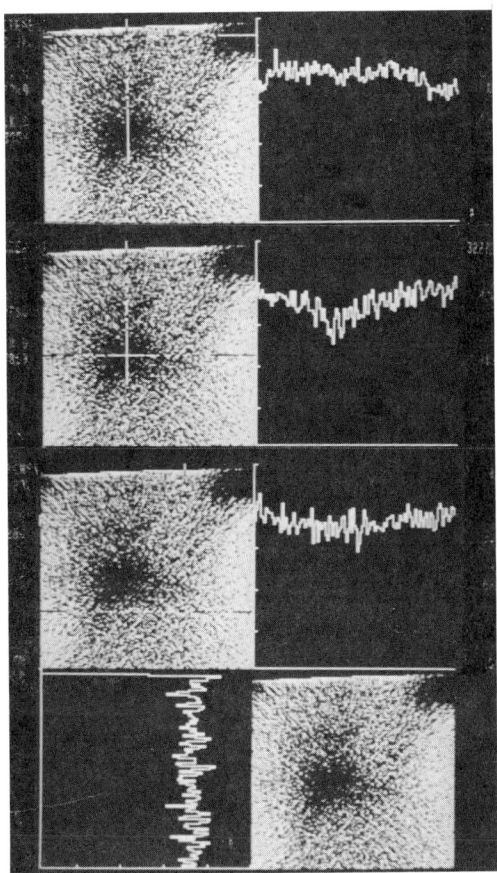

Fig. 4: Density profiles of selected locations on the 6 mm thick tile. The profiles are drawn along the lines superposed on the image and cross the low-density regions, A and C (cf. upper right graph) and B (middle right graph). The bottom image is shown with two density profiles outside the A, B, C regions, the horizontal (lower right graph) and vertical (lower left).

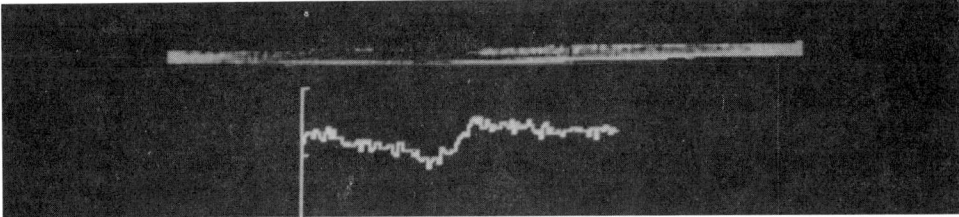

Fig. 5: CT image of the cross-sectional CT scan (i.e., taken along scan line shown in Fig. 4, middle image). The gray scale presenting the densities in the image is thresholded to cover only about 3% of the maximum density. The horizontal density profile across a part of the image (graph below) shows the extension of the low-density region. The scale of the graph covers 8% of the density range ($\sim$3.7-4.1 g/cm$^3$) which allows one to show both the low density B region and a long range, small ($\sim$1%) density gradient along the tile cross section.

regions and 3.67 g/cm³ in their centers); these two regions are separated by an S-shape ribbon of denser material. The high density ribbon is quite narrow in the center of the tile (5 pixels, i.e., 5 mm or less), and it widens towards the edges of the tile to about 10 pixels (9 mm) at one end and to about 22 pixels (21 mm) at the other end (respectively right and left sides in the image in Fig. 7), at positions where it merges with the denser, edge regions (i.e., about 1 cm from the edge of the tile). At the side of this larger widening there is a crack, seen in the CT image as running along the high density ribbon, roughly in the middle, for about 4 cm. The crack width is seen in the CT image as 3 pixels wide, but in reality it is narrower than the width of one pixel. The real width of the crack was determined from the CT image by assessing the decrease in the value of the absorption coefficient in the area enveloping the crack relative to the area without the crack, and is estimated to be about 0.4 mm.

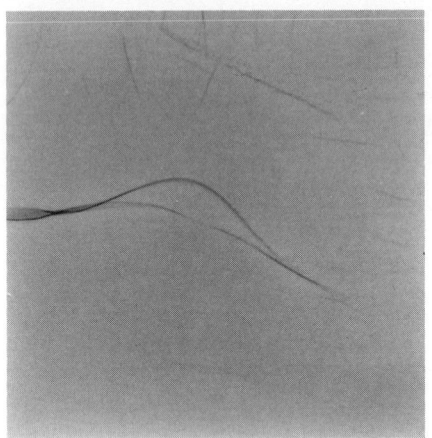

Fig. 6: Low-kV contact radiographic image of the 23 mm thick tile.

 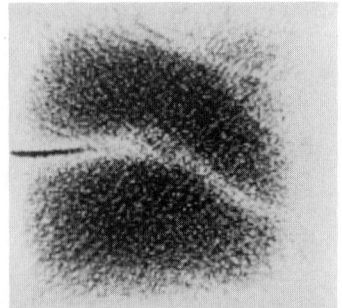

Fig. 7: CT image of the 23 mm thick tile, displayed using a scale that covers a density range 0-4 g/cm³ (left image) and a density range from about 3.7 g/cm³ to 3.95 g/cm³ (right image).

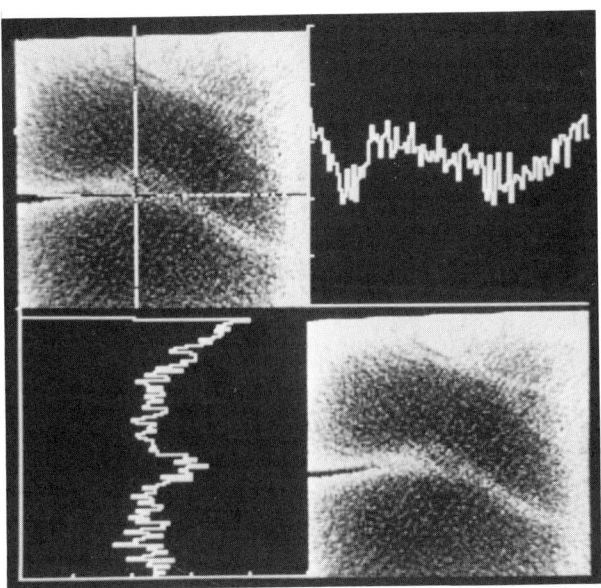

Fig. 8: Density profiles for the CT image presented in Fig. 7. The image is repeated twice, the upper one with crosshairs along which horizontal and vertical density profiles are shown in the upper and a lower graphs respectively. The scale of the graphs covers the range of 20% of the average specimen density, i.e., ~3.3-4.1 g/cm$^3$.

From the CT image presented one cannot determine precisely where the crack ends. However, the low-kV contact radiograph shows this quite well (see Fig. 6). In the CT image the crack is clearly seen as long as the dense 'ribbon' around it is wider than several pixels, i.e., a length of about 4 cm. Where the 'ribbon' narrows to less than a width of 3-4 pixels, the crack cannot be seen, probably because its artificially blurred-out width coincides with the width of the high density 'ribbon'. The 'ribbon' density determined from the CT image is from 3.75 to 3.80 g/cm$^3$ in various places. If the crack extends further than 4 cm through the 'ribbon', then this value is artificially lowered in the CT data by the presence of the crack, and the real density is higher (e.g., if the crack runs through the narrow part of the 'ribbon' then the density in this region would be ~3.9 g/cm$^3$ for a crack width of about 0.2-0.4 mm). To see how far the crack extends and to confirm the above numerical data using CT imaging will require higher spatial resolution at similar contrast.

A second crack is just detectable over the noise in the CT image, in the lower density region above the high density 'ribbon'. A third crack appears to be present near the upper edge of the upper low density region, associated with another high density 'ribbon'.

A comparison of the CT image (Figs. 7,8) with the low-kV contact radiograph (Fig. 6) shows several interesting features. First, the three cracks observed in the CT image are more clearly observed in the radiograph (indicating that a higher CT spatial resolution is desirable). The tomograph locates the cracks in the mid-plane. Second, the low-kV radiograph is not as sensitive to density gradients as the CT image which shows that the cracks lie along high density 'ribbons'. Third, the small cracks in the radiograph are not visible in the CT image. Either these cracks are not located in the midplane or the resolution of the CT image is too low.

CONCLUSIONS

Computed tomography (CT) using a high energy (1.25 MeV) photon source has been employed to image two specimens of sintered alumina ceramic. Because a monoenergetic source was used the absolute values of the densities could be measured to a high accuracy. High contrast (low noise) CT images were obtained permitting density measurement within each pixel to be determined with an accuracy of better than 1%. The average densities in regions encompassing several pixels have been determined with an accuracy of 0.2%. Porous regions and long-range density gradients were observed and quantified. Higher spatial resolution and still better contrast of the CT image could be obtained, however at the expense of the scan time. Good agreement was obtained for the two imaging techniques (CT and contact radiography), although both techniques showed slightly different features in the objects under study and therefore complement each other.

ACKNOWLEDGEMENT

The authors wanted to acknowledge the cooperation of Dr. John Weyand, ALCOA Inc., in providing the samples, and P.W. Reynolds for technical assistance in performing CT scans, and C.J. Allan for his interest and support in this work.

REFERENCES

1. D.R. Johnson, A.C. Schatthauser, V.J. Tennery, E.L. Long (Jr), and R.B. Schultz, "The Ceramic Technology for Advanced Heat Engine Project", Proc. 22nd Automotive Technology Development Contractors, published by the Society of Automotive Engineering, Warrendale, PA, March 1985 (Report ISBN 00-89883-712-2, pp. 369-377).
2. A.R.C. Westwood and J.P. Skalny, J. Advanced Ceramic Materials, 1, 21 (1986).
3. W.A. Ellingson, R.A. Roberts and M.W. Vannier, Proc. XVth Symposium on Nondestructive Evaluation, South-West Research Institute, San Antonio, Texas, 1985.
4. G.T. Herman, "Image Reconstruction from Projections", Academic Press, N.Y. (1980).
5. R.A. Brooks and G. Di Chiro, Phys. Med. Biol. 21, 689 (1976).
6. A.C. Kak, Proc. IEEE 69, 1245 (1979).
7. T. Taylor, W.A. Ellingson and W.D. Koenigsberg, Atomic Energy of Canada Limited report AECL-9005 (1985), and Ceramic Eng. and Science Proc. (in print, 1986).
8. E. Segal and W.A. Ellingson, to be published in the Proc. of the 2nd International Symposium on the Nondestructive Characterization of Materials, Montreal, July 1986, Plenum Publ.
9. P.D. Tonner, G. Tosello, D.S. Hall, L.R. Lupton and B.D. Sawicka, Atomic Energy of Canada Limited report (in preparation).
10. E. Storm and H.I. Israel, Nucl. Data Tables 7, 565 (1970).
11. W.M.J. Veigele, Atomic Data Tables 5, 51 (1970).
12. D.A. Chester, S.J. Riederer, N.J. Pelc, Journal Comp. Assist. Tomography 1, 1 (1977).
13. P.D. Tonner and G. Tosello, Materials Evaluation, 44, 203 (1986).
14. T. Taylor and L.R. Lupton, Nucl. Inst. & Meth., A242, 603 (1986).
15. G.N. Ramachandran and A.V. Lakshminarayan, Proc. Natl. Acad. Sci., USA 68, 2236 (1971).

DISCUSSION

Mr. Notea: I would like to refer to the optimization that you showed at the beginning of your talk. This optimization is correct only for cylindrical objects. Generally the function depends on the shape of the object, and also on the number and shape of the holes and the inclusions inside the object. In the Moscow meeting in 1982 it was shown that the optimum energy might be not for $\mu \cdot x = 2$ but for $\mu \cdot x$ equal 3 or 4.

Ms. Sawicka: I agree that the relation between the noise of the CT image and the product of $\mu$ times x (absorption coefficient times object thickness) depends on the shape of the object. Also the noise varies across the cross section of the object, depending on the object shape, which I neglected in my discussion. But these factors do not change the noise estimate greatly, which means that although the noise estimate from my curves may be not exact in all the cases, it is good, let us say, to within a factor of 2 or 3, and certainly within less than an order of magnitude. On the other hand, the improper choice of the energy in the case of dense objects can increase the noise by orders of magnitude, even many orders of magnitude. If this increase is not too large, up to 2 to 4 orders of magnitude, it can be compensated for by the use of much stronger sources, but there is a limit when this is possible. This is what my curves were supposed to show.

From the Floor: What is the thickness of the section that you are imaging?

Ms. Sawicka: The "in-slice" thickness in our current scanner can be changed depending on the specific application. It is determined by the half maximum of the beam width and therefore by the source size and the aperture of the collimators. For the images that I have shown here, the "in-slice" thickness was about 1 mm for the case of the green alumina pellets, and 2 mm for the CT images of the sintered tiles.

From the Floor: What is the size of your isotopes?

Ms. Sawicka: The geometrical size of the sources?

From the Floor: Yes.

Ms. Sawicka: Sources are delivered packed in small containers. The Co-60 source of required activity consists of many small pellets packed into a cylindrical container, with the circular base having the diameter of 4 mm; in our scanner this circular end is facing the detector. Ir-192 consists of several pellets stacked one behind the other, and the active area is again circular with the diameter of 3 mm. The size of the active area imposes a limit on the beam width used for the CT scans and on the possible "in-slice" thickness.

Mr. William Friedman, Standard Oil: High-aspect ratio objects sometimes produce artifacts inside the objects. Are certain generation scanners better for looking at those types of objects? Does the arrangement of the detectors and the source tend to reduce those problems?

Ms. Sawicka: It seems that the first generation scanners, which have a pencil beam and one detector, have better defined beam geometry and therefore should have less artifacts than scanners that use a fan beam and many detectors. In multidetector systems one has to very carefully suppress the scattering which may cause artifacts. If this is properly done, perhaps there is no preference. Does anybody want to comment on this?

From the Floor: With one detector you have less scattering, so perhaps for a high-aspect ratio object, upon the significance of the scattering (cross-talk between the detectors), a first generation machine might be better. On the other hand, multidetector machines are much faster, so you gain the speed.

Mr. Segal: If you have 10 detectors instead of one, the speed is 10 times better.

Ms. Sawicka: You gain the speed but you do not decrease the artifacts, and, in machines of higher generations, perhaps increase the number of artifacts. The artifacts may arise due to both scattering between the detectors (cross talk) and scattering in the object, and both can be larger in higher generation scanners in comparison to the first generation machines. Probably the second generation scanner would be a good option.

Mr. H. Ringermacher, United Technologies Research Center: What is a typical time of acquisition of all the data you need?

Ms. Sawicka: I waited for this question. (Laughter). How many detectors?

Mr. Ringermacher: I knew you waited for it.

Ms. Sawicka: The acquisition time depends on many factors, namely the noise you allow in the CT image (whether you want a high contrast or a low contrast image), the number of detectors and the source strength you use in your scanner, the number of rays and projections you take to generate the image and the required spatial resolution. The images which I have shown today were measured with high contrast (low noise). In the case of the green pellet, the acquisition time per ray was 8 sec., and the total time was 8 sec times 90 projections times 64 rays, which gave 12.8 h, because one-detector system was used. The first generation scanner is not fast! Using 64 detectors, the acquisition time could have been reduced to 12 minutes. Our source at the time of these measurements was not very strong. We could have used a source about 7 times stronger, which would give you a 2 minute scan. This is assuming you have 64 detectors. If you are using one detector, the time is 64 times longer.

Mr. Ringermacher: In the case of the plate what did it take to do?

Ms. Sawicka: In the case of the plate the image was generated using 224 rays and 360 projections, which--with the use of 224 detectors which is quite possible to be implemented--the acquisition time would be 14 minutes. The source was Co-60, and we could have used Ir-192, which is stronger and would reduce the time by about a factor of 10, which gives less than 2 min. Again, one detector gives you the time...

Mr. Ringermacher: 224?

Ms. Sawicka: Yes, that many times longer. One more remark. The acquisition times that I quoted were required to generate the images with the noise much lower than 1%. In most cases, images having the noise 3 times larger are sufficient, and those can be generated in a time 9 times shorter in comparison to the quoted numbers.

# APPLICATIONS OF FILM TOMOGRAPHY TECHNIQUE FOR ONDE

A. Notea

NDE Lab., Department of Nuclear Engineering
Technion - IIT
32000 Haifa, Israel

INTRODUCTION

Tomography with X- and gamma- rays provides three-dimensional radiographic information on the examined object. The film-based tomography (1,2) generates a summation-image of a surface within the object by continuously combining back projections directly on the film. This method has many attractive features for industrial applications in which cost and simplicity are of primary importance. Some of the features are:

(a) The absence of post processing allows this method to yield an image immediately on development of the film.

(b) Conventional radiologists need a short training time to master the technique as most components and concepts are familiar to them: e.g. radiation sources, films, screens, collimators, filters, processing units, viewers, exposure, contrast, resolution.

(c) Purpose oriented system optimized for a certain range of products, may be built with costs much less than digital computing transaxial tomography (CT).

(d) Tomographic images of surfaces of critical areas within the examined object are directly recorded on films curved to match the required shape.

(e) Tomographs on film may be digitized and imaged-processed by commercial system developed for conventional radiographs.

(f) The slice thickness of the recorded surface may be in the order of magnitude of the thickness of the film-emulsion.

(g) The quality of the tomograph is high especially for high- contrast objects (2) and whenever the noise and the dynamic-range of the film do not impose a limitation on the information to be extracted.

In the present study the possibility to record images of surfaces within the object was studied and the response function was modeled.

## METHOD

The method is based on synchronous rotation of the object and the film holder. Both axes of rotation are in the same plane as well as the radiation source. The radiation beam is perpendicular to the rotation axis. The radiation fan beam covers the full width of the object. The thickness of the slice covered by the beam can be varied according to the area of interest in the object. The source-object distance is adjusted so that the cone beam geometry does not introduce significant distortions in the volume under inspection. This method in principle was suggested over forty years ago (3).

In this system every point in the inspected volume of the object corresponds to a point in the generated three-dimensional tomographic image and both points remain relatively stationary. Each point to be imaged is passed by the radiation from all directions (0 to $\pi$). A film placed within the 3D tomographic image records the corresponding surface within the object.

The analytical approach for noise-free continuous summation on the film is given by the following transforms. The spatial distribution of the radiation attenuation in the examined volume may be expressed by

$$g(x,y,z) = \mu_E(x,y,z)\rho(x,y,z) \tag{1}$$

where $\mu$ is the mass attenuation coefficient ($cm^2/g$) for a given energy, and $\rho$ is the material density ($g/cm^3$) at $x,y,z$. The interogate radiation passes the x-y planes in the examined volume and the z axis is parallel to the symmetric axis of rotation.

The g function may be expressed in coordinates defined by the penetrating radiation $s,u,z$ (see Fig. 1). The projection P along the u direction for a certain $\theta$ is solely a function of s, and is given by the Radon transform (4) $[Rg]$ (5).

$$P(s,\theta,z) = \int_0^{E_{max}} S(E) \exp\{-[Rg](s,\theta,z,E)\} dE = \tag{2}$$

$$= \int_0^{E_{max}} S(E) \exp\{-\int_{-\infty}^{\infty} g(s\cos\theta - u\sin\theta, s\sin\theta + u\cos\theta, z, E) du\} dE$$

$$s \geq 0, \ 0 \leq \theta < 2\pi$$

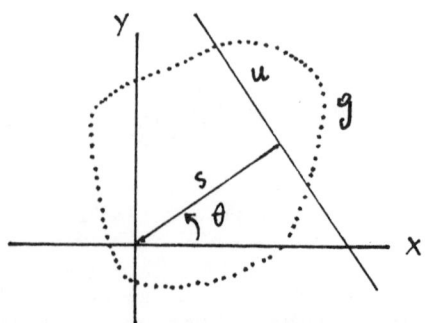

Fig.1. Coordinate Systems.

where S(E) describes the X-ray energy spectrum and the film spectral response.

The image generated by the continuous summation on the film is expressed by the unfiltered integration over $\theta$, i.e. by the back-projection transform $[\mathcal{B}P]$ (5).

$$I(x,y,z) = [\mathcal{B}P](x,y,z) = \int_0^\pi P(s,\theta,z)d\theta . \qquad (3)$$

The tomographic image I differs from that obtained by the CT, as in the later method the $\mathcal{B}$ transform is performed on $[Rg]$ after applying an appropriate filter. The film placed within the the 3D image along a surface $F(x,y,z) = C$ will record a two-dimensional image representing the intersection of I with F.


MEASURED IMAGES

The object used for the demonstration was a ceramic hand-made jar with a defect in the neck area (see fig.2). The tomographic images were generated on Agfa-Gevaert Structurix D7 films with industrial X-ray unit (Andrex, 2.9 mm. focus) at 150 KV and source to jar distance of 310 cm.

The following "cuts" are presented:

i)  horizontal planar tomogram at the neck through the damaged area (location P in fig.3) is presented in fig.4.

ii) planar tomograms at $45°$ in the neck area (location A and B of fig.3) are presented in fig.5.

iii) vertical planar tomograms (parallel to the symmetric axis of the jar) along the diameter and off-center (see fig.6).

iv) vertical cylindrical tomograms generated by two concentric film cylinders whose symmetric axis coincides with that of the jar. The smaller cylinder (C in fig.4) cuts within the neck's wall where the defected area is observed, and through the base (see fig.7a). The larger cylinder (D in fig.3) cuts through the walls of the spherical volume (see fig.7b). As is seen the two bands of the intersections with wall are not parallel i.e. the perimeter of the jar is not circular. Both films were exposed simultaneously.

The resolution was measured using a perspex block 20 by 45 by 200 mm. with fifteen slits of 10 mm. depth and widths: 10;5;3;2;1.5;1.2;1;0.8;0.6; 0.5;0.4;0.3;0.2;0.1 and 0.05 mm. The tomogram reveals the slits down to 0.2 mm., but as is expected (6,7), the image contrast reduces with the width. The limit is achieved with 2.9 mm. focus and 1.19 magnification i.e. about 0.5 mm. geometric unsharpness.

Fig.2. The examined jar

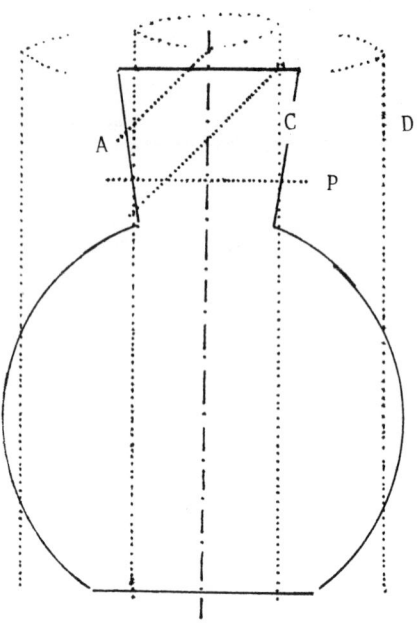

Fig.3. Locations of films.

Fig.4. Horizontal planar tomogram at P, fig.3.

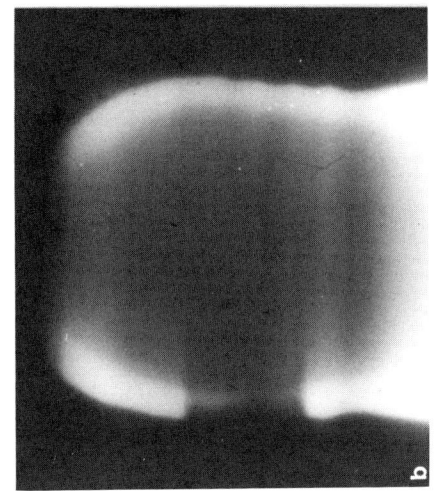

Fig. 5. Planar tomograms at 45° in the neck area
(a) at A, fig.3. (b) at B, fig. 3.

437

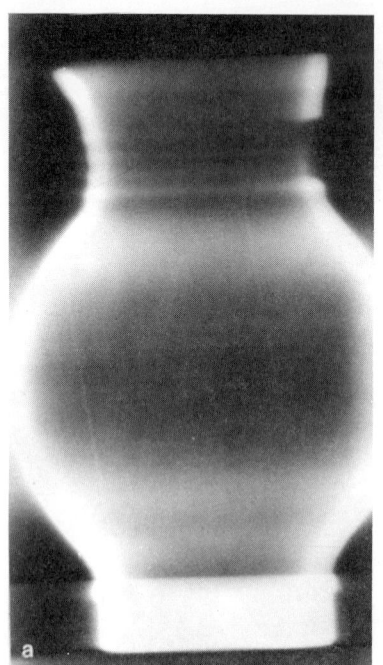 

Fig.6. Vertical planar tomograms

(a) along the diameter  (b) off-center

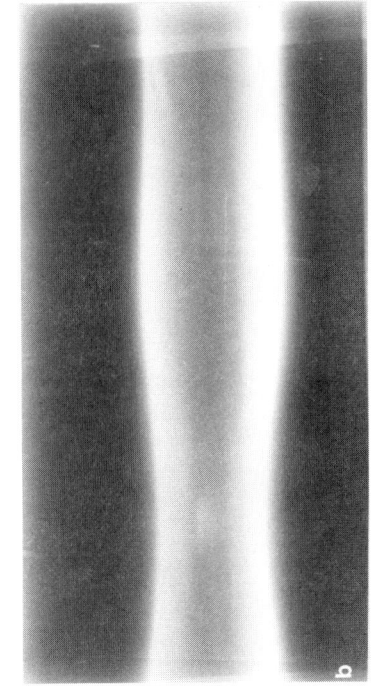
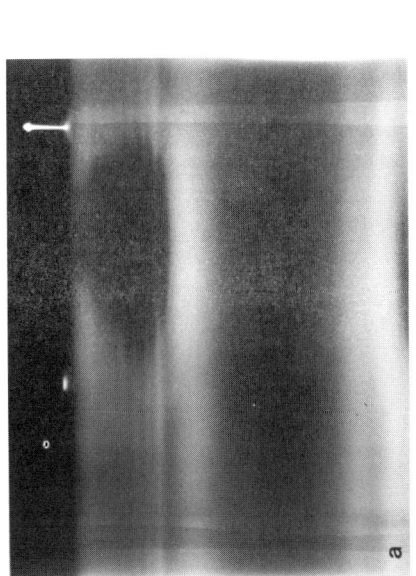

Fig.7. Vertical cylindrical tomograms

(a) cylinder at C in fig.3
(b) cylinder at D in fig.3.

CONCLUSIONS

The study has shown that it is possible to obtain 2D tomographic images of planar and curved surfaces within the object under investigation. Thus critical areas may be examined with films exposed simultaneously to the X-ray beam.

The film tomograph method can yield satisfactory geometrical resolution using X-ray units of small focus and relatively accurate motion mechanism.

In applying digitization and image enhancement techniques, the advantage of photographic film as a detector becomes apparent due to the spatial resolution achieved, which is far better than with CT systems.

ACKNOWLEDGMENT

The author thanks Ing. K. Rudich for his assistance in the system design and the measurements.

REFERENCES

1. A. Notea, "Film-based Industrial Tomography", NDT International, 18, 179-184 (1985).
2. G. Thuesen, A. Lindegaard-Andersen, "Grazing Incidence Tomography", Phys. Med. Biol. 25, No. 6 1049-1058 (1980).
3. W. Watson, "X-ray Apparatus", U.S. Patent 2 196 618 (Apr. 9, 1940).
4. J. Radon, "Uber Die Bestimmung Von Funktionen Durch Ihre Integralwerte Längs Gewisser Mannigfaltigkeiten", Ber. Verh. Saechs. Akad. Wiss. (Leipzig) 69, 262-278 (1917).
5. S. W. Rowland, "Computer Implementation of Image Reconstruction Formulas", in Image Reconstruction from Projections, G. T. Herman, ed., Springer-Verlag (1979).
6. E. Segal, A. Notea, Y. Segal, "Dimensional Information Through Industrial Computerized Tomography", Materials Evaluation 40, 1268-1279 (1982).
7. A. Notea, "Evaluating Radiographic Systems Using The Resolving Power Function", NDT International 16, 263-270 (1983).

# NDE OF POLYMER COMPOSITES USING MAGNETIC RESONANCE TECHNIQUES

Wayne A. Bryden and Theodore O. Poehler

Applied Physics Laboratory
Johns Hopkins University
Laurel, MD 20707

Polymer based materials have become increasingly important in structural applications primarily due to their high strength to weight ratio. As the use of polymer-based composites has increased, so has the need for reliable non-destructive evaluation techniques. In this paper, a new NDE method for these materials is proposed. The technique relies on the observation of an electron paramagnetic resonance (epr) absorption at the site of damage in a polymer. Using applied magnetic field gradients the physical location of damage can be discerned and an image of the damage site can be obtained. This should allow the detection of cracks and delaminations with high resolution, good sensitivity and good contrast.

The foundations for epr spectroscopy rest on the interaction of unpaired electron spins with an applied magnetic field. In zero field the energy levels of the "free" electron are degenerate. As a field is applied the energy levels are Zeeman split with the energy difference increasing linearly with the strength of the field (see Fig. 1(a)). The substance with "free" electrons is placed in a microwave system of fixed frequency and the magnetic field is swept; as the energy difference of the electron states is increased it will at one point match the energy of the applied microwave field. At that point a resonant absorption of microwave energy will result (see Fig. 1(b)). The equation governing the absorption is $\Delta E = g\beta H_o = h\nu$ where $\beta$ is the Bohr magneton, $H_o$ is the resonant magnetic field, $h$ is Planck's constant and $\nu$ is the microwave frequency. The resonance is characterized by the g factor which gives information as to the nature of the free radical, the linewidth $\Delta H_{pp}$ which is informative as to the dynamics of spin relaxation and the integrated intensity I which is proportional to the number of spins. In practice, the magnetic field is modulated, phase sensitive detection is used and the first derivative of absorption is displayed. The result for the simplest case is shown in Figure 2.

It has previously been shown [1,2,3] that epr signals are generated when polymeric materials are damaged. The mechanism for generating free spins is the breaking of polymer bonds. For each bond cleaved, two free radicals are formed. These species by nature are highly reactive and prone to recombination such that the intrinsic free radical has a very short lifetime. However, species that interact with the primary radicals

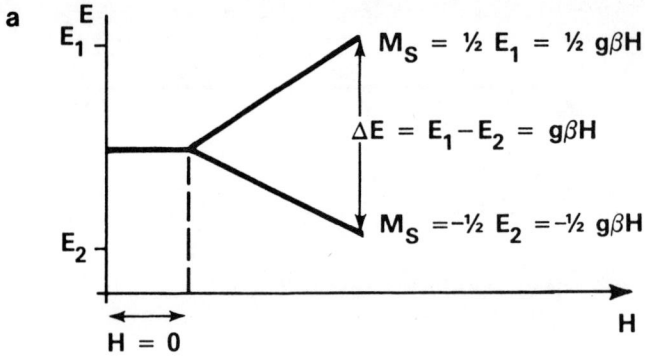
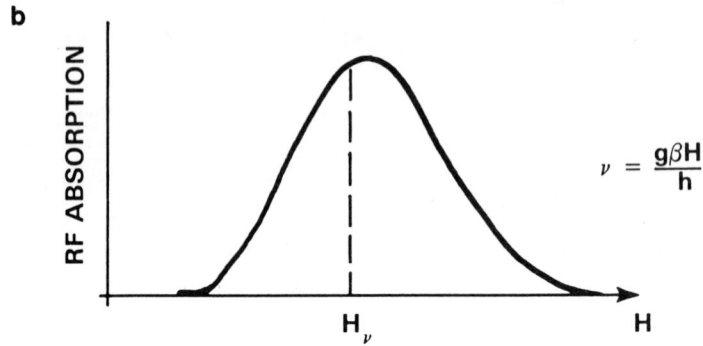

Fig. 1  a) Energy levels of a "free" electron in a magnetic field illustrating Zeeman splitting.
b) Microwave absorption for a free electron as a function of applied field.

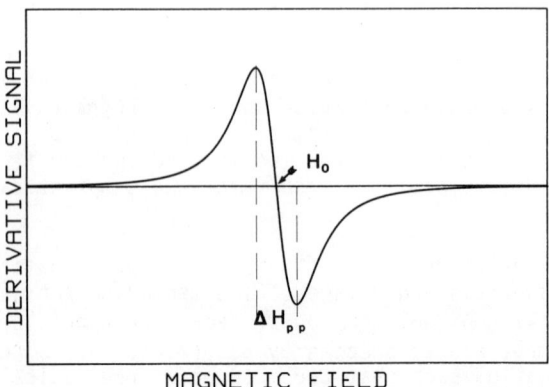

Fig. 2  EPR signal for an ideal free radical with Lorentzian lineshape.

often become paramagnetic and secondary or tertiary free radical species can be observed.

We have previously shown that stable epr signals could be generated in nylon that has been damaged by mechanical stress [2]. It was postulated that the free radicals were trapped by inorganic impurities present in the nylon, due to the high g-value and broadness of the line. This was borne out by later experiments where organic spin traps were added to the fibers. In this case all of the free radicals were preferentially trapped by the strong electron acceptor that was added. This led to a very well conditioned epr signal, with the integrated intensity tracking with the stress on the fibers [3].

These results allowed the quantification of the integrated (over time and space) amount of damage in a polymer sample. This did not give information, on a local scale, as to the topography of the damage. A single large crack would, using this technique, give the same signal as the equivalent number of small cracks. It was therefore necessary to impart spatial sensitivity to the epr signal.

This is achieved by adding magnetic field gradient coils to the instrument. The effect of these coils is to add a linear magnetic field gradient to the sample such that the applied field becomes $H = H_o + Gx$ where $H_o$ is the applied field without the gradient, G is the gradient strength and x is a positional variable. The analogous techniques using nuclear magnetic resonance (NMR) have become widely used in the biomedical community since the first report of NMR imaging by Lauterbur [4]. We feel that epr techniques have significant advantages over NMR techniques in solid samples.

To illustrate the technique, samples were made with various spin phantoms. Small (~ 100 micron) spots of the stable organic free radical DPPH (Fig. 3) were placed on a quartz holder in the epr cavity. The quartz holder was attached to a goniometer to allow accurate values of rotation angle. The dashed line of Figure 4 shows a spectrum taken of a simple two spot phantom without an applied gradient. The separation between the two spots is ~ 3 mm and one spot is about twice as large as the other. This spectrum is similar to Fig. 2 and is what is typically observed for solid state free radicals. The solid line of Figure 4 shows the effect of an applied field gradient of 9.5 G/cm on the sample. The single peak has split into two peaks due to the two sites of spin density. The spectrum shows two clearly resolved lines with intensity ratios of approximately 2:1. The separation in magnetic field between the two peaks can be related back to the spatial separation by using the known field gradient. The separation between the peaks in a sample is dependent on the angle ($\theta$) that the sample plane makes with the field gradient direction. In Fig. 5 the peak separation is plotted as a function of cos $\theta$. The straight line obtained indicates that the projection of spin density on the gradient axis is indeed generated by sample rotation. If a series of rotations is carried out, an image of the spin density in an object can be generated.

The generation of the spin density profiles in a real object can become somewhat complicated by the non-ideal lineshape exhibited by some samples. For example Figure 6 is an epr spectrum, with no gradient applied, of nylon damaged by X-ray irradiation. The complicated lineshape is due to the hyperfine interaction between the magnetic nuclei of the polymer and the electronic spins. It is obvious that the application of a field gradient to this would simply make a more complicated spectrum without giving significant information as to the distribution of damage in the system.

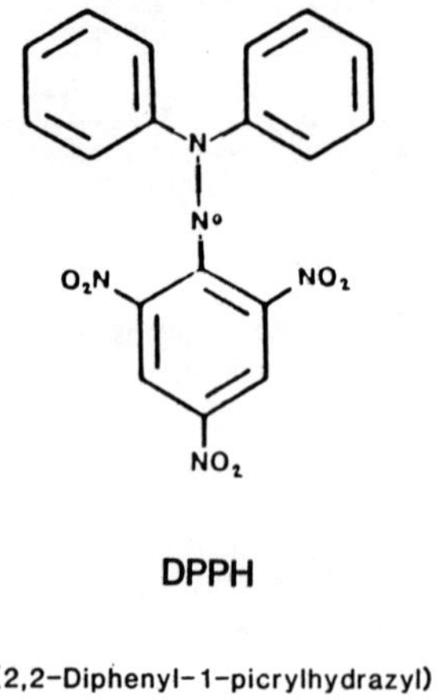

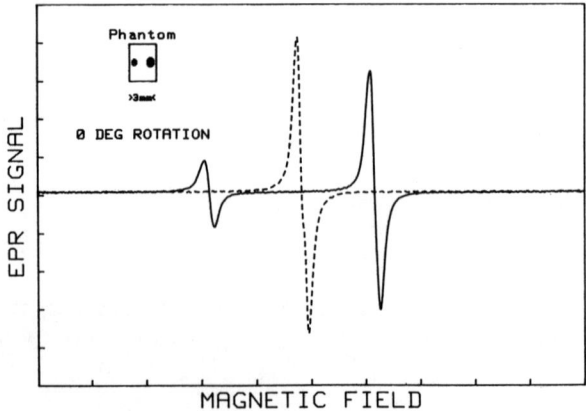

DPPH

(2,2-Diphenyl-1-picrylhydrazyl)

Fig. 3 Molecular structure of the stable organic free radical used to produce spin phantoms.

Fig. 4 EPR spectrum of spin phantom with (-----) and without (- - -) an applied field gradient.

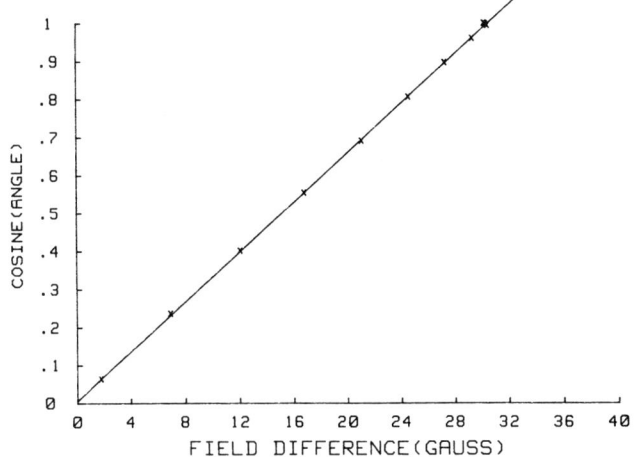

Fig. 5   Angular dependence of peak separation.

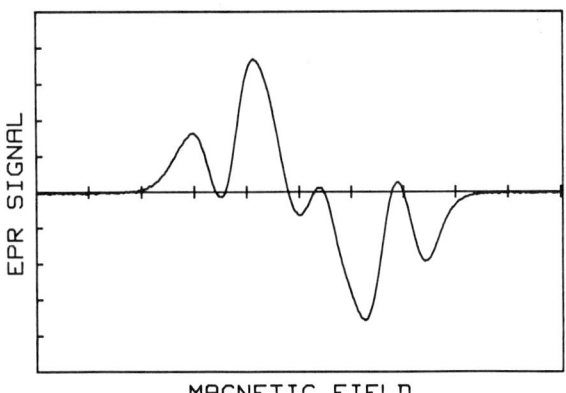

Fig. 6   EPR spectrum of X-ray damaged nylon.

This type of spectral complexity can be handled mathematically by using deconvolution techniques. One technique that is easy and fast to use is based on the convolution theorem of Fourier transforms. Mathematically the spectrum with gradient applied is given by

$$S_g(h) = \int_{-\infty}^{\infty} S_o(h-h') \rho(h') \, dh'$$

where $S_o(h-h')$ is the spectrum obtained in the absence of field gradient and $\rho(h')$ is the spin density in the object. The convolution theorem states that the Fourier transform of the spectrum with applied gradient is equal to the product of the Fourier transforms of the spectrum without the gradient and the spin density i.e.,

$$\mathcal{F}\{S_g(h)\} = \mathcal{F}\{S_o(h)\} * \mathcal{F}\{\rho(h)\}$$

From this it follows that the spin density can be obtained by taking the inverse Fourier transform of the quotient of the Fourier transforms of $S_G(h)$ and $S_o(h)$ i.e.,

$$\rho(h) = \mathcal{F}^{-1}\left\{\frac{\mathcal{F}\{S_g(h)\}}{\mathcal{F}\{S_o(h)\}}\right\}$$

This mathematical technique allows one to readily deal with complex lineshapes and to subtract out the intrinsic linewidth of simply-shaped lines.

After the spin density in the sample has been derived as a function of angle in one plane, an orthogonal set of field gradient coils are employed to measure spin density perpendicular to this plane. All of the projections of spin density are subsequently used in an image reconstruction process to generate a three-dimensional image of damage in a material.

The two classes of image reconstruction methods in the greatest use today are the algebraic reconstruction techniques (ART) [5] and Fourier methods [6]. In the algebraic methods the reconstruction space is divided into nonoverlapping elements and iterative techniques are used to match the sums of the reconstruction elements with the projection data. The Fourier methods make use of the projection slice theorem that states that the one-dimensional Fourier transform of a projection at an angle $\theta$ is equal to the Fourier transform of the original two-dimensional data evaluated along the angle $\theta$ in two-dimensional Fourier space.

The algebraic reconstruction technique has been used with a limited amount of data to generate the two-dimensional plot of spin density of the two spot phantom. This result is displayed in Figure 7. With improved angular resolution and enhanced image processing software the image quality will surely be improved.

In summary, the basis of the epr imaging technique for polymeric and composite materials is the creation of free radicals when a polymer is damaged. These free radicals are trapped by species that are added to the polymer so that a well conditioned signal is obtained. Field gradient coils are employed on an epr spectrometer to map out the spatial extent of damage in the material. Mathematical techniques are employed to lead to a technique that has good sensitivity, high resolution and good contrast.

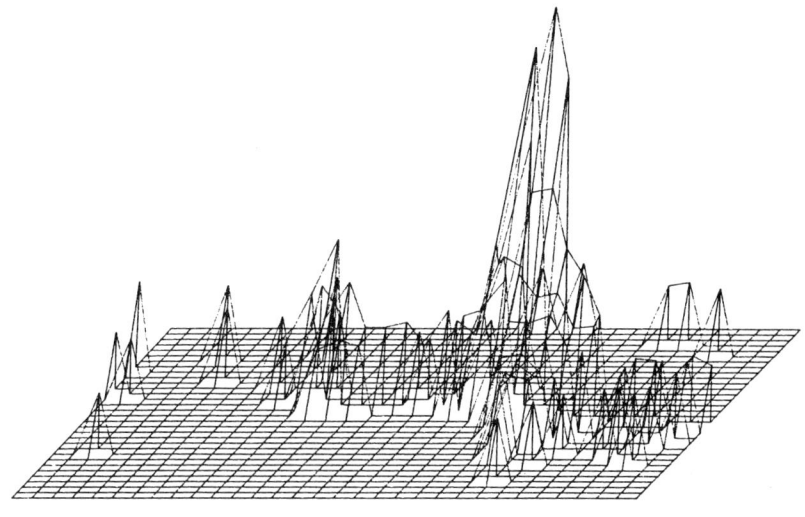

Fig. 7  Image reconstruction using Algebraic Reconstruction Techniques.

ACKNOWLEDGEMENT

This work was supported in part by the Naval Sea Systems Command under Contract No. N00024-85-C-5301.

REFERENCES

1. K. L. DeVries, D. K. Roylance and M. L. Williams, J. Polym. Sci. A1, 8, 237 (1970); K. L. DeVries, B. A. Lloyd and M. L. Williams, J. Appl. Phys., 42, 4644 (1971); T. C. Chiang and J. P. Sibilia, J. Polym. Sci. Polym. Phys. Ed., 10, 2249 (1972); M. Igarashi, J. Polym. Sic. Polym. Chem. Ed., 21, 2405 (1983).
2. W. A. Bryden and T. O. Poehler, in: "Proc. Eleventh World Conf. on Nondestructive Testing, November 1985," Taylor Publishing, Dallas (1985).
3. W. A. Bryden and T. O. Poehler in: "Review of Progress in Quantitative Nondestructive Evaluation 5B," D. O. Thompson and D. E. Chimenti, Eds., Plenum, New York (1986).
4. P. C. Lauterbur, Nature 242, 190 (1973).
5. R. Gordon, IEEE Trans. Nucl. Sci NS-21, 78 (1974).
6. R. M. Mersereau and A. V. Oppenheim, Proc. IEEE 62, 1319 (1974).

PROTON NMR IMAGING OF GREEN STATE CERAMICS


L. B. Welsh, S. T. Gonczy, R. T. Mitsche, and L. J. Bauer

Allied-Signal Engineered Materials Research Center
50 East Algonquin Road
Des Plaines, Ill. 60017-5016

Jay Dworkin and Anthony Giambalvo

FONAR Corporation
Melville, N. Y. 17746


INTRODUCTION

High performance ceramic materials in advanced technology applications are becoming of increasing importance. As a result, the necessity of finding new quantitative non-destructive evaluation (QNDE) methods for ceramics is becoming increasingly apparent. This paper explores the applicability of proton NMR imaging to the QNDE of ceramic materials. While proton NMR imaging is clearly well developed in the area of medical applications (1), only a few experiments have been performed to determine the applicability of this technique to the analysis of ceramic bodies (2). Compared to the NMR imaging of soft tissues for medical applications, the magnetic interactions of protons in solids or semi-solids make high resolution image generation more difficult. These interactions both broaden the proton NMR lines and shorten the spin-spin relaxation times. As a result, larger encoding magnetic field gradients and faster gradient switching are required of a NMR imaging system to produce high resolution, high signal-to-noise ratio images of solids.

In the plastic forming of ceramics, a wax or polymeric binder system is mixed with the dry ceramic powder, which is then molded, extruded, or rolled. After forming the green state ceramic piece, the binder is removed by thermal vaporization or pyrolysis. The physical and chemical homogeneity of the powder/binder mix has effects on the rheology, physical properties, and pyrolysis chemistry of the mix. Variations in these characteristics can produce defects in the ceramic piece during the fabrication steps. It would be valuable to be able to detect such inhomogeneities. While techniques such as computer X-ray tomography produce density maps of the green state ceramics, NMR proton imaging has the potential for producing maps of the proton's chemical state, in addition to maps of the proton density. The chemical map may provide information on the local viscosity of the binder and on the chemical interactions among the binder components and ceramic powders. That type of knowledge is very pertinent to development studies in ceramic processing. In the longer term, NMR imaging for that chemical information may also be usable for process monitoring in a production situation. It is this potential which motivated this preliminary study of NMR imaging of green state ceramics.

GENERATION OF PROTON NMR IMAGES

The images of the different types of ceramic samples were generated on either Fonar Corporation Beta-3000 or Beta-3000M proton NMR imaging systems. The Beta-3000 system uses a permanent magnet to produce the external static magnetic field of 0.3 Tesla (3000 gauss), while the Beta-3000M system uses a non-superconducting magnet to provide a magnetic field of the same strength. The resonant frequency of protons in this magnetic field is 12.7 MHz. Gradient coils produce switchable magnetic fields which are used both to select a region for imaging and to code the locations from which the NMR signals originate. This imaging system uses a processer which reconstructs and scales a digital 256 pixel x 256 pixel image in 8 seconds. An image with 2000 digital intensity levels is displayed at the conclusion of each scan.

Images of the ceramic disks were produced using a two-dimensional Fourier transform multi-slice spin echo technique, using echo delay times of 14 or 28 ms. The gradient multiplier feature of the imager was used to continuously vary the slice thickness down to 2 mm, with an interslice gap of 1 mm. In this technique, the slope of the magnetic field gradient normal to the selected plane of the slice was scaled up from its nominal value of 0.1203 Gauss/cm, which corresponded to a slice thickness of 7 mm for the standard bandwidth radio frequency pulse. The nominal 512 Hz difference between the center frequency of the transmitted rf pulses exciting adjacent slices remained unadjusted, allowing a proportional scaling of the slice separation. A similar scaling of the frequency encoding (read-out) gradient allowed high resolution image acquisition on a 256 x 256 matrix with the nominal pixel size of 1 mm being reduced to as small a value as 0.2 mm.

Scan times varied from 2 to 30 minutes depending on the number of signal averages, the number of phase-encoding levels, and the specific pulse sequence repetition time chosen. Selection of the slices in the sample was accomplished with an oblique multi-slice cursor. This software feature overlays a translatable and rotatable set of parallel lines on a displayed image from an already completed scan. The lines can then be adjusted to choose the orientation and positioning for the slices in the subsequent scan.

SAMPLES

Green state disks were prepared with alpha alumina powder using a two component binder consisting of a 50/50 mixture of a proprietary mix of plasticizer and resin. Disks were prepared with binder contents varying from 5 to 30 wt.%, with most disks having a 25 wt.% binder content. The disks were pressed at $90^{\circ}C$ and 10,000 psi. The sample cylinders to be imaged were prepared by epoxying several disks together. An example of one such sample is shown in Figure 1. In this case, the disks contain 25 wt.% binder. For two disks, glass beads 4 mm in diameter were included as sample "defects" of a known size, in the pattern shown in Figure 1. The disks containing these glass beads were formed by pressing half of the required alumina/binder mix into a thinner disk, pressing the glass beads on the surface in the desired pattern, and then adding the rest of the alumina/binder mix and repressing to form the complete disk.

# PROTON NMR IMAGES OF GREEN STATE CERAMICS

In green state ceramics, proton NMR imaging occurs via the protons contained in the binder. For the green state ceramics discussed here, the use of the two component binder ensures that protons are present in a number of different chemical environments. Protons may also be present in material of differing viscosity, depending on the uniformity of solution of the resin in the platicizer. At sufficiently high binder viscosity, the NMR properties of the binder will be those of a solid material, with broad proton NMR lines and short spin-spin relaxation times ($T_2$). Since medical NMR imaging systems are limited in both the pixel-to-pixel separation in NMR frequency and the shortness of the spin echo delay, obtaining high resolution, high signal-to-noise ratio images can be expected to be difficult.

To illustrate some of the problems that can be anticipated from the binder proton NMR properties, the proton NMR free induction decay (FID) spectrum of the resin/plasticizer solution at 300 MHz is shown in Figure 2. The NMR spectrum is composed of a number of lines, with the aliphatic lines falling near 1100 Hz (from the reference line of water), and the aromatic lines occurring near 3000 Hz. The width of each region is roughly 1200 Hz. For a proton spectrum obtained at 12.7 MHz, the frequency at which the imager operates, the frequency difference of the aliphatic and aromatic regions which scales with the NMR frequency, would decrease to almost 80 Hz. This difference is much less than the pixel frequency difference of 512 Hz. However, the widths of the aromatic and aliphatic regions will probably not decrease very much, as these widths are primarily determined by the proton NMR linewidths, which will not change much with the NMR frequency. As a result, image resolution will be limited to about two pixels if the pixel frequency separation is 512 Hz.

Proton NMR images of green state ceramics have been obtained under a variety of imager conditions with several different samples. The images discussed here are typical of those obtained for the samples. A typical sagittal image of the green state ceramic sample shown in Figure 1, is displayed in Figure 3a. This is an eight scan image with an echo delay

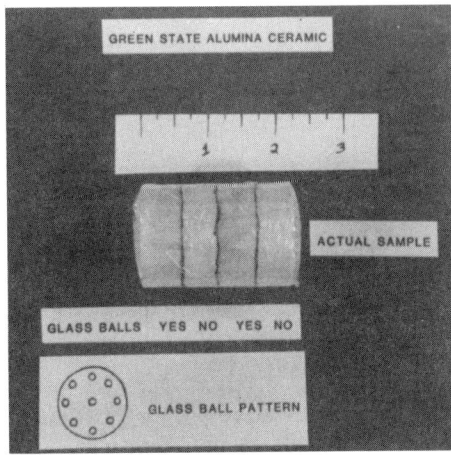

Fig. 1. Photograph of green state alumina disks, epoxied together.

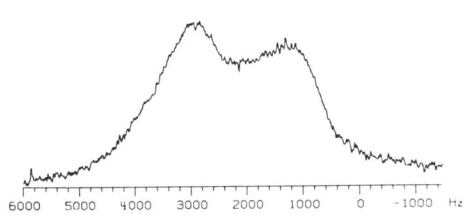

Fig. 2. 300 MHz proton FID NMR spectrum of binder resin dissolved in the plasticizer.

time of 28 ms. The location of the glass beads in two of the four disks is apparent. Also clearly shown are the regions of epoxy between the disks and the presence of several regions of severe sample inhomogeneity. In the upper disk with glass beads, the region between the two halves of the disk can be discerned. A higher resolution axial image of the lower disk is

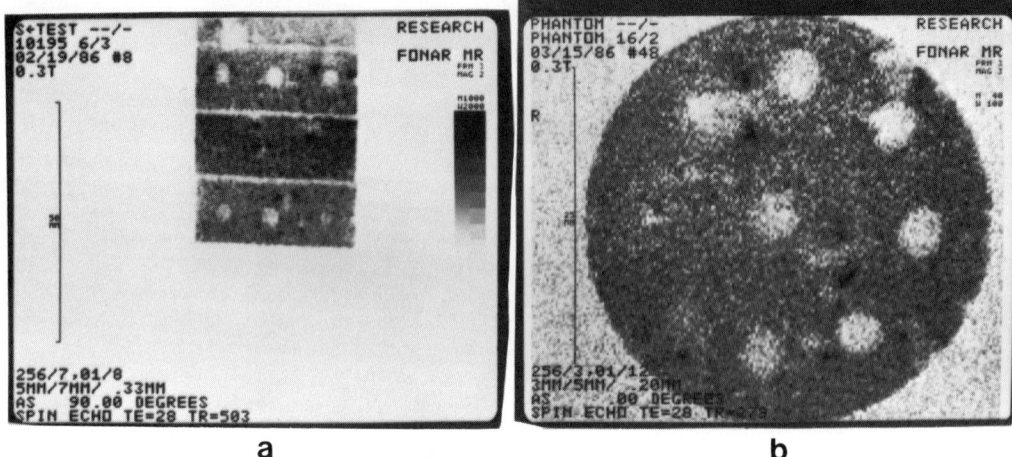

a          b

Fig. 3. Proton NMR images of green state ceramic disks with echo delay of 28 ms and recycle time of 0.5 sec. 1) Sagittal image - 8 scans, 5 mm thick slice near cylinder axis, 0.33 mm x 0.33 mm pixels. b) Axial image - 12 scans, 3 mm thick slice of bottom disk, 0.2 mm x 0.2 mm pixels.

presented in figure 3b. Again, a spin echo delay time of 28 ms was used. The cylinder axis is tilted with respect to the axial slice, so that the glass beads are visible on the right side of the disk, but less so on the left side. Other inhomogeneities are visible in the sample, particularly near the glass beads. This indicates that the technique by which the glass beads were incorporated into the sample introduced additional non-uniform regions.

The utility of multislice proton NMR imaging to the QNDE of green state ceramic materials is displayed in the series of axial images shown in Figure 4. A one scan off-axis sagittal image of the sample is shown in Figure 4a, which displays the positions of the axial slices in the rest of Figure 4 by the position of the multislice cursor. The axial images in Figures 4b through 4h are sequential images of the green state ceramic starting with the slice at the bottom edge of the disk. A total of 25 minutes was required for the generation of the seven interleaved images. The appearance and tracking of the glass beads and other sample nonuniformities through the seven sequential images indicates how the locations of such inhomogeneities or "defects" in the green state ceramic can be located and characterized with a medical NMR imaging system.

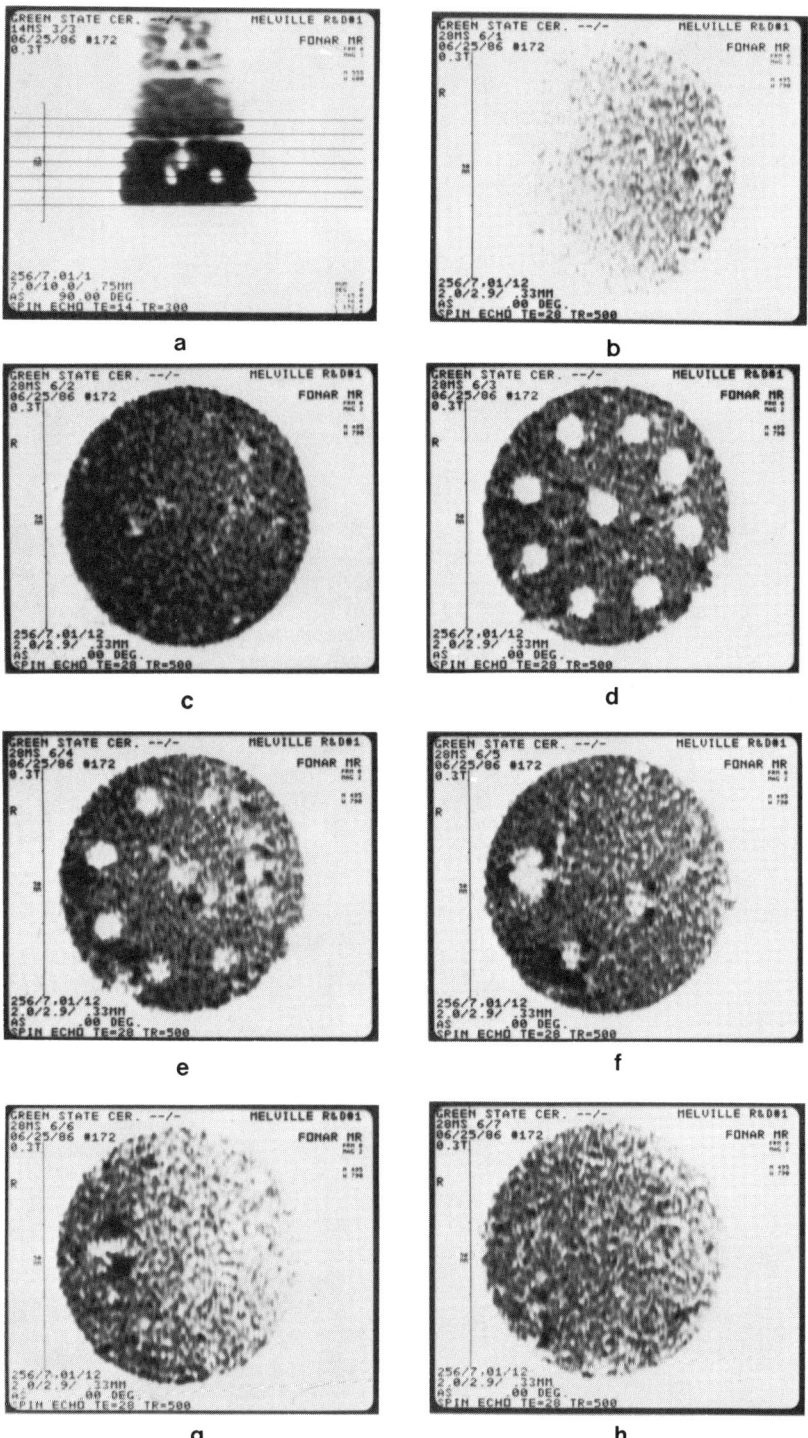

Fig. 4. Multislice proton NMR images of green state ceramic disks with echo delay of 28 ms and recycle time of 0.5 sec. a) Off axis sagittal image of green state ceramic disks showing multislice cursor placement. b - h) Sequential axial images starting from the edge of the sample - 12 scans, 2 mm thick slice, slice center separation of 2.9 mm, 0.33 mm x 0.33 mm pixels.

The origin of the contrast difference or inhomogeneities in the images of these green state ceramic samples is of considerable importance in determining the utility of the proton NMR imaging technique for the QNDE of such materials. Visual inspection of the samples did not indicate the existence of any inhomogeneities or "defects" of the size indicated in the images (exclusive of the glass beads). However, with a two component binder, questions arise as to the uniformity of distribution of the plasticizer and resin, and the degree to which the resin is uniformly dissolved in the plasticizer. Spatial variations in the plasticizer/resin mix uniformity could effect the proton NMR properties, leading to spatial variations in the contrast of the image. To determine if the NMR properties of the binder depended on the exact conditions of component mixing and heat treatment during pressing the green state material, spin echo proton NMR spectra at 300 MHz were generated with different echo delay times for samples of differing homogeneity. Of particular interest is a comparison of a binder sample where the resin has been fully dissolved in the plasticizer, and a sample which is a physical mixture of the resin and plasticizer. Proton NMR spin echo spectra of these samples at different echo delay times are shown in Figure 5. The spectra of the solution are shown in Figure 5a, and the spectra of the physical mixture are shown in Figure 5b. The size of the NMR signal decays very rapidly with increasing echo delay time for the solution, while the signal of the physical mixture decays much more slowly. The change in the relative intensities of the various lines in the spectra with increasing echo delay, indicates that protons in chemically different environments will dominate the NMR signal and hence the image at longer echo delays. These spectra demonstrate that the images are $T_2$ selective, with the contrast possibly arising from plasticizer rich regions of the green state ceramic samples or regions where the resin has been poorly dissolved in the plasticizer.

To determine if the use of shorter spin echo delay times would alter the images of the green state ceramics significantly, proton NMR images

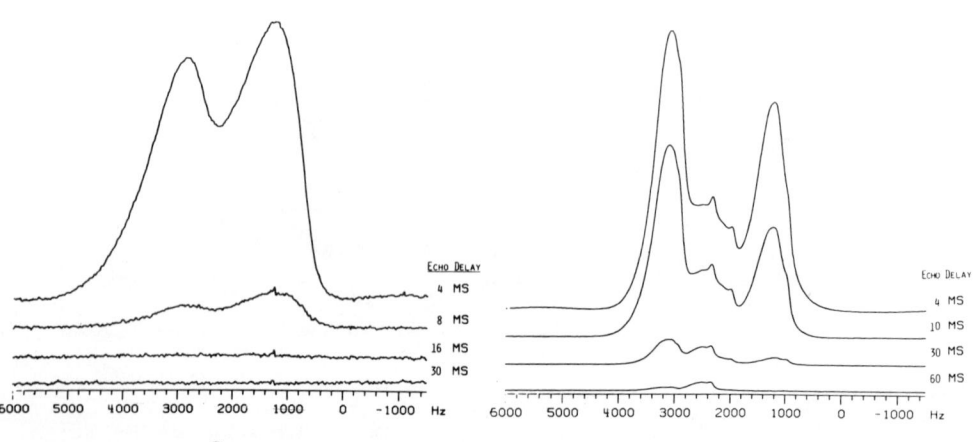

Fig. 5. 300 MHz proton spin echo NMR spectra at the indicated echo delays. a) Solution of resin in plasticizer. b) Physical mixture of resin and plasticizer.

were generated at echo delay times of 14 and 28 ms. The use of a 14 ms echo delay time required the use of thicker image slices and larger pixel size. In Figures 6a and 6b, sagittal image slices at 14 and 28 ms delay are compared. A similar comparison of two axial image slices is given in Figures 6c and 6d. In general, the spatial variation of the image intensity at a 28 ms delay appears to be greater than at 14 ms (exclusive of the glass beads), although, the differences are not dramatic. These results suggest that if an imaging system with a much shorter echo delay were used to image these samples, the resulting images would show much less contrast.

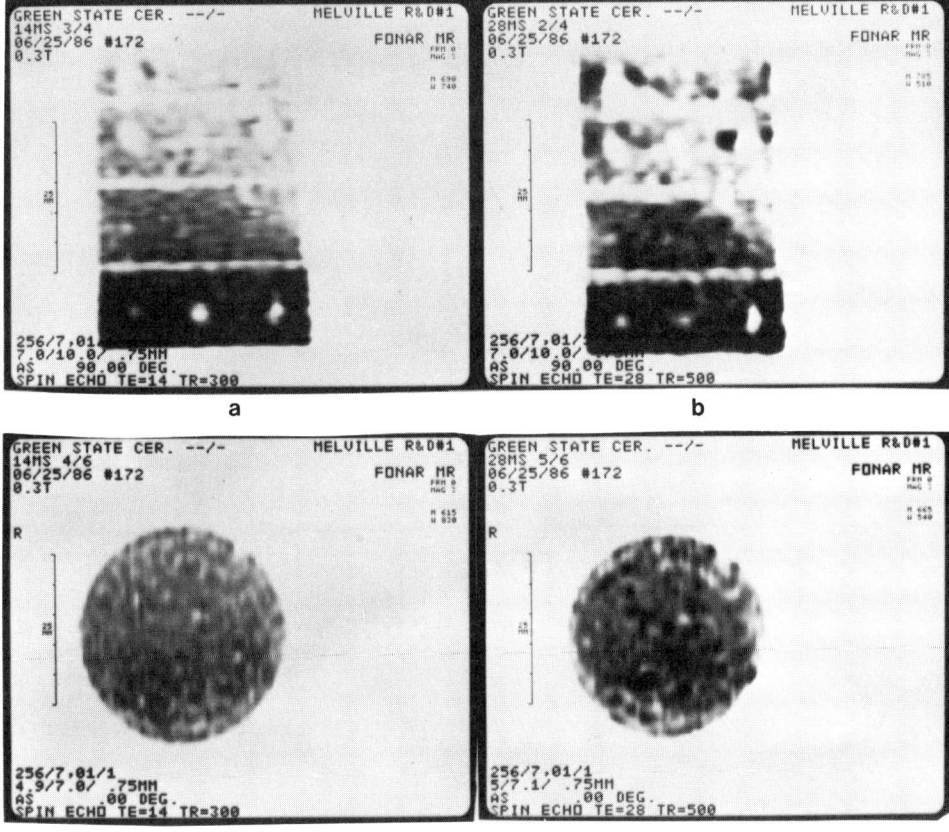

Fig. 6. Single scan proton NMR images of green state ceramic disks using either 14 or 28 ms echo delays with 0.75 mm x 0.75 mm pixels.
a) Sagittal image with 14 ms echo delay - 7 mm thick slice.
b) Sagittal image with 28 ms echo delay - 7 mm thick slice.
c) Axial image with 14 ms echo delay - 4.9 mm thick slice.
d) Axial image with 28 ms echo delay - 5.0 mm thick slice.

CONCLUSIONS

The preliminary proton NMR images of green state alumina ceramics discussed here clearly indicate that proton NMR imaging can develop into a valuable advanced QNDE technique for such materials. Sample nonuniformities of 1 mm diameter or larger appear to be easily detectable, and smaller nonuniformities can probably be detected under certain conditions, using the Fonar Beta-3000 NMR imaging system. The decrease in signal size with increasing echo delay time for the green state ceramics, indicates the need for imager modifications to achieve shorter echo delay times. However, in $T_2$ selective images of the type presented here, longer echo delays may in some cases provide better contrast and easier "defect" detection.

The work of Kupperman, et al. (2), demonstrated the imaging of porosity in green state ceramics by proton NMR of water in the pores. However, Kupperman was not successful in imaging a green state ceramic with a polyvinyl alcohol binder. In contrast, our study demonstrates spatial image variations in a green state ceramic which contained a two component (resin/plasticizer) binder system. Preliminary analysis indicates that inhomogeneous solvation of the resin component by the plasticizer is a possible source of the observed image contrast.

The possibility of imaging binder chemistry variations in a green state ceramic opens up a new tool for ceramic processing studies. If the NMR response of the protons in an organic binder is modified by changes in solvation, chemical crosslinking, phase, composition, and viscosity, it may be possible to map the extent of those changes in the green state ceramic. Similar types of results should also be obtainable in polymer and composite materials.

REFERENCES

1. For a review of proton NMR imaging, particularly for medical applications, see, for example, T. F. Budinger and P. C. Lauterbur, Science 226, 288 (1984), or P. A. Bottomley, Rev. Sci. Instrum 53(9),1319 (1982).

2. D. S. Kupperman, H. B. Karplus, R. B. Poppel, W. A. Ellingson, H. Berger, C. Robbins, and E. Fuller, "Application of NDE Methods to Green Ceramics: Initial Results," Argonne National Lab Report ANL/FE-83-25, Chapter V, p. 22, March 1984.

DISCUSSION

Mr. W. Friedman, Standard Oil: Have you compared these with x-ray C.T.'s and are the NMR images also susceptible to some of the artifacts that occur in more complicated shapes?

Mr. Welsh: We planned to compare with C.T., but don't have the results yet. We obtained the final images just before the conference, but did not have time to perform the C.T. experiments. What was the second part of the question?

Mr. Friedman: Whether you get artifacts if you have complicated flaw shapes.

Mr. Welsh: I think the main problem with artifacts is if you have anything metallic or conductive--

Mr. Friedman: But what about looking at high-aspect ratio things like blades?

Mr. Welsh: We haven't tried that. Images obtained so far are on model systems, to determine image quality from the present system.

From the Floor: Can you explain to me what is it that controls the slice thickness in this?

Mr. Welsh: The strength of the magnetic field gradient across the sample. As you increase the size of that gradient, you will get a thinner slice. What you are basically doing is spatially separating the resonance of protons in nearby spatial locations.

From the Floor: I see how that sort of affects your pixel size, but the slice thickness seems to be something other --

Mr. Welsh: Well, it depends on the bandwidth of the transmitter, but it also depends on the magnetic field gradient. So if you keep the bandwidth fixed and increase the magnetic field in the 2 directions you will get a thinner slice thickness.

Mr. D. Copley, G.E.: What do you see is the prospect for imaging fully-cured polymerical composite materials?

Mr. Welsh: I think, with the current medical imaging systems, that's a very tough problem. I think you generally have to go to multi-pulse methods.

Mr. Copley: Do you think that would be feasible?

Mr. Welsh: Certainly, several groups are putting a lot of effort into it and getting some positive results.

From the Floor: What's your minimum delay time?

Mr. Welsh: Minimum delay time for this imager was 14 milliseconds.

From the Floor: That's an instrumental factor?

Mr. Welsh: Yes. That relates to how fast you can switch the gradients. If you go to a smaller imager, you can probably switch a lot faster. And this is a large imager with a small object.

From the Floor: Do you consider that 1 millimeter or so resolution is sufficient to find what you want in green ceramics?

Mr. Welsh: We turned the question around the other way; that is, we wanted to see what kind of spatial information can be obtained by this technique. Right now, we are working with pixel sizes that are about 0.2 millimeters, and it may be possible to get down to that kind of resolution. For the samples that we imaged, I believe the resolution is closer to a half to 1 millimeter. A year from now, the resolution should be better.

# OPTICAL GENERATION OF COHERENT ULTRAHIGH FREQUENCY SURFACE WAVES

Sheryl M. Gracewski and R. J. Dwayne Miller

University of Rochester
Rochester, New York 14627

INTRODUCTION

It has been demonstrated [1-4] that coherent bulk acoustic waves can be holographically induced by the interference pattern of two intersecting picosecond laser pulses. The coupling between the optical and elastic fields can either be due to electrostriction or optical absorption which produces thermal stresses. The resulting waveform can then be detected by a third time-delayed laser pulse which diffracts off the density grating of the wave field. The elastic response within the laser spot caused by either coupling mechanism has been shown to consist of 1-D counter-propagating pressure waves. From the frequency and attenuation of the diffracted signal, the wavespeed and attenuation coefficient, respectively, of the medium can be obtained.

A similar technique can be used to generate surface acoustic waves (SAW) by interfering the laser pulses at a solid interface. Figure 1 shows a schematic diagram of the proposed technique. Two laser pulses, separated by angle $\theta$ are used to generate the surface acoustic waves. The interference pattern spacing and thus the acoustic wavelength $\Lambda$ are given by

$$\Lambda = \frac{\lambda}{2\sin(\frac{\theta}{2})}$$

where $\lambda$ is the wavelength of the laser light. Therefore the SAW wavelength is continuously tunable (up to approximately 30GHz) by adjusting the angle $\theta$. The diffracted signal of a third laser pulse will again be used to measure acoustic wave amplitude, but in this case, the periodic surface displacements will form the phase grating to diffract the incident signal.

Laser generation of surface waves and specifically generation by optical interference had been proposed as early as 1968 by Lee and White [5]. In [5], Lee and White demonstrated that surface waves can be induced by transient heating due to laser illumination and that the coupling to surface waves of a specific wavelength is enhanced when a spatially periodic mask is placed between the laser and specimen. Since that work other configurations have been implemented. Aindow et al. [6] and researchers referenced therein investigated waves generated by an unfocused laser spot.

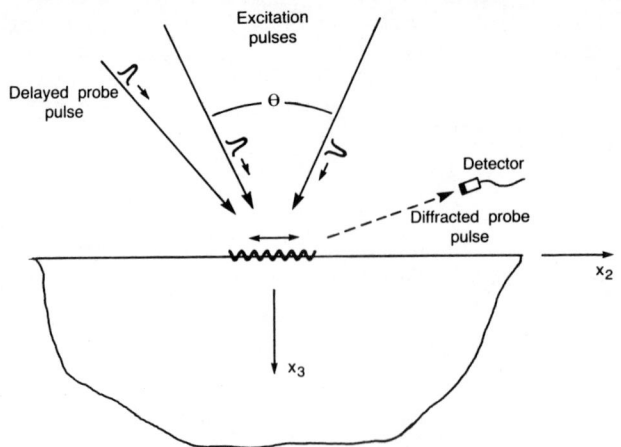

FIGURE 1.  Diagram of the laser generation of coherent SAW.  Interference of two excitation pulses separated by angle θ sets up a spacially harmonic temperature profile which induces counterpropagating surface waves.  A third laser pulse which diffracts off the surface displacement profile is used to detect the wave amplitude.

The use of directivity patterns which increase surface wave amplitude while avoiding surface damage has been investigated by Cielo et al. [7] and Hutchins et al. [8]. Enhancement of SAW generation by using a spacially period mask can be found in the works of Ash et al. [9] and Lee and Jackson [10], though in the latter paper, the device was designed for optical detection rather than SAW generation for NDE. SAW generation by laser interference, but at a lower frequency (30MHz) than proposed here has been demonstrated by Cachier [11]. An approximate calculation was used to estimate surface displacement amplitude.

In this paper, a theoretical formulation for SAW generation by optical interference will be discussed and a solution to the limiting case of heating confined to a region close to the surface will be given. The solution for a step function time dependence will be obtained by integrating time harmonic solutions over frequency.

THERMOELASTIC FORMULATION

The temperature field $T(\underline{x},t)$ caused by the optical interference pattern can be represented by

$$(T-T_o) = \frac{\Delta T_{max}}{2} \left(1 + \cos \frac{2\pi x_2}{\Lambda}\right) e^{-\gamma x_3} H(t) \tag{1}$$

where $H(t)$ is the unit step function, $T_o$ is the initial temperature, $\gamma$ is related to the optical absorption, and the maximum temperature change $\Delta T_{max}$ is on the order of 0.1°C. To justify this equation, various assumptions must hold. The step function time dependence implies an instantaneous temperature rise that does not dissipate with time. The laser pulse (80 psec) and thermal relaxation times are on the order of a tenth or less of the acoustic wave time constant at 10GHz and therefore the application of the temperature field can be considered instantaneous. Also, the diffusion time constant (O(μsec)) is much greater than the observation time (O(nsec)) and therefore diffusion is negligible on the experimental time scale. To neglect intensity variations from fringe to fringe, the generation laser spot size must be large enough to contain many fringes and must be sufficiently larger than the detection laser spot size.

The thermoelastic equations of motion in terms of the displacement components $u_i$ for a homogeneous, isotropic, linear elastic body are

$$(\lambda+\mu)u_{j,ji} + \mu u_{i,jj} - \rho \ddot{u}_i = \alpha(3\lambda+2\mu)(T-T_o)_{,i} \tag{2}$$

and

$$\kappa T_{,kk} = \rho c_v \dot{T} + (3\lambda+2\mu)\alpha T_o \dot{u}_{k,k} \tag{3}$$

where $\lambda$ and $\mu$ are the Lamé constants, $\kappa$ is the thermal conductivity, $c_v$ is the constant volume specific heat, $\alpha$ is the coefficient of linear thermal expansion, and $,_i$ denotes partial differentiation with respect to $x_i$. If temperature changes due to diffusion and elastic strains are negligible, equation (3) becomes

$$\dot{T} = 0$$

and the governing equations (Eqns. (2)) are then uncoupled and can be solved for the displacement components subject to an applied temperature profile.

In the analytical model, boundaries other than the top surface will be neglected so the applied temperature field given by equation (1) will be incident on a solid halfspace. Traction-free boundary conditions on the planar boundary $x_3 = 0$ are given as

$$\tau_{33} = (\lambda+2\mu) u_{3,3} + \lambda u_{2,2} - \alpha(3\lambda+2\mu)(T-T_o) = 0$$

$$\tau_{23} = \frac{\mu}{2} (u_{2,3} + u_{3,2}) = 0 \tag{4}$$

$$\tau_{13} = \frac{\mu}{2} (u_{1,3} + u_{3,1}) = 0 ,$$

where $\tau_{ij}$ are stress components of the stress tensor $\underline{\tau}$. Radiation conditions are applied at $x_3 \to \infty$. The initial conditions are

$$u_i(\underline{x},t) = \dot{u}_i(\underline{x},t) = 0 . \tag{5}$$

Ultimately the solution of this thermoelastic problem subject to the temperature profile given by equation (1) with an exponential $x_3$ dependence is desired. In this paper we consider only the two limiting cases of $\gamma \ll 1/\Lambda$ and $\gamma \gg 1/\Lambda$. For $\gamma \gg 1/\Lambda$, the heating is confined to a small layer of material near the boundary $x_3 = 0$ with thickness much smaller than an acoustic wavelength. For this limit, the displacements within the bulk of the material would satisfy the homogeneous equations of motion

$$(\lambda+\mu)u_{j,ji} + \mu u_{i,jj} - \rho \ddot{u}_i = 0 . \tag{6}$$

The boundary conditions can be approximated by equations (4) with the temperature profile given by equation (1) evaluated at $x_3 = 0$. A similar approximation was used by Lee and Jackson [10] for time-harmonic surface heating. This approximation becomes even more accurate if a transparent constraining layer is added onto the halfspace [12]. The solution to this problem for $\gamma \gg 1/\Lambda$ will be derived in the next two sections.

If $\gamma \ll 1/\Lambda$, the applied temperature field can be considered independent of $x_3$. This situation would apply if the material optical absorption coefficient was small. In this case, the resulting displacement field $\underline{u}_T$ can be written as

$$\underline{u}_T = \underline{u}_B + \underline{u} \tag{7}$$

where $\underline{u}_B$ is the bulk wave solution derived in [4] ($\underline{u}_B$ is the 1-D solution resulting from the temperature profile (Eqn. (1)) applied to an infinite medium) and $\underline{u}$ is the difference solution. The difference solution satisfies the homogeneous equations of motion (Eqns. (6)) subject to boundary conditions at $x_3 = 0$ for the stress components $\tau_{ij}$ given by

$$\tau_{33} = -\tau_{33}^B = C\{(\lambda+2\mu)[1+\cos(\omega_\ell t)\cos(k_o x_3)] + 2\mu[1-\cos(\omega_\ell t)]\cos(k_o x_3)\}$$

$$\tau_{23} = -\tau_{23}^B = 0, \quad \tau_{13} = -\tau_{13}^B = 0 \tag{8}$$

where

$$C = \frac{\alpha(3\lambda+2\mu)}{\lambda+2\mu} \frac{\Delta T_{max}}{2}$$

$$k_o = \frac{2\pi}{\Lambda}, \quad \omega_\ell = k_o C_\ell, \quad C_\ell [(\lambda+2\mu)/\rho]^{\frac{1}{2}}$$

and $\tau_{ij}^B$ denote the stresses of the bulk wave solution. (The term in $\tau_{33}$ independent of $x_2$ must be dropped for a finite solution. Deleting this term will not affect the resulting surface displacement profile.) The problem statement for this difference field is of the same form as the time-harmonic solution derived in the next section. Therefore, once the complete time-harmonic solution is known, the solution for $\gamma \ll 1/\Lambda$ can be obtained.

TIME HARMONIC SOLUTION

Away from the boundaries of the laser spot, intensity variations in the $x_1$ direction are negligible and the motion is plane strain. In this case $\partial/\partial x_1 = 0$ and $u_1 = 0$ so the displacement components can be written in terms of two potentials $\phi$ and $\psi$ as

$$u_2 = \frac{\partial \phi}{\partial x_2} + \frac{\partial \psi}{\partial x_3}, \quad u_3 = \frac{\partial \phi}{\partial x_3} - \frac{\partial \psi}{\partial x_2}. \tag{9}$$

In terms of potentials, the equations of motion (Eqns.(6)) for $\gamma \gg 1/\Lambda$ become

$$\nabla^2 \phi - \frac{1}{c_\ell^2} \ddot{\phi} = 0, \quad \nabla^2 \psi - \frac{1}{c_t^2} \ddot{\psi} = 0 \tag{10}$$

where

$$c_\ell^2 = (\lambda+2\mu)/\rho, \quad c_t^2 = \mu/\rho, \quad \text{and}$$

$$\nabla^2 = \frac{\partial^2}{\partial x_2^2} + \frac{\partial^2}{\partial x_3^2}.$$

The boundary conditions of $x_3=0$ for a time-harmonic applied temperature field at frequency $\omega_o$ are

$$\tau_{23} = \mu(2\phi_{,23} + \psi_{,33} - \psi_{,22}) = 0$$

$$\tau_{33} = (\lambda+2\mu)\nabla^2\phi - 2\mu(\phi_{,22} + \psi_{,23}) = \cos(k_o x_2)e^{-i\omega_o t}. \qquad (11)$$

For the given applied temperature field, the potentials are of the form

$$\phi(x_2,x_3,t) = \Phi(x_3)\cos(k_o x_2)e^{-i\omega_o t}$$

$$\psi(x_2,x_3,t) = \Psi(x_3)\sin(k_o x_2)e^{-i\omega_o t}. \qquad (12)$$

By substitution into the equations of motion, expressions for $\Phi$ and $\Psi$ are obtained as

$$\Phi(x_3) = C_1 e^{i\gamma_\ell x_3} + C_3 e^{-i\gamma_\ell x_3}, \quad \gamma_\ell^2 = \frac{\omega_o^2}{c_\ell^2} - k_o^2$$

$$\Psi(x_3) = C_2 e^{i\gamma_t x_3} + C_4 e^{-i\gamma_t x_3}, \quad \gamma_t^2 = \frac{\omega_o^2}{c_t^2} - k_o^2. \qquad (13)$$

For branches $\text{Re}[\gamma_\ell, \gamma_t] \geq 0$, $\text{Im}[\gamma_\ell, \gamma_t] \geq 0$, the radiation condition at $x_3 \to \infty$ requires $C_3 = C_4 = 0$. Expressions for the constants $C_1$ and $C_2$ are obtained by substitution into the boundary conditions. Equations (11) become

$$\begin{bmatrix} (k_o^2 - \gamma_t^2) & -2i\gamma_t k_o \\ -2i\gamma_\ell k_o & (k_o^2 - \gamma_t^2) \end{bmatrix} \begin{bmatrix} C_1 \\ C_2 \end{bmatrix} = \begin{bmatrix} 1/\mu \\ 0 \end{bmatrix}. \qquad (14)$$

Solution by Cramer's rule yields

$$C_1 = \frac{1}{\mu} \frac{(k_o^2 - \gamma_t^2)}{R}, \quad C_2 = \frac{1}{\mu} \frac{2i\gamma_\ell k_o}{R} \qquad (15)$$

where

$$R = (k_o^2 - \gamma_t^2)^2 + 4\gamma_\ell \gamma_t k_o^2 = 0$$

is the Rayleigh wave equation.

The final solution for the displacement components, obtained by substituting equations (12), (13), and (15) into equation (9) is then

$$u_2 = -\frac{B}{\mu}\sin(k_o x_2)e^{-i\omega_o t}\left\{\frac{k_o(k_o^2-\gamma_t^2)}{R}e^{i\gamma_\ell x_3} + \frac{2\gamma_\ell\gamma_t k_o}{R}e^{i\gamma_t x_3}\right\}$$

$$u_3 = \frac{B}{\mu}\cos(k_o x_2)e^{-i\omega_o t}\left\{\frac{i\gamma_\ell(k_o^2-\gamma_t^2)}{R}e^{i\gamma_\ell x_3} - \frac{2i\gamma_\ell k_o^2}{R}e^{i\gamma_t x_3}\right\}. \tag{16}$$

In Figure 2, the magnitudes of the normalized displacement components $\hat{u}_i = k_o u_i$ along the boundary $x_3 = 0$ are plotted as a function of $\hat{\omega}_o = \omega_o/k_o c_R$. The constant $c_R$ is the Rayleigh wavespeed of the material. Maximum coupling occurs at $\omega_o = k_o c_R$. Note that material damping needs to be added to predict the response for this value of $\omega_o$.

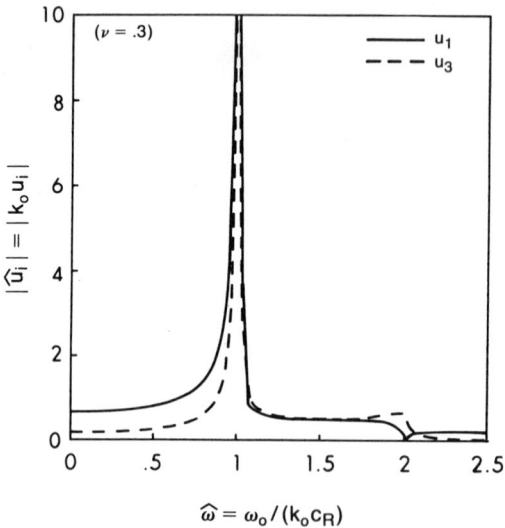

FIGURE 2. Normalized displacement components along $x_3 = 0$ versus the nondimensional frequency $\hat{\omega}_o$ for a time-harmonic excitation.

## STEP FUNCTION TIME DEPENDENCE

To obtain the solution $\bar{u}_i$ for the applied temperature field with a step function time dependence, the time-harmonic solution must be integrated as follows:

$$\bar{u}_2 = \frac{k_o}{2\pi}\frac{B}{\mu}\sin(k_o x_2)\lim_{\beta\to 0}\int_{-\infty}^{\infty}\frac{1}{\beta-i\omega_o}\left[\frac{(k_o^2-\gamma_t^2)}{R}e^{i\gamma_\ell x_3} + \frac{2\gamma_\ell\gamma_t}{R}e^{i\gamma_t x_3}\right]e^{-i\omega_o t}d\omega_o$$

$$\bar{u}_3 = \frac{1}{2\pi}\frac{B}{\mu}\cos(k_o x_2)\lim_{\beta\to 0}\int_{-\infty}^{\infty}\frac{1}{\beta-i\omega_o}\left[\frac{i\gamma_\ell(k_o^2-\gamma_t^2)}{R}e^{i\gamma_\ell x_3} - \frac{2i\gamma_\ell k_o^2}{R}e^{i\gamma_t x_3}\right]e^{-i\omega_o t}d\omega_o. \tag{17}$$

The integrands have poles at $\omega_0 = -i\beta$, $\pm k_0 c_R$ and branch points at $\omega_0 = \pm k_0 c_\ell, \pm k_0 c_t$. Therefore to carry out the integration numerically, the contour of integration must be deformed as shown in Figure 3.

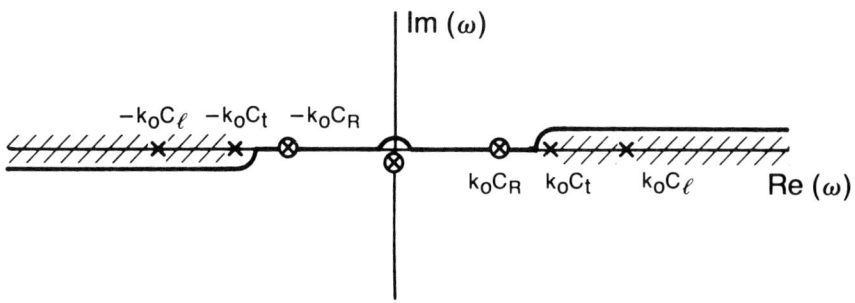

FIGURE 3. Contour of integration for integrals in equations (17).

SUMMARY

In this paper, a method of inducing high frequency SAW waves was discussed and a solution was derived for the case of heating confined to a surface layer small in comparison to an acoustic wavelength. The main advantage of this technique is the ability to generate coherent SAW waves at ultrahigh frequencies and to continuously adjust the acoustic wavelength by changing the angle between the two excitation pulses. Also, because the waves are optically generated, there is no mechanical contact and the required specimen size is small.

REFERENCES

1. R. J. Dwayne Miller, Roger Casalegno, Keith A. Nelson, and M. D. Fayer, Chemical Physics, 72, pp. 371-379, (1982).
2. Keith A. Nelson, R. J. Dwayne Miller, D. R. Lutz, and M. D. Fayer, J. Appl. Phys., 53(2), pp. 1144-1149, (1982).
3. Keith A. Nelson, D. R. Lutz, M. D. Fayer, and Larry Madison, Phys. Rev. B, 24, pp 3261-3275, (1981).
4. Keith A. Nelson and M. D. Fayer, J. Chem Phys, 72(9), pp. 5202-5218, (1980).
5. R. E. Lee and R. M. White, Appl. Phys. Letters, 12(1), pp. 12-14, (1968).
6. A. M. Aindow, R. J. Dewhurst, D. A. Hutchins, and S. B. Palmer, J. Acoust. Soc. Am., 69(2), pp 449-455, (1981).
7. P. Cielo, F. Nadeau, and M. Lamontagne, Ultrasonics, pp. 55-62, (1985).
8. D. A. Hutchins, R. J. Dewhurst, and S. B. Palmer, J. Acoust. Soc. Am., pp. 1362-1369, (1981).
9. E. A. Ash, E. Dieulesaint, and H. Rakouth, Elect. Letters, 16(12), pp 470-472, (1980).
10. R. E. Lee and D. W. Jackson, Ultrasonics Symp. Proc., IEEE Cat. # 75, CHO 994-4SU, pp 126-129, (1975).
11. G. Cachier, Appl. Phys. Letters, 17(10), pp 419-421, (1970).
12. D. A. Hutchins, R. J. Dewhurst, and S. B. Palmer, Ultrasonics, 19, pp. 103-108, (1981).

USE OF A CHIRP WAVEFORM IN PULSED EDDY CURRENT CRACK DETECTION


R. E. Beissner and J. L. Fisher

Southwest Research Institute
San Antonio, Texas 78284


INTRODUCTION

When an electrical conductor containing a surface-breaking crack is subjected to a short pulse of electromagnetic radiation, the reflected field contains transient features related to the depth of the crack. This has been demonstrated in both theoretical calculations [1] and in experiments [2,3] on shallow (0.13 mm to 1.3 mm) slots in a low conductivity titanium alloy. Specifically, these results show that the peak crack signal is delayed in time by an amount proportional to the square of the crack depth, and that the signal decay times also increase with increasing depth.

Implementation of the pulsed eddy current method for crack depth measurement is, however, made difficult by the very short rise and decay times typical of crack depths and materials of interest. As shown in earlier work [1], pulse times of the order of tens of nanoseconds are required to separate the flaw signal from the background response associated with direct coupling of the source and receiver coils. To achieve adequate signal intensity with such short pulse widths, peak power requirements are large, thus complicating the design of an appropriate eddy current probe.

The work reported here is concerned with the use of a chirp waveform, coupled with autocorrelation analysis of the return signal, to synthesize a short, high intensity pulsed measurement. In principle, the advantage offered by the chirp approach over the direct pulsed eddy current method is that the pulse energy is distributed over a much longer time interval, thus reducing peak power requirements. The specific question addressed in this study is whether one can synthesize a pulse short enough for crack depth measurement within the constraints imposed by limited bandwidth and a chirp repetition rate adequate for convenient signal averaging.

THEORY

Our study is based on the two-dimensional crack model shown in Figure 1. To simplify the mathematics, we assume that the incident pulse is spatially uniform and produces a time-dependent magnetic field $H_o(t)$, with direction perpendicular to the plane of the figure, on the crack and conductor surfaces. The first step in the calculation is the determination of the field on the crack face.

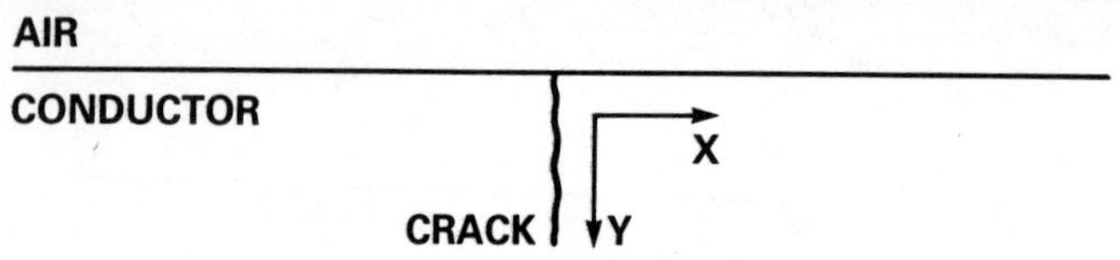

Fig. 1. Two-dimensional model of a crack. A spatially uniform, time-dependent field exists on the surface of the conductor and crack.

Rather than work directly with time-dependent fields, we choose instead to solve this problem first in the frequency domain by taking Fourier transforms with respect to time. If $h_o(\omega)$ is the transform of $H_o(t)$, then the transform of the field in the conductor can be written

$$h(x,y,\omega) = h_o(\omega)\ [e^{iqy} + \psi(x,y,\omega)] \tag{1}$$

with $q = (1+i)/\partial$, where $\partial$ is the skin depth. The first term in the bracket is proportional to the unperturbed field at depth y, and the second term is proportional to the perturbation caused by the flaw.

The procedure for calculating $\psi(x,y,\omega)$ is similar to that reported earlier [1] and need not be described in detail. Briefly, we start with the scalar Helmholtz equation for $\psi$ and transform to the following integral equation by means of Green's theorem:

$$1 - e^{iqy} = \int_o^d K(y,y') \left(\frac{\partial \psi(x,y',\omega)}{\partial x}\right)_{x=0} dy' \tag{2}$$

where d is the crack depth,

$$K(y,y') = i[H_o^{(1)}[(q|y - y'|)] - H_o^{(1)}[q(y + y')]] \tag{3}$$

and $H_o^{(1)}$ is the zero order Hankel function. To solve for the x derivative of $\psi$, we first note that this quantity is proportional to the current density on the crack surface and is therefore singular at the crack tip. With this in mind we make the substitution $y/d = 1 - \gamma^2$ and solve instead for the nonsingular function

$$\gamma \frac{\partial \psi}{\partial x} = \sqrt{1 - y/d}\ \frac{\partial \psi}{\partial x}. \tag{4}$$

To obtain a numerical solution we use a 48-point Gaussian quadrature approximation to the integral in (2) and solve the resulting system of algebraic equations by matrix inversion.

Based on the development presented earlier [1], we next make use of the reciprocity theorem [4] to obtain the following expression for the transform of the flaw signal per unit crack length:

$$\Delta V(\omega) = h_o(\omega) G(\omega) \tag{5}$$

where the transform of the impulse response function is

$$G(\omega) = \frac{2}{\sigma} \int_0^d e^{iqy} \left(\frac{\partial \psi(x,y,\omega)}{\partial x}\right)_{x=0} dy \qquad (6)$$

Plots of the real part of $G(\omega)$ are presented in Figure 2; labels on the curves are crack depths in units of 0.001 inch (0.025 mm).

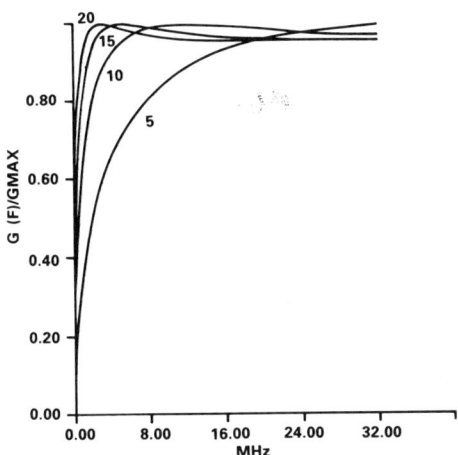

Fig. 2. Real parts of the Fourier transforms of impulse response functions for the geometry shown in Figure 1. Labels on the curves are crack depths in units of 0.001 inch (0.025 mm).

CHIRP/AUTOCORRELATION CALCULATIONS

Figure 2 shows that accurate determination of the time-dependent impulse response functions requires a bandwidth of more than 20 MHz. There is, however, a significant dependence on crack depth at much lower frequencies, which suggests that conventional probes with lower bandwidth might be useful for crack depth measurement by a pulsed eddy current method. The calculations described in this section were undertaken to explore this possibility using the chirp/autocorrelation approach.

To illustrate the principle, let us first consider a damped chirp waveform given by

$$H_o(t) = \begin{cases} \cos^2 \frac{\pi}{2}\frac{t}{T} \sin \omega_m \frac{t^2}{T} & 0 \leq t \leq T \\ 0 & \text{otherwise} \end{cases} \qquad (7)$$

This is a pulse of duration T with angular frequencies from zero to $\omega_m$. Figure 3a is a plot of this function with T = 10 μsec and a maximum frequency of 4 MHz.

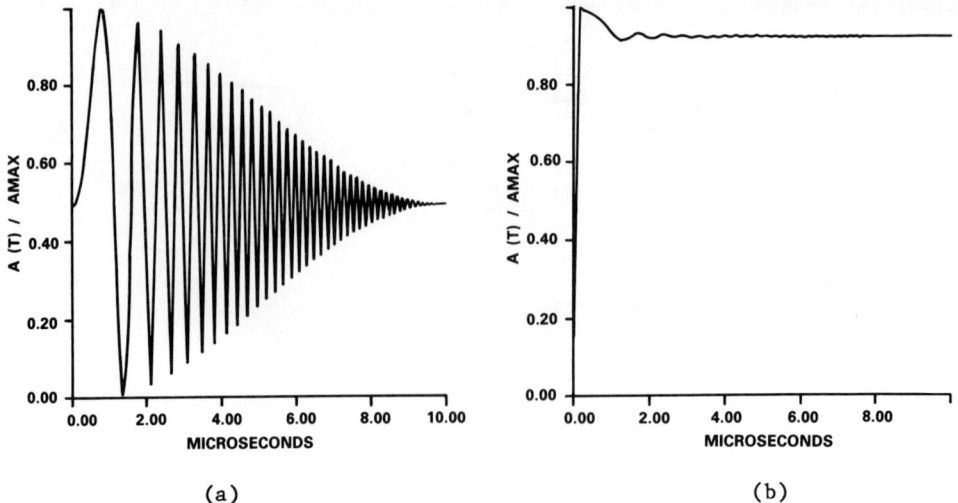

Fig. 3. A chirp waveform (a) and its autocorrelation function (b). Autocorrelation of a 10 μsec chirp with 4 MHz bandwidth synthesizes a short, time-domain pulse.

The autocorrelation function of $H_o(t)$ can be written in terms of $H_o(t)$ or its Fourier transform $h_o(\omega)$ as follows:

$$A(t) = \int_{-\infty}^{\infty} H_o(t')H_o(t+t')dt'$$

$$= \frac{1}{2\pi} \int_{-\infty}^{\infty} |h_o(\omega)|^2 e^{-i\omega t} d\omega \tag{8}$$

Figure 3b is the autocorrelation of the chirp plotted in Figure 3a, and shows how the autocorrelation operation synthesizes a short, time-domain pulse. The motivation for the present study is to see whether depth information can be obtained by a similar operation on the flaw response function given by (5).

Figures 4a and 4b are autocorrelation functions obtained by using $\Delta V(\omega)$ in place of $h_o(\omega)$ in (8), with the transform of (7) used for $h_o(\omega)$ in (5). The pulse duration T is 20 μsec in both cases; the bandwidth is 8 MHz in Figure 4a and 4 MHz in Figure 4b.

From Figure 4 we see two effects of limited bandwidth on the chirp/autocorrelation signal. As the bandwidth is decreased, peak arrival times for flaws of different depths tend to merge and shift to later times. Also, as bandwidth is decreased, differences in decay times become less pronounced. Both of these effects tend to make depth measurement based on temporal response more difficult to achieve with a small bandwidth chirp. On the other hand, a bandwidth of 8 MHz is not unreasonable, and, based on the results shown in Figure 4, should suffice for crack depth estimation. Further support for the idea is provided by experimental results reported elsewhere [3], which show that crack depth information can be derived from low-to-moderate bandwidth data.

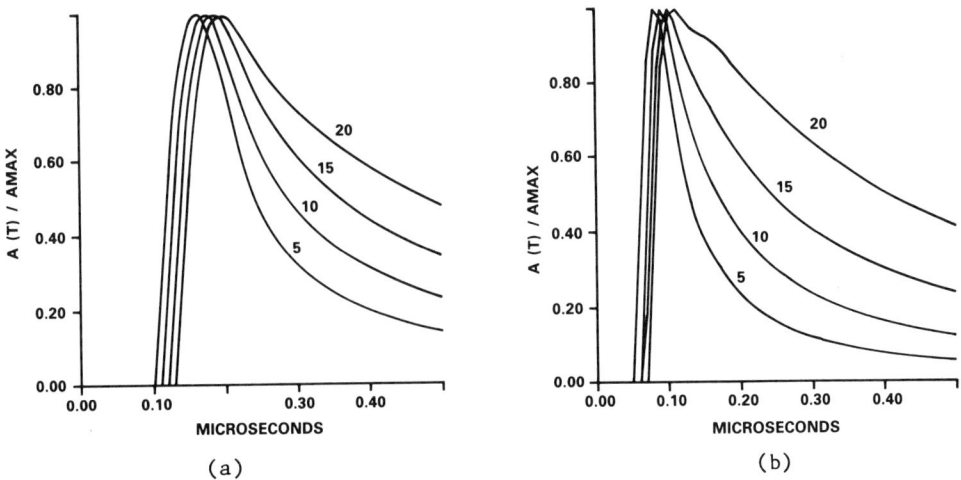

Fig. 4. Autocorrelation functions of the flaw response to chirp waveforms of 20 sec duration. The bandwidth is 8 MHz in (a) and 4 MHz in (b). Labels on the curves are crack depths in units of 0.001 inch (0.025 mm).

CONCLUSIONS

Calculations of crack response to a chirp waveform indicate that crack depth can be estimated from the autocorrelation of the flaw signal. For crack depths of 0.13 mm or greater in low conductivity materials, a bandwidth of about 8 MHz should be adequate. Although only one chirp duration was considered in the calculations, this chirp length (20 μsec) appears to be adequate from the standpoint of depth resolution. The 20 μsec chirp is also long enough to significantly reduce the peak power that would be needed to achieve comparable depth resolution by the direct pulsed eddy current method. Finally, the 20 μsec chirp is short enough to allow several repetitive pulses for signal averaging purposes while operating in a scanning mode. The chirp/autocorrelation synthesis of a pulsed eddy current experiment, with chirp bandwidth of about 8 MHz and duration of about 20 μsec, therefore appears to be a practical approach to the estimation of crack depth.

ACKNOWLEDGEMENT

This work was sponsored by the Center for Advanced Nondestructive Evaluation, operated by the Ames Laboratory, U.S. Department of Energy, for the Air Force Wright Aeronautical Laboratories/Materials Laboratory under Contract No. W-7405-ENG-82 with Iowa State University.

REFERENCES

1. R. E. Beissner and J. L. Fisher, "A Model of Pulsed Eddy Current Crack Detection," in Review of Progress in Quantitative NDE, 5A, D. O. Thompson and D. E. Chimenti, eds., Plenum Press, New York (1986), p. 189.

2. J. L. Fisher and R. E. Beissner, "Pulsed Eddy Current Crack Characterization Experiments," in Review of Progress in Quantitative NDE, 5A, D. O. Thompson and D. E. Chimenti, eds., Plenum Press, New York (1986), p. 199.
3. J. L. Fisher, R. E. Beissner, and G. L. Burkhardt, "Chirp Waveform Simulation in Pulsed Eddy Current Crack Detection," these proceedings.
4. B. A. Auld, "Theoretical Characterization and Comparison of Resonant-Probe Microwave Eddy-Current Testing with Conventional Low-Frequency Eddy-Current Methods," in "Eddy-Current Characterization of Materials and Structures," ASTM STP 722, G. Birnmaum and G. Free, eds., American Society for Testing and Materials, Philadelphia (1981), p. 332.

A NEW METHOD FOR THE MEASUREMENT OF ULTRASONIC ABSORPTION IN

POLYCRYSTALLINE MATERIALS

H. Willems

Fraunhofer-Institut für zerstörungsfreie Prüfverfahren
Universität, Gebäude 37
D-6600 Saarbrücken 11, FRG

INTRODUCTION

In general, the ultrasonic attenuation in polycrystalline materials at room temperature (RT) is described in terms of scattering losses and absorption losses. Ultrasonic scattering is caused by the grain structure of the material whereas the interaction of the ultrasonic wave with lattice imperfections (e.g. dislocations, Bloch walls) leads to energy absorption. Usually, it is impossible to separate the different contributions by using conventional pulse-echo techniques which measure the total ultrasonic attenuation plus some artificial attenuation due to specimen geometry and sound-field divergence. Only if special assumptions on the frequency dependence of each contribution can be made, a separation might be possible /1/. Recently, several techniques have been proposed in order to measure ultrasonic absorption directly. One technique is based on resonance measurements in small cylindrical specimens by exciting standing waves at frequencies below 1.2 MHz /2/. A second technique uses the infrared detection of the heat produced by the ultrasonic absorption /3/, and a third technique is based on ultrasonic diffusion measurements /4/. This paper presents a new method for direct absorption measurements in polycrystalline materials by measuring ultrasonic reverberation. The method is especially appropriate in the case of large scattering contributions usually present in coarse grained steels.

DESCRIPTION OF THE METHOD

The block diagram of the experimental set-up is shown in Fig. 1a. The ultrasonic system used enables burst excitation of the ultrasonic probe, and a logarithmic amplification of the received signals. Ultrasonic pulses with a length of usually 10-20 cycles are insonified into the specimen using contact technique. Initially, the ultrasonic energy is contained within a volume $V_p$ which is determined by the pulse length and the probe aperture. For the frequencies and the probes considered here, $V_p$ is in the order of 1 cm$^3$. After a time $t_1$, depending on the amount of scattering as well as on the volume $V_s$ and the geometry of the specimen, the ultrasonic energy will be equally distributed within the whole specimen volume provided the specimen is not too large. From that moment onwards the amplitude decay of the received ultrasonic signals exibits a considerably smaller slope than for times below $t_1$ (see Fig. 1b). This new slope can be associated with the ultrasonic absorption coefficient $\alpha_A$ characterizing the absorption of ultrasonic energy by intrinsic material properties. The important point is that the ultrasonic scattering, which is necessary to reach the condition of homogeneous energy distribution, obeys the law of energy conservation and, thus, does not contribute to the amplitude decay for $t > t_1$. Obviously, the method cannot be applied to non-scattering (or weakly scattering) materials, because the ultrasonic energy would essentially remain in the initial pulse until complete absorption. The ultrasonic signals after $t > t_1$ may be considered as a kind of reverberation and, therefore, the method is called the reverberation method in what follows. It should be mentioned, that the reverberation signal consists of a (unknown) mixture of longitudinal (L) waves, shear (T) waves and surface (R) waves due to the mode conversion by the scattering processes as well as by the reflections at the specimen boundaries. In fact, the reverberation signal can be picked up with any probe (L, T or R) of the same frequency at any specimen position. Because no definite propagation velocity can be attributed to the reverberation signals, all absorption coefficients measured with this method are given in units of dB/ms.

Obviously, the method is restricted with regard to the specimen volume. In the case of very small absorption coefficients a rough estimation yields for the upper limit of the specimen volume $V_S$

$$V_S = V_P \cdot 10^{\frac{D_0(dB) - D_1(dB)}{20}} \qquad (1)$$

Here, $D_0$ is the dynamic range of the initial ultrasonic signal, $D_1$ is the dynamic range necessary for signal evaluation, i.e. the determination of the slope of the logarithmic signal, and $V_P$ is the initial pulse volume as explained above. With $D_0 = 80$ dB, $D_1 = 20$ dB and $V_P = 1$ cm$^3$ one finds $V_S = 1000$ cm$^3$. Taking into account actual absorption coefficients and the reduced dynamic range of ultrasonic probes at higher frequencies, $V_S$ will be below 100 cm$^3$ in realistic cases. If the specimens are very small, losses due to mechanical damping by the ultrasonic probe have to be considered as can be seen from Fig. 2. Here, $\alpha_A$ was measured as a function of the specimen volume $V_S$ using cylindrical bars. The result is well described by the relation (solid line in Fig. 2):

$$\alpha'_A = \alpha_A + K/V_S \qquad (2)$$

Here, $\alpha'_A$ is the measured absorption coefficient, $\alpha_A$ is the true absorption coefficient and K depends on the ultrasonic probe. Concerning the specimens investigated in this work, $V_S$ was usually of the order of 10 cm$^3$.

Keeping in mind the limitations mentioned above, the method is very appropriate to measure absorption coefficients between 0 db/ms and approximately 400 dB/ms. This range is usually encountered for most steels at room temperature (RT) and frequencies below 20 MHz. Under these conditions the ultrasonic absorption is mainly attributed to the interaction with dislocations /5,6/ and, in ferromagnetic materials also with the magnetic structure /7,8,9,10/. The reverberation method was used to investigate the influence of these parameters on the ultrasonic absorption. Some examples are demonstrated in what follows showing the high sensitivity of the method.

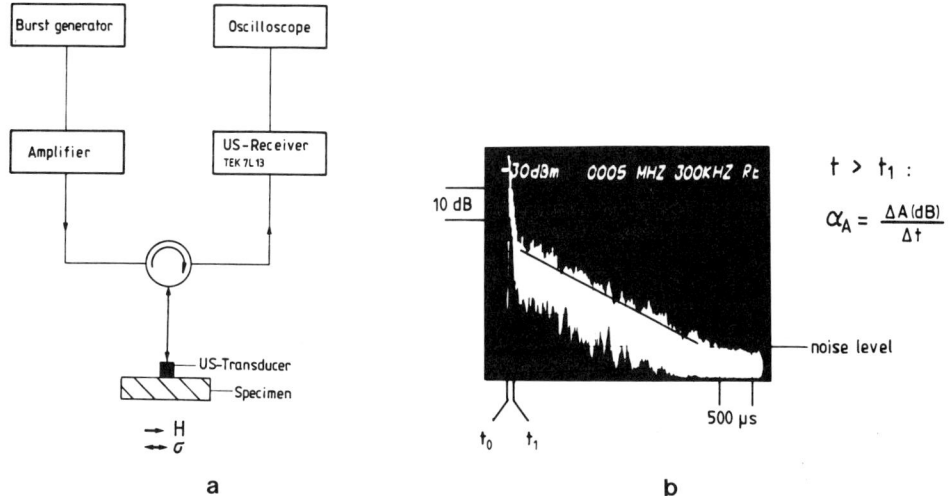

Fig. 1  a) Experimental set-up for ultrasonic absorption measurements
        b) Example of measuring signal

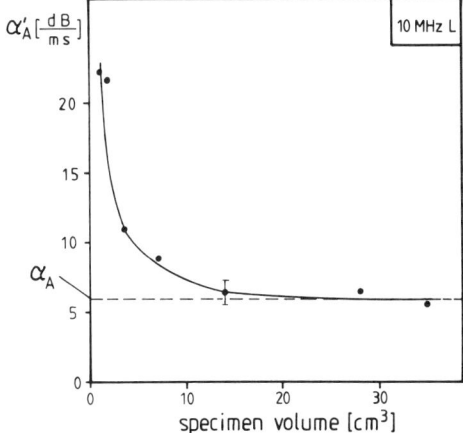

Fig. 2  Influence of specimen volume on the absorption measurement as measured using cylindrical bars

Table 1. Microstructural and ultrasonic data of some investigated specimens. Ultrasonic frequency: 5 MHz (shear wave)

| Specimen | Microstructure | Carbon content (wt%) | Grain size (μm) | $\alpha$ (db/ms) | $\alpha_A$ (db/ms) |
|---|---|---|---|---|---|
| 1 | Ferrite | 0.05 | 122 | 948 | 118.0 |
| 2 | Ferrite/Pearlite | 0.12 | 80 | 753 | 93.8 |
| 3 | Ferrite/Pearlite | 0.20 | 43 | 334 | 85.0 |
| 4 | Pearlite | 0.79 | 30 | 251 | 20.6 |
| 5 | Austenite(Ni-base) | 0.15 | 21 | 552 | 21.9 |
| 6 | Austenite(Fe-base) | 0.07 | 80 | 1240 | 10.0 |

475

EXAMPLES OF APPLICATION

Materials

Some of the investigated materials are described in Table 1, which also displays attenuation coefficients and absorption coefficients measured at a frequency of 5 MHz. It can be seen, that the absorption coefficients are much lower than the corresponding attenuation coefficients which have been determined by conventional techniques. Here, this is due to the fact that the attenuation coefficients are dominated by scattering losses produced by the relatively large grain sizes of the specimens. However, it should be mentioned that in the case of very fine-grained steels the absorption losses might dominate the total attenuation especially at lower frequencies.

Elastic deformation

The absorption coefficient $\alpha_A$ in a bar of austenitic Alloy 800H (X10 NiCrAlTi 32 20) was measured as a function of elastic strain by applying stresses up to one third of the yield strength. As can be seen from Fig. 3, the absorption increases with increasing strain and goes back to the initial value if the stress is released. This behaviour may be explained by the theory of dislocation damping yielding for the absorption due to the interaction of the ultrasonic wave with the dislocations /5/

$$\alpha_A \sim \Lambda \cdot L^4 \cdot f^2 \tag{3}$$

for frequencies far below the resonance frequency of the dislocations. Here, f is the frequency, L is the dislocation length, and $\Lambda$ is the dislocation density. The elastic deformation leads to a bowing-out of those dislocations that are suitably oriented to the applied load. According to equation (3), the resulting increase in the loop length should cause a higher absorption in qualitative agreement with the experimental results.

Plastic deformation

The dislocation density in a polycrystalline material can be strongly enhanced by plastic deformation. Fig. 4 shows the dependence of the ultrasonic absorption coefficient on the plastic strain in Alloy 800H for two different heat treatments. In the as-received condition the material was solution annealed (1130°C/30 min/water), i.e. most of the carbon was in solution. A distinct increase in absorption is observed with increasing plastic deformation leveling off at plastic deformations above 10%. If the material is aged 16h/800°C, precipitation of carbides takes place. During plastic deformation, the carbides act as sources for dislocation multiplication. Consequently, a higher dislocation density is expected during deformation for this material state compared to the as-received state. This is exactly what one observes by means of absorption measurements (Fig. 4) showing that the ultrasonic absorption increases much stronger for the aged material.

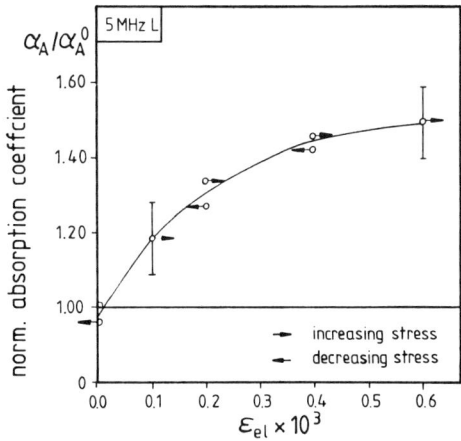

Fig. 3  Normalized ultrasonic absorption coefficient $\alpha_A/\alpha_A^0$ as a function of elastic strain $\varepsilon_{el}$ in Alloy 800H

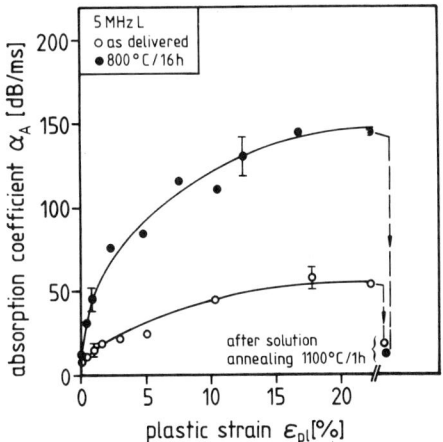

Fig. 4  Ultrasonic absorption as a function of plastic strain $\varepsilon_{pl}$ in Alloy 800H for different heat treatments. After the deformation the specimens were solution annealed.

After the deformation, the specimens were solution annealed allowing the microstructure to recover. This is clearly indicated by a drop of the absorption coefficients to nearly the initial values (Fig. 4). It must be emphasized that the changes in ultrasonic absorption measured here (and also in most cases described below) did not lead to significant changes in the total attenuation coefficients as obtained from the changes in the total attenuation coefficients as obtained from the evaluation of backwall-echo sequences. An example of the change of the reverberation signal due to plastic deformation is shown in Fig. 5c.

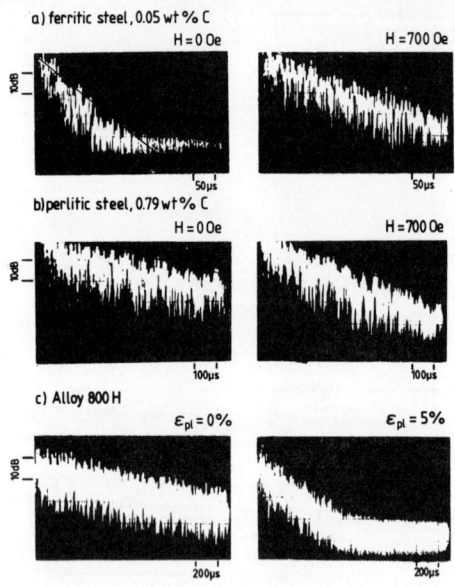

Fig. 5  Examples of ultrasonic reverberation signals showing the influence of magnetic fields (a,b) and plastic deformation (c). Frequency: 10 MHz longitudinal waves

Creep deformation

Dislocation mechanisms play an important role during high-temperature creep (see for example /11/). During early creep one usually observes the generation of dislocation networks forming subgrain structures. Ultrasonic absorption measurements were performed on specimens of Alloy 800H which are creep-tested at 800°C/31 MPa. In order to take non-destructive measurements, the tests which are still in progress were interrupted every 2000 hours /12/. Fig. 6 shows the measured ultrasonic absorption coefficient as a function of creep strain. The absorption coefficients are normalized to the absorption coefficients measured in a reference specimen. At strains of approximately one percent, $\alpha_A$

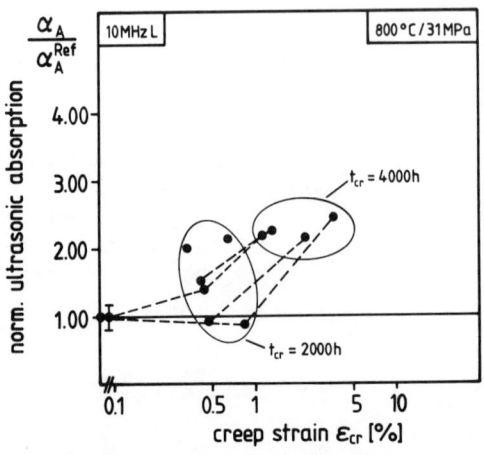

Fig. 6  Ultrasonic absorption in Alloy 800H as a function of creep strain (tests still in progress)

increases by a factor of about two. Around strain values of 1 %, steady state creep starts in the investigated material /13/. Here, TEM-investigations have shown that dislocation networks develop between the range of minimum creep rate and the end of steady state creep under the considered conditions /14/. Thus, we believe that the observed change in ultrasonic absorption is caused by the change in dislocation structure.

Influence of magnetic fields

In addition to the losses caused by dislocation damping, one observes in ferromagnetic materials losses due to the interaction of the ultrasonic wave with the magnetic structure. The ultrasonic wave induces changes of the magnetization by Bloch wall movements as well as by rotation processes. This leads to the generation of micro-eddy currents and hence to the absorption of ultrasonic energy. Theory /7/ yields for the frequency dependence of the ultrasonic absorption coefficient $\alpha_A^M$

$$\alpha_A^M \sim \frac{f^2/f_0}{1+(f/f_0)^2} \cdot \qquad (4)$$

Here, $f_0$ is the frequency at which the penetration depth of the micro-eddy currents becomes comparable to the mean domain size. For rotational processes, $f_0$ is by a factor of about 10 higher than in the case of wall movements because a different magnetic permeability enters into equation (4) /10/. In the magnetic saturated state all magnetic losses should vanish.

Fig. 7 shows the ultrasonic absorption measured in a ferritic steel (specimen 1) and a perlitic steel (specimen 4, see Table 1) as a function of magnetic field strength. $\alpha_A$ drops drastically at very low fields and then slightly decreases up to the highest fields applied. A completely different behaviour at low fields is observed for the perlitic steel (Fig. 7). Here, the absorption increases at low fields, reaches a maximum, and decreases at higher fields similar to the case of the ferritic specimen. The effect becomes more pronounced at higher frequencies. The findings described above were typical for both kind of steel types.

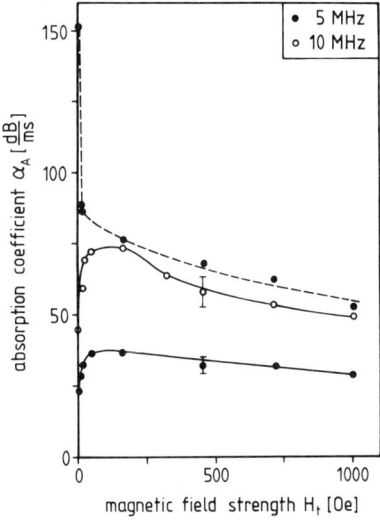

Fig. 7   Ultrasonic absorption as a function of magnetic field. (Dashed line: ferritic steel, 0.05 wt% C, solid lines: perlitic steel, 0.65 wt% C) $H_t$- tangential magnetic field strength

Without applied magnetic field the main contribution to the magnetic part of the ultrasonic absorption should be explained by the interaction of the ultrasonic wave with the Bloch walls. In the ferritic steel the Bloch walls are easily movable thus leading to a high absorption. In contrast, the perlitic structure consisting of small lamellas of ferrite and cementite seems to hinder the movements of the walls. Actually, there is no magnetic contribution to the absorption at zero-field as indicated by the measured absorption at "saturation". With applied field, the Bloch walls become movable. At the same time the Bloch wall density decreases. The combined action of both processes could explain the maximum in Fig. 7. The decrease of absorption measured at higher fields clearly indicates the presence of rotational processes because the Bloch wall density goes to zero in that range. This is confirmed by the fact that the observed decrease follows the decrease of the magnetostriction at higher fields. Examples of the measuring signals with and without magnetic field are shown in Fig. 5a,b.

Frequency dependence

The measured frequency dependence of the ultrasonic absorption between 2 MHz and 15 MHz is shown in Fig. 8 for different steels without magnetic field and with "saturation" field. In the latter case, the absorption should be mainly due to dislocation damping. Then, for the ferritic steel as well as for the perlitic steel and the austenitic steel (Alloy 800H) the frequency dependence seems to be quadratic in agreement with equation (3) but taking into account the measuring error of about ± 10 % a linear dependence cannot be excluded. A linear frequency dependence is found for the ferritic-perlitic steel (specimen 2, see Table 1). The magnetic part of the absorption, i.e. the difference of $\alpha_A$ measured without and with saturation field, depends linearly on frequency for both the ferritic and the ferritic-perlitic steel in the investigated frequency range. Concerning the perlitic steel (Fig. 8a) the absorption is higher at a field strength of 700 Oe than without field. This indicates that there is no complete saturation at 700 Oe as well as that there is no magnetic contribution to the ultrasonic absorption at zero field.

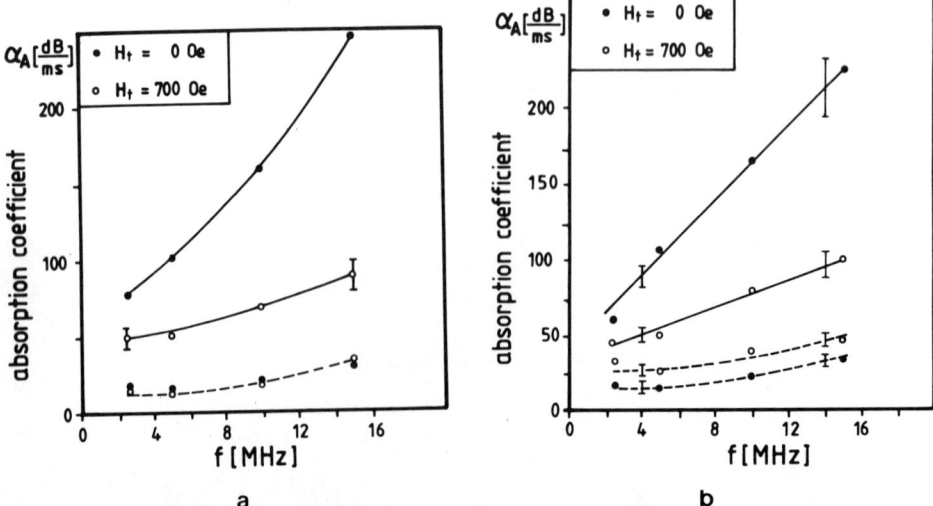

Fig. 8  Ultrasonic absorption in different steels (see Table 1) as a function of frequency without and with magnetic saturation.
a) solid lines: specimen 1, dashed lines: specimen 5
b) solid lines: specimen 2, dashed lines: specimen 4

CONCLUSIONS

The method presented here allows the direct measurement of ultrasonic absorption in polycrystalline materials which scatter ultrasound sufficiently. It can be used with a high sensitivity to study changes of the dislocation structure as well as the magnetic structure of materials, especially steels. This might be important for the nondestructive evaluation of materials properties. Additionally, the proposed method enables a separation of scattering losses and absorption losses when combined with conventional attenuation measurements.

ACKNOWLEDGEMENTS

Part of this work was financially supported by the "Bundesministerium für Forschung und Technologie" of the FRG. The creep-tested specimens were supplied by Dr. U. Hildebrandt from BBC Mannheim, FRG.

REFERENCES

1. K. Goebbels, in Research Techniques in NDT, Vol. IV, ed. R. S. Sharpe, (Academic Press, London, 1980), 87.
2. M. Deka and N. Eberhardt, in Non-Destructive Methods for Material Property Determination, eds. C. O. Ruud and R. E. Green, (Plenum Press, New York, 1984), 135.
3. J. P. Monchalin and J. F. Bussiere, Ibid., 289.
4. C. B. Guo, P. Holler and K. Goebbels, Acustica, 59 (1985), 112.
5. A. Granato and K. Lucke, J. Appl. Phys., 27, (1956), 583.
6. A. B. Bhatia, Ultrasonic Absorption, (Clarendon Press, Oxford, 1967).
7. R. Becker and W. Doring, Ferromagnetismus, (Springer Verlag, Berlin, 1939).
8. G. Simon, Ann. Phys., 1 (1958), 23.
9. W. P. Mason, Rev. Mod. Phys., 25, (1953), 136.
10. H. Franz, Z. Metallkde, 53, (1962), 27.
11. J. Bressers, ed., Creep and Fatigue in High Temperature Alloys, (Applied Science Publishers Ltd, London, 1981).
12. U. Hildebrandt, H. Persch, K. Schneider and H. Willems, The potential of NDT methods for the characterization of creep and fatigue damage in high temperature materials. Paper submitted to: COST 50/COST 501 Conference "High Temperature Alloys for Gas Turbines and other Applications 1986", held in Liege Oct. 6-9, 1986.
13. H. P. Degischer, H. Aigner, R. Danzer and W. Mitter, Evaluation and qualification of uniaxial creep of X10NiCrAlTi 32 20, Ibid.
14. V. Guttman and J. Timm, 2nd progress report, COST 501 project CCR3, JRC Petten (1985).

A SINGLE TRANSDUCER BROADBAND TECHNIQUE FOR LEAKY LAMB WAVE DETECTION

P. B. Nagy*, W. R. Rose**, and L. Adler

Department of Welding Engineering
The Ohio State University
Columbus, Ohio  43210

INTRODUCTION

The introduction of advanced composite materials into many of the new generation aircraft and spacecraft has given rise to a significant increase in all aspects of operational capability. These materials come in many forms; organic and non-organic, fibrous and particulate, and endless combinations of the above which not only give enhanced strength characteristics, but deliver them in very specific design directions. The more common uses are in both primary and secondary aircraft structures, however, considerable effort is also being put into designing composite materials for use in hostile environments such as those destined for use as jet engine turbine blades.

The obvious advantage of these materials is in their high strength and stiffness to weight ratios. The anisotropy of the material allows strength to be utilized in a given design direction without the addition of extra weight resulting from strength in unnecessary directions, as is usually the case in isotropic materials. The use of these materials over their more conventional metallic counterparts can yield dramatic weight savings which can reduce fuel consumption and allow for carrying additional fuel or cargo.

The present and future importance of these materials demands that unique methods be developed for their testing and evaluation. Their inherent anisotropy and non-homogeneity create problems for standard testing procedures, however, through the use of new techniques these can be overcome.

The graphite/epoxy samples used in this work were manufactured from Hercules Corporation AS4/3501-6 Prepreg Tape. Each layer, or ply, in a given sample is made up of a thermo-set epoxy impregnated with unidirectional graphite fibers. The plies in a given laminate can all be in the same direction or the directions can be varied as desired. Due to the two components present in each ply and because of the discrete directions of the fibers, these materials are non-homogeneous and anisotropic by nature.

---

*Permanent Address: Applied Biophysics Laboratory, Technical University Budapest, Budapest, Hungary.

**Permanent Address: Aerospace Maintenance Development Unit, Canadian Armed Forces, Trenton, Ontario, Canada.

## LEAKY LAMB TECHNIQUES

Lamb waves are much better suited for interrogation of thin plates than the more conventional bulk wave techniques, because they propagate along the plates rather than through them. This study is limited to immersion techniques only, when the Lamb modes become "leaky" as they are highly attenuated by reradiating their energy into the liquid on both sides of the plate as they propagate.

Fig. 1 shows the schematic diagram of leaky Lamb wave generation in a thin plate. When a narrow-band tone-burst signal is incident on the plate at an arbitrary angle of incidence, and there is no phase-matching between the compressional wave in the liquid and one of the Lamb modes in the plate, the incident wave is simply specularly reflected. The beam profile is undistorted in this case, as it is shown by digitalized schlieren photography in Fig. 2a. The Lamb modes are highly dispersive, therefore a small change in the carrier frequency can result in phase-matching at the same angle of incidence. Fig. 2b and c show the redistributed reflected and transmitted beam profiles in the case of phase-matching. The slowly decaying leaky fields are the same on both sides of the plate, but the presence of the somewhat reduced specular reflection results in a destructive interference on the front side.

The most accurate technique to observe the Lamb resonances seems to be the detection of the beam split due to destructive interference between the specularly reflected component and the reradiated Lamb wave [1,2]. The main drawback of this method is its sensitivity to mechanical misalignments, especially in the case of highly anisotropic plates when the reradiated field of the leaky Lamb wave is shifted laterally not only in the plane of incidence, but out of it, also [3]. In spite of these difficulties, the outstanding accuracy and sensitivity of this technique ensure its important role as a basis for comparison for other methods. In the following, we shall discuss two broadband, single-transducer techniques of promising features for ultrasonic NDE of composite plates.

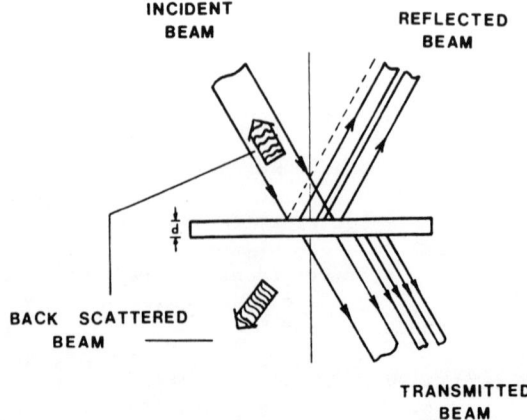

Fig. 1.  Schematic diagram of leaky Lamb wave generation.

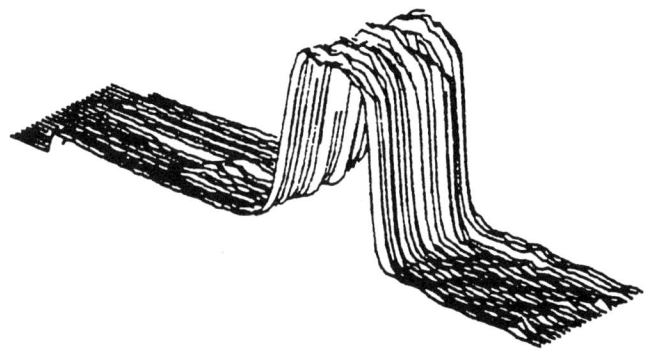

a, Specular reflection when there is no phase-matching.

b, Redistributed reflection when there is phase-matching.

c, Transmission when there is phase-matching.

Fig. 2. Beam profiles by digitalized Schlieren photography.

BACKSCATTERING TECHNIQUE

Backscattering from fiber reinforced composite laminates was first studied by Bar-Cohen and Crane [4]. They found that in directions perpendicular to the fibers of one or more plies the backscattered signal was much higher than elsewhere, which suggests that the scattering inhomogeneity is mainly oriented parallel with the fibers. The obvious possibility that the strong scattering is directly due to the embedded fibers was discarded by Achenbach and Qu [5] on the basis that they are too small in diameter and too closely packed to give rise to strong backscattering in the applied frequency range. The same authors explained the presence of strong backscattering even in seemingly porosity free, flawless plates by the irregularities of the fiber texture [6].

The physical nature of the scattered acoustic wave remained somewhat hidden in these studies, probably because many factors seem to contribute. Our experimental results show that the principal source of backscattering is the leaky Lamb wave generated in the plate. De Billy, Adler, and Quentin [7] found increased backscattering from macroscopically isotropic, but microscopically inevitably inhomogeneous plates at so-called Lamb angles using tone-burst technique. This phenomenon is analogous to the increased backscattering from liquid-solid interfaces at the Rayleigh angle, and can be fully explained by the similarities of the Rayleigh and Lamb waves [8].

In our experiment, we used a broadband ultrasonic transducer in pulse-echo mode. Fig. 3 shows the frequency spectrum of the backscattered signal at a particular fiber orientation and angle of incidence. The spectrum is modulated by random interference between uncorrelated inhomogeneities, therefore the Lamb resonances can not be recognized. Fig. 4 shows the same backscattered spectrum after spatial averaging and deconvolution by the transducer's frequency response. The arrows indicate the Lamb modes as measured by the well proven accurate beam split reflection technique. The peaks of the average backscattered spectrum obviously correspond to these Lamb modes, which indicates the important role of these plate vibrations in the backscattering generation.

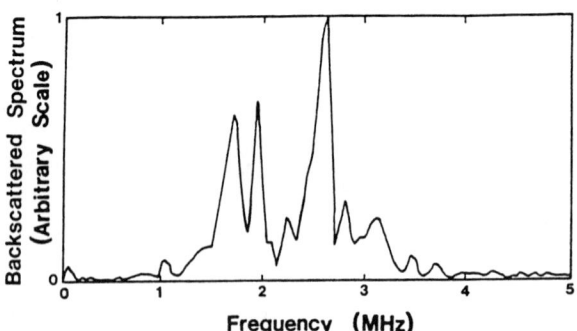

Fig. 3. Frequency spectrum of the backscattered signal from a unidirectional plate (perpendicular to the fibers at 24° angle of incidence).

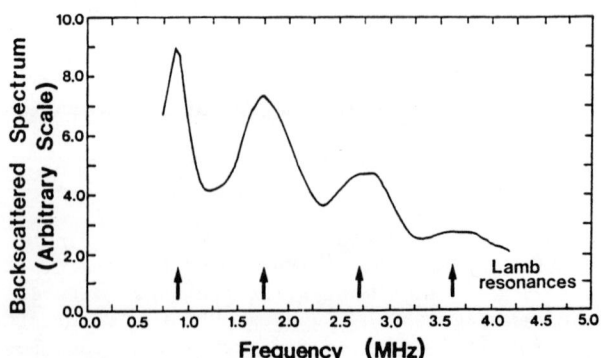

Fig. 4. Spatial averaged frequency spectrum of the backscattered signal (20 points, other parameters are the same as in Fig. 3).

## TRANSMISSION TECHNIQUE

Due to the strong acoustic impedance mismatch between the composite plate and the surrounding liquid, the specular transmission through the sample is usually very weak. At the so-called Lamb resonances, the transmission is greatly enhanced by double mode conversion between the compressional wave in the liquid and the Lamb wave in the plate. Half of the Lamb wave energy leaks back into the liquid on the back side of the plate and propagates parallel with the incident beam. The schematic arrangement for detecting these transmission peaks is shown in Fig. 5.

We use double transmission technique in order to simplify the mechanical alignment and enhance the contrast of the Lamb resonances. A broadband ultrasonic transducer is used to pick up the echo signal from a perpendicular plane reflector. The plate under study is placed somewhere between the transducer and the reflector, and the first double transmitted signal from the reflector is time-gated and frequency analyzed. There is no need for further alignment whatsoever. The lateral displacement of the through-transmitted signal both in and out of the plane of incidence is fully compensated by this double way arrangement, as well as the sharpness of the resonance peaks is further enhanced. It is very easy to change the angle of incidence and plate orientation since the received signal is always well defined and sufficiently strong. Fig. 6 shows the frequency spectrum of the through-transmitted signal for a unidirectional plate at 30° angle of incidence. The arrows indicate the Lamb resonances as measured by the beam split reflection technique.

Fig. 7 shows the dispersion curves for a unidirectional plate at normal fiber orientation. The phase velocity of the Lamb modes was controlled by the angle of incidence. Owing to the outstanding stability of the suggested double transmission technique, the sample can be rotated around any axis normal to the transducer in order to get a similar dispersion curve.

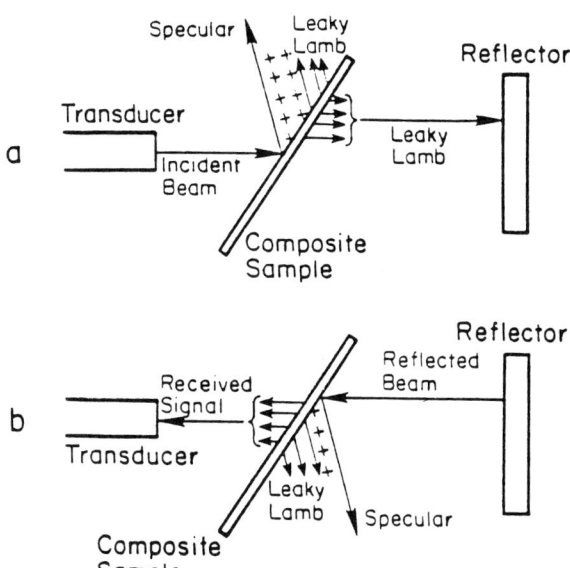

Fig. 5. Schematic diagram of the through-transmission method in (a) forward and (b) backward directions.

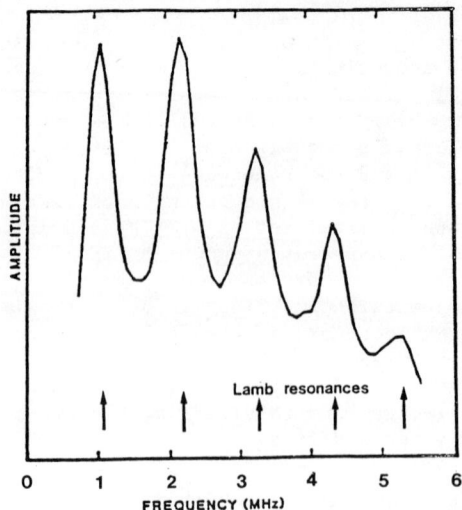

Fig. 6. Frequency spectrum of the through-transmitted signal (unidirectional plate at normal fiber orientation, 30° angle of incidence).

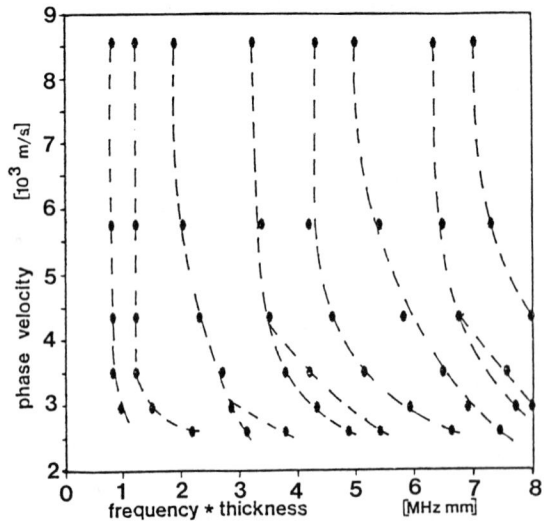

Fig. 7. Dispersion curves for a unidirectional plate at normal fiber orientation.

Polar diagrams can be readily obtained by rotating the sample around any axis normal to the plate. Different modes have very different degrees of anisotropy, and, beside the phase and group velocities, the strength of the mode conversion changes as well as the fiber orientation. We found that the very modes showing the highest (phase velocity) anisotropy are bound to disappear at certain azimuthal angles. Therefore, we chose as an example in Fig. 8 one of the less anisotropic modes which can be followed easily through all azimuthal angles.

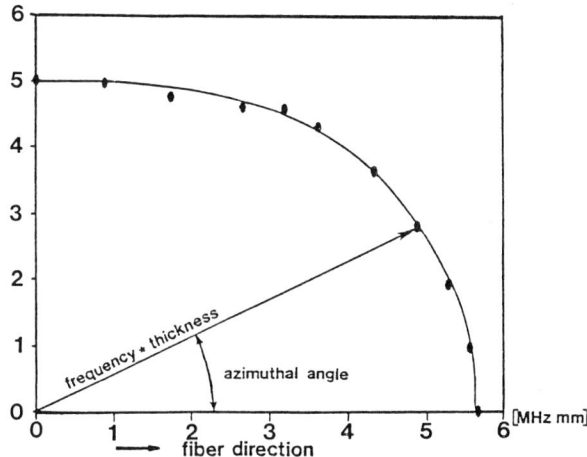

Fig. 8. Polar diagram for a unidirectional plane at 10° angle of incidence.

CONCLUSIONS

We investigated two broadband, single-transducer techniques for Lamb wave interrogation on thin anisotropic plates. The backscattering technique was shown to yield the elastic properties of the plate after extensive spatial averaging, but its main field of application is probably the qualitative or even quantitative characterization of inhomogeneities inside the plate.

A new double way through-transmission technique was introduced to determine the elastic properties of the plate as well. This simple, very stable arrangement yields superior signals with respect to other single-transducer techniques. Both (phase velocity) dispersion and anisotropy data are readily measured by this technique. The elastic properties of the plate can be mapped easily by scanning the sample at high speed. Furthermore, by measuring the amplitude of a certain Lamb resonance in the spectrum of the double through-transmitted signal, we can determine the attenuation coefficient of each mode at any point of the sample.

ACKNOWLEDGEMENT

This work was partially supported by the Canadian Armed Forces.

REFERENCES

1. D. E. Chimenti and A. H. Nayfeh, "Leaky Lamb waves in fibrous composite laminates," J. Appl. Phys. 58, 4531 (1985).
2. Y. Bar-Cohen and D. E. Chimenti, "Nondestructive evaluation of composite laminates," Review of Progress in Quantitative Nondestructive Evaluation Vol. 5B, D. O. Thompson and D. E. Chimenti, eds., (Plenum, New York, 1986) p. 1199.
3. W. R. Rose, S. I. Rokhlin, and L. Adler, "Evaluation of anisotropic properties of graphite/epoxy composite plates using Lamb waves," Review of Progress in Quantitative Nondestructive Evaluation Vol. 6, D. O. Thompson and D. E. Chimenti, eds., (Plenum, New York, in press).

4. Y. Bar-Cohen and R. L. Crane, "Acoustic backscattering imaging of subcritical flaws in composites," Mater. Eval. 40, 970 (1982).
5. J. D. Achenbach and J. Qu, "Backscattering from flaw distributions in composite materials," Review of Progress in Quantitative Nondestructive Evaluation Vol. 5B, D. O. Thompson and D. E. Chimenti, eds., (Plenum, New York, 1986) p. 1179.
6. J. Qu and J. D. Achenbach, "Analytical treatment of polar backscattering from porous composites," Review of Progress in Quantitative Nondestructive Evaluation Vol. 6., D. O. Thompson and D. E. Chimenti, eds., (Plenum, New York, in press).
7. M. de Billy, L. Adler, and G. Quentin, "Measurements of backscattered leaky Lamb waves in plates," J. Acoust. Soc. Am. 75, 998 (1984).
8. L. E. Pitts, T. J. Plona, and W. G. Mayer, "Theoretical similarities of Rayleigh and Lamb modes of vibration," J. Acoust. Soc. Am. 60, 374 (1976).

SIMULTANEOUS MEASUREMENTS OF ULTRASONIC

PHASE VELOCITY AND ATTENUATION IN SOLIDS


N. K. Batra and P. P. Delsanto

Department of the Navy
Naval Research Laboratory, Washington, D.C. 20375-5000

ABSTRACT

Several techniques have been proposed for the ultrasonic characterization of metals, ceramics and composite materials. For highly attenuative materials, e.g., fiber-reinforced composites, either the sound phase velocity, or the attenuation can usually be determined. In this paper, we extend the correlation method for simultaneous measurement of phase velocity and attenuation in liquids, first proposed by Sedlacek and Asenbaum[1], to the case of highly attenuative solids. By using specially designed specimens in the shape of wedges, the path of propagation can be continuously varied. Cross-correlations of pressure amplitude and phase between different points in the field of continuous ultrasonic plane waves allow measurement of attenuation and phase velocity. Experiments performed on several specimens of various polymers confirm the efficiency and reliability of the technique. Corrections due to refraction and other effects are also discussed and evaluated.

INTRODUCTION

Ultrasonic pulse echo systems have been widely used in recent years to measure the attenuation and ultrasonic group velocity with applications for tissue characterization in medicine and material characterization in NDE. Pulse-echo techniques require plane parallel specimens of low attenuation so that multiple echoes can be set up in the specimen. They do not measure absolute attenuation because of errors due to reflection, transmission and surface scattering. Due to a lack of spectral purity, they do not provide us with phase velocity, which is a more relevant quantity for the determination of elastic constants, stresses, texture and material phases in the material. Ultrasonic attenuation due to impedance mismatch, surface roughness and diffraction is generally irrelevant as far as the material characterization is concerned. The more relevant quantities are the attenuation due to absorption, $\alpha_{abs}$ and to scattering, $\alpha_{sc}$. Thus, accurate and absolute values of the attenuation $\alpha_{tot} = \alpha_{abs} + \alpha_{sc}$ and phase velocity can characterize the material almost completely. Simultaneous measurements of these quantities are therefore particularly important for material characterization.

Simultaneous measurements of phase velocity and attenuation can be carried out by continuous wave (CW) ultrasonic spectroscopy.[2] Standing waves are set up in the parallel plate specimen acting as a resonant cavity. The "$Q$" of the resonances gives attenuation, whereas frequency spacing between two consecutive resonances gives the phase velocity. This technique, however, is not applicable to heterogeneous materials or to materials with high acoustic attenuation. In fact, in these cases, multiple resonances are either missing or broad and overlapping making the measurements meaningless, particularly when accurate measurements of acoustic parameters are required to study variations in material properties.

Recently, Sedlacek and Asenbaum[1] measured attenuation and phase velocity in liquids by continuously varying the path of propagation of ultrasonic waves and cross correlating the transmitted and received signals. It is the purpose of the present paper to propose a variation of their technique, which can be applied to the case of highly attenuative solids. By using specially designed specimens in the shape of wedges, the path of propagation can be continuously varied. Cross-correlations between the incident CW and the propagated wave through continuously varying thickness allow absolute measurements of attenuation and phase velocity.

The theory and experimental set-up used for the development of this technique are explained in the next two sections. In the following sections, we illustrate some of the results obtained with our technique for an epoxy specimen.

THEORY

In order to characterize a material, we fabricate a specimen of it in the shape of a wedge and mount it on a rotating turntable in an ultrasonic immersion tank, as shown in Fig. 1. Continuous longitudinal waves are propagated through the sample. We assume that the wedge angle $\theta$ is small and neglect, for the time being, all refraction and multiple reflection effects. The ultrasonic signals, $S_T$ sent by the transmitter $T$, and $S_R$ received by the tranducer $R$, are given by

$$S_T = F_T \; A_0 \cos(\omega t + \phi_0) \tag{1}$$

$$S_R = T_{12} \cdot T_{21} \cdot F_R A_0 \exp[-\alpha_1 (l-y) - \alpha_2 y] \tag{2}$$
$$\times \cos\left[\omega\left(t - \frac{l-y}{v_1} - \frac{y}{v_2}\right) + \phi_0\right]$$

where $F_T$ and $F_R$ are the response factors of the transmitter, receiver and associated electronics; $T_{12}$ and $T_{21}$ are the transmission factors between medium 1 (water) and medium 2 (material) and viceversa; $\alpha_1, \alpha_2, v_1$, and $v_2$ are the attenuation coefficients and phase velocities in the two media; $A_0$ is the initial amplitude of the CW; $l$ is the (fixed) distance between the transducers and $y$ is the length of the propagation path of the CW in the material. By moving the wedge (or the transducers) parallel to the $x$-axis, $y$ can be varied. By choosing the origin of the $x$-axis at the vertex of the wedge, we have

$$y = x \tan\theta \tag{3}$$

where $\theta$ is the angle of the wedge.

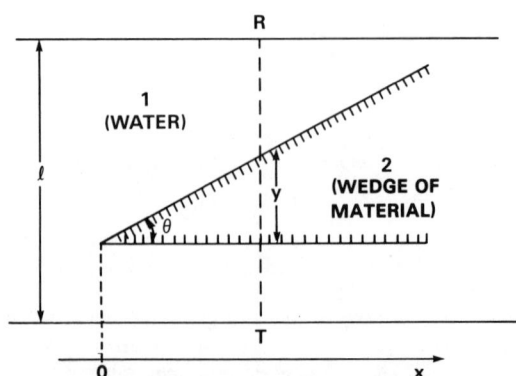

Fig. 1  Arrangement of specimen and transducers (refraction effects ignored).

Through an electronic mixer (phase detector) the two signals $S_T$ and $S_R$ are correlated. The corresponding non-normalized cross-correlation factor is given by

$$\psi = \int_{-T}^{T} S_T(t) S_R(t) \, dt \qquad (4)$$

$$= F_T F_R T_{12} T_{21} A_0^2 \exp[-\alpha_1(l-y) - \alpha_2 y] \cdot \mathcal{I}$$

where

$$\mathcal{I} = \int_{-T}^{T} \cos(\omega t + \phi_0) \cos\left[\omega\left(t - \frac{l-y}{v_1} - \frac{y}{v_2}\right) + \phi_0\right] dt$$

$$= 1/2 \int_{-T}^{T} \cos\left[\omega\left(\frac{l-y}{v_1} + \frac{y}{v_2}\right)\right] dt \qquad (5)$$

$$+ 1/2 \int_{-T}^{T} \cos\left[2\omega t + 2\phi_0 - \omega\left(\frac{l-y}{v_1} + \frac{y}{v_2}\right)\right] dt$$

The cross-correlation time interval $2T$ can be chosen arbitrarily large. If we choose

$$2T \gg 2\pi/\omega \qquad (6)$$

The second integral in Eq. (5) vanishes and we obtain

$$\mathcal{I} \cong T \cos\left[\omega\left(\frac{l-y}{v_1} + \frac{y}{v_2}\right)\right] \qquad (7)$$

It then follows

$$\psi = C \exp(-\alpha y) \cos(by + \phi) \qquad (8)$$

where

$$C = F_T F_R T_{12} T_{21} A_0^2 T \exp(-\alpha_1 l)$$

$$\alpha = \alpha_2 - \alpha_1$$

$$b = \omega(s_2 - s_1) \qquad (9)$$

$$\phi = \omega l s_1$$

$s_i$ is the slowness given by

$$s_i = 1/v_i \qquad (i = 1,2) \qquad (10)$$

A typical experimentally obtained plot of the cross-correlation factor $\psi$, showing the sinusoidal oscillations and exponentially decreasing amplitudes, as predicted by Eq. (8), is shown in Fig. 2. In order to analyze it, we can consider three maxima (or minima) of the cross-correlation factor $\psi_1, \psi_2$, and $\psi_n$ corresponding to the propagation paths $y_1, y_2$, and $y_n$ respectively (Fig. 3). At these points we have

$$by_i + \phi = \pm 2\pi m \qquad (i = 1, 2, n) \qquad (11)$$

where $m$ is an integer and

$$\psi_i = C \exp(-\alpha y_i) \qquad (i = 1, 2, n) \qquad (12)$$

It immediately follows

$$\alpha_2 = \alpha_1 + \log(\psi_1/\psi_n)/(y_n - y_1) \qquad (13)$$

and

$$s_2 = s_1 \pm 2\pi/\omega \Delta y \qquad (14)$$

The sign in Eq. (14) can be easily determined by looking at the direction in which $\psi$ moves when $y$ increases: If $\psi$ moves to the left (right), the sign is positive (negative). From Eq. (13) and (14) one can immediately determine the attenuation $\alpha_2$ and slowness $s_2 = 1/v_2$ of the material, since the corresponding quantities for water are well known.

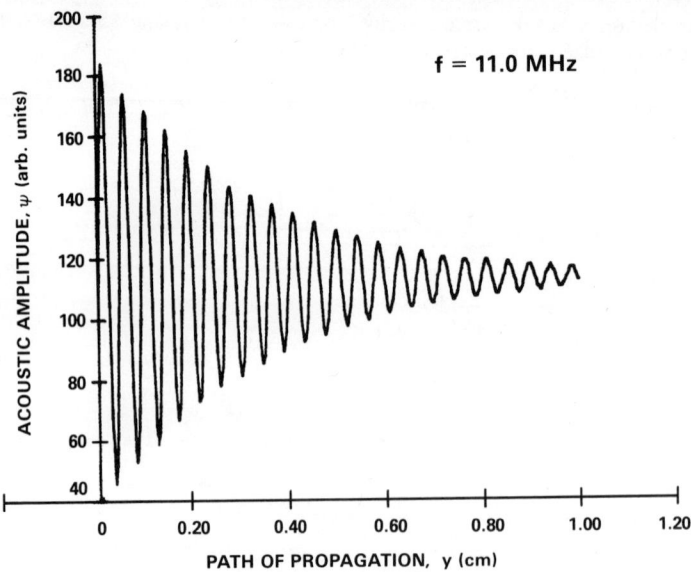

Fig. 2   Typical experimental cross-correlation curve.

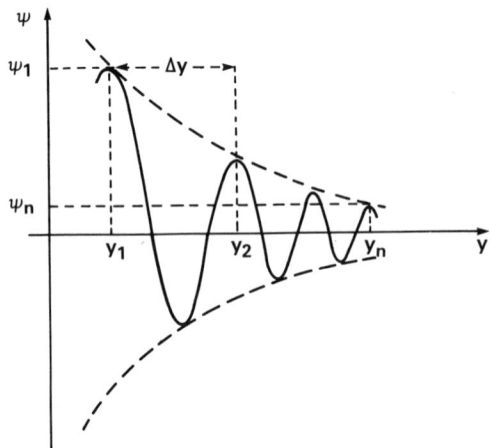

Fig. 3   Graphical representation of the data analysis.

We wish to analyze now the effect of refraction (Fig. 4). One side of the wedge is normal to the incoming wave and, therefore, does not cause refraction. The other side causes a deviation $\theta - \theta_R$ from the original propagation direction. $\theta$ is the wedge angle, assumed to be small, and $\theta_R \leqslant \theta$ is the correspondent refraction angle $\theta - \theta_R$. As a consequence, the receiving transducer must be rotated through an angle. The acoustic path increases by a distance

$$\Delta = (l - l_0 - y)\,[1/\cos(\theta - \theta_R) - 1] \qquad (15)$$
$$\simeq 1/2\,(l - l_0 - y)(\theta - \theta_R)^2$$

As a result, one obtains frequency independent corrections to both the attenuation $\alpha_2$ and the slowness $s_2$

$$\Delta\alpha_2 = \alpha_1\,(\theta - \theta_R)^2/2$$
$$\Delta s_2 = s_1\,(\theta - \theta_R)^2/2 \qquad (16)$$

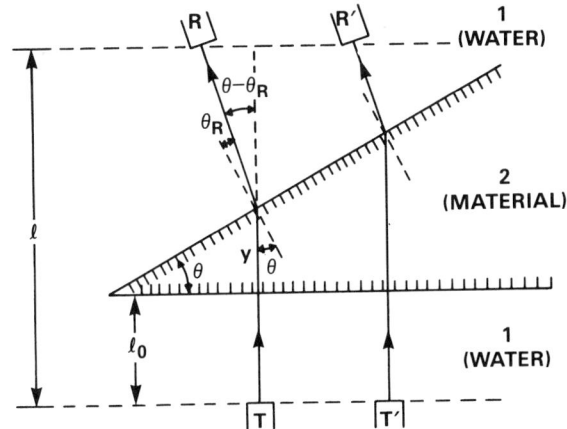

Fig. 4  Arrangement of specimen and transducers (refraction effects included).

which can be easily evaluated if the refraction index of the material is known or by simply measuring $\theta_R$.

Finally, we discuss briefly the effect of multiple reflections. Unless $T_{12} = T_{21} = 1$, there will be multiple reflections of the wave at both sides of the wedge and at $T$ and $R$. A number of waves are thus generated, which add up to the basic signal $S_R$, given by Eq. (2). Assuming $\theta \ll 1$ and considering only multiple reflections within the material wedge, we note that if the wave is reflected $n$ times each side of the wedge, its acoustic path within the wedge is $(2n + 1)y$. Its amplitude is multiplied by a factor $R^{2n}$, where $R$ is the reflection coefficient of the material, assumed to be the same at the two surfaces of the wedge. As a result, we obtain waves with cross correlation factors still given by Eq. (8) but with different values of the parameters. The interference of all these waves creates a slight modulation pattern, which needs to be explicitly taken into account for a better fitting of the experimental curves.

Usually, the factor $R^{2n}$ and the attenuation in the return trips reduce the amplitude of the waves so much that only doubly reflected waves ($n = 1$) need to be considered. Since the attenuation increases with the frequency, the modulation due to multiple reflection becomes negligible at higher frequencies.

EXPERIMENTAL SET-UP

Before starting with the acquisition of the experimental data, the orientation of the sample is adjusted such that the incident beam parallel to the $Z$-axis scans the horizontal surface ($XY$) of the wedge. The transmitting and receiving transducers are on the same bridge and scan together in the $X$-direction and index in $Z$-direction. The $Z$-position of $R$, $Y$-position of $T$ and their polar as well as azimuthal positions can be adjusted independently. The positions of $T$ and $R$ are adjusted using pulse-echo as well as pitch-catch techniques, using a pulse-receiver. The ultrasonic beam is incident normal to the $X-Y$ plane. The position and orientation of the transducer $R$ are adjusted so that the received signal is maximum.

Once the orientations of specimen and transducers are fixed, the experimental set-up is switched to, as shown in Fig. 5. Continuous waves of suitable frequency are divided into two parts by a splitter. One portion of these waves is amplified and excites a broad-band immersion transducer ($T$). The other portion is fed into a phase-sensitive detector (mixer). The dc output of the mixer (proportional to the phase difference between input and output signals) is filtered, digitized and transmitted to a VAX computer, under the control of a Tektronix desk top computer. The filtered signal can also be recorded on a $X-Y$ recorder as a function of the propagation path $y = x \tan \theta$ in the material wedge.

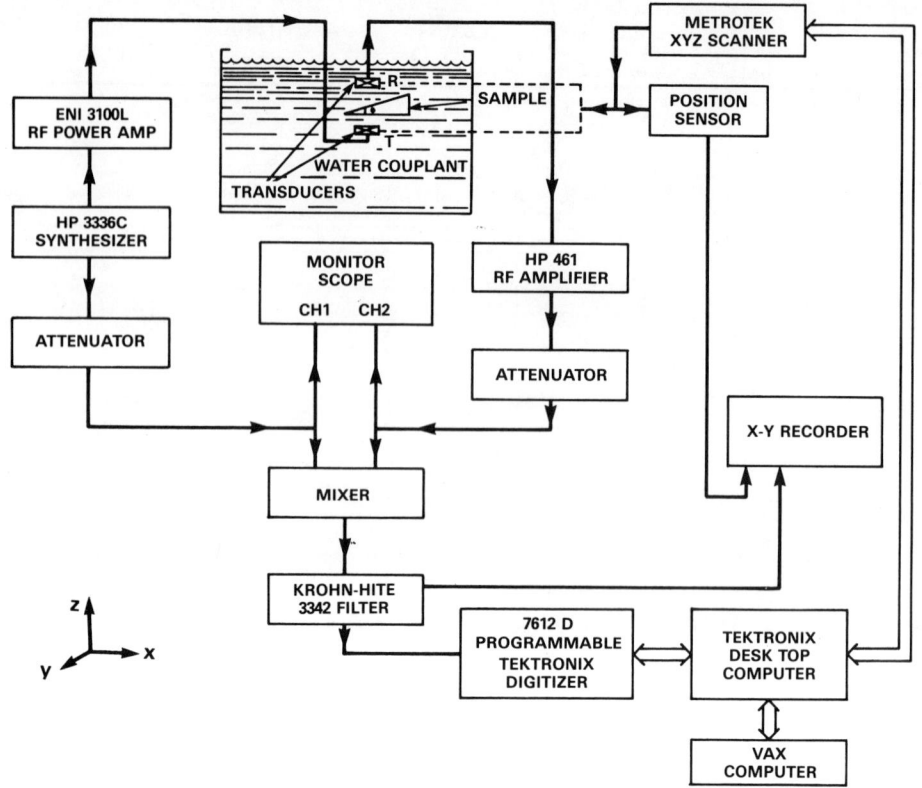

Fig. 5 Experimental set-up.

## RESULTS AND DISCUSSIONS

We have already seen in Fig. 2 a plot of the acoustic amplitude as a function of the propagation path in the specimen at a frequency of 11 MHz. Two more examples are shown in Fig. 6 and 7 (at 9 and 19 MHz, respectively). From Figures 2, 6 and 7 it clear that the wave is exponentially decaying, as predicted by Eq. (8), the damping being caused by the interaction of the ultrasound with the molecules of the material. Figure 6 also shows a slight modulation, due to multiple reflections, as discussed earlier. At higher frequencies (11 and 19 MHz), we see that the modulation becomes negligible. It is also clear that the attenuation $\alpha_{tot}$ increases with the frequency, as expected.

Similar results, obtained at other frequencies, have allowed us to determine through Eqs. (13) and (14), the frequency dependence of the phase velocity $v$ and attenuation coefficient $\alpha_{tot}$. The corresponding plots (Figs. 8 and 9) show that the material investigated (epoxy) has no dispersion in the frequency range considered (9 to 20 MHz) and a linear dependence of the attenuation on the frequency. The phase velocity and the attenuation coefficient were found to be

$$v_2 \ (km/s) = 2.165 \ \pm 0.005 \tag{17}$$

$$\alpha_2 (dB/cm) = (4. \pm 1.) + (2.25 \pm 0.05) \ f \ (MHz) \tag{18}$$

If there are any defects present in the material (either due to density change, cracks or inclusions or velocity changes due to material inhomogeneities or stress), the acoustic wave loses its intensity not only by absorption, but also by scattering. Consequently, the total attenuation will increase as the propagating ultrasound interacts with those regions. Figure 10 shows an example of such a case. It refers to a specimen with a cylindrical hole of diameter 100 mil, normal to the direction of propagation of the wave. As the specimen is scanned, one can obtain

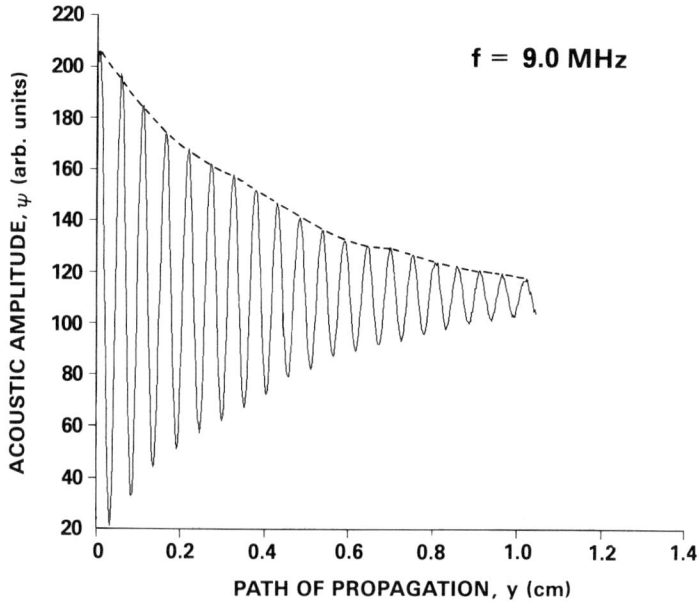

Fig. 6  Experimental cross-relation at 9 MHz. A slight modulation of the exponential envelope, due to multiple reflections is apparent.

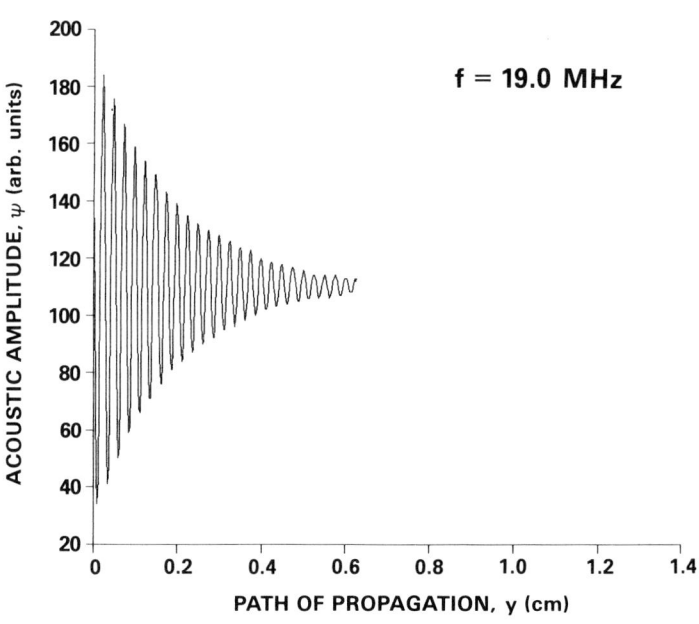

Fig. 7  Experimental cross-correlation curve at 19 MHz. No modulation of the exponential envelope is noticeable.

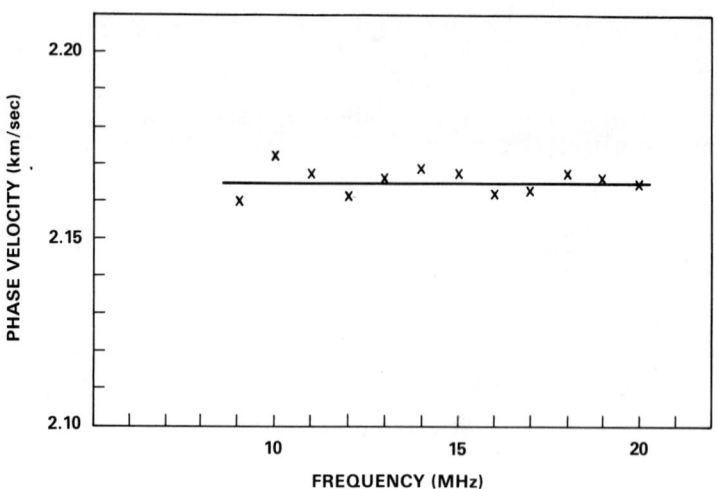

Fig. 8  Dispersion curve for an epoxy specimen.

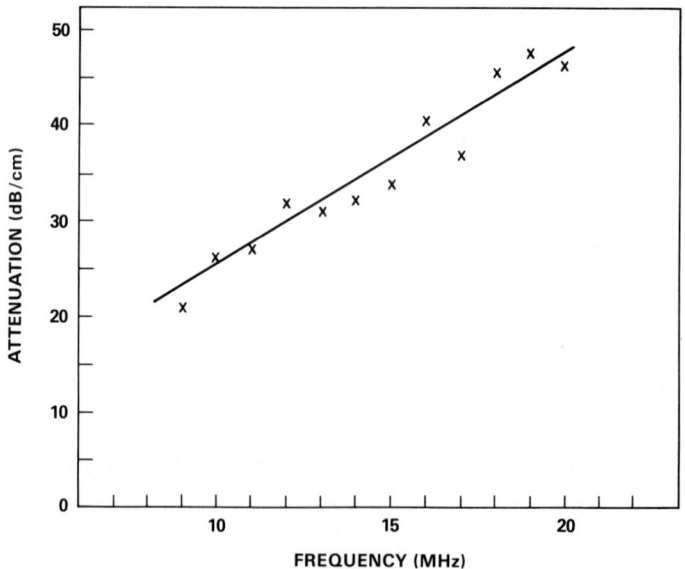

Fig. 9  Attenuation vs. frequency for an eqoxy specimen.

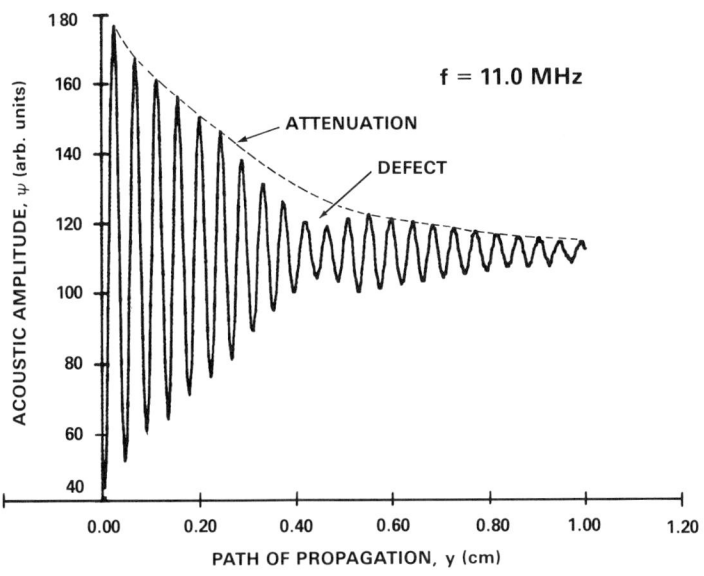

Fig. 10  Cross-correlation curve for a specimen with a defect.

values of the attenuation due to absorption and phase velocity from the initial (or final) part of the curve. The strong deviation of the envelope from exponential at $y \simeq 0.4$ cm, indicates the presence of the flaw in that region. With this technique, one can not only detect the presence of flaws, but also characterize them quantitatively. In fact, if one subtracts the attenuation due to absorption from the total attenuation, one obtains the attenuation due to scattering which can be interpreted as the signature of the flaw.

CONSLUSION

We have presented an ultrasonic technique, which allows the *simultaneous* determination of phase velocity, attenuation and flaws in a material. The technique is based on the cross-correlation between an input CW of given frequency and the corresponding signal transmitted through a wedge specimen of the material. The theory predicts that, for an unflawed homogeneous specimen, the acoustic amplitude vs. propagation path is a sinusoidal curve, with exponentially decaying amplitude. From the experimentally obtained curve one can easily obtain, by fitting the parameters of the theoretical curve, both phase velocity and the attenuation coefficient. If flaws are present, they manifest themselves as deviations from the theoretically predicted exponential curve. By subtracting the experimental from the theoretical curve, one obtains the "signature of the flaw", which can be used for a quantitative characterization.

Corrections due to refraction and multiple reflections do not substantially alter the results of our analysis. They can produce, however, a modulation of the envelope of the cross-correlation curve, which can be easily taken into account through a least-square analysis of the experimental data.

ACKNOWLEDGEMENTS

This project was sponsored by the Ships and Submarine Materials Block Program managed by DTNSRDC.

REFERENCES

1. M. Sedlacek and A. Asenbaum, J. Acoust. Soc. Am. 62, 1420-3 (1977).

2. N.K. Batra and H.H. Chaskelis, Nondestructive Testing Comm. 2, 65-76 (1985).

SHEAR WAVE IMAGING OF SURFACE AND NEAR-SURFACE FLAWS IN CERAMICS

J. D. Fraser and C. S. DeSilets

Precision Acoustic Devices, Inc.
Fremont, CA 94539

B. T. Khuri-Yakub

Ginzton Laboratory, Stanford University
Stanford, CA 94305

INTRODUCTION

A 50 MHz shear wave transducer has been built on a silicon nitride buffer wedge, and operated with a modified C-scan imaging system to form a high frequency shear wave nondestructive testing system for ceramics. We have shown that coupling between the wedge and a sample at 45 degrees incidence angle is adequate for imaging purposes, and can be maintained even on fine-ground surfaces (about 1 micron rms surface roughness), while scanning at linear speeds of at least 25 mm per second, using sucrose solutions of 50% to 60% concentration as the couplant. We have shown that it is possible to build a system which will detect flaws of order 20 microns in size with .4mm resolution to a depth of 7mm, while scanning about 1 square inch per minute. Advantages of this method over the commonly used longitudinal wave techniques include efficient coupling of the sound into the sample, absence of a front-surface echo, and enhanced depth resolution due to the shorter wavelength of shear waves.

DISCUSSION

The basic geometry of the system is shown in Figure 1. A 50MHz lithium niobate shear wave transducer is fabricated on a silicon nitride 45-degree wedge with a slightly convex bottom face, so good contact can easily be made to a substrate. The transducer is mounted in a gimballed housing to allow scanning while tracking the surface of the substrate. A computer controlled, two-dimensional raster scanner, pulse echo electronics, a gated detector and an analog to digital converter complete the setup. Samples with known defects can be scanned, and echoes from the defects can be recorded as RF oscilloscope photographs or as digitally stored amplitudes. These amplitudes can be used to construct line scans and pictures.

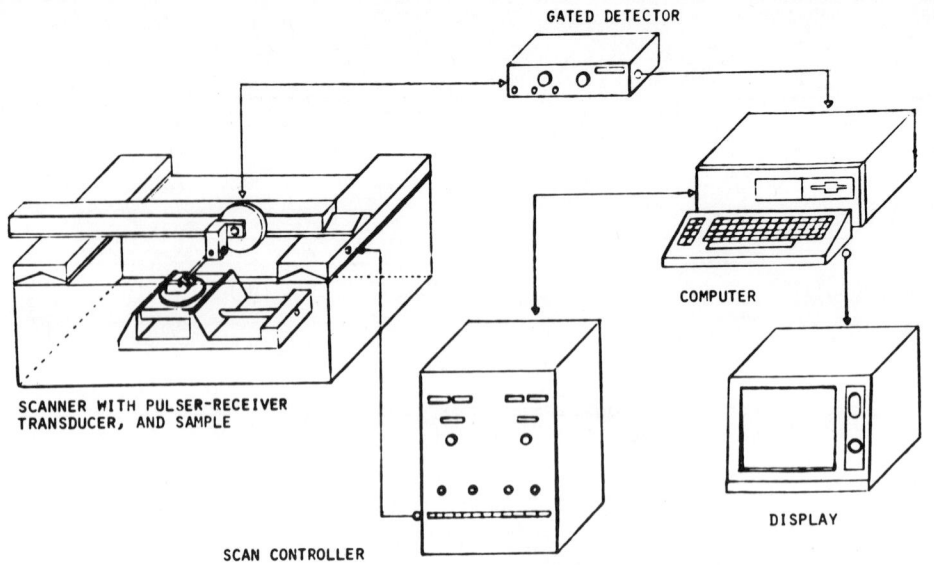

Figure 1. Block diagram of the 50MHz, wedge coupled shear wave imaging system.

    A transducer was fabricated indium bonding a lithium niobate shear oriented crystal to a silicon nitride block, and polishing it to a thickness of 48 microns, yielding a shear wave transducer of resonant frequency 50 MHz. The far side of the block was then lapped and polished at a 45 degree angle, with a slightly convex (approximately 1 meter radius) surface. A top electrode of chrome-gold with a one millimeter diameter defined the acoustic beam, and the transducer was mounted in a brass housing for mechanical support and electrical shielding. The housing included pivot points for the gimballed scanning system which would allow the transducer to move while tracking the surface of the samples.

    During fabrication, the impulse response of the transducer was measured and compared to theoretical values. The result is shown in Figure 2. As can be seen, the correspondence is good, indicating that the transducer was well fabricated. The differences between the two impulse responses are probably due to lack of parallelism between the transducer and the far side of the silicon nitride block.

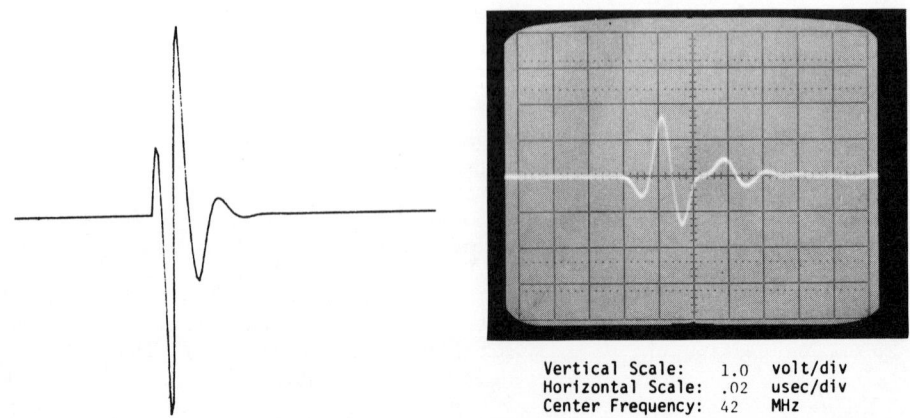

Figure 2. Theoretical vs. experimental impulse response of the 50MHz shear wave transducer.

A set of hot-pressed silicon nitride samples with seeded defects was obtained from Professor Khuri-Yakub at Stanford University. These samples are approximately one inch in diameter and .2 to .3 inch thick. Each sample contains inclusions of known type near its center. The types of inclusions represented are silicon, iron, and carbon. For each type, both 100 micron and 400 micron sizes are represented, and samples are available with both polished and fine ground surfaces.

Experiments were performed to verify efficient coupling of shear waves and detection of defects in the samples by manual coupling of the transducer from the wedge to the sample is by mode conversion to and from longitudinal waves in the water, as shown in Fig. 3. It was found that with careful attention to the cleanliness of the surfaces and the water, and with some wringing of the surfaces together, that efficient coupling could be achieved, and echoes could be detected from some flaws. However, this form of coupling was difficult to maintain, as it tended to weaken if pressure were not maintained or if much movement were attempted, or to lock in position if pressure were maintained. Also, it was suspected from the strengths of the back wall echoes that the coupling was not close to complete. We had predicted theoretically that the coupling efficiency would be approximately 50% if the water layer could be made 1/50 wavelength thick (about .6 micron), but would decay to about 3% if the layer were 1/10 wave thick (3 microns). We could not directly measure the coupling or the water layer thickness, but we could inspect the echoes from the far end of the buffer wedge. As shown in figure 4, there are two principal echoes from the fare end of the wedge, one which reflects from the front surface of the sample, and one which reflects from the back. If the coupling is poor, the first will be much stronger than the second. However, if the coupling is strong, both echoes will be present, and the second echo may be larger than the first. We were able to use this technique to estimate the strength of the coupling.

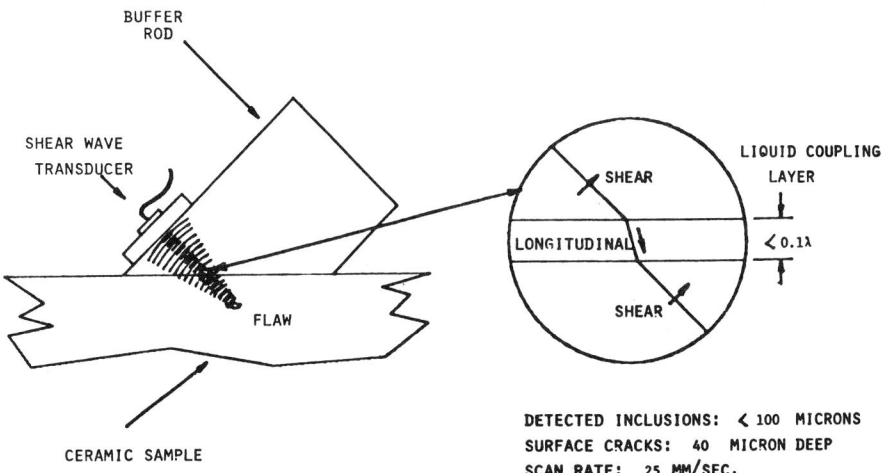

Figure 3. Detail of the coupling mechanism by which shear waves of vertical polarization are transmitted with low loss through a liquid interface.

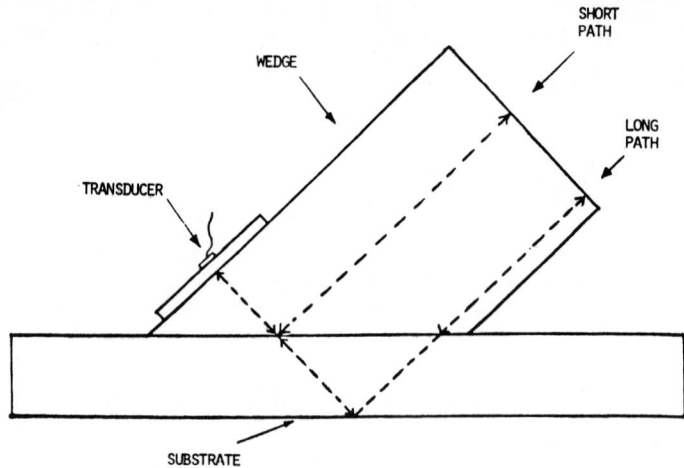

Figure 4. Sound paths for back wall echoes for a wedge transducer with incomplete coupling between wedge and sample.

Due to the difficulty of maintaining good coupling while scanning with plain water, a series of experiments was carried out with sucrose solutions to achieve better performance. Sucrose solutions are known to have extremely high shear viscosities (of order $10^{15}$ centipoise), are easy to prepare and clean up, and are non-toxic. We measured the relative strengths of the two rear-surface echoes as a function of sucrose concentration, both with the transducer wrung to the sample, and while scanning at one inch per second (the maximum scanning rate of our system). We also noted the duration for which coupling would be maintained during continuous scanning without addition of couplant. The results are shown in Table 1 below.

Our experiments have indicated that a scanning speed of at least 25mm per second is feasible. With a lateral resolution of .4mm, this will yield a scan rate of 6 square cm per minute, or .93 square inch per minute. This system could scan from the surface of a sample to a depth of about 6mm at this rate when configured as we envision.

We have done tests with polished and fine ground surfaces on the samples. Our polished samples have a surface roughness of less than .1 micron rms, while our fine ground samples have a surface roughness of

Table 1. Relative coupling effectiveness and persistence for various concentrations of sucrose in water.

| Concentration (weight %) | Scanning Coupling (dB) | Persistence (minutes) |
|---|---|---|
| 0 | -1.3 | 3 |
| 10 | -3.7 | 6 |
| 25 | -4.1 | > 15 |
| 50 | -2.5 | > 15 |
| 60 | -6.0 | 10 |

approximately 1 micron rms. We have seen no difference in the coupling achievable with the two surface finishes. Scanning is more difficult on rough surfaces, however, and the polished wedge tended to lock to the samples even with 50% sucrose solution. We found that increasing the sucrose concentration in our solution to 60% allowed smooth scanning.

There is a strong background signal level arising from backscattering of the sound beam propagating in the wedge material by small flaws in the wedge material itself, both between the transducer and the front face, and in the reflected beam propagating toward the far end of the wedge. Backscattered sound from flaws in the path of the beam propagating toward the far end of the wedge form a serious limitation on the sensitivity of the system, since it arrives at the transducer at the same time as an echo from a flaw in the sample. If the flaws in the wedge material are larger or more numerous than those in the sample, their signals will obscure those from the sample. We compared the system electrical noise to the backscattered signal, and found the degradation to be 16 dB. That is, the minimum detectable signal in the presence of the backscattering from our particular wedge transducer was 16 dB higher than it would have been in the absence of backscattering. Additionally, the receiver signal to noise ratio could easily be improved from the present 15 dB to 5 dB. Thus a total improvement of 25 dB could be achieved in a new system incorporating design changes to eliminate backscattering in the wedge and to reduce the electrical noise to the lowest practical value.

We modified the transducer and the electronics before continuing with the experiments. The ideal solutions, making a new wedge transducer on a defect-free sapphire wedge and designing and building a new receiver, were beyond the time and money allocated to this project. In order to decrease the effects of background noise due to backscattering in the wedge, we chose to operate in what is called a pitch-catch mode. The single transducer electrode was replaced with a pair of electrodes 1.3 mm in diameter and separated by .7 mm. We then used one electrode for the transmitter, and one for the receiver. This kept the transmitted beam separate from the receiver transducer's field of view in the near part of the field, decreasing the strength of the backscattering relative to the desired flaw echoes. Essentially, the lateral resolution was improved, because the system was only sensitive to reflections arising in the region where the two beam patterns overlapped. A 50% sucrose solution was adopted as the standard coupling medium.

We have done scattering calculations, as shown in Figure 5, which show that different types of flaws can be expected to show different echo shapes. An RF echo from a silicon flaw is shown in Figure 6, which shows similarity to the calculation of Figure 5. We expect to be able to use a signal processing to determine the type and size of flaws detected by the system, by automatically comparing features of RF echoes to learned features of echoes from known flaws.

With the modified protocol and equipment, we were able to scan samples reliably, both manually and automatically. We have been able to test the samples for detectability of the flaws, and obtained the results shown in Table 2. Flaws were detected in samples of all three types and both nominal sizes, with adequate signal to noise ratios. From our theoretical results, however, we expect the ratios of signals received from 400 micron flaws to those received from 100 micron flaws to be 20dB for silicon, 16dB for iron, and 22dB for carbon. We do not see that experimentally. The theory is based on well established results, and it is probable that the variance is due to multiple targets within the beamwidth, or clumps of particles, which would act like larger particles. Another possibility is that during the hot pressing process, the seeded defects react chemically with the silicon

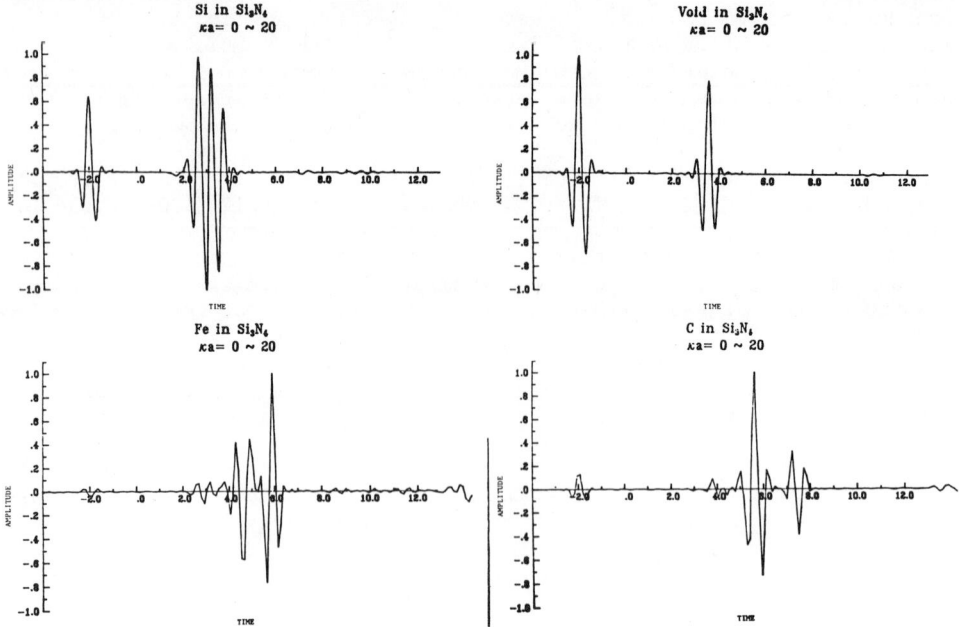

Figure 5. Theoretical impulse responses for various spherical flaws in silicon nitride.

nitride to form a zone of changed material larger than the original seed. In the next phase of the project, detailed tests comparing ultrasonic measurements to destructive test results will be required to establish the actual response to a single flaw of known size.

We expect that further work could produce increases in sensitivity of 10dB due to improved receiver signal to noise ratio, 16dB due to decreased beam diameter when focussing is used, and 16dB due to decreased scattering noise when a sapphire buffer is used. Thus, the total improvement possible should be about 42dB. From these facts, the results of our scattering calculations, and using the signal to noise ratios from 400 micron flaws in

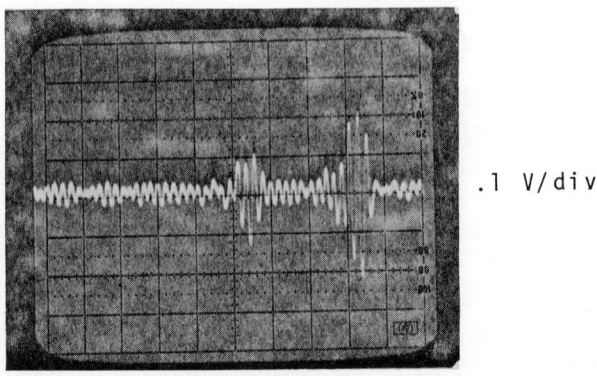

Figure 6. RF echo from a 400 micron silicon inclusion in silicon nitride.

Table 2, we estimate that the flaw sizes detectable by this technique will be approximately 18 microns for silicon, 25 microns for iron, and 20 microns for carbon. We have presumed here that these echoes are from single flaws, since only one target was found in each case, and 400 micron powder is not very susceptible to clumping.

An interesting experiment was done with the digital storage system and a 2.5 mm thick silicon nitride sample which had a calibrated 40 micron deep tight crack. The crack was traversed in a line scan while recording the peak value of the reflected signal for every .25mm of lateral motion. These data are plotted in Fig. 7, showing two distinct echoes; one direct reflection and one resulting from a two-bounce path to the crack. Both echoes have high signal to noise ratios, suggesting that considerably smaller cracks could be detected.

Table 2. Detectibility of Flaws

| Sample Number | Type | Nominal Flaw Size | Target Signal to Noise Ratio | Comments |
|---|---|---|---|---|
| 4 | Si | 400 | 20dB | |
| 1 | Si | 100 | 12dB | Multiple targets |
| | | | 8dB | Clumps? |
| 86 | Fe | 400 | 13dB | |
| 2 | Fe | 100 | 16dB | Multiple targets |
| | | | 10dB | Clumps? |
| 67 | C | 400 | 19dB | Polished |
| 81 | C | 400 | 12dB | Fine ground |
| 27 | C | 100 | 15dB | Clumps? |

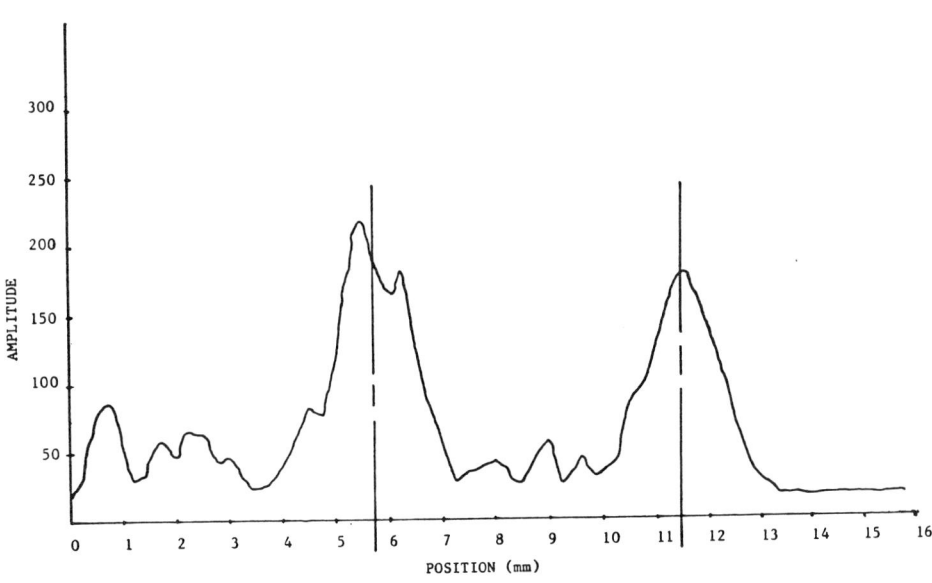

Figure 7. Line scan over a 40 micron tight crack, showing direct and multiple reflection images of the crack.

Future work will be oriented toward building a prototype system capable of demonstrating the commercial utility of our technique for solving real inspection problems. Our intent is to build a system incorporating all of the improvements we have discovered, as well as fast, accurate and convenient scanning hardware, high quality image display and recording equipment, and a signal processing capability for analyzing and classifying the RF echoes from flaws.

During construction of the prototype, we will upgrade the acoustic and electronic aspects of the system to include the following:
1. Sapphire buffer wedge - to eliminate backscattering noise
2. Weak focussing - to improve lateral resolution and sensitivity
3. Lower noise receiver - to improve sensitivity
4. Fast, accurate, and easily set up scanner

We intend to solicit practical samples from manufacturers and users of ceramics, and to demonstrate to them a capability to quickly scan their samples to detect defects, and then to return to the defect locations and analyze the defects by RF signal processing. If the system elicits interest from potential purchasers or inspection service customers, it will be made commercially available.

CONCLUSIONS

We have demonstrated detection of internal inclusions in silicon nitride of several types, with diameters of 400 and 100 microns; and of a tight surface crack 40 microns deep. Reasonable improvements to the system will improve the detection limits to be of order 20 microns. We have also calculated the scattering properties of several types of flaws and have noted the possibility of flaw characterization by processing of the echoes obtained. Scanning a six inch square plate up to one quarter inch thick 39 minutes with lateral resolution of .4mm appears to be feasible.

# A REAL-TIME SAFT SYSTEM APPLIED TO THE ULTRASONIC INSPECTION OF NUCLEAR REACTOR COMPONENTS[a]

T. E. Hall, S. R. Doctor, and L. D. Reid

Pacific Northwest Laboratory
P.O. Box 999
Richland, WA 99352

## INTRODUCTION

In 1982 Pacific Northwest Laboratory began activity under the sponsorship of the U.S. Nuclear Regulatory Commission to implement SAFT technology in a field usable system. The University of Michigan had previously laid the groundwork by performing extensive research related to the development of the SAFT algorithm in the area of ultrasonics and the investigation of ways to improve the computation time [1,2]. The task given PNL was to deploy the results of this research effort by developing an instrument that would perform in-service inspection of nuclear reactor components using the SAFT-UT algorithm.

Initially at PNL, a study was performed to review the status of SAFT as it applied to ultrasonic testing. This established a firm understanding of the SAFT algorithm and investigated a number of parameters related to SAFT, such as expected resolution and system bandwidth effects. The results of this review may be found in a document by Busse, et al. [3].

A number of items needed to be addressed in order to accomplish the deployment task. The physical realities of developing a field system, such as portability and environmental concerns, were important. A pipe scanner needed to be developed to perform the precision scanning necessary for SAFT-UT data collection. The algorithm itself needed to be accelerated to produce a real-time three-dimensional SAFT-UT system. And finally, graphics software must be developed to display the resultant image in a clear and interpretable manner.

The hardware components employed in the SAFT-UT field system are commercially available whenever possible. The purpose for this was to improve the final system reliability and facilitate technology transfer to the commercial realm. Also it allowed internal PNL engineering resources to be focused on developing the SAFT field system algorithm and related software rather than developing unique hardware.

---

(a) Work supported by the U.S. Nuclear Regulatory Commission, under Contract DE-AC06-76RLO 1830; Dr. J. Muscara, NRC Program Monitor.

A significant amount of work has been focused on acceleration of the computationally intensive coherent summation SAFT algorithm. This has been accomplished through advanced signal processing techniques and development of a Real-Time SAFT Processor peripheral device. These efforts have been described in a separate paper [4] and will not be dealt with in depth here. It is essential for the successful deployment of SAFT in a commercial environment to achieve rapid imaging. This allows the operator to make timely judgments with respect to specimen integrity, thus reducing the costs involved with reactor inservice inspection.

This paper will focus on describing configurations developed for the implementation of SAFT-UT on thin-wall materials such as piping. Of primary importance is detection and sizing of vertically oriented defects, such as intergranular stress corrosion cracks (IGSCC), in the primary cooling system of light water reactors.


SAFT-UT IMAGE RECONSTRUCTION

"Synthetic aperture focusing" refers to a process in which the focal properties of a large-aperture focused transducer are generated from an orderly series of measurements over a large area using a small aperture transducer. The processing required to focus this collection of data has been called beam-forming, coherent summation, or synthetic aperture processing.

Inherently, SAFT has an advantage over physical focusing techniques in that it provides a full-volume focused characterization of the inspected area, where traditional physical focusing techniques provide focused data only at the depth of focus of the lens. For the typical data collection scheme used with SAFT-UT, a focused transducer is positioned with the focal point located at the surface of the part to be inspected. This configuration is used to produce a broad, divergent ultrasonic beam in the object under test. As the transducer is scanned over the surface of the object, a series of A-scan lines (rf waveforms) are recorded for each position of the transducer. Each reflector produces a collection of echoes in the A-scan records. If the reflector is an elementary single point reflector, then the collection of echoes will form a hyperbolic surface. The shape of the hyperboloid is determined by the depth of the reflector within the test object and the velocity of sound in the material under test. This relationship between echo location in the series of A-scans and the actual location of the reflectors within the test object makes it possible to reconstruct a processed image from the acquired signals.

If the scanning and surface geometries are well known, it is possible to accurately predict the shape of the locus of echoes for each point within the test object. The process of coherent summation involves shifting the A-scans by a predicted time delay and summing the shifted A-scans. This process may also be viewed as performing a spatial-temporal matched filter operation for each point within the volume to be imaged. Each element is then averaged by the number of points that were summed to produce the final processed value. If the particular location correlates with a locus of A-scan echoes, then all of the values summed will be in phase and produce a high-amplitude result. If the location does not correlate with a locus of predicted echoes, then destructive interference will take place and the spatial average will result in a low amplitude.

## SAFT-UT PULSE-ECHO IMPLEMENTATION

The principles of the elementary single-transducer SAFT-UT pulse-echo configuration are essential to comprehend in order to accurately interpret results of the SAFT-UT process and to understand the more complex multiple-transducer configurations. SAFT-UT pulse-echo is a single-transducer source-receiver scanning configuration (i.e., the same transducer that generates the sound field in the object space also receives the return echo).

In thin materials, where reflections from the back surface need to be considered, multiple sound paths may contribute to the reconstructed image. In the SAFT-UT pulse-echo configuration there are two of these candidate propagation paths that are dominant when observing generally vertical oriented defects. These are illustrated in Figure 1. The part as shown has a thickness of T with an elementary point reflector at depth D from the scanned surface. The transducer scanned has its focus located at the surface, and in general has a non-zero incident angle.

The first path to be considered is the direct path ($P_1$-$P_1$) from the beam entry point to the reflector and back to the entry point, without intersecting the far surface. Analysis shows that data received along this path will be deterministic and represent accurately and uniquely lateral position and depth information [5]. This path represents the sometimes elusive 'tip' signal commonly referred to when imaging vertical-oriented defects such as IGSCC.

The second path ($P_2$-$P_3$-$P_1$) is the propagation path from the entry point to the far surface to the object and back to the transducer. This sound path length may be represented by Eq. 1.

$$P = (D^2+X^2)^{1/2} + [(2T-D)^2+X^2]^{1/2} \qquad (1)$$

Figure 2 shows the family of curves generated when solving Eq. 1, while discretely varying depth D. A definite ambiguity is apparent when observing this diagram. Very little unique object depth information is available from reconstruction of this sound path, and as a result the time-of-flight of the reflected signal will be in general independent of the object depth. From this elementary analysis, it appears that this bounce path would be difficult to incorporate into a SAFT imaging system since the expected propagation data is weakly indicative of the reflector depth. This fact is unfortunate, since this path is commonly encountered in normal ultrasonic data gathered with angle-beam illumination. This is especially true for verti-

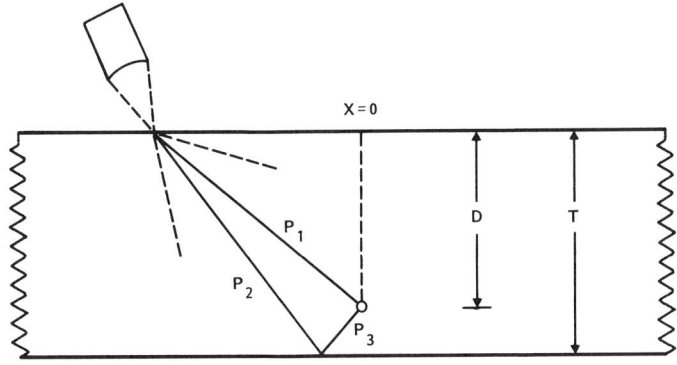

Fig. 1. Single transducer, SAFT-UT, pulse-echo configuration.

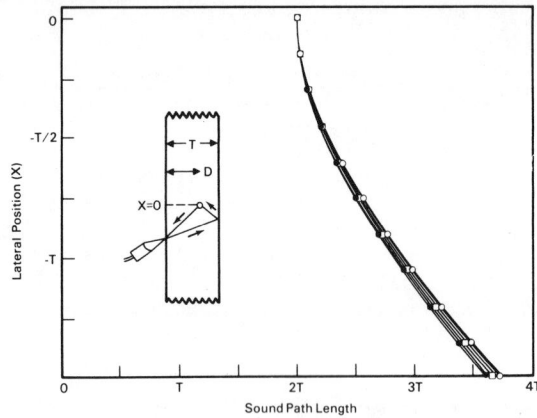

Fig. 2. Length of sound path vs. transducer lateral position for SAFT-UT pulse-echo path $P_1$-$P_2$-$P_3$

cally oriented planar defects. This phenomenon is the cause for the high amplitude 'corner trap' signal that is encountered when scanning this class of defects. It aids in the detection of these defects, but hinders accurately sizing the vertical extent of a given reflector.

Figure 3 shows a SAFT-UT image of a machined vertical sawcut in a stainless steel coupon. The data was collected in the SAFT-UT pulse-echo configuration. The notch, which simulates a vertical defect, was 0.6 inches in length and 0.3 inches deep into the material, and the coupon thickness was 0.585 inches. The image displays the B-scan side view and B-scan end view of the volume inspected. A tip signal has been imaged successfully in this example and the very strong corner reflection is apparent. As predicted, then, it would be difficult to size this defect in the vertical direction if the tip signal was not present. Also an operator may judge that this image represents a single continuous defect because of the duality nature of the indications, or he may judge that two independent reflectors on the same vertical plane are present in the material. Not enough information is available in the data to distinguish between these two cases.

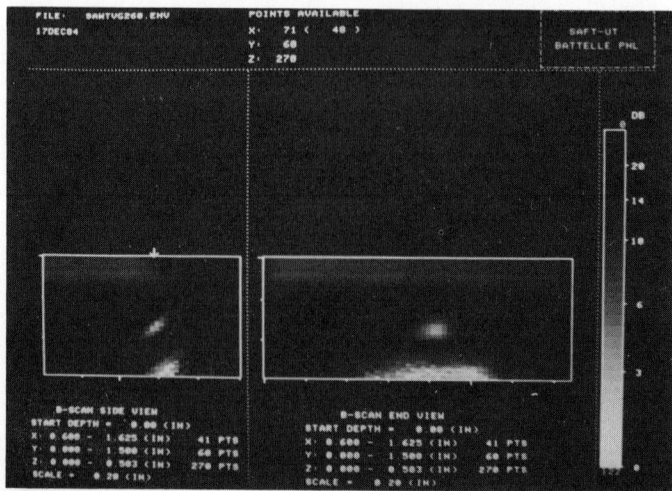

Fig. 3. SAFT-UT pulse-echo image of a semi-circular sawcut (0.6 x 0.3 inches) showing both the B-scan side view on left and B-scan end view on right.

For vertically oriented defects in materials with a finite thickness, the strength of the pulse-echo configuration lies in detection of these defects. The corner-trap echo is typically very strong and is present even in very attenuative materials.

TANDEM SAFT CONFIGURATION (TSAFT)

Tandem SAFT or TSAFT has been implemented to provide a method for characterizing and sizing these vertical defects in materials with a finite thickness. The TSAFT configuration capitalizes on the forward scatter from the object rather than the direct back scatter as in the pulse-echo case. It consists fundamentally of a fixed transmitter that is placed in line with, or in tandem with, a scanned receiving transducer. The transmitter is placed such that the divergent sound beam illuminates the primary object area. The scanned receiver is then translated to receive direct energy reflected from the defect area. As in the pulse-echo configuration, after each pass of the receiver transducer, the transport mechanism is incremented so that a rectilinear pattern is obtained (e.g., around the circumference of a pipe).

In the interest of brevity we will consider here only the most common sound path used in the tandem SAFT configuration. A more complete description of TSAFT multiple sound paths may be investigated by reviewing a previous paper by S.R. Doctor [6]. In the configuration we will consider here, the energy from the illuminating transducer reflects from the far surface prior to striking the object area, and the receiver in turn captures the energy directly reflected from the object. The total path length for a given point reflector, then, may be represented by Eq. 2.

$$P = [X_T^2 + (NT+D)^2]^{1/2} + [X_R^2 + (T-D)^2]^{1/2} \qquad (2)$$

where $X_T$ and $X_R$ are the lateral position of the transmitter and receiver relative to the object plane, T is the thickness of the material, D is the distance of the object point from the near surface, and N is the number of reflections the transmit ray undergoes prior to striking the object point.

A graph may be generated using Eq. 2 that will assist in predicting results when implementing this configuration. Figure 4 shows the family of curves that resulted from performing this exercise. These curves represent the total path length with respect to the receiver lateral position for various object depths (D). It was assumed that the echo from the point reflector was received in the central angle of the receiver. It can be predicted from this graph that the object depth, D, will map to a unique position in the image space and that the redundancy experienced in the pulse-echo configuration will not be present here.

Figure 5 is the tandem SAFT image of the identical sawcut described earlier with the pulse-echo case. Notice that, as predicted, the full extent of the surface of the vertical object is represented. Using TSAFT in this example provides the information to the interpreter so that a judgement may be made concerning the vertical extent of the defect. The image is independent of the tip diffracted signal. Also it is plain to the observer that there is a single vertical defect and not two co-planar defects.

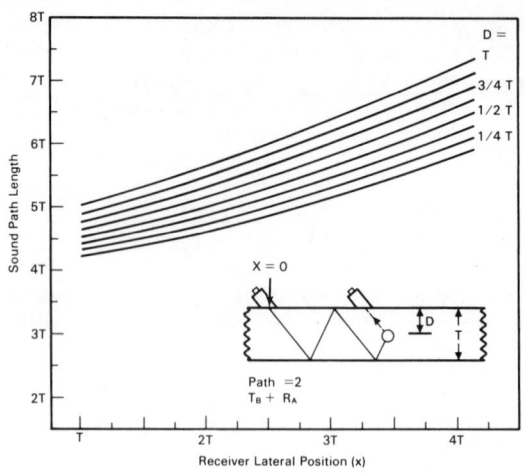

Fig. 4. Length of sound path vs. receiver lateral position for TSAFT configuration.

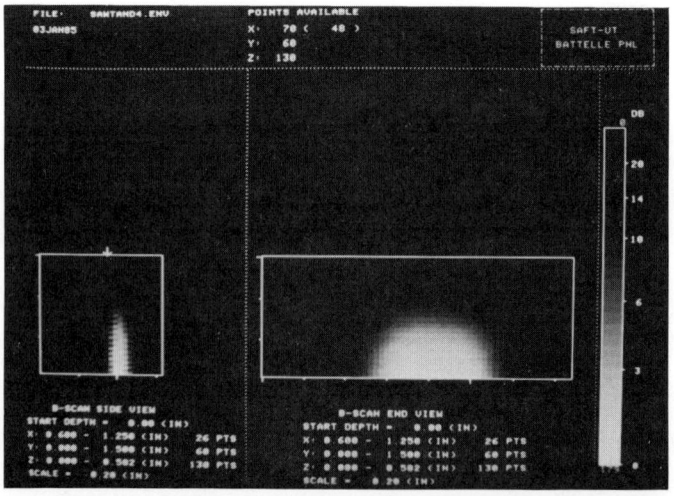

Fig. 5. SAFT-UT tandem image of a semi-circular sawcut (0.6 x 0.3 inches). Both the B-scan side view and B-scan end view are shown.

MODIFIED TANDEM SAFT CONFIGURATION (TSAFT-2)

When considering inspection of certain materials, defects may exist at any depth in the specimen--from small cracks located at the far surface, to deep cracks or fabrication defects that will reflect energy near the scanned surface. A full vertical object plane needs to be uniformly illuminated in this type of specimen in order to accurately determine the nature and size of the defects. Analysis of the tandem SAFT configuration reveals that illumination throughout a vertical object plane is in general not fully uniform. This is due to the fact that the transmitter is stationary and, due to laws of diffraction [7], will exhibit a non-uniform intensity distribution throughout the illumination aperture. This is easily seen by observing Fig. 6(a). Given two points in the object plane, $P_1$ and $P_2$, the rays striking these points from the transmitter arrive at unequal angles $\phi_1$ and $\phi_2$. It can be shown that the amplitude of the signal received by the scanned transducer, along independent paths, is not equal simply because the wave transmittance at the transducer-specimen interface is angle dependent. This phenomenon leads to undersizing when the object of interest extends beyond about half the beam-width (insonification angle).

An alternative tandem technique (TSAFT-2) was implemented to alleviate the illumination deficiency of the TSAFT configuration. This second configuration may be seen in Fig. 6(b). To maintain a constant illumination throughout a given object plane, the transmit transducer is scanned in the opposite direction as the receiver. For each object plane in this case, $\phi_1$ and $\phi_2$ are equal, thus ensuring that the total center ray path length is constant, from the transmitter to receiver. This concept has proven to be valuable when imaging the full vertical-object plane, and in particular it has enhanced the imaging of near-surface defects.

Figure 7 shows a comparison view of two SAFT-UT scans of the same object. Two opposing vertical sawcuts (semi-circular) were placed in a 1.0-inch-thick aluminum coupon. They were inserted such that there was a 0.050-inch separation at the center of the part. These objects were scanned implementing 5.0-MHz contact transducers in the 45-degree shear-mode configuration. The upper image shows the B-scan side view and B-scan end view of results obtained from the TSAFT configuration (fixed transmitter). It can be easily seen that the lower machined defect is imaged, while the upper

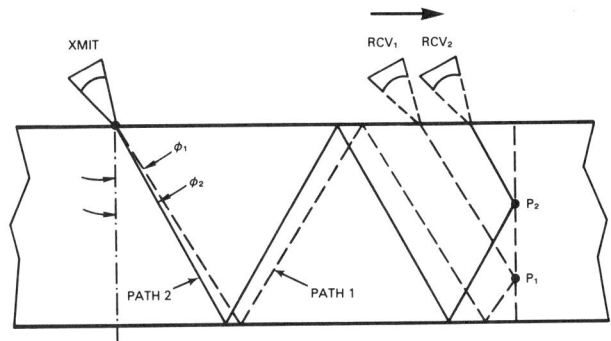

Fig. 6a. SAFT-UT tandem (TSAFT) configuration.

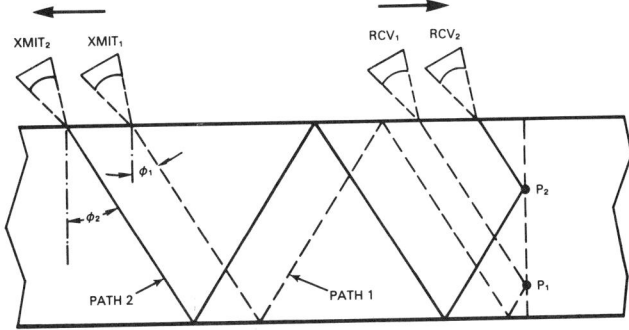

Fig. 6b. SAFT-UT Tandem-2 (TSAFT-2) configuration.

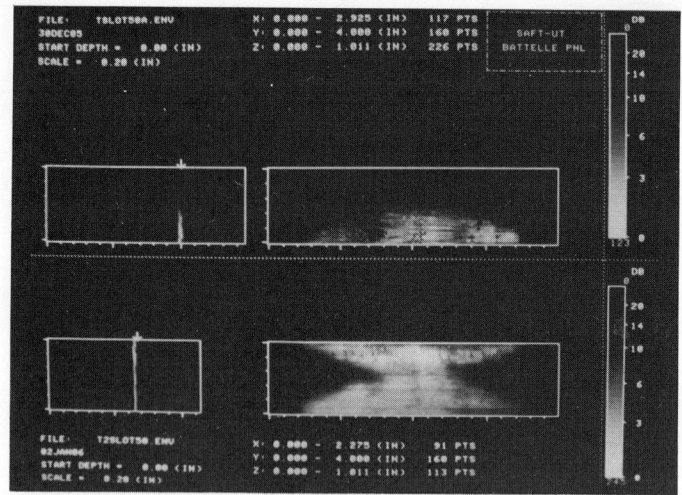

Fig. 7. SAFT-UT tandem image (upper) and SAFT-UT Tandem-2 (lower) image of coplanar opposing sawcuts in 1.0-inch aluminum coupon.

structure is not. This is primarily due to the aperture-limiting effect of the insufficient illumination. The lower image of Fig. 7 shows the corresponding results of the TSAFT-2 configuration. Notice that very good illumination resulted from simultaneously scanning the transmitter with the receiver. Also it may be observed that the image shows uniform reconstruction amplitude throughout the thickness of the specimen.

It may be expected that improved near-surface imaging should result from this technique. And this is the case. Figure 8 shows the B-scan side view and B-scan end view of an image of fabrication defects in the weld region of a carbon steel block. The block was 1.5-inches thick. The image shown is a result of a TSAFT-2 scan using 5.0-MHz contact transducers in a 45-degree shear-mode orientation. Of particular interest in this image is

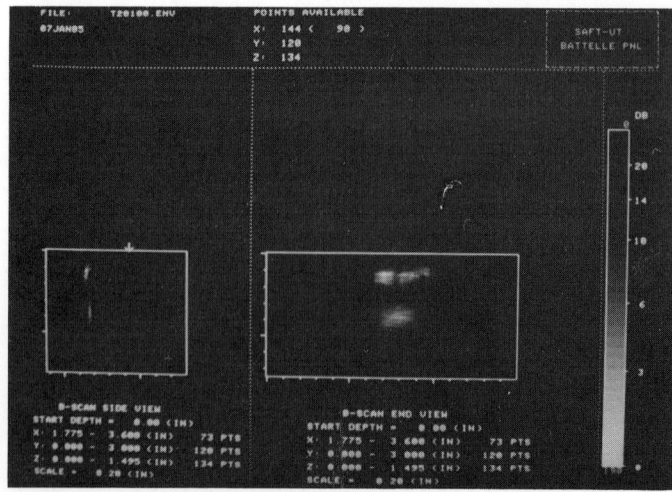

Fig. 8. SAFT-UT Tandem-2 (TSAFT-2) image of lack-of-fusion defects in 1.5-inch-thick carbon steel specimen.

the imaged defect located near the scanned surface. It is located approximately 0.2 inches from the near surface and is approximately 0.8 inches in length. This defect may easily have been overlooked in a volumetric inspection utilizing other configurations. The advantage of TSAFT-2 is the uniform illumination of the complete vertical object plane.

SUMMARY

SAFT-UT has evolved in recent years to become a viable real-time acoustical imaging method. A variety of configurations have been developed to accommodate various geometrical requirements and provide full-volume focusing capability. The single transducer, pulse-echo configuration provides a straight-forward configuration for imaging volumetric defects, and for reliably detecting strong 'corner-trap' signals from vertical defects. The tandem SAFT configurations provide a means for sizing vertical defects and the TSAFT-2 technique, in particular, performs well for imaging the full vertical object plane. These results, along with the development of real-time processing, have moved SAFT-UT from a laboratory-imaging method to a field-usable inspection technique.

REFERENCES

1. S. Ganapathy, B. Schmult, W. S. Wu, T. G. Dennehy, N. Moayeri, and P. Kelly. "Design and Development of a Special Purpose SAFT System for Nondestructive Evaluation of Nuclear Reactor Vessels and Piping Components," Report No. NUREG/CR-4365, Nuclear Regulatory Commission, Washington, D.C. (1985).

2. G. S. Kino, D. Corl, S. Bennett, and K. Peterson, Real Time Synthetic Aperture Imaging System, in: "Ultrasonics Symposium Proceedings," Cat. #80CH-1602-2S7, IEEE (1980).

3. L. J. Busse, H. D. Collins, and S. R. Doctor, "Review and Discussion of the Development of SAFT-UT," Report No. NUREG/CR-3625, U.S. Nuclear Regulatory Commission, Washington, DC (1984).

4. T. E. Hall, S. R. Doctor, and L. D. Reid, Implementation of a Real-Time Ultrasonic SAFT System for Inspection of Nuclear Reactor Components, to be published in Vol. 15 of "Acoustical Imaging" (the 1986 Proceedings of the Acoustic Imaging Conference), Plenum Press, New York (1986).

5. S. R. Doctor, T. E. Hall, S. L. Crawford, and L. P. Van Houten, SAFT-UT Field Experience in: 1985 Pressure Vessels and Piping Conference, PVP-Vol. 98-1, New Orleans (1985).

6. S. R. Doctor, L. J. Busse, S. L. Crawford, T. E. Hall, R. P. Gribble, A. J. Baldwin, and L. P. Van Houten, "Development and Validation of a Real-Time SAFT-UT System for the Inspection of Light Water Reactor Components," Report No. NUREG/CR-4583, Semi-annual Report April to September 1984, U.S. Nuclear Regulatory Commission, Washington, DC (1985).

7. J. Krautkramer and H. Krautkramer, "Ultrasonic Testing of Materials," Third revised edition, Springer-Verlag (1983).

ON THE ULTRASONIC IMAGING OF TUBE/SUPPORT STRUCTURE

OF POWER PLANT STEAM GENERATORS*

Jafar Saniie and Daniel T. Nagle

Illinois Institute of Technology
Department of Electrical & Computer Engineering
Chicago, IL  60616


INTRODUCTION

The corrosion and erosion of steam generator tubing in nuclear power plants can present problems of both safety and economics. In steam generators, the inconel tubes are fit loosely through holes drilled in carbon steel support plates. Corrosion is of particular concern with such tube/support plate structures. Non-protective magnetite can build up on the inner surface of the support plate holes, and allowed to continue unchecked, will fill the gap, eventually denting and fracturing the tube walls. Therefore, periodic nondestructive inspection can be valuable in characterizing corrosion and can be used in evaluating the effectiveness of chemical treatments used to control or reduce corrosion. Presently, we are investigating the feasibility and practicality of using ultrasound in routing testing for gap measurement, for evaluating the corrosion and assessing the degree of denting. The tube/support structure can be modeled as a multilayer, reverberant target, which when tested with ultrasound results in two sets of reverberating echoes [1]. One set corresponds to the tube wall and the other to the support plate. These echoes must be decomposed and identified in order to evaluate the tube/support structure. This report presents experimental results along with a discussion of various measurements and processing techniques for decomposing and interpreting tube/support echoes at different stages of corrosion.

MULTILAYER MODEL AND ECHO CLASSIFICATION

The steam generator tube/support structure is composed of the following thin layers: the inner inconel tubing, the intermediate water gap, and the surrounding steel support plate (see Figure 1). Ultrasonic examination of these inherent thin layers produces multiple reflections (reverberations) which are not readily resolvable. Thus, this section contains the presentation of a model of reverberant echoes and their application to steam generators tube/support plate evaluation.

Through detailed experimentation and extensive computer simulation, an echo classification model was developed and presented in earlier reports [1-3]. This model allows for the classification and identification of the boundaries

---
*This work has been support by EPRI under Contract RP2673-5.

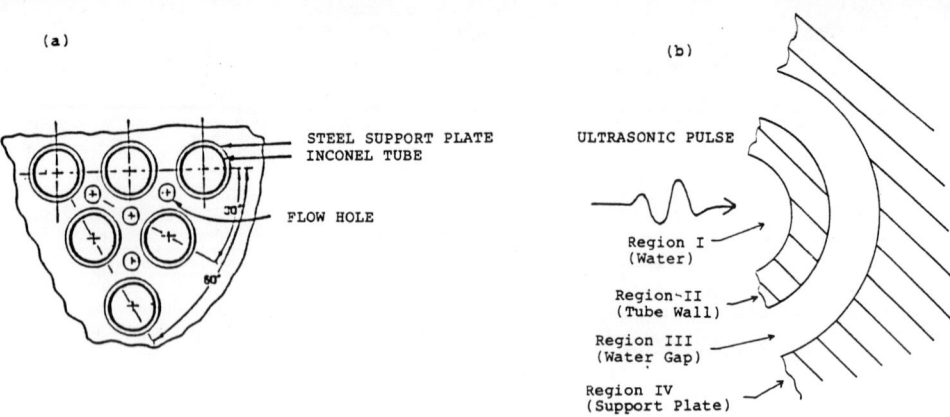

Figure 1. Steam Generator Tube/Support Plate Structure (a) A Cross Section Schematic (b) The Multi-layer Model

of the inconel tube, the water gap, and the steel support plate. Ultrasonic testing of the tube/support plate structure results in two dominant sets of multiple echoes, one set is referred to as class 'a' echoes (tube echoes) and the other as class 'b' echoes (support plate echoes). Class 'a' echoes are defined as the set of all echoes which reverberate continually in the inconel tube, and only once in the water gap. The equation for the measured backscattered signal is:

$$r(t) = s_1(t) + s_2(t) + n(t) \qquad (1)$$

where

$$s_1(t) = \sum_k a_k u(t-2kT_2) \quad : \text{Tube Echoes}$$

$$s_2(t) = \sum_k b_k u(t-2kT_2-2T_3) \quad : \text{Support Plate Echoes}$$

$$a_k = \tau_{12}\tau_{21}\rho_{21}^{k-1}\rho_{23}^k$$

$$b_k = k\tau_{12}\tau_{21}\tau_{23}\tau_{32}\rho_{34}\rho_{21}^{k-1}\rho_{23}^{k-1} ,$$

$u(t)$ = measurement system impulse response.

$\rho_{ij}$ = reflection coefficient between (incident) region i and region j (see Figure 1b).

$\tau_{ij}$ = transmission coefficient between (incident) region i and region j (see Figure 1b).

$T_2$ = represents the traveling time in the tube wall (see Figure 1b).

$T_3$ = represents the traveling time in the water gap (see Figure 1b).

$n(t)$ = All other echoes plus measurement noise.

The component n(t) is the estimation error, where it is dependent upon: (1) the other reverberant low intensity echoes not considered, (2) the amount and type of the debris in the water gap, (3) the magnitude of grain noise in the tube wall, and (4) the degree of deterioration, or roughness, of the boundaries of the tube/support structure.

The general envelope of the two sets of multiple echoes, class 'a' and 'b', are very different from each other. The 'a' echoes are depicted by an exponential decay, where as the number of reverberations increases, the amplitude decreases. However, the 'b' echoes have a maximum not at the beginning of the backscattered signal, but at a point which is dependent upon the reflection coefficients of the boundaries of tube and support plate.

The 'a' echoes and 'b' echoes are dependent on several different variables, which allows for the analysis and evaluation of subsequent boundaries of the tube/support structure. It can be seen from Equation 1, as well as experimentally, that the destruction of tube wall boundaries will result in scattering and in a significant reduction of reflected echoes. However, corrosion of the support plate will have little or no effect on class 'a' echoes since these echoes represent only the reflection characteristics of tube wall. Corrosion manifests itself in the characteristics of class 'b' echoes which allow a proper interpretation of corrosion by examining the reduction and spreading of support plate echoes. The presence of 'a' echoes masks 'b' echoes which necessitates class 'a' and 'b' echo decomposition through a proper experimental set-up coupled with suitable signal processing. The decomposition and further evaluation of these two sets of multiple echoes can provide more detailed information concerning the progress of corrosion in support plates and denting problems on the tube walls.

## EXPERIMENTAL RESULTS & DISCUSSIONS

### Transducer Configuration

For routine ultrasonic evaluation of steam generator tube/support plates, the choice of transducer and transducer apparatus must be practical and proficient. The transducer's position with respect to the target and reflector plays an important role in properly focusing the ultrasonic beam onto the tube wall. The focal point of the transducer must lie on the surface of the tube wall, and provide minimal deviation from the normal entry angle. This condition will decrease differences in the time-of-flight of incident rays, and, therefore, minimize echo distortion.

In our present studies, the transducer used has a center frequency of 20 MHz, a transducer element of 0.035" by 0.016", and a focal length of .5" (cylindrically focussed). The transducer apparatus (See Figure 2) is designed such that: (1) it projects the beam onto the plane of the inconel tube for inspection; (2) it provides mobility in the vertical and rotational directions with the usage of rubber 0-rings; and (3) it provides position stability due to its precise fitting. The reflector used is a plane mirror oriented at a 45 degree angle.

### Verification of Echo Classification

Experimental verification of our classification techniques and their usefulness is investigated in ultrasonically testing the multilayer structure consisting of water, inconel layer (56mils), water gap and steel support plate. Results are shown in Figure 3. The uppermost A-scan in this figure shows reverberation echoes produced by the 56 mil thick flat inconel layer, suspended in a water bath. As expected, the uniformly spaced class 'a' echoes decrease with time as energy leaks from the reverberating wave packet into the surrounding water bath. The center trace shows the effect of positioning

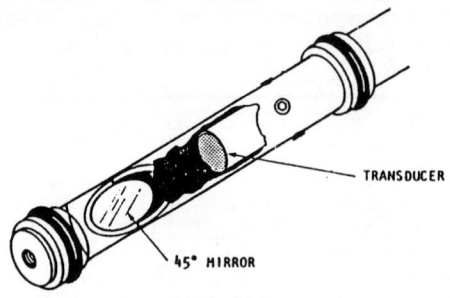

Figure 2. The Transducer Assembly

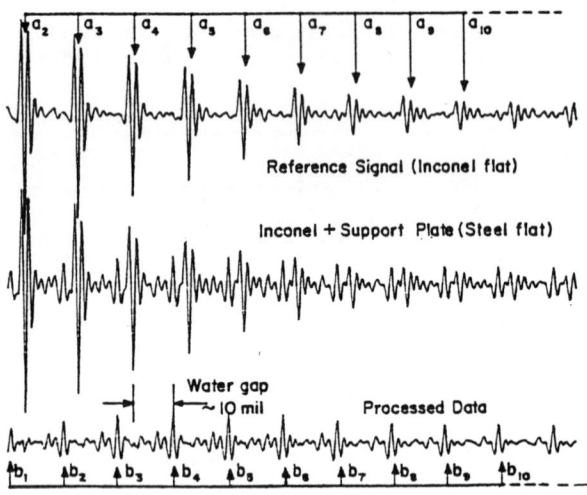

Figure 3. Example of Echo Classification

a steel plate behind the inconel tube in such a way that it leaves a water gap of approximately 10 mils. Although the addition of te support plate introduces class 'b' echoes, their exact location is not easily resolved. By substracting the reference signal (upper trace) from the A-scan of the multilayer structure (center trace), class 'a' echoes can be removed, clearly revealing the locations of class 'b' echoes. The bottom trace of Figure 3 shows the result of this simple processing. In this trace, the inverted 'b' type echoes are easily identified and show the characteristic increase in amplitude with increasing time. Finally, the water gap width is determined from the measured delay between a type 'a' echo and its associated type 'b' echo as shown in the figure.

Denting and Corrosion

The denting of steam generator inconel tubing due to nonprotective magnetite formations has been simulated by using tapered or dented tubes to assess the sensitivity of the ultrasonic system and to develop procedures for automatic pattern recognition and characterization. An inconel tube was deformed by tapering the outside as shown in Figure 4, and the other tube has a partial flat-bottom hole with a depth of 4 mils.

The tapered tube measurements are shown in Figure 5. The unaltered tube thickness is equal to 62.5 mils, and, after tapering, the smallest tube thickness is 37.5 mils. Five equally-spaced measurements were taken on the tapered half of the tube shown in Figure 4. The results (see

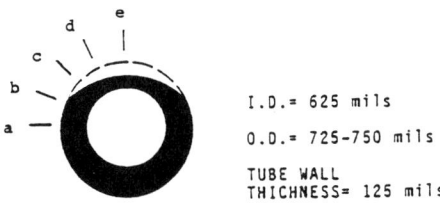

Figure 4. Tapered Tube Schematic

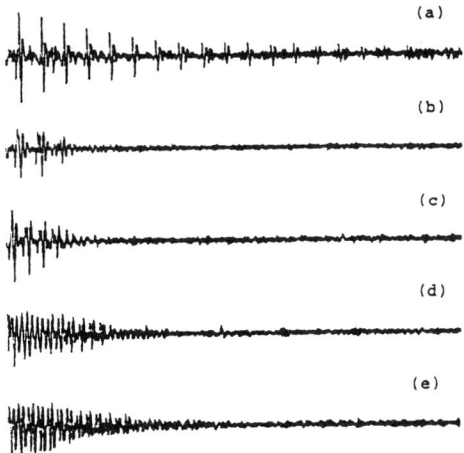

Figure 5. Tapered Tube Experimental Results

Figure 5) reveal changes in various positions indicating different tube thicknesses. Figure 5a is the signal from the untapered tube wall, where all 'a' echoes are present. The signal in Figure 5b indicates a sudden change in the outer tubes wall which causes most of the enrgy to be reflected away from the receiving unit and results in a significant reduction and distortion of the intensity of tube echoes. The signals in Figure 5c-e show the gradual compression of 'a' echoes indicating a reduction in the tube wall thickness. Also, those three signals show sequentially a prolongation of the number of reverberant echoes which implies that the inner and outer boundaries are becoming more parallel and that more echoes are reflected towards the transducer.

The presence of flaws on the tube wall results in a reduction of the intensity of 'a' echoes, as flaws scatter some of the energy away from transducer. The experimental flaw measurement results are shown in Figure 6, where the signals were taken in three different places on the tube; (1) at the undented portion of the tube, (2) at the edge of the hole, and (3) at the center of the hole. The hole is flat bottomed with a depth of 4 mils. Trace 1 represents the tube signal in the absence of dent or flaw. Trace 2 shows severe degradation of the echoes because a large amount of the signal is reflected away from the detectable beam field. The attenuation of the 'a' echoes of Trace 3 is about 2-3 dB down from the 'a' echoes of Trace 1 (reference signal) and may be caused by the surface roughness of the flat-bottom hole. These results confirm the reliability of examining 'a' echoes to

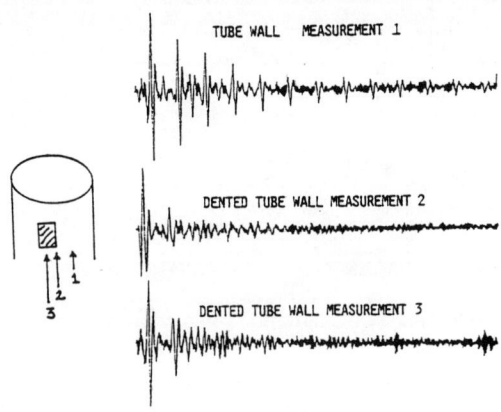

Figure 6. Dented Tube Experimental Results

describe the condition of the tube wall.

Presently, we have a limited number of corroded samples available to us which provide different spectrums of corrosion. A random selection of preliminary ultrasonic measurements of available samples are shown in Figure 7. Traces 1-3 come from the same tube/support sample, and traces 4-5 come from a different sample. Corrosion at the highest degree usually becomes a bridge in the gap which destroys the boundaries as shown in traces 1-3. The effects of corrosion at a lesser degree can be observed in traces 4 and 5. The present results suggest that a variety of echo patterns can be expected depending on the degree of corrosion or denting that exists. It is our objective to investigate a broad range of situations and classify them by their echo characteristics.

Oblique Angle Scanning

The use of oblique angle scanning is a complementary procedure to normal angle scanning that results in an automatic rejection of tube echoes. This enables us to detect support plate echoes without the interference of tube

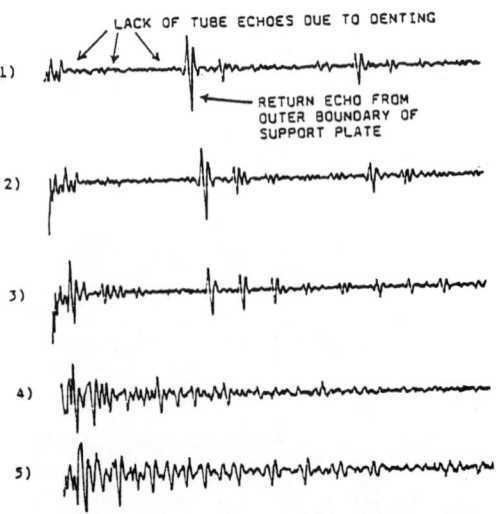

Figure 7. Example of Experimental Measurements Associated with Varying Degrees of Corrosion

echoes. For detecting support plate echoes, it is necessary for the reflecting surface of the support plate to be normal, or near normal, to the incident ultrasonic beam. Any significant deviations in the orientation of the support plate to be normal, or near normal, to the incident ultrasonic beam. Any significant deviations in the orientation of the support plate will result in reflecting the incident ultrasonic beam away from the receiving unit, except when the support plate has a rough surface. Furthermore, flaws in the tube walls are sometimes difficult to detect when using normal scanning, due to the orientation of the flaw, however, with the use of oblique angle scanning flaws are more visible because of the improved viewing geometry.

The results of oblique scanning of the tube/support structures are shown in Figure 8. The data was taken using an inconel tube with a position adjustable steel support plate. For the purpose of demonstration, the support plate was moved to give the best possible reflection. Oblique scanning automatically rejects almost all the tube echoes and provides improved resolution and visibility of the support plate echoes as shown in Figure 8.

It is important to point out that mode conversion is evident with oblique angle scanning. When scanning at an oblique angle there are two waves produced in the tube wall, namely, longitudinal and shear. As a result, the complexity of the reverberation increases significantly. Nevertheless, for a small incident angle the intensity of the shear wave is relatively small and can be ignored.

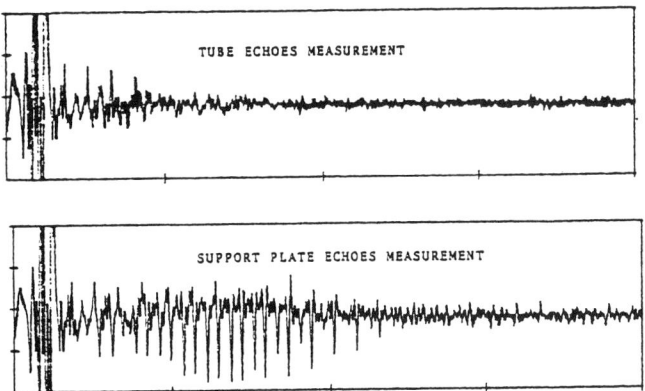

Figure 8. Results of Oblique Angle Scanning

REFERENCES

1. J. Saniie, E.S. Furgason, and V.L. Newhouse, "Ultrasonic Imaging Through Highly Reverberant Thin Layers - Theoretical Considerations." Materials Evaluation, Vol. 40, pp. 115-121, 1982.
2. J. Saniie, "Identification of Reverberant Layered Targets Through Ultrasonic Wave Classification." Review of Progress in Quantitative Nondestructive Evaluation. Eds.: D.O. Thompson and D.E. Chimenti. Plenum Press, pp. 1011-1018, 1984.
3. J. Saniie, "Resolution and Visibility Enhancement of Ultrasonic Reflected Echoes from Targets Hidden by Highly Reverberant Thin Layers." IEEE Ultrasonic Proceedings, pp. 903-907, 1984.

EDDY CURRENT IMAGING FOR MATERIAL SURFACE MAPPING


E. J. Chern[1] and A. L. Thompson

General Electric Company
Aircraft Engine Business Group
Cincinnati, Ohio 45215

INTRODUCTION

    For most nondestructive inspections, quantitative evaluations have to be performed to correlate the measured signals and the desired material properties. In some applications, the relationship between the signals and the material parameters is simple and straightforward. An analytic mathematical function can be easily constructed and solved to describe the interaction. The inverse function also can be readily determined to transform the measured data into the desired information. In some cases, such as defect characterizations, the interaction between the sensing field and the test object is often too complicated for such an approach. Advanced signal analysis techniques with complex assumptions, approximations, and computations are required to interpret the signals. Such an intricate approach is frequently time consuming and beyond general comprehension. Alternative methods that enable the direct correlation of the signals with the test piece are being sought. Imaging techniques which provide a unique capability of correlating NDE signals with component geometry, are gaining in popularity.

    Major manufacturing operations such as material processes, component machining and field service often unavoidably cause geometric marks, stress concentrations, and structural defects. These variances in term induce localized changes in electric conductivity and magnetic permeability of the material. Eddy current are sensitive to these electromagnetic parameters. Thus eddy current NDE techniques are widely used to characterize material properties in various applications.[1-4] Incorporated with a mechanical scanning mechanism, an eddy current image can be constructed to map out material surface properties and identify various features. Eddy current imaging can be a vital technique for signal discrimination, classification, and interpretation.[5,6]

    An analog eddy current imaging system which is capable of real time material surface mapping has been developed and integrated. Eddy current signals and positional information are superimposed by a special designed mixer/amplifier. Outputs from the signal mixer/amplifier are then used to drive an analog X-Y recorder for the generation of 3-D eddy current images.

---

[1]present address: Combustion Engineering, Inc., 1000 Prospect Hill Road, Windsor, Connecticut 06095-0500

Experiments on machining features, residual stresses, and surface defects are studied using the system. The images obtained have demonstrated positive results in mapping and recognizing signatures of the described structural effects. Other potential applications are discussed. A digital system is proposed for data manipulation, image analysis, pattern recognition, and advanced signal processing.

BACKGROUND

Eddy current NDE techniques utilize electromagnetic induction effects to characterize the material properties.[7] As it is well known all electromagnetic phenomena are governed by Maxwell's equations. With appropriate boundary conditions, exact analytic equations and solutions can be obtained to describe the eddy current effects. A mathematical transformation can be obtained to invert the eddy current signal to the desired material parameters. However, in practice, it is often very difficult to define the proper boundary conditions to model the specific eddy current interaction. Eddy current imaging method is an effective approach to correlate eddy current signals with specimen coordinates. Theoretical analysis of the general eddy current phenomena has been discussed by many researchers.[8-11] We shall examine only the theoretical background as it is applied to eddy current imaging.

The fundamental operating principle of the eddy current imaging system is to utilize the eddy current impedance measuring technique to detect the induced changes of the electrical conductivity, magnetic permeability, and geometries due to residual stresses and surface defects in the electrically conductive materials. The impedance change of the sensing coil due to the presence of a pertubation field can be written as

$$dZ = 1/I^2 \int_s (E \times H' - E' \times H) \, ds \qquad (1)$$

where dZ is the variation in eddy current coil impedance, I is the current flow in the coil, E and H are the unperturbed electric field and magnetic field, E' and H' are the perturbed electric and magnetic fields, and s is the area enclosed in the fields. The impedance Z is a complex quantity and is generally witten as $Z=R+jX_L$, the sum of the resistance R and reactance $X_L$. By correlating the variations in coil impedance dZ (= dR + j $dX_L$) with the specimen coordinates, an eddy current image can be generated.

Also, the depth penetration of the eddy current field in the material is inversely proportional to the square root of the conductivity, the permeability and the operating frequency.[12] It can be expressed as

$$\delta = (2\pi\mu\sigma f)^{-\frac{1}{2}} \qquad (2)$$

where $\delta$ is the skin depth, f is the operating frequency, $\sigma$ is the electric conductivity, and $\mu$ is the magnetic permeability of the material. Since eddy current field is divergent, the measurement is an integrated effect to the depth of the field. The material surface up to the desired depth thus can be examined by operating at the specifically selected frequency.

In our particular application of eddy current imaging, only the relative changes in impedance dZ with the scanning X-Y coordinates are monitored. Advanced analysis of the impedance amplitude and phase measurments can provide additional information of the material characteristics. Combining the advanced signal analysis capability with the imaging mechanism, detailed structural images of the material may be shown.

EXPERIMENTS

The block diagram of the laboratory configuration of this eddy current material surface mapping system is shown in Figure 1. The instruments used are a Staveley Nortec NDT-25L Eddyscope, an Aerotech Unidex III controller with ATS406 X-Y scanning axes, a Hewlett-Packard 7045B X-Y recorder, and a specially designed signal mixer/amplifier controller. The functional diagram of this signal mixer/amplifier controller is shown in Fig. 2. The eddy current probes used are standard absolute pencil probes typically with 2 MHz, 1 MHz and 500 kHz in frequencies.

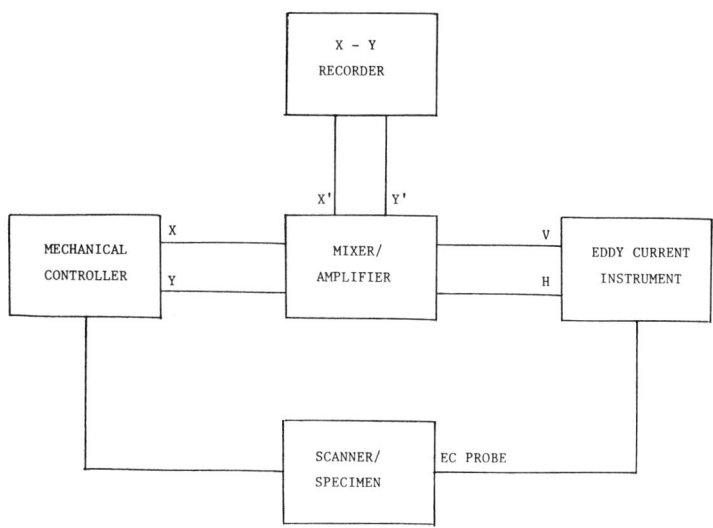

Figure 1. Block Diagram of the Eddy Current Surface Mapping System Configuration

The mechanical scan plan is programmed into the Aerotech controller with numerical control (NC) codes. The scan plan is designed to scan on X axis with the speed of 1 cm/sec and index on Y axis with 0.4 cm increments. The parameters of the NDT-25L eddy current instrument are set by nulling the system at the specimen and calibrating to a liftoff amplitude of 10 V. Experimentally, nulling the system at the specimen surface removes the dc component of the eddy current signal. This procedure enables the monitoring of signal variations with higher sensitivity.

The mechanical X and Y raster scanning voltage signals, and both vertical and horizontal eddy current signals are input to the signal mixer/amplifier assembly. The output signals from the mixer/amplifier, X' and Y' are used to drive the X-Y recorder for imaging plotting. Where X' is X plus Y $\cos\Theta$, Y' is y $\sin\Theta$ plus the arithmetic sum of the two measured impedance components as described in Fig. 2. A three dimensional image is generated in real time while the probe is scanning across the surface of the specimen. In addition to an X-Y recorder which allows the real time tracking of signal variation, a grey scale amplifier can also be used to

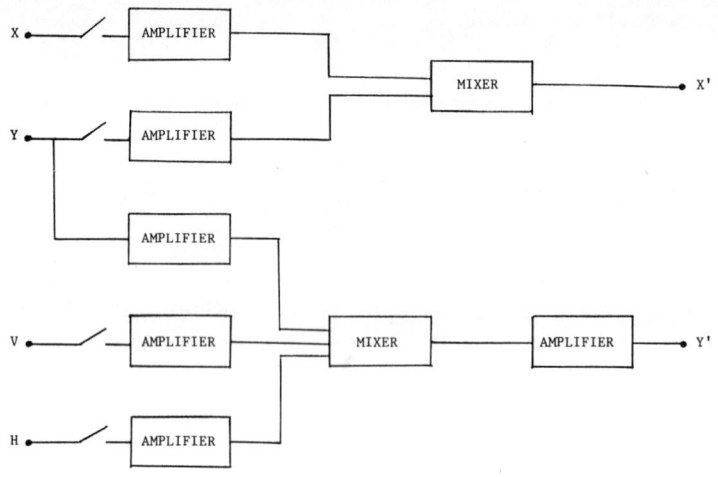

Figure 2. Functional Diagram of the Signal Mixer/Amplifier Controller

present the image. A typical result of the eddy current material surface property mapping image is shown on Figure 3. The surface structures from electrostatic-discharge machined (EDM) notches and machining features are clearly shown.

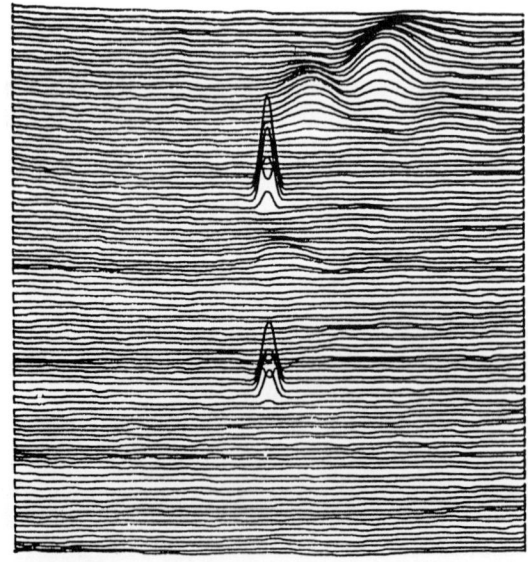

Figure 3. A typical Eddy Current Image from the Material Surface Mapping System.

RESULTS AND DISCUSSION

The results of the experiments are summarized in three technical areas: (i) "electromagnetic microscope"; (ii) surface stress mapping; (iii) defect characterization. Each subject is described and discussed as follows.

Electromagnetic Microscope

The function of a scanning electron micrscope (SEM) is to study the metallurgical structure of the surface within a few thousandths of a centimeter, while the function of an acoustic microscope (A/M) is to study the acoustic structure from a surface dead zone of about two tenths of a centimeter down to a few centimeters for most of the materials. The Electromagnetic Microscope can investigate surface and subsurface regions which cannot be investigated by either SEM or A/M nondestructively. With the use of different frequencies to vary the depth of penetration, various depths of the material electromagnetic properties can be magnified and studied. Figure 4 shows an magnified eddy current image of a 0.015 inch (0.381 mm) deep, 0.030 inch (0.762 mm) long, and 0.003 inch (0.076 mm) wide EDM notch. Figure 5 shows the "electromagnetic" grain boundary of an aluminum specimen with various grain structures.

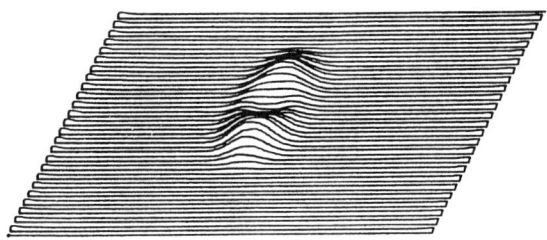

Figure 4. An Eddy Current Image of a 0.015 inch deep, 0.030 inch long, and 0.003 inch wide EDM Notch

Surface Stress Mapping

Localized static or dynamic stress concentrations generated by uneven distribution of applied and residual stresses, can cause fatigue cracks to initiate and propagate. Applied stresses and residual stresses induced changes in material properties such as acoustic and electrmagnetic parameters. Using the eddy current imaging mechanism, the stress distribution across the material surface can be easily mapped out. The localized high stress areas, i.e. potential crack initiation sites, thus can be identified and corrected to prevent the failure. Figure 6 shows an area with residual a stress distribution created by applying external stresses on legs of an H-shaped aluminum specimen.

Defect Characterization

Surface breaking defects such as porosities, fatigue cracks and electrostatic-discharge machined (EDM) notches cause an abrupt change in the secondary electromagnetic field induced in the material by an eddy current sensor. Various surface discontinuities are all different in nature. For example, a fatigue crack which is a "close" surface discontinuity has a higher frequency eddy current response than an "open" EDM notch. The shape and size of the discontinuity can also be depicted by the imaging system.

Figure 5. The "Electromagnetic" Grain Boundary of an
Aluminum Specimen with Many Grain Structires

Fatigue cracks, machining marks, and geometric features thus can be identified and categorized. Figure 7 shows the eddy current image obtained from a 0.006 inch (0.1524 mm) diameter hole on a 5/8 inch (15.875 mm) thick flat specimen. The image of a 0.024 inch (0.610 mm) fatigue crack is shown in Figure 8.

Figure 6. An Image of a Residual Stress Distribution

Figure 7. An Eddy Current Image of a 0.006 inch (0.152 mm) Diameter Hole on a Flat Specimen

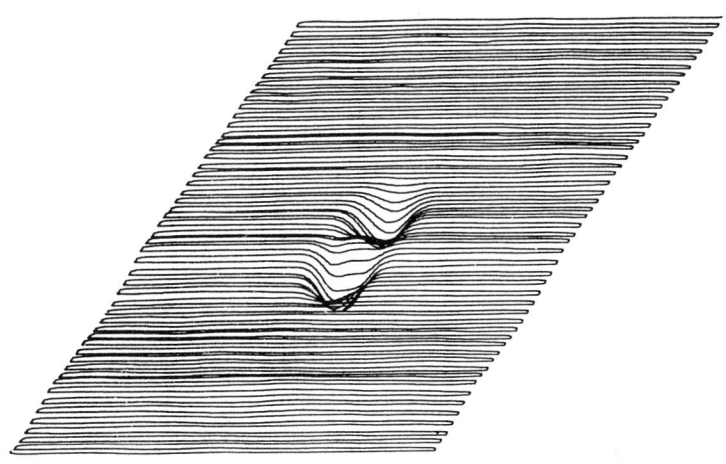

Figure 8. The Eddy Current Image Obtained from a 0.024 inch (0.610 mm) Long Fatigue Crack

The eddy current images can be presented with either the conventional grey scale approach or a continuous line X-Y 3-D approach. The continuous line approach which allows the tracking of the varying signal has the significant advantage over the conventional grey scale approach in which only discrete levels of signal amplitudes can be set and observed. The stress induced changes in conductivity and permeability are more slowly varying in nature, i.e., lower frequency, compared to abrupt continuity changes of surface breaking flaws. The system can discern stress responses from crack responses. This capability enables the prediction and monitoring of the initiation and propagation of surface cracks.

In summary, analog eddy current imaging techniques can undoubtably improve the system detection and evaluation capabilities. However, to fully utilize the capacity of imaging approach, a digital eddy current imaging system is proposed. The data storage and manipulation, imaging processing, signal enhancement, and other advanced analysis capabilities of the digital system will greatly facilitate the potential developments and applications of eddy current imaging.

REFERENCES

1. C. V. Dodd, W. E. Deeds and W. G. Spoeri, Optimizing Defect Detection in Eddy Current Testing, Materials Evaluation, March 1971, 59-63.
2. C. V. Dodd, and W. A. Simpson, Jr., Thickness Measurements Using Eddy Current Techniques, Materials Evaluation, May 1973, 73-79.
3. W. R. Junker and W. G. Clark, Jr., Eddy Current Characterization of Applied and Residual Stresses, in "Review of Progress in Quantitative Nondestructive Evaluation, 2B", ed. Thompson and Chimenti, (Plenum Press, New York, 1983), 1269-1286.
4. E. J. Chern, Automated Eddy Current Hole Measurements, Materials Evaluation, 43, No. 13, (December 1985) 1644-1648.
5. D. C. Copley, Eddy Current Imaging for Defect Characterization, in "Review of Progress in Quantitative Nondestructive Evaluation 2", ed. Thompson and Chimenti, (Plenum Press, New York, 1984) 1527-1540.
6. R. O. McCary, D. W. Oliver, K. H. Silverstein, and J. D. Young, Eddy Current Imaging, IEEE Transactions on Magnetics, MAG-20, No. 5, (September 1984) 1986-1988.
7. R. C. McMaster, The Present and Future of Eddy Current Testing, Materials Evaluation, 43, No. 12, (November 1985) 1512-1521.
8. C. V. Dodd and W. E. Deeds, Analytical Solutions to Eddy Current Probe-Coil Problems, J. Appl. Phys., 39, No. 6, (May 1968) 2829-2838.
9. R. Palanisamy and W. Lord, Finite Element Analysis of Eddy Current Phenomena, Materials Evaluation, 38, No. 10, (October 1980) 39-43.
10. I. Imam, M. V. K. Chari and A. Konrad, Development of Modal Analysis Technique for Eddy Current Nondestructive Evaluation, in "Proceedings of the 13th Symposium on NDE", ed. B. E. Leonard, (San Antonio, TX, April 1981) 480-485.
11. L. Udpa and W. Lord, Diffusion, Waves, Phase and Eddy Current Imaging, in "Review of Progress in Quantitative Nondestructive Evaluation, 1", ed. Thompson and Chimenti, (Plenum Press, New York, 1984) 499-506.
12. D. J. Hagemaier, Eddy Current Standard Depth of Penetration, Materials Evaluation, 43, No. 11, (October 1985) 1438-1442.

IMAGING OF ADVANCED COMPOSITES WITH A LOW-FREQUENCY ACOUSTIC MICROSCOPE

J. D. Fraser and C. S. DeSilets

Precision Acoustic Devices, Inc.
Fremont, CA 94539

B. T. Khuri-Yakub

Ginzton Laboratory, Stanford University
Stanford, CA 94305

INTRODUCTION

We have built the first commercially-made low frequency acoustic microscope, measured its characteristics, and explored its utility for NDE of composites and other aerospace materials. We studied the effects of numerical aperture and frequency on resolution and defect sensitivity. Interesting effects were noted. We determined limits on scanning speed and imaging depth and demonstrated the technical feasibility of the technique for practical problems. We obtained samples of rocket motor casings from the Navy, and structural parts from Navy and Air Force contractors, and demonstrated successful detection of defects in several cases of real interest.

DISCUSSION

The system we constructed is similar to Professor Khuri-Yakub's acoustic microscope in use at Stanford University. The block diagram is shown in Fig. 1. It was designed with special attention to the preamplifier and detector stages, which govern the sensitivity and dynamic range of the instrument. A dynamic range of 48dB, corresponding to 256 levels of gray scale, was desired for high-quality images. Dynamic range was tested as a function of frequency, as shown in Table 1.

Table 1. Receiver Sensitivity and Dynamic Range

| Frequency MHz | Sensitivity dBm | Dynamic Range dB | Equivalent Bits |
|---|---|---|---|
| 1.0 | -47 | 50 | 8.3 |
| 2.0 | -47 | 48 | 8.0 |
| 5.0 | -43 | 42 | 7.0 |
| 10.0 | -32 | 43 | 7.2 |
| 20.0 | -53 | 39 | 6.5 |

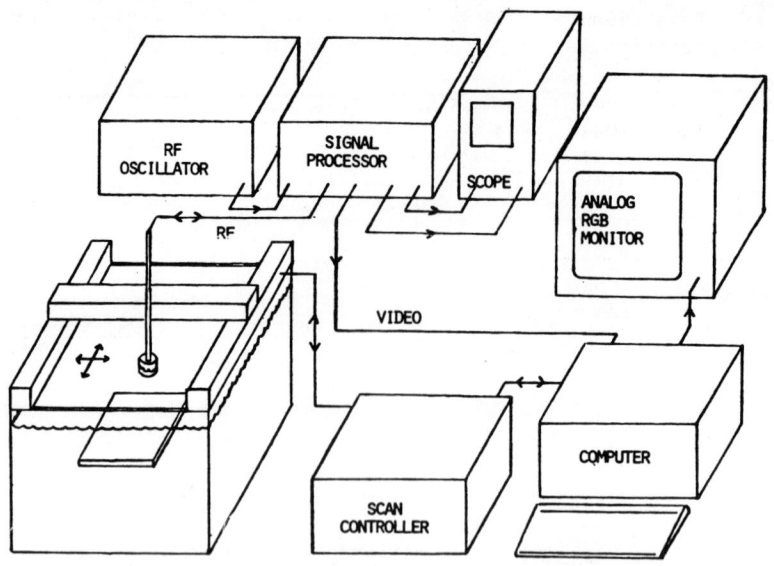

Figure 1. Block diagram of the low-frequency acoustic microscope system.

As can be seen, the performance at low frequencies is fine, while from 5 MHz up the receiver dynamic range is slightly less than desired. The high-frequency dynamic range will be improved in later versions of the system.

To build transducers for this project, P.A.D. used a technology developed for producing sharply focussed transducers for Professor Khuri-Yakub's microscope project at Stanford. The basic construction and performance parameters are those of our single quarter-wave matched medical transducer line: a piezoelectric ceramic element with acoustic matching by a single quarter-wave plate to water and electrical matching by a tuned transformer to 50 Ohms real at the center frequency. The insertion loss was typically 3dB at the center frequency, and the 6dB bandwidth was approximately 50%.

We made the following transducers for this project: 1, 3, and 5MHz f1, and 3MHz f3. We also made use of an existing .5 MHz f1 transducer, and operated the 3 MHz f1 transducer at a third harmonic frequency of 10 MHz. A 3 MHz f.7 transducer was constructed, but was accidentally broken before results could be obtained with it.

P.A.D. obtained or fabricated a variety of samples for evaluation. These included a variety of filament wound and laid up kevlar-epoxy and graphite-epoxy parts, as well as wire and epoxy resolution targets. Some of these samples are described below in Table 2.

Table 2.

| Label | Source | Description |
|---|---|---|
| I | Hercules | 16mm thick graphite, wound |
| II | Navy contractor | 2.7mm graphite cloth layup, seeded |
| III | Navy contractor | 7.5mm graphite bracket, seeded |
| IV | P.A.D. | 9mm thick, made from IV, seeded |
| V | Air Force contr. | 4mm, 24 ply quasi-isotropic graphite |

Sample I, a filament would graphite-epoxy motor case, was obtained from Hercules Corporation for us by China Lake N.W.C. We had problems imaging this sample. When we measured the propagation characteristics, we found that sound would pass through the material, but that the velocity depended strongly on direction. The velocity parallel to the fibers was found to be hard to measure, but greater than 6km/sec, while the velocity perpendicular to the fibers was 3.1km/sec. The individual plies of the fibers were about .75mm thick.

This much velocity dispersion could be expected to create problems in a system which depends for its resolution on coherent propagation of sound waves in a wide range of directions, and the problem would be expected to be worse when the thickness of the plies is comparable to the wavelength of the sound than when it is small compared to the wavelength. A seeded sample, Number II was fabricated from the Hercules graphite material, as shown in Fig. 2. The size of the flat-bottomed holes was 3mm, equal to the cross-fiber wavelength at 1 MHz. The targets were 3mm from the top side of the block, and 6 mm from the bottom side. The thickness of the plies was about 3/4 of a wavelength at a frequency of 3MHz.

Images were made from the top side at 3 MHz, 1 MHz, and .5 MHz, as shown in Fig. 3. The progression from poor image quality to better quality as the frequency decreases is striking, as it is exactly the opposite of what would happen in normal materials. At .5 MHz, 3mm voids are reliably detected 3mm below the surface. However, when imaging was attempted from the bottom side, 6mm from the flaws, nothing was detected. The dispersion and loss experienced in traversing some eight layers of fibers decreased the signal from the flaws below the detectable limit. This means that imaging through the whole thickness of Sample I would have to be done at a much lower frequency, say .1 MHz, and that the resolution would correspondingly be about 15mm, roughly equal to the thickness of the material. This mode of imaging might be effective for detecting some important types of flaws. The minimum flaw size detectable would probably be at least 5mm in lateral extent.

Samples III and IV were obtained by P.A.D. from a Navy contractor. Sample V was a seeded graphite-epoxy plate, laid up from cloth and vacuum bagged. It was 2.7mm thick, and had a series of syntactic foam squares and a series of saran wrap squares inserted at the midplane. A map of the

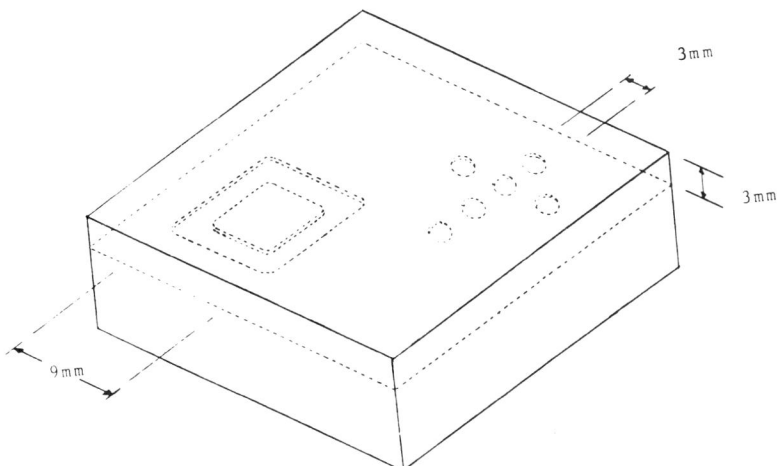

Figure 2. Schematic diagram of Seeded sample laminated from filament wound graphite, with 3mm holes and 9mm square.

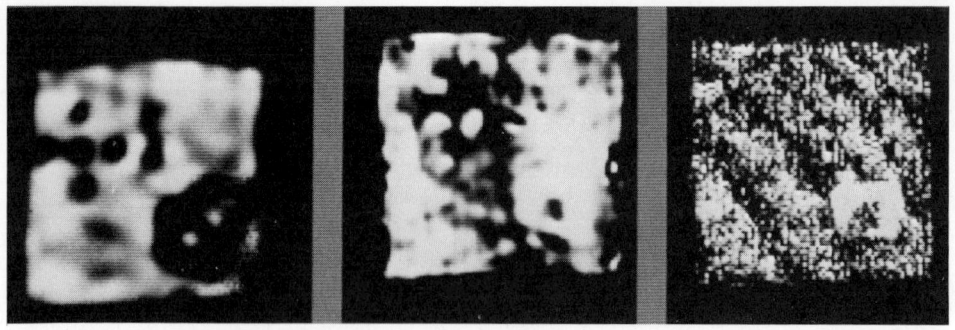

Figure 3. Acoustic images of a seeded, filament would graphite-epoxy sample. From left to right: 3MHz, 1MHz, .5MHz.

sample is shown in Figure 4. The contractor's own x-ray inspection had detected the foam squares with weak contrast, but had not found the saran wrap squares, which are representative of a delaminated but closed region. We imaged this sample at 1 MHz as shown in Figure 5. Both foam and saran squares are easily seen at 3mm size, and 1mm squares show some trace. Additionally, other flaws are seen, which are probably bubbles introduced during the layup. At 3 MHz, the wavelength would be comparable to the thickness of one layer of the graphite cloth, and the image (not shown) included only surface topography.

Sample IV was a thick, sharply curved layup of the same type of cloth as Sample III. Normal ultrasound inspection with weakly focussed transducers and a through transmission-back reflection technique worked on the flat areas, but not on the corner. We imaged this sample with a 1 MHz F1 transducer, taking the image in strips 9mm wide and moving the part between

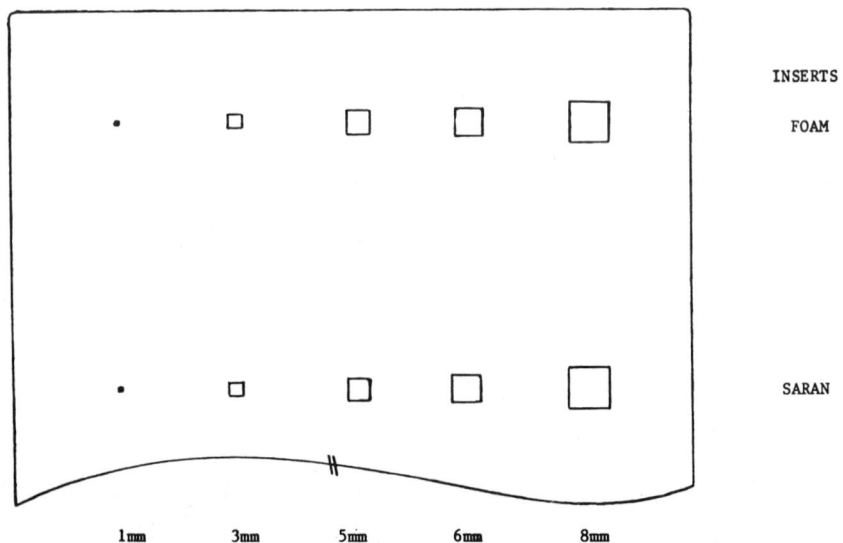

Figure 4. Schematic diagram of seeded graphite-epoxy laminate, showing locations and sizes of flaws.

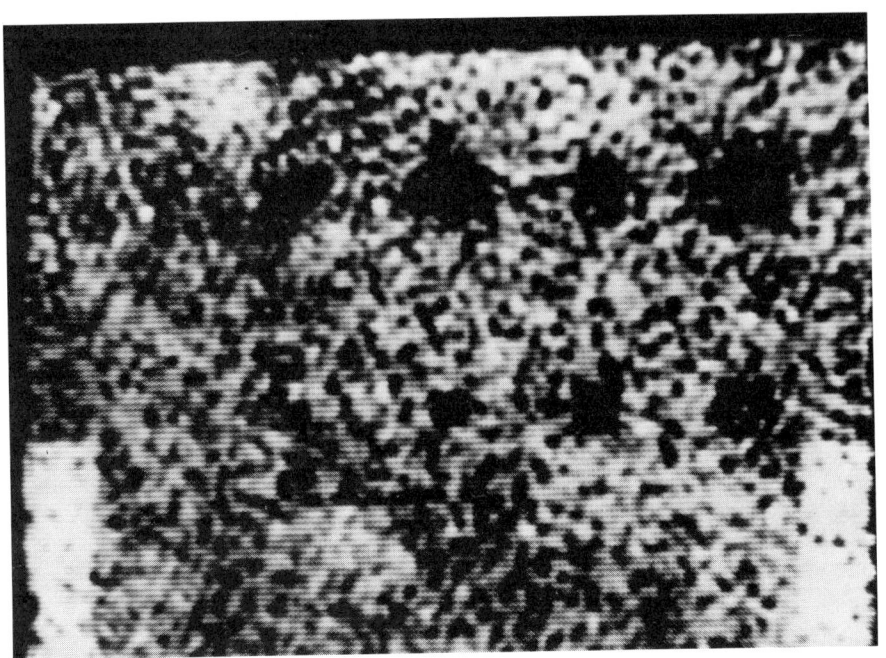

Figure 5. Photograph of the seeded sample shown in Figure 4.

strips to simulate contour following. This sample contained seeded Teflon strips, as shown in Fig. 6. A 1 MHz image is shown in Fig. 7, and shows reliable detection of all the strips, both on the flat and around the corner.

Another, more recently developed, material of interest to aerospace manufacturers is so-called quasi-isotropic graphite-epoxy. This is fabricated from prepreg, with the individual layers of fibers being of the order of .2mm thick, and with layers laid at 0 degrees, 90 degrees, and plus and minus 45 degrees in such a way as to minimize the anisotropy of the

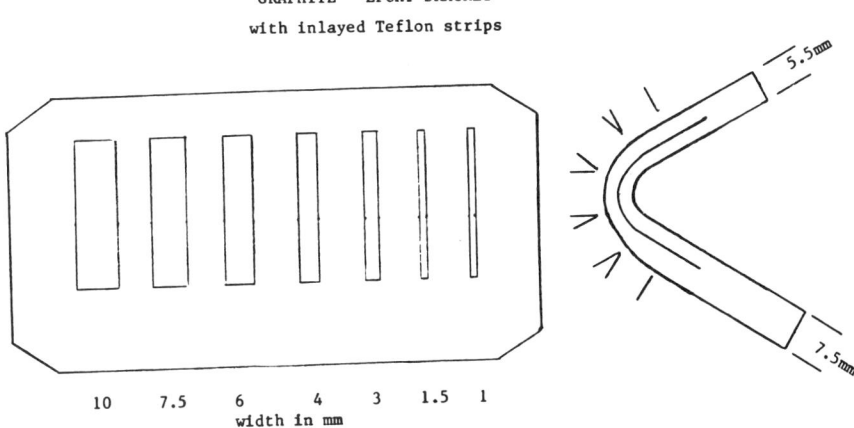

Figure 6. Schematic of seeded graphite-epoxy bracket.

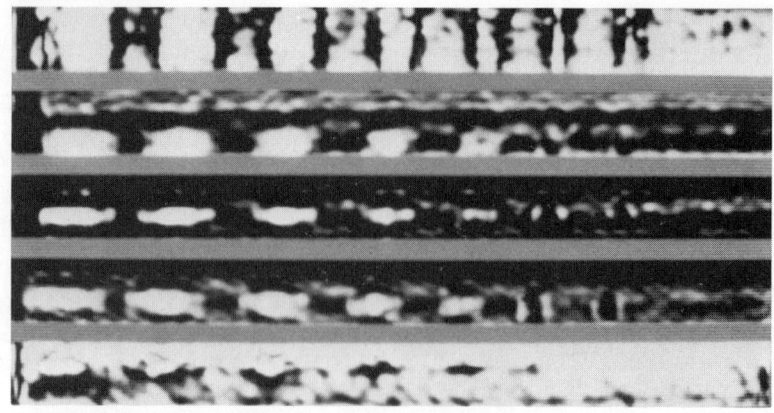

Figure 7. Photograph of the seeded graphite-epoxy bracket of Figure 6, showing detection of teflon strips.

mechanical, and incidentally the acoustic properties. We obtained sample V, a 4mm thick sample made of 24 plies, from an Air Force contractor. This sample had been deliverately damaged by clamping it to a steel plate and dropping a weight onto a 3mm diameter, flat ended pin contacting the sample.

The extremely fine structure of the quasi-isotropic material suggested that is could be imaged at higher frequencies than previous samples, and this turned out to be true. 3 MHz F1 and 10 MHz F1 pictures are shown in Figure 8. At 3 Mhz, the sample is about four wavelengths thick, and the sound easily penetrates to the back surface. Subtle details of the way the material delaminates with depth may be obtainable from the shading of the image, since the signal strength received depends on the interference of the front surface echo with the echo from the delamination surface. The 10 MHz image is also interesting, and shows the extent of the damaged zone with

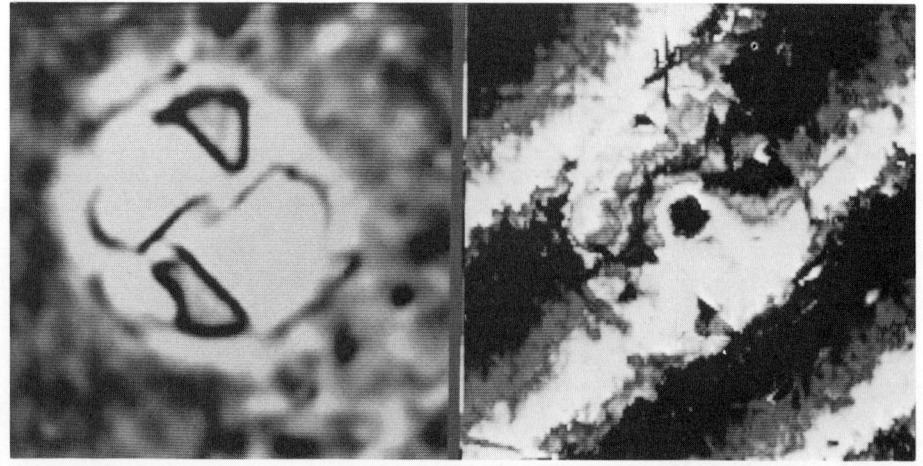

Figure 8. Acoustic images of damaged quasi-isotropic graphite laminate. Left: 3 MHz Right: 10 MHz.

excellent lateral resolution. The ultrasound is probably penetrating all the way to the back surface in this case also.

The theoretical maximum scanning speed of the system was analyzed theoretically, as it was well higher than could be realized with the mechanical scanner at P.A.D. The time per pixel for a single transducer system is equal to the round trip transit time between the transducer and the focal point, which, for samples of longitudinal sound velocity greater than that of the coupling medium (water), is maximum when the transducer is focussed on the surface of the sample. For a typical transducer, such as our 3 MHz F1 transducer, the radius of curvature is 25mm, so the round trip path is 50mm, and at a sound velocity in water of 1.5mm/microsec, this gives a transit time of 33 microsec. For this transducer, the spot size at the focus is about .6mm. Using about 2 pixels per spot size gives a smooth looking picture. This would yield a linear scanning rate of .3mm/33microsec, or 9 meter/sec. The areal scan rate would be 27 square centimeters per second, or 4 square inches per second. This would be comparable to existing mechanical equipment, but with better lateral resolution. However, mechanical scanning at 9 meters per second would be extremely improbable with any mechanical system. The only practical way to achieve the theoretical speed is to use a phased array and scan it electronically. Many of the technical details of electronically scanned, phased array imaging have been worked out during the last ten years by academic and corporate researchers interested in medical applications of ultrasound, and it now appears feasible to build such a system for NDE applications. An additional advantage of the phase array approach is that multiple beams could be managed simultaneously, further increasing scanning speed. This would require additional electronics, but since the transducer, position sensor, and display would remain the same, the cost would increase much less than the speed.

CONCLUSIONS

P.A.D. has demonstrated a Low Frequency Acoustic Microscope instrument and transducers which can be added to a standard mechanical c-scanner to provide the benefits of acoustic microscopy to commercial ultrasonic NDE users. This instrument and these transducers are now available for sale, and several transducers have already been sold. Several companies have also indicated interest in buying the instrument, and we expect sales before the end of 1986.

The utility of the instrument was investigated for imaging kevlar-epoxy and graphite-epoxy structures of filament wound, cloth layup, and quasi-isotropic prepreg construction. It was found to be unusable on filament would kevlar, and to achieve resolution and depth of penetration of 3mm on filament wound graphite. On this laid up cloth part we detected square and strip delaminations as small as 1mm wide at depths of up to 4.5mm with good reliability, at 1 MHz. On quasi-isotropic prepreg parts, impact induced delaminations were clearly visualized. Imaging of this material is possible at frequencies of at least 10 MHz, due to the fine scale of the structure. At 3 MHz, penetration is completely through a 4mm sample, and resolution is good.

The system is most useful for flaw detection in thin, flat or smoothly curved structures such as plates or skins. It is not useful on thick filament wound structures, particularly those made of kevlar. With the use of a phased array and a solid coupling system such as is common in the medical ultrasound business, the acoustic microscope could be a fast, cheap method for inspecting parts either at the manufacturing site or in the field. Scanning speeds of the order of ten square inches per second should be achieved with resolution of .6mm. By using a "paintbrush transducer" connected to a mechanical position detector, a swath of aircraft wing skin, say five inches wide, could be inspected as quickly as two inches per second.

ACOUSTIC MICROSCOPY USING AMPLITUDE AND PHASE MEASUREMENTS

P. Reinholdtsen and B. T. Khuri-Yakub

Edward L. Ginzton Laboratory
W. W. Hansen Laboratories of Physics
Stanford University
Stanford, California 94305

INTRODUCTION

We have built a low-frequency scanning acoustic microscope (SAM) that measures both amplitude and phase. The majority of SAMs simply measure the amplitude of the reflected signal. Measuring the phase gives a great deal more information. For one thing, the phase is very sensitive to height variations. Measuring the phase also gives us the ability to do signal processing on the resulting images, such as removing the effects of surface features from defocused images of subsurface defects.

THE MICROSCOPE

An efficient broadband focused transducer is excited with a tone burst. The transducer also receives the reflected echo from the sample of interest. As shown in Fig. 1, the 3 MHz tone burst is generated

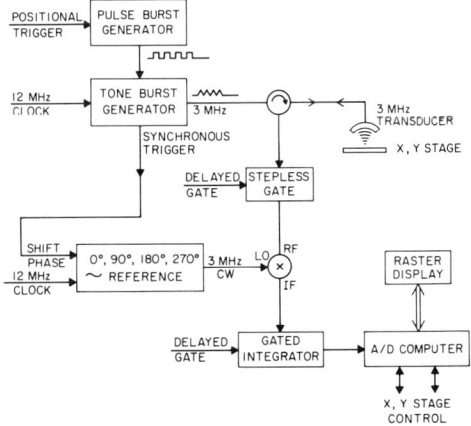

Fig 1.   Schematic of the amplitude and phase acoustic microscope. The real and imaginary parts of the reflected signal are measured from a sequence of four tone bursts with the reference shifted in 90` increments.

543

from a 12 MHz clock. The 12 MHz clock is also used to generate a 3 MHz cw reference whose phase can be shifted by 90° increments relative to the transmitted tone burst. The return signal is then mixed with the reference signal and the result is integrated and digitized. The in-phase component of the reflected signal is obtained by taking the difference of the resulting 0° and 180° components, and the quadrature component is obtained by taking the difference between the 90° and 270° components. This removes any dc offset that may be present in the mixer, gated integrator, or digitizer.

## NUMERICAL DEFOCUSING

A common technique used with SAMs is defocusing the transducer to obtain enhanced subsurface defect detection which consists of moving the transducer closer to the sample of interest. This concentrates more acoustic energy below the surface, leading to greater contrast when subsurface features are present. A defocused image contains information from both surface and subsurface features. If there are features on the sample's surface, such as random roughness, they will show up in the defocused image. It would be useful to be able to distinguish between surface and subsurface effects in order to remove the surface effects from a defocused image. With both amplitude and phase information, it is possible to do this to a large degree by also taking an image with the transducer on focus.

By taking an on-focus image, it is possible to numerically defocus the image by an amount equivalent to the defocused image. Because the on-focus image contains surface information, the numerically-defocused image will contain defocused surface features. The difference image between the numerically- and experimentally-defocused images will contain only subsurface features.

In order to perform this numerical defocusing, we need to do three things: (1) characterize the transducer; (2) determine the effect of the transducer's characteristics on image formation; and (3) determine how these characteristics change as the transducer is defocused.

## TRANSDUCER CHARACTERIZATION

The transducer will be characterized by plane wave decomposition [1] and V(z) inversion [2]. The field produced by a transducer can be decomposed into a superposition of plane waves by taking the two-dimensional Fourier transform of the generated field in a plane perpendicular to the transducer's face.

$$S_{10}(z; k_x, k_y) = \mathcal{F}\{s_{10}(z; x, y)\} \tag{1}$$

where $s_{10}$ is the generated field, $S_{10}$ corresponds to a plane wave with wave number $(k_x, k_y, k_z)$, and

$$k^2 = k_x^2 + k_y^2 + k_z^2 \ .$$

It is easy to relate the plane wave decomposition from one plane to another by multiplying each plane wave component by a phase factor (assuming the water is approximately lossless):

$$S_{10}(z_1; k_x, k_y) = S_{10}(z_2; k_x, k_y) * \exp[i k_z (z_2 - z_1)]. \qquad (2)$$

The field at another plane can be calculated by inverse transforming the propagated plane wave spectrum

$$s_{10}(z; x, y) = \mathscr{F}^{-1}\{S_{01}(z; k_x, k_y)\} \qquad (3)$$

or

$$s_{10}(z; \vec{X}) = \mathscr{F}^{-1}\{S_{01}(z; \vec{\kappa})\} \qquad (4)$$

where $\vec{X} = (x,y)$ and $\vec{\kappa} = (k_x, k_y)$.

The response of a transducer to an incoming plane wave can be related to its outgoing spectrum if the transducer is reciprocal so that

$$S_{01}(z; \vec{\kappa}) \propto k_z S_{10}(z; -\vec{\kappa}) \qquad (5)$$

$S_{01}(z;\vec{\kappa})$ is a response to plane wave with a wave number $(k_x, k_y, -k_z)$.

If the transducer is located over a uniform sample, we can compute the transducer's output as a function of its distance from the sample, if we know the sample's reflection coefficient for plane waves as a function of wave number. This is the famous $V(z)$ curve and is given by

$$V(z) \propto \int_{-\infty}^{\infty} S_{10}(\vec{\kappa}) S_{01}(\vec{\kappa}) R(\vec{\kappa}) e^{2ik_z z} d\kappa \qquad (6)$$

where $R(\kappa)$ is the sample's reflection coefficient.

If we assume a reciprocal transducer and sample that are circularly symmetric, this reduces to

$$V(z) \propto \int_{-\infty}^{\infty} k_z S_{10}^2(\vec{\kappa}) R(\vec{\kappa}) e^{2ik_z z} d\kappa. \qquad (7)$$

With a couple of variable changes, this becomes a Fourier relationship

$$V(z) \propto \int_0^k k_z^2 S_{10}^2(k_2) R(k_2) e^{2ik_z z} dk_z$$

$$\propto \int_0^{k/2} \beta^2 S_{10}^2(\beta) R(\beta) e^{i\beta z} d\beta$$

$$\propto \mathscr{F}\{\beta^2 S_{10}^2(\beta) R(\beta)\} \qquad (8)$$

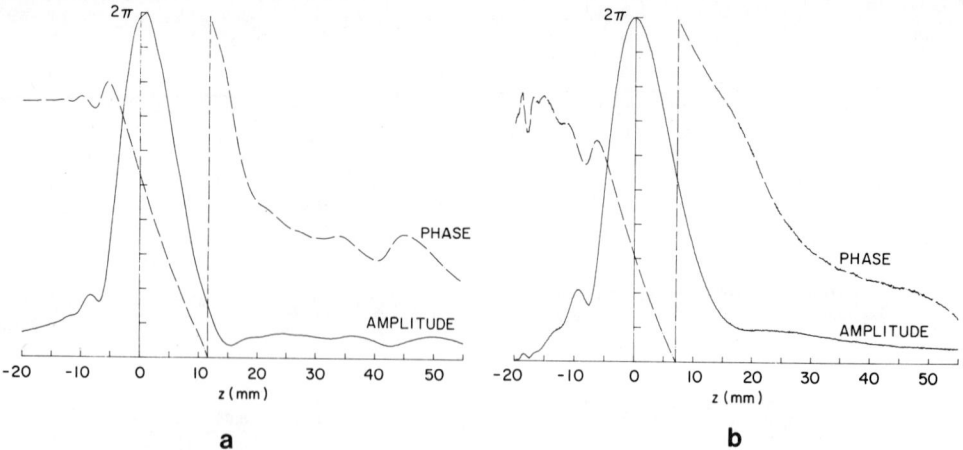

Fig. 2.  (a) Theoretical V(z) for an F2 transducer operating at 3 MHz. The focal length is 32 mm and the diameter is 16 mm. (b) Measured V(z) for a real F2 transducer operating at 3 MHz. Notice the good agreement with the theoretical model.

where $\beta = 2k_z$, and $S_{10}(\beta) = 0$ for $\beta < 0$ and $\beta > k/2$.

We can invert this to obtain the product of the sample's reflectance function and the square of the transducer's spectrum

$$\mathscr{F}^{-1}\{V(z)\} \propto \beta^2 S_{10}^2(\beta) R(\beta) \ . \tag{9}$$

If we know the reflectance function of a reference sample, and take V(z) for an unknown transducer using this sample, we can experimentally determine the transducer's spectrum.

Taking the square root in Eq. 9 leads to a 180° phase ambiguity. If we assume the spectrum is continuous, we can flip the sign of the computed square roots in order to minimize the difference between adjacent points in the spectrum.

Figure 2 shows a theoretical V(z) curve for an F2 transducer (32 mm focal length, 16 mm diameter) operating at 3 MHz. The theoretical model assumed the field at the transducer's face was a truncated converging spherical wave. Notice the good agreement with the experimentally-measured V(z) curve for a real F2 transducer. The curves are different, however, which is the reason behind experimentally characterizing the transducer's spectrum rather than strictly using a theoretical model.

Figure 3 shows the corresponding theoretical plane wave spectrum and the inverted spectrum from the measured V(z) curve. The sharp drop off in amplitude corresponds to the acceptance angle of the transducer. Notice that at a certain point, the inverted spectrum is zero. This was done because beyond that point the phase varied too rapidly to reliably remove the phase ambiguity.

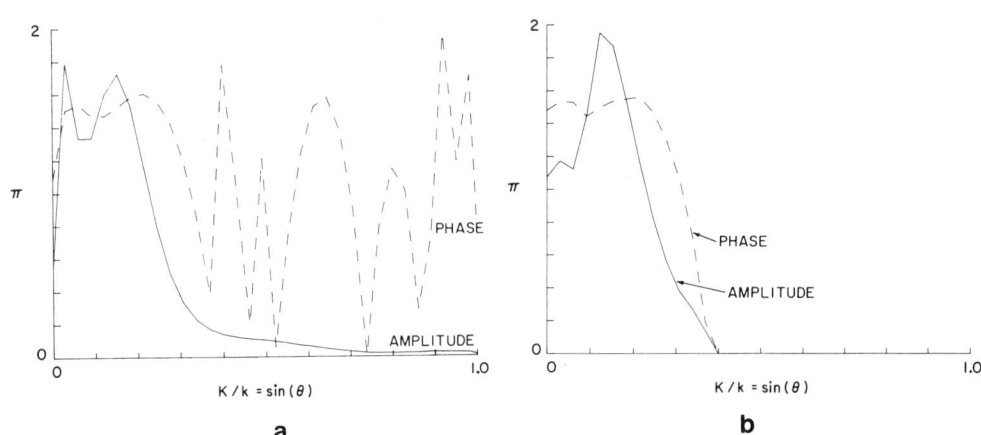

Fig. 3. (a) Theoretical spectrum for the F2 transducer of Fig. 2.
(b) Inverted spectrum for the real F2 transducer of Fig. 2. Again, notice the good agreement with the theoretical model. The spectrum was set to zero because of the algorithm used to remove ambiguity in the phase.

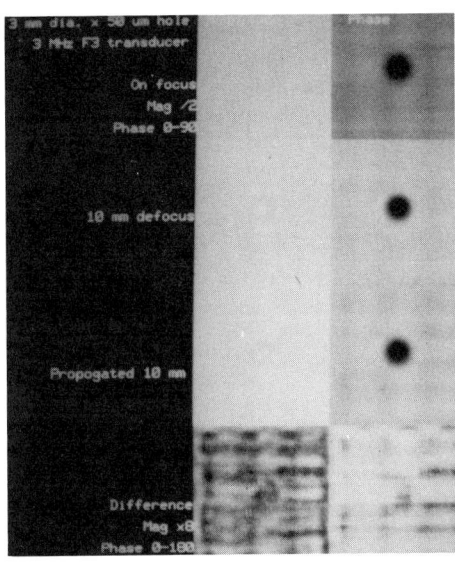

Fig. 4. Acoustic images of fused quartz with a shallow depression 3 mm across and 50 μm deep. Top images: On-focus image with F3 transducer at 3 MHz (amplitude on left; phase on right). Second from the top: Experimentally measured, defocused image (10 mm). Third: Numerically defocused image obtained from on-focus image using theoretical model for transducer's spectrum. Bottom: Difference between experimentally-defocused and numerically-defocused images. The magnitude is multiplied by 10.

Now that we can experimentally characterize the transducer, we need to know what effect the transducer has on image formation. In order to do this, we will make several simplifying assumptions. We assume that the reflected acoustic field at the object's surface is a point-by-point product of the incident field and the object's acoustic field response. Second, the sample is relatively flat compared to the focal depth of the transducer. This is a good approximation if the transducer used has a large F-number (a small aperture). Third, there should be no mode conversions (such as into pseudo-Rayleigh waves). This is true for many composites or other materials with low acoustic velocities, or when the transducer is on focus.

With these assumptions, the measured image is simply the round trip impulse response of the transducer convolved with the object's field response

$$i(z; \vec{X}) = i_0(\vec{X}) * t(z; \vec{X}) . \tag{10}$$

Where the round trip impulse response is simply the product of the transmitting and receiving impulse responses

$$t(z; \vec{X}) = s_{10}(z; \vec{X}) \, s_{01}(z; \vec{X}) . \tag{11}$$

These, in turn, are simply the Fourier transforms of the transmitting and receiving spectra of the transducer, which we experimentally determined

$$s_{10}(z; \vec{X}) = \mathscr{F}^{-1}\{S_{10}(z; \vec{\kappa})\} = \mathscr{F}^{-1}\{S_{10}(\vec{\kappa}) e^{ik_z z}\} .$$

$$s_{01}(z; \vec{X}) = \mathscr{F}^{-1}\{S_{01}(z; \vec{\kappa})\} = \mathscr{F}^{-1}\{S_{01}(\vec{\kappa}) e^{ik_z z}\} \tag{12}$$

If we transform Eq. 13 to the Fourier domain, the convolution becomes a product

$$I(z; \vec{k}) = I_0(\vec{k}) \, T(z; \vec{k}) . \tag{13}$$

By dividing Eq. 13 by itself, for different defocus levels, we can relate images taken at different defocus depths

$$\frac{I(z_1; \vec{k})}{I(z_0; \vec{k})} = \frac{I_0(\kappa) \, T(z; \vec{k})}{I_0(\kappa) \, T(z_0; \vec{k})}$$

$$I_1(z_1; \vec{k}) = I(z_0; \vec{k}) \frac{T(z_1; \vec{k})}{T(z_0; \vec{k})} . \tag{14}$$

This gives us the ability to take an on-focus image and numerically defocus it to any depth, if we know the transducer's spectrum.

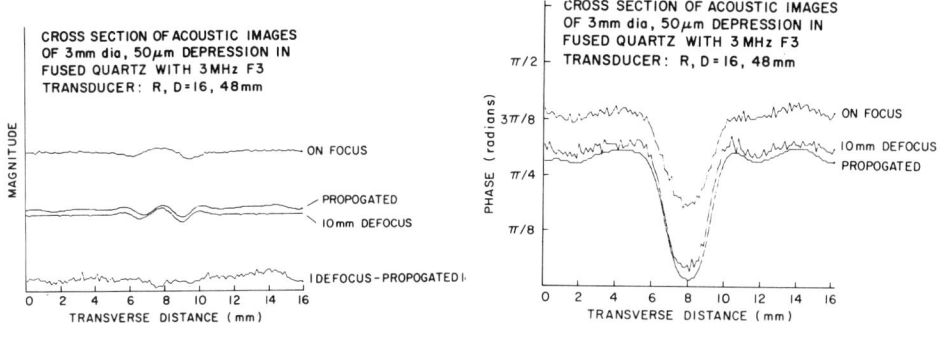

Fig. 5. Line scans through the center of the depression of images in Fig. 4. (a) magnitude. (b) phase.

EXPERIMENTAL RESULTS

A fused quartz sample was made to determine, experimentally, if we could indeed predict a defocused image from an on-focus image. The sample was polished smooth and then a 50 μm depression 3 mm across was drilled into the surface. This was designed to simulate a small phase step.

Some initial results were obtained by imaging with a 3 MHz F3 transducer, and numerically defocusing by using a theoretical model for the transducer's spectrum. The images were taken on focus and with a 10 mm defocus. The results are shown in Fig. 4. Notice the good agreement in both magnitude and phase between the experimentally-defocused and numerically-defocused images. Figure 5 shows single line scans of these images through the center of the defect in order to give a more quantitative indication of the agreement. The difference image demonstrates that surface features can indeed be subtracted from defocused images.

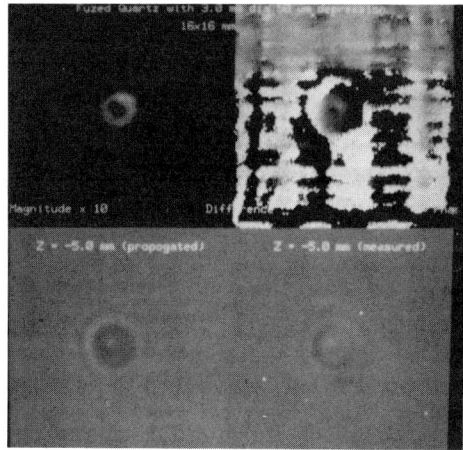

Fig. 6. Acoustic images of fused quartz with a shallow depression using the F2 transducer at 3 MHz. Bottom right: magnitude of experimentally-defocused image (defocused 5 mm). Bottom left: magnitude of numerically-defocused image using spectrum obtained by inverting V(z). Top left: magnitude of difference multiplied by 10. Top right: phase of difference.

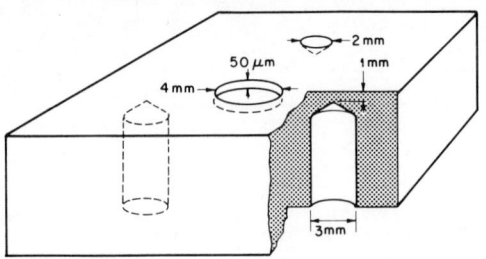

Fig. 7. Schematic of aluminum block with surface and subsurface defects.

The same sample was then imaged using the F2 transducer, whose spectrum had been experimentally determined. Figure 6 shows the results. Again, the numerically-defocused image is close to the experimentally-defocused image.

Next, a sample with both surface and subsurface features was imaged with the F2 transducer. The sample was made of an aluminum plate shown schematically in Fig. 7. It had two holes drilled into the back side to within 1 mm of the front surface. These were the subsurface defects. Two depressions were drilled into the top side that was to be imaged: a shallow V-shaped hole, similar to the top of those drilled into the back side; and a shallow depression similar to the one imaged in the quartz sample.

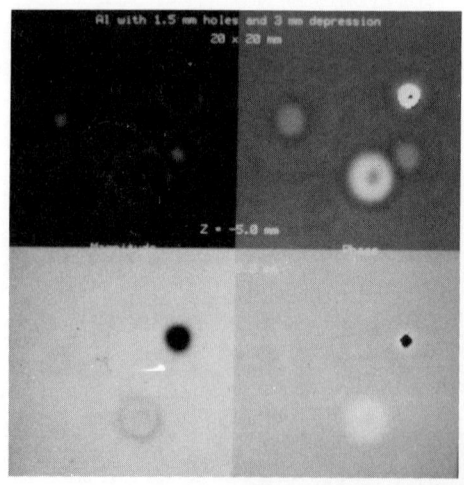
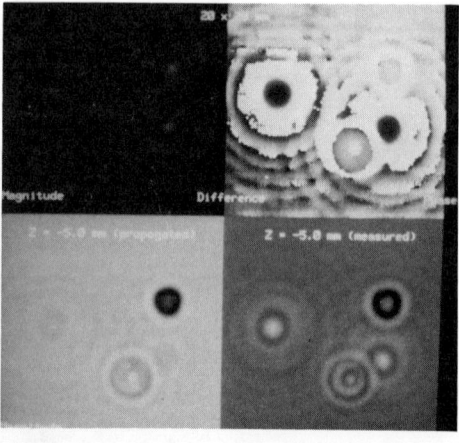

a  b

Fig. 8. Acoustic images of aluminum block using F2 transducer at 3 MHz . (a) Experimentally-measured images. Amplitude is on the left; phase is on the right. Bottom: on-focus image. Top: defocused 5 mm . (b) Bottom right: magnitude of experimentally-defocused image. Notice the increased contrast of subsurface features relative to numerically-defocused image (bottom left). Also, the surface features are very similar. Top images: difference between numerically-defocused and experimentally-defocused images. Note increased contrast of subsurface defects relative to surface defects.

In Fig. 8, we see that in the on-focus image, the surface defects appear clearly, especially in the phase images, whereas the subsurface features are much less visible. In the experimentally-defocused image, the subsurface features are greatly enhanced, much more so than in the numerically-defocused images. This is what we were looking for, because, when the difference is taken, the surface features are suppressed much more than the subsurface features. The subsurface features were somewhat suppressed because they were within the focal depth of the transducer. The shallow surface depression was suppressed much more than the V-shaped hole which extended deeper into the sample.

CONCLUSION

With the low-frequency acoustic microscope we can now do signal processing on images because of the ability to measure both amplitude and phase. We can characterize the angular spectrum of transducers by the inversion of $V(z)$ data. This gives the ability to numerically defocus acoustic images and it enables surface features to be removed from experimentally-defocused images, leading to increased relative contrast of subsurface features.

ACKNOWLEDGMENT

This work was sponsored by the Center for Advanced Nondestructive Evaluation, operated by the Ames Laboratory, USDOE, for the Air Force Wright Aeronautical Laboratories/Materials Laboratory under Contract No. SC-86-102.

REFERENCES

1. D. H. Kerns, J. Acoust. Soc. America 57 (2) (February 1975).
2. K. K. Liang, G. S. Kino, and B. T. Khuri-Yakub, IEEE Sonics & Ultrasonics SU-32 (2), 213 (March 1985).
3. R. A. Lemons and C. F. Quate, Appl. Phys. Lett. 24, 163 (1974).
4. C. F. Quate, A. Atalar, and H. K. Wickramasinghe, Proc. IEEE 67, 1052 (1979).
5. A. Atalar, J. Appl. Phys. 45, 5130-5139 (October 1979).
6. P. Reinholdtsen, W. W. Hipkiss, and B. T. Khuri-Yakub, Review of Progress in Quantitative Nondestructive Evaluation, D. O. Thompson and D. E. Chimenti, Eds., Plenum Press, 263 (1985).
7. See IEEE Trans. on Sonics and Ultrasonics, Special Issue on Acoustic Microscopy SU-32 (2), (March 1985).

ACOUSTIC MICROSCOPY:   MATERIALS ART AND MATERIALS SCIENCE


R.S. Gilmore, R.E. Joynson, C.R. Trzaskos, and J.D. Young

General Electric Company
Corporate Research and Development
P.O. Box 8
Schenectady, New York   12301

INTRODUCTION

Significant progress has been made in acoustic microscopy and other forms of acoustic imaging over the last two decades. Originally introduced by Quate [1], this technology has been established by Weglin [2], Kino [3], Wickramasinghe [4], Bertoni [5], and Quate [6] as a powerful tool for materials characterization and development. The work described here [7] goes beyond that cited: it utilizes time-resolved acoustic signals of much greater bandwidth, and does not rely on $V(z)$ behavior to form images. Instead only the digitized amplitudes of the spatially and temporally resolved acoustic signals are processed and displayed to form the images. Much of the progress reported here is also due to advances in computer display technology. Originally presented as posters, the included figures demonstrate various hardcopy and high-resolution raster displays incorporated in the described acoustic microscope. Keeping in mind the purpose for which each image was intended, it is instructive to compare the image quality that the different displays can produce. Six figures, containing twenty-nine separate images, make up the presentation. In their original display format, each figure was a 30 × 40 in. poster in which the individual images were displayed at the identical magnifications that were initially presented to the acoustic microscopist.

The 512 × 512 pixel CRT images were therefore displayed at 11 × 11 in., and the hardcopy gray-scale recorder images were displayed at an 18 in. width and either an 18 or 36 in. length, depending on their original format, i.e., 1024 × 1024 or 1024 × 2048 pixels. The transducers used in this work produce 1.5 to 2.0 wavelength ultrasonic pulses at center frequencies of 50 MHz, 20 MHz, and 5 MHz. The beams were focused by high-velocity lenses of fused quartz or [111] cut silicon at apertures from F/0.8 to F/7.0, which resulted in single-point resolution spot sizes from 20 to 300 microns. Work reported elsewhere [7] suggests that the resolution for surface wave imaging is determined by the frequency, the surface wave velocity, and the diameter of the entry circle (Figure 1) due to the intersection of the cone of convergence caused by the Rayleigh incident angle and the entry surface of the material. The maximum dynamic range of the images is controlled by the 8-bit, single-pulse (up to 80 MHz) gated peak detector; however, most of the image files were attenuator adjusted to contain maximum amplitude signals between 7 and 8 bits. This was done to avoid loss of detail due to saturation.

The reduction of the 30 × 40 in. posters to the 8 × 10 in. page size in the published volume reduces the detail available from the original magnifications. In addition, more information was lost by the necessity to display the color images in black and white. In spite of these necessary limitations in the published version of the images, the versatility and image quality of the microscope are still apparent. In addition, a careful selection of the samples used to demonstrate surface inspection (Figure 2) displays the artistic possibilities of acoustic microscopy as well as the detail that highly focused direct reflections can provide with respect to surface damage.

RESULTS

The acquisition and display of acoustic images from V(z), the interference between the Leaky-Rayleigh wave, and the direct surface reflection are well documented [1,6]. The use of cylindrically convergent, time-resolved surface wave signals is a more recent development [7] and is possible only at frequencies for which broadband transducers can be fabricated. Figure 1A shows a schematic of the broadband acoustic microscope used in this work. In addition to the microscope schematic, a description of how the signals are set up and gated, and an indication of how deconvolving the acoustic beam from the raw data can improve the resolution of a surface wave image, are shown in Figure 1B.

Although the signal that is directly reflected from the surface contains little if any subsurface information, a directly reflected beam focused on the surface can provide an excellent display of the surface morphology. Figure 2 shows five images of circulated coins, which simultaneously show the surface damage sustained by a coin during circulation, the extraordinary skill of the artists who engrave the dies from which the coins are struck, and finally the clear promise offered by acoustic microscopes for artistic expression.

The inspection of semiconducting devices with an acoustic microscope is normally associated with large-scale integrated devices. Figure 3 shows a set of images acquired during the development of methods to diffusion bond a gate turnoff array onto a silicon-controlled thyristor with a diameter of 77 mm (3 in.). The thickness of [100] cut silicon that was penetrated to produce these images is 0.7 mm (0.028 in.). During the process development, this was one of the first devices in which all but two of the fingers in the gate array were fully developed and attached. The cracking of the spacer ring did not affect the initial successful operation of the device, but it does provide probable failure sites for additional cracking during the thermal cycling that would accompany operation at 5 kV and 2 kA.

In the attempts to achieve better and better resolution, work in acoustic microscopy has tended to the use of higher frequencies and shorter focal lengths. In the inspection of a 0.5 in.-thick titanium plate (Figure 4), the selection of a moderately high frequency, 20 MHz, and a long focus, F/7.0, develops a 0.5 in., −6 dB depth of field that permits 95% of the plate volume to be inspected with one scan. The large number of displayed indications are caused by having seeded the originating billet with nitrided titanium sponge. The 5% of the material obscured during the volume scan by the entry surface reflection was in turn inspected by a surface wave scan, but these images have been deleted for brevity.

The 5 MHz surface wave images in Figure 5 all originate from the 1024 × 2048 data file displayed in Figure 5(1). The sample displayed was originally fabricated from a 0.5 in. thick plate of René 95, a nickel-based superalloy, for use as an eddy current calibration block. The surface was given a mirror polish, but although the original machining marks were smoothed somewhat, they remain in the form of rounded valleys in a front surface mirror. The twelve slots and six holes in the sample show a very different interaction with the surface wave than the polished fly-cutter gouges left by the machinist. The purpose of the two Wiener filter enhanced images [Figure 5(3) and Figure 5(4)] is to demonstrate that the images do contain 1/3 surface wavelength resolution, which is 0.008 in. (0.2 mm) at 5 MHz. These images also demonstrate software that permits a 256 × 256 image segment containing the unknown discontinuity to be windowed from the 1024 × 2048 image file and processed to display the enhanced shape of the discontinuity. This of course permits the unknown target to be analyzed at a small fraction of the time and computer memory that would be required to process the larger file.

Most of the acoustic images showing grain size textures are made of as-cast materials, such as the 1% silver, 99% copper sample shown in Figures 6(1) to 6(4). The as-cast structures usually show grains with somewhat random orientations, which in turn produce contrast in the acoustic image that is controlled by the elastic anisotropy of the grains and the particular orientation that each grain has with respect to the imaged surface. Since the metal has not been deformed, the grain boundaries are usually straight lines and the interior is imaged at a single shade of gray when the parent metal is single phase. For the Cu99-Ag1.0 sample shown, Figure 6(4) shows some subgranular structure. In the forged titanium textures shown in Figures 6(5) and 6(6), the grain boundaries are much less distinct. However, a clear gradient in structure can be seen from the bottom to the top of both Figure 6(5) and the upper display of the same data file in Figure 6(6). These types of acoustic images can therefore be used to establish the uniformity of the microstructure in a large, highly stressed part such as a titanium compressor spool for an aircraft engine.

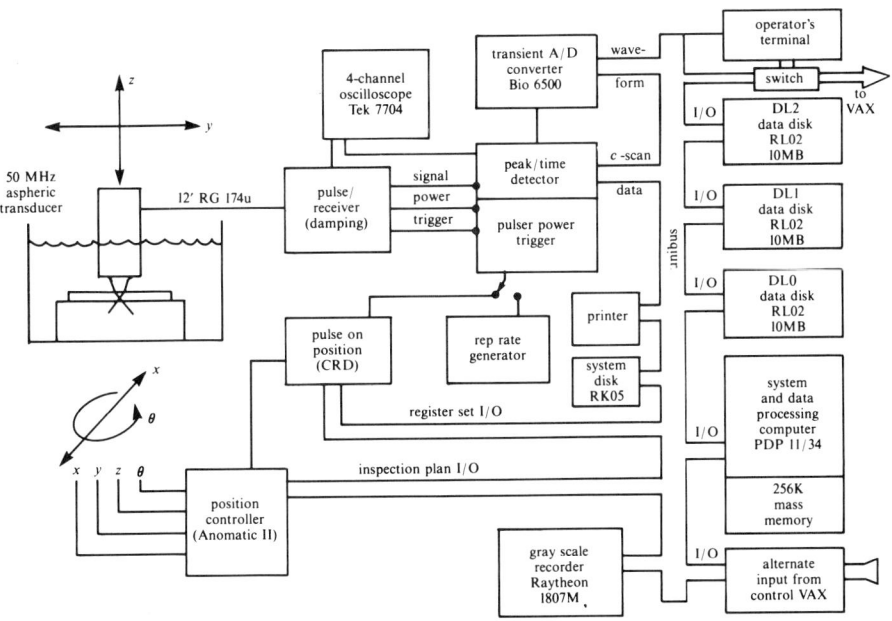

Figure 1A. Expanded view of GE-CRD's ultrasonic microscope also shown in Figure 1B(1).

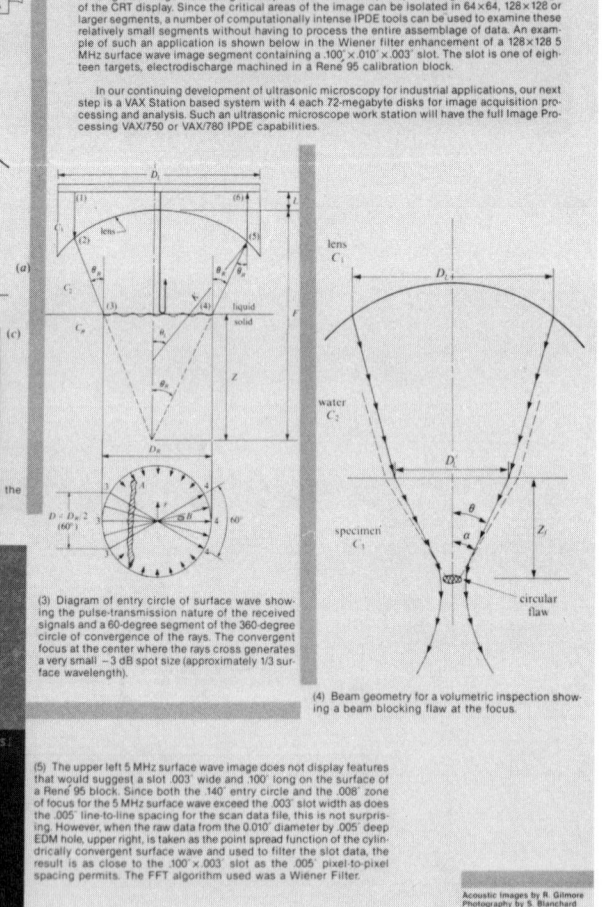

Figure 1B. Description of the ultrasonic microscope and surface and volume imaging.

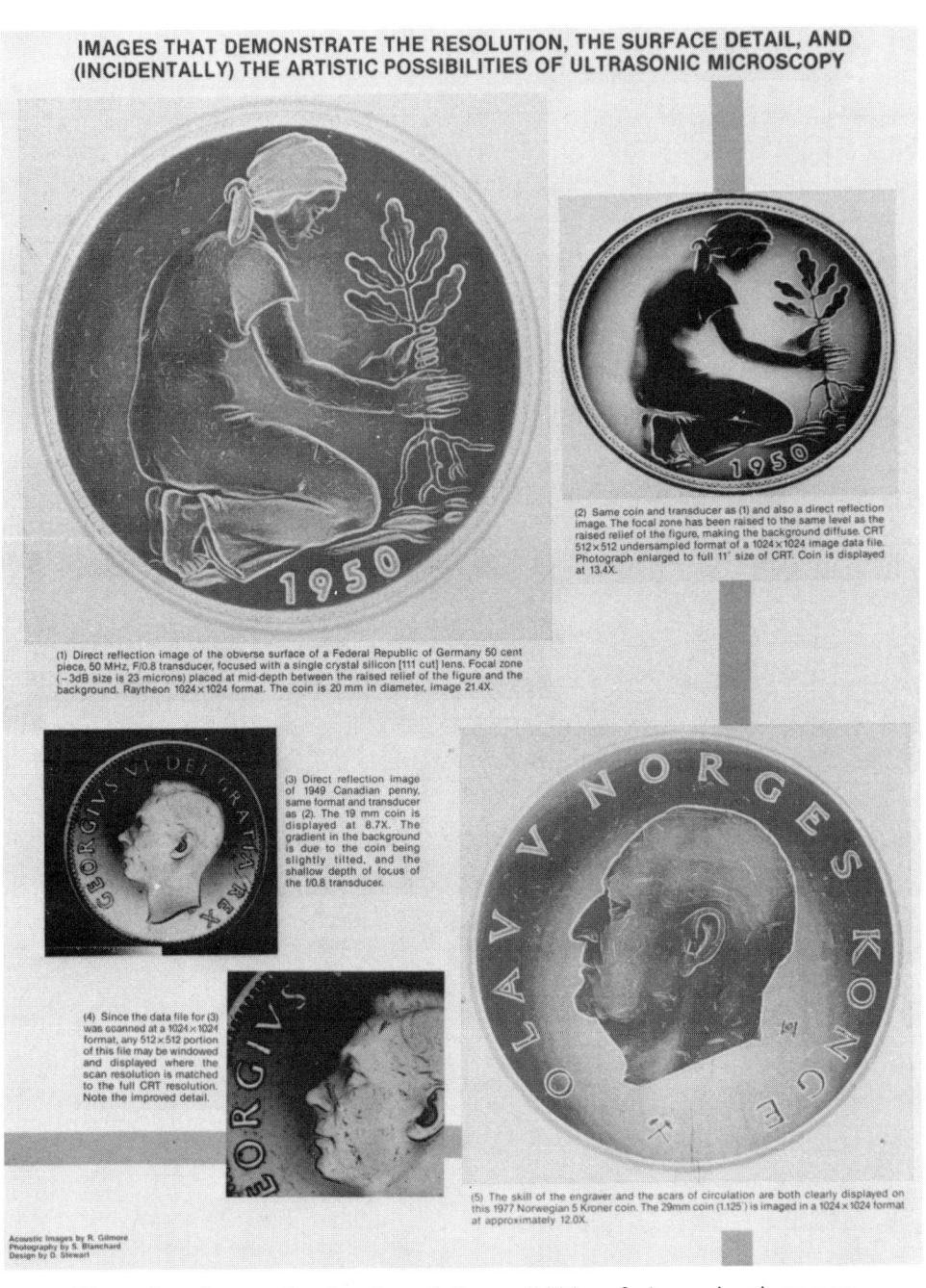

Figure 2. Images showing the artistic possibilities of ultrasonic microscopy.

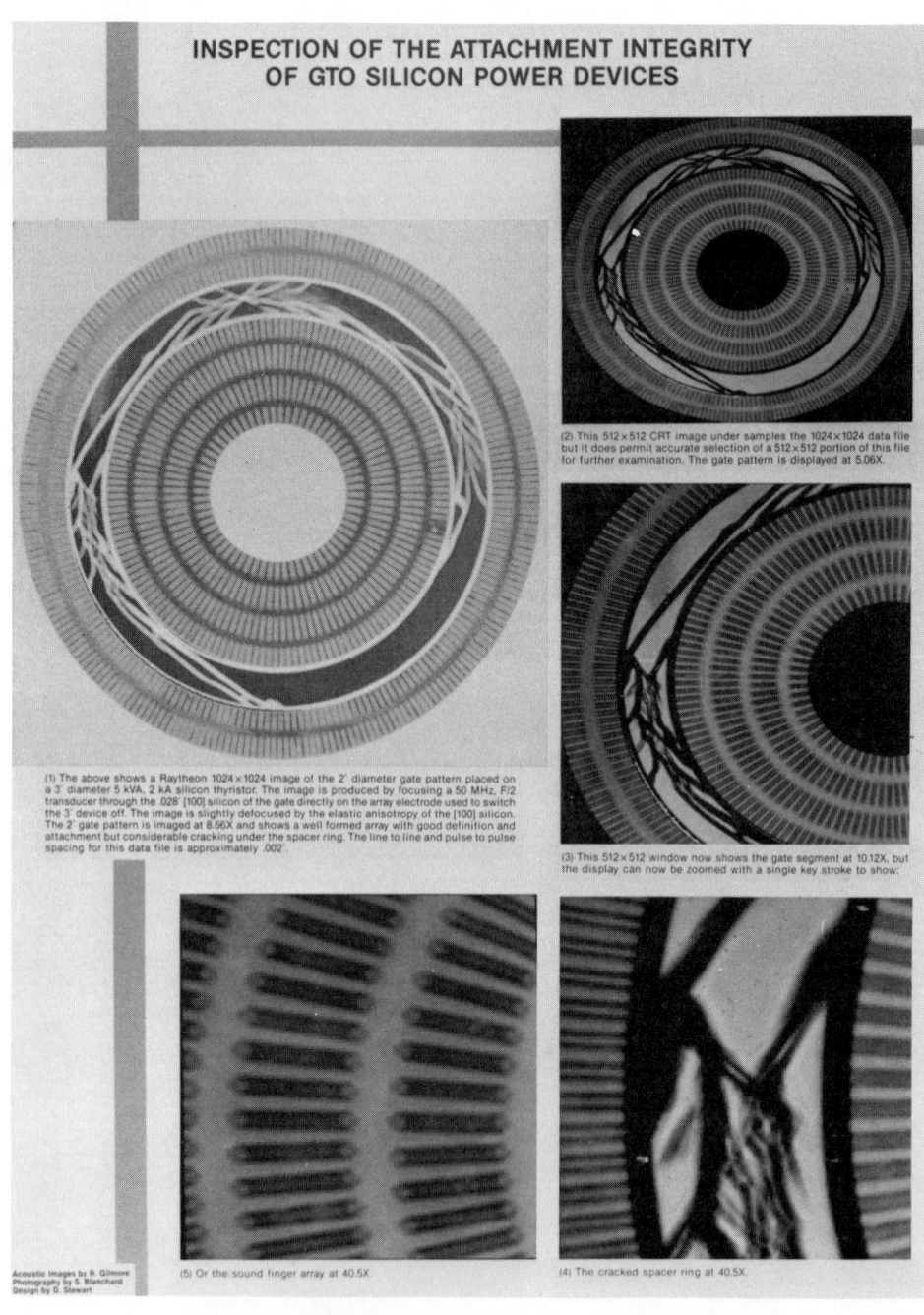

Figure 3. Inspection of the attachment integrity and details of the gate array in GTO silicon power devices.

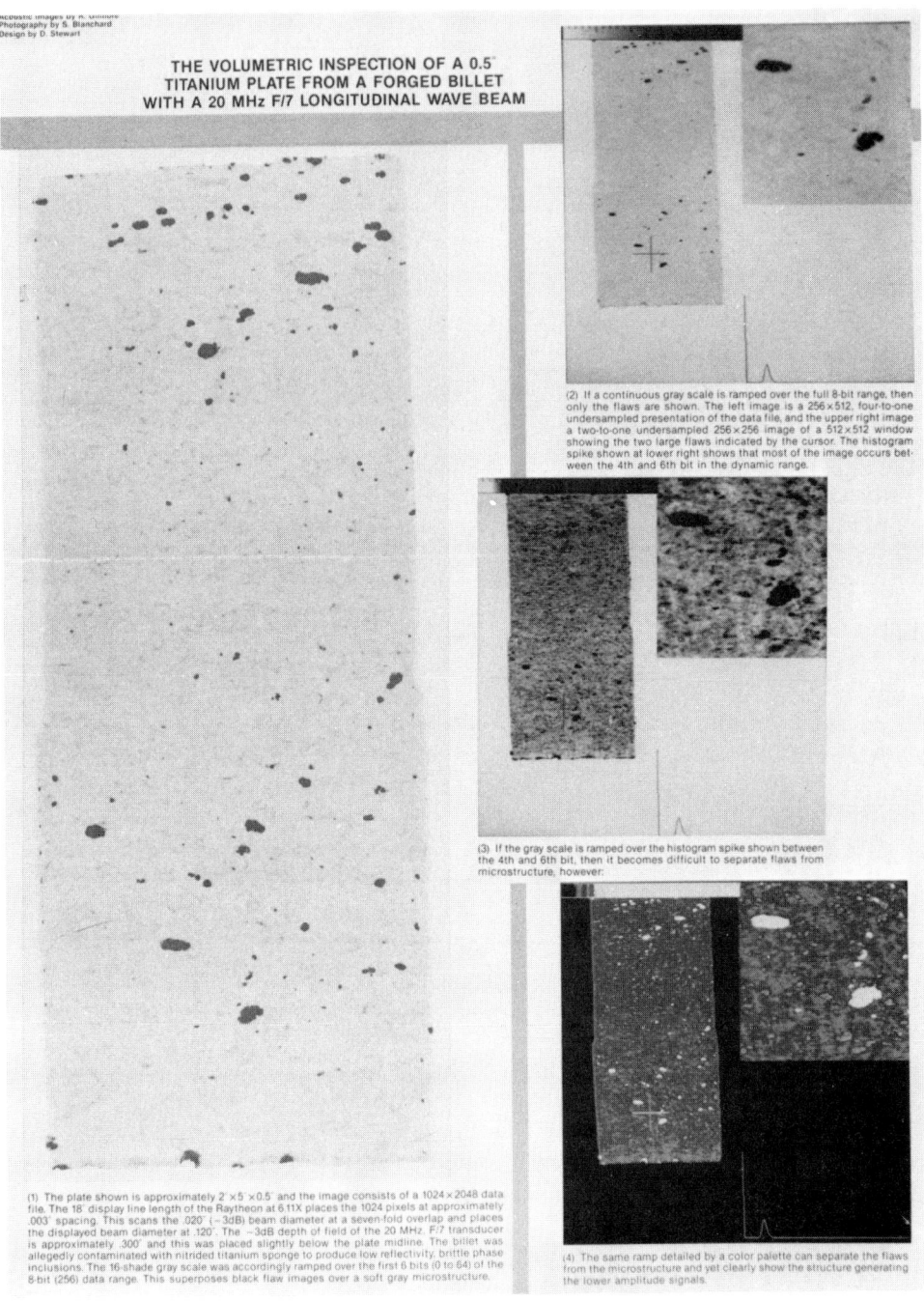

Figure 4.  20 MHz, F/7 inspection of a forged titanium plate.

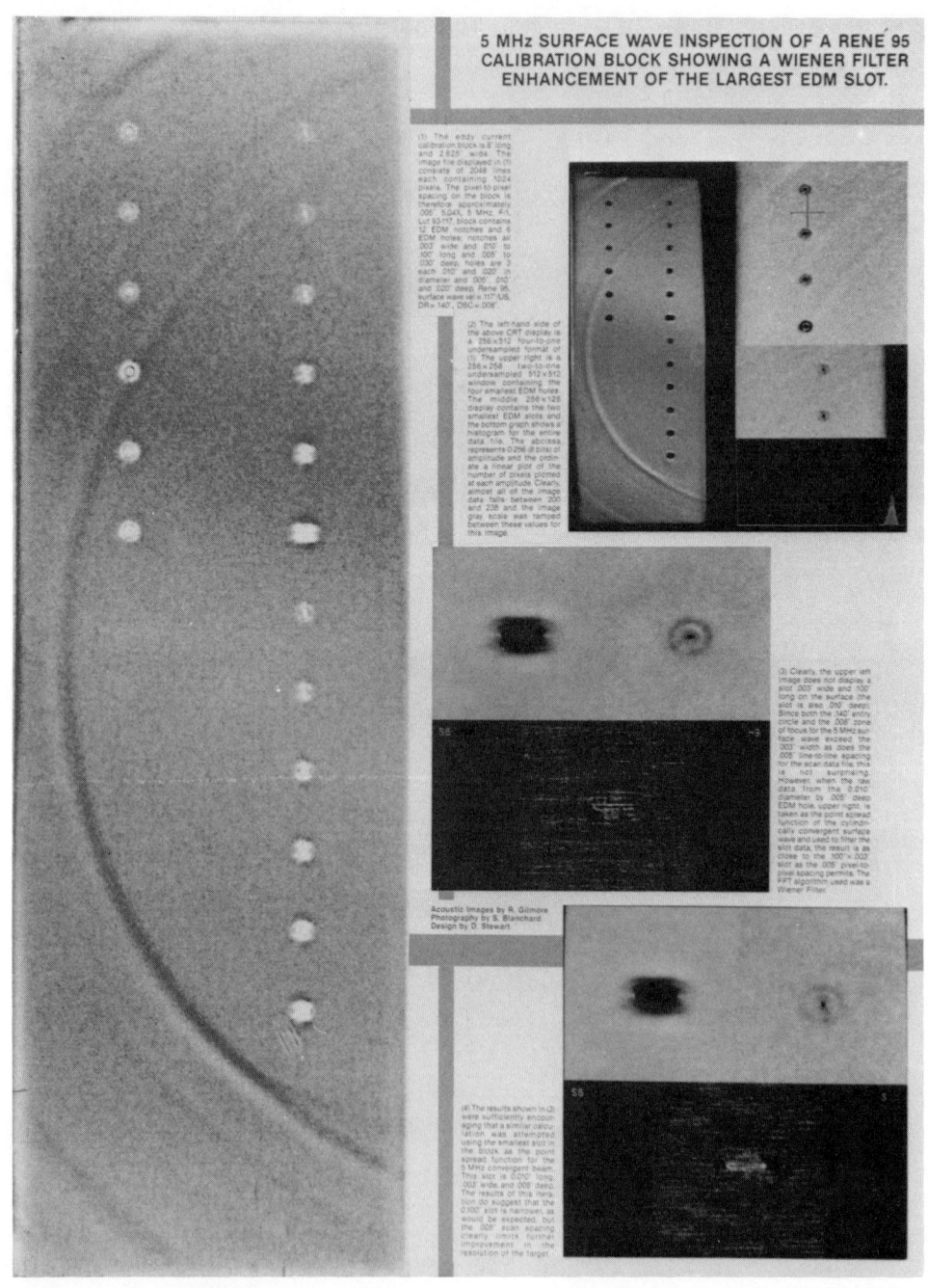

Figure 5. 5 MHz surface wave inspection of the nickel-based superalloy, René 95.

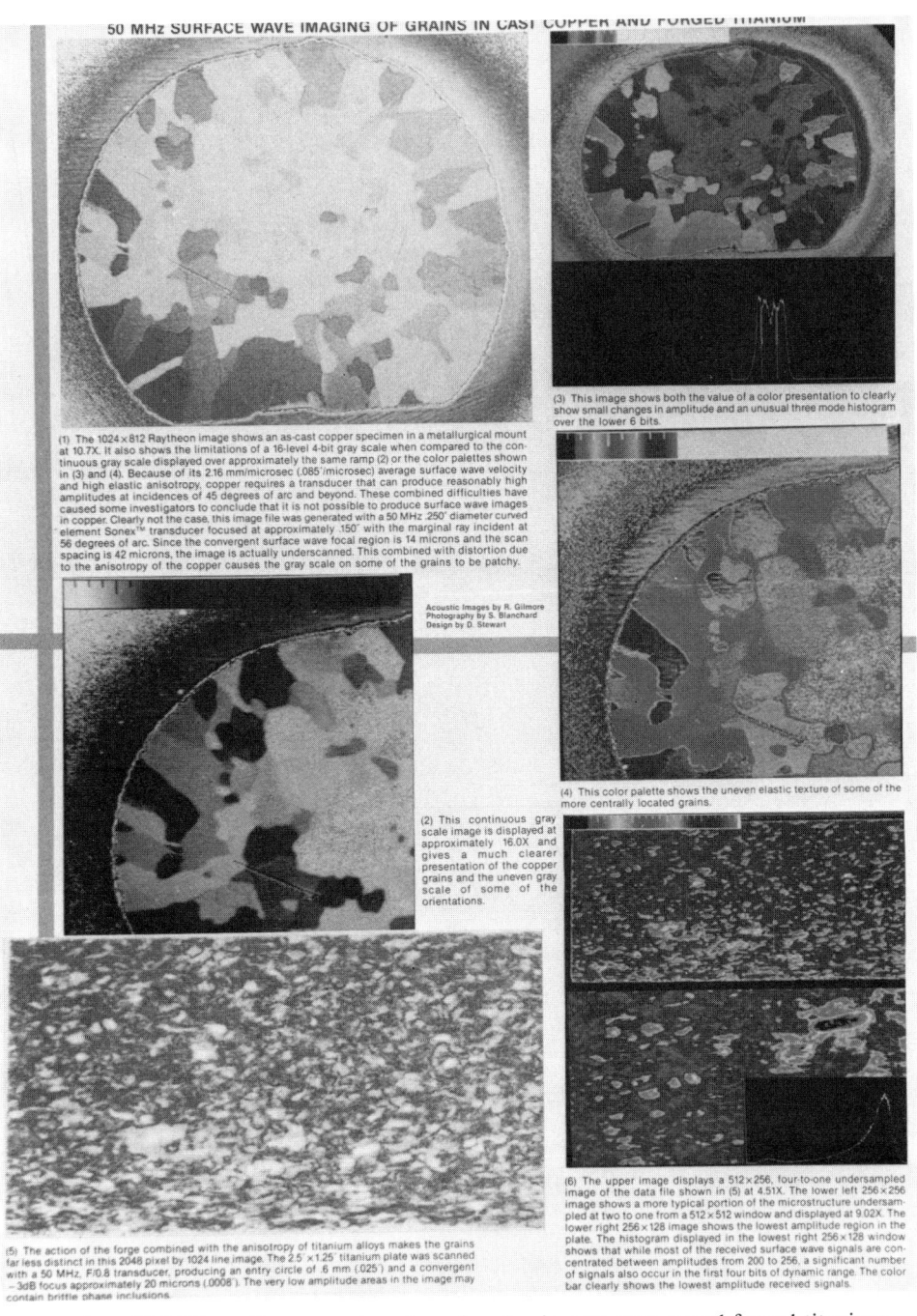

Figure 6. 50 MHz surface wave imaging of grains in cast copper and forged titanium.

## REFERENCES

[1] R.A. Lemons and C.F. Quate, "Acoustic Microscopy by Mechanical Scanning," Appl. Phys. Lett. *24,* 165 (1973).

[2] R.D. Weglin and R.G. Wilson, "Characteristic Materials Signatures by Acoustic Microscopy," Electron. Lett. *14,* 352 (1978).

[3] G.S. Kino, "Fundamentals of Scanning Systems," in *Scanned Image Microscopy,* ed. E.A. Ash, Academic Press, London (1980).

[4] H.K. Wickramasinghe, "Contrast and Imaging Performance in the Scanning Acoustic Microscope," J. Appl. Phys. *50,* 664 (1979).

[5] H.L. Bertoni and T. Tamir, "Unified Theory of Rayleigh-Angle Phenomena for Acoustic Beams at Liquid-Solid Interfaces," Appl. Phys. Lett. *2* (1973).

[6] C.F. Quate, "Microwaves, Acoustics and Scanning Microscopy," in *Scanned Image Microscopy,* ed. E.A. Ash, Academic Press, London (1980).

[7] R.S. Gilmore, K.C. Tam, J.D. Young, and D.R. Howard, "Acoustic Microscopy From 10 to 100 MHz for Industrial Applications," Phil. Trans. R. Soc. London, *A320,* 215-235 (1986).

ULTRASONIC INVERSION: A DIRECT AND AN INDIRECT METHOD

C.A. Chaloner and L.J. Bond

Royal Ordnance, Explosives Division
Westcott, Aylesbury, Buckinghamshire, UK
and University College, London

INTRODUCTION

A formal yet widely applicable definition of inversion is difficult to find; this has led to Flax et al [1] to comment 'Inverse scattering means many things to many people'. However, the inverse problem, viewed with particular reference to defect characterisation, can be regarded as gaining information on the features of an unknown or concealed body which can be made to cause a disturbance in an interrogating field. Thus imaging can be classed a non parametric inversion and indeed imaging and inversion processes have been shown to be mathematically equivalent under certain conditions [2]. Solutions to an inverse problem can be divided into two groups: direct and indirect. The direct method involves mathematical operations (usually transforms) for which the experimental data are the input and the interpretation of those data is the output. Indirect inversion, on the other hand, means finding the best fit between experimental data and a previously assumed theoretical model, and is usually an iterative process.

This paper examines two inversion methods. The 1-D Born technique is a direct inversion algorithm utilising frequency information in the intermediate range (approx 0.5 to 2.5 ka, $k = 2\pi$/wavelength, a = flaw radius) which has been found to be robust in the presence of high noise levels.

An indirect inversion method has also been examined and initial results with bandlimited noisy synthetic data and experimental samples have been encouraging. The technique involves parameterising the features of the flaw about which information is required and then searching the parameter space for the best fit between flaw and ideal data.

INVERSION: PROBLEM AND SOLUTION

The inverse problem can be illustrated schematically as

Inverse Problem

observed disturbance $\xrightarrow{\text{infer}}$ model of object and

Forward Problem

model of object ─determine→ disturbance produced

More formally most inverse problems can be reduced to solving an integral equation of the type

$$g(\tau) = \int_0^\infty K(\nu,\tau)\, p(\nu)\, d\nu \qquad 0 \leq \tau < \infty$$

for $p(\nu)$ from measurements of $g(\tau)$, where $K$ is the kernel relating the governing parameters $\tau$ and $\nu$; i.e. a Fredholm integral equation of the first kind (an equivalent matrix formulation can also be used). Thus the objective of inverse methods can be viewed as the determination of one or more of the parameters in the governing equations or system of equations of some process. The problem of solving such equations is basically ill-conditioned and this causes difficulties in practical inversion techniques.

The potential method of solution to the inverse problem must accommodate four important considerations.

## Existence

Before solution parameters are calculated it must be considered whether in fact a solution exists within the limitations of the inversion model assumptions. Although in some cases existence conditions can be established [4] it is also possible that the inversion algorithm will produce an 'answer' within a very limited model which is unrelated to the true solution. It may, therefore, be necessary to ensure that all possible solutions are included in the model.

## Uniqueness

The solution of the inverse problem is not unique for cases involving experimental data which are incomplete or inexact (noisy) since incomplete data must result in an incomplete solution. Thus experimental data inversion cannot be regarded as a deterministic problem. However, methods do exist for establishing the significance of a particular solution [5,6] within a solution set. It may be the case that the common features (if any) in the range of possible solutions may be sufficient information. Alternatively, further assumptions about the physical situation or further experimental measurements could be made to narrow the class of possible solution. The considerations of existence, requiring a model with a large possible solution set, and uniqueness, requiring limitations on the solution set so that meaningful information can be obtained, must be balanced against each other.

## Stability

A problem is defined as stable if the solution depends continuously on the data. If small perturbations in the data (such as noise) cause large changes in the solution, the problem is unstable or ill-conditioned. In fact, the majority of inverse problems are ill-conditioned [7]. The stability of a particular solution can be measured [8] and the problem is often dealt with by considering the smoothness of the model.

Solution Construction

This is the aspect of inversion which receives most attention and is obviously very dependent on the physics of the situation and the forward models available. What is required is an algorithm which finds a solution to a specified precision within a finite number of iterations. In order to solve an inverse problem the related forward problem must be fully understood and it is often the case that the forward problem must be solved in conjunction with the inverse problem [9] and these must be uncoupled in some way, say by a simplifying approximation of the forward problem.

The method of solution is dependent on the initial formulation of the problem (for example, matrix, integral or differential) and include trial and error techniques, approximations, iterative procedures, transforms, and exact analytical solutions.

BORN INVERSION

The basic technique for obtaining suitable ultrasonic signals for 1-D Born Inversion is well documented elsewhere [10,11]. Therefore, only the major differences between our experimental/development protocol and other published works will be outlined. The transducers used were a specially constructed lead metaniobate probe and a commercial contact probe. The samples were diffusion bonded Titanium alloy or maraging steel blocks containing spheroidal or ellipsoidal voids.

The Born Inversion gives good sizing results for strong scatterers such as voids in metals although it is derived for weak scatterers. The inversion was therefore examined using exact scattering data from spherical voids and experimental data. The technique operates on the real part of the back-scattered frequency spectrum which has been time-shifted such that the flaw centre corresponds to the time origin. A comparison of the real part of the time-shifted spectra for experimental data [11], exact theoretical data [12] and Born theoretical data [13], which have each gone through the same signal processing procedure shows good correspondence.

Analysis of the back-scattered time domain signal from a weak scatterer shows reflections from the front and back of the inclusion as well as other contributions, not fully resolved due to the signal being obtained from a bandlimited frequency spectrum . A similar plot for a void also shows a secondary signal due to creeping wave circumnavigating the flaw surface.

It is expected, therefore, that the Born Inversion will work reasonably well for voids in elastic media which support a creeping wave of detectable magnitude. However, it should be noted that for a weak scatterer the signal path difference between front and back face is proportional to 4a, (a = flaw radius), whereas for a void where the creeping waves are tangentially launched and re-radiated the path difference is $(2+\pi)a$. This is a 28% change in pathlength if the longitudinal host velocity and creeping wave velocity are approximately equal. In fact, if strong and weak scattering ideal data are treated in an equivalent manner in the inversion algorithm the strong scatterer does give larger radius predictions although not as significantly different as expected from path and velocity differences (see Table 1).

However, for small voids, low signal to noise ratios or voids with rough uneven surfaces it is unlikely that creeping waves will be detected. The effect on the inversion of degrading the creeping wave contribution, however, is not serious. For example, reducing the secondary impulse amplitude by over 80% causes a change of less than 8% in the Born radius prediction

The second major problem was that of accurately obtaining the time shift required to locate the flaw centroid and various methods were investigated to this end.

(i) Area function: This method examines the cross-sectional flaw area normal to the direction of wave propagation which is a maximum at the flaw centroid [14].

(ii) Maximum flatness: Using ideal Born data it is found that if the Born predicted radius is calculated for a large number of time shifts, a pattern is produced where the correct time shift is located where the function is maximally flat [15].

(iii) Minimisation of imaginary part (MOIP): It can be shown that for an ideal Born scatterer only the real part of the back-scattered spectrum exists if the flaw centroid corresponds to the time origin. Owing to noise, the imaginary part of the experimental data spectrum remains finite; however, the integral of this (imaginary part of spectrum against frequency) tends to a minimum at the correct time shift.

(iv) Low-frequency examination: Low-frequency information in the range $ka < 0.5$ can be used to predict the required time shift, however, it was found that experimental data was inaccurate in this region.

It was found that for good data, i.e. frequency spectra covering adequate bandwidth (approximately 0.5 to 2.5 ka) and containing no anomalous signals, the three methods of time-shifting predicted time shifts usually within one or at most two resolution points of the measurement system. The correspondence or otherwise was then used to classify the data as suitable or unsuitable for further processing.

On inversion of experimental data according to the Born equation, a further good/bad classification was carried out. Because the experimental system is bandlimited the characteristic function is a smoothed step function and the radius location therefore has to be estimated using:

(a) area under function/peak of function
(b) distance corresponding to the point that is 50% of the peak value.

Radius predictions by these two methods which did not agree within 10% were rejected, being of inadequate bandwidth or too noisy. Table 1 shows the time shifts and corresponding radius predictions for some data from spherical voids.

MONTE CARLO/HEDGEHOG PROTOCOL

In this section the Monte Carlo/Hedgehog search for ultrasonic defect sizing is described and the reasons for choosing this indirect trial and error type routine detailed. The method was first proposed by Valyus [3] and is commonly used in seismology.

The information required from the inversion must first be parameterised. For ultrasonic defect sizing these parameters could be flaw radius, density and flaw longitudinal and shear wave velocity, thus characterising

Table 1 - Timeshifts and corresponding radius predictions for synthetic and experimental data

| Input data | Timeshift range (µs) | Radius area/peak (µm) | Radius 50% peak (µm) | Nominal radius (µm) | Comment |
|---|---|---|---|---|---|
| Ideal Born data | 1.85 | 197 | 188 | 200 | No noise<br>0-3 ka bandwidth |
| Ideal Born data | 1.85 - 1.854 | 199 - 201 | 192 | 200 | 10 dB S/N<br>0-3 ka bandwidth |
| Ideal Born data | 1.78 - 1.88 | 120 - 129 | 142 - 158 | 200 | 10 dB S/N<br>1-3 ka bandwidth |
| Ideal Void data | 0.8 | 212 | 203 | 200 | No noise<br>0-3 ka bandwidth |
| Ideal Void data | 0.8 - 0.803 | 209 - 210 | 198 - 203 | 200 | 10 dB S/N<br>0-3 ka bandwidth |
| Ideal Void data | 0.74 - 0.79 | 109 - 140 | 138 - 148 | 200 | 10 dB S/N<br>1-3 ka bandwidth |
| Expt data Void in Ti alloy | -0.08 - -0.1 | 195 - 202 | 193 - 198 | 200 | Accept result<br>0.4-2.2 ka bandwidth |
| Expt data Void in Ti alloy | -0.13 - -0.14 | 238 - 240 | 229 - 237 | 300 | Accept result<br>0.35-2 ka bandwidth |
| Expt data Void in Ti alloy | -0.09 - -0.18 | 113 - 188 | 148 - 182 | 200 | Reject result<br>0.85-2.2 ka bandwidth |

the flaw composition and size. A four dimensional discretised parameter space is therefore set up which should contain all the potential solutions. The space can be examined using any conventional searching technique, in practice the Monte Carlo is used and a random point chosen. At each point, the theoretical scattering function for the given parameter is calculated. The synthetically generated data is then compared to experimental data from a flaw of unknown characteristics. The degree of correspondence between the two data sets is calculated by standard techniques (least squares best fit) and recorded. In the Monte Carlo another random point is chosen and the process repeated.

The Hedgehog, however, utilises the information obtained from the correspondence calculations; if the point on the parameter space is determined to be a 'good fit' to the experimental data then the nearest neighbours in that space are also examined. If any of these are found to be 'good fits' then their nearest neighbours are examined. This process continues until all the points adjacent to a point within the 'good fit' region have been examined. The algorithm then reverts to the Monte Carlo and another random point is chosen for comparison with experimental data. The process continues until all regions of the parameter space have been examined.

The advantages of the Hedgehog routine are:

1) It can locate and describe the minimum regions (i.e. best fit regions) of a multivariate function.
2) Unstable solutions appear as isolated points. The point distribution can be regarded as a measure of solution stability.
3) Any a priori information available about the defect to be characterised can be used to restrict the parameter space. Further, the parameter space can, if necessary, be extended virtually indefinitely (the limitation being the computing power available) to encompass all possible solutions.
4) The algorithm can be used in conjunction with any forward model which can be adequately parameterised or indeed a combination of models (say one for cracks, one for volumetric defects). The Hedgehog is only limited by the forward models available.
5) Experimental deviations from the ideal can be incorporated in the forward model.
6) When zones of interest/minimum regions are located a finer mesh spacing (i.e. finer spacing between points in the discretised parameter space) can be generated for more detailed examination.

The disadvantages are:

1) The algorithm can be computationally demanding depending on the complexity of the forward model. However, a data base of signals can be built up which would reduce computing time considerably.
2) The problem of existence can only be dealt with by expanding the parameter space.
3) Additional processing is required to distinguish between different minimum regions within a parameter space.

Results

Experiments were first performed with noisy bandlimited synthetic data to examine the performance of the algorithm under controlled conditions. This confirmed that the least square calculation was unaffected by white, gaussian, uncorrelated noise even at 3 dB signal-to-noise and that a bandwidth of 0.5 to 1.8 ka still give good sizing results. However, the algorithm was degraded by loss of low-frequency data such that a bandwidth of 1 to 2.5 ka proved inadequate.

Initial results with (fairly noisy) experimental data are encouraging and are shown in Table 2. A data comparison is shown in Fig. 1 and an example of the graphical output of the Hedgehog program is shown in Fig. 2.

CONCLUSIONS

The Born algorithm works for types of data for which it is not theoretically derived. The bandwidth requirements of the Born are on the limits of commercially available probes. However a protocol for matching flaw size to probe bandwidth has been developed and has been used to measure the quality of data input to the inversion.

An indirect inversion method (Hedgehog) has also been developed and initial experimental results with fairly poor quality data have been encouraging. Evaluation of the Hedgehog with synthetic data has shown it to be robust in the presence of uncorrelated Gaussian noise and limited bandwidths. Indirect methods can utilise all the available information about an unknown scatterer and do not involve approximations or simplifications of the forward model and the consequent inaccuracies.

Table 2 - Hedgehog inversion results on synthetic data and real data from diffusion bonded samples

| Sample data | Nominal radius ($\mu$m) | Best prediction ($\mu$m) | Range of first 4 predictions ($\mu$m) | Bandwidth (ka) |
|---|---|---|---|---|
| Ti spherical void | 250 | 260 | 260-290 | 0.1 -2.5 (10 dB S/N) |
| (Synthetic data) | 250 | 260 | 260-290 | 0.5 -2.5 (10 dB S/N) |
| Ti-6Al-4V Spherical void | | | | |
| Face A | 200 | 200 | 180-210 | 0.4 -2.02 |
| C | 200 | 200 | 190-240 | 0.4 -2.08 |
| D | 200 | 200 | 180-210 | 0.4 -2.05 |
| Maraging Steel Spherical void | | | | |
| Face A | 300 | 280 | 270-300 | 0.4 -1.65 |
| D | 300 | 300 | 290-320 | 0.45-1.65 |
| A | 200 | 250 | 230-260 | 0.3 -1.11 |
| C | 400 | 440 | 430-460 | 0.6 -2.5 |

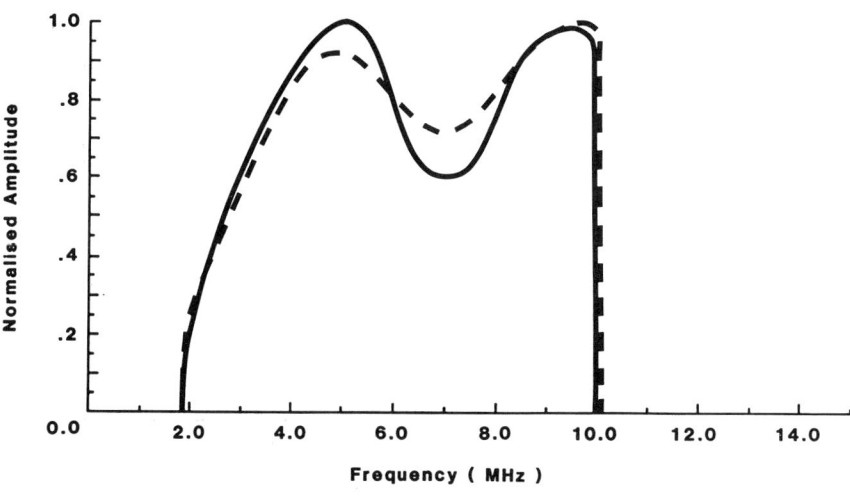

– – – – Ideal Data      ———— Expt. Data

Fig. 1  Comparison of the backscattered magnitude spectra for ideal and experimental data from a 200 $\mu$m radius void in Ti-alloy

569

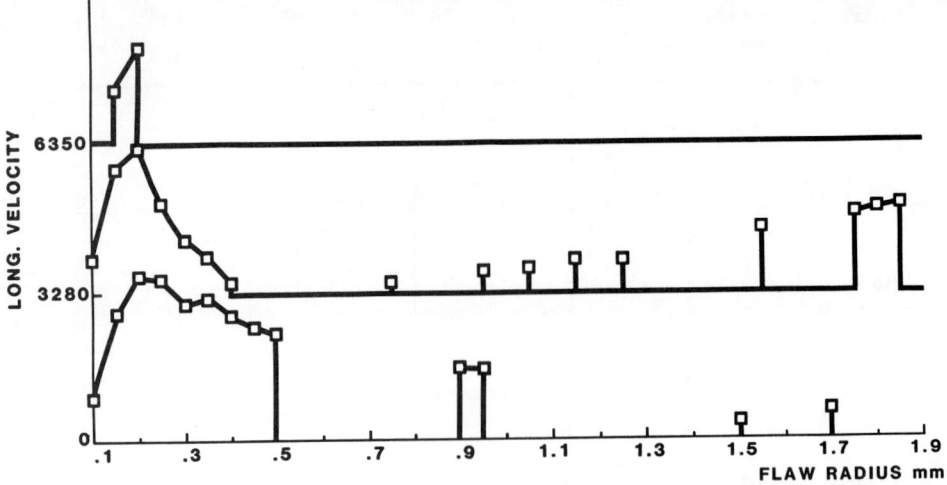

Fig. 2   Output from Hedgehog program showing the degree of correspondence between ideal and experimental data for various flaw radii and compositions (long. vel.) (only good correspondence points shown)

ACKNOWLEDGEMENT

This work has been carried out with the support of Royal Ordnance, Explosives Division and the Procurement Executive, Ministry of Defence.

REFERENCES

1. L. Flax, G.C. Gaunaurd, H. Uberall, Theory of resonance scattering, in: "Physical Acoustics", Vol. 15, W.P. Mason and R.N. Thurston, eds., Academic Press (1981).
2. R.B. Thompson, K.M. Lakin, J.H. Rose, A comparison of the inverse Born and imaging techniques for reconstructing flaw shapes, in: "Proceedings IEEE Ultrasonic Symposium", B.R. McAvoy, ed., IEEE, (1981).
3. V.P. Valyus, Determining seismic profiles from a set of observations, in: "Computational Seismology", V.I. Keilis Borok, ed., Consult Bureau (1972).
4. R.L. Parker, Ann. Rev. Earth Planet. Sci., 5, 35 (1977).
5. G.E. Backus, F. Gilbert, Geophys. J.R. Astr. Soc., 13, 247 (1967).
6. C.R. Smith, W.T. Grandy, "Maximum entropy and bayesian methods in inverse problems", Reidel Pub. Co., (1985).
7. J.G. McWhirter, E.J. Pike, J. Phys. A: Maths. Gen., 11, 1729 (1978).
8. P.C. Sabatier, J. Geophys., 43, 115 (1977)
9. A.J. Devaney, IEEE Trans. on Sonics and Ultrasonics, SU-30, 355, (1983).
10. R.K. Elsley, R.C. Addison, "Dependence of the accuracy of the Born inversion on noise and bandwidth", Proc. of the DARPA/AFWAL Review of Prog. in QNDE", AFWAL-TR-81-4080, Air Force Wright Aeronautical Lab., (1981).
11. C.A. Chaloner, L.J. Bond, IEE Proc., Part A, 133, (1986), (In press).
12. C.F. Ying, R. Truell, J. Appl. Phys., 27, 1086, (1956).
13. J.E. Gubernatis, E. Domany, J.A. Krumhansl, M. Huberman, J. Appl. Phys., 48, 2812, (1977).

14. J.H. Rose, J.M. Richardson, "Time domain Born approximation", Proc. of the DARPA/AFWAL Review of Prog. in QNDE", AFWAL-TR-81-4080, Air Force Wright Aeronautical Lab., (1981).
15. R.C. Addison, "Ultrasonic test bed for quantitative NDE", Report No. AFWAL-TR-82-4075, Air Force Wright Aeronautical Lab., (1982).

# IMAGING OF FLAWS IN SOLIDS BY VELOCITY INVERSION

Jack K. Cohen, Norman Bleistein and Frank G. Hagin

Center for Wave Phenomena
Colorado School of Mines
Golden, Colorado

We describe the application of a method for ultrasonic imaging of flaws in solids. These methods greatly extend earlier work along these lines at Rockwell and the Langenberg group in Germany, see [1,2,3,4,5,6,7,8,9,10]. The new inversion methods allow reflector imaging and parameter estimation in progressively more complex media with progressively more realistic source/receiver configurations. This research has been carried out in the context of seismic exploration. However, the problems are sufficiently similar that these more realistic models have direct counterparts in nondestructive testing [11,12,13,14,15,16,17]. In particular, both problems are high frequency inverse scattering problems. High frequency means that the wavelengths are much smaller (by a factor of three or more) than the other length scales of the problem.

An essential element of all inversion algorithms is "back propagation" with respect to some reference propagation speed. In the earlier work, used by both the Rockwell program and by Langenberg, the reference speed was constant and the flaw was required in the "far field." We soon developed techniques that do not restrict us to far field inversion. For constant background propagation speed, the evolution of this method was described in [11,12,18]. More recently, we have developed methods that allow for a background speed that depends on one variable (stratified reference) [13,14] and on three variables [15,17]. In the latter two papers, we also extend the method to the case of separated source and receiver and non-planar datum surfaces. We present solutions formulas for the case of one source, multi-receiver array; one receiver, multi-source array; fixed offset between source and receiver. The solutions are in the form of integrals, leading to summations in the discrete processing. They are extremely stable and tend to diminish the noise in the data.

Thus, the flaws and inclusions being imaged by the new methods no longer need be small and in the far field, but need only be a few (three or more) wave lengths from the sources and receivers.

When tested on seismic data sets, the computer algorithms developed to implement our theories use a few tens of minutes of CPU time on a machine such as an IBM 3034 or a Cyber 7600. A test case on a Cray ran in under a minute. Seismic data sets typically consist of 400-800 data traces with 1024 points per trace. The inversion algorithms produce between 250,000 and 400,000 output data points for such sets. In NDT, the data sets tend to be at least an order of magnitude smaller in size, as is the output. Thus, we anticipate CPU times that are two orders of magnitude smaller than those for the seismic experiments.

Figures 1 through 4 demonstrate our methods and their computer implementation in the context of seismic exploration. Figure 1A depicts a model in which the background sound speed in the right half of the figure has one dimensional variation. Figure 1B depicts synthetically generated backscattered data for this model. Each vertical line represents the record of the response to an impulsive source at that transverse spatial location. The horizontal deviation from a straight line depicts the amplitude. The objective of such an inversion is to accurately depict the flank of the intrusion ("salt dome" in the seismic literature) on the left of Figure 1A. A constant background sound speed algorithm could not successfully image that flank because it would not properly account for refractions at each of the horizontal interfaces when inverting the data. A $c(z)$ (one dimensional variation) background speed was used for this model. The result is shown in Figure 1C. The flank is successfully imaged.

The background speed used for this case was not the discontinuous speed of the input model, but a piecewise linear speed $c(z)$, connecting the values of sound speed at each of the interfaces. Thus, the robustness of the method with respect to errors in true propagation speed is also demonstrated. Furthermore, the data was not "true amplitude" data but did have accurate phase information for the model. Thus, the output also demonstrates the value of the method as an imaging technique when only phase or "time-of-flight" information is retained.

We remark that, in a nondestructive testing context, this algorithm would be useful in examining a sample with a flat surface immersed in a fluid the sources and receivers placed in the fluid. With this immersion technique, multi-directional wave propagation in the solid is more easily achieved than with a transducer in contact with the solid's surface. The $c(z)$ algorithm allows one to account for refractions at the fluid-solid interface. It also would allow for inclusion of a laminar structure into the background propagation speed. It is this algorithm that will provide the starting point for inversion of NDE data from an immersed solid.

Figure 2A is a model for which the $c(z)$ algorithm will not suffice. Backscatter data for this model was generated under the assumption of 1000 ft/sec increments in sound speed across each interface. A $c(x,z)$ background sound speed was used to invert the data. To test the method, a background sound speed that was a close approximation of the "true" sound speed was employed. The result is shown in Figure 2B. The missing portions of the reflectors are the result of the limited aperture of the experiments on the upper surface: reflected rays from the outer portions of the reflectors propagate outside the model data set. Within the aperture, the reflectors are all satisfactorily imaged.

As a comparison, Figure 2C shows the inversion with respect to a constant background, namely the sound speed in the first region. While the first reflector is properly imaged, the lower reflectors are not. This demonstrates the need for the more complex background sound speed. For true amplitude data, we could use this first result to estimate the sound speed below the first reflector and then properly image the second reflector, as well. We would then estimate the sound speed below that reflector thereby obtaining the full complex sound speed used to produce Figure 2B. By this method, one could recurse through a medium. The practicality of this type of recursion has not yet been tested.

The algorithm used to generate the output of Figure 2 can also be used

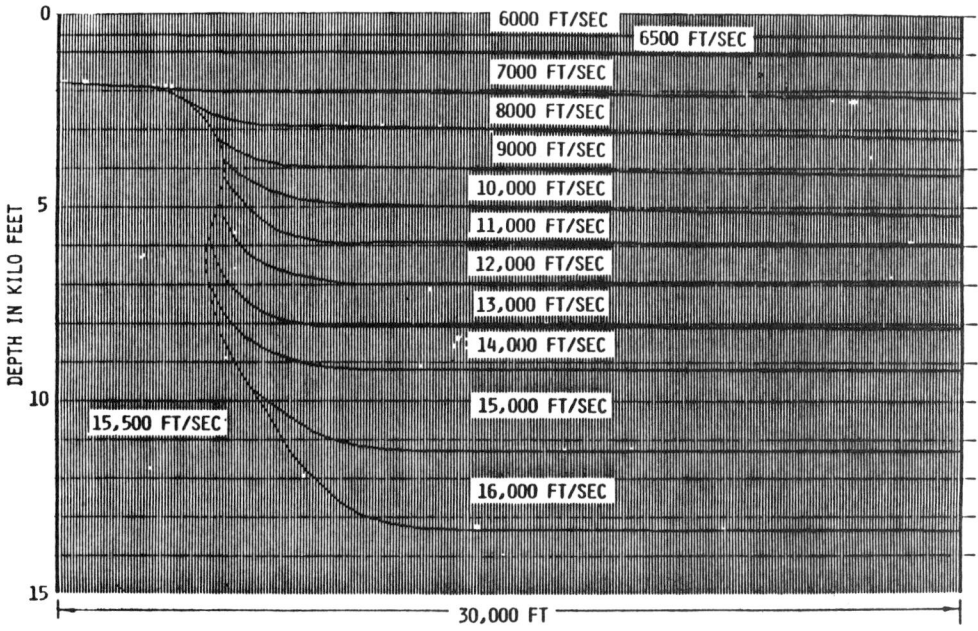

FIGURE 1A

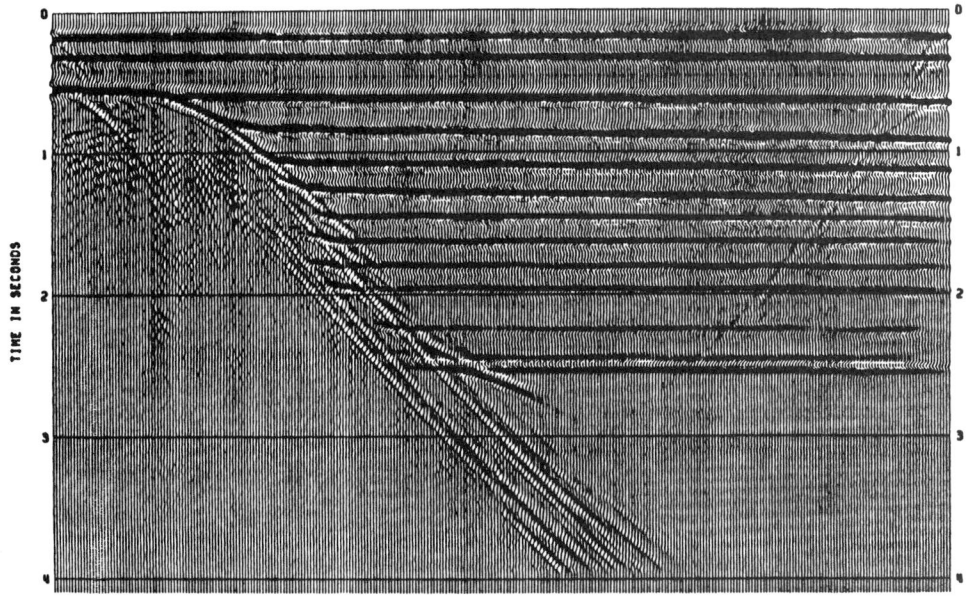

FIGURE 1B

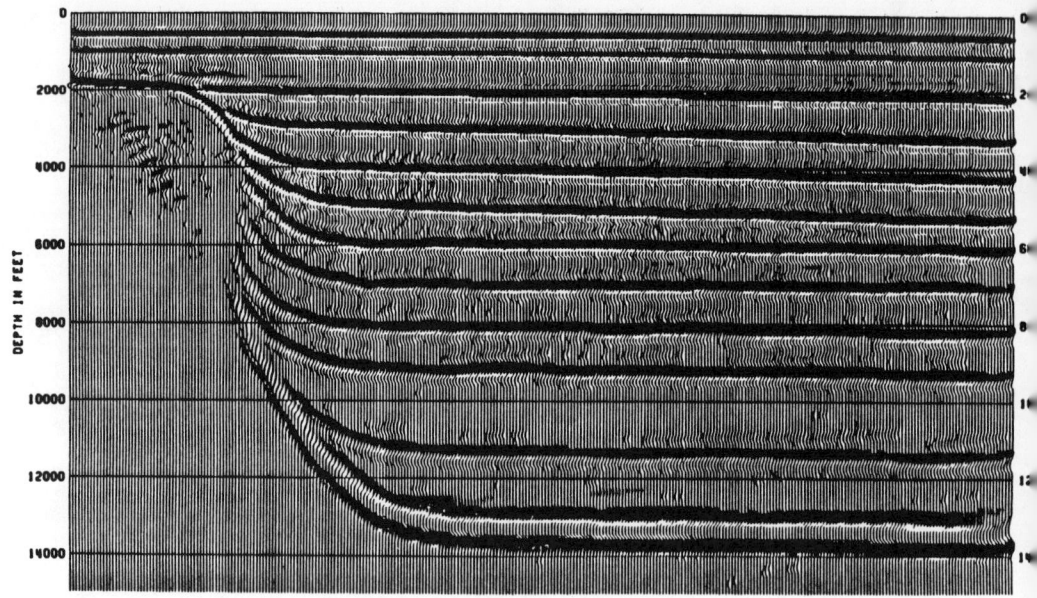

FIGURE 1C

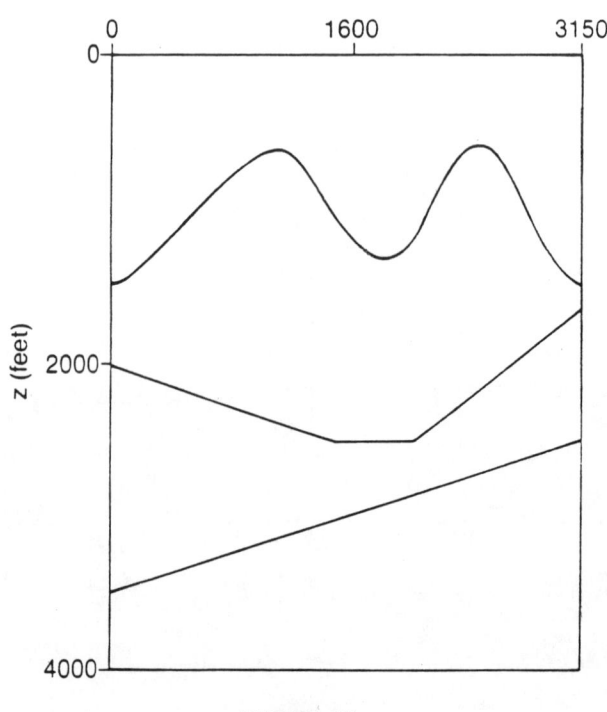

FIGURE 2A

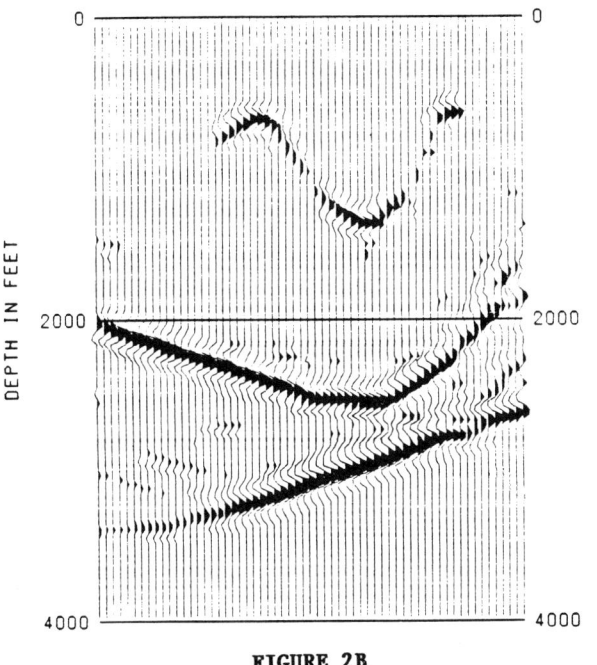

**FIGURE 2B**

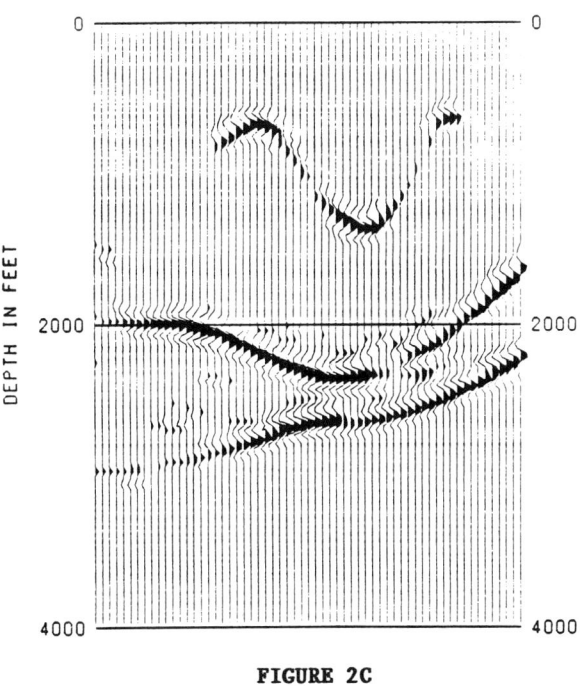

**FIGURE 2C**

with data for which the amplitude has not been accurately preserved (e.g., when automatic gain control has been employed). In this case, the location of the reflectors with respect to the background propagation speed is accurately reproduced but parameter estimation is no longer feasible except possibly at the crudest level of relative magnitude.

In the context of nondestructive testing, a $c(x,z)$ inversion of this type would be the method of choice for a sample with an irregular upper surface such as the one in this example. Again we would model the experiment in which the part is immersed in a fluid. Without a $c(x,z)$ inversion algorithm, one could not hope to take account of the lensing effect of the outer boundary.

The example of Figures 3A-C is another example of an intrusion in an otherwise stratified medium. Figure 3A shows the model, also showing the normal incidence geometrical optics rays at one reflection surface. The blanks along that surface depict regions from which the normal incidence rays do not emerge at the upper surface within the range of the experiments. Figure 3B demonstrates an inversion with a $c(x,z)$ algorithm, while Figure 3C depicts the inversion with a $c(z)$ algorithm. Both inversions produce almost the same image of the flanks of the intrusion. However, the latter method inaccurately images the reflector directly below the saltdome, while the former method depicts portions of that reflector properly located. The gaps

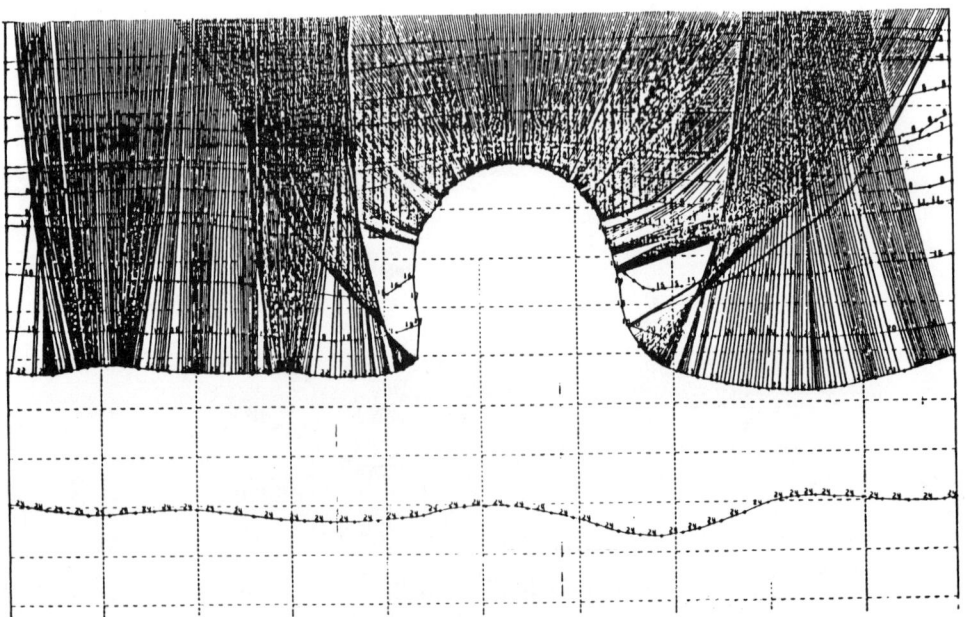

FIGURE 3A

in the lowest reflector arise from rays that pass out of the data field or from rays that passed through a caustic on their propagation path to the surface. The latter have not yet been incorporated into this algorithm although they are a topic of current research.

The last example demonstrates a finite offset inversion algorithm with a constant background. In this case it was assumed that the source/receiver pair were separated by a fixed distance. Figure 4A shows the model, Figure 4B shows the fixed offset data with each trace plotted at the midpoint between source and receiver; Figure 4C shows the output of the inversion algorithm and demonstrates the validity of this method.

This constitutes a first demonstration of the implementation of our theory in the case of a separated source and receiver. Clearly, this has its direct counterpart in nondestructive testing to experiments in which the source and receiver are separated. In this case, a source/receiver pair would be moved across the domain containing the test object. Similarly, we have a theory for the case in which one source and an array of receivers is employed or the opposite case of one receiver and an array of sources.

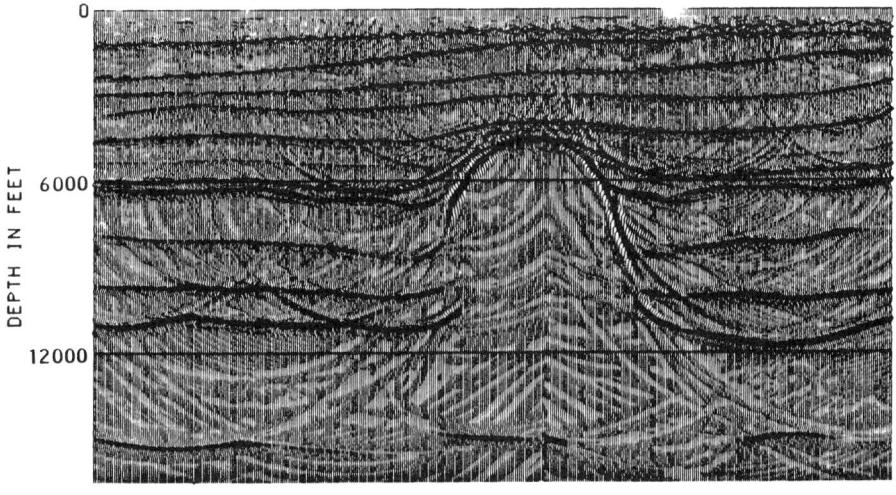

FIGURE 3B

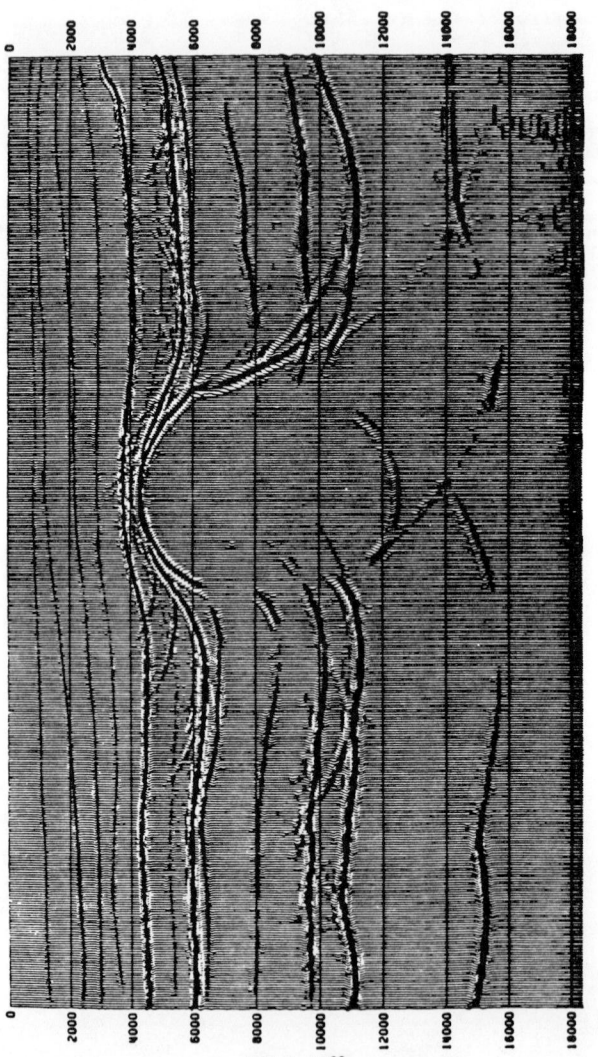

FIGURE 3C

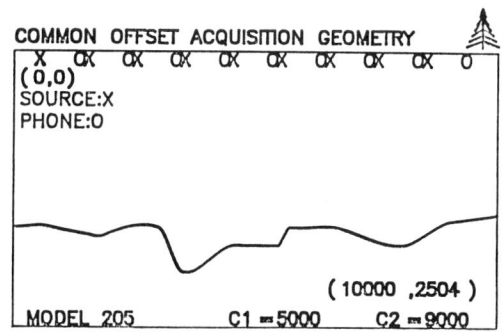

EACH SOURCE (X) CONNECTED TO THE RECEIVER (O) AT ITS RIGHT, 800 FT. SEPARATION.

**FIGURE 4A**

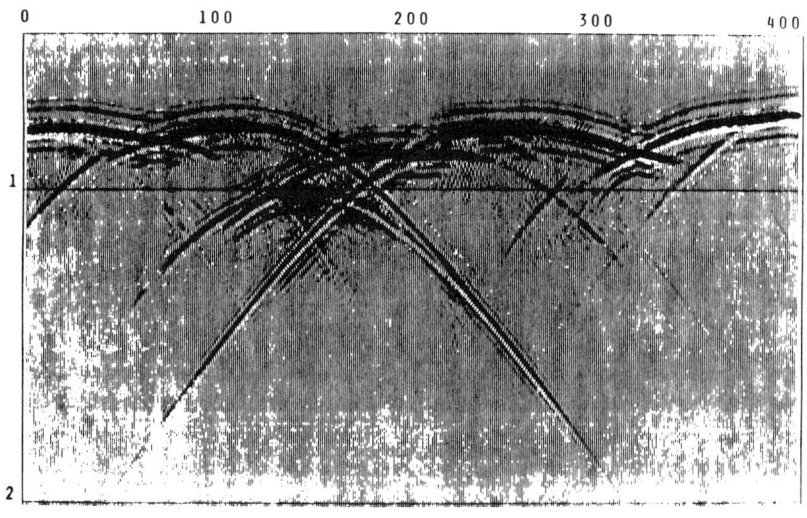

**FIGURE 4B**

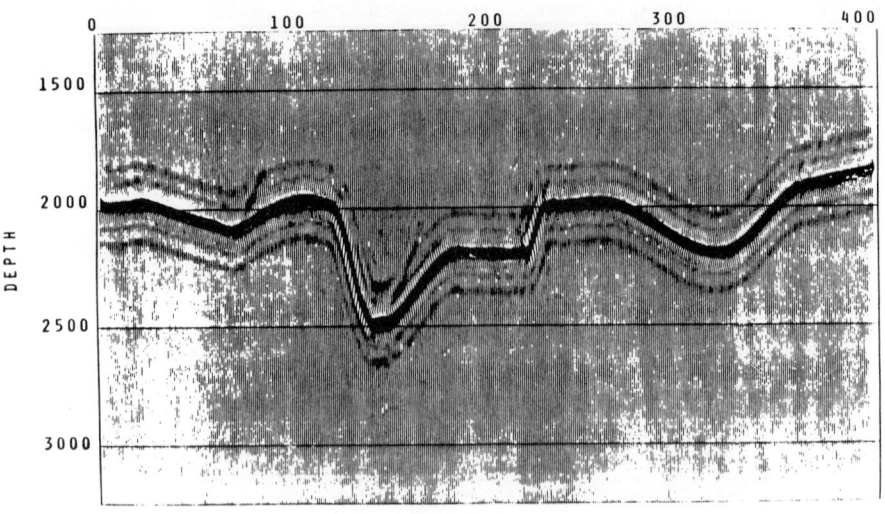

FIGURE 4C

## REFERENCES

1. N. Bleistein and J. K. Cohen, Application of a new inverse method to non-destructive evaluation: Res. Rep., Rockwell International Science Center (1977).
2. N. Bleistein and J. K. Cohen, Application of physical optics inverse scattering to non-destructive testing: Res. Rep. MS-R-8007, Denver Res. Inst. Rep. (1980).
3. N. Bleistein and J. K. Cohen, Progress on a mathematical inversion technique for non-destructive evaluation: Wave Motion $\underline{2}$, 75-81 (1980).
4. N. Bleistein and J. K. Cohen, Progress on a mathematical inversion technique for non-destructive evaluation II: Res. Rep., Rockwell International Science Center (1980).
5. J. K. Cohen and N. Bleistein, A note on non-destructive detection of voids by a high frequency technique: Res. Rep. MS-R-7813, Denver Res. Inst. Rep. (1978).
6. J. K. Cohen, N. Bleistein, and R. K. Elsley, Non-destructive detection of voids by a high frequency inversion technique: Res. Rep., Rockwell International Science Center (1978).
7. P. Höller, K. J. Langenberg, and V. Schmitz, Mathematische verfahren zūr lösung des inversen beugungproblems: Joint report of Fachgebiet Theoretische Electrotechnik der Gesamthoschule Kassel, Ghk-TET and Fraunhofer Institute für zerstörungsfreie Prüfverfahren, FhG-Izfp (1984).
8. K. J. Langenberg, M. Fischer, M. Berger, and G. Weinfutter, Imaging performance of generalized holography: Fachgebiet Theoretische Electrotechnik der Gesamthoschule Kassel, Ghk-TET (1984).
9. K. J. Langenberg, D. Brück, and M. Fischer, Inverse scattering algorithms, $\underline{\text{in}}$: "New procedures in nondestructive testing, Proceedings of the German-U.S. workshop, Fraunhofer Institute, Saarbrücken, Germany, August 30 - September 3, 1982", P. Höller, ed., Springer Verlag, New York (1985).

10. V. Schmitz and P. Höller, Reconstruction of defects by ultrasonic testing using synthetic aperture procedures, in: "Review of Progress in Quantitative Nondestructive Evaluation, 4A", D. O. Thompson and D. E. Chimentai, eds., Plenum, New York (1985).
11. J. K. Cohen and N. Bleistein, Velocity inversion procedure for acoustic waves, Geophysics 44, 1077-1085 (1979).
12. N. Bleistein and J. K. Cohen, The velocity inverse problem – Present status, new directions, Geophysics 47, 1499-1511 (1982).
13. N. Bleistein and S. H. Gray, An extension of the Born inversion method to a depth dependent reference profile, Geophys. Prosp. 33, 999-1022 (1985).
14. J. K. Cohen and F. G. Hagin, Velocity inversion using a stratified reference, Geophysics 50, 1689-1700 (1985).
15. J. K. Cohen, F. G. Hagin, and N. Bleistein, Three dimensional Born inversion with an arbitrary reference, Geophysics 51, n.8 (1986).
16. N. Bleistein, J. K. Cohen, and F. G. Hagin, Computational and asymptotic aspects of velocity inversion, Geophysics 50, 1253-1265 (1985).
17. N. Bleistein, On the imaging of reflectors in the earth: Res. Rep. CWP-038, Center for Wave Phenomena, Colo. Sch. of Mines (1985).
18. N. Bleistein, J. K. Cohen, and F. G. Hagin, Two-and-one-half dimensional Born inversion with an arbitrary reference, Geophysics 52, n.1 (1987).

CHARACTERIZATION OF FLAW SHAPE AND ORIENTATION USING ULTRASONIC

ANGULAR SCANS

D. K. Hsu, S. J. Wormley, and D. O. Thompson

Ames Laboratory, USDOE
Iowa State University
Ames, IA 50011

INTRODUCTION

To exploit theoretical advances in elastic wave inverse scattering, an automated multiviewing ultrasonic transducer system and the associated signal processing algorithms have been developed at the Ames Laboratory for the reconstruction of the size, shape, and orientation of volumetric flaws [1]. The flaw sizing algorithm is based on elastic wave inverse scattering theories in the long and intermediate wavelength regime [2,3] and the three-dimensional reconstruction algorithm finds the equivalent ellipsoid that best fits the flaw sizes in the various viewing directions [4,5]. The original multiprobe system consists of six peripheral transducers equally spaced in a circle surrounding one transducer at the center. The peripheral transducers may be tilted at an angle toward the center to increase the aperture and can also be translated along their respective axes to allow an equalization of the acoustic propagation time. The axis of the aperture cone is normally placed perpendicular to the part surface. The flaw sizing procedure was a one-dimensional inverse Born algorithm to determine the flaw's centroid-to-tangent plane distances for a number (normally 13 or 19) of pulse-echo or pitch-catch scattering directions within a finite aperture cone. The flaw sizes are then used as inputs to a nonlinear least squares regression program to yield a complete geometric reconstruction in the form of three semi-axes and three Euler angles of the best-fit ellipsoid. Using this system, successful reconstructions have been obtained for both oblate spheroidal (disk-like) and prolate spheroidal (rod-like) inclusions and voids. The readers are referred to a complete description of the system in Ref. 1.

Recently, efforts have been devoted to the assessment of the reconstruction reliability as a function of the aperture size and the signal-to-noise ratio of the flaw waveforms [6]. Of particular interest is the effects of flaw orientation on the reconstruction reliability [7]. Computer simulations of the reconstruction errors were made for flaws untilted and tilted with respect to the viewing aperture. For the same aperture size, the reconstruction errors were much greater for a tilted flaw, consistent with experimental observations. With the viewing aperture situated normal to the part surface, a flaw tilted with respect to the surface may afford a very low leverage for reconstruction due to limited surface area coverage by the wavefront tangent planes, small signal-to-noise ratio, and possibly also the presence of flash point interferences in the scattering amplitude spectrum. Examples are edge-on views of a disk-like flaw or end-on views of a rod-like flaw. As a result, a larger aperture containing the same number of scattering directions (kept as a constant for speed considerations) may not be sufficient in improving the reconstruction reliability. Besides, the aperture size has practical limits in a single-side access inspection situation. A more advantageous approach is to tilt the interrogation aperture to compensate for the particular flaw orientation and to restore the leverage for a reliable reconstruction. To do so, some prior knowledge about the flaw shape and orientation is required. In this work an angular scan method is developed in which the flaw shape and orientation are estimated from azimuthal and polar scans of the flaw signal amplitude. Based on such preliminary determination of flaw shape and orientation, an aperture orientation may then be chosen so that the aperture axis is perpendicular to the flaw surface where the total curvature is a minimum. The angular scans and the judicial choice of the aperture configuration for data acquisition have several advantages: the signal-to-noise ratios of the flaw waveforms are improved, the flash point interference phenomena are avoided, and the symmetry planes determined in the angular scans allow two-dimensional cross-sectional reconstructions in the principal planes of the ellipsoid-like flaws.

RECONSTRUCTION OF TILTED FLAWS

To assess the reconstruction reliability for tilted flaws using the multiviewing transducer system, two flaws were studied: a short section of copper wire (160μm dia. x 400μm) tilted 45° in a thermoplastic host and a 400 x 200μm oblate spheroidal void in titanium tilted 30°. Using a data acquisition aperture perpendicular to the part surface, the usual reconstruction procedure failed to yield good results for both tilted flaws. However, when the aperture was tilted to compensate the flaw orientation, excellent reconstructions were obtained for these tilted flaws. These experiments clearly demonstrated the advantages of conducting the construction with an aperture perpendicular to the flattest part of the flaw surface. Because the orientation of the flaw under investigation is generally unknown, an angular scan plan was developed to obtain this information.

ANGULAR SCAN METHOD

In the high frequency regime the principles of geometrical optics apply and the amplitude of the ultrasonic signal backscattered from a flaw should be proportional to $\sqrt{\rho_1\rho_2}$ where $\rho_1$ and $\rho_2$ are the principal radii of curvature at the point where the wavefront makes contact with the flaw surface. The high frequency regime is characterized by $ka \gg 1$, where k is the wavevector and a is some characteristic size of the flaw. The product $\rho_1\rho_2$ is refered to as the total curvature (or Gaussian curvature) of the flaw surface at the contact point. For an ellipsoid it can be shown that $\sqrt{\rho_1\rho_2} = A_x A_y A_z / r_e^2$ where $A_x$, $A_y$ and $A_z$ are the semiaxes of the ellipsoid and $r_e$ is the center-to-tangent plane distance for the scattering direction [8]. The value of $r_e$ depends on the sizes of the semiaxes as well as the orientation of the flaw [5]. (The signal amplitude also depends on the impedance mismatch between the flaw and host material; but, for a given flaw, it is simply a constant multiplicative factor.) Based on this, one would expect that flaw shape and orientation information can be deduced from the angular dependence of the backscattered flaw signal amplitude in the large ka limit. We shall first describe the angular scan method and then discuss its applicability in the comparison with the experimental data.

Computer Simulation

To illustrate the flaw shape and orientation estimation, we computed the high frequency signal amplitude $(A_x A_y A_z/r_e^2)$ for a prolate spheroid with $A_x=A_z=80\mu m$, $A_y=250\mu m$, and with the major axis of the prolate pointed at an azimuthal angle of 120° and a polar angle of 45° in the laboratory coordinates fixed on the sample. Figure 1 shows azimuthal scans at five polar angles: 0, 15, 30, 45, and 60°. The distance from the origin to a point on the curve represents the amplitude of the computed flaw signal. As can be seen, a plane of mirror symmetry exists at an azimuthal angle of 120° (or 300°). A polar scan at this azimuthal angle, shown in Fig. 2(a), shows a peak at $\alpha=45°$ and $\beta=300°$; thus revealing the tilt angle of the prolate flaw. Here we use the notations $\alpha$ and $\beta$ for the polar and azimuthal angles, respectively. A second scan in a plane containing the direction of maximum signal ($\alpha=45°$, $\beta=300°$) and perpendicular to the plane of mirror symmetry is shown in Fig. 2(b). (The second scan requires changing the polar angle $\alpha$ and the azimuthal angle $\beta$ simultaneously). The second scan shows a constant signal amplitude; thus confirming the prolate spheroidal shape of the flaw. For the convenience of discussion, we shall call the plane of mirror symmetry containing the normal to the part surface the vertical sagittal plane (or VSP). The VSP bisects the flaw and is normal to the part surface. The plane of the second scan perpendicular to the VSP and also bisecting the flaw, is called the perpendicular sagittal plane (or PSP), as shown in Fig. 3.

The angular scan pattern of a general ellipsoid characterized by semiaxes $A_x=400\mu m$, $A_y=200\mu m$, $A_z=100\mu m$ and Euler angles $\theta=37°$, $\phi=54°$, and $\psi=0°$ is shown in Fig. 4. Pictorially, the long axis of the ellipsoid lies along an azimuthal angle of 144° (or 324°) and the tip in the second quadrant is tilted out of the figure. The symmetry of the pattern in Fig. 4 shows that the VSP is at an azimuthal angle of 144° (or 324°). The VSP and PSP scans are shown respectively in Figs. 5a and b. As can be seen, the VSP curve shows a peak at the expected 37° and, because

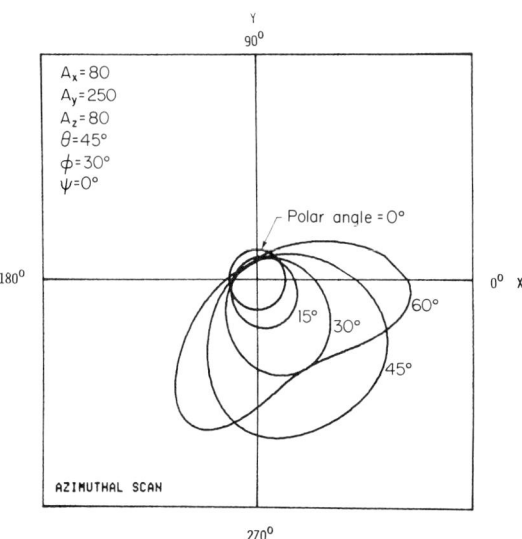

Fig. 1. Computed signal amplitude of a tilted prolate flaw as a function of azimuthal angle at five different polar angles. $A_x$, $A_y$, and $A_z$ are the semi-axes and $\theta$, $\phi$, and $\psi$ are the Euler angles specifying the flaw orientation.

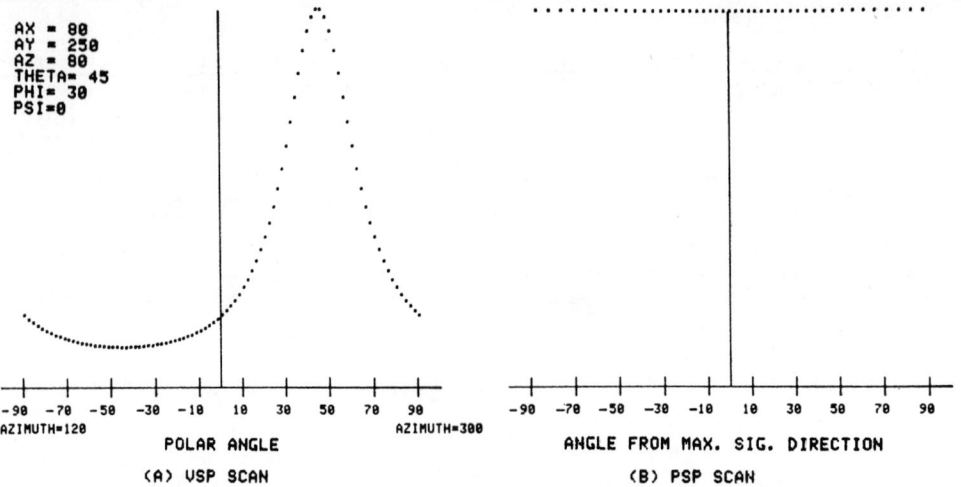

Fig. 2. a) Scan in the vertical sagittal plane revealing the 45° tilt angle of the flaw. b) Scan in the perpendicular sagittal plane revealing the prolate shape of the flaw.

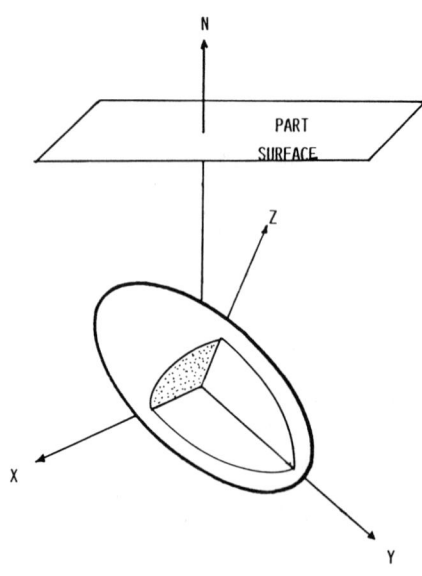

Fig. 3. yz plane is the vertical sagittal plane (containing the normal to the part surface). xz plane is the perpendicular sagittal plane.

$A_x > A_y$, the width of the VSP peak is broader than the PSP peak. If this were an oblate spheroid with $A_x = A_y > A_z$, then the widths of the VSP peak and the PSP peak would have been equal.

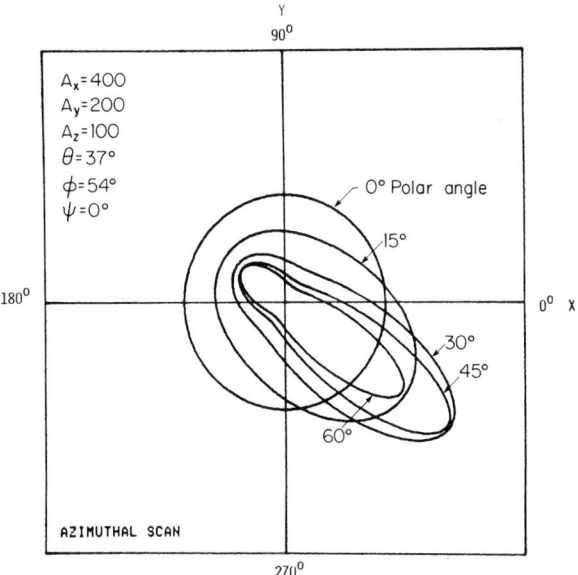

Fig. 4. Computed angular scan pattern of a general ellipsoid.

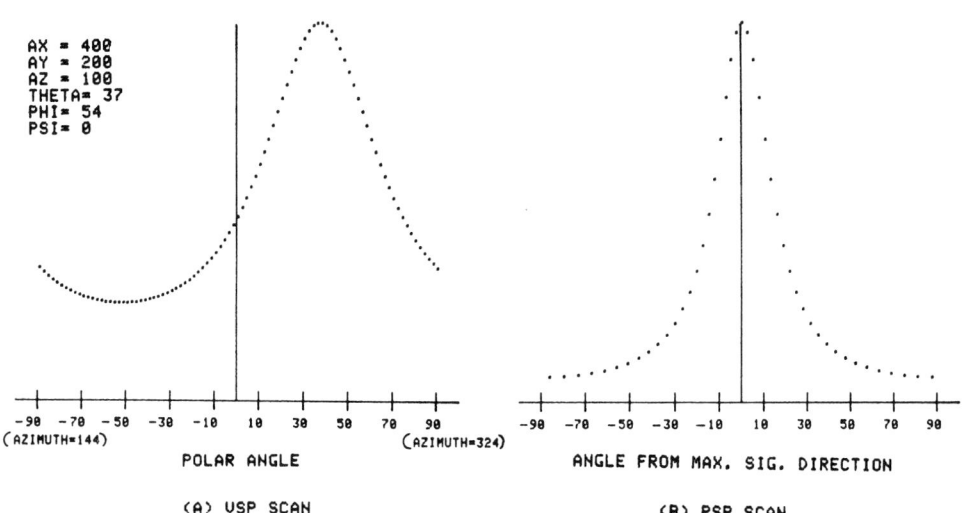

Fig. 5. a) VSP scan showing a broader peak at a polar angle of 37°.
b) PSP scan showing a narrower peak because $A_x > A_y$.

Using computer simulation, the angular scans of a large number of flaw shapes and orientations have been studied. It was noticed that spheroidal flaws of any orientation always have a VSP. A VSP also exists for general ellipsoids if the third Euler angle $\psi$ is zero. The angular scan of a general ellipsoid with a nonzero $\psi$ possesses no mirror symmetry. Further investigations are needed to extract its orientation information and to distinguish it from nonellipsoidal flaws. Based on this rather extensive simulation study, a flow chart, as shown in Table 1, has been made for characterizing the shape and orientation of ellipsoidal flaws.

COMPARISON WITH EXPERIMENTS

To compare with the predicted signal amplitude contours in angular scans, we obtained experimental data on a tilted oblate spheroidal void in titanium. The sizes of the flaw are $A_x=A_y=400\mu m$ and $A_z=200\mu m$. The azimuthal and polar angles of its z-axis (normal to the "flat" surface) are respectively 255° and 30°. Experimental data in Fig. 6(a) show that the VSP occurs near the expected azimuthal angle of 255° and that the signal amplitude at this azimuth first increases with polar angle and then decreases after the polar angle exceeds the tilt angle 30° of the oblate flaw. The frequency spectrum of the transducer used in the angular scans is such that ka ranges approximately from 0.5 to 3. As a comparison, the computed high frequency limit signal amplitude contours are shown in Fig. 6b. Although the computed results are made for the large ka limit, they clearly display qualitatively the same features as the experimental results. It should also be noted that the experimental data represent the raw flaw signal amplitude without correcting for diffraction and interface refraction effects, except

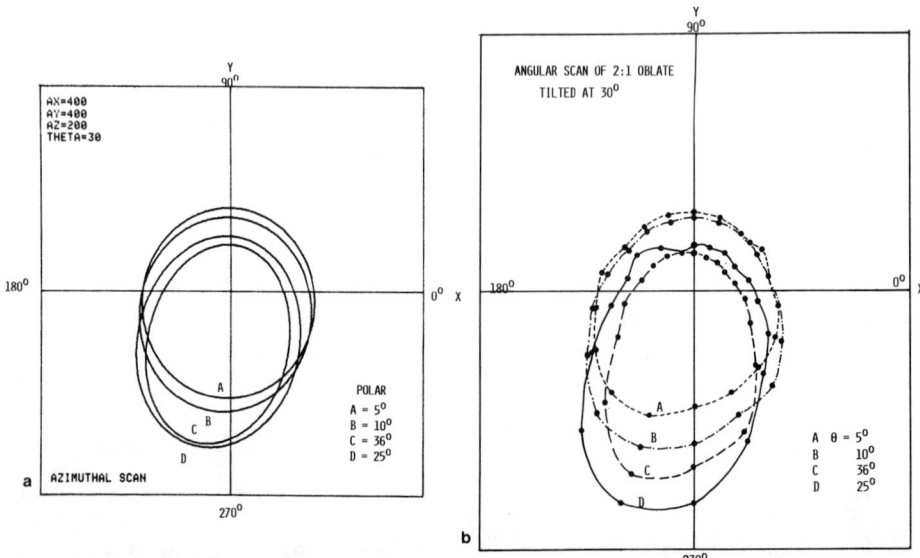

Fig. 6. a) Experimental angular scan results for an oblate spheroidal void in titanium tilted 30° with respect to the part surface. b) Computed signal amplitude contours for the titled oblate spheroid.

Table 1. Flow chart for flaw shape characterization based on angular scans of the flaw signal amplitude.

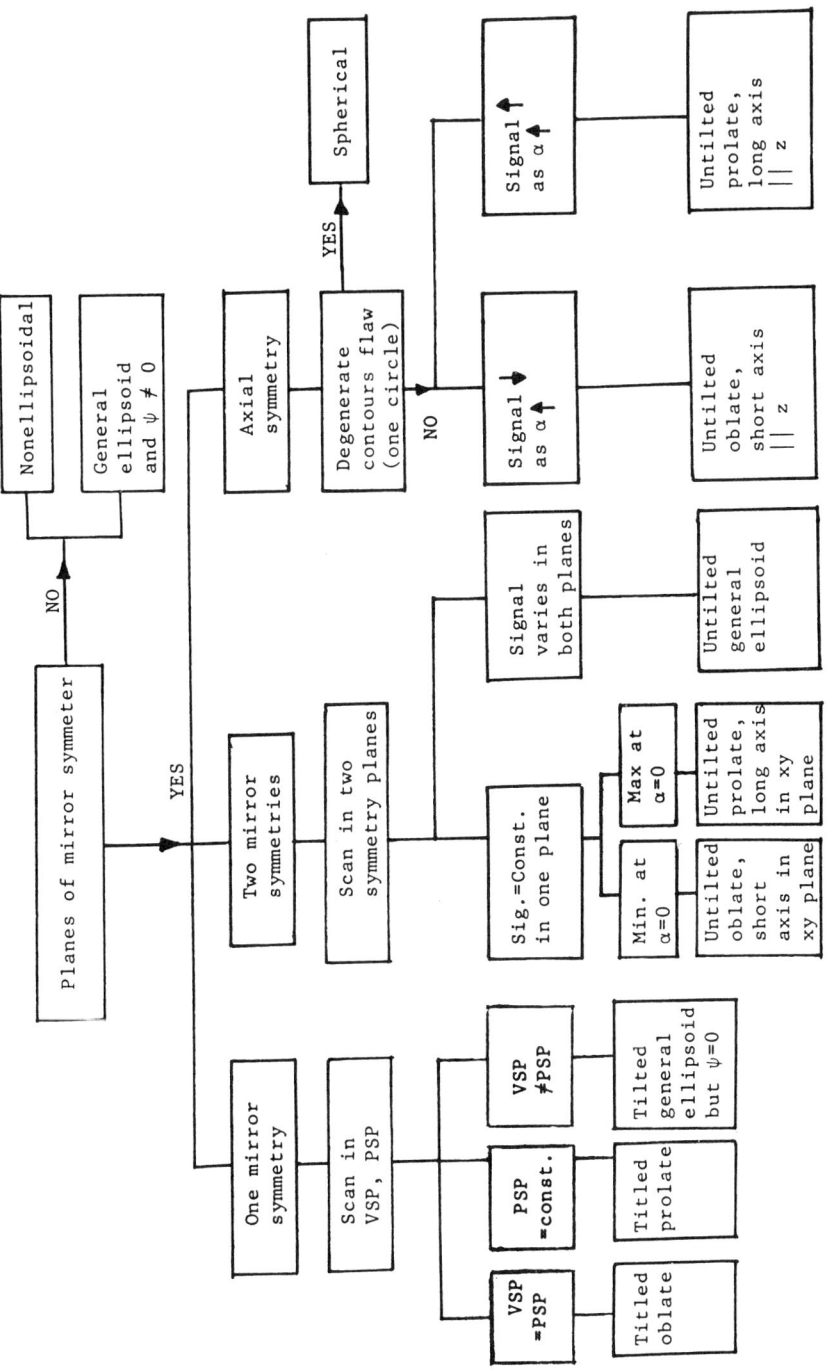

Note: Abbreviated notations in this table are: VSP-vertical sagittal plane, PSP = perpendicular sagittal plane, $\psi$ is the third Euler angle, $\alpha$ is the polar angle, xyz is the laboratory system of coordinates, and z is perpendicular to the part surface.

that the acoustic propagation time was equalized for the various look angles. The important conclusion to be drawn from this comparison is that the symmetry and Gaussian curvature of the flaw surface are such strong features that they are observable even in the intermediate frequency regime. Furthermore, the flaw signal amplitude is dominated by the front surface δ-function of the impulse response, which contains most of the high frequency components.

To investigate the angular dependence of the flaw signal amplitude (the front surface echo strength) more quantitatively, a polar scan was made in the VSP of the 2:1 oblate and the results are shown in Fig. 7. The experimental flaw signal was processed with the measurement model algorithm [9] to correct for diffraction and refraction effects in order to extract the front surface echo strength of the impulse response function the experimental results are shown in Fig. 7 as crosses. As a comparison, the dashed line in Fig. 7 represents Opsal's calculated results [10] bandlimited by the transducer response used in the experiment. Although the experiment and the calculation show some discrepancy, the trend supports the conclusion reached above.

In Fig. 8 we compare the experimental and computed angular scan contours of a copper wire inclusion tilted 45°. The experimental amplitude contours clearly differ from the computed results by having a bulge in the second quadrant. This difference, in fact, was caused by a strongly

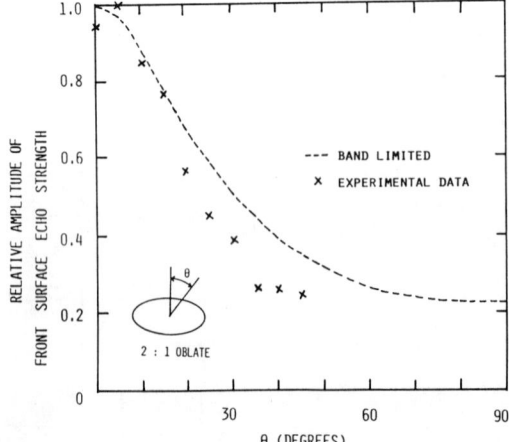

Fig. 7. Comparisons of experimental front surface echo strength for the 2:1 oblate spheroidal void and the computed results with transducer band limiting effects included.

reflecting flat facet on one end of the copper wire (cut by a wire cutter, see micrograph inset). This shows that the angular scan contours are sensitive to detailed surface features of the flaw. A PSP scan perpendicular to the axis of the wire section is shown in Fig. 9. As expected, the signal amplitude is a constant. In contrast, the PSP scan of the oblate spheroid tilted 30° (also shown in Fig 9) shows a peak.

CONCLUSIONS

We have deomonstrated that angular scans of the flaw signal amplitude cannot only reveal considerable flaw shape and orientation information, but also may be used to position the finite aperture of the multiviewing apparatus for a spatial data acquisition pattern that greatly improves the reconstruction reliability. The angular scan method has therefore provided an alternative approach for flaw reconstruction using the multiviewing transducer system. Instead of determining the full flaw characterization (size, shape, and orientation) by relying on the iterative fitting of the sizing data to the best-fit ellipsoid, the approximate shape and orientation of the flaw may be first determined with angular scans of the flaw echo amplitude prior to any sizing measurements. It should be stressed that the angular scan method is not limited to any particular flaw sizing inversion algorithm. Instead, it is a preliminary step that provides useful information to make the subsequent sizing and reconstruction more stable and reliable. Future work will be directed toward extracting more quantitative information such as aspect ratios and developing reconstruction methods based on the Gaussian curvature of the flaw surface.

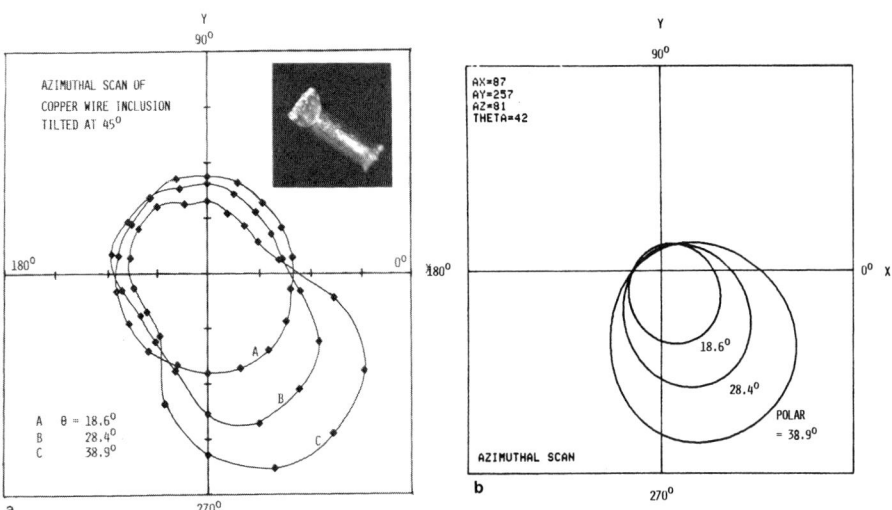

Fig. 8. a) Experimental angular scan contours of a copper wire section titled 45°. b) Computed angular scan contours for the same flaw.

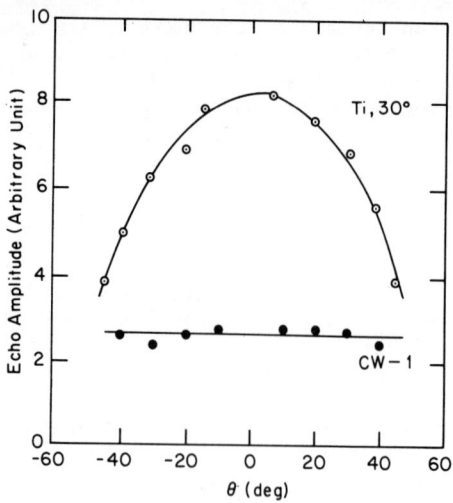

Fig. 9. PSP scans of the copper wire inclusion and the oblate spheroidal void in titanium. Here θ is the angle measured from the maximum signal direction.

ACKNOWLEDGEMENT

The Ames Laboratory is operated for the U.S. Department of Energy by Iowa State University under Contract No. $-7405-ENG. This work was supported by the Director of Energy Research, Office of Basic Energy Sciences.

REFERENCES

1. D. O. Thompson, S. J. Wormley, and D. K. Hsu, "Apparatus and technique for reconstruction of flaws using model-based elastic wave inverse ultrasonic scattering", to be published in Rev. Sci. Instrum., Dec., 1986.
2. J. H. Rose and J. A. Krumhansl, J. Appl. Phys. 50, 2951 (1979).
3. J. M. Richardson and K. Fertig, in Proc. DARPA/AF Review of Progress in Quantitative NDE, AFWAL-TR-80-4078 (1980), p. 528.
4. J. H. Rose, R. K. Elsley, B. R. Tittmann, V. V. Varadan, and V. K. Varadan, in Acoustic, Electromagnetic, and Elastic Wave Scattering--Focus on the T-Matrix Approach, Edited by V. K. Varadan and V. V. Varadan (Pergamon, NY, 1980), p. 605.
5. D. K. Hsu, J. H. Rose, and D. O. Thompson, J. Appl. Phys. 55, 162 (1984).
6. S. J. Wormley and D. O. Thompson, "Error sensitivity of long and intermediate wavelength flaw reconstruction", in Review of Progress in Quantitative NDE, 4A, D. O. Thompson and D. E. Chimenti, Eds., (Plenum Press, NY, 1985), pp. 203-211.
7. S. J. Wormley, D. K. Hsu, and D. O. Thompson, "The effects of flaw orientation and finite aperture on model based reconstruction using multiprobe transducers", ibid, 5A, pp. 529-539 (1986).
8. D. O. Thompson and S. J. Wormley, "Absolute magnitude of front surface reflections in ultrasonic measurements", ibid, 3A, pp. 385-393 (1984).
9. R. B. Thompson and T. A. Gray, J. Acoust. Soc. Am. 74, 1279 (1983).
10. J. L. Opsal, J. Appl. Phys. 58, 1102 (1985).

# RECONSTRUCTION OF THE ELECTROMAGNETIC WAVEFIELD FROM SCATTERING DATA

James H. Rose

Center for NDE
Iowa State University
Ames, IA   50011

## INTRODUCTION

Inverse scattering theory concerns itself with determining the properties of a scatterer (e.g., the spatial variation of the scatterer's dielectric constant and conductivity) from measured scattering data. Two general approaches to this problem exist. The first approach is a direct variational method. One starts by computing the scattering amplitude for some assumed properties of the scatterer. The resulting theoretical scattering amplitude is compared with the measured scattering data. If these results differ, then one varies the properties of the assumed scatterer and recomputes the scattered field for the new properties. This process is truncated when the measured and computed fields agree to a specified accuracy. The properties of the assumed scatterer are then supposed to coincide to within some accuracy with the unknown scatterer which generated the measured data.

The second approach is more direct. One develops linear integral (or differential) equations whose input is the scattered field [1,2]. The solution of these equations then either yields the properties of the scatterer directly or after simple post-processing.

Both approaches have been successful in solving the inverse problem for one-dimensional scatterers; i.e., scatterers whose properties depend only on a single spatial variable. Examples are layered or spherically symmetric scatterers. Both approaches suffer great difficulties for scatterers whose properties vary in a general three-dimensional way. In the first method the computational effort of solving the direct problem often becomes prohibitive. The second approach is not generally fully developed for the three-dimensional case. Nevertheless, approximations to it such as the inverse Born approximation and optical imaging indicate that good progress can be made. In particular, the exact general equations which govern the reconstruction of three-dimensional scatterers are currently unknown for acoustics, electromagnetics and elastodynamics.

The purpose of this paper is to propose a candidate for such a general reconstruction equation in the case of electromagnetics. The structure of the paper is as follows. First, we briefly review the current situation. Then we introduce some of the elements of scattering theory. Finally, we derive the new equation.

Review

Recently, Newton [3] has developed the exact equations which govern three-dimensional inverse scattering for Schrödinger's equation. Newton's approach proceeds in two steps. First, he showed that the wavefield everywhere in space (including the interior of the scatterer) could be deduced from a linear integral equation whose input is the far-field scattering amplitude. Second, the potential is recovered straightforwardly from the wavefield using wavefront conditions.

Recently, Rose, Cheney, and DeFacio [4] have shown that the integral equation used by Newton is also valid for a wide range of scalar wave equations. For example, this linear integral equation relates the wavefield to the scattering amplitude for the acoustic wave equation. (Note, however, that it is currently unknown if the equation has unique solutions in this case). The derivation of the integral equation in Ref. [4] is very general and relies only on such features of the scattering process as linearity, causality and the far-field decay of the wavefield.

The form of the derivation makes it clear that the same integral equation (modified slightly) will also hold for linear, hyperbolic vector wave equations. In this report, the derivation is generalized for electromagnetic scattering. A similar derivation is also possible for elastodynamics.

## Elements of Electromagnetic Scattering Theory

The propagation of electromagnetic waves in a linear medium is governed by [2]

$$\vec{\nabla} \times (\mu^{-1}(\omega,\vec{x}) \vec{\nabla} \times \vec{\tilde{E}}(\omega,\vec{x})) - k^2 n^2(\omega,\vec{x}) \vec{\tilde{E}}(\omega,\vec{x}) = 0. \tag{1}$$

Here $\omega$ is the angular frequency, k is the wavevector of light in free space, and $\omega = k$, since we choose the speed of light c=1. The electric field is denoted by $\vec{\tilde{E}}(\omega,\vec{x})$ and the magnetic permeability by $\mu$. Finally,

$$n^2(\omega,\vec{x}) = \varepsilon(\omega,\vec{x}) + \frac{4\pi i\, \sigma(\omega,\vec{x})}{\omega}. \tag{2}$$

Here $\varepsilon(\omega,\vec{x})$ denotes the dielectric permittivity and $\sigma(\omega,\vec{x})$ the conductivity. In general, $\mu$, $\varepsilon$ and $\sigma$ are tensors.

The geometry for the scattering experiment is described. The scatterer has finite spatial extent. It is enclosed by some finite region of space called R which also contains the origin of coordinates. Within the region R the material properties ($\mu$, $\varepsilon$ and $\sigma$) vary as functions of $\vec{x}$ and $\omega$ and may, in fact, be discontinuous. However, they are assumed to be bounded functions. Exterior to the scattering region the conductivity is zero and $\varepsilon$ and $\mu$ take on constant frequency independent values, e.g., $\varepsilon$ and $\mu$ could be chosen to take on their free space values exterior to R. Thus, the geometry presupposes a finite (possibly metallic) object embedded in an infinite otherwise homogeneous space. Finally, the entire scattering region is surrounded by a large ball S which is centered about the origin of coordinates.

The derivation to be given is clearest in the time-domain. The time-domain electromagnetic wave equation is obtained by taking the Fourier transform of the quantities in Eq. (1) with respect to ω. For example,

$$\vec{E}(t,\vec{x}) = \frac{1}{2\pi} \int_{-\infty}^{\infty} d\omega e^{-i\omega t} \vec{\tilde{E}}(\omega,\vec{x}). \tag{3}$$

Exterior to the region R the wave equation reduces to

$$\nabla \times \nabla \times \vec{E}(t,\vec{x}) - \frac{\partial^2}{\partial t^2} \vec{E}(t,\vec{x}) = 0. \tag{4}$$

Here we have set the speed of light in the surrounding material equal to one and used the fact that $c^{-2} = \varepsilon\mu$.

The scattering experiment is assumed to proceed as follows. At early times, a transversely polarized plane wave delta pulse is incident propagating in the $\hat{e}$ direction with polarization $\hat{\alpha}$. This incident field is a solution of Eq. (4) and is given explicitly by

$$E_i^0(t,\hat{e},\hat{\alpha},\vec{x}) = \delta(t-\hat{e}\cdot\vec{x}) \hat{\alpha}_i. \tag{5}$$

Here $\hat{e}$ and $\hat{\alpha}$ are unit vectors which are indicated by carets. Note $\hat{e}\cdot\hat{\alpha}=0$ since $E_i^0$ is transversely polarized. Tensor notation will be used from here on in denoting the component of field quantities.

The incident pulse propagates until it strikes the scattering region where it interacts with the scatterer in a complicated way. Finally, at late times the field propagates outward from the region R and is measured on the surface of the large sphere S which is centered about the origin. These measurements are the scattering data.

The scattering data can be defined in terms of the field $E_i^+(t,\hat{e},\hat{\alpha},\vec{x})$ which result from the incident field $E_i^0$. We take $x = |\vec{x}|$ large and define the impulse response functions $R^\perp$ and $R''$ by

$$E_i^+(t,\hat{e},\hat{\alpha},\vec{x}) = \delta(t-\hat{e}\cdot\vec{x}) \hat{\alpha}_i \tag{6}$$
$$+ \frac{R^\perp(t-x,\hat{e},\hat{\alpha},\hat{x})}{x} \hat{y}_i^\perp + \frac{R''(t-x,\hat{e},\hat{\alpha},\hat{x})}{x} \hat{y}''_i + o(x^{-1})$$

Here $\hat{x} \equiv \vec{x}/x$ denotes the direction of scattering, while $\hat{y}^\perp$ and $\hat{y}''$ denote transverse polarization directions on the surface of S. The unit vector $\hat{y}''$ lies in the plane of scattering and has $\hat{x}\cdot\hat{y}''=0$, while $\hat{y}^\perp \equiv \hat{x} \times \hat{y}''$.

Integral Equation Derivation

The idea behind the derivation is quite simple. The field everywhere in space can be computed given the initial data (i.e., the wavefield at very early times) and the properties of the scatterer $\varepsilon$, $\mu$, and $\sigma$. The result is formally denoted by $E_i^+(t,\hat{e},\hat{\alpha},\vec{x})$. However, the field can also be computed given the final data (i.e., the wavefield at very late times which is known from the scattering data) and $\varepsilon$, $\mu$ and $\sigma$. When these two ways of computing the field are equated, the integral equation is obtained. Perhaps surprisingly $\varepsilon$, $\mu$ and $\sigma$ do not appear, and a relation strictly between the wavefields and impulse response functions results.

The derivation rests on the following assumptions concerning the time-domain electromagnetic wave equation. (1) The solution of the wave equation which yields the scattering data in Eq. (6) is unique. This rules out the existence of solutions which decay more rapidly than $x^{-1}$ at large x. (2) The signal (wave train) received at an observation point on the surface of the sphere S is assumed to decay sufficiently rapidly to zero outside a time interval which contains $t=x$. A closely related further assumption is that the wavefield approaches zero in any bounded region of space as $t \to +\infty$. (3) It is assumed that the amplitude of the scattered field decays as $x^{-1}$ and varies continuously as a function of angle for sufficiently large x.

We start with the second method of computing $E_i^+(t,\hat{e},\hat{\alpha},\vec{x})$. Namely, we compute it for all $\vec{x}$ and t from the scattering data. First we define $E_i^{+sc} = E_i^+ - \delta \hat{\alpha}_i$. Then we re-express the scattering data in Eq. (6) as
$(\vec{x} \to \infty)$

$$E_i^{+sc}(t,\hat{e},\hat{\alpha},\vec{x}) = \sum_P \int_{-\infty}^{\infty} d\tau x^{-1} \delta(t-\tau-x) R^P(\tau,\hat{e},\hat{\alpha},\hat{x}) \hat{y}_i^P + o(x^{-1}). \quad (7)$$

Here the sum is over the polarization $P = (\perp, \|)$. We note that the Dirac delta contributes only when $t-\tau=x$, so that without loss of generality we may restrict the $\tau$ integration to $t-\tau$ positive.

Next we will express the outgoing spherical wave, $x^{-1}\delta(t-\tau-x)$, as a sum of plane waves $\delta(t-\tau-\hat{e}\cdot\vec{x})$ via the formula

$$x^{-1}[\delta(s+x)-\delta(s-x)] = (2\pi)^{-1} \frac{d}{ds} \int_{S^2} d^2e' \, \delta(s-\hat{e}'\cdot\vec{x}). \quad (8)$$

We use (8) in (7) with $s>0$:

$$E_i^{sc}(t,\hat{e},\hat{\alpha},\vec{x}) = -(2\pi)^{-1}(d/dt) \int_{-\infty}^{\infty} d\tau \int_{S^2} d^2e' \sum_P \delta(t-\tau-\hat{e}'\cdot\vec{x}) R^P(\tau,\hat{e},\hat{\alpha},\hat{x})$$
$$+ o(x^{-1}). \quad (9)$$

Here $S^2$ denotes the two-sphere and the integration is over all angles. We see from this expression that the Dirac $\delta$ contributes only when $\tau = t-\hat{e}'\cdot\vec{x}$. Since we know from (3) that this must be the same as $\tau = t-x$, it follows that the Dirac $\delta$ contributes only when $\hat{e}'=\hat{x}$. This implies that $R^P(\tau,\hat{e},\hat{\alpha},\hat{x})$ which appears on the right-hand side can be replaced with $R^P(\tau,\hat{e},\hat{\alpha},\hat{e}')$ in Eq. (9). The result is

$$E_i^+(t,\hat{e},\hat{\alpha},\vec{x}) = \delta(t-\hat{e}\cdot\vec{x})\hat{\alpha}_i - (2\pi)^{-1}(d/dt) \int_{-\infty}^{\infty} d\tau \int_{S^2} d^2e' \sum_P$$
$$\delta(t-\tau-\hat{e}'\cdot\vec{x})\hat{y}_i^P R^P(\tau,\hat{e},\hat{\alpha},\hat{e}') + o(x^{-1}). \quad (10)$$

Thus, the entire $\vec{x}$ dependence on the right-hand side resides in the Dirac delta, a result needed below.

Equation (10) is a rewriting of the final (i.e., scattering) data as a weighted sum of plane waves. Now we need to answer the following question. What solution of the wave equation corresponds to the final data defined by Eq. (10)? By construction $E_i^+$ is such a solution.

However, a different representation of the same solution can be obtained from the right-hand side of Eq. (10). In order to state this clearly, a second type of solution is needed for the electromagnetic wave equation. Namely, we define $E_i^-(t,\hat{e},\hat{\alpha},\vec{x})$ to be that solution which for sufficiently <u>late</u> times is given by $\delta(t-\hat{e}\cdot\vec{x})\hat{\alpha}_i$. Thus, the solution $E_i^-$ is defined by its behavior at late times (i.e., by its final data).

Now consider the final (late time) data defined by the right-hand side of Eq. (10). Each Dirac delta plane wave, $\delta(t-\tau-\hat{e}'\cdot\vec{x})\hat{y}_i^P$, can be considered to have evolved from a solution $E_i^-(t-\tau,\hat{e}',\hat{y}_i^P,\vec{x})$. Consequently, the whole right-hand side of (6) is seen to have evolved from the solution

$$E_i^+(t,\hat{e},\hat{\alpha},\vec{x}) - (2\pi)^{-1}(d/dt) \sum_P \int_{-\infty}^{\infty} d\tau \int_{S^2} d^2 e' \; E_i^-(t-\tau,\hat{e}',\hat{y}_i^P,\vec{x}) R^P(\tau,\hat{e},\hat{\alpha},\hat{e}'). \quad (11)$$

Here we have used the fact that the solutions of a linear wave equation can be superimposed to form a new solution.

Both $E_i^+$ and the expression given in (11) are solutions of the electromagnetic wave equation which evolve into the same scattering data for sufficiently late times. Since the solutions of the wave equation are assumed to be unique, these solutions can be equated to give our basic result

$$E_i^+(t,\hat{e},\hat{\alpha},\vec{x}) = E_i^-(t,\hat{e},\hat{\alpha},\vec{x}) - (2\pi)^{-1}(d/dt) \int_{-\infty}^{\infty} d\tau \int_{S^2} d^2 e' \sum_P$$

$$R^P(\tau,\hat{e},\hat{\alpha},\hat{e}') \; E_i^-(t-\tau,\hat{e}',\hat{y}_i^P,\vec{x}). \quad (12)$$

We have shown that subject to a few mild assumptions, Eq. (12) holds for electromagnetic scattering. We note that the scalar wave analogue of Eq. (12) is the basis of Newton's exact solution of the inverse problem for Schrödinger's equation. Substantial work is underway to see if Eq. (12) can be made to yield unique solutions for $E_i^+$ given the scattering data for electromagnetic scattering.

ACKNOWLEDGEMENT

This work is supported by the NSF university/industry Center for NDE at Iowa State University.

REFERENCES

1. K. Chadan and P. C. Sabatier, <u>Inverse Problems in Quantum Scattering Theory</u>, (Springer-Verlag, NY, 1977).
2. R. G. Newton, <u>Scattering Theory of Waves and Particles</u>, (Springer-Verlag, NY, 1982), 2nd Edition.
3. R. G. Newton, Phys. Rev. Lett., <u>43</u>, 541 (1979).
4. J. H. Rose, M. Cheney, and B. DeFacio, Phys. Rev. Lett., <u>57</u>, 783 (1986).

UNIFORM FIELD EDDY CURRENT PROBE:  EXPERIMENTS AND INVERSION

FOR REALISTIC FLAWS*

J. C. Moulder,[1] P. J. Shull,[1] and T. E. Capobianco[2]

[1]Fracture and Deformation Division
[2]Electromagnetic Technology Division
National Bureau of Standards
Boulder, Colorado 80303

INTRODUCTION

Uniform field eddy current (UFEC) probes operate by interrogating flaws with a spatially uniform electromagnetic field. Their use in quantitative NDE is particularly attractive because theoretical models of the field-flaw interaction are greatly simplified by the assumption of a uniform field. This in turn leads to much simpler inversion protocols for determining flaw sizes from measurements.

A theory for the interaction of a uniform interrogating field with three-dimensional surface flaws in the limit of small skin depth ($a/\delta \gg 1$) has been developed by B. A. Auld and his co-workers at Stanford University [1-3]. Last year, E. Smith reported the first quantitative measurements on fatigue cracks and EDM notches with an "essentially" uniform field eddy current probe, demonstrating excellent agreement between theory and experiment [4].

We report here the results of an extensive series of measurements to evaluate in detail the use of uniform field eddy current probes for quantitative NDE. As described in a companion paper [5], we designed and built two UFEC probes that operate in the frequency ranges of 0.5 to 4 and 2 to 8 MHz. The probes were calibrated with either cylindrical or part-spherical recesses formed by electrical-discharge machining.

We used these probes to study a number of surface-connected, semi-elliptical EDM slots and actual fatigue cracks in a variety of materials, including 7075 Aℓ, Ti-6Aℓ-4V, and Haynes 188 alloys. Because of the limited space available, we report only the results for high-frequency measurements on EDM notches and fatigue cracks in Ti-6Aℓ-4V. Flaws ranged in length from 0.5 to 3.0 mm and in depth from 0.25 to 1.5 mm.

All measurements and calibrations were performed with an automatic network analyzer. We compare measured flaw signals with the predictions of Auld's uniform field theory, and the limits of applicability of the theory are determined. We also explore the accuracy with which flaw depth can be determined using a simple least-squares inversion algorithm.

---
*Contribution of the National Bureau of Standards; not subject to copyright in the United States.

# THEORY

The general ΔZ theory for the interaction of a spatially uniform magnetic field with two- and three-dimensional flaws has been developed by Auld et al. [1-3]. In the limit of small skin depth, the change in probe impedance caused by a flaw can be represented as

$$\Delta Z = \frac{c}{\sigma} \frac{H^2}{I^2} \left[ \Sigma^0 + (1+i) \frac{c}{\delta} \Sigma^1 + \frac{i\Delta u c}{\delta^2} \Sigma^1 \right], \quad (1)$$

where H/I is the magnetic field strength per unit current, $\sigma$ is the conductivity, 2c is the flaw length, $\Delta u$ is the flaw width, $\delta$ is skin depth. $\Sigma^0$ and $\Sigma^1$ are shape factors that depend only on a/c (a is flaw depth); they must be calculated for the particular flaw geometry. The three terms in Eq. 1 correspond roughly to resistive losses at the crack corners, the wall impedance of current flowing over the flaw surfaces, and Faraday induction due to the volume enclosed by currents encircling the flaw [2].

It is clear from Eq. 1 that the flaw signal ΔZ is proportional to the square of the magnetic field strength and inversely proportional to the conductivity of the workpiece. ΔZ depends on frequency only through the skin depth $\delta = \sqrt{2/\mu\omega\sigma}$. ΔZ depends strongly upon the flaw length, but depends on flaw depth only through $\Sigma^0$ and $\Sigma^1$.

The $\Sigma$'s appropriate to rectangular and semi-elliptical flaws were calculated numerically last year at Stanford University and published in reference [3]. This year it was discovered that the values for $\Sigma^0$ for semi-elliptical flaws published last year are in error for values of a/c > 0.35. These were recalculated this year by Steve Jefferies at Stanford and the new values are shown in Table 1. For use in the inversion of flaw signals, the numerically calculated values of $\Sigma^0$ and $\Sigma^1$ were fit to polynomials by a nonlinear least-squares procedure, giving

$$\Sigma^0 = -3 - 1.35 \left(\frac{a}{c}\right) + 0.66 \left(\frac{a}{c}\right)^2 \quad (2)$$

and

$$\Sigma^1 = -0.02 + 2.56 \left(\frac{a}{c}\right) + 1.11 \left(\frac{a}{c}\right)^2 - 1.70 \left(\frac{a}{c}\right)^3. \quad (3)$$

Table 1. Values for the shape parameter $\Sigma^0$ used in Eq. 1. Courtesy of S. Jefferies, Stanford University.

| a/c | $\Sigma^0$ |
|---|---|
| 0.1 | -3.13 |
| 0.2 | -3.25 |
| 0.34 | -3.37 |
| 0.5 | -3.52 |
| 0.75 | -3.64 |

CALIBRATION

For quantitative measurements calibration is extremely important since, to calculate ΔZ for a particular probe, it is necessary to know H/I. For air core probes, one may calculate this quantity using the results of Dodd and Deeds for a vertical coil [6] or Burke for a horizontal coil [7]. But for ferrite core probes like the NBS UFEC probe, analytical expressions for the field strength are not available and the field strength must either be measured or calculated numerically. We chose to use a procedure first suggested by B. A. Auld [8] and used last year by Smith [4].

The method consists of measuring the impedance change ΔZ caused by a cylindrical recess of depth greater than the skin depth and calculating H/I from the expression

$$\Delta Z = i \left(\frac{H}{I}\right)^2 \frac{2}{\sigma} \frac{Ad}{\delta^2} \qquad (4)$$

where A is the surface area of the recess and d is its depth. This formula was adapted by Auld from a microwave perturbation theory solution for a smooth dimple in the surface of a perfect conductor [8].

Smith found that to obtain agreement of calculated and measured flaw signals, the numerical factor of 2 in Eq. 4 had to be replaced by a factor of 1 [4]. We confirmed his result this year in our measurements on cylindrical recesses in Ti-6Aℓ-4V ($\sigma$ = 5.9 x $10^5$ S/m). For a recess 0.78 mm in diameter and 0.83 mm deep (d/δ > 2), the values of H/I calculated using the factor of 1 in Eq. 4 gave excellent agreement of calculated and experimental ΔZ values for a series of EDM notches. These results are shown in Fig. 8 of reference [5]. For another recess 0.77 mm dia. x 0.40 mm deep (d/δ ≈ 1), a factor of 1.2 in Eq. 4 gave good agreement of calculated and measured flaw signals.

Since these recesses had to be quite deep to meet the criterion d/δ > 2, the discrepancy between Eq. 4 and measurement may be related to the fact that the cylinders very little resemble a smooth dimple. To approximate more closely the theoretical assumptions, we made recesses in the shape of a spherical cap in the same titanium alloy. Calibration with a recess that was deeper than the skin depth (1.46 mm surface diameter, 0.89 mm radius, and 0.37 mm depth) gave excellent agreement of theory and experiment for the same EDM slots when a factor of 2 was used in Eq. 4 and with the appropriate volume substituted for Ad. For a spherical cap approximately the same depth as the skin depth (0.67 mm surface diameter, 0.42 mm radius, and 0.19 mm depth), a factor of 4 gave the best results. Thus, we find that Eq. 4 may be generalized to

$$\Delta Z = i \left(\frac{H}{I}\right)^2 \left(\frac{CV}{\sigma\delta^2}\right) \qquad (5)$$

where V is the volume of the recess (either cylindrical or spherical) and C is an empirical shape factor whose value depends on the shape and depth of the recess relative to δ, as described above.

Recently, R. Collins and his colleagues at University College London have derived analytical expressions for the electric and magnetic fields inside axisymmetric surface flaws (including cylinders and spherical caps) interrogated with a uniform field in the small skin depth limit [9, 10]. This raises the possibility of calculating ΔZ exactly for cylindrical and spherical-cap recesses.

Another possible calibration method is to measure ΔZ for an EDM notch of known dimensions, then to determine H/I by forcing the calculated flaw signals into agreement with experimental signals. Results of calibrating the NBS UFEC probe using these different methods are shown in Figure 1. Agreement is excellent among the different calibration procedures, except at the lowest frequencies, where $d/\delta < 2$ for two of the artifacts.

EXPERIMENT

As described in the companion paper [5], all measurements were made with a computer-controlled automatic network analyzer that measured probe impedance at 401 discrete frequencies distributed uniformly over the frequency range of 2 to 8 MHz. Each measurement was the result of signal averaging 64 times; no fewer than five independent measurements on each flaw were smoothed with a ten-point running average and then averaged to obtain the final result. A measurement consisted of measuring the impedance of the probe on and off the flaw over the stated frequency range, and then computing the vector difference of the two impedances.

The probe was positioned by a computer-controlled x-y positioner. Liftoff in the z direction was controlled with a manual micropositioner; tilt in the x-z and y-z planes could be independently adjusted. As described in reference [5], the probe was extremely sensitive to changes in tilt or liftoff during the measurements and it was necessary to control these parameters very closely.

A series of five semi-elliptical EDM slots were prepared in a specimen of Ti-6Aℓ-4V 7.5 cm wide, 21 cm long, and 1 cm thick. They ranged in length from 0.5 to 2.5 mm and in depth from 0.33 to 1.05 mm; their exact dimensions are given in Table 2 of reference [5]. The fatigue cracks that we studied were grown in similarly sized specimens of Ti-6Aℓ-4V by W. Rummel and co-workers [11]. They were grown in cantilevered bending. The lengths and estimated depths of these flaws are shown in Table 2. Lengths were measured optically during growth of the cracks; depths were estimated from the lengths using the expected aspects ratio of $a/c = 0.67$. Three additional fatigue cracks in a Ti-6Aℓ-4V specimen 3.5 cm x 23 cm x 0.6 that were measured are also shown in Table 2 (MM2A-C).

RESULTS

The magnitude and phase of ΔZ for four of the semi-elliptical EDM slots are shown in Fig. 2 and 3, where they are compared with calculated signals obtained using the spherical-cap calibration method to determine H/I. Corresponding results obtained using a ccylindrical recess to calibrate the probe are shown in Fig. 8 of reference [5].

For both calibration methods the agreement between the calculated and measured magnitude of ΔZ is excellent for the larger flaws (within ±5 percent). For frequencies less than 3 MHz, theory and experiment diverge somewhat owing to the fact that $d/\delta$ for the calibration recesses fell below 2.0 at these frequencies. The effect is more pronounced in the case of the cylindrical recess calibration. The agreement between theory and experiment for the phase of ΔZ is better for the spherical calibration method, and this is the calibration method we use in the remainder of this paper.

For the smallest flaw shown in Figure 2, NBS15D, theory and experiment diverge for frequencies below 7 MHz, but at this frequency $a/\delta$ for this flaw is only 1.6. This shows that the limits of applicability of the theory extend to $a/\delta = 1.6$, which is less conservative than the usual estimate of

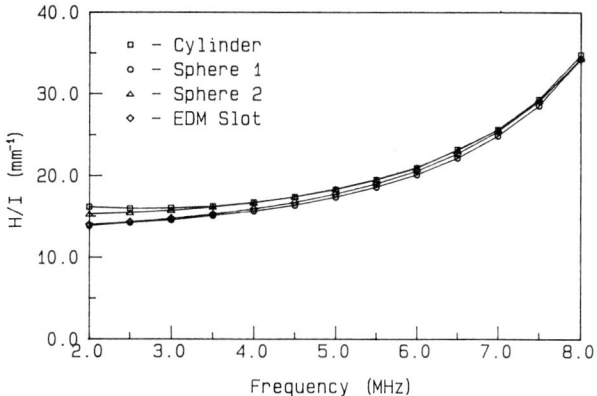

Fig. 1.  Calibration curves of magnetic-field intensity per unit current for uniform field eddy current probe, obtained using three different types of artifacts.

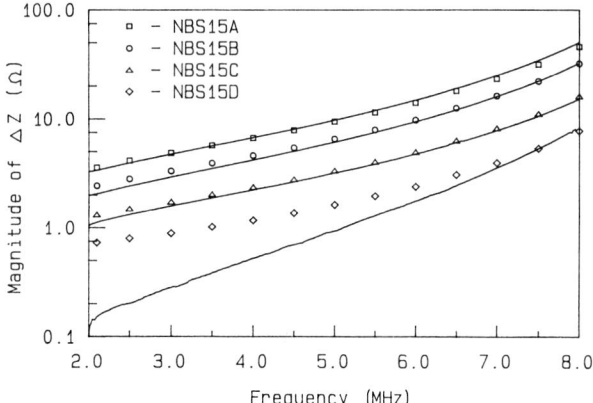

Fig. 2.  Measurements (solid lines) of the magnitude of the flaw signals from four different semi-elliptical EDM slots in Ti-6Aℓ-4V compared with theoretical predictions (symbols).

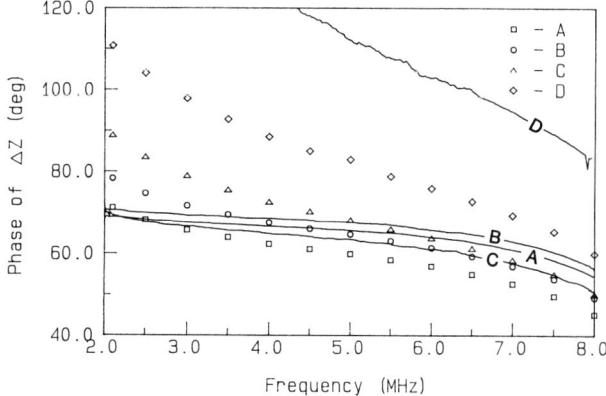

Fig. 3.  Measurements (solid lines) of the phase of the flaw signals from four different semi-elliptical EDM slots in Ti-6Aℓ-4V compared with theoretical predictions (symbols).

Table 2. Ti-6Aℓ-4V Fatigue Cracks

| Specimen ID | Length (mm) | Estimated Depth (mm) |
|---|---|---|
| NBSF1A | 2.87 | 0.96 |
| NBSF1B | 2.50 | 0.83 |
| NBSF2A | 0.43 | 0.14 |
| NBSF2B | 0.42 | 0.14 |
| NBSF3A | 0.99 | 0.33 |
| NBSF3B | 0.98 | 0.33 |
| NBSF4A | 1.37 | 0.46 |
| NBSF4B | 1.51 | 0.50 |
| NBSF5A | 1.25 | 0.45 |
| NBSF5B | 2.08 | 0.69 |
| MM2A | 3.18 | 1.06 |
| MM2B | 1.52 | 0.51 |
| MM2C | 0.76 | 0.25 |

would have required operating the probe at 15 MHz. The excellent agreement between theory and experiment is very encouraging, because no adjustable parameters were used in calculating flaw signals.

A comparison of measured and calculated flaw signals for five fatigue cracks in Ti-6Aℓ-4V is shown in Figure 4. Flaw dimensions used for the calculations were the optically measured lengths and estimated depths shown in Table 2. Flaw width was arbitrarily set to 1 μm. Changing the crack width to 10 μm increased $\Delta Z$ by only 3-4 percent. As shown in Figure 4, the agreement between theory and experiment is excellent for these fatigue cracks, which ranged in length from 1.0 to 1.5 mm. For the smallest fatigue cracks we examined (NBSF2A and B, results not shown) the condition $a/\delta > 1.6$ could not be attained.

A similar comparison for four of the larger fatigue cracks (2c = 2.0-3.2 mm) is shown in Figure 5. For these four flaws the measured flaw signal was consistently lower than the flaw signal calculated using estimated flaw sizes. The difference in depth necessary to account for the discrepancy ranged from 32 to 42 percent, as determined by using the inversion procedure described below. These differences are discussed more fully in the next section.

INVERSION

With the exception of the puzzling results shown in Figure 5, the agreement between theory and experiment is excellent for both EDM slots and fatigue cracks. This suggests that a simple inversion procedure may be devised to determine the flaw sizes from eddy current measurements. Possible approaches to the complete inversion problem have been considered before [2]. In principle, it is possible to derive both the length and depth of a fatigue

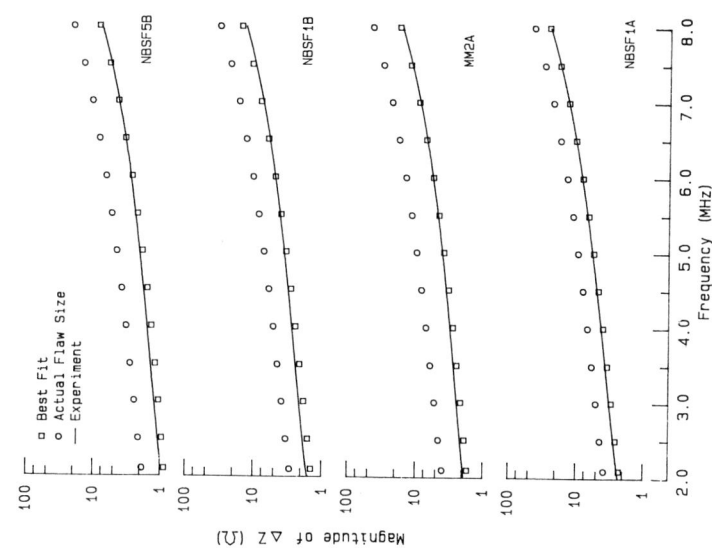

Fig. 5. Comparison of measured and predicted flaw signals for four fatigue cracks in Ti-6Aℓ-4V. All cracks were greater than 2.0 mm long.

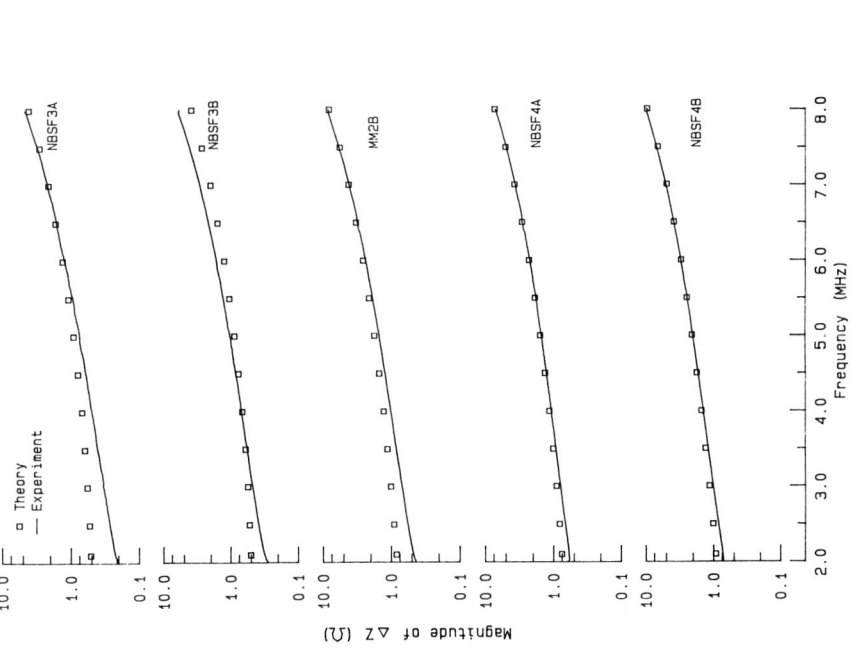

Fig. 4. Comparisons of theory and experiment for five fatigue cracks grown in Ti-6Aℓ-4V. All cracks were less than 1.5 mm long.

crack from measurements at three or more frequencies. However, as shown in Eq. 1, $\Delta Z$ depends strongly on flaw length, but only indirectly on flaw depth through the dependence of $\sum^0$ and $\sum^1$ on a/c. Rather than attempting a full inversion, we assumed that the length of the flaw could be determined easily with a conventional, nonuniform field probe using the imaging concepts suggested by Auld to determine flaw length from the shape of the flaw profile obtained from a scan down the length of the flaw [12].

If the flaw length is known, it is quite simple to invert the measurements to obtain flaw depth. This is the approach we took. Flaw length was assumed to be the actual length of the flaw in the case of the EDM notches or, in the case of fatigue cracks, the optically measured length. Flaw depth was then determined by minimizing the root-mean-square difference between measured and calculated flaw signals. This was accomplished on a portable microcomputer using a commercial spread sheet program to demonstrate the simplicity of the inversion procedure.

Results of using this inversion method to estimate the depths of the EDM notches we measured are shown in Fig. 6a. In all but one case the predicted depths are within ±0.1 mm of the actual depth. Results for the inversion of the fatigue crack measurements are shown in Fig. 6b. For the smaller flaws (a < 0.5 mm), the inversion results are also excellent, but for the larger flaws the discrepancies first shown in Fig. 5 are evident.

The nature of the discrepancy in the inversion results is such that measured flaw signals underpredict the flaw depths by up to 42 percent. Because this can have serious implications when using eddy current methods to size flaws, we broke open four flaws, three of which exhibited serious discrepancies, to determine their actual sizes. The results are listed in Table 3; they show that in all cases the actual flaw dimensions were close to the estimated dimensions. The flaw lengths listed in Table 3 under the heading "NBS Prediction" were determined by etching the specimen surface and then measuring the flaw length with the specimen in an unloaded condition, using a traveling microscope. Flaw depths were then calculated by using this length as an input to the inversion procedure described before. This method

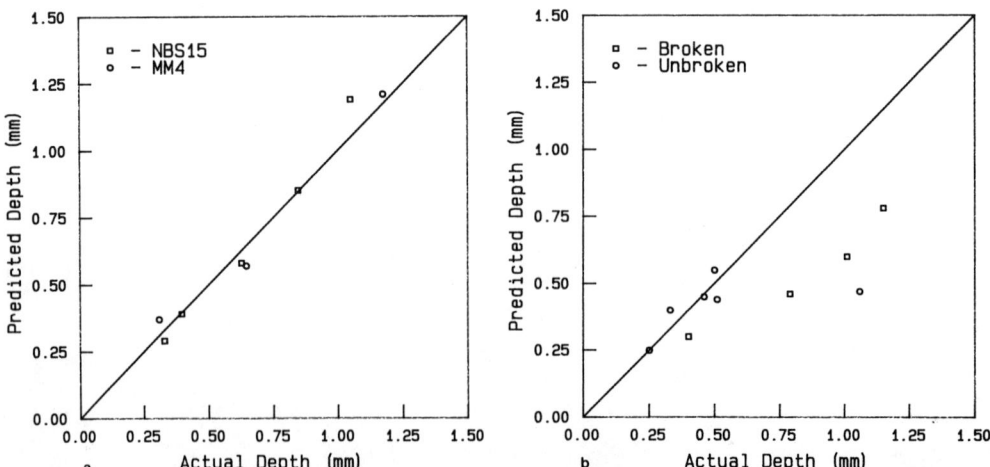

Fig. 6. Inversion results for (a) semi-elliptical EDM slots and (b) fatigue cracks in Ti-6Aℓ-4V. Flaw depths predicted by the inversion procedure are plotted against the actual depths, or, in the case of unbroken fatigue cracks, the estimated depth.

Table 3. Ti-6Aℓ-4V Fatigue Crack Results

| Specimen ID | MM Estimate 2c (mm) | a (mm) | NBS Prediction 2c (mm) | a (mm) | Actual 2c (mm) | a (mm) | Percent Difference (Depth) |
|---|---|---|---|---|---|---|---|
| NBSF1A | 2.87 | 0.96 | 2.54 | 0.78 | 2.98 | 1.15 | -32% |
| NBSF1B | 2.50 | 0.83 | 2.07 | 0.60 | 2.62 | 1.01 | -41% |
| NBSF5A | 1.25 | 0.42 | 1.25 | 0.30 | 1.31 | 0.40 | -25% |
| NBSF5B | 2.08 | 0.69 | 1.57 | 0.46 | 2.16 | 0.79 | -42% |

underestimated the flaw depths by up to 42 percent. For one of the flaws that was broken (NBSF5A), the predicted depth was only 0.1 mm less than the actual depth; this is considered acceptable agreement.

The discrepancy in inverted flaw depths is difficult to explain. The excellent agreement between theory and experiment for EDM slots and smaller fatigue cracks would seem to indicate the theory is valid as long as the condition $a/\delta > 1.6$ is met. There is no reason to expect the theory to break down for fatigue cracks, since we found good agreement in some cases. It is difficult to ascribe the problem to measurement errors, since the discrepancy arises for larger flaws, where the signal-to-noise ratio is highest. It also seems unlikely that random errors would result in uniformly underestimating flaw sizes. One possible explanation that is consistent with the observations is that closure forces on the crack faces caused partial contact, short-circuiting the eddy currents flowing around the flaw. That this should only occur for the longer cracks is consistent with the fact that the longer cracks had a greater stress intensity factor during growth. This could cause greater plastic deformation at the crack tip, leading to greater closure forces. But, before this can be considered a valid explanation of the observed discrepancies, a more conclusive experiment needs to be performed demonstrating that $\Delta Z$ changes when the crack is opened under load.

CONCLUSIONS

A UFEC probe was designed and fabricated at NBS and used to study a series of semi-elliptical EDM slots and fatigue cracks in Ti-6Aℓ-4V. From this study the following conclusions emerge:

Calibration of a UFEC probe can be carried out with any of three calibration artifacts -- cylindrical recess, spherical cap, or EDM slot. Empirical shape factors needed to calculate H/I were determined for cylindrical and spherical recesses.

Measurements on all EDM slots and those fatigue cracks less than 1.5 mm long were in excellent agreement with the predictions of Auld's uniform field theory for $a/\delta > 1.6$.

For larger fatigue cracks, measurements underestimated flaw depth by up to 42 percent. Crack closure forces acting on the longer cracks is one possible explanation that is consistent with experimental observations.

A simple method for inverting UFEC measurements to obtain flaw depth when the length is known was demonstrated. Predicted depths were within ±0.1 mm in the absence of possible closure effects.

ACKNOWLEDGMENTS

This work was sponsored by the Center for Advanced Nondestructive Evaluation, operated by the Ames Laboratory, USDOE, for the Air Force Wright Aeronautical Laboratories/Materials Laboratory under Contract Number W-7405-ENG-82 with Iowa State University. Fatigue cracks were provided by W. Rummel, B. Christner, and D. Long of Martin Marietta Denver Aerospace. We thank W. Rummel for numerous helpful discussions on eddy current methods and for his continuing interest in this work. We are grateful to S. Jefferies of Stanford University for contributing new calculations of the shape factor used in Eq. 1. We especially thank B. A. Auld of Stanford University for continuing encouragement and for providing many useful insights into the theoretical model and the calibration methods used here. We also thank E. Smith of Lockheed-Sunnyvale for helpful discussions on the experiment. We were aided in this study by the skilled programming of S. Ciciora and by the careful measurements contributed by C. Cherne.

REFERENCES

1. B. A. Auld, F. G. Muennemann, and D. K. Winslow, J. Nondestructive Evaluation 2, 1 (1981).
2. B. A. Auld, F. G. Muennemann, and M. Riaziat, Quantitative modeling of flaw responses in eddy current testing, in: "Research Techniques in Nondestructive Testing, Vol. VII," R. S. Sharpe, ed., Academic Press, London (1984).
3. B. A. Auld, S. Jefferies, J. C. Moulder, and J. C. Gerlitz, Semi-elliptical surface flaw EC interaction and inversion: theory, in: "Review of Progress in Quantitative Nondestructive Evaluation 5A," D. O. Thompson and D. E. Chimenti, eds., Plenum Press, New York (1986).
4. E. Smith, Application of uniform field eddy current technique to 3-D EDM notches and fatigue cracks, in: "Review of Progress in Quantitative Nondestructive Evaluation 5A," D. O. Thompson and D. E. Chimenti, eds., Plenum Press, New York (1986).
5. P. J. Shull, T. E. Capobianco, and J. C. Moulder, Design and characterization of uniform field eddy current probes, these proceedings.
6. C. V. Dodd and W. E. Deeds, J. Appl. Phys. 39, 2829 (1968).
7. S. K. Burke, J. Phys. D: Appl. Phys. 19, 1159 (1986).
8. B. A. Auld, Stanford University, Stanford, CA 94305, private communication.
9. R. Collings, D. Mirshekar-Syahkal, and D. H. Michael, Proc. R. Soc. Lond. A393, 159 (1984).
10. R. Collins, D. H. Michael, D. Mirshekar-Syahkal, and H. G. Pinsent, J. Nondestructive Evaluation 5, 81 (1985).
11. W. D. Rummel, B. K. Christner, and D. L. Long, Comparative responses from cracks and EDM slots as measured by eddy current methods, these proceedings.
12. B. A. Auld, G. McFetridge, M. Riaziat, and S. Jefferies, Improved probe-flaw interaction modeling, inversion processing, and surface roughness clutter, in: "Review of Progress in Quantitative Nondestructive Evaluation 4A," D. O Thompson and D. E. Chimenti, eds., Plenum Press, New York (1985).

INVERSION OF EDDY CURRENT DATA AND THE RECONSTRUCTION

OF FLAWS, PART 1: ACQUISITION OF DATA

Jeff C. Treece, John M. Drynan, John A. Nyenhuis, and Charles D. Beaman

Purdue University
School of Electrical Engineering
West Lafayette, IN 47907

INTRODUCTION

Measuring the eddy currents in a material induced by an exciting field can provide useful information about the shape of the material. Several methods of nondestructive evaluation using eddy currents do not utilize a uniform exciting field over the area of interest [1]. When a non-uniform exciting field is used, the presence or absence of a flaw in the material is detected. However, some applications require more specific information about the size and shape of the flaw. If reconstruction of the flaw is required, current mathematical algorithms [2,4] require that the magnetic field due to eddy currents induced by a uniform exciting field be accurately measured. The magnetic fields can be measured by placing small inductive pickup coils in the vicinity of the material. Several different frequencies can be used to take advantage of the skin depth effect in conductors. Low frequencies can be used to look for flaws relatively deep beneath the surface; high frequencies can be used to look for "shallow" flaws.

In this project our goal is to develop a system to gather data for quantitative reconstruction of flaws in stainless steel tubing. Our sets of data are to be used by Sabbagh Associates for the the purpose of high resolution inversion of the flaws and calibration of their computer model. A number of sensors are needed to span the inner perimeter of the tube. Sensors must be oriented to measure primarily the axial (parallel to tube axis) and radial (perpendicular to tube axis) magnetic field components [3]. The computer model requires that twenty five frequencies be used. A sensor spacing of approximately 0.03" in a 0.8" ID tube is required.

EXPERIMENTAL PROCEDURE

Refer to Figure 1 for a diagram of the probe assembly. A solenoidal exciting coil of approximately 4" length and 0.4" diameter was used to produce an electric field which was uniform and in the axial direction near the sensors. Eight small

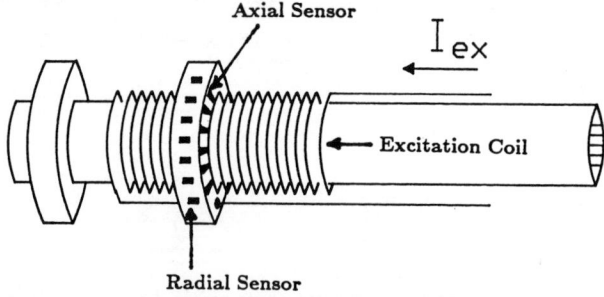

Fig. 1 Probe assembly for axial and radial field measurements.

sensors fabricated using a photolithographic technique were encapsulated around the perimeter of the exciting coil. Axial and radial magnetic fields were measured by suitable orientation of the sensors around the exciting coil.

Multi-channel signal conditioning circuitry was developed to amplify and measure the signals from several sensors simultaneously. Phase-sensitive detection was used to perform phase and magnitude measurement. Signals from the sensors were first preamplified, then fed into the phase detector circuitry. The phase detector produced two DC outputs, one which was proportional to the in-phase field component, and one which was proportional to the quadrature component. The two DC outputs were measured by analog-to-digital converters.

A lab computer was used to control the data acquisition by stepping the function generator through the different frequencies, selecting the proper gain setting on the mixer circuit, positioning the probe by turning a stepping motor, and recording the A/D converter voltages. The set of data was stored on floppy disk and later transferred to a larger computer for analysis.

ELECTRONICS AND SENSORS

The computer model requires using 25 frequencies in the range of 100KHz to 5MHz. The required spatial resolution limits the sensor dimensions to about 1mm x 1mm. This small sensor size limits the available signal. Amplification must be sufficient to obtain a measurable signal from a flaw anywhere in the wall of the tube. Crosstalk and other noise must be kept small enough not to interfere with the measurements.

Phase shift in the electronics must be characterized so that the data can be properly rotated to extract true in-phase and quadrature information. The computer model developed by Sabbagh Associates utilizes the in-phase, as well as the quadrature, information. A small phase error can cause significant measurement error in the smaller of the two signals.

To improve noise performance, we preamplied the signal from the inductive sensors before passing the signals to the phase detecting electronics. The preamplifier, based on the MC1733 video amplifier, gave a gain of about 40dB. The phase detector was a circuit based on the MC1596 balanced modulator IC. The overall gain of the system (peak-to-peak input to DC output) was selectable between 88dB and 108dB. Mixing signals were the voltage driving the exciting coil and the trigger signal from the signal generator. This trigger signal was 90 degrees out of phase with the exciting coil voltage. A lowpass filter at the output of the MC1596 extracted the phase information.

The sensors we used were small inductive pickup coils. Ideally such a coil is perfectly flat so that there is no pickup of orthogonal fields. The first sensors we used were hand wound from magnet wire. These sensors had a very low source impedance, so there was negligible signal loss at the preamplifier input and thermal noise was low. It was very difficult to control the geometry of the hand wound coils. The hand wound coils also were larger than the spatial resolution desired. The next coils we used were fabricated using a photolithographic technique. These sensors had a consistent geometry and were flat, and had a source impedance of approximately 100 Ohms.

Sensors oriented to sense the axial field pick up a large background field resulting from magnetic coupling to the exciting coil. This background signal limits the dynamic range of the measured signal by causing the amplifier to saturate at a low value of the exciting field strength. This problem can be overcome by using a bridge arrangement to subtract the background signal from the sensor before amplification. A large "background coil" is used to sense the average axial field inside of the tube. The background signal is subtracted from the sensor signal at the input to a differential amplifier.

Figure 2 presents a block diagram of our eight-sensor data acquisition system. We accomplish the spatial resolution by placing eight sensors on the probe and making three runs of data acquisition. Successive runs have slightly different azimuthal rotation angles. The data from these successive runs are used to obtain measurements spaced on a uniform grid near the flaw. The eight signals from the sensors are fed into a preamplifier box. Eight coaxial cables connect the preamplifiers to the phase detectors. Each of the eight phase detectors is built on a printed circuit board, and has an amplifier, an in-phase mixer, and a quadrature mixer. Each board has two gain settings, selectable by computer. A phase detector board produces two DC outputs. The sixteen DC outputs from the eight mixer boards are read by an A/D converter. A program was written to control all aspects of taking data. A photograph of the experimental setup appears in Figure 3.

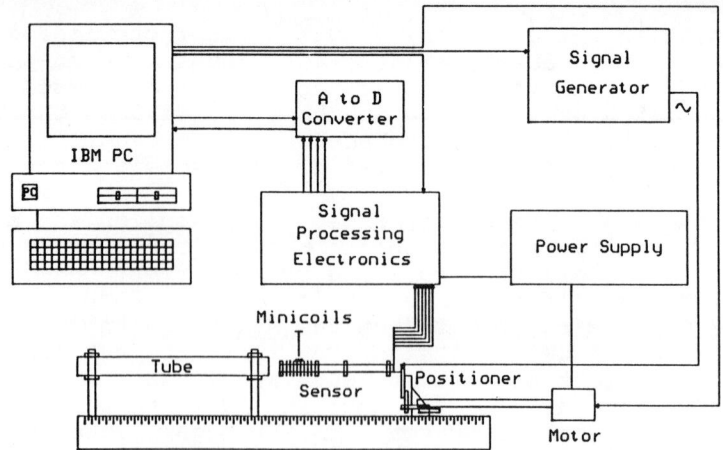

Fig. 2   Block diagram of data acquisition system showing: lab computer, function generator, electronics, and positioning motor.

Fig. 3   Photograph of experimental apparatus showing tube mounted on optical rail, electronics, probe, and lab computer.

EXPERIMENTAL RESULTS

We used our system to obtain axial and radial field data from several electro-discharge-machined (EDM) flaws in the 0.8" ID tube ranging in dimensions from 10 mils to 200 mils. The peak-to-peak flaw signal induced in the sensor ranged from about 25 nanovolts to a few microvolts. This resulted in DC signal outputs in the range of about 10 mV to 10 V. We present some of this data, and compare the data with calculations from a computer model developed by Sabbagh Associates.

Figure 4 shows a small flaw on the inner wall of the tube with dimensions 0.203" long by 0.019" deep by 0.022" wide. In Figure 5 we present our measured data for a frequency of 500 KHz. Figure 6 shows the corresponding computer model data.

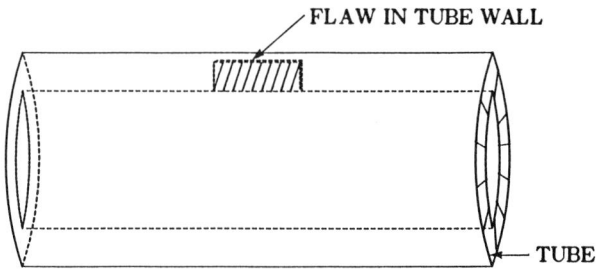

Fig. 4  Test flaw on inner surface of tube wall.

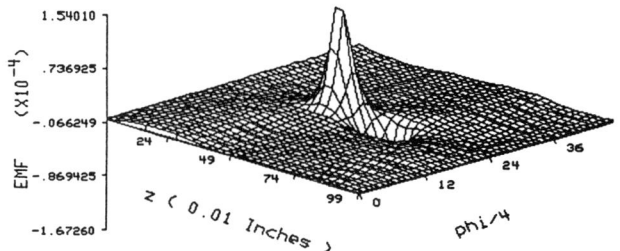

Fig. 5  Experimental data obtained from flaw of Figure 4.  The plot represents the EMF induced in a radial sensor at 500 KHz, that is 90 degrees out of phase with the exciting coil current, beneath a tube surface spanning 1 inch in the axial direction by 192 degrees in the azimuthal direction.

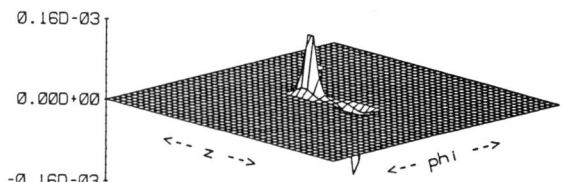

Fig. 6  Computer model calculations predicting the observed EMFs in Figure 5.  The computer calculations represent the theoretical EMF induced at 500 KHz beneath a surface of tube spanning 1 inch in the axial direction by 360 degrees in the azimuthal direction.

The data plot represents the quadrature field measured beneath a section of tube 1 inch long in the axial (z) direction, and spanning 192 degrees in the azimuthal (phi) direction.  We have normalized the DC outputs to correspond with the EMF calculated with the computer model.  We notice a very close qualitative agreement; the experimental and the model data plots both exhibit the same characteristics.  The signal magnitude is also very close to the expected; thus there is at least some quantitative agreement.

Measurements were also made using axial field sensors. The flaw used for the axial experiment measured 0.200" long by

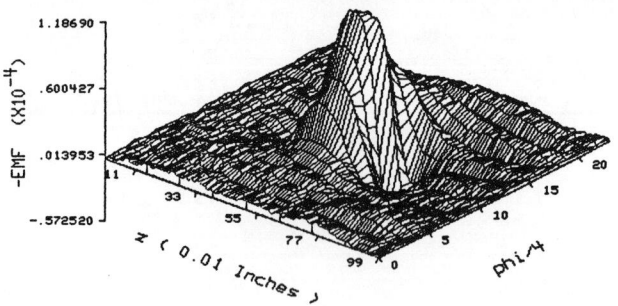

(a) Unimproved axial sensor.

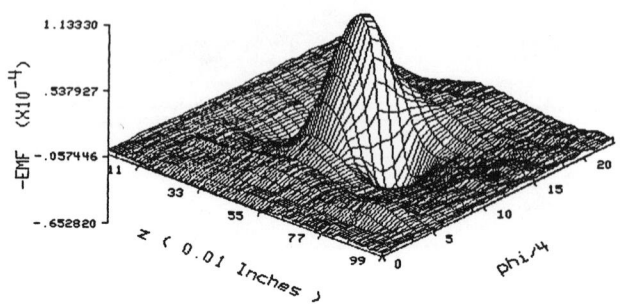

(b) Axial sensor using background reduction.

Fig. 7  Quadrature axial data at 500KHz spanning 1 inch by 96 degrees.

0.039" deep by 0.020" wide. Two sets of measurements were made; one using the background coil method discussed above, and one using the unimproved axial signal. Two plots of the axial magnetic field that is 90 degrees out of phase with the exciting coil current at 500 KHz appear in Figure 7. Figure 7(a) was obtained using the unimproved axial sensor, and Figure 7(b) was obtained using the background-reducing coil. The improvement in signal-to-noise ratio is noticeable in the Figure 7(b). The axial measurements presented here agree qualitatively with the computer model calculations.

CONCLUSIONS

Our goal of flaw inversion appears to be very feasible based on the qualitative and quantitative agreement between measured data and computer model data. Further development of the inversion algorithm by Sabbagh Associates is expected to result in high resolution inversion of the flaws. Measurement of the axial signal was sufficiently accomplished by using a bridge circuit to subtract the background signal; this technique improved the signal-to-noise ratio. We noticed that the flaw signal from the radial sensor was, in general, about two to three times as large as the corresponding (for the same exciting field) axial signal. This is partly due to the fact that the sensor is much closer to the wall of the tube [3]. Flaw fields were successfully measured at different frequencies by small sensors in the presence of a uniform exciting field.

Several improvements to the system would probably improve performance and result in better data. Better cable shielding and grounding might help eliminate stray RF pickup, resulting in smaller background signals and less noise. It would also be beneficial to somehow eliminate the effect of variation in the tube diameter. Natural variations in the wall of the tube give false signals which might be interpreted as flaws. The voltage from the axial background coil could be used to monitor the exciting field strength. This might provide information about variations in the tube diameter. Perhaps changing the probe design would keep the sensors a more uniform distance from the wall of the tube. It might be possible reduce the effect of tube variations by using a bridge-type sensor arrangement. Also, digital filtering or number processing might improve the signals from the electronics for use by the computer model.

In summary, we have taken eddy current measurements to be used for high-resolution reconstruction of flaws in a stainless steel tube. Magnetic field measurements were made at many frequencies with small sensors in the presence of a practically uniform exciting field. Sensor design and electronics were developed to make the acquisition of large amounts of data feasible. Plots of the magnetic fields were compared with model calculation plots generated by Sabbagh Associates, and were found to show good promise of reconstruction.

ACKNOWLEDGEMENTS

This work was sponsored through Sabbagh Associates, Inc., Bloomington, IN. Work at Sabbagh Associates was supported by Department of Energy Contract No. DE-AC02-83ER80096. We thank Harold and David Sabbagh for providing computer simulations of the magnetic fields.

REFERENCES

1. H.L. Libby, "Introduction to Electromagnetic Nondestructive Test Methods," John Wiley & Sons, Inc., (1971).

2. H.A. Sabbagh and L.D. Sabbagh, IEEE Transactions on Magnetics, Mag-22, 282, (1986).

3. J.A. Nyenhuis, J.C. Treece, and J.M. Drynan, Review of Progress in Quantitative Nondestructive Evaluation, 5A, 241, (1986).

4. L.D. Sabbagh and H.A. Sabbagh, "Inversion of Eddy Current Data and the Reconstruction of Flaws Part 2: Inversion of Data," Presented at the 1986 Review of Progress in Quantitative Nondestructive Evaluation Conference, San Diego, CA.

INVERSION OF EDDY CURRENT DATA AND THE RECONSTRUCTION OF FLAWS

PART 2: INVERSION OF DATA

L. David Sabbagh and Harold A. Sabbagh

Sabbagh Associates, Inc.
2634 Round Hill Lane
Bloomington, IN 47401

INTRODUCTION

Our multifrequency reconstruction algorithm is based on a rigorous electromagnetic model for eddy-current interactions with flaws in conducting tubes. The theoretical model was originally developed for nonferrous (stainless steel) tubes and was published in the IEEE Transactions on Magnetics [1]. The reader is encouraged to read [1] in order to understand the electromagnetic details of the model and algorithm; this will allow us to concentrate on applying the algorithm to laboratory reconstructions.

The electromagnetic analysis leads to certain integral equations which give the components of the perturbed magnetic induction field in terms of the two-dimensional Fourier transforms of the conductivity distribution of the anomalous region at each of $N_r$ layers within the cylinder wall. The objective is to invert these equations, i.e., to determine the conductivity distribution, given the measured induction field as input data. Most of the experimental work has used the radial (r) component of the induction field as the data, but toward the end of the project we began to receive some axial (z) data. The z data has not been inverted yet.

The data is acquired by measuring the perturbed induction field at several frequencies, and then writing down an equation for each frequency. If there are $N_f$ frequencies, then we arrive at a system of $N_f$ equations in $N_r$ unknowns, in each couple (m,h). In the laboratory experiments to be described later, we used 25 frequencies ($N_f$ = 25), starting at 100 kHz and ending at 5 MHz. The number of layers, $N_r$, was equal to ten.

In our experiments we measure the radial component of the induction field as our input data. This introduces a slight complication in the inversion, because the transfer function that corresponds to the radial field has a zero on the axis (n,0) in Fourier space. This makes it impossible to invert the system. Therefore, we divide the transfer function by jh (analytically, of course) and integrate the data with respect to z. This produces a new, but equivalent, system that does allow inversion along the Fourier axis (m,0).

INVERSION VIA CONSTRAINED LEAST SQUARES (CLSP)

The unknown conductivities are normalized in physical space so that $-1 \leq \sigma_k(\phi,z) \leq 0$, where a value of -1 indicates the presence of an anomaly at $(\phi,z)$ and a value of 0 indicates the absence of an anomaly. In transform space this inequality translates into the constraint

$$|\tilde{\sigma}_k(m,h)| \leq |\tilde{\sigma}_k(0,0)|. \tag{1}$$

We define the support of a function as the region in space in which the function takes on nonzero values. In terms of the support the conductivity anomaly satisfies the following constraint in transform space

$$|\tilde{\sigma}_k(0,0)| \leq support. \tag{2}$$

This, together with (1), implies that the unknown vector has a bounded norm

$$\|\tilde{\sigma}(m,h)\| \leq M, \tag{3}$$

where M is known whenever the support can be estimated.

In order to enforce the norm constraint, (3), on the least-squares solution we use a Levenberg-Marquardt parameter, together with a Newton iteration [2-3]. The constraints on each component of the solution vector, as in (2) and (3), can be satisfied by solving a Least Distance Problem [3-4].

EXTRAPOLATION IN THE FREQUENCY DOMAIN (EXFD)

The system matrix is a low-pass filter in the two-dimensional Fourier domain (m,h). This means that the matrix elements go to zero as (m,h) goes to infinity. Therefore, it is impossible (or at least extremely difficult) to invert the system to obtain the solution vector $\tilde{\sigma}_k(m,h)$ for large values of (m,h). It is usually desirable to have these high frequency components of $\tilde{\sigma}_k$ available, because they provide the resolution in the reconstructed conductivity anomaly. Therefore, some form of 'signal extrapolation' in the Fourier domain is necessary to recover the inherent resolution of the reconstructed signal. This is a common problem in deconvolution, and has received much attention in recent years in a variety of areas, ranging from image reconstruction and enhancement to 'super-resolving' radar antennas.

The possibility of signal extrapolation (which in mathematical terminology is called analytic continuation) is fully reported in [5].

EXPERIMENTAL RESULTS

Overview

There are two issues involved in the reporting of our experimental results: (1) how well does the model data agree with the laboratory data, both quantitatively and qualitatively, and (2) given the laboratory data, how well can the flaws be reconstructed.

The laboratory experiments were conducted on a 304 stainless steel tube whose outer diameter measured 7/8 inch, with a wall thickness of 0.048 inch. This is a standard tube for nuclear reactor applications, and has other applications as well. This class of stainless steel is nominally nonmagnetic, and has an electrical conductivity of $1.4 \times 10^6$ S/m. Defects of certain orientations and dimensions were machined into the tube by the process of electro-discharge machining (EDM).

The model and reconstruction algorithm are capable of predicting and utilizing the complete magnetic induction field vector, i.e., the radial (r), azimuthal ($\phi$) and axial (z) components. These components are sensed by appropriately oriented microloop sensors.

In order to test the validity of the model, we generated model data for several flaws, in particular, flaws C and H for the radial sensor and flaw H for the axial sensor. We must remember that it is easy to generate the real and imaginary components of the model data at the various frequencies; however, since it is difficult to determine an exact 90° phase shift experimentally, the real and imaginary laboratory data are subject to phase errors as well as measurements errors.

Data Agreement for Radial Sensor

Since flaw C was used to calibrate the reconstruction technique, Figs. 1a and 1b show how well the model data and the laboratory data agree. We see that confidence about our radial model. By qualitative agreement, we mean that the peaks and valleys occur in the same places and have the same shapes; by quantitative agreement, we mean that the peak to peak values are in good agreement as are the rates of increase and decrease.

The agreement between the model data and the laboratory data for flaw H was similar to that of flaw C. We spot tested this agreement at several frequencies since we didn't need the model data at all 25 frequencies.

Integrated Data

As we mentioned previously, there is a need to integrate the laboratory data to produce a new, but equivalent, system so that we can invert the radial field at (m,0). Figures 2a and 2b show the agreement of the equivalent model data for flaw C and the integrated laboratory data for 300 kHz. One set of data must be multiplied by -1 to get the phase agreement.

Inversion of Radial Data

Since the resolution of the model and the laboratory data are different, we need to process the latter to make it consistent with the model so that the processed data can be presented to the inversion algorithm in the proper manner. The model has the following properties;

1. Its resolution is 1° in $\phi$ and 4 mils in z,

2. It is set up for 360 X 256 points at each of 25 frequencies,

3. The system matrix is even in m and odd in h; thus it is zero whenever h = 0.

The laboratory data has the following properties:

1. Its resolution is 4° in $\phi$ and 10.3125 mils in z,

2. It is set up for 24 X 100 data points at each of 25 frequencies.

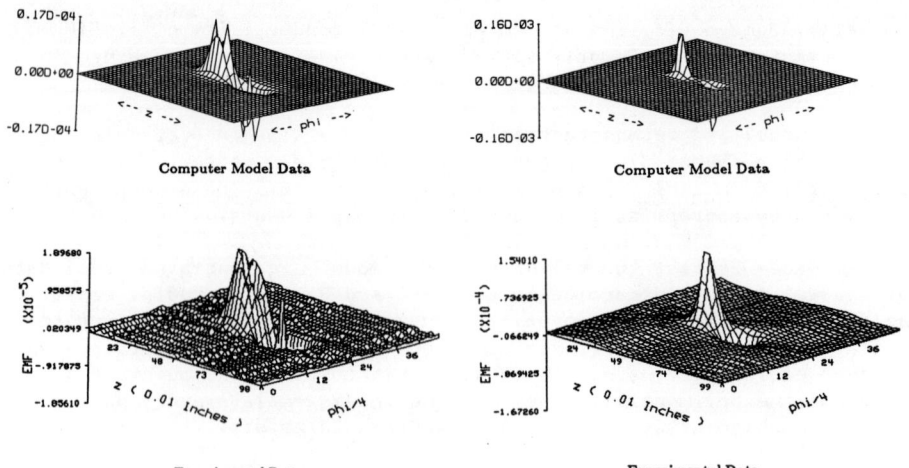

Fig. 1a. Real part of model data and laboratory at 500 kHz.

1b. Imag part of model data and laboratory at 500 kHz.

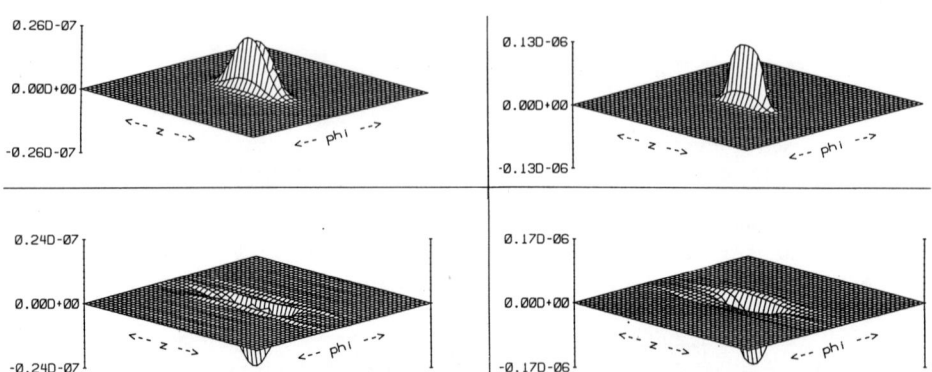

Fig. 2a. Real part of integrated model and laboratory data at 300 kHz.

2b. Imag part of integrated model and laboratory data at 300 kHz.

There are three steps in processing the radial laboratory data to make it consistent with the model. Step 1 is to interpolate from the laboratory resolution to the model resolution. We use a two dimensional quadratic interpolation scheme for this step. Step 2 is to extrapolate so that the laboratory data from step 1 is extended to cover the same range as the model. We use an extrapolation technique that incorporates known information such as the bandwidth (generated from model) and the model data is zero at $(m,0)$, $\forall m$. Step 3 is to integrate (with respect to the z direction) the resulting data from step 2 for the reasons stated before. We use a simple trapezoidal integration scheme.

The reconstruction algorithm consists of two parts as explained previously. The first part is to solve a CLSP for certain pairs $(m,h)$. In these experiments, we considered $\{(m,h): 0 \leq m \leq 35, 0 \leq h \leq 15\}$. The processed laboratory data is then presented to CLSP and the output is presented to the second part algorithm, EXFD. EXFD is an iterative technique that switches between the frequency domain and the real domain and uses the FFT. In these experiments, EXFD was iterated ten times. The output of EXFD is then presented in a graphical manner so that the flaw can be visualized.

Figure 3a gives a gray scale visualization of the original flaw (OFC) at level 1. The gray scale visualization of OFC would be identical at levels 2, 3, and 4. Levels 5--10 should all be blank. Figures 3b, 3c, 3d and 3e give a gray scale visualization of the reconstructed flaw C (RFC) at levels 1, 2, 3 and 4, respectively. The gray scale visualization for RFC at levels 5--10 are all blank.

The gray scale visualization of levels 1--8 for the original flaw H (OFH) is the same as in Figure 3a since OFC and OFH differ only in their depth. Figures 4a, 4b, 4c, 4d and 4e give a gray scale visualization of the reconstructed flaw H (RFH) at levels 1, 2, 3, 4 and 5, respectively. Levels 6--10 are all blank.

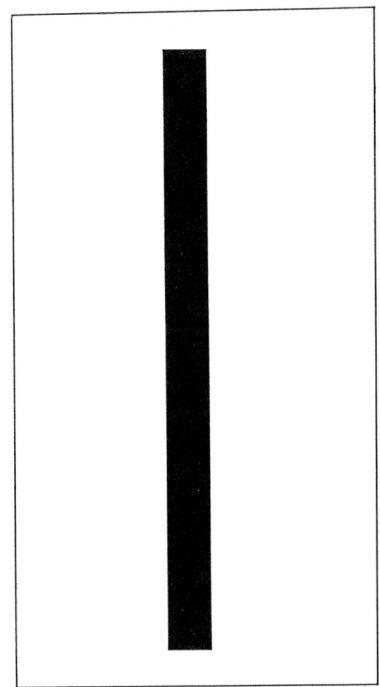
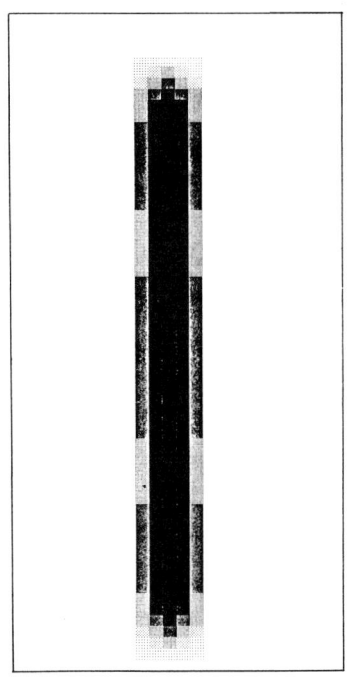

Fig. 3a. Gray scale image for flaw c at levels 1 - 4.

3b. Gray scale image for reconstructed flaw c at level 1.

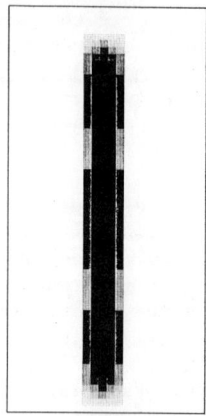 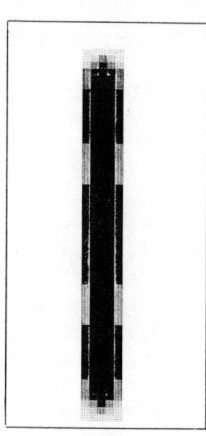

Fig. 3c. Gray scale image for reconstructed flaw c at level 2.

3d. Gray scale image for reconstructed flaw c at level 3.

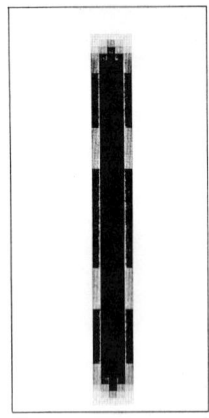 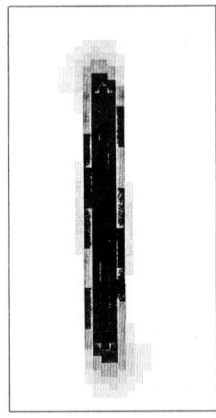

Fig. 3e. Gray scale image for reconstructed flaw c at level 4.

4a. Gray scale image for reconstructed flaw h at level 1.

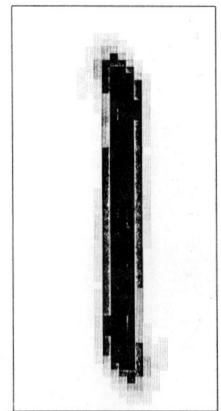 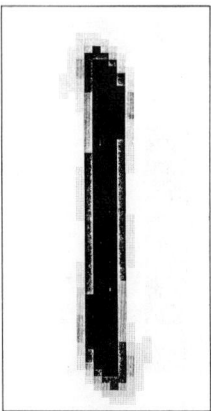

Fig. 4b. Gray scale image for reconstructed flaw h at level 2.

4c. Gray scale image for reconstructed flaw h at level 3.

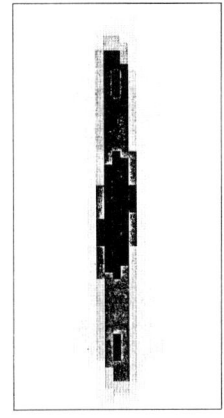 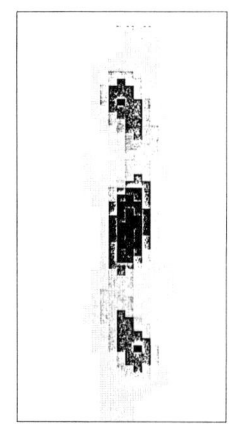

Fig. 4d. Gray scale image for reconstructed flaw h at level 4.   4e. Gray scale image for reconstructed flaw h at level 5.

Since flaw C was used to calibrate the model, we expect the results for its reconstruction to be quite good. It is possible that if we had kept more pairs (m,h) in CLSP and/or used more than 10 iterations in EXFD, the reconstruction of flaw C would have been even more faithful. However, there is a point where the computational results from CLSP are no longer valid due to the mathematical nature of inverse problems. Using more iterations, while time consuming on our conventional computer, would have resulted in a clearer picture.

The results from the reconstruction of flaw H were quite good through five levels. Again, due to the mathematical nature of inverse problems, the ability to get fine resolutions at depths is a difficult problem.

ACKNOWLEDGEMENT

This work was supported by the Department of Energy under Contract No. DE-AC02-83-ER80096 with Sabbagh Associates, Inc.

REFERENCES

1. H. A. Sabbagh and L. D. Sabbagh, "An Eddy-Current Model for Three-Dimensional Inversion", IEEE Transactions on Magnetics, Vol Mag-22, No. 4, (July 1986), 282-291.
2. Gene H. Golub and Charles F. Van Loan, Matrix Computations, (The Johns Hopkins University Press, Baltimore, MD).
3. Charles L. Lawson and Richard J. Hanson, Solving Lease Squares Problems, (Prentice Hall, Inc., Englewood Cliffs, NJ).
4. H. A. Sabbagh and L. D. Sabbagh, "Development of a System to Invert Eddy-Current Data and Reconstruct Flaws: Part 2, Multifrequency Approach", International Advances in Nondestructive Testing, 10, (1984).
5. L. D. Sabbagh, H. A. Sabbagh and J. S. Klopfenstein, "Image Enhancement Via Extrapolation Techniques: A Two Dimensional Iterative Scheme and A Direct Matrix Inversion Scheme", Review of Progress in Quantitative Nondestructive Evaluation 5A, ed. D. O. Thompson and D. E. Chimenti, (Plenum Press, New York, 1986), 473-483.

DISCUSSION

From the Floor: What dictated the choice of 10 levels in the r direction?

Mr. Sabbagh: The tube we were using is 48 mills, and so we chopped them 10 levels, which means each r level is approximately five mills.

From the Floor: I mean, does that choice enter into the compilation itself? I mean, is there a break-even point in subdividing?

Mr. Sabbagh: No. Just depends how fine you want to go in the r direction.
Now, by r equaling 10 levels, that means the least squares problems we solved for (each case) has 10 unknowns. Since we are working with 25 frequencies, I was solving a 25, actually, I made everything real, so that is doubled, thus we were solving a 50 by 10, least squares problem. But, no, it only depends how much resolution you want.

From the Floor: You might consider, say, if you decided to have 10 levels, to concentrate on the number of those close to the even diameter of your tube and make them a little bit more spaced out at the edge.

Mr. Sabbagh: That's high on our priority list of things to do. That's right. I agree.

A PULSED LASER/ELECTROMAGNETIC ACOUSTIC TRANSDUCER APPROACH TO ULTRASONIC SENSOR NEEDS FOR STEEL PROCESSING

G. A. Alers

Magnasonics, Inc.
215 Sierra Drive, SE
Albuquerque, NM 87108

H. N. G. Wadley

Metallurgy Division
National Bureau of Standards
Gaithersburg, MD 20899

INTRODUCTION

Many of the traditional NDE techniques of the past are today being investigated for their potential role as process control sensors for materials processing [1]. Ultrasonics appears one of the most promising because of its ability to penetrate opaque bodies and allow determination of microstructure variables (such as grain size), process variables (such as internal temperature distribution) and detect internal discontinuities (cracks, pores and inclusions). A key problem with the traditional approaches to ultrasonic measurements is the need to contact the body being probed with piezoelectric transducers. These transducers are fragile, require couplants, and fail when exposed to temperatures of more than a few hundred degrees Celsius. Their use during processing may thus require practices that unacceptably interfere with the process.

Pulsed lasers, as a remote ultrasonic generation methodology, have been available for several years. Their use would overcome all the difficulties above if a similar remote ultrasonic receiver were available. The obvious candidate for this would be a laser interferometer, but these are of insufficient sensitivity for applications which involve unpolished, poorly reflecting surfaces. An alternative noncontact (but no longer remote) receiver could be an electromagnetic acoustic transducer (EMAT). These have the potential to be engineered to survive the process environment, and if carefully designed, to match the characteristics of a laser ultrasonic source.

This laser/EMAT approach is being evaluated as part of a collaborative American Iron and Steel Institute (AISI)/NBS sensor development program at NBS. The sensor described here is primarily needed to detect internal discontinuities within steel bodies at temperatures up to 1200°C but numerous other uses could be envisaged including time-of-flight tomography for determining internal temperature distribution.

## LASER GENERATED ULTRASOUND

The generation and subsequent propagation of ultrasound by the absorption of an intense laser pulse is now a relatively well understood process [2]. Research at Hull University and Harwell in England during the early 1980's indicated that two modes of laser generated ultrasound may occur. For low absorbed fluxes, the surface where absorption occurs never exceeds its melting temperature and the source of ultrasound is then a transient dilatation. The stresses associated with this dilatation are for the most part below the elastic limit, and this mode of generation is therefore referred to as thermoelastic. At higher fluxes, the surface temperature rise is capable of exceeding the vaporization temperature. Atoms leave the surface at high velocity imparting a momentum to the substrate which is the source of ultrasound. This mode of generation is referred to as ablation.

It is possible to model the thermoelastic source using heat flow theory. For simple bodies such as isotropic elastic half-spaces and infinite plates, this theoretical description of the source can be coupled with the bodies dynamic elastic Green's tensor to quantitatively predict the temporal waveform of the ultrasonic signal [2,3,4]. The displacement signals at epicenter and various locations on the plate surface are shown in ref. 3. The response of an EMAT is proportional to surface velocity and so the velocity at the longitudinal arrival at epicenter is a positive pulse whose half width is approximately that of the source duration (in the absence of grain scattering/absorption) whilst at the shear arrival it is a negative pulse.

The ablative source is not so amenable to quantitative modelling because we do not know the mass or velocity of material ablated, or how this varies with time during optical absorption. However, if it is assumed that the source is mechanically equivalent to a momentum of arbitrary strength, temporal waveforms of arbitrary amplitude may then readily be evaluated [2]. The principle difference, as seen by an EMAT, to a thermoelastic source is that the epicenter longitudinal arrival is now a bipolar pulse with an initially negative direction. For this source, the epicenter shear arrival remains a unipolar pulse but of opposite sign to that of the thermoelastic source.

In practice, it is not possible to produce pure ablation without accompanying thermoelastic contributions to the source. The relative contributions of each to the emitted ultrasound is yet to be fully resolved, and will vary with optical flux. However, the interesting possibility exists of identifying a range where only longitudinal waves propagate because of cancellation of the shear components due to thermoelastic and ablative sources.

The displacement amplitude of a laser generated ultrasonic signal is proportional to the absorbed optical energy in the thermoelastic region. Thus, the velocity amplitude will be proportional to the absorbed optical power. This should be maximized if we wish to maximize the EMAT signal:noise ratio. At NBS, a Q-switched Nd:YAG laser is available with a 25 ns pulse duration and variable energy per pulse up to 800 mJ (average power of 32 MW). 175 mJ, 1 mm diameter pulses were used for the results reported here.

## EMAT ULTRASONIC RECEIVERS

Although the EMAT is technically a noncontact ultrasonic transducer, it must be held close to the surface before the electromagnetic induction

process can operate with a useful efficiency. For application to
inspections of hot steel, any small air gap at the surface of the metal
acts as a good thermal insulator and greatly reduces the heat transfer
into the transducer structure. In addition to being physically capable of
operating in a high temperature environment, the EMAT has several other
features that can be exploited to advantage. By simply changing the
direction of the magnetic field used, the transducer can be made to detect
either longitudinal or shear waves. Also, by choosing the shape of the
coil of wire inside the EMAT, the angle of incidence of the waves to which
it is sensitive can be controlled. Since the electromagnetic induction
process can be made to operate over a wide range of frequencies, EMATs are
fundamentally broadband devices. Their frequency of operation is usually
set by the electronic receiver circuits used to amplify the electrical
signals. For hot steel, the best frequency of operation is determined
primarily by the laser source spectrum, the ultrasonic attenuation
properties of the steel and only secondarily by the structure of the EMAT
and its associated electronic circuits.

An important problem to be overcome when designing an EMAT for
operation on hot steel is to be sure that the electrical resistivity of
the hot steel is not so high that the electromagnetic skin depth becomes
comparable with the wavelength of the ultrasonic waves involved. For
ordinary metals with resistivities near 10 $\mu$ohm-cm, this criterion is
easily satisfied for frequencies less than 10 MHz. However, hot steel has
a resistivity between 100 and 200 $\mu$ohm-cm so the frequency response may be
attenuated above 1 MHz for longitudinal waves and 0.3 MHz for shear waves.

## EMAT DESIGN

Since an EMAT consists of a coil of wire held close to the surface
of the part being inspected plus a large magnetic field to flood the
region around the coil, two problems must be addressed to enable hot steel
bodies to be interrogated. First, the coil must be constructed with high
melting point wires and a layer of thermal insulation between the coil and
the steel. Second, the magnet must be designed for supplying a large
magnetic field to the total area of the EMAT coil and it must be protected
from the heat generated by the hot steel.

The most direct approach to supplying a magnetic field to the EMAT
coil is to use a large electromagnet with pole pieces designed to
concentrate the magnetic flux at the EMAT coils as shown in Figure 1(a).
Here, the EMAT coils can be located in either a normal or tangential field
direction and hence receive either longitudinal or shear waves. (Position
A in a tangential field is suitable for longitudinal waves while Position
B, in a normal field, is suitable for shear waves.) Such an electromagnet
was constructed but it proved to be quite massive and difficult to move in
a scanning apparatus even though it generated magnetic fields in the 4 to
5 koe range at an EMAT coil with a 1/2 inch diameter. Figures 1(b) and
(c) show magnet configurations based on samarium cobalt permanent magnets
located very close to the EMAT coil itself in order to maximize the fields
at the coil wires. Both of these structures were assembled out of 1/4
inch thick by 1/2 inch square slabs of SmCo and fields in the range of 3
to 4 koe were measured at the EMAT coil positions. These permanent magnet
structures were quite light and manueverable and thus could be easily
scanned over large areas. Their reduction in efficiency caused by
slightly smaller magnetic fields could be tolerated in exchange for this
simplification in mechanical structure.

The EMAT coils were made thermally stable by housing their copper
wires in small (1/16" diameter), four-hole ceramic thermocouple tubes.

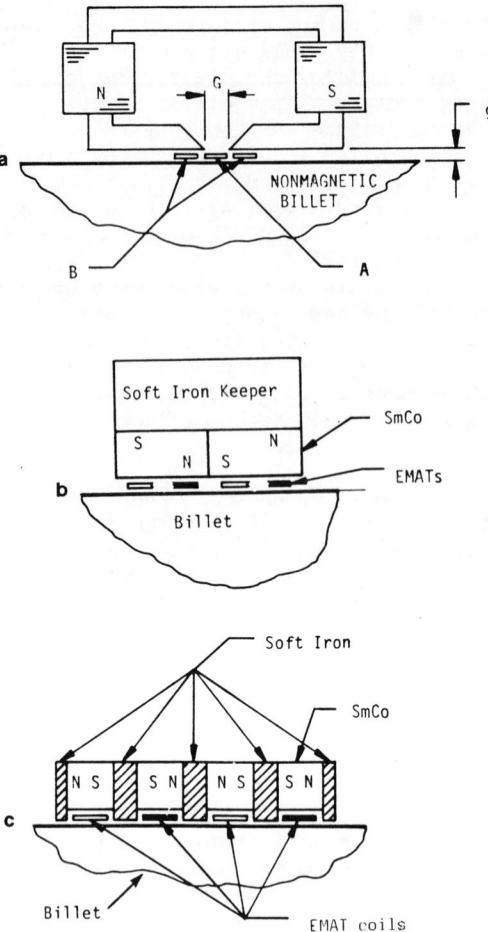

Fig. 1. Magnet configuration needed to operate EMATs. (a) Electromagnet for exciting or detecting either longitudinal or shear waves. (b) Permanent magnet arrangement for exciting or detecting shear waves. (c) Permanent magnet design for excitation or detection of longitudinal waves.

Since the duration of the exposure to high temperature was intended to be short in order to protect the permanent magnets, it was deemed unnecessary to use more expensive, high resistance wires such as platinum for the EMAT coils. By using thermocouple tubes with 1/16 inch diameters, the minimum lift-off or separation between the EMAT coil and the sample was 1/32 inch (0.8 mm). In the regions at the ends of the thermocouple tubes where the copper wires were unprotected, the wires were coated with a high temperature ceramic cement that served to thermally and electrically insulate each wire from its surroundings.

Since it was of interest to make time-of-flight measurements and because the attenuation of high frequency shear waves in hot steel is higher than longitudinal waves [5], the high temperature EMAT designed to detect longitudinal waves propagating directly through the thickness dimension of the steel samples received the most attention. Thus, the comb magnet construction technique shown in Figure 1(c), was used in all the experiments reported here. All the wires in each section of the EMAT coil between the pole pieces conducted current in the same direction but this direction reversed in the adjacent pole piece gap where the magnetic field was also in the opposite direction. In this way, the entire face of the EMAT would respond in phase to a plane longitudinal wave striking the face along its normal direction. This method of assembling the permanent magnet structure allows the pole pieces to conduct heat directly to the samarium cobalt. Thus, the ability of the probe to operate on hot objects depended critically on the length of time it took to heat the permanent magnets to a temperature at which their performance was jeopardized. The manufacturer of the magnets used in these EMATs set the upper temperature limit for his product at 330°C for continuous operation. During the tests on hot steel blocks, a thermocouple on the samarium cobalt showed that it required over 1-1/2 minutes for the temperature of the magnets to exceed 250°C. This 90 second time interval was ample time for collecting all the necessary ultrasonic information. Therefore, it was not necessary to add cumbersome water cooling apparatus at this stage.

RESULTS

In accord with the ablative mechanism of generating ultrasonic waves, very large ultrasonic signals were observed with the longitudinal wave EMAT shown in Figure 1(c) when it was positioned on the face of a sample directly opposite to the point at which the laser beam struck the sample. When a shear wave sensitive EMAT was used (Figure 1(b)), shear waves were detected only when the EMAT was displaced from epicenter along the back side of the sample. As expected from EMAT theory [6], shear waves at an angle of 30 degrees relative to the surface normal were easily detected. One surprising result was the observation of a second large shear wave signal leaving the surface of optical impact at an angle of 60 degrees relative to the surface normal. This may be a manifestation of the thermoelastic contribution to the source.

A very important performance characteristic of the longitudinal wave EMAT used here was the amount of air gap or lift-off that could be tolerated between the sample and the front face of the EMAT. Figure 2 shows how the output signal from the EMAT decreased as a function of this separation distance. This graph shows that the addition of up to 0.02 inches of thermally insulating material between the pole pieces and the hot object would not cause a significant reduction in sensitivitiy but would probably make a dramatic improvement in the ability of the structure to withstand exposure to high temperatures. The fact that the drop in signal strength is described by an exponential function is consistent with the periodic structure of this form of magnet. The rate of drop could be decreased (to achieve higher sensitivity at large air gaps) by using thicker permanent magnets to increase the separation distance between pole pieces.

A second important performance characteristic of the longitudinal wave EMAT used here was the frequency bandwidth that could be achieved with that specific EMAT design and amplifier circuit. Figure 3 shows the waveforms observed when the longitudinal wave EMAT shown in Figure 1(c)

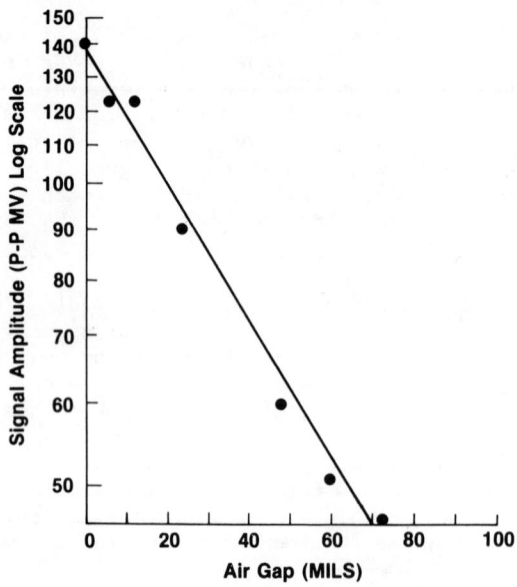

Fig. 2. Dependence of the EMAT receiver probe sensitivity on the thickness of the air gap between the probe and the surface of the sample.

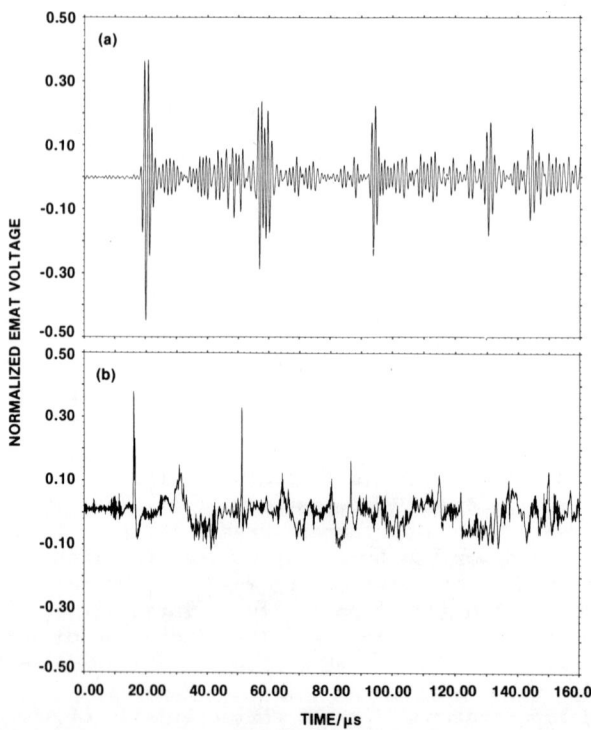

Fig. 3. Ultrasonic longitudinal waveforms observed with different electronic tuning circuits connected to the EMAT coil. (a) Band pass filter combined with a tuning capacitor across the EMAT coil. (b) No tuning.

was used to receive directly transmitted longitudinal waves generated by the pulsed laser. When narrow band filters were used in the amplifier stages, the waveform shown in Figure 3(a) was observed. Note that the noise received prior to the ultrasonic wave arrival is very small and the electrical signal is a tone burst containing many cycles of oscillation. Figure 3(b) shows the effect of removing all the filters and tuning capacitors from across the EMAT coils to achieve a large bandwidth. The noise level prior to the arrival of the acoustic wave is much higher than in the previous cases. However, it appears that the arrival time could be measured to an accuracy greater than 0.1 µsec (or 0.6%).

In order to demonstrate laser excitation and EMAT reception of ultrasonic waves in hot steel, two large steel block samples were heated in a large furnace at 980°C (1800°F). One sample was made of 304 stainless steel and hence underwent no ferromagnetic transition as it cooled off. This sample was a 4-inch diameter cylinder in which the sound wave propagated along the 4-inch long axis of the cylinder. The second sample was a 6-inch by 6-inch square slab of 1018 steel arranged so that the sound wave propagated parallel to the 4-inch thickness dimension. For this sample, a transition from the nonmagnetic to magnetic state would be expected when the temperature passed through the Curie temperature of 770°C (1418°F). These hot blocks were positioned such that the focal point of the laser beam was at the center of the front surface of the sample. The EMAT was then rested lightly against the back surface directly opposite the laser impact point. Thermocouples were held against the sample surface and inserted into the EMAT structure at the location of the samarium cobalt permanent magnets in order to monitor the local temperatures while the steel cooled and the EMAT heated up.

Figure 4 shows examples of the waveforms observed on the 4-inch long, stainless steel cylinder as its surface temperature fell from 752°C to 322°C. The directly transmitted longitudinal wave signal and three reverberations can be easily distinguished from the background noise. Most of this noise is probably from ultrasonic signals reflected by the side walls because the noise is at the EMAT frequency and increases with time after the laser impact. Note that the time interval immediately following the laser pulse and prior to the arrival of the first, longitudinal wave signal is very quiet at all temperatures. As the sample became more cool, the acoustic noise between the reverberation signals increased. This may be explained by mode conversion at the side walls plus a dramatic lowering of the attenuation for shear waves formed by mode conversion as the temperature of the steel became lower.

Figure 5 shows the waveforms observed on the 4-inch thick slab of 1018 steel as the surface temperature dropped from 810°C (1490°F) through the Curie temperature to 618°C (1144°F). At the highest temperature studied, the ultrasonic longitudinal wave signals were very well defined and appear similar to the signals observed on stainless steel. Below the Curie temperature, the magnetic field from the pole pieces of the comb-type EMAT flowed directly into the steel in the immediate vicinity of the pole piece so that the tangential field at the EMAT coil became greatly reduced. Therefore, the sensitivity to longitudinal waves fell and the directly transmitted longitudinal wave which should arrive at about 20 µsec was no longer the dominant signal observed. Instead, a late arriving pair of signals appear which are probably shear waves that have reached the EMAT by reflecting from the side walls of the sample. Since these shear waves have reflected from the side of the sample, they must impinge on the EMAT at an angle and would be detected if a periodic normal magnetic field existed in the EMAT structure. Such a field is actually present in the immediate vicinity of the pole pieces of the comb-type EMAT

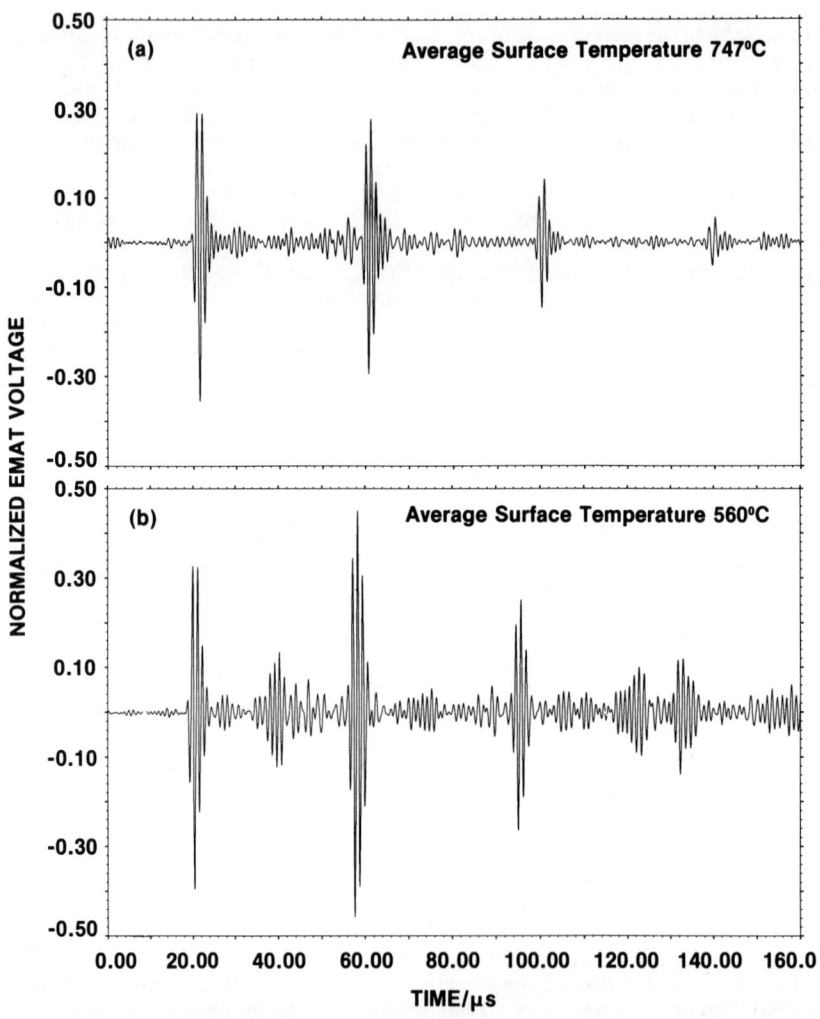

Fig. 4. Waveforms on a 4-inch thick stainless steel block.

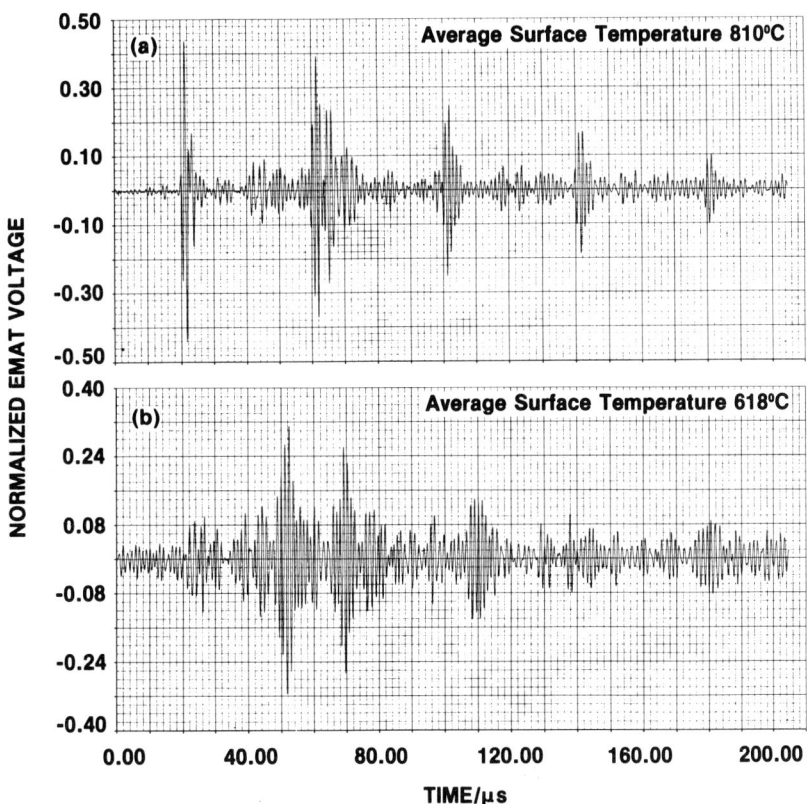

Fig. 5. Waveforms on a 4-inch thick 1018 steel block.

when it is in contact with a sample that is ferromagnetic. An analysis of the geometrical dimensions of the EMAT used in these experiments indicates that the EMAT would be sensitive to shear waves that approach the face of the probe at an angle of about 48 degrees relative to the surface normal. Such an angle is reasonably consistent with the sound wave paths reflected by the side walls of the sample.

CONCLUSIONS

1. A lightweight, easily scanned EMAT receiver probe has been constructed out of heat resistant materials and samarium cobalt permanent magnets to withstand intermittent use at high temperatures. It has been optimized for use with a pulsed laser ultrasonic source.

2. This probe was able to detect 1 MHz ultrasonic pulses generated by a focused 175 millijoule laser pulse with a signal-to-noise ratio of 40:1 (32 dB) on a mild steel block with a surface temperature of 810°C (1490°F). Similar results were also obtained on a stainless steel sample.

3. No degradation of performance was observed after the probe had been repeatedly held in contact with hot steel for time intervals that ranged from 60 to 90 seconds for each contact even though no active water cooling was employed. It now appears relatively straightforward to devise EMAT's for semi-continuous use at temperatures up to 1300°C.

ACKNOWLEDGEMENTS

The help and assistance of F. Mauer and J. Martinez is gratefully acknowledged.

REFERENCES

1. H. N. G. Wadley, Journal of Metals, Oct. 1986.

2. C. B. Scruby, R. J. Dewhurst, D. A. Hutchins, and S. B. Palmer, Research Techniques, in: "Nondestructive Testing V," ed., R. S. Sharpe, p. 281, Academic Press, Oxford (1982).

3. H. N. G. Wadley, C. K. Stockton, J. A. Simmons, M. Rosen, S. D. Ridder, and R. Mehrabian, Review of Progress in QNDE, $\underline{1}$, p. 421, Plemum Press, New York (1982).

4. H. N. G. Wadley, J. A. Simmons, and C. Turner, Review of Progress in QNDE, $\underline{3B}$, p. 683, Plenum Press, New York (1984).

5. E. P. Papadakis, et al., J. Acous. Soc. Am. $\underline{52}$, p. 855 (1972).

6. C. F. Vasile and R. B. Thompson, Ultrasonics Symposium Proceedings, IEEE Cat. No. 77CH1264-ISU, p. 84 (1977).

DISCUSSION

From the Floor:  Have you considered using the new permanent magnet materials to achieve more efficient transducers?

Mr. Alers:  Yes.  The EMATs I described used samarium cobalt magnets already available in our laboratory.  Much better materials with about one and a half times the energy product are not available.

From the Floor:  I found that a neodymium-iron-boron alloy gave about 25 percent more field.  However, this result depends on the composition and the geometry.

Dr. R. B. Thompson:  The neodymium iron alloys have a much lower Curie temperature.

Mr. Alers:  That's right.  In any case, we may have to spray some water on the back.

Dr. C. Fortunko:  I just wanted to make a few comments.  One is that samarium cobalt doesn't go up to 360 degrees.  I think it has a phase transition at about 200 or so degrees centigrade.

Mr. Alers:  The manufacturer's literature advertises continuous operation of some samarium cobalt compositions at 350° C.

From the Floor:  What are the highest frequencies that your EMAT can pick up this wave?

Mr. Alers:  Well, that would have to be determined experimentally, so I can't answer your question.  Personally, I doubt if we can even get 5 MHz through the steel.  Therefore, I think it's academic whether the EMAT will respond to frequencies higher than 5 MHz or not.

Mr. S. Rokhlin:  I understand that the thickness of your sample was roughly three inches.

Mr. Alers:  A four-inch thickness.

Mr. Rokhlin:  What response was obtained?

Mr. Alers:  We observed perfectly good signal-to-noise ratios after three or four bounces, which would correspond to an acoustic path of more than 12 inches in this case.

Mr. Rokhlin:  It was not clear how the lift-off was minimized.  How were you able to fix the gap under the transducer?

Mr. Alers:  The graph of signal versus lift-off indicates that the rate of decrease of signal with separation is about what would be expected for an array of magnets with about one quarter of an inch spacing between the pole pieces.  This agrees with the dimensions on the inside of the EMAT used.  If you made an EMAT with a half inch magnet array its signals would fall by a factor of two in 60 mils of lift off instead of the 30 mils we observed.  Our experience indicates that the normal air gap under an EMAT in light rubbing contact is much, much less than these characteristic dimensions of 30 to 60 mils.

Mr. Bernie Tittmann, Rockwell Science Center: Your radiation patterns on both the transmitter and the receiver are fairly broad, so what does that say about the need to carefully align the transmitter beam with the detector beam? Maybe the problem of scanning is not nearly as severe as one might believe.

Mr. Alers: We are operating with pretty big dimensions, so alignment is not very critical. The real reason we looked into the angular dependence was there may be a lot of energy going off at some angle and we could therefore optimize the receiver for that angle and achieve sensitivity. However, one must be careful not to walk into a big trap of mode conversion and loss of the entire signal.

Mr. Sachse, Cornell: The other day, there was a talk here about a wheel with a transducer built into its center. How do your signals compare to the kind of signals that those people get?

Mr. Alers: They didn't show any pictures of their raw signals which would allow us to compare the signal-to-noise ratios obtained by the two techniques. I think we are getting comparable signal-to-noise ratios. However, they were looking at three quarter inch diameter flat bottom hole reflectors while we were just doing straight transmission. Therefore, there is still a lot to be done to compare the two techniques.

DEVELOPMENT AND COMPARISON OF BEAM MODELS FOR TWO-MEDIA

ULTRASONIC INSPECTION

B. P. Newberry, R. B. Thompson, and E. F. Lopes

Ames Laboratory, USDOE
Iowa State University
Ames, IA  50011

INTRODUCTION

This paper reports on an effort to model the radiation pattern of a submerged ultrasonic transducer exciting a beam which is incident on a liquid-solid interface. The important aspects of this process are the diffraction of the beam as it propagates in the liquid and solid media, focussing of the beam due to a lens at the transducer face and/or the curvature of the interface, and aberrations induced by refraction at the interface.

Ray tracing techniques have commonly been used to describe focussing and aberrations introduced in ultrasonic beams due to refraction and reflection at surfaces [1]. These methods generally ignore beam spreading due to diffraction. If aberrations are neglected, simple formulae are available to predict the effects of diffraction on the axial fields of unfocussed piston sources [2] and the full fields of Gaussian sources [3] after refraction through a planar or cylindrically curved liquid-solid interface. Presented here are two models which treat both diffraction and aberration effects as an ultrasonic beam passes through a planar or cylindrical interface. The approximate Gauss-Hermite model is presented as a working computational tool. The accuracy of its predictions are evaluated by comparison to those of the more exact and computationally intensive Green's function model.

GAUSSIAN-HERMITE MODEL

The Gauss-Hermite (G-H) beam model is based on an expansion of the radiation field in a complete set of orthonormal Gauss-Hermite functions. The use of this type of solution for transducer radiation fields was proposed by Cook and Arnoult [4]. Thompson and Lopes [5] combined this description of the diffraction of a propagating sound beam with a ray tracing model for the refraction of the fields at an interface to produce a hybrid model which accounts for both diffraction and aberrations.

In the fluid, the G-H formalism expresses the velocity potential, $\phi$, as the sum of eigenfunctions

$$\phi(x,y,z) = \sum_{mn} C_{mn} \psi_m(x,z) \psi_n(y,z) e^{j(\omega t - kz)} \qquad (1)$$

where the $C_{mn}$ are constant complex coefficients and the beam is directed along the z-axis. The eigenfunctions, $\psi$, are composed of transversely varying Gaussian exponential and Hermite polynomial factors, along with axially varying amplitude and phase terms (see Ref. [5] for a complete description). These eigenfunctions satisfy a reduced wave equation in which a term of the order $\partial^2\psi/\partial z^2$ has been dropped [4]. This approximation is equivalent to the Fresnel approximation and should be good for the well collimated beams often used in ultrasonic NDE.

The coefficients, $C_{mn}$, can be determined by employing orthogonality relationships [6], provided that the potential $\phi$ is known on some plane, such as the plane containing the transducer face,

$$C_{mn} = e^{-j(\omega t - kz_0)} \int_{-\infty}^{\infty} dx \int_{-\infty}^{\infty} dy \psi_m^*(x,z_0) \psi_n^*(y,z_0) \phi(x,y,z_0). \qquad (2)$$

This integral may be evaluated exactly for a Gaussian distribution in the initial plane, but must be numerically integrated for a piston or other more complicated source. One benefit of this method is that, in principle, any type of source may be treated with equal ease.

Ray Tracing Through interface

Figure 1 illustrates the geometry of the procedure for treating refraction at the liquid-solid interface. Suppose a cylindrically curved liquid-solid interface (bold arc) is illuminated by a transducer whose central ray is given by the central solid line. An incident field in the fluid may be defined, via the transducer radiation pattern, on a surface just before the interface (dashed arc), assuming no influence by the solid. A transmitted plane is defined perpendicular to the refracted central ray of the beam. One would predict a set of virtual fields on this transmitted plane which, were the plane fully embedded in the solid, would produce the actual radiation field in the solid. A ray tracing analysis is used to relate the complex amplitudes of the incident and transmitted virtual fields. The fields on the transmitted plane can be used to generate a new set of coefficients for the G-H functions in order to describe the propagation of the beam into the solid.

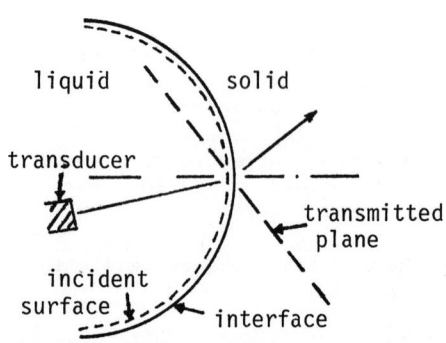

Fig. 1. Geometry of interface transmission computation.

Actually, the scalar G-H solution is not rigorous in the solid media. However, for longitudinal waves, it should be a good approximation to the behavior of scalar elastic displacement potential. For transverse waves the problem is more severe. But, nonetheless, for many cases in which one component of the vector displacement potential is dominant, the scalar solution may be an accurate representation. For example, a well collimated beam passing at a moderate angle through a planar or gently curved interface should be a valid geometry for modeling shear waves.

Numerical Results - Axicon Transducer

Previously, the G-H model has been applied to beams generated by planar, spherically focussed and cylindrically focussed piston and Gaussian transducers. In order to demonstrate the applicability of the G-H model to general types of transducers, some results are presented here for a conically focussed, or axicon, transducer. This type of transducer has received some attention due to an apparent extended depth of focus [7-9]. The geometry of the axicon is shown in Fig. 2. The probe is characterized by a radius, "a", a cone angle, "$\alpha$", and a frequency, "f".

Figures 3a and b show the axial profile of an axicon transducer radiating into water at 2.5 MHz. For this case, $\alpha$ = 5.7° and a = 2.0 cm. The G-H prediction shown in Fig. 3a was obtained with a 65 x 65 term expansion, and is compared with the calculations of Dietz [7], shown in Fig. 3b. In Fig. 3b, the dashed line corresponds to a numerical integration of the Rayleigh diffraction integral and the solid line represents an approximation using the Method of Stationary Phase (MSP). The G-H model agrees well with the numerical result of Dietz, except in the very near field where convergence is sensitive to the number of terms taken in the expansion. The G-H expansion is also seen to be superior to the MSP approximation. Figures 3c,d,e show the full fields of an axicon with f = 2.25 MHz, $\alpha$ = 4.3°, and a = .635 cm. In Fig. 3c, the transducer is radiating into water. In Figs. 3d,e, the fields are shown after passage through a plane water-steel interface for 45° refracted shear waves and 70° refracted longitudinal waves, respectively. Note that for the 45° T wave case, the beam retains a good focal region, whereas for the 70° L case the beam shape has been degraded severely by aberrations.

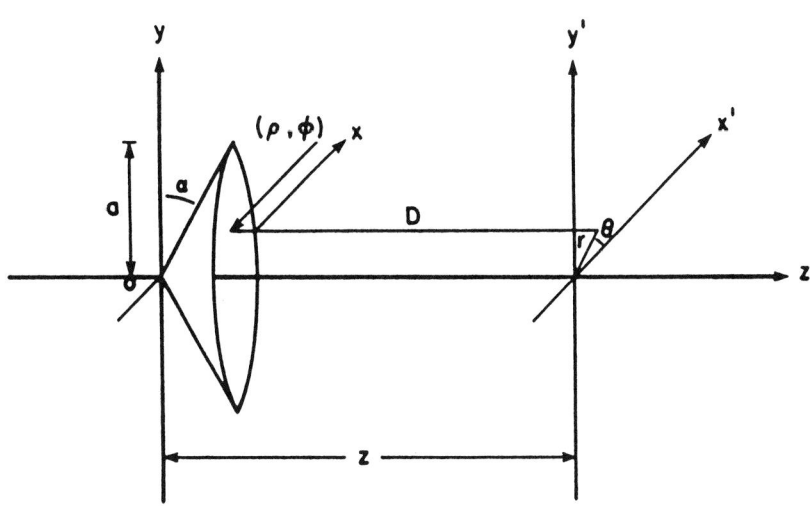

Fig. 2. Geometry of axicon having radius a, cone angle $\alpha$, and frequency f.

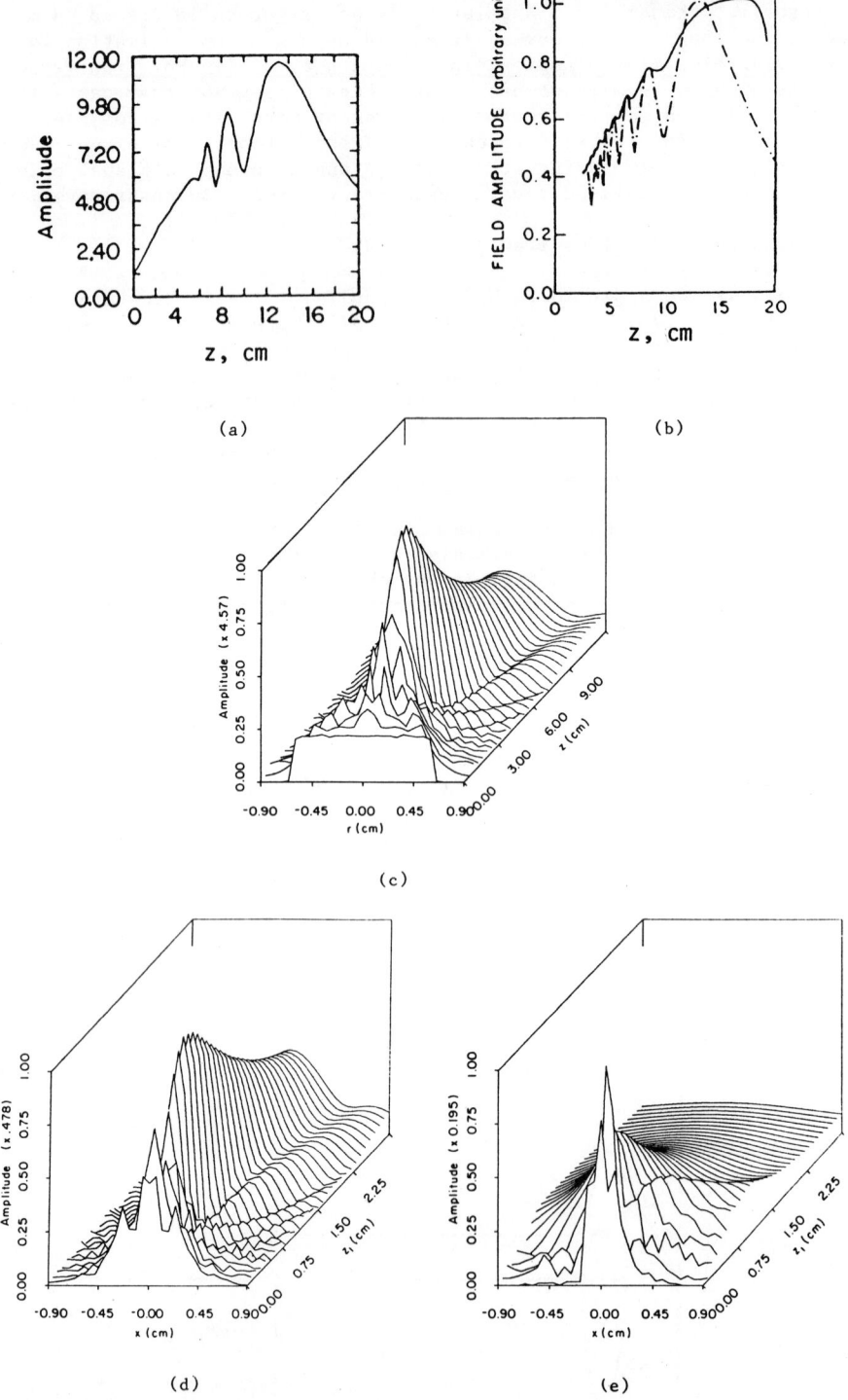

Fig. 3. Axicon Radiation: a) axial profile in water - G-H; b) axial profile in water - Dietz (dashed-numerical calculation, solid - MSP); c) full field in water; d) 45° T wave through water-steel interface ($C_L$ = .596 cm/μs, $C_T$ = .3235 cm/μs, standoff in water = 1.27 cm); e) 70° L wave through water-steel interface (standoff in water = 1.27 cm).

## GREEN'S FUNCTION MODEL

In order to assure the proper application of the Gauss-Hermite beam model, it is necessary to perform either an experimental or theoretical validation of the theory. Experimental data has been compared to the G-H model with favorable results for a variety of different cases [10]. However, there are limitations to this approach due to uncertainties regarding the degree to which the surface motion of actual probes corresponds to the "piston" assumption. Here, a more rigorous theoretical approach has been developed in order to provide a touchstone against which the G-H model may be evaluated.

Consider the semi-infinite volume, V, depicted in the left half of Fig. 4, which has a bounding surface, S. It is known that the time-harmonic displacement at a point in V can be represented as an integral over S of the form [11,12].

$$u_m(x) = - \int_S [\tau^G_{ijm}(x-X)u_i(X) - u^G_{im}(x-X)\tau_{ij}(X)] n_j dA(X) \quad (3)$$

where standard tensor index notation has been employed. In Eq. (3), $u_i$ and $\tau_{ij}$ are the displacement and stress conditions at the surface S, $n_j$ is the outward normal and $u^G_{im}$ is the free-space Green's displacement tensor given by

$$u^G_{im} = (\rho\omega^2)^{-1}[-G_L(R) + G_T(R)]_{,im} + \mu^{-1} G_T(R) \delta_{im} \quad (4)$$

where

$$G_\beta(R) = (1/4\pi R)e^{ik_\beta R}, \quad (\beta = L, T) \quad (5)$$

$$R = |x-X|. \quad (6)$$

The L and T subscripts denote properties of longitudinal and shear waves, respectively. The corresponding Green's stress tensor, $\tau^G_{ijm}$, is obtained by substitution of Eq. (4) into Hooke's Law. From the physical point-of-view, $u^G_{im}$ and $\tau^G_{ijm}$ may be thought of as the displacement and stress fields that would be radiated to the field point x by a localized body force $f_m$ applied at the source point X, as sketched on the right hand side of Fig. 4.

If the surface S is considered to be a liquid-solid interface, then Eq. (13) provides a formal solution for the field in the solid provided that the boundary fields, $U_i$ and $\tau_{ij}$, on the solid surface due to a transducer beam illumination can be determined. The approach taken by the Green's function (GF) model then, is to determine these boundary fields by an appropriate approximation and evaluate the integral numerically.

The determination of the boundary fields has been based on the use of the G-H beam model, which is rigorously correct, within the Fresnel approximation, to describe the radiation pattern in the fluid. However, any appropriate transducer radiation model will work as well. First, the incident field is determined on the fluid side of the interface in the same manner as depicted in Fig. 1. Then, to calculate the actual

fields at each point on the surface, the transducer beam is assumed to
be locally a plane wave and the surface is assumed to be locally planar.
Thus, everywhere on the surface the continuity of stress and displacement
can be approximately introduced through the classical theory of plane
wave incidence on a plane boundary between two media. The boundary fields
are equal to those of the transmitted wave in this form of the Kirchhoff
approximation. As the interval in the numerical integration procedure
is reduced, the error in these approximations is decreased. One drawback
to this method is that a small integration step is required to resolve
the phase variations over the surface. The computation time is therefore
quite lengthy.

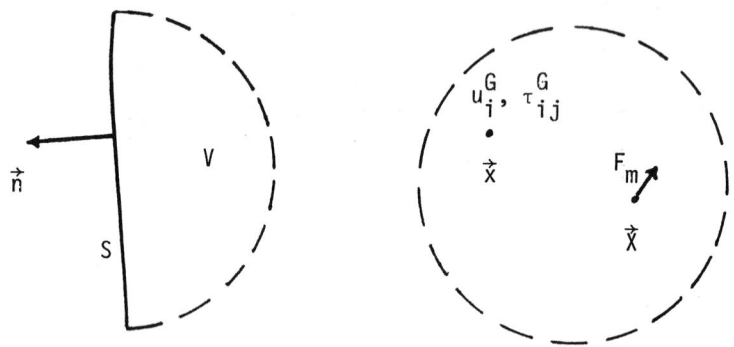

Fig. 4. Schematic geometry of Green's function formulation.

NUMERICAL RESULTS

Results of the Green's function model are compared here with the
predictions of the Gauss-Hermite model for an unfocussed piston illuminating a cylindrical interface. For this case, the transducer radius
"a" is .635 cm, the frequency "f" is 5 MHz, the radius of curvature of
the interface "R" is 7.62 cm, and the standoff in water "$z_o$" is for all
cases 5.6 cm. The material properties used are those for water and fused
silica: $V_{water}$ = .15 cm/µs, $V_L$ = .597 cm/µs, $V_T$ = .376 cm/µs, and $\rho_{solid}$=
2.2 gm/cm$^3$. All results shown are for longitudinal waves.

Figure 5 shows the axial radiation fields for refracted angles of
0, 5, 15 and 30 degrees, where the refracted angle is that of the beam
axis. The amplitude has been normalized to that at the transducer face,
and has been divided by the transmission coefficient of the central ray.
In Fig. 5a, the GF prediction is seen to have a slightly lower amplitude
in the focal region than does the G-H profile. However, a case was run
(dashed line) in which the Fresnel approximation was applied to the spherical
wave functions of the GF code. The result agrees well with the G-H theory,
suggesting that the difference between the G-H and GF models is due to

the presence of the Fresnel approximation in the G-H model. The remaining plots demonstrate the good agreement of the G-H model with the GF predictions at higher angles.

In Fig. 6, two profiles are shown which are taken transverse to the central axis in a plane perpendicular to the cylindrical axis of the interface. The first is at normal incidence and the second is at a refracted angle of 30°. For both cases the profile is taken at a distance in the solid, $z_1$, of 2.5 cm. At normal incidence this corresponds to the geometrical focal point of the interface. The agreement is seen to be excellent at normal incidence and at 30° the G-H model approximates the structure of the profile very well.

SUMMARY

The Gauss-Hermite model for transducer radiation through an interface has been shown to be a versatile tool for modeling various types of transducers. Furthermore, when compared to a more rigorous model, it has proven to be remarkably accurate for the cases studied. The Green's function model will be very useful in further evaluation of the G-H and other models for more extreme cases and for shear wave cases.

ACKNOWLEDGEMENT

This work was sponsored by the Electric Power Research Institute under research project RP2687-1.

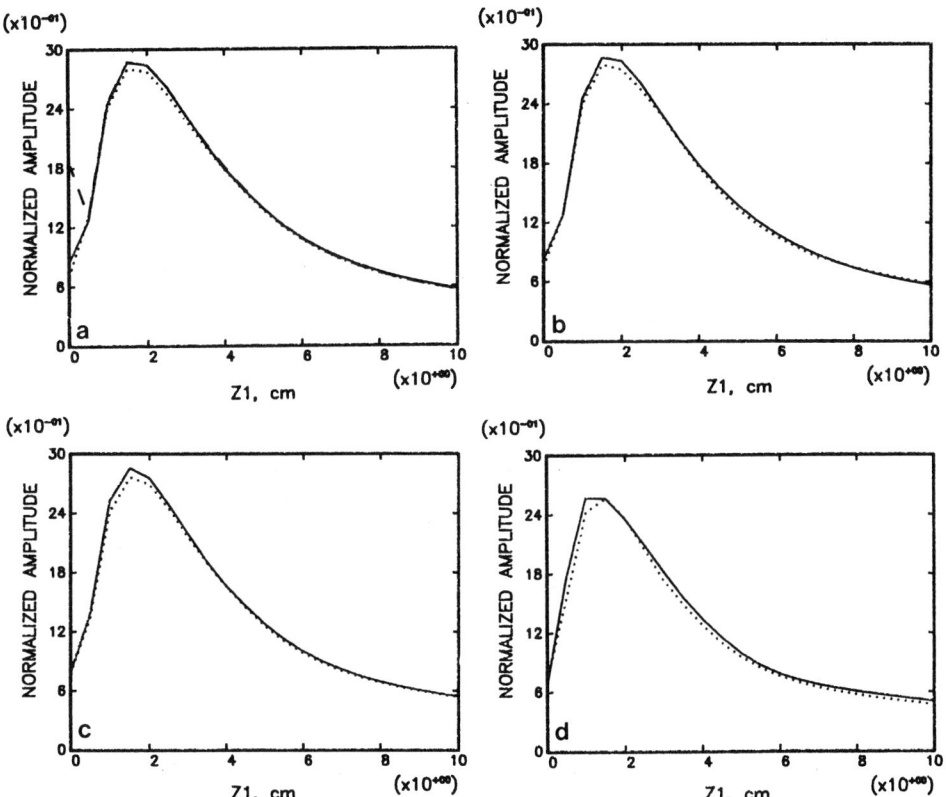

Fig. 5. Axial profiles in solid (solid-G-H, dotted-GF, dashed-GF w/Fresnel approx.) a) $\theta_1=0°$, b) $\theta_1=5°$, c) $\theta_1=15°$, d) $\theta_1=30°$.

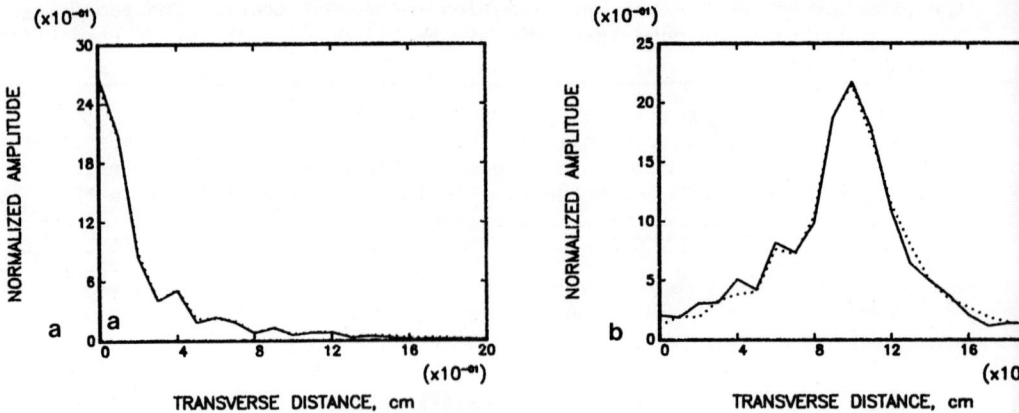

Fig. 6. Transverse profiles in solid (solid-GH, dotted-GF) a) $\theta_1=0°$, b) $\theta_1 = 30°$.

REFERENCES

1. C. Fiedler, S. Meng, and L. Adler, Review of Progress in Quantitative NDE, 5, D. O. Thompson and D. E. Chimenti, Eds., (Plenum Press, NY, 1986), p. 1697.
2. R. B. Thompson and T. A. Gray, Review of Progress in Quantitative NDE, 2, D. O. Thompson and D. E. Chimenti, Eds., (Plenum Press, NY, 1983), p. 567.
3. R. B. Thompson and E. F. Lopes, J. Nondestr. Eval., 4, 107 (1984).
4. B. D. Cook and W. J. Arnoult, J. Acoust. Soc. Am., 59, 9 (1976).
5. R. B. Thompson and E. F. Lopes, Review of Progress in Quantitative NDE, 5, D. O. Thompson and D. E. Chimenti, eds., (Plenum Press, NY, 1986), p. 117.
6. M. Abramowitz and I. A. Stegun, Handbook of Mathematical Functions, (U.S. Dept. of Commerce, Washington, DC, 1964), Ch. 22.
7. D. K. Dietz, IEEE Trans. Sonics & Ultrasonics, SU-29, 128 (1982).
8. M. S. Patterson and F. S. Foster, ibid, 83.
9. C. B. Burckhardt, H. Hoffmann, and P. A. Grandchamp, J. Acoust. Soc. Am., 54, 1628 (1973).
10. B. P. Newberry, T. A. Gray, E. F. Lopes, and R. B. Thompson, Review of Progress in Quantitative NDE, 5, D. O. Thompson and D. E. Chimenti, Eds., (Plenum Press, NY, 1986), p. 127.
11. J. E. Gubernatis, E. Domany and J. A. Krumhansl, J. Appl. Phys., 48, 2804 (1977).
12. J. D. Achenbach, K. Viswanathan, and A. Norris, Wave Motion, 1, 299 (1979).

DISCUSSION

From the Floor: Your Green's Functions were vector functions; is that right?

Mr. Newberry: Tensor functions.

From the Floor: How did you make the comparison to the scalar? What components did you pick? Did you pick the magnitude of the tensor?

Mr. Newberry: The result of the Green's Function model is a vector displacement in the solid, and the result of the Gauss-Hermite is a scalar displacement potential. We made an approximation of quasi-plane waves so we can deduce the displacement from the displacement potential.

From the Floor: So you calculated the displacement potential from your vector?

Mr. Newberry: We calculated a displacement from the displacement potential in the Gauss-Hermite model.

Mr. R. L. Ludwig, Colorado State University: Did you look at your beam model, where you tried to model a liquid-solid interface -- at the possibility of using a liquid as a first approximation to a couplant?

Mr. Newberry: I'm not sure I understand the question.

Mr. Ludwig: Well, what I had in mind was to use a fairly small layer of liquid so that you can actually use it as a couplant, as the first approximation for a couplant. Is this a possibility? Can you account for it?

Mr. Newberry: I think eventually we can. We (haven't reached) that problem where we just have a thin layer. We hope to extend this method to using wedges which would have a couplant in between. We haven't done that yet, no.

Mr. Ludwig: Thanks.

Mr. Fraser: Thank you.

# TRANSDUCER MODELS FOR THE FINITE ELEMENT SIMULATION
# OF ULTRASONIC NDT PHENOMENA

R. L. Ludwig, D. Moore and W. Lord

Electrical Engineering Department
Colorado State University
Fort Collins, Colorado 80523

## INTRODUCTION

Numerical studies of realistic ultrasonic NDT situations in their most general form assume arbitrarily shaped defects excited by pulsed transducer waves all subject to the specimen's external boundaries. The combination of three features, transmitter model, ultrasound/defect interaction, and receiver model makes a thorough quantitative evaluation extremely difficult. Although a numerical finite element code capable of handling ultrasonic wave propagation and scattering has been developed [1,2] the overall system remains incomplete without apropriate transducer models.

Reviewing some of the existing transducer models, such as the layered device structure [3,4] and the equivalent electronic circuits by Mason and others [5], one observes that all these models are one-dimensional approximations of three dimensional transducer behavior. Diffraction losses resulting in ultrasonic beam spread, nonuniform field variation across the aperture and, if the transducer is directly contacted to a solid, mode conversion are not included. In this paper a simple approach to incorporating the transducers within the finite element formulation is investigated.

## TRANSMITTER MODEL

The basis for the transmitter model is the homogeneous elastic wave equation

$$(\lambda+\mu) \nabla \cdot \bar{u} + \mu \nabla^2 \bar{u} - \rho \ddot{\bar{u}} = 0 \tag{1}$$

To incorporate the transmitter excitation over a finite aperture a displacement vector is chosen such that

$$\bar{u}(x_1 y_1 t) = \delta(x) w(y) f(t-\phi(y)) \hat{x} \tag{2}$$

where $\delta(x)$ ensures application of the displacement vector on the surface of the specimen, $w(y)$ specifies a window function, which for the present study is assumed to be rectangular, $f(t-\phi(y))$ accounts for the pulsed excitation function with the phase delay $\varphi(y)$ allowing for beam steering. It should be emphasized that f(t) can be any realistic transducer function, but for the present purpose the following expression is chosen

$$f(t) = (1-\cos\omega_0 t)\cos 3\omega_0 t \tag{3}$$

with $\omega_0 = 2\pi \times 10^6 s^{-1}$. Incorporating (3) in a two dimensional, plane strain, finite element model allows the simulation of a line source, $w(y)=\delta(y)$, acting on a half space with an otherwise stress-free boundary condition $(\bar{T} \cdot n = 0)$. The wave-front construction of the line source response can be obtained by displaying the displacement vector as a function of time. This is shown for the x component in Fig. 1. Adding several line sources over a finite aperture permits the simulation of a strip-like transducer based on the assumption that precise elastic, electric field coupling phenomena as well as couplant variation is neglected. The resulting x-components of the displacement vector are given in Fig. 2. Comparing Fig. 1 with Fig. 2 reveals the focusing effect of the finite aperture.

RECEIVER MODEL

In order to obtain signals that are directly comparable to practical A-scan measurements, a receiver model has been developed assuming the form

$$S(t) = \frac{1}{2N}\{a_1 \sum_{i=1}^{N} b_i u_{xi}(t) + a_2 \sum_{i=1}^{N} c_i u_{yi}(t)\} \tag{4}$$

where

$a_1 = 1$:  Couplant variation for longitudinal components

$a_2 = 2$:  Couplant variation for shear components

$b_i, c_i = 1$:  Amplitude factors for shear and longitudinal components

$N = 11$:  Number of nodal points forming the aperture.

Applying (4) to model a shear receiver, an aluminum block with a fixed L-wave transmitter was used as shown in Figure 3. The receiver was scanned across the back wall in increments of 0.05 inches as indicated in Figure 3. The combined signals, as predicted by (4) and the finite element code are shown in Figure 4. For illustration purposes the incident L-wave, as produced by the fixed transmitter over the finite aperture, is also included. In this display one can observe the change in amplitude both for the incoming L-wave and S-wave. For three distinct receiver locations a comparison between numerical predictions and practical measurements is shown in Figure 5. Comparing the predictions with the experiments (top line) shows good qualitative agreement. However, for more complex situations such as scattering or diffraction by defects this agreement can generally not be expected owing to the two dimensional nature of the numerical model.

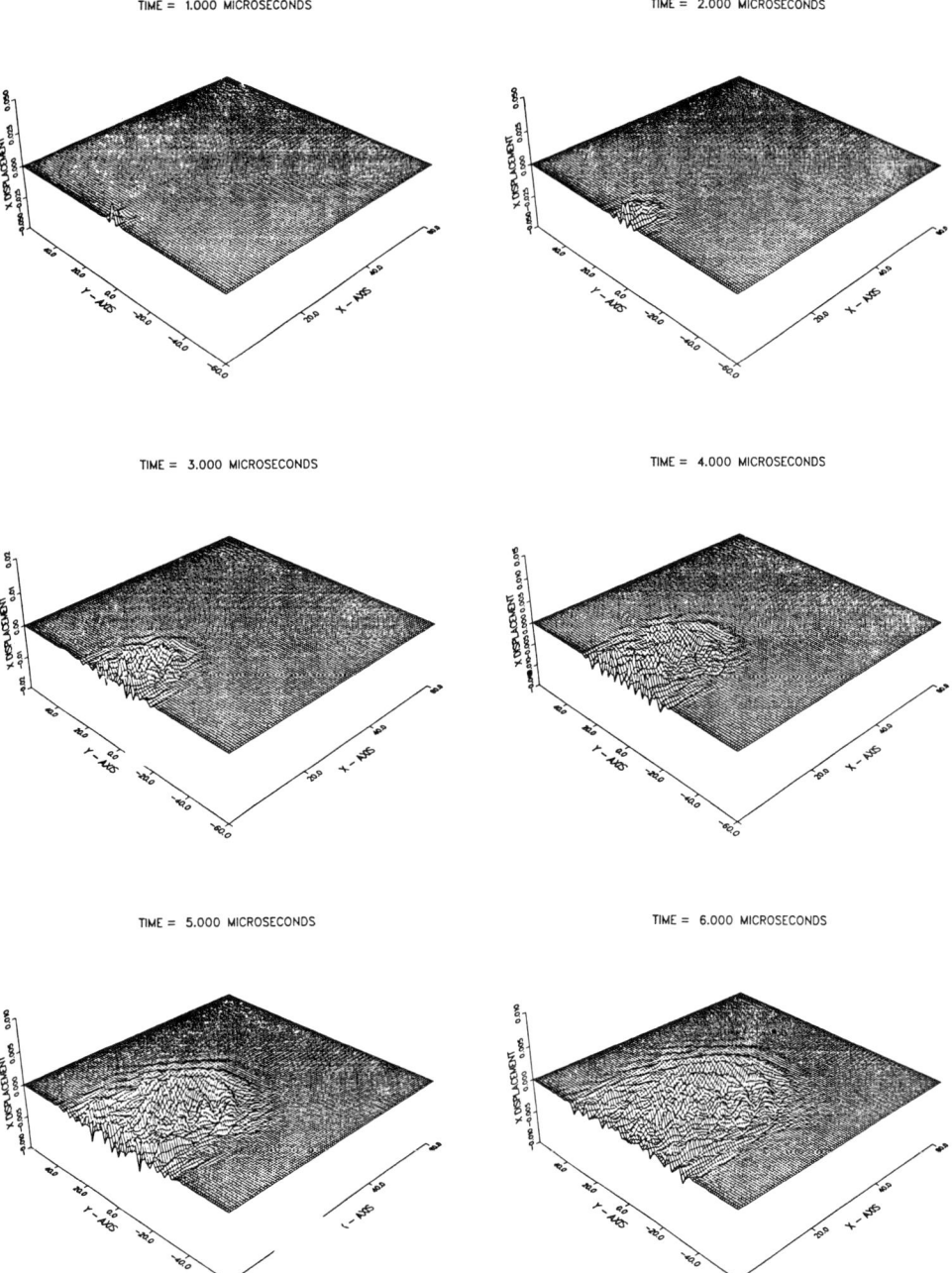

Fig. 1. Time evolution for the x-component of the displacement vector excitated by a line source.

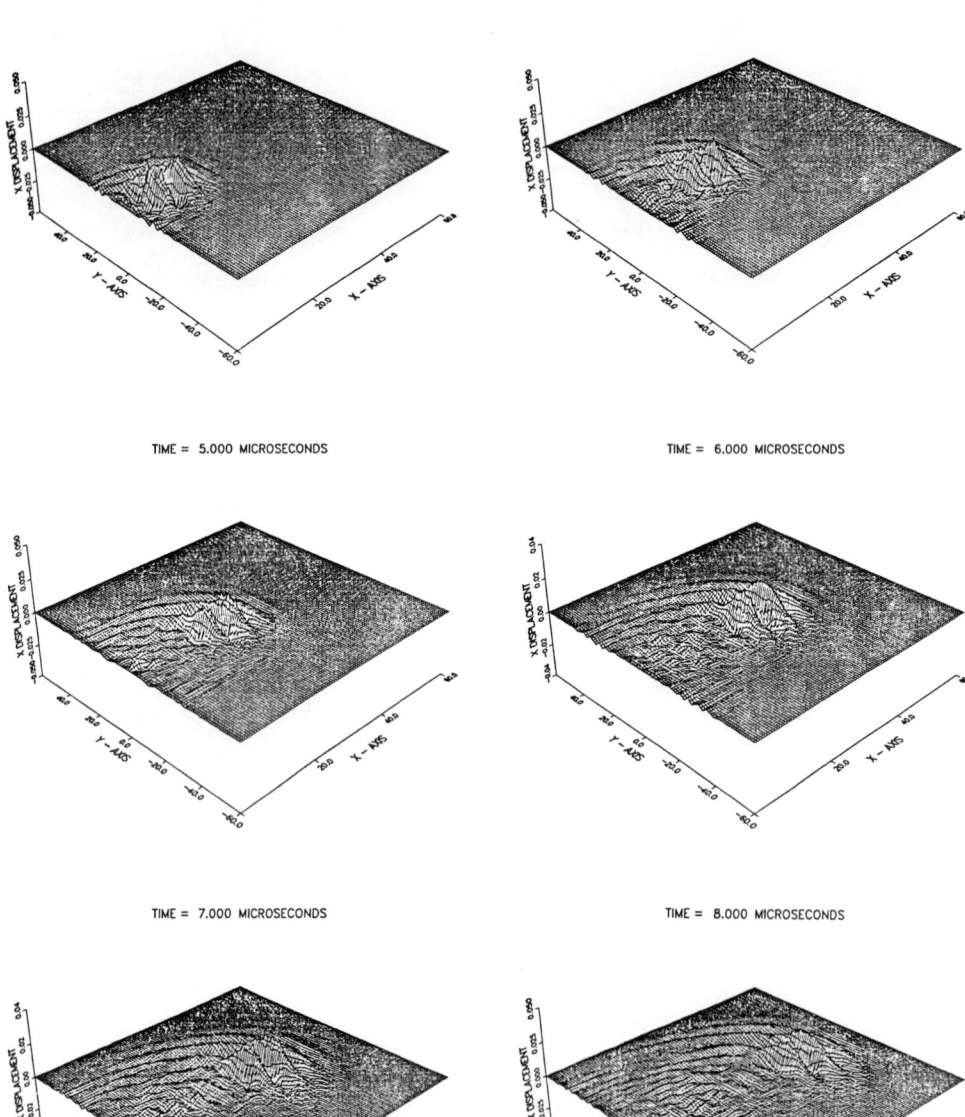

Fig. 2. Combined effect of 9 line sources forming a finite aperture.

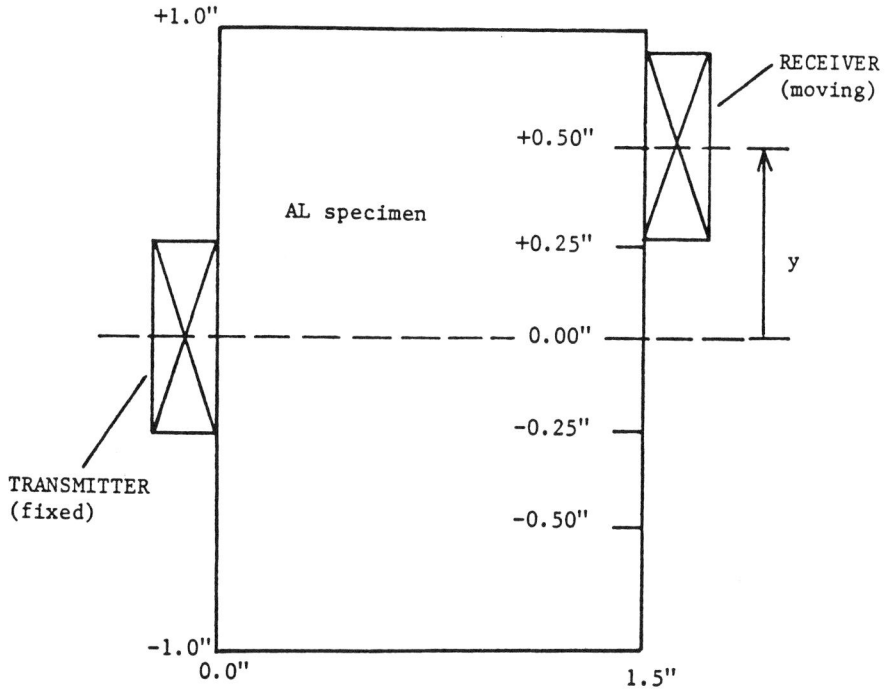

Fig. 3. Test arrangement.

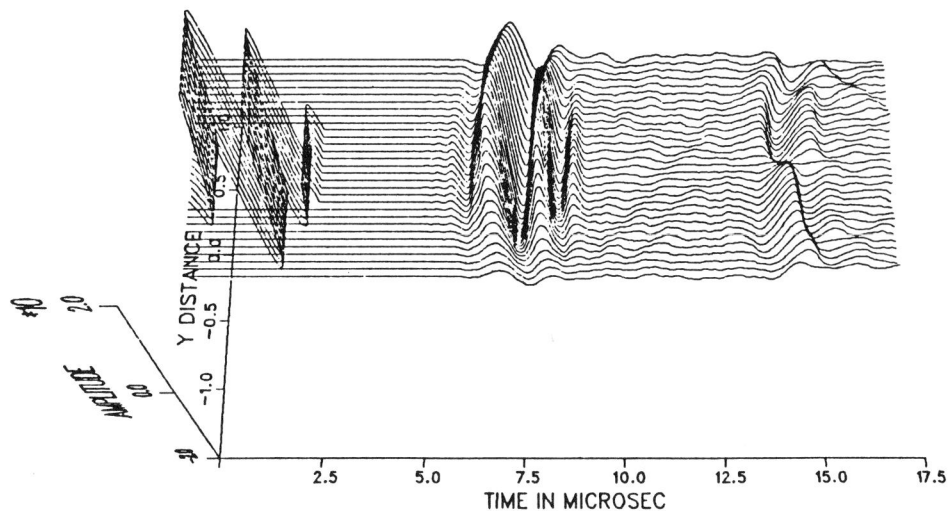

Fig. 4. Combined receiver signals.

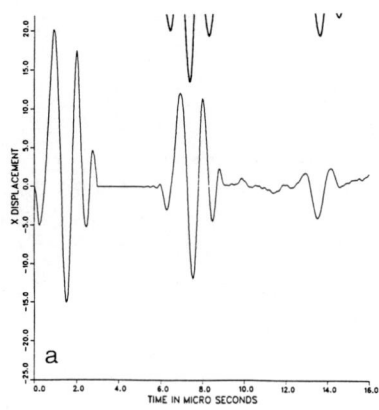
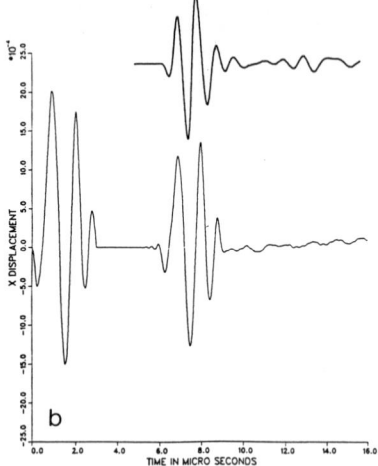
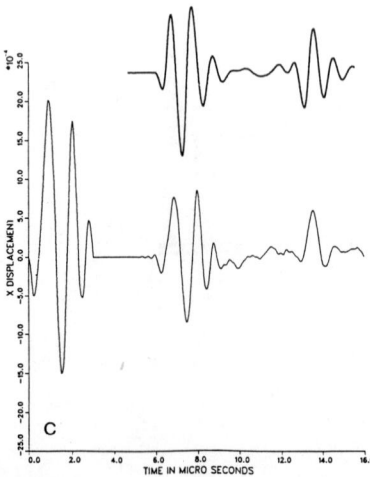

Fig. 5. Experimental A-scan (top line) versus numerical predictions. Receiver location at a) y=+.25″, b) y=0.0, c) y=−0.25″.

CONCLUSIONS

The application of finite element modeling to ultrasonic nondestructive evaluation problems requires transducer models which both mimic reality and also lend themselves to incorporation into numerical code. This paper describes transmitter and receiver models which meet these requirements. Unfortunately exact quantitative confirmation of these models cannot occur until a full three dimensional code has been developed.

ACKNOWLEDGMENTS

Financial support for this work has been provided by the Electric Power Research Institute under Project RP 2687-2. CYBER 205 computer time has been provided by Colorado State University and the National Science Foundation.

REFERENCES

1. R. Ludwig and W. Lord, A Finite Element Formulation for Ultrasonic NDT Modeling, Review of Progress in Quantitative NDE, Vol. 4A, D. O. Thompson and D. E. Chimenti, Eds., (Plenum Press 1985).
2. R. Ludwig and W. Lord, Developments in the Finite Element Modeling of Ultrasonic NDT Phenomena, Review of Progress in Quantitative NDE, Vol. 5A, D. O. Thompson and D. E. Chimenti, Eds., (Plenum Press, 1986).
3. J. Krautkraemer and H. Krautkraemer, Ultrasonic Testing of Materials, Springer Verlag (1977).
4. M. G. Silk, Ultrasonic Transducers for Nondestructive Testing, Adam Hilger (1984).
5. W. P. Mason, Electromechanical Transducer and Wave Filters, Van Nostrand (1948).

TRANSDUCER RADIATION MODELING FOR ULTRASONIC

INSPECTION PURPOSES

David D. Bennink, Anna L. Mielnicka-Pate
D. O. Thompson, and R. B. Thompson

Center for NDE
Iowa State University
Ames, Iowa 50011

INTRODUCTION

One of the main goals of ultrasonic inspection is to determine the absolute scattering response from a given reflector (i.e., defect). In the literature there are a number of reported successful approaches for evaluation of the absolute scattering response from spheroidal inclusions and voids [1]. These approaches were based on a measurement model that accounts for transducer diffraction effects and the scattering and propagation through liquid-solid interfaces. Approximate analytic diffraction corrections were developed for some experimental configurations, and consequently the scattering responses were deconvolved from received ultrasonic signals in an absolute sense. However, one of the important conditions for the accuracy of the deconvolution process is the proper modeling of the individual transducer diffraction characteristic. There is evidence [2,3] based upon on-axis pressure studies and C-scan profiles that considerable discrepancies can occur between individual transducers of the same diameter and nominal frequency. The main purpose of this paper is to study the acoustic characteristics of a set of three unfocused, immersion, piezoelectric, pulse-echo transducers. The analysis of transducer modeling was performed by correlating the vibrating piston theory with experimental results.

THEORETICAL MODELING

In this work the ultrasonic transducer is considered to act as a two-port device: one port being the electrical connection characterized by the voltage across and the current into the transducer, and the second port being the plane flush with the transducer face and characterized by the volume velocity and average pressure across it. It is only these terminal variables--voltage, current, velocity, and pressure--that are of interest and the internal details of the transducer are assumed unknown. In order to derive an expression for the received signal, it is necessary to determine the relationships between the terminal variables when the transducer acts as either a generator (transmitting) or receiver of acoustic waves.

## Transmitting Transducer

The transmitting transducer is characterized by the normal velocity profile, $v_T(\vec{r}_T,f)$, generated across the transmitting plane, $S_T$, as shown in Fig. 1. The transmitting plane is defined as the plane flush with the face of the transmitting transducer. It is assumed that this normal velocity profile can be separated into two terms

$$v_T(\vec{r}_T,f) = v_T(f) t(\vec{r}_T) \tag{1}$$

where $v_T(f)$ is a complex scale factor, dependent on the spectrum of the input voltage [$V_T(f)$] but independent of position, and $t(\vec{r}_T)$ is a real shape function, possibly dependent on frequency but independent of input voltage. In taking $t(\vec{r}_T)$ to be real-valued, it is assumed that the phase remains constant across the transmitting plane. If the transducer operates as a linear device, then we can define a transmitting response function

$$\beta_T(f) = \frac{v_T(f)}{V_T(f)} \tag{2}$$

so that the velocity profile can be written as

$$v_T(\vec{r}_T,f) = [\beta_T(f) t(\vec{r}_T)] V_T(f) \tag{3}$$

where the terms inside the bracket characterize the transmitting transducer.

A spatial imaging technique based upon data taken in the Fresnel zone can be used to determine $t(\vec{r}_T)$ directly [3]. Such an approach, however, would require large amounts of data storage should $t(\vec{r}_T)$ prove to vary with frequency. A second approach, and the one taken in this paper, is to choose a functional form for $t(\vec{r}_T)$ and to vary the functional parameters for the best fit between theory and experiment. It is assumed that a form for $t(\vec{r}_T)$ can be found that satisfactorily predicts the radiated field for a proper choice of the parameters involved.

## Receiving Transducer

The receiving transducer is characterized by its response to the weighted average of the pressure $P_R(\vec{r}_R,f)$, existing across the receiving plane, $S_R$, as shown in Fig. 2:

$$\bar{P}_R(f) = \frac{1}{A_R} \int_{S_R} P_R(\vec{r}_R,f) r(\vec{r}_R) dS_R \tag{4}$$

where $r(\vec{r}_R)$ is a sensitivity function, possibly dependent on frequency, and $A_R$ is the area over which $r(\vec{r}_R)$ does not equal zero. The receiving plane is defined in the same manner as the transmitting plane. It is often convenient to evaluate Eq. (4) using the pressure that would have existed had the transducer not been present. This approach has already proven useful in correcting ultrasonic measurements for diffraction effects. If the transducer operates as a linear device, then we can define a receiving response function

$$\gamma_R(f) = \frac{V_R(f)}{\bar{P}_R(f)} \tag{5}$$

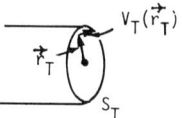 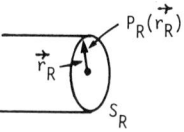

Fig. 1. Transmitting transducer's characterization.   Fig. 2. Receiving transducer's characterization.

so that the received voltage can be calculated as

$$V_R(f) = \frac{\gamma_R(f)}{A_R} \int_{S_R} p_R(\vec{r}_R, f) r(\vec{r}_R) dS_R \tag{6}$$

Although methods may be available for determining $r(\vec{r}_R)$ directly, the same approach is used in this paper for $r(\vec{r}_R)$ as for $t(\vec{r}_T)$.

Linear Input-Output Model

An input-output model can be defined for the entire ultrasonic measurement system relating the received and transmitted voltages as

$$V_R(f) = H(f) V_T(f) \tag{7}$$

where $H(f)$ is the transfer function for the transmitting-receiving transducer system

$$H(f) = \frac{\text{voltage at receiving transducer}}{\text{voltage at transmitting transducer}} \tag{8}$$

Defining $T\{\ldots\}$ as an operator representing the transformation of the source field, $v_T(\vec{r}_T, f)$ on $S_T$, to the received field, $p_R(\vec{r}_R, f)$ on $S_R$, that is,

$$p_R(\vec{r}_R, f) = T\{\rho c v_T(\vec{r}_T, f)\} \tag{9}$$

where the fluid density $\rho$ and acoustic velocity $c$ have been introduced to preserve dimensionality, then the received voltage is, from Eq. (6)

$$V_R(f) = \frac{\gamma_R(f)}{A_R} \int_{S_R} T\{\rho c v_T(\vec{r}_T, f)\} r(\vec{r}_R) dS_R \tag{10}$$

Substituting Eqs. (3) and (10) into Eq. (8) and realizing that $T\{\ldots\}$ is a spatial transformation only results in

$$H(f) = \frac{\rho c \beta_T(f) \gamma_R(f)}{A_R} \int_{S_R} T\{t(\vec{r}_T)\} r(\vec{r}_R) dS_R \tag{11}$$

where $T\{t(\vec{r}_T)\}$ contains all the effects of propagation, transmission, and scattering. The details involved in modeling $T\{t(\vec{r}_T)\}$ are generally complicated and only approximate solutions are often available for even the simplest cases. The measurement system schematics used in the input-output modeling approach are shown in Fig. 3.

Once the shape function, the sensitivity function, and the transmitting and receiving response functions are known for all frequencies of

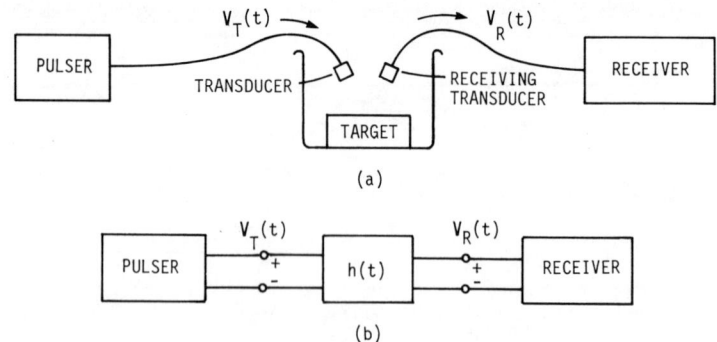

Fig. 3. Ultrasonic measurement system schematic (a), and its input-ouput model (b).

interest, it is possible to determine the ultrasonic transfer function provided a method is available for evaluating $T\{t(\vec{r}_T)\}$. The measurement model [1] provides a particularly useful approach to $T\{t(\vec{r}_T)\}$ for the case of a small spherical reflector located on the acoustic axis of both the transmitting and receiving transducers and in the same fluid medium, as shown in Fig. 4. Generalizing the results presented for a pulse-echo measurement to a pitch-catch one, the transfer function for this configuration can be written as

$$H(f) = 2\pi \frac{\rho c \beta_T(f) \gamma_R(f)}{jkA_R} A(\alpha_R, \beta_R) C_T(z_F) C_R(z_R) e^{-jk(z_F + z_R)} \quad (12)$$

where

$$C_T(z_F) = \left(\frac{jk}{2\pi}\right) e^{jkz_F} \int_{S_T} t(\vec{r}_T) \frac{e^{-jkR_F}}{R_F} dS_T \quad (13)$$

and

$$C_R(z_R) = \left(\frac{jk}{2\pi}\right) e^{jkz_R} \int_{S_R} r(\vec{r}_R) \frac{e^{-jkR_R}}{R_R} dS_R \quad (14)$$

Likewise, $A(\alpha_R, \beta_R)$ is the scattering amplitude evaluated for the coordinate angles of the receiving transducer's axis. By looking at the variation of $H(f)$ with either $z_F$ or $z_R$, it is possible to determine the "best fit" parameters for an estimated $t(\vec{r}_T)$ or $r(\vec{r}_R)$.

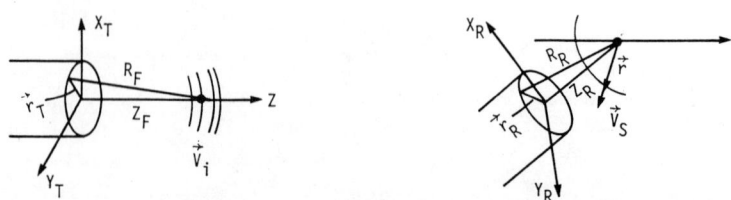

Fig. 4. Geometry for a spherical reflector.

For this paper the experimental results were correlated with piston radiator theory, which defines the shape and sensitivity functions as follows:

$$t(\vec{r}_T) = \begin{cases} 1, & |\vec{r}_T| \leq a_T(f) \\ 0, & \text{otherwise} \end{cases} \tag{15}$$

and

$$r(\vec{r}_R) = \begin{cases} 1, & |\vec{r}_R| \leq a_R(f) \\ 0, & \text{otherwise} \end{cases} \tag{16}$$

Although these assumed shape and sensitivity functions are not physically realistic, they are often used because of their simplicity and it would be convenient if they provide satisfactory agreement with experimental results. In addition, since all experimental data were taken in pulse-echo measurements, it is assumed that $a_T(f) = a_R(f) = a(f)$. For this choice of shape function and sensitivity function, the pulse-echo response from a small spherical reflector becomes

$$H(f) = 2 \frac{\rho c \beta(f) \gamma(f)}{jka^2(f)} A(0,\pi) C^2(z) e^{-2jkz} \tag{17}$$

where

$$C(z) = \left(\frac{jk}{2\pi}\right) e^{jkz} \int_0^{a(f)} \frac{e^{-jk\sqrt{z^2+r^2}}}{z^2+r^2} 2\pi r \, dr \tag{18}$$

$$= 1 - e^{-jk(\sqrt{z^2+a^2(f)}-z)}$$

The magnitude of $C(z)$ is thus given by

$$|C(z)| = 2\left|\sin \frac{1}{2} k(\sqrt{z^2 + a^2(f)} - z)\right| \tag{19}$$

or, from Eqs. (7) and (17)

$$|C(z)| = |B(f)| \sqrt{|V_R(f,z)|} \, e^{\alpha(f)z} \tag{20}$$

where $B(f)$ is a complex scale factor, $V_R(f)$ the received voltage at axial distance Z, and $\alpha(f)$ the attenuation in water. Equation (20) is valid when the attenuation is small compared to the wave number. The parameter of variation is a frequency-dependent effective radius $a(f)$.

EXPERIMENTAL INVESTIGATIONS

The objective of the experimental investigations was to measure axial profiles on the basis of Eq. (20) for a set of three transducers. These profiles will be compared with theoretical profiles described by Eq. (19).

Measurement Procedures

Three ultrasonic pulse-echo commercial transducers were investigated. These transducers were piezoelectric, of the unfocused immersion type with a center frequency of 5MHz. Transducers referred to as #1 and

#2 had nominal diameters of 1/2 inch and transducer #3 a nominal diameter of 1/4 inch. The axial profiles were measured using a 1-mm diameter, spherical reflector located in water on the acoustic axis of the transducer. An automatic scanning device was used for transducer positioning. The investigated positions consisted of between 50 and 70 points per profile for each frequency of interest with the spacing varied to achieve at least 20 near-field points. The axial profiles were extracted from the Fast Fourier Transform of the received voltage within the range of 3MHz to 8MHz with a spacing of 0.1MHz.

In order to compare the theoretical results described in Sec. 2, the experimental axial profiles were "curve-fitted" with Eq. (19), using the active radius a(f) as a functional parameter. Because the scale factor B(f) in Eq. (20) depends on several unknown quantities, it was also necessary to determine the best-fit value of B(f) for each active radius value chosen. Three different approaches were used for the curve fitting: the first approach was based on fitting the entire range of measured distances, the second approach involved curve-fitting the measured data over the first maximum, and the third approach involved curve-fitting the measured data over the second maximum. The maxima were numbered increasingly as z approaches zero from infinity.

The curve-fitting procedure was based on minimizing the accumulated error between the theoretical and measured values, defined as

$$\bar{E} = \frac{1}{N} \sqrt{\sum_{i=1}^{N} (V_{TH} - V_M)^2} \cdot 100 \qquad (21)$$

where N is the number of positions included in the curve-fitting procedure, $V_{TH}$ is the theoretical value of the on-axis profile (according to Eq. [19]), and $V_M$ is the scaled experimental value of the on-axis profile (according to Eq. [20]).

RESULTS

An example of the measured and calculated axial profiles for one of the transducers is shown in Fig. 5. In this figure only three representative frequencies (out of 51 calculated) are shown to demonstrate the details of the individual curve-fitting results. As expected, the on-axis zeros predicted by theory show up as minima in the experimental data. Also, the maxima tend to decrease in height as z approaches zero. This tendency appeared consistent across transducers and could be the result of several factors, including variations from planar and circular apertures, phase shifts across the transducer element, or discrepancies between the actual and assumed shape and sensitivity functions.

Those active radii that resulted in the least curve-fitting error for the three approaches used are shown in Fig. 6. Both transducers #1 and #2 display active radii that are relatively constant across frequency for curve-fitting based on the entire curve or the second maximum but not for curve-fitting based on the first maximum. Transducer #3, on the other hand, displays active radii that are not consistent across frequency, regardless of the curve-fitting approach used. The average value of the active radii across frequency is also larger than the nominal value for transducer #3, whereas it is lower than the nominal for both transducers #1 and #2. Although the experimental data agree well with the piston radiator theory in general, it is clearly a better approximation for certain transducers and may not even be valid for some.

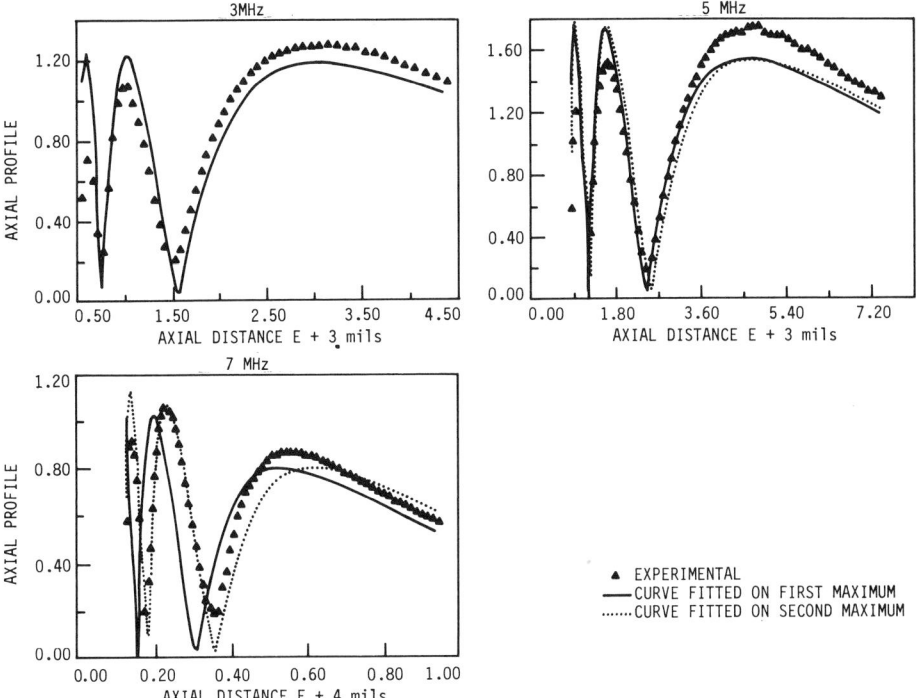

Fig. 5. Axial profiles obtained for transducer #1 (5 MHz; 0.5 inch). Curve-fitting results plotted using scale factors based on all data values.

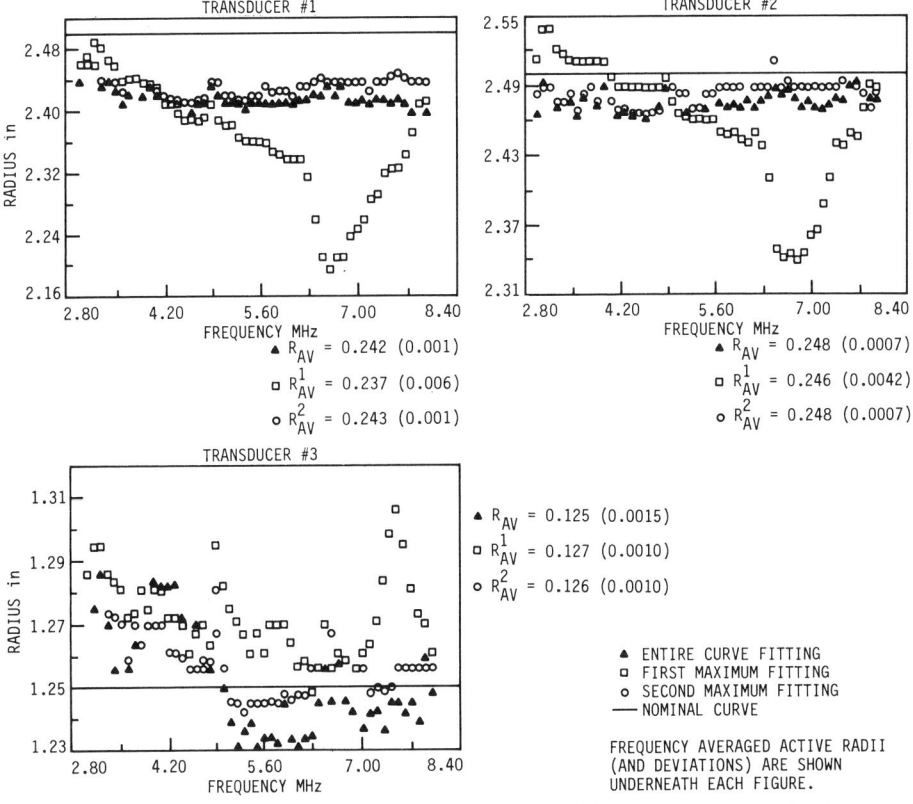

Fig. 6. Active radius calculated using different curve-fitting approaches.

The accumulated error, defined in Eq. (21), is shown in Fig. 7. In addition to the errors at a particular frequency, averaged errors across investigated frequencies are indicated underneath each curve. These results demonstrate that the error between the experimental data and the theoretical data can often be significantly reduced by using a curve-fitted active radius. Fitting the entire curve provided the best results but the second-maximum fitting gave nearly equivalent results. This is particularly interesting because the second maximum approach requires substantially less measurement and computation (about a 70% reduction in both). It is also interesting to note that, although significant variations are shown in the active radius values for transducer #3, the curve-fitting does not significantly affect the accumulated error. All three transducers show the smallest accumulated errors for frequencies slightly above the center frequency.

CONCLUSIONS AND DISCUSSION

This paper developed a technique for approximating the characteristics of ultrasonic transducers based on axial profiles from a small spherical reflector. Results were presented for three piezoelectric, immersion transducers assumed to act as piston radiators. All three transducers behaved in a manner that could be approximated to varying degrees by the piston radiator theory. Transducers #1 and #2 showed active radii almost constant across frequency for two of the curve-fitting approaches, suggesting the use of the average active radius value for characterizing these transducers. The active radius values for transducer #3, however, were not as consistent. Further investigations are needed to determine the causes for the discrepancies that were observed between the theoretical and experimental results.

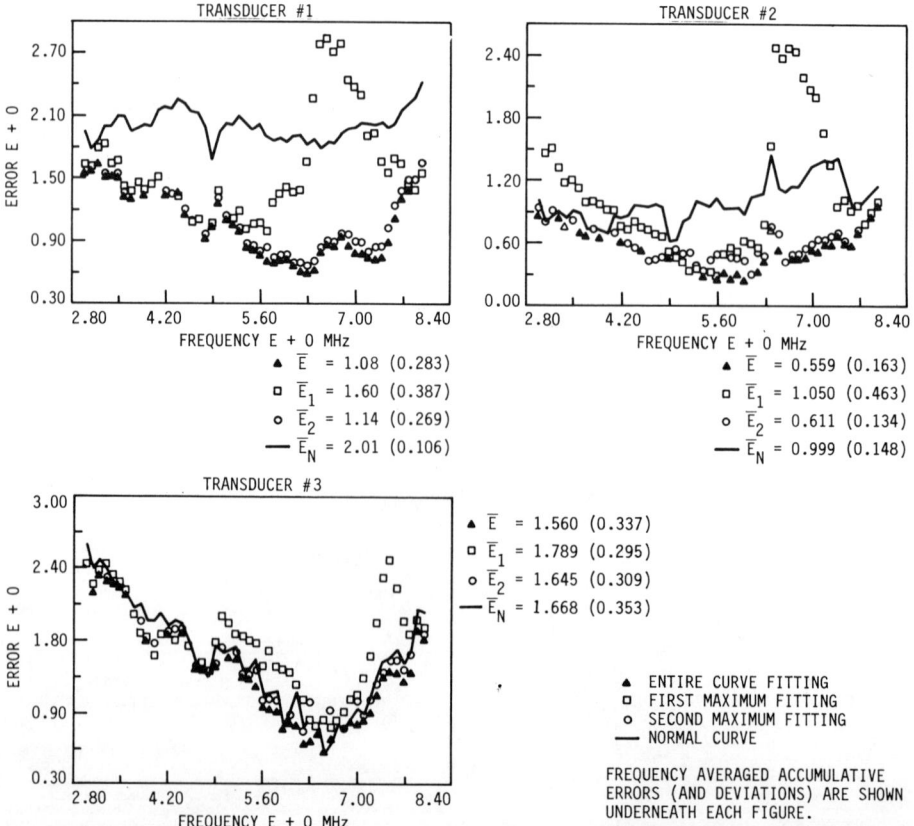

Fig. 7. Accumulative error in transducer active radii calculations.

ACKNOWLEDGMENTS

The work was sponsored by the Center for NDE and Iowa State University, Ames, Iowa. The authors would like to recognize the assistance of S. J. Wormley and T. A. Gray in designing the experiments.

REFERENCES

1. R. B. Thompson and T. A. Gray, "A model relating ultrasonic scattering measurements through liquid-solid interfaces to unbounded medium scattering amplitudes," J. Acoust. Soc. Am. 74(4) (1983).

2. D. J. Chwirut and G. D. Boswell, "The evaluation of search units used for ultrasonic reference block calibrations," NBS report 781454.

3. K. M. Lakin and A. Fedotowsky, "Characterization of NDE transducers and scattering surfaces using phase and amplitude measurements of ultrasonic field patterns," IEEE Trans. on Sonics and Ultrasonics, SU-23(5) (1976).

4. R. B. Thompson and T. A. Gray, "Analytic Diffraction Corrections to Ultrasonics Scattering Measurements," Review of Progress in Quantitative NDE, Vol. 2A, Edited by D. O. Thompson and D. E. Chimenti, p. 567-585 (1983).


DISCUSSION

From the Floor: You have fairly significant assumptions built in those transducer models. One of them is that you have a uniform radiation field. We have done experimental work in direct sensing techniques from the transducer surface in order to find how diffraction behaves, what is the orientation of the wave, and then to analyze the scattering phenomenon as the wave propagates through a structure.
  The other item is that the absolute amplitude of the radiation pattern is the unknown function which the model needs to consider. Your method considers only data that are normalized with respect to the shape and are not absolute values. However, the amplitude often plays a big role when one deals with sizing of cracks and other measurements. Can you make a comment?

Ms. A. Mielnicka-Pate, Iowa State University, Ames: In the initial stage of our research we did not concentrate on the absolute amplitude when modeling the transducer radiation. Our first objective was to compare the scaled spatial (in particular axial) field distribution with the theory of the rigid disk of unknown "active" diameter. We have demonstrated that there is very good agreement between experimental and theoretical data. However, we agree that the absolute amplitudes are important. Consequently, we intend to measure constants $\beta$ and $\Gamma$ using some kind of transducer calibration techniques.
  Your comment regarding the surface velocity distribution addresses a very important aspect of our transducer modeling. We do not know how sensitive the radiated field is to the shape of the velocity distribution, which resulted in good agreement of the theory with the scaled experimental data. The next step in our analysis would be to determine the sensitivity of the spatial distribution of the radiated field to the change of the velocity shape function. However, we presently do not know what this sensitivity might be.

Mr. Fraser: Actually, given the realities of transducers, it's probably impressive that you can use the frequency-dependent diameter to model that and get reasonably good results.

From the Floor: John, I want to ask a question. I have a problem with this way of treating a transducer. This approach could be fine for a transducer that is driven by a harmonic of a constant frequency. However, most of the transducers that we use are driven by some kind of a transient signal. Simplistically, one looks at the radiation that's being generated as two components; the first component is generated by the front face, and there's a second component due to the rim of the transducer. So I find the idea of an effective radius a very disturbing one.

Ms. Mielnicka-Pate: All experimental data were measured for impulses. We performed the discrete Fourier transform on impulses measured at 25 locations along the main axis. Next we analyzed 52 frequency components extracted from the spectra of those impulses. I have just presented results for only three of those frequency components. In fact, that agreement was excellent for most of the 52 spectral components that we extracted from measured spectra.

From the Floor: That's correct. But your theory is a continuous wave theory. You showed the Green function to start. I'm not certain that the theory really applies to the experiment.

Ms. Mielnicka-Pate: We used frequency domain representation for the analysis of an acoustic impulse propagating away from a transducer. The Green's function integrated over the surface of the transducer provided the axial pressure distribution which agreed remarkably well with our scaled measured data. Moreover, we matched the measured data with the theoretical values using the "active" diameter as a parameter. We have modeled the transducer on the basis of data measured at the main axis of the transducer. The reason was that there is an analytic solution to the Green's function integrals for on-axis locations. Therefore, only on-axis locations were investigated in order to avoid any approximation errors.

However, as reported, one probe did not agree very well with the rigid disk on-axis theory. In addition, the impulse scattered from the flat bottomed hole received by this probe was significantly different than received by other probes. It is possible that this probe had some non-symmetry in the radiation field, or its acoustical axis did not overlap with the geometrical normal. We intend to investigate in more detail its spatial radiation characteristic in order to understand the results that we obtained on the axis.

TECHNIQUE FOR GENERATION OF UNIPOLAR ULTRASONIC PULSES

D. O. Thompson and D. K. Hsu

Ames Laboratory, USDOE
Iowa State University
Ames, IA 50011

INTRODUCTION

Substantial progress has been made in recent years in the development of inverse elastic wave scattering theories for use in ultrasonic nondestructive evaluation (NDE). These include theories that are applicable in different ultrasonic frequency ranges and include formulations in various approximations [1-15]. It is by application of these inverse scattering solutions to ultrasonic inspection results that quantitative measures of the size, shape, and orientation of a flaw can be determined.

Implementation of the inverse scattering solutions mentioned above, however, has been limited. One of the principal reasons for this limitation is the narrow bandwidth of commercial ultrasonic instrumentation relative to the ultrasonic bandwidths needed to properly employ the inverse results over a range of flaw sizes. Addison et al. [16] have shown that the minimum bandwidth needed to size a flaw with radius a using the simplest, 1-D inverse Born scattering solution must extend from $ka=0.5$ to $ka=2.0$ to achieve sizing accuracies of 20%. $k$ is the magnitude of the wave vector and is given by $2\pi/\lambda$ where $\lambda$ is the ultrasonic wavelength. This requirement is considerably more demanding than the capabilities of most commercially available transducers. Addison et al. also showed that inattention to this specification will result in size estimates that are either too small or too large depending upon whether there is a deficiency of the low or high frequency content in the interrogating ultrasonic spectrum.

In order to utilize the inverse theories and to accommodate a reasonable range of flaw sizes that are unknown to the investigator, it is evident that considerable improvement in ultrasonic system bandwidths must be obtained. Unipolar pulses provide an attractive way to approach this objective. Features of this pulse have been described by a number of authors [17,18,19]. Current practice, however, seems to be limited to the pitch-catch mode. In this configuration one transducer is connected to a low impedance electronic pulser and is used to transmit the signal while a second transducer is connected to a high impedance amplifier and is used to receive the signal. This configuration, however, is much too limited for NDE applications in which the preponderance of applications requires the pulse-echo mode in which the same transducer is used for transmitting and receiving.

The purpose of this paper is to report results that describe a new way to generate broadband, unipolar pulses in pulse-echo operation using conventional planar transducers. A review of requirements is first given; this summary is then followed by a description of a transmit/receive switch that has been developed for pulse-echo operation and some results that have been obtained with it. These results are then discussed and compared with results obtained with conventional instrumentation.

EXPERIMENTAL PROCEDURE

Review of Requirements

The production of broadband, unipolar stress pulses using a piezoelectric generator in a pulse-echo mode requires consideration of both the transmit and receive aspects of the process. It is well known [20] that a step function voltage pulse applied to an untuned planar piezoelectric transducer with perfectly matched backing will produce a unipolar stress pulse that shows similarities to, but in fact is not, the time derivative of the applied voltage pulse. The "effective" differential character arises because the charge layers produced on opposite faces of the piezoelectric by the applied voltage pulse produce stress pulses that are replicas of the pulse, but which are of opposite polarities and which are separated in time by an amount equal to one transit time through the piezoelectric. In the receive mode, the piezoelectric transducer generates a voltage signal that faithfully represents the unipolar stress pulse under open circuit (no load) conditions. However, the features of this reproduction are modified considerably by values of the parameter $R_L C_0 \omega$ in which $R_L$ is the impedance of the external circuit used to measure the signal, and $C_0 \omega$ is the reciprocal of the transducer source impedance. If $R_L C_0 \omega \ll 1$, it follows that the open circuit voltage signal is differentiated by the measuring circuit, whereas if $R_L C_0 \omega \gg 1$, the measuring circuit does not significantly perturb the open circuit signal generated by the piezoelectric. In the latter case, then, the received signal will also appear as a unipolar pulse.

Examples of the discussion given in the previous paragraph are given schematically in Fig. 1 for two different applied voltage pulses and for both the transmit and receive modes. Figure 1a shows that the expected transmitted stress pulse is unipolar if the applied voltage pulse is a step function, and that the measured, or received, signal is either unipolar or bipolar depending upon whether $R_L C_0 \omega \gg 1$ or $R_L C_0 \omega \ll 1$. Figure 1b shows the expected behavior for the transmitted and received pulse for the commonly used spike applied voltage pulse. In this case the stress pulse is bipolar, and depending again upon the relative values of $R_L C_0 \omega$, the received pulse is either bipolar if $R_L C_0 \omega \gg 1$ or the commonly observed tripolar pulse if $R_L C_0 \omega \ll 1$. It should be emphasized that this figure is only schematic; the actual shape of the unipolar stress pulse is expected to vary according to the rise time of the applied voltage pulse and the transit time of the back surface stress pulse through the piezoelectric. Furthermore, the idealized examples assume that there are no diffraction or frequency dependent attenuation losses associated with pulse propagation that would tend to distort the shapes of both the stress pulses and the received signals when viewed in the time domain.

Transmit/Receive Switch

From the above descriptions, it is clear that a special transmit-receive switch is needed in order to generate and receive unipolar pulses in a pulse-echo mode. The switch must provide a way to connect

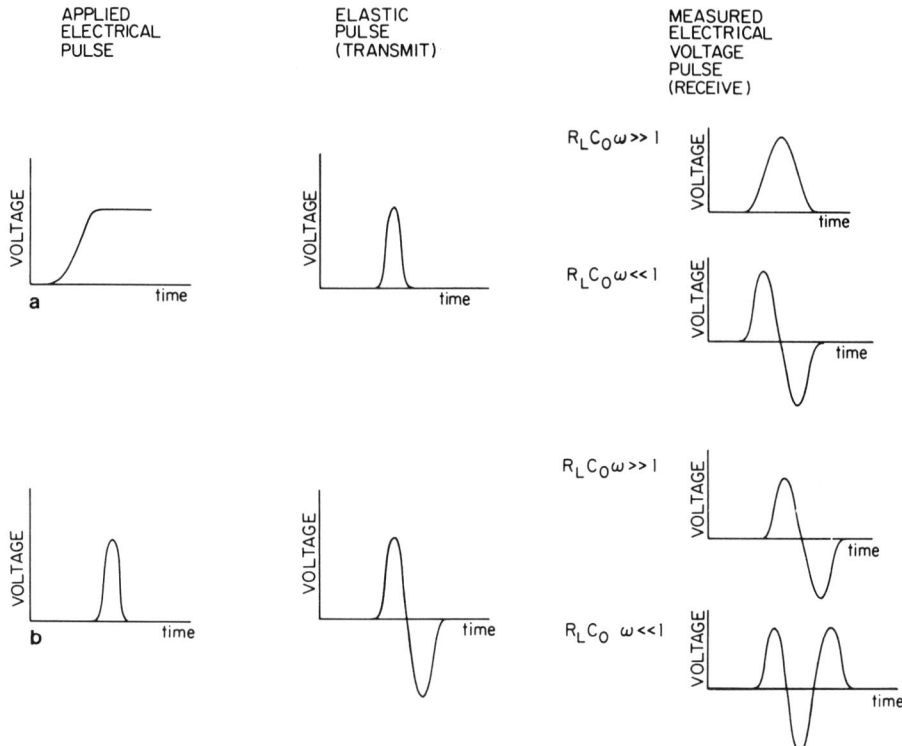

Fig. 1. Transmitted and received pulse shapes for both a step function excitation voltage and a spike excitation voltage.

the transducer electronically to a low impedance applied voltage pulser during the transmit cycle and to a receiver with high input impedance during the receive cycle. Figure 2 shows a recently developed transmit-receive switch that has these properties. The low impedance pulsing transmitter is shown at the left, the transducer T with internal impedance $1/jC_0\omega$ in the center, and an FET with high input impedance and low output impedance on the right. $R_1$ is nonessential to the operation of the circuit for single transducer operation. It is included because the circuit is used in a multiplexing mode in which the transmitter is switched electronically to alternate transducers [21]. The only purpose of this resistance is to keep the circuit input at ground potential when the transmitter is switched to other circuits. The diodes $D_1$ provide a low impedance conducting path during the transmit cycle of the pulser. During the receive cycle of the pulse echo mode, these diodes effectively block the low impedance transmitter from the transducer because of the relatively low signal levels. The inclusion of the diodes $D_1$ is therefore essential since the impedance in all current paths from the transducer must be included in the calculation of the load impedance $R_L$ presented to the transducer. Diodes $D_2$ are included as a protective device to shield the FET (or following preamplifier) from the "main bang" of the transmitter. Resistor $R_2$ is included as a parallel load resistor for the transducer so that the diodes $D_2$ do not simply short the main transmitter pulse to ground during the transmit cycle. A value of $R_2 = 270\Omega$ has been used to date. This value is not critical, but must be several times larger than the input impedance of the transducer for all operating frequencies. Received signals in the pulse echo mode are coupled to the high impedance FET through the network $R_3C_1$. Two conditions apply to the choice of these values although, as before, the actual values are not very critical. In one of these the $R_3C_1$ value is selected to produce the desired time constant and to set the low frequency end of the signal pass band. In the case shown, the time constant is 2.4μsec with a low frequency pass limit of 0.4MHz. Secondly, the impedance of this network should be chosen so that the total load impedance presented to the transducer is 10-20 times the source impedance of the transducer. Since the diodes are essentially nonconducting in the receive mode and the FET possesses a very large input impedance, the only substantive contributors to the transducer load impedance are $R_2$, $C_1$, and $R_3$. A value for $R_3$=10K ohms has been found satisfactory for work to date. The transmit-receive switch is completed with an FET buffer that is connected to a following broadband preamplifier (Comlinear). The values of resistances used in the FET produce an output impedance of about 120Ω. Lower output impedances can be obtained using other values for $R_4$ and $R_5$. It should be noted that the use of an FET is optional and depends upon circuit arrangements of the following electronics.

Results

The transmit-receive switch described in the previous section has been used to generate unipolar pulses with standard, commercially available piezoelectric transducers. All results given in this section were obtained using a water immersion technique and a flat metal reflector that could be placed at prescribed distances from the transducer. Panametrics transducers with 10 and 15 MHz center frequencies and 1/4 inch diameters were used. It is important that the transducers contain only an internal piezoelectric element and no internal tuning elements. The voltage pulses applied to the transducers were step function pulses generated by a Hewlett Packard 214A generator. After reflection from the metallic reflector, received signals were taken from the transmit-receive

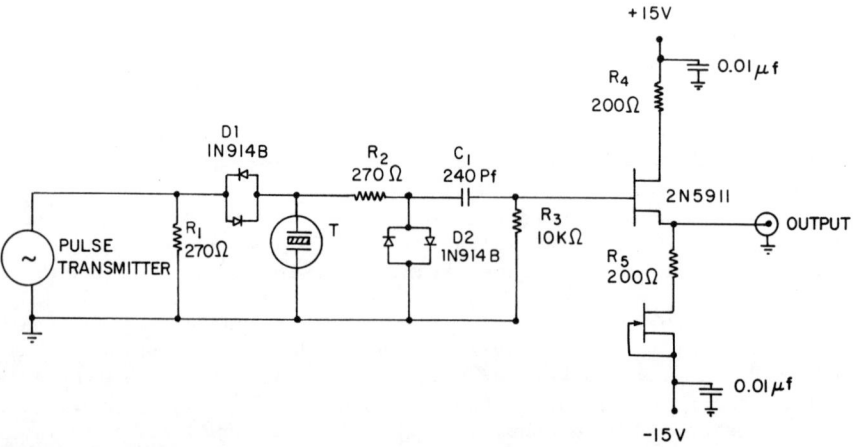

Fig. 2. Schematic circuit diagram for the transmit-receive switch used in producing the unipolar ultrasonic pulse.

switch shown in Fig. 2, amplified in a Comlinear preamplifier with 3dB points at 0 and 150 MHz, digitized in a Tektronix 7912 A/D converter, and displayed. Results were obtained using the transmit-receive switch both with and without the FET buffer.

Figures 3a and 3b show results obtained for the 10 MHz, 1/4 inch, immersion transducer in which the FET buffer in the transmit-receive switch was included. Figure 3a shows the received time domain pulses taken at several distances between the transducer face and the metallic reflector, and Fig. 3b shows the spectral analysis of the time domain pulses at the same distances. The results in Fig. 3a show that, at close distances, acceptable unipolar pulses are produced, and that with increasing separation, the unipolar pulse degrades toward bipolar behavior. The small, sharper spikes on the positive side of the negative-going pulse appear to be associated with internal reverberations in the piezoelectric due to imperfect backing. It would appear from the results in Fig. 3b that the unipolar degradation observed in Fig. 3a is due to the loss of low frequency content because of diffraction. The results in Fig. 3b also show the signal enhancement at 10MHz due to the quarter wave plate that is present in the immersion transducers.

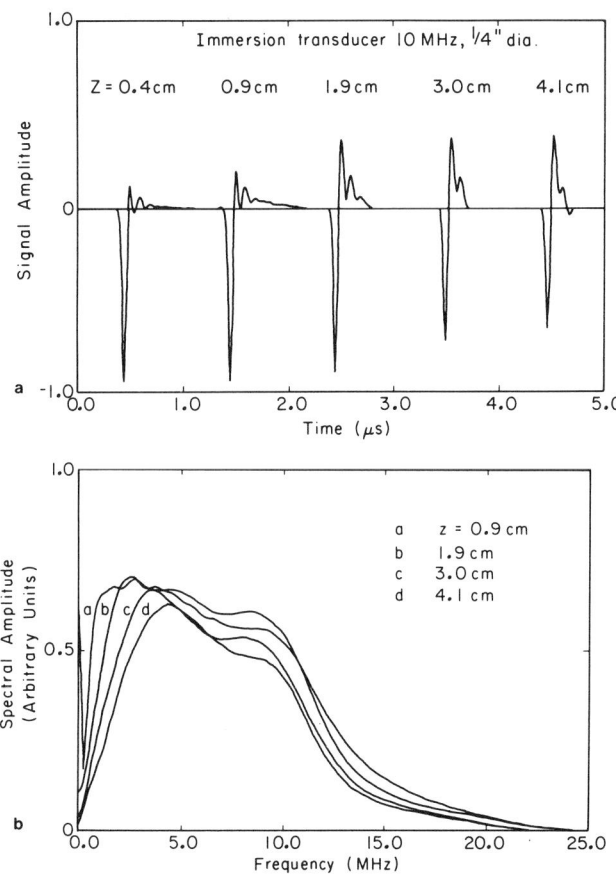

Fig. 3. a) Received ultrasonic pulses in the pulse-echo mode at various transducer-reflector distances using a 10 MHz, 1/4 inch diameter immersion transducer and with the FET in Fig. 2 included.
b) Frequency spectra of the ultrasonic pulses shown in Fig. 3a.

671

The frequency spectra obtained for a contact transducer used in an immersion mode are shown in Fig. 4a. Even though this transducer is not intended for immersion applications and does not possess a matching quarter wave plate, the omission of this plate does not appreciably affect the generation of the unipolar pulse. The results shown in Figs. 3b and 4b show a degradation in the low frequency content of the unipolar pulse signal with separation between the transducer and the reflecting plate. As noted, this effect is to be expected in terms of diffraction losses. If this be true, application of diffraction corrections to the various frequency spectra should reduce all the spectral results to a common curve for a particular transducer. Figure 4b shows the results obtained when diffraction corrections are applied to the results of Fig. 4a. It will be seen that a universal curve results which is the true spectral response of the transducer for the applied voltage step function condition. Similar results have been obtained for all the other cases examined.

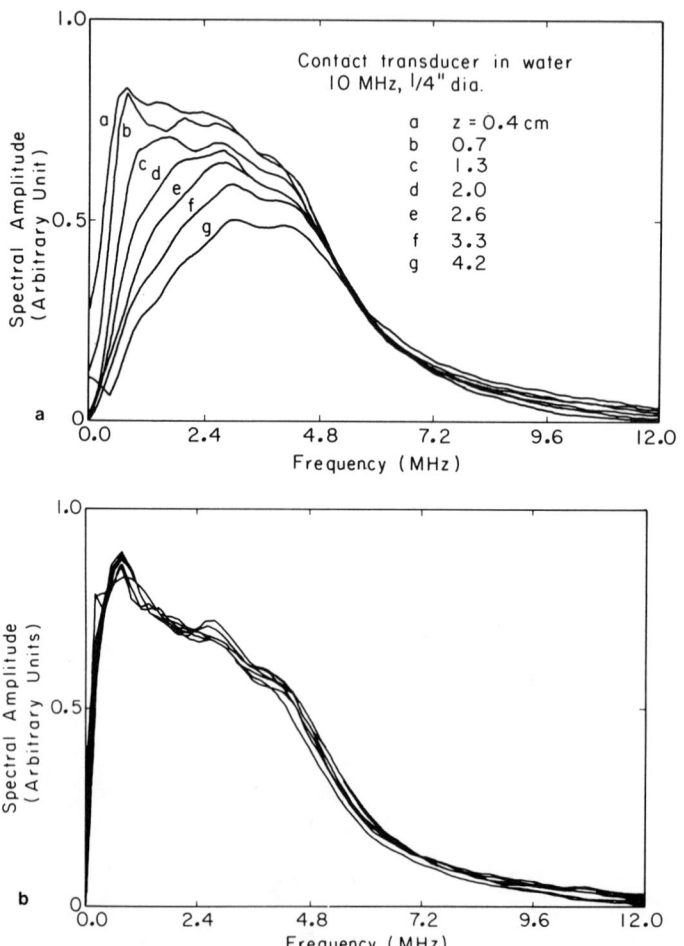

Fig 4. a) Frequency spectra of ultrasonic pulses in the pulse-echo mode at various transducer-reflector distances using a 10MHz, 1/4 inch contact transducer in the immersion mode. The FET buffer in the transmit-receive switch is bypassed. b) Intrinsic frequency spectrum of the unipolar pulse obtained by applying diffraction corrections to the spectra in Fig. 4(a).

Although it is possible to correct analytically for diffraction losses as shown above, the corrections do not restore lost energy into the interrogating ultrasonic beam. A principal consequence of this loss is a degraded signal/noise ratio at the lower end of the frequency spectrum. Considerations are in progress aimed at reducing the diffraction losses through the use of long focal length lenses. Figure 5 shows the results obtained using a plane wave, 15 MHz, 1/4 inch transducer both with and without a 69 centimeter focal length lens. The lens in this case was cast from epoxy to the desired radius of curvature and then cut to an edge thickness of 0.012 inches using a diamond saw. For test purposes, the lens was bonded to the transducer face with glycerin. The top part of the figure shows spectral results obtained at three separations with no lens and the bottom part shows the same results with the lens in place. It is evident that the lens action helps to provide beam collimation and suppression of diffraction losses in this case. Further studies to select a more ideal collimating lens are in progress.

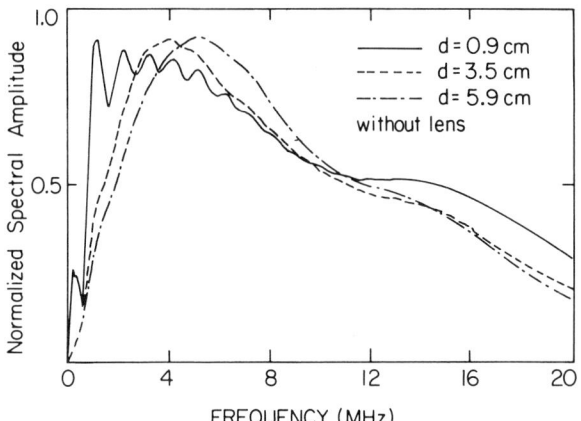

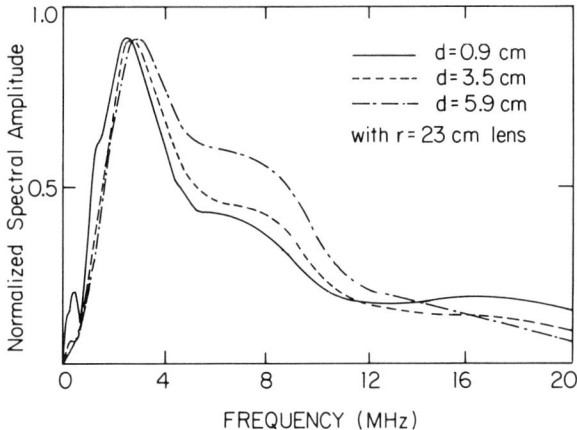

Fig. 5. Frequency spectra of ultrasonic pulse-echo signals at three transducer-reflector distances with the 69cm focal length lens (bottom figure) and without the lens (top figure). The spectra are normalized to have the same peak value. The transducer is a 15 MHz, 1/4 inch diameter immersion transducer.

## DISCUSSION OF RESULTS

As noted in the Introduction, one of the principal purposes of this work was to develop a convenient way to obtain broadband ultrasonic pulses that can be used with pulse-echo measurement techniques and with a variety of more or less standard transducers. The transmit-receive switch that has been developed appears to satisfy this purpose. It can be assembled and used separately or it can easily be incorporated into existent voltage pulser units. Two limitations have been found that need to be noted. First, the transducer used must not contain an inductive tuning element and secondly, the applied voltage pulser used as a driver must be capable of providing sufficient current so that the step function pulse applied to the transducer is sharply formed. Further, it has been found that the top frequency limit of the transducer's bandwidth in this mode of operation is effectively defined by the high frequency roll off of the piezoelectric resonance. The low frequency pass band limit is determined by the values of components in the transmit receive switch.

The magnitude of the transducer bandwidth improvement obtained with the current transmit-receive switch is shown in Fig. 6. Curves are shown in this figure for the pulse echo response of another 10MHz, 1/4 inch immersion transducer with the reflector placed at 6 cm from the transducer and for four different values for the resistance $R_3$ shown in Fig. 2. The values for the resistance are 50Ω, 390Ω, 10KΩ, and 47KΩ for curves 1, 2, 3, and 4, respectively. No diffraction corrections have been applied to these results. It is apparent that the improvement in bandwidth is very significant across this range. It is also apparent that increases in $R_3$ above 10KΩ produce only small changes in the overall bandwidth and at the expense of lengthened time constants.

One of the tests that must be applied to the extended bandwidth transducer as an indicator of its flaw sizing capability is to compare the measured scattering amplitude with theoretical predictions for a well-defined ultrasonic target. Such a comparison is shown in Fig. 7 for a machined ellipsoidal void that was diffusion bonded in a titanium block.

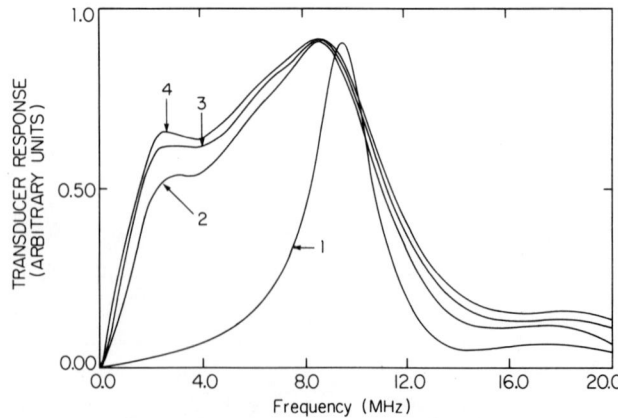

Fig. 6. Frequency spectra of a 10 MHz, 1/4 inch diameter immersion transducer placed 6 cm from a flat reflector and operated in the pulse-echo mode using the transmit-receive switch of Fig. 2. Curves 1, 2, 3, and 4 correspond, respectively, to $R_3$=50, 390, 10K, and 47KΩ.

The ellipsoidal void was oblate in shape, and the nominal values of the semiaxes were 400μ, 400μ and 200μ. Opsal [22] has calculated the scattering amplitude for this target; his results are given by the dashed line. The solid curve shows the experimental results obtained using the unipolar pulse in the pulse-echo mode. These curves are given in absolute units and include diffraction and attenuation corrections. The agreement between the curves is considered very satisfactory. It is evident that it would have not been possible to obtain results that agree this well with theoretical predictions had the low impedance response curve of Fig. 6 been used.

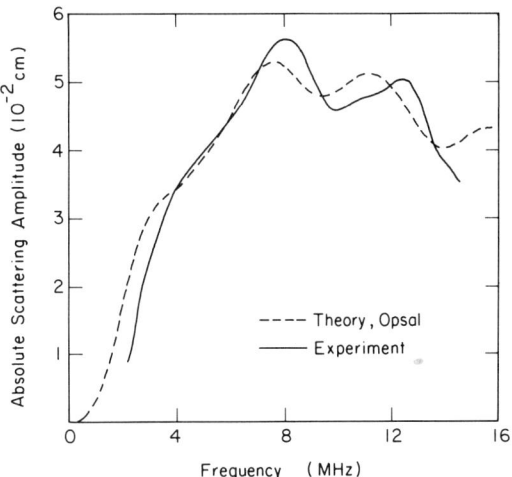

Fig. 7. Absolute scattering amplitude of an oblate spheroidal cavity in titanium with semiaxes of 400 and 200μm. The dashed curve is the theory and the solid curve is the experimental result obtained with the unipolar pulse in the pulse-echo mode.

ACKNOWLEDGEMENT

The Ames Laboratory is operated for the U.S. Department of Energy by Iowa State University under Contract No. W-7405-ENG-82. This work was supported by the Director of Energy Research, Office of Basic Energy Sciences.

# REFERENCES

1. J. M. Richardson, in 1978 Ultrasonics Symposium Proceedings (IEEE, New York, 1978), pp. 759-766.
2. W. Kohn and J. R. Rice, J. Appl. Phys. 50, 3353 (1979).
3. J. E. Gubernatis, J. A. Krumhansl, and R. M. Thomson, J. Appl. Phys. 50, 3338 (1979).
4. J. E. Gubernatis and E. Domany, J. Appl. Phys., 50, 818 (1979).
5. B. Budianski and J. R. Rice, J. Appl. Mech. 45, 453 (1978).
6. M. T. Resch, J. C. Shyne, G. S. Kino, and D. V. Nelson, in Review of Progress in Quantitative NDE 1, D. O. Thompson and D. E. Chimenti, Eds., (Plenum Press, NY, 1982), p. 573.
7. B. T. Khuri-Yakub, G. S. Kino, K. Liang, J. Tien, C. H. Choa, A. G. Evans, and D. B. Marshall, Ref. 6, p. 601.
8. J. M. Richardson and K. Fertig, in Proceedings DARPA/AF Review of Progress in Quantitative NDE, AFWAL-TR-80-4078 (1980), p. 528.
9. J. M. Richardson, in Proceedings DARPA/AFML Review of Progress in QNDE, AFML-TR-78-205 (1979), pp. 332 and 340.
10. N. Bleistein and J. K. Cohen, in Proceedings DARPA/AFML Review of Progress in QNDE, AFML-TR-78-55 (1978), pp. 73-80.
11. J. H. Rose and J. A. Krumhansl, J. Appl. Phys. 50, 2951 (1979).
12. E. Domany, K. E. Newman, and S. Teitel, in Proceedings DARPA/AFML Review of Quantitative NDE, AFWAL-TR80-4078 (1979), p. 341.
13. J. H. Rose and J. L. Opsal, in Review of Progress in Quantitative NDE 1, D. O. Thompson and D. E. Chimenti, Eds., (Plenum Press, NY, 1982), p. 573.
14. V. G. Kogan and J. H. Rose, in Review of Progress in Quantitative NDE 2, D. O. Thompson and D. E. Chimenti, Eds., (Plenum Press, NY, 1983), p. 1141.
15. J. H. Rose, R. K. Elsley, B. R. Tittman, V. V. Varadan, and V. K. Varadan, in Acoustic, Electromagnetic, and Elastic Wave Scatteirng--Focus on the T-Matrix Approach, V. K. Varadan and V. V. Varadan, Eds., (Pergamon, NY, 1980), p. 605.
16. R. C. Addison, R. K. Elsley, and J. F. Martin, in Review of Progress in QNDE, 1, D. O. Thompson and D. E. Chimenti, Eds., (Plenum Press, NY, 1982), pp. 251-261.
17. F. Yu, D. B. Ilic, B. T. Khuri-Yakub, and G. S. Kino, in IEEE Ultrasonic Symposoium Proceedings, 1979, pp. 284-288.
18. R. B. Thompson, K. M. Lakin, and J. H. Rose, in IEEE Ultrasonics Symposium Proceedings, 1981, pp. 930-935.
19. J. F. Muratore, H. R. Carleton, and H. Austerlitz, in IEEE Ultrasonics Symposium Proceedings, 1982, pp. 1049-1053.
20. B. A. Auld, Acoustic Fields and Waves in Solids, Vol. 1, (Wiley, 1973).
21. D. O. Thompson and S. J. Wormley, in Review of Progress in QNDE, 4, D. O. Thompson and D. E. Chimenti, Eds., (Plenum Press, NY, 1984), pp. 287-296.
22. J. L. Opsal, J. Appl. Phys. 58, 1102 (1985).

# ACOUSTIC TRANSDUCERS AND LENS DESIGN FOR ACOUSTIC MICROSCOPY

C-H. Chou and B. T. Khuri-Yakub

Edward L. Ginzton Laboratory
W. W. Hansen Laboratories of Physics
Stanford University
Stanford, California 94305

## INTRODUCTION

The transducer-lens system is one major component in the performance of an acoustic microscope. The design criteria for the various types of applications of an acoustic microscope are different. For surface imaging applications, it is desired to have a small spot size and low sidelobe level. For materials characterization and subsurface imaging applications (such as subsurface crack imaging), it is required to have high surface wave excitation efficiency. Several researchers addressed the problem of surface imaging.[1-5] Also, some work has been done to investigate materials properties indirectly by using so-called V(z) curves.[6-8] The resolution obtained in the latter case is always much worse than that of the corresponding lens. The direct measurement of surface wave velocity by using conventional transducer-lens systems also gives poor spatial resolution because it suffers from the interference of the specularly-reflected signal with the surface wave component caused by the low efficiency of surface wave excitation.[9] In order to increase the surface wave excitation efficiency, we first modified the design of the standard longitudinal transducer-lens system. Furthermore, we worked out a novel configuration, i.e, the shear transducer-lens system. This gives very high surface wave excitation efficiency and anisotropic acoustic beam distribution. Therefore, it can be used for direct measurements of materials properties in different directions with much less defocus than in the case of conventional transducer-lens systems. This gives the potential of many new applications of materials characterization with excellent spatial resolution.

## LONGITUDINAL TRANSDUCER-LENS SYSTEM

A typical transducer-lens system is schematically depicted in Fig. 1. For surface imaging applications, the acoustic beam generated by the longitudinal transducer propagates through the buffer rod and is focused at the surface of the sample. In order to obtain a low sidelobe level, the buffer rod length is chosen to correspond to S = 1 where

$$S = \lambda l/a^2 \tag{1}$$

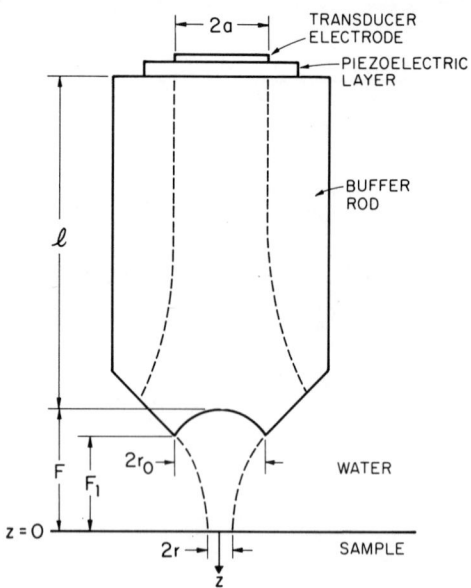

Fig. 1. Schematic diagram of typical transducer-lens system.

where $\lambda$ is the wavelength of the longitudinal wave in the buffer rod, $l$ is the length of the buffer rod, and $a$ is the radius of the transducer.

Bringing the lens closer to the sample (defocusing) induces surface acoustic waves, which give quantitative information about the materials properties, such as surface wave velocity and residual stress. Therefore, it is very important to excite surface waves efficiently for materials characterization. To evaluate the efficiency of surface wave excitation, we will look at the impulse response of the transducer-lens system in the time domain.

The theoretical calculation of the time domain response of a transducer-lens system is described in reference 10. Figure 2a shows the theoretical result of the time domain response of a longitudinal transducer-lens system with the buffer length corresponding to $S = 1$ at the center frequency of 50 MHz when defocused $-1.8$ mm. The diameter of the lens is 4.5 mm and the F-number is 1.65. The sample is hot pressed silicon nitride. It is clear that the received signal consists of the specularly-reflected signal, which comes first, and the surface wave component, which arrives second, and that the surface wave signal is weaker than the specularly-reflected signal. Figure 2b shows the corresponding experimental result which agrees with the theoretical result very well.

In order to increase the relative amplitude of the surface wave signal, it would be helpful to minimize the acoustic illumination at the center region of the lens. One way to do this would be to place the lens in the near field of the transducer, at a location where the on-axis field strength is at a minimum, such as for $S = 0.5$.

Figures 3a and 3b show the theoretical and experimental results for the time domain response of a longitudinal transducer-lens system that we constructed with a buffer length corresponding to $S = 0.6$, a center

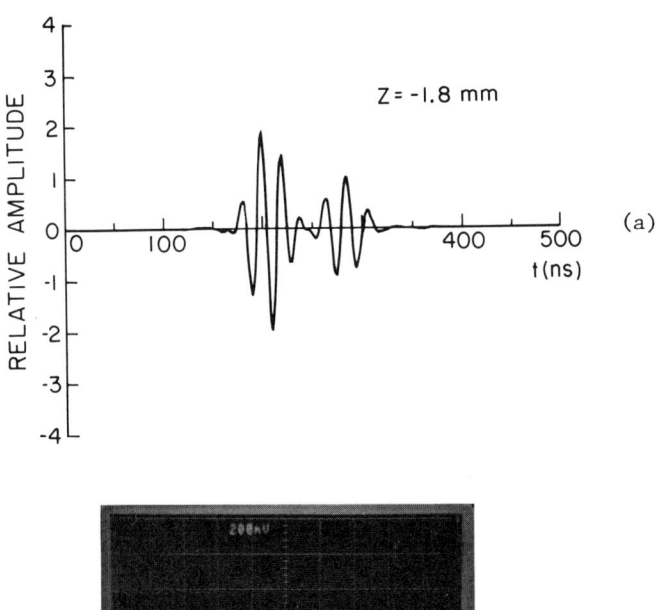

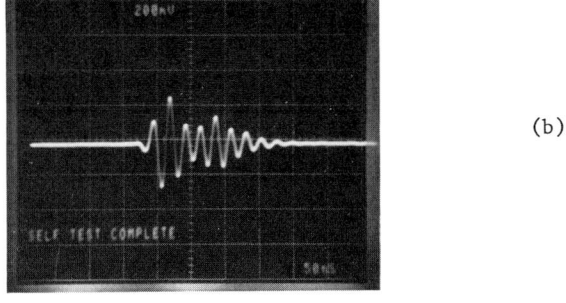

Fig. 2. The time domain response of a longitudinal transducer-lens system. S = 1 at $f_0$ = 50 MHz , Z = 1.8 mm , Lens Diameter = 4.5 mm , F-number = 1.65 , Sample: $Si_3N_4$. (a) Theoretical. (b) Experimental.

frequency of 40 MHz , and a defocusing distance of -1.8 mm . The lens diameter is 3.5 mm , the F-number is 1.65 , and the sample is also hot pressed $Si_3N_4$. The surface wave excitation efficiency, with respect to the specular reflection, in this case has been increased by about 12 dBs compared to that of the transducer-lens system of Fig. 2.

SHEAR TRANSDUCER-LENS SYSTEM

A new configuration of a transducer-lens system is a shear transducer-lens system. In this case, we use a shear polarized transducer on the buffer rod instead of the longitudinal transducer. The shear wave propagates through the buffer rod to the lens-water interface and is mode converted into a longitudinal wave in the water. Since the incident angle at different locations of the interface varies, the longitudinal transmittance is a function of r/F , where r is the radial distance from the center of the lens, and F is the focal length of the lens. Figure 4 shows the theoretical calculation of the transmittance function at the lens-water interface of a shear transducer-lens system with a buffer rod of fused quartz.[11] In this calculation, we assume that the transducer is radially polarized and the lens is in the far field of the transducer so that the incident shear wave is uniformly distributed. Considering the transmittance function as the equivalent lens illumination, we calculated the time domain response as we did for

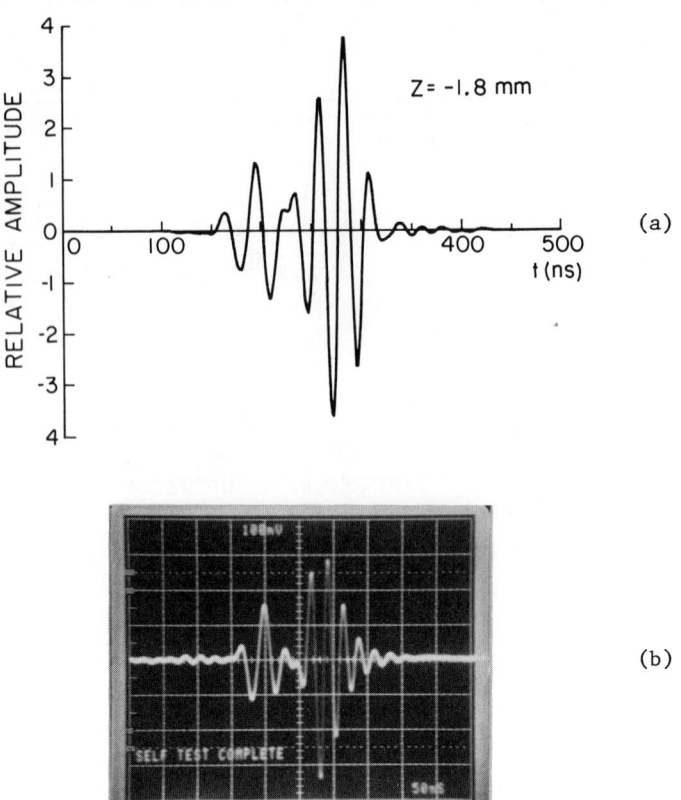

Fig. 3. The time domain response of a longitudinal transducer-lens system. S = .6 at $f_0$ = 40 MHz , Z = -1.8 mm , Lens Diameter = 3.5 mm , F-number = 1.65 , Sample: $Si_3N_4$. (a) Theoretical. (b) Experimental.

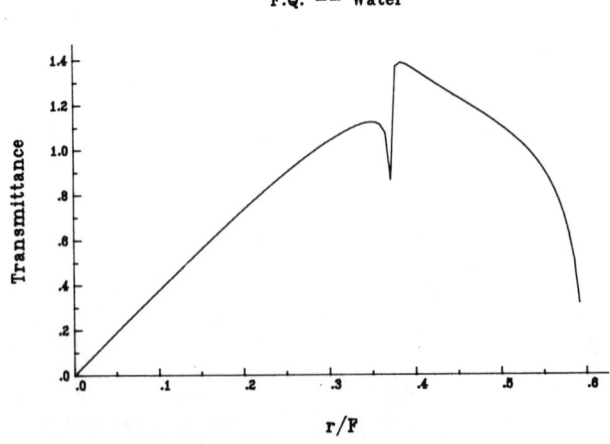

Fig. 4. Transmittance at lens-water interface of a shear transducer-lens system. Buffer: fused quartz.

the longitudinal transducer lens. The result is shown in Fig. 5 for a shear transducer-lens system with defocusing of -1.8 mm operating at a center frequency of 50 MHz. The diameter and F-number of the lens are 2 mm and 1.65, respectively. The sample is still hot pressed $Si_3N_4$.

Figure 5 shows clearly that the specularly-reflected signal is negligible compared to the surface wave component in this case. This allows us to obtain a clean surface wave with little defocusing, which is very important for materials characterization with high spatial resolution as we will see in the next session.

In practice, it is easier to construct a linearly-polarized shear transducer. Figure 6 shows the theoretical result of the acoustic field distribution in the aperture plane for a linearly-polarized shear transducer-lens system. It can be seen that the acoustic field is anisotropic in this case.

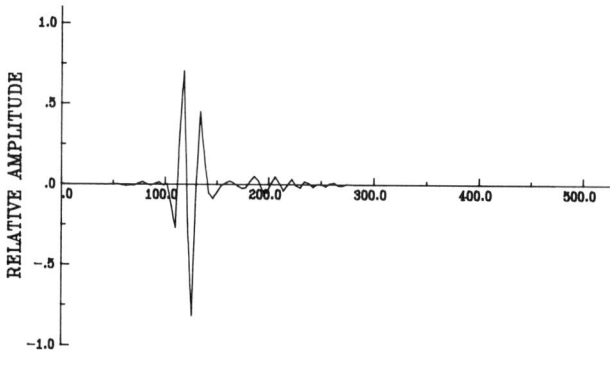

Fig. 5. The theoretical time domain response of a radially-polarized shear transducer-lens system. $f_0$ = 50 MHz, Z = -1.8 mm, lens diameter = 2 mm, F-number = 1.65, Sample: $Si_3N_4$.

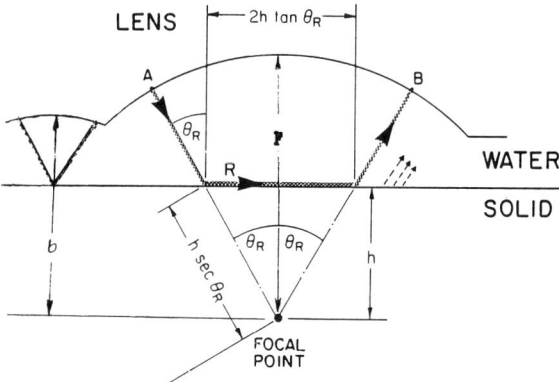

Fig. 6. Theoretical result of acoustic fields distribution in an aperture plane of a linearly-polarized shear transducer-lens system.

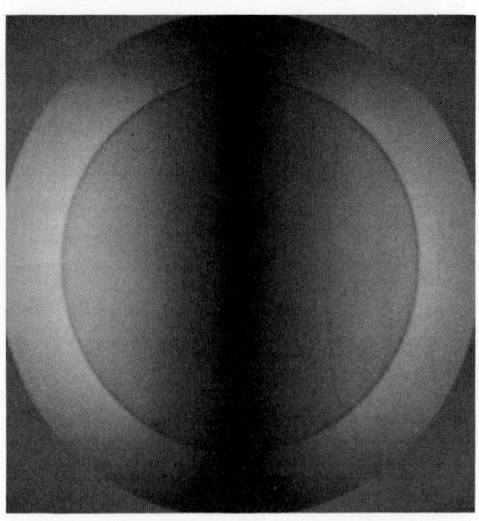

Fig. 7. Schematic diagram of the configuration of a lens for the surface wave velocity perturbation measurement by a shear transducer-lens system.

APPLICATIONS

To take advantage of the "clean" surface wave excitation and the anisotropic property of the shear transducer-lens system, many novel applications can be expected, such as measuring the surface wave velocity, residual stress, and anisotropy of materials as well as film thickness, and subsurface crack depth with good spatial resolution. Here we will give some examples of the potential applications of the shear transducer-lens system.

In order to measure the surface wave velocity of materials, a reference signal is needed. Figure 7 shows the schematic diagram of the configuration of the lens for surface wave velocity perturbation measurement. In this configuration, we introduced a cylindrical lens with weak focusing to provide a reference signal. It is designed to focus at the surface of the sample when the main lens is defocused by -1.5 mm. The F-number of this cylindrical lens is 2.5. The corresponding focal depth is 0.67 mm at 50 MHz. The maximum half-angle of the lens is below the Rayleigh angle of the samples we will work with. Therefore, the received signal by this lens has no surface wave component.

By using this configuration, we measure the phase change of the signal received by the main lens with respect to the signal received by the cylindrical lens. The phase variation caused by the surface wave velocity perturbation along the sample can be expressed by[11]

$$\Delta\phi = 2\omega h (\Delta V_R/V_R) \tan\theta_R / V_R \qquad (2)$$

where $V_R$ is the surface wave velocity and $\theta_R$ is the Rayleigh angle of the sample. Therefore, if we measure the phase change, we will be able to determine the surface wave velocity perturbation.

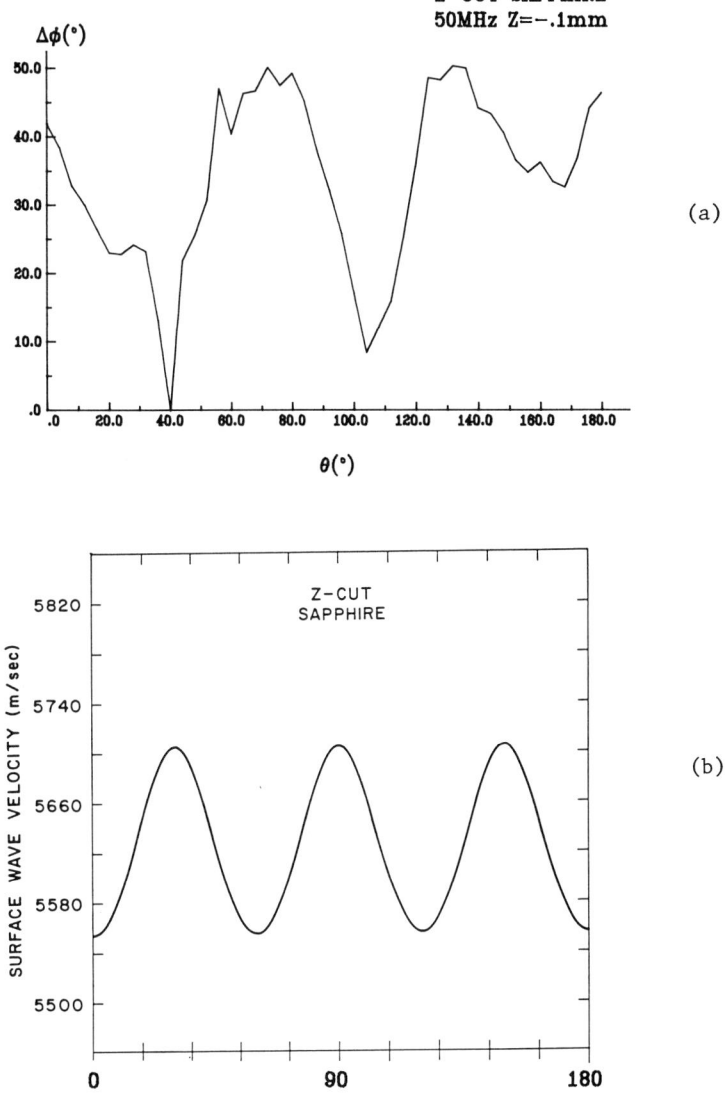

Fig. 8. (a) Measured slowness curve of Z-cut sapphire by a shear transducer-lens system with a defocus of .1 mm. (b) Theoretical prediction of the surface wave velocity slowness curve of Z-cut sapphire.

Figure 8a shows the preliminary results of the surface wave velocity slowness curve measurement of a Z-cut sapphire sample by the shear transducer lens, as described above. The operating frequency is 50 MHz. The defocusing distance is -.1 mm. Figure 8b is the theoretical prediction. The experimental result shows the six-fold crystal symmetry which is in agreement with theory.

Figure 9 shows the measured phase change across a gold film step with a thickness of 2000 Å deposited on a glass substrate by the same transducer-lens system at 50 MHz with a -.2 mm defocusing distance.

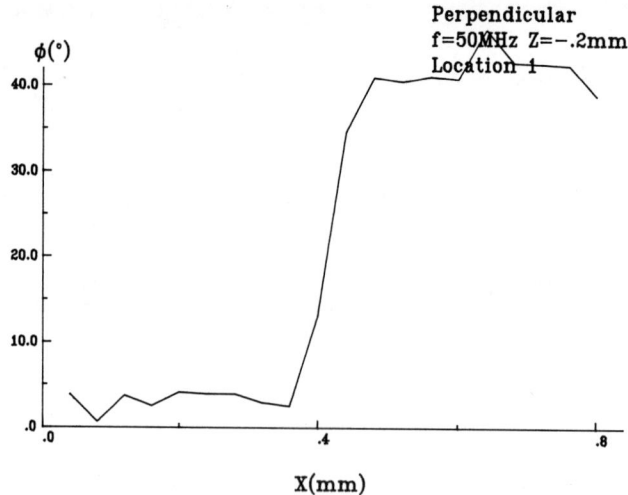

Fig. 9. Measured phase change along a film step by a shear transducer-lens system with a defocus of .2 mm. Sample: glass substrate with deposited 2000 Å Au film.

The theoretical phase change due to the existence of the film is 43°, which agrees with the experimental result of 39° very well. The spatial resolution is .1 mm in this case, which is 3.3 λ in water.

CONCLUSION

We have developed both a longitudinal transducer-lens system and a shear transducer-lens system for providing high efficiency of surface wave excitation. The novel configuration of the shear transducer-lens system makes us able to measure surface wave velocity in different directions with good resolution, which is very useful for materials characterization. The work presented here is still preliminary. There is much work to be done for improvement. Many applications are expected.

ACKNOWLEDGMENT

This work is supported by the Department of Energy under Contract No. DOE DE-FG03-84ER45157.

REFERENCES

1. C. F. Quate, A. Atalar, and H. K. Wickramasinghe, Proc. IEEE 67, 1092-1113 (1979).
2. A. J. Miller, "Applications of Acoustic Microscopy in Semiconductor Industry," in Acoustical Imaging, Vol. 12, 67-78 (Edited by E. A. Ash and C. R. Hill, Plenum Press, NY, 1982).
3. F. Faridian, Proc. IEEE Ultrasonics Symp., 759-762 (1985).
4. M. Nikoonahad and E. A. Ash, Proc. IEEE Ultrasonics Symp., 557-560 (1984).
5. A. Atalar, "Acoustic Reflection Microscope," Ph.D. Dissertation, Stanford University (1978).

6. K. K. Liang, G. S. Kino, and B. T. Khuri-Yakub, IEEE Trans. on Sonics & Ultrasonics SU-32, 213-224 (1985).
7. J. Kushibiki and N. Chubachi, IEEE Trans. on Sonics & Ultrasonics SU-32, 189-212 (1985).
8. C-H. Chou and G. S. Kino, "The Evaluation of V(z) in a Type II Reflection Microscope," accepted for publication, IEEE. Trans. on UFFC (1986).
9. K. K. Liang, "Precision Phase Measurement in Acoustic Microscopy," Ph.D. Dissertation, Stanford University (March 1985).
10. C-H. Chou, B. T. Khuri-Yakub, and G. S. Kino, "Lens Design for Acoustic Microscopy," accepted for publication, IEEE Trans. on UFFC (1986).
11. B. A. Auld, Acoustic Fields and Waves in Solids, Vol. 2, Chapter 9 (John Wiley & Sons, NY, 1973).

FIELD MAPPING AND PERFORMANCE CHARACTERIZATION OF COMMERCIAL

EDDY CURRENT PROBES*


T. E. Capobianco

National Bureau of Standards
Electromagnetic Technology Division
Boulder Colorado 80303

INTRODUCTION

The reliability of eddy current nondestructive evaluation (NDE) depends not only on correct operator procedure but also on probe quality. With the increasing emphasis on detecting smaller flaws, the logical question is, "How can probe quality be measured quantitatively?" Electrical parameters are commonly measured to determine proper impedance matching of the probes to detection instruments. While this is important, impedance matching by itself is not necessarily an assurance of satisfactory eddy current probe performance.

Previous attempts to define eddy current probe performance have used artifact standards such as plates or blocks with slotted holes or saw cuts. While this may be a good starting point, there are many unresolved questions concerning the reproducibility and size certification of manufactured defects and their reliability in simulating real fatigue cracks. Additional complications arise when considering the effects on the measured signal of the ratio of the slot depth to the skin depth of the induced eddy currents. Widely varying measurements of a single slot, i.e. data scatter, can make determination of probe performance difficult. This can result from a variety of problems ranging from poor operator technique to loosely fitting bolt hole probes. Thus, artifact standards travel with a lot of baggage which may not be immediately apparent. Their popularity can be attributed to simplicity in use and conceptual similarity to the real thing.

During the past three years, NBS in Boulder has been conducting a program aimed at understanding and quantifying probe performance and developing a characterization method based on measurement of a commonly recognized physical quantity. The issue of probe performance characterization is of enough concern to the U. S. Army that they have sponsored part of this work in an effort to develop a military standard for this purpose. This paper discusses the development of such a performance characterization method for use in a U.S. Army standard.

---

*Publication of the National Bureau of Standards, not subject to copyright.

## SAMPLES, ARTIFACTS, AND APPARATUS

The probes used for this study are commercially made, manual bolt hole probes in three sizes, 4.76, 6.35, and 7.94 mm (3/16, 1/4, and 5/16 in). These were acquired from two Air Force bases and included probes that were categorized as either good or defective based on their performance in actual NDE inspections. The defective characteristics reported include low sensitivity, noisy behavior, broken probe stems, poorly shaped probe heads, and inappropriate stem materials. The NBS program focused on the sensitivity problem and looked for evidence of "noisy behavior." The remaining complaints dealt with physical attributes which can be controlled by properly specifying purchases from vendors.

Also included with one of the probe sets was a "calibration block", an artifact standard, which is a 6.35 mm thick (1/4 in) plate of 7075 aluminum with slotted holes ranging in size from 4.76 to 19.05 mm (3/16-3/4 in) in 1.59 mm (1/16 in) increments. The slots are parallel to the hole axes and the depths are 0.48 ±0.075 mm (0.019 ±0.003 in), the width of the slots is 0.38 ±0.025 mm (0.015 ±0.001 in). Two tests were performed for each probe using this piece. One measured the change between the probe impedance in air and its impedance in the hole, but off the slot. The second measured the impedance change from "in the hole, off the slot," to "in the hole on the slot."

Another impedance change test was devised using two different conductivity materials. For this test, three tapered holes were spark cut in 19.05 mm (3/4 in) thick specimens of 7075 T6 aluminum and Ti-6Al-4V titanium alloy. The holes were tapered to accommodate dimensional variations within each probe size.

Impedance and inductance measurements were taken with a commercial low frequency impedance analyzer and voltage measurements were made with a lock-in amplifier. Probe fields were mapped using an apparatus that has been described in two earlier papers [1,2]. Briefly, it consists of a very small pickup coil which is moved around by a computer controlled, two-axis positioner. The eddy current probe being mapped is excited with an ac current and is held stationary. A computer logs pickup coil voltages as a function of position relative to the probe via a lock-in amplifier. In a variation of this setup, the bolt hole probe is rotated while the field mapping pickup coil is held stationary. The results from this rotating method are reported here. All measurements were made at 100 kHz, the nominal operating frequency of the sample probes.

## TEST METHODS EVALUATED

Because of the problems related to artifact standards, the aim of the NBS program has been to find a commonly recognized electrical or magnetic quantity which can be related to how well a probe responds to a defect without having to use a defect, either real or manufactured, for the characterization measurement.

### Field Mapping

The first approach taken was to examine the magnetic field produced by representative probes. The field mapping data were taken by rotating the eddy current probe while the field mapping sensor rested lightly on the face of the probe (Fig. 1). Figure 2 shows field profiles of two bolt hole probes with evidence of either misaligned coils, ferrite core inhomogeneities or both.

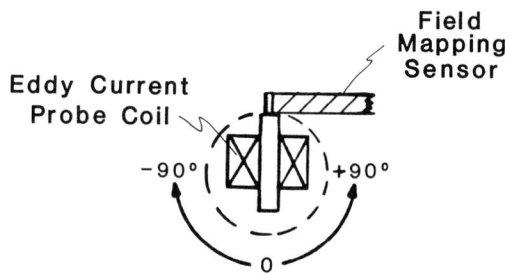

Fig. 1. Field Mapping Configuration.

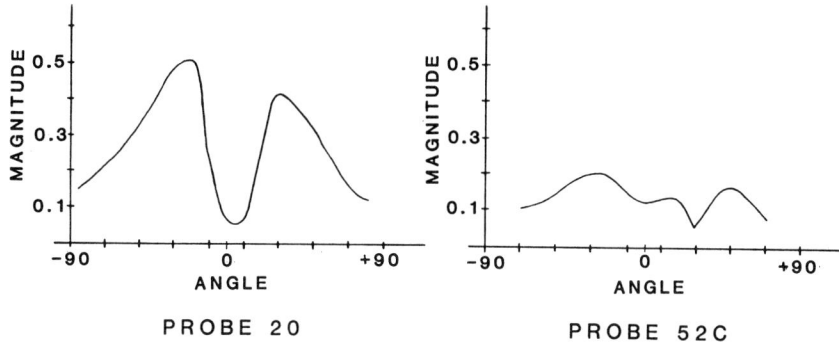

Fig. 2. Asymmetric Field Profiles of Two Bolt Hole Probes.

X-ray examinations of the ferrites were not able to resolve texture inhomogeneities. Coil tilt has been observed visually and on radiographs, but has not yet been quantified and correlated to the field maps.

Deviations of the field profiles from the expected symmetric distribution do not necessarily indicate a defective probe; however, some probes with very asymmetric profiles produced strong responses to tests of other electrical parameters. Conversely, some probes with symmetric distributions produced weak responses. The pertinent indicator seemed to be magnitude of the peak field, or the average peak value in the asymmetric cases.

Electrical Parameters

Other electrical characteristics were examined. These included the probe voltage induced by the ac field of a calibrated Helmholtz electromagnet, the impedance and frequency of the probe's self resonance, the probe impedance change ($\Delta Z$) between air and in contact with aluminum, inductance in air and on aluminum, probe Q, and probe $\Delta Z$ measured between aluminum and titanium. Of all these methods, a few are potentially useful predictors of eddy current probe performance. These include $\Delta Z$ in slotted holes, $\Delta Z$ between air and aluminum, $\Delta Z$ between titanium and aluminum, inductance in air or on aluminum. Figures 3, 4, and 5 compare the results of several characterization methods for three sets of bolt hole probes. The probe order in all three figures was determined by the magnitude of $\Delta Z$ on the slotted holes.

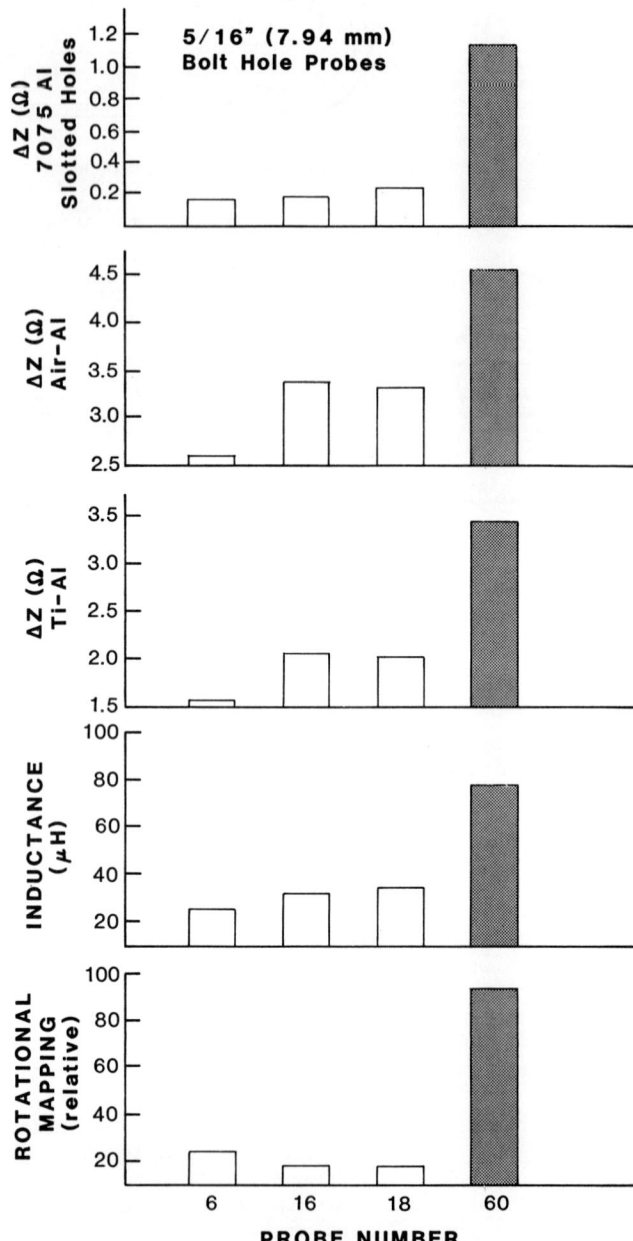

Fig. 3. Characterization method comparisons for 5/16" (7.94 mm) bolt hole probes. □ Denotes probes with complaint of low sensitivity. ▨ Denotes probes without complaint of low sensitivity.

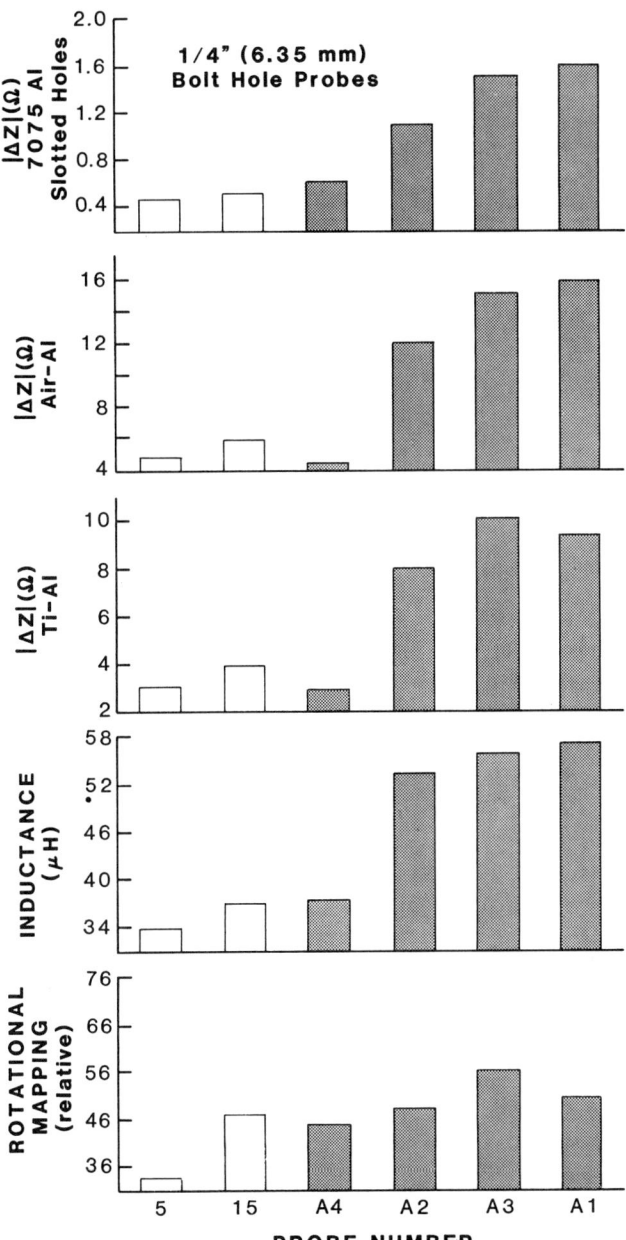

Fig. 4. Characterization method comparisons for 1/4" (6.35 mm) bolt hole probes. □ Denotes probes with complaint of low sensitivity. ▦ Denotes probes without complaint of low sensitivity.

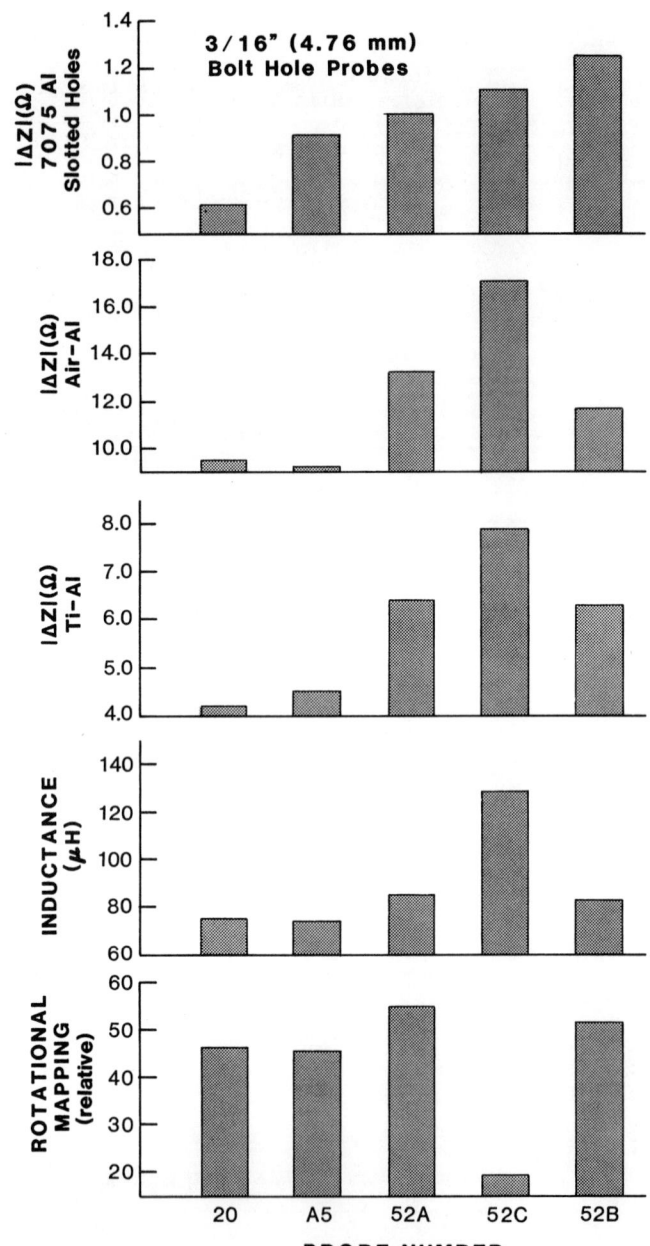

Fig. 5. Characterization method comparisons for 3/16" (4.76 mm) bolt hole probes. ☐ Denotes probes with complaint of low sensitivity. ▨ Denotes probes without complaint of low sensitivity.

DISCUSSION

One feature which becomes apparent from Figs. 3, 4, and 5 is that different test methods rank probe sensitivity differently. This, in part, is the result of circumstances particular to each measurement method. For example, during the measurements with the slotted artifact standard the probes fit rather loosely in the corresponding holes. In order to reduce the scatter observed when several readings were taken in the same hole with the same probe a tight fit was required and it was necessary to shim each probe in the hole. The probe was shimmed until the impedance variations seen while rotating the probe around the hole (off the slot) were reduced to ±0.05 ohms. Measurements of $\Delta Z$ on the same slot with the same probe became reproducible to within 5% in most cases and often within 1%. But the best $\Delta Z$ reproducibility obtained for certain probes was greater than 10% no matter how they were shimmed. This would appear to confirm the complaints by some users that some probes are noisier than others.

Measurements of the change in probe impedance on two materials of different conductivity were made. There are several advantages of this method over $\Delta Z$ measurements made on slotted holes. First, conductivity is not a difficult measurement to make; therefore, any test blocks can be well characterized to assure uniform test conditions. Second, the use of tapered holes and thick blocks minimizes data scatter from loosely fitting probes and edge effects. Third, the measurement is independent of the relationship between the eddy current skin depth and the depth of a notch, so probes of different operating frequencies can be tested on the same blocks. In addition, there is no need to recertify slot depths to account for wear. Finally, the measurement is reported in ohms and, as a result, performance criteria can be set in units independent of any particular eddy current instrument.

Inductance measurements seem to correlate well with the conductivity measurements, but enough differences exist between the two that they should be considered complementary rather than competing.

Measurements of $\Delta Z$ between air and aluminum also show good correlation with other measurements and the inclusion of this test adds the advantage of redundancy with a minimum of complications.

The field mapping measurements track the electrical parameter measurements in most cases. The exceptions, however, are so glaring that they suggest we have a long way to go to fully understand the whys of eddy current probe performance. Yet, on the plus side, unusual results from electrical measurements are often accompanied by unusual field distributions (probe 52C, Figs. 2 and 5), emphasizing the usefulness of having an additional diagnostic tool.

CONCLUSIONS

Several electrical parameters can be used to differentiate eddy current probes with poor sensitivity from those with good sensitivity. They are measurements of the probe inductance in air, impedance in air, and impedance on aluminum and titanium. Minimum values of the impedance change between air and aluminum and between titanium and aluminum and the probe inductance can be set as the performance criteria for a particular type of probe. These electrical quantities can be easily measured in a production or repair facility using standard electronic instruments with specified calibration procedures. The units used for reporting are commonly recognized, such as ohms and henries. This avoids the situation where probe performance is measured with a variety eddy current instruments and the units reported are arbitrary and relative to a particular type or brand of test instrument.

Additionally, the results suggest that the amount of data scatter in repeated $\Delta Z$ measurements of an individual probe may indicate noisy probe operation but further investigation is needed in this area.

ACKNOWLEDGEMENTS

This work was supported by the NBS Office of Nondestructive Evaluation and the U. S. Army Materials Technology Laboratory. L. Dulcie and T. Zink assisted with the measurements.

REFERENCES

1. F. R. Fickett and T. E. Capobianco, "Magnetic Field Mapping with a SQUID Device", Proceedings of the Review of Progress in Quantitative Nondestructive Evaluation, 4:401 (Plenum, NY, 1985).
2. T. E. Capobianco, F. R. Fickett, and J. C. Moulder, "Mapping of Eddy Current Probe Fields", Proceedings of the Review of Progress in Quantitative Nondestructive Evaluation, 5:705 (Plenum, NY, 1986).

# DESIGN AND CHARACTERIZATION OF UNIFORM FIELD EDDY CURRENT PROBES[*]

P. J. Shull,[1] T. E. Capobianco,[2] and J. C. Moulder[1]

[1]Fracture and Deformation Division
[2]Electromagnetic Technology Division
National Bureau of Standards
Boulder, CO 80303

## INTRODUCTION

Current practice in the design of eddy current probes calls for an optimum balance between detection sensitivity and false rejection of test pieces. Naturally, designers have emphasized improving the sensitivity and signal-to-noise ratio of eddy current probes to achieve these goals. As progress has been made in mathematical modeling of eddy current flaw signals, it has become possible to consider designing probes for specific functions and workpieces. It is now recognized that probes of different design might be required for detection and inversion. The study described here is one of the first instances in which an eddy current probe was designed specifically for inversion. This has involved balancing an entirely different set of constraints than those that must be optimized for a probe intended for detecting flaws.

The theory we use for inversion is one developed by B. A. Auld and his co-workers at Stanford University for the interaction of a uniform field with a three-dimensional flaw [1]. The use of a uniform field to interrogate flaws greatly simplifies the calculation of flaw responses. Quantitative comparisons of experimental measurements with the predictions of this theory were first reported in 1985 by Smith [2]. He used an eddy current probe described as having an "essentially uniform" field distribution and found excellent agreement between measured and predicted signals.

In this paper we describe our efforts to design a uniform field eddy current (UFEC) probe optimized for quantitative inversion of flaw signals using the uniform field theory. We characterized the probe's field uniformity by two-dimensional field mapping and studied its sensitivity to liftoff, tilt, and the proximity of edges. Measurement methods for flaws are described and illustrated with results on a series of semi-elliptical EDM slots in Ti-6Aℓ-4V. A companion paper describes in detail extensive experiments on both real and simulated surface flaws, the calibration procedures that were used, and inversion results [3].

---

[*]Contribution of the National Bureau of Standards; not subject to copyright in the United States.

PROBE DESIGN

Auld's theoretical model [1] for the interaction of a uniform magnetic field with a three-dimensional flaw makes several assumptions, and this imposes certain constraints on the probe design. First, the interrogating magnetic field must be uniform in an area that is larger than the flaw. It is difficult to estimate the degree of field uniformity required for adequate agreement between theory and experiment, but we sought to achieve a variability no greater than 10 percent over the active region of the probe. Second, the theory assumes $a/\delta \gg 1$, where a is the crack depth and $\delta$ is the skin depth. Usually, $a/\delta = 2$ is assumed to be adequate to meet this criterion. For small flaws in low-conductivity materials, this requires high-frequency operation of the probe. A third constraint, that the probe operate far below its self-resonant frequency, was found necessary to control measurement precision. Initial measurements with prototypes showed that the scatter in flaw signal measurements was acceptable as long as the phase angle of the probe impedance $\theta \geq 80°$.

The necessity for the probe to have a large area of uniformity, a strong magnetic field, and high-frequency operation produce competing effects. Increasing the number of turns on the probe increases the magnetic-field intensity between the pole tips and thus improves probe sensitivity. Unfortunately, this increases the probe's inductance, which lowers the resonant frequency. Enlarging the uniform-field area decreases the field's strength and, thereby, the probe's sensitivity and resonant frequency.

To induce a uniform flow of current across the flaw, an appropriate size and uniformity of the field are essential. In operation, the flaw is aligned parallel to the magnetic field, so that the area of uniform magnetic field must be longer in the field direction than the largest flaw to be measured. The area of uniformity must also be wider than the diameter of the calibration recesses, which were about 0.8 mm in diameter. Any increase in these dimensions relaxes the degree of accuracy necessary in positioning the probe. We assumed that the pole-tip spacing would have to be only slightly larger than the longest flaw to be measured; this fixed the size of the ferrite. To produce a large area of uniformity, we shaped the poles of a horseshoe-shaped ferrite, as shown in Fig. 1. The curving, chamfered edges of the pole tips were produced with a conical grinding tool.

The theoretical requirement that $a/\delta \gg 1$ was used to establish the upper and lower frequency bounds for the probe. To achieve $a/\delta = 2$ for the lowest conductivity material we intended to test (Ti-6Aℓ-4V, $\sigma = 5.9 \times 10^5$ S/m) and the smallest flaw that we wished to measure (a = 0.33 mm) required operation of the probe at 15 MHz. We chose as a design goal a self-resonant frequency for the probe of 18 MHz, slightly higher than the required operating frequency. Assuming a lead capacitance

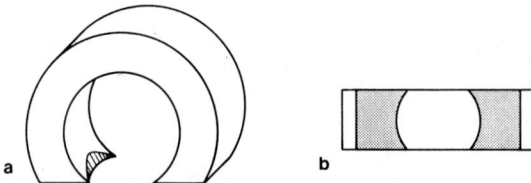

Fig. 1. Schematic of the UFEC probe showing a) the horseshoe-shaped ferrite probe body with shaped and chamfered pole tips and b) a plan view of the probe's footprint.

of 30 pF and a resonant frequency of 18 MHz, we calculated that the probe would require an inductance of 100 µH. To estimate the number of turns that would be required to obtain this inductance, we used an equation for the inductance of a ferromagnetic toroid with a narrow gap:

$$L = C \frac{\mu_0 \mu N^2}{\mu_0 (2\pi r_0 - l_g) + \mu l_g} \qquad (1)$$

where L is inductance, $\mu_0$ is the permeability of free space, µ is the permeability of the ferrite, N is the number of turns, $r_0$ is the mean radius of the toroid, $l_g$ is the length of the gap, and $C = 6.4 \times 10^{-5}$ is a geometrical factor. We arrived at an estimate of 57 turns for the ferrite we used ($\mu_r$ = 1500).

Since we wished to be able to study a wide variety of materials with conductivities ranging over two orders of magnitude, we fixed a low-frequency operating limit for the probe of 0.5 MHz by determining the frequency required to achieve a/δ = 2 for the smallest flaw (a = 0.27 mm) in the highest conductivity material, 7075 Aℓ ($\sigma = 1.87 \times 10^7$ S/m).

When we began to fabricate and test probes, we found that 57 turns on the ferrite did not give a strong enough magnetic-field intensity to produce adequate flaw signals. Furthermore, we had underestimated the amount of lead capacitance. By increasing the number of turns to 70, we were able to obtain adequate flaw signals (ΔZ = 10-60 Ω at 8 MHz) and a resonant frequency of 12.5 MHz. However, at frequencies below 2 MHz, the flaw signals became smaller and the signal-to-noise ratio fell below 10. In the end, we found it necessary to build two probes to cover the desired frequency range. NBS I, the low frequency probe, operated between 0.5 and 4 MHz; NBS II, the high frequency probe, operated from 2 to 8 MHz. Mechanical and electrical parameters for the two probes are given in Table 1.

To determine the size and uniformity of the probe's magnetic field, the magnetic field in the area between the pole tips was mapped in air, using an apparatus that has been described before [4]. The field-mapping setup consisted of a ten-turn pickup coil, 0.4 mm diameter and 0.25 mm long, mounted on a two-axis, computer-controlled positioner. Signals from the pickup coil were detected with a two-phase lock-in amplifier, also under computer control. The coil was aligned to measure the tangential component of the probe's field, i.e., the y component of the field in the x-y plane, as shown in Fig. 2.

Table 1. Mechanical and Electrical Parameters of UFEC Probes NBS I and NBS II.

|  | NBS I | NBS II |
|---|---|---|
| ID (mm) | 4.6 | 4.6 |
| OD (mm) | 9.5 | 9.5 |
| Width (mm) | 3.2 | 3.2 |
| Number of turns (AWG 44) | 100 | 70 |
| $\mu_r$ | 1500 | 1500 |
| Inductance (µH) | 421 | 112 |
| $\omega_0$ (MHz) on Ti | 4.5 | 12.5 |
| $\omega_0$ (MHz) in air | 3.9 | 8.5 |
| Quality Factor (Q) | 19.5 | 13.3 |

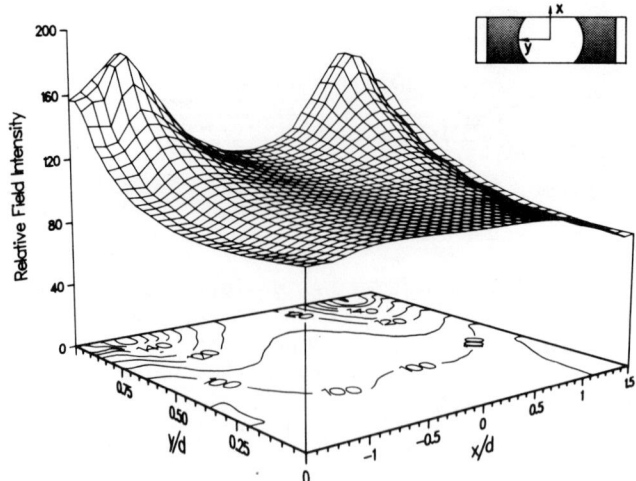

Fig. 2. Two-dimensional map of the relative magnetic-field intensity between the poles of the NBS UFEC probe. One-half of the area between the two poles is mapped from the center of the probe in the foreground to the tip of the pole in the background.

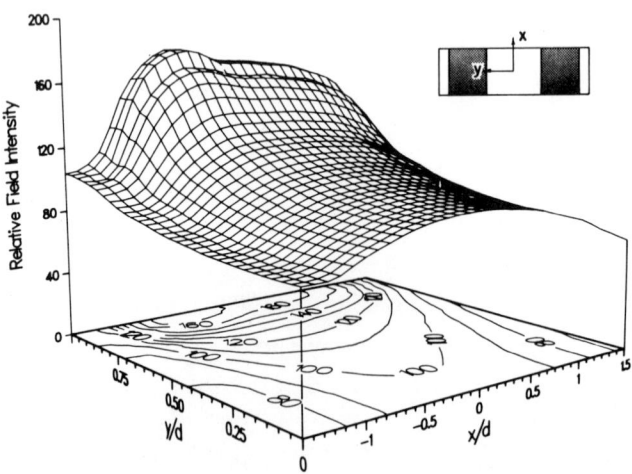

Fig. 3. Two-dimensional map of the relative magnetic-field intensity between the poles of a probe with unshaped pole tips.

For the field-mapping experiments, the probes were driven with a 25-mA ac signal at 100 kHz. With reference to Fig. 2, the probe was scanned in a raster along the x direction, starting at the center of the probe, and incrementing in the y direction until the pickup coil reached the tip of the probe. Only one-half of the probe was scanned. The results of such a scan of the UFEC probe are shown in Fig. 2 as a three-dimensional plot of the relative magnetic-field intensity, normalized to 100 at the center of the probe. The scan coordinates are normalized by d, which is half the distance between the probe tips. Contours of constant field intensity are plotted on the floor of the field map in Fig. 2. The contour plot shows that there is a broad region in the center of the probe where the field is constant to within 10 percent. For contrast, Fig. 3 shows a similar field map for a probe that did not have its feet shaped and chamfered. The superior uniformity of the design with shaped pole tips is obvious.

We also measured the x component of the probe's magnetic field along the z axis. These measurements revealed a very strong gradient of the field along this direction, and a relatively large amount of flux leakage inside the horseshoe. By distributing the windings over the entire body of the probe and chamfering the tips, the flux leakage was reduced, thereby maximizing the field in the plane of the probe's feet.

EXPERIMENT

The UFEC probe was connected in a four-terminal arrangement to an automatic network analyzer, which measured the vector impedance of the probe at 401 discrete frequencies equally distributed over the frequency range of 2-8 MHz. For brevity, only measurements with the high frequency probe, NBS II, will be discussed here. The network analyzer was controlled by a laboratory computer, which also controlled an x-y scanner that positioned the probe. The head of the scanner had an additional three degrees of freedom to aid in precise positioning of the probe: a micropositioner to control motion in the z axis, and a platform that could be tilted in the x-z and y-z planes. We found it extremely important to maintain constant liftoff and tilt during the measurements.

Flaw signals were measured by recording the impedance of the probe over the entire frequency range of interest with the probe centered over the flaw, and then again with the probe displaced laterally, at least 10 mm from the flaw. The vector difference of the on- and off-flaw impedances was then calculated at each frequency and stored. The network analyzer was capable of performing signal averaging during a measurement, and we found it helpful to average 64 times in our measurements.

RESULTS AND DISCUSSION

Fig. 4 shows an example of the results for one measurement on a semi-elliptical EDM notch in Ti-6Al-4V, 2.0 mm long, 0.85 mm deep, and 0.2 mm wide. The magnitude of $\Delta Z$ ranges from 2.3 $\Omega$ at 2 MHz to 31.9 $\Omega$ at 8 MHz. The phase of $\Delta Z$ is fairly constant at low frequencies, but begins dropping as the probe's resonant frequency is approached. To achieve higher precision, at least five such independent measurements were made on each flaw, smoothed with a running ten-point average, and then averaged to obtain the final result, illustrated in Fig. 5. We also calculated the standard deviation of both the magnitude and phase of $\Delta Z$, as illustrated in Fig. 6. The standard deviation of the magnitude of $\Delta Z$ varies from 6 m$\Omega$ at 2 MHz to 2.4 $\Omega$ at 8 MHz. The standard deviation of the phase of $\Delta Z$ remains approximately constant at about 0.8°.

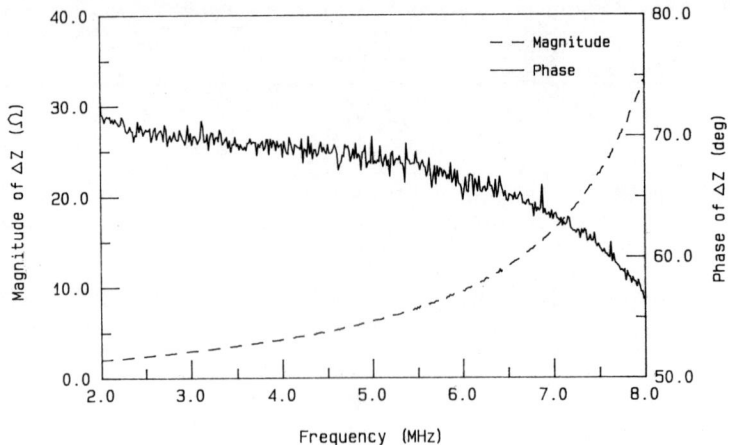

Fig. 4.   ΔZ measurement for a semi-elliptical EDM slot in Ti-6Aℓ-4V (NBS15B).

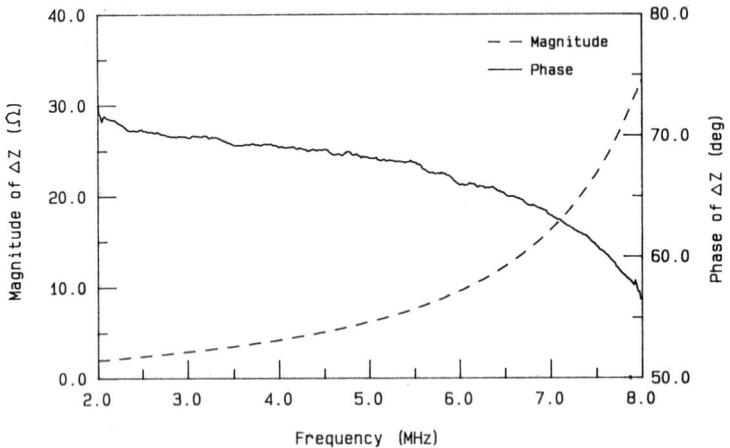

Fig. 5.   Average of five independent ΔZ measurements made on a semi-elliptical EDM slot in Ti-6Aℓ-4V.

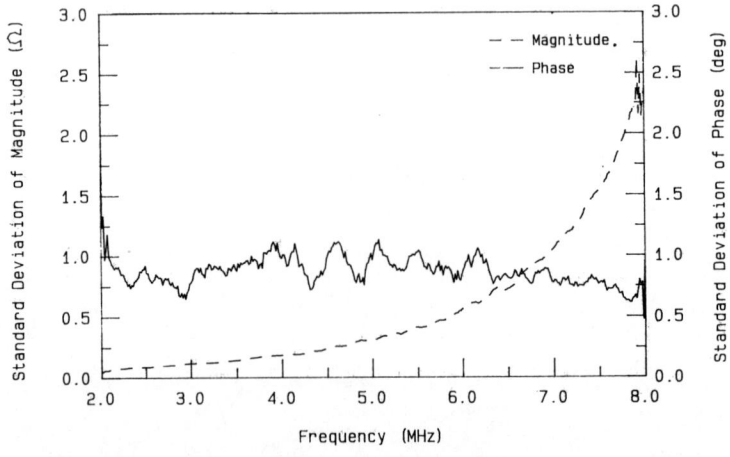

Fig. 6.   Standard deviation for five independent ΔZ measurements made on a semi-elliptical EDM slot in Ti-6Aℓ-4V.

Fig. 5 shows that the flaw signal increases with frequency, following the resonance curve of the probe. This phenomenon is well known, and follows from the fact that the current I, as measured at the input terminals of the probe, remains constant as resonance is approached, while current through the coil decreases, owing to the effects of lead capacitance, thus effectively increasing $H/I_L$ [1]. Since the signal increases as resonance is approached, it might seem best to operate at frequencies close to resonance. But, as Fig. 6 shows, the standard deviation also increases with the flaw signal, which has an adverse affect on the precision of the measurements. We found that by operating far enough below resonance that the phase of the probe impedance, Z, was greater than 80 degrees, an acceptable precision resulted.

Sensitivity of the UFEC probe to liftoff and tilt were studied by measuring flaw signals as the liftoff and tilt were varied. The same flaw shown in Figs. 4-6 was used for this study. Liftoff was varied from 0.1 mm to 0.5 mm during a series of measurements, being careful to keep the liftoff the same in both the on- and off-flaw locations. The flaw signal, ΔZ, remained constant, within experimental error, for all values of the liftoff. A similar experiment was performed while the tilt was changed over a range of a few degrees. The results of this experiment also revealed no change in the flaw signal. This was somewhat surprising, since we had experienced considerable difficulty in achieving good reproducibility of measurements unless we carefully controlled these two parameters.

The cause of the liftoff sensitivity was revealed in another experiment where we measured the signal ΔZ produced by a small change in liftoff. The probe was left in one location, away from any flaw, and ΔZ was measured for a change in height of 10 μm. This signal was measured as a function of height above the test piece. The results are shown in Fig. 7, which shows that there is a very steep gradient in the liftoff signal as a function of height in the first 0.25 mm above the surface. Thereafter, the liftoff signal remains constant, within experimental error. But the size of the liftoff signal is very large: 4.6 ohms/μm. This means that changes in the height of the probe above the surface as small as a few micrometers can give signals as big as typical flaw signals. This demonstrates clearly that the liftoff must be held to very close tolerances to obtain reliable data.

The effect of an edge on UFEC probe measurements was studied with the probe axis in two different orientations: parallel and perpendicular to the edge of a specimen. In the perpendicular orientation, the effects of the edge became apparent when the center of the probe was about 18 mm from the edge. It was slightly more sensitive in this orientation than in the parallel orientation, where effects were not noticeable until the probe was about 15 mm from the edge.

A "no-flaw" measurement was made with the probe in the same position, away from a flaw, to determine the noise level. For the magnitude of ΔZ, the amplitude of the noise varied from 0.025 Ω at 2 MHz to 0.35 Ω at 8 MHz. For a flaw 1.5 mm long, this noise level would yield a signal-to-noise ratio of approximately 10 at 2 MHz and 90 at 8 MHz.

Finally, in Fig. 8, we show the results of a series of measurements on five semi-elliptical EDM slots in Ti-6Aℓ-4V. The dimensions of the flaws are given in Table 2. Experimental results in Fig. 8 are shown as solid lines; the symbols represent theoretical predictions based on Auld's uniform field theory. The probe was calibrated with a cylindrical recess, as described in more detail in [3]. We found excellent agreement between theory and experiment for the three larger flaws (A-C). For the smaller

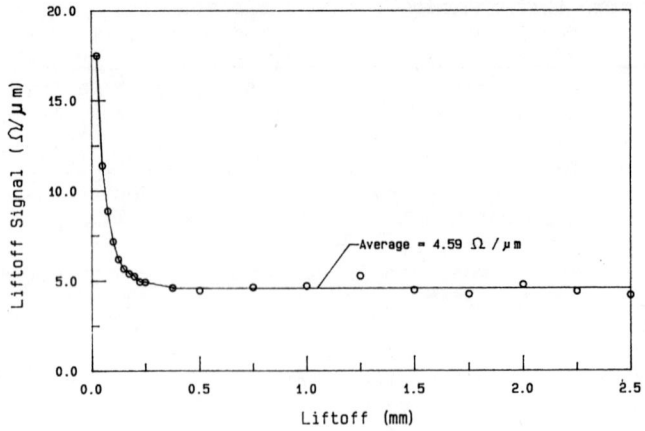

Fig. 7.   UFEC probe liftoff response on Ti-6Aℓ-4V.

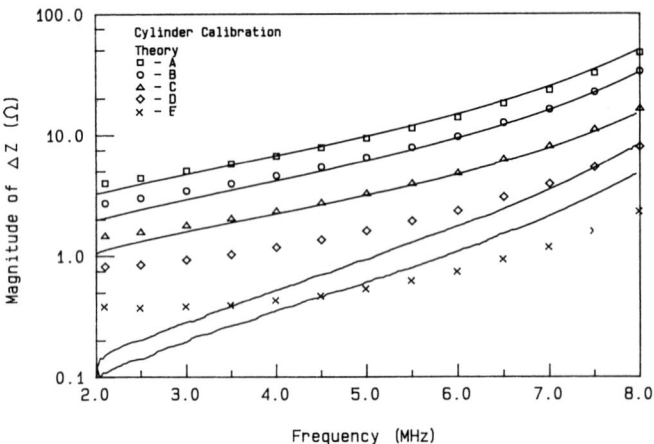

Fig. 8.   Comparison of measured and calculated flaw signals for five semi-elliptical EDM slots in Ti-6Aℓ-4V. Flaw dimensions are listed in Table 2.

Table 2. Semi-Elliptical EDM Slots in Ti-6Aℓ-4V.

| Specimen ID | Length (mm) | Width (mm) | Depth (mm) |
|---|---|---|---|
| NBS15A | 2.48 | 0.16 | 1.05 |
| NBS15B | 2.01 | 0.20 | 0.85 |
| NBS15C | 1.60 | 0.12 | 0.63 |
| NBS15D | 1.18 | 0.12 | 0.40 |
| NBS15E | 0.61 | 0.10 | 0.33 |

flaws, D and E, theory and experiment diverged significantly, but this was because, for flaw D, $a/\delta$ is 1.6 at 7 MHz and decreases with frequency from that point. For the smallest flaw, E, $a/\delta$ never exceeds 1. This suggests that even these small flaws could be measured if the probe could be operated at 15 MHz.

CONCLUSIONS

We have discussed the constraints and trade-offs involved in designing a uniform field eddy current probe intended for quantitative flaw inversion. Two probes were fabricated to cover the frequency ranges of 0.5-4 MHz and 2-8 MHz. We found that shaping the pole tips of the ferrite probe improved the size and uniformity of the field's spatial distribution, as shown by the results of field mapping. We achieved a uniformity within ten percent over an area of 2.4 x 2.2 mm.

The probe was found to be relatively insensitive to the amount of tilt or liftoff as long as it remained constant during the measurement. On the other hand, the probe was extremely sensitive to changes in either tilt or liftoff during the measurements. A signal of 4 $\Omega$ resulted from only 1 $\mu$m of vertical motion. This means that extremely precise control of probe positioning is necessary.

The probes were used successfully for quantitative measurements on surface-connected flaws in Al and Ti alloys. The size of the uniform field enabled us to measure flaws as long as 3 mm. The smallest flaws that could be accurately measured were 1 mm long and 0.4 mm deep. Smaller flaws would require a higher frequency of operation to achieve $a/\delta > 1.6$, a limit that was determined empirically in this study.

ACKNOWLEDGMENTS

This work was sponsored by the Center for Advanced Nondestructive Evaluation, operated by the Ames Laboratory, USDOE, for the Air Force Wright Aeronautical Laboratories/Materials Laboratory under Contract Number W-7405-ENG-82 with Iowa State University. We are grateful to S. Ciciora for his assistance in computer programming and to C. Cherne and T. Zinc for assistance with experimental measurements.

REFERENCES

1. B. A. Auld, F. G. Muennemann, and M. Riaziat, Quantitative modelling of flaw responses in eddy current testing, in: "Research Techniques in Nondestructive Testing, Vol. VII," R. S. Sharpe, ed., Academic Press, London (1984).
2. E. Smith, Application of uniform field eddy current technique to 3-D EDM notches and fatigue cracks, in: "Review of Progress in Quantitative Nondestructive Evaluation 5A," D. O. Thompson and D. E. Chimenti, eds., Plenum Press, New York (1986).
3. J. C. Moulder, P. J. Shull, and T. E. Capobianco, Uniform field eddy current probe: experiments and inversion for realistic flaws, these proceedings.
4. T. E. Capobianco, F. R. Fickett, and J. C. Moulder, Mapping of eddy current probe fields, in: "Review of Progress in Quantitative Nondestructive Evaluation 5A, D. O. Thompson and D. E. Chimenti, eds., Plenum Press, New York (1986).

ASSESSMENT OF EDDY CURRENT PROBE INTERACTIONS WITH

DEFECT GEOMETRY AND OPERATING PARAMETER VARIATIONS

Ward D. Rummel, Brent K. Christner, and Donald L. Long

Martin Marietta Aerospace
Denver Aerospace
Denver, Colorado 80201

INTRODUCTION

Many current generation aerospace designs are based on the application of linear elastic fracture mechanics to establish critical flaw sizes and inspection intervals based on the expected service loads and life cycles. These fracture critical designs require the application of highly reliable nondestructive inspection procedures during production and maintenance to detect defects which would result in system failure if allowed to enter or remain in service.

Eddy current inspection procedures are being increasingly used for the inspection of fracture critical hardware before entering service and subsequently during in-service or overhaul inspections due to its demonstrated capabilities for flaw detection and suitability for automation. However, the inspection technique has been improperly applied in many instances due to the lack of supporting theory, insufficient performance data, and the cost associated with experimental validation. Experimental validation of an inspection technique is time consuming and costly and must be repeated for each new application or change in parameters. The difficulty associated with experimental validation has lead to considerable attention and effort being directed toward determining the critical characteristics of NDE application and to the generation of supporting theory to facilitate the predictive modeling of NDE flaw detection capabilities. The work described in this paper is a continuation of previous work completed on assessing the characteristics of eddy current probes as applied to flaw detection in production applications [1,2,3,4].

APPROACH

Haynes 188, titanium, and aluminum test specimens containing laboratory grown fatigue cracks and electrical discharge machined (EDM) notches were prepared in cooperation with the National Bureau of Standards (NBS) for use in eddy current probe / flaw interaction studies being conducted by NBS, Martin Marietta and other investigators [1,4,5].

The work described herein was performed using a Nortec SP02065 500 KHz to 2 MHz shielded, differential surface probe and Nortec NDT-25L eddy current instrument to characterize the response obtained by scanning fatigue cracks and EDM notches in the Haynes 188 specimens while varying operating frequency, defect size, and probe orientation. The SP022065 probe has a differentially wound split ferrite, 4.75 mm in diameter and 50 mm long. All scanning was performed using a high accuracy computer controlled x-y scanner. The NDT-25L instrument was nulled on the Haynes 188 alloy specimens and the phase angle was adjusted to align lift-off response in the horizontal channel to minimize the contribution of lift-off in the vertical (amplitude) channel. The vertical channel signal was stored and recorded using a digital oscilloscope. The fatigue crack and EDM notch defects were scanned in a single line coincident with the axis of the defects with the probe starting position such that the eddy current field was unaffected by the defect. The scan speed was 0.1 inch/minute with the probe passing directly over the defects finishing in a position with the defect outside the eddy current field. The analog output from the NDT-25L was sampled every 8 msec. or .0008 inch scan distance.

The Haynes 188 alloy used in conducting this investigation is a cobalt base alloy having an electrical resistivity (70 deg F) of 92.2 microhm-cm. The fatigue cracks and EDM notches in the specimens used during the investigation are described in TABLE I.

TABLE I
DESCRIPTION OF SPECIMEN DEFECTS

| FLAW PAIR | FATIGUE CRACKS LENGTH (mm) | DEPTH (mm) | EDM NOTCHS LENGTH (mm) | DEPTH (mm) |
|---|---|---|---|---|
| 1 | 2.67 | 0.41 | 2.39 | 0.58 |
| 2 | 1.52 | 0.30 | 1.62 | 0.52 |
| 3 | 0.53 | 0.10 | 0.60 | 0.19 |

The defects listed in TABLE I were scanned at 500 KHz, 1 MHz, and 2 MHz with the probe orientated such that the flaws were perpendicular to the split in the probe ferrite (Fig. 1.a.) and again with the flaws parallel with respect to the split in the probe ferrite (Fig.1.b.).

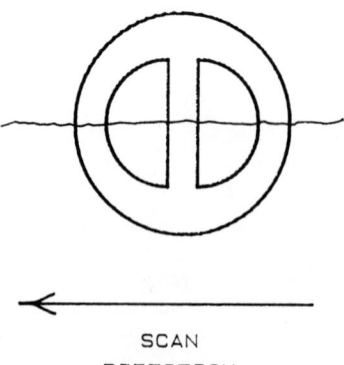

Figure 1a. Scan with split in probe ferrite perpendicular to defect axis.

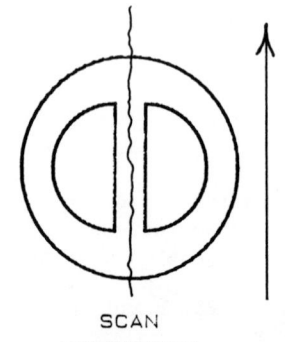

Figure 1b. Scan with split in probe ferrite parallel to defect axis.

EDM VERSUS CRACK RESPONSE

The fatigue cracks and EDM notches listed in TABLE 1 were selected to give three different crack/EDM pairs with roughly similar dimensions. The defects were scanned at 500 KHz, 1 MHz, and 2 MHz with the probe passing directly over and parallel to the defects with the probe oriented to the defect as shown in Fig. 1.a. The vertical channel of the NDT-25L instrument was monitored and plotted for each of the crack/EDM pairs at the three frequencies. The resulting plots for the 2.39 mm EDM and 2.69 mm crack are shown in Figs. 2 through 4.

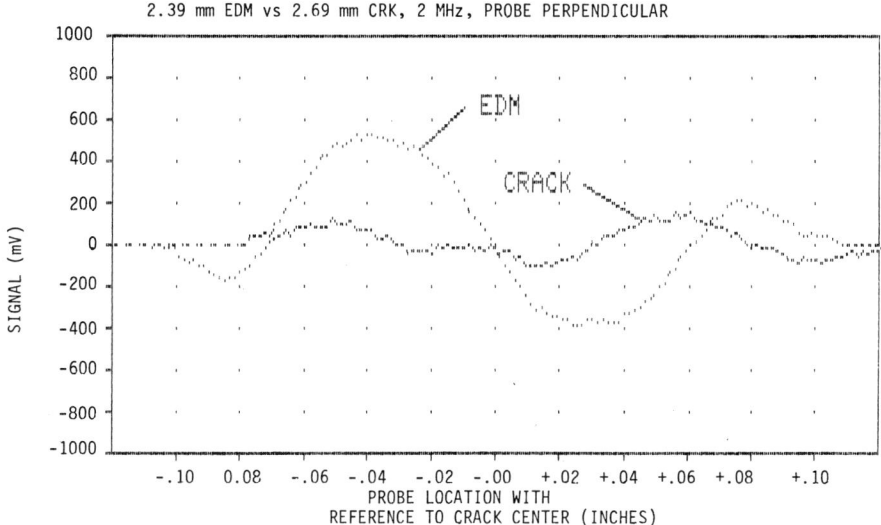

Figure 2.   Results from 2.39 mm EDM and 2.69 mm Crack Scanning at 2 MHz.

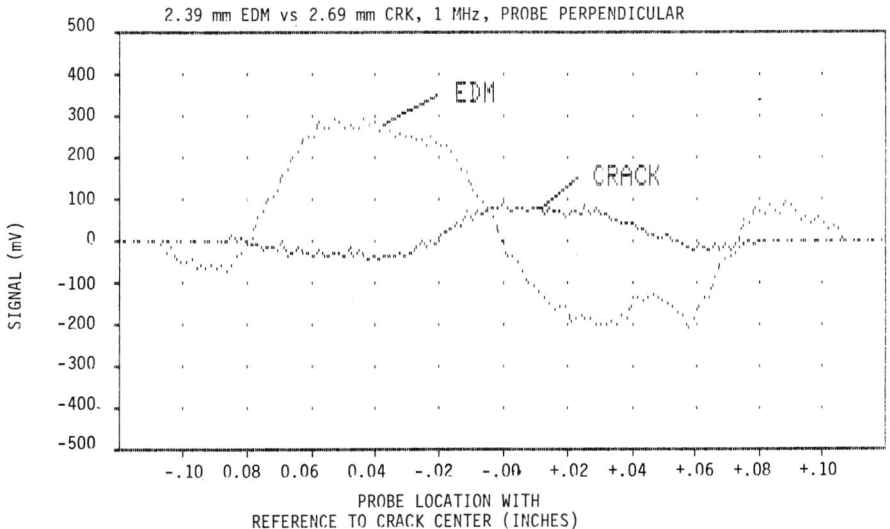

Figure 3.   Results from 2.39 mm EDM and 2.69 mm Crack Scanning at 1 MHz.

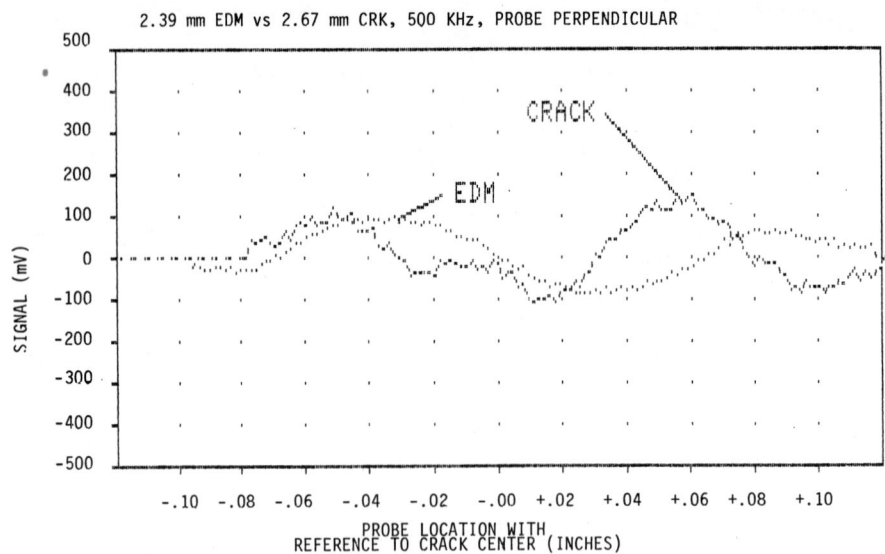

Figure 4. Results from 2.39 mm EDM and 2.69 mm Crack Scanning at 500 KHz.

The plots illustrate that the difference in response between the EDM notches and cracks decreased as a function of frequency. At 2 MHz, the maximum response obtained from scanning the EDM notches was from 4 to 5 times greater than the response obtained from scanning the similar sized fatigue cracks. At 1 MHz, the difference in response decreased to a 2 to 3 times greater response from the EDM notches. At 500 KHz, the responses from the fatigue cracks and EDM notches of similar sizes were roughly equal. It is theorized that as the eddy current field depth of penetration increases with decreasing frequency and becomes large with respect to the flaw depths, the resulting response from the defects becomes less influenced by the width of the defects. The EDM notches used in this investigation had widths ranging from 0.112 to 0.146 mm. The faces of the fatigue cracks were in partial or full contact approximating a width of 0.

The 1.62 mm EDM and 1.52 mm crack demonstrated the same decrease in differenece in amplitude as the operating frequencey decreased. The small 0.53 x 0.10 mm fatigue crack was not detectable (distiquishable from the background noise) by the probe and instrument set-up used for this investigation at the frequencies evaluated. The responses from the inspections of this defect were therefore not plotted.

RESPONSE AS A FUNCTION OF FREQUENCY

The EDM notches in the specimens were used to evaluate the decrease in response associated with a decrease in the probe operating frequency. The EDM notches were scanned at 500 KHz, 1 MHz, and 2 MHz with the probe passing directly over and parallel to the axis of the defects with the probe oriented to the defect as shown in Fig. 1.a. The vertical channel of the NDT-25L instrument was monitored and plotted for each EDM at the three frequencies. The resulting plots for the 2.37 mm EDM are shown in Fig. 5.

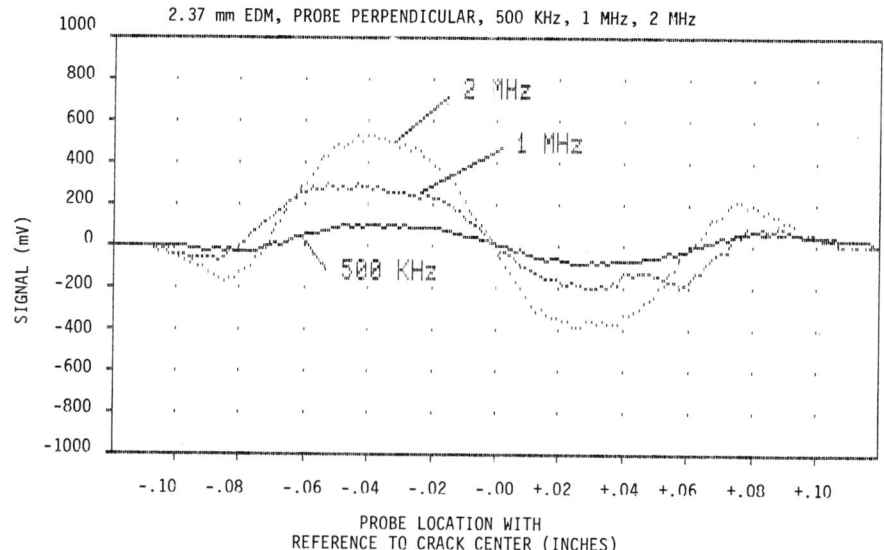

Figure 5. Results from 2.37 mm EDM Scanned at 500 KHz, 1 MHz and 2 MHz.

There was a clear and, in most instances, a proportional decrease in maximum response with a decrease in operating frequency for all three EDM sizes. The length of the response signature increased with EDM length but was not a direct function of the EDM length dimension. In all cases, the response signature was "S" shaped from positive to negative as the differentially wound probe passed over the defect from one side of the ferrite to other accounting for the change in direction of absolute amplitude.

The probe movement distance between the maximum positive amplitude and the maximum negative amplitude was nearly equal for all three defect sizes and frequencies and was roughly equal to 1/3 the probe diameter.

RESPONSE AS A FUNCTION OF PROBE ROTATION

The absolute amplitude and response signature as a result of scanning a defect are influenced by the orientation of the defect with respect to the probe coils when using a split ferrite, differentially wound probe. To evaluate this effect the three EDM defects were scanned at 500 KHz, 1 MHz and 2 MHz with the split in the ferrite oriented perpendicular to the defects (Fig. 1.a.) and with the split in the ferrite oriented parallel to the defects (Fig. 1.b.). The resulting amplitude plots as a function of probe position for the 500 KHz scans are shown in Figs. 6 and 7.

The "S" shaped response signature obtained when the probe is orientated such that the split in the ferrite is perpendicular to the axis of the defects, as described above, changes to a one directional response if the probe is rotated 90 deg and with the exception of the 2.39 mm EDM notch the absolute amplitude of the positive deflection was reduced significantly.

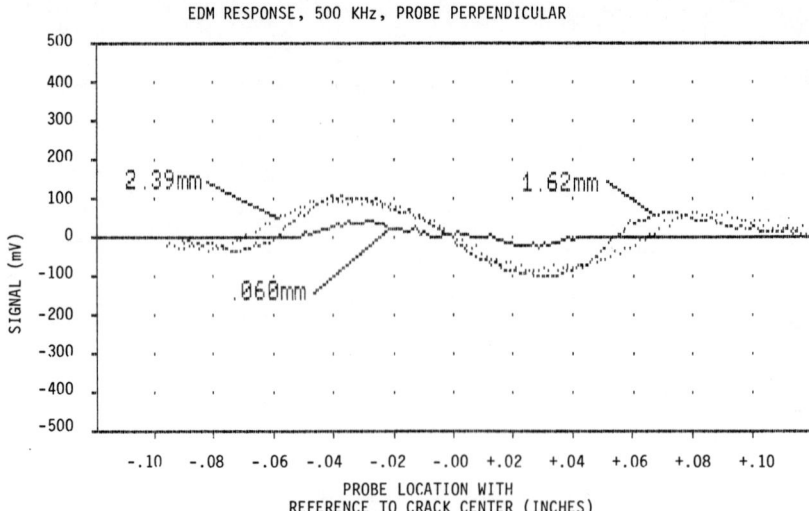

Figure 6. Results of EDM Scans at 500 KHz with flaws oriented perpendicular to the split in the probe ferrite.

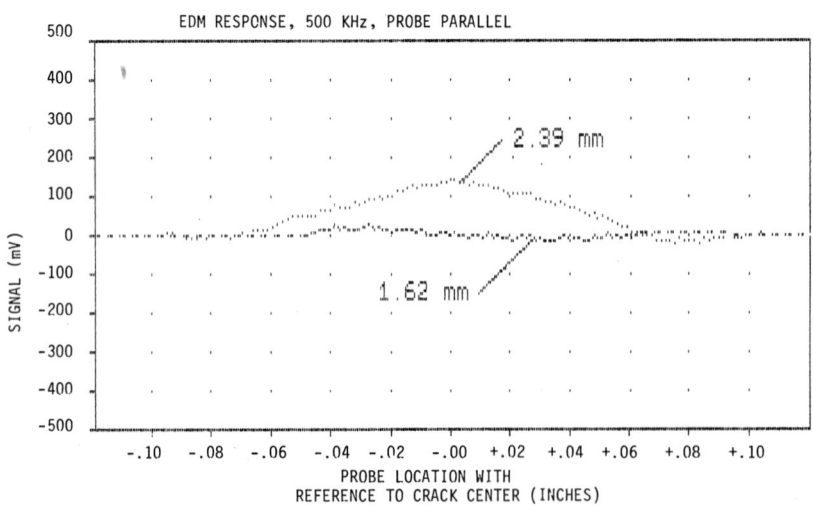

Figure 7. Results of EDM Scans at 500 KHz with the flaws oriented parallel to the split in the probe ferrite.

The response obtained from the 2.39 mm EDM notch after rotating the probe 90 deg increased over the absolute value of the positive deflection obtained with the probe oriented such that the split in the ferrite was perpendicular to the axis of the defect. The response from the 0.60 mm EDM notch became undistiquishable from the background noise after rotating the probe 90 deg at 2 MHz. At 500 KHz the response from both the 1.62 and 0.60 mm EDM notches became undistinquishable from the background noise after the probe was rotated 90 deg. The loss in detection of the two smaller cracks as the probe was rotated indicates that inspection reliability or probability of detection for defects that are small with respect to the eddy current field size is a function of probe orientation using differentially wound type probes and must be considered when designing and engineering eddy current inspection processes for the detection of small flaws.

To illustrate how the response changes as the probe is rotated, the 2.39 mm EDM was scanned repetitively rotating the probe in 30 deg increments between scans. The resulting amplitude plots are shown in FIGURE 9. The "S" shaped signature became more skewed with each increment until the probe was rotated the full 90 deg at which point all negative deflection in response was eliminated. There was also an increase in absolute value of the positive deflection with each increment in probe rotation.

The responses obtained from the two EDM notches smaller than the probe diameter also exhibited the skew in the "S" shaped response as the probe was rotated. However, the absolute amplitude from the smaller flaws decreased as described above rather than increased as did the responses from the EDM notch larger than the probe diameter.

CONCLUSIONS

Eddy current probe / flaw interaction assessments such as those described in this paper offer the opportunity to increase the knowledge base required to extend the theoretical modeling procedures that exist for aiding the engineering assessment and predictive analysis of eddy current inspection performance and reliablity in critical production NDE applications.

The work described in this paper has shown that defect width contributes significantly to the eddy current response from a defect at high operating frequencies. The EDM notches evaluated in this investigation produced significantly greater amplitudes than did similar sized fatigue cracks at an operating frequency of 2 MHz. The difference in response between the two defect types became insignificant as the operating frequency was reduced to 500 KHz.

The response obtained from all the defect types and sizes evaluated decreased as the operating frequency was decreased. It most cases the reponse decreased proportionally to the decrease in frequency.

The response and detectability of flaws smaller than the eddy current probe diameter are functions of the probe orientation with respect to the axis of the flaws when using differentially wound type probes. This knowledge must be incorporated in the design and engineering of eddy current inspection processes for the detection of small defects.

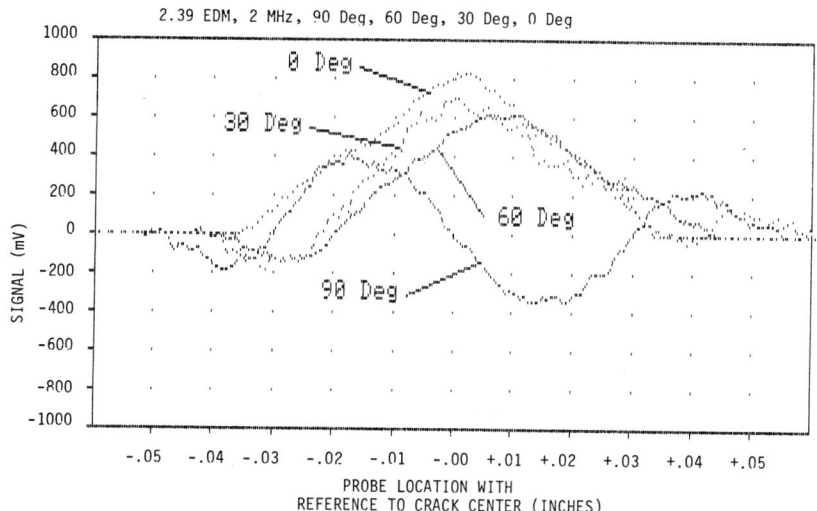

Figure 8. Results of 2.39 mm EDM Scans rotating probe 30 deg. between scans.

ACKNOWLEDGEMENTS

The work described in this paper was sponsored by the Center for Advanced Nondestructive Evaluation, operated by the Ames Laboratory, USDOE, for the Air Force Wright Aeronautical Laboratories / Materials Laboratory under Contract No. W-7405-ENG-82 with Iowa State University.

REFERENCES

1. D. O. Thompson and D. E. Chimenti, eds., <u>Review of Progress in Quantitative Nondestructive Evaluation</u>, (Plenum Press, New York).
2. Ward D. Rummel, Richard Rathke and Robert Schaller, "Eddy Current Test Samples, Probes and Scanning System", in <u>Review of Progress in Quantitative Nondestructive Evaluation 3A</u>, ed. by D. O. Thompson and D. E. Chimenti (Plenum Press, New York, 1983) pp. 561-568.
3. Ward D. Rummel, Brent K. Christner and Steven J. Mullen, "Assessment of the Effects of Scanning Variations and Eddy Current Probe Type on Crack Detection", in <u>Review of Progress in Quantitative Nondestructive Evaluation 4</u>, ed. by D. O. Thompson and D. E. Chimenti (Plenum Press, New York, 1984) pp. 1319-1326.
4. Ward D. Rummel, Brent K. Chrisner and Steven J. Mullen, "The Influence of Calibration and Acceptance Criteria on Crack Detection and Discrimination by Eddy Current Techniques", in <u>Review of Progress in Quantitative Nondestructive Evaluation 5A</u>, ed. by D. O. Thompson and D. E. Chimenti (Plenum Press, New York, 1985) pp. 929-946.
5. J. C. Moulder, J. C. Gerlitz, B. A. Auld and S. Jefferies, "Semi-Elliptical Surface Flaw Eddy Current Interaction and Inversion: Theory and Experiment", in <u>Review of Progress in Quantitative Nondestructive Evaluation 5A</u>, ed. by D. O. Thompson and D. E. Chimenti (Plenum Press, New York, 1985) pp. 383-402.

# EFFECTS OF SHIELDING ON PROPERTIES OF EDDY CURRENT PROBES WITH FERRITE CUP CORES

S. N. Vernon and T. A. O. Gross[*]

Naval Surface Weapons Center, Silver Spring, MD 20903-5000

[*]T. A. O. Gross, Inc., Lincoln, MA 01773-4116

## INTRODUCTION

In eddy current inspection the ability to detect small defects depends on the sensitivity of the system and on the relative sizes of the probe and the defect. To detect defects on the opposite surface the probe radius should be at least as great as the thickness of the material. This limits the sensitivity to small defects that can be achieved by decreasing the probe size. Assuming the instrumentation is a given, further sensitivity can be achieved by improving the sensitivity of the probe itself.

Probe sensitivity depends on the coupling between the probe and the test material. Coupling is illustrated by the normalized impedance diagram (Fig. 1). Normalized reactance ($X_n$) is the ratio of the reactance of the probe in contact with the material ($\omega L_m$) to the reactance of the probe in air ($\omega L_o$). Normalized resistance is the difference between the resistance of the probe on the material and the resistance in air divided by $\omega L_o$. Sensitivity can be described in terms of the coupling coefficient $K$ where $K^2 = 1 - X_\alpha$. The quantity $X_\alpha$ is the value of the normalized impedance where the extrapolated impedance curve intersects the reactance axis [1] (see Fig. 1). The coupling coefficient is proportional to the percentage of the total flux which links with the test material.

A number of factors affect the coupling coefficient. The most obvious is the separation between the probe and the test material (lift-off) as illustrated in Figure 1. Even with zero lift-off a coupling coefficient of 0.89 ($X_\alpha = .2$) would be excellent. Probe type [1] and distribution of windings [2] also affect probe sensitivity. Shielding can improve coupling by forcing into the test material some of the flux that would otherwise link within the coil itself.

This paper describes the effects of shielding on the sensitivity, or coupling, of a ferrite cup, or pot, core probe. Two types of shielding were investigated: self-shielding which results from skin effects in adjacent turns of the wire forming the coil and imposed shielding consisting of highly conductive metal on those surfaces of the cup which are not in contact with the test material. Bailey [3] and Ellsberry and Bailey [4] have reported on the characteristics of ferrite cylindrical core probes with imposed shielding. Gross [5] has investigated the effects of torroidal core probes with imposed shielding both on field intensity as a function of distance from the probe face and on coupling. The cup core probe was selected for this investigation because even without shielding it appears to offer better coupling than either cylindrical or torroidal core probes.

SELF-SHIELDING

The normalized impedance diagrams shown in Figure 2 illustrate the shielding that can be achieved using thick wire*. The inner curve was generated by a probe having 15 turns of AWG #32 wire which has a diameter of 0.2 mm. It should be noted that the normalized impedance values for a carbon/carbon 3-D weave material ($\rho \simeq 900$ $\mu\Omega$ cm) fall on the same curve as the values for aluminum ($\rho \simeq 6$ $\mu\Omega$), as would be expected from theory. [6]. The lower curve was generated with a probe having 15 turns of #20 wire (diameter of 0.8 mm). At about 16 KHz the slope of the curve begins to change and the coupling improves with increasing frequency.

The curves in Figure 3 show those values of the probe inductance in air, normalized with respect to that inductance at 1 KHz, which are associated with each point on the lower portions of the impedance diagrams for the two probes. The impedance diagram for the #32 wire probe shows no change in slope and no change in the inductance of the probe in air. In contrast, the impedance curve for the #20 wire probe has a change in slope and the normalized inductance begins to decrease at a frequency slightly lower than the frequency at which the change in slope in the impedance curve becomes evident.

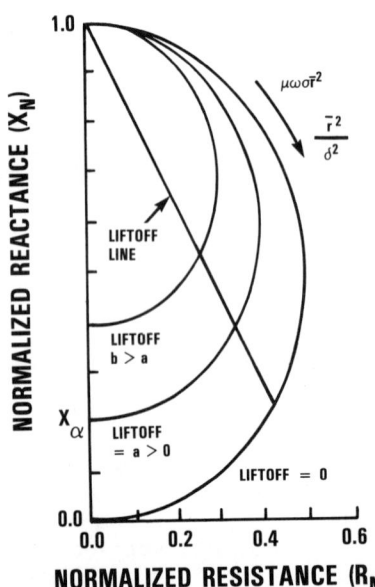

Fig. 1. Normalized impedance diagram showing the effects of lift-off, conductivity ($\sigma$), frequency ($\omega$), mean coil radius ($r$), and skin depth ($\delta$) for an idealized case.

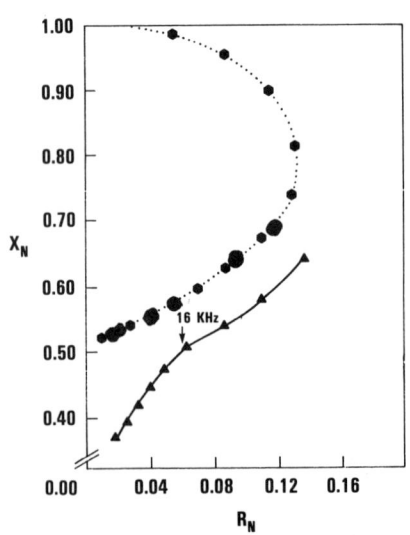

Fig. 2. Impedance diagrams for a probe wound with #32 wire (....) and for a #20 wire probe (--). Test materials: aluminum, $\rho$ = 6 $\mu\Omega$ cm (● and ▲); carbon/carbon, $\rho$ = 900 $\mu\Omega$ (●).

---

* All the probes described in this paper had an outside diameter of 3.6 cm. Similar results were obtained with probes as small as 0.8 cm.

The inductance of the probe in air is a measure of the mutual inductance among the turns in the coil. It is our hypothesis that as the frequency increases to a point where the skin depth in the wire approaches the radius of the wire, the flux associated with each turn in the coil no longer penetrates adjacent turns; there is mutual shielding in the coil. This reduces the mutual coupling and the inductance of the coil is slightly reduced. This mutual shielding has two effects. First, the flux that was linking within the coil is pushed into the surrounding ferrite and the test material, thereby increasing the coupling coefficient. Second, the mutual coupling and, consequently, the coil inductance, is slightly reduced. This reduction in coil inductance is not directly reflected in the change in slope of the impedance curve because the reactance is normalized with respect to both the frequency and the inductance. If the change in slope were a direct reflection of the decrease in probe inductance, the slope would change in the opposite direction.

Figure 4 shows the impedance curves and corresponding normalized inductance curves for a probe made with #22 wire (diameter = 0.6 mm) as well as those for the #20 wire probe. The change in slope in the impedance curve for the #22 wire probe appears at about 40 KHz. The inductance, again, begins to decrease at a slightly lower frequency.

The frequency at which the skin depth in copper is equal to the wire radius is given in Table I for the three gauges of wire investigated. For #20 and #22 wire, the changes in slope were observed at 16 KHz and 40 KHz respectively. The curves were generated by increasing the frequency in octave steps, consequently the frequency at which the slope change occurs cannot be identified more accurately than about ± 75%. Given this uncertainty, there is good agreement with the values given in the table - except for the #32 wire. The effect on the curves for the #22 wire probe is less than the effect on the #20 wire. It is reasonable that the thinner the wire the less the shielding effect. #32 wire is simply too thin to prove an observable effect.

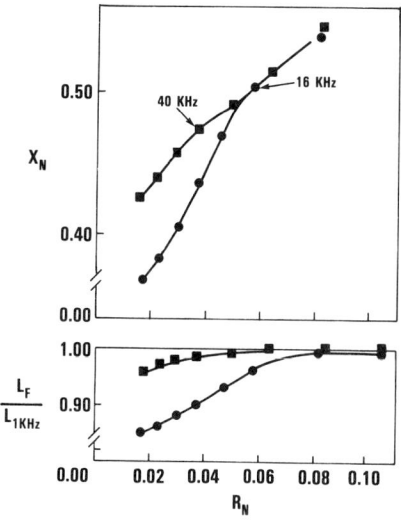

Fig. 3. Impedance diagrams and values of normalized inductance $L_f/L_{1 KHz}$) associated with each impedance value for #32 wire probe (●) and #20 wire probe (▲).

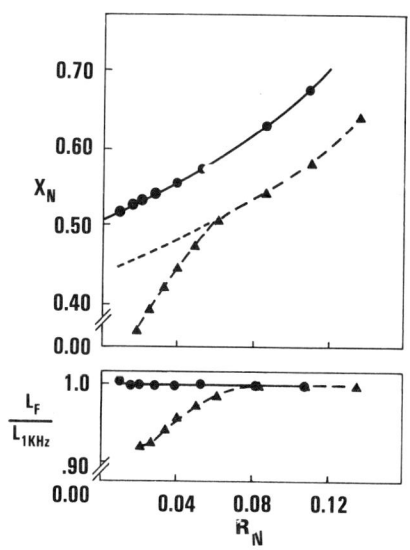

Fig. 4. Impedance diagrams and associated normalized inductance values for #20 wire probe (●) and for #22 wire probe (■).

TABLE I

Frequency at which the skin depth in Cu = wire radius

| WIRE GAUGE | WIRE RADIUS (in.) | FREQ. AT WHICH SKIN DEPTH = RADIUS (KHz) |
|---|---|---|
| 32 | 0.004 | 422 |
| 22 | 0.0125 | 43 |
| 20 | 0.016 | 26 |

Figure 5 shows the #20 wire probe impedance curves for aluminum, tantalum, titanium alloy, and carbon/carbon. The curve for each material deviates from the unshielded curve between about 16 and 40KHz. The deviation is much less obvious in the upper portion of the curve, but it is present. It is generally assumed that the impedance curve is independent of the resistivity of the test material, as was the case for the #32 wire probe curve.

The benefits of increased sensitivity resulting from self-shielding effects can be applied to the inspection of any material. In the inspection of aluminum, for example, frequencies above 40 KHz are normally used. Since aluminum components inspected by eddy current methods are generally quite thin, smaller probes are used than those considered here. As a consequence the curve for aluminum could correspond to the curve for carbon/carbon where the benefits of self-shielding occur in the useable [1] mid range of the curve.

If a very small probe is required it may not be possible to use thick wire and still have enough turns to provide the impedance required by the instrumentation. Similar shielding effects can be achieved with imposed shielding.

IMPOSED SHIELDING

Shielding, in the form of metal sheets or electro-deposited metal, can be placed on any or all of the five surfaces: a, b, c, d, and e, shown in Figure 6. The method for the electro-deposition of copper is described in reference 5. Experimentation has indicated [5] that the shielding should be electrically insulated from the ferrite. Copper sheeting having a thickness of 0.005 inches is convenient to use on large probes. Since it is difficult to accurately cut small pieces of the sheeting, it may be more convenient to use electro-deposition for small probes. Any high conductivity metal will suffice. Silver offers some improvement but may not be worth the additional cost. It is important that there be a gap in the circumference of the shielding to avoid the shorted-turn effect. [5]

The effects typical of imposed shielding are shown (Fig. 7) for the #32 wire probe. The solid curve is the unshielded impedance diagram for both carbon/carbon and aluminum. The dashed line is the aluminum impedance curve for the same coil placed in a ferrite cup core with copper deposited on the 5 surfaces. The dotted line is the corresponding curve for carbon/carbon. Probe sensitivity, or coupling, begins to show an improvement over the unshielded probe at about 16 to 32 KHz. Below this frequency the coupling is worse for the shielded than for the unshielded probe. This effect was found to be independent of number of turns and probe size. The reason for this effect is not well understood.

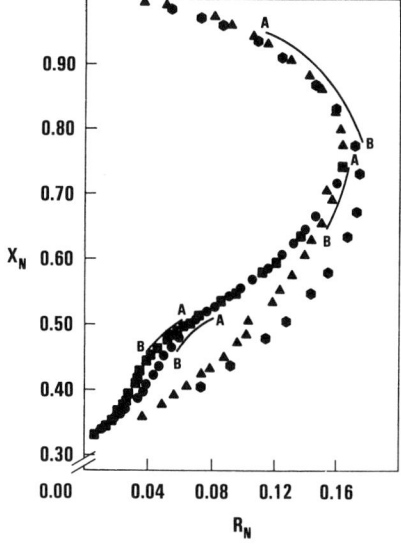

Fig. 5. Impedance diagrams for #20 wire probe on aluminum (■) tantalum (● $\rho$ = 13 $\mu\Omega$ cm), titanium alloy (▲ $\rho$ = 171 $\mu\Omega$ cm) and carbon/carbon (⬢).
A = 16 KHz, B = 40 KHz

Fig. 6. Cross-section of a ferrite cup core showing surfaces that can be shielded.

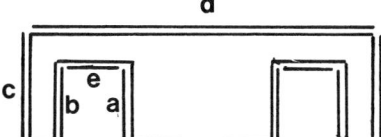

Again the benefits of shielding can be achieved for the inspection of high conductivity metals by using a smaller probe at frequencies above 16 KHz.

Some further increase in coupling can be achieved by adding imposed shielding to probes which already provide self-shielding. Figure 8 illustrates this effect on the impedance curve of the #20 wire probe. The normalized inductance of the shielded probe in air begins to decrease immediately, agreeing with the evidence that shielding has an effect, albeit an unwanted one, at very low frequencies.

Reference 5 reports the effects of imposed shielding on the intensity of the field as a function of distance, both along the probe axis and normal to the axis. Within a distance of one radius normal to the probe face, the field was stronger for the shielded probe. At a distance of five radii the field was stronger for the unshielded probe.

A cursory investigation indicated (Figure 9) that when the windings filled the cup of the core, shielding only surface b resulted in a less negative effect on coupling at low frequencies and improved coupling at a lower frequency.

CONCLUSIONS

A significant improvement in probe sensitivity can be achieved through the use of shielding. Self-shielding results from the use of larger diameter wire when the operating frequencies are greater than the frequency at which the skin depth in the wire is greater than the radius of the wire.

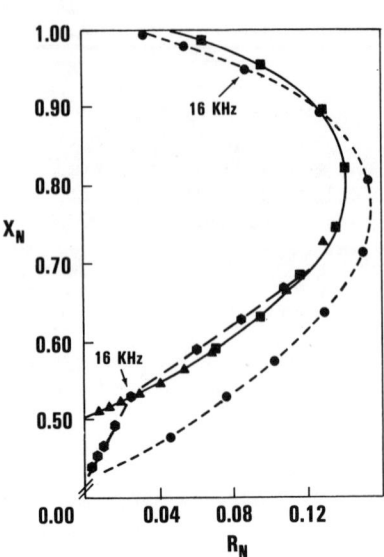

Fig. 7. Impedance diagrams for the #32 wire probe: ■ unshielded on carbon/carbon, ▲ unshielded on aluminum, ● shielded on c/c, ⬢ shielded on Al.

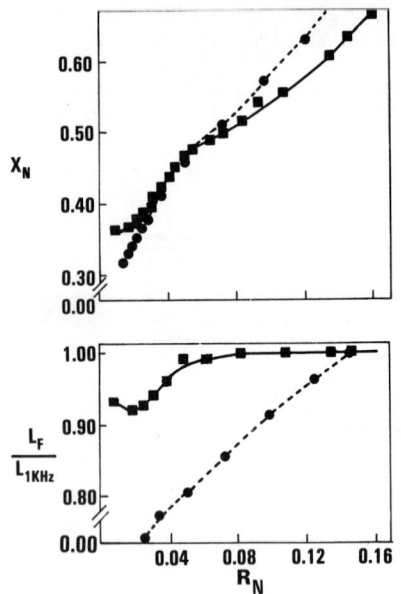

Fig. 8. Aluminum impedance diagrams and associated values of normalized inductance for the #20 wire probe with no imposed shielding (■) and with imposed shielding (●).

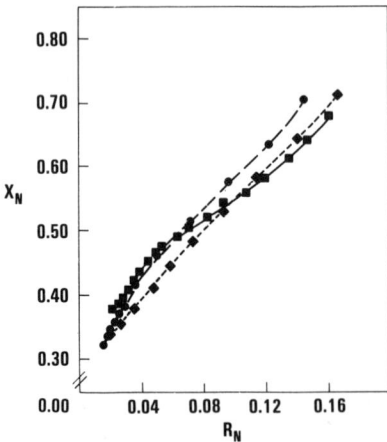

Fig. 9. Impedance diagrams for the #20 wire probe with no imposed shielding (■), with deposited shielding on 5 surfaces (●), and with shielding only on surface b (◆).

Theoretical modeling should take into account the effects of self-shielding. In the absence of self-shielding the coupling coefficient is independent of the resistivity of the test material. For a given probe normalized impedance depends on the quantity $\mu\sigma\omega r^2$ where $\mu$ is the relative permeability, $\sigma$ is the conductivity, $\omega$ is the angular frequency and r is the mean radius of the coil. When the frequency and wire size are such that there is a self-shielding effect, the normalized probe impedance has an additional dependency on material conductivity and frequency. The impedance associated with the material affects the flaw response, so skin effects in the wire should be considered in flaw inversion modeling.

Imposed shielding can increase the sensitivity of probes that are too small to contain the necessary number of turns of larger diameter wire.

REFERENCES

1. Vernon, S.N., "Probe Properties Affecting the Eddy Current NDI of Graphite Epoxy," Review of Progress in Quantitative Nondestructive Evaluation, Vol 5B, Donald O. Thompson and Dale E. Chementi, Ed., Plenum Press, New York, (pub 1986), pp 1113-1123.
2. Smith, S.H. and C.V. Dodd, "Optimization of Eddy Current Measurements of Coil toConductor Spacing," Materials Evaluation, Vol. 33, No. 12, Dec 1975, pp 279-283, 292.
3. Bailey, D.M., "Shielded Eddy Current Probes," Materials Evaluation Evaluation, Vol. 41, Jun 1983, pp 776-777.
4. Elsberry, R.T. and D.M. Bailey, "Characterization of Shielded Eddy Current Probes," Materials Evaluation, Vol. 44, No. 8, July 1986, pp 984-988.
5. Gross, T.A.O., "Final Report: Eddy Current Inspection of Graphite Epoxy Composites," Final Report, Naval Surface Weapons Center Contract No. N60921-85-C-0257, 7 Apr 1986.
6. Libby, H.L., Introduction to Electromagnetic Test Methods (New York: John Wiley and Sons, Inc., 1971), pp 122-179.

PICKUP COIL SPACING EFFECTS ON EDDY CURRENT

REFLECTION PROBE SENSITIVITY[†]

        T. E. Capobianco and Kun Yu*

        National Bureau of Standards
        Electromagnetic Technology Division
        Boulder Colorado

        * Institute of Physics
        Chinese Academy of Sciences
        Beijing China

INTRODUCTION

    Differential probes have existed for many years and have been produced in a variety of configurations. The common feature of these diverse probes is that they detect an ac magnetic field gradient, but it is the desired direction of gradient detection that results in the numerous design variations [1,2,3]. The probes commonly used in a production shop for nondestructive testing (NDT) are invariably wound on ferrite cores because of the increased sensitivity that results from the use of high magnetic permeability core materials. However, recent advances in the theory of flaw-field interactions have stimulated interest in the use of air core probes [4,5]. The use of air core coils in the detector helps to minimize the complexity of the calculations and leaves the experimenter with very adequate tools for verification studies. This theoretical work has been a critical element in the development of quantitative eddy current measurements.

Designing an air core differential probe for optimum sensitivity involves geometric as well as electrical considerations. This experiment addresses the problem of optimizing pickup coil position inside the excitation coil.

EXPERIMENT

The probe used for this experiment consists of an air core excitation coil surrounding two air core pickup coils. All three coils have vertical axes with respect to the flat test plate. The probe cross section and electrical schematic are shown in Fig.1, and Table 1 gives the dimensions of the coils.

The pickup coils were potted in paraffin and the probe response was measured for three different pickup coil separations. To clarify the three separation distances let r equal the outside radius of a pickup coil, R be

---

†Publication of the National Bureau of Standards, not subject to copyright.

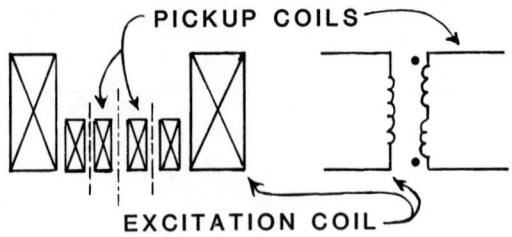

Fig. 1. Probe Cross Section and Electrical Connections.

Table 1  Probe Coil Dimensions

|                | Excitation Coil       | Pickup Coils           |
|----------------|-----------------------|------------------------|
| Turns          | 150                   | 10                     |
| Inner Diameter | 4.57 mm (0.180 in)    | 0.203 mm (0.008 in)    |
| Outer Diameter | 5.79 mm (0.228 in)    | 0.431 mm (0.017 in)    |
| Length         | 2.79 mm (0.110 in)    | 0.254 mm (0.010 in)    |

the inside radius of the excitation coil and S be the separation distance, the distance between pickup coil centers. The first configuration was with the pickup coils adjacent to each other at the center of the excitation coil so that $S = 2r$. In the second arrangement, we placed the coils at the 1/4 and 3/4 positions on the excitation coil inside diameter which resulted in $S = R$. In the third separation, the coils were placed at the inside edges of excitation coil, where $S = 2R - 2r$ (Fig. 2). At each separation distance a set of four measurements was taken on the test plate, then the pickup coil separation was changed. This was accomplished by melting the paraffin out of the center of the excitation coil, repositioning the pickup coils and potting the pickup coils with paraffin again.

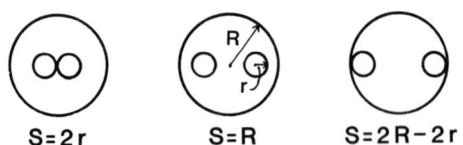

Fig. 2. Pickup Coil Separation Distances.

The test plate consisted of four semi-elliptical electrical discharge machined (EDM) notches in a 6061 aluminum plate. The notch dimensions are given in Table 2. The operating frequency was 60 kHz, resulting in a skin depth of 0.43 mm, and the ratio of flaw depth to skin depth was greater than 1 for all EDM notches measured.

At each pickup coil separation the probe was scanned over four EDM notches. The scan direction was perpendicular to the long axis of each EDM notch and the probe was oriented so that the pickup coils were aligned with the scan direction. The voltage and phase of the pickup coils were measured with a lock-in amplifier and recorded as a function of probe position on the flaw. The flaw signal was calculated from the maximum and minimum values of the notch scan, i. e. the peak-to-peak amplitude of the pickup coil output.

Table 2  Semi-Elliptical EDM Notch Dimensions

| EDM Notch | Length mm (in) | Width mm (in) | Depth mm (in) | Area mm$^2$ (in$^2$) |
|---|---|---|---|---|
| A | 3.698 (0.145) | 0.190 (0.007) | 0.704 (0.028) | 2.05 (0.003) |
| B | 3.818 (0.150) | 0.200 (0.008) | 0.954 (0.038) | 2.86 (0.004) |
| C | 4.146 (0.163) | 0.254 (0.010) | 1.254 (0.049) | 4.05 (0.006) |
| D | 4.202 (0.165) | 0.218 (0.009) | 1.436 (0.057) | 4.74 (0.007) |

Figure 3 shows a typical plot of voltage vs. position for an EDM notch scan and the corresponding flaw signal.

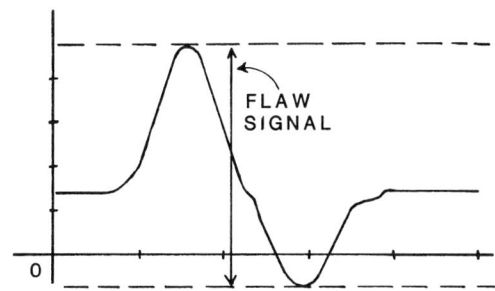

Fig. 3.  Typical Notch Scan and Flaw Signal.

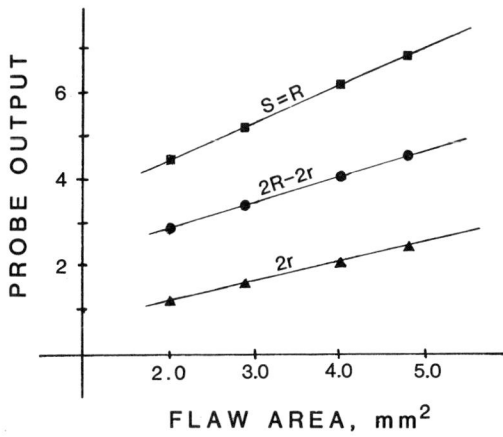

Fig. 4.  Linear Relationship of Flaw Signal to Flaw Area.

723

One of the unique features of this class of probes is the linear relationship of the flaw signal to the flaw area [3]. In the case of a rectangular notch, the flaw area would be the notch length times the notch depth. We took advantage of this effect and used the slope of the the signal vs. area plot as the indication of probe sensitivity. The effectiveness of a particular separation distance was evaluated using this sensitivity criterion.

SENSITIVITY RESULTS

The data show a stong dependence of probe sensitivity on the location of the pickup coils in the excitation field (Fig. 4). When the separation distance was small, S = 2r, the probe showed the least sensitivity. The probe calibration for this spacing was 0.34 $\mu V/mm^2$. When the separation distance was large, S = 2R-2r, the probe provided a stronger response and the calibration was 0.55 $\mu V/mm^2$. The location providing the greatest sensitivity was the 1/4 and 3/4 points of the excitation coil inside diameter, where S = 2R. This position resulted in a calibration of 0.97 $\mu V/mm^2$.

PICKUP COIL UNBALANCE

During a flaw scan using a differential probe in the orientation reported here, a smooth bipolar output can be expected from the pickup coils when they are symmetrically placed in the excitation field and the two coils are electrically balanced. For the probe used in these measurements, the voltages induced in the pickup coils were found to differ by 11.5% when the coils were placed in a calibrated ac magnetic field; thus the coils are unbalanced. This condition causes two noticeable deviations from ideal

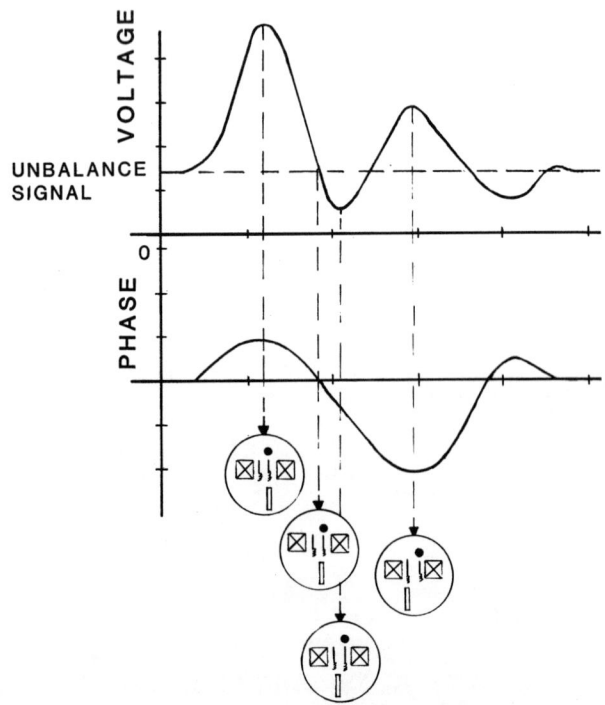

Fig. 5.  Effects of Pickup Coil Unbalance.

behavior (Fig. 5). First, when the probe is positioned off the flaw a voltage is present proportional to the unbalance of the pickup coils. Second, the bipolar signal typically expected does not start at zero and is not symmetric. Fortunately, the peak-to-peak amplitude remains linearly proportional to the flaw size in spite of the unbalanced voltage. The disadvantage is that this unbalanced signal imposes a lower limit on the smallest measurable flaw signal because of the problems involved in trying to extract a small voltage change from a large background.

CONCLUSIONS

For circular pickup coils in a differential configuration, the most sensitive position tested is at the 1/4 and 3/4 positions on the excitation coil inside diameter. In many cases, pickup coil windings are designed to fill the space available inside the excitation coil which forces the coil centers to these locations automatically.

Unbalanced pickup coils do not affect the linear characteristics of the probe but will limit the smallest measurable flaw signal.

REFERENCES

1. J. H. Smith, C. V. Dodd, L. D. Chitwood, "Multifrequency Eddy-Current Inspection of Seam Weld in Steel Sheath," Oak Ridge National Laboratory Report ORNL/TM-9470, April 1985.

2. R. E. Beissner, Gary L. Burkhardt, and Felix N. Kusenberger, "Exploratory Development on Advanced Surface Flaw Detection Methods", Final Report, Contract No. F33615-82-C-5020, Air Force Systems Command, June 1984.

3. T. E. Capobianco, J. C. Moulder and F. R. Fickett, "Flaw Detection with a Magnetic Field Gradiometer," Proceedings of the 15th Symposium on Non-destructive Evaluation, p. 15, (NTIAC, San Antonio, 1986).

4. B. A. Auld, S. Jefferies, J. C. Moulder, J. C. Gerlitz, "Semi-Eliptical Surface Flaw EC Interaction and Inversion: Theory," Proceedings of the Review of Progress in Quantitative Nondestructive Evaluation 5:383 (Plenum, NY, 1986).

5. J. C. Moulder, J. C. Gerlitz, B. A. Auld, S. Jefferies, "Semi-Eliptical Surface Flaw EC Interaction and Inversion: Experiment," Proceedings of the Review of Progress in Quantitative Nondestructive Evaluation 5:395 (Plenum, NY, 1986).

ACKNOWLEDGEMENTS

This work was sponsored by the NBS Office of Nondestructive Evaluation. J. Moulder supplied us with the flaws and flaw sizes.

EDDY CURRENT PROBE PERFORMANCE VARIABILITY

G. L. Burkhardt, R. E. Beissner, J. L. Fisher

Southwest Research Institute
6220 Culebra Road
San Antonio, Texas  78284

INTRODUCTION

Eddy current testing is used extensively by the Air Force for nondestructive inspection of many aircraft structural components. Although the reliability and consistency of inspections depends to a large extent on the characteristics of the eddy current probes used, no adequate specifications or certification methods presently exist for assuring probe performance. Because of the variability in probe performance, a need exists for establishing a means to control probe performance characteristics which will eventually lead to improved test results.

The primary objective of the work reported in this paper was to determine the degree of variability in eddy current probe performance for a group of probes representative of those presently in use by the Air Force for inspecting aircraft structural components. By presenting the data in terms of probability distributions, an assessment can be made of how many probes would be rejected if acceptance criteria were based on requiring probe performance to be within a desired range. A determination of the percentage of rejectable probes based on the probes presently in use would allow the acceptance range to be set to a reasonable value without rejecting an excessive number of probes. An additional objective was to provide a comparison between probe performance on artificial flaws (slots) and a fatigue crack as an initial step for estimating probe performance on cracks based on performance on slots.

EXPERIMENTAL PROCEDURE

Thirty nonshielded and thirty shielded probes were tested. These probes were obtained from many different Air Force bases and are representative of probes typically in routine use. Probe coil diameters were limited to less than approximately 1/8 in., since these would be most commonly used for small flaw detection. Typical probes are shown in Figure 1.

The probes were characterized by measuring their responses to four slots in a prototype Air Force (AF) general purpose eddy current standard, NSN 6635-01-092-5129, P/N 7947479-10. The slots measured 1 in. long with depths of 0.05, 0.02, 0.01, and 0.005 in. A piece of 0.0025-in. thick mylar tape was positioned on the standard for measuring the probe response

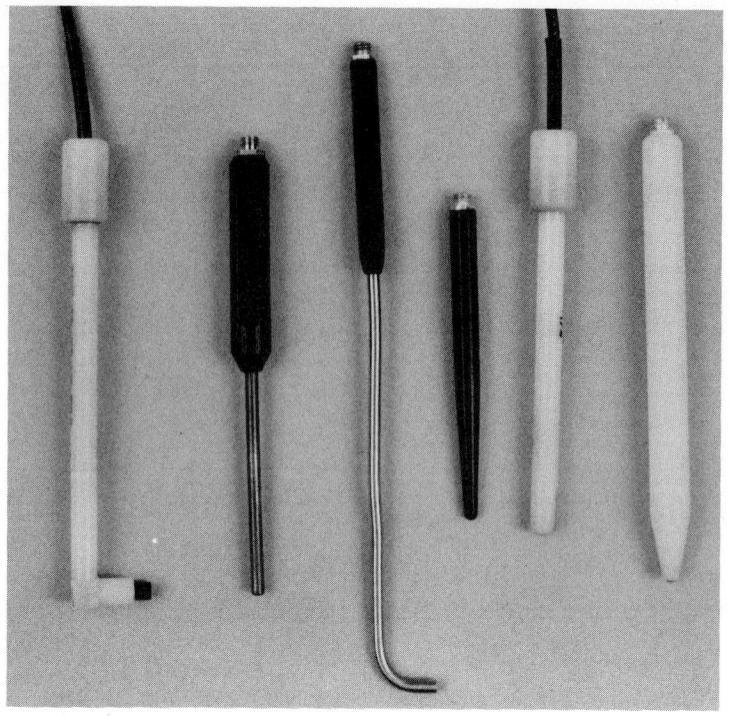

Fig. 1. Typical eddy current probes evaluated in this program

to liftoff. The probe responses were also measured for a fatigue crack (0.050 in. long by approximately 0.012 in. deep). Both the standard and the fatigue crack specimen were aluminum.

A block diagram of the laboratory setup is shown in Figure 2. A Hewlett Packard 4194A impedance/gain-phase analyzer was used to measure the impedance characteristics of each probe as it was scanned over the specimens. Impedance data were taken at 200 kHz since this frequency is commonly used by newer Air Force eddy current equipment for inspection of aluminum structures.

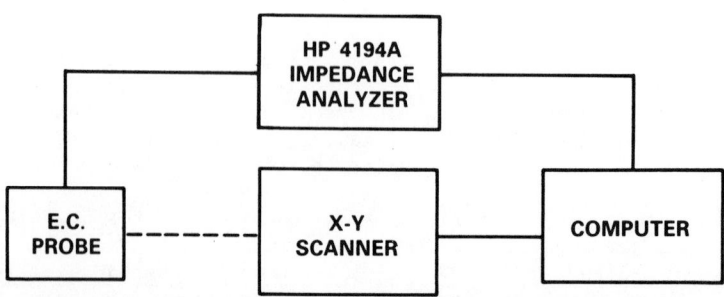

Fig. 2. Block diagram of data acquisition system

Each probe was spring-loaded against the specimen using a double cantilever spring arrangement which assured that the probe was always perpendicular to the test specimen regardless of the amount of spring deflection. The probes were scanned by a precision scanning system driven by high-resolution microstepper motors. Impedance measurements were digitized at 0.01-in. increments as the probes were being scanned. Both the scanning system and the impedance analyzer were controlled by a desktop laboratory computer. The digitized data were transferred to the computer for analysis and storage.

The probe scan path is shown in Figure 3. The probe was first scanned over the center of each slot in the standard. Since the slots were much longer than the probe diameter, it was not necessary to position the probe exactly in the center of the slot length. After scanning each probe over the slots, it was moved onto the tape to generate a change in liftoff. The probe was then moved onto the fatigue crack specimen and scanned over the crack. Because of the relatively small size of the crack, probe positioning was more critical, and it was not possible to obtain the maximum crack response in a single scan. Therefore, a raster scan was used, as shown in Figure 3. This resulted in multiple scans across the crack in increments spaced 0.005 in. apart until the maximum crack response was obtained.

Measurements of the following parameters are reported in this paper:

(1) Flaw Response: A typical probe response to the four slots and liftoff is shown on an impedance plane plot in Figure 4. Note that the liftoff response is in a different direction from the flaw responses. In a typical eddy current inspection, the liftoff response is minimized by adjusting the instrument to respond only to the components of signals which are perpendicular in direction to the liftoff response. A similar approach was used in this program. A computer routine was used to calculate the signal component perpendicular to the liftoff direction. The probe impedance component perpendicular to the liftoff direction is plotted in Figure 5 as a function of probe position with respect to the slots. The flaw response is the maximum impedance change obtained from each flaw. This measurement was made on each of the slots as well as the fatigue crack.

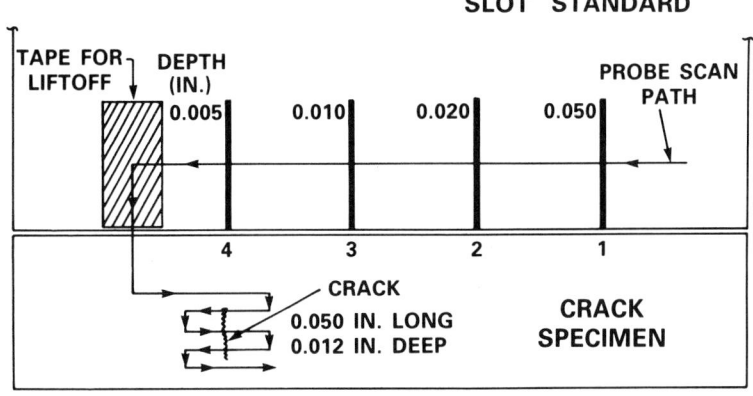

Fig. 3. Scan path for slot standard and fatigue crack specimen

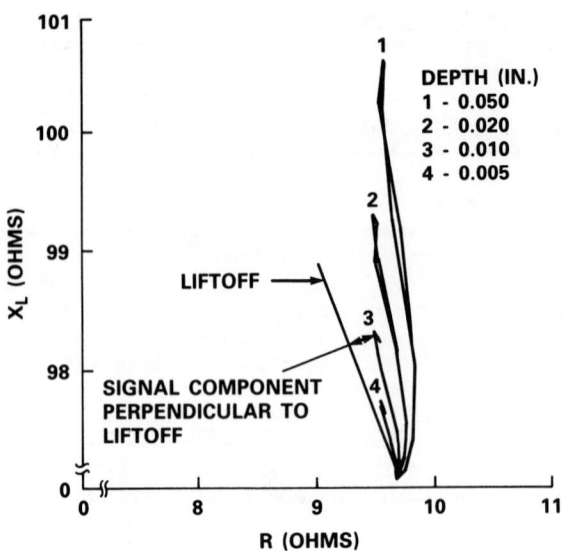

Fig. 4. Impedance plane plot for typical eddy current probe response from slots 1-4 (1 in. long with depths of 0.05, 0.02, 0.01, and 0.005 in. respectively) and from 0.0025-in. liftoff

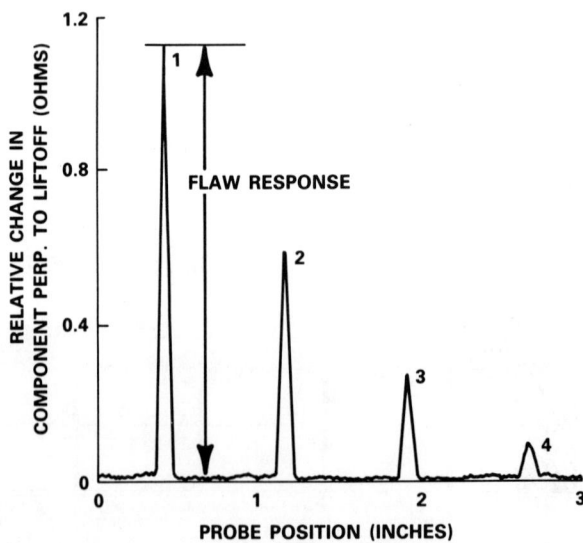

Fig. 5. Change in probe impedance component perpendicular to liftoff vs. position for slots 1-4 (1 in. long with depths of 0.05, 0.02, 0.01 and 0.005 in. respectively)

(2) Absolute Probe Impedance: The total impedance of each probe (without regard to the liftoff direction) was measured at a position in the scan away from any flaw. The probe resonant frequency was also checked to make sure it was not close to the operating frequency.

EXPERIMENTAL RESULTS AND DISCUSSION

For determining the variability in probe performance, the data are presented as histograms showing distributions of the number of probes vs. flaw response (probe impedance change from a flaw). A Gaussian curve has also been fitted to the data, and the mean and standard deviation data are shown. The Gaussian curve was used for convenience; the use of other curves that may provide a better fit to the experimental data was beyond the scope of the project.

The distribution of flaw responses from EDM slot No. 3 (1 in. long x 0.010 in. deep) is shown in Figure 6 for the group of thirty shielded probes. The data from this slot were selected because they were similar to the data from the fatigue crack and would be more representative of smaller flaws. The shapes of the distributions from the other slots are similar except that the distributions are shifted to higher impedance values for the deeper slots and to smaller values for the shallower slots. The probe impedance changes from slot No. 3 are shown on the horizontal axis. The width of each box in the histogram represents an impedance change of 0.025 ohm. The percentage of the total number of probes having an impedance change from the flaw within the range shown by each box is represented by the height of the box. For example, 16.7% of the group of thirty probes had an impedance change from slot No. 3 within the range of 0.225 to 0.250 ohm, as shown by the height of the single box representing this range on the horizontal scale. The percentage of probes having impedance changes within a range represented by more than one box can be obtained simply by adding the number of probes represented by all the boxes in that range.

It is apparent that a wide variation in flaw response exists, as shown in Figure 6. The flaw responses ranged from a minimum of 0.080 ohm to a maximum of 0.533 ohm, or a variation by approximately a factor of 7. The mean (or average) response was 0.242 ohm. Two-thirds of the probes had flaw responses within the range of 0.150 to 0.300 ohm. If acceptance criteria for probes were set to include only this range, for example, then 67% of the probes would be accepted and 33% would be rejected.

Data from slot No. 3 for the group of thirty nonshielded probes are shown in Figure 7. Here the variation in flaw response is from 0.044 to 0.280 ohm or approximately a factor of 6. This is about the same amount of variation as with the shielded probes, but the responses are shifted to smaller values, showing that the nonshielded probes generally give a smaller response to the flaw.

Overall trends in the flaw response data indicate that the responses to each of the four slots and the fatigue crack varied by a factor of 6 to 7 for each probe type (shielded and nonshielded). In each case, this variation could be reduced to a factor of 2 by rejecting approximately one-third of the probes.

Relationship Between Slot and Crack Responses

In order to use a slot in a standard to set up an eddy current probe and instrument for an inspection, the slot response must be representative of the response obtained from a crack in the size range anticipated. In

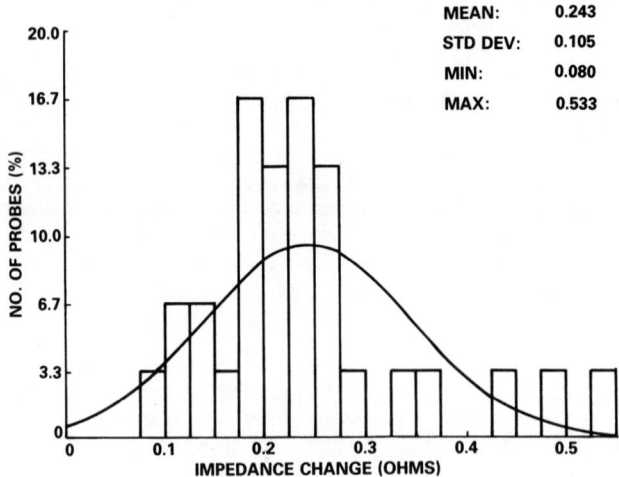

Fig. 6. Distribution of flaw response impedance change from slot No. 3 (1 in. long x 0.01 in. deep) for shielded probes

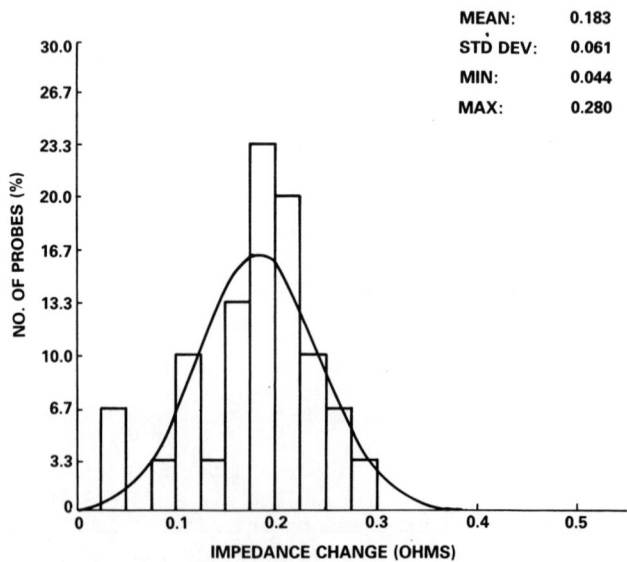

Fig. 7. Distribution of flaw response impedance change from slot No. 3 (1 in. long x 0.01 in. deep) for nonshielded probes

this program, data were obtained from four slots of different depths and one fatigue crack. Although the data from a single crack are too limited to allow an adequate correlation to be made between slots and cracks of different sizes, some valuable conclusions can still be drawn.

As mentioned in the previous section, the data from slot No. 3 (1 in. long x 0.01 in. deep) more closely represented the 0.05-in. long x approximately 0.012-in. deep fatigue crack than the data from the other slots. The relationship between the flaw responses from slot No. 3 and those from the fatigue crack is shown in Figure 8 for both the shielded and nonshielded probes. Here, the impedance change from the crack is plotted as a function of the impedance change from slot No. 3 for each probe. The line drawn on the plot represents a one-to-one correspondence between the crack and slot data. For example, if the data point for a probe falls on this line, then the response for that probe is the same for the crack as it is for the slot. The data show the same wide variation in responses as shown previously in the distributions. However, the slot and crack responses for most of the data are reasonably equivalent since the data points fall close to the line representing a one-to-one correspondence. The shielded probe data generally fall above the line while the nonshielded probe data are generally grouped below the line. This indicates that the shielded probes tend to give a somewhat greater response from the crack as compared to the slot while the nonshielded probes give a smaller response from the crack.

These data provide a degree of confidence that the response of a probe to a crack can be approximated by its response to a slot. Although additional data are needed for other crack sizes, this provides a preliminary indication that probe response acceptance limits can be established on slots in Air Force standards such as the one used here.

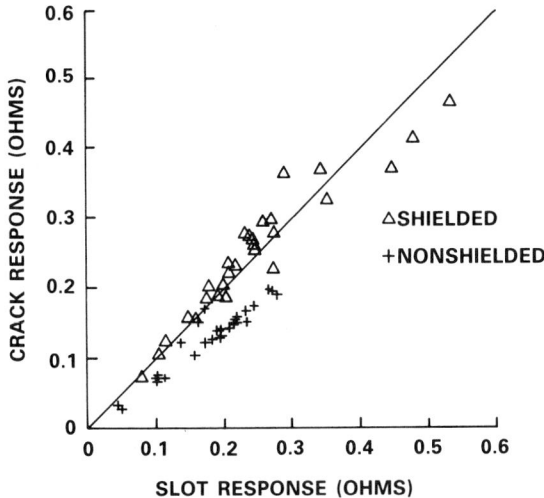

Fig. 8. Response from crack (0.050 in long x approximately 0.012 in. deep) vs. response from slot No. 3 (1 in. long x 0.010 in. deep) for shielded and nonshielded eddy current probes. The line at 45 degrees represents a one-to-one correspondence between slot and crack responses.

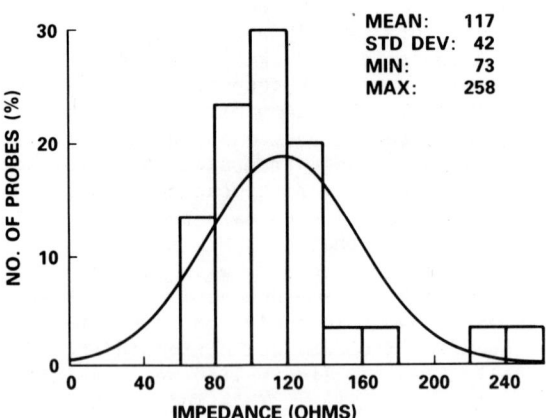

Fig. 9. Distribution of absolute probe impedance from shielded probes at 200 kHz

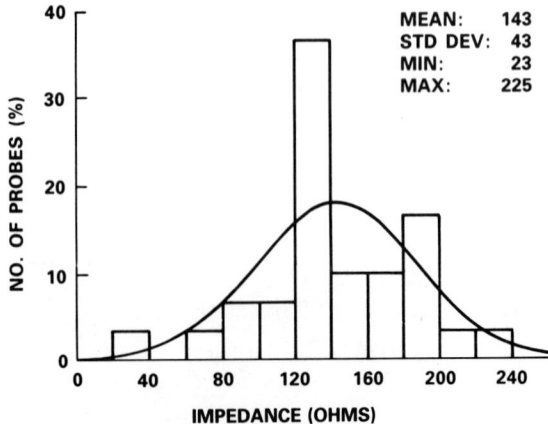

Fig. 10. Distribution of absolute probe impedance from nonshielded probes at 200 kHz

Absolute Probe Impedance

In addition to the impedance change from a flaw, another parameter of importance is the absolute probe impedance. This measurement is significant because eddy current instruments require that probes have an impedance within a certain range or the instrument will not balance properly and will not function with that probe. Also, since the impedance becomes very large at the resonant frequency of a probe, it is not desirable to operate a probe at a frequency close to resonance.

For all of the probes, the resonant frequency was well above 200 kHz and was not a significant factor in the impedance measurements. The distributions of probe impedances at 200 kHz for the shielded and nonshielded probes are shown in Figures 9 and 10, respectively. The variation in impedance for the shielded probes ranged from 73 to 258 ohms while the nonshielded probe impedances ranged from 23 to 225 ohms. For the shielded probes, 73% had impedances between 80 and 140 ohms while only 50% of the nonshielded probes had impedances in this range. Acceptance criteria for probe impedance would require evaluation of instrument specifications to determine the range of impedance values over which the instruments would balance.

CONCLUSIONS

For the probes tested in this program, the flaw responses from the shielded probes were generally greater than those from the nonshielded probes. The responses within each of these probe groups varied by a factor of 6 to 7 for each of four slots (1 in. long with depths ranging from 0.005 in. to 0.05 in.) and a fatigue crack (0.05 in. long x approximately 0.012 in. deep).

By rejecting approximately one-third of the probes, the variation in flaw responses for the shielded probes or for the nonshielded probes could be reduced to a factor of 2.

A relatively good correspondence was obtained between the probe responses to the 0.05-in. long x approximately 0.012-in. deep fatigue crack and the 1-in. long x 0.01-in. deep slot. The crack signal was generally slightly larger than the slot signal for the shielded probes and slightly smaller than the slot signal for the nonshielded probes.

The absolute probe impedance values varied by a factor of 3.5 for the shielded probes and by a factor of almost 10 for the nonshielded probes.

A first step toward obtaining more consistent probe performance could be to establish probe acceptance criteria based on (1) the impedance change from slots in a standard such as the Air Force general-purpose eddy current standard and (2) the absolute probe impedance.

ACKNOWLEDGMENTS

The authors wish to thank Mr. David Jones for assistance with the experimental setup and data acquisition. Support for this work was provided by the San Antonio Air Logistics Center/MMEI, Kelly Air Force Base, San Antonio, Texas.

CAPACITIVE ARRAYS FOR ROBOTIC SENSING

M. Gimple and B.A. Auld

Edward L. Ginzton Laboratory
Stanford University
Stanford, CA. 94305

INTRODUCTION

Electromagnetic arrays have been used effectively for many years in optimizing the responses of antenna systems. The basic principles that make arrayed antennas work and make them easy to control can also be applied to near field electromagnetic array sensors. The array factor allows for flexibility in sensor geometry. Firstly, by exciting only a portion of an array in a sequential fashion one can physically scan and interrogate a region of a sample without having to move the sample or the probe head itself. Secondly, the field configurations can be altered by selectively exciting electrodes of an array. Also, the information received can be selected by combining electrodes to form different effective receiver geometries. Thirdly, array configurations allow for real-time analog signal processing. For instance, one can perform pattern matching by choosing the spatial resolution of the probe to match the spatial resolution of the desired feature.

The electromagnetic basis of the sensor allows for multi-parameter sensing. First, one can measure distance of the probe from an object by measuring probe electromagnetic coupling with the sample object. Second, the existence and size of flaws can be determined by measuring changes in voltage versus current characteristics at the probe terminals. Thirdly, simple surface features such as edges can be located by using differentially connected probe pairs. Finally, material properties such as dielectric constant and conductivity can be extracted by measuring changes in capacitances and resistances in known geometry samples.

Electromagnetic sensor arrays come in two versions: capacitive and inductive. Capacitive probes are the focus of this paper. Inductive probes are discussed in a companion paper [1] by researchers at SRI.

SAMPLE MATERIALS

Capacitive probes can be used to investigate the properties and structure of both metals and dielectrics. For metals only surface features can be extracted. Charges which accumulate at the surface blind the capacitive probe to interior structures. Dielectrics fortunately do not have this problem. Both surface and interior features can be examined.

PROBE CAPABILITIES

Reference [2] discusses different capacitor array configurations and the capabilities that one can get from such sensing systems. In this paper we demonstrate four such basic capabilities with slight modification:

>Distance Ranging
>Edge Detection
>Response Optimization (Field Adaptation)
>Pattern Matching (Filtering)

BASIC PROBE ELEMENT

The basic probe element is shown in Fig. 1(a). It is essentially a parallel plate capacitor that has been opened up such that the two plates lie in the same plane against a common substrate. The electric field lines rather than being parallel (uniform) are now elliptical (non-uniform). The probe is operated by applying a voltage to one electrode and measuring the current to ground from the second electrode (receiver). Interrogated samples are placed in the lower half space. Changes in measured current reflect changes in the sample-probe system configuration. A metal sample, which is grounded, when placed close to the electrodes will shunt current around the receiver electrode to ground and thus lower the output signal. Dielectrics on the other hand will enhance the output for they increase the capacitive coupling between the sensing electrodes, without shunting any significant current to ground.

For a single port eddy current device we know [3] that the change in impedance is given by

$$\Delta Z = \frac{j\omega\mu_o}{I^2} \int \left( \left\{\frac{\partial}{\partial z}\phi\right\} \phi' - \phi \left\{\frac{\partial}{2z}\phi'\right\} \right) dxdy \qquad (1)$$

where $\phi$ is the magnetostatic potential ($H = \nabla\phi$) and the integration is performed over the surface of the mouth of the flaw. The primed notation refers to those quantities that exist in the presence of the flaw. The unprimed quantities refer to flaw absence. For the capacitive device described above we have a dual relationship, that is,

FIELD PENETRATION

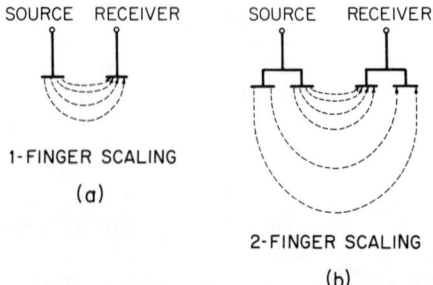

Fig. 1. Elementary capacitive probes. They are used for distance ranging.

$$\Delta y = \frac{j\omega\varepsilon_o}{v^2} \int \left( \psi \frac{\partial \psi'}{\partial z} - \psi' \frac{\partial \psi}{2z} \right) dxdy \qquad (2)$$

$\psi$ is the electrostatic potential ($E = -\nabla\psi$) and the integration is performed over the surface of the test piece. Solutions for simple geometries can be extracted directly. For more complex geometries approximations and numerical techniques must be employed.

Figure 1(b) shows a similar configuration to that in Fig. 1(a) except that the source now consists of two contiguous, elementary electrode fingers excited simultaneously and the receiver consists of two contiguous, elementary electrode fingers whose currents outputs are added together. As shown in the figure this allows for deeper field penetration and higher sensitivity at a given distance away.

Experiments demonstrating this effect were conducted using these probes with metallic samples. The electrodes that were used were 40 mils wide and 475 mils long. They were separated by 50 mils. A full array of electrodes were employed (8 to 9 fingers) even though only a subset was actively involved in the measurements. The remaining electrodes were explicitly grounded. Experimental results are shown in Fig. 2. In order that the two geometries could be compared on a similar basis both outputs were nomalized to a 0dB level at the zero vertical distance position. Note that the curve for the two-finger scaled probe dominates over that for the 1-finger scaling case. This is an indication that the fields do penetrate further into the interrogation region when the effective electrode widths are increased. Also note that for extremely small vertical distances the sensitivity of the probes to distance (slope) is small. This is due to the fact that at small distances the metal sample is so close to the source electrodes that it completely shields the receiving electrode from any interaction.

COMPLEX ELECTRODE CONFIGURATIONS

The configurations of Fig. 1 are sensitive to both vertical distances and horizontal displacements. Separating the two effects from each other is not readily done with a two electrode probe. Thus, there

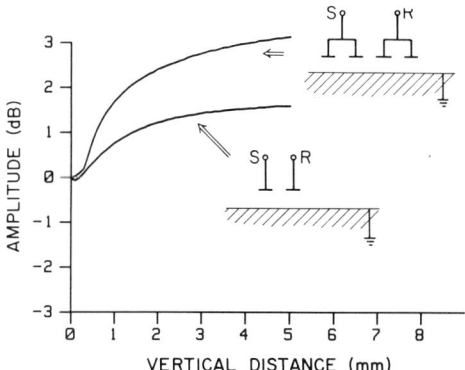

Fig. 2. Normalized distance ranging output for the two elementary capacitive probes.

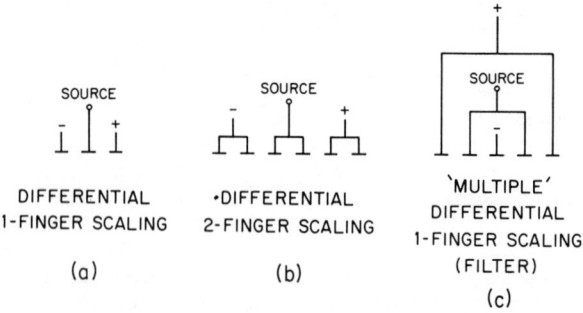

Fig. 3. Three complex electrode probe configurations. They are used to extract lateral surface features.

is a need to adopt a probe geometry having a greater number of source/receiver electrodes. Figure 3 displays electrode configurations that make use of multiple source/receiver electrodes, with differential electronics to eliminate signals from common mode effects. Changes that affect both sides of the differential pair similarly (such as changes in vertical distances) are nulled. Only changes that disturb parts of the probe fields will result in non-zero outputs. Figure 3(a) is the simplest differential configuration. It is essentially two elementary probes connected in tandem with opposite polarity.

Experimental results for a grounded metal, slotted sample are shown in Fig. 4. The slots are .75 inches wide and 10 mils and 5 mils deep, respectively. The sample is scanned in a direction perpendicular to the sample slots. The probe is oriented such that the scan direction is perpendicular to electrode fingers. The peaks and valleys in the signal correspond to the edge components of the slots. A point to note is that the response to the deeper slot is larger than the response to the shallower slot. Thus, the probe is sensitive to slot depths. Also note that the response for a given depth slot is not symmetric for each edge (i.e. the size of the peak does not equal the size of the valley). This is due to the fact that the differential electronics is not balanced to zero. This purposeful asymmetry in gain paths allows us to measure output signal polarity from the signal amplitude alone.

Figure 5 displays results for a dielectric sample. The sample used is Delrin ($\varepsilon_r = 3.7$). The three slots are .625 inches wide and 20

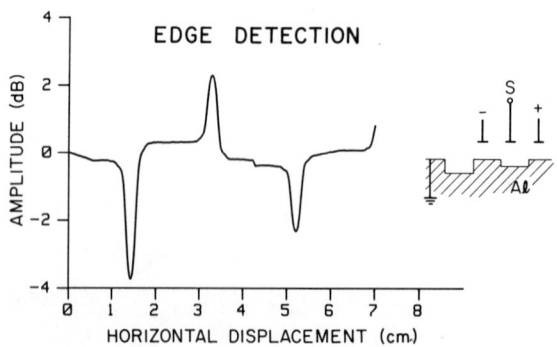

Fig. 4. Horizontal scan of a slotted Al sample with a 1-finger scaled differential probe. The slot depths are 10 mils and 5 mils, respectively. The sample is explicitly grounded.

740

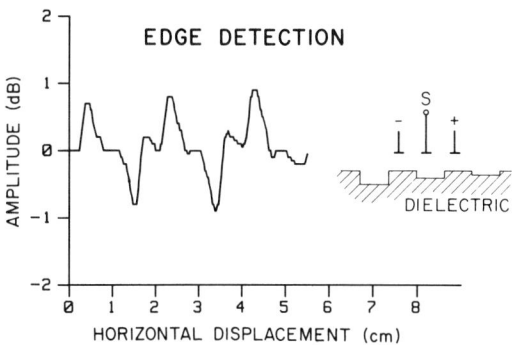

Fig. 5. Horizontal scan of a slotted Delrin sample with a 1-finger scaled differential probe. The slot depths are 20 mils, 10 mils, and 5 mils, respectively.

mils, 10 mils, and 5 mils deep, respectively. Note, that in the dielectric case a peak (valley) in the output signal occurs where a valley (peak) would occur with a metal sample. Also, one should note that the probe's sensitivity to each of the dielectric slots is the same (each peak and valley is of approximately the same size). We infer from this that the probe's sensitivity reaches saturation at a point under 5 mils deep. Further investigation is needed here.

Figure 3(b) displays a configuration similar to Fig. 3(a) save for the fact that 2-finger scaling is employed. One should expect better sensitivity but lower horizontal resolution than in the 1-finger scaling case. The lower horizontal resolution is due to the fact that an edge disturbance starts to cause a mismatch as it starts passing over the receiver electrodes and reaches maximum mismatch when the edge is centered on the source electrode. Since 2-finger scaling probes are physically larger, then the peaks and valleys in the signals will be wider. Experimental results are displayed in Fig. 6.

The third differential probe consists of two single differential probes connected in tandem with the opposite polarity (see Fig. 3(c)). The purpose of this probe configuration is to look at two features (edges) simultaneously that are separated in space (filtering). When

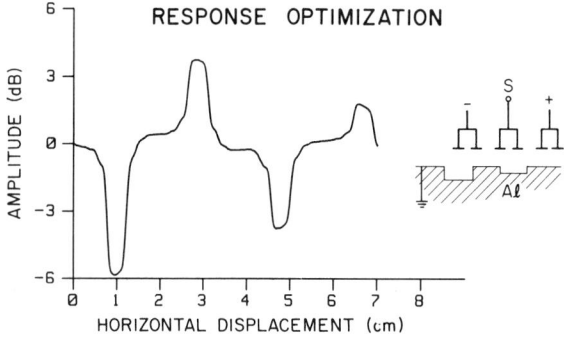

Fig. 6  Horizontal scan of a slotted Al sample with a 2-finger scaled differential probe. The slot depths are 10 mils and 5 mils, respectively. The sample is explicitly grounded.

the spatial separation of the sample edges matches the spatial separation of the source electrodes then the output signal will be the sum of two edge signals overlapping giving a large peak or valley. At any other slot width the output signal tends to separate into two smaller individual signals. The degree of separation is dependent not only on the slot width itself but also on the width of the impulse response of the probe. Figure 7 displays experimental results for this probe scanned over three slots of widths of .180 inches (matched) .125 inches, and .500 inches, respectively in an explicitly grounded metal sample.

To apply capacitive probes to quantitative nondestructive evaluation and intelligent robot sensing it is necessary to quantitatively interpret the probe signals. This requires the development of probe interaction modeling based on Eq. (2). In investigating the behavior of long arrays, one can first approximate the probe as though it is of infinite extent. For an infinite array the analytical solution for the electric fields in the interrogating half-space without any samples present is given by [4]. The solutions are in the form of summations of Legendre polynomials. We are investigating the area of using perturbation methodologies to extract the field expressions that occur when sample objects are present with and without surface features such as edges and slots. This is an area of future work, and will be extended using finite difference calculations for the various probe geometries discussed above.

CONCLUSION

Four basic capabilities of capacitive arrays have been demonstrated. Also it has been demonstrated that capacitative arrays can be effective for both metals and dielectrics.

Future plans involve developing more analytical modeling capacity in order to produce better probes and to eliminate the need for recalibration for new sample geometries. Considerations for future probes include extending array geometries into 2-D and incorporating microelectronic preamplification in close proximity to the measuring electrodes.

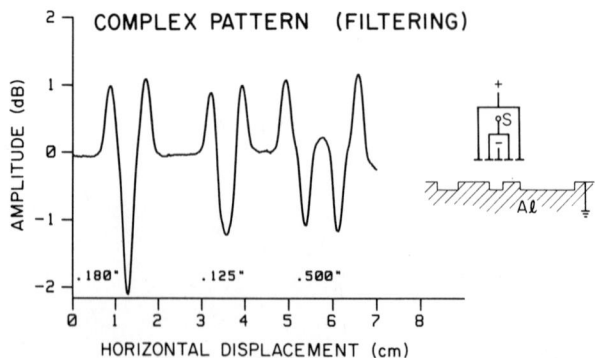

Fig. 7  Horizontal scan of a slotted Al sample with a spatial filter probe. The slots are 20 mils deep and are .180 inch (matched), .125 inch, and .500 inch wide, respectively. The sample is explicitly grounded.

ACKNOWLEDGEMENTS

We would like to thank A.J. Bahr and A. Rosengreen at SRI for their close collaboration. The work done here has been sponsored by the Air Force under Contract No. F49620-84-C-0095.

REFERENCES

1. A. J. Bahr and A. Rosengreen, in: these proceedings.
2. B. A. Auld, J. Kenney, and T. Lookabaugh, "Electromagnetic Sensor Arrays-Theoretical Studies", in: Review of Progress in Quantitative Nondestructive Evaluation 5A, D. O. Thompson and D. E. Chimenti, eds., (Plenum Press, New York, 1985).
3. B. A. Auld and S. Jefferies; J. C. Moulder and J. C. Gerlitz, "Semi-Elliptical Surface Flaw EC Interaction and Inversion Theory", in: Review of Progress in Quantitative Nondestructive Evaluation 5A, D. O. Thompson and D. E. Chimenti, eds., (Plenum Press, New York, 1985).
4. D. P. Morgan, "Surface-Wave Devices for Signal Processing", (Elsevier Science Publishers, Amsterdam, 1985) 335-361.

INDUCTIVE SENSOR ARRAYS FOR NDE AND ROBOTICS

A. Rosengreen and A. J. Bahr

SRI International
Menlo Park, California 94025

INTRODUCTION

The objectives of this research program are to develop and evaluate electromagnetic sensor arrays for use in NDE and robotics. The work at SRI has focused on the use of inductive sensors; parallel work at Stanford University [1] has emphasized capacitive sensors. The previously reported wire-wound coil sensor [2], consisting of a drive coil and several smaller pickup coils, uses a technology that is not well suited for constructing an array. Furthermore, the spatial resolution of such a sensor is limited by the practical size of the pickup coils. In this paper we describe the development of a sensor array that uses printed-circuit technology to overcome these limitations. We expect that printed-circuit technology will simplify the fabrication of an array, permit precise replication of the array elements, reduce the minimum achievable size of individual array elements to improve spatial resolution, and allow the construction of two-dimensional arrays.

PRINTED-LOOP SENSOR ARRAYS

Printed-circuit technology is widely used for fabricating planar circuit patterns and can produce line widths and spacings as small as a few thousandths of an inch. The use of this technology for inductive sensor arrays requires that the multi-turn coils used previously be replaced with single-turn printed loops. Although small single-turn loops reduce the sensor signal, they significantly improve spatial resolution.

A potentially useful feature of printed-circuit technology is that the circuit boards can be mounted in a vertical position with respect to the test sample to facilitate external connection to the sensor array (Fig. 1). The cross-sectional side view in Fig. 1(a) shows two vertical wire-wound drive coils and a circuit board containing the printed loops sandwiched between the coils. The relative position of the loops is shown in Fig. 1(b).

The principle of operation of the printed-loop sensor is the same as that of an eddy-current reflection probe; that is, the sensor output is proportional to any change that occurs in the mutual coupling between the drive coil and pickup loops. The single magnetic flux line in Fig. 1(a) schematically represents this mutual coupling.

We built and tested a vertical sensor array with two of the loops differentially connected. However, even when the sensor was scanned over the edge of a flat metal sample, we found that the signal produced was very weak

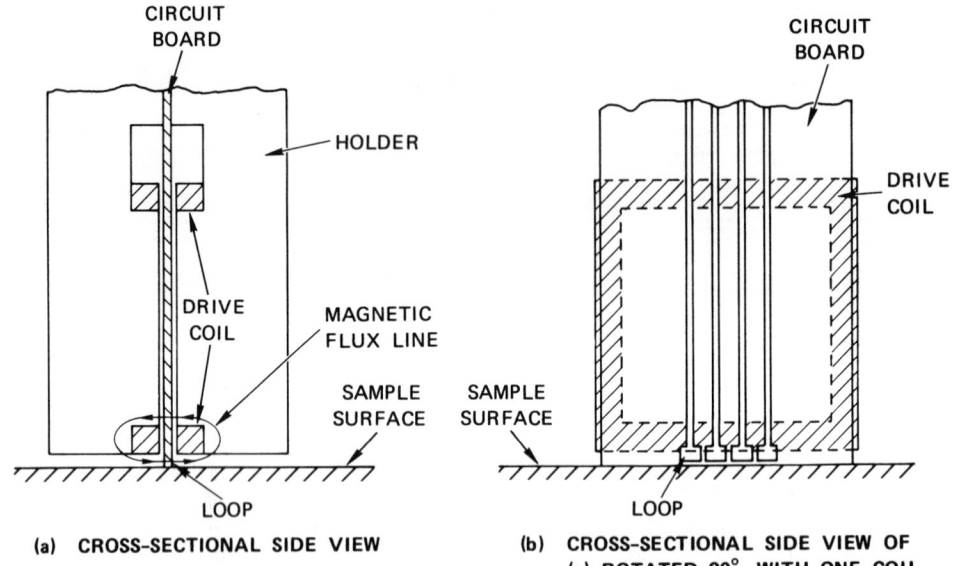

Fig. 1. Vertical printed-loop sensor array.

and could be only marginally detected because the magnetic fields responsible for the mutual coupling between the drive coils and the printed loops were not tightly coupled to the sample.

To improve the sensitivity of the sensor, we converted the vertical sensor array to a horizontal sensor array with a single drive coil (Fig. 2). As shown in Fig. 2(a), the single drive coil with 5 x 6 turns and the circuit board are located parallel to the sample surface. The circuit board is attached to the drive-coil holder, thereby placing the array of printed loops between the drive coil and the test sample at a position close to the center of the drive coil. The array of the four printed loops, two of which are shown in Fig. 2(b), fits well within the 0.5-in. x 0.5-in. opening of the drive coil. Each loop, formed by a 0.005-in.-wide printed conductor, has inside dimensions of 0.03 in. x 0.05 in. The center-to-center spacing between neighboring loops is 0.07 in. To reduce the electromagnetic pickup in the lines that connect the loops to the soldering pads, the entire board is wrapped with a 0.004-in.-thick copper foil so that only the loops are directly exposed to the magnetic field of the drive coil. A 0.002-in.-thick film of Kapton provides insulation between the loops and the copper foil. The copper foil typically touches the sample surface as shown, thus resulting in a lift-off distance of 0.006 in.

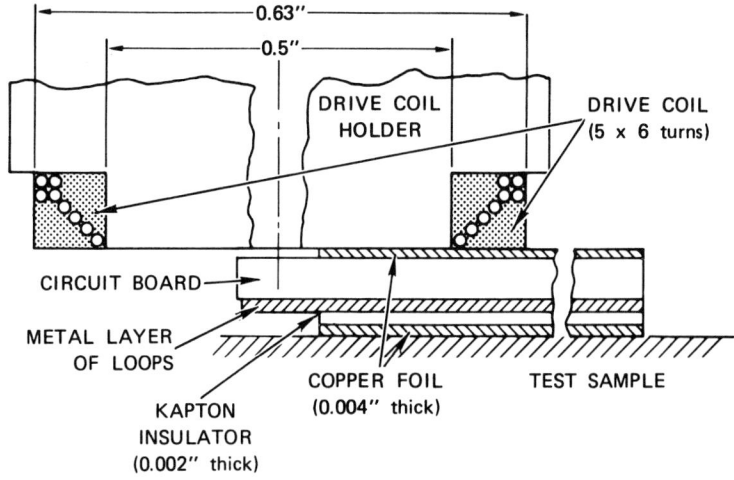

(a) CROSS-SECTIONAL SIDE VIEW

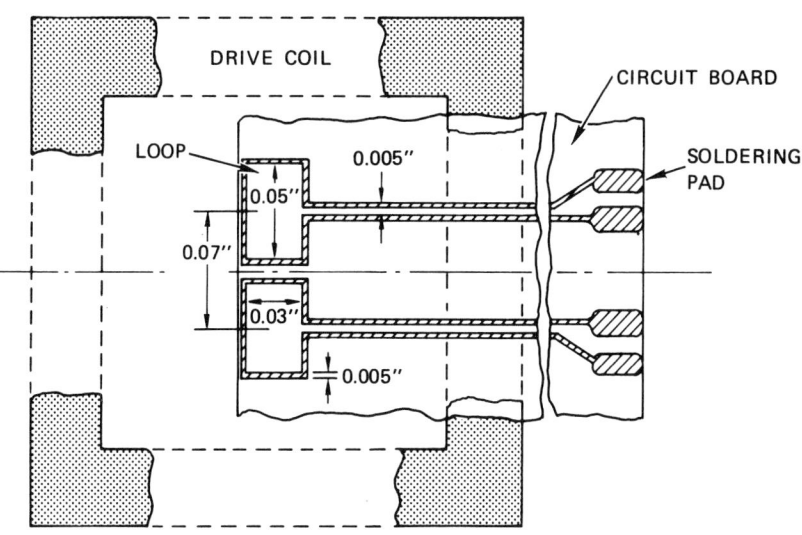

(b) BOTTOM VIEW OF DRIVE COIL AND PRINTED LOOPS

Fig. 2. Horizontal printed-loop sensor array. (Not to scale)

EXPERIMENTAL RESULTS

In the spatial-resolution and sensitivity experiments discussed below, the sensing loops were connected differentially as indicated in Fig. 3, which also shows the measurement setup. The drive coil is excited by a stable 200-kHz signal, with a level such that the current in the coil is a few mA. The signals from the loops are sent to a differential amplifier via two potentiometers that are used to balance the sensor pair. The differential signal is detected by a lock-in amplifier that provides a full-scale 10-V output. This output is captured by the data acquisition system described in [2]. Although a lock-in amplifier in the detection system is not essential, we used it for these experiments because of our initial uncertainty about the sensitivity of the single-loop sensors.

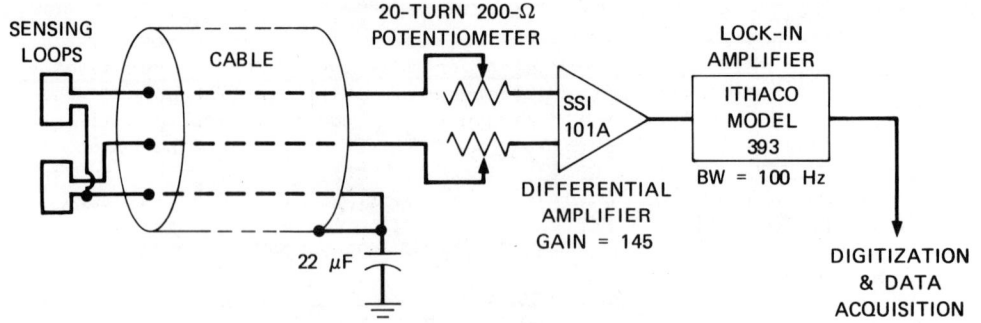

Fig. 3. Measurement setup.

Figure 4 shows the improved spatial resolution of the printed-loop sensor over the wire-coil sensor described in [2]. Figure 4(a) compares the signals detected by a printed-loop and a wire-coil sensor when scanned over a wide slot (0.125-in. wide and 0.125-in. deep) in an aluminum plate. Because it has a center-to-center spacing of 0.070 in., the printed-loop sensor is able to resolve the slot, whereas the wire-coil sensor, which has a center-to-center spacing four times larger, exhibits a response that is determined by the dimensions of the sensor and not those of the slot. Figure 4(b) shows the signal detected by a printed-loop sensor when scanned over a narrow EDM slot (0.010-in. wide and 0.040-in. deep) in an aluminum plate. As expected, the basic resolution of the differentially connected printed-loop sensor is about 0.070 in.

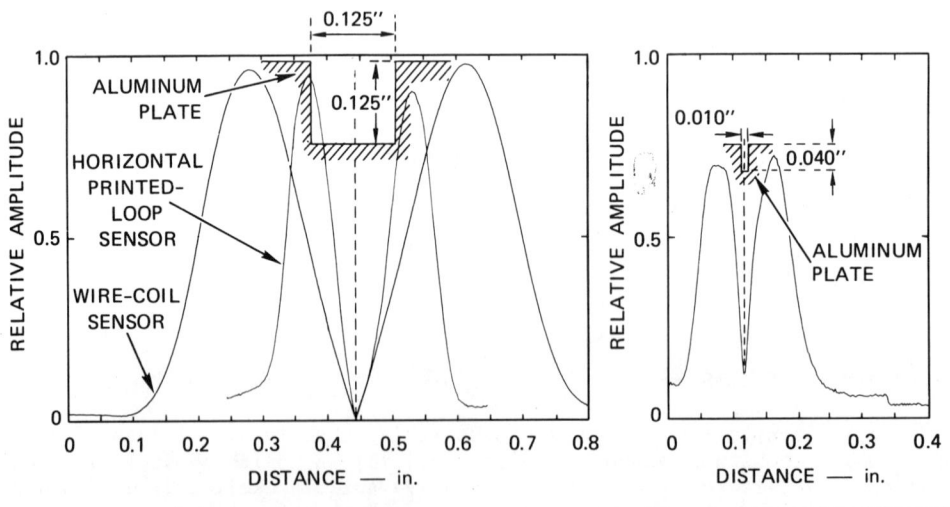

(a) WIDE-SLOT RESPONSE        (b) NARROW-SLOT RESPONSE

Fig. 4. Spatial resolution of printed-loop sensor.

We evaluated the sensitivity of the horizontal printed-loop sensor by scanning small steps in an aluminum plate. Figure 5 shows the response to a

0.002-in. step. The full scale of this figure corresponds to a 30-µV output from the differential amplifier. The voltage signal-to-noise ratio is larger than 100. This ratio is determined not only by the bandwidth, but also by the common-mode rejection of the differential amplifier. The common-mode rejection governs how large a signal can be applied to the drive coil before the common mode begins to mask a weak signal. In this case, the common-mode rejection was measured to be 60 dB. The drive current used to obtain the results shown in Fig. 5 was 3 to 4 mA. For the signal level present in Fig. 5, the exact shape of the skirts of the curve depended on how the probe was aligned because there was some common-mode interference.

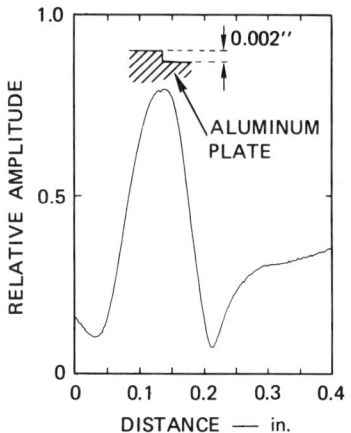

Fig. 5. Step response of horizontal printed-loop sensor.

When the loops are connected so that the signals from each loop add together rather than subtract, the sensor becomes very sensitive to lift-off. Using this sum mode, the probe becomes a proximity sensor. Figure 6 shows the results of varying the lift-off distance. In Fig. 6(a), curves of the lift-off signal at the output of the preamp are drawn as functions of lift-off distance for the two cases where either two or four loops are connected in series. As can be seen, increasing the number of connected loops increases the sensitivity. Connecting either loops 2 and 3 or 1 and 4 made no difference in the detected signal. The output voltage appears to be simply the sum of the voltage from each loop, and there is no evidence of interaction between the loops. At 0.4 in. to 0.5 in. away from the sample, the lift-off voltage saturates. Using this saturation voltage to normalize the curves in Fig. 6(a) produces the curves shown in Fig. 6(b). These curves show that the range of lift-off distance over which the test piece exerts an influence on th detected signal is about 0.1 in. for both combinations of loops. Controlling this range of influence will probably require changing both the drive-field configuration and the loop interconnections. At present only one drive coil is used, but it may be possible to use an array of several localized drive coils to control the ranging effect.

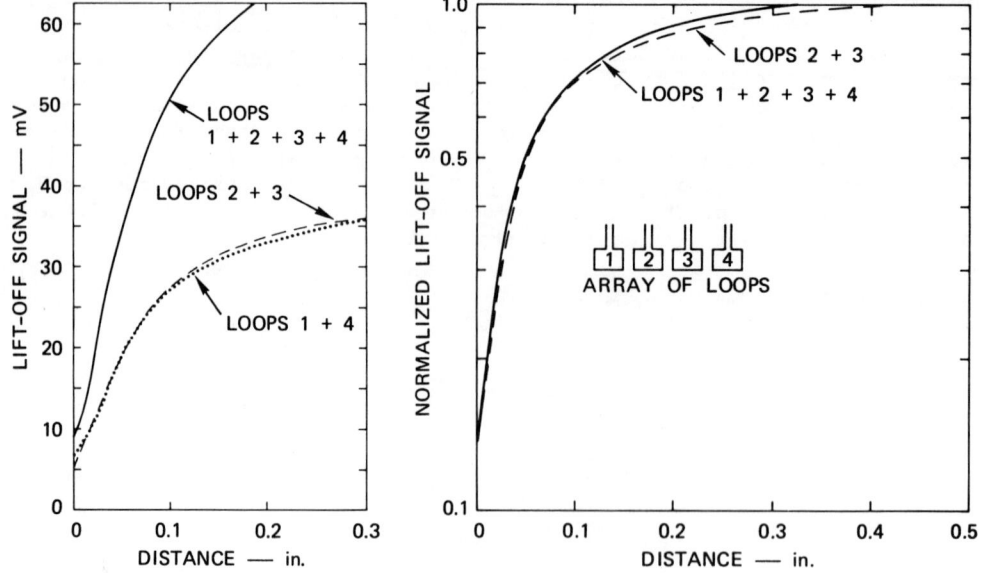

Fig. 6. Lift-off signal vs. distance (sum mode).

SUMMARY

We have demonstrated that small single-turn printed loops can be used as sensors with sufficient sensitivity to be useful in NDE and robotics and that printed-circuit techniques facilitate the fabrication of arrays of small loops to provide electronic scanning with high spatial resolution. Our future research plans include developing a model for inductive sensor arrays; designing and building a horizontal-loop array with vertical connections; demonstrating electronic scanning in one, and perhaps two, dimensions; and exploring the possibilities of arraying drivers as well as sensors.

ACKNOWLEDGMENT

The authors would like to thank Professor B. A. Auld and Mr. M. Gimple of Stanford University for many helpful discussions, and Mr. M. R. Cutter of SRI for constructing the sensor array and assisting with the measurements. This work was sponsored by the Air Force Office of Scientific Research under Contract No. F49620-84-K-0011.

REFERENCES

1. M. Gimple and B. A. Auld, Capacitive Arrays for Robotic Sensing, in these Proceedings.

2. A. J. Bahr, Electromagnetic Sensor Arrays--Experimental Studies, in "Review of Progress in Quantitative Nondestructive Evaluation," Vol. 5A, D. O. Thompson and D. E. Chimenti, eds., Plenum Press, New York (1986), pp. 691-698.

OPTICAL RANGE FINDER

G. Q. Xiao, D. B. Patterson, and G. S. Kino

Edward L. Ginzton Laboratory
W. W. Hansen Laboratories of Physics
Stanford University
Stanford, California 94305

INTRODUCTION

Recently, a great deal of interest has been shown in making accurate range measurements with good transverse definition. This capability makes it possible, in machine vision systems, to extract geometrical shape information from the images. In robot position sensing, it is important to determine the absolute distance instead of distance change so that noncontinuous measurements can be made without the need for calibration at start-up. A third application of great importance is to measure the shape and size of machined parts with a noncontacting sensor.

Research in the area of distance measurements has been carried out for many years using several different techniques.[1] One technique is to use interferometry, which requires a high-quality laser. The method yields excellent range accuracy, but suffers from phase wraparound; thus, without additional complexity, such as the use of two laser frequencies or the employment of extremely high-frequency modulation, it is not possible to make noncontinuous distance measurements.[2] An alternative technique is to use a triangulation; this method does not usually provide very good transverse resolution and is attractive only when the accuracy requirement is on the order of 100 μm or greater.

In this paper, we describe a new optical technique which is based upon the type II microscope.[3,4] With this system, we have made absolute distance measurements with a range accuracy of 2 μm and a transverse definition of 10 μm at a working distance of 15 cm . The system is stable, easy to align, and largely insensitive to the tilt and roughness of the object.

THEORY AND EXPERIMENTAL SET-UP

A schematic diagram of the optical range finder is shown in Fig. 1. The objective lens focuses the laser beam to a spot of the order of 5 μm diameter, so that most of the beam passes through a 10 μm diameter pinhole. A camera lens placed beyond the pinhole focuses the

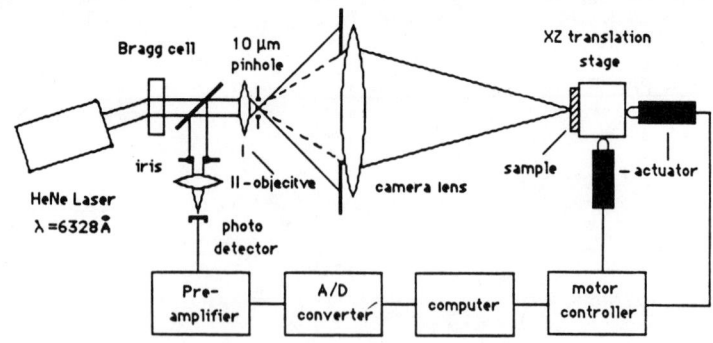

Fig. 1. Experimental arrangement of optical range finder.

light onto a point on the object. An image of this point is, in turn, focused on the pinhole. The reflected light passes back through the pinhole and is detected by the photodetector. When the object is located exactly at the focus of the transmitted light, the maximum amount of light passes through the pinhole. Otherwise, when the system is partially defocused, the amount of light passing through the pinhole is reduced. We therefore expect the signal output from the detector to have a strong dependence on the position of the object.

Because a focused beam is employed and the phase changes between the different rays comprising the beam are very small, the phase fluctuations of the laser are unimportant; therefore, a semiconductor laser, or even an incoherent source, can be used. Furthermore, the employment of a focused beam implies that the transverse definition of the system is excellent.

It will be noted that the system shown is deceptively simple. In confocal scanning microscopy, it is common to employ two pinholes so as to eliminate the reflected light from the transmitting pinhole, or to use a collimated input beam to the objective lens instead of a pinhole.[3,4] In both cases, the light passes through the pinhole only once. The optical alignment procedure required for an optical range finder, which has a much longer working distance than the microscope, is extremely difficult with such a two-pinhole system. Thus, our system uses the single pinhole technique in which the light beam passes through the pinhole twice. It is easy to align and very stable. In this case, the light reflected back from the 10 µm pinhole is eliminated by focusing the beam to a spot size of 5-6 µm.

We assume a lens with a pupil function $P(\theta)$ where $\theta$ is the angle between the ray from the pupil plane to the focal point and the axis. By following the derivation of Richards and Wolf[5] for vector fields, with some minor changes, it can be shown that the transverse electric field associated with a plane wave focused by the lens to a point on axis a distance $z$ beyond the focal plane is of the form

$$D(z) = \int_0^{\theta_0} \frac{(1 + \cos\theta)\sin\theta}{(\cos\theta)^{1/2}} e^{jkz\cos\theta} P(\theta) \, d\theta \qquad (1)$$

It will be noted that there is a $(\cos\theta)^{1/2}$ term in the denominator rather than in the numerator, as it is in Wolf's theory, in order to conserve power at a flat exit plane from the lens.

The integrand of Eq. (1) expresses the amplitude and phase of the plane wave components of the E field at an angle $\theta$ to the axis. When a focused beam is reflected from a plane mirror, its image will be a distance $2z$ away from the focus. In our apparatus, the reflected image is refocused onto the pinhole in front of the detector. Thus, it is the field on axis at the focal plane of the objective lens which is imaged at the pinhole. This field is of the form

$$V(z) = \int_0^{\theta_0} \frac{(1 + \cos \theta) \sin \theta}{(\cos \theta)^{1/2}} e^{2jkz \cos \theta} P(\theta) R(\theta) d\theta \qquad (2)$$

where the output signal from the detector is proportional to

$$I(z) = |V(z)|^2 \qquad (3)$$

In these expressions, $R(\theta)$ is the reflection coefficient of the plane reflector, k is the wave number $(2\pi/\lambda)$, f is the focal length of the lens, and $\sin \theta_0$ is its numerical aperture, where $\sin \theta_0 = a/f$, and the radius of the lens is a. With uniform excitation, [$P(\theta) = 1$ for $\theta < \theta_0$], $ka \gg 100$, and a plane reflector [$R(\theta) = 1$]. An approximate expression for $|V(z)|$ has been derived by Liang et al[6]

$$|V(z)| = \left| \frac{\sin kz(1 - \cos \theta_0)}{kz(1 - \cos \theta_0)} \right| \qquad (4)$$

This expression accurately predicts the shape of the central lobe; however, it fails to account for asymmetries in the sidelobes. The depth of focus is given by the 3 dB points of the central lobe in Eq. (4).

$$(\Delta z)_{3 \text{ dB}} = \frac{0.443 \lambda}{1 - \cos \theta_0} \qquad (5)$$

For small $\theta_0$, the numerical aperture (N.A.) of the lens is given by the relation [N.A. = $\sin (\theta_0) \approx \theta_0$]. With $\cos \theta_0 \approx 1 - \theta_0^2/2$, we can write Eq. (5) in the form

$$(\Delta z)_{3 \text{ dB}} \approx \frac{0.886 \lambda}{(\text{N.A.})^2} \qquad (6)$$

We used a 4 mW He-Ne laser with a wavelength of 6328 Å in these experiments. A semiconductor laser could have been employed equally well; we chose to work with the gas laser initially only because of its availability and the greater ease of working with visible light in the early experiments. We found it convenient to use a Bragg cell for amplitude modulation of the incident light; with a semiconductor laser, we could directly modulate the laser itself. Since the objective lens is not specially coated for the He-Ne wavelength, there is still some light reflected back from the objective lens. To reduce the amount of the reflected light reaching the detector, we simply place an iris diaphragm

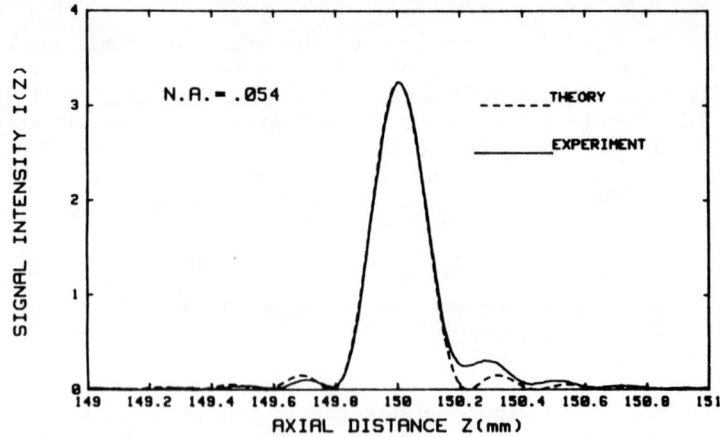

Fig. 2. The dependence of the detected signal amplitude on object distance for a mirrored surface.

before the photodetector. An iris diaphragm placed between the pinhole and the objective lens reduced the numerical aperture of the reflected light. This increased the spot size of the returning one-way spatial filter. The object was moved by two actuators in the axial direction z and the transverse direction x . The whole system was controlled by a computer.

RESULTS

To demonstrate the usefulness of this system for obtaining accurate distance measurements, we used a mirror as the reflecting object and scanned it in the axial direction z . Figure 2 shows the dependence of the signal on the position of the object. The numerical aperture of the optical system is 0.054 , the corresponding depth of focus is 188 µm , and the spot size is about 10 µm . The experimental curve fits the theoretical curve very well, except for some discrepancy in the side-lobes, which is believed to be caused by aberrations in the optical system. By determining the position of the peak of the curve, we can determine the position of the object very accurately.

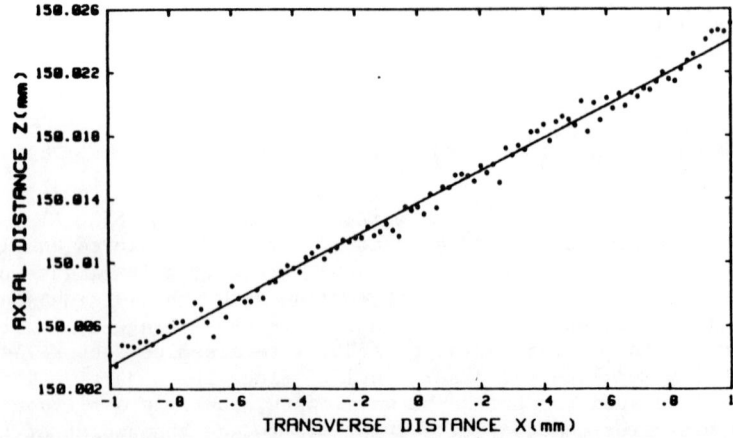

Fig. 3. One-dimensional scan over a tilted mirror.

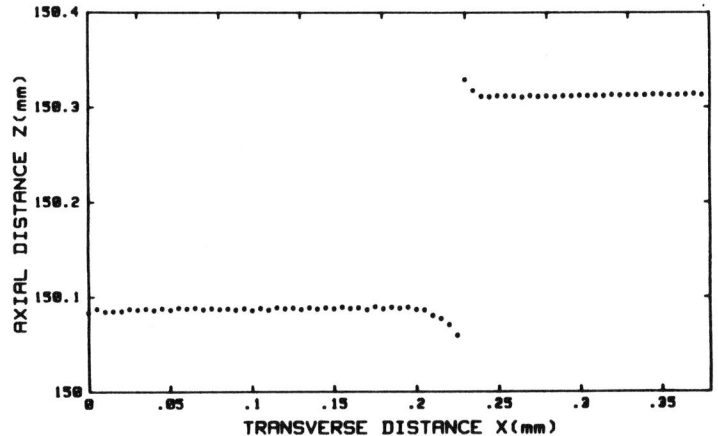

Fig. 4. One-dimensional scan over a 220 µm mirror step.

To determine the sensitivity of our measurement technique, we scanned the focused beam over a tilted mirror, as shown in Fig. 3. The straight line represents the mirrored surface and the dots are experimental results. The estimated accuracy is about 2 µm. As we tilted the mirror, the height of the central lobe decreased and the depth of focus increased. This is because less light could be collected by the camera lens, so the effective numerical aperture decreased.

The results of a subsequent scan over a 220 µm step on a specular reflector are shown in Fig. 4. The transverse spacing between adjacent points is 5 µm. Except for the overshoot near the step, the accuracy of the range measurement is about 2 µm. The overshoot is due to shadowing and interference effects which are illustrated in Fig. 5. The focused light is reflected by both top and bottom surfaces. Most of the light reflected from the bottom surface is blocked by the step, except that reflected back at a small angle. The interference between the two beams shifts the center peak and causes the overshoot effect.

This system is not only suitable for measurements on smooth surfaces, but also for measurements on rough surfaces. This is because when the light is focused to a point on a rough surface, the reflected light will be scattered in all directions, but an image of this point

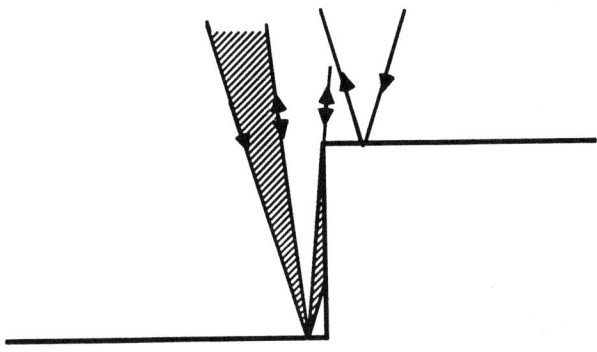

Fig. 5. Illustration of shadowing and interference effects.

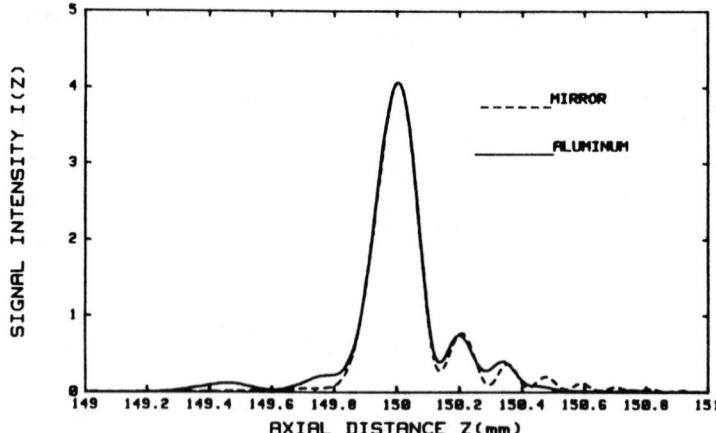

Fig. 6. Dependence of the detected signal amplitude on the object distance for a raw aluminum sample.

can still be obtained at the pinhole. We pay a price for this convenient result - the light reaching the pinhole is much reduced in intensity; the advantage is that the alignment of the rough surface is relatively uncritical. To demonstrate this effect, we scanned a sample of sheet aluminum over the surface, as supplied by the manufacturer. The results we obtained are shown in Fig. 6. The central lobe of this curve is unchanged from the central lobe obtained using a mirror. As expected, the detected signal from this rough surface is much weaker than that for a smoother surface, but it is still strong enough to be easily detected with a narrowband receiving system. The results obtained were insensitive to the tilt of the object, which is essential for the measurement of ordinary machined parts.

CONCLUSIONS

We have demonstrated an optical range finder with a range accuracy of 2 μm and a transverse definition of 10 μm at a working distance of 15 cm . The device, based upon the type II microscope principle, is very stable, easy to align, and suitable for measurement over both smooth and rough surfaces.

ACKNOWLEDGMENT

This work was supported by the Stanford Institute for Manufacturing and Automation, and by the Air Force Office of Scientific Research under Contract No. AFOSR 84-0063C.

REFERENCES

1. T. C. Strand, Opt. Engineering 24, 033 (1985).
2. C. C. Williams and H. K. Wickramasinghe, "Optical Ranging by Wavelength Multiplexed Interferometry," accepted for publication, J. Appl. Phys.
3. C. J. R. Sheppard, Scanned Image Microscopy, Edited by E. A. Ash (Academic Press, London, 1980).

4. T. R. Corle, C-H. Chou, and G. S. Kino, "Depth Response of Confocal Optical Microscopes," accepted for publication, Opt. Lett.
5. B. Richards and E. Wolf, Proc. R. Soc. London A. 253, 358 (1959).
6. K. K. Liang, G. S. Kino, and B. T. Khuri-Yakub, IEEE Trans. on Sonics and Ultrasonics, Spec. Issue on Acoustic Microscopy SU-32 (2), (March 1985).

THE EFFECT OF OXYGEN ON THE ION-ACOUSTIC SIGNAL GENERATION PROCESS*

F. G. Satkiewicz, J. C. Murphy, J. W. Maclachlan, and
L. C. Aamodt

The Johns Hopkins University, Applied Physics Laboratory
Laurel, MD  20707

INTRODUCTION

Thermal wave imaging (TWI) describes a family of methods for materials characterization based on temperature changes induced by an external source. The source can be a laser [1,2,3] or an electron [4,5,6] beam modulated to produce time-varying surface and bulk temperatures in the specimen. Recently, ion sources have been used for excitation [7,8,9] and share some imaging features in common with laser and electron sources. All three types of sources have the ability to detect buried defects in opaque solids and to locate tightly closed cracks [10]. However, the fundamental physical mechanisms of signal generation and image contrast vary to some extent based on the source and, in particular, on the physics of ion-acoustic signal generation process. Non-thermal acoustic wave generation by particle beam-specimen interactions may play a role in this process along with the thermal generation processes familiar from the case of laser excitation. In a continuing effort to study acoustic generation mechanisms employing an ion beam source, the question of sensitivity to surface chemistry arose. Since our apparatus was equipped to conduct experiemnts under $O_2$ partial pressures, the effect of $O_2$ on ion-acoustic generation was a natural choice. Oxygen is one of the most common impurities encountered in materials development, and its presence in materials such as turbine blade coatings raises questions about propagation through microstructures containing different phases. Demonstration of acoustic contrast associated with oxygen would thus be of some interest for NDE. This paper is directed at improved understanding of the signal generation processes for ion-acoustic imaging and investigates the effect of oxide films on acoustic signal generation by ions. A dependence of the acoustic signal amplitude on oxide coverage is observed which varies with specimen material and shows a kinetic variation with the time of film formation and erosion.

EXPERIMENTAL

Ion-acoustic measurements were made using a SIMS (Secondary Ion Mass Spectrometer) apparatus modified to include beam blanking and beam steering capability. Figure 1 is a block diagram of the system. A piezoelectric transducer attached to the underside of the sample detected the specimen elastic response to square-wave modulation of the ion beam current. The modulation frequency varied over a range between 100 Hz and 250 kHz and the magnitude and phase of the acoustic signal were determined as a function of frequency. Specimen current (Faraday current) could be measured simultaneously with the acoustic signal. Beams of $Ar^+$, $Xe^+$, and $Ne^+$ ions at primary beam voltages ranging from 1 to 10 keV could be used in this apparatus but in these experiments only $Xe^+$ and $Ar^+$ ions were employed. The ability to choose the specific ion mass had allowed investigation of the presence of acoustic generation by direct momentum transfer from beam to specimen [11]. Both focused (beam diameter = 300 microns) and unfocused beams were used and no discernible differences in the parametric dependence of the response on frequency, ion type or specimen material could be correlated with focusing.

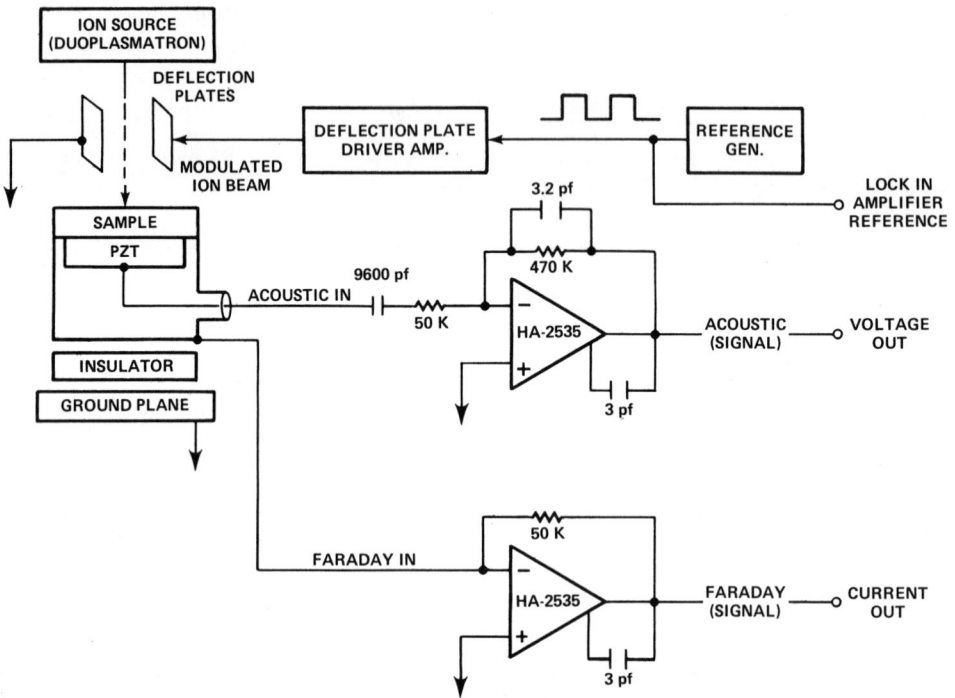

Fig. 1 Block diagram of ion-acoustic system.

The specimens consisted of bulk metal and single crystal silicon substrates several centimeters on a side ranging from 0.25 mm to 20 mm thick. They were covered by surface oxide films prepared by both anodization and thermal oxidation in the case of Ta or thermal oxidation alone for the other metals and silicon. The thermal oxidation was accomplished by heating under known partial pressures of oxygen. The samples included Al, Mg, Ti, Ta, V, W, Mo, and crystalline Si as summarized in Table I. The acoustic signal amplitudes were measured as a function of time during film formation and film removal. Measurement of changes in the specimen current were also made under identical experimental conditions. For selected samples, SIMS measurements of secondary ion flux were made under the same conditions used for the ion-acoustic studies.

TABLE 1   Samples Used for $O_2$ Enhancement Studies

|  |  |
|---|---|
| Mg-alloy(Al,Mn,Zn) | 2.4 mm × 12.7 mm disc |
| Al alloy(Mg,Mn,Fe,Si,Cu) | 1.6 mm × 14 mm × 13 mm rectangle |
| Si single crystal | 0.38 mm × 8 mm × 25 mm rectangle |
| Ti(CP) | 0.64 mm × 16 mm × 21 mm rectangle |
| Steel(Mn) | 2.1 mm × 8 mm × 13 mm segment of washer |
| Ta(pure) | 1.5 mm × 13.2 mm × 13.2 mm rectangle |
| Anodic $Ta_2O_5$(purple) on Ta | a. Film: ~ 350Å |
|  | b. Substrate: 0.13 × 14 × 15 mm |
| Anodic $Ta_2O_5$(copper) on Ta | a. Film: ~ 1400Å |
|  | b. Substrate 0.1 × 14 × 13 mm |

Coupling to the PZT housing was made with DAG 154 (Acheson Colloids Co., Port Huron, Michigan)

RESULTS

The changes in the magnitude and phase of the acoustical signal, for samples presputtered to remove native oxide films, were measured as a function of time following introduction of $O_2$ at a pressure of approximately $10^{-5}$ Torr into the specimen chamber. Figure 2 shows the results for aluminum. Following the admission of $O_2$, the acoustic signal increased until finally reaching a stationary value. The ratio of the final signal amplitude to that measured before oxygen is admitted is defined as the enhancement ratio, R. R values for both acoustic and Faraday current signals vary with the specimen material as shown in Table II. The time to reach equilibrium is dependent on the material. No systematic study of the effect of oxygen partial pressure on formation time has yet been made.

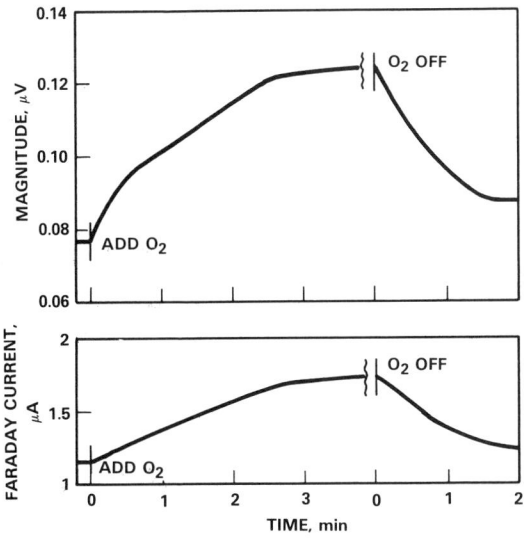

Fig. 2  Variation in ion-acoustic response during sputtering of Al alloy with 7 keV ar$^+$ in the presence and absence of oxygen. The modulation frequency is 89 kHz.

TABLE II  Comparison of the Effect of Oxygen on the Acoustical Response and Ion Yield of Materials Sputtered with 6-7 keV Ar$^+$; 40 kHz (Blanking)

| Material | Acoustic $\dfrac{S(O_2)}{S(\text{No } O_2)}$ | $\dfrac{i_F(O_2)}{i_F(\text{No } O_2)}$ | SIMS $\dfrac{I_{X+}(O_2)}{I_{X+}(\text{No } O_2)}$ | $\dfrac{i_F(O_2)}{i_F(\text{No } O_2)}$ |
|---|---|---|---|---|
| Mg Alloy | 2.5 | 2.47 | 1.7 | 1.74 |
| Al Alloy | 1.16 | 1.5 | 2.9 | 1.17 |
| Si(SC) | - | - | 12 | 1.12 |
| Ti | 1.02 | 1.07 | - | - |
| Fe(Steel) | ND | ND | 196 | 1.07 |
| Ta | 1.09 | 1.10 | 91 | 1.09 |
| (Ta$_2$O$_5$/Ta)#1 | 1.14 | 1.11 | - | - |
| (Ta$_2$O$_5$/Ta)#2 | 1.16 | 1.09 | - | - |

The frequency dependence of the acoustic signal and the Faraday current was measured before oxygen was admitted and in the equilibrium region. Figure 3 summarizes the experimental results for the ion-acoustic signal for aluminum and magnesium using Xe$^+$ ions as the source. For both specimens there is a low frequency region where the acoustic signal varies as $f^{-1}$ and where little oxygen enhancement is seen. In a second region at higher frequencies there is a significant enhancement. The enhancement ratio at high frequencies is $R_{Mg} = 2.5$ and $R_{Al} = 1.16$. Magnesium exhibits the largest enhancement of any of the metals studied. Figures 4(a) and (b) show $R_f$ for both metals for Ar$^+$ and Xe$^+$ ions, respectively. The variation of the specimen current with frequency is not shown in these figures but was found to be independent of frequency for the range of frequencies used in this work, revealing no dependence of the kind observed in the acoustic case. The origin of the frequency dependence for the ion-acoustic signal has not been established but it is noteworthy that a similar frequency dependence has been observed in electron-acoustic imaging and is correlated with the visibility of lateral structures such as grain boundaries or closed cracks [10]. There it was suggested that two generation mechanisms for acoustic waves in solids exist because of the presence of two frequency regimes.

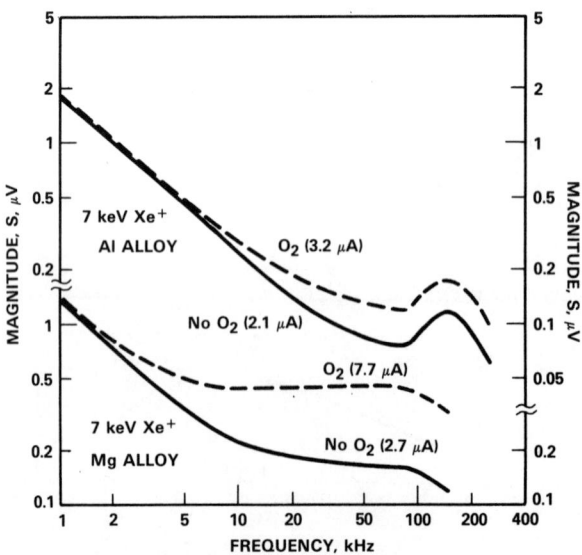

Fig. 3  Frequency dependence of ion-acoustic response in the presence and absence of oxygen during sputtering of Mg and Al alloys with 7 KeV Xe$^+$.

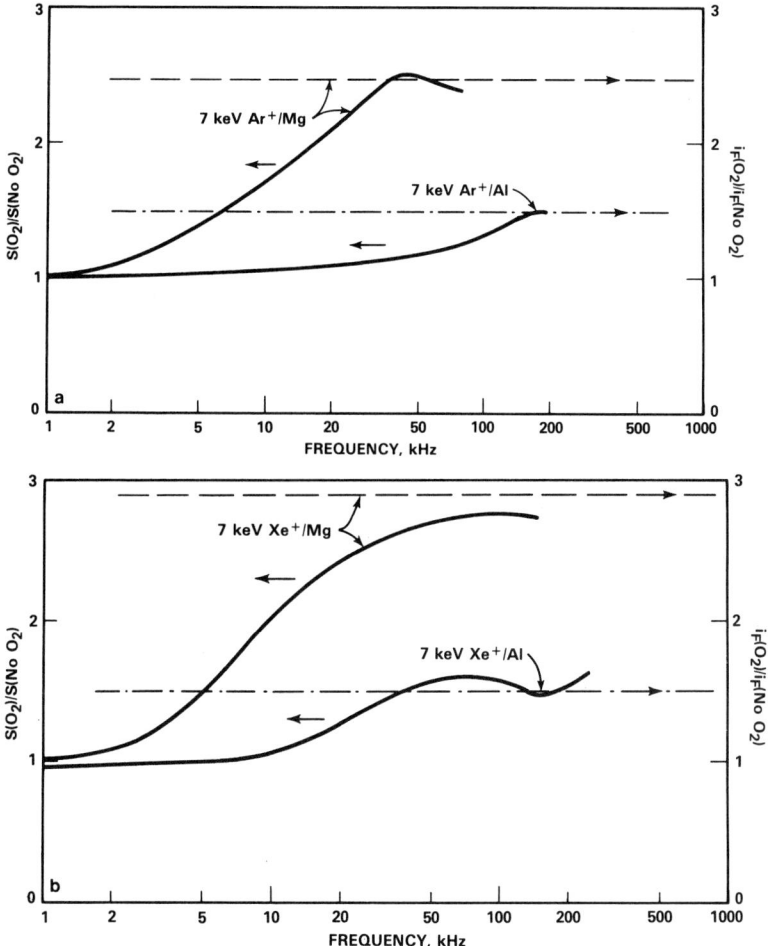

Fig. 4  Frequency dependence of enhancement ratio, R, for Mg and Al alloys sputtered with (a) 7 keV Ar$^+$ ions and (b) 7 keV Xe$^+$ ions.

Measurements were also made on anodized Ta specimens covered initially by thick oxide films nominally 37 nm and 150 nm thick. The variations of acoustic signal and specimen current with time are shown in Fig. 5(a) and (b) for the 37 nm and 150 nm specimens, respectively, each for several different areas on the specimen. The second areas of the 37 nm specimen sputtered did not show the initial enhancement observed in the other areas studied and suggests the possibility of changes in surface chemistry caused by either condensation of sputtered neutrals or the chemisorption of $O_2$ introduced during the enhancement step in the first profile. Under non-oxidizng conditions, the ion beam erodes the oxide film, finally breaking through after some minutes of exposure. At the interface, both the acoustic signal and specimen current drop. For the conditions of Fig. 5, the breakthrough time was independently estimated using SIMS by monitoring the $16_O+$ signals. The values corroborated the earlier results. After exposing the Ta substrate, $1.5 \times 10^{-5}$ Torr oxygen was added to the target chamber. A rapid rise occurred in both the acoustic and specimen current signals. For the second $Ta_2O_5/Ta$ sample with a thick anodic film, the initial response is different than that for the thinner film. The rapid surface

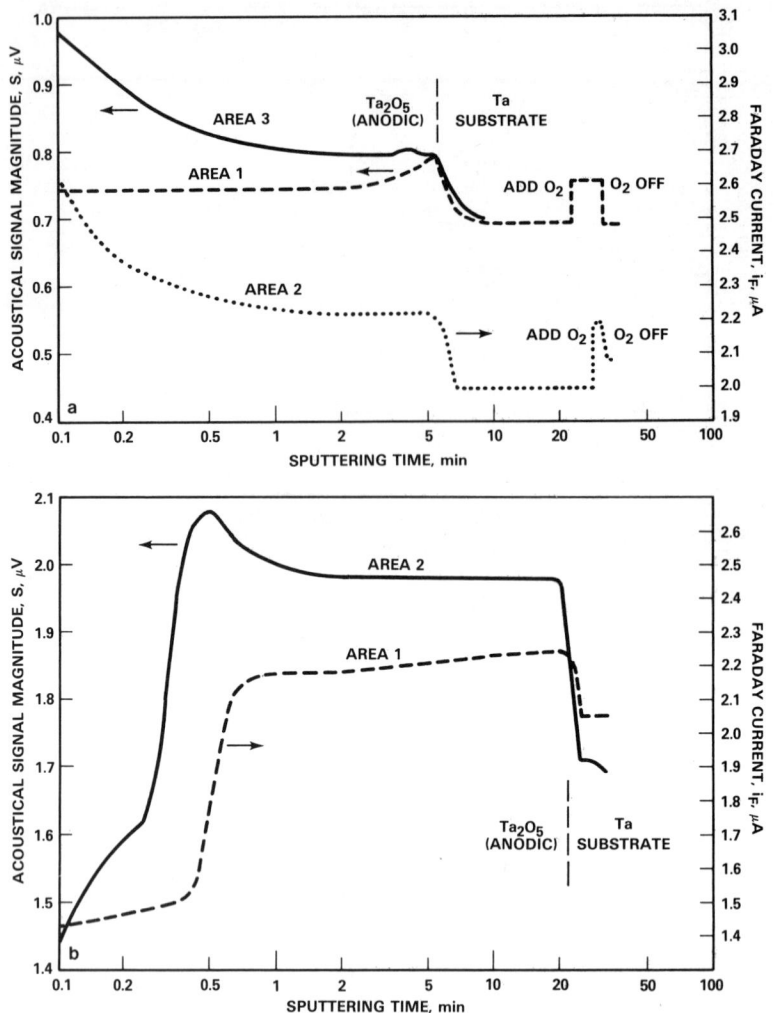

Fig. 5 Variation in ion-acoustic signal and specimen current with time for different areas on $At_2O_5$ film on Ta. (a) Sample 1 with film thickness 37 nm, (b) Sample 2 with film thickness 150 nm.

enhancement is not evident; in addition, there appears to be an offset in the rise times of both the acoustic signal and specimen current responses. One conclusion from these measurements is that enhancement of the acoustic signal does not depend strongly on the oxide film thickness and hence not on the mechanical properties of the film. This distinguishes it from the sensitivity shown for the case of laser-acoustic generation where the mechanical film properties are important in controlling the shape and amplitude of the acoustic response. The absence of any dependence of the signal on the mechanical properties of the layer is significant since in experiments on ultrasound generation by laser heating, the stress boundary conditions at the specimen surface and the presence of constraining surface layers control the shape and amplitude of the acoustic response.

DISCUSSION

Unlike laser (non-ablative applications) and electron beam sources, heavy ion beams cause erosion of the specimen. At the same time, ions are implanted below the surface with the deposition of energy and momentum distributed with depth. As these time-dependent processes occur, acoustic generation proceeds by various mechanisms in different regions in the sample. Thus ejecta could contribute to the signal. Since the surface condition can affect the sputtering process and perhaps the acoustical process, the sputtering behavior of the materials used was examined. The positive ion spectra of the Mg alloy, Al alloy, Si, steel, and Ta samples were obtained both with and without oxygen enhancement. The rise and decay of the $MO^+$ signal (where M is a metal) was obtained. In addition, the Mg, Al, and Si spectra showed a large and somewhat comparable population of multiple-charged ions ($MO^{2+}$, $MO^{3+}$). The silicon acoustic change with oxygen is much less than that seen for either Mg and Al suggesting that there is not a direct correlation between sputtering efficiency and acoustic signal generation. Of course, the sputtering process also involves production of negative ions and a large number of neutrals so that a full picture of the correlation between the two processes must take all components of the sputtering into account. One possible explanation for the increase in the acoustic signal with oxygen (especially for Mg and Al) is that the sputtering rates for the oxides are substantially less than that for the metals. Thus less energy is removed as ejecta and is then available for acoustic wave generation on implantation. Since the change is more prominent at higher frequencies, this suggests a possible non-thermal enhancement process.

The acoustic signal and specimen current are correlated in many of the experiments described in this paper. The main area where correlation does not exist is for oxidized material at low frequency. Several possible origins of the current changes have been considered in light of prior work. Benninghoven and Wiedmann [12] studied the reaction of $O_2$ on Mg and show positive and negative ion spectra of clean and oxidized surfaces. Both polarity ion populations rise dramatically on an oxidized surface; however, the rise for negative ions is greater and a shift in specimen current would be expected. On the other hand, Muller and Benninghoven [13] studied the reaction of $O_2$ on Ti and showed a net increase in positive ions on an oxidized metal surface. Since the sputtering rate of $TiO_2$ is less than Ti, an increase in specimen current and a decrease in the acoustic signal would be expected. The former is seen but not the latter. Again, no account was made of the secondary ion distributions nor of the true ion yields.

CONCLUSIONS

Acoustic signals generated in solids by modulated ion beams had been shown to be able to image surface and near subsurface regions of the specimen. However, the mechanisms leading to acoustic wave generation are uncertain. Non-thermal generation mechanisms had been expected to be important since current theories of sputtering would support a strong role for momentum transfer from the ion beam to the specimen. Experimentally, little evidence for these mechanisms has been found. This paper has investigated ion-acoustic signal generation for certain metals and for silicon crystals covered with oxide films and finds that the films enhance the magnitude of the acoustic signals observed. Results obtained using different ion species and through measurement of the frequency and time dependence of the oxidation/erosion process of film formation suggest that the mechanical properties of the films are not responsible for the observed enhancement. However, the nature of the oxygen attachment to the surface may influence the signals. Also, the detailed interpretation of specimen current changes and acoustic signal changes are future tasks.

ACKNOWLEDGEMENT

This was was supported by the U. S. Dept. of the Navy, Naval Sea Systems Command under Contract No. N00024-85-C-5301.

REFERENCE

1. Y. H. Wang, R. L. Thomas, and G. F. Hawkins, Appl. Phys. Lett. 32, 538 (1978).
2. D. Fournier and A. C. Boccara, The Mirage Effect in Photothermal Imaging, in: "Scanned Image Microscopy," E. A. Ash, ed., Academic Press, London (1980).
3. J. C. Murphy and L. C. Aamodt, Appl. Phys. Lett. 38, 196 (1981).
4. G. S. Cargill, III, Nature 286, 691 (1980).
5. A. Rosencwaig, Elec. Lett. 16, 928 (1980).
6. J. C. Murphy, J. W. Maclachlan, R. B. Givens, F. G. Satkiewicz, and L. C. Aamodt, Generation of Ultrasound by Laser, Electron and Ion Probes and Its Application to the Characterization of Materials in: "Ultrasonics International '85 -- Conference Proceedings," Butterworth Scientific, Ltd., U.K. (1985).
7. F. G. Satkiewicz, J. C. Murphy, L. C. Aamodt, and J. W. Maclachlan, Review of Progress in Quantitative Nondestructive Evaluation 5A, 455 (1986).
8. K. Kimura, K. Nakanishi, A. Nishimura, and M. Mannami, Jap. J. Appl. Phys. 24, L449 (1985).
9. D. N. Rose, H. R. Turner, and K. O. Legg, 4th Int'l. Topical Meeting on Photoacoustic, Thermal and Related Sciences, Ville d'Esterel, Quebec, Aug. 4-8, 1985.
10. J. C. Murphy, J. W. Maclachlan, and L. C. Aamodt, IEEE Trans. Ultrason. Ferroelectrics, Freq. Cont., UFFC-33, 529 (1986).
11. F. G. Satkiewicz, et al, unpublished work, 1986.
12. A. Benninghoven and L. Wiedmann, Surf. Sci. 41, 483 (1974).
13. A. Muller amd A. Benninghoven, Surf. Sci. 41, 493 (1974).

DISCUSSION

Mr. H. K. Wickramashinghe: With what accuracy can you determine the end point in the oxide erosion process and locate the interface between the surface film and the substrate?

Mr. Murphy: That is difficult to answer now because the SIMS work is not being done simultaneously with the ion-acoustic studies at this time. We hope to better answer that question in the relatively near future when data from two simultaneous measurements are available. Based on current estimates for oxide sputtering rates, it looks as if the interfaces can be determined relatively accurately on the time scale that I showed in the figure. A better number will have to wait for the next set of experiments.

APPLICATION OF ADDITIVE REGIONAL KALMAN FILTERING

TO X-RAY IMAGES IN NDE

John P. Basart, Yi Zheng, and Edward R. Doering

Center for NDE
Iowa State University
Ames, Iowa 50011

INTRODUCTION

One of the time consuming procedures in inspecting parts by x-ray film is the identification of a flaw. Low contrast films of dense objects especially cause problems. A radiologist must have considerable experience in identification in order to keep the examination time relatively small, but also keep the reliability high. Our objective in this project is to develop a computer procedure that will sufficiently enhance flaws in an image in a manner that will reduce the time it takes a human to locate and identify a flaw.

Factors limiting the quality of an X-ray image are image unsharpness, quantum fluctuation, film grain and film contrast [1,2]. The unsharpness caused by scattered radiation reduces the image contrast. The quantum fluctuation caused by random emission and absorption of X-ray quanta smears or masks the contrast. The film grain and contrast limit the recorded information capacity. A coarse-grained image conveys less detail than one of fine grain. In this paper, we discuss a method for enhancing the image by reducing the fluctuation due to disturbances, such as quantum fluctuation and granularity, etc. The main tool used is the Kalman filter. The basic idea is to estimate a pixel optimally in an image using a given pixel and its near neighbors. One advantage of a Kalman filter is that it incorporates information about every aspect of the process. It can include a model of the process that generated the desired information, a model of the noise added to this process, a model of the measurement system, and a model of the noise within the measurement system. In addition, there can be multiple models representing multiple processes at any one, or all, of these stages. Another advantage of the Kalman filter is that it can distinguish between stochastic processes that have strongly overlapping spectra. Ordinary spectral filters are of limited benefit under such conditions. When processing noisy images one often finds that the noise, system, and signal processes overlap in frequency.

Three steps are involved when implementing our method of filtering. They are 1) image segmentation, 2) image modeling, and 3) Kalman filtering. Each of these procedures will be explained. The results of filtering a low contrast flaw in an x-ray image will be discussed at the end of the article.

## SEGMENTATION

Segmentation is an image classification procedure. Autoregressive modeling, which we incorporate in our method, requires stationarity. Generally, the stationarity assumption is not true for the processes in an image over the whole image and this violation will cause blurred edges and reduced contrast in a filtered image. Therefore segmentation is necessary to find regions in which the statistics, mean and variance, are stationary.

An image is segmented by partitioning it with respect to local mean and local spatial activity of the image [3,4]. Spatial activity is defined as the rate of change of spatial luminance from one pixel to another. It is related to the concept of variance. The formula used to calculate the spatial activity is called the masking function. Regions of stationary mean and stationary variance can be found by segmenting an image by local means and by the masking function, respectively. With these two segmentations in hand, they can be combined to produce new segments that are wide-sense stationary.

Local means are found by a window of running average. A $(2n+1) \times (2n+1)$ window is selected in one corner of the image. All pixels within the window are averaged. This average is assigned to the center pixel. The window is then moved and the process is repeated. The mathematical expression for the local mean is

$$m_n(i,j) = \frac{1}{(2n+1)^2} \sum_{p=i-n}^{i+n} \sum_{q=j-n}^{j+n} z(p,q) \qquad (1)$$

where $z(p,q)$ is the image intensity at pixel $p,q$. After calculation of the mean for all windows, a file of the local means is set aside for later use.

The next step is to determine the masking function for the image. The masking function is defined by

$$M_r(i,j) = \sum_{p=i-r}^{i+r} \sum_{q=j-r}^{j+r} e^{-\|(x,y) - (p,q)\|} [\frac{1}{q} \sum_{p=0}^{7} D_{pqn}] \qquad (2)$$

where $\|(x,y) - (p,q)\|$ is the Euclidean distance between points $(x,y)$ and $(p,q)$, $(x,y)$ is center pixel of a window, $(p,q)$ is any other point in the window, and D is the difference in intensity between a pixel adjacent to $(p,q)$ and the pixel at $(p,q)$. The difference, D, is summed over all pixels adjacent to $(p,q)$. The average of these differences is weighted exponentially by the distance from $(p,q)$ to $(x,y)$. After the masking function is calculated for all the windows, it is recorded in a file.

The next step is to use the local means and masking function to segment the image. A cluster seeking procedure, somewhat similar to the K-means cluster seeking algorithm [5], is used to cluster local means and masking functions. It differs from the standard K-means cluster seeking algorithm in that the thresholds of the distance between the cluster center are given for simplicity. Each local mean and masking function is assigned to a certain cluster. All combinations of local mean clusters and masking clusters form wide-sense stationary regions which we desire.

MODELING

After completing the segmentation, the process in each segmented region is represented by a p-order AR process:

$$s(k) = \sum_{n=1}^{p} \Phi_n \cdot s(k-n) + w(k) \tag{3}$$

$$z(k) = \mathbf{H} \cdot \mathbf{s(k)} + v(k) \tag{4}$$

where $z(k)$ is the measurement of intensity at a pixel, $v(k)$ is an additive measurement white-noise sequence, $s(k)$ is a "true image" process, $w(k)$ is a residual sequence, $\mathbf{H}$ is a $(1 \times m)$ measurement vector and $\mathbf{s(k)}$ is an $(m \times 1)$ vector of $s(k)$. $v(k)$ and $w(k)$ are independent and uncorrelated with $E[v(k)] = 0$, $E[w(k)] = 0$, $E[v(k)w(h)] = 0$, $E[v(k)v(k-h)] = R \cdot \delta(h)$, and $E[w(k)w(k)] = Q \cdot \delta(h)$. $\Phi$'s are coefficients to be estimated. There are a number of ways to estimate $\Phi$'s such as maximum likelihood or least squares approaches [6]. "Marquardt's compromise" [7,8] and Yule-Walker equation [6] methods are often used in practice. We estimate the $\Phi$'s by solving the Yule-Walker equation

$$\mathbf{r_0} \cdot \mathbf{\Phi} = \mathbf{r_1} \tag{5}$$

where

$$\mathbf{\Phi} = [\Phi_1 \; \Phi_2 \; \Phi_3 \; \cdots \; \Phi_p]^T$$

$$\mathbf{r_1} = [r_1 \; r_2 \; r_3 \; \cdots \; r_p]^T$$

$$\mathbf{r_0} = \begin{vmatrix} r_0 & r_1 & \cdots & r_{p-1} \\ r_1 & r_0 & & \\ & & & \\ r_{p-1} & \cdots & & r_0 \end{vmatrix}.$$

The r's are autocorrelation coefficients of $s(k)$. Given the measured $z(k)$'s and the variance R of $v(k)$, the r's can be found by taking the expectation of (4). The semi-positive definite property of the r's must be considered when the r's are calculated [9].

After a state-space form of (3) is obtained [10], we are ready to apply the Kalman filter.

KALMAN FILTERING

The Kalman filter is an optimal filter that can separate two or more stochastic process. The Kalman filter theory and applications can be found in many sources [11,12,13].

Since all quantities required for Kalman filtering have now been found, the optimal estimates of pixels are obtained by the following recursive procedure:

1. Enter the recursive loop with the initial values of the a priori estimated (n×1) vector $s(k|k-1)$ and its error covariance matrix $P(k|k-1)$

2. Compute the Kalman gain

$$K(k|k) = P(k|k-1) \cdot H^T \cdot (H \cdot P(k|k-1) \cdot H^T + R)^{-1} \qquad (6)$$

3. Estimate a pixel

$$s(k|k) = s(k|k-1) + K(k|k) \cdot (z(k) - H \cdot s(k|k-1)) \qquad (7)$$

4. Compute the error covariance matrix

$$P(k|k) = (I - K(k|k) \cdot H) \cdot P(k|k-1) \qquad (8)$$

5. Predict

$$s(k+1|k) = \Phi \cdot s(k|k) \qquad (9)$$

$$P(k+1|k) = \Phi \cdot P(k|k) \cdot \Phi^T + Q \qquad (10)$$

The process is repeated for the next pixel $z(k+1)$ from step 2 until all pixels are processed. One should be careful that the Kalman equations are simplified due to the scalar modeling.

RESULTS

By applying the above procedure to low-contrast X-ray images, we have produced enhanced images. One example of a processed image is shown here. It is an 88×88 pixel subimage of an X-ray image of a casting. There is a flaw located near the center area of the image. The flaw is not obvious in the original image which is very dense and has low contrast. The contour map of the original image is shown in Fig. 1(a). The variance of the disturbance fluctuation measured from a flat area in the original image is 1.6. The result from filtering is shown in the contour map in Fig. 1(b). Since the dynamic ranges of the images are too small (about 20 to 30), histogram equalization with an exponential transformation function was applied to both the original and filtered images. Ruled surface plots of the original and the transformation results are shown in Fig. 2(a) and Fig. 2(b), respectively. The dynamic range in Fig. 2 has increased to 128. In the original image, the flaw region is broken into many spikes which make flaw detection difficult. The filtered image shows a bigger concentration of intensity within a region that can be defined by a single boundary. Compared with the original image, the filtered one has a lower and smoother background. Thus the flaw in the filtered image is easily detected now.

The experiment was done on an ISU VAX 11/780 computer. The CPU time for running the Kalman filter part was about 8 minutes (88×88 pixels) using a moving window. The modeling and filtering were applied to a 7×7 moving window and the 3×3 pixels in the center of the window were saved each time.

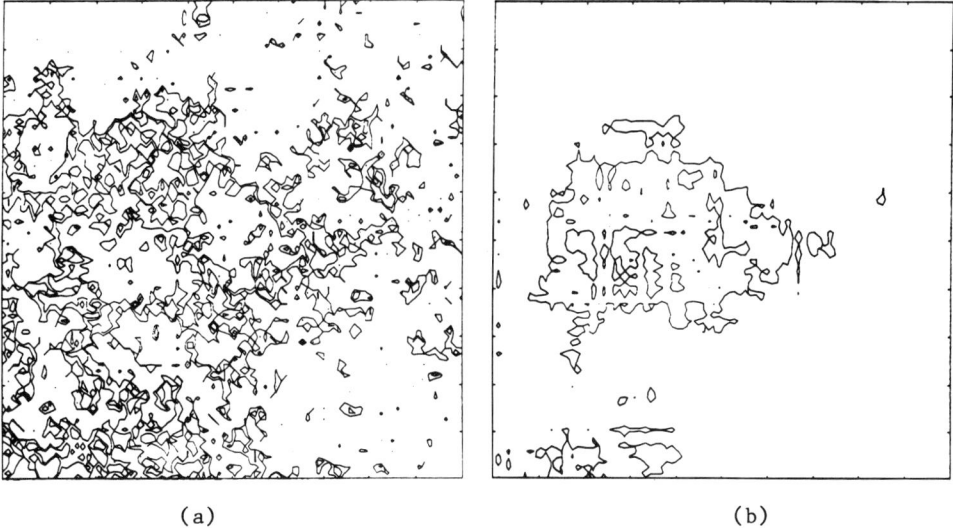

Fig. 1. Contour plots of intensity before (a) and after (b) filtering. Contour levels are the same in both plots. The flaw in the filtered map (b) clearly stands out above the background.

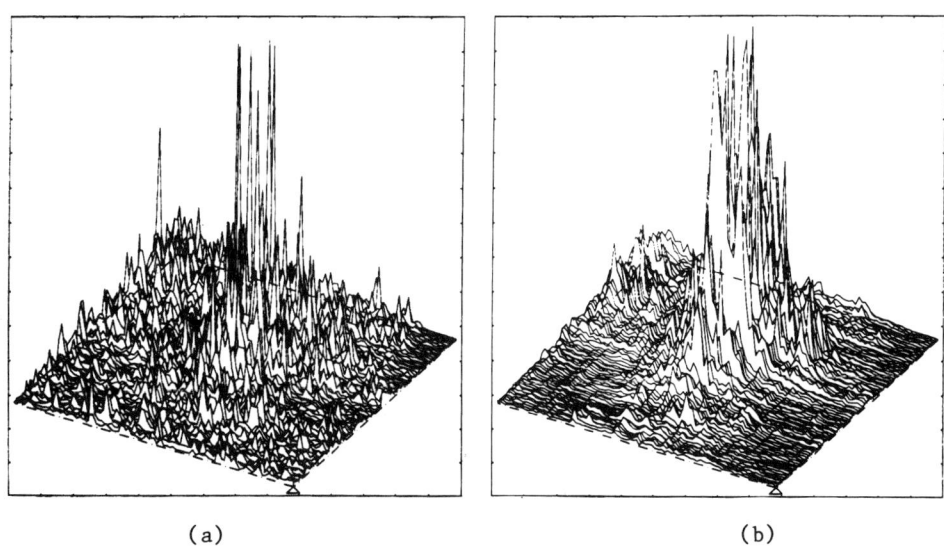

Fig. 2. Ruled surface plots before (a) and after (b) filtering. In the filtered map (b), the background noise is lowered and smoothed, and the power in the flaw is more centralized and less "spikey" than in the original map.

ACKNOWLEDGEMENT

This work was supported by NSF grant No. AST-8217135 and the NSF University/Industry Center for NDE at Iowa State University.

REFERENCES

1. R. Halmshaw, Information theory applied to industrial radiography, in: Research Techniques in Nondestructive Testing, vol. II, R. S. Sharpe, ed., Academic press, New York (1973).

2. R. Halmshaw, Industrial Radiology: Theory and Practice, Applied Science Publishers Ltd, England (1982).

3. B. J. Schachter, L. S. Davis, A Rosenfeld, Pattern Recognition, 11, 19 (1979).

4. S. A. Rajala, R. J. P. de Figueiredo, IEEE Trans. A.S.S.P. 29, 1033 (1981).

5. J. T. Tou, and R. C. Gonzalez, Pattern Recognition Principles, Addison-Wesly, Reading, MA (1974).

6. G. E. P. Box, and G. M. Jenkins, Time Series Analysis: Forecasting and Control, rev. ed., Holden-day, San Francisco, CA (1976).

7. D. W. Marquardt, J. Soc. Ind. Appl. Math., 11, 431 (1963).

8. A. Pankratz, Forecasting with Univariate Box-Jenkins Models, John Wiley, New York (1983).

9. W. A. Fuller, Introduction to Statistical Time Series, John Wiley & Sons, Inc, New York (1976).

10. A. Bovas and J. Ledolter, Statistical Methods for Forecasting, John Wiley & Sons, Inc., New York (1983).

11. R. E. Kalman, Trans. ASME, 82, 35 (1960).

12. R. E. Kalman, and R. S. Bucy, Trans. ASME 83, 95 (1961).

13. R. G. Brown, Introduction to Random Signal Analysis and Kalman Filtering, John Wiley, New York (1983).

# AN IMAGE SEGMENTATION ALGORITHM

# FOR NONFILM RADIOGRAPHY

Zane W. Bell

Oak Ridge Y-12 Plant
Martin Marietta Energy Systems, Inc.
Oak Ridge, Tennessee 37831

## INTRODUCTION

Radiographic inspection is a commonplace task at the Oak Ridge Y-12 Plant.* It is a labor intensive operation that requires human beings to make subjective decisions that are precise and consistent with previous decisions. These qualities of the inspection procedure have made it desirable to seek to automate both the irradiation of the part under test and the interpretation of the resulting radiographs.

Radiographs for automatic interpretation are acquired by an X-ray sensitive scintillating lightpipe that illuminates a charge-coupled device (CCD) camera. The CCD camera is interfaced to an image processor and the pixels in the camera are digitized at video rates. The image is available for storage and/or processing in the image processor as it is acquired.

There are two distinct phases in the automated inspection of such radiographs. These can be loosely described as image segmentation and pattern classification. The segmentation phase is concerned with dividing the radiograph into regions of interest and regions of boredom. The regions of interest consist of those areas in the image that contain the shadows of mass excess or mass defect, while the regions of boredom are marked by the signature of noise on an otherwise uneventful average mass density. The classification problem is one of characterizing discovered defects (the term "defect" will be used to denote any deviation from the average mass density) as to their shape, size, radiographic density, or the like. This paper will be limited to a discussion of an algorithmic approach to the implementation of the segmentation phase.

The image generated by the CCD camera must be processed before segmentation can be attempted. The raw data are simultaneously corrected for the exponential variation of detected intensity with density, for the variation of the source intensity over the surface of the lightpipe-CCD assembly, and for the X-ray conversion efficiency of the camera system by subtracting the logarithm of a reference image from the raw data image. The effects of the scattering of photons in the part under test are

---

*Operated by the U.S. Department of Energy by Martin Marietta Energy Systems, Inc., under contract DE-AC05-84OR21400.

removed, after the subtraction, by deconvolving a calculated scattering distribution.

At this point in the processing, the image pixel values represent mass defect or excess. It is now the job of the segmentation algorithm to distinguish real defects from a noisy background. The steps involved in this process are averaging with edge retention, soft thresholding, and cluster acceptance. All of these steps are designed to be implemented in an image processor; only certain computations requiring floating-point operations are performed in a host computer. In the sections that follow, each of the steps in the segmentation algorithm and their effects are described. In the concluding section, possible improvements to the algorithm are discussed.

AVERAGING WITH EDGE RETENTION

The noise present in radiographs after the initial processing makes it desirable to remove the small, grain-like artifacts that are not significant. Simple spatial averaging is a rapid and easily implemented procedure to accomplish this task. Figure 1 shows the histogram of pixel values before (Figure 1a) and after (Figure 1b) averaging over a 5 by 7 region.

Being equivalent to a low-pass filtering operation, averaging tends to blur the edges of real defects and invalidate subsequent size estimates. To counteract this side effect, an edge detection algorithm could be employed to dynamically reduce the area over which the average is performed. Since most image processors are designed to perform operations on an entire image during each processing cycle, this solution is not viable.

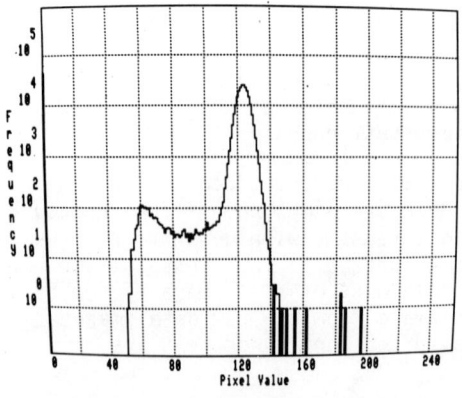
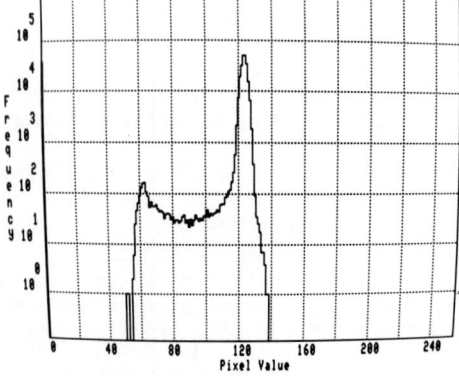

Fig. 1. a) Histogram before averaging.   b) Histogram after averaging.

A better idea is to use an edge detection algorithm to locate edges in an image and, after averaging, to copy original (unaveraged) data into the averaged data in the edge regions. This technique has the advantages of retaining the original sharpness of defect edges and allowing the image processor to perform almost all the computation.

Edge detection is accomplished with omnidirectional Sobel operators[1], which give a measure of the gradient in the vicinity of a pixel. Large gradients are indicative of edges; small gradients are indicative of plateau or valley regions of pixel value. The qualitative criteria "large" and "small" are quantified in terms of the histogram of pixel values occurring in the image.

After averaging, a histogram of pixel values is accumulated by the image processor and transferred to host memory. The most frequently occurring pixel value is assumed to approximate the mean value of the normal distribution describing the noisy background, and a least squares fit is performed over an eleven pixel interval centered on the mean value to determine the parameters of the distribution. The values of the gradient, as estimated by the Sobel operators, are then compared to what might be expected from random variables described by the computed normal distribution. Operator values exceeding three standard deviations are accepted as indicating an edge pixel. Since the image processor operates on an entire image, and the output of its arithmetic unit is another image, the result of this edging is a mask image whose values are set to 255 (all eight bits set) to indicate an edge pixel and to zero to indicate any other pixel. An example of an edge mask is shown in Figure 2.

SOFT THRESHOLDING

One of the cues humans use in scene analysis is gray level thresholding. Objects are distinguished from each other and from a background on a basis of the relative amount of light received by the eye. Since humans operate in an analog world, thresholding decisions tend toward fuzzy decision boundaries rather than toward sharp boundaries. The soft thresholding technique used in the inspection process at Y-12 resulted from an effort to mimic human behavior.

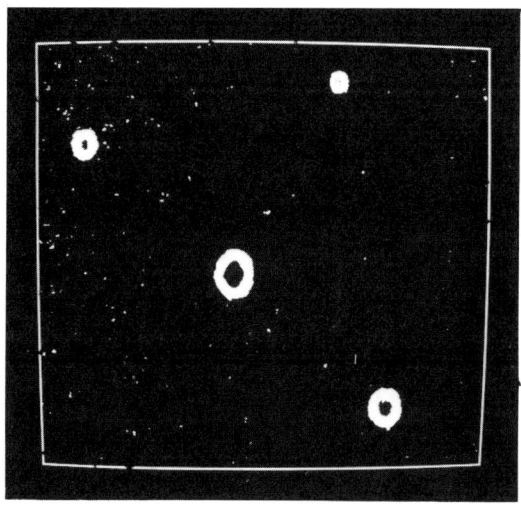

Fig. 2. Edge mask image.

The segmentation problem in radiography is concerned with distinguishing foreground objects from a grainy background. The problem is somewhat ill-conditioned in that there is no specific shape for which the radiograph is to be searched. Consequently, template matching schemes must be discarded because of the tremendous number of templates that must be tested. Thus, the problem may better be stated in the reverse: Distinguish the background from non-background.

As seen from Figure 1, the background is well described by a normal distribution with a calculable mean and variance. A hard thresholding algorithm due to Chow and Kaneko[2], used local histograms (rather than the histogram of the entire image) to estimate the position of the threshold. The use of local thresholds suppresses the effects of the background intensity varying slowly over the image. However, such algorithms suffer from the effect shown in Figure 3. The proximity of $P_1$ to $P_2$ causes a large fraction of $P_1$ to be lost, and estimates of the parameters of $P_1$ are poor at best.

An improvement over local hard thresholding is the development of a local measure that assigns a pixel to the foreground or the background according to the likelihood that it is a member of the background. Such a measure is the conditional probability that a pixel whose value is x belongs to the background, $p(bkg|x)$.

This conditional probability can be calculated in terms of quantities measurable from the histogram of the image using Bayes' Theorem,

$$p(bkg|x) = \frac{p(x|bkg) \cdot p(bkg)}{p(x)}$$

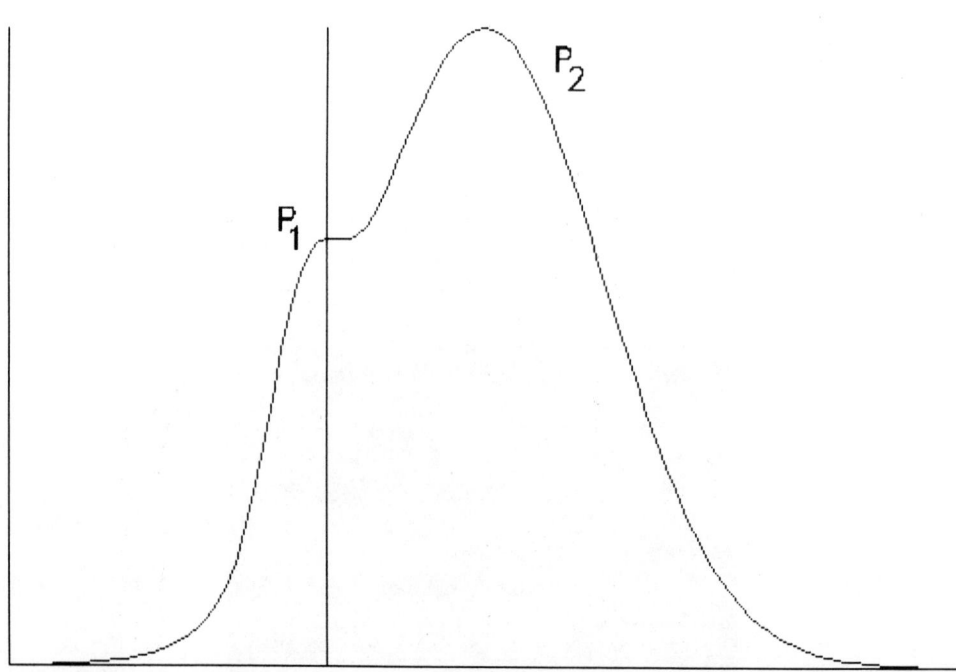

Fig. 3. Effects of hard threshold on peaks in close proximity.

where p(x|bkg) is the probability of a pixel value x occurring given that
it is part of the background, p(bkg) is the a priori probability of the
occurrence of a background pixel, and p(x) is the a priori probability of
the occurrence of a pixel whose value is x.

The conditional probability p(x|bkg) is calculated from the
normalized least squares fit (as mentioned in the previous section) of a
Gaussian to the region of the histogram containing the most frequently
occurring pixel value. The a priori probability p(x) is simply the
normalized value of the histogram at the pixel value x, and p(bkg) is
given by the ratio of the integral of the Gaussian integrated over all
pixel values to the total number of pixels in the image.

If p(x|bkg) is written as $A \cdot \exp\{-[(x-\mu)/\sigma]^2/2\}$, then p(bkg|x) is
given by

$$p(bkg|x) = \frac{A \cdot e^{-\frac{1}{2}\left(\frac{x-\mu}{\sigma}\right)^2}}{p(x)}.$$

This is just the ratio of the fitted Gaussian to the actual histogram, and
the result agrees with intuition.

The local aspect of the algorithm is implemented by dividing the
radiograph into 16 sub-images and acquiring the "sub"-histogram of each
sub-image in addition to the histogram of the entire image. Under the
assumption that the most likely pixel value in the entire image is near
the mean background value, a Gaussian is fitted to that region of the
histogram of the entire image. The search for the mean background level
in each sub-histogram is then restricted to the background region ($\mu \pm 3\sigma$)
of the histogram of the entire image. This is necessary because the
unlucky placement of a large defect completely in a sub-image can produce
more defect area than background. Note that this technique assumes that
each image is mostly background.

For each sub-image, p(bkg|x) is computed as described above and
pixels are assigned to the foreground or background according to the Monte
Carlo method. For each pixel, p(bkg|x) is compared to a random number.
If the random number is the larger, then the pixel is assigned to the
foreground. Otherwise it is consigned to the background. This process is
accomplished in the image processor in a single cycle by implementing the
random numbers as an image and using the conditional arithmetic
capabilities of the processor.

Figure 4 shows a radiograph after the averaging and edging process.
Five circular defects were machined into the test piece although only four
are evident in the image. Figure 5 shows the results of soft thresholding
displayed as a region-of-interest mask. Note that the five defects have
survived and many small speckles have also been generated. These are
caused by noise in the original data and by chance in the Monte Carlo
decision process. The checkerboard appearance is caused by the division
of the image into sub-images.


CLUSTER ACCEPTANCE

Figure 5 shows many small, and probably false, defects in addition to
the five real defects. The analysis of this figure would present a
prohibitive cost in terms of time because of the false alarms. For this
reason it is necessary to discriminate against clusters not meeting some
minimum size requirement.

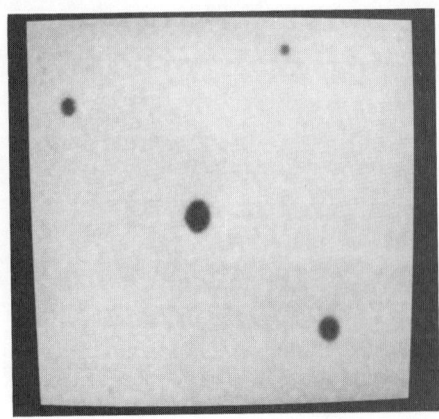

Fig. 4. Averaged image.

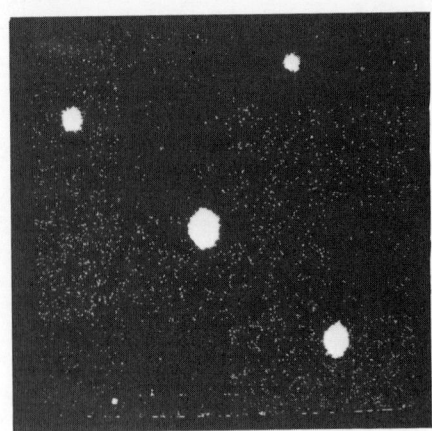

Fig. 5. Region-of-interest masks.

A simple way to accomplish this is to take advantage of the fact that there are only two pixel values present in the region-of-interest mask image (Figure 5). Those pixels corresponding to defects, or suspected defects, are indicated by the value 255. All pixels to be disregarded are represented by the value zero. If a simple average were performed over each 5-by-7 pixel region centered at each pixel in Figure 5, then only 35 discrete values could possibly occur in the resultant image. These values would correspond to the center pixel being surrounded by an integral number, up to 3 (of pixel)s. Thus, the average of mask image results in an image whose values are indicative of the "surroundedness" of each pixel in the mask.

Figure 6 shows the result of using a 5-by-7 averaging on the mask image of Figure 5 and retaining only those pixels that are surrounded by at least 20 other pixels. The image is free of noise defects and only the five real defects remain. Note also that the edges of the real defects are sharpened by this step in the processing because some of the edge fuzziness in the mask image is due to statistical fluctuations in the random number image.

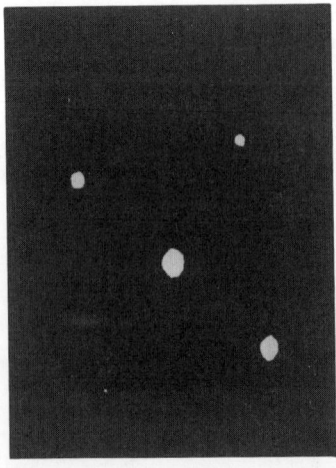

Fig. 6. Results after cluster acceptance step.

CONCLUSION

The image processing algorithm described above has been shown to produce clean results with actual radiographs. The computation is relatively simple, with the greater burden of work falling to the image processor. The host computer is needed only to perform least squares fits which must be done in floating point. Even this task, however, can be migrated onto an intelligent image processor, thus freeing the host to continue with the task of defect characterization.

The Monte Carlo process of soft thresholding brings the probability of defect detection smoothly to zero as the pixel values in the defect approach the background level. This is preferable to a hard threshold which abruptly brings the sensitivity to zero. Thus, there remains a small, albeit rapidly decreasing, chance of detecting a small defect even in the presence of an overwhelming background.

The use of a weighted coin toss to classify pixels is a method that attempts to be correct on average. This means that a fraction of pixels that belong to the foreground are misclassified and that defects consisting of only a few pixels can lose an appreciable amount of their size. This potentially causes the difficulties that the cluster acceptance criterion may no longer be met, and, if it is met, the apparent size of the defect is incorrect. These effects can be reduced by maintaining a "probability image" that gives the probability of retention of each pixel and using that probability to normalize the apparent sizes of the discovered defects.

There are, undoubtedly, several aspects of the segmentation algorithm that could be improved. The current method of averaging uses a simple unweighted average, which, in the frequency domain, results in poor high frequency attenuation. Filters custom designed for the noise present in the radiographs could be designed to improve the final results, although this would become tedious if it were necessary to have different filters for each radiography system.

From the previous discussion, it is apparent that the Monte Carlo selection process does not include correlations among nearby pixels with high probability of being foreground. Isolated and clustered pixels are treated identically, and correlations are considered at the cluster acceptance stage. Inclusion of correlations at the thresholding stage could, possibly, ease or eliminate the acceptance procedure. This is currently under active investigation.

References

1. Rafael C. Gonzalez and Paul Wintz, "Digital Image Processing," Addison-Wesley, Reading MA (1977), pp. 337-338.
2. C. K. Chow and T. Kaneko, Comp. and Biomed. Res. $\underline{5}$, 388 (1972).

APPLICATIONS OF DIGITAL IMAGE ENHANCEMENT TECHNIQUES FOR IMPROVED

ULTRASONIC IMAGING OF DEFECTS IN COMPOSITE MATERIALS

Brian G. Frock and Richard W. Martin

University of Dayton
Research Institute
Dayton, Ohio 45469

ABSTRACT

Several standard digital image enhancement techniques have been applied to digitized ultrasonic C-scan images of defects in graphite/epoxy composites. Features in the computer enhanced images are much more distinct than those in the original images. In some cases, cracks which are too close together to be visually resolved in the original images are clearly resolved in the enhanced images. Noisy images have been improved by first smoothing the original data and then edge enhancing the smoothed data. The resulting sharpened edges improve the visualization of features in the images. The application of digital image enhancement techniques has allowed the imaging of defects in composite materials at frequencies as low as 3.5 MHz. This use of these lower frequencies permits better imaging of defects in thick composite materials.

INTRODUCTION

Ultrasonic C-scan images which have not been created from digitized and stored data often fail to display all of the information which is available. This is due primarily to the necessity of setting the breakpoints between grey levels prior to scanning rather than after the scan has been completed. Digitizing and storing the data overcomes these problems because the number of grey levels and their threshold values are selected after the scanning is completed. This allows the image to be displayed as often as desired and with an almost unlimited number of different choices of threshold values. More importantly, digital image enhancement techniques which have been developed and used in other disciplines [1-3] can now be applied in the field of ultrasonic NDE.

In this paper we demonstrate how some of the more common digital image enhancement techniques can be applied to the ultrasonic NDE of graphite/epoxy composites to improve the visualization and spatial resolution of image features of interest.

SCANNING AND DISPLAY SYSTEMS

A high-precision, computer controlled scanning system was developed for the acquisition, storage, enhancement, and display of ultrasonic images of up to 512 x 512 points in size. Control of the system is accomplished by an LSI-11/23 microcomputer with an RT-11 operating system through a high speed IEEE-488 DMA interface to a data acquisition module and a stepper motor controller. A schematic diagram of the system and additional information are available in prior publications [4-5].

The video display system is capable of displaying from 2 to 16 colors or levels of grey-scale representing the signal amplitude. The user has the option of selecting equal percent of range thresholding, equalized histogram thresholding, or the user may specify the threshold values. The number of breakpoints can also be selected by the user.

DIGITAL IMAGE ENHANCEMENT TECHNIQUES

With the exception of histogram equalization and edge sharpening by blurred image subtraction, the image enhancement techniques presented in this paper were implemented through the use of a moving 3 pixel by 3 pixel mask. The numbering system for the individual elements in the mask is illustrated in Fig. 1a.

Mean Value Filter

The simplest of the techniques is the mean value filter [6] in which the amplitude of the central pixel in the mask is replaced by the mean value of the amplitudes of the nine pixel elements within the mask. This type of filter is used mostly for noise removal in cases where the noise severely interferes with visual interpretation of image features, and for preprocessing to remove low level noise in images which will be subjected to high frequency enhancements for improved edge detection and visualization.

Edge Enhancement by First Differences

There are a large number of directional edge enhancement techniques described in the literature [2,3,7-11]. Most of the techniques use an approximation to the first derivative in which the amplitude of the central pixel in the original image is replaced by the sum of first differences of pixel amplitudes within the mask. The first difference mask [3] which we used for vertical edge enhancements is given in Fig. 1b. If the difference in pixel amplitudes across the vertical region enclosed by the mask is large and positive, the value of the corresponding central pixel in the transformed image will be large and positive. If the difference is large and negative, the value of the corresponding central pixel in the transformed image will be large and negative. Not only are vertical edges enhanced, but they are also directionally "shaded", thus creating a three-dimensional effect. The mask [3] used in this paper for horizontal edge enhancement is given in Fig. 1c. Horizontal edges in the resultant image are both sharpened and shaded.

Images which are both visually pleasing and very useful for feature identification can be created by adding the vertical edge enhanced image to the horizontal edge enhanced image. The resultant image has sharpened edges and directionally dependent shading which creates a three-dimensional effect. The mask for this transformation is given in Fig. 1d.

| 2 | 3 | 4 |
|---|---|---|
| 9 | 1 | 5 |
| 8 | 7 | 6 |

a

| -1 | 0 | 1 |
|---|---|---|
| -2 | 0 | 2 |
| -1 | 0 | 1 |

b

| -1 | -2 | -1 |
|---|---|---|
| 0 | 0 | 0 |
| 1 | 2 | 1 |

c

| -2 | -2 | 0 |
|---|---|---|
| -2 | 0 | 2 |
| 0 | 2 | 2 |

d

Fig. 1.  Enhancement masks: (a) Numbering system; (b) First difference vertical edge enhancement; (c) First difference horizontal edge enhancement; (d) Sum of "b" and "c".

Edge Enhancement by Blurred Image Subtraction

Castleman [1] has demonstrated that when the blurred image is subtracted from the original image, the higher spatial-frequency image which results has sharpened edges with bands on both sides of the edges. As Hall, et al. [11] point out, this banding aids the visual detection of the edges and improves feature identification. No directional shading is produced by this technique, and thus the image has no three-dimensional appearance.

Histogram Equalization

Histogram equalization [2] sets the grey level thresholds for image display such that each grey level has approximately the same number of pixel elements. Thus, most of the grey level thresholds are placed in the regions of the image histogram where most of the pixel amplitudes occur. This greatly increases the probability that amplitudes of pixels on different sides of an edge will be placed in different grey levels, and increases the probability that the edge will be visible.

SAMPLE

The sample used for this study is a 16 ply graphite/epoxy specimen with a $[90_4/0_4]_s$ fiber orientation. Its physical dimensions are 6.4 cm long by 2.5 cm wide by 0.25 cm thick. The sample contains matrix cracks in both of the 90 degree plies and also in the 0 degree plies (see Fig. 2).

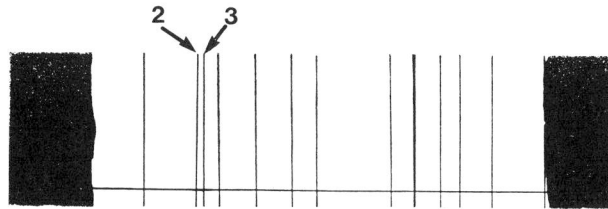

Fig. 2.  X-ray radiograph of cure crack sample.

DATA COLLECTION

The data were collected using normal incidence ultrasonic immersion C-scanning techniques with focused transducers. Three different transducers with the following parameters were used: (1) 25 MHz center frequency, 0.64

cm diameter, 2.54 cm focal length; (2) 10 MHz center frequency, 1.27 cm diameter, 7.6 cm focal length; and (3) 3.5 MHz center frequency, 1.27 cm diameter, 5.1 cm focal length. For each transducer data were acquired with the transducer focused on the front surface of the sample and also with the transducer focused on the back surface of the sample (not all of the data are presented here).

In all cases the RF echoes were pre-amplified, then further amplified, rectified and low-pass filtered (5 MHz cutoff) with a MATEC Broadband Receiver. An electronic gate was centered over the rectified and filtered back surface echo and the averaged value of the gated signal was digitized and stored during scanning. The step sizes for all scans were 0.013 cm by 0.013 cm.

Near noiseless images were generated by using small amounts of attenuation in the pre-amplifier stage and little gain in the Broadband Receiver stage. A noisy image was generated by using more attenuation in the pre-amplifier stage. The signal which was subsequently amplified at the Broadband Receiver stage had a much lower signal-to-noise ratio and produced a noisy image.

IMAGES

Typical binary C-scan images are presented in Figs. 3a and 3b. Data for these images were collected with the 25 MHz transducer focused on the front surface of the sample. Since the amplitude of the back surface echo was used to produce the image, discontinuities appear as white areas on a black background. Some of the cracks are readily visible in Fig. 3a where the threshold was set at 50% of the range. When the threshold level was raised to 70% of the range (Fig. 3b), more of the cracks became visible, but other features also appeared.

Better visual images of the sample are shown in Figs. 3c and 3d. All sixteen grey levels are used for these images rather than just two as was the case for Figs. 3a and 3b. The image in Fig. 3c is displayed using an equal percent of range thresholding, whereas the image in Fig. 3d is displayed in an equalized histogram threshold format. Clearly, more visual information is made available to the viewer through the use of the equalized histogram display format. The very narrow vertical and horizontal lines which are visible in Fig. 3d are the result of surface irregularities caused by the bleeder cloth during the fabrication process.

Comparison of the images in Fig. 3 with the X-ray radiograph in Fig. 2 reveals that not all of the cracks in the sample have been imaged by the ultrasonic interrogation. This is due to the very short acoustic depth of field of the transducer used for the data collection. Those cracks in the upper 90 degree plies are sharply imaged, while those in the lower 90 degree plies are out of focus and are too blurred to be visually identified.

An enhanced version of the images of Fig. 3 is shown in Fig. 4. This image was created by first smoothing the data with a mean value filter and then subtracting the mean value filtered (blurred) version from the original image. Several characteristics are evidenced in this image. First, the cracks are more precisely defined than in the previous images. Second, the low spatial-frequency variations in image intensity have been removed, producing an image which is much more uniform in intensity than the previous images. Third, the surface texture is visible over the entire image. Finally, there are some dark bands on both sides of most of the cracks which aid the eye in the detection of the cracks.

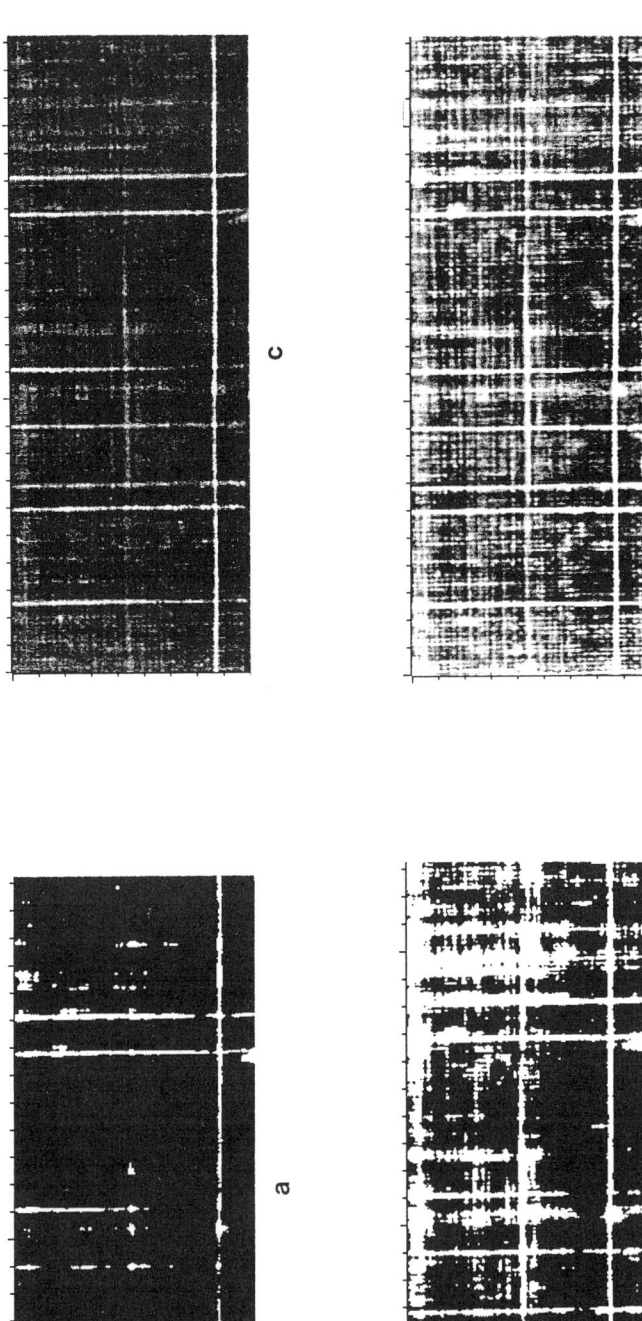

Fig. 3. Images of cure crack sample using 25 MHz transducer: (a) Binary image with threshold at 50% of range; (b) Binary image with threshold at 70% of range; (c) Sixteen grey-level image, equal percent of range thresholding; (d) Sixteen grey-level image, equalized histogram thresholding.

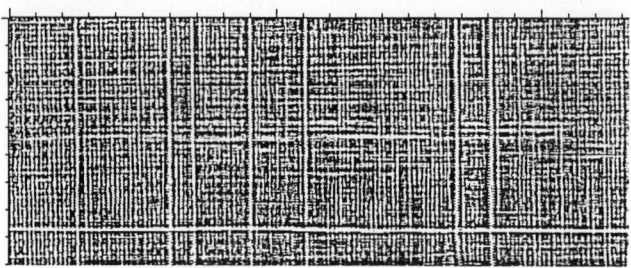

Fig. 4. Edge enchanced version of images in Fig. 3 using blurred image subtraction technique.

Images generated from data acquired with the 10 MHz transducer focused on the back surface of the sample are presented in Fig. 5. Cracks in all of the plies are visible because of the longer acoustic depth of field of this transducer. The unenhanced data is displayed in a 16 grey level equalized histogram format in Fig. 5a. Note in particular the cracks labeled 2 and 3 in Fig. 5a and in the X-ray radiograph of Fig. 2. Crack number 2 is in the bottom 90 degree ply layer while crack number 3 is in the top 90 degree ply layer. An edge enhanced version of the image in Fig. 5a is presented in Fig. 5b. This image was generated by summing the vertically and horizontally edge enhanced version of the data which was used to generate the image in Fig. 5a. The image in Fig. 5b has a pseudo three-dimensional appearance which produces the illusion that the cracks are elevated above the background. The two cracks (labeled 2 and 3) are also more visible in this image than in the original image.

A noisy 10 MHz image of the sample is shown in Fig. 5c. In this image the cracks labeled 2 and 3 are blurred together and are not clearly discernable. Fig. 5d shows the same data after three consecutive passes with a mean value filter followed by horizontal and vertical edge enhancements. This image is the sum of the vertically and horizontally edge enhanced versions of the smoothed image. The pseudo three-dimensional appearance is present, and cracks 2 and 3 are visually resolvable.

Images generated by using the 3.5 MHz transducer to insonify the sample are presented in Fig. 6. Data for Fig. 6a were acquired with the transducer focused on the front surface of the sample, while the data for Fig. 6b were acquired with the transducer focused on the back surface of the sample. Outlines of the cracks are fuzzy, and cracks 2 and 3 are not resolvable. There is, however, some evidence in Fig. 6c that the two cracks are resolvable when the images of Fig. 6a and 6b are suitably combined. Generation of this image from the two previous images required multi-stage processing techniques. First the data for Fig. 6b was subtracted from the data of Fig. 6a. Then, vertically and horizontally edge enhanced versions of the subtraction image were generated. Finally, the vertically and horizontally edge enhanced images were added together to create the image in Fig. 6c. The cracks are more distinct in this image, and cracks 2 and 3 are visually resolvable.

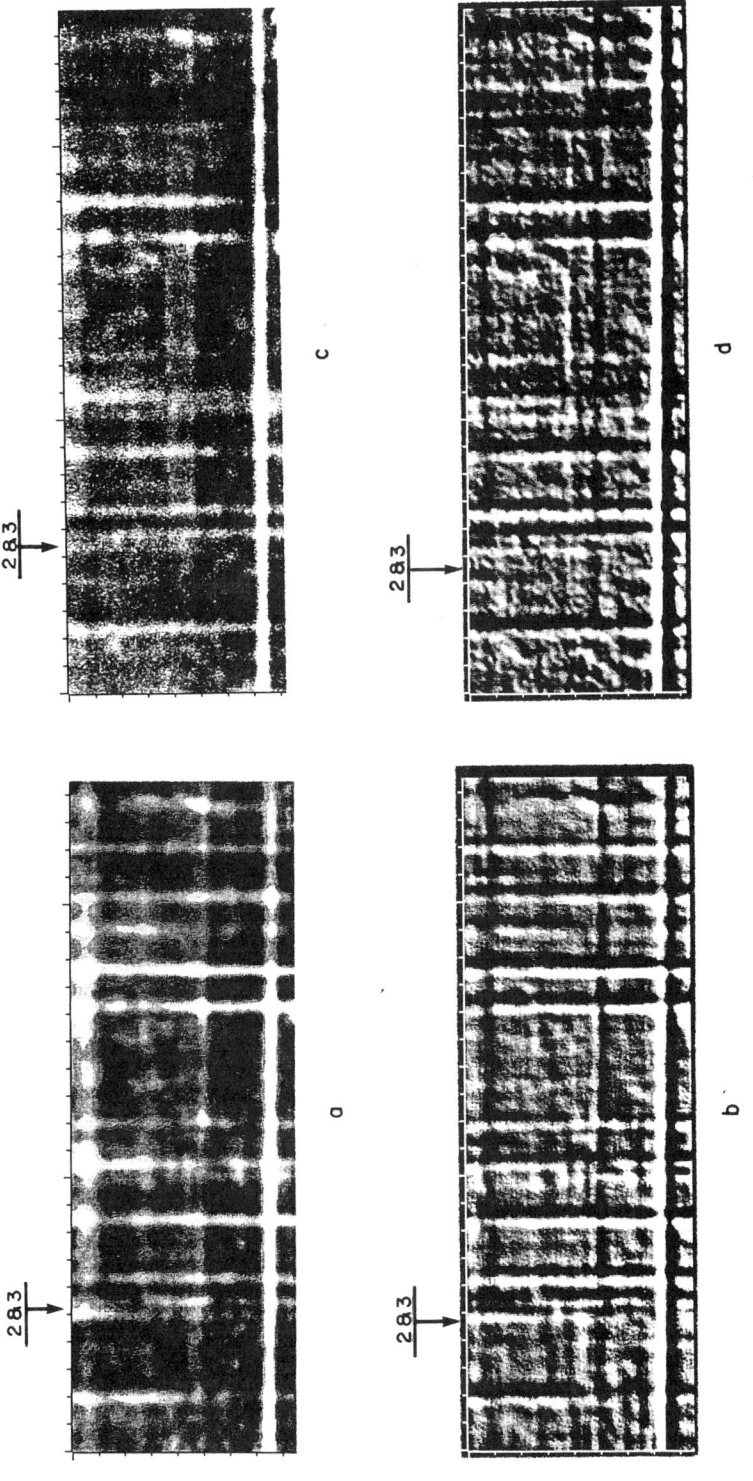

Fig. 5. Images of cure crack sample using 10 MHz transducer: (a) Low noise image; (b) Sum of first difference vertical and horizontal edge enhanced versions of "a"; (c) Noisy image; (d) Sum of first difference vertical and horizontal edge enhanced versions of "c" after mean value filtering.

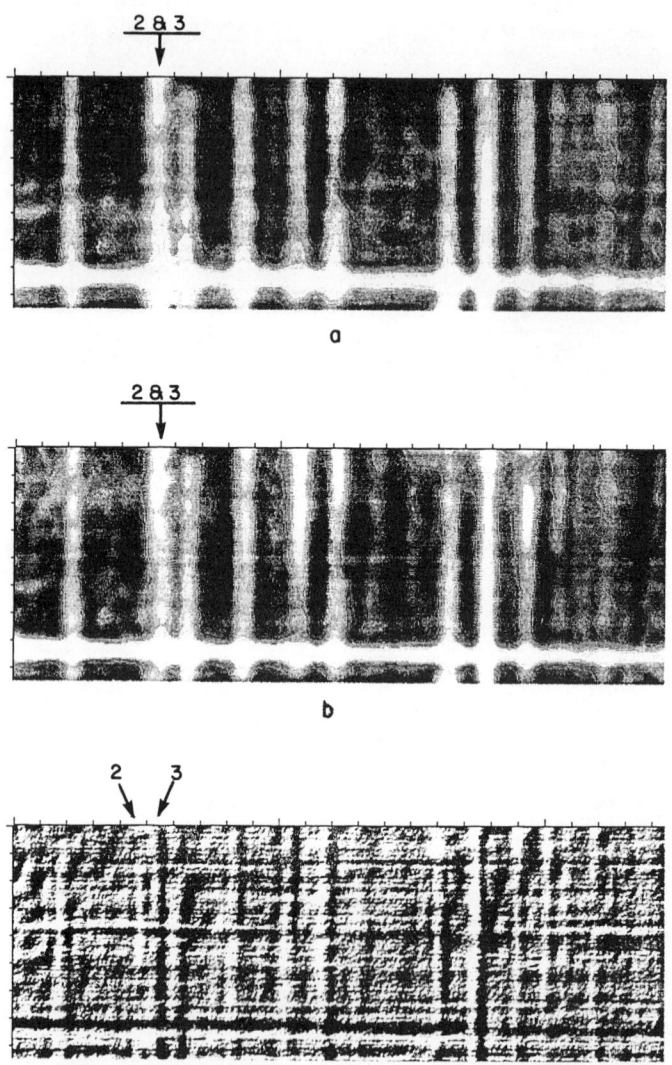

Fig. 6. Images of cure crack sample using 3.5 MHz transducer: (a) Transducer focused on front surface; (b) Transducer focused on back surface; (c) Sum of first difference vertical and horizontal edge enhanced versions of subtraction image ("a" minus "b").

SUMMARY

We have demonstrated how some very simple digital image processing techniques can be used to significantly improve the visibility of features in images generated from ultrasonic data. Further, we have demonstrated how some of these techniques can be applied to significantly improve the spatial resolution of features in images generated from data acquired at frequencies as low as 3.5 MHz. This is very important for the inspection of graphite/epoxy composite because of the large amounts of ultrasonic energy scattering which occur at temporal frequencies above 5 MHz.

ACKNOWLEDGMENTS

This research was sponsored by the AFWAL Materials Laboratory under Contract Number F33615-83-C-5036. The authors acknowledge Dr. Thomas J. Moran and Mr. Robert J. Andrews for their support and encouragement in the area of image enhancements. The authors also thank Mr. Mark Ruddell for his data collection efforts.

REFERENCES

1. K. R. Castleman, "Digital Image Processing," Prentice-Hall Inc., Englewood Cliffs, New Jersey, (1979).
2. R. C. Gonzalez and P. Wintz, "Digital Image Processing," Addison-Wesley Publishing Company, Reading Massachusetts, (1977).
3. H. C. Andrews and B.R. Hunt, "Digital Image Restoration," Prentice-Hall Inc., Englewood Cliffs, New Jersey, (1977).
4. T. J. Moran, R. L. Crane and R. J. Andrews, High Resolution Imaging of Microcracks in Composites, Materials Evaluation, Vol. 43, No. 5, (April 1985), pp 536-540.
5. R. W. Martin and R. J. Andrews, Backscatter B-Scan Images of Defects in Composites in: "Review of Progress in Quantitative Nondestructive Evaluation," Vol. 5, Plenum Press, (1986), pp 1189-1198.
6. J. Lee, Digital Image Enhancement and Noise Filtering by Use of Local Statistics, IEEE Transactions on Pattern Analysis and Machine Intelligence, Vol. PAMI-2, No. 2, (1980), pp 165-168.
7. L. S. Davis, A Survey of Edge Detection Techniques, Computer Graphics and Image Processing, 4, (1975), pp 248-270.
8. W. Frei and C. Chen, Fast Boundary Detection: A Generalization and a New Algorithm, IEEE Transactions on Computers, Vol. C-26, No. 10, (1977), pp 988-998.
9. A. Rosenfeld and M. Thurston, Edge and Curve Detection for Visual Scene Analysis, IEEE Transactions on Computers, (May 1971), pp 183-190.
10. J. E. Hall and J. D. Awtrey, Real-Time Image Enhancement Using 3 x 3 Pixel Neighborhood Operator Functions, Optical Engineering, Vol. 19, No. 3, (1980), pp 421-424.
11. E. L. Hall, R. P. Kruger, S. J. Dwyer, D. L. Hall, R. W. McLaren and G. S. Lodwick, A Survey of Preprocessing and Feature Extraction Techniques for Radiographic Images, IEEE Transactions on Computers, Vol. C-20, No. 9, (1971), pp 1032-1044.

# AN IMPROVED DEFECT CLASSIFICATION ALGORITHM BASED ON FUZZY SET THEORY

M. Carkhuff and S. S. Udpa

Electrical Engineering Department
Colorado State University
Fort Collins, Colorado 80523

## INTRODUCTION

The characterization of defects in materials constitutes a major area of research emphasis. Characterization schemes often involve mapping of the signal onto an appropriate feature domain. Defects are usually classified by segmenting the feature space and identifying the segment in which the feature vector is located. As an example Udpa and Lord [1] map differential eddy current impedance plane signals on to the feature space using the Fourier Descriptor approach. Doctor and Harrington [2] use the Fisher Linear Discriminant method to identify elements of the feature vector that demonstrate a statistical correlation with the nature of the defect. Mucciardi [3] uses the Adaptive Learning Network to build the feature vector. In all these cases defect classification is typically accomplished by categorizing the mapped feature vectors using Pattern Recognition methods employing either distance or likelihood functions [4].

Clustering algorithms which employ distance functions have gained in popularity in recent years owing to the fact that apriori knowledge of the statistical distribution of the feature vectors is not necessary. A number of clustering algorithms have been developed to cater to a wide variety of applications. Among these, the K-Means algorithm represents one of the most widely used due to its simplicity and the need to specify only the number of clusters. However, one of the disadvantages associated with the algorithm lies in the inability of the algorithm to discard feature vectors that lie in the "gray areas" between clusters. This paper presents a more robust clustering procedure capable of flagging such stray feature vectors for further analysis and interpretation.

## CLUSTERING ALGORITHMS

Clustering algorithms group the elements of a data set into a number of subsets or clusters based on an appropriate performance criterion. A popular procedure currently used for defect classification is the K-means clustering algorithm [4]. This algorithm sorts a given N-member data set X into K mutually exclusive and exhaustive clusters based on minimization of a performance index J, where

$$J_j = \sum_{x \varepsilon S_j(\ell)} ||x - z_j(\ell+1)||^2 \quad j=1,2,\ldots,K \tag{1}$$

$\ell$ denotes the iteration number, and $S_j(\ell)$ denotes the set of samples whose cluster center is $z_j(\ell)$.

A drawback of the K-means procedure is that its performance deteriorates with the introduction of "ambiguous" feature vectors; i.e., stray points lying in the gray areas between established clusters. Since the K-means algorithm includes every data point in cluster calculations, ambiguous data can cause erroneous cluster center computation and cluster "loosening". Figure 1 shows an example of such effects. Figure 1a illustrates the results of applying the K-means procedure (for K=2) with no ambiguous data. In Fig. 1b, two stray data points have been introduced. Cluster loosening, shifting of cluster centers and misclassification of the stray data result. These effects are undesirable in defect classification. A procedure capable of recognizing and excluding ambiguous feature vectors from cluster calculation can contribute to improvement in the defect classification scheme. In addition the ambiguous vectors should be flagged for later examination and interpretation.

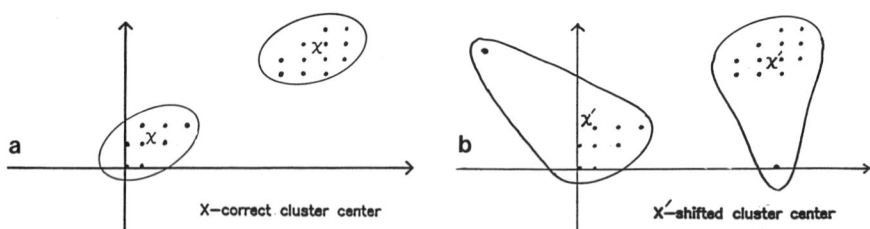

Fig. 1. K-Means clustering with a) no ambiguous feature vectors and b) two ambiguous feature vectors.

An improvement upon the K-means procedure is the Fuzzy K-means clustering algorithm [5]. The two are related in that both utilize distance minimization for cluster calculations. However, unlike the standard K-means algorithm, the fuzzy K-means procedure assigns relative or "fuzzy" classifications to all members of the data set; membership of data points to clusters is not a strict assignment. Membership of a data point j to each of K clusters is defined by a set of K values called membership values, $u_{ij}$, $i=1,2,\ldots,K$ which describes the "probability" or "degree" of belonging of that point with respect to each cluster. Membership values, like probabilities, are restricted to numbers between 0 and 1, with higher membership values indicating higher probability of belonging. An additional constraint is that the sum of membership values for each data point must equal 1,

$$\sum_{i=1}^{K} u_{ij} = 1 \quad j=1,2,\ldots,K \tag{2}$$

implying that the union of the K clusters represents the entire membership space.

Before the fuzzy K-means algorithm is discussed in detail, the idea of membership values will be further illustrated with an example.

Figure 2 shows a 22 point data set mapped onto a 2-dimensional feature space. Points 1-8 belong to cluster 1; points 9-20 belong to cluster 2; and points 21 and 22 are stray data points belonging to neither cluster. It is assumed here that only two clusters exist; i.e., points 21 and 22 cannot be clusters themselves.

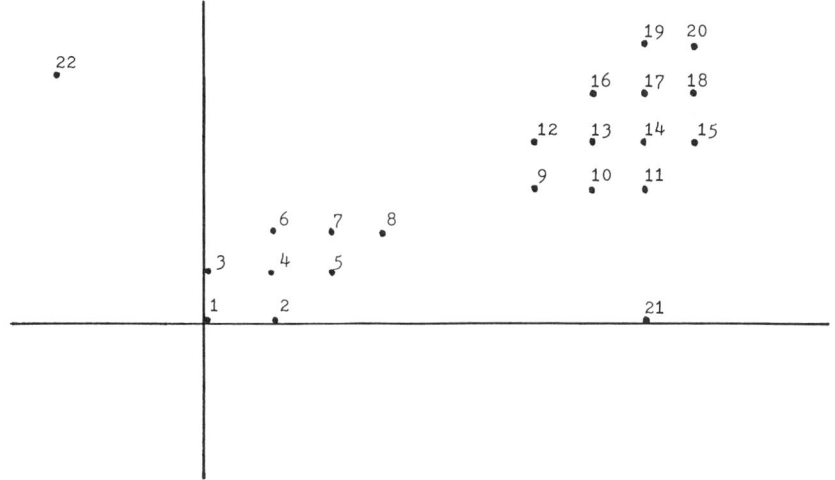

Fig. 2. Sample data set.

Table 1 lists the membership values relating each point to clusters 1 and 2. Membership values $u_{1m}$ (m=1,2,...,8) and $u_{2n}$ (n=9,10,...,20) indicate that points 1-8 and 9-20 belong to clusters 1 and 2, respectively. Membership values $u_{1m}$ (m=9,10,...,20) and $u_{2n}$ (n=1,2,...,8) indicate that points 1-8 and 9-20 do not belong to clusters 2 and 1, respectively. Mid-range membership values (.3<u<.7) relate points 21 and 22 to both clusters. These mid-range values indicate that points 21 and 22 should not be classified as belonging to either cluster. Clearly, the relative values of the membership function could be used as a measure to exclude ambiguous data points from known clusters.

Table 1. Membership values for data of Fig. 2.

| Point | $U_1$ | $U_2$ | Cluster | Point | $U_1$ | $U_2$ | Cluster |
|---|---|---|---|---|---|---|---|
| 1  | .96  | .04  | 1 | 12 | .05 | .95 | 2 |
| 2  | .99  | .01  | 1 | 13 | .01 | .99 | 2 |
| 3  | .98  | .02  | 1 | 14 | 0.0 | 1.0 | 2 |
| 4  | 1.0  | 0.0  | 1 | 15 | .02 | .98 | 2 |
| 5  | .99  | .01  | 1 | 16 | .01 | .99 | 2 |
| 6  | .99  | .01  | 1 | 17 | .01 | .99 | 2 |
| 7  | .98  | .02  | 1 | 18 | .02 | .98 | 2 |
| 8  | .93  | .07  | 1 | 19 | .03 | .97 | 2 |
| 9  | .09  | .91  | 2 | 20 | .04 | .96 | 2 |
| 10 | .04  | .96  | 2 | 21 | .44 | .96 | - |
| 11 | .03  | .97  | 2 | 22 | .70 | .30 | - |

The fuzzy K-means clustering algorithm is based on minimization of the objective function

$$\sum_{i=1}^{K} \sum_{j=1}^{N} (u_{ij})^m |x_j - z_i|^2 \qquad (3)$$

It should be noted that this function is similar to the performance index J minimized in the standard K-means procedure. However, the objective function includes the weighting factor $(u_{ij})^m$, $(m>1)$. This weighting factor suppresses the effect of data which have low membership values in cluster calculations.

Differentiation of Eq. 3 with respect to $u_{ij}$ and $z_i$ yields

$$z_i = \frac{\sum_{j=1}^{N} (u_{ij})^m x_j}{\sum_{j=1}^{N} (u_{ij})^m} \qquad (4)$$

$$u_{ij} = \frac{\left(\frac{1}{|x_j - z_i|^2}\right)^{\frac{1}{m-1}}}{\sum_{\ell=1}^{K} \left(\frac{1}{|x_j - z_\ell|^2}\right)^{\frac{1}{m-1}}} \qquad (5)$$

These equations are solved iteratively.

The fuzzy K-means algorithm involves the following steps:

Step 1.  The number of clusters K and initial membership values must be specified.

Step 2.  Solve equation 5 to determine cluster centers.

Step 3.  Solve equation 6 for new membership values.

Step 4.  If $u_{ij} < e$ (e is a specified error threshold) the procedure is terminated. Otherwise, go to Step 2.

Classification of data is performed using the final set of membership values calculated.

Although the fuzzy K-means procedure does not force a strict classification upon each data point in the set (as the standard K-means procedure does), and suppresses the effect of stray data on cluster determination, it does not exclude the ambiguous data from calculation of clusters. This paper presents an algorithm which not only identifies the stray feature vectors but also excludes them from cluster determination.

K/FUZZY K-MEANS ALGORITHM

This algorithm combines the K and fuzzy K procedures into a single process, which is outlined below.

Step 1.  The number of clusters K is determined by a maximin distance algorithm [4]. To be considered a valid grouping, a cluster must contain 3 or more data points.

Step 2. Implement the K-means procedure until it converges.

Step 3. Implement the fuzzy K-means algorithm, using the final K-means cluster centers (rather than membership values) to initiate the procedure.

Step 4. Calculate the product of each data point's membership values. Discard the jth point if

$$\prod_{i=1}^{K} u_{ij} > b \qquad (6)$$

where b is a specified threshold value.

Step 5. If a point or points were discarded in the previous step, go to Step 2. Otherwise, continue.

Step 6. Compare each data point's classifications as determined by the K-means and by the fuzzy K-means procedures. If the two classifications agree for each and every point, the procedure is terminated. Otherwise, go to Step 2.

Results

In order to assess the performance of the algorithm, the procedure was tested using feature vectors derived from eddy current impedance plane signals using the Fourier descriptor method. although this method is described elsewhere [1] a brief description of the approach is presented here for completeness.

Consider a closed contour u( ) as shown in Fig. 3.

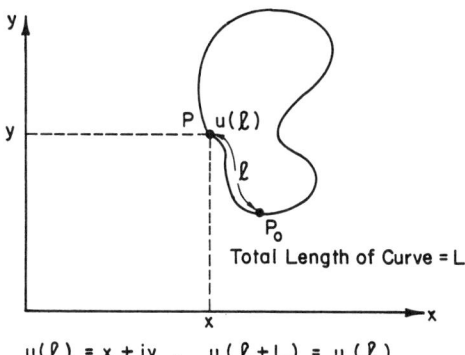

$u(\ell) = x + iy$,  $u(\ell + L) = u(\ell)$

Fig. 3. Closed contour u(1).

Starting at an arbitrary point P on the curve, if one traces its entire length L, one comes again to the starting point P. Therefore, u( ) is a periodic function. Since any periodic function can be expanded into a Fourier series, we have

$$u(1) = \sum_{n=-\infty}^{\infty} C_n \exp(jn(\frac{2\pi}{L})) \qquad (7)$$

where

$$C_n = \frac{1}{L} \int_0^L u(\ )\exp(-jn(\frac{2\pi}{L})\ ) \qquad (8)$$

are the Fourier series coefficients. In practice the infinite summation is truncated since a finite number of Fourier series coefficients allows a sufficiently accurate reconstruction of the original curve u( ).

The Fourier series coefficients $C_n$ can be transformed to descriptors $D_{mn}$ which are invariant to rotation, translation and scaling of the original curve. One such transformation is

$$D_{mn} = \frac{C_{1+m}^n \, C_{1-n}^m}{C_1^{m+n}} \qquad \begin{array}{l} m=1,2,3,\ldots \\ n=2,3,4,\ldots \end{array} \qquad (9)$$

The transformed coefficients $D_{mn}$ are called the Fourier Descriptors of the function u(1). It has been shown [1] that only eight Fourier Descriptors are needed to represent even the most complex impedance plane trajectories. Using the procedure outlined in [1], eight-dimensional feature vectors were determined for each signal. A representative set of test data is listed in Table 2, along with the classifications as determined by the standard K means procedure and by the K/Fuzzy K procedure.

The data set consists of 55 feature vectors. Vectors 1 through 50 represent actual defects belonging to 5 different classes. Vectors 51 through 55 are ambiguous vectors belonging to none of the 5 classes.

The K-means algorithm correctly determines membership of the 50 valid vectors. However, the stray vectors are assigned to the clusters to which they are closest; these are considered misclassifications.

The improved algorithm also correctly determines membership of the first 50 vectors. Moreover, it correctly discards these stray vectors, excluding them from cluster calculations. This results in 100% correct classification of this data set.

The effect of excluding stray feature vectors from calculation of clusters is shown in Table 3. Standard deviations of clusters (a measure of cluster "looseness") are tabulated for each of the K-means clusters and K/Fuzzy K clusters. The K-means clusters are generally much "looser" than those of the improved procedure.

CONCLUSIONS

Test results indicate that the K/Fuzzy K algorithm correctly identified ambiguous feature vectors and excluded them from cluster determination. This resulted in fewer misclassifications and significant "tightening" of clusters. The discarded vectors were saved for later interpretation. The procedure is simple and robust. Stray data points are detected in the first pass through the algorithm and very few iterations are required for convergence. In addition, the cluster-seeking algorithm determines the number of clusters K and consequently no apriori knowledge of the data is necessary.

Table 2. Data set classified by K-means and by K/Fuzzy K algorithms.

| Vector # | | | | | | | | | Defect type Classification K-Means | K/Fuzzy K |
|---|---|---|---|---|---|---|---|---|---|---|
| 1 | 0.886 | 0.380 | 0.637 | 0.629 | 0.741 | 0.233 | 0.416 | 0.676 | 1 | 1 |
| 2 | 0.874 | 0.616 | 0.708 | 0.622 | 0.801 | 0.292 | 0.503 | 0.278 | 5 | 5 |
| 3 | 0.875 | 0.356 | 0.694 | 0.602 | 0.801 | 0.160 | 0.514 | 0.227 | 4 | 4 |
| 4 | 0.856 | 0.992 | 0.801 | 0.606 | 0.863 | 0.284 | 0.695 | 0.779 | 3 | 3 |
| 5 | 0.833 | 0.255 | 0.989 | 0.207 | 0.848 | 0.113 | 0.978 | 0.987 | 2 | 2 |
| 6 | 0.886 | 0.392 | 0.632 | 0.630 | 0.749 | 0.238 | 0.402 | 0.676 | 1 | 1 |
| 7 | 0.887 | 0.404 | 0.632 | 0.634 | 0.749 | 0.213 | 0.393 | 0.677 | 1 | 1 |
| 8 | 0.885 | 0.373 | 0.639 | 0.631 | 0.743 | 0.241 | 0.414 | 0.676 | 1 | 1 |
| 9 | 0.886 | 0.393 | 0.637 | 0.627 | 0.746 | 0.227 | 0.410 | 0.677 | 1 | 1 |
| 10 | 0.886 | 0.335 | 0.650 | 0.601 | 0.755 | 0.173 | 0.416 | 0.724 | 1 | 1 |
| 11 | 0.855 | 0.337 | 0.654 | 0.592 | 0.749 | 0.174 | 0.426 | 0.725 | 1 | 1 |
| 12 | 0.886 | 0.337 | 0.648 | 0.602 | 0.755 | 0.177 | 0.425 | 0.725 | 1 | 1 |
| 13 | 0.886 | 0.330 | 0.652 | 0.602 | 0.752 | 0.178 | 0.420 | 0.724 | 1 | 1 |
| 14 | 0.886 | 0.340 | 0.651 | 0.595 | 0.753 | 0.167 | 0.433 | 0.725 | 1 | 1 |
| 15 | 0.875 | 0.581 | 0.695 | 0.657 | 0.804 | 0.311 | 0.460 | 0.256 | 5 | 5 |
| 16 | 0.875 | 0.589 | 0.699 | 0.654 | 0.803 | 0.300 | 0.456 | 0.254 | 5 | 5 |
| 17 | 0.879 | 0.632 | 0.683 | 0.676 | 0.821 | 0.270 | 0.434 | 0.258 | 5 | 5 |
| 18 | 0.873 | 0.593 | 0.706 | 0.652 | 0.801 | 0.306 | 0.485 | 0.257 | 5 | 5 |
| 19 | 0.874 | 0.591 | 0.705 | 0.644 | 0.803 | 0.297 | 0.469 | 0.274 | 5 | 5 |
| 20 | 0.873 | 0.578 | 0.707 | 0.642 | 0.797 | 0.303 | 0.493 | 0.272 | 5 | 5 |
| 21 | 0.874 | 0.597 | 0.703 | 0.645 | 0.808 | 0.293 | 0.461 | 0.274 | 5 | 5 |
| 22 | 0.873 | 0.588 | 0.707 | 0.635 | 0.801 | 0.288 | 0.479 | 0.273 | 5 | 5 |
| 23 | 0.873 | 0.588 | 0.705 | 0.639 | 0.800 | 0.294 | 0.488 | 0.273 | 5 | 5 |
| 24 | 0.873 | 0.477 | 0.693 | 0.600 | 0.804 | 0.224 | 0.552 | 0.220 | 4 | 4 |
| 25 | 0.873 | 0.452 | 0.702 | 0.610 | 0.777 | 0.317 | 0.603 | 0.221 | 4 | 4 |
| 26 | 0.872 | 0.451 | 0.704 | 0.574 | 0.784 | 0.248 | 0.568 | 0.216 | 4 | 4 |
| 27 | 0.874 | 0.435 | 0.701 | 0.616 | 0.777 | 0.292 | 0.606 | 0.219 | 4 | 4 |
| 28 | 0.877 | 0.468 | 0.684 | 0.647 | 0.799 | 0.268 | 0.419 | 0.220 | 4 | 4 |
| 29 | 0.883 | 0.492 | 0.669 | 0.662 | 0.813 | 0.226 | 0.455 | 0.271 | 4 | 4 |
| 30 | 0.876 | 0.473 | 0.692 | 0.613 | 0.798 | 0.231 | 0.501 | 0.272 | 4 | 4 |
| 31 | 0.882 | 0.497 | 0.669 | 0.656 | 0.816 | 0.205 | 0.453 | 0.273 | 4 | 4 |
| 32 | 0.877 | 0.451 | 0.689 | 0.628 | 0.798 | 0.241 | 0.488 | 0.249 | 4 | 4 |
| 33 | 0.868 | 0.985 | 0.749 | 0.680 | 0.874 | 0.268 | 0.564 | 0.701 | 3 | 3 |
| 34 | 0.867 | 1.000 | 0.748 | 0.677 | 0.875 | 0.271 | 0.566 | 0.702 | 3 | 3 |
| 35 | 0.865 | 0.956 | 0.762 | 0.661 | 0.861 | 0.308 | 0.588 | 0.702 | 3 | 3 |
| 36 | 0.866 | 0.955 | 0.752 | 0.681 | 0.862 | 0.320 | 0.556 | 0.704 | 3 | 3 |
| 37 | 0.866 | 0.948 | 0.762 | 0.663 | 0.860 | 0.313 | 0.594 | 0.698 | 3 | 3 |
| 38 | 0.863 | 0.932 | 0.765 | 0.658 | 0.845 | 0.339 | 0.582 | 0.689 | 3 | 3 |
| 39 | 0.863 | 0.938 | 0.766 | 0.653 | 0.849 | 0.329 | 0.604 | 0.689 | 3 | 3 |
| 40 | 0.862 | 0.932 | 0.765 | 0.656 | 0.847 | 0.335 | 0.599 | 0.691 | 3 | 3 |
| 41 | 0.863 | 0.948 | 0.765 | 0.652 | 0.851 | 0.324 | 0.605 | 0.689 | 3 | 3 |
| 42 | 0.834 | 0.246 | 0.979 | 0.228 | 0.856 | 0.100 | 0.954 | 0.986 | 2 | 2 |
| 43 | 0.833 | 0.242 | 0.988 | 0.210 | 0.848 | 0.080 | 0.970 | 0.986 | 2 | 2 |
| 44 | 0.834 | 0.255 | 0.983 | 0.214 | 0.854 | 0.096 | 0.963 | 0.986 | 2 | 2 |
| 45 | 0.830 | 0.250 | 1.000 | 0.175 | 0.840 | 0.083 | 1.000 | 0.985 | 2 | 2 |
| 46 | 0.831 | 0.265 | 0.995 | 0.183 | 0.843 | 0.049 | 0.985 | 0.989 | 2 | 2 |
| 47 | 0.833 | 0.276 | 0.988 | 0.207 | 0.852 | 0.057 | 0.963 | 0.990 | 2 | 2 |
| 48 | 0.833 | 0.265 | 0.984 | 0.213 | 0.854 | 0.053 | 0.961 | 0.989 | 2 | 2 |
| 49 | 0.831 | 0.260 | 0.995 | 0.193 | 0.842 | 0.046 | 0.977 | 0.990 | 2 | 2 |
| 50 | 0.832 | 0.257 | 0.989 | 0.203 | 0.849 | 0.075 | 0.972 | 0.988 | 2 | 2 |
| 51 | 0.878 | 0.832 | 0.682 | 0.734 | 0.849 | 0.296 | 0.480 | 0.363 | 5* | DISCARDED |
| 52 | 0.976 | 0.070 | 0.076 | 0.616 | 0.301 | 0.559 | 0.135 | 0.472 | 1* | DISCARDED |
| 53 | 0.859 | 0.962 | 0.784 | 0.625 | 0.851 | 1.000 | 0.645 | 0.707 | 3* | DISCARDED |
| 54 | 0.858 | 0.985 | 0.979 | 0.504 | 0.839 | 0.972 | 0.168 | 0.363 | 3* | DISCARDED |
| 55 | 1.000 | 0.148 | 0.699 | 0.652 | 0.805 | 0.083 | 1.000 | 0.989 | 2* | DISCARDED |

*-Misclassification

Table 3. Standard deviations of K-Means and K/Fuzzy K clusters.

STANDARD DEVIATIONS OF CLUSTERS

| Cluster # | K-Means | K/Fuzzy K |
|---|---|---|
| 1 | 0.227e+00 | 0.532e-02 |
| 2 | 0.140e+00 | 0.113e-01 |
| 3 | 0.222e+00 | 0.378e-01 |
| 4 | 0.337e-01 | 0.337e-01 |
| 5 | 0.648e-01 | 0.167e-01 |

ACKNOWLEDGMENT

This work was supported by the Electric Power Research Institute under contract RP 2673-4.

REFERENCES

1. S. S. Udpa and W. Lord, Materials Evaluation, 42, 1136 (1984).
2. P. G. Doctor, et al., Pattern Recognition Methods for Classifying and Sizing Flaws Using Eddy-Current Data, in "Eddy Current Characterization of Materials and Structures," G. Birnbaum and G. Free, eds., ASTM, Philadelphia (1981).
3. A. N. Mucciardi, "Elements of Learning Control Systems with Applications to Industrial Processes," Proc. of the IEEE Conf. on Decision and Control, IEEE, New Orleans (1982).
4. J. J. Tou and R. C. Gonzalez, "Pattern Recognition Principles,"] Addison-Wesley, Reading (1974).
5. M. P. Windham, IEEE Trans. on Pattern Recog. and Mach. Intel., Vol. PAMI-4, 357 (1982).

FREQUENCY MODULATED (FM) TIME DELAY-DOMAIN THERMAL WAVE TECHNIQUES,
INSTRUMENTATION AND DETECTION: A REVIEW OF THE EMERGING STATE OF
THE ART IN QNDE APPLICATIONS

Andreas Mandelis

Photoacoustic and Photothermal Sciences Laboratory
Department of Mechanical Engineering, University of Toronto
Toronto, Ontario M5S 1A4, Canada

INTRODUCTION

In this work, the concept of the frequency modulation (FM) technique
is applied to thermal wave systems. Heyser [1] introduced this technique
in the field of acoustical measurements of loudspeakers and named it
time delay spectrometry (TDS). Through its implementation and long-term
use in acoustic engineering, TDS has been shown to outperform any other
time selective technique with respect to noise rejection and nonlinearity
suppression from measurements of systems with linear behavior [2]. The
time delay technique, based on a linear frequency sweep of the excitation
function, has been specifically compared to the impulse response transformation method and the wide-band random noise method [3] and has been
proven to have superior measurement dynamic range properties. The present
first application of TDS to a thermal wave system and subsequent comparison
with a pseudo-random noise method has shown that our FM technique is
superior to the more conventional types of excitation and signal analysis,
consistently with conclusions presented in Ref. [3].

THEORETICAL BACKGROUND

The excitation function in a TDS system is characterized by an excellent
dynamic range. This is a direct consequence of its nature as a minimum
phase system [4]. As such, the instantaneous value of the frequency-like
quantity $f_i(t)$ is given by [1]

$$f_i(t) = (\frac{\Delta f}{T})t + f_c \tag{1}$$

where $\Delta f = f_2 - f_1$ is the carrier signal modulation bandwidth, $f_c = \frac{1}{2}(f_2+f_1)$
is the average carrier frequency, T is the total sweep period, and
$S = \Delta f/T$ is the (constant) sweep rate (in Hz/s). Assuming the excitation
function to be a cosinusoidal carrier wave, an FM thermal wave system
input will be given by

$$X(t) = A(t)\cos[\phi_i(t)] \tag{2}$$

where A(t) is the amplitude modulation (AM) function, usually chosen
to be constant. $\phi_i(t)$ can be shown to be [5]

$$\phi_i(t) = (\pi S)t^2 + \omega_c t + \phi_o \tag{3}$$

where $\phi_o \equiv \phi_i(0)$ is the input phase at t=0.

The experimental conditions chosen for thermal wave measurements are [6-8]: $\phi_o=0$, $f_i=0$, and $\phi_i(T+\delta t)=\phi_i(\delta t)$ for $\delta t \to 0$; and A=constant. These correspond to a linear sawtooth frequency sweep between DC and $f_2=f_{max}$ with multiple fast repetitions of the sweep process every period T (chirp). The signal generated using the linearly swept wave X(t) of eq. (2) is a special case of the more general class of Phase-angle Modulated systems [9]. It is in this context that TDS thermal wave excitation can be regarded as similar to a Frequency Modulated (FM) wave, with the time integral of the applied swept wave

$$m(t) = 2\pi \int_o^t (\frac{\Delta f}{T}) q \, dg \qquad (4)$$

in eq. (3) acting as the FM wave modulating the baseband signal $f_c$. In either case the instantaneous frequency $f_i(t)$ is the sum of the time-varying component and the unmodulated carrier wave, $f_c$.


EXPERIMENTAL

The first reported photothermal wave system with FM Time Delay Domain optical excitation is a recent photothermal deflection spectroscopic (PDS) apparatus assembled in our laboratory [7] and shown in Fig. 1. We investigated the performance of the PDS apparatus using a blackbody reference sample (anodized aluminum) in water, as well as thin quartz films of variable thickness. A fast beam position detector was fabricated using a pinhole-photodiode arrangement with a 34 ns response time. The excitation beam from a 2 W $Nd^{3+}$:YAG pump laser was expanded over the sample in order to facilitate the (one-dimensional) theoretical interpretation of the data. Frequency modulation of the pump beam intensity was effected using an HP 3325A Synthesizer/Function Generator. The system output was registered as a photo-voltage whose amplitude was proportional to the spatial deflection of the He-Ne probe beam due to the Mirage effect [10]. All the necessary Frequency and Time Domain functions were calculated via a Nicolet Scientific Corp. Model 660A dual channel FFT analyzer.

The magnitude and phase of the complex transfer function H(f) of the blackbody/water interface is shown in Fig. 2(i). These data were taken between DC ($f_i=0$) and 1280 Hz (= $f_2$ in eq. (1)) with T=0.41 sec and a sweep rate S=3.122 kHz/sec applied to the acousto-optic modulator. Correlation and spectral processing, averaged over 1000 frequency sweeps with 1024 data points per sweep, required approx. 6-7 min. This time can be reduced, however, to be as low as 1 min., corresponding to a minimum number of ca. 200 sweeps/average. This time is by far shorter than the time required to obtain the same information dispersively using lock-in detection as in Fig. 2 (ii). The reliability of the data shown in Fig. 2(ii) is, furthermore, inferior to that of Fig. 2(i), as the mean of only 20 samples per average was taken over 14 data points. From this comparison it was concluded that the superior speed and reliability of the FM Time Delay system make it a very attractive candidate for thermal mapping or depth-profiling applications in environments requiring fast turn-around, such as industrial quality control laboratories. Using a raised Hanning FIR filter (i.e. cosine to the fourth power window [2]), the input autocorrelation function $R_{xx}(\tau)$ was found to be extremely narrow on the time scale of the PDS experiments and could be accurately approximated by the Dirac delta function. It thus follows [5] that the input-output cross-correlation function is equal to the unit impulse response of the system; this was also verified experimentally [7].

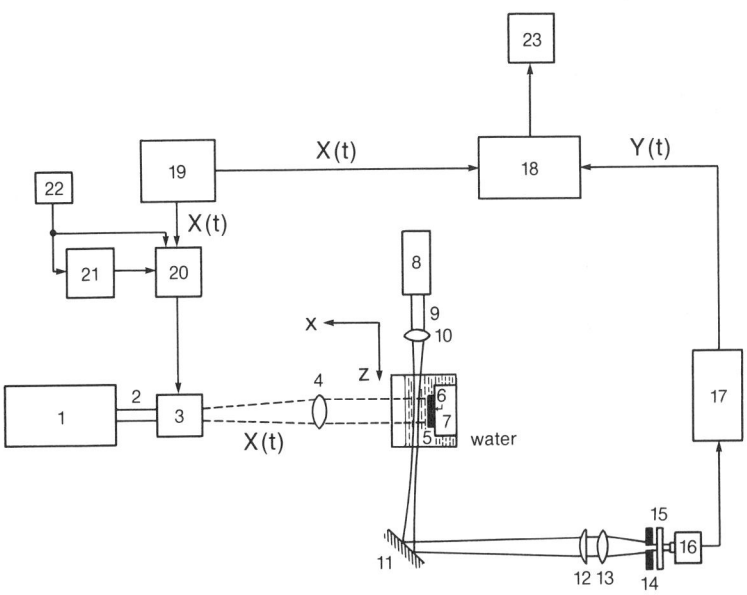

Fig. 1. FM time delay mirage effect spectrometer; 1: $Nd^{3+}$:YAG pump laser; 2: cw 1.06-μm beam; 3: acousto-optic (A/O) modulator; 4: alignment lens; 5: water; 6: sample; 7: sample holder; 8: He-Ne probe laser; 9: 632.8-nm probe beam; 10: focusing lens; 11: optical lever reflector mirror; 12, 13: lenses; 14: 50-μm-diam pinhole; 15: He-Ne beam interference filter; 16: fast rise-time photodiode; 17: wide bandwidth preamplifier; 18: dual channel FFT analyzer; 19: synthesizer/function generator; 20: A/O modulator driver; 21: A/O driver power amplifier; 22: A/O modulator power supply; 23: computer memory storage. X(t) and Y(t) are identified with input and output system functions, respectively.

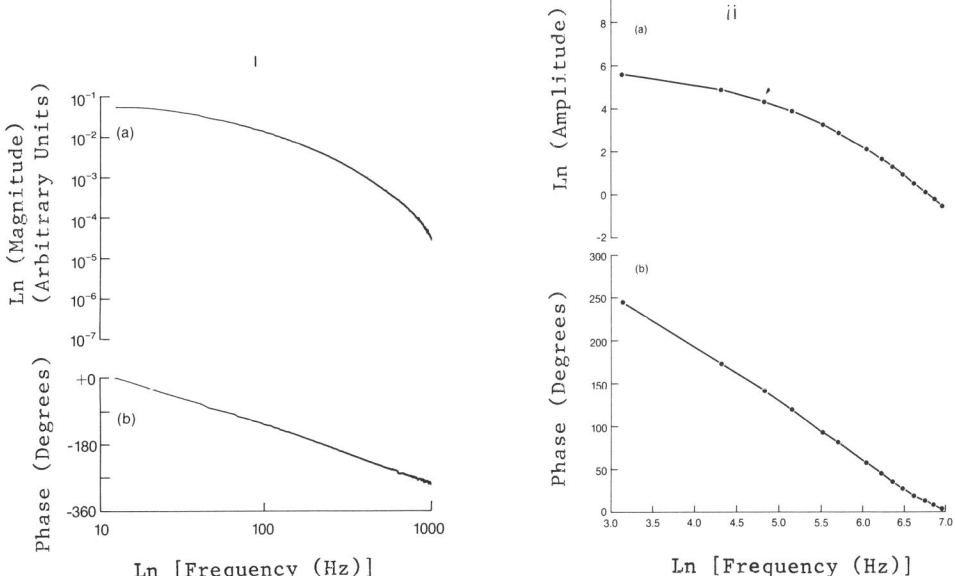

Fig. 2. i) Magnitude (a) and phase (b) of the complex transfer function H(f) of the blackbody/water interface at 10 μm probe beam offset, using the FM Time Delay Domain PDS apparatus. ii) Same data obtained using lock-in detection.

The spectrometer was further used to measure the response from thin microscope quartz slide layers in direct contact with the backing material (anodized aluminum support). A single slide cut in many pieces was used for these experiments, to assure material uniformity. Each piece was etched in 50% HF: 50% $H_2O$ down to the desired thickness. Fig. 3 shows a superposition of the impulse responses for two different thicknesses, 30 μm (curve a) and 100 μm (curve b). The cross-correlation functions show similar features, i.e. an increased peak delay time, a broadened FWHM and an increased trough time delay $\tau_{min}$ with increasing thickness. In each case data were taken at beam offset positions which maximized the PDS output at the detector.

The secondary oscillations on both wings of the main pulse in Fig. 3 were found to be consistent with thermal energy arrivals at the sample surface after multiple reflectins at the sample-backing interface. The delay time $\Delta\tau$ between two successive peaks corresponded roughly to twice the thermal transit time $\tau_{transit}=l^2/\alpha_2$ through the bulk of the sample. Similar effects have been predicted theoretically by Burt [11] in fluids excited by pulsed lasers, and have been observed experimentally in liquids and solids by Tam et al. [12,13].

A theoretical model for the impulse response of the PDS system was also presented by us in Ref. [7], where we calculated the peak delay time $\tau_o$ from the heat conduction Green's Function of the composite system:

$$\tau_o(l) = \frac{L_o^2}{6\alpha_3}[1 + (\alpha_3/\alpha_2)^{1/2}(l/L_o)]^2 \quad (5)$$

where $\alpha_1$, $\alpha_2$, $\alpha_3$ are the thermal diffusivities of the backing material, quartz layer and water, respectively; and $l$, $L_o$ are the quartz layer thickness and probe beam offset at the beam-waist, respectively. A fit of (5) to the data gave a value for $\alpha_2$ in good agreement with the published value [14]. Further experimentation with silicon wafer samples, on which 1 μm thick field $SiO_2$ oxides were grown thermally, showed that the FM

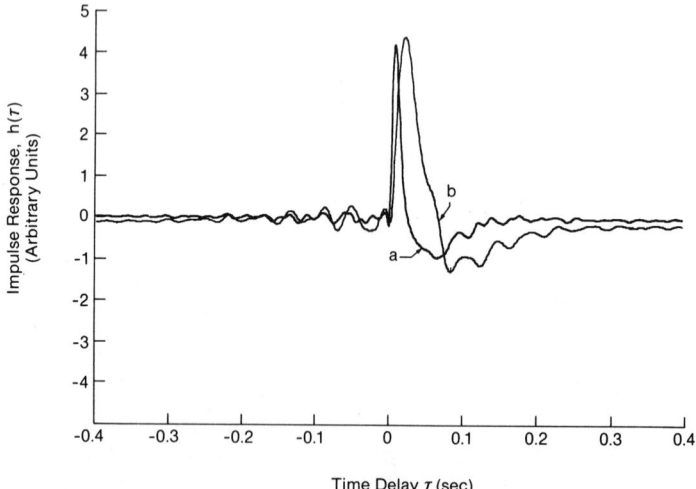

Fig. 3. Impulse response functions from quartz layers of thickness 30 μm (a) and 100 μm (b). $\tau_o^a$=4.69 ms, $\tau_{FWHM}^a$=6.04 ms, $\tau_{min}^a$=32.81 ms; $\tau_o^b$=10.94 ms, $\tau_{FWHM}^b$=28.6 ms, $\tau_{min}^b$=42.97 ms.

Time Delay Domain technique was quite sensitive to the presence of such oxides on the silicon surface [7]. These results were deemed promising for the future of the technique as a non-destructive semiconductor probe capable of replacing the pulsed laser excitation conventionally used [12], the two main advantages of the FM method being a) its much higher pulse tolerance threshold on sensitive materials such as those used in optoelectronic and microelectronic applications; and b) its higher dynamic range than that of the impulsive excitation.

We [8] further made a detailed comparison between the FM Time Delay and the Pseudo-Random Binary Sequence (PRBS) method [15] of optical excitation and Mirage effect response. Fig. 4 shows a comparison of the auto-correlation functions $R_{xx}(\tau)$ and $R_{yy}(\tau)$ of inputs and outputs, respectively, of the two techniques. In Fig. 4b the secondary peaks of the PRBS input autocorrelation function are clearly seen at the onset of the second multiple of the frequency band spanned by the PRBS pseudo-period. These spikes are also present in the PRBS output autocorrelation, albeit much more broadened and of much lower magnitude. A comparison of $R_{xx}(\tau)$ between Figs. 4a and 4b shows that the PRBS function is more broadened than the FM Time Delay function on the time scale of the experiment. Therefore, it is expected that the PRBS $R_{xx}(\tau)$ convolution with the impulse response (i.e. the input/output cross-correlation function) will be somewhat broader than the PRBS impulse response function $h(\tau)$, a fact borne out by the experiments. On the other hand, the narrow $R_{xx}(\tau)$ of the FM Time Delay spectrometer is a closer representation of a Dirac delta function than the PRBS counterpart and produces essentially identical lineshapes between $h(\tau)$ and $R_{xy}(\tau)$.

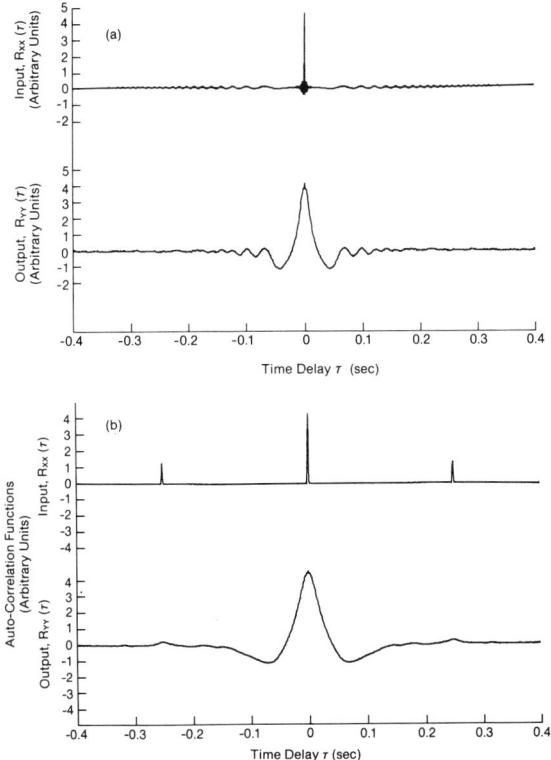

Fig. 4. Autocorrelation functions of system input and output: (a) FT Time Delay spectrometer; (b) PRBS spectrometer.

Fig. 5 is a comparison between the coherence functions obtained from the signal inputs and outputs for the two techniques. The coherence function is a most sensitive indicator of the quality of the relation between input and output. The superior performance of the FM Time Delay spectrometer is unequivocally exemplified in this figure. Essentially no correlation can be found above 600 Hz for the PRBS method, while a strong relation between input and output well beyond 1 kHz is observed for the FM Time Delay system. The dips in the coherence functions are due to non-system related signal sources, such as line ripple and multiples of 60 Hz. These sources are completely deterministic at well-defined frequencies and they do not appear in the statistics of the coherence function. The coherence of the PRBS system exhibited large discrete sawtoothed band components with peaks and valleys of rapidly varying functional quality of the relationship between input and output. This resulted in a poor signal-to-noise ratio of the transfer function $H(f)$, as seen in Fig. 6. This figure indicates the degree of dynamic range superiority of the FT Time Delay spectrometer to that of the device operating with a PRBS excitation. The exceptional quality of the FM Time Delay spectrometer transfer function is intimately related to the quality of the impulse response, whose Fourier transform the transfer function is.

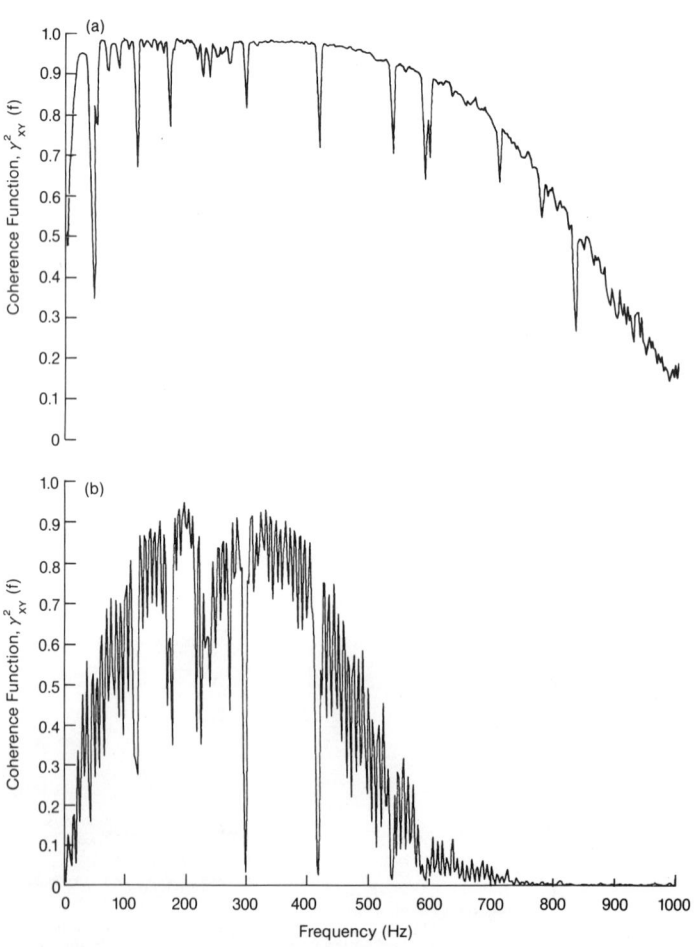

Fig. 5. Coherence functions: (a) FT Time Delay spectrometer; (b) PRBS spectrometer.

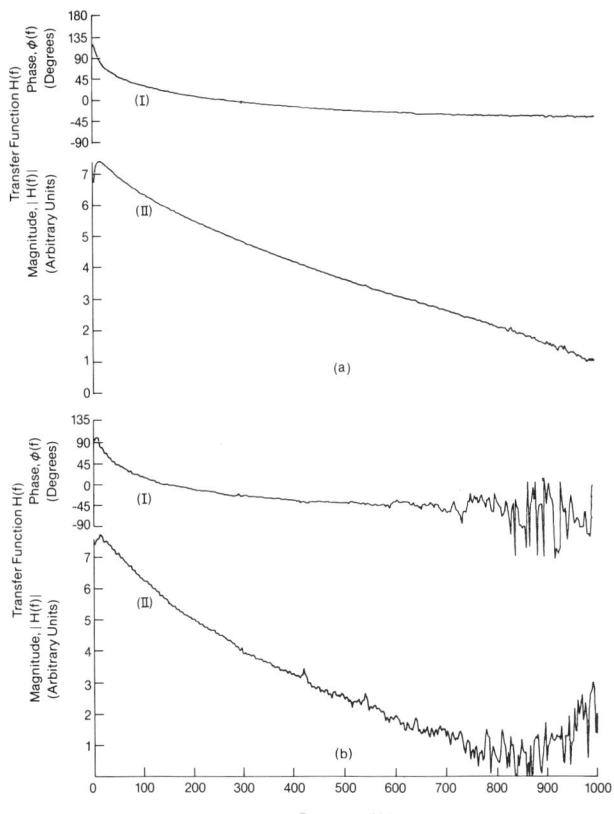

Fig. 6. Transfer function channels of the mirage effect system; (a) FT Time Delay spectrometer; (b) PRBS spectrometer. (I): phase; (II): magnitude.


CONCLUSIONS

1. FT Time Delay and Pseudo-Random thermal wave methods are both suitable for nondestructive depth profiling of materials thermally and spectroscopically.

2. Advantages over frequency-domain methods:
   a) Much faster data acquisition and frequency response processing than lock-in detection.
   b) Better layer separation in time delay than dispersive thermal wave phase lag measurements.

3. Advantages over pulsed laser methods:
   a) Much higher thermal destruction/degradation tolerance threshold due to sequential deposition of laser energy.
   b) Higher dynamic range of the FT Time Delay method.

4. FT Time Delay is superior to PRBS in spectral function processing and signal quality. It holds great promise in NDE of delicate and sensitive materials, such as microelectronic and optoelectronic structures, probing of poor thermal conductors, ceramic subsurface flaw imaging, etc.

# REFERENCES

1. R. C. Heyser, J. Audio Eng. Soc. 15, 370 (1967).
2. H. Biering and O. Z. Pedersen, Brüel Kjaer Tech. Rev. 1, 5 (1983)
3. J. Trampe Broch, Brüel Kjaer Tech. Rev. 4, 3 (1975).
4. R. C. Heyser, Monitor-Proc. IREE, 67 (March 1976).
5. A. Mandelis, IEEE Trans. UFFC, Sept. 1986 (in press).
6. A. Mandelis, Rev. Sci. Instr. 57, 617 (1986).
7. A. Mandelis, L.M.-L. Borm and J. Tiessinga, Rev. Sci. Instru. 57, 622 (1986).
8. A. Mandelis, L.M.-L. Borm and J. Tiessinga, Rev. Sci. Instr. 57, 630 (1986).
9. S. Haykin, in "Communication Systems", Wiley, New York (1978).
10. W. B. Jackson, N. M. Amer., A. C. Boccara, and D. Fournier, Appl. Opt. 20, 1333 (1981).
11. J. A. Burt, J. Phys. D: Appl. Phys. 13, 1985 (1980).
12. H. Sontag and A. C. Tam, Appl. Phys. Lett. 46, 725 (1985).
13. A. C. Tam and C. K. N. Patel, Appl. Opt. 18, 3348 (1979).
14. A. Rosencwaig, in "Photoacoustics and Photoacoustic Spectroscopy" Chemical Analysis, vol. 57, Wiley, New York (1980).
15. Y. Sugitani, A. Uejima and K. Kato, J. Photoacoust. 1, 217 (1982).

# AN A PRIORI KNOWLEDGE BASED WIENER FILTERING APPROACH TO ULTRASONIC SCATTERING AMPLITUDE ESTIMATION

Steve Neal and D. O. Thompson

Ames Laboratory
Iowa State University
Ames, IA  50011

## INTRODUCTION

The Wiener filter is currently used in the ultrasonic scattering amplitude estimation problem as a means of desensitization during deconvolution [1,2,3]. The work summarized here focuses on a Wiener filtering approach which incorporates a priori flaw and noise information. It will be shown that this approach leads to improved scattering amplitude estimates and improved radius estimates.

## BACKGROUND

A frequency domain equation describing the flaw characterization experiment for a single transducer operating in the pulse-echo mode can be written as [2,4]

$$F(\omega) = R(\omega)A(\omega) + n(\omega) \tag{1}$$

where

F = Fast Fourier transform of the measured time domain signal
R = "Known" measurement system frequency response
A = Flaw scattering amplitude
n = Total noise = acoustic noise + electronic noise

The scattering amplitude estimation problem involves estimating $A(\omega)$ given that $F(\omega)$ has been measured in the presence of noise, $n(\omega)$. The simplest approach to the estimation problem would be to ignore the noise and deconvolve the system response, $R(\omega)$, out of the measured signal, $F(\omega)$. The resultant scattering amplitude estimate will be referred to as the experimental result and can be written as

$$\hat{A}(\omega) = \frac{F(\omega)R^*(\omega)}{|R(\omega)|^2} \tag{2}$$

where $\hat{A}(\omega)$ indicates an estimate of the scattering amplitude and $R^*(\omega)$ is the complex conjugate of $R(\omega)$. In order to desensitize the deconvolution as $R(\omega)$ goes to zero, a positive constant is added to the denominator of Eq. (2). The resultant scattering amplitude estimate, written as

$$\hat{A}(\omega) = \frac{F(\omega)R^*(\omega)}{|R(\omega)|^2 + Q^2} \quad (3)$$

represents the most widely used form of the Wiener filter as applied to the scattering amplitude estimation problem [1,4].

STATISTICAL FILTER

A form of the Wiener filter which incorporates a priori flaw and noise information will now be considered [2,4]. The filter can be derived from a statistical approach where $A(\omega)$ and $n(\omega)$ are assumed to be uncorrelated, Gaussian random variables. Further, $n(\omega)$ is assumed to have a zero mean ($E[n(\omega)]=0$), and $A(\omega)$ is assumed to have, in general, a non-zero mean ($E[A(\omega)]\neq 0$). Derivation of the filter proceeds by maximizing the probability that the scattering amplitude estimate, $\hat{A}(\omega)$, is equal to the correct scattering amplitude, $A(\omega)$, given that $F(\omega)$ has been measured in the presence of noise. This approach results in the maximum likelihood estimate of $A(\omega)$ given by the following frequency domain equation

$$\hat{A} = \frac{\frac{|R|^2 \text{Var}(A)}{E[n^2]}}{\frac{|R|^2 \text{Var}(A)}{E[n^2]} + 1} \left\{ \frac{FR^*}{|R|^2} \right\} + \frac{1}{\frac{|R|^2 \text{Var}(A)}{E[n^2]} + 1} \left\{ E[A] \right\} \quad (4)$$

where $\text{Var}(A(\omega))$ = the ensemble variance of $A(\omega)$ and $E[n^2(\omega)]$ = the ensemble average noise power spectrum.

Equation (4) represents a weighting term formulation of the Wiener filter. The two bracketed terms in Eq. (4) represent the experimental result (Eq. (2)) and the ensemble mean scattering amplitude, respectively. Each of the bracketed terms is multiplied by a weighting term. Note that the weighting terms add to unity and are controlled by a single ratio, $|R|^2 \text{Var}(A)/E[n^2]$. The filter determines the maximum likelihood estimate of $A(\omega)$ by optimally weighting the experimental result versus the ensemble mean scattering amplitude.

Consider the behavior of the filter relative to variations in each of the parameters which make up this ratio. Outside of the system bandwidth (i.e., as $|R(\omega)|^2$ goes to zero) the weighting term on the experimental result goes to zero and the weighting term on the ensemble mean goes to one. Thus, at frequencies where no experimental information is available, $\hat{A}(\omega)$ is determined primarily by the ensemble mean, $E[A(\omega)]$. In the strength of the bandwidth (i.e., as $|R(\omega)|^2$ becomes larger) the weighting terms shift the emphasis from the ensemble mean to the experimental result.

At frequencies where the flaw characterization experiment is dominated by noise (i.e., $E[n^2(\omega)]$ is large relative to $|R(\omega)|^2 \text{Var}(A(\omega))$) the noisy experimental result is de-emphasized and $\hat{A}(\omega)$ is determined primarily by $E[A(\omega)]$. At frequencies where the experiment is less noisy (i.e., $E[n^2(\omega)]$ is smaller) more emphasis is placed on the experimental result and less emphasis is placed on the ensemble mean.

Behavior of the filter with respect to variations in Var($A(\omega)$)
is more subtle. A narrow distribution of flaws (i.e., small Var($A(\omega)$))
implies that $A(\omega)$ for most of the flaws in the ensemble will vary only
slightly from the ensemble mean, $E[A(\omega)]$. Thus, for a narrow distribution of flaws, $\hat{A}(\omega)$ is determined primarily by $E[A(\omega)]$. A broad distribution of flaws (i.e., large Var($A(\omega)$)) implies that $A(\omega)$ for many of
the flaws in the ensemble will vary greatly from $E[A(\omega)]$. Thus, for
a broad distribution of flaws, more emphasis is placed on the experimental
result and less emphasis is placed on the ensemble mean.

RESULTS

The statistical filter (Eq. (4)) was compared to the desensitization
filter (Eq. (3)) with Q equal to 10% of $|R|_{max}$. Comparisons were made
of scattering amplitude estimates and associated radius estimates which
were determined from noise corrupted, simulated flaw signals. Radius
estimates were determined from scattering amplitude estimates by utilizing
the one-dimensional inverse Born sizing algorithm [6]. Noise in the
simulated signals was attained by measuring backscattered signals from
a stainless steel specimen with an average grain size of 22.5µm and
from an aluminum specimen with 2% porosity. For the measurement system
utilized, the stainless steel specimen and the aluminum specimen represent
a high frequency noise source and a low frequency noise source, respectively.
The average noise power spectrum was determined from the backscattered
signals from each specimen thus establishing an estimate of $E[n^2(\omega)]$
for each noise type. Measurement of backscattered noise signals, estimation
of noise power spectra, and generation of simulated flaw signals are
discussed in Ref. 4.

A priori flaw information was simulated by assuming a Gaussian
distribution on the radius of an ensemble of spherical voids. For the
results presented in this paper, a mean radius of 200µm and a standard
deviation on the radius of 30µm were assumed. Spherical void radii
were generated at random out of the indicated Gaussian distribution.
For each spherical void size (i.e., for each generated radius), a noise
corrupted flaw signal was created and a scattering amplitude estimate
and an associated radius estimate were determined.

Average results (e.g., average percent errors on radius estimates)
were determined by averaging the errors over 10 generated flaw sizes
at each signal to noise ratio (S/N). Signal to noise ratios were varied
[4] from 3 to 10. As used here, S/N is defined as the square root of
the ratio of the maximum value of the flaw signal power (i.e., the maximum
of $|R(\omega)A(\omega)|^2$) to the maximum value of the average noise power (i.e.,
the maximum of the estimate of $E[n^2(\omega)]$).

The first three figures represent results in which the noise is
taken from the stainless steel specimen. Figure 1 shows the mean square
error of the scattering amplitude estimate versus S/N. The mean square
error was calculated by averaging the square of the difference between
$\hat{A}(\omega)$ and $A(\omega)$ over the measurement system bandwidth. The statistical
filter shows a reduced mean square error at all signal to noise ratios
relative to the desensitization filter.

Figure 2 shows the magnitude of the percent error of the radius
estimates versus S/N. As expected, improved scattering amplitude estimates
(Fig. 1) have resulted in improved radius estimates. This figure shows
that the magnitude of the percent error for the statistical filter is
reduced by a factor of approximately 2 relative to the magnitude of
the percent error for the desensitization filter.

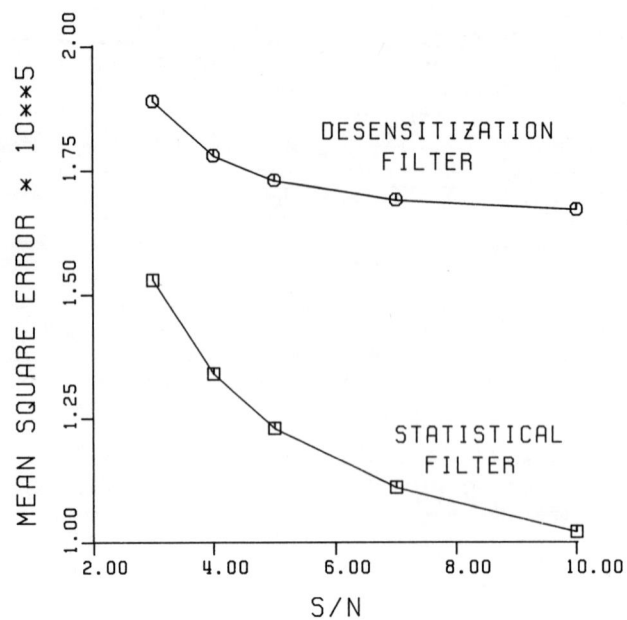

Fig. 1. Scattering amplitude estimate mean square error vs. S/N. Noise utilized in the simulated flaw signals was taken from the stainless steel.

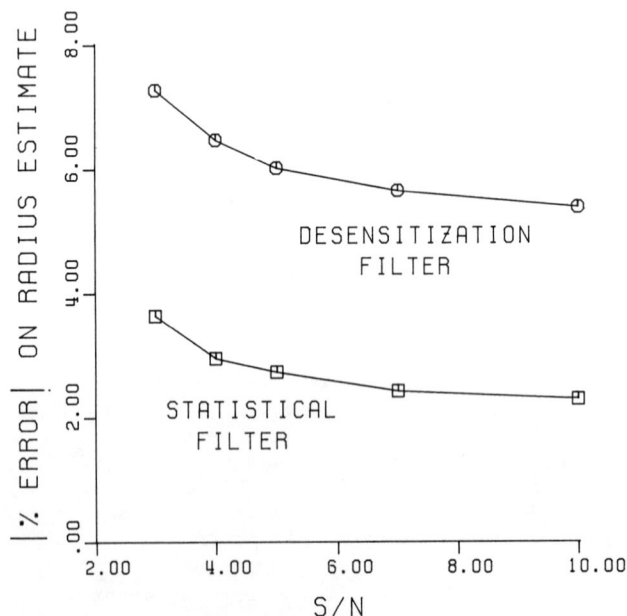

Fig. 2. Magnitude of the radius estimate percent error vs. S/N. Noise utilized in the simulated flaw signals was taken from the stainless steel.

810

Figure 3 shows the percent error in radius estimates versus S/N. Again, the statistical filter shows significant improvement relative to the desensitization filter. The statistical filter forces the average percent error to zero while the desensitization filter tends to undersize the flaws on the average (as indicated by the negative error) by approximately 6%. Note that knowledge of the average percent error alone is not a sufficient comparison tool since an average error of zero could be achieved by greatly oversizing half of the time and greatly undersizing half of the time. Thus, consideration of either the magnitude of the percent error (as in Fig. 2) or the variance of the percent error (as in Table 1) is important.

Motivation for considering noise taken from the aluminum specimen is as follows. The desensitization filter requires the post processing steps of truncation at low frequency followed by extrapolation to zero frequency [6]. For the case of noise with its strength at low frequency, such as the noise taken from the aluminum specimen, extrapolation becomes very difficult. The statistical filter does not require truncation followed by extrapolation. The statistical filter uses the ensemble mean scattering amplitude to reach below the system bandwidth to zero frequency. Figure 4 shows the magnitude of the percent error of the radius estimates versus S/N where the noise is taken from the aluminum specimen. This figure shows that even for the case of low frequency noise, the statistical filter is successful in estimating the scattering amplitude to zero frequency through a weighted average of the experimental result and the ensemble mean scattering amplitude. Note that the extrapolation difficulty is reflected in the increased magnitude of the percent error for the aluminum noise case (Fig. 4) versus the stainless steel noise case (Fig. 2).

Table 1 shows a comparison of a number of parameters for the desensitization filter (top row of numbers) versus the statistical filter. While each data point on the graphs represents the average result for

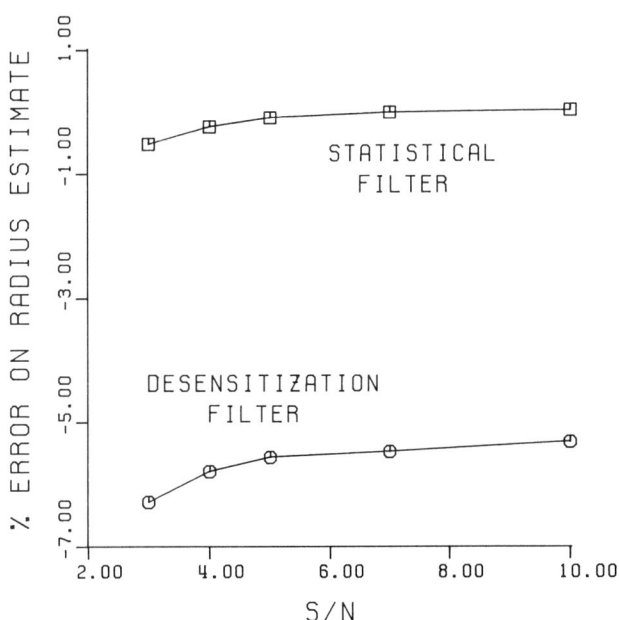

Fig. 3.  Radius estimate percent error vs. S/N. Noise utilized in the simulated flaw signals was taken from the stainless steel.

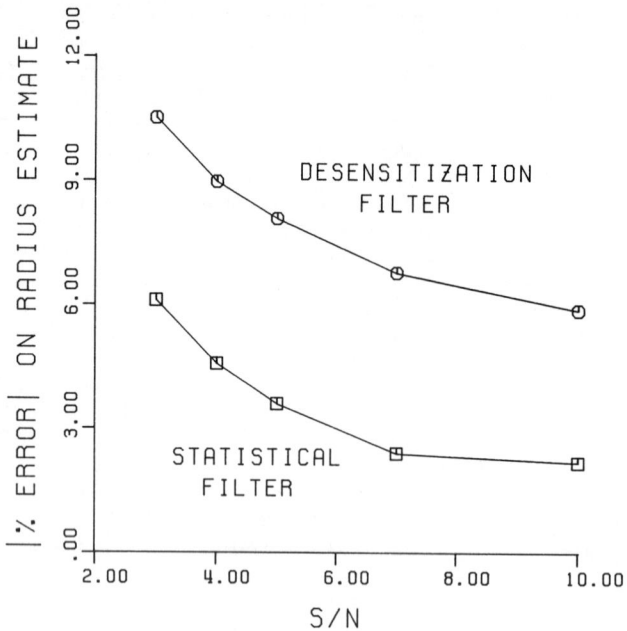

Fig. 4. Magnitude of the radius estimate percent error vs. S/N. Noise utilized in the simulated flaw signals was taken from the aluminum.

10 flaws, each number in the table represents the average result for 200 flaws. This greater number of averages was achieved by averaging over five signal to noise ratios, two types of noise, three breadths of radius distributions, and over ten flaws at each condition. For each parameter, the table shows improved results for the statistical filter relative to the desensitization filter.

Table 1. Comparison of the statistical filter to the desensitization filter.

| Filter | \|% Error\| on Radius Estimate | % Error on Radius Estimate | Variance on Radius Estimate % Error | Mean Square Error * $10^5$ $\hat{A}(\omega)$ vs. $A(\omega)$ |
|---|---|---|---|---|
| Desensitization | 5.60 | -4.60 | 32.1 | 1.94 |
| Statistical | 2.86 | +0.53 | 13.2 | 1.19 |

SUMMARY

In summary, three points are to be noted. First, as indicated by the graphical results and by Table 1, the statistical filter shows improved scattering amplitude estimates and improved radius estimates in comparison to the desensitization filter. Radius estimates were determined with the one-dimensional inverse Born sizing algorithm; however, it is expected that improved scattering amplitude estimates will result in improved radius estimates independent of the inverse sizing algorithm utilized. Second, unlike the desensitization filter, the statistical filter has a statistical base which could be useful in establishing the quality of scattering amplitude estimates and radius estimates. Finally, given the statistical filter's utilization of a priori flaw and noise information, the filter fits in well with NDE procedures for detection and classification of flaws which employ such a priori information.

ACKNOWLEDGEMENT

The Ames Laboratory is operated for the U.S. Department of Energy by Iowa State University under Contract No. W-7405-ENG-82. This work was supported by the Director of Energy Research, Office of Basic Energy Sciences.

REFERENCES

1. Y. Murakami, B. T. Khuri-Yakub, G. D. Kino, J. M. Richardson, and A. G. Evans, "An application of Wiener filtering to nondestructive evaluation", Appl. Phys. Lett. 33(8), (1978).
2. R. K. Elsley, J. M. Richardson, and R. C. Addison, "Optimum measurement of broadband ultrasonic data", in 1980 Ultrasonic Symposium Proceedings, IEEE, New York, (1980).
3. R. B. Thompson and T. A. Gray, "A model relating ultrasonic scattering measurements through liquid-solid interfaces to unbounded medium scattering amplitudes", J. Acoust. Soc. Am. 74(4), (1983).
4. S. Neal and D. O. Thompson, "An examination of the application of Wiener filtering to ultrasonic scattering amplitude estimation", in, Review of Progress in Quantitative NDE, 5, D. O. Thompson and D. E. Chimenti, Eds., (Plenum Press, NY, 1986), p.737.
5. C. F. Ying and R. Truell, "Scattering of a plane longitudinal wave by a spherical obstacle in an isotropically elastic solid", J. Appl. Phys. 27(9), 1086 (1956).
6. R. B. Thompson and T. A. Gray, "Range of applicability of inversion algorithms", in, Review of Progress in Quantitative NDE, 1, D. O. Thompson and D. E. Chimenti, Eds., (Plenum Press, NY, 1982), p.233.

# SIGNAL PROCESSING OF LEAKY LAMB WAVE DATA FOR DEFECT IMAGING IN COMPOSITE LAMINATES

Richard W. Martin
University of Dayton Research Institute
Dayton, OH 45469-0001

D. E. Chimenti
Materials Laboratory
Wright-Patterson Air Force Base, OH 45433-6533

## INTRODUCTION

Inspection of composite laminates with Leaky Lamb waves (LLW) has been shown to hold promise of improved reliability and increased sensitivity to important defects [1]. Conventional scanning with the LLW has the possible disadvantage that the method is sensitive not only to internal structure, but also to small variations in plate thickness, which are indistinguishable from elastic property changes. To circumvent this potentially irrelevant sensitivity, a technique has been developed [2] whereby such variations can be selectively ignored, while retaining sensitivity to important defects or material property variations. The method consists of applying frequency modulation to the usual tone burst RF signal and exploiting detailed knowledge of the Lamb wave spectrum of composites [3] to discriminate between significant defects or property changes and small thickness variations in the plate. The current work extends and expands this analog signal processing scheme by performing the analysis on digitized data, permitting a much more general and flexible approach which will be described.

## LEAKY LAMB WAVE GENERATION

In Fig. 1, a beam of ultrasound propagates to the sample and strikes the plate at an incident angle, generating plate waves satisfying the condition $\underline{k}_f \cdot \hat{p} = k_{pl}$, where $\hat{p}$ is a unit vector in the plane of the plate. If conditions of beam width and frequency are favorable, the reradiated field appears distorted and displaced in the direction of propagation of the plate wave. Most of the energy is then contained in the two shaded regions. "N" denotes null zone, and "LW" indicates leaky wave. The transmitted wave below the plate is omitted for clarity.

Exploiting the fact that in certain regions of incident angle the LLW spectrum is quite regular [3], we frequency modulate the RF tone burst with a triangular waveform at low audio rates, about 5 to 20 Hz. At the output of the gated integrating amplifier we then have a time-varying signal whose power is contained principally in the fundamental and whose frequency equals twice the product of the modulation frequency with the number of plate modes subtended by the modulation bandwidth. This signal can then be further processed digitally to extract information about how the sample under study

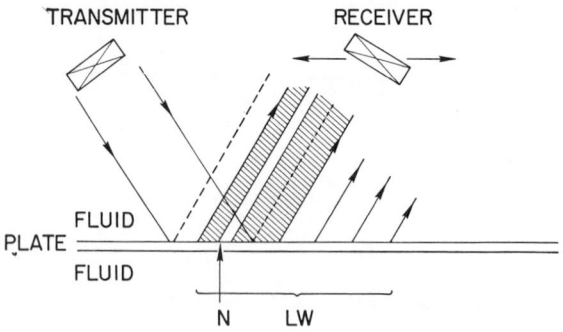

Fig. 1. Schematic of Leaky lamb wave experiment.

behaves elastically. The simplest method consists of filtering this signal and integrating the result with respect to time. If the time constant is chosen to be suitably short, the method can be employed in a C-scan type arrangement, where both transducers are scanned raster fashion across the sample. The result of this signal coding and analysis technique permits discrimination between small elastic property or plate thickness variations and larger excursions indicative of defects [2]. A more sophisticated version of this method is the spectral processing method to be described in this paper.

DATA ACQUISITION SYSTEM

In Fig. 2, a sweep generator B sets the repetition rate for the system by triggering both RF generator A and the digitizer. The sweep generator generates a 56 ms ramp causing the RF generator A to sweep from 1.3 to 4.0 MHz. Pulse generator A determines the period (425 microseconds) and the duration of the tone bursts (10-30 microseconds) by controlling a diode switch. Therefore, 132 tone bursts are generated during the 56 ms sweep period. The received signal is amplified, gated and integrated. The resulting signal of Lamb wave modes is then bandpass filtered at 400 Hz on the high frequency end for anti-aliasing and at 25 Hz on the low frequency end to remove effects of the 56 ms sweep period and D.C. components. The low frequency cutoff is required to remove D.C. and low frequency (<20 Hz) components from the reflected signal which degrade the resolution of the spectra calculations described in later paragraphs. The Lamb wave signals are digitized over the entire 56 ms sweep period at a rate of 2064 samples per second with a total of 128 points recorded. All data points are then sent to a LSI 11/23 microcomputer for data processing and image generation.

TEST SAMPLE

The characterized samples used in this study are unidirectional graphite/epoxy laminates. One is a 32-ply, AS4/3501-6 panel and the second is a 24-ply, T300/CG914 panel. They have simulated defects embedded at layers 8, 16, and 24 or 6, 12, and 18, respectively. Delaminations are simulated with circular teflon wafers and porosity with 40 micron diameter microballoons.

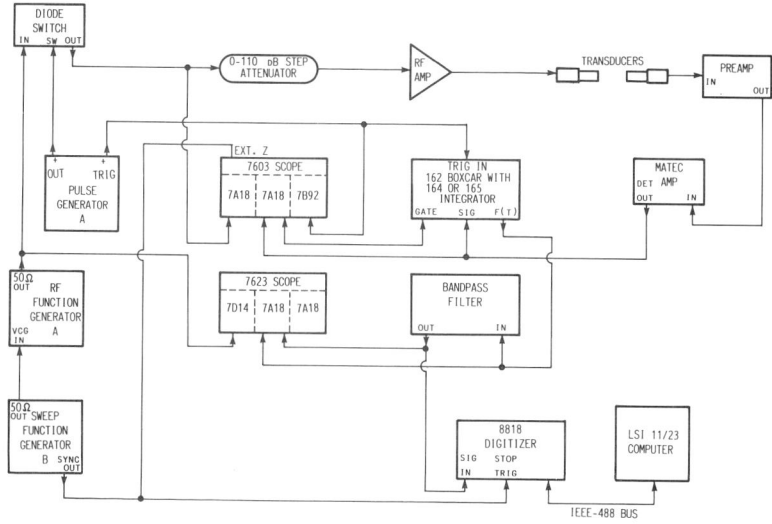

Fig. 2. Block diagram of swept frequency Lamb wave instrumentation.

## LAMB WAVE DATA SPECTRAL PROCESSING

Data processing software first scales the 128 digitized Lamb wave mode data points, buffers the data with zeros, and then calculates a 256 point FFT. In this case, only the 0-200 Hz frequency range of the spectrum is of interest. In the general case, the spectral frequency range of interest depends upon the modulation frequency introduced by the sweep, the RF frequency range and the thickness of the panel under test. Figs. 3a and 4a are plots of digitized Lamb wave modes, while Figs. 3b and 4b are plots of the resulting FFT spectra of these modes. The modes of Fig. 3a occur in a defect-free area of the panel. This type of plot can be interpreted two ways. First, as a spectral plot of the reflected signal from the panel, it identifies the swept ultrasonic frequencies of the Lamb wave modes as indicated by the frequency scale of the RF frequencies ($\omega/2\pi$). However, Fig. 3a depicts only the relatively low frequency detected envelope (modulation) of the RF frequencies and not the RF cycles themselves. Secondly, it is a plot of the time-dependent digitized signal over the 56 ms sweep period (top scale) representing the modulated, low frequency components of the reflected ultrasonic signal. The FFT of this time-dependent signal is shown in Fig. 3b with a predominant peak at around 140 Hz on the FFT frequency scale ($\nu/2\pi$). In contrast, Fig. 4a is a plot of the Lamb wave modes measured over a delamination located at layer 18 of the 24-ply panel and has fewer minima than did the plot from the defect-free area. Because it has fewer modes, the predominant frequency of the layer-18 defect is lower than that of the defect-free material, as shown in the FFT plot of Fig. 4b. The number of modes is fewer with the beam over the delamination because of the mechanical decoupling effect of this defect. The Lamb wave therefore samples an effectively thinner panel than in the defect-free area where the sound energy propagates through the entire thickness of the panel. Defects located closer to the surface produce correspondingly fewer modes and therefore have lower peak frequencies, shifting the entire frequency band toward the low end of the FFT spectrum.

To present the data in a C-scan image format, each FFT spectrum must be reduced to a single value and located on an x-y position grid. Of the many ways to represent the FFT data, the peak frequency and median frequency

817

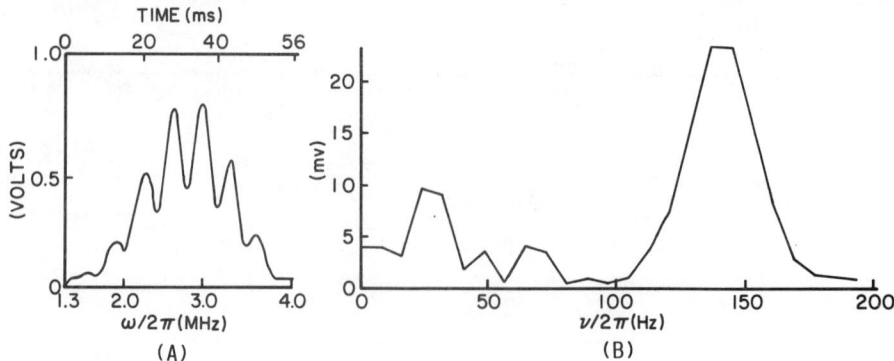

Fig. 3. (a) Plot of Lamb wave modes of a defect-free area showing envelope of detected and filtered ultrasonic signal. (b) Spectrum of the signal shown in (a).

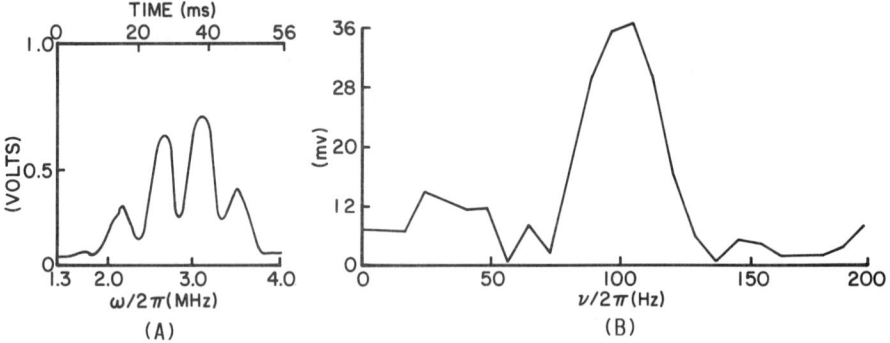

Fig. 4. (a) Plot of Lamb wave modes of a layer-18 defect showing envelope of detected and filtered ultrasonic signal. (b) FFT spectrum of the signal shown in (a).

methods were considered to be the best. On Figs. 3b and 4b the peak frequency method works well, but in the general case it does not. This is because the presence of multiple defects, porosity or changes in the fiber/resin ratio can all cause multiple peaks in the FFT spectrum which may be of nearly equal amplitude. However, we found that even in the presence of multiple peaks, a corresponding shift in the center of the FFT spectrum still occurred. This observation led us to examine the defect discrimination possibilities of the median frequency representation.

To make the analysis slightly more concrete, we introduce the following definitions. Let $Q(\omega,\underline{x})$ be the position-dependent Lamb wave spectrum function, as in Figs. 3a or 4a, where $\omega$ is the angular radio frequency and $\underline{x}$ is a vector position on the sample surface. If the RF tone burst is frequency modulated, the function $Q(\omega,\underline{x})$ will be time dependent and have the periodicity of the modulating function. Considering $Q(\omega(t),\underline{x})$ as a function of time, we may perform a Fourier transform on it to obtain a pseudo-frequency spectrum of this implicit time-dependent function,

$$\hat{Q}(\nu,\underline{x}) = \left| \int Q(\omega(t),\underline{x}) e^{-2\pi i \nu t} dt \right|, \qquad (1)$$

where $\nu$ is the pseudo frequency, generally in the audio range. With this result we may now consider several methods to represent the data. The peak frequency $\nu_{peak}$ is simply the value of pseudo frequency at which $\hat{Q}(\nu_{peak}, \underline{x})$ is a maximum. This representation presents the problems indicated above with regard to multiple defects. Band limiting, discussed below, consists of summing the spectral components within a specified pseudo-frequency band,

$$\int_0^{\nu_{max}} F(\nu_o, \delta\nu) \hat{Q}(\nu, \underline{x}) \, d\nu, \quad (2)$$

where $F(\nu_o, \delta\nu)$ is a unit window function centered at $\nu_o$ and having width $\delta\nu$, and $\nu_{max}$ is the aliasing frequency of the FFT. In the median frequency representation, the pseudo frequency is sought which bisects the area under the FFT spectrum, such that one half the area lies above and one half below that frequency. That is,

$$\int_0^{\nu_{1/2}} \hat{Q}(\nu, \underline{x}) \, d\nu = 1/2 \int_0^{\nu_{max}} \hat{Q}(\nu, \underline{x}) \, d\nu, \quad (3)$$

where $\nu_{1/2}$ is the median pseudo frequency.

The median frequency gives the best representation of frequency shifts in the FFT spectra and therefore relative defect depth and response when the mode structure varies across the panel. An important advantage of median frequency processing is that no previous knowledge of the location and type of defects in the panel, nor the resulting mode spectra it produces are required to obtain a high quality image with accurate relative depth information. The depth information is retained so long as the frequency difference representing defects close to the same depth is greater than the resolution of the FFT. In the case presented, the FFT resolution is 8 Hz, but it could be improved to 4 Hz by doubling the digitizing rate, recording 256 data points, and performing a 512 point FFT. This would double the depth resolution and also retain defect depth information of flaws near the plate boundaries. An important advantage of spectral processing of Lamb wave data in general lies in the ability of this scheme to permit a very wide dynamic range of defect response while preserving sensitivity to small or weak flaws.

TEST RESULTS

Fig. 5 is the result of a swept frequency tone burst scan with median frequency processing on the 24-ply test panel. The top row of defects is located between layers 6 and 7 and consists of (from left to right) 1, 1/2 and 1/4-inch diameter delaminations, porosity (middle), and a ply cut (right). The middle row consists of the same defects as the first row except that all defects are located between layers 12 and 13 within the panel. The bottom row of defects, located between layers 18 and 19, consists of porosity (near the middle) and three delaminations on the right side of 1/4, 1/2 and 1-inch diameter, respectively. In this image, increasing median frequency is proportional to increaingly darker shades of gray proportional to defect depth. Black represents the highest median frequency (130 Hz) and occurred in defect-free areas of the panel. Layer-18 defects (bottom row) appear dark gray and represent a median frequency of 90 Hz. Layer-12 defects (middle row) appear medium gray and represent a median frequency of 75 Hz, while layer-6 defects (top row) appear light gray to white and represent a median frequency of 48 Hz. Note that objects in the same row (same depth) appear the same shade except for some of the less concentrated porosity areas down the middle which appear dark gray. Also appearing dark gray are long horizontal streaks which probably represents small changes in the fiber/resin ratio in those areas. Fig. 6 is presented as a comparison to Fig. 5 to show the results of a normal incident C-scan

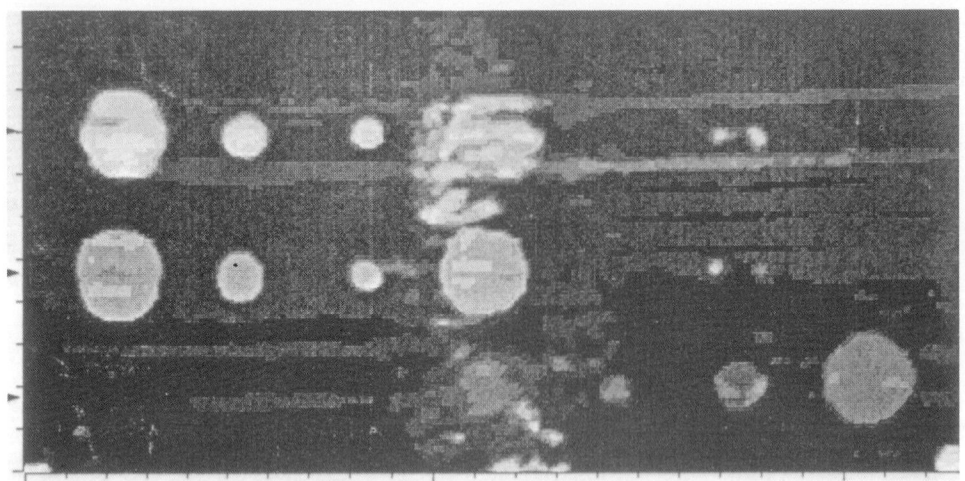

Fig. 5. C-scan image utilizing median frequency processing of the Lamb wave mode spectrum. Increasing median frequency is proportional to increasingly darker shades of gray level. Frequency is proportional to defect depth. Black represents defect-free areas while progressively lighter shades of gray indicate defects progressively closer to the surface.

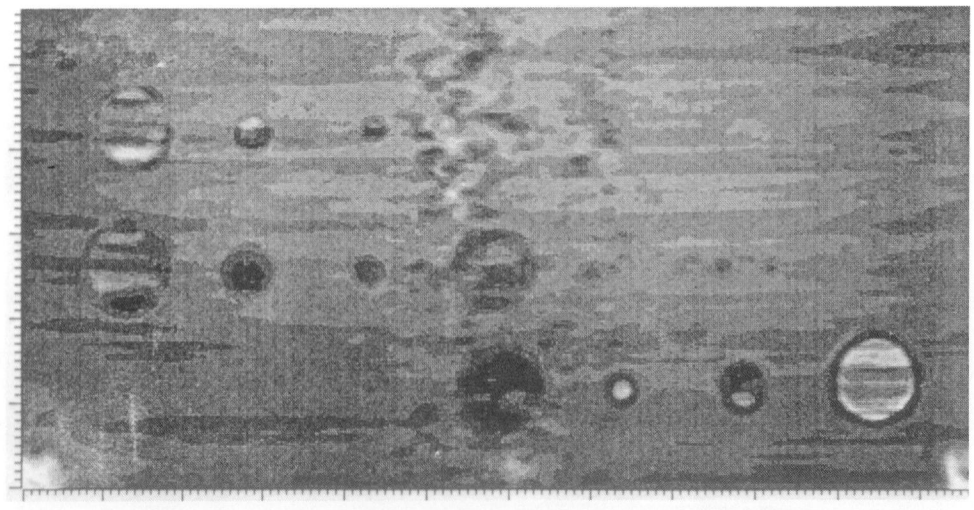

Fig. 6. Normal incidence C-scan image of same panel as in Fig. 5 utilizing an unfocused 2.2 MHz transducer gated on the back surface.

image of the same sample. The C-scan was produced using a 2.2 MHz unfocused transducer and was gated on the back surface. Most of the delaminations were clearly shown, but the porosity and ply cuts are not as sharp and of course no depth information is available.

Since defects of different depths are represented by different frequencies, processing software was also developed to selectively image defects at a particular depth. First, a frequency band is selected which is believed to best represent defects at the desired depth. Then that band in the FFT is integrated (digital bandpass filtered) and used as the value for the C-scan image. Fig. 7 is an image produced by integrating the frequency band that represents defect-free material. This produces an image where defect-free material will appear dark. The frequency band selected was 100-160 Hz and was determined by selecting a band around the frequency of the large peak in Fig. 3b. Fig. 8 was produced by selecting a frequency band around the large peak in Fig. 4b and bandpass filtering (81-120 Hz), thus producing an image representing defects at layer 18. The defects appear dark in Fig. 8 and very good discrimination was achieved in selectively imaging only defects at that particular depth.

To demonstrate the discrimination capabilities of our method, spectral processing was applied to a 32-ply test panel whose thickness was not constant. For comparison purposes, the image of Fig. 9 was generated by measuring the depth of a single null in a 4-MHz constant frequency tone burst. Many of the defects are obscured due to small thickness changes in the plate. Then, the image of Fig. 10 was generated by swept frequency tone burst and spectral processing techniques. The spectral processing technique utilized here was selective imaging of the frequency band which includes defects at all depths. The defects appear dark and all other panel material a lighter shade of gray. Digital bandpass filtering of the 24-137 Hz frequency band of the FFT was selected. Defects shown in the 32-ply panel are similar to those in the 24-ply panel. Each row of defects is at a different depth and includes (from left to right) delaminations, ply cuts, and porosity. All defects were imaged in Fig. 10. Porosity (on the right side) and a ply drop, running across the middle of the panel, are clearly imaged in Fig. 10, but not visible in the non-modulated LLW image in Fig. 9.

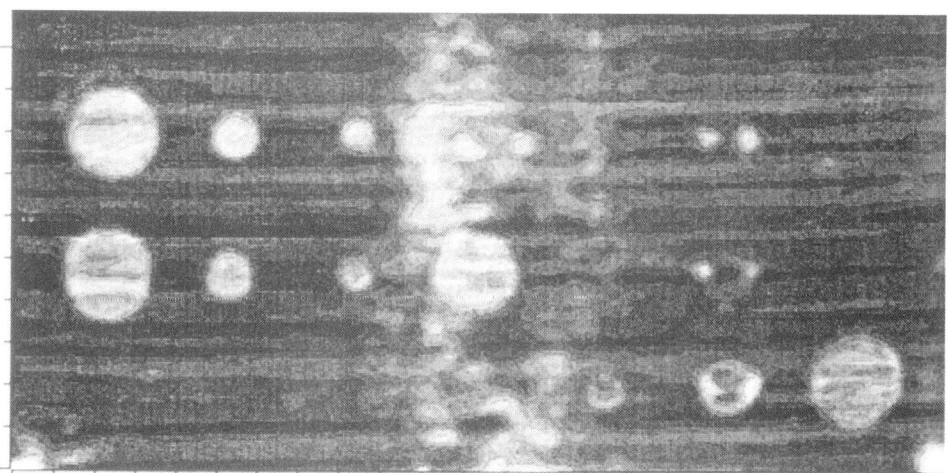

Fig. 7. Image formed by digital bandpass filtering of the 110-160 Hz band of the Lamb wave mode spectrum. An example spectrum is shown in Fig. 3b. Defect-free areas of the panel appear darkest while all defect areas appear lighter in shade.

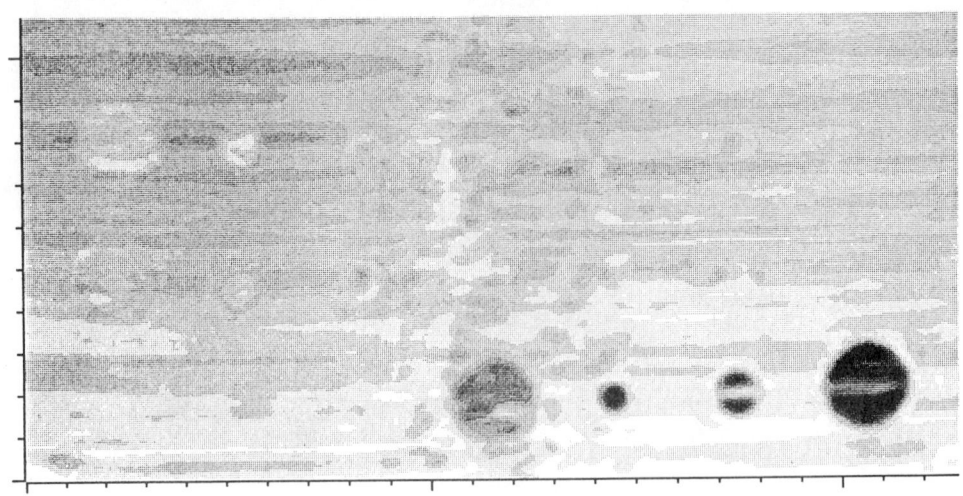

Fig. 8. Selective depth image formed by digital bandpass filtering of the 81-120 Hz band of the Lamb wave mode spectrum. An example spectrum is shown in Fig. 4b. Defect areas of the panel at layer 18 appear darkest while all other areas appear lighter in shade.

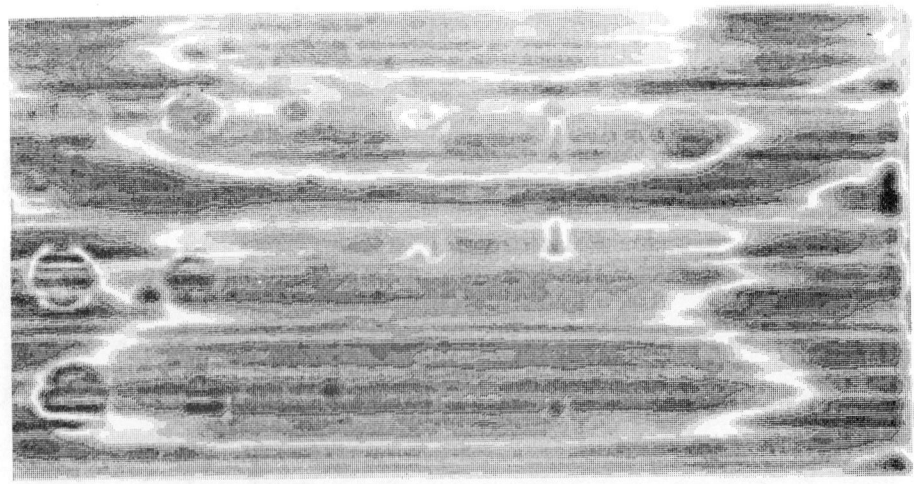

Fig. 9. Lamb wave image of a 32-ply panel generated by measuring the depth of a single null in a 4 MHz, constant frequency tone burst.

CONCLUSIONS

Swept frequency Lamb wave techniques have been shown to provide reliable detection of simulated delaminations, porosity and ply cuts in unidirectional laminates. Also, spectral processing techniques avoid the problems associated with tracking the amplitude of a specific null because the entire signal from several modes is utilized. Median frequency processing of the mode spectra provides a good general purpose processing technique for defect detection and relative depth determination, since median frequency is proportional to depth. Finally, LLW techniques allow the use of low frequency unfocused transducers while providing good resolution and complete depth penetration.

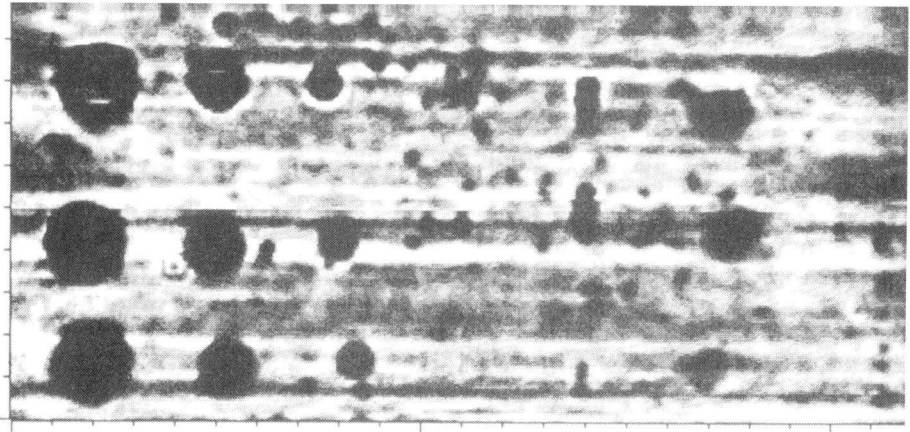

Fig. 10.  Image generated using swept frequency tone burst and spectral processing techniques to selectively image the frequency band which would include defects at all depths (24-137 Hz). Defects all appear dark and all other panel material a lighter shade of gray.

ACKNOWLEDGMENTS

This work was performed at the AFWAL Materials Laboratory at Wright-Patterson Air Force Base, Ohio and supported in part under Contract Number F33615-83-C-5036. The authors would like to acknowledge Y. Bar-Cohen of the Douglas Aircraft Company for providing the samples used in these experiments, and Mark Ruddell of UDRI for expert technical assistance in the performance of the experiments described in this paper.

REFERENCES

1. Y. Bar-Cohen and D.E. Chimenti, *Review of Progress in QNDE 5*, D.O. Thompson and D.E. Chimenti, eds., Plenum Press, New York, (1986), p 1199.
2. D.E. Chimenti and Y. Bar-Cohen, 1985 IEEE Ultrasonic Proceedings, B.R. McAroy, ed., New York, (1986), p 1028,.
3. D.E. Chimenti and A.H. Nayfeh, *J. Appl. Phys. 58*, (1985), 4531.
4. R.W. Martin and R.J. Andrews, "Backscatter B-Scan Images of Defects in Composites," *Review of Progress in Quantitative Nondestructive Evaluation*, Vol. 5B, (1986), pp 1189-1198.

A HIGHLY INTERACTIVE SYSTEM FOR PROCESSING LARGE VOLUMES

OF ULTRASONIC TESTING DATA

H. L. Grothues, R. H. Peterson, D. R. Hamlin, K. S. Pickens

Southwest Research Institute
San Antonio, Texas

INTRODUCTION

Automated ultrasonic testing (UT) of big structures poses particular problems related directly to economics and productivity. Generally, UT examinations on these large structures are performed with multiple channels to reduce scan time and collect data from various orientations. The amount of resulting data also is quite large. Traditional approaches have relied on up-front gating and signal thresholding to reduce the amount of data recorded. This has been a practical approach, as the capability of data processing and recording devices has also been limited. Even with the incorporation of computer technology, most systems performing UT of large structures still operate on this same data acquisition principle. General purpose computer configurations lack the performance to provide any substantial improvement in data analysis. Computer resources have been focused on number crunching, data summary, and data comparison using general criteria such as signal amplitude and sound path location. In practice, examiners use this type of system to identify areas of concern and then perform "re-looks" while observing the instrument A-scan display. Years of experience are then applied in interactive analysis of the A-scans for final resolution. For the particular area of concern, as much additional information as possible is collected (e.g., different angles and orientations) to provide information crucial to the final disposition. If the system collected the proper data and was capable of presenting these data in a meaningful format, this manual "re-look" procedure would not be necessary.

DISCUSSION

The Enhanced Data Acquisition System (EDAS) is designed for automated multichannel UT of large components. Its goals are to

- Provide rapid disposition of indications
- Be usable in low signal-to-noise ratios
- Require less stringent calibration requirements
- Increase productivity of the UT crew

In addition, EDAS must be able to quickly collect the proper type of data in sufficient quantity to present easy-to-interpret displays.

The EDAS philosophy is based on using the best probes for the particular examination rather than attempting sophisticated enhancements to less than adequate data. Along with that, the system provides features directed toward typical problems encountered in the UT of large components: multi-channel operation to reduce examination time, realtime display of test data to confirm proper UT system operation during the examination, development of compressed images in realtime to assist in methodical data evaluation, high-speed interaction to enable searches of data to help correlation of results, and a mechanism for annotating and linking different sets of results for analysis and reporting.

To achieve productivity, EDAS is composed of two subsystems: one dedicated to data acquisition and one to data analysis. Data transfer between the subsystems is via high-density laser disk. Separation of acquisition and analysis activities permits data acquisition to occur on subsequent components while data analysis is being performed. EDAS uses the basic ultrasonic C-, B- and A-scan displays as well as side and end views of the material volume as data analysis tools. Since data evaluation is based on visual image analysis, the system must record full waveforms at close intervals. For EDAS, waveform length can vary from 256 to 4096 points, and typically are collected at a scan interval of 0.05 inch. The power of image analysis is that even an inexperienced analyst can detect the predictable pattern of a reflector in an ultrasonic B-scan. In the B-scan image all A-scans are of value in data analysis. Since this is the case, waveforms are continually recorded based on position--no amplitude thresholding is employed.

ACQUISITION SUBSYSTEM

EDAS is a multichannel system composed of a replication of independent channels, as shown in Figure 1. Each channel functions under the control of a channel processor--a microprocessor responsible for data acquisition, generation of data for realtime display, generation of data for 3-view display, and general housekeeping tasks. Each channel interfaces with a high-speed signal averager and a commercially available UT instrument. System interface with the scanner mechanism is accomplished through a position processor responsible for triggering data acquisition, recording position location, and ensuring scan constraints are met. Data flow from each of the channels and the position processor is orchestrated by a command processor responsible for passing data to the realtime display and storage of all data on the laser disk. The data acquisition system provides functions for: diagnostics, calibration, parameter entry, 3-view display, realtime display, and A-scan/B-scan presentation.

In the process of acquiring data, the channel processor is responsible for developing summary displays to be used in data analysis. These displays (termed 3-view) are developed in realtime and written to disk as part of the examination record. The information is for immediate use at the data analysis subsystem. The purpose of the 3-view display is to assist the analyst in locating relevant areas and to provide a logical framework for the analysis process. It assists in reducing the amount of information to be reviewed.

To cope with the speed and volume of information being recorded by the system, EDAS provides a realtime color display for the examiner. This display enables simple review of UT data for all channels during each scan and permits comparison of data patterns between successive scans and between different channels. The realtime display also provides the operator with confidence that the system is performing as expected.

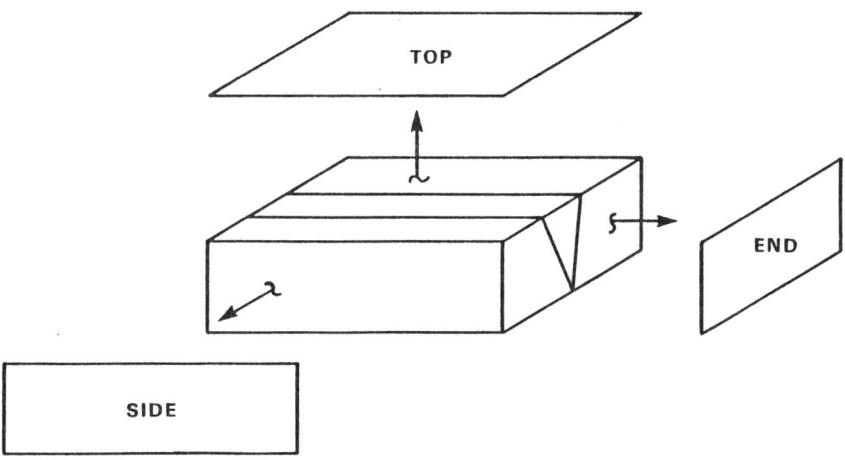

Fig. 1. Volumetric projections (3 views)

To provide further detail, the acquisition subsystem makes available a combination B-scan/A-scan presentation for localized use by the examiners in performing interactive operation and investigation in resolving anomalies observed during the examination. The data acquisition subsystem is not intended to be used for data analysis activities.

Operation of the data acquisition subsystem is through a high-performance color graphics workstation incorporating a window-based mouse-driven user interface. All system functions are displayed as push-buttons on the graphics display in front of the operator. Selection is made by "pointing and pressing" with the mouse--making the system easy to use and not restricting operation through a series of hierarchical menus or complicated commands. The basic acquisition configuration is capable of supporting up to seven channels simultaneously at a scan speed of 6 inches per second.

ANALYSIS SUBSYSTEM

The analysis subsystem consists of a high-performance color graphics workstation, a color hardcopy unit, and a laser disk drive. It can be located separately from the acquisition subsystem, since it operates independently of it. In performing data analysis, EDAS provides a highly interactive, response-oriented environment for the operator. The system uses a series of command display screens and data display windows. User interface with the analysis subsystem is very similar to that of the acquisition subsystem with controls displayed of the graphic screen as push-buttons and operator selection accomplished through "pointing and pressing" with the mouse. This type of operator interface enables freestyle application of the system. The operator is free to use the system in the manner best suited to his or her methods. System response is near realtime for any operator command. Since the system is highly interactive, this enables the operator to perform analysis and recall at will with no observable penalty for repeating steps. The powerful multitasking system and window-based user interface enables selection and simultaneous arrangement of displays to assist the analyst in understanding the data.

Generally speaking, the operator would start analysis with the 3-view summary display generated from the entire set of scan data during the examination. The 3-view display, illustrated in Figure 2, is a volumetric projection (top, side, and end views) of the peak signal data acquired from the inspected volume. A 3-view display is produced for each channel. All three views are properly aligned in a single image so that positional correlation can be made. The image is scaled in coordinates of the component, with signal amplitude mapped to a color scale adjustable by the analyst. The analyst uses this image to identify indications where more detailed information is required for proper disposition. Areas of interest are defined directly from this image through use of a box-type cursor which appears on each of the views simultaneously. The size of the box cursor is adjustable for each of the three coordinate axes, and its location is controlled through manipulation of the mouse. For the volume specified by the cursor boundaries, corresponding 3-view images for other channels and basic B-scan and A-scan data can be retrieved through push-button selection by the analyst.

EDAS handles all coordinate transformations between channels as well as coordinate determination between images automatically--the operator is free to concentrate on interpretation of the data. In particular, EDAS offers powerful enhancements in data analysis by providing proper images for scan-to-scan and channel-to-channel correlations at a speed which is in direct response to the analyst's request. The process is much like switching between color slides in a projector, but with the additional capability for the analyst to retain and arrange multiple images on the display as desired. Throughout this process, the analyst can: sequence through B-scans for a selected channel, review successive A-scans for echo dynamics and predictable sound-path variations, and switch to corresponding images for other channels. In each of these images, the analyst may identify indications and electronically mark them on the image. The EDAS software keeps track of these operator notations and the references to the corresponding data. Where the analyst determines correlation of data from different ultrasonic channels, these data are logically linked and recorded for retrieval and reporting purposes.

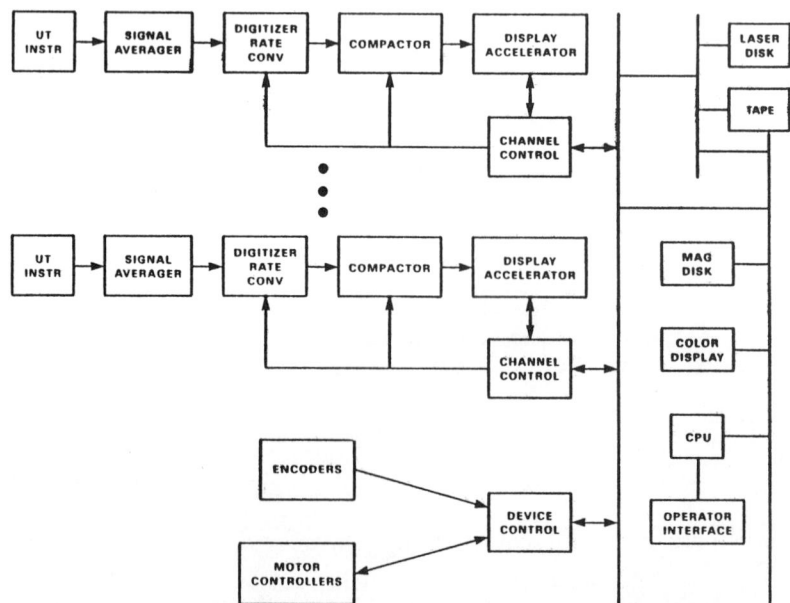

Fig. 2. Acquisition subsystem block diagram

The analysis subsystem software provides a dynamic capability that permits the operator to interactively control the display of recorded A-scans so that waveform dynamics can be observed and evaluated. This permits experienced examiners who are unfamiliar with image interpretation to bridge the gap between traditional A-scans and image analysis, thus taking advantage of their full range of experience.

EDAS is not limited to image display; it offers capabilities to extract quantitative information from the images so that complete analysis and reporting can be performed. The ability to efficiently handle enormous amounts of data in a nonpredefined manner makes EDAS a realistic tool for experienced ultrasonic examiners.

SUMMARY

EDAS is designed for use with both conventional and multibeam UT techniques. The power of the system is in its ability to efficiently handle large volumes of ultrasonic data in a highly responsive, interactive manner. Keith Pickens, in a companion paper, discusses the development of image processing and artificial intelligence applications that are potential enhancements to the system.


DISCUSSION

Mr. Green: I know it's not your work, so maybe it's an unfair question but why are you doing these experiments? You don't have a problem with (inaudible)?

Mr. Peterson: This work was a more fundamental study of the interactions between hydrogen and defects. In terms of applications, the positron annihilation technique has promise as a nondestructive method of evaluating hydrogen embrittlement in steels.

Mr. Green: And gives you the basic fundamentals of what you can do and can't do?

Mr. Peterson: Correct.

Mr. Jeff Eberhard, G.E.: Can you say a little bit more about the experimental configuration, how the positrons get in, and all that kind of stuff?

Mr. Peterson: Two different radioactive isotopes were used in this work, one of which was sodium-22. That source consisted of sodium-22 chloride deposited in a narrow trench on a lucite block. The source is placed on one side of the sample. A gamma ray detector is placed on the other side of the sample to detect the annihilation photons. The signals from the detector are fed into a multi-channel analyzer and then into a computer system. The resultant peak consists of the number of counts as a function of energy. The peak is centered about 511 keV, which is the energy that the annihilation photons would have if the electron in the annihilation event was stationary. As the changes in the spectra produced by defects are very subtle, computer analysis is required to extract meaningful parameters from the raw data.

Mr. R. Green: Are there restrictions or some restriction on the thickness of the specimen?

Mr. Peterson: Yes, mainly on the lower end. The sample should be thick enough to effectively stop all of the positrons. This minimizes the background contribution from positrons annihilating in the detector.

Mr. Ron Smith, Harwell: You are going about the very small changes you need to measure with the positron system. Could you say why you use the R parameter rather than, say, an S parameter, which is the peak width parameter?

Mr. Peterson: S, P or P/W parameters give an indication of the number of positrons annihilating in defect traps. The R parameter analysis considers both the peak and wing portions of the spectra in order to detect changes in the predominant positron trap.

From the Floor: You mentioned the proton screening of the dislocation. What does that mean?

Mr. Peterson: Hydrogen (protons) and positrons are both attracted to defects where the crystal structure is relaxed. A proton at a dislocation will repel any positrons since they both have positive charge. Consequently, the positrons will be more likely to annihilate in the matrix.

ULTRASONIC FLAW DETECTION USING A TIME SHIFTED MOVING AVERAGE

David A. Stubbs and Bob Olding

Systems Research Laboratories
2800 Indian Ripple Road
Dayton, Ohio 45440

INTRODUCTION

The Retirement For Cause (RFC) inspection system uses both eddy current and ultrasonics for the real-time inspection of gas turbine engine components. The ultrasonic inspection module uses a squirter technique to couple the ultrasound to the engine part. The current flaw detection requirement for the ultrasonic system is 0.020 inch diameter, mal-oriented, penny-shaped, internal voids and extensive testing has shown that the squirter technique is comparable to immersion for the detection of these type defects [1,2]. The signal-to-noise ratio is nearly the same for both techniques with only occasional water noise signals occuring in the squirter technique. Additionally, on complex geometries, low amplitude reflections from the nozzle are sometimes present. A typical scan of a bore of an engine part (see Figure 1) has approximately 40,000 A-scans with each A-scan having 300-1000 digitized points; thus even an occasional noise signal can add up to many false indications over the course of an entire scan. For example, using the above numbers a noise signal 0.01% of the time would result in 2000 "flaw indications" in one scan. To help reduce the number of noise-induced flaw indications, an algorithm has been developed to allow the system to distinguish between a signal from an actual defect and one induced by stray reflections, electrical, or water noise.

The trained human observer has relatively little difficulty distinguishing an actual flaw signal from other noise. For an engine part rotating about the transducer some attributes of a true defect signal are: 1) it is present for more than one A-scan, 2) its amplitude increases and then decreases as the defect passes through the ultrasound, and 3) the temporal postion of the signal progresses regularly as the defect moves past the ultrasound. The photograpnhs in Figure 2 show reflections from a 0.032 inch diameter, penny-shaped void. Each photograph represents a 0.5 degree movement of the engine part. The temporal shifting of the flaw signal is evident. A standard averaging technique could make use of the fact that the defect signal is present for more than one A-scan, while the presence of the noise signal is more or less random. But since the defect signal shifts in time the averaged amplitude of the signal would be severely degraded. With the type of flaw detection requirements that the RFC system has, no degradation of the signal amplitude can be tolerated. It would be desirable to incorporate one or both of the other two observations noted above along with a standard averaging technique. For the RFC system an averaging technique has been developed that ultilizes both 1) and 3) above.

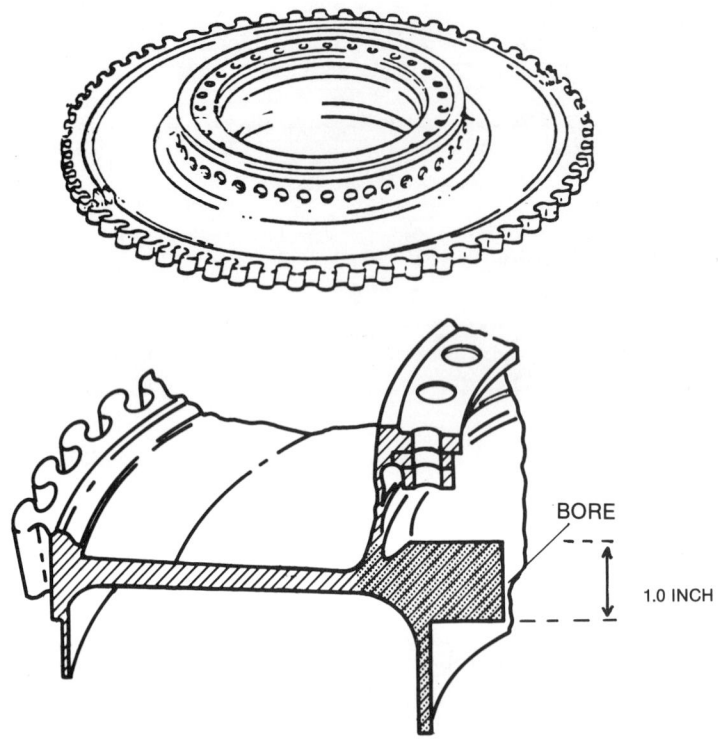

Figure 1. - A typical gas turbine engine part. The ultrasonic scan would inspect the shaded area.

It is easily seen that the temporal shifting of a defect signal is due to the defect moving closer to (or away from) the transducer as the part rotates. If this shifting is constant and fairly independent of flaw depth, then an algorithm that makes use of this shift could be implemented into an automatic inspection system for real time inspections. Early in the RFC program empirical data showed that the amount of shift of the flaw signal in time was nearly constant as a function of angular movement of the engine part. Additionally, data showed that this shift was the same regardless of the flaw depth. These results are easily verified analytically and can be seen in Figures 3 and 4. Figure 3 shows a simple approximation to the path taken by the ultrasound when inspecting the bore of an engine part. The transducer is offset from the center of the part so that a 45 degree shear wave is created in the part. The path segment labled "A" is the portion of the ultrasound path that changes as the part rotates. Figure 4 shows graphs of the metal-path-length of the ultrasound for detection of defects at depths of 0.5, 1.0, and 1.5 inches respectively. The horizontal axis is the angular movement of the part (labelled "$\alpha$" in Figure 3). The point on each curve labled "maximum amplitude" is the angle where the defect intersects the "center line" of the transducer beam. Note that the constant shift only occurs because of the 45 degree propogation of the ultrasound.

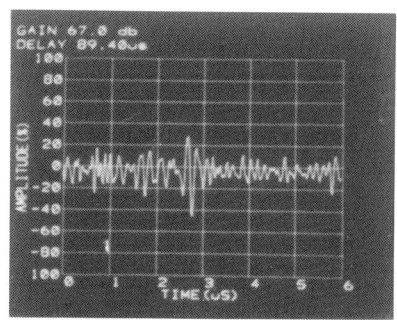

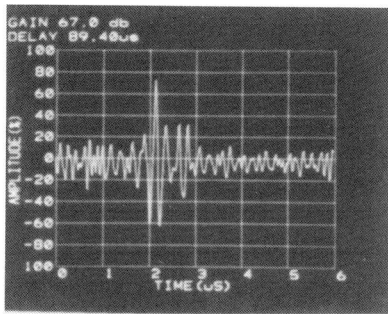

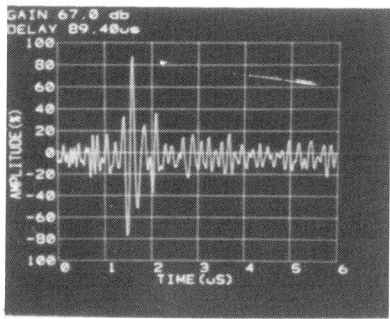

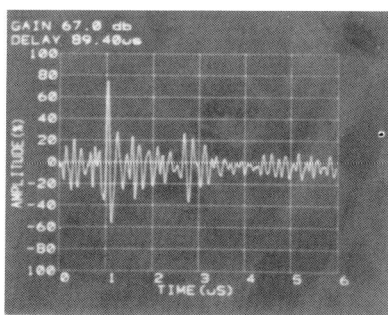

Figure 2. - The four photographs show the reflection from a 0.032 inch diameter, mal-oriented, penny-shaped void 0.6 inches below the surface of the part. Each photograph represents a 0.5 degree rotation of the part. The temporal shift of the signals is easily seen.

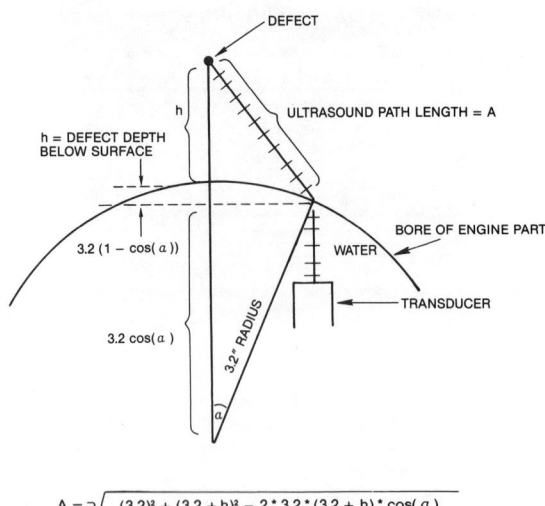

$$A = \sqrt{(3.2)^2 + (3.2 + h)^2 - 2 \cdot 3.2 \cdot (3.2 + h) \cdot \cos(\alpha)}$$

Figure 3. - A ray tracing of the ultrasonic path into an engine bore of 3.2 inch radius. The path length "A" changes as the part rotates.

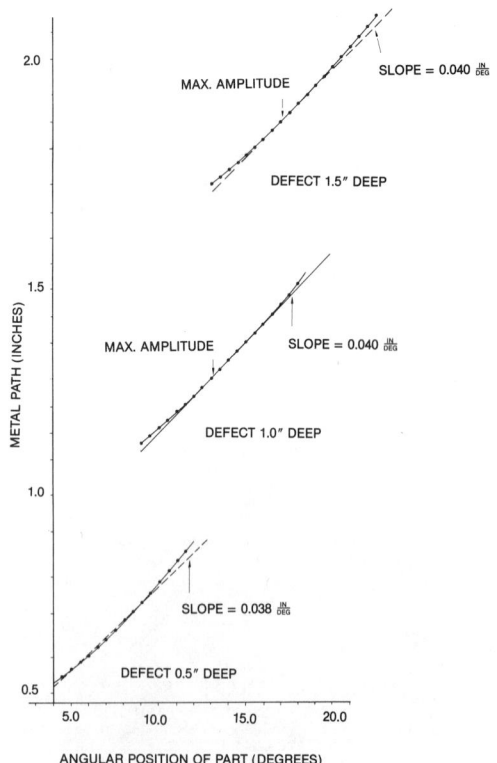

Figure 4. - These three curves show the change in ultrasonic metal-path length as the part is rotated for defects of 0.5, 1.0, and 1.5 inches below the surface. Note that a large segment of each curve is nearly linear and the slope of all these linear regions is nearly the same.

The previously discussed results allowed for the creation of an algorithm that takes into account the temporal shifting of the defect signal while averaging successive A-scans. This algorithm is called a "time shifted moving average"; the moving average refers to the fact that as each new A-scan is received the oldest A-scan is removed from the averaged result. It must be stressed that it is necessary for all of the conditions previously discussed to hold true in order to implement a time shifted moving average algorithm. If the temporal shift was depth dependent extensive testing would be necessary within the algorithm to classify the temporal shift according to depth. If the amount of shift could only be calculated (and was not a constant) then a substantial amount of computer time would be necessary to calculate how much to shift the data. Both of these conditions would sufficiently slow down the algorithm to prevent real time scanning using the averaging technique.

The time shifted moving average was to be implemented using a high level language without the use of an array processor. Because of these constraints, the goal was to discover an algorithm that required a minimum number of arithmetic operations, and to limit the operations to addition and subtraction since the microprocessor used could perform these relatively quickly. The microprocessor could also move blocks of memory quickly relative to the manipulation of individual bytes or words. Memory was not a problem and could be sacrificed if the result was an increase in performance of the algorithm. Besides performing the moving average to reduce noise, the algorithm would also have to perform threshold detection on the averaged data.

Central to the moving average algorithm is a buffer which will be referred to as the accumulator. The accumulator holds the summation of the most recent A-scans. If the moving average was expected to average, for example, five A-scans then the accumulator would contain the summation of the five most recent. Each location (the A-scans are digitized at 50 MHz so each location represents digitized A-scan data twenty nano-seconds apart) within the accumulator can hold 16 bits worth of information. The A-scans have only eight bits of resolution so the accumulator can contain the summation of up to 256 A-scans without overflowing.

Before an A-scan is added, the accumulator is shifted a specified number of locations to the right. This shift takes into account the change in time-of-flight of any flaw indications represented in the A-scan. The amount shifted is directly proportional to the expected change in time-of-flight. The accumulator is shifted by performing a block move so that the first word in the accumulator (the first digitized datum point in the A-scan) now appears at location N+1 where N is proportional to the time shift. Location 1 through N-1 are then set to zero. Once shifted, the current A-scan is simply added to the accumulator. Location one of the A-scan is added to location one of the accumulator, then location two, etc.

Once the current A-scan is added, the oldest A-scan summed in the accumulator must be subtracted. This requirement demands that all incoming A-scans must not only be added to the accumulator but also be saved in memory so they may be later subtracted. This is accomplished by keeping a FIFO queue of A-scans whose length is equal to the number of A-scans maintained in the accumulator. Because the accumulator has been shifted several times since the oldest A-scan was first added, the first location of the saved A-scan (A-scan(n)) must be subtracted from a several-times shifted location (A-scan(n+total_shift)). This location equals the number of A-scans averaged times the number of locations shifted per A-scan.

As each subtraction is completed, the result is compared to a threshold limit. However, the result does not yet represent an averaged point within the A-scan since it has not been divided by the number of A-scans present within the accumulator. To avoid the time consuming operation of division for every location, the threshold value is initially multiplied by the number of A-scans accumulated.

The above process of adding A-scans to a shifted accumulator, storing the A-scans in a FIFO queue and subtracting the oldest A-scan from the accumulator is repeated as new A-scans are obtained for processing.

The algorithm was developed and tested. To summarize, the steps in the algorithm are:

1) Read the A-scan data from the digitizer.

2) Shift the data in the accumulator buffer by one shift.

3) Add the new A-scan to the accumulator buffer.

4) Subtract the oldest A-scan from the accumulator buffer.

5) Search through the accumulator buffer for a peak above a threshold level that could signify a flaw.

6) Go to step 1.

Through efficient software design this algorithm executes the time shifted moving average algorithm fast enough to allow real time processing of the ultrasonic data even though a high level language was used. Figure 5 shows a curve of the effective "rep rate" - the number of A-scans processed per second, as a function of the A-scan buffer size. Using a spatial sampling (number of A-scans per inch) sufficient to detect the 0.020 inch diameter penny-shaped voids, the scan rate for a 0.5 inch thick bore is approximately 1.25 inches per second when using the time shifted moving average.

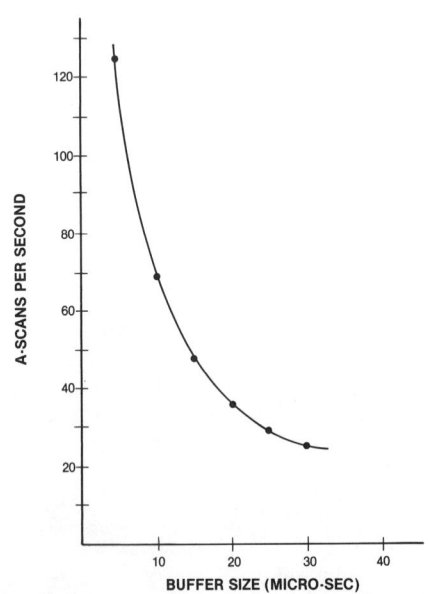

Figure 5. - The effective A-scan rep rate of the time shifted moving average as a function of buffer size.

An example of the power of the time shifted moving average is shown in Figures 6 and 7. Figure 6 shows four rectified, low-pass filtered (video) A-scans of a 0.032 inch diameter, mal-oriented, penny-shaped void 0.60 inches below the surface. A-scans number 2 and 3 also have a large water noise spike in them. The time shifted moving average algorithm was applied to these four A-scans and produced the averaged A-scan shown in Figure 7. The water noise signal amplitude has been reduced by about a factor of six whereas the void signal amplitude has not been reduced at all. This is a very significant reduction of the noise signal and occurs even though the noise was present in two of the four A-scans.

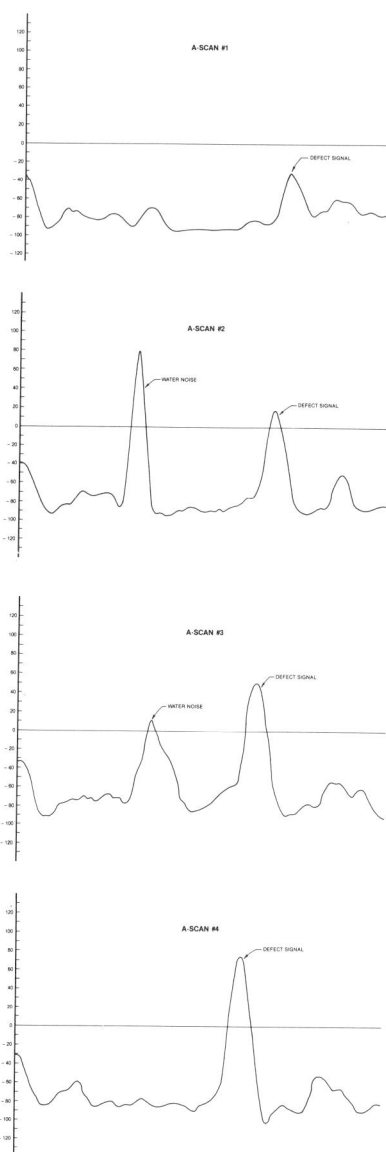

Figure 6. - These four photographs show video signals from a 0.032 inch diameter void 0.6 inches below the surface. Each photograph represents a 0.5 degree rotation of the part containing the void. Notice the large water noise spike in photographs 2 and 3.

837

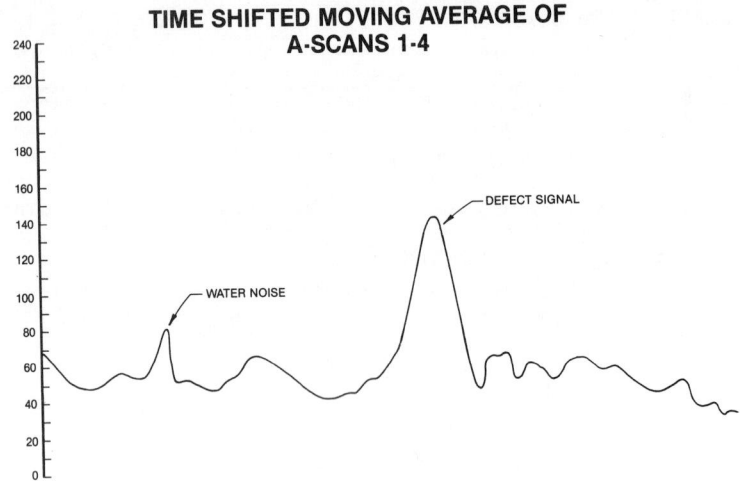

Figure 7. - The resultant ultrasonic signal after being processed through the time shifted moving average algorithm. Note the reduction of the noise signal.

SUMMARY

An algorithm that makes use of the temporal shifting of the ultrasonic signal from an actual defect as the defect moves past the transducer has been developed. This algorithm is very effective in reducing the amplitude of unwanted random noise signals while preserving the amplitude of actual defect signals. This algorithm has been implemented into the RFC ultrasonic inspection system and allows for the real time detection of worst case mal-oriented, penny-shaped voids in gas turbine engine components.

ACKNOWLEDGEMENT

This work was conducted under Air Force Contract F33615-81-C-5002.

REFERENCES

1. G. M. Light, R. A. Cervantes, and W. R. Van der Veer, "Review of Progress in Quantitative Nondestructive Evaluation", D. O. Thompson and D. E. Chimenti, editors, Plenum Press, New York (1984), 1359.
2. G. M. Light, W. R. Van der Veer, D. A. Stubbs, and W. C. Hoppe, "Review of Progress in Quantitative Nondestructive Evaluation", D. O. Thompson and D. E. Chimenti, editors, Plenum Press, New York (1986), 885.

AN UPDATE ON AUTOMATIC POSITIONING, INSPECTION, AND SIGNAL
PROCESSING TECHNIQUES IN THE RFC/NDE INSPECTION SYSTEM

Ray T. Ko, Wally C. Hoppe, David A. Stubbs,
Donald L. Birx, Bob Olding, and Gary Williams

NDE Systems Division
Systems Research Laboratories, Inc.
Dayton, OH 45440-4696

INTRODUCTION

This paper updates several techniques developed for the Retirement For Cause (RFC) Nondestructive Evaluation (NDE) Inspection System eddy current inspection module. Techniques to be discussed include:

1. Scallop Centering - development of an automatic scallop centering routine makes scallop inspections reliable.

2. Soft Survey Mode - improvements have been made for fast peak detection.

3. Method 2 Select Mode - a fine flaw detection technique based on the acquired waveform.

4. Antirotation Window Inspection - a frequency select mode has been established for detecting flaws in antirotation windows.

5. Scaling of Flaw Depth - a scaling factor has been developed, based on Phase I Reliability Test data, which converts flaw signal amplitude into estimated flaw depth.

SCALLOP CENTERING

Scallops are common geometries on gas turbine engine disks, with most disks having from 36 to 48 scallops of a given size. Physically, the scallops are shaped in a semi-circular arc with a certain angle between the two sharp corners and the geometric center of the region, as shown in Figure 1. In most cases, the scallop diameter is greater than 0.500-inch.

To center an eddy current probe in the scallop, the operator first must align the disk on the fixture such that the laser alignment line is at the bottom of one of the scallops. This assures that the initial probe placement will be within the 0.250-inch off-center tolerance required for the centering algorithm to work. Centering is actually performed 0.020-inch above the scallop region, using a rotary eddy current probe with a diameter 0.010-inch less than the scallop diameter. The probe must be centered

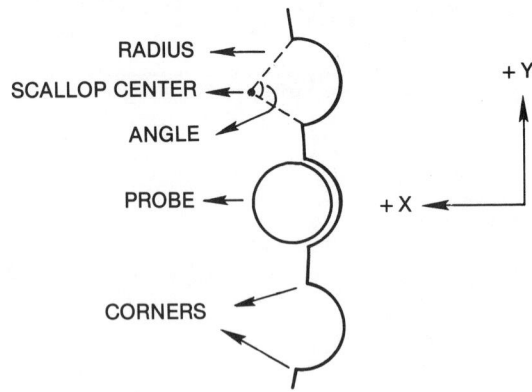

Fig. 1. Top view of scallop geometry

within 0.001-inch for a reliable inspection. Experiments showed that as the probe moves around the scallop region, a strong signal occurs when the probe is near a corner or edge (two strong signals if there are two corners). When the probe is at the geometric center of the scallop, an equal amplitude with a specific time delay between two corner signals is observed, as seen in Figure 2. To automatically center the probe above the scallop, and satisfy both of the centering criteria (equal amplitude and a specific time delay) the mechanical axis, amount of movement, and direction requirements must be provided to the mechanical controller. The automatic procedure that was developed contains four steps, each of which relates to a specific mechanical axis. The amount of movement and direction are deduced from a waveform, or two waveforms at different positions, as shown in Table 1.

The Y-zoning is simply a method of moving the probe from a single peak to a double peak region as shown in Figure 3. Each time the probe rotates through the +X axis a sync pulse is produced which is the trigger source for producing a waveform. Experiments showed that if the probe is far off-center in the +Y direction, but not farther than the probe radius, the upper edge (+Y) gives a strong signal that appears in the first-half section of the waveform. The lower edge (-Y) gives a signal in the second-half section of the waveform. Thus, the necessary direction of movement can be predicted based on the location of the signal peak in the waveform; a fixed movement along the Y axis is executed until the double peaks appear in both sections of the waveform.

Table 1. Scallop Centering Procedure

| Step | Procedure | Mechanical Motion |
|---|---|---|
| 1. | Y-zoning | Move probe into the double corner signal region |
| 2. | Y-centering | Move probe into the equal amplitude region |
| 3. | X-centering | Move probe into a specific time-delay region |
| 4. | Y-centering | Move probe back to the equal amplitude region |

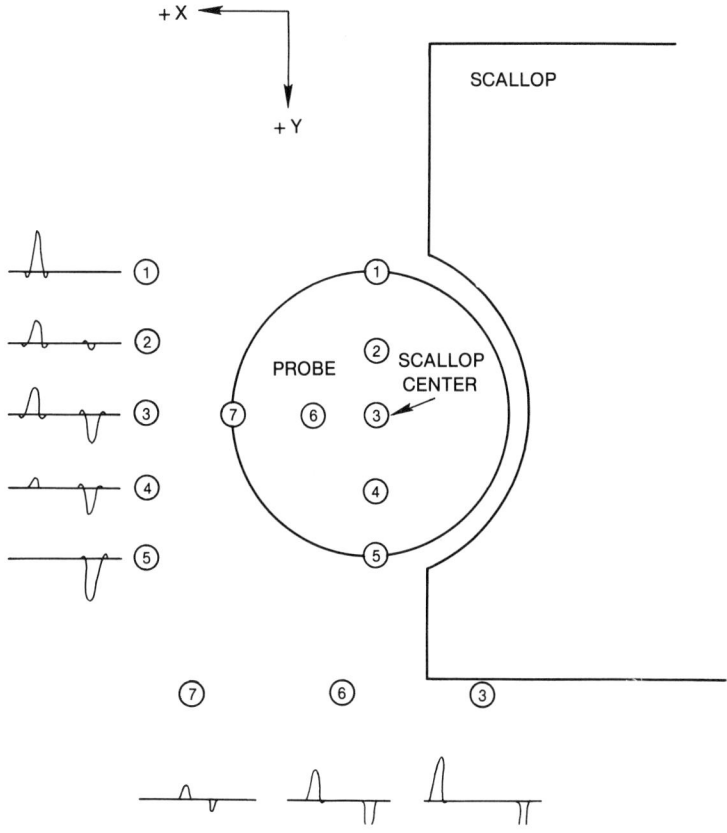

Fig. 2. Scallop signals along the machine X and Y axes

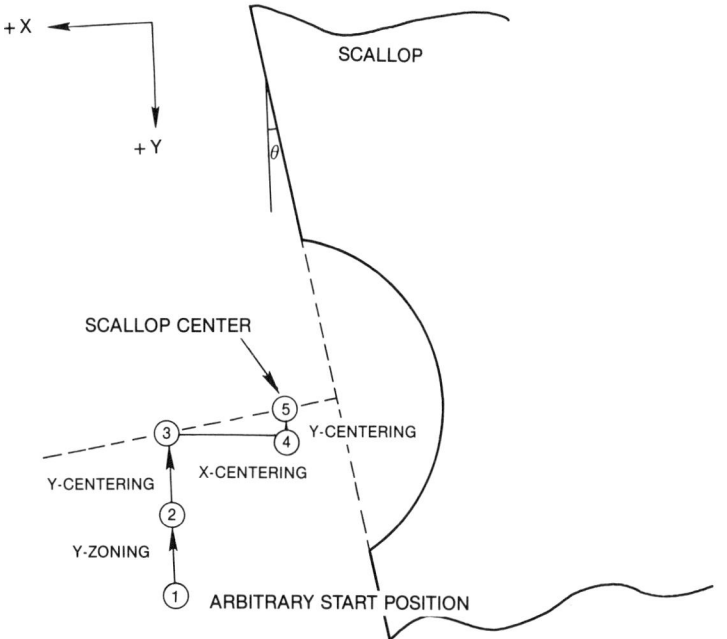

Fig. 3. Mechanical motions in scallop centering

The Y-centering uses the amplitude difference between the double peaks to predict the amount of movement and direction along the Y axis. The amount of movement is the product of current amplitude difference and a scaling factor. This factor is a ratio of the distance to the amplitude difference, and is calculated from two waveforms at different positions of known distance. The direction of movement along the Y axis (+ or -) is determined by the waveform section with the higher amplitude. The initial predicted movement usually places the probe in the equal amplitude region; if not, the Y-centering routine continues executing until the equal-amplitude criteria is satisfied.

The X-centering uses the time delay between double peaks to predict the amount of movement and direction along the X axis. Experiments showed that the time delay between the two peaks increases proportionally as the probe moves along the X axis (starting at the open end and moving toward the scallop). The scaling factor is deduced by comparing the time delay at two different positions of known distance. The amount of movement is the product of the difference between the current time delay and the specified time delay, and the scaling factor. The direction of movement depends upon the relationship of the current time delay to the specified time delay (lesser or greater). The specified time delay controls the position of the probe along the X axis, and is a constant for a given scallop size.

Upon completion of the three centering steps, the probe should remain at the centered position and the waveform should satisfy the centering criteria. However, experiments showed that after the X-centering was completed, the amplitude difference of the two peaks changed slightly. This is because the line between the two corners of the scallop is not necessarily parallel to the machine X axis, as seen in Figure 3. To achieve consistent centering, the Y-centering and X-centering must be executed repeatedly until the criteria are met. Usually, an additional Y-centering step is sufficient to achieve 0.001-inch precision along both axes.

SOFT SURVEY MODE

"Soft" survey is a new method for accomplishing the survey mode of inspection. In the old survey mode method, the sampler monitored the signal for threshold excursions. When an excursion occurred, the sampler sent an interrupt to the mechanical controller, which stored the seven axis positions at which the interrupt was sent. An unusually large amount of noise meant a correspondingly large number of positions being stored in the controller, which resulted in an "interrupt overflow" error condition.

In the new soft survey, raw data is read out of the sampler's memory into the inspection module computer during the scan. The computer performs simple signal processing in software, hence the name "soft" survey. The signal processing consists of a four-point digital filter which is centered on the center frequency expected from the flaws, and a simple peak correlation routine in which peak threshold excursions are found, and closely spaced peaks are combined into one peak. This processing reduces noise, and reduces the number of indications from a single flaw. In soft survey, threshold excursions do not generate interrupts. Instead, the computer keeps track of the location at which the peak occurs in the buffer, along with the buffer number. This is the equivalent of recording two axis positions (scan axis and index axis). At the end of each buffer of data, the sampler sends an end-of-buffer interrupt to the mechanical controller, which stores the related seven axis positions. The length of the scan is determined by the maximum number of positions the mechanical controller can store, which is approximately 300 for seven axis positions. The two survey modes are compared in Table 2.

Table 2. Comparison of Two Survey Modes

| | Item | Old Survey | "Soft" Survey |
|---|---|---|---|
| 1. | Interrupt Source | Threshold Excursion | End of Buffer |
| 2. | Number of positions stored | Seven axis positions for each excursion | Two "axis" positions |
| 3. | Correlation | After scan | During scan |
| 4. | Advantages | -- | Peak detection<br>Digital filter<br>Tolerant to noise<br>Tolerant to closely spaced indications |

Although a "peak excursion overflow" condition can occur, this should happen only in extremely noisy situations, and probably would indicate a problem in the scan, the disk, or the probe. Digital filtering makes the soft survey mode more tolerant to noise, and the correlation routine makes soft survey more tolerant to extremely rough or gouged surfaces which may produce many closely spaced indications. The soft survey is also very fast, allowing maximum scanning speeds to be used in the RFC/NDE Inspection System.

METHOD 2 SELECT MODE

In the Method 1 Select Mode, the inspection module computer monitored the entire waveform for threshold excursions at the suspected flaw area identified during the survey mode. If both the signal-to-noise ratio and threshold level were not high enough, false calls could be produced. To reliably detect small flaws, which usually give small signals, Method 2 Select Mode was established.

Method 2 is a window-threshold technique in which detection depends not only on the threshold, but also on the waveform shape. Experiments showed that when an EC probe (double-ended differential coil) scans across a flaw area, the resultant waveform has three common characteristics: 1) Zero crossing; 2) higher amplitude than average background; and 3) predictable slope (dependent upon each scan plan). Usually, the background does not satisfy conditions (2) and (3). Thus, when a buffer of data is read out of the sampler's memory into the computer, the Method 2 routine first looks through the entire waveform for the zero-crossing index. When the crossing is found, a maximum peak-to-peak amplitude searching routine is executed within a fixed time window which is centered about the zero crossing index. The maximum amplitude is then compared to the threshold to determine if it is a flaw or not. Method 2 is also very fast, allowing maximum scanning speeds in the RFC/NDE Inspection System. Method 1 and Method 2 are compared in Table 3.

Table 3. Comparison of Method 1 and Method 2 Select Modes

| | Item | Method 1 | Method 2 |
|---|---|---|---|
| 1. | Flaw determination criteria | Amplitude | Amplitude, zero crossing, and waveform slope |
| 2. | Advantages | -- | Tolerant to noise, detects small flaws |

Data from the Phase I Reliability Test, in which Method 2 was used, indicated a 90-95% confidence level for detecting surface flaws in the 0.005-inch to 0.010-inch range in flat plates (web/bore surface), boltholes, and rivet holes. For example, a 0.003-inch deep fatigue crack in a rivet hole was reliably detected using Method 2, with the threshold at 40 millivolts and background at 34 millivolts.

ANTIROTATION WINDOW INSPECTION

Antirotation windows (ARWs) are critical geometries on seals which are used to join two gas turbine engine disks together. The shape, number, and grouping of ARWs vary depending upon the particular seal, as shown in Figure 4. The edges of the ARWs are sources of strong geometry signals which often have a higher amplitude than flaw signals, thus making a conventional amplitude inspection technique very difficult, as shown in Figure 5. Therefore, a frequency select mode was developed which uses the frequency difference as a flaw determination criteria [1].

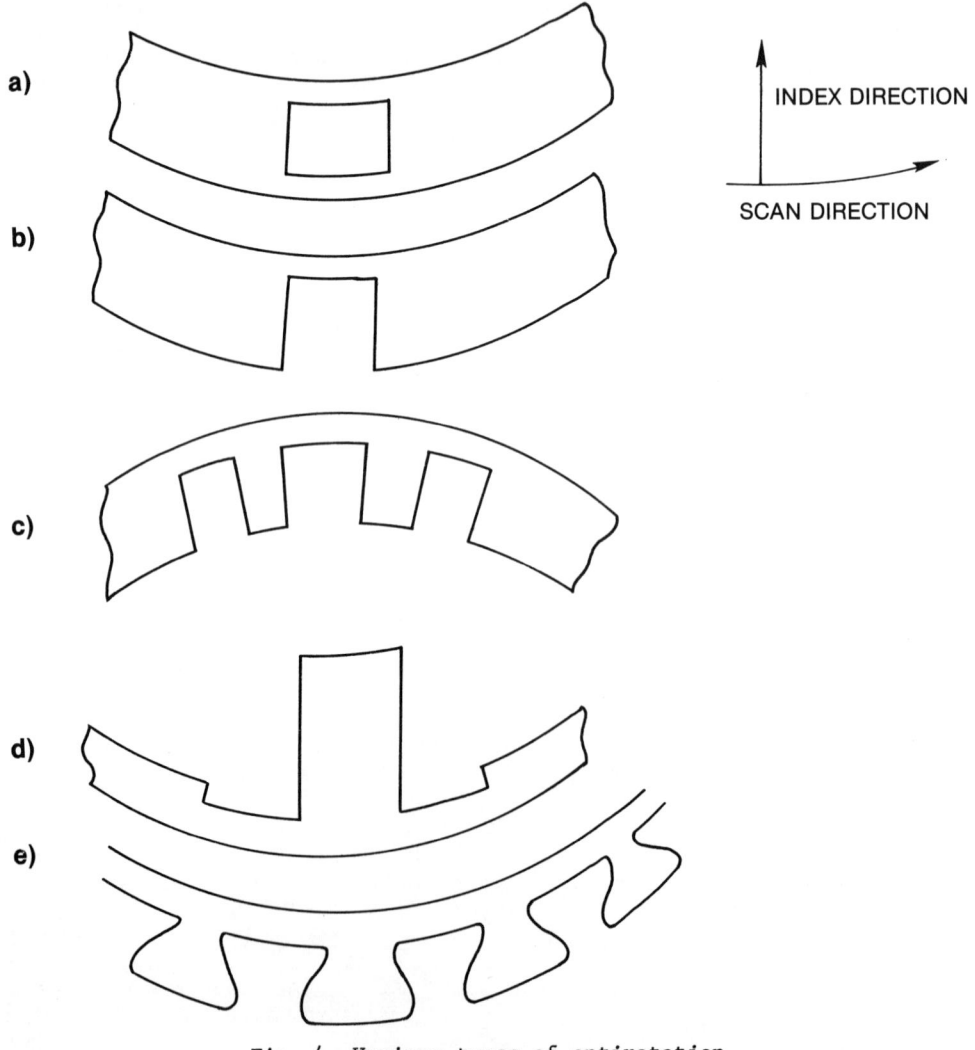

Fig. 4. Various types of antirotation windows, (a), (b), (c); antirotation tang (d); live rim (e)

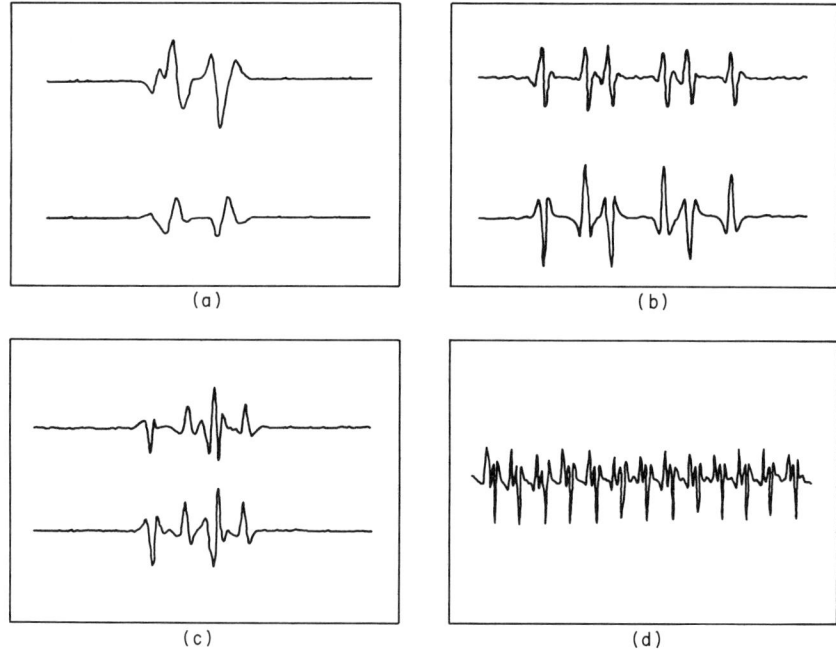

Fig. 5. Antirotation window signals: (a) top waveform is from vertical channel and indicates a flaw; bottom waveform is from horizontal channel; (b) shows waveform from different type of antirotation window. Antirotation tang signal (c) indicates a flaw. Live rim signal (d) shows vertical channel only.

Prior to applying the frequency select mode on an ARW, a dimensioning routine is executed to locate the position of the ARW edges. When the frequency select mode is executed, the probe scans across the critical ARW area, or the entire window if desired, and waveforms are collected from both vertical and horizontal channels. The waveforms are sorted into sections, with each section corresponding to a specific edge. A Fast Fourier Transform (FFT) routine then converts all the sections of the time-domain waveform into the frequency domain spectrum, and the mean frequency of each spectrum is calculated. The mean frequency difference between the channels is compared to the frequency difference threshold for flaw determination, as seen in Figure 6. Plots of the frequency difference vs. index direction positions are shown in Figure 7. Plot (a) shows a flaw-free window, plot (b) shows a flawed window, and plots (c) and (d) show flawed windows with the flaws at different positions along the index axis.

The frequency select mode has been applied to antirotation tang and live rim inspections, with encouraging results in both cases.

SCALING OF FLAW DEPTH

When a flaw is detected in the Method 2 Select Mode, the flaw size is recorded as length by amplitude. A study of the Phase I Reliability Test data revealed a linear relationship between the amplitude data and the flaw depth. However, a comparison of the amplitude-vs.-depth data plots for different specimen types (rivet hole, bolthole, flat plate) showed that the slopes differed by a magnitude of 4 for each specimen type.

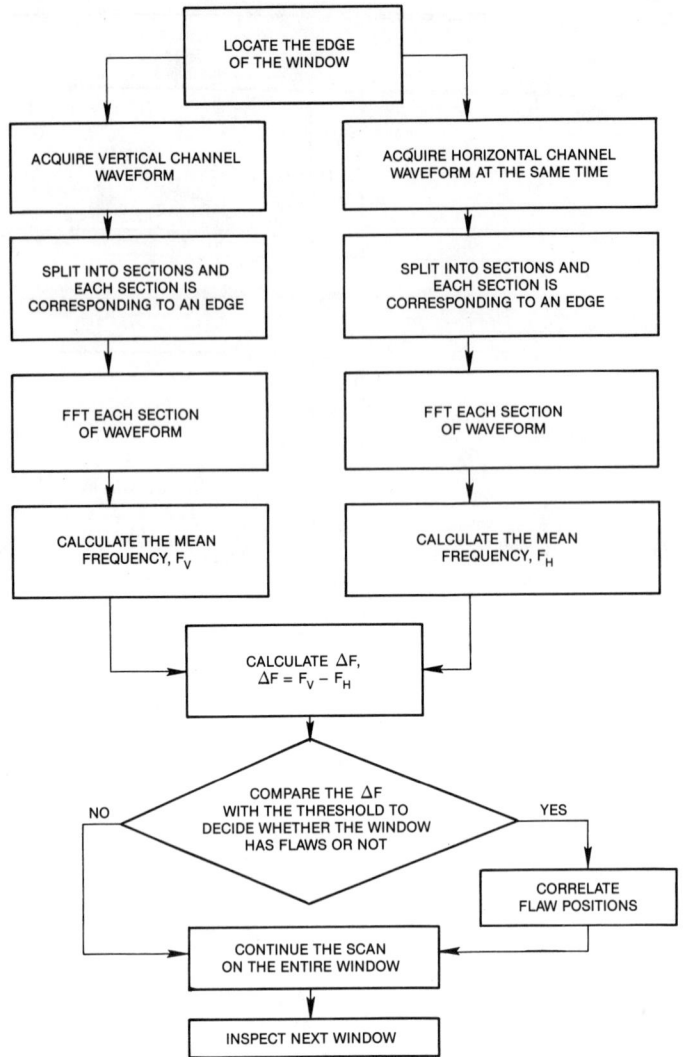

Fig. 6. Flow chart of frequency select mode

To rationalize the difference in slope, the specimen scan plans were studied in terms of gain and frequency factors. A higher gain setting results in higher amplitudes for flaws. The gain setting is controlled by the "desired gain" parameter during calibration, and the notch size in the calibration block. If the desired gain is higher, the calibrated gain will be higher and the detected flaw amplitude would be increased. However, if the notch size in the calibration block is larger, the signal from the notch is larger and, subsequently, a lower gain would be sufficient to calibrate the signal to the desired amplitude. Thus, a larger notch gives a lower gain and the detected flaw signal is lower.

The second factor is the frequency effect. The expected frequency of the flaw signal can be calculated by dividing the coil size by the scan speed. For the rotary probe, if the probe is rotated at a constant rate (say, 1500 rpm), the frequency of the signal depends on the probe diameter. For surface probes, since the scan axis is the turntable, the frequency of the flaw signal depends on the turntable rotation speed, and the distance

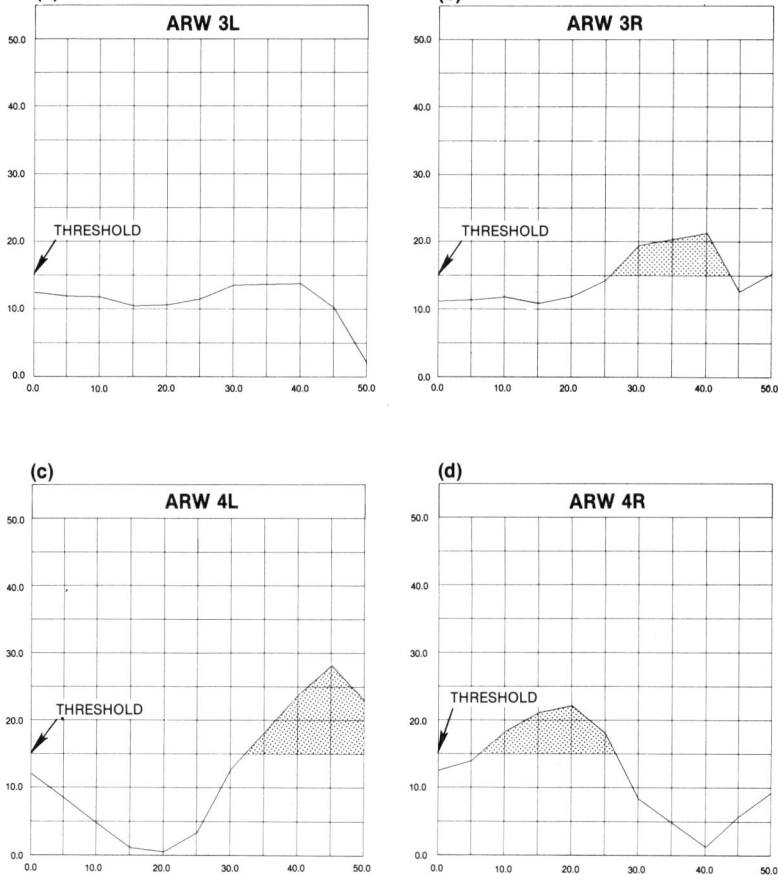

Fig. 7. Antirotation window inspection results, frequency difference vs. index distance: (a) flaw-free window at one side; (b) flawed window on the other side; (c) and (d) flawed window on both sides at different index positions.

from the probe to the center of the turntable. When the window size is smaller than the peak-to-peak time, the higher frequency will have higher amplitude.

Based on these factors, the flaw amplitude is normalized by the gain and frequency factors. This normalized value is called the scaled flaw depth. The gain factor is the "desired gain" parameter in the scan plan, and the frequency factors are window size (in Method 2 Select Mode), probe diameter or the distance from the probe to the center of the turntable, and probe rotary speed or turntable rotary speed. This general formula applies to rotary probes of any size and surface probes at any location. The proportional constant is deduced by making the slope of scaled-depth-to-flaw-depth equal to one from the rivet hole data. No scan plan changes are necessary to utilize this scaling factor.

Plots of the scaled depth vs. flaw depth for the three different specimen types are shown in Figure 8. The straight lines are the least-squares best-fit lines for data from each specimen. The slope is 1.03 for the

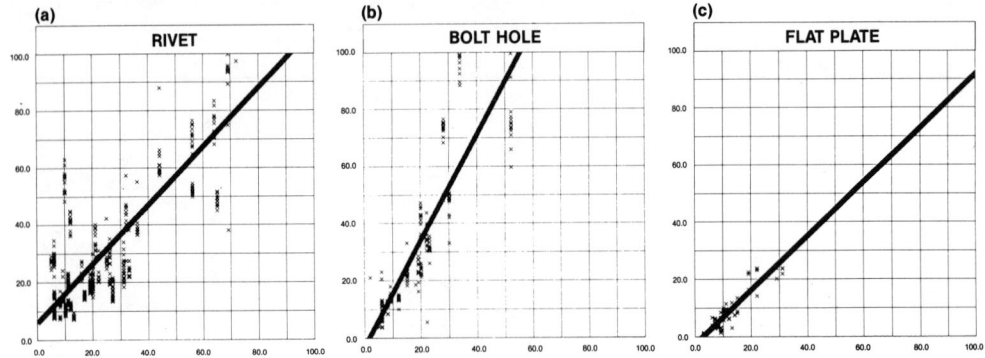

Fig. 8. Scaled depth vs. flaw depth: (a) rivet hole (b) bolthole and (c) flat plate

rivet hole data, and 0.94 for the flat plate data. However, the slope is 1.86 for the bolthole data. The current scaling of the flaw depth utilizes the amplitude of the flaw signal and is deconvoluted by the gain and frequency factors with a general normalization formula. The scaled depth of the flaw using this formula is within 100% of the flaw depth. Further development of this scaling factor, or sizing technique, will examine in more detail the effects of high pass and low pass filters on the flaw signals, and the effects associated with flaw location (i.e. at a chamfer, edge, or inside a hole). Additional sizing studies will be conducted using antirotation window data in which the frequency difference between both channels is used as a flaw indication.

SUMMARY

Improvements made in the automatic positioning, inspection, and signal processing algorithms in the RFC/NDE Inspection System have made possible the inspection of difficult geometries, including scallops and antirotation windows. Moreover, the flaw detection techniques implemented in the new survey and select modes have resulted in the reliable detection of small flaws. Finally, a scaling factor for estimating flaw depth has been formulated, marking the first step toward a precise sizing technique.

ACKNOWLEDGEMENTS

This work was conducted under USAF contract number F33615-81-C-5002 of the Air Force Wright Aeronautical Laboratories/Materials Laboratory.

REFERENCE

1. Wally Hoppe and Dave Stubbs, "Automatic Eddy Current Inspection of Antirotation Windows in F100 Engine Compressor Air Seals", in <u>Review of Progress in Quantitative Nondestructive Evaluation</u>, <u>4</u>, D. O. Thompson and D. E. Chimenti, eds., (Plenum Press, New York, 1985).

MULTIPARAMETER METHODS WITH PULSED EDDY CURRENTS

C. V. Dodd and W. E. Deeds[*]

Metals and Ceramics Division
Oak Ridge National Laboratory[†]
Oak Ridge, Tennessee 37831

ABSTRACT

Multiparameter methods have been used for a number of years to distinguish certain material properties from others that may be varying in the same eddy-current inspection. Usually the measured data are the magnitudes and phases of the eddy currents at several fixed frequencies. Alternatively, the necessary data can be obtained from pulsed eddy currents by measuring the pulse heights at various times or the times to reach various pulse heights. Such data can be used to analyze the pulse into various Fourier components, but that is time consuming and unnecessary. The raw data (for example, the pulse heights at various times) can be used as variables in polynomial approximations to the various properties in exactly the same way as has been used with the multifrequency, multiparameter method. This approach has several advantages, including simpler equipment, ability to use higher frequencies, and less modification required for different inspection problems.

INTRODUCTION

The greatest problem in eddy-current nondestructive evaluation is to distinguish particular sample properties from others that may be unimportant but are capable of strongly affecting the eddy-current readings. A popular approach is to use the extra information available from tests at several frequencies to eliminate the unwanted variables. Pulsed eddy currents can also give the additional information necessary to discriminate among various sample properties. An extensive bibliography of early work with pulsed eddy currents has been given by Libby.[1]

___

[*]Adjunct Research Participant from the University of Tennessee, Knoxville.

[†]Operated by Martin Marietta Energy Systems, Inc., for the U.S. Department of Energy under Contract DE-AC05-84OR21400.

EXTRACTION OF THE PULSE INFORMATION

The most common method of obtaining the necessary information from the pulse response has been to pass the coil output signal through a filter that can separate the pulse into various orthogonal function components. Alternatively, the pulse shape can be recorded and then analyzed with a Fast Fourier Transform (FFT) computer routine. The filter or FFT outputs can then be correlated with various sample properties or "mixed" to minimize the effects of unwanted variables.

The purpose of this paper is to show that it is not necessary to filter or FFT the pulse output before correlating it with the sample properties. Indeed, the additional steps involved in the filtering or FFT calculation can introduce additional noise in the data and decrease the accuracy of the correlation. The filtering method may also require extensive and expensive hardware changes for different inspection applications.

ORNL DIGITAL CORRELATION METHOD

A simpler approach is to use the digitized pulse shape directly for input data to the correlation process. For example, the pulse heights can be measured at various preset times, or the times to reach certain pulse heights can be measured. Figure 1 shows a plot of voltages measured at certain preselected times during a pulse, and Fig. 2 shows a block diagram of an instrument to measure the pulse heights at the preset times. The TRACK AND HOLD units are usually called sample and hold (S/H) modules, and each measures the pulse height at the time the correct signal comes from the computer. Figure 3 shows a plot of the times at which the pulse height reaches certain preselected voltages, and Fig. 4 shows a block diagram of an instrument to measure the times to the preselected pulse heights. Each voltage comparator stops a timer when the pulse voltage passes a computer-determined reference voltage.

Present S/H modules become less effective at frequencies above a few megahertz, as do filters in multifrequency equipment. Therefore, time-to-pulse height modules based on voltage comparators are more effective at frequencies higher than a few megahertz. On the other hand, if the lift-off becomes too great, no reading will be obtained from the voltage comparator, whereas the S/H circuit loses resolution only slowly.

Whichever method is used, the digitized data can then be used as the variables in a polynomial approximation to the desired property. The coefficients in the polynomial are determined so as to give the best fit (in a least-squares sense) to the given property. This is exactly the same process as has previously been used in the multifrequency, multiparameter method [2], except that pulse heights (or times-to-pulse heights) are used as measured variables instead of the magnitudes and phases at various fixed frequencies.

The pulse method has a number of advantages: the equipment is much simpler and cheaper, the changes required for different inspection problems can usually be made with simple software commands rather than hardware changes, and it is possible to work at much higher frequencies than those at which present equipment can make accurate phase measurements. This last advantage is particularly important for inspecting very poor conductors or very small specimens. For inspecting ferromagnetic

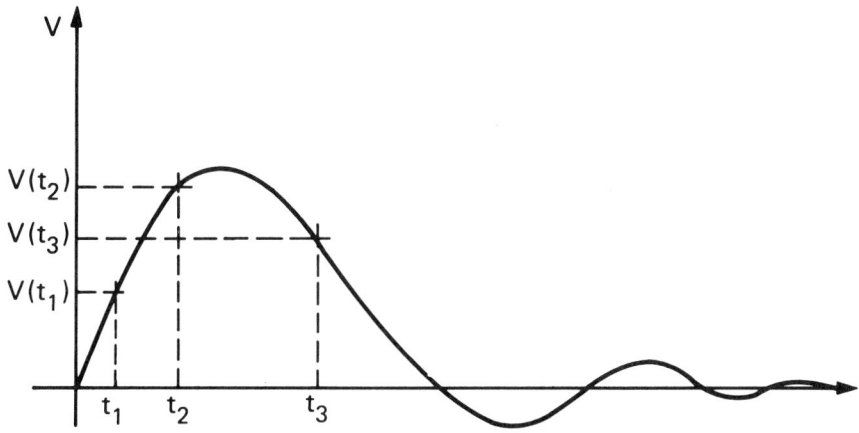

Fig. 1. Voltage at various times in a pulse.

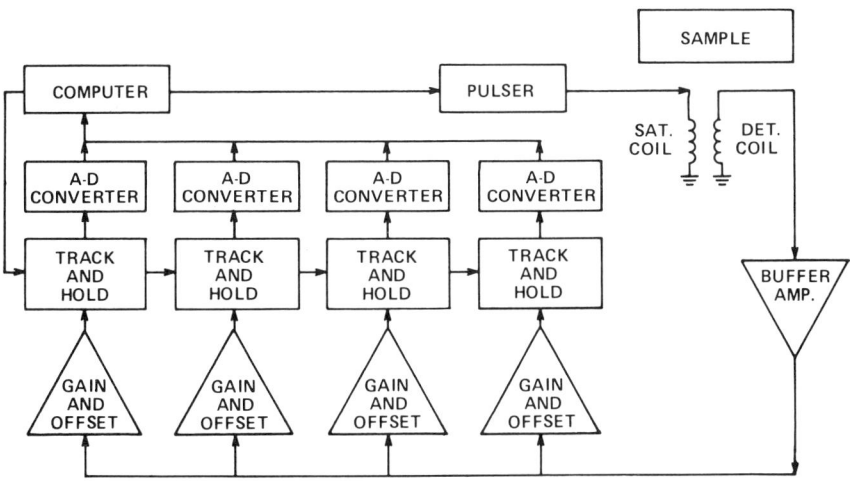

Fig. 2. Pulse amplitude instrument.

materials there is the additional advantage that a large driver pulse can be used to saturate the material as well as generate eddy currents for nondestructive evaluation.

In the past, pulse equipment has not been able to equal the accuracy of the best multifrequency equipment. However, recent electronic modules are capable of producing, measuring, and digitizing pulses with an accuracy, speed, and reproducibility that make the pulse method as accurate as the multifrequency method, while remaining much simpler. To change operating frequencies for different applications of a multi-frequency system usually requires extensive changes of hardware, such as oscillators, amplifiers, and filters, whereas any changes that might be needed with a pulsed system, such as times for measuring the pulse height, can usually be made with software instructions. Ordinarily, only the coil design needs to be optimized, and that is normally necessary for any eddy-current inspection.

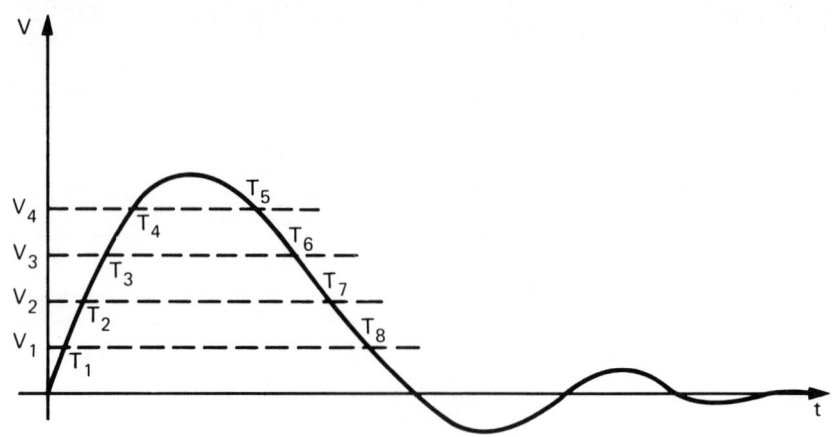

Fig. 3. Times to various voltages in a pulse.

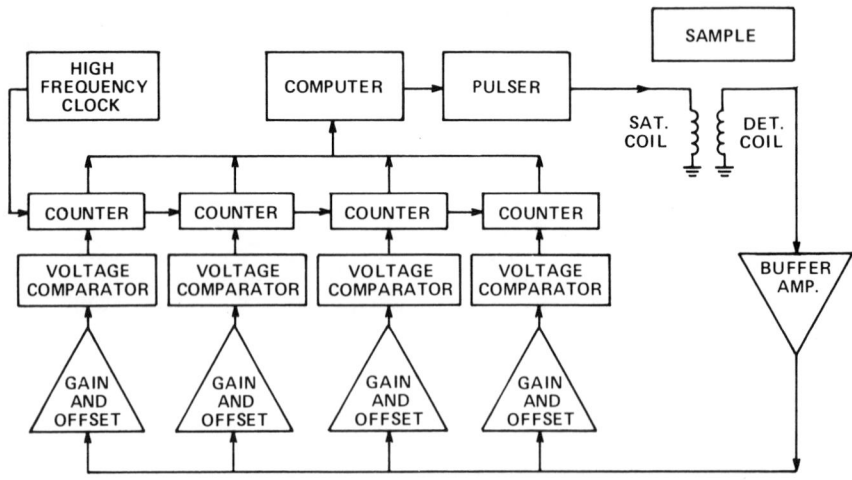

Fig. 4. Pulse time interval instrument.

APPLICATION TO THIN-WALL TUBING

At Oak Ridge National Laboratory we have used a pulsed system to make several inspections that would be difficult with conventional equipment. One inspection was for internal flaws in stainless steel tubing with an outside diameter of 3.56 mm (0.140 in.) and wall thickness of 0.13 mm (0.005 in.).

The optimum operating frequencies for the small, thin tubing were too high for our phase-sensitive detectors to give accurate readings, but a short duration pulse with a very small coil was able to make accurate and reliable measurements of flaws that were less than 10% of the wall thickness and located on the opposite side of the tube wall. Figure 5 shows scans of electrodischarge machined (EDM) notches with

depths of 25, 51, 8, 13, and 18 μm (0.001, 0.002, 0.0003, 0.0005, and
0.0007 in.) from left to right; the top trace is for flaws on the inside
of the tube, the lower trace for flaws on the outside, where the probe
was located. Note that the flaw depths are reliably indicated regardless of the location of the flaw, because the flaw size polynomial can
compensate for such extraneous variables. Figure 6 shows flaw readings
obtained in a scan across an area of the tube with EDM notch depths of
25, 51, 76, 76, 51, and 25 μm (0.001, 0.002, 0.003, 0.003, 0.002, and
0.001 in.) from left to right, the first three being on the near side
and the last three being on the far side of the 127-μm-thick (0.005-in.)
stainless steel tube wall.

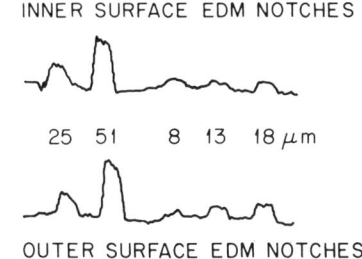

Fig. 5.  Scan of machined notches with depths of 25, 51, 8, 13, and
18 μm on the opposite side (upper trace) and near side (lower
trace) of a stainless steel wall 0.13 mm thick.

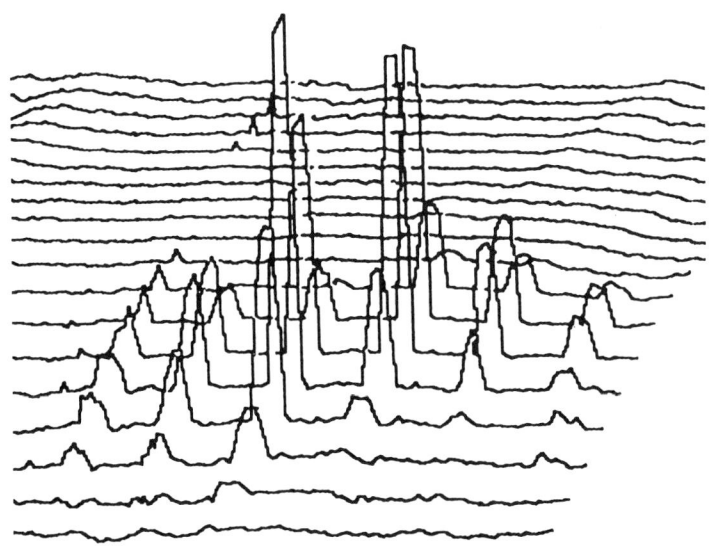

Fig. 6.  Flaw depth scan of stainless steel tube with 3.56-mm outside
diameter and 0.13-mm wall thickness. From left to right the
EDM notches have depths of 0.03, 0.05, and 0.08 mm on the
outside and 0.08, 0.05, and 0.03 mm on the boreside of the
tube wall.

The conventional reflection-type coil had opposing twin pickup coils with mean radii of 0.25 mm (0.010 in.) inside a driver coil of mean radius 0.50 mm (0.020 in.). The pulse rise and fall times were approximately 10 ns, and the maximum pulse rate was 10 MHz. Eight test readings could be taken per pulse, but it was found that four data points per pulse were sufficient to give very good defect sensitivity and lift-off rejection, if taken at the proper parts of the pulse. In fact, the percentage accuracy was at least as good as that obtainable with conventional multifrequency equipment measuring magnitudes and phases at three frequencies. Of course, the latter equipment could not even operate at frequencies high enough to be effective for such thin, small-diameter tubing.

REFERENCES

1. H. L. Libby, "Introduction to Electromagnetic Nondestructive Test Methods," Wiley-Interscience, New York (1971).
2. W. E. Deeds and C. V. Dodd, International Advances in Nondestructive Testing, 8, 317—33 (1981).

By acceptance of this article, the publisher or recipient acknowledges the U.S. Government's right to retain a nonexclusive, royalty-free license in and to any copyright covering the article.

AN EFFICIENT TECHNIQUE FOR STORING

EDDY CURRENT SIGNALS

S. S. Udpa

Electrical Engineering Department
Colorado State University
Fort Collins, Colorado 80523

INTRODUCTION

Preventive maintenance schemes often involve periodic inspection of plant and equipment in industry. Evaluation of the integrity of the plant at regular intervals provides data not only for the detection of defects in their incipient stages but also for monitoring the evolution in the growth of defects that were considered benign in previous inspections. Evolution in the nature of defects is tracked by comparing the signals obtained with test data obtained and recorded from previous tests. Consequently the procedure entails storage of a large amount of data during each inspection. As an example, the Nuclear Regulatory Commission mandates periodic inspection of steam generator tubing in nuclear power plants using eddy current techniques. The signals are recorded on a multichannel analog tape recorder with one of the tracks dedicated for storing the voice of the operator who records the details of the test conditions. A major disadvantage of this approach lies in the voluminous nature of the data necessitating the use of a large amount of storage media. In addition it is difficult to address and access specific data records.

An alternative method involves sampling the signal and recording the digitized signal. The method results in lower distortion if appropriate sampling frequencies and word lengths are used to represent the signal. In addition, it is relatively simple to access individual records if suitable headers are attached to them. However, the method shares the disadvantage of requiring significant storage media since no attempt to compress the information in the stored signal is made. This paper presents a simple technique capable of compressing the information contained in a signal significantly. Results demonstrating the potential of the method are also presented.

SIGNAL MAPPING AND COMPRESSION

The method described in this paper shares a philosophy that has been used to compress speech signals extensively [1]. The objective is accomplished in two stages as shown in Figure 1. The first stage involves mapping of the signal onto an appropriate parameter space using a suitable model. The parameters are coded efficiently to achieve additional compression and stored. Signals are reconstructed using the coded coefficients in the model.

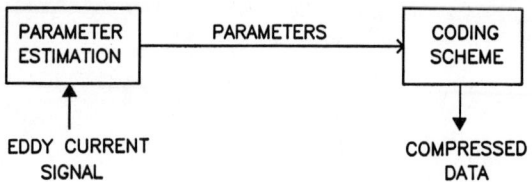

Fig. 1. Mapping and compression scheme for storing signals.

The degree of compression and the quality of reconstruction hinge on the appropriateness of the model relative to the signal. As an example, ultrasonic signals can be represented using Autoregressive models. Eddy current impedance plane trajectories can be modelled using Fourier descriptors or Circular Autoregressive models. In all these cases the process of mapping yields a compact representation containing a minimal amount of redundant information. This paper demonstrates the feasibility of using this general approach by examining a specific application involving eddy current signals and using the Fourier descriptor method for mapping.

FOURIER DESCRIPTORS

The use of Fourier descriptors for the representation and classification of impedance plane trajectories has been reported elsewhere [2,3]. However, for the sake of completeness, a brief description of the method follows.

Consider a closed curve as shown in Figure 2. If we define the contour function $u(\ell)$ as

$$u(\ell) = x(\ell) + jy(\ell) \tag{1}$$

where $(x(\ell), y(\ell))$ represent the coordinates of a point in the impedance plane $\ell$ arc length units away from an arbitrary starting point $P_0$, then

$$u(\ell) = u(\ell+L) \tag{2}$$

where L is the total length of the closed curve. Equation 2, which is valid since the curve is closed, indicates that the function $u(\ell)$ is periodic with period L. Consequently $u(\ell)$ can be expanded in a Fourier series

$$u(\ell) = \sum_{n=-\infty}^{\infty} C_n \exp(\frac{j2\pi n\ell}{L}) \tag{3}$$

where

$$C_n = \frac{1}{L} \int_0^L u(\ell) \exp(\frac{-j2\pi n\ell}{L}) d\ell \tag{4}$$

Compression is achieved by computing only a limited number of coefficients, $C_n$. These coefficients can be substituted in a finite version of the sum in equation (3) to resynthesize a curve which approximates the original curve closely.

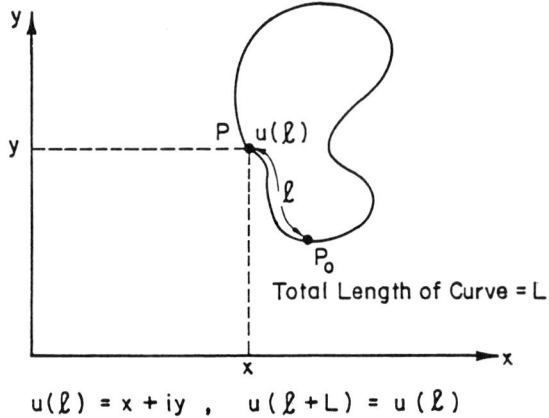

$$u(\ell) = x + iy, \quad u(\ell+L) = u(\ell)$$

Fig 2. Closed curve showing contour function $u(\ell)$.

$$u(\ell) \simeq \sum_{n=-M}^{M} C_n \exp(\frac{j2\pi n \ell}{L}). \quad (5)$$

The quality of reconstruction is directly related to the number of coefficients (2M+1) used in the resynthesis equation with the trade off being the degree of compression. Smaller values of M lead to a higher degree of compression at the expense of the quality of reconstruction.

The coefficients cannot be computed directly using Equation 4 since we have only a sampled version of $u(\ell)$. Instead the trajectory is approximated by a polygon of m sides with the vertices located at the sample points as shown in Figure 3 and the following expression used for calculating the coefficients [3,4].

$$c_n = \frac{L}{4\pi^2 n^2} \sum_{k=1}^{m} (b_{k-1}-b_k) \exp(\frac{-j2\pi n \ell_k}{L}), \quad n=0 \quad (6)$$

where

$$\ell_k = \sum_{i=1}^{k} |v_i - v_{i-1}|, \quad k>0, \quad \ell_0=0$$

and

$$b_i = \frac{v_{k+1}-v_k}{|v_{k+1}-v_k|}$$

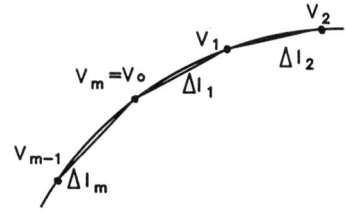

Fig. 3. Polygon of m sides approximating the curve.

## CODING

Additional compression is achieved by coding the Fourier coefficients efficiently. However, coding schemes using the conventional floating point format cannot be employed since the dynamic range of the coefficients is very large. This is due to the fact that the trajectories are very smooth and consequently the coefficients decay at a rate faster than $1/n^2$. Recognizing the large dynamic range, the coefficients are coded using the mantissa/exponent form with 3 bits reserved for the exponent. The number of bits used for the mantissa was then varied and the quality of reconstruction evaluated.

## RESULTS

In order to assess the performance of the approach, the method was first applied to a lemniscate of Bernoulli synthesized by using the expression

$$r^2 = a^2 \cos 2\theta \tag{7}$$

in a polar coordinate system.

Figure 4 shows the quality of reconstruction obtained by using progressively smaller number of bits in the mantissa. It is clear that the quality deteriorates rapidly when fewer than 8 coefficients are used. The method was then used to compress information contained in an eddy current signal obtained from a test arrangement consisting of an inconel tube with a through wall hole defect located at the center of a support plate as shown in Figure 5 with d=0.0. The trajectory obtained represents a composite signal consisting of the support plate and defect signals. Figure 6 shows the quality of reconstruction as a function of the number of bits in the mantissa. Figure 7 shows results obtained with a signal due to a through wall hole defect in an inconel tube. In all the cases it is seen that the performance is satisfactory when the number of bits used is 8 or larger.

The signal shown in Figure 6 was obtained by sampling each channel at the rate of 2000 samples per second using an A/D converter with 12 bit resolution. Fourier coefficients of the trajectory containing 300 points were then computed and coded. If we use 16 Fourier coefficients and employ 8 bits in the mantissa the total number of bits required is 456. This implies that a compression ratio of 15.8 has been achieved. Higher compression ratios can be achieved for simpler trajectories.

Figures 8 and 9 show results of attempts at comressing information by undersampling the original signal and using fewer bits. It is clear that the performance of the method proposed in this paper is superior. This is due to the fact, that in the process of mapping the signal to the parameter space most of the redundant information contained in the signal is eliminated.

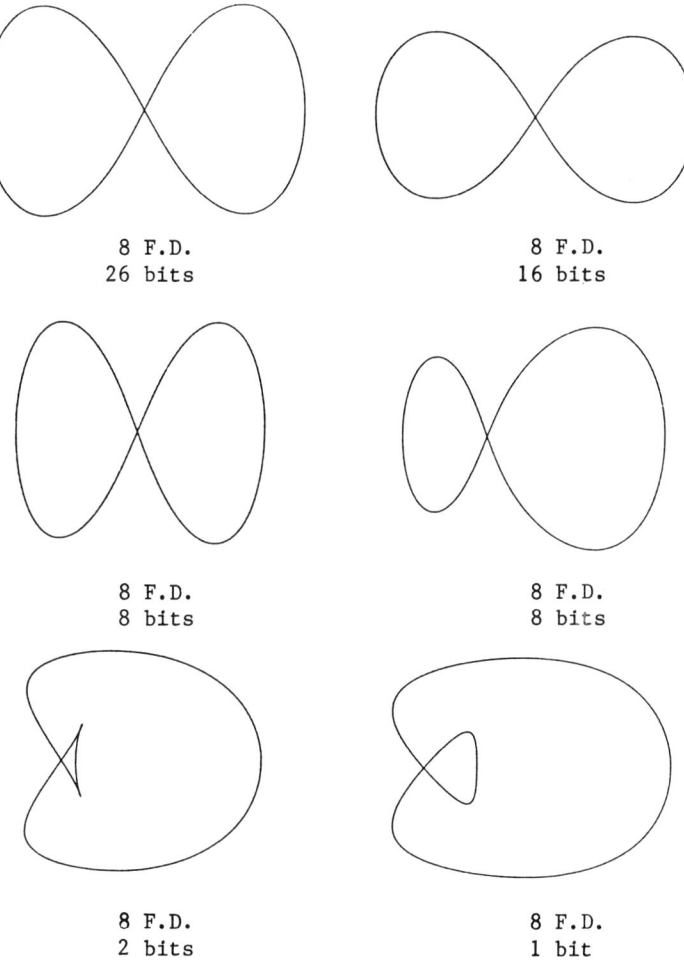

Fig. 4. Reconstructed signal obtained by using 8 Fourier coefficients and varying number of bits in the mantissa.

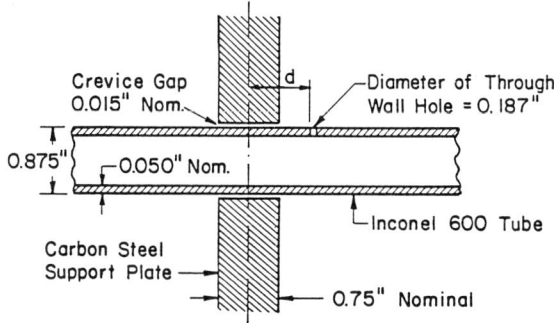

Fig. 5. Defect and support plate arrangement used to generate eddy current signal.

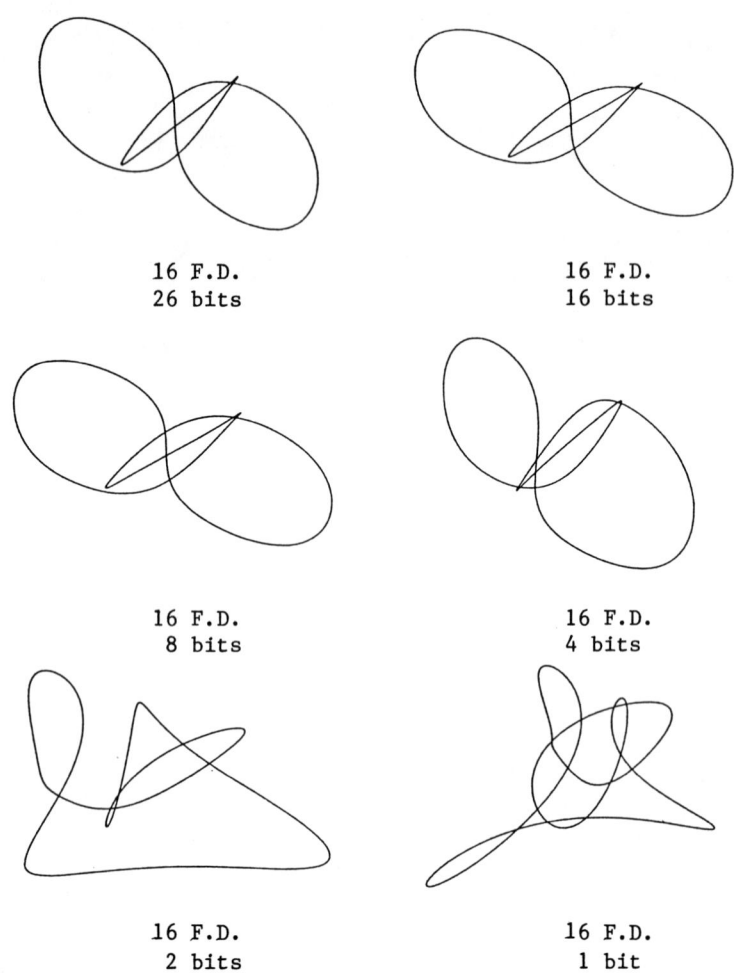

Fig. 6. Signals reconstructed using 16 Fourier coefficients and varying number of bits in the mantissa.

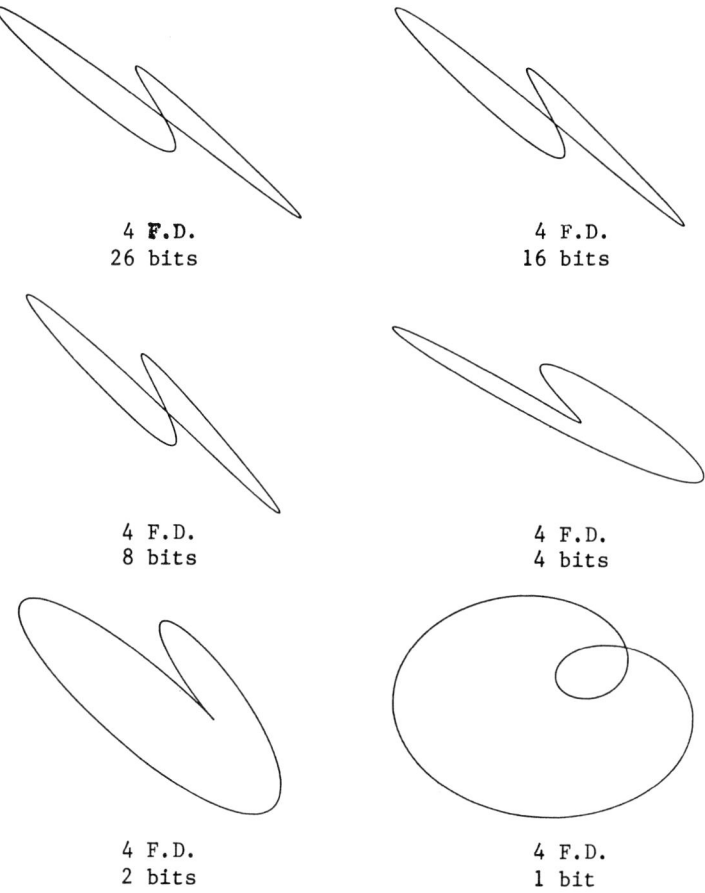

Fig. 7. Signals reconstructed using four coefficients and varying number of bits used in the mantissa.

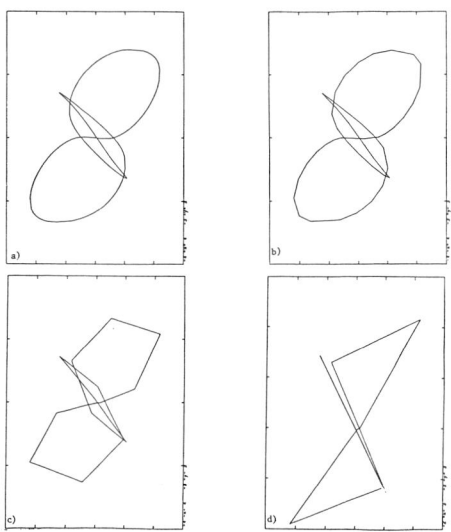

Fig. 8. Eddy current impedance plane track sampled at a) 2000 samples/ second, b) 660 samples/second, c) 300 samples/second and d) 150 samples/second.3

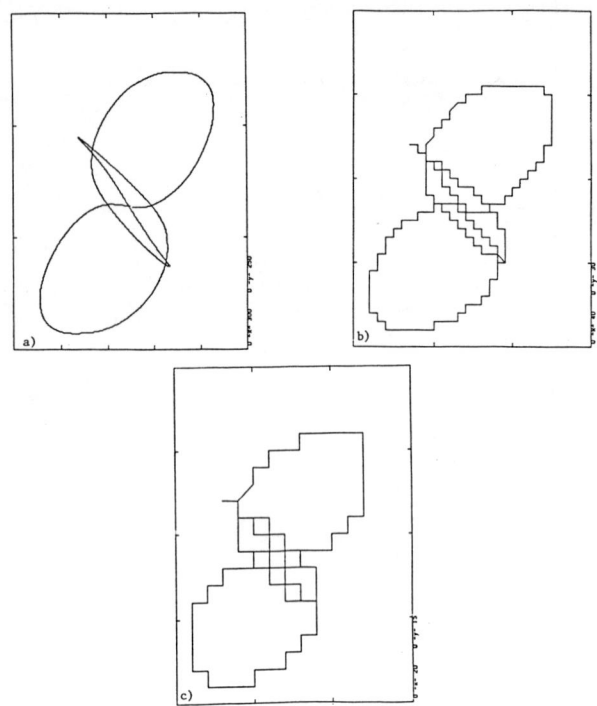

Fig. 9. Signal obtained by quantizing the impedance plane trajectory, using a) 8 bits, b) 5 bits, c) 4 bits.

CONCLUSIONS

A simple technique involving a two step procedure has been presented. Significant compression can be achieved using the method with little loss in quality resulting in considerable savings in the amount of storage media required. Access to individual records can be made easy by attaching headers to the record.

ACKNOWLEDGMENTS

The support of the Electric Power Research Institute for carrying out this work under Contract RP 2673-4 is gratefully acknowledged. Thanks are also due to Russell Bridgewater who performed most of the computational work.

REFERENCES

1. B. S. Atal and S. L. Hanauer, J. Acoustic Soc. of America, 50, 637 (1971).

2. S. S. Udpa and W. Lord, Materials Evaluation, 42, 1136 (1984).

3. W. Lord and S. S. Udpa, Fourier Descriptor Classification of Eddy Current Probe Signals, in "Review of Progress in Quantitative NDE," D. O. Thompson and D. E. Chimenti, Eds., Plenum Press, New York (1983).

4. E. Persoon and K. S. Fu, IEEE Trans. on Systems, Man and Cybernetics, SMC-7, No. 3 (1977).

IMAGE PROCESSING AND ARTIFICIAL INTELLIGENCE FOR DETECTION AND

INTERPRETATION OF ULTRASONIC TEST SIGNALS*

Keith S. Pickens, John C. Lusth, Pamela K. Fink,
Karol K. Palmer, Ernest A. Franke

Southwest Research Institute
San Antonio, Texas

INTRODUCTION

Detection of flaws is an important industrial concern. For example, aircraft and nuclear-power reactor owners and regulatory authorities need effective means of detecting flaws that could pose a threat to public safety. Operators of costly equipment require information on service-induced flaws to be able to make run-or-retire decisions. As the cost of parts and concerns for public safety increase, the importance of flaw detection and size estimation has likewise escalated.

Ultrasonic nondestructive evaluation (NDE) is one of the primary tools in the inspection for flaws (such as voids, nonmetallic inclusions, and cracks). Because ultrasonic testing (UT) uses acoustic waves for detection, it can inspect the interior of a thick material and reach inaccessible surfaces. Detection of a flaw, however, is only one step in the process. Flaw sizing is equally important and becoming more so with the increasing concern about lifetime prediction.

Current UT technology does not provide adequate sizing information in all cases. At Southwest Research Institute (SwRI), this problem was addressed by the development of the patented SLIC (shear and longitudinal waves to inspect for cracks) multibeam technology, which greatly extends ultrasonic inspection accuracy [1-4].

The importance with which the development of the SLIC technology must be viewed is evidenced by the recently announced result of an international round-robin test. The round robin was conducted to measure the efficiency of organizations and NDE methods for correctly detecting and characterizing flaws contained in test blocks representing nuclear-power reactor pressure-vessel components. These test blocks were examined by more than thirty of the world's foremost organizations who individually applied their own best technology and efforts. The results speak for themselves. When practiced by an SwRI expert, the SLIC technology outperformed every other technique and international organization.

---

*Funded by the Southwest Research Institute Internal Research Program

The limiting factor in the wide application of SLIC technology to industrial problems is, ironically, the large amount of information it generates. An analyst requires a high level of expertise to understand and interpret a SLIC module signal. The lack of skilled analysts has limited the application of SLIC technology to the many problems for which it can provide a solution.

OVERVIEW

The technical approach to the recognition and interpretation of the SLIC multibeam-multipulse signals involves three major subject elements-- UT data acquisition, image analysis, and analysis of image features by an expert system. These elements separately represent known and established technology within their limited domain. The strength of this system lies in the synthesis of these three different fields to form an interdisciplinary solution to a difficult problem.

One of the most productive areas of work in artificial intelligence (AI) has involved the production of programs with expert-level performance. These systems operate in their limited domains by incorporating the knowledge of a human expert, usually in the form of decision rules. Rules embody the relationships that a human expert uses in a formal statement of his knowledge. Thus, an expert system provides access to this specific expertise in a consistent and reproducible manner and on a much larger scale.

The expert system is provided with a description of the key features extracted from ultrasonic signals received during an examination. To facilitate extraction of key features, the examination data are organized as an image. The UT signals do not form an image in the classic sense, but it is possible to treat them as a two-dimensional image by using time as one axis. Forming an "image" allows the powerful and well-developed tools of image processing to be applied to extract a description of the signals for analysis by the expert system.

The acquisition of UT signals to form an "image" (see Figure 1) is the basis upon which the image processing and expert system rest. Computer-based data-acquisition systems allow the necessary large-scale signal acquisition. The data acquisition system technology used with the expert system is described in another paper in these proceedings [5].

Image Enhancement

Image processing [6-7], used to extract a description of the signals, reduces the amount of data to be analyzed by the expert system by two orders of magnitude. This reduction greatly increases overall system performance. The image processing itself involves three major steps. First, noise from the image formed by the SLIC signals is filtered out which enhances clarity of the image. Second, the characteristics that define the features of interest are accentuated. Third, these features are analyzed to determine the characteristics that describe the SLIC signals. The characteristics thus determined are then passed to the expert system for analysis.

Noise Reduction

The first step in processing the image is noise reduction. This is accomplished in the system by convolution with two digital filters. The digital filters take advantage of the fact that design of the SLIC module and the accompanying data acquisition process result in signals that form

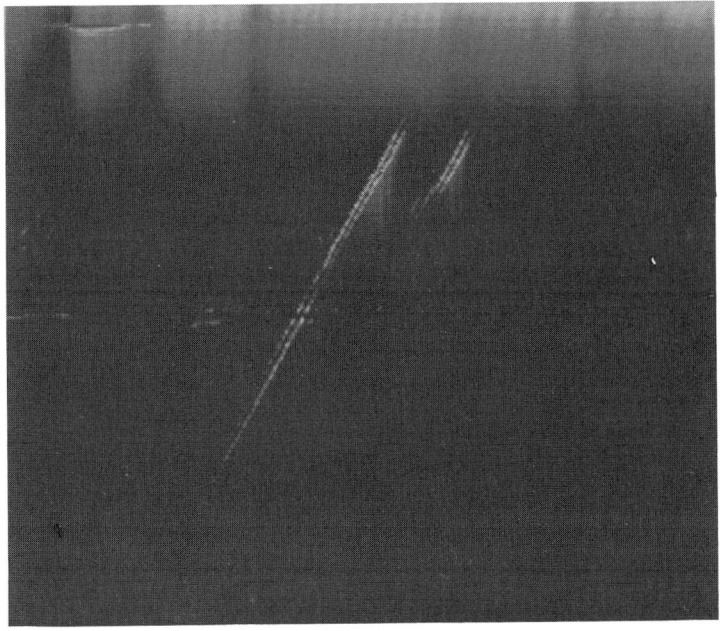

Fig. 1. Image formed by a SLIC-50 transducer before processing

lines of known slope. One filter emphasizes lines with positive slopes; the other filter emphasizes lines with negative slopes; and both filters reduce the background noise. Since different filters are used for the positive and negative slope lines, the image-processing module tries both and selects the one that produces the "brightest" image.

Conversion to a Binary Image

To identify the lines, each pixel in the image is placed in one of two groups. One group contains pixels within the line; the other contains pixels not in a line. This binary image can then be used to build a set of parameters for the lines. To convert the enhanced image to binary, a threshold must be selected. A histogram of pixel intensities shows a large number of background pixels with low intensity and a smaller number of brighter pixels that form the enhanced lines (see Figure 2).

A threshold value must be selected that will suppress the background while leaving the line pixels. If the threshold selected is too low, the binary image will contain extraneous (noise) pixels that can distort the features, introduce extra features, and increase processing time. If too high a threshold is used, portions of lines or some entire lines may be suppressed. The program determines the threshold based on an empirically determined number of bright pixels. By basing the threshold on the number of pixels rather than a given level, good results can be obtained on a wide range of images.

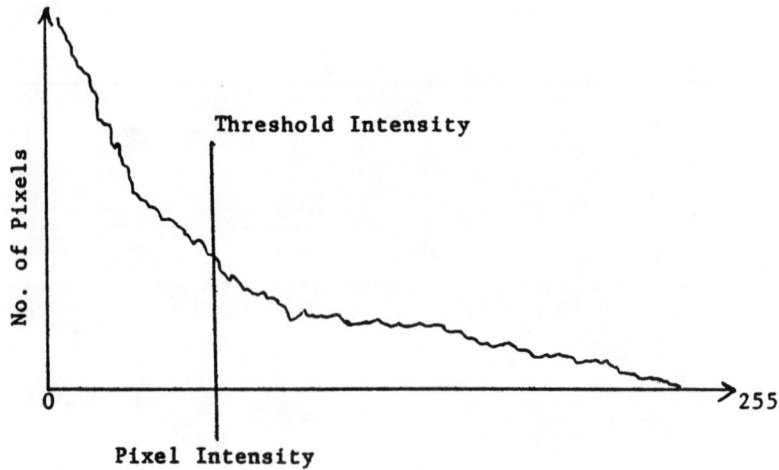

Fig. 2. Typical histogram of pixel intensities after digital filtering

Blob Labeling

After a binary image is obtained, the next step is to group connected pixels into discrete groups, referred to as "blobs." This is done by scanning the binary image pixel by pixel and constructing a new image in which each pixel is associated with a unique blob.

When a binary image point is found, the adjacent points that have already been scanned are examined to determine if any of them belong to a blob. If so, the pixel under consideration is labeled with that same blob identifier. The scan pattern is from top to bottom and left to right for images with positive slope lines on the monitor, and from right to left for negative slope images. This scan pattern was selected to reduce the possibility of fragmenting lines into multiple blobs.

During blob labeling, the number of pixels in each blob is counted. When labeling is completed, small blobs with fewer than four pixels are discarded.

Feature Extraction

The last section of the image processing module scans the labeled blob image and accumulates data to compute parameters identifying lines.

After these data are accumulated, final line parameters are computed by least squares regression. A file is then written to disk containing (for each line):

- length of line
- slope of line
- x, y coordinates of the line start point and its intensity
- x, y coordinates of the line end point and its intensity
- x, y coordinates of the maximum intensity.

Figure 3 shows the computed lines overlaid on a SLIC image.

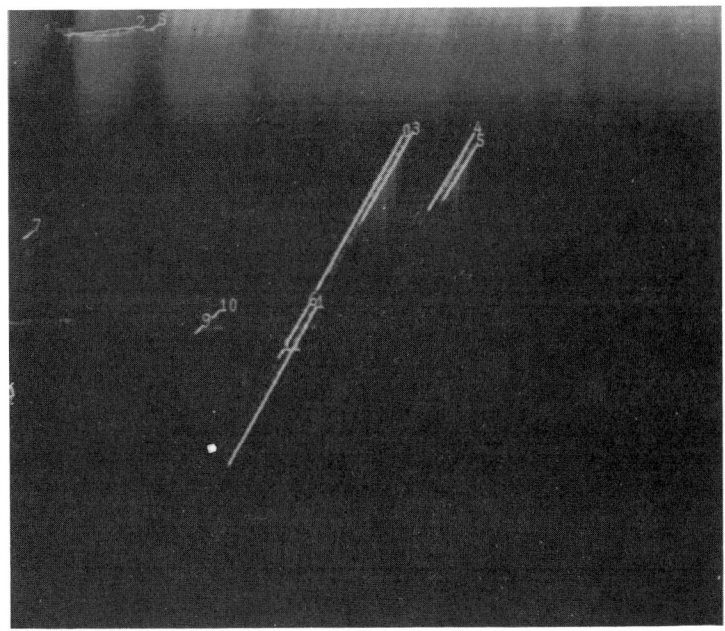

Fig. 3. Lines identified by the image-processing module. Fig. 1 shows the same image before processing.

Artificial Intelligence Module

The artificial intelligence [8-10] module interprets the image parameters produced by the image-processing module. From the image parameters received, this module must distinguish between data which represents actual artifacts and data which is the result of interference or noise. The analysis is implemented in three steps: line merging, artifact identification, and crack depth computation.

Prior to implementation of these modules, the merging process was mainly considered a trivial task requiring no expertise. The expertise, however, that was believed to lie in the merging process (which seems simple, if not intuitive, to a human) was found to be more complex than expected. Findings such as this are not unusual when attempting to automate an intuitive process.

The line merging program detects and corrects two types of problems existing in the data produced by the image-processing module. First, the method used by the image-processing module tends to break line segments into two or even more segments. These segments must be reconnected in order to achieve a proper representation of the data resulting from the SLIC test. Second, the input data sometimes consisted of under-filtered video signals; a single signal appears to be multiple, closely spaced signals. These signals must be blended into the meaningful line segments. The merging program uses two sets of rules to determine if either of these situations exists.

Line segments that actually belong to the same original line tend to have certain attributes in common. For example, the endpoint of one line segment will tend to be close to the endpoint of the adjacent segment, the orientation of the two-line segments will tend to be the same (i.e., the line segments will be close to parallel), and intensity of the points along the two-line segments should produce a smooth change from low intensity to high intensity and back to low intensity.

Thus, three rules are used to determine whether line segments have been broken apart. First, the endpoints of two-line segments must lie within a certain distance (d1) of each other. Second, both endpoints of the shorter line segment must lie within a certain distance (d2) of an extension of the longer line segment. Both distances d1 and d2 are currently constant values which were empirically determined. In future implementations, these distances may be adaptively determined.

The third rule used to determine whether line segments have been broken apart involves the intensities of the various points on the line segments. When graphed, intensities along a typical line segment would form a curve with a single peak. Therefore, the intensities along the line segment that would result from the merge must also form a curve with a single peak.

Under-filtered video signals appear in the image as line segments which are parallel to and lie very close to another line segment. The merging routine uses two rules for identification of echoes. First, the two-line segments must overlap. Second, both endpoints of the shorter line segment must lie within a certain distance of an extension of the longer line segment.

The merge program tests each line segment against every shorter line segment for broken segments and echoes. Segments are tested beginning with the longest and working in descending order by length. This process is reiterated until no more broken segments or echoes can be identified.

For both broken segments and echoes, the line segments are merged by extending the longer line segment to the projection of the far endpoint of the shorter segment. The shorter segment is then deleted. The points of greatest intensity of the two-line segments are compared, and the one with the larger value is preserved.

From the line segments resulting from the merging program, a third program finds pairs of line segments that meet the criteria for a flaw. Currently, the criteria for cracks detected by the SLIC-50 method with their origin at the cladding-to-base metal interface are implemented. The criteria for other SLIC-50 cracks and for SLIC-40 cracks will be implemented in the future.

There are two rules currently implemented for determining whether a pair of line segments have the characteristics of a crack detected by the SLIC-50 method. The first rule dictates that the slopes of the two segments be nearly equal. The second rule involves a check against the known geometry of cracks within the clad sample. The depth of a crack can be measured in two ways. The first method is simply to measure the distance between the longer line segment and the maximum intensity point on the shorter line segment. The second method relates the location of the cladding-to-base metal interface to the location of the shorter line segment in the image. This assumes the origin of the crack is at the cladding-to-base metal interface. These two rules are applied to the line segments in descending order starting with the longest segment.

In an early implementation, the first pair of segments found that had the characteristics of a crack was determined to be the actual crack. This method is adequate for images in which there is very little noise and the crack is the most significant feature present. However, in more realistic scenarios, varying amounts of noise as well as spurious artifacts, such as a weld prep, may appear in the image. Therefore, all pairs having the characteristics of a crack are found. The "best" crack is then chosen from among these based on a set of rules that distinguish a true crack from other artifacts. Currently, two rules have been implemented for this set.

The first of these rules is based on the observation that for SLIC-50 images the "top" line segment of the pair of line segments representing the crack is consistently longer and brighter than any line segment representing noise or interference. Therefore, the longest and brightest line segment is initially assumed to be the "top" line segment of a crack. If no matching "bottom" line segment exits for this line, it is disqualified as the "top" line segment of a crack.

The second rule of this set is the last rule to be executed. It selects, from the set of all line-segment pairs that pass the criteria of all the previous rules, the pair which represents the largest crack. This rule's success lies in the fact that the amount of noise present in an image is empirically found to decreases with depth. Therefore, line segments found deeper in the image are more likely to represent a crack than to be the result of noise or interference.

Finally, if a crack is present, a third program determines the depth of the crack. The depth is calculated by measuring the distance between the first line segment and the maximum intensity point of the second segment. This distance is then converted from pixels to millimeters by a constant scaling factor determined through a calibration.

To date, this system has been tested against two sets of images. The first set has of five scenarios. Each scenario consists of a flaw and the corresponding image or images containing relatively clear images of both line segments constituting the crack. These scenarios also have some noise. The system successfully identifies and measures the crack in all five of these scenarios.

The second set also consists of five scenarios. These have much fainter images of the line segments constituting the cracks. The "bottom" lines of the pairs are especially faint, in some cases less bright than the surrounding noise. They also contain more noise than the first set. The system successfully identifies and measures the cracks in three of these five scenarios.

For the other scenarios where the system does not correctly identify the crack, it chooses a "bottom" line segment that is actually noise. In both cases the reported crack depth was greater than that of the actual crack. The first of these scenarios represents the case where a small crack is present. The correct "bottom" line segment was never eliminated from the list of segments having the characteristics of a crack. The noise is chosen over the correct segment only because it lies deeper in the image. In this case, it is believed that a rule that addresses the intensities of the second segments along with their position in the image may resolve this problem. In the second of these two scenarios, the "bottom" segment of the line pair lies among a large cluster of line segments caused by the signals received from the edge of the sample. It is suspected that in this case it would be very difficult for even the human expert to identify the crack. Further consultation with the expert is planned for this scenario.

CONCLUSION

The combination of image processing techniques with an expert system has thus far proven to be extremely successful. "Real world" signals with both noise and artifacts have been handled. At this writing, the expert system uses only a few rules, yet it can deal with the entire class of SLIC-50 signals. Extension to other members of the SLIC family and to conventional UT can be accomplished by the addition of new rules to the expert system.

REFERENCES

1. "Ultrasonic Satellite-Pulse (Observation) Technique for Characterizing Defects of Arbitrary Shape," U.S. Patent No. 4299128, November 1981.
2. "Ultrasonic Multibeam Inspection Technique for Detecting Cracks in Bimetallic or Coarse-Grained Materials," U.S. Patent No. 4435984, March 1984.
3. "Multibeam Satellite-Pulse Observation Technique for Characterizing Cracks in Bimetallic or Coarse-Grained Components," U.S. Patent Application Serial No. 588898.
4. "Detection of Surface Cracks in Bimetallic Structures - A Reliability Evaluation of Six Ultrasonic Techniques," published in the Conference Proceedings of DARPA/AF Review of Progress in Quantitative NDE, University of Colorado, Boulder, Colorado, August 2-7, 1981.
5. H. L. Grothues, D. R. Hamlin, R. H. Peterson, and Keith S. Pickens, "A Highly Interactive System for Processing Large Volumes of Ultrasonic Testing Data," to be published.
6. Ramakant Nevatia, <u>Machine Perception</u>, New Jersey, Prentice-Hall, Inc., 1982.
7. Dana H. Ballard and Christopher M. Brown, <u>Computer Vision</u>, New Jersey, Prentice-Hall, Inc., 1982.
8. Avron Barr and Edward Feigenbaum (eds), <u>The Handbook of Artificial Intelligence</u>, Vols. 1-3, Los Altos, California, William Kaufmann, Inc., 1981.
9. Nils Nilsson, <u>Principles of Artificial Intelligence</u>, Palo Alto, California, Tioga Publishing Co., 1980.
10. Patrick Henry Loinston, <u>Artificial Intelligence</u> (2nd ed.), Reading, Massachusetts, Addision-Wesley, 1984.

# AN EXPERT SYSTEM FOR ULTRASONIC MATERIALS CHARACTERIZATION AND NDE

Ming-Shong Lan and Richard K. Elsley

Rockwell International Science Center
P.O.Box 1085, Thousand Oaks, CA 91360

## INTRODUCTION

Science, engineering, and manufacturing all depend on accurate measurements. These measurements can be made either by a human or by an automated system. In both NDE and materials characterization, there are numerous evaluations and decisions which must be made based on the experience and judgment of an operator or engineer. Current automated systems are not capable of making these judgments. Instead, typically, the operator or engineer evaluates the results after the measurements have been made. Expert systems provide a method for building the expertise of the human into the measurement apparatus, thereby causing all decisions made during the measurement process to be made with the skill of expert operators.

A human operator can have expert level skill or less than expert skill. A number of differences in approach and performance can be observed between an expert human, a less than expert human, and conventional automated measurement systems. Table 1 lists a number of steps that a measurement process can include and whether or not these steps are typically performed by each of the three types of operators. All measurement systems must perform at least steps 1 and 2. Most automated measurement systems do not go much beyond these two. Human operators usually add a number of steps that evaluate the validity of the data and results. The difference between an expert and a nonexpert is whether and how well he performs these steps. Finally, only the expert will decide what measurements ought to be performed or will discover new methods.

Expert systems technology provides a means of implementing, in an automated system, the qualitative and judgmental reasoning used by experts and has been widely applied[1] to problems that do not involve direct processing of sensor data. In addition, a growing number of expert systems are addressing the interpretation of sensor data.[2,3]

The objective of this project is to develop an automated measurement system that will perform ultrasonic measurements and provide expert interpretation of them without the need for an operator who is himself an expert in these measurements. The resulting system is a hybrid that uses the methods and tools of expert systems to flexibly manipulate the symbolic aspects of the problem and uses numerical algorithms for experiment control, data acquisition and signal processing.

Table 1. Steps of the measurement process that are performed by experts, non-experts and automated systems

| Step | Expert | Non-expert | Automated system |
|---|---|---|---|
| 1. Measurement of the raw data | yes | yes | yes |
| 2. Calculation of the result | yes | yes | yes |
| 3. Verification that the apparatus is working correctly | yes | sometimes | sometimes |
| 4. Direct estimation of the accuracy of the raw data | yes | sometimes | sometimes |
| 5. Error propagation | yes | sometimes | |
| 6. Inference of validity of data from nondata features | yes | sometimes | |
| 7. Evaluation of validity and usefulness of the results | yes | sometimes | |
| 8. Selection of apparatus for the measurement | yes | sometimes | |
| 9. Selection of appropriate measurement methods | yes | | |
| 10. Discovery of new measurement methods | yes | | |
| 11. Should the measurement have been made at all? | yes | | |

In operation, MCES first presents the operator with a menu of the properties that it can measure. He selects one or more and MCES evaluates each of a number of measurement methods in turn to determine applicability. This process includes asking the operator if he can provide required external information. Once MCES has selected applicable methods, the operator is instructed to place the ultrasonic transducer on the sample in a desktop water tank. When this has been done, MCES fires the transducer, measures the echoes and does the required signal processing and calculations. The results are presented to the operator along with heuristic judgments as to the accuracy of the results.

During this project, several new measurement methods were discovered, and a method for automatic method discovery was developed. Work is in progress to implement this and the other steps from Table 1 into the system.

MEASUREMENT METHODS

Figure 1 illustrates several requirements for a measurement to be performed successfully. First, for a given method to be applicable, certain conditions must be true about the specimen. For example, a method may require that the specimen have two flat parallel surfaces. The method may also require that certain data be obtained from sources external to the measurement process. For example, it may be necessary to know the thickness of the specimen in order to calculate its velocity. Finally, it may be necessary to make measurements on reference specimens in order to obtain calibration information. Once it is decided that a method is applicable, the actual measurements are performed. This involves operating the ultrasonic pulser/receiver and data acquisition equipment and performing signal processing on the results. Finally, the calculations that give the desired answers are performed and the data and results evaluated.

In the initial version of MCES, each method is stored as a single entity, including the measurements that must be made, the series of calculations used and the operator interfacing required to get external data. An approach in which the components that make up all of the methods are stored separately and MCES links them together to form new methods in response to the operator's requests has been developed (see Method Discovery section below).

ULTRASONIC MEASUREMENTS

Figure 2 shows the measurement geometry used for this preliminary version of MCES. Normal incidence, unfocussed waves are incident on the sample. Echo $u_1$ returns from the front surface of the specimen. If the specimen has a parallel back surface, echoes $u_2$ and $u_3$ may also be present. However, due to noise considerations, echo $u_3$ may not be detectable.

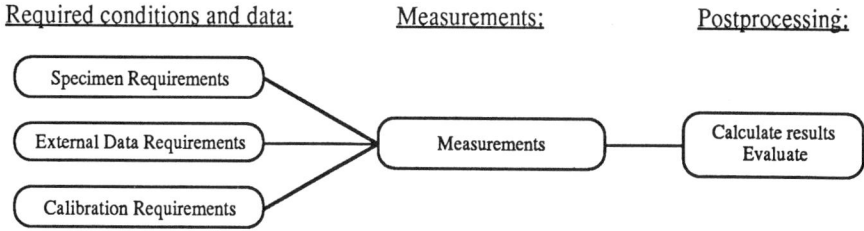

Fig. 1. A measurement method includes conditions and data required for its use, execution of the measurements, calculation of results and evaluation of the data and results.

For some measurement methods, only a front surface echo is required, although these tend to be less accurate because of their dependence on absolute amplitude measurements. For some methods, a measurement from a reference specimen of known properties (echo $u_0$) is required. The expert system must take all of these considerations into account.

The features of the measured signals that are used for interpretation can be simple ones such as the peak amplitude and arrival time of pulses, or the frequency spectrum may be used as a signature of the properties of the specimen. This initial version uses the amplitude and arrival time of the video pulses of the echoes. More sophisticated features can readily be added.

## DESCRIPTION OF MCES

### Architecture

Unlike conventional expert systems, MCES performs not only symbolic (rule-based) computations, but also performs data acquisition and numeric computations (signal processing). This is accomplished via the architecture shown in Figure 3. Symbolic processing is performed on an LMI Lambda 2X2 computer whose native language is LISP. The software is written in KEE (Knowledge Engineering Environment) by Intellicorp, Inc., which is a software tool that provides an object oriented programming environment and a production system that supports both forward and backward chaining mechanisms.[4] Numeric processing (data acquisition and signal processing) is performed in a VAX 11/780 using ISP, a high-level signal processing language developed by Rockwell International. Ultrasonic data acquisition is performed under control of the VAX using a Data Precision D/6000 transient recorder and Panametrics ultrasonic transducers and electronics. The symbolic processor controls the system and contains a knowledge base, an inference engine, and an operator interface.

Fig. 2. Ultrasonic measurements. Normal incidence, unfocussed ultrasound pulses produce one or more echoes.

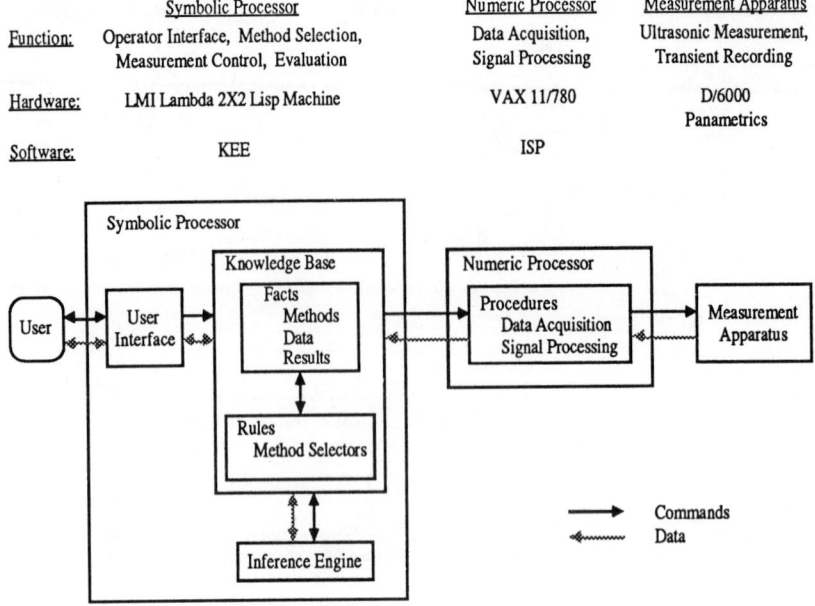

Fig. 3. Architecture of the MCES system.

## Knowledge Base

The knowledge base (see Figure 3) includes several classes of facts and a set of rules. The facts include properties of reference materials, features extracted from the measurements, the current state of knowledge of the specimen at any time during the measurement process, and the procedures to be used in executing the methods. The rule set includes rules for selecting applicable methods and will soon include rules for detecting invalid data and evaluating the accuracy of the results.

The measurement methods included in MCES are listed in Table 2. MCES can measure ultrasonic velocity, ultrasonic attenuation, density, and specimen thickness. These quantities were chosen because they are the building blocks that will be used to determine a range of other material properties in the next version of the system. The system knows two methods for measuring each of the four measurable quantities.

Table 2. MCES Measurement Methods

| Method | Quantity Measured | Geometry Required | Echoes Required | Data Required | Expected accuracy |
|---|---|---|---|---|---|
| SV1 | velocity | - | $u_1, u_0$ | density | Medium |
| SV2 | velocity | parallel faces | $u_1, u_2$ | thickness | Good |
| SA2 | attenuation | parallel faces | $u_1, u_2, u_0$ | thickness | Medium |
| SA3 | attenuation | parallel faces | $u_1, u_2, u_3$ | - | Good if atten is low |
| ST1 | thickness | parallel faces | $u_1, u_2, u_0$ | density | Comparable to SV1 |
| ST2 | thickness | parallel faces | $u_1, u_2$ | velocity | Comparable to SV2 |
| SD1 | density | - | $u_1, u_0$ | velocity | Comparable to SV1 |
| SD2 | density | parallel faces | $u_1, u_2, u_0$ | thickness | Comparable to SV1 |

For each method, the knowledge base contains rules that describe the requirements for its applicability (third, fourth and fifth columns of Table 2). These requirements are encoded in method selector rules (see Figure 3) and are used to determine which methods can be used for a given specimen. Also included are the a-priori accuracy estimates (last column of Table 2). They are derived from the experience of an expert and are used to determine the order in which methods are applied (best first) if more than one method is applicable. For each method selected, a sequence of operations needs to be performed to execute the method. This sequence is implemented in the method selector rules.

The diverse sources of data that are used by MCES are stored in objects (units) in KEE for use in the calculation of results and their evaluation. The collection of measured features are stored in an object that contains attributes (slots) including the values of measured features, quantitative estimates of their uncertainty, and a qualitative measure of their validity. Data derived from the operator dialog are stored in objects that describe the material and specimen properties. The properties of the reference specimen are stored in another object.

Knowledge Acquisition

The knowledge base used by MCES was developed through a series of interviews between an expert and a knowledge engineer. This knowledge includes measurement methods, operational procedures, signal processing methods and calculation methods. The project involved three people: a knowledge engineer (MSL) familiar with the required programming techniques and also familiar with the basics of ultrasonics; Expert #1 (RKE), who is the source of the expertise, and Expert #2 (LAA), against whose measurements MCES was tested. The structure of the system evolved as experience with the knowledge base accumulated.

Operation of MCES

MCES offers the operator a menu of properties that it can measure. After he has chosen one or more properties, MCES applies the method selector rules to determine if any methods are applicable to this specimen. This process includes asking the operator if the specimen has the required geometry and if he can provide the necessary external data. If one or more methods are applicable, the measurements are begun. MCES instructs the operator to position the specimen and the transducer. It then instructs the VAX computer to perform the data acquisition and signal processing.

For each method in the knowledge base, companion ISP procedures in the VAX acquire data and perform the signal processing necessary to identify the required pulses and extract their features. The signal processing consists of searching the measured waveform for the appropriate ultrasonic pulses and then extracting from the pulses the features needed to calculate the required results. Two types of features can be extracted. The first and simplest type is scalar features of the pulses, such as peak amplitude and arrival time. The second approach to feature extraction is to measure the full frequency dependence of the measured quantities. This approach can provide additional information, but requires more extensive computation and more subtle interpretation. This first version of MCES uses only scalar features extracted from the video envelope of the pulse. Their simplicity makes them preferable to frequency-dependent features and their accuracy is often quite adequate. Future work will incorporate frequency-dependent methods as well.

The features extracted from the measured data by ISP are then passed to KEE for result calculation and evaluation. The properties selected by the operator are reported to him, after which he can choose to perform other measurements.

Data Evaluation

A human expert performs two types of evaluation of the measurement results. One is a quantitative a-posteriori estimate of the accuracy of the result based on measured accuracies of the data. The other is a qualitative, often heuristic, evaluation of whether anything is seriously wrong with the result and whether it may be of questionable accuracy. Approaches to these methods of evaluation have been developed and are currently being implemented.

Each piece of numeric data in the system consists of numbers:

$x_i$, the expected value of the quantity,

$\sigma_i^2$, the variance of $x_i$,

$v_i$, the number of degrees of freedom associated with the measurement, and

$c_i$, a qualitative confidence factor.

The a-posteriori accuracy estimate will be performed by propagating through the calculations estimates of the variance of each variable and of the number of degrees of freedom of that estimate. For a function $f(\{x_i\})$, the variance $\sigma_f^2$ of $f$ in terms of the variances $\sigma_i^2$ of the $x_i$ is given by

$$\sigma_f^2 = \sum_i (\partial f/\partial x_i)^2 \cdot \sigma_i^2 \qquad (1)$$

and the number of degrees of freedom $v_f$ of $f$ in terms of the number of degrees of freedom $v_i$ of the $x_i$ is given by

$$v_f = \sigma_f^4 / \sum_i [(\partial f/\partial x_i)^4 \, \sigma_i^4 / v_i] \qquad (2)$$

In order to use this method, estimates of $\sigma_i^2$ and $v_i$ for the measured features and the operator supplied data are required. For measured features, several repetitions of the measurement can be made quickly. From them, $\sigma_i^2$ and $v_i$ can be estimated. For user-supplied data, the user will be asked if he knows $\sigma_i^2$, and if so, how sure he is of it. If he does not know $\sigma_i^2$, a reasonable value from the knowledge base will be substituted, assuming one degree of freedom. If he gives a value of $\sigma_i^2$, a value of $v_i$ will be supplied based on his degree of certainty of the value.

The confidence factors $c_i$ can take on four values:

Good data,
Adequate, but perhaps not highly accurate,
Suspect,
Probably bad.

The confidence factors for experimental data will be determined from several sources: unusual characteristics in the data (such as the presence of unexpected pulses, unexpected feature values and lower than expected signal to noise ratios), or a-priori prejudice on the part of the expert. The confidence factors will be propagated through the calculations as follows. The confidence factor resulting from a calculation will be equal to the worst confidence factor of any of the inputs to the calculation. This is a pessimistic philosophy, but remember that the purpose of the confidence factor is not to say quantatively how accurate the measurement is. The variance and degrees of freedom serve this purpose. The purpose of the confidence factor is to give qualitative warning that something unusual has happened.

## Discovery of New Measurement Methods

As mentioned above, the current version of MCES groups the measurements, calculations, and external information requests that compose a method into one entity. During codification of the knowledge, it was observed that if these components were rearranged, new measurement methods could be created. As a result, methods for measuring the density of a specimen or the thickness of a plate purely ultrasonically were discovered. Neither expert #1 nor #2 was previously aware of these methods, although in retrospect any expert would probably have discovered them if he had asked himself the right questions. This points up one of the advantages often cited for the knowledge codification process: that thinking systematically about the knowledge often yields new insights.

An approach was therefore developed to allow the version of MCES currently under development to discover new methods. In this approach, the knowledge base contains tables of all raw data that can be measured and all equations which express useful relationships between measured data and calculable quantities. When the operator requests measurement of a given quantity, the expert system searches in turn all equations which contain that quantity. It then chains through all other equations, attempting to establish one or more methods that link available measurements and external data to the desired quantity. This capability would discover all of the methods that MCES currently contains.

RESULTS

A test was performed in which four specimens of widely varying acoustic properties were measured by MCES and by Expert #2. The specimens were made of Lucite, aluminum, Inconel and beryllium. They were chosen for their wide range of density and ultrasonic velocity. MCES and the human expert used methods and apparatus of their own choosing. Hence, the methods used were not identical. The results were evaluated in terms of the accuracy of the measurements and how long the measurements took to perform. Handbook values (not measured on the same specimens) were also compared where appropriate. The results are presented in Table 3. They are grouped by methods that are approximately equivalent.

The time required for MCES to make measurements consists primarily of the user dialog and the manual positioning of the specimen in the water tank. Typical time is 3 minutes per measurement. The human operator required 1 hour to measure velocity and attenuation for one sample. This time was 75% setup and 25% measurement and analysis. These numbers are not directly comparable because different methods were used, but they clearly show the speed advantage of the expert system.

Velocity measurements: The more accurate method (SV2) gave results that were 1.5-5% higher than handbook values, probably due in part to a systematic error in the simple signal features used. Method SV1, which is generally less accurate because it depends critically on absolute amplitude measurements, had errors of 7-15%. The human expert's results were within 2% of handbook values.

Attenuation measurements: Attenuation is inherently more difficult to measure accurately than velocity. Methods SA2 and SA3 gave values that were within a factor of two of one another and of reasonable magnitude. This represents an acceptable level of agreement The expert's measurements were of the same order of magnitude in two cases, but were clearly invalid (negative attenuation) in another. Hence, in this case, MCES's measurements appear to be more stable than the expert's.

Thickness and density measurements: At the beginning of the project, Expert #2 did not know a method for ultrasonically measuring these quantities. Hence, no measurements are reported. The accuracy of MCES's measurements are directly related to the accuracies of methods SV1 and SV2, on which they are based, and are in the 5-12% range.

CONCLUSIONS

The Materials Characterization Expert System (MCES) is designed to ultrasonically measure a number of material properties. It knows several methods for measuring each one. The user is presented with a list of properties and is asked to select one or more for MCES to measure. MCES then carries on a dialog with the user to determine which of the methods are applicable in this case and which are likely to give the most accurate results. It then performs the measurements, analyzes the data, and reports the results to the user.

Table 3. Results comparing human expert and MCES.

| Measurement | Method | Material: Lucite | Aluminum | Inconel | Beryllium |
|---|---|---|---|---|---|
| Velocity: | SV1 | 2.92 | 6.75 | 5.15 | 15.11 |
| (mm/µs) | SV2 | 2.80 | 6.40 | 5.90 | 13.51 |
|  | Expert (1) | 2.72 | 6.31 | 5.62[2] | 12.62 |
|  | Handbook | 2.68 | 6.32-6.50 | 5.72 | 12.90 |
| Attenuation: | SA2 | 0.57 | 0.092 | 0.091 | 0.194 |
| (dB/mm) | SA3 | 0.59 | 0.143 | 0.172 | 0.265 |
| (at 5 MHz) | Expert (3) | 0.75 | 0.19 | 1.85[4] | (5) |
| Thickness: | ST1 | 5.97 | 10.50 | 2.53 | 10.09 |
| (mm) | ST2 | 5.47 | 9.84 | 2.81 | 8.61 |
|  | Expert (6) | - | - | - | - |
|  | Micrometer | 5.72 | 9.96 | 2.90 | 9.02 |
| Density: | SD1 | 1.276 | 2.87 | 7.23 | 2.14 |
| (g/cc) | SD2 | 1.22 | 2.84 | 7.01 | 2.04 |
|  | Expert (6) | - | - | - | - |
|  | Direct Meas. | 1.171 | 2.691 | 8.028 | 1.827 |

Notes:
1. Expert's method similar to SV2, except visual pulse overlap is used to measure time delay.
2. 15 MHz transducer used.
3. Expert calculated attenuation as function of frequency and read off value at given frequency.
4. Measured at 15 MHz; cannot be compared to 5 MHz measurements.
5. Measurement gave unreasonable (negative) value. Expert advised against its use.
6. Expert did not know a method for making this measurement.

Developing MCES involved the following: 1) development of a knowledge representation for ultrasonic measurement methods and data; 2) codification of ultrasonic measurement knowledge; 3) discovery of new ultrasonic measurement methods; 4) integration of symbolic data processing, numeric data processing and ultrasonic measurements in one system; 5) automatic selection of measurement methods; 6) automatic identification of ultrasonic pulses in a measured signal and extraction of features from them. Work currently in progress to improve MCES includes: 1) automatic evaluation, by both quantitative and heuristic methods, of the quality of measured data and the calculated results; and 2) automatic discovery of new measurement methods.

ACKNOWLEDGEMENTS

This paper is dedicated to the memory of Lloyd Ahlberg, colleague and friend. This work was supported by the Rockwell International Independent Research and Development Program.

REFERENCES

1. **Building Expert Systems**, F. Hayes-Roth, D. A. Waterman, D. B. Lenat, eds., Addison-Wesley, 1983.
2. L. D. Erman, F. Hayes-Roth, V. R. Lesser, D. R. Reddy, "The Hearsay-II Speech-Understanding System: Integrating Knowledge to Resolve Uncertainty", **Computing Surveys** 12 (2) June 1980, pp. 213-253.
3. H. P. Nii, E. A. Feigenbaum, "Rule-based Understanding of Signals", **Pattern-Directed Inference Systems**, D. A. Waterman, F. Hayes-Roth, eds., Academic Press, 1978, pp. 483-501.
4. KEE Software Development System Reference Manual, Document R-2.0-1.0, Intellicorp, Menlo Park, CA, March 1985.

DEVELOPMENT OF AN EXPERT SYSTEM FOR ULTRASONIC FLAW CLASSIFICATION

Lester W. Schmerr, Jr., Ken E. Christensen, and Stephen M. Nugen

Center for NDE
Iowa State University
Ames, IA 50011

INTRODUCTION

The complete characterization of a flaw requires information about the flaw type (crack, void, inclusion, etc.), flaw size, and orientation. Here we are only concerned with the determination of the flaw type so that the appropriate sizing algorithms can be chosen. This type of classification problem using ultrasonic waves is very suitable for employing the tools and techniques of artificial intelligence [1,2]. Adaptive learning methods, for example, have in the past been employed to train a flaw classification module so that it can distinguish between cracks and volumetric flaws [3]. Some of the limitations of this approach, however, have been due to the empirical nature of the features used for classification and the difficulty of understanding and adjusting the decision-making process when errors occur.

In contrast, we have chosen to develop a classification scheme in the form of a rule-based expert system where the features used by the system for classification come from model-based fundamental knowledge, and where the rules are made explicit and modifiable. In this paper we will describe the nature of the expert flaw classification system we are building and demonstrate its use with some ultrasonic data. As currently constituted, the domain of knowledge of the system is highly constrained. The flaw classifier is concerned with distinguishing between single isolated volumetric and crack scatterers. The design of the system, however, is such that these constraints are not essential.

SYSTEM OVERVIEW

As outlined in Fig. 1, the expert flaw classification system, FLEX (Flaw Expert), consists of essentially four major components: 1) a user-interface that allows the visual display and manipulation of ultrasonic data, 2) a set of tools that allow a user to manipulate and modify the rules of the system, 3) a module called FEAP (for FEAture Processing), and 4) a module called FLAP (for FLAw Processing). FEAP and FLAP are being designed as two separate, cooperating expert systems.

Feature Processing (FEAP)

It is the job of FEAP to take the preprocessed ultrasonic data from a given experiment, and determine confidence factors associated with each feature being used by the system. These confidence factors are numbers in the range [-1,1], where -1 indicates complete certainty that a feature is not present, 1 indicates complete certainty that a feature is present, and numbers in between indicate the degree of certainty or uncertainty (see Appendix). Both FEAP and FLAP manipulate these confidence factors according to the conventions and methods developed by Shortliffe and Buchanan for the MYCIN project [4]. FEAP also determines the percentage of the ultrasonic data sets, if there are more than one, in which there is positive evidence (positive confidence factors) for each feature. Currently, FEAP assumes that the data it uses has had non-flaw dependencies removed through the application of the measurement model of Thompson and Gray [5]. This preprocessing is done so that we can rationally evaluate features characteristic of the flaw type only. FEAP uses a combination of fundamental knowledge of flaw scattering properties, and heuristic knowledge based on our familiarity with actual experimental data. For example, a Kirchhoff model of how a flat crack behaves at normal incidence indicates [6] that in the frequency domain we can expect a linearly increasing amplitude with increasing frequency (solid line in Fig. 2a). A more exact numerical model of the scattering process verifies this linear behavior, but indicates

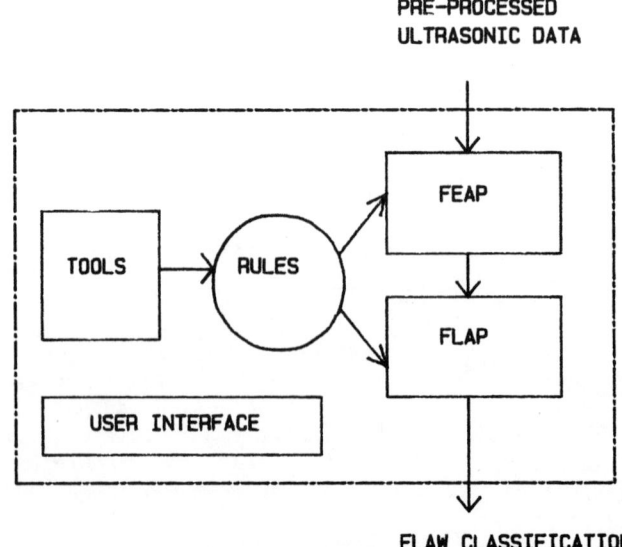

Fig. 1. Outline of components of FLEX.

that a modulation also exists (dotted line in Fig. 2a). In principle,
therefore, we would expect to seek such a characteristic feature over
the entire bandwidth of our experimental ultrasonic system. In reality,
however, experimental results (see Fig. 2b) show the presence of this
feature only up to about the center frequency of the transducer. This
type of heuristic knowledge is then factored into our actual search
for, and evaluation of, this feature.

Currently, the module FEAP is in an early stage of development.
We have outlined a set of decision trees, for extracting all the features
and their associated confidence factors, and are now in the process
of automating that extraction process. However, because of the modular
nature of the system, flaw classification evaluations are still possible
by having a human operator replace FEAP and provide the necessary confidence factors to FLAP. Below we will outline a sample application of
FLEX to some ultrasonic data which does the feature evaluation in just
that manner.

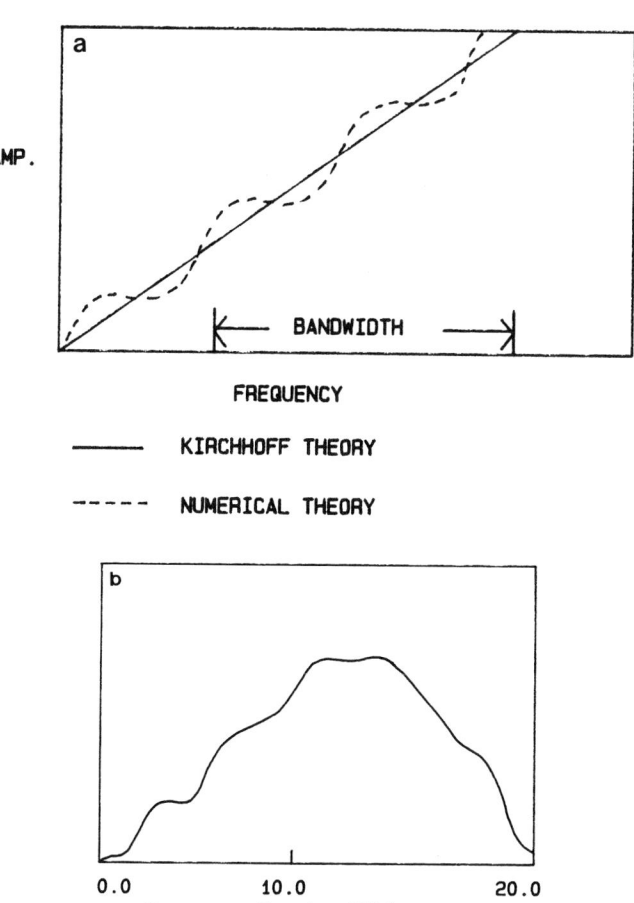

Fig. 2.  a) Theoretical response of a crack at normal incidence. b)
400 micron crack in IN100.

Flaw Processing (FLAP)

Once the feature evaluation process is complete, FLAP evaluates the evidence produced by FEAP according to an explicit set of rules and reaches a conclusion as to flaw type. Currently, we are using nine features and a corresponding set of nine rules to perform the classification. These nine features (see Table 1) were chosen because analytical, numerical, and approximate models of the scattering process show that they can be expected to help distinguish between single, isolated volumetric and crack-like scatterers. For more complicated geometries (flaws near a surface, etc.) and flaw types (porosity, microcracking, etc.), these features would have to be modified and/or supplemented by other flaw features. As indicated in Table 1, both time and frequency domain features are used in the evaluation process. Having multiple domains has been found, in our test cases, to be particularly useful for handling noise and other experimental system inaccuracies.

Table 1. Features used in FLEX for classification of an isolated flaw as to flaw type (volumetric or crack) and the corresponding domain in which the feature is defined.

| Feature | Domain |
|---|---|
| 1. Positive-Leading-Edge-Pulse | Time |
| 2. Flash-Points | Time |
| 3. Rayleigh-Wave | Time |
| 4. Creeping-Wave | Time |
| 5. Ringing | Time |
| 6. Linearly-Increasing-Amplitude | Frequency |
| 7. Plateau-With-Shallow-Nulls | Frequency |
| 8. Decreasing-Amplitude-With-Deep-Nulls | Frequency |
| 9. Sharp-Nulls | Frequency |

The flaw classification rules used by FLAP exist in two forms: external and internal. The external form of the rules is designed for English-like readability and easy modification by non-programmers. The external rules are in the form of (if... then ... ) types of conditional statements where the "if" part of the rule is called the antecedent part of the rule and the "then" part is the consequent. A typical external rule in FLAP is:

(Rule 204
(if    ringing is detected in the trailing response of at least 50 percent of the flaw signals in the time domain)
(then  there is weak belief in the accumulating evidence supporting the determination of a volumetric flaw)

There are four important pieces of such a rule: 1) the feature it applies to, 2) the percentage of flaw signals where this feature exists, 3) the qualification (e.g., weak belief) on the strength of evidence associated with this feature, and 4) the type of evidence (accumulating or non-accumulating). The percentage value is used as a threshold to decide if the evidence is sufficient to warrant invoking the rule. This gives us, in a simple manner, a way to account for uncontrollable uncertainties in the system, such as noise, and to minimize their influence on the flaw classification process. Similarly, the use of phrases such

as "weak belief" in our rules allows us to control the "weight" of the
evidence associated with that rule. Note that, unlike adaptive learning
systems, such weights are explicit in their meaning and modifiable in
clear English text. The use of accumulating and non-accumulating evidence
allows us to merge the existence of strong features (non-accumulating
evidence), where the existence of a single data set with this feature
is sufficient to add this evidence into our conclusion, and less strong
features (accumulating evidence) that are indicative of flaw type, but
where an average value over all data sets is taken as the net evidence
for this feature. Actually, our feature evidence is categorized into
three classes:

(1) Sufficient. Example: The unique response of linearly increasing
amplitude in the frequency domain is sufficient evidence to conclude
that the flaw type is a crack. This type of evidence is considered
to be non-accumulating.

(2) Necessary. Example: We always expect to see a negative leading
edge pulse in the time domain for cracks. If positive evidence exists
for a positive leading edge pulse, we record negative accumulating evidence
for cracks.

(3) Indicative. Example: Ringing, or resonance, in the time
domain is indicative of a volumetric flaw. When a net positive evidence
is found for this feature, we record positive accumulating evidence
for a volumetric flaw.

The external rules of FLAP are automatically translated into internal
form as part of the rule modification tools of FLEX. This internal
rule form is designed for simple evaluation by the program. For example,
the above Rule 204 would translate to:

```
(RULE 204
       (IF (>= (FEAT-PC (QUOTE RINGING)) 50))
       (THEN (SETQ ACC_VOL_EVD
             (CONS (* 0.3 (ZERO-CLIP (FEAT-CF (QUOTE RINGING))))
                   ACC_VOL_EVD))))
```

Once all the external rules are translated into such internal forms,
it is the task of FLAP to evaluate these rules and reach a conclusion
based on the confidence factors and percentage values provided by FEAP
(or an equivalent human operator). This part of FLAP is a simple inference
engine whose actions can be summarized as follows:

Rule evaluation consists of two phases: evaluating the rule antecedent
and evaluating the rule consequent. Recall the rule antecedent is the
conditional or "if" clause. If the conditional is true, then the consequent or "then" clause of the rule is evaluated for side effects, i.e.,
making an entry into one of the evidence lists supporting the two hypotheses. At the current time, FLAP selects every rule for evaluation without
regard to order or weight. A rule is said to "fire" if the antecedent
is true.

Another capability of FLAP is its ability to explain the results
of its classification. We have chosen to implement this feature of
the system in the form of an audit trail which provides a summary of
the rule firings and the evidence (or lack of it) which caused the particular conclusions to be reached. An example of such an audit trail for
a specific problem will be given in the next section.

The conclusion of the system is in the form of the total evidence
(confidence factors) for both volumetric and crack-like flaws). Typically,
there may be positive evidence for both types of flaws. If, however,
there is a wide enough variation in this evidence, a firm conclusion
can be reached.

FLEX is currently being developed on a Symbolics 3670 system using
Symbolics extended Common Lisp. The user interface, which employs bit-
mapped graphics and a mouse, is necessarily system dependent. The overall
architecture, knowledge representation scheme, and inference strategy
are not system dependent. Our expectation is that production-oriented
implementations will be possible on a variety of 32-bit microcomputers.

AN EXAMPLE FLAW CLASSIFICATION

To see some of the elements of the behavior of this system, we
have given in diagram 1 an outline of the application of FLEX to the
classification of an artificial 400μm radius crack placed in a sample
of IN100. Thirteen different time domain waveforms and corresponding
frequency domain results were available from this sample through the
use of the multi-viewing transducer system developed at Iowa State by
Dr. D. O. Thompson and his co-workers [7]. Each of the waveforms or
spectra corresponded to a different "look-angle" at this flaw. In diagram
1, we have followed the behavior of this sytem by examining in detail,
for this particular example, one of the nine rules, Rule 202, and its
consequences. Diagram 1 shows the external form of this rule and the
internal form that this rule is translated into by the Translator.
For this particular example, a human operator, using the visual display
features of FLEX, examined all thirteen look-angles and provided estimates
of the confidence factors associated with each feature. This data was
fed into the inference procedure of FLAP and the conclusion shown was
drawn. As mentioned previously, there is typically evidence for both
flaw types, as we see here. However, we also see that the difference
in confidence values is strong enough so that one could, with moderate
confidence, conclude that this was a crack.

By invoking the explanation facility, we can see the reason why
each rule did or did not fire. In the case of Rule 202, we found flash-
points in 100 percent of the look-angles so the threshold of 50 percent
was exceeded and the rule fired. The value of 0.68 given is the average
confidence factor given for this feature over the thirteen look-angles.
This value is multiplied by 0.5 to factor in the weight of the evidence
(moderate belief) for this feature (see Appendix). Following this explanation in diagram 1, we see a tabulation of all the non-accumulating and
accumulating evidence from all the rules for this example. Rule 202
is seen to provide an entry in the accumulating crack evidence list
(ACC-CRK_EVD) as it should. Finally, we note that the total confidence
factors given in the conclusion are just the "sum" of all the accumulating
or non-accumulating evidence for crack or volumetric, where the "sum"
is labelled with an M superscript to indicate it is actually carried
out according to the methods developed for the MYCIN expert system [4]
(see also the Appendix).

SUMMARY AND CONCLUSIONS

We have shown some of the major elements of FLEX that are currently
operational. As mentioned, the automating of the feature extraction
portion of the system (FEAP) is a particularly challenging task that
we are now undertaking. Extension of this system to do intelligent
signal preprocessing and post-classification flaw sizing is also possible
because of the very modular nature of the system. However, probably

more important from a practical standpoint, is the ability of the system to handle different testing needs and classification problems. The design of the system will also take into account this important necessity.

ACKNOWLEDGEMENTS

We would like to express our appreciation to Dr. D. O. Thompson for the use of data taken on the multiviewing transducer system and to S. J. Wormley for his time and help in the initial "shakedown" of FLEX. We also appreciate the assistance of Dr. T. A. Gray in providing crack data for use in the system. This work was supported by the NSF University/Industry Center for NDE at Iowa State University.

APPENDIX

Confidence Factor Notes

1. Confidence Factors (CF's) are not the same as probabilities. Particularly, a CF of N for conclusion X does not imply a CF of (1-N) for conclusion not-X. For this reason, CF's are usually calculated and manipulated with heuristics developed as part of the expert system.

2. The CF for a particular feature evaluation can be regarded as the difference between the belief that the feature is present and the belief that the feature is not present. That is,

$$CF = MB - MD$$

MB = Measure of Belief.        $0 =< MB =< +1$
MD = Measure of Disbelief.     $0 =< MD =< +1$
CF = Confidence Factor         $-1 =< CF =< +1$

3. For uncertain judgements, we partition the range of CF [-1,+1] into nine non-overlapping subranges:

```
-1.0 =< CF =< -0.9 ------> Certain disbelief
-0.8  < CF =< -0.6 ------> Strong disbelief
-0.6  < CF =< -0.4 ------> Moderate disbelief
-0.4  < CF  < -0.2 ------> Weak disbelief

-0.2 =< CF =< +0.2 ------> Uncertainty

+0.2  < CF  < +0.4 ------> Weak belief
+0.4 =< CF  < +0.6 ------> Moderate belief
+0.6 =< CF  < +0.8 ------> Strong belief
+0.8 =< CF =< +1.0 ------> Certain belief
```

These subranges have two purposes:

    a. When explaining a conclusion to an end user, map the CF values to the applicable partition name.

    b. When getting uncertain judgements from users, map them to the CF value which is the midpoint of the applicable partition. For example, if a user's confidence in his/her evaluation of ringing is "moderate belief", then assign a value of +0.5 to the CF.

4. The $\overset{M}{+}$ operation for summing accumulating evidence (see the note in Diagram 1) is borrowed from MYCIN [4]. For example, in adding two positive confidence factors CF1 and CF2, the sum is defined as:

$$CF1 \stackrel{M}{+} CF2 \equiv CF1 + (1-CF1)*CF2$$

Note that this sum is independent of the order of adding CF1 and CF2.

REFERENCES

1. S. M. Weiss, and C. A. Kulikowski, <u>A Practical Guide to Designing Expert Systems</u>, Rowman and Allanheld, Totowa, NJ, 1984.
2. E. B. Hunt, <u>Artificial Intelligence</u>, Academic Press, NY, 1975.
3. M. F. Whalen, L. J. O'Brien, and A. N. Mucciardi, Proceedings of the DARPA/AFML Review of Progress in Quantitative NDE, AFWAL-TR-80-4078, 1980.
4. B. G. Buchanan, and E. H. Shortliffe, <u>Rule-Based Expert Systems</u>, Addison-Wesley, Reading, MA, 1984.
5. R. B. Thompson, and T. A. Gray, J. Acoust. Soc. Am., 74, 1297-1290, 1983.
6. Chien-Ping, Chiou, Unpublished Master's Thesis, Iowa State University, Ames, IA  50011, 1986.
7. D. O. Thompson, and S. J. Wormley, in, <u>Review of Progress in Quantitative NDE</u>, D. O. Thompson and D. E. Chimenti, Eds., (Plenum Press, NY, 1985), pp. 287-296.

Diagram 1. An example flaw classification of an artificial circular crack in IN100.

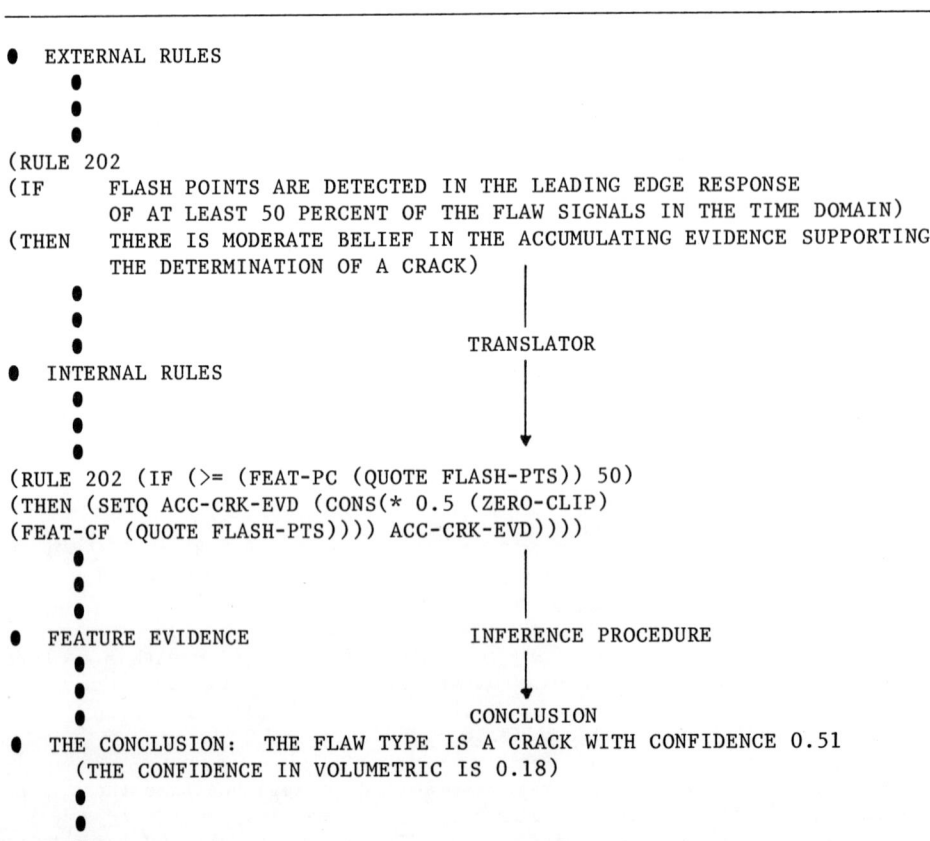

- EXTERNAL RULES

(RULE 202
(IF      FLASH POINTS ARE DETECTED IN THE LEADING EDGE RESPONSE
         OF AT LEAST 50 PERCENT OF THE FLAW SIGNALS IN THE TIME DOMAIN)
(THEN    THERE IS MODERATE BELIEF IN THE ACCUMULATING EVIDENCE SUPPORTING
         THE DETERMINATION OF A CRACK)

                                    TRANSLATOR

- INTERNAL RULES

(RULE 202 (IF (>= (FEAT-PC (QUOTE FLASH-PTS)) 50)
(THEN (SETQ ACC-CRK-EVD (CONS(* 0.5 (ZERO-CLIP)
(FEAT-CF (QUOTE FLASH-PTS)))) ACC-CRK-EVD))))

- FEATURE EVIDENCE           INFERENCE PROCEDURE

                                    CONCLUSION
- THE CONCLUSION: THE FLAW TYPE IS A CRACK WITH CONFIDENCE 0.51
  (THE CONFIDENCE IN VOLUMETRIC IS 0.18)

- EXPLANATION FACILITY ⟶ AUDIT TRAIL
  •
  •
  RULE 202 FIRED.  FLASH-PTS DETECTED IN 100.0 PERCENT OF THE
     SIGNALS (50 PERCENT NEEDED), THE RESULTING CONTRIBUTION
     TO ACC-CRK-EVD IS (0.5 * 0.68 = 0.34).
  •
  •
  (NAC-CRK-EVD 0.0)
  (ACC-CRK-EVD (0.175 0.108 0.34))
  (NAC-VOL-EVD NIL)
  (ACC-VOL-EVD (0.03 0.075 0.081)
                    M       M
- NOTE:  0.51 = 0.175 + 0.108 + 0.34
                    M       M
         0.18 = 0.030 + 0.075 + 0.081

SIIA: A KNOWLEDGE-BASED ASSISTANT FOR THE

SAFT ULTRASONIC INSPECTION SYSTEM(a)

R. B. Melton, S. R. Doctor, T. T. Taylor, R. V. Badalamente

Pacific Northwest Laboratory
P.O. Box 999
Richland, Washington, 99352

INTRODUCTION

SIIA(b) is a knowledge-based system designed to assist in making the operation of the Synthetic Aperture Focussing Technique (SAFT) Ultrasonic Inspection System more reliable and efficient [1]. This paper reports on our effort to develop a prototype version of SIIA to demonstrate the feasibility of using knowledge-based systems in nondestructive evaluation (NDE).

One of our prime motivations for developing SIIA is to provide a means for insuring that the SAFT system is used correctly and consistently and to assist in interpreting the results of a SAFT inspection. Our initial formulation of the problem was to develop a system to assist in the interpretation of the images resulting from a SAFT inspection. As we started to identify the structure of the inspection problem, however, we realized that a more effective application of the knowledge-based system technology would be to develop a system that is in essence an on-line procedure generator that guides a user through a SAFT inspection. Such a system assists in proper setup of the inspection equipment for each of the steps in a SAFT inspection and in interpreting the inspection results for each step.

The first section of the paper describes the structure of the problem and our conceptual design of the knowledge-based system. The next section describes the current state of the prototype SIIA system and relates some of our experiences in developing the system. The final section discusses our plans for future development of SIIA and the implications of this type of system for other NDE techniques and applications.

---

(a) This work was supported by the United States Department of Energy under contract DE-AC06-76RLO 1830.

(b) SIIA stands for the SAFT Image Interpretation Assistant. Our original intention was to build a system to assist in interpreting inspection results. As we explored the problem our objectives were reformulated as described in the paper.

The SAFT Inspection Problem

The SAFT ultrasonic inspection system has been developed, under U.S. Nuclear Regulatory Commission funding, for inspecting primary pressure boundary weldments in nuclear reactors or similar facilities. The system scans an area of material and produces a three-dimensional view of the entire volume of material scanned. Flaws and other reflectors are interpreted by an operator viewing the three-dimensional image or cross-sections thereof.

In practice there are three modes of inspection for the SAFT system: Normal-Beam, Pulse-Echo, and Tandem Mode scanning. In normal-beam scanning the ultrasonic beam is normal to the inspecting surface. This mode is generally used to characterize the weldment by locating the weld-root and the counter-bore regions. In the case of ferritic materials in pressure-vessels it might also be used for the first attempt at flaw detection.

The second step in scanning is to perform a pulse-echo inspection. In this mode a shear or longitudinal wave is used at a 45 or 60 degree angle. A single transducer is used as both transmitter and receiver. The objective of pulse-echo scanning is always oriented to flaw detection and characterization. If the quality of data is high enough then a decision about the presence or absence of a flaw may be made using the pulse-echo data. Otherwise, a tandem mode scan is performed.

In tandem mode scanning two transducers are used, one for transmitting, the other for receiving. There are three different configurations for the transducers. In addition the operator must choose between shear and longitudinal waves at a 45 or 60 degree beam angle. Like the pulse-echo scan the tandem mode scan is intended to detect and characterize flaws.

In all three modes the operator must choose the appropriate transducer(s) center frequency, bandwidth, and diameter. In some cases he must also choose the type of transducer, contact vs. booted-shoe, or the coupling technique, immersion vs. direct contact.

In each of the inspection steps described above the operator must make a number of choices in setting up and performing the inspection. In addition to those mentioned above he must also decide what type of SAFT processing if any will be done on the data. The operator's choices are determined by the characteristics of the specific inspection he is performing. The most important parameters are the type of material being inspected and the components of the weldment (i.e. a pipe welded to a valve or a nozzle to pressure vessel weld).

Conceptually the SAFT inspection problem breaks down into two components: procedural and interpretive knowledge of how to perform and interpret an inspection, and description of the physical objects that are combined to represent a specific inspection situation.

Procedurally we have broken the problem into subproblems or steps. The three primary steps are the normal-beam, pulse-echo and tandem mode scans. Within each primary step there are secondary steps of setup, determination of desired transducer characteristics, initial transducer selection, transducer checkout with respect to signal-to-noise ratio, the scanning itself, and data interpretation.

In the setup step we must decide what type of SAFT processing is appropriate -- line-SAFT, full-SAFT, or to not use SAFT processing. In addition for normal beam scanning we must decide whether the inspection objective is weld characterization or flaw detection.

In determining transducer characteristics we primarily consider the type of material being inspected and for ferritic welds the type of cladding, if any. In doing so we determine a desired transducer center frequency, bandwidth, and diameter. We follow this by matching desired transducer characteristics with actual transducer descriptions and select a specific transducer for the inspection.

Before the scan is performed the system requests a check of signal to noise ratio. In this step the back surface signal is compared to the overall noise from grain structure and other material characteristics. If the ratio is less than 6dB then a different transducer is selected and the check performed again. This process continues until an acceptable signal to noise ratio is achieved.

The Current Status of SIIA

We have chosen to implement SIIA on a Symbolics Lisp Machine using the KEE knowledge-based system development software (a). KEE provides the ability to describe the problem in terms of object-class hierarchies with procedural attachments, lisp functions, and If-Then rules. It provides both forward and backward chaining as control strategies for the rule bases. For a discussion of frame based knowledge representation systems such as KEE see [2]. For an introductory level coverage of knowledge-based systems see [3].

For the purposes of developing the SIIA prototype we have concentrated on representing the structure of the physical components of the problem and then on the normal beam scanning step without data interpretation. We have created the base level structures for the rest of the problem, but have not supplied the procedural know-how.

The physical components related to the problem are described in object-class hierarchies. For example, Fig. 1 shows a graphical representation of the object-class hierarchy for materials. Starting at the left a KEE unit for materials is shown. This unit contains slots for descriptive parameters common to all materials of interest. In this case these parameters are grain-shape, grain-size, and sensitization. Moving to the right we see the next level breaks down into ferritic steels and stainless steels. There is no further breakdown of ferritic steels, but stainless steels breaks down further into cast and wrought stainless steels. At the lowest level, connected by dashed lines, we have specific instances of materials. For example A533B is a specific ferritic steel and SS304 is a specific wrought stainless steel.

In a similar manner other physical entities are described including reflectors, primary system components, and transducers. The object class hierarchies have been defined for each of these classes. As with materials these definitions begin with a generic description and proceed with increasing detail to specific descriptions. For example, primary system components are broken down into vessels, components, and pipes. Vessels are further broken into pressure vessels, pressurizers, and steam generators. Components are broken into elbows, pumps, valves, and a catch all other category. There is no further breakdown of pipes.

Flow of control in solving the problem centers around a description of the "inspection problem". Figure 2 shows the menu that is used to collect

---

(a) Symbolics is a trademark of Symbolics, Inc. KEE is a trademark of Intellicorp, Inc.

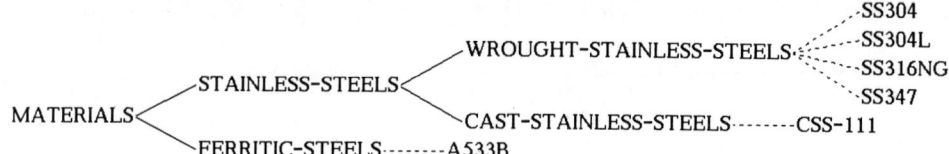

Figure 1. Object-Class hierarchy for description of Materials

```
Inspection Problem Description
Problem Name: TEST-1
Problem Description: Pipe-Pipe Pipe-Component Component-Component Pipe-Vessel Component-Vessel Vessel
Type of Reactor: PWR BWR
Plant System Containing Weldment: High-Pressure-Injection Low-Pressure-Injection Main-Steam-Line Return-Lines
Exit
```

Figure 2. Initial problem description menu. User enters problem name from keyboard. Other items are entered by selecting an item using a "mouse". In this illustration the selected values are shown in boldface.

information about the problem. This basic information determines what other information to collect. For example if we are inspecting a weld on a pressure vessel then we only have one material to consider, whereas if we are inspecting a weld between a pipe and a pump we need to ask what type of material each piece is made of. Associated with each inspection problem is one or more "inspection worksheets". The worksheet is filled in with intermediate information that further describes the problem or is derived from information about the problem. For example the desired transducer characteristics are entered into the worksheet when they have been determined. A problem may have one or two worksheets associated with it. A problem has two worksheets when two separate scans will be required to complete the inspection. This occurs primarily with bimetallic welds, such as cast to wrought stainless steel, where material characteristics require that two different transducers be used.

Finally associated with each worksheet is an "inspection procedure." The inspection procedure contains the information required by the operator to proceed with the scan. A specific transducer is identified along with the type of SAFT processing to be performed and other information such as beam angle and wave propagation mode. A portion of the generic form for an inspection procedure is shown in Figure 3.

Knowledge Representation

Knowledge is represented in the system through the object-class hierarchy descriptions and through the use of If-Then rules. If-then rules set reference parameters associated with objects in the system. For example in determining how many procedures will be required one of the rules looks at the type(s) of material joined by the weld. If one is wrought stainless steel and the other cast stainless steel then the rule concludes that two inspection procedures will be required.

**(Output) The INSPECTION-PROCEDURES Unit in SIIA Knowledge Base**

Member slot: **NORMAL-BEAM-INSPECTION-OBJECTIVE** from **INSPECTION-PROCEDURES**
  Inheritance: **OVERRIDE.VALUES**
  ValueClass: (ONE.OF FLAW-DETECTION PROFILING)
  Cardinality.Max: 1
  Comment: "The purpose for the normal-beam scan."
  Values: PROFILING

Member slot: **NORMAL-BEAM-SCANNING-MODE** from **INSPECTION-PROCEDURES**
  Inheritance: **OVERRIDE.VALUES**
  ValueClass: (ONE.OF NO-SAFT LINE-SAFT FULL-SAFT)
  Cardinality.Max: 1
  Comment: "The scanning mode to be used in normal-beam."
  Values: NO-SAFT

Member slot: **NORMAL-BEAM-TRANSDUCER** from **INSPECTION-PROCEDURES**
  Inheritance: **OVERRIDE.VALUES**
  ValueClass: **TRANSDUCERS**
  Cardinality.Max: 1
  Comment: "The transducer selected for use in normal beam scanning."
  Values: Unknown

Member slot: **PULSE-ECHO-BEAM-ANGLE** from **INSPECTION-PROCEDURES**
  Inheritance: **OVERRIDE.VALUES**
  ValueClass: (ONE.OF 45 60)
  Cardinality.Max: 1
  Comment: "The beam angle desired in the material being inspected"
  Values: Unknown

Member slot: **PULSE-ECHO-MODE** from **INSPECTION-PROCEDURES**
  Inheritance: **OVERRIDE.VALUES**
  ValueClass: (ONE.OF SHEAR LONGITUDINAL)
  Cardinality.Max: 1
  Comment: "The transducer mode to be used in pulse-echo scanning"
  Values: Unknown

Member slot: **PULSE-ECHO-TRANSDUCER** from **INSPECTION-PROCEDURES**
  Inheritance: **OVERRIDE.VALUES**
  ValueClass: **TRANSDUCERS**
  Cardinality.Max: 1
  Comment: "The transducer selected for use in pulse-echo scanning."
  Values: Unknown

Member slot: **TANDEM-MODE** from **INSPECTION-PROCEDURES**

Figure 3. A portion of a generic inspection procedure form showing the information for normal beam and pulse-echo scanning. Default values are shown for the normal beam inspection objective and scanning mode.

Figure 4 shows the "external" form of a rule that determines desired transducer characteristics for a ferritic steel pressure vessel weld with multiple wire cladding. The rule is entered into the system in a quasi natural language form. The KEE system parses this form and resolves references to objects in the system and their parameters.

```
(Output) The EXTERNAL.FORM Slot of the CLADDING-RULE-3 Unit
```
Own slot: EXTERNAL.FORM from CLADDING-RULE-3
   Inheritance: SAME
   ValueClass: (LIST in kb KEEDATATYPES)
   Avunits: (RULEPARSE in kb RULESYSTEM2)
     Facet Inheritance: UNION
   Values: (IF
         ((?WORKSHEET IS IN CLASS INSPECTION-WORKSHEETS) AND
         (THE INSPECTION-PROBLEM OF ?WORKSHEET IS ?PROBLEM)
         AND (THE TYPE-OF-REACTOR OF ?PROBLEM IS PWR) AND
         (THE SYSTEM OF ?PROBLEM IS PRESSURE-VESSEL) AND
         (THE INSPECTION-TECHNIQUE OF ?PROBLEM IS
         DIRECT-CONTACT) AND
         (THE CLADDING OF
         (THE WELDMENT-TO-BE-INSPECTED OF ?PROBLEM) IS
         MULTIPLE-WIRE)) THEN
         ((THE DESIRED-TRANSDUCER-CENTER-FREQUENCY OF
         ?WORKSHEET IS 2.25) AND
         (THE DESIRED-TRANSDUCER-DIAMETER OF ?WORKSHEET
         IS 0.375)))

Figure 4. A typical rule from the SIIA system. The rule is shown in a quasi-natural language form. It is entered into the KEE system in this form.

### User Interface

In operation the user starts the system by using the mouse to activate the process of "generate an inspection procedure." The system begins by asking the user to fill in the menu previously shown in figure 2 that describes the basic inspection problem. Based on the values indicated on the menu a second menu is presented that asks for more specific information about the problem.

The system then forward chains through rule bases corresponding to the steps in solving the problem as described earlier. If necessary two worksheets are created. During the forward chaining through the rule bases displays can be activated that show the user which rules are being considered and when they are fired. This is primarily useful for debugging but not for day-to-day operation. In the final version of the system we will likely not use these displays, but provide an indicator of what step the system is working on to show the user that something is happening.

When an inspection procedure has been filled in for the current inspection step the signal-to-noise ratio for the chosen transducer must be checked as described above. At this point the user is asked to make a single point measurement and report the signal-to-noise ratio to the system. If it is acceptable then the system will clear the user to perform the scan, if not then it will determine a new transducer and ask for another signal to noise ratio measurement.

### Interpretation of Scanning Results

At this point we have not implemented any rulebases for interpreting the results of a scan. We are planning that the first version of this part

of the system will advise the user on what to look for in color displays of
scan results and ask him questions about what he sees. Based on the users
responses the system will recommend other data displays that might be useful
and attempt to determine whether there is a flaw displayed in the images.
If there is a flaw it will assist the user in characterizing the flaw.

Complexity of SIIA

As it stands today the SIIA knowledge base consists of approximately
130 KEE units. Included in this count are 25 rules. As a given problem
is solved from four to six additional units are created describing the
problem and its associated worksheet(s) and inspection procedure(s).
Figure 5 shows a graph of most of the current knowledge base. Notice
that the rules are broken into subsets corresponding to the steps in
performing a SAFT inspection. At this point there are only rules related
to general problem description and normal beam scanning.

CONCLUSIONS

Designing and implementing SIIA has been a valuable experience.
From an expert systems point of view it is an interesting problem because it is fairly complex and required the use of a variety of types
of knowledge to solve the problem. From an NDE point of view it has
caused us to consider what role knowledge-based systems should have in
the NDE inspection process.

For the most part the NDE community has not developed detailed
procedures for optimized inspection or procedures for analyzing the results of inspections. There are ASME code or other requirements that
provide procedures for making a weld and tell how often to inspect it,
but generally there is very little that tells how best to inspect it.

Even with an advanced computer based system such as SAFT, the margin
for misapplication still exists. Furthermore, the vast amount of data
generated by such a system can be overwhelming and will increase inspection
times unless optimized analysis procedures (based on expert knowledge)
are employed. We were motivated then to find a way to guide a SAFT user
through the proper use of the system in order to produce consistent high
quality results and to reduce the time to perform a thorough analysis
of the data. One option is to require extensive training and qualification for SAFT operators. This, however, is an expensive option. We
are optimistic that by integrating a knowledge-based system with the
rest of the SAFT system that we can achieve the same result with lower
cost to the end user.

In determining the structure of the SAFT inspection problem we
realized that there are many parallels with conventional ultrasonic inspection and with other techniques such as eddy current inspection.
As with SAFT the other techniques require that the proper transducer
be used and that other aspects of the inspection be set up properly.
The other techniques also require that data be properly interpreted in
the context of the specific inspection. Knowledge-based systems can
assist with consistent solution to all of these problems. We expect
to see more knowledge-based systems for NDE in the future. This will
help reduce the sensitivity of NDE inspections to variation between individual inspectors.

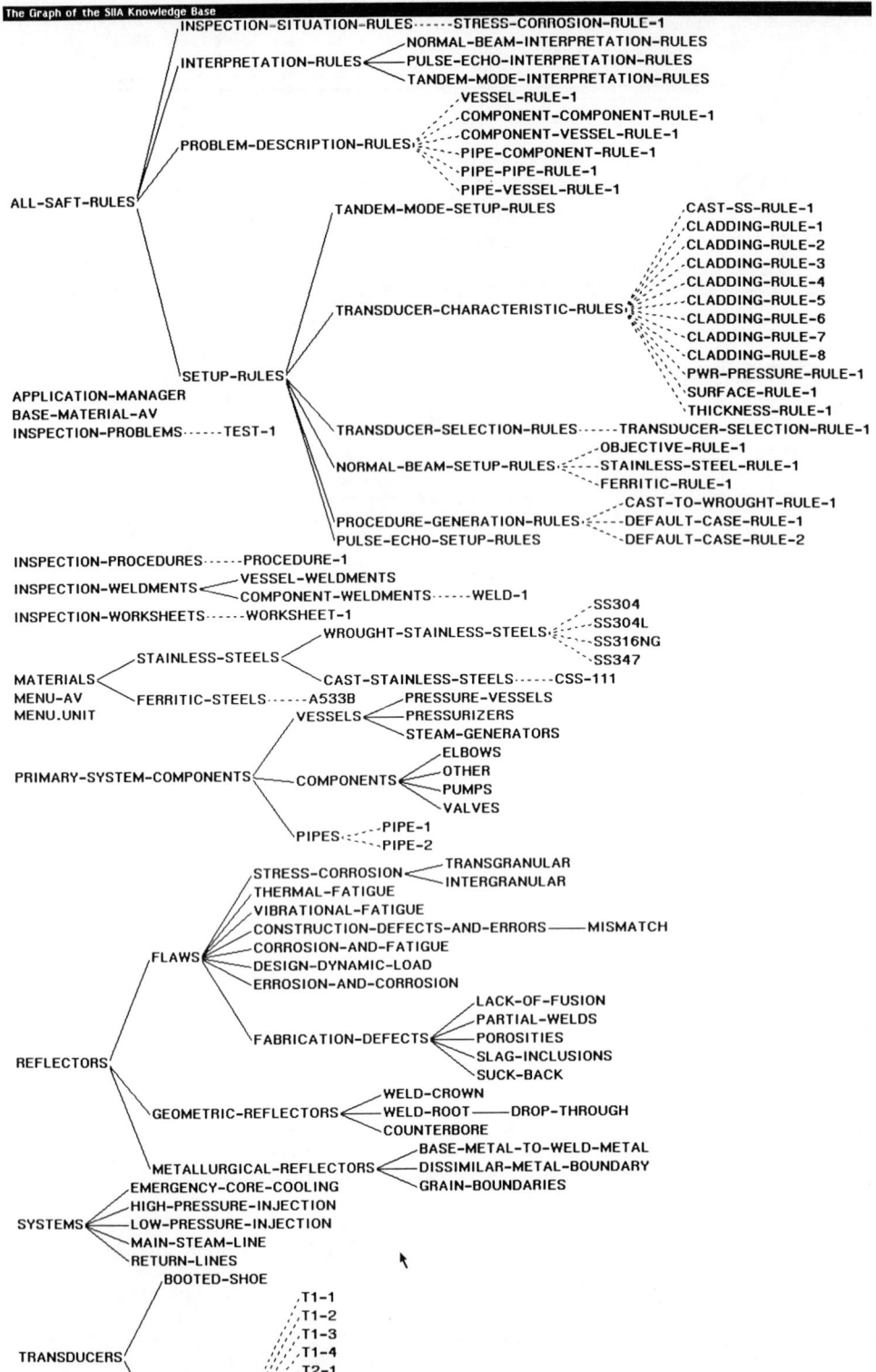

Figure 5. A graphical representation of the SIIA knowledge base. Solid lines indicate subclasses, dashed lines indicate members of a class or subclass, i.e. specific instances.

REFERENCES

1. T. E. Hall, S. R. Doctor and L. D. Reid, A Real-Time SAFT System Applied to the Ultrasonic Inspection of Nuclear Reactor Components, these Proceedings.
2. R. E. Fikes and Thomas P. Kehler, "The Role of Frame-Based Representation in Reasoning", ACM Communications, $\underline{28}$, 904-920 (1985).
3. D. A. Waterman, "A Guide to Expert Systems", Addison-Wesley, Reading, Massachusetts (1985).

# AN AI APPROACH TO THE EDDY CURRENT DEFECT CHARACTERIZATION PROBLEM

L. Udpa and W. Lord

Electrical Engineering Department
Colorado State University
Fort Collins, Colorado 80523

## INTRODUCTION

Conventional eddy current NDT methods rely for their operation on the interaction of quasi-static electromagnetic fields with flaws in the specimen under test. The physics of such interactions are described completely by a parabolic diffusion equation

$$\nabla \times (\frac{1}{\mu} \nabla \times A) = -\sigma \frac{\partial A}{\partial t} + J_s \tag{1}$$

derived from the quasi-static form of Maxwell's equations. This precludes the use of methods such as holography and tomography in the analysis of data from eddy current probes [1]. Ideally, one would desire an analytical closed form solution of equation (1) in terms of the material parameters $\mu(\bar{r})$ and $\sigma(\bar{r})$, so that one has a direct method for solving the inverse problem or imaging problem. The nature of the defect characterization problem in eddy current NDT and the difficulties involved in the analytical modeling of realistic test geometries are described at length in [2,3]. Simulation of nonlinear, practical problems with arbitrary defect shapes are generally done using numerical techniques such as the finite difference and finite element methods. However, the major drawback of numerical models is lack of a closed form solution, which renders them inadequate for directly solving the inverse problem. This paper presents a new approach to the general inverse problem in NDT of defect imaging, that uses the finite element model iteratively for estimating the test specimen parameters $\{\underline{\sigma}\}$ as shown in Figure 1.

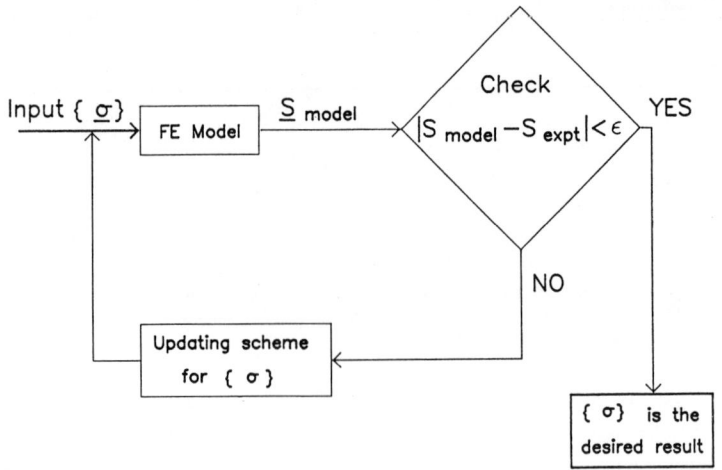

Fig. 1. Algorithm for the parameter estimation problem.

The method, based on artificial intelligence techniques, reduces the general inverse problem to a tree or state space search. One of the commonly used problem solving techniques in artificial intelligence is the method of heuristic search where one searches for the optimal solution in a space of all potential solutions. The method typically consists of two steps

i) State space representation of problem
ii) Search procedure

In order to understand the formulation and implementation of the algorithm, the following definitions are useful.

DEFINITIONS AND NOTATIONS

A _state_ or a _node_ is defined as a possible configuration of relevant parameters in the problem domain. The set S of all possible states of the problem is called the problem _state space_. A state transformation operator O is a rule which transforms a node 'a' in S to a node 'b' in S. 'b' is called the _successor_ of 'a'. The generation of all successors of a node is referred to as the _expansion_ of the node. The starting state of the system is called the _root node_ or _initial node_ and the goal node is a node that satisfies some prescribed termination criteria. The _tree representation_ of the problem state space is given by the quadruple [4] {S,T,I,G} where S is the problem state space, T is the set of state transformation operators, I is the set of initial nodes and G is the set of goal nodes. The problem solution is then obtained using a tree search procedure. In most problems, however, the state space to be searched is explosively large and one has to use _heuristics_, defined as a strategy that reduces the search effort. The following two sections describes the tree-representation of the eddy current imaging problem and the search procedure used.

PROBLEM REPRESENTATION

The tree representation of a problem state space is completely specified by identifying precisely the four sets in the quadruple defined in the previous section. Heuristic information about the problem can also be built into the general scheme.

## Problem State Space

A typical test specimen is a three dimensional object but this paper considers an axisymmetric test geometry shown in Figure 2 where an eddy current probe moves inside a hollow tube with axisymmetric defects in the tube wall. Discretizing the test object by a two dimensional matrix of rectangular cells, the states of the problem are identified by matrices of the form

$$s = \begin{bmatrix} p_{11} & p_{12} & \cdots & p_{1n} \\ p_{21} & p_{22} & \cdots & p_{2n} \\ p_{mn} & p_{m2} & \cdots & p_{mn} \end{bmatrix} \quad (2)$$

where the row and column indices of an element select a cell in the object and the value of the element is either 0, indicating presence of a flaw, or 1 indicating no flaw in the cell. The state s, then represents the discrete spatial distribution of material properties in the test specimen. As shown in Figure 3, the discretization of the tube wall is not uniform but conforms to the finite element mesh that models the problem.

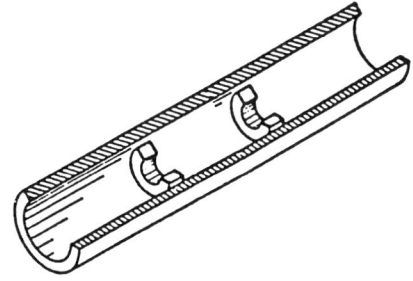

Fig. 2. Axisymmetric eddy current NDT geometry.

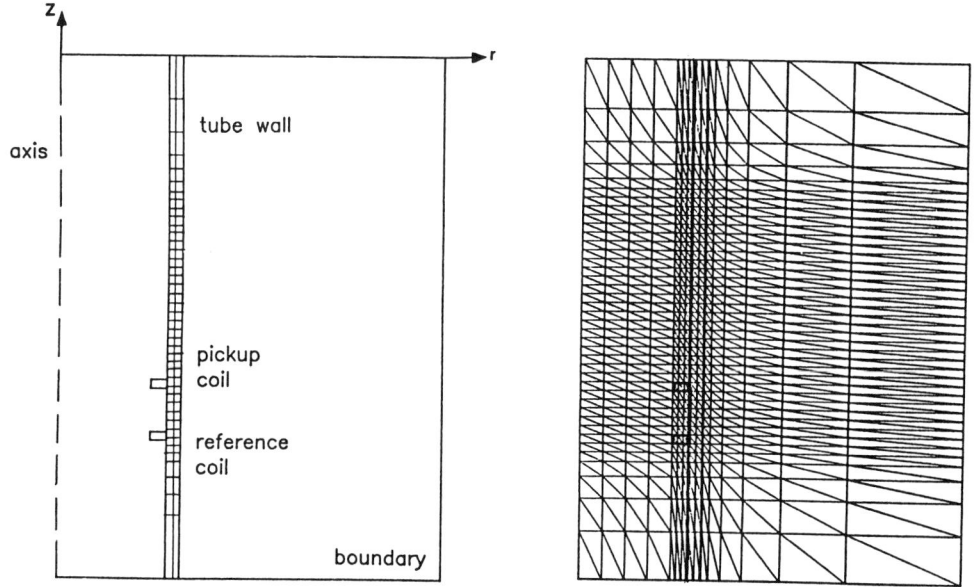

Fig. 3. Two dimensional discretization of tube wall corresponding to the mesh used in the finite element model.

## State Transformation Operators

Since the search space for a discretization of m=n=10 consists of $2^{100}$ states, it is required to introduce heuristics to constrain the search to a reduced subspace. These constraints can be incorporated into the state transformation rules so as to limit the expansion of a node to a select few successors. In the first constraint the defect boundary, defined by a two dimensional sequence of edges as explained in Figure 4, is considered to be a two dimensional discrete time Markov process indexed by the depth into the material in units of cells. The transition probability matrices characterizing the Markov process is chosen to allow only smoothly varying boundaries and transitions leading to sharply varying boundary sequences are prohibited. Figure 5 contains a very simple choice of transition probability matrices. The second constraint is based on the assumption that the crack/defect grows only narrower with depth. These constraints result in a considerable reduction of the search space by defining T such that the fanout factor for a node expansion is 6 as seen in Figure 6.

Fig. 4. Defect boundary, defined by sequences of edges, modelled by a Markov process.

$$P_{12} = \frac{1}{3}\begin{bmatrix} 1 & 1 & 1 \end{bmatrix}$$

$$P_{23} = \frac{1}{3}\begin{bmatrix} 1 & 1 & 1 & 0 & 0 \\ 0 & 1 & 1 & 1 & 0 \\ 0 & 0 & 1 & 1 & 1 \end{bmatrix}$$

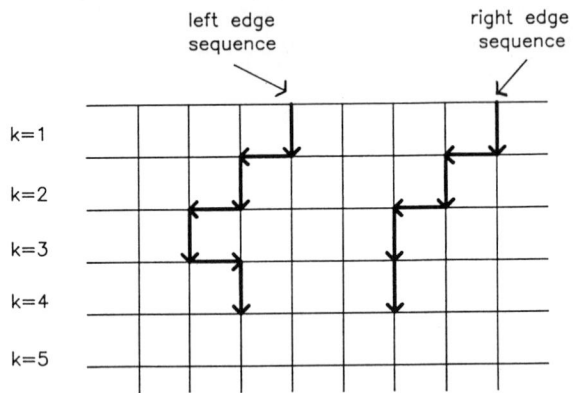

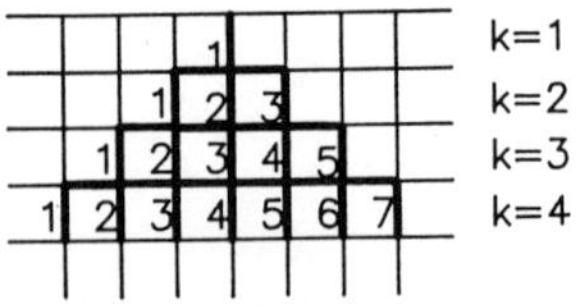

$$P_{34} = \frac{1}{3}\begin{bmatrix} 1 & 1 & 1 & 0 & 0 & 0 & 0 \\ 0 & 1 & 1 & 1 & 0 & 0 & 0 \\ 0 & 0 & 1 & 1 & 1 & 0 & 0 \\ 0 & 0 & 0 & 1 & 1 & 1 & 0 \\ 0 & 0 & 0 & 0 & 1 & 1 & 1 \end{bmatrix}$$

Fig. 5. Transition probability matrices.

## Initial Node

Restricting the class of defects to surface breaking cracks, the initial node is defined by the spatial extent of the flaw at the surface. It is represented by a two dimensional, unit length Markov chain $\{x_1^L, x_1^R\}$

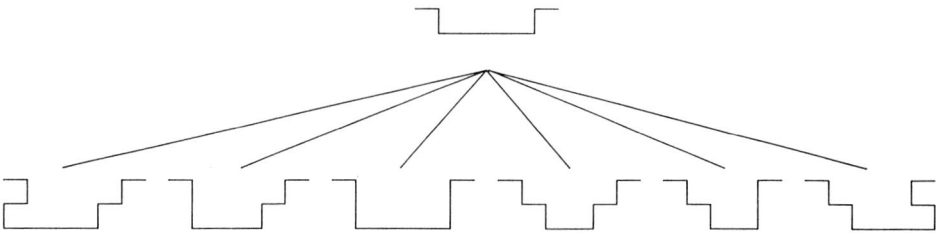

Fig. 6. Expansion of a node into its successors.

corresponding to the left and right edges of the surface opening. The start node is determined using characteristic features in the experimental signal. For instance, in Figures 7 the phase plots of the eddy current probe signal for three defect widths, exhibit a discontinuity close to edges of the defect and is used for estimating the initial node.

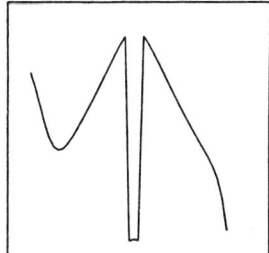 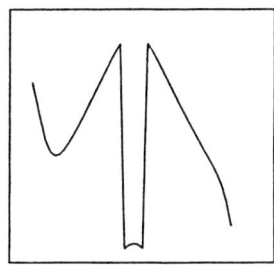 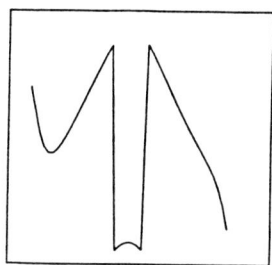

Fig. 7. Eddy current phase signal for three different defect widths.

## Goal Node

The goal node of the problem is defined by the two dimensional depth profile generated by the left and right edge sequences such that the corresponding defect produces a signal at the surface that is 'closest' to the experimentally measured signal.

## SEARCH PROCEDURE

A simple A* type algorithm [4] is used to search for the goal node. The search is conducted in two phases, namely the node generation phase and the node evaluation phase. The first few steps of the basic search procedure are given below.

1. Define $S_1 = [\{x_1^L, x_1^R\}]$ where $x_1^L$ and $x_1^R$ indicate the unit length, left and right edge sequences. The corresponding defect obtained by labeling all the cells in the layer K=1 and in between $x_1^L$ and $x_1^R$ as belonging to defect (air) is input to the FE model. The node $S_1$ is evaluated by computing $e(S_1)$ = 'distance' between model output and input signal. If $S_1$ is not goal node store $S_1$ and $e(S_1)$.

2. Expand node $S_1$. Define

$$S_2^k = [\{x_{2,i}^L, x_{2,j}^R\}, \{x_1^L, x_1^R\}]$$

i=1,2,3; j=1,...i; k=1,2,...6

Compute $e(S_1^k)$, k=1,2,...6

Check for goal node.

3. Select the most promising (smallest e value) unexpanded node and go to 2.

Some of the evaluation functions used in the algorithm were the $\ell_2$ norm and the $\ell_\infty$ norm.

## RESULTS

The method was first tested on defects simulated using the FE model. Simple defects input to the system and the results of the algorithm are shown in Figure 8. The complete tree expansion with the cost of each node using $\ell_2$ and $\ell_\infty$ norms is shown for the simple rectangular defect. A more complex shaped defect input to the system and the corresponding result is seen in Figure 9. Results of implementing the algorithm on some experimental signals from axisymmetric rectangular defects are given in Figure 10.

## CONCLUSIONS

Results obtained so far do indicate the potential of the method as a possible imaging tool. The approach used is very general and can be applied to variety of problems such as eddy current NDT or ultrasonic NDT by employing the appropriate FE model. The imaging procedure depends on extensive use of the numerical model, and is computationally intensive, but when a forward problem is best described by a numerical model the approach presented here seems to be the only method for solving the inverse problem.

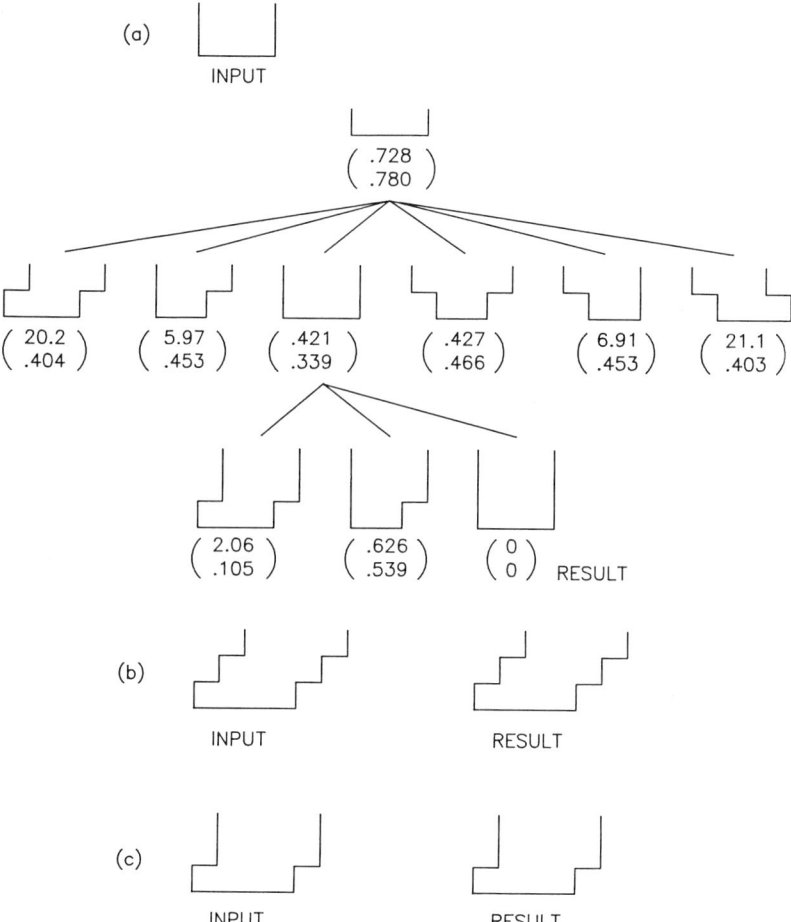

Fig. 8. Eddy current simulation results on some simple defects.

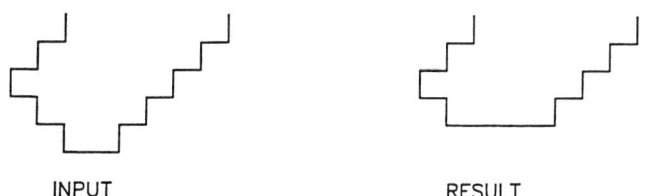

Fig. 9. Eddy current simulation results on a complex defect profile.

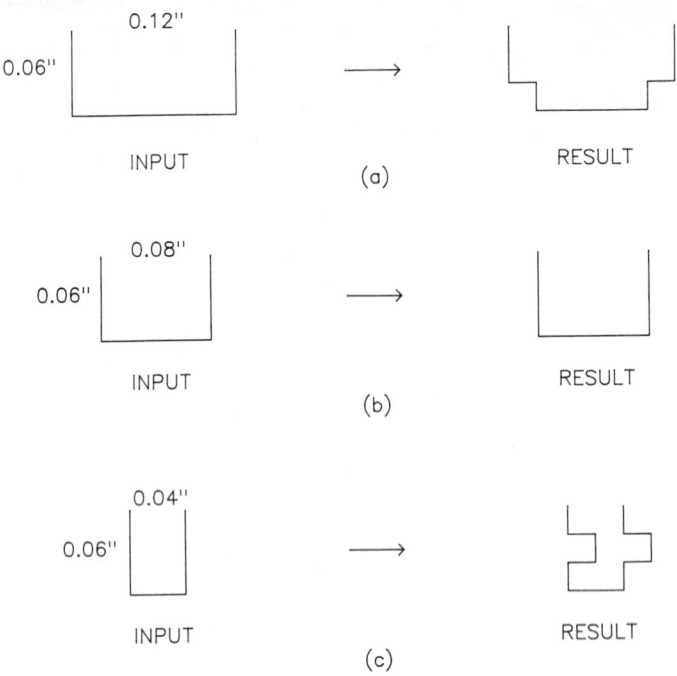

Fig. 10. Results of implementation on experimental eddy current signals from three rectangular defects.

REFERENCES

1. L. Udpa and W. Lord, Diffusion, Waves, Phase and Eddy Current Imaging, in: "Review of Progress in Quantitative NDE," Vol. 4, D. O. Thompson and D. E. Chimenti, Eds., Plenum, New York (1985) pp. 499-506.

2. L. Udpa and W. Lord, A Discussion of the Inverse Problem in Electromagnetic NDT, in: "Review of Progress in Quantitative NDE," Vol. 5, D. O. Thompson and D. E. Chimenti, Eds., Plenum, New York (1986) pp. 375-382.

3. W. Lord and L. Udpa, Imaging of Electromagnetic NDT Phenomena, in: "Review of Progress in Quantitative NDE," Vol. 5, D. O. Thompson and D. E. Chimenti, Eds., Plenum, New York (1986) pp. 465-472.

4. E. Rich, "Artificial Intelligence," McGraw-Hill, New York (1983).

DESIGN AND OPERATION OF A DUAL-BRIDGE ULTRASONIC INSPECTION

SYSTEM FOR COMPOSITE MATERIALS

David C. Copley

General Electric Company
Aircraft Engine Business Group
Cincinnati, Ohio

INTRODUCTION

The accepted method of inspecting composite aircraft and engine components is ultrasonic testing with a C-scan presentation of results. Within the last few years, analog recording systems have been superseded by digital data acquisition and analysis. Programmable control of mechanical scanning has improved upon earlier machines which could test only simple flat and circular shapes. The equipment described here was designed to inspect complex profiles such as aircraft engine vanes and ducts. A unique dual-bridge design with synchronized eleven-axis motion control was used to achieve this. The success of the system resulted from careful definition of initial design, and from the use of high-level software to supervise the otherwise lengthy task of programming complex scanning profiles.

BASIC DESIGN CRITERIA

Mechanical and electrical design requirements are defined by the type of material and component to be tested, and from the nature and critical size of expected defects. In this case, the components were various aircraft engine parts made from carbon-fiber reinforced polyimide. The material is prone to microvoids in the plastic matrix as a result of gases produced in the curing reaction. These microvoids affect the matrix-dominated properties such as cross-ply tensile and interlaminar shear strengths. It is desirable to measure void content down to 1% by volume and delaminations down to 6mm diameter.

Detection and measurement of microvoids is done by an ultrasonic through-transmission method as desribed by Stone and Clark [1]. New curves of attenuation coefficient versus void content were derived empirically for the polyimide material. However, a pulse-echo technique is also needed in order to distinguish between microvoids and discrete delaminations or inclusions, as well as to measure the depth of discrete defects. For this reason the design was configured to perform both types of test.

The size and configuration of the engine components also dictated certain design decisions. A number of parts have a double skin, with an internal cavity. This cavity had to be kept full of water during through-transmission tests, a requirement which could be fulfilled easily for an immersion test, but was almost impossible for a squirter configuration.

Another factor in favor of the immersion system was the more compact gimbal design which could be used, facilitating access inside small components. One should recognize that squirter systems are preferable to immersion systems under many other circumstances; the inherent collimation of the ultrasonic beam imposes less restrictions on mechanical positioning, and they eliminate the difficulty of immersing large sandwich structures.

The most stringent requirements were produced by components with compound curvatures. Components with parallel faces (single or double skin) do not cause deviation of the normal-incidence sound beam, and the principal requirement is that the transmitter and receiver are co-axial, with the ultrasonic beam being within $2^o$ of normal to the surface. This could be achieved using mechanical coupling through a rigid yoke, though this would be very cumbersome for large components. However, components with non-parallel sides, such as airfoil sections, refract the ultrasonic beam, requiring independent manipulation of the transmitting and receiving transducers (Figure 1). Three conceptual designs were outlined which could accomplish the required motions. The dual robot configuration (Figure 2a) has been successfully used for inspection of complex configuration parts. The use of commercial robots such as the Unimation Puma can eliminate some of the design and programming effort, but they cannot operate in the immersion environment. The remaining two possibilities are a single bridge with rotary bearing (Figure 2b), and a dual bridge design (Figure 2c). Both of these require control of eleven axes. The rotary bearing design provides a degree of mechanical coupling between the transducers, but it is restricted in range of transducer spacing and in the ability to straddle a large component. The dual bridge design provides more flexibility, and mechanically it is a simple extension of conventional single-bridge designs. The final choice of the dual bridge design was also influenced by the very short time available to procure the equipment: its commonality with existing designs resulted in a shorter procurement cycle.

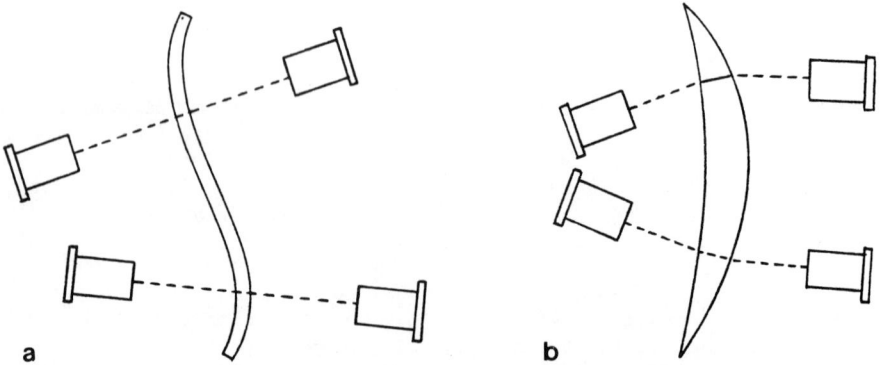

Fig. 1. Transducer alignment needed for scanning of (a) parallel-sided, and (b) airfoil section components.

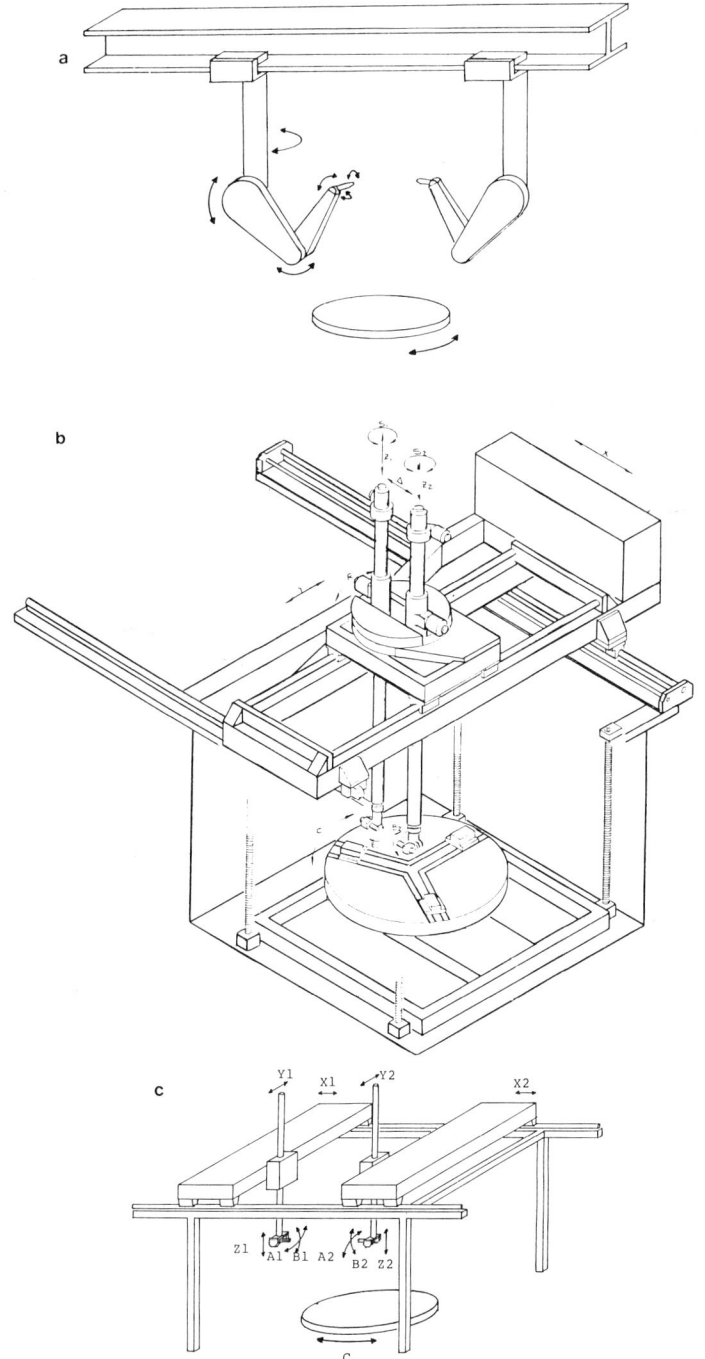

Fig. 2. Mechanical Configurations to Scan Complex Curvatures.
 a) Dual Robot (13 Controlled Axes).
 b) Rotary Bearing (11 Controlled Axes).
 c) Dual Bridge (11 Controlled Axes).

EQUIPMENT DESCRIPTION

Based on the above considerations, a procurement specification was written, covering hardware and software requirements. The limiting constraint on mechanical accuracy was the need to maintain the through transmitted signal constant within 1dB. For the transducers normally used (5MHz, 0.25" diameter), this means that the focal points must be coincident laterally within 1.2mm. This can be accomplished by a combination of gimbal angular accuracy (within $0.1^o$), and linear accuracy (within 0.4mm) in the Y and Z directions.

The detailed design and manufacture of the system were performed by California Data Corporation. Configuration is shown in Figure 3, and Figure 4 is a view of the hardware. The DEC PDP 11/73 host computer runs under the RSX 11M operating system with timesharing of the motion control, data acquisition, display, and editing functions. Mass storage consists of an 80 MByte Winchester, with a tape drive for archival storage and an 8" floppy disc drive. Motion of the six linear axes is accomplished by DC motors with encoder feedback of rotary shaft position. Positioning of the five rotary axes is controlled by stepping motors. The Krautkramer-Branson KB6000 ultrasonic instrument operates under computer control with up to eight channels active. The multiple channels are used to increase the dynamic range beyond the limits of a single channel, and in pulse-echo mode allow "sectioning" of material into depth slices. An output on each channel of signal amplitude (7-bit) and time-of-flight (12-bit) is passed to the data acquisition system. During scanning, the active channels are sampled typically at a 1.2mm interval. Data from all channels are stored on disc, and one channel is displayed in near real time. Typical scans occupy 1 to 5 MBytes of disc storage. Color hard copy is provided by a Quadram inkjet printer and by a Nicolet plotter.

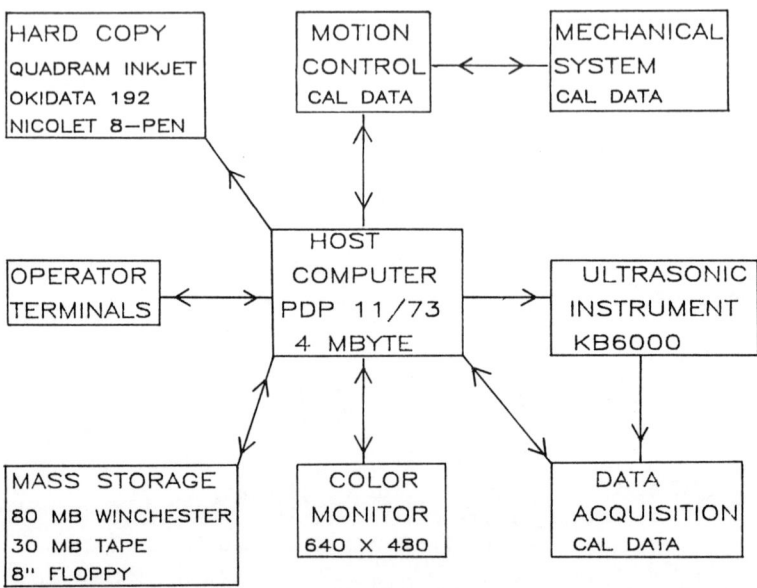

Fig. 3. Equipment Configuration

Fig. 4. View of System Hardware.

OPERATING PRACTICE

Inspection of a part is performed by execution of a command file which contains instructions for motion control, instrument setup, and data acquisition parameters. These functions may be included in the command file or called as subroutines. Prompts are displayed on the operator terminal to instruct the operator to perform any steps not under program control. At present these steps are part loading and unloading, initial alignment of transducers to a reference plane, and checking of calibration signal amplitude. Software routines are currently being written to perform these last three functions.

Programming of scan profiles is usually done by a teach process. A component of the type to be tested is fixtured in the tank. The mechanical axes of one bridge are moved under manual control to selected points on the scan profile, and the coordinates of each point are taught by a menu selection. The spacing at which points are taught depends on the curvature, a 25mm spacing is typical. When the points have been taught for one bridge, the positions of the second bridge are generated by the transformations:-

$$a2 = a1 \qquad b2 = b1$$

$$X2 = X1 + R\cos(a1)\cos(b1)$$

$$Y2 = Y1 - R\sin(a1)$$

$$Z2 = Z1 - R\sin(a1)\cos(b1)$$

Where R is the separation between gimbal pivot points and the axis definitions are as shown in Figure 2c.

A utility program then generates the interpolation between the points. The output is a motion control file which is either incorporated in, or called as a subroutine from, the master command file.

The usual data presentation is a color-mapped amplitude C-Scan for each active channel, as shown in Figure 5. The color lookup table can be defined in either color or grayscale, with control over the slope and offset of the table. Image analysis functions include zoom to magnify a selected area, position reports of cursor location, and histogram presentation of amplitude distribution in a selected area. Estimation of void content using attenuation / void content curves is inconvenient using the standard linear amplitude presentation. To simplify this process, a post-processing program was written to provide logarithmic output. This program also extends the dynamic range by combining up to four linear channels which have different amplification levels. The post-processing program acts effectively as a software logarithmic amplifier with a dynamic range of 80dB. For components of constant cross-sectional thickness, a color look-up table is then defined with color thresholds corresponding to 1% intervals in void content.

A more desirable way of incorporating the attenuation / void content relationship is to make thickness measurements at each data point as a part of the scan, and then calculate the local attenuation coefficient at that point. In theory, this may be implemented by a measurement of time delay between front and back surface echoes, or by a pulse-echo location of each surface from the closest transducer. Either method will present some practical difficulties on complex part configurations.

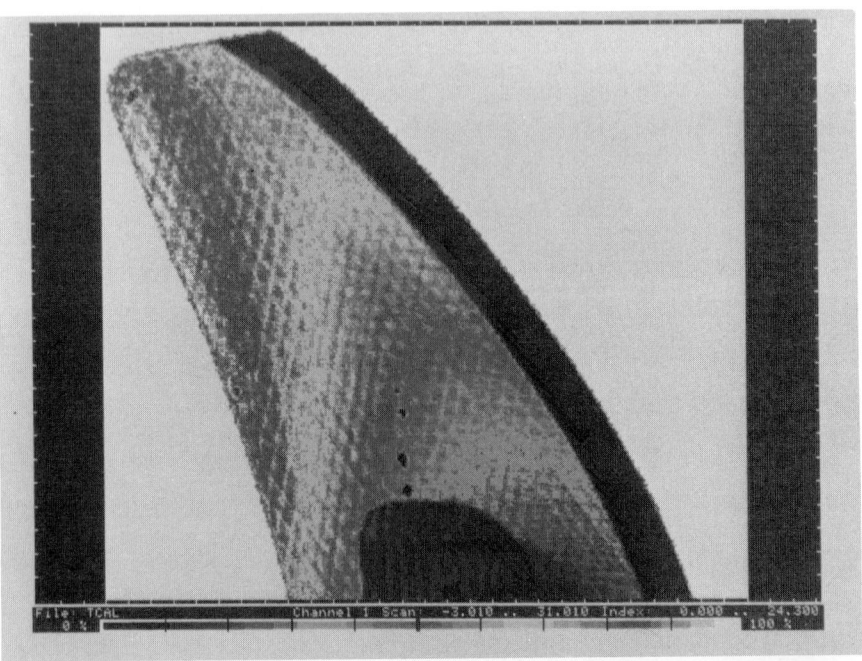

Fig. 5. Amplitude C-Scan of Composite Propulsor Blade.

OPERATING EXPERIENCE

Experience to date has been favorable, with unscheduled downtime below 5%. Most problems have been mechanical in origin, with software and electrical problems being minimal. Time to inspect components has been reduced by a factor of between two and five, compared with previous semi-automated equipment, with the greatest improvement being on more complex components. At present, more time is spent performing initialization and data analysis than is spent scanning parts. For this reason, future developments will be directed towards off-line data analysis, and the automation of the normalization and calibration routines. Another area where development is planned is the generation of scan profiles from geometric data obtained either from computer-aided design, or from an optical scan of the component.

SUMMARY AND CONCLUSIONS

The dual-bridge design has proved to be an extremely flexible and effective configuration for inspecting complex parts. The selection of a well-established computer and operating system (PDP 11/73 and RSX11M) resulted in a smooth introduction of the equipment and ready availability of expert help. Because of time limitations, the design was conservative, using existing modular designs and software wherever possible. Not suprisingly, most of the problems encountered were with the previously untested parts of the system. Improvements are still needed in the areas of interpretation of scan data, and the generation of scan profiles from geometric data.

REFERENCES

1. D.E.W. Stone and B. Clarke. Ultrasonic Attenuation as a Measure of Void Content in Carbon Fiber Reinforced Plastics. Nondestructive Testing. $\underline{8}$, 137 (1975).

USING A SQUIRTER TO PERFORM PULSE-ECHO ULTRASONIC INSPECTIONS

OF GAS TURBINE ENGINE COMPONENTS: THE PROS AND CONS

David A. Stubbs

Systems Research Laboratories
2800 Indian Ripple Road
Dayton, Ohio 45440

INTRODUCTION

The Air Force's Retirement For Cause (RFC) ultrasonic system uses a low pressure water squirter system to couple the ultrasound to the engine part undergoing inspection. From an overall system point-of-view, there are many advantages in the use of the squirter as compared to the use of a standard immersion tank. Foremost of the advantages are the ease of use and ease of maintenance. However, from an NDE point of view (the reliable detection of small flaws) the squirter technique has several disadvantages. The squirter complicates the inspection process by adding factors such as a dynamic water column serving as the couplant, additional size, and many reflecting surfaces to the already difficult task of detecting flaws in the complex shapes of gas turbine engine components. The details of these problems and their solutions are discussed in this paper.

ADVANTAGES

One of the goals of the RFC ultrasonic inspection module is to detect a 0.020 inch diameter, mal-oriented, penny-shaped, internal void in gas turbine engine components. Previous reports [1,2] have shown the squirter technique capable of detecting these small defects. Figure 1 shows a photograph of the squirter and Figure 2 shows photographs of two rf waveforms from a small side drilled hole. The left photograph in Figure 2 shows the reflection using an immersion system and the right photograph shows the signal obtained using a squirter. Note the similiar signal to noise ratios. Based on data such as this the decision was made to use the squirter in the RFC production inspection system. Using the squirter proved very advantageous from an overall system point of view. The total RFC system consists of five eddy current inspection stations and two ultrasonic stations and by using a squirter the mechanical manipulators are nearly identical for both the eddy current and ultrasonic systems (see Figure 3). This commonality proved very beneficial in terms of development work and system maintenance. Another benefit of using a squirter is that a large immersion tank is unnecessary, thus the maintanence tasks associated with a tank (periodic cleaning and rustproofing, constant refilling, etc.) are eliminated. A third advantage is that the water flow through the squirter removes air bubbles from the face of the transducer that would cause errant

Figure 1 - This photograph shows the squirter used in the RFC ultrasonic system. The entire squirter is made of acrylic and is approximately three inches long.

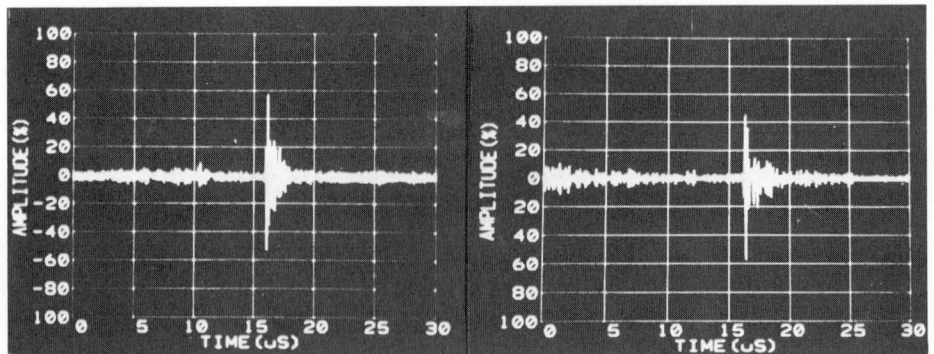

Figure 2 - Both photos show a signal from a 0.020 inch diameter, side drilled hole 0.75 inches below the surface of the bore of a F-100 engine disk. The left signal is from an immersion system. The right signal is from the squirter system.

Figure 3 - A view of the RFC facility at Kelly AFB in San Antonio, TX. Three eddy current inspection stations are on the left. One ultrasonic station is on the right. Notice the similiarity between the two types of stations. This similiarity is a direct result of using a squirter for the ultrasonic inspection.

signals (or no signals) during the inspection process. The RFC systems are designed to be fully automated inspection systems - it would not have been convenient to have an operator wipe the face of the transducer before every inspection as is typically done in an immersion system. Finally, because the part fixture is up on the inspection station rather than down in a tank (see Figure 4), the operator has a much easier task of loading and unloading the engine parts. Since the weight of some of the parts of the F100 engine exceed forty pounds this is not a trivial advantage.

DISADVANTAGES

Acknowledging that nothing in life is free, it is not surprising that along with the many advantages of using a squirter there also come some disadvantages. Much data were gathered supporting the equality of the squirter technique with the immersion method. These data showed the signal-to-noise ratios and the frequency content to be the same for both methods. However, all of these data were gathered under static conditions on a small sample of test specimens. In the production inspection process the engine parts are rotated resulting in linear scan speeds of one to five inches per second. Additionally, all engines parts are not made the same. And finally, in a production enviornment the alignment and stability of the mechanical manipulators cannot be maintained as precisely as in a laboratory enviornment. These conditions result in increased noise in the ultrasonic signal.

Figure 4 - An F-100 engine component undergoing an inspection. The mounting fixture is waist high and easily accessed by sliding open the acrylic splash guard.

One type of noise resulting from the use of a squirter is random water noise. This occurs when the water stream splashes over an edge of the engine part and is the most troublesome when the squirter is near the top or bottom of a bore region (see Figures 5 and 6). A second type of water noise occurs when the inner web region fills up with water and spills down into the stream of water coming from the squirter. Both of these water noise conditions produce random spikes in the ultrasonic signal. A typical bore inspection requires 40,000 A-scans; thus even occasional noise spikes can add up to an unacceptable number over the course of an entire scan. Two approaches have been used to help overcome the water noise. The first involves positioning the squirter nozzle very close to the surface of the part which decreases the length of the water column and helps reduce the splashing. A typical standoff distance between the part surface and the nozzle tip that substantially reduces the water noise is 0.050 inches. The RFC mechanical manipulators have positional resolution of 0.0001 inches so a standoff of 0.050 inches is easily maintained. The second solution utilizes the temporal shifting of the reflection from a true defect signal to an advantage in a software averaging algorithm [3]. Through the combined use of both of these techniques a fairly noise-free ultrasonic signal can be obtained during a production type inspection.

The other predominant type of noise is the presence of unwanted reflections. These reflections usually come from the ultrasound reflecting off the part surface and the squirter nozzle as shown in Figure 7. To help reduce these reflections the orifice of the nozzle was made as wide as possible and the length of the nozzle was shortened. These efforts substantially reduced the frequency of occurence and the amplitude of the reflections when present. The use of the time shifted averaging algorithm mentioned earlier [3] also helps reduce the amplitude of the reflections. However, in most inspections it is the presence of these reflections that determines the flaw detection threshold level and thus the minimum size flaw that can be detected using the squirter.

Some of the tools that are used to help reduce the occurence and amplitude of the noise signals arising from the use of the squirter have been discussed. It has been found that there is one other means of reducing the noise. All of the engine part inspections are executed from a "scan plan". These scan plans control the movement of the mechanical manipulator, the setup of the data aquisition instruments, and the signal processing algorithms. The mechanical movements are derived using the engine part blue prints. By careful positioning of the squirter standoff the amplitude of the unwanted reflections can be reduced to below the desired threshold levels. The nozzle has been designed so that the optimum standoff is usually in the 0.050 inch range that is also desirable to reduce water noise. The exact positioning is very critical. Figure 8 shows the increase in the amplitude of a reflection when the squirter is mis-positioned by only 0.030 inches. Fortunately, this level of positioning is well within the accuracy and repeatability ranges of the mechanical system. Unfortunately, the engine parts themselves sometimes vary by more than this. To compensate for the part variation each inspection incorporates a "dimensioning" algorithm that measures the variation in the engine's part dimensions.

There is one additional disadvantage of using a squirter. The present inspection requirements for some of the engine parts require the inspection of regions with complicated geometries. In many cases the size of the squirter prevents the scan plan writer from positioning the squirter in the most efficient scanning position. Figure 9 shows a typical inspection situation. It would be desirable to inspect the bore of this part from the top and bottom sides as well as from the bore because the top and bottom

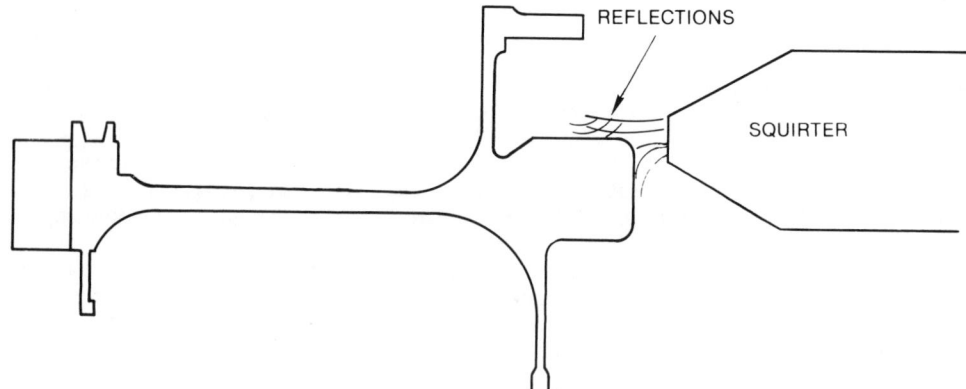

Figure 5 - Unwanted reflections occur when the ultrasound is reflected from the turbulent water splashing over an edge. Also notice how the water can pool in the groove behind the bore and then spill over the bore edge.

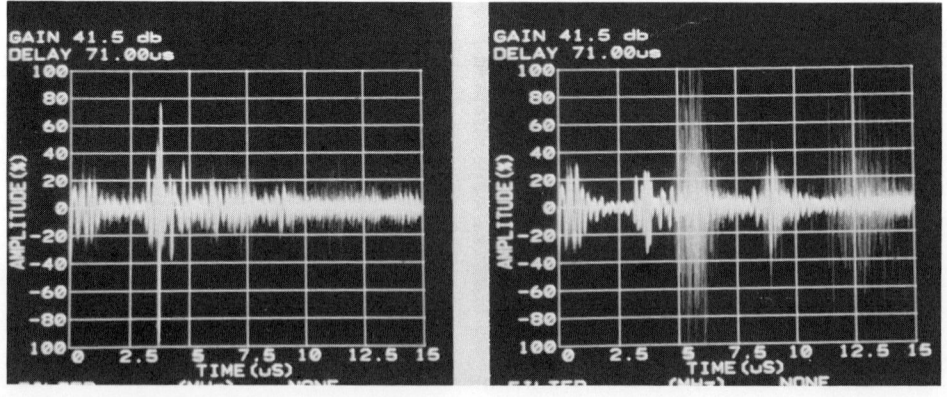

Figure 6 - The left photo shows the signal from a side drilled hole without water noise. The right photo shows the side drilled hole signal and the water noise that occurs as the squirter is moved near the top of the bore. Both photos have a two-second duration exposure

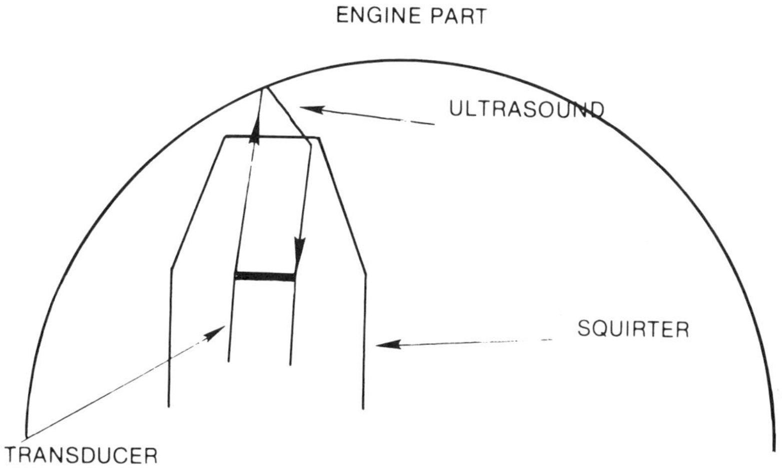

ENGINE PART

ULTRASOUND

SQUIRTER

TRANSDUCER

Figure 7 - This simple drawing illustrates how the ultrasound reflects off the engine part and the squirter nozzle to produce unwanted reflections.

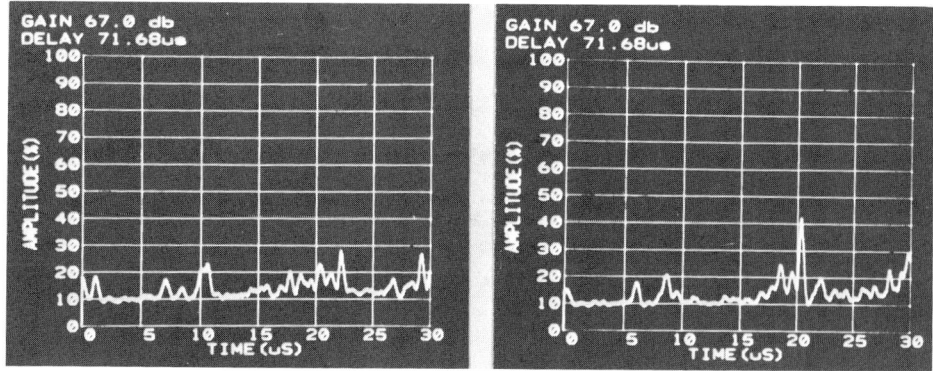

Figure 8 - These two photos show a video signal of the bore region with no defect present. In the left photo the squirter is correctly positioned. In the right photo the squirter is 0.030 inches too far away from the bore. Note the unwanted reflection at 20 u-sec

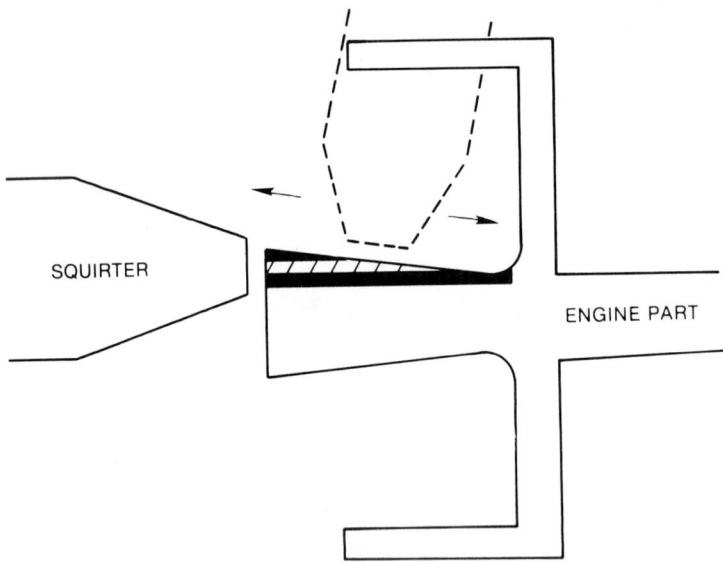

Figure 9 - This illustration shows the disadvantage of the squirter's size in trying to scan various geometries.

corner regions produce noisy signal due to water splashing. But the presence of the "L-shaped" arms prevent the scanning from the top and the bottom. In this case the scan plan must scan the entire region from the bore side. To reduce the effect of the water noise near the corner the scan is broken into several zones where each zone has a different scan depth. Although this is inefficient, it allows the complete coverage of the bore region in this engine part.

SUMMARY

The RFC ultrasonic inspection system uses a squirter technique to couple the ultrasound to the engine part. There are both advantages and disadvantages in using a squirter with most of the advantages being on an overall system level and the disadvantages being the increased noise level in the ultrasonic signal. Through the use of careful squirter design, signal processing algorithms in software, and careful mechanical positioning of the squirter the noise level can be reduced to an acceptable level - in this case a level low enough to allow the detection of 0.020 inch diameter, mal-oriented, penny shaped voids. The final result is that the squirter can be used in a production inspection mode at a sensitivity that is equivalent to the level of sensitivity of an immersion system.

ACKNOWLEDGEMENT

This work was conducted under Air Force Contract F33615-81-C-5002.

REFERENCES

1. G. M. Light, R. A. Cervantes, and W. R. Van der Veer, Evaluation of Captured Water Column Technology for Advanced Ultrasonic Sizing Techniques in: "Review of Progress in Quantitative Nondestructive Evaluation", D. O. Thompson and D. E. Chimenti, editors, Plenum Press, New York (1984), 1359.

2. G. M. Light, W. R. Van der Veer, D. A. Stubbs, and W. C. Hoppe, Evaluation of Captured Water Column Technology for Advanced Ultrasonic Sizing Techniques in: "Review of Progress in Quantitative Nondestructive Evaluation", D. O. Thompson and D. E. Chimenti, editors, Plenum Press, New York (1986), 885.

3. D. A. Stubbs and B. Olding, "Ultrasonic Flaw Detection Using a Time Shifted Moving Average", Published elsewhere in this volume.

AN APPARATUS FOR THE AUTOMATED NONDESTRUCTIVE TESTING OF ELECTRO-

EXPLOSIVE DEVICES USING THE ELECTROTHERMAL TRANSIENT TECHNIQUE

Harper L. Jacoby, Keith S. Pickens, John F. Tyndall,
Steve P. Clark, Robert A. Baker, and Ronald H. Peterson

Southwest Research Institute
San Antonio, Texas

INTRODUCTION

An electroexplosive-device (EED) tester was developed by Southwest Research Institute (SwRI) for the U.S. Army Munitions and Chemical Command under contract number DAAK10-84-C-0206. The tester was designed to nondestructively evaluate (NDE) eight different electrothermal devices at production rates up to 45 parts per minute. The tester automatically removes the device to be tested from an input tray, places it in a test fixture, performs an electrothermal transient test, and places the device into an accept/output tray or a reject bin, depending on the quantitative decision made by an analysis computer. The entire NDE is done without hazard to the operator in a self-contained, purged enclosure. The accept/reject criteria are operator-selectable and are applied through use of Boolean statements. Test results and measurement accuracy are designed to be equal to those achieved using laboratory instrumentation.

The tester also allows producers of such devices to evaluate or discern changes of materials or manufacturing processes in a timely manner. Stored test results and summary information may be used to develop a large statistical database for analysis of electrothermal parameters.

ELECTROTHERMAL TRANSIENT TESTING

An EED consists of a bridgewire with an explosive material in intimate contact with the wire. The wire and explosive are encased in a hermetically sealed container with at least one projecting electrical lead. When a sufficient level of electrical current is introduced into the wire, the temperature and resistance of the wire increase until the explosive ignites. Detonation of the devices triggers a variety of diverse mechanisms such as ordnance, safety equipment, space flight hardware, and defensive countermeasures; thus, failure or imprecise detonation can have extreme implications far beyond the unit price of the item.

Thermal transient testing of EEDs has been done on a limited basis for many years using a manually balanced Wheatstone bridge. Because the testing was labor-intensive and somewhat hazardous, a need for an automatic test station had become apparent. Automated nondestructive electrothermal transient testing of EEDs is accomplished by applying a constant-current, step-function pulse, somewhat below the normal firing current, to the device. Instrumentation measures and records the electrical response to the current pulse. This response describes the physical status of the device. Figure 1 depicts an idealized waveform output during a thermal transient test [1]. The various signal features characterizing the response of the device are identified. Poor bridgewire welds, inadequate contact with the explosive, and other defects are evidenced by anomalies in the electrical responses.

The EED tester consists of four subsystems: (1) mechanical, (2) mechanical control, (3) measurement, and (4) analysis and control. A pictorial representation of the EED tester is shown in Figure 2.

MECHANICAL SUBSYSTEM

The mechanical subsystem (shown in Figure 3) consists of four distinct groups: (1) input X/Y table and pick-and-place arm, (2) rotary table and inspection station, (3) output X/Y table and pick-and-place arm, and (4) reject bin. A tray of devices to be tested is placed onto a platter assembly which is then placed on the input X/Y table. A similar platter of empty trays is placed on the output X/Y table. The input X/Y table moves the device to be tested beneath the pickup point of the pick-and-place arm. The pick-and-place arm removes the device from the input tray using a pickup tube with an applied vacuum. The pick-and-place arm then moves the device to the rotary table where the device is loaded into a holding fixture. The rotary table indexes until the device is beneath the test fixture at the inspection station. The test fixture clamps down to apply four pogo-pin contacts to the device to be tested, and the electrothermal transient test is conducted. The test fixture unclamps at the conclusion of the test. The device is then rotated incrementally until it is beneath the pickup point of the output pick-and-place arm. If the device was judged acceptable, the output pick-and-place arm picks the device from the holding fixture. The arm rotates to place the device into an empty tray location which has been properly positioned by the output X/Y table. If the device is unacceptable, it continues in the rotary table until it is ejected from the holding fixture into a reject bin by an air burst. The actual movement of the device is not performed in a serial fashion as described. Each activity is occurring concurrently during a machine cycle. The operator must replace the output platter when it becomes full, empty the reject bin if its capacity is reached, and feed full platters onto the input stage when it is empty. Status lights on the console alert the operator to perform these functions as they become necessary.

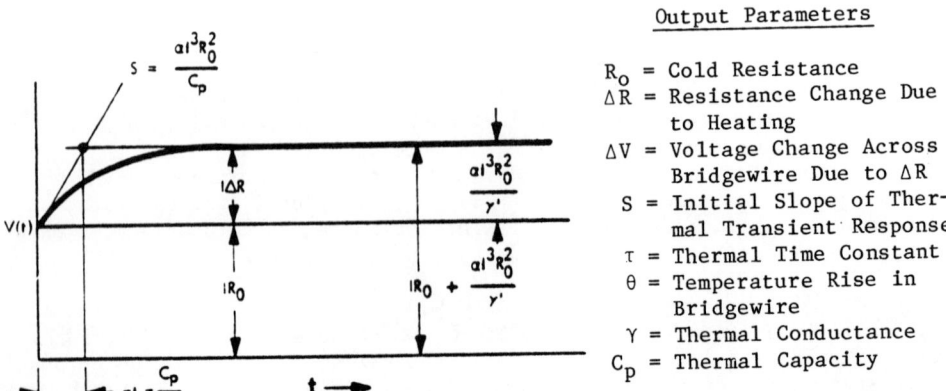

Output Parameters

$R_o$ = Cold Resistance
$\Delta R$ = Resistance Change Due to Heating
$\Delta V$ = Voltage Change Across Bridgewire Due to $\Delta R$
$S$ = Initial Slope of Thermal Transient Response
$\tau$ = Thermal Time Constant
$\theta$ = Temperature Rise in Bridgewire
$\gamma$ = Thermal Conductance
$C_p$ = Thermal Capacity

Fig. 1. Thermal transient output waveform

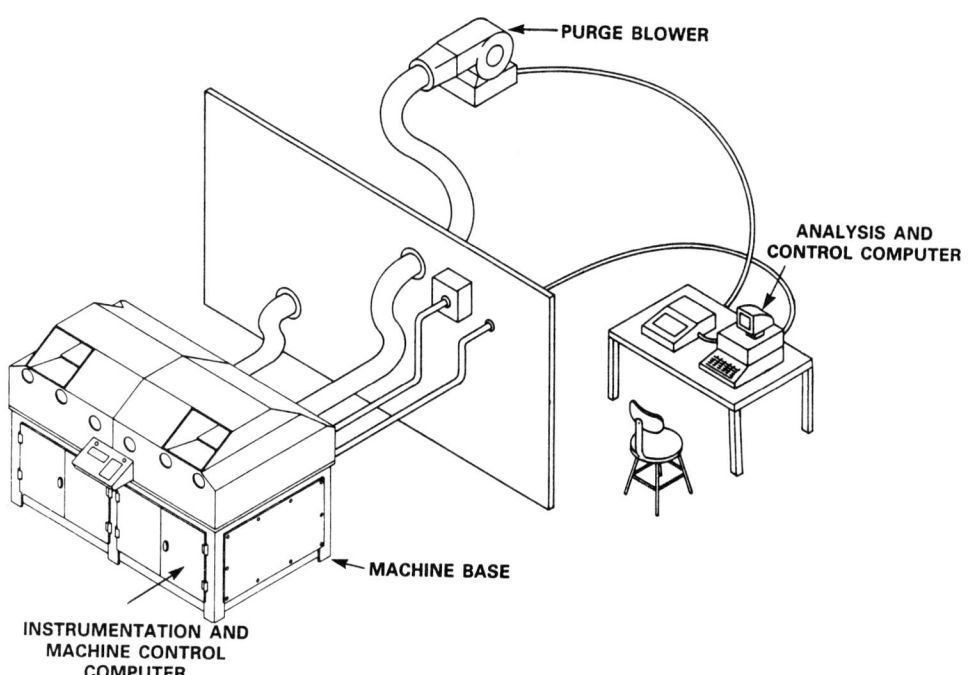

Fig. 2.  EED tester system

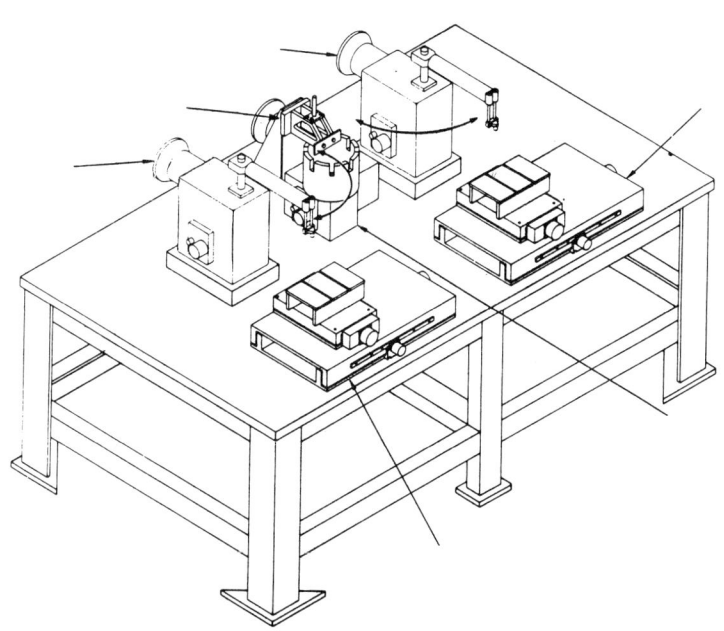

Fig. 3.  EED tester mechanical layout

The machinery elements are mounted onto the machine base or table. The machine base is enclosed in a lexan cover purged and ventilated to prevent accumulation of explosive dust. The cover also provides a shield for the operator in the event a device is detonated during the testing process. Access doors allow the operator to load or unload trays of devices.

MECHANICAL CONTROL

The mechanical control subsystem uses an STD BUS computer to control the various mechanical elements and assess status sensor feedback during operation. The mechanical control computer (MCC) also passes commands from the analysis and control computer to the measurement-subsystem current pulser and provides a communications link to the operator's console. The MCC provides control and synchronization for four stepper motor axes, three DC motor axes, various air control solenoids, and pressure or vacuum sensing switches. It also coordinates the application of vacuum for pickup and assessment of sensory input to detect faults. All components of the mechanical control subsystem are commercially available and configured for this application.

MEASUREMENT SUBSYSTEM

The thermal transient measurement subsystem consists of three major elements: (1) measurement circuit, (2) constant-current pulser, and (3) waveform recorder.

In providing an automated thermal transient tester, one principal difficulty was automating the measurement process. For laboratory or limited-number sample testing, the EED was placed in a Wheatstone bridge network. The operator would manually balance the bridge to determine the initial resistance of the device, using a low-level current pulse. A higher level test current would then be applied to induce the thermal transient. The voltage imbalance across the balanced bridge would be recorded and used to derive additional parameters to assess the quality of the device.

To achieve the production rate of 45 parts per minute, instrumentation was required to automate the measurement process. It was determined that the bridge measurement process would be maintained in order to reduce the dynamic range requirements of the waveform recorder used to record the transient waveform. Several methods were initially considered to balance the bridge circuit before applying the test current. A prototype circuit was developed using a field-effect transistor (FET) as the balance arm. The FET acted as a voltage-controlled resistor. In this configuration, however, the FET did not exhibit a linear response over a sufficient dynamic range. The final circuit configuration employed a resistance ladder network to balance the bridge to a coarse degree, with an FET to balance to a finer degree. The final circuit design is unique in that it employs two methods that have been used individually to balance a bridge network, but to the author's knowledge had previously not been used in combination.

Measurement Circuit

Figure 4 is a simplified schematic of the measurement circuit developed for the EED tester. The circuit sequentially performs five main functions:

- Coarse balance of bridge

- Fine balance of bridge

- Measurement of the initial or "cold" resistance of the device under test

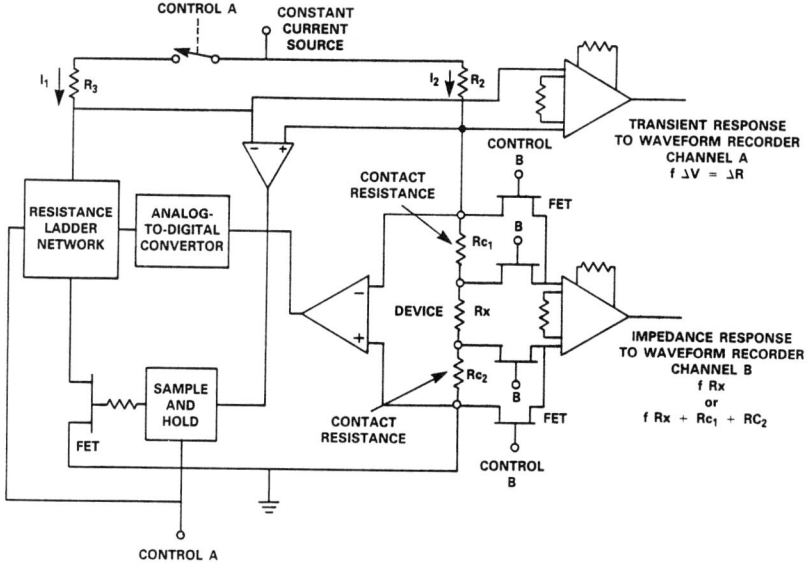

Fig. 4. Simplified schematic of EED tester measurement circuit

- Measurement of the contact resistance between the device under test and the test contacts
- Measurement of the electrothermal transient response.

Coarse Balance

During the coarse balance portion of the measurement cycle, a precise low-level current is applied only to the unknown resistance arm of the bridge network. The voltage developed across the device, including the contact resistance, is amplified and sent to an analog-to-digital (A/D) convertor. The A/D uses this voltage to develop a digital representation of the unknown resistance. The digital outputs of the A/D drive mercury-wetted relays with precision resistors across the contacts. By opening or closing the relay contacts of this ladder network, a resistance is developed that balances the bridge to 1 part in 4096 (12 bits). After the coarse-balance process, the current is applied to the entire bridge network. However, a residual voltage does remain across the bridge. This voltage would represent an unknown quantity of offset to the transient response waveform without the fine-balance sequence.

Fine Balance

Fine balance of the bridge is done by an FET operating as a voltage variable resistor. The gate voltage is controlled by a differential amplifier that measures the voltage unbalance when the bridge is excited by a low-level current. The output of the differential amplifier drives the FET gate through a sample-and-hold circuit, and the output voltage seeks a value that causes the FET to balance the bridge. The sample-and-hold then retains that voltage for the duration of the test. The FET is used to make only small changes to the reference arm resistance; and, therefore, the small nonlinearity of the FET resistance is inconsequential in the final measurements. The coarse- and fine-balance processes require only milliseconds to complete.

Initial Resistance

The initial resistance of the device under test is determined by use of a four-wire measurement technique. Contact with the device is made with four spring-loaded contact pins. Two of the contact pins are used to inject current into the device while the remaining two are connected to the high-impedance input of an instrumentation amplifier. The voltage developed across the device, due to the current flow, is amplified and sent to a channel of the waveform recorder. The recorded voltage value is used by the analysis computer to calculate the initial resistance of the device. The initial resistance value is determined using the same precise low-level current used for the balance process.

Contact Resistance

The quality of the connection with the device using the spring-loaded contacts can be degraded by dirt, oxide film, mechanical wear, and so forth. In the interest of accuracy and precision, the measurement circuit includes provisions for measuring the contact resistance. This is accomplished by switching the input sense lines of the amplifier used to measure the initial resistance of the device. For a short period of time, the FET transistors are used to switch the input sense to the contacts injecting the current into this device. The voltage output of the amplifier during this period is proportional to the device resistance plus the contact resistance with the device. The analysis computer employs the derived values for device resistance and device plus contact resistance to determine if the contact resistance has exceeded a predetermined limit. If this occurs, the machine operator is alerted to the need for corrective measures.

Electrothermal Transient Response

Once the bridge is balanced and resistances are monitored, a step-function constant-current pulse if applied to the bridge. A differential amplifier measures the time-dependent voltage developed by the bridge as the EED resistance increases due to heating. This waveform is digitized to 12 bits and sent to the computer for analysis.

Constant-Current Pulser

The constant-current pulser used to deliver the low-level balance current and the test current were developed specifically for this application. The constant-current pulser may be programmed and "fired" under manual or computer control. The test current amplitude and pulse duration are adjustable. The current amplitude may be set by the operator without applying the test current to a device. The pulser provides test currents of 1 milliamp to 2000 milliamps using three ranges. The low-level balance current and balance cycle duration are fixed by internal adjustment. The current output of the pulser is very precise and stable, as variations in currents supplied can directly influence the accuracy of the measurements. The current pulser is also designed to settle very rapidly (within about 4 microseconds) to the programmed current value. The pulser delivers a precise current to loads ranging from 0.1 to 10 ohms.

Waveform Recorder

The waveform recorder is a standard Nicolet Model 4094 with an IEEE-488 computer interface. The analysis and control computer communicates directly with the recorder to program certain functions and to offload stored waveforms after the thermal transient test cycle. The recorder is triggered by a signal from the current pulser to begin recording and captures two channels of waveform data during the testing process. The

12-bit waveform data are then passed to the analysis and control computer for processing.

ANALYSIS AND CONTROL SUBSYSTEM

The analysis and control subsystem uses a standard Motorola VME/10 desktop computer with a built-in 40-megabyte disk drive. The computer is to be installed in an "office grade" environment, separate from the machine base. An RS 232 serial communications interface provides communications between the analysis and control computer and the machine control computer. Peripherals include a 200-CPS printer and an IEEE-488 interface to the waveform recorder. The waveform recorder is installed in the machine base, and Hewlett Packard 488 BUS extenders provide increased communications distance.

The analysis and control computer provides four distinct processes for the system: (1) operator interface, (2) data acquisition, (3) accept/reject decisions, and (4) report generation.

The operator interface has two levels. At the machine operator level, the system is a menu-driven facility that allows a semiskilled operator to conduct tests. At the system administrator level, the standard operating system allows the user to perform file manipulation. Structured ASCII test-control files may be edited at this level. Test-control files contain device-specific information such as reject criteria, test current levels, tray matrices, equipment setup parameters, and descriptive phrases for operator menus. These test-control parameters can be changed only at the administrator level.

The data acquisition module downloads instrumentation parameters from the test-control parameters. The waveform recorder, constant-current pulser, and machine control computer must be programmed or "set up" prior to testing. During testing, this module reads in the waveform from the recorder, processes the data to pass to the accept/reject module, and passes the accept/reject decision to the machine control computer. In addition, if a machine fault occurs, this module passes the fault notice to the operator interface module.

The accept/reject module evaluates the test data received from the acquisition process. The accept/reject decision is based on the evaluation of a set of Boolean statements contained in the test control file. The statements are constructed from standard arithmetic and logical operators such as +, -, *, /, "and," "or," "less than," "greater than," numeric constants, and the electrothermal parameters generated from the waveform. There may be up to ten Boolean statements for each device test set. Each statement must evaluate "false" before a device is accepted. If a statement evaluates "true," the statement number is listed in a report as the cause for device rejection.

The report generation module can generate several report formats based on the level of information desired by the operator. The report can be as brief as a statistical summary of an entire test lot or a listing of each of the electrothermal parameters for each device tested. Reports may be generated and listed on the printer during the testing process or saved on disk and printed later. As the tests are conducted, the printer has sufficient speed to list all of the parameters from each device to provide a realtime output of test results.

SUMMARY

The system developed by SwRI and described in this paper provides the capabilities necessary to collect a large statistical database. From the statistical database, the parameter values for each device type can be selected from the distribution curves that allow an elevated confidence in detonator performance and provide process control feedback to the manufacturer.

REFERENCE

1. L. A. Rosenthal, "Electrothermal Equations for Electroexplosive Devices", NOVARD Report 6684, U. S. Navy Ordnance Laboratory, Silver Spring, Maryland, (1959).

AN AUTOMATED REAL-TIME IMAGING SYSTEM FOR INSPECTION OF COMPOSITE
STRUCTURES IN AIRCRAFT


D. R. Hamlin, B. M. Jacobs, R. H. Peterson,
and W. R. Van der Veer

Southwest Research Institute
San Antonio, Texas


INTRODUCTION

The use of composite structures in military aircraft has grown significantly in recent years. As an example, the F-15 is approximately 2 percent composite by weight compared to the more recent AV-8B which is approximately 31 percent by weight. Projections for the use of composite materials in future aircraft design indicate that this trend will continue. A significant portion of the increased use of composites involves essential aircraft structural components--wings, stabilators, control surfaces, etc. Consequently, regular inspection of these structures at military installations will become increasingly necessary to help assure structural integrity.

A significant number of currently established NDE inspection procedures for composites in military aircraft are based upon manual ultrasonic (UT) techniques. Using these techniques, the examiner detects and sizes defects such as disbonds and delaminations by monitoring and interpreting A-scan waveform signals on a UT instrument display screen. Manual probe manipulation permits maximum scanning flexibility and also permits the ultrasonic signal response to be optimized by the examiner using manual motions not possible with mechanized scanners. However, these examinations require that the examiner be responsible for instrument calibration, probe manipulation, signal interpretation, and documentation of inspection results. Other important concerns are the inspector's ability to determine completeness of coverage during the examination and the data reviewer's ability to validate instrument calibration and completeness of coverage, confirm signal interpretation, and compare current inspection results to those obtained during previous inspections.

Considerable effort has been expended over the last seven years to develop an inspection system for aircraft composite and bonded structures that provides manual scanning simultaneously with automatic recording of ultrasonic and inspection probe location data. The final goal of this effort was to establish a system that (1) is compatible with current manual inspection procedures, (2) records parameter and inspection data and produces images similar to those obtained during production testing, and (3) increases inspection quality by providing real-time scan coverage and processed data displays for the examiner while generating a comprehensive documentation record for the data reviewer. One primary advantage of having a system that meets these requirements is the availability of permanently recorded data presentations that can be used to identify changes in flaw configuration and size over a period of aircraft operation.

An earlier development program verified the methodology for automatic and simultaneous recording of ultrasonic inspection and position information using three prototype systems evaluated at military facilities. More recently, technical activity required to develop an inservice inspection system providing all of the capabilities itemized above has resulted in the Automated Real-Time Imaging System (ARIS).

OBJECTIVES

The ARIS technology is based on the principle of simultaneously recording ultrasonic inspection data along with the ultrasonic search unit position during a manually scanned inspection; the position is determined using an acoustic triangulation approach. The current production ARIS technology addresses the needs of both the inspector and the data reviewer: simplicity of setup, ease of calibration (downloading of prerecorded parameters), real-time display of coverage and processed data to improve inspection quality, flexibility, and portability, and complete documentation of all calibration and processing parameters, coverage and processed data results. With automated data collection features, the system must collect, store, process, recall and display large amounts of data conveniently. Convenient use of data is the key ARIS technical benefit. ARIS images acquired during different stages of aircraft life can be compared with each other and with production images to monitor flaw initiation and growth.

The principal objectives of the technical approach were to develop:

(1) A transportable system based on modular assemblies ruggedized for shipment as airline luggage while also providing convenient assembly and setup.

(2) A high productivity system with features such as:

  (a) An electronic template for defining component inspection boundaries to guide the operator in manipulating the probe assembly while defining completeness of coverage and processed data results in realtime.

  (b) Simplified operational software based on the use of high level commands of a type familiar to the average inspector.

  (c) Remote display and control capability for convenient inspection system interaction.

  (d) An adaptable search unit assembly configured for use with standard transducer types and transducer coupling systems.

  (e) Archival storage of all parameter (including instrument calibration) and inspection data to facilitate post-examination retrieval and review.

(3) Establish an affordable and producible system that is based for the most part on commercially available components.

(4) A flexible system with features such as:

    (a) A processor-controllable ultrasonic instrument compatible with inspection requirements of advanced composite and bonded structures.

    (b) A modular position locating assembly.

    (c) Modular software adaptable to existing flaw detection and characterization methods.

SYSTEM DISCUSSION

The completed ARIS consists of the components itemized in Figure 1. All components (including the mobile cart) are designed to be conveniently transportable and are mounted in ruggedized enclosures which can be shipped as airline luggage. The system can be quickly and easily assembled at the inspection site (no tools are required) and is designed for use by one examiner to facilitate inspection of components which are remote from the control unit, as depicted in Figure 1.

System control is accomplished using dual microprocessors and a programmable read-only memory (PROM) based software package. The control software is menu driven and permits operator activation of four major modules, as shown in the structural hierarchy of Figure 2. These modules provide the following functional capabilities:

The system checkout module permits the operator to test proper operation of each system component. Selected components can be operated independently and readouts accessed to determine component operational status.

The parameter generation module permits the operator to generate operator parameters, UT instrument calibration parameters, and inspection template data and store this information on diskette for subsequent recall.

The acquisition module permits data acquisition and realtime processing and display. Three modes of operation are supported. Inspection data can be stored on diskette for subsequent analysis.

The post-processing module permits data review and analysis functions using previously stored examination data.

The operator controls the system by activating selected functional capabilities using a hierarchy of control menus. For each menu displayed, the operator is presented with a limited set of options corresponding to the type of function being performed. In each menu, selections are presented using easily understood language. The keyboard control key associated with any option can be easily deduced by noting the option number presented on the menu list. The menu hierarchy can be traversed in both directions and provides an orderly sequence by which to schedule and control activities.

To initiate an examination, the operator can recall all inspection parameters from diskette. These include the UT instrument calibration parameters and the inspection region electronic template data. The electronic template defines areas on the component surface requiring examination and guides the operator in manipulating the inspection probe. The template is oriented to the aircraft component using predefined target

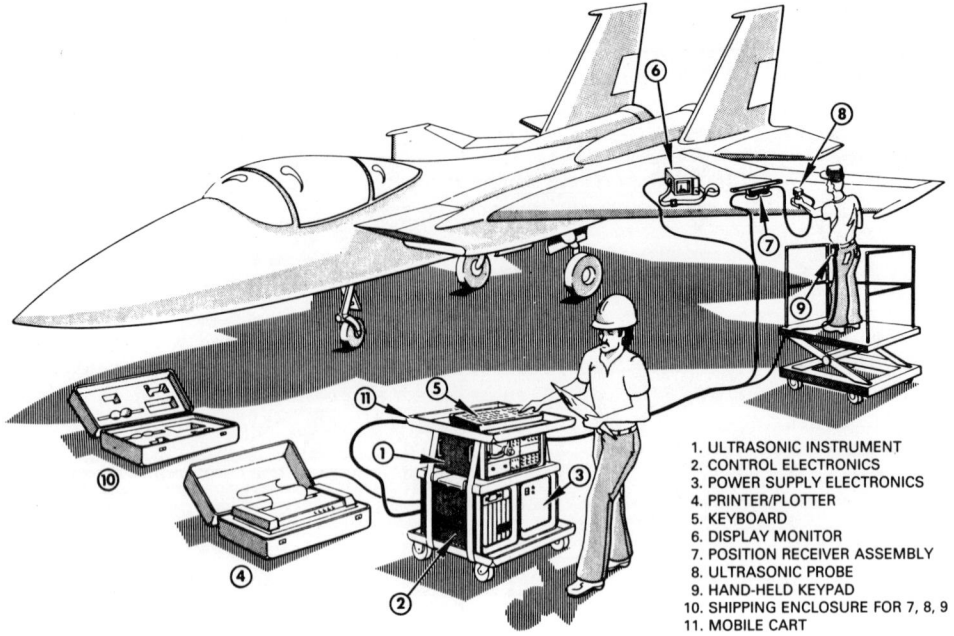

Fig. 1. Conceptual Drawing of System Hardware Configured for Remote Inspection Operation

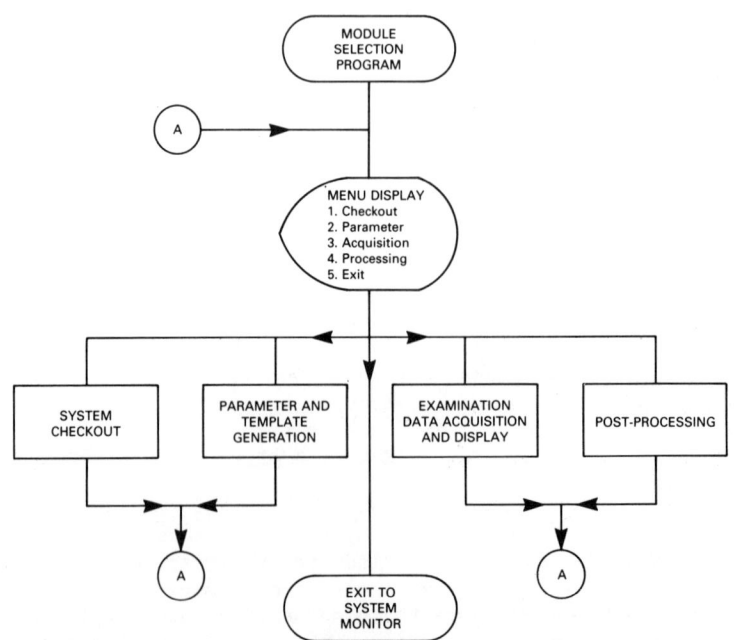

Fig. 2. Operator Menu Hierarchy

points (landmarks such as rivets, fasteners, etc.) The operator procedurally defines target points permitting computer display of the electronic template on the screen. Remote inspections can be performed by utilizing a hinged camera which swings in front of the UT instrument screen to transmit the A-scan display to the monitor. The hand-held keypad allows the operator to select either the A-scan camera display or realtime coverage/template display for presentation on the monitor. The UT instrument (KB-USD1) can also be controlled remotely using the keypad.

During an examination, the coverage/template display is updated in realtime to demonstrate coverage corresponding to operator manipulation of the transducer over the inspected region. A cursor is displayed on the screen to show the operator where the transducer is located in respect to the template defined inspection boundary. In addition to inspection coverage information, two types of realtime ultrasonic data processing are provided. The operator may perform GO/NO-GO (threshold level) processing using either positive or negative threshold violation or Gray-scale (or optional color) processing (16 levels) using either time, depth, or amplitude data. Examinations can be performed using pulse-echo or through-transmission UT inspection techniques. Data can be analyzed during the examination by pausing the real-time processing. The inspection data can be hardcopied (scaled or 1:1) and stored on diskette for subsequent recall and additional analysis. Post-processing functions permit the operator to perform a detailed evaluation of the data, including the capability to locate indications of interest on the actual component surface using either a 1:1 map overlay technique or a triangulation technique using predefined target point locations and arc lengths as defined on a tabular printout.

The system uses specially designed inspection probe devices to facilitate assembly and changeout of components (search units, delay tips, target point locators, boot assembly, handles, etc.), and incorporates several optional handles and a stylus/holder unit which are designed to minimize operator fatigue. The optional components and assembly steps are depicted in Figure 3. No tools are required for assembly or disassembly. The through-transmission yoke assembly shown in Figure 4 features spring-loaded arms and gimbal mounts for each transducer to permit surface compliance. A polyurethane boot assembly is available for examination of components with rough surfaces.

SYSTEM VALIDATION

System operation was validated using the test plate depicted in Figure 5. This plate contains a series of bottom-drilled holes varying in size (0.25" to 2.125") and depth (0.10" to 0.40"). An ARIS display of this plate is shown in Figure 6 and represents a C-scan type presentation in which depth in material is shown using color modulation. The C-scan display area is represented using a collection of 0.111"-square elements (pixels) which account for the step shape boundaries of each hole. The color scale on the left side of the figure defines the color versus depth relationship.

Figure 7 shows an ARIS color scale by depth display for a carbon epoxy composite region of a T-38 wing tip. The part consists of an outer skin which has been bonded to a corrugated type stiffener. The area shown in the C-scan type presentation represents two skin-only regions (running horizontally top and bottom) and a bonded skin/stiffener region (center). Two disbonds are shown as abrupt pattern changes in the central red band running across the image. The skin thickness varies from 0.035" to 0.045" (left to right in the image) while the skin and bonded stiffener region varies in thickness from 0.070" to 0.078" (left to right in the image).

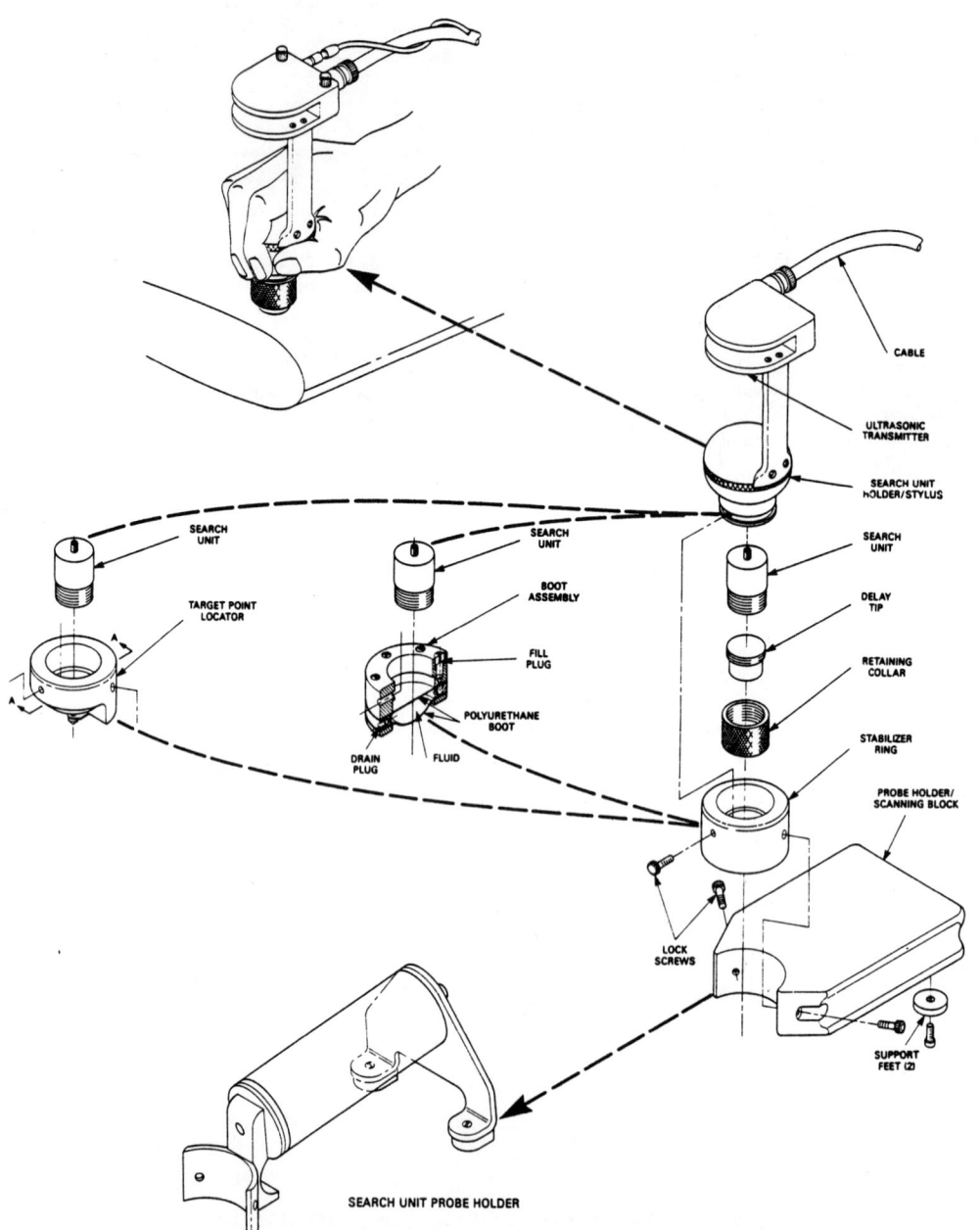

Fig. 3. Specially Configured Inspection Hardware

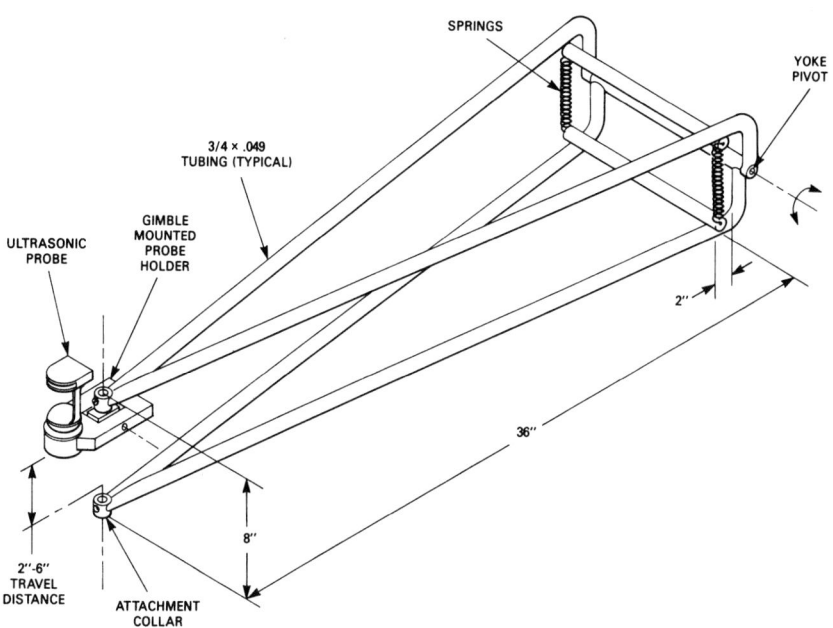

Fig. 4. Through-Transmission Yoke

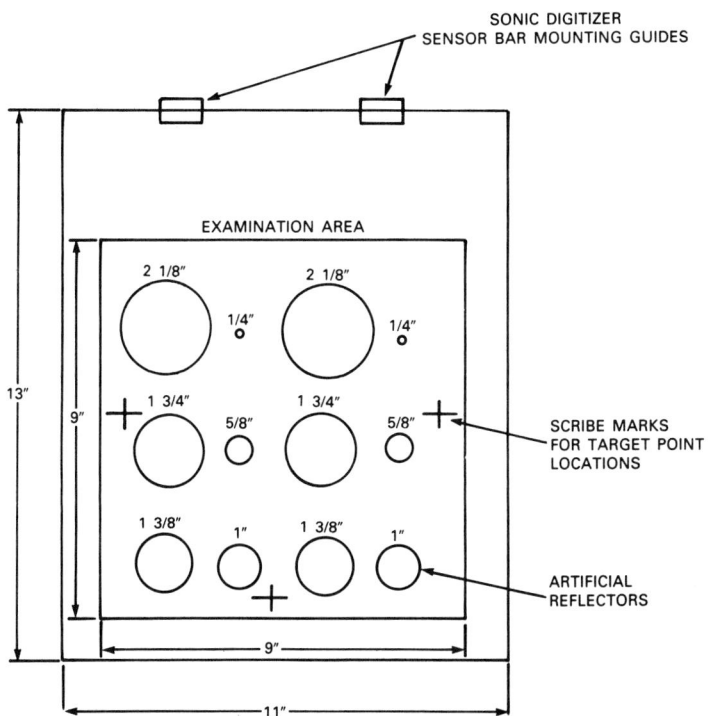

Fig. 5. ARIS test plate hole pattern. Hole diameters vary as shown while the depths are mirrored (reversed) about the vertical centerline.

Figure 8 shows an ARIS GO/NO-GO threshold display for a through-transmission examination of the aluminum skin to honeycomb structure on a T-38 horizontal stabilizer. Numerous disbonds or core damage areas are shown. The examination was performed using the through-transmission yoke assembly.

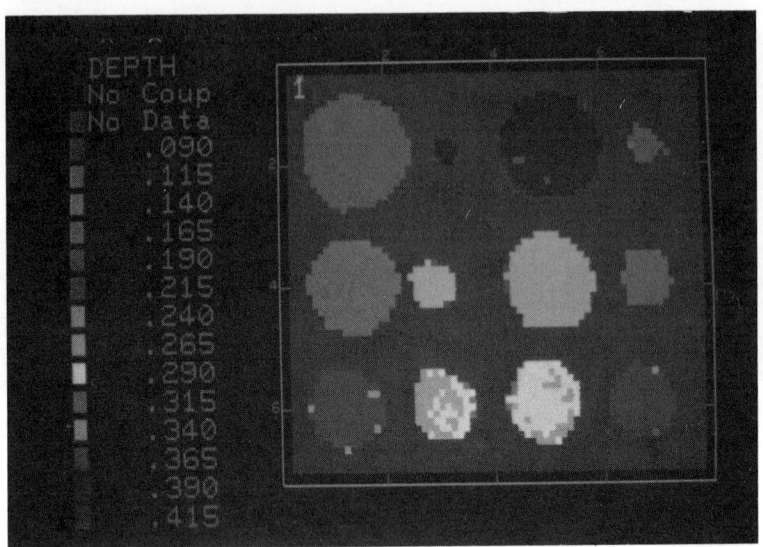

Fig. 6. ARIS pulse-echo test plate examination results using an L-scan presentation in which color represents hole depth within the material.

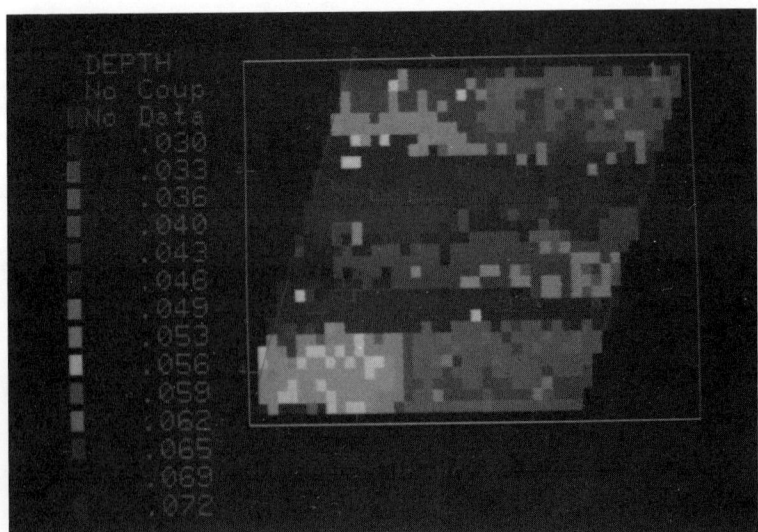

Fig. 7. ARIS pulse-echo examination results of a carbon epoxy region from a T-38 wing tip containing disbonds.

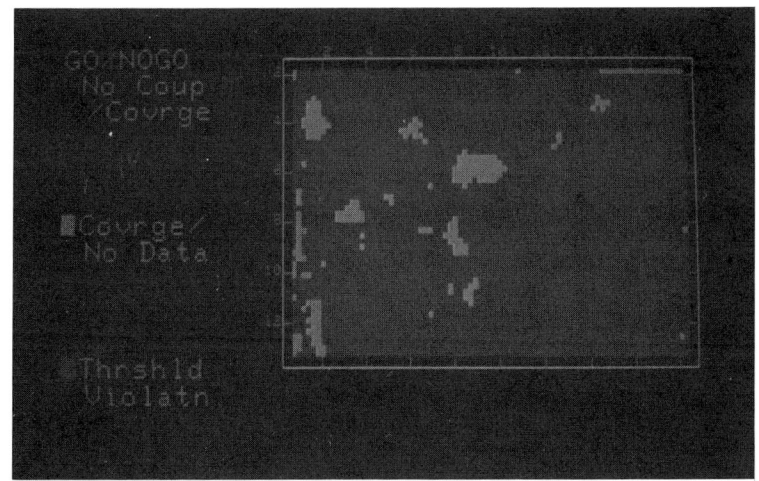

Fig. 8. ARIS through-transmission examination results of a T-38 horizontal stabilizer consisting of an aluminum skin bonded to a honeycomb core. Several disbonds and core damage areas are present.

CONCLUSION

ARIS was designed to permit routine day-to-day examinations of aircraft composite structures using established manual ultrasonic techniques while providing automated data recording, processing, and analysis functions. The system provides realtime coverage and processed data displays, thus permitting higher quality examinations while providing comprehensive documentation to permit more effective data review. The system is currently scheduled for a series of field site evaluations on a variety of aircraft and composite structure components. Results from the field trials will be reported in future publications.

SYSTEM OF INSPECTION ASSISTED BY MICROPROCESSOR*

J.L. Arnaud, M. Floret, and D. Lecuru

Aerospatiale, Central Laboratory
92150 Suresnes, France

PRESENTATION OF THE STUDY

In spite of significant advances in the field of automatic inspection (robotization, motorization), there are still numerous cases where the cost of such facilities cannot be justified due to the low production rates, or to the fact that they are not easily applicable due to the shape of the parts or to the environment (on-site maintenance inspection or inspection during manufacture in particular areas of composite parts).

In both of these cases, these inspections must be carried out manually by an operator, and many questions arise concerning the traceability of such operations (have all the parts been inspected? Were all the settings correct? Was the operator's interpretation of the results correct?).

The system developed by AEROSPATIALE was so designed as to gather around a microprocessor all the functions required to ensure reliability of the inspection and to relieve as much as possible the operator of all the phases where interpretation or positioning errors could occur.

Thus, in order to demonstrate that with a limited investment, the reliability, rapidity and performance of a manual inspection can be equalled, a SIAM (System of Inspection Assisted by Microprocessor) has been set up to tackle an arduous task: the maintenance inspection of aircraft joints (figure 1) for which very stringent requirements have to be satisfied:

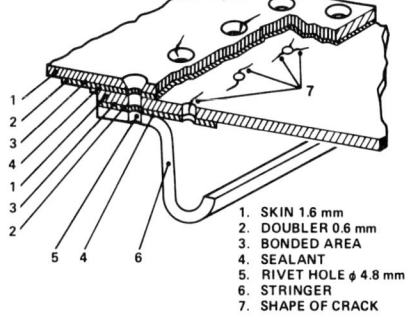

Fig. 1: Location of the cracks searched for on the assemblies.

1. SKIN 1.6 mm
2. DOUBLER 0.6 mm
3. BONDED AREA
4. SEALANT
5. RIVET HOLE $\phi$ 4.8 mm
6. STRINGER
7. SHAPE OF CRACK

---

* This development has been awarded the Golden Medal for Innovation NDT Diploma of the 6th International conference on NDT in STRASBOURG 1986.

- inspection of more than 30,000 rivets in 24 h,

- guarantee that no rivet has been missed on the 500 meters of joints to be inspected,

- length of defect detected with a 95% confidence level: less than 5 mm,

- false alarm rate lower than 1%,

- management of data,

- easy handling for on-site inspection.

All the functions (measurements, positioning, management) required to guarantee the reliability of this kind of manual inspection are presented below.

SELECTION OF THE NDT METHOD

The method selected is not the one which provides the highest detection level, rather it is the method best suited to the problem (search for cracks around rivets, figure 1), which can be easily applied to maintenance. Ease of utilization led us to select the eddy current method.

Two differential probes were used (figure 2); they were specifically designed to provide a solution to this problem. They consist of a central transmitter, two receivers separated by a distance equal to rivet centerline spacing and two lift-off compensation coils.

Such a probe arrangement allows two kinds of deviation to be obtained in the impedance plane; the deviation on the x axis corresponds to resistance variation (rivets), while the deviation on the y axis corresponds to inductance variation (cracks). The value of the latter deviation does not depend on paint thickness and rivet presence (figure 3).

Since the crack may start either on the external sheet (second rivet line) or on the internal sheet (first rivet line), the two probes operate at different frequencies (2.9 kHz and 13.5 kHz, respectively). They are connected to an EC 3000 (HBS) apparatus provided with 4 outputs for each of the two probes.

This type of system allows artificial defects, approximately 2 mm long, to be detected on each inspection line.

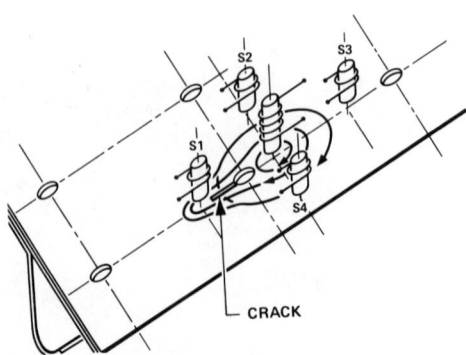

Fig. 2: Design of eddy current probe

Fig. 3: Display of the impedance plane

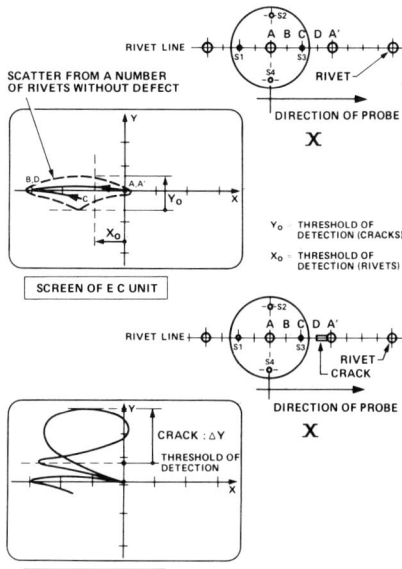

The use of a manual method on a large number of rivets may, however, lead to non-detections due to differences between artificial defects and actual cracks (orientation, shape) and, on the other hand, to detection misses due to human errors (fatigue, incorrect signal interpretation, missed inspection sites).

Therefore, even if the performance requirements of the method seem to be satisfied, its reliability for the whole structure remains to be guaranteed.

SCANNING MODE

To make sure that all the rivets have been inspected, it is necessary to set up a scanning system which provides the position of sensors. The conventional solution of an automated system using stepper motors has been rejected, since it results in a large-size fixture lacking operating flexibility where the coupling of the probes cannot be guaranteed for all local shape irregularities. Therefore, a manual scanning system had to be selected; the movements of the system are repeated by a linear position sensor along which the probes are moved.

The position sensor is a standard model and consists of a bar on which slides a magnet where the probe holder is fitted. Whenever an electric pulse is sent into the bar, it is subjected to a twisting torque at the magnet position. This torsion generates an acoustic wave which propagates in the bar. The time lapse between the exciting pulse and the received acoustic wave is used to determine the position of the magnet and thus that of the scanning system (figure 4).

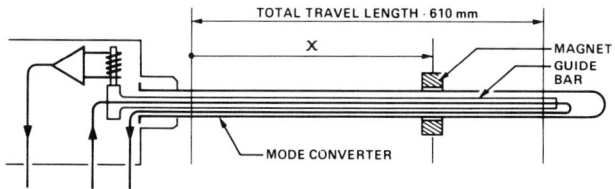

Fig. 4: Schematic of the position probe

945

This system has many advantages as regards precision (1/20,000 of total travel), transmission speed, ruggedness, light weight. Futhermore, flexible or rigid models up to 30 meters long exist. For adaptation to the fuselage, the overall a length of the system is 1 meter, which corresponds to the spacing between two frames (24 rivets on each line). The scanning system thus defined is very easy to use (figure 5) and it allows the fuselage to be segmented into 500 inspection zones which may be readily identified on a symbolic representation.

THE COMPUTERIZED SYSTEM DEVELOPED

The system (APPLE II) carries out the EC channel acquisition and probe position management functions. It also checks the calibration, displays the 4 EC channels (as a function of position) and stores them on a 10-Mbit disk capable of storing the results for 7 complete fuselages (figures 6 and 7).

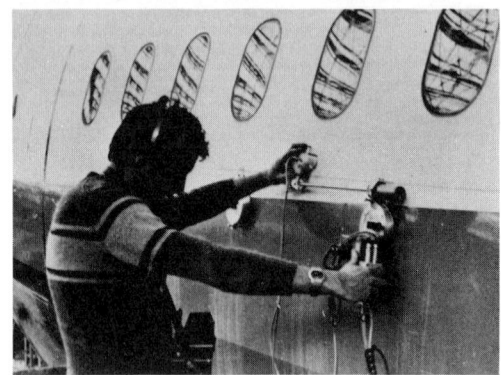

Fig. 5: Overall view of the postion probe

Fig. 6: Overall view of the ground SIAM system

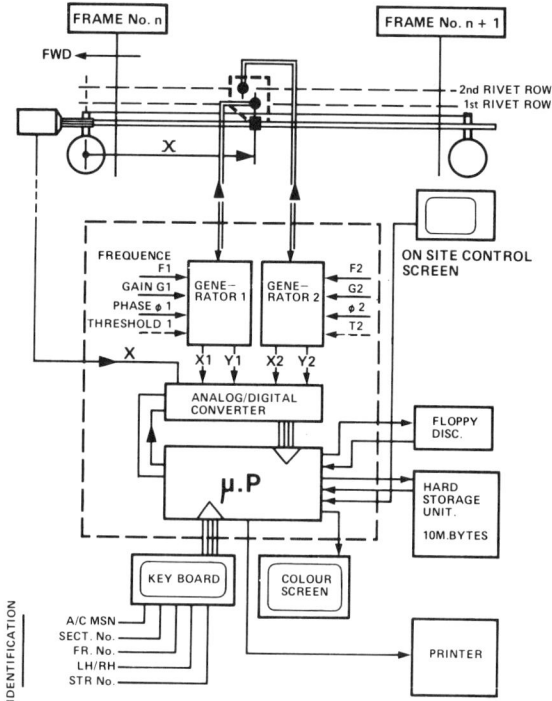

Fig. 7: SIAM block diagram

The study first consisted in structuring the various phases of the inspection and linking them through menus, and then presenting the results in a straightfoward, easy-to-use manner.

As regards the identification of the areas to be inspected, the various fuselage sections are represented symbolically (figure 8). The interframe section subjected to the inspection is identified by blinking of the display and is cleared once inspection is completed. If any anomaly is detected, it is delineated by a thick line of a different color.

The EC data for each interframe section are represented as shown on figure 9. Chanels V1 and V3 represent the responses generated by the rivets, channels V2 and V4 correspond to the response generated by the defects (in this case, standard defects from 2 to 7 mm long). Figure 11 represents the display of a defect-free inspection area (correct inspection of the area can be ascertained by counting the "rivet" indications).

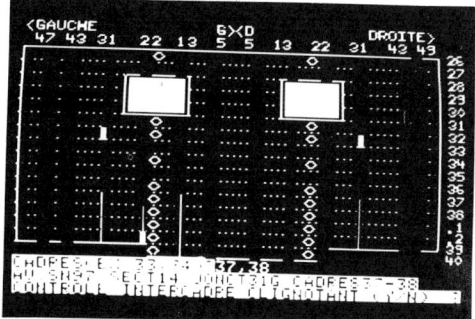

Fig. 8: Symbolic representation of a section during inspection

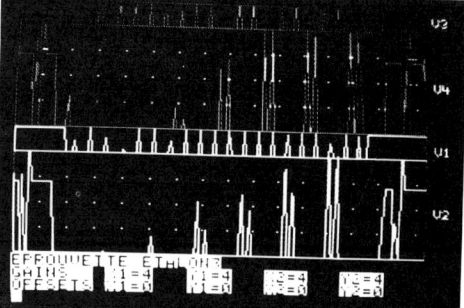

Fig. 9: Display of the EC data on a test specimen

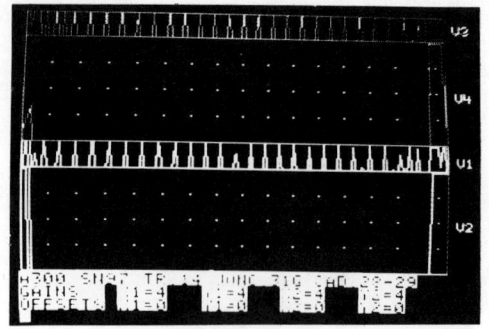

Fig. 10: Display on a defect-free area

VALIDATION OF THE SYSTEM

The SIAM has been mainly designed with a view to rapid detection and inspection management, rather than for assessing the detected defects. Since the probability of cracks is low, the defects detected and identified by the SIAM should be small in number and it should be easy to assess them accurately by a more accurate method which may be slower (i.e. dual-inspection principle: fast detection, then accurate assessment). However, to validate the complete system as regards performance and reliability, it was necesary to know the confidence level of detection of actual defects.

This validation was performed on specimens representative of the actual structure in which 750 fatigue cracks were created on one or the other of the rivet lines. After 3 inspections, the specimens were opened, and this allowed curves for probability of detection by SIAM to be plotted as a function of the actual defect length (figure 11).

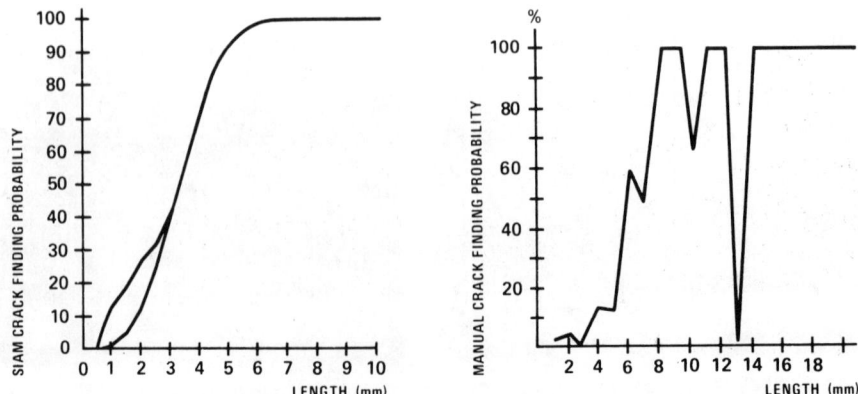

Fig. 11: Probability of detection on real cracks with SIAM

The curve was then fitted using the WEIBULL law, which allows the definition of a 95% confidence level detection length of approximately 4.5 mm for a detection threshold close to 2 mm, and a very high reliability level was achieved since all defects longer than 4.5 mm were detected and no false alarm was trigged.

RESULTS OBTAINED ON AN AIRCRAFT

Once the performance and realibility had been established, the feasibility of actual scale inspection on a fuselage had to be demonstrated.

For this purpose, four aircraft were partly inspected, i.e. 65,000 rivets were inspected. The inspection rate increased from 510 rivets/hour to 1400 rivets/hour from the first to the fourth aircraft. The task was conducted by a team of 3 operators.

After a period of time required for running up the system and for team adaptation, the inspection conducted on the fourth aircraft complied with the requirements of the specification which stipulates a total inspection time of 24 hours.

The indications detected during each inspection, which determined the gross false alarm rate, were due to rivets of different materials (aluminium, monel, standard rivets being in titanium) or to doubler edges or local incidents (highly offset rivets).

After a rapid check of these local indications (visual inspection or conductivity measurements), the net false alarm rate becomes zero.

CONCLUSIONS

In a situation where inspection automation would be difficult and costly, the concept of computerized assistance to manual inspection proves straightforward, easy to use, while provising a high level of rapidity, reliability and performance for a moderate investment cost.

AEROSPATIALE has also adapted this concept to the ultrasonic inspection of composite structures in manufacture, since it had been noticed that although 90% of the surfaces (flat sections of composite parts) were inspected by automatic facilities (pools, sprays), the remaining 10% of the surfaces, which correspond to shape irregularities (strengtheners, bent sections), were still inspected manually, which occupied approximately 50% of the entire inspection time, since the inspection raised many reliability problems.

Both in manufacture and maintenance, the SIAM concept allows the inspection cost to be reduced while still guaranteeing high quality and reliability.

A SEMI-AUTOMATIC SYSTEM FOR THE ULTRASONIC MEASUREMENT OF TEXTURE

S. J. Wormley and R. B. Thompson

Center for NDE
Iowa State University
Ames, IA  50011

INTRODUCTION

   The texture (preferred grain orientation) of rolled metal plates influences a number of important mechanical properties such as their ability to be plastically formed into complex shapes. X-ray diffraction techniques can characterize texture in great detail but are unsuitable for real time process control. Furthermore, x-rays only sample the properties of a near surface layer, whereas the average properties throughout the thickness may be of greater interest. This paper describes an alternate texture characterization approach based on ultrasonic measurements of the anisotropy of plate wave velocities. Relationships have recently been established between the macroscopic elastic constants of a rolled metal plate and the coefficients of an expansion of the crystallite orientation distribution function (CODF) in terms of generalized Legendre functions [1]. It has also been shown that these coefficients can be determined from velocity measurements of ultrasonic plate modes [2,3]. Here a system is described which implements these ideas in a semi-automated fashion as would be required for process control applications. The measurement system consists of two sets of EMAT transducers and associated electronics, one for $SH_0$ mode measurements and the other for $S_0$ mode measurements. Each set consists of one transmitter and two receivers, separated by a fixed distance and placed at a variable angle with respect to the rolling direction of the plate. The pair of received signals are digitized and processed to determine the coefficients $W_{400}$, $W_{420}$ and $W_{440}$, which can, in turn, be used to make first order predictions of pole figures. These steps are reviewed in detail and future directions are discussed.

THEORY

   The rolled metal plate is modeled as a continuum, having macroscopic orthotropic symmetry (three mutually perpendicular mirror planes). The degree of preferred orientation of its crystallites are quantified by a crystallite orientation distribution function (CODF) represented by $w(\xi,\psi,\phi)$, where the arguments are Euler angles describing the orientation of crystallites with respect to the sample axes [1]. It is often convenient to expand the CODF as a series of generalized Legendre functions, $Z_{\ell mn}$, as defined by [4]

$$w(\xi,\psi,\phi) = \sum_{\ell=0}^{\infty} \sum_{m=-\ell}^{\ell} \sum_{n=-\ell}^{\ell} W_{\ell mn} Z_{\ell mn}(\xi) e^{-im\psi} e^{-in\phi}. \qquad (1)$$

For cubic crystallites, symmetry dictates that the lowest order independent coefficients are $W_{000} = 1/2\sqrt{2}\,\pi^2$ (a normalization constant), and $W_{400}$, $W_{420}$, and $W_{440}$. Following the Voigt procedure for averaging elastic constants, the polycrystalline elastic constants, $C'_{IJ}$, of the orthorhombic plate may be expressed in terms of these four $W_{\ell mn}$ coefficients and the single crystal elastic constants, $C^o_{IJ}$ [1]. Typical results are

$$C'_{44} = C^o_{44} + c^o\,[\,1/5 - 16/35\,\sqrt{2}\,\pi^2(W_{400} - \sqrt{(5/2)}\,W_{420})] \qquad (2)$$

$$C'_{55} = C^o_{55} + c^o\,[\,1/5 - 16/35\,\sqrt{2}\,\pi^2(W_{400} + \sqrt{(5/2)}\,W_{420})] \qquad (3)$$

$$C'_{66} = C^o_{66} + c^o\,[\,1/5 + 4/35\,\sqrt{2}\,\pi^2(W_{400} - \sqrt{70}\,W_{440})], \qquad (4)$$

with similar relationships available for the remaining six independent elastic constants.

These elastic constants, and hence the $W_{\ell mn}$, can be inferred from measurements of ultrasonic wavespeeds. For thin plates, Lee, Smith, and Thompson [3] have shown that

$$\rho\{SH_o^2(45°) - \tfrac{1}{2}[SH_o^2(0°) + SH_o^2(90°)]\} = \frac{16c^o\pi^2\sqrt{35}}{35} W_{440} \qquad (5)$$

$$\rho\{\frac{S_o^2(0°) + S_o^2(90°)}{2} - S_o^2(45°)\} = \frac{16c^o\pi^2\sqrt{35}}{35} W_{440} \qquad (6)$$

$$\rho\{S_o^2(0°) - S_o^2(90°)\} = -\frac{32\sqrt{5}\pi^2 c}{35}(1 + \frac{2C_{12}^o}{C_{11}^o}) W_{420} \qquad (7)$$

$$\tfrac{\rho}{2}\{SH_o^2(45°) + \tfrac{1}{2}[SH^2(0) + SH_o^2(90)]\} = C_{44}^o + c^o\{1/5 + \frac{4\sqrt{2}\,\pi^2}{35} W_{400}\} \qquad (8)$$

where $c^o = c^o_{11} - c^o_{12} - 2c^o_{44}$.

Here $SH_o(\theta)$ represents the velocity of the fundamental, horizontally polarized shear mode and $S_o(\theta)$ represents that of the fundamental symmetric Lamb mode, sometimes referred to as the extensional mode.

VELOCITY MEASUREMENT APPROACH

Equations (5)-(8) show that, in order to determine the coefficients $W_{400}$, $W_{420}$, and $W_{440}$, it is necessary to measure the velocity of two ultrasonic plate modes, the $S_o$ and the $SH_o$ at three propagation directions (0°, 45°, and 90° with respect to the rolling direction). For each mode, this is accomplished with a three probe system, as shown in the system block diagram in Fig. 1. One transducer is used as a transmitter

HARDWARE SYSTEM

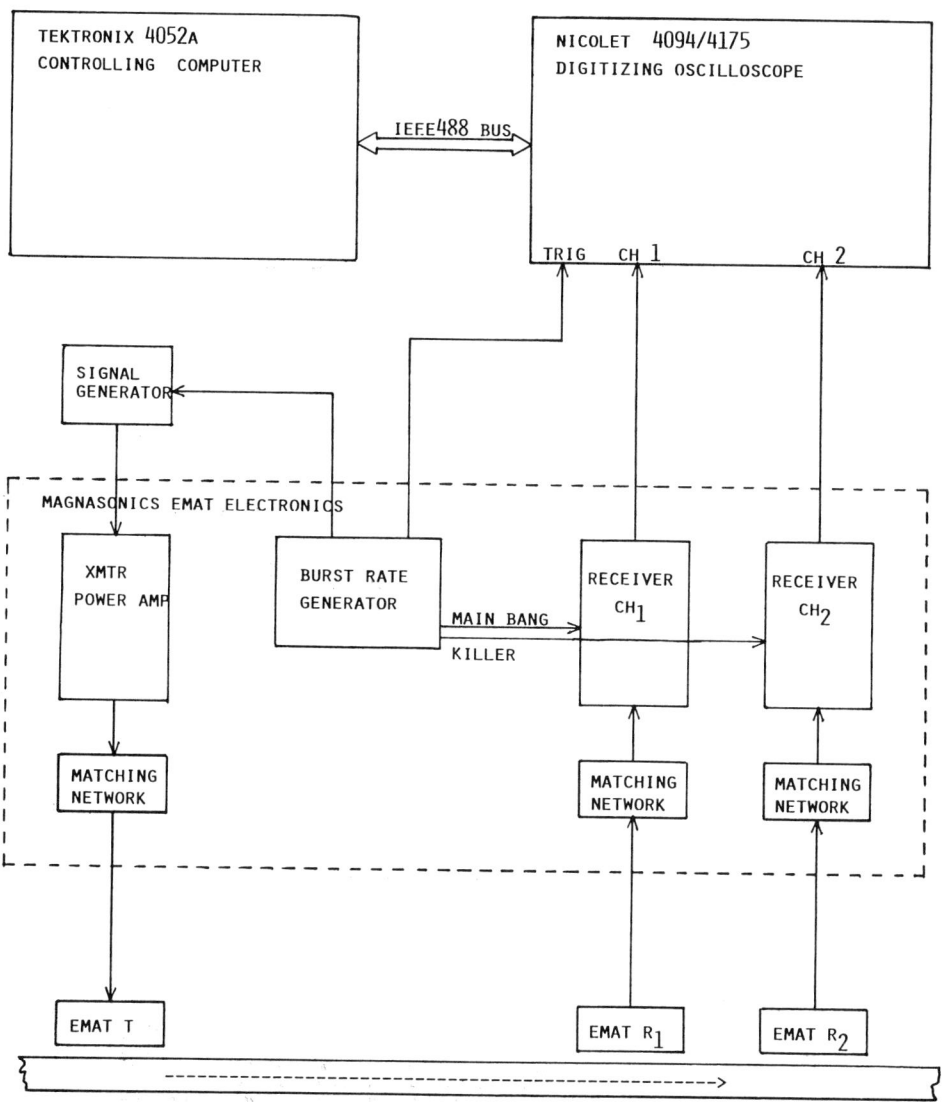

Fig. 1. Block diagram of the hardware used to measure velocity as a function of orientation with respect to the rolling direction in a plate.

and two act as receivers so that absolute velocities can be obtained. EMAT probes were selected because of their couplant free operation and ability to operate in high temperature environments as would be desired in process control applications. The ultrasonic velocity is determined by measuring the time difference between the two received RF bursts by digital cross-correlation techniques. Because many points on the waveform contribute to the answer, a precise time difference measurement is obtained.

Figure 2 shows the hardware system currently implemented. The Magnasonics EMAT Electronics (bottom) and associated signal generator (top) provide the driving, matching, and receiving functions for the EMAT probes. The Nicolet Digitizing Scope (middle) can digitize 8K points per channel with a spacing as small as 2ns per sample point, capturing a complete burst of RF in each channel. A Tektronix 4052A computer (not shown) controls data acquisition and signal processing. Due to memory restrictions, only 512 data points are used per channel with a sample spacing of 5ns per sample point. Use of a more modern computer in the near future will reduce this restriction.

MEASUREMENT PROCEDURE

Given the plate being inspected, the rolling direction is determined from physical inspection. Were this not possible, it could be determined from the ultrasonic data. The velocities of the $S_o$ and the $SH_o$ modes are then determined at the three propagation directions using the cross-correlation procedures. The difference in time is obtained to an accuracy of one sample interval; in this case 5ns. Effective distances between the receiver EMAT probes are known from initial calibrations experiments.

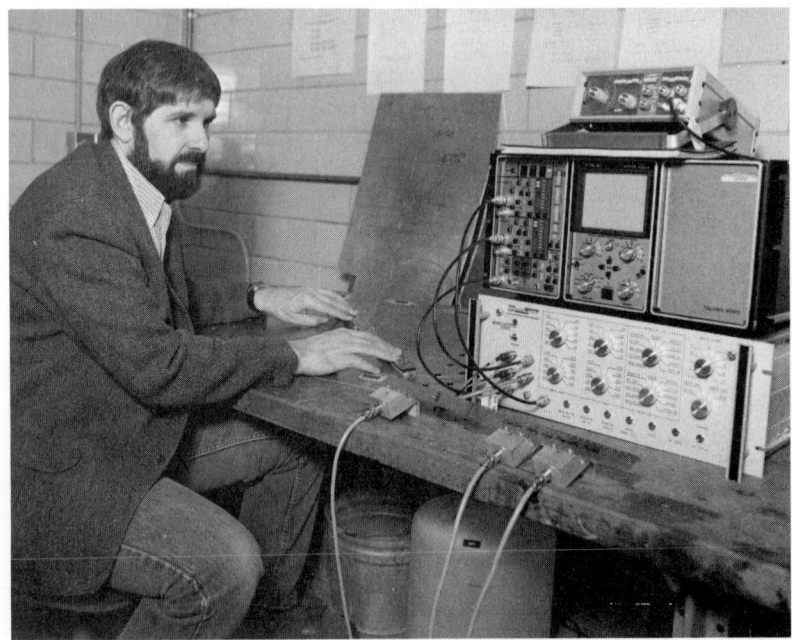

Fig. 2. A semi-automatic system for the ultrasonic measure of texture. Not shown is the computer used for data processing and calculation of CODF.

Therefore, velocity can be calculated directly from the time differential. The CODF coefficients $W_{400}$, $W_{420}$, and $W_{400}$ are then calculated using Eqs. (5)-(8).

MEASUREMENT RESULTS

The system has been evaluated on a set of samples which have been independently characterized by a manually operated ultrasonic system based on the same principles [3-5] but using a time-average interval counter for the velocity determination. The repeatability of the velocity measurement was first examined. For the SH mode, measurements with 512 samples per waveform at 5ns intervals and with signal averaging of 16 repetitions showed a repeatability of ±5ns as would be expected. Based on a receiver separation of 10cm, this corresponds to a velocity measurement precision of approximately 0.02%.

Table I compares the values obtained with the digital system for the coefficient $W_{440}$ to those obtained manually [5,6]. Also shown are values obtained on a few samples by neutron diffraction [7]. With the exception of one anomalous point, the agreement between the two ultrasonic systems is within 15%, with better performance obtained for most cases. When available, the neutron data is also in good agreement. The error of 15% is considerably greater than can be explained on the basis of the measurement precision and further work is required to define all contributing factors. Nevertheless, these initial results are quite encouraging.

Table I. Comparison of CODF $W_{440}$ measurements.

| Material | Manual $W_{440}$ | Auto $W_{440}$ | % DIFF | Neutron $W_{440}$ |
|---|---|---|---|---|
| 6061-T6 AL #1 | -2.79E-4 | -2.58E-4 | -7.5 | ---- |
| 6061-T6AL #2 | 9.96E-4 | 8.98E-4 | -9.8 | ---- |
| 6061-T6 AL #3 | 9.96E-4 | 7.03E-4 | -27.0 | ---- |
| ALCOA AL 629° | 5.02E-3 | 4.75E-3 | -5.4 | 4.9E-3 |
| ALCOA AL 640° | 3.50E-3 | 3.28E-3 | -6.3 | ---- |
| ALCOA AL 675° | 3.03E-3 | 2.92E-3 | -3.6 | 3.3E-3 |
| 304 Stainless #1 | -2.80E-4 | -3.13E-4 | 11.6 | ---- |
| 304 Stainless #2 | 2.71E-4 | 3.10E-4 | 14.3 | ---- |

The results for $W_{400}$ and $W_{420}$ are not yet as satisfactory. For $W_{420}$, the required $S_o$ mode signals have very poor signal-to-noise ratios, which have recently been shown to be a result of improper tuning. It is anticipated the good results will be obtained when the tuning is corrected, as has been demonstrated for manual systems [5,6].

For $W_{400}$, the absolute measurement procedure appears to be a problem. Whereas Eqs. (5)-(7) only require relative measurements, Eq. (8) demands an absolute comparison of measured velocities to theoretical predictions based on Voigt averages of elastic constants. Not only are the experimental difficulties greater in the determination of absolute velocity, but the theoretical foundation is also weaker. The Voigt procedure is generally more accurate in predicting relative variations of elastic constants than their absolute values. When Eq. (8) was applied to experimental data, the predictions of $W_{400}$ were not physically realistic. Future efforts will be directed at other approaches, e.g., those based on the dispersion of higher order plate modes [8], to determine this coefficient.

FUTURE DIRECTIONS

It is planned that the development of this system will be completed during the next year. A number of minor technical problems uncovered by these initial tests will first be addressed. Included will be improvements in EMAT tuning to improve signal-to-noise, correction of some probe construction flaws, and use of a computer with the capacity to process more data points and thereby gain greater time precision.

Of a more fundamental nature, alternative procedures for the determination of $W_{400}$ will be sought. From Eqs. (2)-(4), it can be seen that knowledge of the relative values of $C'_{44}$, $C'_{55}$, and $C'_{66}$ would allow determination of $W_{400}$ as well as $W_{420}$ and $W_{440}$. Such information can be obtained from the dispersion of higher order SH modes [8], and the application of the digital system to this measurement will be investigated.

Once the full set of coefficients are determined, ultrasonic pole figures can be predicted [9]. In some cases, only one of the coefficients may be adequate for process control [6]. In either event, the speed of the ultrasonic technique with respect to x-ray or neutron measurements make it a strong candidate for practical texture control applications.

ACKNOWLEDGEMENT

This work is supported by the NSF university/industry Center for NDE at Iowa State University.

REFERENCES

1. C. M. Sayers, "Ultrasonic velocities in anisotropic polycrystalline aggregates", J. Phys. D 15, 2157-2167 (1982).
2. R. B. Thompson, J. F. Smith, and S. S. Lee, Nondestructive Evaluation of Microstructure for Process Control, H.N.G. Wadley, Ed., ASM, Metals Park, OH, 1986, pp. 73-99.
3. S. S. Lee, J. F. Smith, and R. B. Thompson, "Inference of crystallite orientation distribution function from the velocity of ultrasonic plate modes", Proceedings of the 2nd International Symposium on Nondestructive Characterization of Materials, J. F. Bussiere, Ed., (Plenum Press, NY, in press).
4. R. J. Rose, J. Appl. Phys., 37 (1966), p. 2069.
5. R. Bruce Thompson, S. S. Lee, and J. F. Smith, "Angular dependence of ultrasonic wave propagation in a stressed orthorhombic continuum: Theory and application to the measurement of stress and texture", J. Acoust. Soc. Am. 80(3), Sept. 1986, pp. 921-931.
6. A. V. Clark, Jr., A. Govada, R. B. Thompson, G. V. Blessing, P. P. Delsanto, R. B. Mignogna, and J. F. Smith, "The use of ultrasonics for texture monitoring in aluminum alloys", these proceedings.
7. R. J. Fields, National Bureau of Standards, private communication.
8. J. F. Smith, G. A. Alers, P. E. Armstrong, and D. T. Eash, "Separation and characterization of stress levels in metal sheet and plate: I. Principles of initial test", J. Nondestr. Eval. 4, 157-163 (1984).
9. J. F. Smith, R. B. Thompson, D. K. Rehbein, T. J. Nagel, P. E. Armstrong, and D. T. Eash, "Illustration of texture with ultrasonic pole figures, ibid.

MODEL-BASED ULTRASONIC NDE SYSTEM QUALIFICATION METHODOLOGY


T. A. Gray, R. B. Thompson and B. P. Newberry

Center for Nondestructive Evaluation
Iowa State University
Ames, IA  50011

INTRODUCTION

The use of computer models of ultrasonic NDE inspections is a convenient and cost-effective alternative and/or companion to experimental reliability trials used for qualifying the detection reliability of a given inspection system applied to a given inspection task.  In addition, the use of such models permits qualification of ultrasonic inspection of new component designs even before such components exist.  This paper presents the current status of the implementation of a model-based software package for these system qualification applications.  A brief overview of the model elements and assumptions will be followed by a discussion of the detection system qualification methodology and, finally, by model-predicted qualification results with associated experimental data.

MODEL OVERVIEW

Accurate prediction of ultrasonic detection reliability which could be obtained in an automated scan of a component requires a number of modeling considerations.  First, there must be the capability to predict the ultrasonic response from a flaw of a given size and depth as modified by the scan configuration, by the geometric and material properties of the component, and by random variability both of the defect's position with respect to scan lines and of the flaw's orientation relative to some nominal configuration.

In the current implementation, the ultrasonic beam and its modification due to passage through a bicylindrical liquid-solid interface is approximated by a Gaussian profile model which is matched (axial amplitude and 1/e beam width) to the central lobe of the far-field radiation pattern of a piston source[1].  Scattering from defects is incorporated either via a numerically exact model[2] to longitudinal elastic wave scattering from spherical scatterers (voids or inclusions) or by an elastodynamic Kirchhoff approximation[3] for longitudinal or shear wave scattering from circular flat cracks.  These models are then incorporated using a measurement model[4] which relates the scattering amplitude of a defect to the corresponding ultrasonic waveform as measured in a practical inspection configuration.  The resulting scan simulation model has been found both theoretically and experimentally to predict accurate backscatter results for cases where beam aberrations can be neglected[1]; where the defects are in the far-field of the probe[1] and are small with respect to the transverse variations of the ultrasonic beam[4]; and, for the case of cracks, where the inspection is close to specular[3]. Theoretical and experimental tests have shown, though, that the Kirchhoff approximation works well at scattering angles up to 60 degrees from normal to the crack face[5]. Although the model, as described above, does not apply to simulation of all possible ultrasonic inspection scenarios, it does have significant applicability in the aircraft engine industry, for example, and is a convenient "shell" into which other scattering and beam models can be easily incorporated.

Next, the appropriate model capability for system qualification is prediction of probability of detection (POD). A block diagram of the POD model concept in general, and in the area of system qualification in particular, is shown in Fig. 1. In the current implementation, POD calculation is based upon Rician fading statistics[6], which is commonly used in the analysis of detecting radar signals in the presence of noise. This implementation is described in more detail in Ref. 7. Briefly, to predict a POD value for a given size crack at a certain depth, the scan model is first used to calculate the video signal amplitudes for a variety of defect orientations and positions relative to the scan lines. These scan amplitudes are then fit to a simple analytic function of the defect's orientational and positional variables in order to reduce the computational burden of subsequent numerical integrations used to compute POD values. This function, along with RMS noise and threshold amplitudes, are then used to generate a POD value for the given size and depth of defect, assuming uniformly random orientational and positional degrees of freedom of the defect, whose limits were specified in the scan simulation.

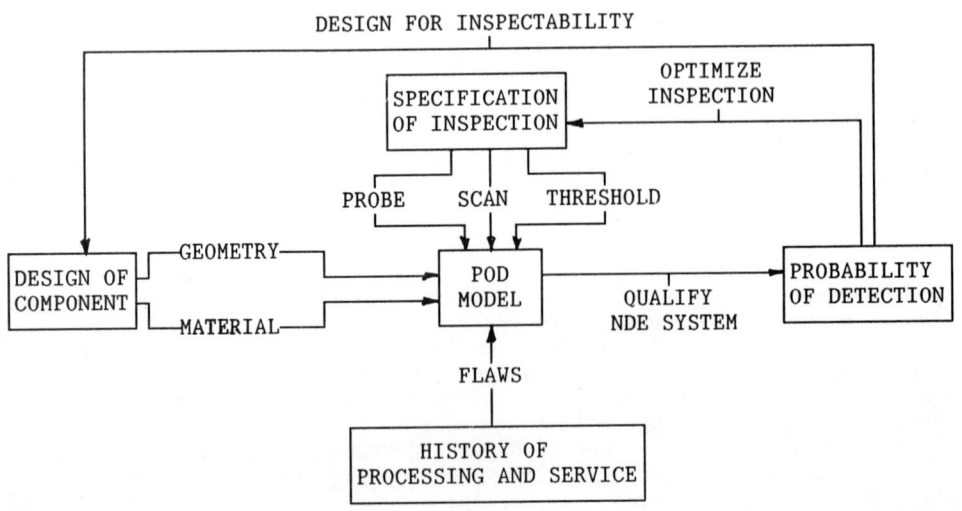

Fig. 1. Diagram of POD model and its application to NDE system qualification and optimization and to computer-aided-design for inspectability.

## QUALIFICATION METHODOLOGY

This section will describe the application of the POD modeling approach discussed above to performing ultrasonic inspection system qualification. The approach chosen here is to qualify a detection system by determining the minimum detectable flaw size (MDF) at any position in the component under test. The MDF is defined according to a predetermined POD value; e.g., assuming a defect is detectable if POD > 0.9. Other qualification criteria, such as false accept and false reject probabilities, can also be determined from the model POD values. Since this section is intended only to be a methodology discussion, other qualification parameters than MDF will not be addressed here.

As in the case of an experimental reliability trial, the first step in model-based qualification of a detection system is to "acquire" scan data from a defect at a given depth in a component using a specified scan plan and configuration. In this application, the component is characterized by its material properties (density, acoustic velocities, ultrasonic attenuation) and its geometry (surface radii of curvature) local to the area where the scan is performed. The defect is specified according to its type (sphere or crack), size, location and (in the case of a crack) orientation relative to the component surface. The scan itself is defined according to the ultrasonic probe characteristics (radius, focal length and bandwidth) and configuration (water path and orientation angles with respect to the component surface) and by the scan plan local to the defect position. The scan plan is defined by a discrete mesh of points (i.e. discrete motion in both the scan and index directions). To model typical scan plans, which have a continuous scan with a discrete index, the mesh size in the scan direction can be set to an arbitrarily small step size (e.g., corresponding to the pulser repitition rate). Finally, a reference RF waveform, such as the ultrasonic reflection from a planar surface using the same probe and other hardware as will be used for inspection, is input to the model.

Based upon the scan configuration, component specifications, and defect characteristics described above, the measurement model is used to compute the amplitudes of the RF signals and/or video envelopes corresponding to the various defect sizes, orientations, and positions with respect to the scan plan. In the current implementation, these calculations are not performed to simulate a scan of an entire component, but, rather, are concerned with the variability of signal amplitudes that could occur for a single grid element in the scan plan. For a discrete scan in a single dimension, at any position of the ultrasonic probe, a defect of a given size, depth, and orientation is most reliably detected at that probe position if the defect lies within one-half of the scan increment of that probe position. If the defect is farther than one-half of a scan increment away, its measured ultrasonic signal would be larger in amplitude, and so more detectable, at the next (or previous) probe position. Thus, the scan simulation algorithm emulates a scan by considering the probe to be at a fixed position while the defect location is varied relative to that probe location.

Since determination of a POD value requires intensive numerical integration, it is not computationally feasible to rely upon the full measurement model to determine the ultrasonic signal amplitudes for all defect positions and orientations needed by the integration mesh (note, this is a 4-D integration for a 2-D scan where the defect has two orientational degrees of freedom). Therefore the measurement model-predicted amplitudes are fit in a least-squares sense to a simple 8-parameter analytic function of the positional and orientational degrees of freedom of the defect, which has been found in practice to reproduce, essentially exactly, the measurement model-predicted variability of signal amplitudes over modest ranges of a defect's variability. In practice, when a POD calculation is to be performed, the measurement model is used to compute the signal amplitudes at three values of each of the defect's positional and orientational degrees of freedom, for a total of 81 points. On a VAX 11/780 computer, the measurement model portion of the calculation takes approximately 90 cpu-seconds to determine the 81 signal amplitudes, while the POD calculation using the simple analytic model takes about 10 cpu-seconds based upon 4096 integration points (8 signal amplitudes for each positional and orientational degree of freedom of the defect), which highlights the improved computational speed of this simple function approach.

Next is the calculation and processing of the POD values. The result of this step consists of POD values for each defect size and depth specified in the scan simulation. A sample POD versus flaw size curve is shown in Fig. 2. Also indicated in this figure are several system qualification parameters of importance, MDF (minimum detectable flaw size), false-accept (FA) and false-reject (FR) probabilities. In the figure, the MDF determination is based upon a POD > 0.90 criterion, which will be the qualification technique described in the remainder of this paper. Based upon such MDF determinations at a number of defect locations in a component, it is then possible to determine MDF = critical-defect-size contours throughout the component, and so qualify the inspection. Similarly, since FA and FR can be determined from a POD curve, as is indicated in Fig. 2, similar analyses could be performed using these as qualification figures of merit.

EXAMPLES

This section will illustrate the model-based qualification protocol along with experimental comparisons, where available. First, experimental and model predicted scan profiles and associated POD's for spherical voids below a cylindrical interface will be compared. This will be followed by a sample application of the model package to NDE system qualification.

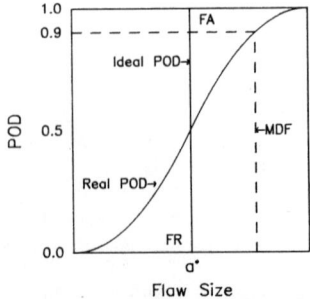

Fig. 2. Typical POD vs flaw size curve, showing various system qualification parameters.

The first example to be presented, which addresses detectability of spherical voids below a cylindrical surface, will illustrate the technique of POD determination from model derived scan data. The particulars of this example are:

>Sample:
>  Fused quartz plate
>  2.54cm (1 in.) thick
>  7.62cm (3 in.) cylindrical curvature
>Defects:
>  Spherical voids (bubbles)
>  Various sizes, depths
>Probes:
>  0.635cm (0.25 in.) diameter
>  Unfocussed
>  5, 10, 15 MHz center frequency
>Scan:
>  Normal incidence
>  Continuous circumferential (cylindrical) scan
>  Discrete axial index
>Threshold:
>  50% DAC for #1 FBH

The scan model "calibration" was effected by using the back-reflected signal, at normal incidence, from the back of the quartz sample through the 1 inch dimension. Since the scan is continuous in the circumferential direction and the flaws are isotropic scatterers, the only random degree of freedom of the defects is their axial position relative to the scan lines. Figure 3 shows a typical comparison of experimental and model-predicted curves of video amplitude as a function of axial position of the defect relative to the probe

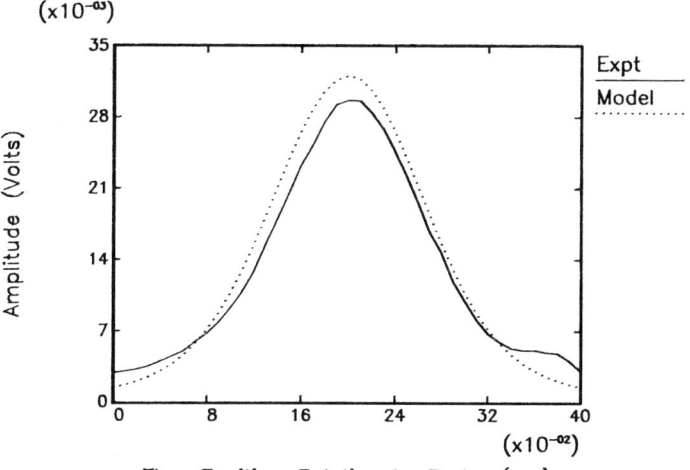

Fig. 3. Comparison of experimental and model-predicted scan profiles for a spherical void below a cylindrical interface.

position. As is evident, the model does quite well at predicting the experimental signal amplitudes.

Based upon the simulated and experimental scan results just mentioned, POD estimates can be determined. In this example, ultrasonic scattering noise is negligible, so the POD for a given size and depth spherical void is just the probability that its noise-free ultrasonic signal will exceed the detection threshold. Thus, if $y_e$ denotes the range of axial defect positions for which the ultrasonic signals exceed the threshold, and if $y_s$ is the axial scan index, then the POD is merely $y_e/y_s$ (if $y_e < y_s$, otherwise POD=1). For a given size and depth of defect, Fig. 4 shows the variation in POD as a function of the scan index mesh size. The model clearly produces accurate estimates of POD values for this simple inspection case.

Next, we consider the utility of the model in a more general NDE system qualification context. The specifics of this example are:

    Component:
      Annular forging, WASPALLOY[8]
      5.715cm (2.25 in.) thick
      8.89cm (3.5 in.) I.D., cylindrical curvature
      12.07cm (4.75 in.) O.D., cylindrical curvature
    Defects:
      Flat, circular cracks
      Alligned (+/- 5 deg.) with forging flow lines
      Various sizes, depths
    Probe:
      0.635cm (0.25 in.) diameter
      Unfocussed
      10 MHz center frequency

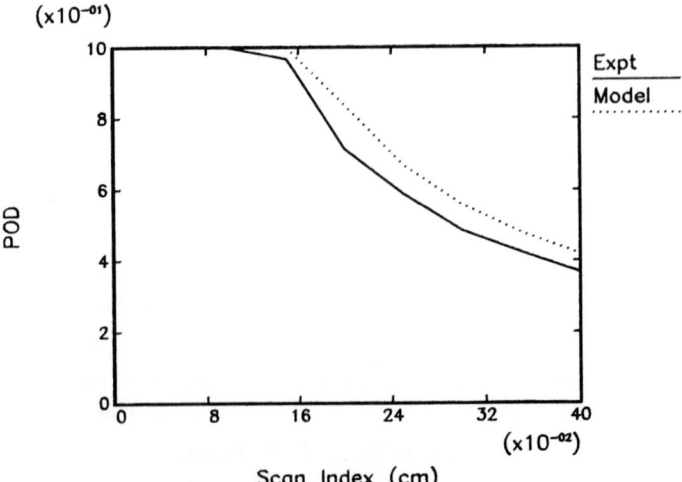

Fig. 4. Comparison of experimental and model-predicted POD vs axial scan increment for a spherical void below a cylindrical surface.

Scan:
  Normal incidence through O.D.
  Continuous circumferential (cylindrical) scan
  Discrete axial index = 0.254 cm (0.10 in.)
Threshold:
  50% DAC for #1 FBH

As in the previous example, a normal incidence, back surface reflection was used as a reference signal for the scan model. Here, though, the defects have orientationally dependent scattering. Moreover, due to the presence of forging flow lines, the defects are assumed to have a nominal orientation that depends upon the flaw's position in the component. Figure 5 indicates the scan model validity by comparing the ultrasonic RF waveform obtained at normal incidence to a #3 FBH interrogated through a planar surface of the forging to its model-simulated counterpart. Evidently, the model, which approximates the FBH as an open, circular flat crack, accurately represents the experimental data.

For system qualification, it is assumed that the key "figure of merit" to be ascertained is minimum detectable flaw size (MDF) as a function of location in the forging. Furthermore, it will be assumed that "detectable" means that POD > 0.9. (Figure 2 indicates this qualification parameter.) Also, since the detection criterion is that an ultrasonic indication exceeds 50% of the amplitude of a #1 FBH, qualification in this example is relative to that threshold type. In order to address the problem of nominal defect orientation in the forging, it was necessary to make some assumption as to forging flow line topography. It was assumed that the flow surfaces were surfaces of revolution (axially symmetric) whose axial cross-sections were concentric parabolas of the form

$$z = a - 2h\left(\frac{y}{w}\right)^2$$

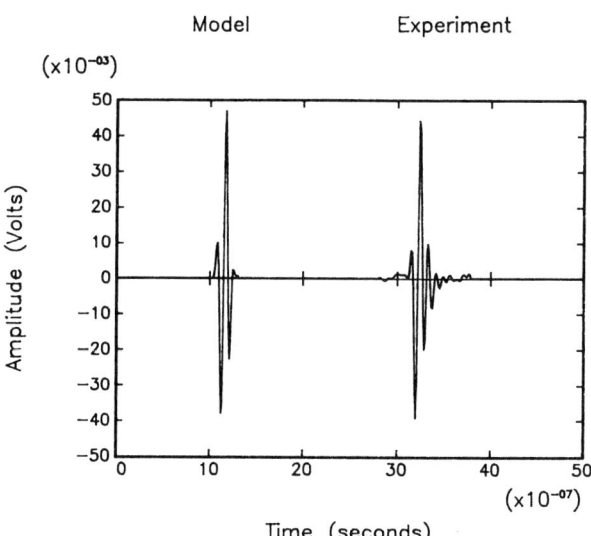

Fig. 5. Comparison of experimental and model-predicted RF waveforms from a #3 FBH in WASPALLOY forging.

where w is the forging thickness (2.25 in.), h is the radial thickness (O.D. minus I.D.), y is axial position through the forging relative to the plane of axial symmetry, and z is radial distance. Based upon this ad hoc assumption of flow line shape, the nominal crack orientation depends solely upon the defect's axial position in the forging, and not upon its depth below the cylindrical component surfaces.

Figure 6 illustrates the qualification concept for this inspection scenario. Shown are contours of constant MDF in the cross-section of the forging. The inner contour (i.e., closest to the centerline of the forging) represents a MDF of 0.01905cm radius (0.015 in. diameter) with successive contours in 0.01905cm increments. As can be seen, the MDF becomes progressively larger (i.e., detectability gets worse) as one nears the edges of the component. This is due to the combined effects of a normally incident ultrasonic beam relative to the component O.D. and the increased tilt of the cracks as one approaches the sample edges (due to the assumed flow lines). This result would highlight, for example, the inadequacy of the assumed scan plan for reliable 0.015 inch defect detectability throughout the component. Of course, if the flow lines are in fact known ahead of time, the scan plan could be adjusted to articulate the probe relative to the part surface so that the refracted ultrasonic beam is more favorably alligned with the nominal defect orientations. In addition, the component could be scanned from the I.D. and edges as well as through the O.D. In such a case, MDF contours at any point in the component could be calculated based upon the maximum POD obtained from the total part scan. The simple example described here does, though, illustrate the technique.

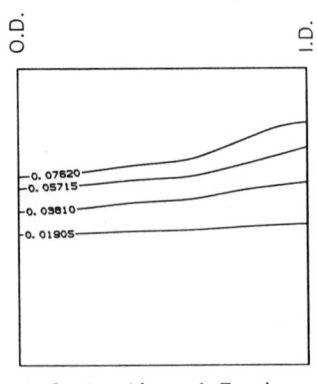

← Center Line of Forging →

Fig. 6. Contours of constant minimum detectable flaw size (MDF) for WASPALLOY forging inspection system qualification example.

SUMMARY

This paper has described the current implementation of a model-based software package and its application to qualification of ultrasonic NDE system qualification applicable to inspection of components with bicylindrical curvature which contain small spherical or crack-like defects and for which abberations in the ultrasonic beam, due to oblique incidence at a curved surface, e.g., are negligible. Several model upgrades to increase its range of applicability are currently under development. First, new scattering model approximations are being considered to relax the limited choice of defects. Work in this area is directed toward enlarging the class of "small" defects which can be accurately modeled and toward modeling inspectability of "large" defects. In addition, the possibility of incorporating experimental scattering data (such as could be available through an enginnering database) is being considered. An enhanced ultrasonic beam model, based upon a Gaussian-Hermite function approach[9], is near completion. This approach allows explicit incorporation of beam abberations and can, with suitable modification, be applied to beam propagation in various classes of anisotropic media, which will find future applicability in modeling of inspection of graphite-reinforced epoxy composite components.

Another area where enhancements are being considered is in the method for extracting POD values from the model data. Such approaches as log-normal, etc. statistics will be investigated. Also considered will be probability distribution functions other than the uniform distribution to describe defect variability. The possibility of extracting confidence levels (to predict "90/95" detectability, e.g.; i.e., POD = 0.90 with 95% confidence) is under study, as well. Future work will also address the use of other qualification criteria, such as false reject probabilities. The goal of these modeling tasks is to develop a system qualification tool with general applicability which can subsequently be used as the "kernel" of a package to be incorporated in computer-aided-design activities to enable specification of component inspectability at the design stage.

ACKNOWLEDGEMENT

This work is supported by the NSF university/industry Center for NDE at Iowa State University.

REFERENCES

1. R. B. Thompson and E. F. Lopes, J. Nondestr. Eval., 4, 1984, pp. 107-133.
2. C. F. Ying and R. Truell, J. Appl. Phys., 27, 1956, pp. 1086-1097.
3. L. Adler and J. D. Achenbach, J. Nondestr. Eval., 1, 1980, pp 87-100.
4. R. B. Thompson and T. A. Gray, J. Acoust. Soc. Am. 74(4), 1983, pp. 1279-1290.
5. T. A. Gray and R. B. Thompson, in Review of Progress in Quantitative Nondestructive Evaluation, vol. 4A, D. O. Thompson and D. E. Chimenti, eds., Plenum Press, New York, 1985, pp. 11-17.
6. M. Schwartz, Information Transmission, Modulation and Noise, McGraw-Hill, New York, 1970, pp. 361-372, 444-451.
7. T. A. Gray and R. B. Thompson, in Review of Progress in Quantitative Nondestructive Evaluation, vol. 5A, D. O. Thompson and D. E. Chimenti, eds., Plenum Press, New York, 1986, pp. 911-918.
8. WASPALLOY forging sample courtesy of Allison Gas Turbine Engine division of General Motors Corporation.
9. R. B. Thompson and E. F. Lopes, in Review of Progress in Quantitative Nondestructive Evaluation, vol. 5A, D. O. Thompson and D. E. Chimenti, eds., Plenum Press, New York, 1986, pp. 117-125.

CRACK SIZING BY THE TIME-OF-FLIGHT DIFFRACTION METHOD, IN THE LIGHT OF
RECENT INTERNATIONAL ROUND-ROBIN TRIALS, (UKAEA, DDT AND PISC II)

Graham J. Curtis

AERE Harwell, UK and US Naval Academy, Annapolis

Building 521, Materials Physics & Metallurgy Division
AERE Harwell, Oxon, OX11, ORA, UK

INTRODUCTION

In 1980-81, Harwell developed a mini-computer controlled multi-probe defect detection and sizing system(1) based on the ultrasonic time-of-flight/diffraction principle introduced by Silk(2). This system proved to be capable of fully automatic data collection from the PWR girth-weld simulation Plates 1 and 2 in the Defect Detection Trials of 1981-82. The speed of collection and subsequent analysis was such that a report on the defects found could be filed within 48 hours. The mode of operation adopted simulated minimum time of access to the defects, and was intended to define that dimension of a defect which has greatest significance, ie the through-thickness dimension.

In 1984, for the PISC II Trial, the approach adopted changed to emphasise the three-dimensional location and sizing capabilities of the time-of-flight/diffraction method. Data collection and analysis became highly interactive and the mode of operation simulated NDE at the manufacturing stage of a pressure vessel.

The purpose of this paper is to indicate the defect through-thickness sizing capability of TOFD achieved in the 1981-82 Defect Detection Trials and the defect mapping capability achieved in the 1984 PISC II Trial.

ACOUSTIC DESIGN ELEMENTS OF THE AUTOMATED TOFD+SAFT SCANNER

The design features of the 8-sender, 8-receiver line array scanning head, together with its control instrumentation has been discussed previously(1). It is perhaps useful to reiterate here the fundamental features, in order that what follows will be readily appreciated:

Accuracy of Depth Determination

Figure 1 depicts a sender-receiver pair of probes straddling a symmetrically located defect. The time-of-flight between the probes, T, is given by:

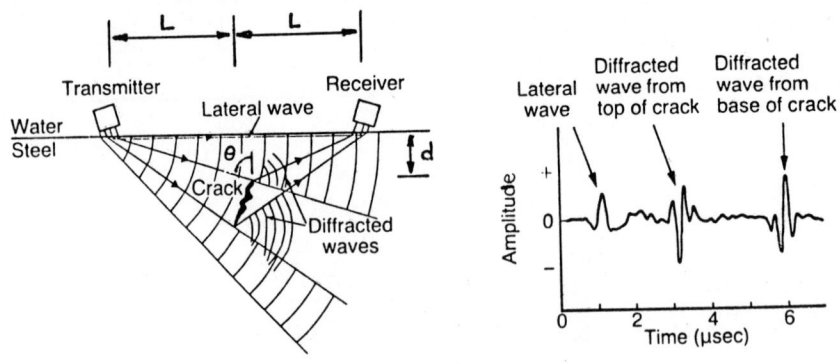

Fig 1  Basis of the Time-of-Flight/Diffraction Technique.

$$T = \frac{2}{C}\left[L^2 + d^2\right]^{\frac{1}{2}} \qquad 1$$

where C = velocity of compressional waves in the metal.

Whence the error, δd in determining the depth is related to the error in the time-of-flight, δt, by:

$$\delta d = \frac{C}{2\cos\Theta}\delta t \qquad 2$$

A target specification of ±1mm was set for the depth determination imposing an upper limit of 80° on Θ for a δt of 0.05μs.

## Optimization of the Diffracted Amplitude

Silk(3) examined the angular dependence of the amplitude of diffraction from a vertical crack upon the angle of incidence and found it to be greatest at approximately 60°. The theory due to Temple(4) largely corroborates this.

## Choice of Working Angles

For maximum diffracted amplitude, the angle for the axis of the refracted beam should be 60°, and for a sizing accuracy of ±1mm, the maximum angle in the refracted beam should not exceed 80°. The latter also corresponds to the upper half-maximum diffracted amplitude angle. The lower half-maximum diffracted amplitude angle is approximately 45°, so this was taken to be the lowest bound on the useful angular range.

## Choice of Ultrasonic Operating Frequency

A balance needs to be struck between using as high a frequency as possible, to increase the accuracy of location and resolution and as low a frequency as possible to both decrease the affects of attenuation in the austenitic cladding layer present on the specimens and increase the precision of the analogue-to-digital converter used. The compromise frequency adopted was 5MHz. The strobe frequency of the digitiser was chosen to be 20MHz.

The linear probe array required to cover the ASME XI Code weld area

The restricted range of working refraction-angle for a pair of probes, dictated by the need to maximise both the diffraction signal amplitude and the accuracy of measurement, makes it necessary to have a linear array of 8 sending transducers and 8 receiving transducers to cover the cross-sectional area of weld in the ASME XI Code of Inspection. The design of the array is described elsewhere (1), Fig 2, however, shows the array in a probe-trailer resting on the surface of the PISC II Trial Plate 2. (The PISC II Plate is 1500 mm square and 300 mm thick). Acoustic coupling is achieved by water immersion. The probe trailer is designed to respond to the changing contour of the specimen, but is constrained to follow the tracking direction without skewing. As the photograph shows the probe trailer is connected to an x-y scanning frame driven by stepping motors. It was found possible to drive the scanning head, under computer control, around the surface of the specimen, in a complex raster, and still have it return to its start location to an accuracy of ±2mm.

DATA COLLECTION

An HP1000 minicomputer controls the entire scan procedure via CAMAC and Harwell 6000 series modules and the storage of accumulated data on magnetic tape. The scan takes place automatically after the head has been calibrated and located at some appropriate starting point relative to the datum mark on the specimen. Scan rates depend upon the number of probe-pair combinations used and the amount of averaging required. With just a single probe-pair, B-scans can be produced on an associated TV monitor with a scan speed of 30mm per minute. In the search mode, which uses 40 combinations out of the possible 64, the scan speed drops to 3mm per minute. A block diagram of the inspection electronics can be found in ref (1). The scanning carriage carries the 8 individual transmitter drivers, each capable of delivering a 400 volt fast rising pulse. The scanning-head itself carries the 8 individual receiver charge amplifiers. Further amplification takes place locally to the scanning frame. The 8 digitisers are accommodated locally to the control logic and the mini-computer in a room some 30 metres away.

Fig 2   The Multi-Probe Time-of-Flight/Diffraction Scanner

DATA INTERPRETATION

By transferring sequential A-scan data into adjacent columns of a frame store memory, the inspection data is presented as a B-scan. The amplitude of the ultrasonic signal modulates the brightness of the image. Any unevenness in the surface of the specimen leads to variation in the length of the water path between the probes and the surface as the scanning head passes across it. Not only is the locus of the lateral wave signal in the B-scan made bumpy, but the effect is translated through the whole duration of each B-scan element, distorting the shape of any defect signals present. To remove these distortions the trace is flattened. The flattened B-scan still retains the effect of beam spread. To determine the true length of a defect, the "tails" at the ends of the defect are removed by synthetic aperture focus-processing (SAFT) of the B-scan (see ref 1 for details of the processing procedures).

All depth measurements are made on B-scans that have been flattened. The B-scan presentation is carried out on a specially programmed PDP11/64 + I2S image processing package. The operator locates a cursor against any feature, presses a button and the cursor changes its shape to indicate the response of a point located at that particular depth. At the same time the depth of the feature below the surface, (ie the cladding-ferritic interface if the inspection is from the clad-side of the specimen), appears on an associated VDU screen, together with the location relative to the start of the scan. The cursor has been found to be very useful in defining the length of defects which are near to the surface where the performance of SAFT processing is not so good. It has also been found to be valuable when determining the location of planar defects in the dimension normal to their plane. For this type of defect, scanning across it produces a B-scan with a parabolic signal-locus for the top and the bottom.

THE GEOMETRY OF DDT PLATES 1 AND 2 AND THE PICS II PLATE 2

The DDT Plates 1 and 2 and the PISC II Pate 2 were approximately 1500mm square and composed of two 750mm wide, 300mm thick A33B ferritic plates, butt-welded together. The weld profile was double "U" and the finished plate was austenitically clad to a depth of 8-10mm. In the case of DDT Plate 1, recesses were milled into the side-walls prior to welding to accommodate rectangular pieces prepared from fatigue-crack specimens. The affect was to simulate lack-of-wall fusion defects. In DDT Plate 2 and PISC II Plate 2 controlled defects of more natural morphology were induced by contaminants in the weld metal; eg planar cracks, branched cracks and slag-lines. To simulate dangerous planar cracks, whole pieces of unbroken fatigue-crack specimens were welded into recesses in the side walls of the weld.

THE BASIC ACCURACY OF DEFECT THROUGH-THICKNESS MEASUREMENT USING TOFD

The best indication of this is obtained from the inspection data for DDT Plate 1. The defects had well controlled perimeters, and yielded very clear, unambiguous TOFD B-scans. The B-scan of Fig 3 shows how clearly the signals obtained from the tops and bottoms of the 15 defects in the lower part of the weld, stand-out against the background of noise produced by grain-scatter. Figure 4a shows the degree of correlation achieved between the through-thickness dimension defined by TOFD and that defined by destructive examination (6). The correlation coefficient is 0.98. The sizing error is on average 2mm. It is useful to compare this figure with a similar one shown in Fig 4b which is composed from all the

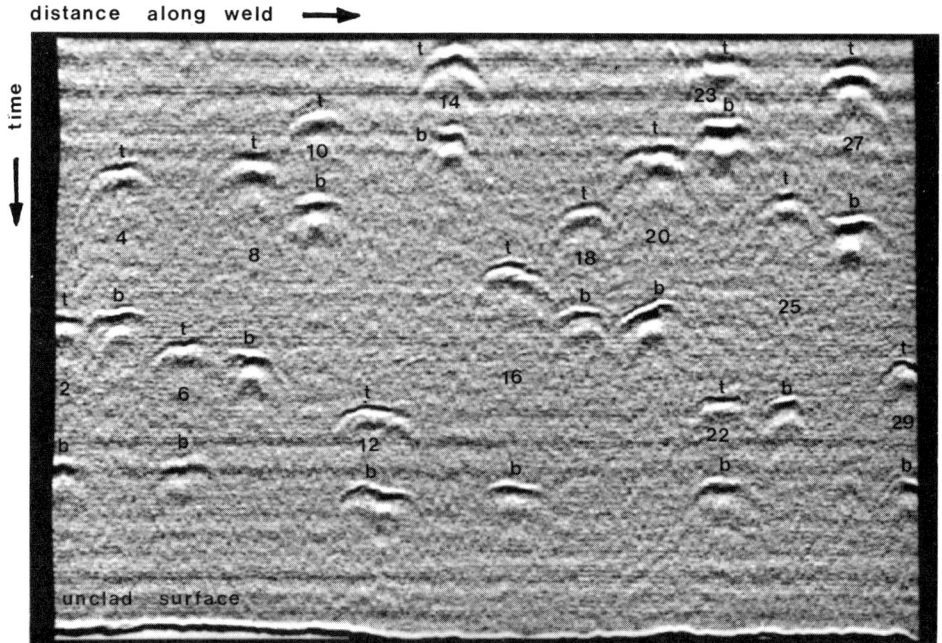

Fig 3  DDT Plate 1: A TOFD B-scan showing signals from the tops and bottoms of 15 defects

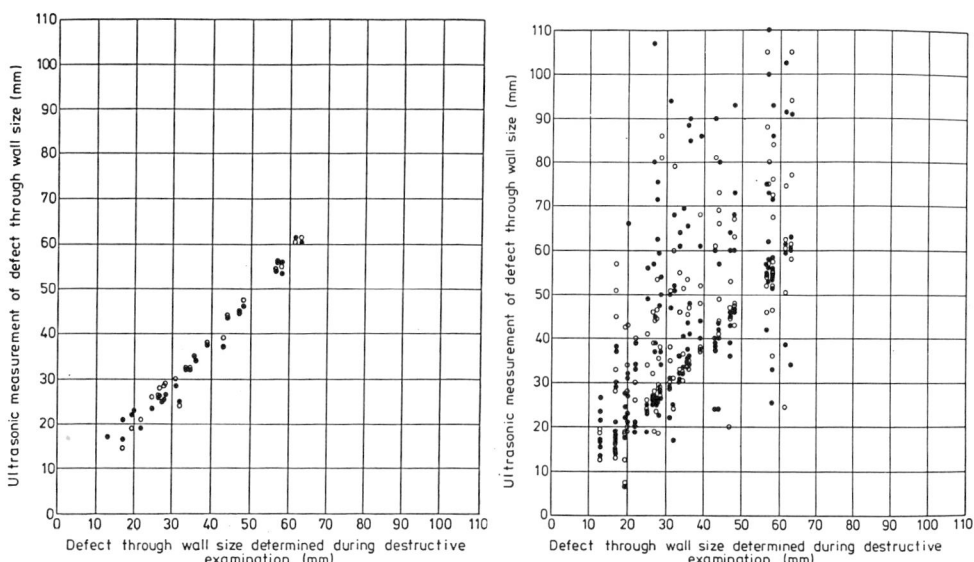

Fig 4a  DDT Plate 1: Correlation of TOFD through-thickness sizing with destructive defect sizing

measurements made by the teams taking part(6). This figure indicates the spread of accuracy possible from the various techniques currently in use in Europe compounded by the skill of the operators.

Currently the best indication of this is obtained from the inspection data from PISC II Plate 2, although it must be stated that truly destructive data has yet to be produced. The "destructive reference data" has been obtained from radiographs and ultrasonic c-scans carried out on small blocks containing each defect, which were excised from the Plate(5).

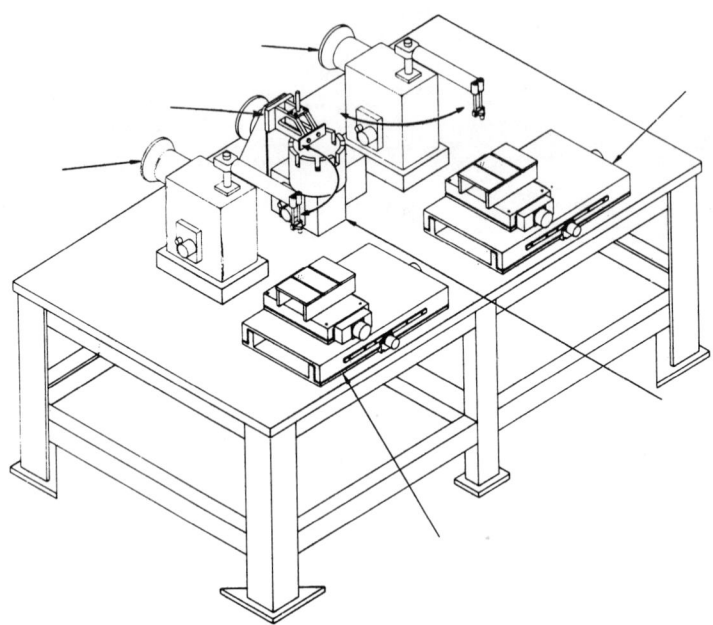

Fig 4b  DDT Plate 1: Correlation of all submitted through-thickness sizing with destructive defect sizing

To characterise a defect, two mutually perpendicular, linear TOFD and SAFT, B-scans are produced through the centroid of the defect. From these, two sectional drawings of the defect - elements are produced. Fig 5a shows the drawings produced for Defect 2 in PISC II Plate 2 and Fig 5b shows one of the SAFT processed TOFD, B-scans used to define the location of the elements of Defect 2, in the through-thickness/weld-length plane of the weld. Superimposed upon the drawing is the box-like perimeter for the defect defined by the semi-destructive examination.

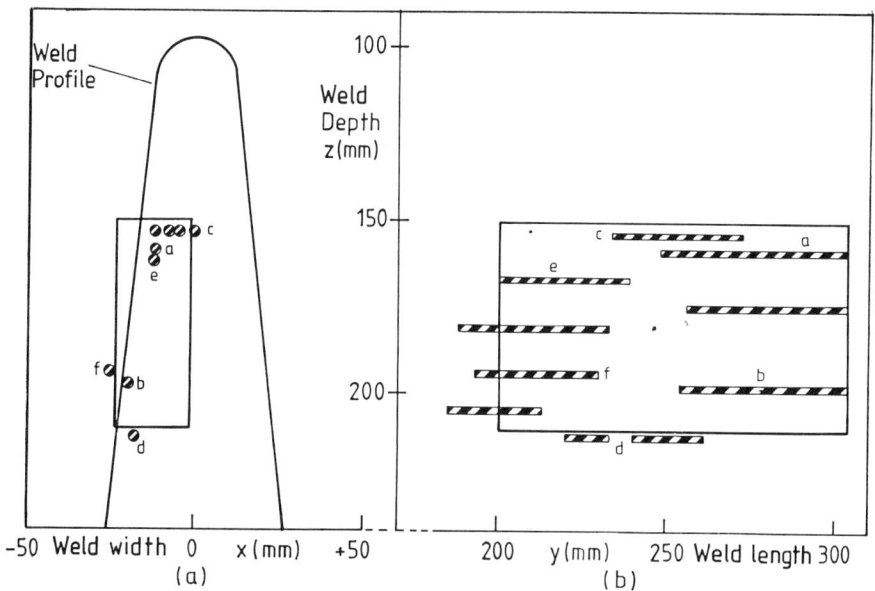

Fig 5a  PISC II Plate 2:  Sectional drawings for defect 2 derived from TOFD and SAFT measurements; (a) end elevation  (b) side elevation

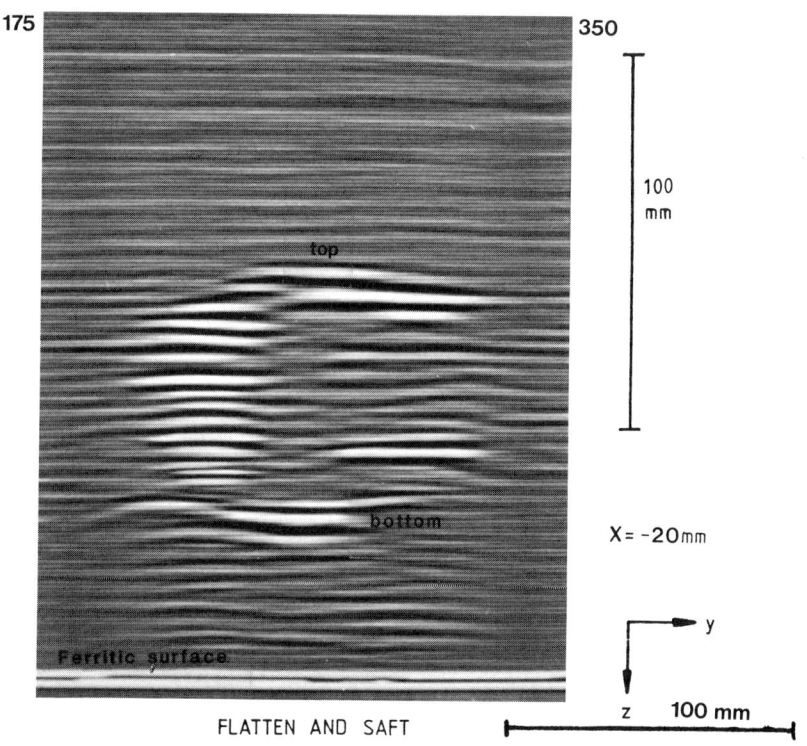

Fig 5b  PISC II Plate 2:  Raw and SAFT Processed TOFD B-Scans of Defect 2

CONCLUSIONS

The TOFD and SAFT technique is currently one of the most accurate defect sizing methods in the armory of NDE methods. It also offers the facility to produce accurate, detailed, three dimensional maps of the components of defect. It, therefore, gives the non-destructive evaluator the facility to characterize as well as to overall size defects.

REFERENCES

1. G. J. Curtis and B. M. Hawker, "Automated time-of-flight studies of the Defect Detection Trial Plates 1 and 2", British Journal of NDT, September 1983, pp. 240-248.
2. M. Silk, "Sizing crack-like defects by ultrasonic means", Res. Techns. in NDT, ed. R. S. Sharpe, Vol. 3, Ch. 2, Ac. Press, 1977.
3. M. Silk, "Transfer of ultrasonic energy in the diffraction technique for crack sizing", Ultrasonics 17 (3), 113, 1979.
4. J. A. G. Temple, "Time-of-flight inspection with compression waves: cracks and slag-lines", AERE TP Report 932, 1981.
5. E. Borloo and S. Crutzen, "Destructive examination of the PISC II RRT Plates", PISC II Report No. 3, June 1985. Published by Joint Res. Centre, Ispra Centre, Varese, Italy.
6. J. A. G. Temple, "Sizing capability of automated ultrasonic time-of-flight diffraction in thick section steel and aspects of reliable inspection in practice", UKAEA Report AERE-R 11548, Nov. 1984.

DISCUSSION

Dr. R. B. Thompson, Ames Lab: A question that always arises, with respect to tip diffraction techniques, is the influence of closure or other things on the strength of that tip signal relative to the background. What sort of studies have you performed relating to that issue?

Dr. Curtis: Well, we have done experiments in the fracture lab to show the effect of compressive and tensile stresses. I have been away for a year so I'm not sure of the latest data. Perhaps Andrew Temple, who is in the audience, can bring me up to date.

Dr. Temple: We have done two types of studies: The first showed that the signal from a crack under compressive stress tended to saturate with increasing compressive stress. Typically it saturates at about 12 dB below the uncompressed crack signal. In the second studies, we have shown there's still saturation with possibly a lower -more signal velocities.
In principal a lot depends on the stage of growth of the crack, the compressive stress and the actual (morphology) of the crack, but nevertheless, in practice, you can still see a crack signal even if you compress it right up to almost the yield. It's a question then of whether you lose 12 or 15 dB, whether the signal-to-noise accommodates that.

From the Floor: I guess your time-of-flight message is very similar to the Aloc method developed by IZP Saabrucken. They also used time-of-flight measurements in order to size defects. We are also involved in the P.I.S.C.

Mr. Mohammed Behravesh, EPRI: Clearly, the results you have shown demonstrate the effectiveness of this technique. Do you have any information as to how well inspectors are able to repeat those measurements?

Dr. Curtis: Yes and no. What I have been showing you is really the results obtained from the use of a prototype system on round-robin test plates, so we are somewhat remote from the field. All I can say is that we have 18 year-old-girls doing the analysis and they do it quite well. I imagine the average Ph.D. chimpanzee could do it quite well, too.
  (Laughter)

From the Floor: I have two questions. My first question is: In your experimental setup, you have both transmitter and receiver in a water bath. Have you thought of using contact transducers for that? Is the quality the same or better?

Dr. Curtis: Yes, it's about the same. We have done about $6 million worth of development on time-of-flight and it's been used with grease or oil contact, with water, with wheel probes, with EMATs, the whole thing.

From the Floor: My second question is that you have a two-transducer configuration. Have you thought of using one transducer as both transmitter and receiver?

Dr. Curtis: We haven't, really. We have seen at least one other set of researchers use a single system, and we can see that it's possible to do if there is a specimen configuration that demands it. So yes, we have considered it, but most of our experience is with getting the optimum sensitivity, which is with a separate send-receive setup.

From the Floor: So does the size of the transducer affect the quality of the B-scan?

Dr. Curtis: It does. As with any ultrasonic test, the amplitude of the signal depends upon the diameter of the probe and a number of other parameters. In the system I have shown here, we used half-inch probes for everything except the widest-spaced pairs of probes, where we double up to an inch. So, the data you've seen here from cracks deep in the weld has been obtained with an inch diameter probe working at 5 Megahertz.

From the Floor: I was wondering, is there some way of finding a frequency that gives you an optimum tip-diffracted wave that you can look at? How do you go about finding that frequency?

Dr. Curtis: Well, Andrew Temple, the theoretician, might have some comments, but my practical experience is that we would have liked to go to, say, 10 or 20 Megahertz. But working through the austenitic cladding layer on the pressure vessel simulation specimens tends to be a dominant feature that controls the frequency-dependent amplitude of the tip diffracted signal.
  But maybe, Andrew, you have a comment.

From the Floor: Well, I was just wondering if there was some kind of an optimal frequency that you can look at. Instead of going to a higher frequency, maybe even going to a lower gives you a better amplitude of detection of the diffracted wave. Is there some theory behind that?

Dr. Curtis: Well, only that which relates to attenuation in the material.

From the Floor: Well, other than just attenuation of the material?

Dr. Curtis: Not that I know of.

RESULTS OF THE PHASE I RELIABILITY TEST

ON THE RFC/NDE EDDY CURRENT STATION

Ray T. Ko

NDE Systems Division
Systems Research Laboratories, Inc.
Dayton, Ohio 45440-4696

INTRODUCTION

The Phase I Reliability Test was successfully conducted on the RFC/NDE Inspection System in November, 1985, at Systems Research Laboratories, Inc. The objective of the test was to ensure the RFC/NDE Inspection System's ability to reliably inspect the crack-critical geometric features on representative F100 engine parts. Three types of EC reliability specimens (representing gas turbine engine disk geometries) were inspected: flat plates (web/bore surface); 0.316-inch diameter boltholes (with 45° chamfers on both sides of the specimen holes); and 0.177-inch diameter rivet holes. Thirty specimens of each type were used, with each containing either a number of fatigue cracks at various orientations and locations, or no cracks at all, as illustrated in Figure 1. The specimens were fabricated to satisfy the crack depth category requirements shown in Table 1.

The specimens were mounted on special fixtures, with rivet hole and bolthole specimens mounted six at a time, and flat plates one at a time, as shown in Figure 2. The fixtures clamped to the EC station rotary table in

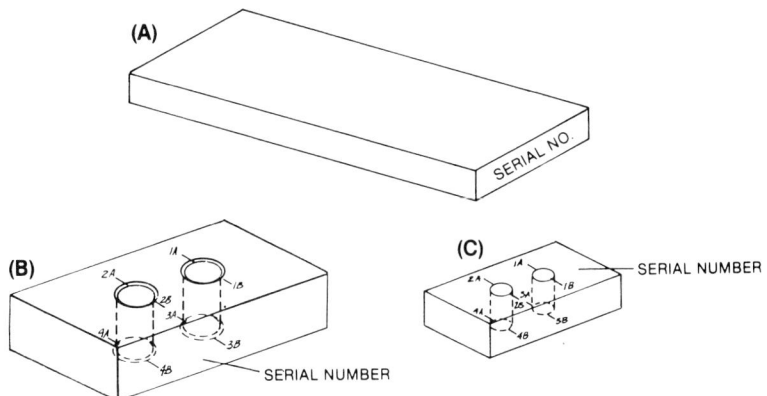

Fig. 1. Reliability specimens: (A) flat plate; (B) bolthole; (C) rivet hole. Flaws were in the top and/or bottom of flat plates, and at any of the eight positions in the holes.

Table 1. Specimen Crack Depth Categories

| Category | Depth (A) |
|---|---|
| I | (A) < 0.010-in. |
| II | 0.010-in. ≤ (A) < 0.020-in. |
| III | 0.020-in. ≤ (A) < 0.050-in. |

a manner similar to an engine disk being clamped. Prior to the test, SRL received six rivet hole specimens, six bolthole specimens, and three flat plate specimens, containing cracks of each category, with which to develop specimen scan plans.

SCAN PLAN DEVELOPMENT

Scan plans for the reliability specimens utilized the same inspection routines as used for engine parts. Some modifications evolved during development of the specimen scan plans, including:

1. Development of the Method 2 flaw detection technique.
2. Implementation of back-and-forth scan motion for flat plates.
3. Development of a consistent phase calibration for rotary probes.
4. Use of two linear axes for centering accuracy.
5. Modification of the clamping mechanism on the specimen fixtures.

Method 2 Flaw Detection

Unlike the Method 1 flaw detection technique which applied a threshold level across an entire waveform, Method 2 was designed to look for signal variations inside a time window and compare them to the threshold [1]. These variations include zero crossing, and the shape and amplitude of the signal. Since the probes utilize a double-ended differential coil, flaw signals usually contain at least one zero crossing. Also, flaw signals usually have a slope that can be calculated from the scan speed and coil size. In most cases, noise signals do not satisfy all of these conditions and are therefore not seen as flaws. For example, Method 2 can reliably detect a 0.003-inch deep fatigue crack in a rivet hole, with the threshold

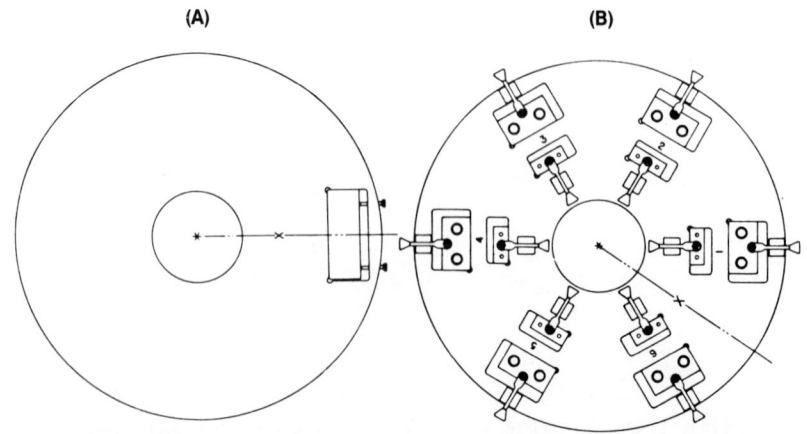

Fig. 2. Specimen fixtures for (A) flat plates, and (B) bolthole and rivet hole specimens

set at 40 millivolts and a peak noise amplitude of 34 millivolts. Method 1 and Method 2 techniques are compared in Table 2.

Back-and-Forth Scan

Development of the flat plate scan plans presented several challenges. While the engine disk scan plan incorporated a continuous rotation of the turntable to inspect flat surfaces, this type of mechanical scan could not be applied to the specimens for the several reasons:

1. The dimensional variations between specimens and the fixture made the step-down movement very difficult, as shown in Figure 3.
2. The probe springs were very easily damaged as the probe traveled over the specimen edge.
3. The edges of the fixture slot and specimens gave strong geometry signals.
4. The aluminum fixture, holding specimens of inconel, titanium, and waspaloy, presented material-related problems.
5. The roughness of the aluminum fixture surface produced relatively large signals compared to those from the specimens.

A back-and-forth scanning motion, shown in Figure 4, was developed to scan the specimens, instead of using a continuous rotation of the turntable. This motion consisted of a $15°$ rotation of the turntable, a reverse rotation back to the starting point, and an index move toward the center of the turntable; this sequence was repeated until the suspected flaw zone was covered. Although the back-and-forth motion solved the five problems stated above, it had two side effects. First, the soft survey peak-detection routine could not be used with the back-and-forth motion, since soft survey depends upon a continuous rotation of the turntable. Second, the turntable could not reach a desired constant speed before it ramped down within the $15°$ scan range, resulting in an inferior scan condition for specimens compared to that for real engine disks. However, small flaws on the flat plates were still reliably detected.

Consistent Phase Angle

A consistent phase angle is very important in achieving reliable flaw detection. Phase calibration of rotary eddy current probes can be done in at least two ways, as described in Table 3. The first method requires the probe to center above a calibration bolthole, then move down into a flaw-free area where the lift-off signal is adjusted to horizontal. The second method is to place the probe beside a calibration block at a certain lift-off and, using the resultant strong lift-off signal as a reference, rotate the signal to the desired phase angle. This beside-the-block method is very repeatable and less sensitive to lift-off variations than the first method. However, the phase angle acquired beside the block usually differs from that acquired when the probe is inside a bolthole. This difference is easily determined, and compensated for, by comparing the two phase angles and inserting the appropriate constant value in the scan plan. Thus, the phase angle used for a bolthole inspection is the angle determined in the

Table 2. Comparison of Method 1 and Method 2 Techniques

| Item | Method 1 | Method 2 |
|---|---|---|
| Flaw determination criteria | Amplitude | Amplitude, zero crossing, and waveform slope |
| Advantages | -- | Tolerant to noise, can detect small flaws |

Table 3. Two Phase Calibration Methods for Rotary Probes

| Inside the Bolthole | Beside the Block |
|---|---|
| 1. Center above the hole | 1. Position beside the block |
| 2. Move to flaw-free area | 2. Make lift-off signal horizontal |
| 3. Make lift-off signal horizontal | 3. Rotate angle by certain degree |

beside-the-block method, offset by the predetermined constant. Although both phase methods are acceptable, the second method was chosen for its repeatability.

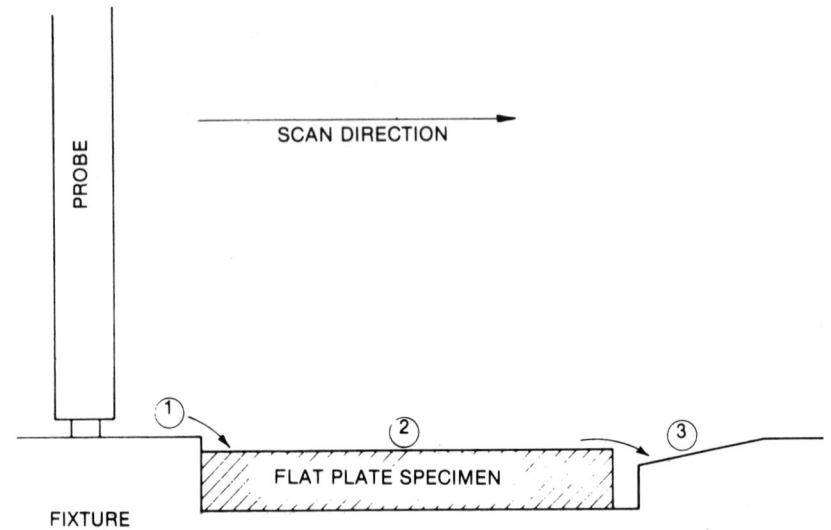

Fig. 3. Side view of the step-down movements for flat plate inspections. The probe moves from the fixture (position 1) down to the specimen (position 2) and back to the fixture via a ramp (position 3).

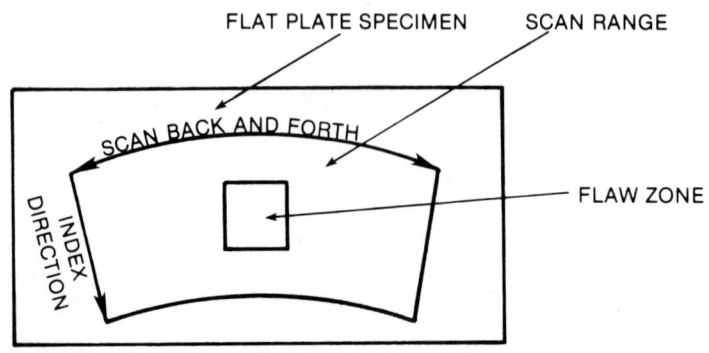

Fig. 4. Top view of the back-and-forth scan pattern on a flat plate specimen

Accurate Centering

Prior to inspecting a bolthole, a centering routine moves the probe to within 0.0010-inch of geometric center of the bolthole. Normally, the turntable and a linear axis are used for centering, but this method does not always give the desired accuracy due to the limited resolution of the turntable. The turntable resolution is $0.01°$, and the corresponding arc length of this angle becomes larger as the radius increases, as shown in Figure 5. Since the center of the bolthole specimen is 7.85-inches from the center of the turntable, the corresponding arc length for a $0.01°$ arc is 0.0013-inch, which is outside the given probe centering tolerance of 0.0010-inch. This problem was solved simply by using two linear axes for centering, since the linear axes have a resolution of 0.0001-inch. This technique not only resulted in very good centering on the chamfered bolthole specimens, but also provided a solid foundation for reliably detecting small flaws in the chamfer region of the boltholes.

Reliable Clamping

Finally, the fixture was modified to provide a reliable clamp on the specimens. Originally, rubber stoppers were inserted into holes located at the corners of the specimens to hold the specimens in place. However, the pressure of the stopper against the side of the specimen tended to raise the specimen off the fixture, and the specimens often loosened and popped out during scans. To remedy this situation, an adjustable rubber-tipped spring-loaded clamp was used to clamp the bolthole and rivet hole specimens, and side-drilled holes with thumb-screws were incorporated to hold the flat plate specimens in place, as shown in Figure 2. To keep the flat plate specimens from rising, the contact surface of the screw was rounded.

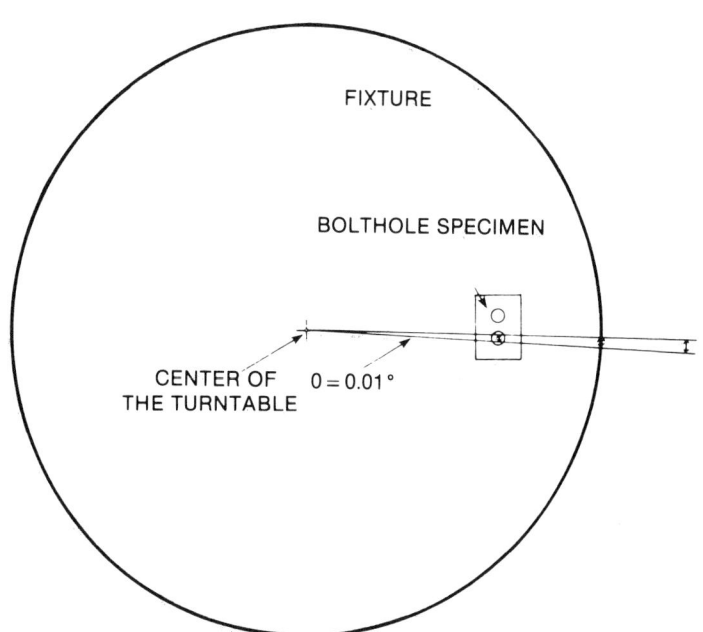

Fig. 5. Bolthole centering accuracy at various axial locations

TEST MATRIX

System performance was assessed in terms of the five performance characteristics outlined in Table 4. Based on those parameters, several types of tests were developed, as described in Table 5. The test matrix shown in Table 6 was devised to accomplish and track all the tests performed on the various specimens.

Table 4. Performance Characteristics

| Characteristic | Definition |
|---|---|
| 1. Capability | Basic flaw detection capability |
| 2. Repeatability | Multiple inspection effects |
| 3. Reproducibility | Effects of changes (instrument, part loading) |
| 4. Variability | Effects of human parameters (operator) |
| 5. Reliability | Composite effect of multiple inspections, instrumentation changes, human parameters |

Table 5. Reliability Test Types

| Test Type | Description | Load Specimen | Side (Top/bottom) | Orientation (Degree) |
|---|---|---|---|---|
| A | Base Capability | Yes | Top | 0 |
| B | Five Repeats | No | Top | 0 |
| C | Probe Change | No | Top | 0 |
| E | Load Change | Yes | Top | 0 |
| F | Load Change | Yes | Top | 0 |
| G | Position Change | Yes | Top | 180 |
| H | Position Change | Yes | Bottom | 180 |
| I/J | Operator Changes | Yes | Top | 0 |

NOTE: Test Type D was planned third probe change but was not accomplished. Test Type H did not apply to the flat plate specimens.

Table 6. Test Matrix for Reliability Specimens

| Specimen | Test Type | | | | | | | | |
|---|---|---|---|---|---|---|---|---|---|
| | A | B | C | E | F | G | H | I | J |
| A1 to A6 | AA | BA1 to BA5 | CA | EA | FA | GA | HA | IA | JA |
| B1 to B6 | AB | BB1 to BB5 | CB | EB | FB | GB | HB | IB | JB |
| C1 to C6 | AC | BC1 to BC5 | CC | EC | FC | GC | HC | IC | JC |
| D1 to D6 | AD | BD1 to BD5 | CD | ED | FD | GD | HD | ID | JD |
| E1 to E6 | AE | BE1 to BE5 | CE | EE | FE | GE | HE | IE | JE |

NOTE: Only one specimen was used for flat plate inspections. Test Type H did not apply to flat plates.

RESULTS

The test data indicated a 90-95% confidence level for detection of surface flaws in the 0.005-inch to 0.010-inch range in the flat plate, bolthole, and rivet hole specimens. An encouraging aspect of this was the strong correlation between the apparent vs. actual flaw data.

Figures 6 through 11 show the cumulative results of the tests on position change (Types G, H, and A), probe change (Types A and C) and repeated inspections (Type B). Of particular interest were the variability test results, which indicated the system is largely unaffected by operator changes, position changes, and probe changes.

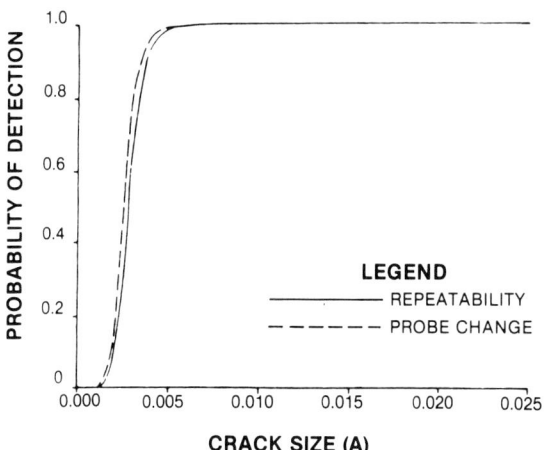

Fig. 6. Flat plate (web/bore surface) specimen - cumulative test types test results

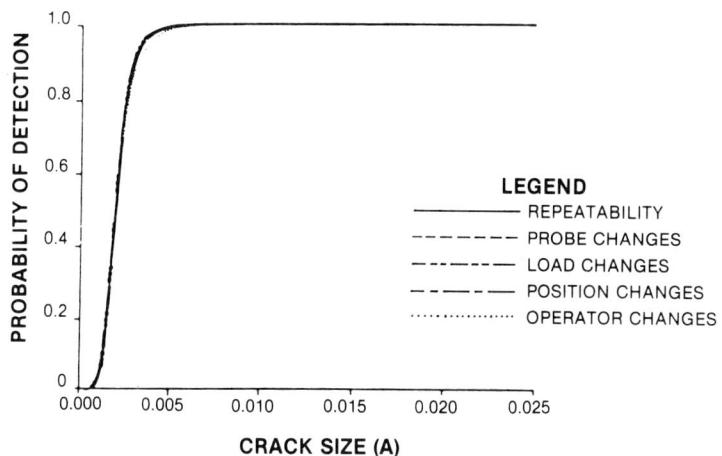

Fig. 7. Bolthole specimen - cumulative test types results

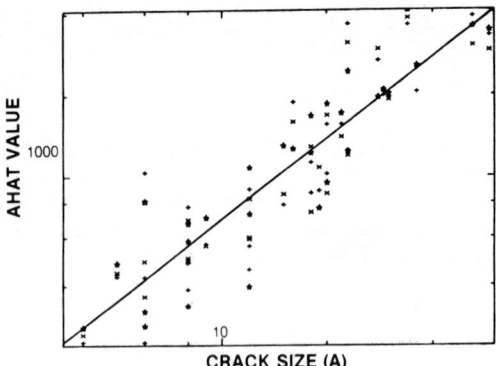

Fig. 8. Bolthole specimen - position change test data

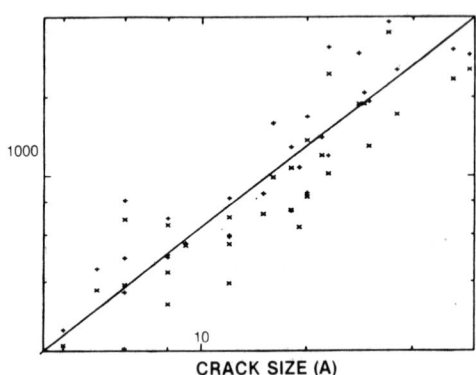

Fig. 9. Bolthole specimen - probe change test data

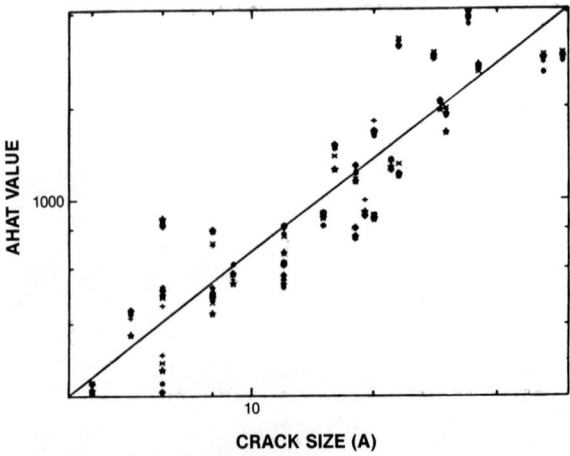

Fig. 10. Bolthole specimen - 5 repeated inspections test data

984

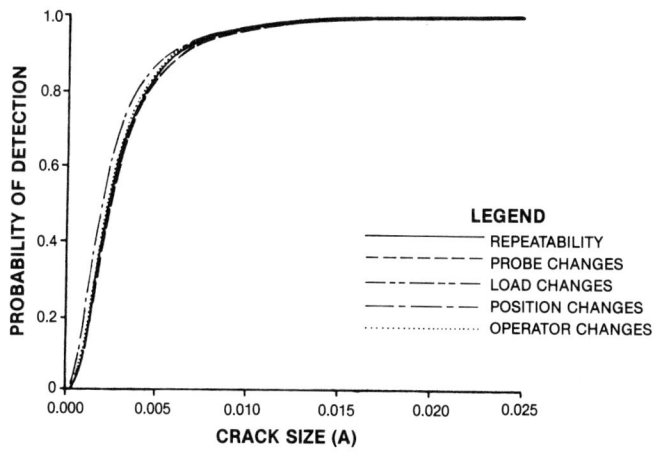

Fig. 11. Rivet hole specimen - cumulative test types results

ACKNOWLEDGEMENT

This work was performed under USAF contract number F33615-81-C-5002 of the Air Force Wright Aeronautical Laboratories/Materials Laboratory.

REFERENCE

1. Ray T. Ko, Wally C. Hoppe, David A. Stubbs, D. L. Birx, B. Olding, & G. Williams, "An Update on Automatic Positioning, Inspection, and Signal Processing Techniques in the RFC/NDE Inspection System", in: Review of Progress in Quantitative NDE, Vol. 6, (to be published in 1987).

ANALYSIS OF THE RFC/NDE

SYSTEM PERFORMANCE EVALUATION EXPERIMENTS

A. P. Berens

University of Dayton Research Institute
Dayton, Ohio 45469-0001

INTRODUCTION

The RFC/NDE System comprises automated eddy current and ultrasonic inspection stations developed by Systems Research Laboratories for Air Force depot level inspections of aircraft engines. The system performance is being evaluated through a series of experiments whose objectives are to quantify crack detection capability as a function of crack size in various engine components. This paper presents the eddy current results from the first (preliminary) phase of this evaluation.

The RFC/NDE eddy current system makes accept/reject decisions by comparing peak voltage, $\hat{a}$, to a pre-defined threshold value. The $\hat{a}$ values from specimens with known crack sizes were recorded and provided the basis for estimating the probability of detection, POD(a), as well as for quantifying the effects of repeated measurements, probe changes, load changes, crack orientation, and operator changes. Analysis of the $\hat{a}$ values also validated the use of the cumulative lognormal distribution as an appropriate model for the POD(a) function.

REVIEW OF $\hat{a}$ vs a ANALYSIS METHOD

Automated NDE systems make accept/reject decisions based on the analysis of a response to an induced inspection stimulus. In most current automated systems, this analysis produces a single numerical value, $\hat{a}$, which is compared to a threshold, $\hat{a}_{TH}$. The magnitude of $\hat{a}$ is influenced by many factors related to both the inspection system and the geometry and material state of a flaw. However, the property of interest for damage tolerance analysis is the characteristic flaw size, a, of a crack growth equation. Therefore, the inspection response, $\hat{a}$, is modeled only as a function of a with all other factors contributing to a random component whose statistical properties determine the probability of detection. That is, if

$$\hat{a} = f(a) + e \tag{1}$$

then

$$POD(a) = P\{\hat{a} > \hat{a}_{TH}\}$$
$$= P\{e > \hat{a}_{TH} - f(a)\} . \tag{2}$$

Since â vs a data from inspections of representative cracks can be used to estimate both f(a) and the statistical properties of e, such data can also be used to estimate the POD function. Further, uncertainty in the POD function can be quantified by a confidence bound which is calculated from the statistical properties of the parameter estimates [1].

The POD function will be a cumulative lognormal distribution if the relationship between â and a is given by

$$\ln \hat{a} = b_0 + b_1 \ln a + e, \tag{3}$$

where e is normally distributed with zero mean and a variance of $S^2$. That is, equations (2) and (3) imply that

$$POD(a) = \Phi \left[ \frac{\ln a - \frac{1}{b_1}(\ln \hat{a}_{TH} - b_0)}{S/b_1} \right] \tag{4}$$

where $\Phi(Z)$ is the standard normal cumulative distribution.

Analysis of â vs a data from several eddy current inspection reliability experiments has shown that equation (3) is often (but not necessarily always) an acceptable model for the â vs a relationship. The analysis of the RFC/NDE eddy current reliability data was first directed toward testing hypotheses regarding the applicability of equation (3) and then toward estimating the POD(a) functions.

## PRELIMINARY RFC/NDE EDDY CURRENT RELIABILITY EXPERIMENTS

A preliminary evaluation of the RFC/NDE system was performed prior to its delivery to the Air Force. The objectives of these preliminary experiments were to demonstrate that the system was reasonably capable of meeting specifications and to determine the effect (if any) of five potential sources of scatter in â measurements. Real fatigue cracks in three specimen types were used in these preliminary tests. The specimens simulated engine rivet holes, bolt holes, and the web/bore surface. Each set of specimens contained at least 30 fatigue cracks and at least 30 more inspection opportunities without cracks. Crack size, a, in these experiments is crack depth and the â value is peak voltage.

By performing repeat measurements under controlled experimental conditions, specific contributions to total scatter in â values can be isolated [1]. Accordingly, repeat inspections of each set of specimens were made to quantify the effects of five potential sources of variability: a) repeatability - five measurements without removing specimens from the fixtures; b) transducers - repeat inspection using a second transducer; c) specimen loading - specimens were removed from fixture, reloaded, and re-inspected; d) operator - independent inspections by three operators; and e) crack orientation - three methods of loading the rivet and bolt hole specimens in the fixtures. Only some of these tests were performed on the web/bore surface specimens due to the length of time required to inspect these specimens. It was agreed before the start of testing that system capability for a specimen type would be estimated from the first inspection of the specimens.

Further details on these tests can be found in two other papers in these proceedings [2,3]. The following paragraphs summarize the results for each of the specimen types.

Rivet Hole Specimens

Figure 1 presents a plot of $\ln \hat{a}$ vs $\ln a$ for the rivet hole specimens. The crack detection threshold, $\hat{a}_{TH}$, was set at 140 mv for these inspections as indicated by the horizontal line. Ten cracks were not detected in this inspection. These cracks had depths of 1, 3, 3, 4, 6, 7, 7, 8, 9, and 16 mils. Four extra indications were recorded but it is not known whether or not these extra indications are false calls.

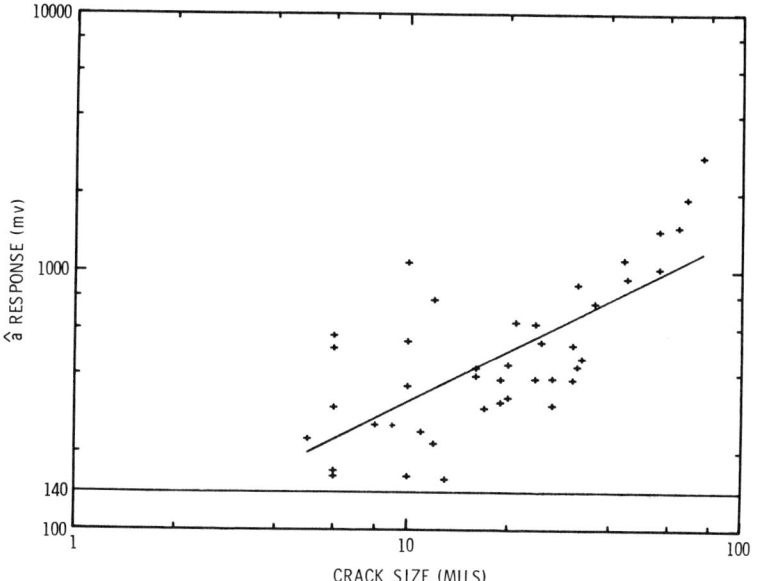

Fig. 1. $\hat{a}$ vs a for the rivet hole specimens.

The linear least squares fit to the $\ln \hat{a}$ vs $\ln a$ data is also shown in Figure 1 and it was judged to adequately fit the data. Deviations (residuals) of individual $\ln \hat{a}$ values from the regression line were obtained and analyzed for normality. Figure 2 displays a plot of the residuals on a form of normal distribution paper. Deviation from linearity in Figure 2 indicates non-normality. The Shapiro-Wilkes test, which is based on the correlation coefficient between the residuals and the normal scores, failed to reject the hypothesis that the $\ln \hat{a}$ values are normally distributed about the $b_o + b_1 \ln a$ line. Thus, the cumulative lognormal model was judged to be an acceptable model for the POD(a) function. Figure 3 presents the estimated POD function and its lower 95 percent confidence bound for the rivet hole specimens.

Figure 4 presents the $\hat{a}$ vs $a$ test results for five repeat inspections during which the specimens were not removed from the fixtures. The scatter reflected in the $\hat{a}$ values for a particular crack in this figure represents the optimum repeatability of the system for this specimen type during Phase I. The scatter was quantified by pooling the standard deviations of log $\hat{a}$ values from each of the cracks. The pooled estimate of the standard deviation for the repeatability measurements was 0.050. Assuming a lognormal distribution of $\hat{a}$ values, this degree of repeatability would imply approximately a 5 percent coefficient of variation, i.e., 95 percent of $\hat{a}$ values for a particular crack would be within ± 10 percent of the average.

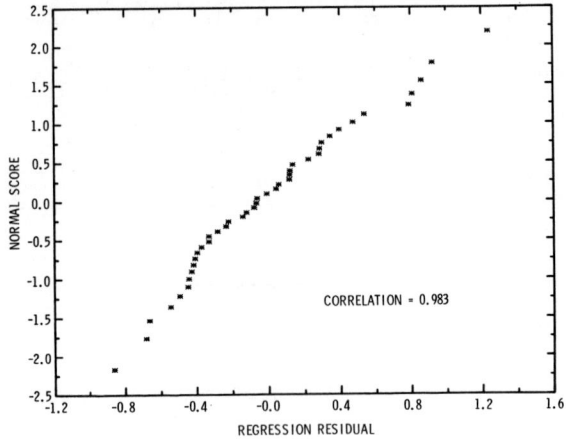

Fig. 2. Normal probability plot of ln â residuals for the rivet specimen.

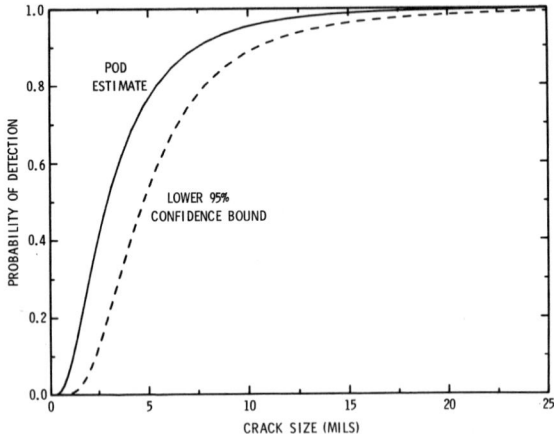

Fig. 3. POD(a) and lower 95% confidence bound for the rivet hole specimens.

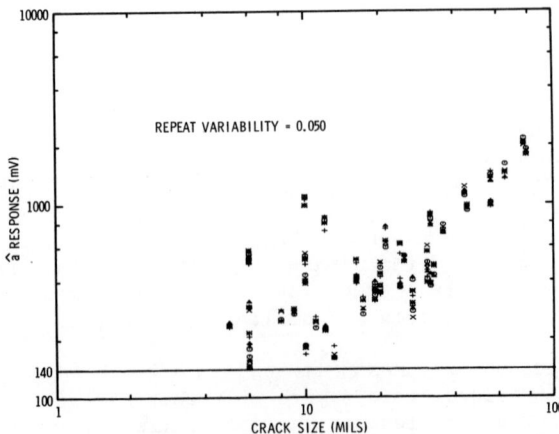

Fig. 4. â vs a for five repeat inspections under identical conditions.

Similar analyses were performed on the repeat â measurements when inspections were repeated with all but a single factor held constant. Table 1 presents these pooled estimates of the standard deviation of ln â values. The degree of scatter exhibited by multiple loadings of the specimens in the test machines and by different operators is about equal to the minimum scatter as measured by the repeatability standard deviation. The â scatter introduced by using a different transducer, however, was significantly greater than the minimum. Changing the orientation of the crack with respect to the transducer also resulted in a significant increase in â scatter.

Table 1. Coefficient of Variation for Sources Isolated during Preliminary RFC/NDE Eddy Current Reliability Evaluation

|  | RIVET HOLE | BOLT HOLE | WEB/ BORE |
|---|---|---|---|
| REPEATABILITY | 0.050 | 0.043 | 0.017 |
| TRANSDUCER | 0.112 | 0.134 | 0.177 |
| LOADING | 0.056 | 0.048 |  |
| OPERATOR | 0.055 | 0.056 |  |
| CRACK ORIENTATION | 0.104 | 0.129 |  |

Bolt Hole Specimens

The â vs a data obtained from the inspections of the bolt hole specimens are presented in Figure 5. All except a one mil deep crack in one of the specimens were detected. Extra indications (possibly false calls) were recorded on eight of the specimens during the characterization inspections. The linear ln â vs ln a relationship was judged to be acceptable and the Shapiro-Wilks tests could not reject the normality hypothesis for the deviations of ln â values about the straight line. Figure 6 presents the estimated lognormal POD(a) function and its lower 95 percent confidence bound for the bolt hole specimens.

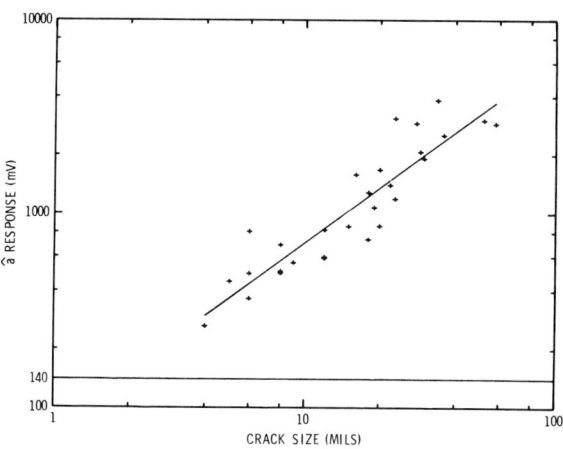

Fig. 5. â vs a for the bolt hole specimens.

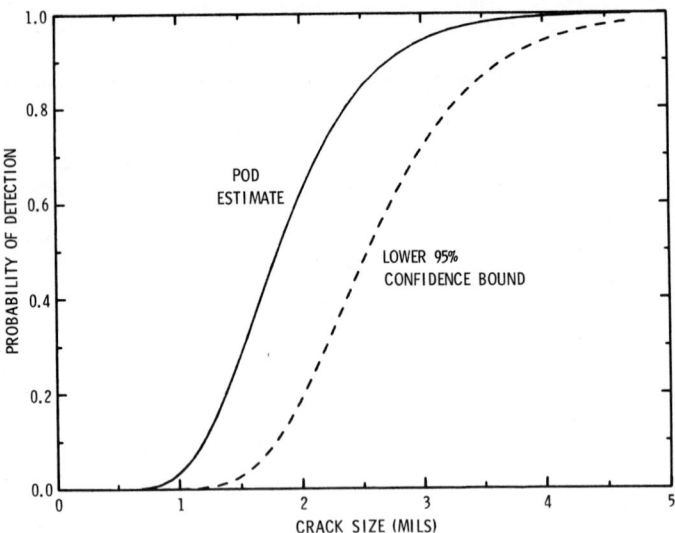

Fig. 6. POD(a) and lower 95% confidence bound for the bolt hole specimens.

The results of the tests to measure the scatter resulting from controlled factors are presented in Table 1. These results are essentially equivalent to those obtained from the rivet hole specimens. Repeat loading and different inspectors do not increase the scatter in â values and, hence, do not degrade the POD(a) function. Different transducers, however, do increase the scatter in â.

Web/Bore Specimens

Figure 7 presents the â vs a data obtained from the specimens which simulated inspections of an engine disk web/bore. Two cracks with depths of one and three mils were not detected in this experiment. Ten extra indications were recorded. Again the ℓn â vs ℓn a relationship was judged to be linear and nomality of the deviations of ℓn â values from the straight line could not be rejected by the Shapiro-Wilkes test. The lognormal POD(a) function and its lower 95 percent confidence bound for these data are presented in Figure 8.

Since scanning the web/bore specimens proceeded slowly, the experimental matrix was greatly reduced for this specimen type. Only one set of repeatability measurements were taken and only the variability due to different transducers was obtained. The results of these experiments are presented in Table 1.

DISCUSSION

Equation (4) implies that the more scatter in â values for a fixed crack depth, the flatter the POD(a) function. This is also easily realized from the â vs a plots. A consistent trend that has been present for â vs

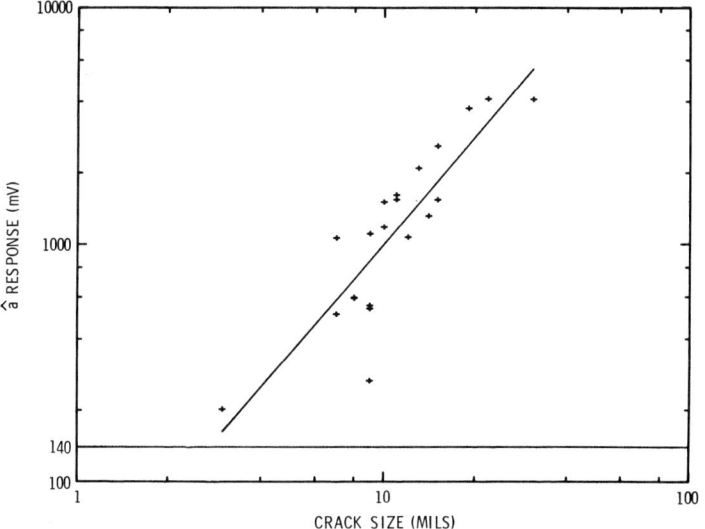

Fig. 7.  â vs a for the web/bore specimens.

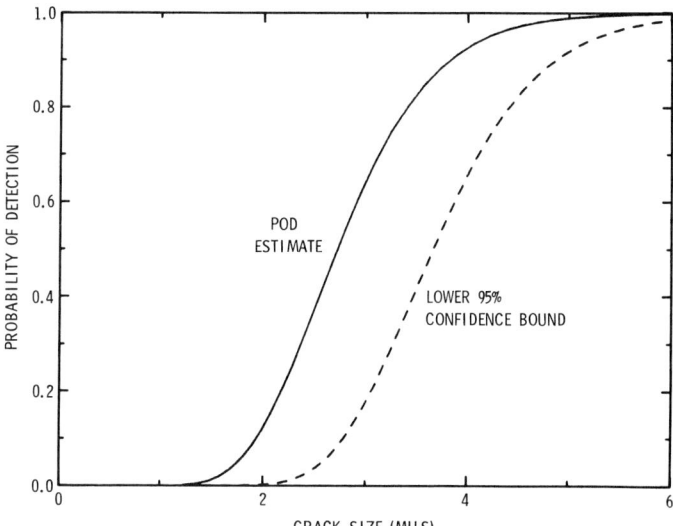

Fig. 8.  POD(a) and lower 95% confidence bound for the web/bore specimens.

a data from automated systems is that the biggest cause of this scatter is due to variation associated with different cracks.  The data of this study agrees with this trend.

The random error term, e, of equation (3) can be partitioned into components and the contributions of the individual components can be estimated from properly conducted experiments.  In the repeat experiments of this study, one factor at a time was varied and the components of the total variation were estimated.  Table 1 lists the components due to the factors that were varied. Estimates of the variation due to cracks were 0.73, 0.32, and 0.26 for the rivet hole, bolt hole, and web/bore specimens, respectively. Thus, the crack-to-crack variation was much larger than that due to the other factors.

The POD(a) function of Figures 3, 6, and 8 included only the crack-to-crack and repeatability variation. Of the factors measured, only the transducer variability could significantly contribute to the total. For example, if it is assumed that a bolt hole is to be inspected and the transducer to be used will be chosen at random, then the proper S to be used in equation (4) would be given by

$$S^2 = (0.32)^2 + (0.13)^2$$

or S=0.35. On the other hand, if the bolt hole was to be inspected by the transducer used to generate Figure 6, then S = 0.32 is the proper measure of the scatter.

Perhaps it should also be noted that POD(a) functions were calculated for each of the experimental conditions. Resulting differences in the POD(a) functions were judged to be negligible in the retirement-for-cause application.

CONCLUSION

The conclusions drawn from these Phase I tests of the RFC/NDE system are:

a) The lognormal model was appropriate for these highly automated inspections.

b) The significant causes of variability in transducer response, â, were the cracks, the transducers, and the crack orientation with respect to the transducer.

c) The reloading of the specimens and the changing of operators had no effect on the transducer response (and, thus, no effect on POD).

d) The test matrix for the Phase II evaluation tests will be greatly reduced.

ACKNOWLEDGMENT

This work was sponsored by the Materials Laboratory of the Air Force Wright Aeronautical Laboratories under Contract F33615-81-C-5002 through a Systems Research Laboratories subcontract.

REFERENCES

1. Berens, A.P., and Hovey, P.W., "Flaw Detection Reliability Criteria," AFWAL-TR-84-4022, Air Force Wright Aeronautical Laboratories, Wright-Patterson Air Force Base, Ohio, April 1984.
2. Shambaugh, R.L., "Design and Execution of the RFC/NDE System Inspection Reliability Tests," these proceedings.
3. Ko, R.T., "RFC Phase I Eddy Current Reliability Tests Results," these proceedings.

APPLICATION OF THE RFC/NDE SYSTEM TESTING RESULTS


C. G. Annis, Jr. and
T. Watkins, Jr.

Pratt & Whitney
Engineering Division, South
West Palm Beach, FL

INTRODUCTION

The evaluation of the RFC/NDE System has produced capability characteristics in the format of â vs. a or apparent cracksize vs. actual cracksize. Although crack sizing has long been a major concern of NDE practitioners and theoreticians, the analysis procedure which permits straightforward conversion of these â vs. a to POD vs. a, or Probability of Detection vs. cracksize, is a recent development [1]. This paper will compare these two descriptions of NDE system capability by comparing their influence on the Retirement for Cause (RFC) process.

The life cycle of a gas turbine component which is fatigue limited can be conceptualized as being comprised of two phases: (1) an initiation phase, during which material undergoes cyclic loading and thus accumulates fatigue damage; and (2) a propagation phase as the initiated crack actively grows until it is either detected and removed during nondestructive evaluation or reaches critical dimensions and fails in service. This simplified representation is illustrated graphically in Fig. 1, where cracksize is plotted as a function of time (and therefore accumulated fatigue cycles). The figure shows that cycles required to initiate a crack vary according to some statistical distribution; once initiated, cracks proceed according to the laws of fracture mechanics [2,3,4].

The Probabilistic Life Analysis Technique, PLAT [5] embodies a Monte Carlo simulator; it is a statistical description of the gas turbine life cycle. It uses distributions of the independent variables rather than the more familiar single-value input functions. Because the input can assume a range of values, the simulator output is also multivalued and is presented in statistical terms, such as expected outcome or rates of occurrence. To do this, the simulator selects from distributions of the independent variables and combines their effects according to the physical laws of the system being modeled. Each "pass" through the simulator will result in one outcome based on one sample from each of the controlling variables. After many -- sometimes tens of thousands of passes -- these individual results are collected and analyzed statistically.

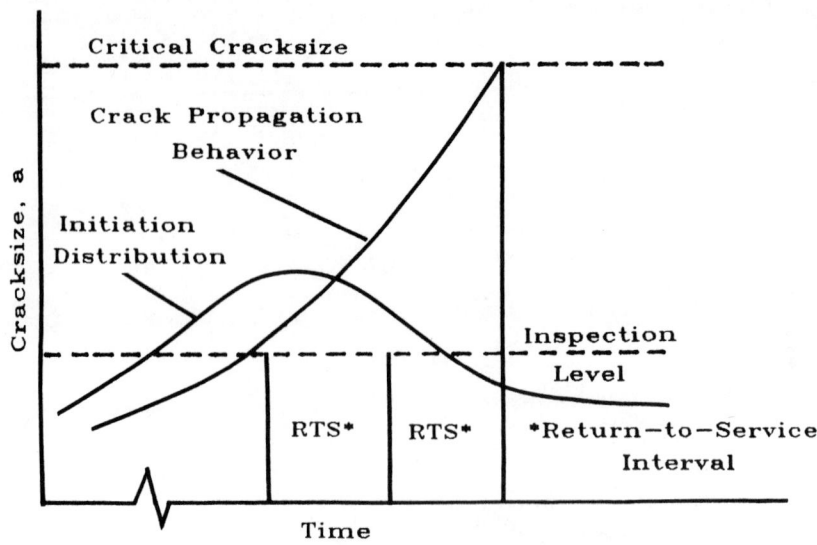

Figure 1. Simplified Damage Tolerant Design.

The PLAT is an application of statistical methods to component life analysis. Previous life analyses were based on a deterministic criterion with the worst case establishing the usable life for all parts. If the worst case occurred only once in a thousand times, then 99.9 percent of the parts (gas turbine disks in this case) were being retired prematurely. PLAT is a methodology for establishing the statistical behavior of all the disks which comprise an entire fleet of engines. Instead of single-valued functions, PLAT input consists primarily of information about statistical distributions of life-controlling parameters. These include Initial Material Quality (the distribution of expected sizes of microstructural anomalies such as voids and includions), crack initiation behavior in the form of a stress vs. cycles (s-N model), mission severity, stress variability, and a transition model of behavior from IMQ through crack initiation, to the propagation phase. Also required is a life prediction model, describing the expected longevity of a component after a crack has initiated. Finally, A Nondestructive Evaluation/Return-to-Service model is required.

## $\hat{a}$ VS. a ANALYSIS

Actual cracksize (a) and indicated cracksize ($\hat{a}$) -- often an eddy current signal voltage or fluorescent penetrant brightness -- are recorded for each observation and a threshold level $\hat{a}_{th}$ is determined. This threshold is considered to be the smallest signal which represents a crack; any signal below this level is defined as "noise".

Figure 2 illustrates the approach to be used in defining POD(a) as a function of cracksize by considering the probability that a crack will appear large enough to be detected [5]. Of course this probability is itself a function of actual cracksize. The basic model for $\hat{a}$ vs. a data is given by:

$$\hat{a} = f(a) + c + e \tag{1}$$

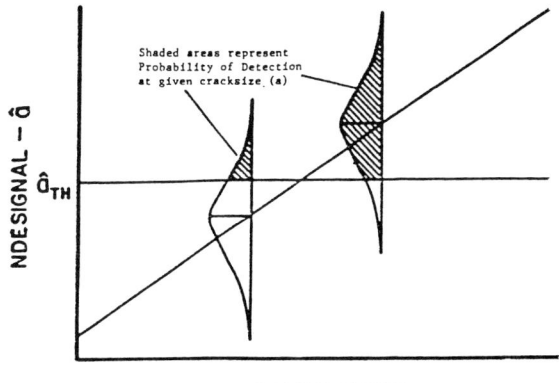

Figure 2. Conversion of Residuals in Log-Log Plane to Probability of Detection vs. a Curve.

where f(a) represents the overall trend in â as a function of a, c represents flaw to flaw variation, and e represents the variation from inspection to inspection of the same flaw. The function f(a) is fixed, while the variables c and e are random variables (the variance components of the model) with means of zero.

There are many data analysis methods based on the foregoing equation, and the appropriate method depends on the form of f(a). One method is to convert f(a) to a linear relationship through transformations of â and a. For example, ln(a) and ln(a) are often observed to be linearly related. The basic equation is then given by:

$$\ln(\hat{a}) = \alpha + \beta \ln(a) + c + e \qquad (2)$$

Again, c and e are random variables with means of zero, but not identical with those in equation (1). Assuming these variance components to have normal distributions leads to the POD function:

$$POD(a) = P(\hat{a} > \hat{a}_{th}) \qquad (3)$$

$$= P(\ln(\hat{a}) > \ln(\hat{a}_{th})) \qquad (4)$$

$$= 1 - \Phi\left(\frac{\ln(\hat{a}_{th}) - (\alpha + \beta \ln(a))}{S}\right) \qquad (5)$$

where $S^2$ is the total variance and $\Phi$ is the standard normal distribution function. Since the area under the standard normal curve is one:

$$POD(a) = \Phi\left[-\left(\ln(\hat{a}_{th}) - (\alpha + \beta \ln(a))S\right)\right] \qquad (6)$$

Dividing numerator and denominator by $\beta$ then gives:

$$= \Phi\left(\frac{\dfrac{\beta \ln(a)}{\beta} - \dfrac{\ln(\hat{a}_{th})}{\beta} + \dfrac{\alpha}{\beta}}{\dfrac{S}{\beta}}\right) \qquad (7)$$

$$= \Phi\left(\frac{\ln(a) - \frac{\ln(\hat{a}_{th}) - \alpha}{\beta}}{\frac{S}{\beta}}\right) \quad (8)$$

which is observed to be a lognormal distribution in a with mean and standard deviation given by:

$$\mu = \frac{\ln(\hat{a}_{th}) - \alpha}{\beta} \quad (9)$$

$$\sigma = \frac{S}{\beta} \quad (10)$$

For simplicity and computational efficiency, we may wish to approximate this lognormal distribution by a log logistic distribution [5] with the same mean and standard deviation. Estimates of the scale and location parameters (t and w) of the log logistic distribution are given by:

$$t = \frac{\pi}{\sigma\sqrt{3}} \quad (11)$$

$$w = \exp[-\mu t] \quad (12)$$

The form of the resulting POD model is then:

$$POD = \frac{(wa^t)}{1 + (wa^t)} \quad (13)$$

which is computationally more straightforward than a lognormal distribution.


PLOTTING â VS. a DATA ON THE POD VS. a CURVE

Cracks of the same physical size (equal a) often exhibit different probabilities of detection because they appear to be different sizes (have different â). Here, we assume that all cracks which appear to be the same size (i.e., have equal â) have equal probabilities of detection, regardless of their actual size. We can now map any point in the â vs. a plane to a corresponding point in POD vs. a space. Although they are not used computationally, individual â vs. a observations can now be plotted in the POD vs. a plane using several simple assumptions. Examining Fig. 3, we choose a particular observation (the circle) in the â vs. a plane. Given no other information than the apparent crack size and our linear regression â vs. a model (equation 2), we would assign a "most probable" cracksize (a*) based on the mean regression line (shown). We could assign a "most probable" POD (POD*) based on the POD vs. a regression (equation 11) and a*. This process is represented schematically by the short dashed line. We do, however, know the true value of a. Our â vs. a observation can now be represented in the POD vs. a plane as shown by the circle in the right hand plot. The point is plotted at its actual cracksize, a, and the POD* associated with an average apparent cracksize of a*.

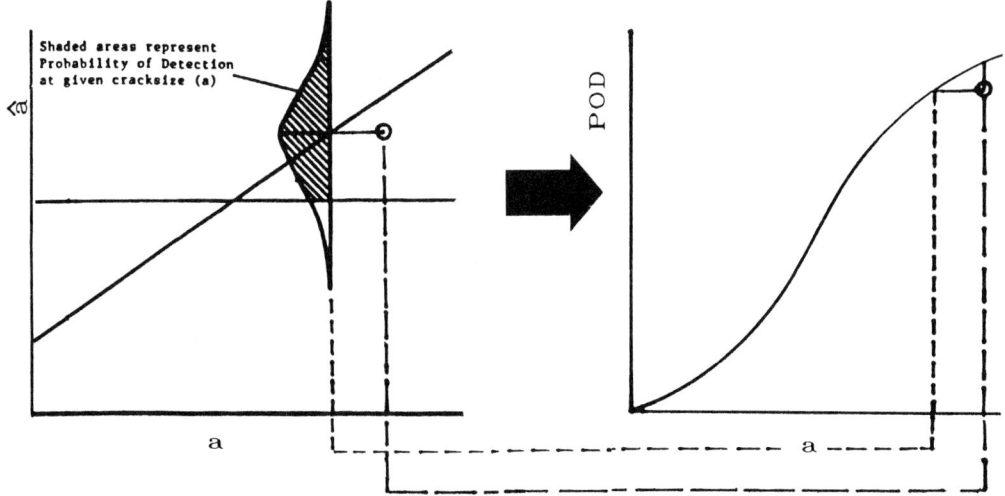

Figure 3.  Plotting â vs. a data as POD vs. a.

CONFIDENCE BOUNDS FOR POD VS. a CURVES

Dr. A. P. Berens [5] has adapted a method described by Cheng and Iles [6] for calculating confidence bounds on cumulative distributions. Confidence bounds on the mean POD are then given by:

$$POD(a) = \Phi(Z_L), \text{ where} \tag{14}$$

$$Z_L = \hat{Z} - \sqrt{\frac{\lambda}{n}\left(\frac{\hat{Z}^2}{2} + \frac{(x-\bar{x})^2}{SSX} + 1\right)} \tag{15}$$

and

$$SSX = \sum_{i=1}^{n} x_i^2 - \frac{\left[\sum_{i=1}^{n} x_i\right]^2}{n} \tag{16}$$

In the above equations, n is sample size, $\lambda$ is the pth percentile of a $\chi^2$ distribution with two degrees of freedom, and

$$\hat{Z} = \frac{X - \mu}{\sigma} \tag{17}$$

Figure 4 illustrates the â vs. a regression. The corresponding POD vs. a plot, Fig. 5, presents the mean capability, as determined using the log logistic approximation (equation 13) with the 95% confidence limit (equation 14). Data are plotted as described in the preceding section.

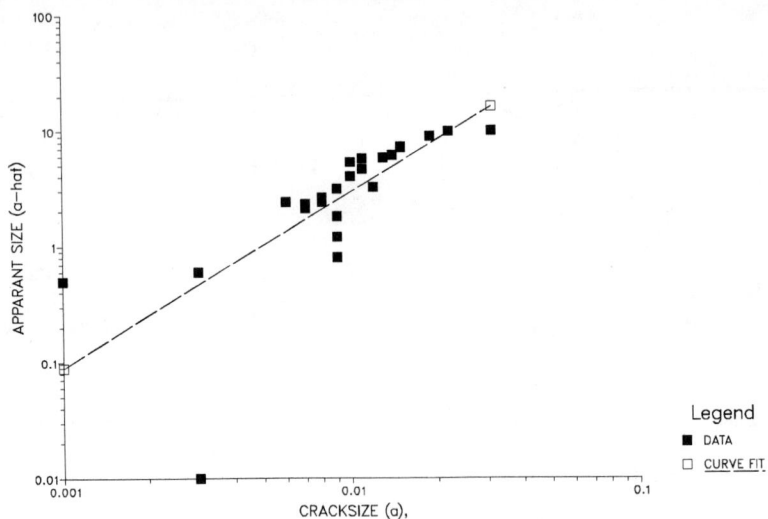

Figure 4.  $\hat{a}$ vs. a Regression in the Log-Log Plane.

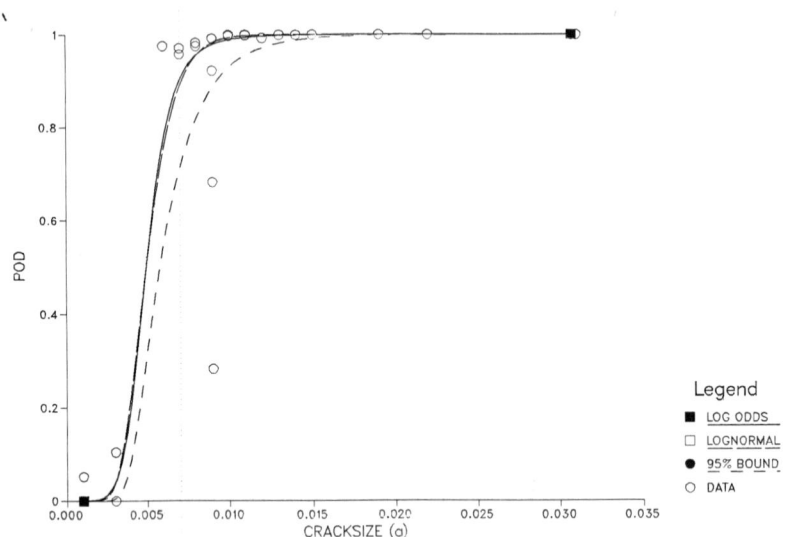

Figure 5.  POD vs. a Plot with 95% Lower Bound.

COMPARISON OF $\hat{a}$ VS. a AND POD VS. a

Since $\hat{a}$ vs. a and POD vs. a are different descriptions of the same phenomenon, they should produce identical results when used in PLAT analysis. This is intuitively obvious. The investigation was conducted because the Monte Carlo simulation is different for each, and therefore the study would uncover any unexpected differences. Simulating NDE performance using $\hat{a}$ vs. a is straightforward: knowing the true size, a, the distribution of $\hat{a}$ is sampled, and an individual $\hat{a}$ is determined. If this value exceeds the threshold cracksize, $\hat{a}_{th}$, then it is "detected", and the simulator proceeds accordingly. The simulation of POD vs. a is another matter. Knowing true size, a, the POD vs. a function is evaluated to determine a POD. A random variable uniformly distributed on the 0,1 interval is generated and compared with POD. If POD is greater than this number, the crack is "detected".

To evaluate any differences in performance, the life cycle of a hypothetical advanced compressor disk lug attachment was simulated using the PLAT. This case was realistic in all respects except two: the crack propagation life was inadequate, and the NDE behavior represented worst-case capability. This was done deliberately. If adequate propagation margin and NDE capaility had been simulated, there would be zero failures. The more realistic design therefore would illustrate nothing. By simulating a less-than-optimum situation, we are able to observe failures and removals, and therefore have a basis of comparison.

RESULTS

The failures using each NDE representation are plotted vs. inspection interval and presented in Fig. 6. A similar plot illustrating NDE removals is presented in Fig. 7. As can be seen, these are essentially (but not identically) equal. The small differences are caused by the random nature of a Monte Carlo simulator. The influence of different random number streams has been investigated and quantified and reported [5].

CONCLUSIONS

NDE capability can be described using $\hat{a}$ vs. a and POD vs. a, and they result in the same failure and replacements rates when used in a simulation of NDE behavior.

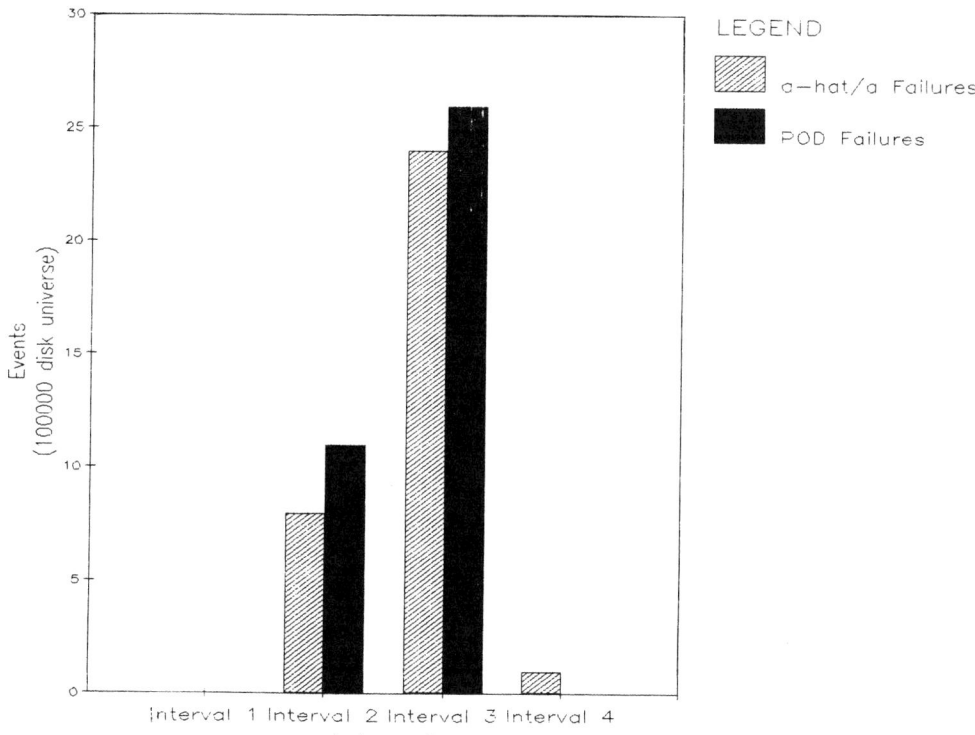

Figure 6.  Comparison of POD vs. a and $\hat{a}$ vs. a Models -- Predicted Failures.

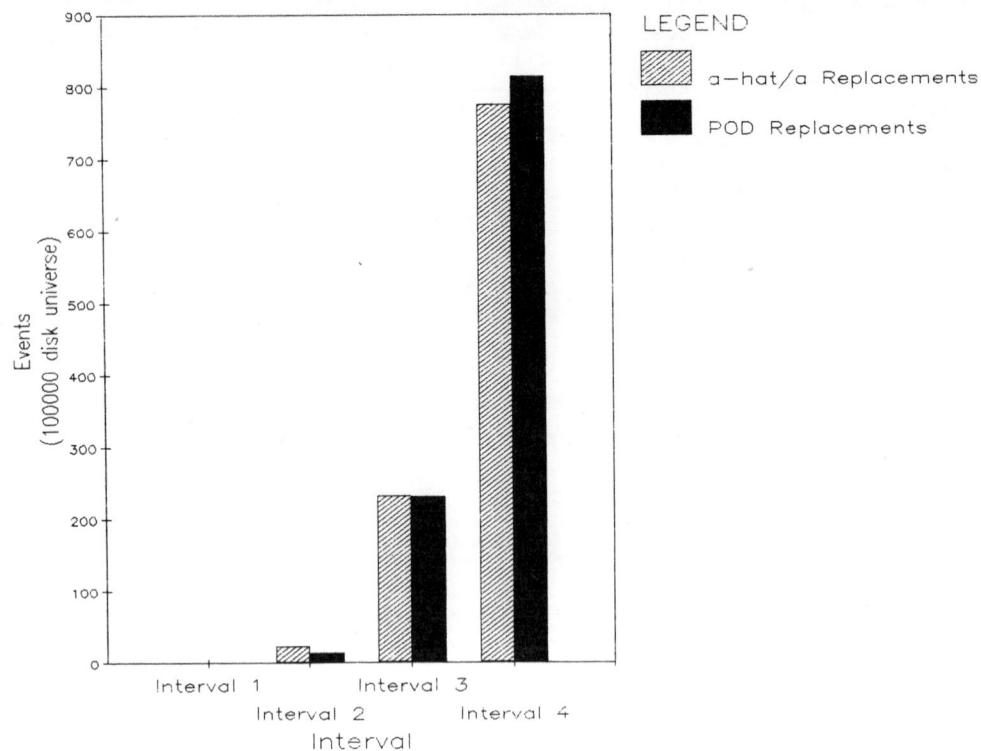

Figure 7. Comparison of POD vs. a and â vs. a models -- Predicted Replacements.

The mean POD and associated confidence limit are easily determined from the â vs. a behavior.

Individual â, a observations can be plotted as POD, a points for illustrative purposes.

REFERENCES

1. A. P. Berens and P. W. Hovey, Flaw Detection Reliability Criteria, UDR-TR-83-137, University of Dayton Research Institute, for Air Force Wright-Aeronautical Laboratories, Wright-Patterson Air Force Base, Ohio 45433.
2. J. N. Yang, G. C. Salivar and C. G. Annis, Jr., Statistics of Crack Growth in Engine Materials -- Volume 1: Constant Amplitude Fatigue Crack Growth at Elevated Temperatures, AFWAL-TR-82-4040, Contract No. F33615-80-C-5189, (July 1982).
3. J. N. Yang, G. C. Salivar and C. G. Annis, Jr., Statistics of Crack Growth in Engine Materials -- Volume 2: Spectrum Loading and Advanced Techniques, AFWAL-TR-82-4040, Contract No. F33615-80-C-5189, (June, 1983).
4. J. M. Larsen, B. J. Schwartz and C. G. Annis, Jr., Cumulative Damage Fracture Mechanics Under Engine Spectra, AFML-TR-77-4159, Contract No. F33615-77-C-5153, (January, 1980).
5. T. Watkins, Jr. and C. G. Annis, Jr., Engine Component Retirement For Cause: Probabilistic Life Analysis Technique, AFWAL-TR-85-4075, Contract No. F33615-80-C5160, (July, 1985).
6. R. C. H. Cheng and T. C. Iles, "Confidence Bands for Cumulative Distribution Functions of Continuous Random Variables", Technometrics, Volume 25, No. 1, (1983), 77-86.

STATUS OF ADVANCED UT SYSTEMS FOR THE NUCLEAR INDUSTRY


M. Behravesh, M. Avioli, G. Dau, S-N. Liu

Electric Power Research Institute
Palo Alto, California  94303

INTRODUCTION

An advanced ultrasonic testing (UT) system is a configuration of hardware that includes some type of computer. The computer may be hardwired to perform specific functions or have appropriate software. It may typically be used for data acquisition, signal processing, image generation, pattern recognition and data analysis. Additionally, advanced systems have data storage and are, therefore, different from the standard transducer-pulser/receiver systems that rely on human filtering and written documentation of the filtered data.

More and more utilities and service vendors are using or are considering an advanced UT system for preservice (PSI) and inservice (ISI) inspections. A major utility has purchased five Amdata IntraSpect Ultrasonic Imaging systems. Another has bought and is using a NES/Dynacon's UDRPS system. Science Application Inc's Ultra Image III, along with Infometric's TestPro software, is being used for turbine rotor bore inspection. IntraSpect imaging technology is being combined with pattern recognition capabilities by a leading service vendor. New and potential users of advanced UT systems are increasing in numbers every day. What these systems offer is so attractive that regulatory agencies consider their deployment an "intelligent choice" for inspection.


BACKGROUND

The most significant difference between nuclear power plant UT inspection and, say, aircraft inspection, is the radiation background that is present and increases with years of service. Inspection personnel must always consider the ALARA guidelines (As Low As Reasonably Achievable) for radiation. The human factors include biological hazard and psychological stress. Under these pressures, the data processing capabilities of the human brain are also limited. This is why UT inspection data, when collected manually, is usually in summary form.

On the economic side, inspection costs are tied directly to the radiation dose rate at the inspection site. Allowable radiation dose per quarter is regulated and oftentimes results in shortages of qualified personnel to perform UT ISI in nuclear plants. This has economic impact via the deployment of limited personnel to do a job that has not changed in size or scope. The longer the down time, the higher the cost of purchasing replacement power to counter the loss of generating capacity.

Advanced systems were developed with the goals of reduced radiation exposure and increased reliability. These systems use scanners for high-density coverage of components, and since some of them are mechanized and remotely controllable, radiation exposures to personnel can greatly be reduced. Digital recording and processing of data facilitate repetition of stored scan patterns, data logging for data review, and application of signal processing techniques for noise reduction.

Advanced Systems

The number of systems becoming commercially available is growing each year. The NDE managers of utilities, the end users of these systems, are often faced with the decision as to "What system is right for my inspection problem?" "Is an advanced UT system a cost effective way to go?" To help this group, the Electric Power Research Institute (EPRI) has initiated a project whose end product will be a Utility NDE Managers' Guide to Advanced UT Systems. A short summary of the available data to date will be presented here. Tables are used to give an immediate overview of capabilities.

There are two broad classifications of advanced UT systems. These are feature-based systems and image-based systems. A feature-based system is very similar in concept to a radar system for aircraft. Parameters, or "features" are extracted from the ultrasonic signals accumulated during an inspection. These features are believed to contain information about the reflector that produced the original ultrasonic echo. As in radar, processing of echo features can be used to identify the reflector. Imaging systems use arrays of ultrasonic signals obtained from surface scanning to create "pictures" of reflectors located within a material. The purpose of the images is to assist an inspector in obtaining information on the spatial characteristics of his inspection problem.

EPRI is currently examining systems that combine feature-based and image-based capabilities into, as they are conceived, feature-enhanced imaging systems. Considerations are being given to the use of Personal Computers (PC's) for the hardware and software needs of these systems. Tables I and II cover systems that are in use today. Table I lists the hardware that comprises these systems. The software and software function capability of these systems are listed in Table II. The systems mentioned are not restricted to Tables I and II hardware and software. Table III describes commercially available components that may be used as alternatives and/or enhancements. Several UT systems that are still under development are listed in Table IV.

Qualification

A major concern of the utility NDE manager is the compliance of the system he chooses with NRC guidelines. The USNRC has issued guidelines that define minimum acceptable performance requirements. Qualification tests have been developed to evaluate advanced UT systems against the established criteria defined by the NRC. Systems that have met all of the USNRC requirements for IGSCC detection and sizing are listed in Table V.

TABLE I

COMMERCIALLY AVAILABLE AUTOMATED UT SYSTEMS/SERVICES -- HARDWARE

| | INTRASPECT 98 | P-SCAN | SDL-1000 | SMART UT | UDRPS | ULTRA IMAGE III | ZIP-SCAN |
|---|---|---|---|---|---|---|---|
| VENDOR - SERVICE GROUP | AMDATA INC./CE | INDEPENDENT TESTING LABS. & UNIVERSAL TESTING LABS. INC. | SIGMA RESEARCH INC. | GENERAL ELECTRIC COMPANY | NES / DYNACON SYSTEMS INC. | ULTRA IMAGE INTERNATIONAL | U.S. TESTING COMPANY INC. |
| AUTOMATED SCANNER | AMAPS | AWS-4 | SDL-1003 | ALARA-1 OR AMAPS | ALARA-1, AMAPS OR VERSASCAN | ALARA-1 OR AMAPS | MIMIC |
| SEMI-AUTOMATED SCANNER | | MWS-2 | SDL-1004 | GENERAL DYNAMICS | VERSASCAN | ULTRA IMAGE STD. MINATURE | |
| GUIDE TRACK & MOUNTING | FLEX-STEEL WITH MAG. WHEELS | STEEL TRACK WITH TIMING BELT | FLEX-STEEL WITH LOCK-ON CARRIAGE | RIGID TRACK OR FLEX-STEEL & MAG. WHEELS | RIGID TRACK FLEX-STEEL & MAG. WHEELS OR PLASTIC BELT | RIGID TRACK OR FLEX-STEEL & MAG. WHEELS | FLEX-CHAIN & NON-METALLIC WHEELS |
| SCANNER CONTROLLER | COMPUTER INTERFACED & JOYSTICK | COMPUTER INTERFACED | COMPUTER INTERFACED | CONTROLLER W/ AUTO OR MANUAL OR COMPUTER INTERFACED | COMPUTER INTERFACED & JOYSTICK | CONTROLLER W/ AUTO OR MANUAL OR COMPUTER INTERFACED | COMPUTER INTERFACED |
| TRANSDUCER | CONTACT OR BOOTED | CONTACT | CONTACT & IMMERSION | CONTACT OR BOOTED | CONTACT, BOOTED OR MULT. (6) ARRAY | CONTACT OR BOOTED | CONTACT |
| UT COMPONENTS | METROTEK | P-SCAN | SDL-1000 | ULTRA IMAGE III | ANY COMMERCIAL UNIT | ULTRA IMAGE III | ZIPSCAN |
| DATA STORAGE | HARD DISK & FLOPPY & CASSETTE | MAGNETIC TAPE CASSETTE | HARD DISK & FLOPPY | HARD DISK & FLOPPY & VHS OF A-SCAN OR RF | HARD DISK, 9-TRACK OR OPTICAL DISK | HARD DISK & FLOPPY | HARD DISK & FLOPPY |
| DATA DISPLAY | RGB COLOR CRT | B/W CRT (2 EA.) | RGB COLOR & B/W CRT | RGB COLOR & B/W CRT | RAMTEK COLOR CRT | RGB COLOR & B/W CRT | B/W CRT |
| HARDCOPY UNIT | TECHTRONIX & COLOR POLAROID | EPSON FX 80 | B/W & COLOR POLAROID | VHS & COLOR POLAROID | RAMTEK VERSATEC & TECHTRONIX COLOR | TECHTRONIX 4632/4634 & MATRIX COLOR POLAROID | 3M VRG 4000 |
| COMPUTER - DATA PROCESSING | HP 9836C | P-SCAN PROCESSOR | SDL-1001 | ULTRA IMAGE III | HP-1000 & AP 400 ARRAY PROCESSOR | ULTRA IMAGE III Z-80 80286/8087 | LSI-11/73 |

## TABLE II

### COMMERCIALLY AVAILABLE AUTOMATED UT SYSTEMS/SERVICES -- SOFTWARE

| VENDOR - SERVICE GROUP | INTRASPECT 98<br>AMDATA INC./CE | P-SCAN<br>INDEPENDENT TESTING LAB & UNIVERSAL TESTING LAB | SDL-1000<br>SIGMA RESEARCH INC. | SMART UT<br>GENERAL ELECTRIC COMPANY | UDRPS<br>NES/DYNACON SYSTEMS INC. | ULTRA IMAGE III<br>ULTRA IMAGE INTERNATL | ZIP-SCAN<br>U.S. TESTING COMPANY INC. |
|---|---|---|---|---|---|---|---|
| **SCAN MOTION** | | | | | | | |
| AXIAL | ••• | ••• | ••• | ••• | ••• | ••• | ••• |
| CIRCUMFRENTIAL | ••• | ••• | ••• | ••• | ••• | ••• | ••• |
| SKEW | ••• | ••• | ••• | ••• | ••• | ••• | ••• |
| **DATA RECORDING** | | | | | | | |
| AMPLITUDE | ••• | ••• | ••• | ••• | ••• | ••• | ••• |
| POSITION | ••• | ••• | ••• | ••• | ••• | ••• | ••• |
| ARRIVAL TIME | ••• | ••• | | ••• | ••• | | ••• |
| RF WAVEFORM | ••• | | | • | ••• | | ••• |
| RECTIFIED WAVEFORM | | ••• | | ••• | ••• | ••• | ••• |
| INSTRUMENT SETTINGS | ••• | ••• | ••• | ••• | ••• | | |
| **DATA DISPLAY** | | | | | | | |
| A-SCAN | ••• | ••• | ••• | ••• | ••• | ••• | ••• |
| B-SCAN | ••• | ••• | ••• | | ••• | ••• | ••• |
| C-SCAN | ••• | ••• | ••• | ••• | ••• | ••• | ••• |
| 3-D ISOMETRIC | • | | ••• | | ••• | | |
| PROJECTED VIEWS | | ••• | | | ••• | | |
| TIME-OF-FLIGHT | ••• | | ••• | ••• | ••• | | ••• |
| FLAW MAP | | | | | ••• | | |
| ZOOM | ••• | | ••• | ••• | ••• | ••• | |
| DAC CURVES | ••• | | | | ••• | | |
| **AVERAGING** | | | | | | | |
| TEMPORAL AVERAGING | • | | | | ••• | | |
| SPATIAL AVERAGING | ••• | | | | ••• | • | |
| **SYNTHETIC APETURE** | | | | | | | |
| HOLOGRAPHY | | | ••• | | | | |
| SAFT | • | | | | • | | ••• |
| **ARTIFICIAL INTELLIGENCE** | | | | | | | |
| ALN | | | | | | | |
| FEATURE EXTRACTION | • | | | | • | | |
| PATTERN RECOGNITION | | | | | • | | |
| **SIZING TECHNIQUES** | | | | | | | |
| TANDEM TECHNIQUE | ••• | ••• | | | ••• | | |
| MODE CONVERSION TECHNIQUE | ••• | ••• | | | ••• | | |
| CREEPING WAVE | ••• | ••• | | | ••• | | ••• |
| CRACK TIP DIFFRACTION | ••• | | | | ••• | | ••• |

••• FULL CAPABILITY
• PARTIAL CAPABILITY

TABLE III

COMMERCIALLY AVAILABLE COMPONENTS SUPPORTING AUTOMATED UT SYSTEMS

| SYSTEM | MANUFACTURER | FUNCTION / DESCRIPTION |
|---|---|---|
| ALARA-1 | VIRGINIA CORP | An automated mechanical scanner and scanner controller that can be interfaced to a computer. Mechanical system provides skew motion as well as standard scan motion. Rigid track and guide assembly. |
| ALN | ADAPTRONICS/ GEN.RES.CORP. | Adaptive Learning Network (ALN) allows training of the system to distinguish key features of the UT signal unique to specific defects or conditions and assists the inspector in confirming defect or geometry conditions. |
| AMAPS | AMDATA SYSTEMS INC. | An automated mechanical scanner and scanner controller that can be interfaced to a computer. Configured for hard shoe or booted transducer assembly. Flex-steel, wrap around track with magnetic wheels. |
| ROBBI | KRAFT WERK UNION | An automated UT inspection system designed primarily for flaw sizing and not significantly used for initial inspection and detection. System is presently being reconfigured for full inspection use. |
| SDL-1000 | SIGMA RESEARCH INC. | An automated UT inspection system capable of A, B, C and 3D scan imaging but principally designed for ultrasonic holography inspection and high resolution imaging of previously detected indications. |
| SUTARS | SOUTHWEST RESEARCH INSTITUTE | An ultrasonic tracking system used with manual UT to integrate scan position with UT data. Tracking sensors strap to pipe to provide position data for hand held transducer. |
| TEST PRO | INFOMETRICS INC. | IBM PC based hardware and software allows integration of all automated UT components and assories to provide fully automated UT systems. PC is a value-added component to current automated UT systems. RF waveforms are recorded; artificial intelligence and pattern recognition algorithms are built into the software. Provides additional analysis capability using standard PC software such as Lotus 1-2-3. |

TABLE IV

NOT YET COMMERCIAL AUTOMATED UT SYSTEMS AND TECHNOLOGIES

| SYSTEM | DEVELOPERS | FUNCTION / DESCRIPTION |
|---|---|---|
| ALOK | IzFP & KRAFT WERK UNION | Amplitude detection with time-of-flight signatures are used to detect flaw conditions. System uses standard UT transducers but is being upgraded to incorporate phased array transducer and will be capable of ALOK, ultrasonic holography and line SAFT imaging. |
| CUDAPS | EPRI & ADAPTRONICS GEN.RES.CORP | Computerized Ultrasonic Data Acquisition & Processing System (CUDAPS) incorporates ALN systems to provide a complete automatic UT system. Major components are ALN-4060 Flaw Descriminator, LSI-11 Microprocessor and AMAPS scanner controlled by an ALN-4033 Scan Controller. System has been configured at the EPRI NDE Center. |
| PVIS | EPRI & SIGMA RESEARCH, INC. | Pressure Vessel Imaging System (PVIS) is an ultrasonic holography inspection system for heavy section steel inspection. The system can preform A,B,C,3D and holographic imaging and integrate multiple images into final 3D images. System under evaluation at NDE Center. |
| SAFT | NRC, BATTELLE-NORTHWEST & CE | Synthethic Aperture Focused Technique (SAFT) is a UT technology which provides focused UT images in the complete inspection volume. A transducer or point UT source is scanned over the volume surface and resulting UT data is analyzed to provide very high resolution focused UT image data at all points in the inspection volume. |
| SECTOR SCAN | EPRI & VINTEK, INC. | Real-time sector scan UT imaging uses a phased array or mechanically scanned transducer to provide real-time L and S wave images in a sector B-scan mode. Technology is similar to commonly used medical real-time sector scan units. |

TABLE V

ADVANCED SYSTEMS SATISFYING IGSCC DEMONSTRATION REQUIREMENTS

| System | Detection | Sizing |
|---|---|---|
| ALARA 1 | X | X |
| ANL 4060 | X | |
| CUDAPS | X | |
| Intraspect | X | X |
| P-Scan | X | X |
| Sutars | X | |
| Robie | | X |
| Ultra Image III | X | X |
| UDRPS | X | X |
| Zipscan | X | |

Cost-Effectiveness

Another concern of the utility NDE manager is the cost-effectiveness of inspections with advanced systems. Benefits most frequently cited by experienced utility users, service groups, and vendors are:

| | |
|---|---|
| Better, and more complete coverage | Post review/analysis |
| Traceable results | Better reliability |
| Permanent recording | Low-radiation exposure |
| Real-time results | Immediate sizing |
| Fewer over/under calls | Documented data |
| Data comparison | Potentially lower cost |

Many of the above benefits have significant long-term cost payoffs that are only realizable over a long time and are not immediately measurable. Each advanced system has different features and components which its marketing literature vigorously promotes. It is not always easy to identify the components or feature most important to a particular utility need. If a significant repair cost can be avoided because of the use of a UT system with advanced data display capabilities, it makes little difference if the system also possesses advanced pattern recognition capabilities. Both capabilities are equally important, but certainly not needed by all utility end users and/or applications.

To obtain cost information on automated/advanced UT inspection, a survey of the utilities, vendors, and service groups who have used such systems was conducted. Table VI summarizes the findings of this survey.

A total of 29 utilities and inspection service groups were surveyed. Twelve groups or 41% responded; however, several had no relevant data which gave an effective response of 21%. Those responding represent a total of more than 120,000 inches of welds that have been inspected by advanced UT--about 1,900 equivalent 20-inch diameter pipe welds. The data incldues GEDAS, IntraSpect, P-Scan, and UDRPS inspection results.

Recognizing that inspections with advanced UT systems provide benefits not readily available with manual UT, the initial equipment cost must be scaled accordingly to obtain the effective inspection cost. Specifically, to obtain the effective cost ratio of advanced UT to manual UT inspection (AUT/MUT), the initial AUT/MUT cost ratio must be divided by the benefit ratio. The average initial AUT/MUT cost ratio was 2.12-- advanced UT inspection initially cost 2.12 times more than equivalent manual UT inspection; the average advanced UT to manual UT benefit ratio was 3.93. Therefore, the average effective AUT/MUT cost ratio is 0.64 and hence the effective cost of advanced UT inspection comes out to be less than two-thirds of the cost of equivalent manual UT inspection.

Technology Transfer

Ultrasonic NDE technology has advanced a great deal within the last two years. Advances have been so rapid that it is often difficult for a utility NDE manager to keep informed of all of the available inspection systems and their capabilities. The terms and concepts used by the developers and manufacturers of the equipment may leave the end user in a state of apprehension when a problem appropriate to an advanced system arises. EPRI is supporting efforts whose goals are to aid the utility NDE manager by providing him with a guide for selection and use of advanced systems and workshops to disseminate the latest information.
MNA/3060RPTSAS.

TABLE VI

AUT/MUT COST QUESTIONNAIRE SUMMARY 7/10/86

| Utility, Vendor or Service Group | 1 | 2 | 3 | 4 | 5 | 6 | 7 | WEIGHTED AVERAGE & TOTALS |
|---|---|---|---|---|---|---|---|---|
| 1. Type of automated UT system: (GEDAS,Intraspect,P-Scan,UDRPS & GESMART UT) | | | | | | | | |
| a) Approximate inches of weld inspected | 9,000 | 1,500 | 35,000 | 10,000 | 500 | 67,000 | 7,190 | 130,190 |
| b) Estimated cost compared to equivalent manual UT inspection (MUT) | 3.00 | 0.90 | 2.50 | 1.00 | 1.90 | 2.00 | 2.50 | 2.14 |
| 2. In your recent experience for an average UT inspection, what is the: | | | | | | | | |
| a) % AUT Inspection | 90 | 7 | 83 | 50 | 0 | 90 | ? | 83 |
| b) % MUT Inspection | 10 | 93 | 17 | 50 | 100 | 10 | ? | 17 |
| c) Total number of welds inspected | 25 | 40 | 120 | 100 | 200 | 130 | ? | 116 |
| d) Total number of weld inches inspected | 1,750 | 1,500 | 6,000 | 5,000 | 7,500 | 8,380 | ? | 6,855 |
| e) Estimated cost to the utility per weld inch | 28.00 | ? | 20.83 | 50.00 | ? | 61.00 | ? | 46.02 |
| f) Average % repeat inspection including sizing | 10 | 5 | 5 | 10 | 1 | 20 | ? | 14 |
| g) Average % sizing | 10 | 4 | 5* | 7.5 | 1 | 2 | ? | 4 |
| h) Average radiation level at weld site (mR/hr) | 250 | 150 | 20* | 50 | 30 | 500 | ? | 415 |
| i) Average inspector (Level I/II) exposure per outage | 1,600 | 1,000 | 150 | 600 | 1,200 | 1,200 | ? | 1,169 |
| j) Average cost per mR for total outage | 8.00 | ? | 83.00 | ? | ? | 9.23 | ? | 9.08 |
| 3. Estimated AUT/MUT cost ratio | 3.00 | 1.00 | 2.50 | 1.10 | 1.90 | 2.00 | 2.50 | 2.15 |
| 4. Average MUT inspection speed: (weld inches/10 hr. shift) | 140 | 70 | 200 | 240 | 300 | 80 | ? | 132 |
| 5. Average AUT inspection speed: (weld inches/10 hr. shift) | 140 | 95 | 400 | 240 | ? | 140 | ? | 222 |
| 6. Average number of welds inspected per outage | 40 | 100 | 75 | 60 | 300 | 130 | ? | 102 |
| 7. Average number of weld inches inspected per man per outage | 550 | 500 | 500 | ? | ? | 160 | ? | 302 |
| 8. Benfits | | | | | | | | |
| 9. AUT/MUT benifits ratio | 4.00 | 2.00 | 2.00 | ? | 1.10 | 5.00 | 2.00 | 3.82 |
| Effective AUT/MUT cost ratio (item 1b divided by item 9) | 0.75 | 0.48 | 1.25 | ? | 1.73 | 0.40 | 1.25 | 0.56 |

\* These are for DOE work and may not be representative.

Utility response -- 3 out of 16 with 1 N.A.
Vendor response -- 9 out of 13 with 4 N.A.
Total response -- 41%
Effective response -- 24%

DISCUSSION

Dr. Berens: I have one question. I think in terms of NDE reliability in terms of either finding them or sizing them. I wonder if you could address that question as to how well we do that in the electric (inaudible) region.

Dr. Behravesh: Well, the reliability studies are being conducted by the regulators and under their contracts, and we have been sort of staying a bit away from that because of the fact that we know that we are not as effective in NDE as we want to be. Most of our concentration in the nuclear industry is in the area of improving the techniques and demonstrating the capabilities of those techniques. Knowing that reliabilities are not as high as we want them to be, we are not, basically, concentrating much on that. So I don't have any reliability numbers for you.

Dr. Berens: Is Gerry Posakony here? I think it's only fair to Mohammed to mention that we did invite representatives of the NRC to present a paper on NDE reliability, and they are in the process of performing experiments and they didn't feel they were performing one that had any results. So, that was a question that was kind of unfair.

Dr. Don Thompson, Ames Lab: Did you take into account the cost-benefit ratio (of the manual reading), whether it was about 50, 55 percent, and take into account the cost of the automation? How do you actually come out in dollars?

Dr. Behravesh: Those were actually in dollars, if you assume that the cost of manual examination is $1, the effective cost of an automated examintion, based on those numbers, was 60 cents.

Dr. R. B. Thompson: That must be based on some assumed lifetime of the equipment, how long you amortize the initial costs.

Dr. Behravesh: That is weighted by the amount of actual examination that is performed, up to a given limit.

Dr. R. B. Thompson: How long does that have to last to pay for itself?

Dr. Behravesh: The equipment pretty much can pay for itself in one outage. Outages are the periods where the plant is shut down for maintenance and examination, and examinations are quite expensive. Typical examination runs into several hundred thousand dollars. So if you were thinking of one system costing a hundred thousand dollars, the price is somewhat insignificant. Not that it does not come into consideration, but it's insignificant.

Dr. Berens: Maybe we can continue this afternoon. Supposed to have someone from Battelle Northwest on our panel this afternoon.

THE DETECTION OF CRACKS UNDER INSTALLED FASTENERS BY MEANS

OF A SCANNING EDDY-CURRENT METHOD

David Harrison

Ministry of Defence (PE)
Royal Aircraft Establishment
Farnborough, Hants.  GU14 6TD
United Kingdom

INTRODUCTION

As a consequence of metal fatigue, cracks can develop and grow in operational aircraft. Periodic inspections must be made in order to detect and repair them before they reach a dangerous length. Cracks which grow from holes are a significant problem for aircraft since the wings and fuselage can contain many thousands of fasteners. Since it is impractical to remove them all, inspection must be made with them installed.

A self-contained automated instrument, the EDDISCAN, has been built to detect and size these cracks. It works on a scanning principle and is based on a microprocessor which controls all aspects of the systems operation, including analysis and display of results. (A photograph of the instrument is shown in Fig 1.) Preliminary tests show that, under laboratory conditions, it can detect the presence of simulated radial cracks as small as 0.2 mm long beneath the heads of fasteners.

PRINCIPLES OF OPERATION

A section of aircraft skin riveted to a stiffener is illustrated in Fig 2. This is a typical example of the situation in which small radial cracks can grow under the heads of installed fasteners. In order to detect these cracks, a small C-shaped ferrite core with a coil on the outer leg is rotated around the axis of the fastener as shown in Fig 2. If it passes over a sub-surface crack, then the reflected impedance of the coil changes. By measuring the impedance repeatedly at different positions as the coil is rotated, a map or image of the impedance variations can be recorded and by suitable analysis, the presence of cracks can be identified.

A block diagram of the instrument is shown in Fig 3. It can be seen that all aspects of its performance are controlled through a microprocessor by means of programmed instructions stored in a 16K EPROM. The coil is rotated at constant speed by a stepper motor. The input signal to the coil is a 1 kHz square wave which is generated as a digital data stream and converted to analogue form by a D/A converter (DAC1). The voltage across the coil is amplified by the pick-up amplifier, converted to digital form by an A/D converter (ADC) and stored in memory (8K RAM). The observed signal consists of four main AC components. These are due to (a) the impedance of the coil in vacuo, (b) the reflected impedance of the specimen, (c) the reflected impedance of the fastener (and hole) and (d) the reflected impedance of any cracks. Components (a)

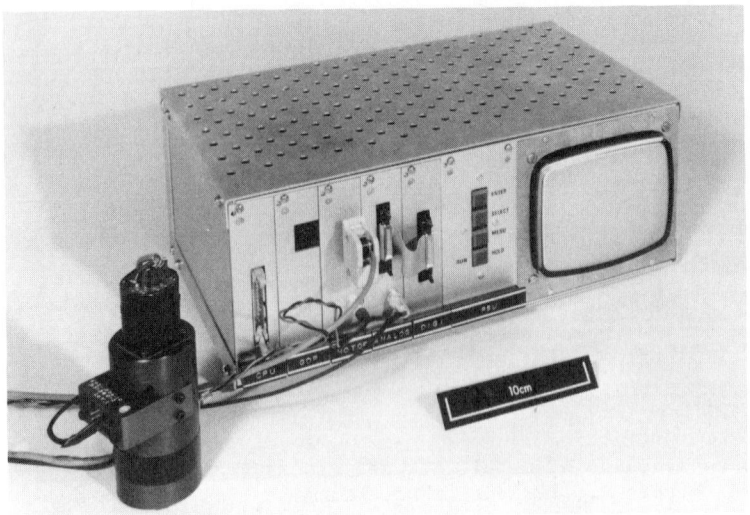

Fig. 1. A photograph of the EDDISCAN instrument

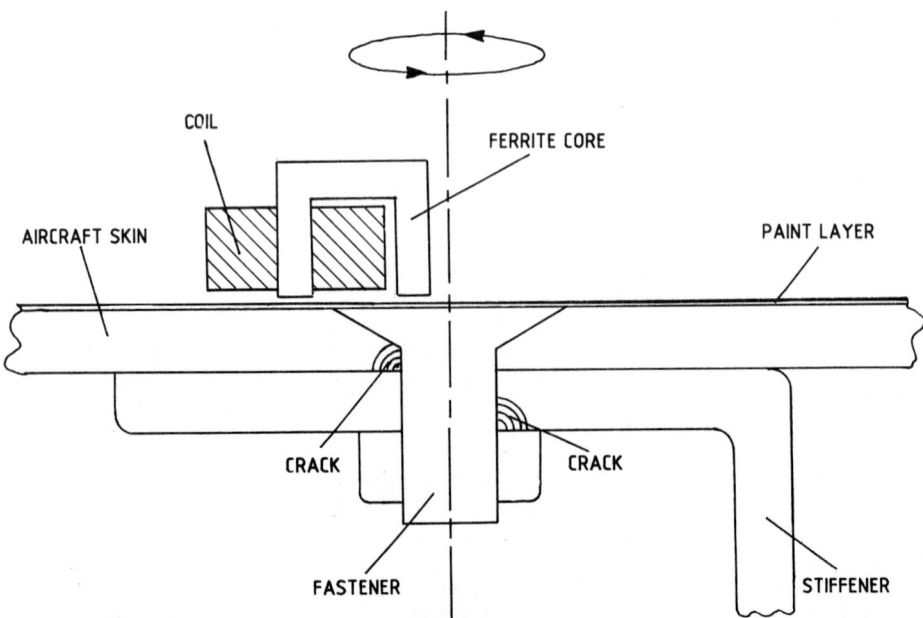

Fig. 2. Schematic diagram illustrating the application of the principles of scanning eddy-current methods to the detection of typical small radial cracks growing beneath the head of a rivet.

and (b) are relatively large and contain no useful information since they do not vary throughout each revolution of the coil. They must be removed so that components (c) and (d) can be amplified. They are cancelled by using a hardware/software negative feedback loop to generate a balancing signal through DAC2. Thus at this stage RAM memory contains a digital representation of the variation of components (c) and (d) during one complete revolution of the coil.

Component (c), the reflected impedance of the fastener, is dependent on the position of the coil relative to the fastener. Because the axis of rotation of the coil may not coincide with the axis of the fastener, this impedance may vary considerably during each revolution. By identifying and analysing (c), the relative positions of the axes can be calculated and used to generate a graphical centering display. By observing this, an operator can rapidly adjust the transducer so that it is concentric with the fastener. Under these circumstances, (c) is reduced to zero leaving only (d) in memory, the signal due to any cracks. Digital processing and analysis programs then analyse this data and display results in the form of text and graphics on a VDU which is driven by a graphics display processor.

By taking this approach to the design, it is possible to make considerable reductions in the amount of hardware required. Analogue circuitry is reduced to a minimum since most of the traditional analogue functions can now be done by the microprocessor. Shifting the boundary between hardware and software in this way inevitably involves a significant increase in the level of effort on software development. However, this approach is justified not only by increased flexibility but also by the ability to do things that are not possible using conventional analogue techniques.

PRELIMINARY PERFORMANCE

The VDU screen layout is represented in Fig 4. The screen is divided into four quadrants, each displaying a separate function. Top left is the centering display. As described above, the position of the transducer relative to the axis of the fastener is displayed on the screen in the form of a small cross and a set of axes. The cross is programmed to move in unison with the transducer and when it is at the origin of the display then the transducer is coaxial with the fastener. With this aid, an operator can rapidly center the transducer. Bottom left is a display of the raw impedance measurements which is updated each revolution of the transducer. Bottom right shows the real and imaginary components of the same data after it has been processed and digitally filtered. Finally, top right shows an impedance plane display of the processed data in which the real and imaginary parts are plotted against each other.

Although development of the software for this instrument is not yet complete, in order to obtain a preliminary estimate of its performance, some measurements have been made using test specimens which contained spark-eroded notches to simulate real defects. These results are presented in Fig 5 in the form of photographs taken of the screen during measurements of four different specimens. In each case the specimen consisted of six aluminium alloy plates, each 2.5 mm thick, held together with a 5.0 mm diameter non-ferrous fastener. Figs 5(A) to 5(D) show respectively the instruments response to (a) a 1.0 mm defect in the top layer, (b) a 1.5 mm defect in the second layer, (c) a 0.6 mm defect in the top layer and a diagonally opposed 1.5 mm defect in the second layer and (d) no defects at all. In Figs 5(A), 5(B) and 5(C) the effect of the defects can clearly be seen on the real and imaginary traces in contrast to the traces of Fig 5(D) which are for no defects. Furthermore, since the phase of the signal depends on the depth of the crack, cracks in the

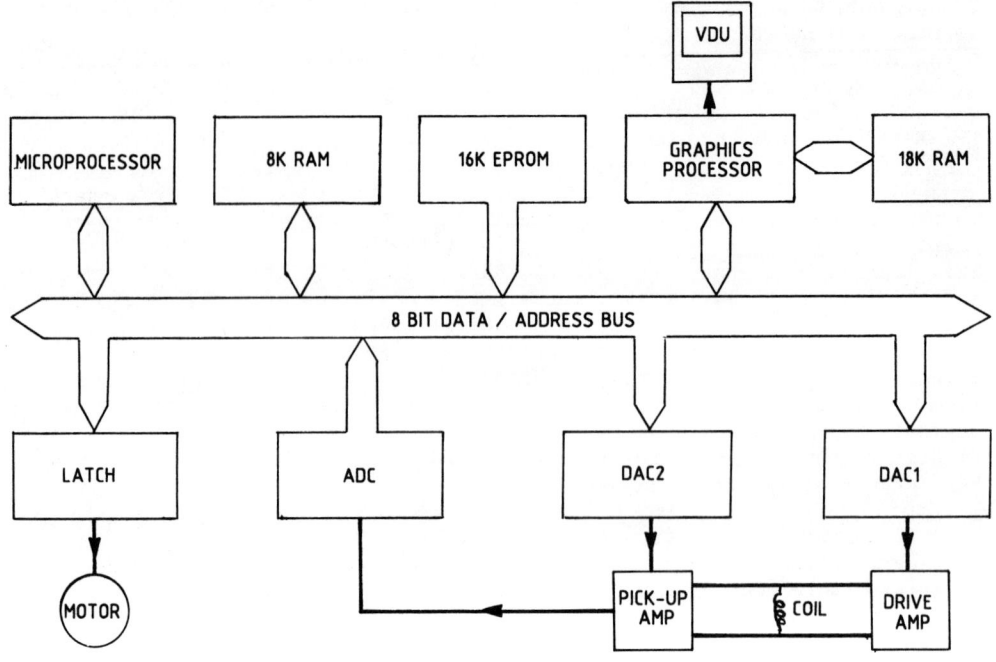

Fig. 3. Block diagram of the electronic hardware

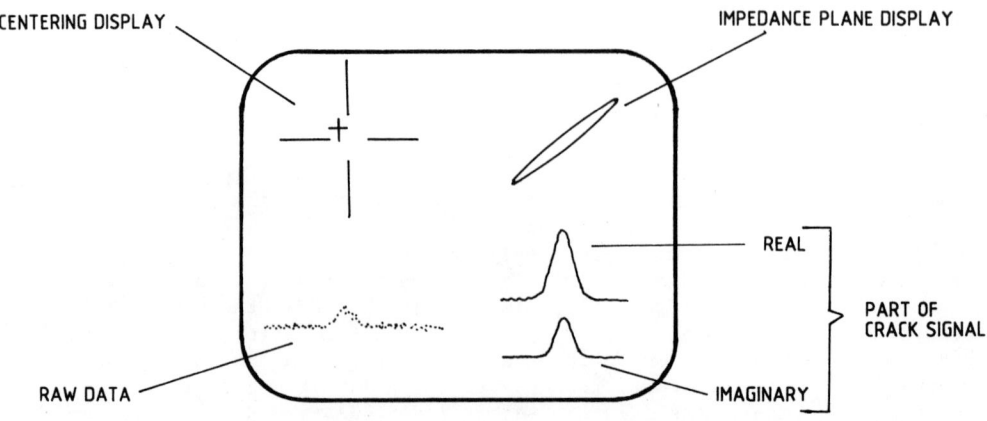

Fig. 4. Schematic diagram of the graphical display of results presented on the screen.

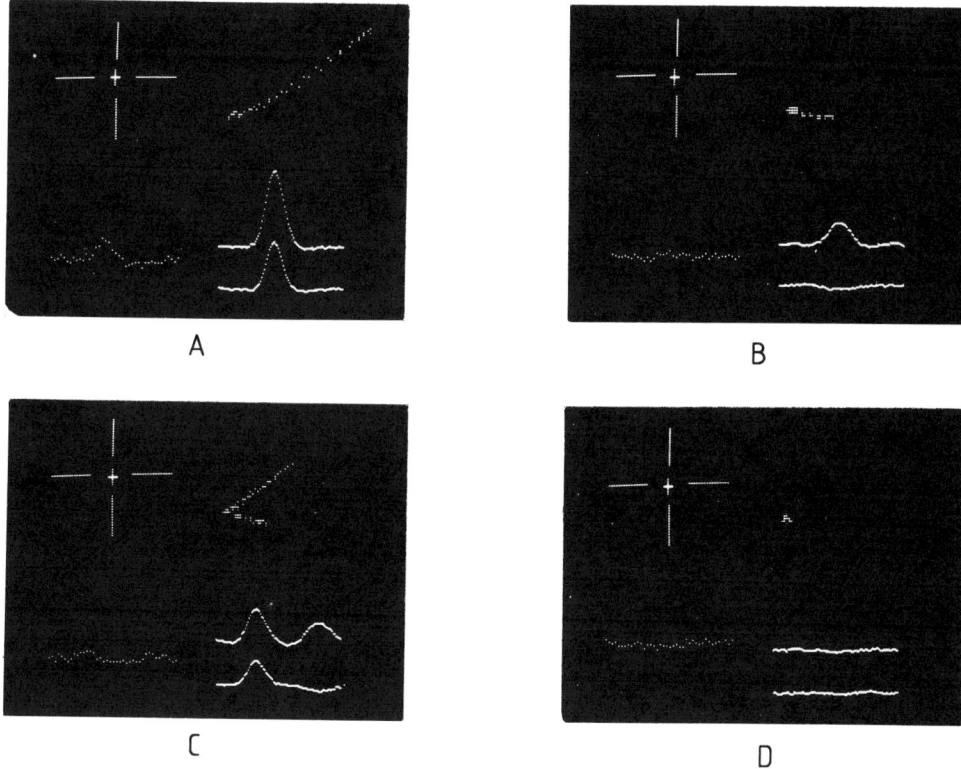

Fig. 5. Photographs of the screen taken during the measurement of various sized notches: (A) 1.0 mm in the skin, (B) 1.5 mm in the second layer, (C) 0.6 mm in the skin and 1.5 mm in the second layer and (D) no defects.

top and second layers cause distinct loci in the impedance plane and thus can be distinguished as is evident from Fig 5(C).

FUTURE DEVELOPMENT

With the instrument in its present form the effect of cracks in the top layer as small as 0.2 mm can be observed by eye. Analysis programs based on pattern recognition are being written which will automatically process the measurements, estimate the size, angular position and depth of any cracks, and generate a comprehensive display of the results. In addition, it is planned to make slight modifications to the transducer that will enable it to work with ferrous fasteners.

REAL DEFECTS FOR VERIFICATION OF

ULTRASONIC TESTING MODELS

John R. Bower

Babcock & Wilcox, Lynchburg Research Center
P.O. Box 11165
Lynchburg, VA  24506-1165

INTRODUCTION

Most ultrasonic inspections are based on idealized calculations of ultrasonic responses. The adequacy of a proposed inspection is determined by the accuracy of the calculations and the model of the assumed flaw. For generalized cases, the calculations can be quite difficult.

Direct experimentation is obviously the best source of data to compare with theoretical or computer results. A large amount can be carried out on blocks with drilled holes and saw cuts. But drilled holes and saw cuts are not real inclusions and cracks, which have irregular shapes and varied surface conditions. They do not provide the confidence needed for evaluation of complicated situations. As a result, blocks have been built at great expense incorporating real fatigue cracks and lack of fusion. These give valuable information, but they are limited to special cases. Slag inclusions or porosity samples must be destructively analyzed to evaluate the actual condition, which is expensive and destroys the test block. It is not practical to provide the quantity and variety of real flaw data that would illuminate theoretical results and allow empirical modeling.

The problem of verification of internal flaws can be solved by using transparent blocks. Glass test blocks have been made and used in imaging internal wave patterns[1]. These blocks are not easy to build or modify in an ordinary lab, but ice is easy to work with.

PRODUCTION OF TEST BLOCKS

The production of test blocks requires the ability to make large blocks of clear ice. If water is simply placed in a large container in a freezer, the resulting ice will be opaque because of the trapped air bubbles. After the top freezes over, it will rupture as freezing progresses, and the container may also rupture. Commercial clear ice is produced by having a continuous stream of bubbles in the water. This prevents the dissolved gas concentration from reaching the bubble nucleation point and keeps the top from freezing.

In the absence of a compressed air supply for a bubbler, an alternate technique may be used. First, deionized water is degassed by boiling.

Boiling under vacuum, at room temperature, is faster and easier than boiling over heat, at atmospheric pressure. It works as well. When degassed water is frozen, about half the volume will freeze as clear ice before bubbles begin to form. At that point, the unfrozen portion is emptied out and replaced by freshly degassed precooled water. During freezing, the top is kept clear of ice by covering the top of the container with insulation.

Figure 1 shows a block grown in this manner without changing the water when bubble formation began. During freezing, the gas concentration in the remaining liquid increases until bubbles are nucleated essentially simultaneously over the freezing surface. A few begin sooner, but most of the bubbles which appear outside the central area are reflected or refracted views of the central mass.

During the process, various defects can be introduced. Surface connected cracks are easiest. A partially completed block is like a bowl. It can be emptied and left to cool to freezer temperature. If room temperature water is added to the bowl, the thermal stress will cause cracking, and the water will then proceed to freeze into a solid block with surface cracks to the depth of the original bowl wall thickness. Internal porosity can be created by allowing the ice to freeze past the point where bubbles begin to form, although this porosity is uncontrolled, and if the water has been thoroughly degassed, the first bubbles to form become nucleation centers and grow into very long thin bubbles as freezing progresses.

Figure 1: Bubble Growth in Ice

Internal cracks and controlled porosity are also simple to make. As a first step, a piece of ice with a suitable crack is selected or a container of non-degassed water is frozen. Its porosity will vary continuously from the outside to the center. The frozen blocks are easily sliced up with a fine jet of water.

A suitably porous piece and/or a cracked piece are then cooled to freezer temperature. At the same time, a block of clear ice is being frozen. While the freezing is progressing, the water is emptied and the defect samples are pressed against the wet ice. Because they are well below freezing, they will quickly adhere to the ice. The cold water is returned to the vessel, freezing continues and the flaws are embedded. Figure 2 shows a test block with an embedded planar porosity region and surface connected cracks.

SCALING MEASUREMENTS

The speed of sound in ice is 3980 m/sec for longitudinal waves and 1990 m/sec for shear waves. In steel, the corresponding values are 5850 and 3230 m/sec. The velocity ratios are 0.680 and 0.616, respectively. These ratios enter into several relations useful in scaling differences in ultrasonic responses from defects in ice and steel.

Figure 2: Planar Porosity and Surface Connected Cracks

The far field spread of a beam from a transducer (or of a reflection from a defect) is determined by the ratio of transducer (or defect) size to wavelength. This means that to get identical results at any frequency it is necessary to scale the transducer and defect sizes by the velocity ratios. In many cases it may be more desirable to keep the sizes the same in steel and ice and reduce the frequency used in ice by the velocity ratio. In either case, the distances would be the same and the transit time in steel will be reduced by the velocity ratio. With this scaling, the same defect in the two materials will give the same distance-amplitude curve and the same variation of response with beam angle.

    Angled beams are commonly used in inspection of steel, using lucite wedges. The angles used are determined from the relation $(\sin A_1)/V_1 = (\sin A_2)/V_2$, where $A_1$ and $A_2$ are angles in the two materials and $V_1$ and $V_2$ are the wave velocities. Shear waves in steel are commonly generated by mode conversion of longitudinal waves in lucite, whose wave velocity is 2680 m/sec. A longitudinal wave in lucite, incident on steel at the critical angle of 27 degrees, will produce longitudinal waves in the steel parallel to the surface. At greater angles of incidence, only shear waves can be generated. At this critical angle, the shear wave angle is 34 degrees, so it is possible to generate shear waves of any greater angle with no longitudinal wave interference.

    For longitudinal waves in lucite on ice, the critical angle is 42 degrees. At this angle, shear waves will be generated at 30 degrees.

    In both cases, because the ratio of shear to longitudinal velocities is so close (0.55 in steel, 0.50 in ice), any mode conversion process in either material will be very similar to the process in the other. However, the shear velocity in ice is lower than the longitudinal velocity in lucite, so it is not possible to create angled shear beams in ice much beyond 45 degrees using lucite wedges. If high angle shear beams are required they can be generated with a shear transducer.

REFERENCE

1.  K.G. Hall, Mat. Eval. 42, 922 (1984).